Advanced Structured Materials

Volume 170

Common engineering materials are reaching their limits in many applications, and new developments are required to meet the increasing demands on engineering materials. The performance of materials can be improved by combining different materials to achieve better properties than with a single constituent, or by shaping the material or constituents into a specific structure. The interaction between material and structure can occur at different length scales, such as the micro, meso, or macro scale, and offers potential applications in very different fields.

This book series addresses the fundamental relationships between materials and their structure on overall properties (e.g., mechanical, thermal, chemical, electrical, or magnetic properties, etc.). Experimental data and procedures are presented, as well as methods for modeling structures and materials using numerical and analytical approaches. In addition, the series shows how these materials engineering and design processes are implemented and how new technologies can be used to optimize materials and processes.

Advanced Structured Materials is indexed in Google Scholar and Scopus.

Holm Altenbach · Arkadi Berezovski ·
Francesco dell'Isola · Alexey Porubov
Editors

Sixty Shades of Generalized Continua

Dedicated to the 60th Birthday of Prof. Victor A. Eremeyev

Springer

Editors
Holm Altenbach
Institut für Mechanik, Fakultät für Maschinenbau
Otto-von-Guericke-Universität Magdeburg
Magdeburg, Sachsen-Anhalt, Germany

Francesco dell'Isola
Università degli Studi dell'Aquila
L'Aquila, Italy

Arkadi Berezovski
Tallinn University of Technology
Tallinn, Estonia

Alexey Porubov
Institute of Problems of Mechanical Engineering
Saint Petersburg, Russia

ISSN 1869-8433 ISSN 1869-8441 (electronic)
Advanced Structured Materials
ISBN 978-3-031-26185-5 ISBN 978-3-031-26186-2 (eBook)
https://doi.org/10.1007/978-3-031-26186-2

This Springer imprint is published by the registered company Springer Nature Switzerland AG
The registered company address is: Gewerbestrasse 11, 6330 Cham, Switzerland

Preface

Victor Anatolyevich Eremeyev is a brilliant scientist with an international reputation who has enriched the science with many achievements in the field of mechanics of deformable solids with the main focus on *Generalized Continua* and the theories of one-dimensional and two-dimensional structural elements like beams, rods, plates and shells and celebrates his 60th birthday in March 2023.

Victor Anatolyevich Eremeyev

This anniversary is reason that numerous friends and colleagues from more than 10 countries have dedicated scientific works to him, which have been summarized in this book. It is easy to see that some of the works are directly related to the publications of the jubilarian. Others originate from the scientific environment or from the circle of friends, which, however, do not always have a direct connection to the publications of Victor Eremeyev. This also explains why the book title was chosen by the editors - there are different aspects of the generalized continuum models.

Victor Eremeyev was born on March 7th, 1963 in Piliponovka/Vinnytsia region (Soviet Union, now Ukraine). He studied at the Department of Mechanics and Mathematics of the State University in Rostov-on-Don (former Soviet Union) from 1980 to 1985 and successfully graduated (with distinction). From 1985 until 1989 he had a fellowship for Candidate of Science in Physics and Mathematics at the Division of Elasticity of Rostov State University. In 1990, he was awarded PhD (Candidate of Physics & Mathematics) at Rostov State University. The thesis title was *The stability of two-phase nonlinear thermo-elastic bodies*. In 2004, he finished his second thesis *Mechanics of two-phase bodies with microstructure under finite deformations* and was awarded Doctor of science of Physics & Mathematics at the

Institute of Problems of Mechanical Engineering of Russian Academy of Sciences (Saint-Petersburg, Russian Federation).

In 1996, he was appointed as Associate Professor (Docent) at the Southern Federal University in Rostov-on-Don. From 2003 he started his international carrier. During the first period he was a research associate at the Institute of Fluid-Flow Machinery of the Polish Academy of Sciences in Gdańsk (Poland). He was in close contact with Prof. Wojciech Pietraszkiewicz. Together they make some fundamental publications highly cited by the scientific community, among them the paper *On natural strain measures of the non-linear micropolar continuum* published in the International Journal of Solids and Structures (**46**(3-4), pp. 774-787) cited 229 times (20.11.2022).

Later he was a invited as a research associate to Germany. He spent several year at the Chair of Engineering Mechanics at the Martin-Luther-Universität Halle-Wittenberg and Otto-von-Guericke-Universität Magdeburg. During this period he organised together with one of the editors (H.A.) a Euromech Colloquium in 2011 (EM 527 - Shell-like Structures – Nonclassical Theories and Applications) and two Courses at the CISM (International Centre for Mechanical Sciences) in Udine, Italy (2011 - Generalized Continua - from the Theory to Engineering Applications, 2014 - Shell-like Structures - Advanced Theories and Applications). The proceedings of the Colloquium and the two Courses were published by Springer. Among his activities was also the publication of his most cited paper (304 times/20.11.2022) *On generalized Cosserat-type theories of plates and shells: A short review and bibliography* prepared together with J. & H. Altenbach and published in the Archive of Applied Mechanics (**80**(1), pp. 73-92).

Leaving Germany, in 2015 Victor Eremeyev was appointed as a professor (PRz) at Politechnika Rzeszowska im. Ignacego Lukasiewicza in Rzeszow (Poland), later (2017) as a professor (PG) and finally as a full professor at Politechnika Gdańska in Gdańsk (Poland). During this period he collaborated with several Polish scientists, among them V. Konopińska-Zmysłowska, T. Wiczenbach, J. Chróścielewski and K. Wiśniewski. Since 2020 he has the position of Professore associato at the Università degli Studi di Cagliari in Cagliari (Italy).

Victor Eremeyev's scientific interests are very diverse. They include but not limited to

- Nonlinear elasticity. Infinitesimal stability of solids undergoing stress-induced phase transitions were studied;
- Plates, shells and beams. Within six-parameter theory of shells phase transformation in thin-walled structures were studied, the material symmetry group was introduced. The five-parameter theory of shells was applied to thin structures made of functionally graded materials and foams. Layer-wise approach was used for modelling of solar panels;
- Generalized media including micropolar and micromorphic media, stress and strain gradient elasticity and strain gradient fluids. Infinitesimal stability of micropolar solids was studied. Material symmetry group was discussed for micropolar, micromorphic, and strain gradient continua. Acceleration waves and ellipticity conditions were studied. Elastic networks were modelled as a Cosserat continuum;

- Surface elasticity. The effective properties for plates and shells with surface stresses were discussed. Surface waves in media with surface/interface microstructure were discussed.
- Electro-magnetic-mechanical coupling. The effects of piezo-, flexoelectricity, piezo- and flexomagneticity were discussed for applications in mechanics of composites and thin-walled structures;
- Application of mathematical methods in mechanics. Using the calculus of variations and functional analysis the existence and uniqueness of weak solutions of some boundary-value problems of six-parameter shells, media with surface elasticity, and strain gradient continua were proved. For example, a remarkable application of anisotropic Sobolev's space to well-posedness analysis of pantographic metamaterials was given.

He published more than 270 scientific papers indexed in Scopus and the list of co-authors looks like *Who is who in Mechanics*: from Russia L. Zubov, A. Nasedkin, S. Lurie, L. Igumnov, N. Morozov, A. Porubov, E. Ivanova, E. Vilchevskaya, A. Freidin, from Germany W.H. Müller, E. Abali, K. Naumenko, M. Lazar, from Italy F. dell'Isola, I. Giorgio, L. Placidi, E. Reccia, G. Rosi, A. Cazzani , A. Luongo and E. Turco, from France J.-F. Ganghoffer, C. Boutin, S. Naili, from UK G. Mishuris, from Colombia L. Lebedev and L. Rendón, from USA D. Steigmann and M. Cloud, from Belarus G. Mikhasev, from Greece E. Aifantis and from Poland T. Sadowski, F. Stachowicz, A. Skrzat, and M. Malikan. In addition, he wrote and edited 21 books in Russian, English and Spanish. During the last years he was invited by several universities and research institutes: Institute of Fluid-Flow Machinery of the Polish Academy of Sciences in 2003, 2005, 2008, 2009 (Gdańsk, Poland), Martin-Luther-Universität Halle-Wittenberg in 2007, 2008, 2009 (Halle, Germany), National University of Colombia in 2008, 2012 (Bogota, Colombia), Politechnika Lubelska in 2011, 2012 (Lublin, Poland) and International Research Center for Mathematics & Mechanics of Complex Systems (M&MoCS) in 2014 (Cisterna di Latina, Italy). Since 2015 he served as a visiting professor at Sapienza University of Rome, Italy (2015, 2017), at Université Paris-Est Créteil Val de Marne, France (2015, 2019), at Universita Degli Studia di Genova, Genoa, Italy (2016), at Università degli Studi di Cagliari, Cagliari, Italy (2017) and at Università degli Studi di Sassari, Alghero, Italy (2017, 2018).

Victor Eremeyev's scientific activities were awarded in 2000 and 2008 with the prize from Elsevier Science Ltd. for the best paper published in the journal Prikladnaya Matematika i Mekhanika (Journal of Applied Mathematics and Mechanics):

- Yeremeyev VA, Zubov LM (1999) The theory of elastic and viscoelastic micropolar liquids, Journal of Applied Mathematics and Mechanics **63**(5):755–767.
- Eremeyev VA, Freidin AB, Sharipova LL (2007) The stability of the equilibrium of two-phase elastic solids, Journal of Applied Mathematics and Mechanics (PMM) **71**(1):61–84.

In 2018, he got the International Prize "Tullio Levi-Civita" for the Mathematical and Mechanical Sciences and in 2021-2022 the Royal Society Wolfson Visiting Fellowships. In 2021, he was awarded the professorship from the President of Poland, he is included in the list of the World's Top-cited Scientists (World's Top 2% Scientists List, 2020-2022, https://elsevier.digitalcommonsdata.com/datasets/btchxktzyw).

Victor Eremeyev is an active member in the following scientific societies or organizations: American Mathematical Society, International Research Center on Mathematics & Mechanics of Complex Systems (M&MoCS) and the Polish Society of Theoretical and Applied Mechanics. In addition, he is a member of the following editorial boards: Acta Mechanica, Archive of Applied Mechanics, Computer Modeling in Engineering & Sciences, Mathematics and Mechanics of Complex Systems, Mathematics and Mechanics of Solids, Mechanics of Composite Materials, Symmetry, Technische Mechanik, Nanomechanics Science and Technology: An International Journal, Zeitschrift für Angewandte Mathematik und Mechanik, Problems of Strength and Plasticity, World Journal of Mechanics and Mechanical Bulletin of Perm National Research Polytechnic University. He was appointed as a guest editor of Zeitschrift für Angewandte Mathematik und Mechanik (2009, 2010, 2011, 2014), International Journal of Engineering and Science (2014), Continuum Mechanics and Thermodynamics (2015, 2019, 2020), Mathematics and Mechanics of Solids (2015, 2021), Symmetry (2020, 2021), Mathematics (2020), Nanomaterials (2020) and Mechanics Research Communcations (2021). He acted as a co-editor and member of the scientific committee of Gabrio Piola edition which was published in 2014 (Volume I) as "The complete works of Gabrio Piola, commented English translation(Eds Francesco dell'Isola, Giulio Maier, Umberto Perego, Ugo Andreaus, Raffaele Esposito, Samuel Forest), Advanced Structured Materials, Vol 38, Springer".

Last, but not least, it is essential to highlight his personal charisma. He easily finds a common ground with different people regardless of their age, gender, nationality, education, or key scientific field. His discussion style is distinguished by respect for opposing viewpoints. He possesses the rare ability to highlight the main idea and formulate a problem with complete mathematical rigor. Pure science does not cover the entire area of his interests. Like any thinking man, he is interested in art, literature, music, and social issues.

Magdeburg,	*Holm Altenbach*
Tallin,	*Arkadi Berezovski*,
Rome,	*Francesco dell'Isola*,
St. Petersburg,	*Alexey Porubov*
	January 2023

Contents

List of Contributors

Bilen Emek Abali
Uppsala University, Division of Applied Mechanics, Department of Materials Science and Engineering, Uppsala, Sweden, e-mail: bilenemek@abali.org

Reza Afshar
Uppsala University, Division of Applied Mechanics, Department of Materials Science and Engineering, Uppsala, Sweden, e-mail: reza.afshar@angstrom.uu.se

Sergei M. Aizikovich
Laboratory of Functionally Graded and Composite Materials, Research and Education Center "Materials", Don State Technical University, 1 Gagarin sq., 344000, Rostov-on-Don, Russian Federation,
e-mail: saizikovich@gmail.com

Holm Altenbach
Chair of Engineering Mechanics, Institute of Mechanics, Faculty of Mechanical Engineering, Otto von Guericke University Magdeburg, Universitätsplatz 2, 39106 Magdeburg, Germany, e-mail: holm.altenbach@ovgu.de

Igor V. Andrianov
Chair and Institute of General Mechanics, RWTH Aachen University, Eilfschornsteinstraße 18, Aachen, D-52062, Germany, e-mail: igor.andrianov@gmail.com

Artem Antonov
Mechanical Engineering Research Institute of Russian Academy of Sciences, 85 Belinskogo str., Nizhny Novgorod 603024, Russian Federation,
e-mail: artem.antonov@autorambler.ru

Harm Askes
The University of Twente, Faculty of Engineering Technology, Twente, 7500 AE Enschede, The Netherlands, e-mail: h.askes@utwente.nl

Marcus Aßmus
Chair of Engineering Mechanics, Institute of Mechanics, Faculty of Mechanical Engineering, Otto von Guericke University Magdeburg, Universitätsplatz 2, 39106 Magdeburg, Germany, e-mail: marcus.assmus@ovgu.de

Vladimir V. Balandin
National Research Lobachevsky State University of Nizhny Novgorod, 23 Gagarin Avenue, Nizhny Novgorod 603022, Russian Federation,
e-mail: rustydog2007@yandex.ru

Emilio Barchiesi
Università degli Studi di Sassari, Dipartimento di Architettura, design e urbanistica (DADU), Alghero, Università degli Studi dell'Aquila, International Research Center for the Mathematics and Mechanics of Complex Systems (M&MoCS), L'Aquila, Italy & Universidad de Lima, Instituto de Investigación Científica, Santiago de Surco, Peru, e-mail: ebarchiesi@uniss.it

Artem V. Basalin
National Research Lobachevsky State University of Nizhny Novgorod, 23 Gagarin Avenue, Nizhny Novgorod 603022, Russian Federation,
e-mail: artem.basalin@mech.unn.ru

Aleksandr A. Belov
National Research Lobachevsky State University of Nizhny Novgorod, 23 Gagarin Avenue, Nizhny Novgorod 603022, Russian Federation,
e-mail: belov_a2@mech.unn.ru

Petr A. Belov
Institute of Applied Mechanics RAS, Moscow, Russian Federation,
e-mail: belovpa@yandex.ru

Arkadi Berezovski
Department of Cybernetics, School of Science, Tallinn University of Technology, Ehitajate tee 5, 19086 Tallinn, Estonia, e-mail: arkadi.berezovski@cs.ioc.ee

Nikolay Bessonov
Institute for Problems in Mechanical Engineering RAS, Bolshoy pr., 61, V.O., St. Petersburg, 199178, Russian Federation, e-mail: bnm@ipme.ru

Mircea Bîrsan
Faculty of Mathematics, University Duisburg-Essen, Thea-Leymann Str. 9, 45127 Essen, Germany & Alexandru Ioan Cuza University of Iaşi, Department of Mathematics, Blvd. Carol I, no. 11, 700506 Iaşi, Romania,
e-mail: mircea.birsan@uni-due.de

Anatoly M. Bragov
National Research Lobachevsky State University of Nizhny Novgorod, 23 Gagarin Avenue, Nizhny Novgorod 603022, Russian Federation,
e-mail: bragov@mech.unn.ru

Antonio Cazzani
University of Cagliari, Marengo St. 2, 09123 Cagliari, Italy,
e-mail: antonio.cazzani@unica.it

Sergei V. Chapek
The Smart Materials Research Institute, Southern Federal University, 178/24 Sladkova Street, 344090, Rostov-on-Don, Russian Federation, e-mail: chapek@sfedu.ru

Alessandro Ciallella
Department of Civil, Construction-Architectural and Environmental Engineering (DICEAA) & International Research Center on the Mathematics and Mechanics of Complex Systems (M&MoCS), University of L'Aquila, Italy,
e-mail: alessandro.ciallella@univaq.it

Francesco D'Annibale
Department of Civil, Construction-Architectural and Environmental Engineering (DICEAA) & International Research Center on the Mathematics and Mechanics of Complex Systems (M&MoCS), University of L'Aquila, Italy,
e-mail: francesco.dannibale@univaq.it

René de Borst
The University of Sheffield, Department of Civil and Structural Engineering, Sheffield, United Kingdom, e-mail: r.deborst@sheffield.ac.uk

Francesco dell'Isola
Department of Civil, Construction-Architectural and Environmental Engineering (DICEAA) & International Research Center on the Mathematics and Mechanics of Complex Systems (M&MoCS), University of L'Aquila, Italy,
e-mail: rancesco.dellisola@univaq.it

Dionisio Del Vescovo
Department of Mechanical and Aerospace Engineering, SAPIENZA Università di Roma, Rome & International Research Center on the Mathematics and Mechanics of Complex Systems (M&MoCS), University of L'Aquila, Italy,
e-mail: dionisio.delvescovo@uniroma1.it

Margarita Dementeva
Institute of Mechanics, Chair of Continuum Mechanics and Constitutive Theory, Technische Universität Berlin, Einsteinufer 5, 10587 Berlin, Germany,
e-mail: m.dementeva@tu-berlin.de

Mikhail A. Drepin
Institute for Problems in Mechanical Engineering of the Russian Academy of Sciences, 199178, Bolshoy pr. V.O., 61, St. Petersburg, Russia,
e-mail: mihail.drepin@gmail.com

Dmitrii S. Dudin
Institute of Continuous Media Mechanics of the Ural Branch of Russian Academy of Science, Academician Korolev St., 1, Perm 614013, Russian Federation,
e-mail: dudin.d@icmm.ru

Sinan Eraslan
The University of Sheffield, Department of Mechanical Engineering, Sheffield, S1 3JD, United Kingdom, e-mail: seraslan1@sheffield.ac.uk

Vladimir Erofeev
Mechanical Engineering Research Institute of Russian Academy of Sciences, 85 Belinskogo str., Nizhny Novgorod 603024, Russian Federation,
e-mail: erof.vi@yandex.ru

José Manuel Torres Espino
Universidad de Lima, Instituto de Investigación Científica, Santiago de Surco, Peru,
e-mail: 20181896@aloe.ulima.edu.pe

Roberto Fedele
Politecnico di Milano, Dipartimento di Ingegneria Civile e Ambientale (DICA), Milan & Università degli Studi dell'Aquila, International Research Center for the Mathematics and Mechanics of Complex Systems (M&MoCS), L'Aquila, Italy,
e-mail: roberto.fedele@polimi.it

Andrey R. Filippov
National Research Lobachevsky State University of Nizhny Novgorod, 23 Gagarin Avenue, Nizhny Novgorod 603022, Russian Federation,
e-mail: filippov@mech.unn.ru

Samuel Forest
Mines Paris PSL University, Centre des Matériaux, CNRS UMR 7633, France
e-mail: samuel.forest@ensmp.fr

Alexander B. Freidin
Institute for Problems in Mechanical Engineering RAS, Bolshoy pr., 61, V.O., St. Petersburg, 199178, Russian Federation, e-mail: alexander.freidin@gmail.com

Ksenia Frolova
Institute for Problems in Mechanical Engineering RAS, Bolshoy pr., 61, V.O., St. Petersburg, 199178, Russian Federation, e-mail: fkp@ipme.ru

Gregor Ganzosch
Institute of Mechanics, Chair of Continuum Mechanics and Constitutive Theory, Technische Universität Berlin, Einsteinufer 5, 10587 Berlin, Germany,
e-mail: ganzosch@tu-berlin.de

Flavien Ghiglione
Mines Paris PSL University, Centre des Matériaux, CNRS UMR 7633,
e-mail: flavien.ghiglione@minesparis.psl.eu

Ivan Giorgio
Department of Civil, Construction-Architectural and Environmental Engineering (DICEAA) & International Research Center on the Mathematics and Mechanics of Complex Systems (M&MoCS), University of L'Aquila, Italy.
e-mail: ivan.giorgio@univaq.it

Inna M. Gitman
The University of Twente, Faculty of Engineering Technology, Twente, 7500 AE Enschede, The Netherlands, e-mail: i.m.gitman@utwente.nl

Evgenii A. Gladkikh
Perm National Research Polytechnic University, Perm, Russian Federation, e-mail: gladkih.ea@mail.ru

Mikhail E. Gonov
National Research Lobachevsky State University of Nizhny Novgorod, 23 Gagarin Avenue, Nizhny Novgorod 603022, Russian Federation, e-mail: gonov_mikhail@mech.unn.ru

Vasily Govorukhin
Vorovich Institute of Mathematics, Mechanics and Computer Science, Southern Federal University, 344090, Milchakova Street 8a, Rostov-on-Don, Russian Federation, e-mail: vngovoruhin@sfedu.ru

Elena F. Grekova
Institute for Problems in Mechanical Engineering of the Russian Academy of Sciences, 199178, Bolshoy pr. V.O., 61, St. Petersburg, Russia, e-mail: elgreco@pdmi.ras.ru

Polina Grigoreva
Institute for Problems in Mechanical Engineering RAS, Bolshoy pr., 61, V.O., St. Petersburg 199178, Russian Federation, e-mail: polina.grigoreva239@gmail.com

Mikhail A. Guzev
Institute for Applied Mathematics of the Far Eastern Branch of the Russian Academy of Sciences, Vladivostok & Perm National Research Polytechnic University, Perm, Russian Federation, e-mail: guzev@iam.dvo.ru

Leonid A. Igumnov
National Research Lobachevsky State University of Nizhny Novgorod, 23 Gagarin Avenue, Nizhny Novgorod 603022, Russian Federation, e-mail: igumnov@mech.unn.ru

Elena A. Ivanova
Peter the Great St.Petersburg Polytechnic University, Dept. of Theoretical Mechanics and Mathematical Physics, Polytechnicheskaya, 29, 195251, Saint-Petersburg & Institute for Problems in Mechanical Engineering of Russian Academy of Sciences, Bolshoy pr. V.O., 61, 199178, Saint-Petersburg, Russian Federation, e-mail: elenaivanova239@gmail.com

Faizullah Jan
Gdańsk University of Technology, G. Narutowicza St. 11/12, 80-233 Gdańsk, Poland, e-mail: faizullah.jan@pg.edu.pl

Zia Javanbakht
Mechanical Engineering and Industrial Design, School of Engineering and Built Environment, Griffith University, 1 Parklands Dr, Southport 4215, Queensland, Australia, e-mail: zia.javanbakht@gmail.com

Simon Jeanne
National Graduate School of Engineering and Research Center, CAEN, France, e-mail: simon.jeanne@ecole.ensicaen.fr

Mikhail Karyakin
South Mathematical Institute, Vlkadikavkaz & Southern Federal University, Rostov-on-Don, Russian Federation, e-mail: karyakin@sfedu.ru

Ilya E. Keller
Institute of Continuous Media Mechanics of the Ural Branch of Russian Academy of Science, Academician Korolev St., 1, Perm 614013, Russian Federation, e-mail: kie@icmm.ru

Anna Knyazeva
Institute of Strength Physics and Materials Science of Siberian Branch Russian Academy of Sciences, 2/4, pr. Akademicheskii, Tomsk, 634055, Russian Federation, e-mail: anna-knyazeva@mail.ru

Alexey M. Kolesnikov
Southern federal university, Milchakova St. 8a, Rostov-na-Donu, 344090 & Don state technical university, Gagarin square 1, Rostov-na-Donu, 344000, Russian Federation, e-mail: Alexey.M.Kolesnikov@gmail.com

Alexander G. Kolpakov
SysAn (System Analysis in Engineering), A. Nevskogo str., 12a, 34, Novosibirsk, 630075, Russian Federation, e-mail: algk@ngs.ru

Elena I. Komarova
National Research Lobachevsky State University of Nizhny Novgorod, 23 Gagarin Avenue, Nizhny Novgorod 603022, Russian Federation, e-mail: komarova@mech.unn.ru

Violetta Konopińska-Zmysłowska
Gdańsk University of Technology, G. Narutowicza St. 11/12, 80-233 Gdańsk, Poland, e-mail: viokonop@pg.edu.pl

Aleksandr Yu. Konstantinov
National Research Lobachevsky State University of Nizhny Novgorod, 23 Gagarin Avenue, Nizhny Novgorod 603022, Russian Federation, e-mail: konstantinov@mech.unn.ru

Evgenii V. Kozhevnikov
Perm National Research Polytechnic University, Perm, Russian Federation, e-mail: kozhevnikov_evg@mail.ru

Ilya I. Kudish
ASME Fellow, ILRIMA Consulting LLC, Millersburg, MI 49759, USA,
e-mail: ilyakudish@gmail.com

Marcin Kujawa
Gdańsk University of Technology, G. Narutowicza St. 11/12, 80-233 Gdańsk, Poland, e-mail: mark@pg.edu.pl

Polina A. Lapina
Laboratory of Functionally Graded and Composite Materials, Research and Education Center "Materials", Don State Technical University, 1 Gagarin sq., 344000, Rostov-on-Don, Russian Federation, e-mail: polina_azarova86@mail.ru

Nguyen Le
Institute of Geophysics, Vietnam Academy of Science and Technology, Hanoi City, Cau Giay District, Bulding A8, 18 Hoang Quoc Viet Street, Vietnam,
e-mail: dinhnguyen081017@gmail.com

Anna Leonteva
Mechanical Engineering Research Institute of Russian Academy of Sciences, 85 Belinskogo str., Nizhny Novgorod 603024, Russian Federation,
e-mail: aleonav@mail.ru

Tse-Chun Liao
National Cheng Kung University, 1 University Road, Tainan, Taiwan 70101, R.O.C.,
e-mail: 3951242a@gmail.com

Yiming Liu
Technische Universität Berlin, Chair of Continuum Mechanics and Constitutive Theory, Institute of Mechanics, Berlin, Germany, e-mail: liuym0507@icloud.com

Sergey A. Lurie
Institute of Applied Mechanics RAS, Moscow, Russian Federation,
e-mail: salurie@mail.ru

Alexey Malkhanov
Mechanical Engineering Research Institute of Russian Academy of Sciences, 85 Belinskogo str., Nizhny Novgorod 603024, Russian Federation,
e-mail: alexey.malkhanov@gmail.com

Angela Madeo
Chair of Continuum Mechanics, Faculty of Architecture and Civil Engineering, TU-Dortmund, August-Schmidt-Str. 8, Dortmund, Germany,
e-mail: angela.madeo@tu-dortmund.de

Vladimir S. Metrikin
National Research Lobachevsky State University of Nizhny Novgorod, 23 Gagarin Avenue, Nizhny Novgorod 603022, Russian Federation,
e-mail: v.s.metrikin@mail.ru

Gennadi Mikhasev
Belarusian State University, 4 Nezavisimosti Ave., 220030 Minsk, Belarus,
e-mail: mikhasev@bsu.by

Anil Misra
Civil, Environmental and Architectural Engineering Department, The University of Kansas, 1530 W. 15th Street, Lawrence, KS 66045-7609, USA & International Research Center on the Mathematics and Mechanics of Complex Systems (M&MoCS), University of L'Aquila, Italy, e-mail: amisra@ku.edu

Alexandr Morozov
Institute of Mechanics, Chair of Continuum Mechanics and Constitutive Theory, Technische Universität Berlin, Einsteinufer 5, 10587 Berlin, Germany
e-mail: morozov@tu-berlin.de

Wolfgang H. Müller
Institute of Mechanics, Chair of Continuum Mechanics and Constitutive Theory, Technische Universität Berlin, Einsteinufer 5, 10587 Berlin, Germany,
e-mail: wolfgang.h.mueller@tu-berlin.de

Dmitry V. Nazarenko
Laboratory "Intelligent Electrical Agricultural Machines and Complexes", Don State Technical University, 1 Gagarin Square, 344000, Rostov-on-Don, Russian Federation, e-mail: dmnazar@mail.ru

Nelli Nazarenko
Institute of Strength Physics and Materials Science of Siberian Branch Russian Academy of Sciences, 2/4, pr. Akademicheskii, Tomsk, 634055, Russian Federation,
e-mail: nnelli@ispms.ru

Patrizio Neff
Lehrstuhl für Nichtlineare Analysis und Modellierung, Faculty of Mathematics, University Duisburg-Essen, Thea-Leymann Str. 9, 45127 Essen, Germany,
e-mail: patrizio.neff@uni-due.de

Sergey A. Nesterov
Southern Mathematical Institute, Vladikavkaz Scientific Center of Russian Academy of Sciences, 22 Markusa Str., 362027 Vladikavkaz, Russian Federation,
e-mail: 1079@list.ru

Andrey L. Nikolaev
Laboratory for Mechanics of Biomaterials, Don State Technical University, Rostov-on-Don 344000, Russian Federation, e-mail: andreynicolaev@eurosites.ru

Yuri V. Pavlov
Institute for Problems in Mechanical Engineering of Russian Academy of Sciences, 61 Bol'shoy, V.O., Saint Petersburg 199178, Russian Federation,
e-mail: yuri.pavlov@mail.ru

Luca Placidi
Faculty of Engineering, International Telematic University Uninettuno, Rome & International Research Center on the Mathematics and Mechanics of Complex Systems (M&MoCS), University of L'Aquila, Italy
e-mail: luca.placidi@uninettunouniversity.net

Vladimir Polyanskiy
Institute for Problems in Mechanical Engineering RAS, Bolshoy pr., 61, V.O., St. Petersburg 199178, Russian Federation, e-mail: vapol@mail.ru

Vladimir V. Poplygin
Perm National Research Polytechnic University, Perm, Russian Federation,
e-mail: poplygin@bk.ru

Alexey V. Porubov
Institute for Problems in Mechanical Engineering, Bolshoy 61, V.O., Saint-Petersburg, Russian Federation, e-mail: alexey.porubov@gmail.com

Sergey I. Rakin
Siberian Transport University, Dusi Koval'chuk Str., 191, Novosibirsk, 630049, Russia Federation, e-mail: rakinsi@ngs.ru

Leonardo A. Perez Ramirez
Chair of Continuum Mechanics, Faculty of Architecture and Civil Engineering, TU-Dortmund, August-Schmidt-Str. 8, Dortmund, Germany,
e-mail: leonardo.perez@tu-dortmund.de

Wojciech Pietraszkiewicz
Faculty of Civil and Environmental Engineering, Gdańsk University of Technology, G. Narutowicza St. 11/12, 80-233 Gdańsk, Poland, e-mail: wpietrasz@wp.pl

Evgenii P. Riabokon
Perm National Research Polytechnic University, Perm, Russian Federation,
e-mail: riabokon.evgenii@gmail.com

Wihelm Rickert
Institute of Mechanics, Chair of Continuum Mechanics and Constitutive Theory, Technische Universität Berlin, Einsteinufer 5, 10587 Berlin, Germany,
e-mail: rickert@tu-berlin.de

Gianluca Rizzi
Chair of Continuum Mechanics, Faculty of Architecture and Civil Engineering, TU-Dortmund, August-Schmidt-Str. 8, Dortmund, Germany,
e-mail: gianluca.rizzi@tu-dortmund.de

Evgeniy V. Sadyrin
Laboratory for Mechanics of Biomaterials, Don State Technical University, Rostov-on-Don 344000, Russian Federation, e-mail: e.sadyrin@sci.donstu.ru

Jaime Heman Espinoza Sandoval
Universidad de Lima, Instituto de Investigación Científica, Santiago de Surco, Peru, e-mail: jhespino@ulima.edu.pe

Bekir Cagri Sarar
University of L'Aquila, International Research Center on Mathematics and Mechanics of Complex Systems, L'Aquila, Italy, e-mail: bekircagri.sarar@graduate.univaq.it

Leah L. Sharipova
Institute for Problems in Mechanical Engineering RAS, Bolshoy pr., 61, V.O., St. Petersburg, 199178, Russian Federation, e-mail: sleah07@gmail.com

Denis N. Sheydakov
Southern Scientific Centre of the Russian Academy of Sciences (SSC RAS), Chekhov Ave. 41, 344006 Rostov-on-Don, Russian Federation, e-mail: sheidakov@mail.ru

Milad Shirani
Department of Mechanical Engineering, University of California, Berkeley, CA 94720, USA, e-mail: milad_shirani@berkeley.edu

Sergei Shubin
Institute for Problems in Mechanical Engineering of Russian Academy of Sciences, V.O., Bol'shoy pr. 61, 199178 St. Petersburg, Russian Federation, e-mail: shubinsn@gmail.com

Lukasz Smakosz
Gdańsk University of Technology, G. Narutowicza St. 11/12, 80-233 Gdańsk, Poland, e-mail: luksmako@pg.edu.pl

Yury O. Solyaev
Institute of Applied Mechanics RAS, Moscow, Russian Federation, e-mail: yos@iam.ras.ru

David J. Steigmann
Department of Mechanical Engineering, University of California, Berkeley, CA 94720, USA, e-mail: dsteigmann@berkeley.edu

Mezhlum Sumbatyan
Vorovich Institute of Mathematics, Mechanics and Computer Science, Southern Federal University, 344090, Milchakova Street 8a, Rostov-on-Don, Russian Federation, e-mail: masumbatyan@sfedu.ru

Czesław Szymczak
Gdańsk University of Technology, G. Narutowicza St. 11/12, 80-233 Gdańsk, Poland, e-mail: szymcze@pg.edu.pl

Kai-Wen Tan
National Cheng Kung University, 1 University Road, Tainan, Taiwan 70101, R.O.C., e-mail: n66091427@gs.ncku.edu.tw

Chuong Anthony Tran
Università degli Studi dell'Aquila, Dipartimento di Ingegneria Civile, Edile-Architettura, Ambientale (DICEAA) & Università degli Studi dell'Aquila, International Research Center for the Mathematics and Mechanics of Complex Systems (M&MoCS), L'Aquila, Italy, e-mail: tcanth@outlook.com

Emilio Turco
Università degli Studi di Sassari, Dipartimento di Architettura, design e urbanistica (DADU), Alghero & Università degli Studi dell'Aquila, International Research Center for the Mathematics and Mechanics of Complex Systems (M&MoCS), L'Aquila, Italy, e-mail: eturco@uniss.it

Francisco James León Trujillo
Universidad de Lima, Instituto de Investigación Científica, Santiago de Surco, Peru, e-mail: fleon@ulima.edu.pe

Vyacheslav Tsybulin
Vorovich Institute of Mathematics, Mechanics and Computer Science, Southern Federal University, 344090, Milchakova Street 8a, Rostov-on-Don, Russian Federation, e-mail: vgcibulin@sfedu.ru

Mikhail S. Turbakov
Perm National Research Polytechnic University, Perm, Russian Federation, e-mail: msturbakov@gmail.com

Andrey S. Vasiliev
Laboratory of Functionally Graded and Composite Materials, Research and Education Center "Materials",Don State Technical University, 1 Gagrina sq., 344001, Rostov-on-Don, Russian Federation, e-mail: andre.vasiliev@gmail.com

Alexander O. Vatulyan
Institute of Mathematics, Mechanics and Computer Sciences named after I.I. Vorovich, Southern Federal University, 8a Mil'chakova Str., 344090 Rostov-on-Don & Southern Mathematical Institute, Vladikavkaz Scientific Center of Russian Academy of Sciences, 22 Markusa Str., 362027 Vladikavkaz, Russian Federation, e-mail: aovatulyan@sfedu.ru

Elena N. Vilchevskaya
Institute for Problems in Mechanical Engineering of the Russian Academy of Sciences, Bol'shoy pr. 61, V.O., 199178 St. Petersburg, Russian Federation, e-mail: vilchevska@gmail.com

Andrey I. Volkov
National Research Lobachevsky State University of Nizhny Novgorod, 23 Gagarin Avenue, Nizhny Novgorod 603022, Russian Federation, e-mail: pmptmvgavt@yandex.ru

Ivan A. Volkov
National Research Lobachevsky State University of Nizhny Novgorod, 23 Gagarin Avenue, Nizhny Novgorod 603022, Russian Federation,
e-mail: pmptmvgavt@yandex.ru

Sergei S. Volkov
Laboratory of Functionally Graded and Composite Materials, Research and Education Center "Materials", Don State Technical University, 1 Gagrin sq., 344000, Rostov-on-Don, Russian Federation, e-mail: fenix_rsu@mail.ru

Yun-Che Wang
National Cheng Kung University, 1 University Road, Tainan, Taiwan 70101, R.O.C., e-mail: yunche@ncku.edu.tw

Karol Winkelmann
Gdańsk University of Technology, G. Narutowicza St. 11/12, 80-233 Gdańsk, Poland, e-mail: karolwin@pg.edu.pl

Mingxiu Xu
University of Science and Technology Beijing, Department of Applied Mechanics, Beijing, P.R. China, e-mail: xumx@ustb.edu.cn

Oksana V. Yavruyan
Institute of Mathematics, Mechanics and Computer Sciences named after I.I. Vorovich, Southern Federal University, 8a Mil'chakova Str., 344090 Rostov-on-Don & Southern Mathematical Institute, Vladikavkaz Scientific Center of Russian Academy of Sciences, 22 Markusa Str., 362027 Vladikavkaz, Russian Federation, e-mail: oyavruyan@sfedu.ru

M. Erden Yildizdag
University of L'Aquila, International Research Center on Mathematics and Mechanics of Complex Systems, L'Aquila, Italy & Istanbul Technical University, Faculty of Naval Architecture and Ocean Engineering, Istanbul, Turkey
e-mail: yildizdag@itu.edu.tr

Victor O. Yurov
Institute of Mathematics, Mechanics and Computer Sciences named after I.I. Vorovich, Southern Federal University, 8a Mil'chakova Str., 344090 Rostov-on-Don & Southern Mathematical Institute, Vladikavkaz Scientific Center of Russian Academy of Sciences, 22 Markusa Str., 362027 Vladikavkaz, Russian Federation, e-mail: vyurov@sfedu.ru

Leonid M. Zubov
Southern Federal University, Russian Federation, e-mail: lmzubov@sfedu.ru

Chapter 1

Effects of 3-D Printing Infill Density Parameter on the Mechanical Properties of PLA Polymer

Reza Afshar, Simon Jeanne, and Bilen Emek Abali

Abstract This study presents some results on the mechanical behavior of polylactide (PLA) material, produced using the fused deposition modeling (FDM) additive manufacturing technique. We investigate the effect of infill density on the mechanical properties of PLA specimens. We used tensile specimens, prepared according to ISO 527-2 standard, and tested them by a universal testing machine with analysis by means of digital image correlation (DIC) method. The results in terms of UTS and nominal strain at break of PLA material are presented. They demonstrate a significant impact of infill density on material behavior of PLA specimens, as expected. Yet the effect is nonlinear that is indeed valuable to understand. As infill density increases, from 10% to 100%, the nominal strain at break decreases from about 2.1% to 1.2%, respectively. In other words, the material becomes more ductile by decreasing the infill density of PLA material, which is possible to justify with an effect of the microstructure created by the infill density. There is a transition of this observed behavior, from being more ductile to more brittle, by increasing the infill density of the PLA specimens.

Key words: 3-D printing · Infill density · Digital image correlation · PLA polymer · Ultimate tensile strength

Reza Afshar · Bilen Emek Abali
Uppsala University, Division of Applied Mechanics, Department of Materials Science and Engineering, Uppsala, Sweden,
e-mail: reza.afshar@angstrom.uu.se, bilenemek@abali.org

Simon Jeanne
National Graduate School of Engineering and Research Center, CAEN, France,
e-mail: simon.jeanne@ecole.ensicaen.fr

H. Altenbach et al. (eds.), *Sixty Shades of Generalized Continua*,
Advanced Structured Materials 170, https://doi.org/10.1007/978-3-031-26186-2_1

1.1 Introduction

Additive manufacturing increases the design freedom in such a way that engineers may create more sophisticated geometries with internal substructures. The manufacturing is layer-by-layer production such that the CAD model is prepared by using a so-called slicer software. In this way, not only standard topologies but also internally void structures are possible. Innovative design concepts [1]-[7] has been suggested in the literature by means of advanced fabrication techniques [8]-[12]-[14], for historical remarks, we refer to [15].

Layer-by-layer fabrication approach allows to create sophisticated structures with internal holes. One famous feature is "infill density" or "infill ratio", where the outer shell is printed by a chosen wall thickness and inside is filled with a given ratio by means of a repeating microstructure. We demonstrate this feature in Fig. 1.1 by using a (rectangular) grid as the microstructure. This feature is often used since its generation is automatized in slicer programs such as Cura, Slic3r, and similar. Apparently, this feature reduces the weight significantly yet the mechanical response changes as well. Despite the common desire that the stiffness is simply reduced linearly by the porosity, experimental observation declares that the relation is more challenging [16]. By incorporating a micrustructure at the same length scale as the geometry itself, infill density may generate a multiscale material with an isotropic behavior [17]-[19] as well as possible novel mathematical models as presented in [20]-[23].

The structure of this paper is as follows: First, the concept of additive manufacturing (AM) with specific focus on fused deposition modeling (FDM) technique is briefly presented. In Sect. 1.3, the experimental procedure for obtaining the ultimate tensile strength (UTS) of PLA specimens is explained in details. Then in Sect. 1.4, the results in terms of ultimate tensile strength and nominal strain at break of PLA material are presented. Finally, the main outcome and its connection to microstructure related mechanical response is discussed.

1.2 Additive Manufacturing

Additive manufacturing (AM) is based on constructing the structure layer-by-layer by adding material. As opposed to the subtractive methods like machining parts by removing material, it uses just the amount of the material reducing waste production. Moreover, it allows to create more sophisticated or even with inner holes design

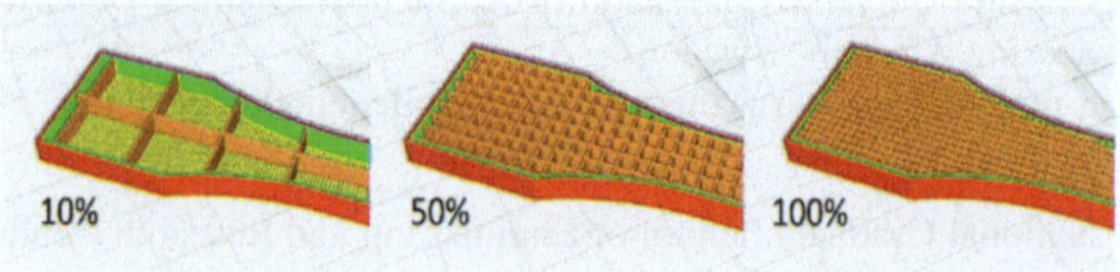

Fig. 1.1 Different infill densities of specimens, namely 10%, 50%, and 100%.

parts. In other words, AM increases the design freedom such that this manufacturing techniques is expected to broaden the design perspective in the near future.

Various methods exist for polymers and metals. Among the methods used for polymers, herein, we concetrate on the method of fused deposition modeling (FDM). The process is to fuse layers of materials together according to a defined blueprint to create an object. For this purpose, an FDM 3D printer was used, which allows the production of three-dimensional objects by depositing molten polymer. Therefore, only thermoplastic polymers are possible to 3D print with FDM printers. Made of renewable sources, PLA is often used in FDM 3D printers. PLA (polylactic acid) is a type of polyester synthesized from fermented plant starch (sugarcane or corn or similar). Therefore, during melting and printing, no toxic gases are released. Its processing is fairly simple and it is fully recyclable.

There are various stages to be carried out in order to arrive at the final part: model the object, choose the parameters, prepare the print code, and finally manufacture the structure. Especially determination of process parameters is key for a successful 3D print. Previous research on FDM based manufactured materials, such as PLA, show that infill density, layer height and infill pattern have the highest impact in mechanical properties of finished parts [26].

1.3 Materials and Methods

In this study, PLA is used as material for printing and the ISO 527-2 standard is chosen for the geometry of 3D printing specimens, for a discussion of this standard to be used in 3D printing, we refer to [27]. The drawing of the specimen is depicted in Fig. 1.2 and its specification is compiled in Table 1.1.

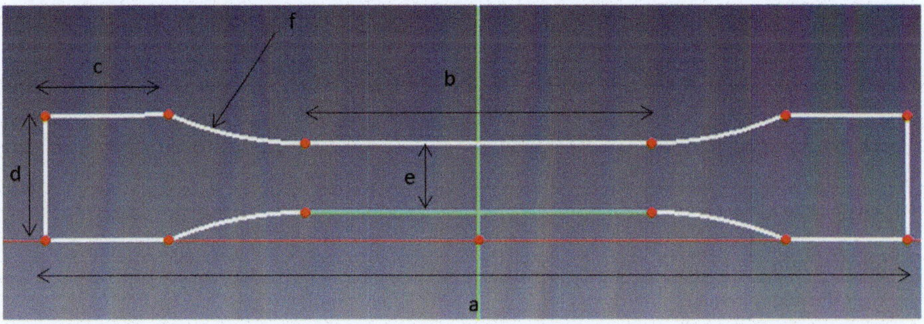

Fig. 1.2: Tensile test specimen geometry and its specifications based on the ISO 527-2 standard.

Table 1.1: Specimen geometric dimensions according to the ISO 527-2 standard with the notation as seen in Fig. 1.2.

Description	Notation	Value
Total length	a	150.00 mm
Tab width	d	21.70 mm
Thickness	t	3.80 mm
Gauge length	b	60.00 mm
Gauge width	e	12.00 mm
Tab length	c	21.40 mm
Filet radius	f	R60.00 mm

1.3.1 Fused Deposition Modeling (FDM)

As the slicer software, we have employed open-source Ultimaker Cura 4.13.1. We stress the importance of process parameters, hence, for securing the repeatibility of this study, we compile them in Table 1.2 used for creating the G-code. As an FDM 3D printer, we have used Ender 3 v2 equipped with a bed-leveler and hot-end. The firmware has been replaced to Klipper running on Raspberry Pi 3 with an interface of Octoprint for interpreting G-code. We use Klipper libraries to control pressure for increased accuracy on curvatures as well as bed-leveling in the configuration file. All specimens were printed with the same parameters, all contours are avoided in 3D printing in order to secure layers along the unidirectional configuration of the structure. All layers were unidirectional with 0° orientation, as visible in Fig. 1.3. For printing the structure with a unidirectional manner, we used line patterns, which steers the layers and thus generates the fibers—fiber is used to describe line pattern in 3D printing jargon. These fibers were not connected to the endpoints, the parameter for connecting infill lines was switched off.

Table 1.2: Process parameters of 3D printing.

Parameter	Value
Layer height	0.2 mm
Wall thickness	1.2 mm
Top/bottom thickness	0.8 mm
Nozzle temperature	210 °C
Build plate temperature	60 °C
Print speed	50 mm/s

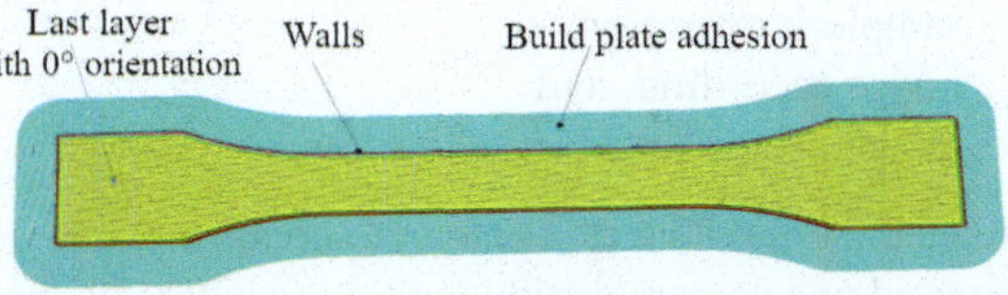

Fig. 1.3 Optimized setup, the travel lines are along the outer surface of the specimen.

1.3.2 Unidirectional Tensile Tests

Prior to mechanical tests, all specimens were preserved at 40°C in a dehydrator against the moisture uptake. Uniaxial tensile tests were performed with a Shimadzu AGS-X universal (electromechanical) testing machine. The tensile equipment and the test set-up is shown in Fig. 1.4. Experiments were conducted by steering the displacement with a ramp speed of 1 mm/min. This choice is justified by additional tests (not shown here) in order to ensure that the material is purely elastic as well as inertia does not play a role in this speed. So the quasi-static experiment leads to elastic results. The strain field was obtained by using digital image correlation (DIC) method (also called digital speckle photography), which is explained in what follows.

1.3.3 Digital Image Correlation Method

The DIC method is a robust non-contact technique for measuring material deformation [28]-[30]. Digital speckle photography is well suited for measurements on small objects [31]. An open source 2D DIC Matlab code ncorr [32] was used for image analysis in this study. The basic principle of DIC is to record a series of images of the sample during deformation, and then to calculate the degree of displacement from an image analysis. The strain field is then calculated from the displacement field numerically. Any DIC implementation consists of three main steps:

Fig. 1.4 Unidirectional testing machine equipment in the Ångström Laboratory, Uppsala University.

- sample preparation,
- image recording, and
- image processing.

Since the uniaxial test is known to create a planar deformation for isotropic materials, we used one camera setting leading to a 2D DIC analysis. For a better image analysis, the sample surface is enriched with a random texture before the measurement. This layer has no specific orientation and is non-periodic [33]. Such a layer, thin enough not to add an additional stiffness to the structure, is mainly achieved by spray painting. By spraying a black color on a white background, randomly, black and white speckles are generated. In this way, the software determines the speckles and calculate the displacement based on this texture. For taking photos, a digital camera with a constant focus and manual settings in the laboratory (diffuse, white) light was used. The image processing starts by dividing the image into subsets. The displacement is then calculated separately for each subset. The basic principle of DIC is to use cross correlation to identify the same subset in two subsequent images recorded during loading. The correlation function is a function of 2D displacement. The location of the peak of the correlation function gives the position of the deformed subset. An iterative cross-correlation algorithm is needed for more complex deformation fields. Deformation of the subset is accounted for by introducing a shape function that translates the pixel coordinates in the reference subset into coordinates in the image after deformation. The grey values are interpolated between the pixels in the image to achieve sub-pixel accuracy [33].

1.4 Results and Discussions

Postprocessing was performed by ncorr MATLAB code leading to values of the ultimate tensile strength (UTS) and Young's modulus. In total 48 specimens were printed, prepared, and tested. Different infill density of PLA samples, from 10% to 100%, was studied experimentally. The grid infill pattern was selected and the part oriented so that the grid is in the direction of the sample, to be depicted in Fig. 1.1.

As expected, the results show a significant impact of infill density on material behavior of PLA specimens. First, as infill density increases, from 10% to 100%, the nominal strain at break decreases from about 2.1% to 1.2% respectively. That means the material behavior becomes more brittle by increasing the infill density of PLA material. In other words, the introduced microstructure at the same length scale of the geometry, causes an increased ductility in the elastoplastic behavior. Especially for fatigue properties, this fact may be of importance to study further, whenever large deformations are sought after. Secondly, the strength of the samples, here defined as the maximum stress at failure, also increases, from about 19 MPa to 30 MPa, as infill density increases. It is due to the stiffness increase, where more material indeed enables to carry more load with increasing the infill density.

In order to make a comparative study of strength of specimens with different infill densities, the nominal strain of 1% is selected and the corresponding stresses for each infill density are obtained. The strain at 1% is roughly the threshold where we observe a deviation of linearity in the stress-strain curves. This deviation indicates a plastic deformation. Specially for the curves with the lower infill density this fact is more obvious in Fig. 1.5. Therefore, we compare the stress values at 1% after a normalization process. By dividing each stress value at 1% strain to the corresponding stress for 100% infill density, we obtain normalized stress values in each experiment. All values are compiled in Table 1.3. Moreover, we add the PLA consumption by emphasizing that the infill density is justified by weight optimization. This PLA consumption is given in meter and "saved PLA" is introduced by relating to the consumption in 100% infill density case. One should note that the corresponding stress value for 90% infill density, was higher than the stress for 100% infill density. Hence, the value of normalized stress for 90% infill density is left blank. A similar behavior is reported in [26], where fracture behavior of PLA samples have been investigated.

Furthermore, we use normalized stress values and calculate the loss of strength in order to demonstrate with saved PLA values as a function of infill density in Fig. 1.6. We may call economy of PLA instead of saved PLA in percent. In this way we observe a one-to-one relation between mechanical behavior and used material

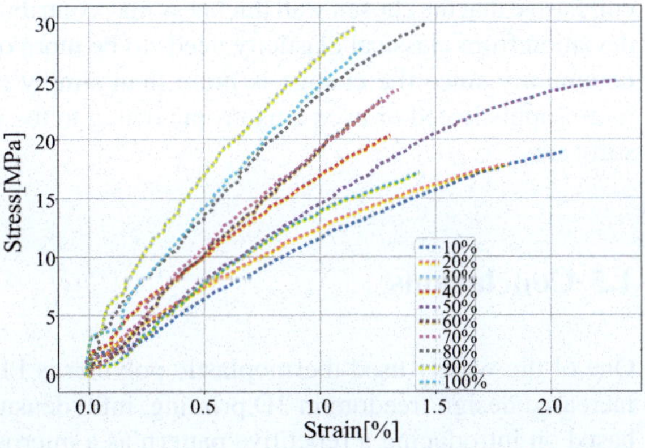

Fig. 1.5 Stress-strain curves for varying infill density for the same specimen structure out of PLA.

Table 1.3: Stress values and PLA material consumptions for different infill densities (*: Stress at 1% strain divided to stress for 100% infill density; **: PLA consumption relative to 100% infill density).

Infill density	**10%**	**20%**	**30%**	**40%**	**50%**	**60%**	**70%**	**80%**	**90%**	**100%**
Maximum stress in MPa	19.50	18.50	18.70	21.20	25.40	25.90	26.80	31.90	30.3	29.50
Stress at 1% strain in MPa	11.87	12.60	14.10	17.10	14.70	19.20	19.50	23.80	27.10	25.20
Normalized stress* in %	53	50	44	32	42	24	23	6	-	-
PLA consumption in m	1.56	1.81	2.06	2.31	2.56	2.82	3.07	3.32	3.57	3.75
Saved PLA** in %	58	52	45	38	32	25	18	11	5	-

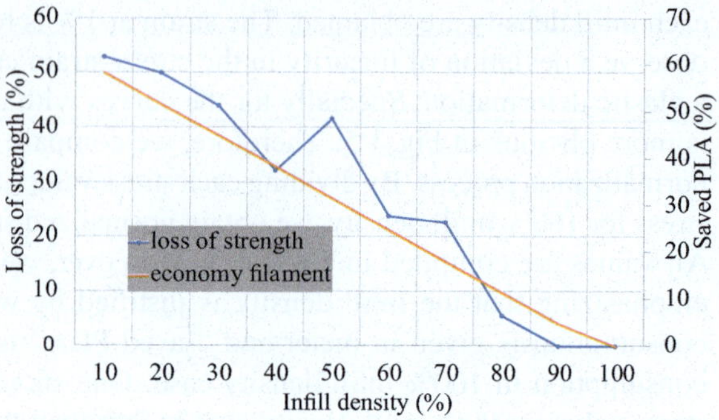

Fig. 1.6 Relation between loss of strength and saved PLA for different infill densities.

amount. According to Fig. 1.6, there is a clear compromise between the strength and amount of saving PLA material. For example, for the case of 10% infill density, there is a saving of PLA material by around 60% with compromise of strength about 53%. Despite the fact that the stress at the same strain level is demonstrating a linear relation, the maximum stress obtained from the UTS is showing a nonlinear relation in Table 1.3. Possible explanation of this mismatch is the brittle or ductile deformation varying because of the infill density. We leave a quantification to later studies, but emphasize that the chosen wall thickness may contribute as well. Mechanical response deviation from classical elasticity needs to be more openly discussed in 3D printing community since the change is more than simply the Young's modulus drop but more sophisticated or even anisotropic owing to the infill shape chosen in the slicer software.

1.5 Conclusions

One of the widely used thermoplastic polymer is PLA in 3D printing. Because of increased design freedom in 3D printing, infill density is also in standard use that is based on introducing a repetitive pattern as a microstructure and reducing amount of material inside of the structure. By experimentally investigating the change of material behavior owing to infill density, we have found out

- a nonlinear relation between UTS and infill density,
- a linear relation between stress at 1% representing an elastic deformation regime,
- a transition of material behavior, from being more ductile to more brittle, by increasing the infill density.

Indeed these results are for PLA material; however, they may be representative for other plastics as well.

References

[1] dell'Isola F, Lekszycki T, Pawlikowski M, Grygoruk R, Greco L (2015) Designing a light fabric metamaterial being highly macroscopically tough under directional extension: first experimental evidence, Zeitschrift für angewandte Mathematik und Physik **66**(6):3473-3498. DOI 10.1007/s00033-015-0556-4

[2] Turco E, dell'Isola F, Cazzani A, Rizzi NL (2016) Hencky-type discrete model for pantographic structures: numerical comparison with second gradient continuum models, Zeitschrift für angewandte Mathematik und Physik **67**(4):1-28. DOI 10.1007/s00033-016-0681-8

[3] Solyaev Y, Lurie S, Ustenko A (2019) Numerical modeling of a composite auxetic metamaterials using micro-dilatation theory, Continuum Mechanics and Thermodynamics **31**(4):1099–1107. DOI 10.1007/s00161-018-0730-y

[4] Turco E (2018) Discrete is it enough? The revival of Piola-Hencky keynotes to analyze three-dimensional elastica, Continuum Mechanics and Thermodynamics **30**(5):1039–1057. DOI 10.1007/s00161-018-0656-4

[5] Seppecher P, Alibert JJ, Lekszycki T, Grygoruk R, Pawlikowski M, Steigmann D, Giorgio I, Andreaus U, Turco E, Gołaszewski M, Rizzi N, Boutin C, Eremeyev VA, Misra A, Placidi L, Barchiesi E, Greco L, Cuomo M, Cazzani A, Corte AD, Battista A, Scerrato D, Zurba IE, Rahali Y, Ganghoffer JF, Müller W, Ganzosch G, Spagnuolo M, Pfaff A, Barcz K, Hoschke K, Neggers J, Hild F (2019) Pantographic metamaterials: an example of mathematically-driven design and of its technological challenges, Continuum Mechanics and Thermodynamics **31**(4):851-884. DOI 10.1007/s00161-018-0689-8

[6] Kwon J, Evans K, Ma L, Arnold D, Yildizdag ME, Zohdi T, Ritchie RO, Xu T (2020) Scalable electrically conductive spray coating based on block copolymer nanocomposites, ACS applied materials & interfaces **12**(7):8687-8694. DOI 10.1021/acsami.9b20817

[7] Giorgio I (2021) Lattice shells composed of two families of curved Kirchhoff rods: an archetypal example, topology optimization of a cycloidal metamaterial, Continuum Mechanics and Thermodynamics **33**(4):1063-1082. DOI 10.1007/s00161-020-00955-4

[8] Thompson MK, Moroni G, Vaneker T, Fadel G, Campbell RI, Gibson I, Bernard A, Schulz J, Graf P, Ahuja B, Martina F (2016) Design for additive manufacturing: Trends, opportunities, considerations, and constraints, CIRP Annals **65**(2):737-760. DOI 10.1016/j.cirp.2016.05.004

[9] Vangelatos Z, Yildizdag ME, Giorgio I, dell'Isola F, Grigoropoulos C (2021) Investigating the mechanical response of microscale pantographic structures fabricated by multiphoton lithography, Extreme Mechanics Letters **43**:101202. DOI 10.1016/j.eml.2021.101202

[10] dell'Isola F, Giorgio I, Pawlikowski M, Rizzi NL (2016) Large deformations of planar extensible beams and pantographic lattices: heuristic homogenization, experimental and numerical examples of equilibrium, Proceedings of the Royal Society A: Mathematical, Physical and Engineering Sciences **472**(2185), 20150,790. DOI 10.1098/rspa.2015.0790

[11] Seppecher P, Spagnuolo M, Barchiesi E, Hild F, Lekszycki T, Giorgio I, Placidi L, Andreaus U, Cuomo M, Eugster SR, Pfaff A, Hoschke K, Langkemper R, Turco E, Sarikaya R, Misra A, De Angelo M, D'Annibale F, Bouterf A, Pinelli X, Misra A, Desmorat B, Pawlikowski M, Dupuy C, Scerrato D, Peyre P, Laudato M, Manzari L, Göransson P, Hesch C, Hesch S, Franciosi P, Dirrenberger J, Maurin F, Vangelatos Z, Grigoropoulos C, Melissinaki V, Farsari M, Müller W, Abali BE, Liebold C, Ganzosch G, Harrison P, Drobnicki R, Igumnov L, Alzahrani F, Hayat T (2019) Advances in pantographic structures: design, manufacturing, models, experiments and image analyses, Continuum Mechanics and Thermodynamics **31**(4):1231-1282. DOI 10.1007/s00161-019-00806-x

[12] Giorgio I, Spagnuolo M, Andreaus U, Scerrato D, Bersani AM (2021) In-depth gaze at the astonishing mechanical behavior of bone: A review for designing bio-inspired hierarchical metamaterials, Mathematics and Mechanics of Solids **26**(7):1074-1103. DOI 10.1177/1081286520978

[13] Öchsner A, Altenbach H (Eds) (2015) Mechanical and Materials Engineering of Modern Structure and Component Design, Advanced Structured Materials, Vol 70, Springer, Cham. DOI 10.1007/978-3-319-19443-1

[14] Öchsner A, Altenbach H (Eds) (2016) Machining, Joining and Modifications of Advanced Materials, Advanced Structured Materials, Vol 61, Springer, Singapore. DOI 10.1007/978-981-10-1082-8

[15] dell'Isola F, Eremeyev VA (2018) Some Introductory and Historical Remarks on Mechanics of Microstructured Materials, In: Advances in Mechanics of Microstructured Media and Structures (Eds dell'Isola F, Eremeyev V), pp 1-20, Advanced Structured Materials, vol. 87, Springer International Publishing, Cham. DOI 10.1007/978-3-319-73694-5_1

[16] Özen A, Abali BE, Völlmecke C, Gerstel J, Auhl D (2021) Exploring the role of manufacturing parameters on microstructure and mechanical properties in fused deposition modeling (FDM) using PETG, Applied Composite Materials **28**(6):1799-1828. DOI 10.1007/s10443-021-09940-9

[17] Altenbach H (2005) Modelling of anisotropic behavior in fiber and particle reinforced composites, In: Multiscale Modelling of Damage and Fracture Processes in Composite Materials (Ed Sadowski T), CISM International Centre for Mechanical Sciences, Vol 474, pp 1-62. Springer, Vienna. DOI 10.1007/3-211-38102-3_1

[18] Nordmann J, Aßmus M, Altenbach H (2018) Visualising elastic anisotropy: theoretical background and computational implementation, Continuum Mechanics and Thermodynamics **30**(4):689-708. DOI 10.1007/s00161-018-0635-9

[19] Altenbach H, Eremeyev VA, Igumnov LA (2021) Multiscale Solid Mechanics: Strength, Durability, and Dynamics, Advanced Structured Materials, Vol 141, Springer, Cham. DOI 10.1007/978-3-030-54928-2

[20] Abali BE, Barchiesi E (2021) Additive manufacturing introduced substructure and computational determination of metamaterials parameters by means of the asymptotic homogenization, Continuum Mechanics and Thermodynamics **33**:993-1009. DOI 10.1007/s00161-020-00941-w

[21] Aydin G, Yildizdag ME, Abali BE (2022) Strain-gradient modeling and computation of 3-d printed metamaterials for verifying constitutive parameters determined by asymptotic homogenization, In: Theoretical Analyses, Computations, and Experiments of Multiscale Materials (Eds Giorgio I, Placidi L, Barchiesi E, Abali BE, Altenbach H), Advanced Structured Materials, Vol 175, pp 343-357, Springer, Cham. DOI 10.1007/978-3-031-04548-6_16
[22] Golaszewski M, Grygoruk R, Giorgio I, Laudato M, Di Cosmo F (2019 Metamaterials with relative displacements in their microstructure: technological challenges in 3D printing, experiments and numerical predictions, Continuum Mechanics and Thermodynamics **31**(4):1015-1034. DOI 10.1007/s00161-018-0692-0
[23] Aydin G, Sarar BC, Yildizdag ME, Abali BE (2022) Investigating infill density and pattern effects in additive manufacturing by characterizing metamaterials along the strain-gradient theory, Mathematics and Mechanics of Solids **27**(10):2002-2016. DOI 10.1177/10812865221100978
[24] Juritza A, Yang H, Ganzosch G (2019) Qualitative investigations of experiments performed on 3D-FDM-printed pantographic structures made out of PLA, In: Abali B, Altenbach H, dell'Isola F, Eremeyev V, Öchsner A (Eds) New Achievements in Continuum Mechanics and Thermodynamics, Advanced Structured Materials, Vol 108, pp 197-209, Springer, Cham. DOI 10.1007/978-3-030-13307-8_15
[25] Özen A, Abali BE, Völlmecke C, Gerstel J, Auhl D (2021) Exploring the role of manufacturing parameters on microstructure and mechanical properties in fused deposition modeling (FDM) using PETG, Applied Composite Materials **28**(6):1799-1828. DOI 10.1007/s10443-021-09940-9
[26] Milovanović A, Golubović Z, Trajković I, Sedmak A, Milošević M, Valean E, Marsavina L (2022) Influence of printing parameters on the eligibility of plane strain fracture toughness results for PLA polymer, Procedia Structural Integrity **41**:290-297. DOI 10.1016/j.prostr.2022.05.034
[27] Özen A, Auhl D, Völlmecke C, Kiendl J, Abali BE (2021) Optimization of manufacturing parameters and tensile specimen geometry for fused deposition modeling (FDM) 3D-Printed PETG, Materials **14**(10):2556. DOI 10.3390/ma14102556
[28] Peters W, Ranson W (1982) Digital imaging techniques in experimental stress analysis, Optical Engineering **21**(3):213,427. DOI 10.1117/12.7972925
[29] Chu T, Ranson W, Sutton MA (1985) Applications of digital-image-correlation techniques to experimental mechanics, Experimental Mechanics **25**(3):232-244. DOI 10.1007/BF02325092
[30] Vendroux G, Knauss W (1998) Submicron deformation field measurements: Part 2. improved digital image correlation, Experimental Mechanics **38**(2):86-92. DOI 10.1007/BF02321649
[31] Jernkvist L, Thuvander F (2001) Experimental determination of stiffness variation across growth rings in picea abies, Holzforschung **55**:309-317. DOI 10.1515/HF.2001.051

[32] Blaber J, Adair B, Antoniou A (2015) Ncorr: open-source 2d digital image correlation matlab software, Experimental Mechanics **55**(6):1105-1122. DOI 10.1007/s11340-015-0009-1

[33] Sutton MA, Orteu JJ, Schreier H (2009) Image Correlation for Shape, Motion and Deformation Measurements: Basic Concepts,Theory and Applications, Springer, New York. DOI 10.1007/978-0-387-78747-3

Chapter 2
Advance Approximate Analytical Solutions of the Contact Problem for an Inhomogeneous Layer

Sergei M. Aizikovich, Polina A. Lapina, and Sergei S. Volkov

Abstract The paper considers contact problems on the shear of the surface of an elastic layers that are inhomogeneous with depth and lie on a non-deformable and elastic base. Contact problems are reduced to solving integral equations. An analysis of the proximity of the kernel transforms of integral equations depending on the parameters of the problem is carried out. The solutions of integral equations are constructed by asymptotic methods, including the Wiener-Hopf method and the bilateral asymptotic method. The closeness of contact stresses is investigated. It is obtained, that when the value of the shear modulus at the lower boundary of the layer exceeds the value of the shear modulus at the layer surface by more than e^2 times, the divergence of contact stresses for layers on a non-deformable and elastic base does not exceed 5 percent.

Key words: Contact problem · Inhomogeneous material · Exponential shear modulus · Layer on the non-deformable base · Layer on the elastic base

2.1 Introduction

Inhomogeneous coatings of various structures have various applications and are of interest to researchers. The choice of adequate mathematical models for the calculation of inhomogeneous materials remains an important issue in modern mechanics [1]-[9].

The paper analyzes the equivalence of solutions for two models of a layer inhomogeneous with depth. The solution of the problems is carried out on the example of contact problems on the pure shear of the surface of an elastic layers inhomogeneous with depth on a non-deformable and elastic base. The exponential law of

Sergei M. Aizikovich · Polina A. Lapina · Sergei S. Volkov
Laboratory of Functionally Graded and Composite Materials, Research and Education Center "Materials", Don State Technical University, 1 Gagarin sq., 344000, Rostov-on-Don, Russian Federation,
e-mail: saizikovich@gmail.com, polina_azarova86@mail.ru, fenix_rsu@mail.ru

H. Altenbach et al. (eds.), *Sixty Shades of Generalized Continua*,
Advanced Structured Materials 170, https://doi.org/10.1007/978-3-031-26186-2_2

shear modulus change was chosen for the analysis, that is also considered in various studies [10]-[18]. The choice of the exponential law made it possible to obtain the kernels of the integral equations in an analytical form. Models of layers on a non-deformable and elastic base are close in a certain range of problem parameters. In this range, the proximity of solutions to contact problems is studied. The problems were solved approximately analytically using asymptotic methods using the simplest approximations of the kernel transforms of integral equations.

2.2 Statement of the Problem of a Shear of the Surface of an Inhomogeneous Layer

Let us consider the contact problem of pure shear of the surface of an inhomogeneous layer lying on a non-deformable or an elastic base by a strip punch with a flat base. The equation of the theory of elasticity in the case of antiplane deformation without taking into account the forces of friction has the form

$$\frac{\partial \sigma_{zx}}{\partial x} + \frac{\partial \sigma_{zy}}{\partial y} = 0, \tag{2.1}$$

where stresses $\sigma_{zx}(x,y), \sigma_{zy}(x,y)$ and deformation $w(x,y)$ are related by the relations

$$\sigma_{zx} = \mu(y)\frac{\partial w}{\partial x}, \quad \sigma_{zy} = \mu(y)\frac{\partial w}{\partial y}. \tag{2.2}$$

where $\mu(y)$ is shear modulus of a layer inhomogeneous by the vertical coordinate.

In the case of a layer lying on a non-deformable base (Fig. 2.1a), the boundary conditions of the contact problem have the form

$$w(x,0) = \varepsilon \quad |x| \leq a, \qquad w(x,H) = 0 \quad |x| < \infty \tag{2.3}$$

$$\sigma_{yz}(x,0) = \begin{cases} -\varphi(x) & |x| \leq a \\ 0 & a < |x| < \infty \end{cases} \tag{2.4}$$

where ε is the value, to which the punch is shifted, a is the half-width of the contact area, $\varphi(x)$ are contact stresses under the punch to be determined, H is the layer thickness.

A layer lying on an elastic base (Fig. 2.1b) is represented by an inhomogeneous half-space, the shear modulus of which is given as a piecewise function

$$\begin{cases} \mu(y) & 0 \leq y \leq H \\ \mu_\infty & H < y < \infty \end{cases} \tag{2.5}$$

where μ_∞ is elastic base shear modulus. In this case, the boundary conditions of the contact problem consist of conditions (2.3), (2.4) on the layer surface, to which the

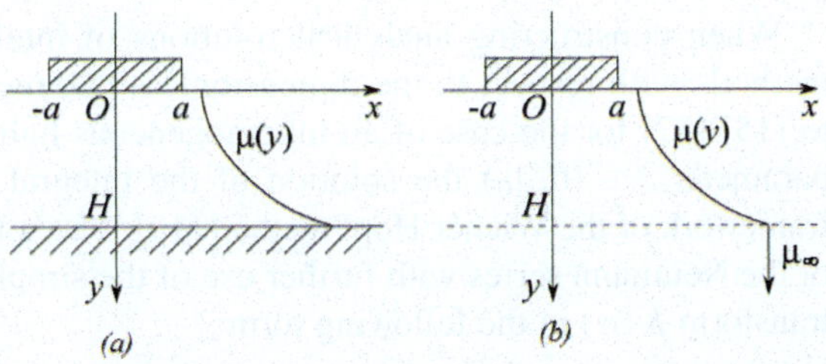

Fig. 2.1 Layer models on a non-deformable (a) and elastic (b) base.

condition of decreasing solution at infinity is added

$$\lim_{\sqrt{x^2+y^2}\to\infty}\left\{w,\frac{\partial w}{\partial x}\right\}=0. \tag{2.6}$$

2.3 Integral Equations of Contact Problems

The paper considers the case when the shear modulus of an inhomogeneous layer changes according to the exponential law

$$\mu(y)=\mu_0 e^{2dy},\quad 0\le y<\infty, \tag{2.7}$$

where μ_0 is the value of the shear modulus on the upper surface of the inhomogeneous layer, d is the value that characterizes the rate of change in the shear modulus with depth.

In this case, in problem (b), the shear modulus of the elastic base is $\mu_\infty=\mu_0 e^{2dH}$, and the shear modulus of the elastic base is e^{2dH} times larger than the shear modulus of the layer on the surface ($\mu_\infty/\mu_0=e^{2dH}$). For the exponential law of the shear modulus change, it is possible to construct solutions to contact problems in an analytical form. Using the integral Fourier transform, the solutions of problems are reduced to solving integral equations for unknown contact stresses.

For the problem (a), the integral equation in dimensionless form has the form

$$\int_{-1}^{1}\varphi(\xi)\mathrm{d}\xi\int_{-\infty}^{\infty}K(\alpha)e^{i\alpha\frac{\xi-x}{\lambda}}\mathrm{d}\alpha=2\pi a^{-1}\varepsilon\mu_0,\quad |x|\le 1,\quad \lambda=\frac{H}{a}, \tag{2.8}$$

$$K(\alpha)=\frac{1}{d_0+V\coth V},\quad V=\sqrt{d_0^2+\alpha^2},\quad d_0=Hd \tag{2.9}$$

Integral equation (2.8) is an integral equation of the Fourier convolution type of the first kind with a difference kernel with respect to unknown contact stresses $\varphi(\xi)$. The transform $K(\alpha)$ has the following asymptotic properties

$$K(\alpha)=|\alpha|^{-1}+O(\alpha^{-2})\text{ at }|\alpha|\to\infty,\quad K(\alpha)=K(0)+O(\alpha^2)\text{ at }|\alpha|\to 0, \tag{2.10}$$

When constructing analytical solutions of integral equation (2.8), asymptotic methods with respect to the dimensionless parameter λ are used in a similar way to [15]-[17] for the case of an inhomogeneous half-space. For small values of the parameter $\lambda \in (0, \lambda_0)$ the solution of the integral equation is constructed in the framework of the Wiener-Hopf method [19]. The solution is written as the zero term of the Neumann series with further use of the simplest approximation of the kernel transform $K(\alpha)$ in the following form

$$K(\alpha) = \frac{1}{\sqrt{A^2+\alpha^2}} \frac{B^2+\alpha^2}{C^2+\alpha^2}, \quad K(0) = \frac{1}{A}\frac{B^2}{C^2} \tag{2.11}$$

For large values of the parameter $\lambda \in (\lambda_0, \infty)$, as in [20], the solution of the integral equation is written as a double functional series in powers of the parameter λ.

In the case of problem (b), the integral equation for unknown contact stresses according to [18] is written in the same form as (2.8), (2.9), taking into account the notation

$$K(\alpha) = \frac{1}{|\alpha|} L(\alpha) \tag{2.12}$$

The function $L(\alpha)$ is obtained analytically and has the form

$$L(\alpha) = \frac{(|\alpha| - d_0)\sinh V + V\cosh V}{(|\alpha| + d_0)\sinh V + V\cosh V}, \quad V = \sqrt{d_0^2+\alpha^2}, \quad d_0 = Hd \tag{2.13}$$

The function $L(\alpha)$ given by expression (2.13) has the following asymptotic properties

$$L(\alpha) = 1 + O(|\alpha|^{-1}) \text{ at } |\alpha| \to \infty, \quad L(\alpha) = L(0) + O(|\alpha|) \text{ at } |\alpha| \to 0. \tag{2.14}$$

When solving the integral equation of problem (b), the bilateral asymptotic method [21]-[24] is used. The kernel transform is approximated by the following product

$$L(\alpha) = L_{\Pi}^{N}(\alpha) \equiv \prod_{i=1}^{N} \frac{\alpha^2 + A_i^2}{\alpha^2 + B_i^2}; (B_i - B_k)(A_i - A_k) \neq 0 \text{ where } i \neq k \tag{2.15}$$

Based on the proposed approximation, an approximate analytical solution of the integral equation is constructed, which is effective over the entire range of values of the dimensionless geometric parameter λ.

2.4 Numerical Analysis

Figures 2.2-2.4 show the kernel transforms of integral equations for two models of a layer that is inhomogeneous with depth at different values of the parameter $d_0 = 2, 1, 0.5$, which corresponds to cases where the shear modulus at the lower boundary of the layer (the shear modulus of the elastic base in problem b) is e^4, e^2, e times greater than the shear modulus at the layer surface respectively. For model (a), the graphs show functions $|\alpha|K(\alpha)$, for model (b), the graphs show functions $L(\alpha)$. As the α grows, the graphs $|\alpha|K(\alpha)$ for the model (a) and the graphs $L(\alpha)$ for the model (b) begin to coincide and tend to the value 1 at $\alpha \to \infty$. We consider $d_0 > 0$, which means that the shear modulus increases with depth. At $d_0 \to +0$, the inhomogeneous layer transforms into a homogeneous one. In this case, as the parameter d_0 decreases, models (a) and (b) cease to be similar, which is confirmed by the graphs of the transforms in Figs. 2.2-2.4 .

For values of the shear modulus parameter d_0, at which the relative difference between the kernel transforms $|\alpha|K(\alpha)$ and $L(\alpha)$ of the integral equations of problem (a) and problem (b) is small, the contact stresses are constructed. Note that thin layers are of interest from the point of view of applications. Figures 2.5 and 2.6 show the relative contact stresses $a\varepsilon^{-1}\mu_0^{-1}P^{-1}\varphi(\xi)$ for different values of the parameters d_0 and λ. Here P is the force acting on the punch. In the case of the model (a), the solution is constructed by the Wiener-Hopf method, which corresponds to the case of small values of the dimensionless parameter $\lambda \in (0, \lambda_0)$ (thin layers). In the case of the model (b), the solution is constructed by the bilateral asymptotic method.

Numerical results confirm the coincidence of contact stresses for the two layer models (on a non-deformable and elastic base) in the case when the problem parameters are such that the models are physically close to each other. Thus, at $d_0 \geq 1$, when the value of the shear modulus at the lower boundary of the layer exceeds the value of the shear modulus at the upper boundary of the layer by more than a e^2 times, the contact stresses constructed for the case of a layer on a non-deformable base are close to the contact stresses constructed for the case of a layer on an elastic base, and the error between solutions does not exceed 5 percent.

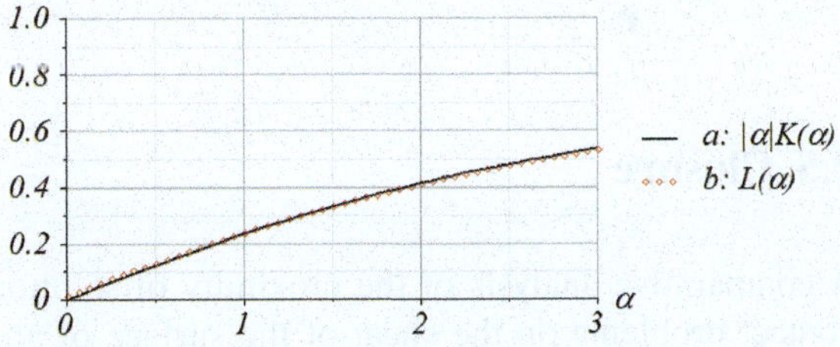

Fig. 2.2 $|\alpha|K(\alpha)$ for model (a) and $L(\alpha)$ for model (b) at $d_0 = 2, \mu_\infty/\mu_0 = e^4$, the solid black line - problem (a), red dots - problem (b).

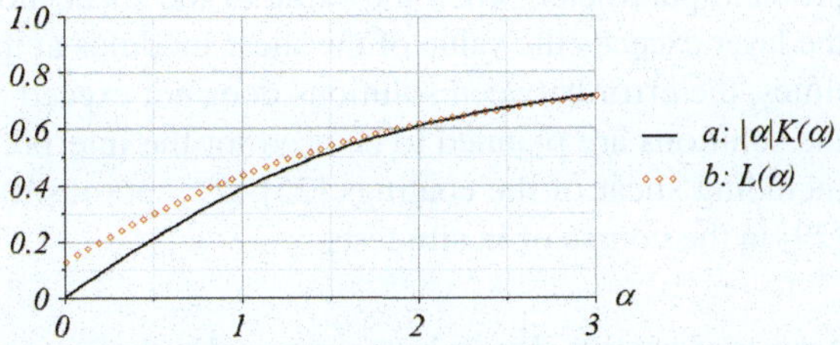

Fig. 2.3 $|\alpha|K(\alpha)$ for model (a) and $L(\alpha)$ for model (b) at $d_0 = 1, \mu_\infty/\mu_0 = e^2$, the solid black line - problem (a), red dots - problem (b).

Fig. 2.4 $|\alpha|K(\alpha)$ for model (a) and $L(\alpha)$ for model (b) at $d_0 = 0.5$, $\mu_\infty/\mu_0 = e$, the solid black line - problem (a), red dots - problem (b).

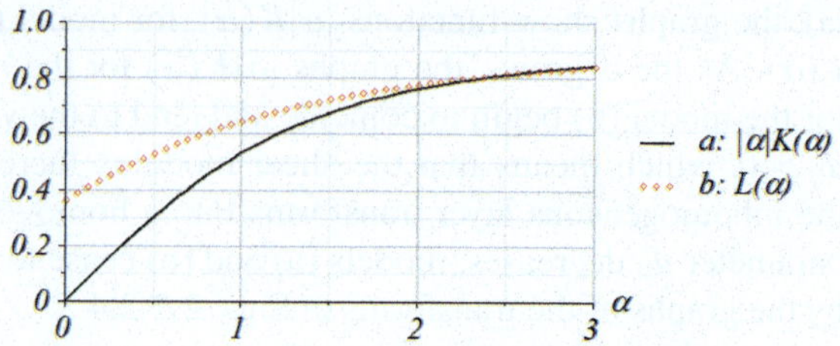

Fig. 2.5 Relative contact stresses $a\varepsilon^{-1}\mu_0^{-1}P^{-1}\varphi(\xi)$ at $d_0 = 1$, $\lambda = 1$, the solid black line - problem (a), the solid red line - problem (b).

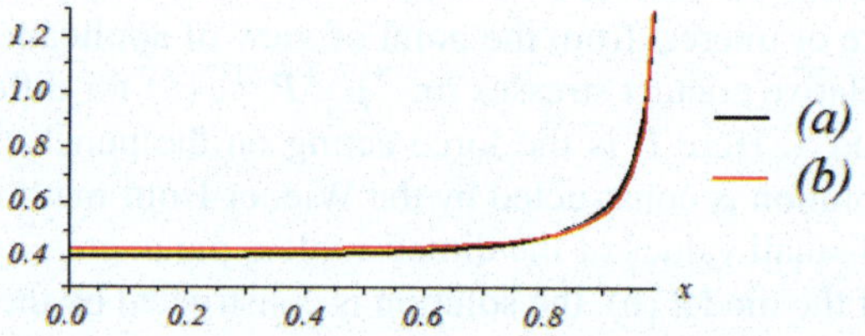

Fig. 2.6 Relative contact stresses $a\varepsilon^{-1}\mu_0^{-1}P^{-1}\varphi(\xi)$ at $d_0 = 2$, $\lambda = 0.5$, the solid black line - problem (a), the solid red line - problem (b).

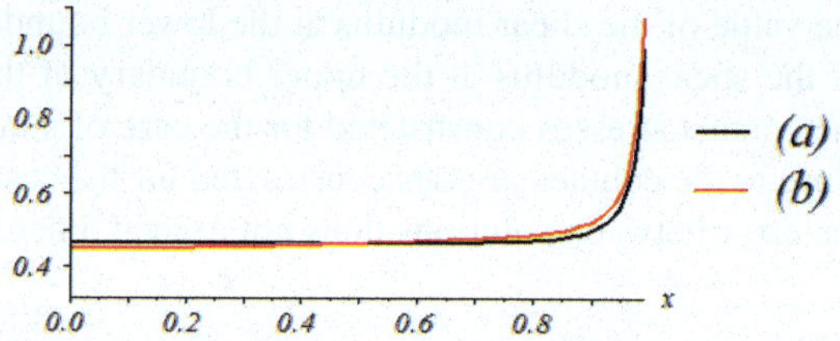

2.5 Closure

A comparative analysis of the proximity of solutions and the equivalence of two contact problems on the shear of the surface of an elastic layers inhomogeneous with depth on a non-deformable and elastic base is carried out. For the values of the problem parameter, when the value of the shear modulus at the lower boundary of the layer exceeds the value of the shear modulus at the layer surface by more than e^2 times, the error between solutions does not exceed 5 percent. In the further studies the solutions are planned to be used for the interpretation of the experimental data, including shear of the coatings [25]-[27], porous materials [28] and scaffold walls [29] in the course of in situ tests.

Acknowledgements The study was supported by the Russian Science Foundation grant No. 22-19-00732.

References

[1] Shiraki Y, Usami N (Eds) (2011) Silicon-Germanium (SiGe) Nanostructures: Production Properties and Applications in Electronics, Woodhead Publishing, Cambridge, UK.

[2] Miyamoto Y, Kaysser WA, Rabin BH, Kawasaki A, Ford RG (Eds) (1999) Functionally Graded Materials: Design, Processing and Applications, Springer, New York, NY, USA.

[3] Gibson RE (1967) Some results concerning displacements and stresses in a non-homogeneous elastic half-space, Geotechnique **17**(1):58–67.

[4] Awojobi AO (1976) On the hyperbolic variation of elastic modulus in a non-homogeneous stratum, Int J Solids Struct **12**:739–748.

[5] Altenbach H, Eremeyev VA (2009) Eigen-vibrations of plates made of functionally graded material, Comput Mater Continua **9**(2):153–177.

[6] Chen P, Chen S (2012) Contact behaviors of a rigid punch and a homogeneous half-space coated with a graded layer, Acta Mechanica **223**:563–577.

[7] Liu T-J, Wang Y-S, Zhang C-Z (2008) Axisymmetric frictionless contact of functionally graded materials, Arch Appl Mech **78**:267–282.

[8] Ke L-L, Wang Y-S (2007) Two-dimensional sliding frictional contact of functionally graded materials, Eur J Mech A/Solids **26**:171–188.

[9] Sadyrin EV (2022) Correlating the mechanical properties to the mineral density of brown spot lesion in dentine using nanoindentation and X-ray microtomography, In: Altenbach H, Eremeyev VA, Galybin A, Vasiliev A (Eds) Advanced Materials Modelling for Mechanical, Medical and Biological Applications, Advanced Structured Materials, Vol 155, pp 389–398, Springer, Cham.

[10] Guler MA, Erdogan F (2004) Contact mechanics of graded coatings, Int J Solids Struct **41**:3865–3889.

[11] Selvadurai APS, Singh BM, Vrbik J (1986) A Reissner-Sagoci problem for a non-homogeneous elastic solid, J Elast **16**:383–391.

[12] Selvadurai APS, Katebi A (2013) Mindlin's problem for an incompressible elastic half-space with an exponential variation in the linear elastic shear modulus, Int J Eng Sci **65**:9–21.

[13] Tokovyy Y, Ma C-C (2015) An analytical solution to the three-dimensional problem on elastic equilibrium of an exponentially-inhomogeneous layer, J Mech **31**(5):545–555.

[14] Selvadurai APS, Katebi A (2016) The Boussinesq–Mindlin problem for a non-homogeneous elastic halfspace, Zeitschrift für angewandte Mathematik und Physik **67**:68.

[15] Zelentsov VB, Lapina PA, Mitrin BI, Kudish II (2019) An antiplane deformation of a functionally graded half-space, Continuum Mechanics and Thermodynamics **34**:909-920.

[16] Zelentsov VB, Lapina PA, Mitrin BI, Eremeyev VA (2020) Characterization of the functionally graded shear modulus of a half-space, Mathematics **2020**(8):644.

[17] Zelentsov VB, Lapina PA, Zagrebneva AD (2022) Method for determining the parameters of the exponential shear modulus of a functional-gradient half-space, In: Altenbach H, Eremeyev VA, Galybin A, Vasiliev A (Eds) Advanced Materials Modelling for Mechanical, Medical and Biological Applications, Advanced Structured Materials, Vol 155, pp 441-460, Springer, Cham.
[18] Aizikovich SM (1978) Shear by a stamp of an elastic inhomogeneous half-space of a special form, Izvestiya AN SSSR MTT (5):74–80.
[19] Noble B (1958) Methods Based on the Wiener-Hopf Technique for the Solution of Partial Differential Equations, Pergamon Press, London, UK.
[20] Aleksandrov VM, Belokon' AV (1967) Asymptotic solution of a class of integral equations and its application to contact problems for cylindrical elastic bodies, J Appl Math Mech **31**(4):718–724.
[21] Aizikovich SM (1990) An asymptotic solution of a class of coupled equations, J Appl Math Mech **54**(5):719–724.
[22] Aizikovich SM, Vasiliev AS (2013) A bilateral asymptotic method of solving the integral equation of the contact problem of the torsion of an elastic half-space inhomogeneous in depth, J Appl Math Mech **77**(1):91–97.
[23] Volkov S, Aizikovich S, Wang Y-S, Fedotov I (2013) Analytical solution of axisymmetric contact problem about indentation of a circular indenter into a soft functionally graded elastic layer, Acta Mechanica Sinica **29**(2):196–201.
[24] Vasiliev AS, Volkov SS, Sadyrin EV, Aizikovich SM (2020) Simplified analytical solution of the contact problem on indentation of a coated half-space by a conical punch, Mathematics **8**(6):983.
[25] Sadyrin E, Karotkiyan R, Sushentsov N, Stepanov S, Zabiyaka I, Kislyakov E, Litvinenko A (2020) Mechanical properties derived by spherical indentation of TiN coating deposited by a method combining magnetron sputtering and arc evaporation, In: Parinov IA, Chang S-H, Long BT (Eds) Advanced Materials, pp 85-95, Springer, Cham.
[26] Burlakova VE, Tyurin AI, Drogan EG, Sadyrin EV, Pirozhkova TS, Novikova AA, Belikova MA (2019) Mechanical properties and size effects of self-organized film, Journal of Tribology **141**(5):051601.
[27] Zhilin DA, Lyanguzov NV, Nikolaev LA, Pushkariov VI, Kaidashev EM, Kaidashev VE (2014) Photoelectric properties of MSM structure based on ZnO nanorods, received by thermal evaporation and carbothermal syntesis, Journal of Physics Conference Series **541**(1):012038.
[28] Sadyrin EV, Mitrin BI, Yogina DV, Swain MV (2021) Preliminary study of distribution of mechanical properties and mineral density by depth of liquid saturated carious dentine, In: IOP Conference Series: Materials Science and Engineering **1029**(1):012056.
[29] Mitrin BI, Chapek SV, Sadyrin EV, Swain MV (2021) Mechanical properties and failure mechanisms of 3D-printed PLA scaffolds: A preliminary study, In: IOP Conference Series: Materials Science and Engineering **1029**(1):012074.

Chapter 3
The Direct Approach for Plates Considering Hygrothermal Loading and Residual Kinetics

Marcus Aßmus, Zia Javanbakht, and Holm Altenbach

Abstract The direct approach for plates is based on Cosserat continuum theory, or rather on the Cosserat surface. The same principles as used in classical continuum mechanics are applied whereby a restriction to two-dimensional body manifolds is imposed. At the same time independent rotational degrees of freedom are introduced. Pavel Andreevich Zhilin then proposed a reduced treatment that is widely used in engineering, especially in composite mechanics. However, the approach still lacks important extensions to include significant quantities influencing the mechanical behaviour. These quantities result from moisture exposure, temperature changes and initial tense in the material. We here delineate extensions based on physical justifications while discussing three-dimensional causation and two-dimensional impact. A resulting set of equations for the combined loading is derived and discussed.

Key words: Generalized plate theory · Hygroscopic impact · Thermal effects · Residual kinetics

3.1 Introduction

We consider a rectangular plate of length L_1, width L_2 and constant thickness H with arbitrary supports at four edges, cf. Fig. 3.1. The plate is subjected to moisture exposure and temperature change. Within the plate, fields exist that result from

---

Marcus Aßmus · Holm Altenbach
Chair of Engineering Mechanics, Institute of Mechanics, Faculty of Mechanical Engineering, Otto von Guericke University Magdeburg, Universitätsplatz 2, 39106 Magdeburg, Germany,
e-mail: marcus.assmus@ovgu.de, holm.altenbach@ovgu.de

Zia Javanbakht
Mechanical Engineering and Industrial Design, School of Engineering and Built Environment, Griffith University, 1 Parklands Dr, Southport 4215, Queensland, Australia,
e-mail: zia.javanbakht@gmail.com

H. Altenbach et al. (eds.), *Sixty Shades of Generalized Continua*,
Advanced Structured Materials 170, https://doi.org/10.1007/978-3-031-26186-2_3

residual quantities, namely residual forces and moments. Such influences are to be treated in the context of the direct approach for homogeneous thin-walled structures. A first attempt was successfully utilized in [1] where a three-layered plate was considered with reduced influence quantities. We here sharpen our reflections and reduce ourselves to a single layer.

The direct approach for plates is strongly associated with the ideas of Zhilin, cf. [2], even if other scientists have had similar thoughts before, cf. Cosserat & Cosserat [3] and Ericksen & Truesdell [4]. We here reduce our considerations to an initially plane (undeformed) surface on which in analogy to the procedure at classical three-dimensional continua, kinematic and kinetic measures are defined and constitutive relations are introduced. The resulting theory is an analogy to a generalized plate theory in accordance with the works of Kirchhoff [5], Reissner [6], and Mindlin [7]. The generalization is meant in the spirit of [8], i.e., where in-plane, out-of-plane and transverse shear state of a plate may superimpose.

We are considering a monolithic conventional material with isotropic and purely elastic material behaviour. Thereby, we restrict ourselves to the small deformation setting. We furthermore consider

- the thermo-static case ($\Delta\dot{\theta} = 0 \wedge \dot{\square} = 0 \; \forall \, \square \in \{v_\alpha, w, \varphi_\alpha\}$),
- the hygro-static case ($\Delta\dot{\mu} = 0 \wedge \dot{\square} = 0 \; \forall \, \square \in \{v_\alpha, w, \varphi_\alpha\}$), and what we call
- the residual-static case ($\dot{\square}^0_{\alpha\beta} = 0 \forall \square \in \{N, L\} \wedge \dot{\square} = 0 \; \forall \, \square \in \{q^0_\alpha, v_\alpha, w, \varphi_\alpha\}$)

for the entire plate at any time. Herein, θ, μ, $N^0_{\alpha\beta}$, $L^0_{\alpha\beta}$, q^0_α, v_α, w, φ_α are the degrees of freedom, which we will specify in more detail in the further course. Their consideration results in a description for two-dimensional body manifolds which is similar to the description of classical three-dimensional Cauchy continuum in a hygrothermal environment with residual stresses. With this analogy we formulate a generalized plate theory for hygro-thermo-elastostatics with preload.

Notation: Throughout present treatise, a direct tensor notation is used. First- and second-order tensors are denoted by lowercase and uppercase bold letters, e.g., $\boldsymbol{a}$ and $\underset{\sim}{\boldsymbol{A}}$, respectively. Fourth-order tensors are designated by uppercase calligraphic letters, e.g. $\mathcal{A}$. Sets groups and spaces are written in blackboard bold letters, e.g. $\mathbb{Sym}$. If index notation is required, we will make use of implicit summation over indices 1 to 3 in case of latin letters, and over indices 1 to 2 in case of greek letters. Furthermore, $\cdot$, $:$, $\otimes$, and $\times$ are the scalar product, the double scalar product, the dyadic product, and the cross product. We additionally introduce the three-dimensional $\boldsymbol{\nabla}_3 = \boldsymbol{e}_i \, {}^{\partial\square}/_{\partial x_i}$ and the two dimensional Hamiltonian $\boldsymbol{\nabla}_2 = \boldsymbol{e}_\alpha \, {}^{\partial\square}/_{\partial x_\alpha}$. With its help the operations divergence $\boldsymbol{\nabla} \cdot \square$ and gradient $\boldsymbol{\nabla}\square$ of a tensor $\square$ can be specified compactly. Also, the symmetric part of a gradient $\boldsymbol{\nabla}^{\mathrm{sym}}\square = {}^1/_2[\boldsymbol{\nabla}\square + \boldsymbol{\nabla}^\top\square]$ can be given. For detailed penetrations of these operations we refer to [9] or [10].

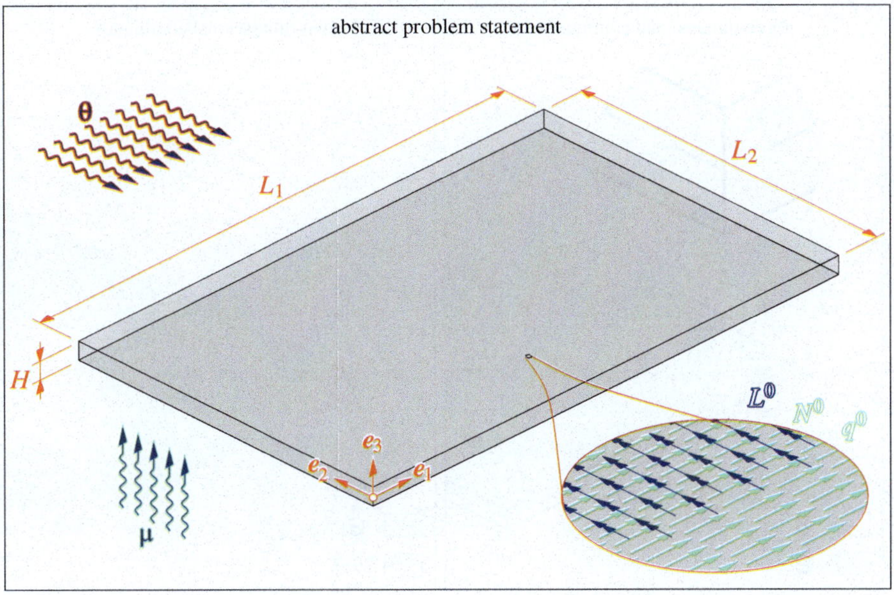

Fig. 3.1: A plate exposed to moisture (μ) and temperature (θ) as well as as kinetic quantities (N^0, L^0, q^0) which are already present in the interior of the unloaded state.

3.2 Frame of Reference

In plate theory it is generally accepted that the plate can be represented by a deformable plane. Such considerations can traced back to Germain [11], Lagrange [12], Navier [13], Poisson [14], and of course Kirchhoff [5], among others (for further discussion see [15]). Hence, all contemplations are reduced to a two-dimensional problem due to the slenderness of the body considered. For this reason, one selects a surface $\mathfrak{S}$ related to planar dimensions. It is reasonable to choose the mid surface due to material and geometrical symmetries in transverse direction. We reduce our concern to coplanarity of all material points of this surface, i.e. initially plane surfaces. For present treatise we furthermore assume that each material point is an infinitesimal rigid body with five kinematic degrees of freedom, cf. Fig. 3.2 at the right-hand side. As will be described in more detail later, these are three translations and two rotations. Thus, we introduce a plate theory as proposed by Zhilin [2]. We directly enter the topic by introducing the degrees of freedom. In present case these are three translations $\boldsymbol{a}$ and two rotations $\boldsymbol{\psi}$, which components are visualized in Fig. 3.2 on the ride-hand-side.

$$\boldsymbol{a} = \boldsymbol{v} + w\boldsymbol{n} \qquad \text{with}\, \boldsymbol{v} = v_\alpha \boldsymbol{e}_\alpha, \tag{3.1}$$

$$\boldsymbol{\psi} = -\varphi_2 \boldsymbol{e}_1 + \varphi_1 \boldsymbol{e}_2. \tag{3.2}$$

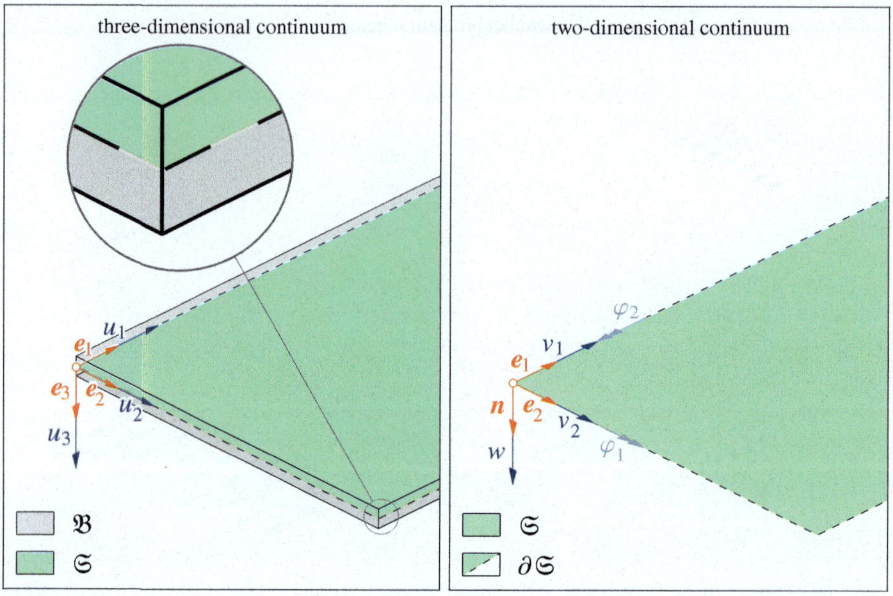

Fig. 3.2: Mid surface $\mathfrak{S}$ in context of a three-dimensional body $\mathfrak{B}$ with degrees of freedom assigned.

Herein, our tripod is still a set of orthogonal unit vectors with longitudinal directions $\boldsymbol{e}_\alpha$ and one direction perpendicular to the plane spanned by the longitudinal ones, called $\boldsymbol{n}$. However, rotations around the surface normal remain unconsidered since they have no physical justification due to the fact that the in-plane shear stiffness of classical engineering structures is much larger that the bending stiffness, cf. [2]. However, instead of employing the rotational vector $\boldsymbol{\psi}$, we introduce the vector $\boldsymbol{\varphi} = \varphi_\alpha \boldsymbol{e}_\alpha$, which is related via $\boldsymbol{\varphi} = \boldsymbol{\psi} \times \boldsymbol{n}$.

Typical deformation measures are the membrane strains $\underset{\sim}{\boldsymbol{G}}$, the curvature changes $\underset{\sim}{\boldsymbol{K}}$, and the transverse shear strains $\boldsymbol{g}$. In a geometrically linear framework, we can define these measures as follows.

$$\underset{\sim}{\boldsymbol{G}} = \boldsymbol{\nabla}^{\text{sym}} \boldsymbol{v} \;= G_{\gamma\delta} \boldsymbol{e}_\gamma \otimes \boldsymbol{e}_\delta, \tag{3.3}$$

$$\underset{\sim}{\boldsymbol{K}} = \boldsymbol{\nabla}^{\text{sym}} \boldsymbol{\varphi} \;= K_{\gamma\delta} \boldsymbol{e}_\gamma \otimes \boldsymbol{e}_\delta, \tag{3.4}$$

$$\boldsymbol{g} = \boldsymbol{\nabla} w + \boldsymbol{\varphi} = g_\beta \boldsymbol{e}_\beta. \tag{3.5}$$

The decoupling of the constitutive equations is reasonable due to

- coplanarity of all material points of the surface considered,
- geometrical symmetry in thickness direction,
- material symmetry in thickness direction, and
- isotropic material behaviour.

We can now introduce a decoupled, which means the in-plane, out-of-plan and transverse shear behaviour not influencing each other, strain energy function of the deformable plane surface in terms of a Taylor series limited to quadratic terms where positive definiteness holds true.

$$W(\underset{\sim}{G},\underset{\sim}{K},\boldsymbol{g}) = \frac{1}{2}\left(\underset{\sim}{G}:\mathcal{A}:\underset{\sim}{G}+\underset{\sim}{K}:\mathcal{D}:\underset{\sim}{K}+\boldsymbol{g}\cdot\underset{\sim}{Z}\cdot\boldsymbol{g}\right) \tag{3.6}$$

Therein, we have introduced the constitutive tensors, whereby $\mathcal{A} = A_{\alpha\beta\gamma\delta}\boldsymbol{e}_\alpha\otimes\boldsymbol{e}_\beta\otimes\boldsymbol{e}_\gamma\otimes\boldsymbol{e}_\delta$ is the fourth-order in-plane stiffness tensor, $\mathcal{D} = D_{\alpha\beta\gamma\delta}\boldsymbol{e}_\alpha\otimes\boldsymbol{e}_\beta\otimes\boldsymbol{e}_\gamma\otimes\boldsymbol{e}_\delta$ is the fourth-order out-of-plane stiffness tensor, and $\underset{\sim}{Z} = Z_{\alpha\beta}\boldsymbol{e}_\alpha\otimes\boldsymbol{e}_\beta$ is the second-order transverse shear stiffness tensor. They depend on the material and geometrical properties. In the isotropic case, they depend on two material parameters only and the thickness of the plate. The kinetic measures arising from this potential.

$$\frac{\partial W}{\partial \underset{\sim}{G}} = \mathcal{A}:\underset{\sim}{G}, \qquad \frac{\partial W}{\partial \underset{\sim}{K}} = \mathcal{D}:\underset{\sim}{K}, \qquad \frac{\partial W}{\partial \boldsymbol{g}} = \underset{\sim}{Z}\cdot\boldsymbol{g}. \tag{3.7}$$

Hence we have found dual quantities, i.e. the in-plane forces $\underset{\sim}{N}$, the moments $\underset{\sim}{L}$, and the transverse shear forces $\boldsymbol{q}$.

$$\frac{\partial W}{\partial \underset{\sim}{G}} = \underset{\sim}{N} = N_{\alpha\beta}\boldsymbol{e}_\alpha\otimes\boldsymbol{e}_\beta, \tag{3.8}$$

$$\frac{\partial W}{\partial \underset{\sim}{K}} = \underset{\sim}{L} = L_{\alpha\beta}\boldsymbol{e}_\alpha\otimes\boldsymbol{e}_\beta, \tag{3.9}$$

$$\frac{\partial W}{\partial \boldsymbol{g}} = \boldsymbol{q} = Q_\alpha\boldsymbol{e}_\alpha. \tag{3.10}$$

The constitutive equations are then given as linear mappings.

$$\underset{\sim}{N} = \mathcal{A}:\underset{\sim}{G}, \tag{3.11}$$

$$\underset{\sim}{L} = \mathcal{D}:\underset{\sim}{K}, \tag{3.12}$$

$$\boldsymbol{q} = \underset{\sim}{Z}\cdot\boldsymbol{g}. \tag{3.13}$$

The isotropic stiffness tensors can be given in following form [16].

$$\mathcal{A} = 2Bh\ \mathcal{P}_1^{\mathfrak{S}} + 2Gh\ \mathcal{P}_2^{\mathfrak{S}}, \tag{3.14}$$

$$\mathcal{D} = 2B\frac{h^3}{12}\mathcal{P}_1^{\mathfrak{S}} + 2G\frac{h^3}{12}\mathcal{P}_2^{\mathfrak{S}}, \tag{3.15}$$

$$\underset{\sim}{Z} = \kappa Gh\ \underset{\sim}{P}. \tag{3.16}$$

Herein we have introduced the compression modulus of the surface $B = {}^{Y}/_{2(1-\nu)}$ [16] with Young's modulus Y, Poisson's ratio ν and a correction factor $0 < \kappa \le 1$ accounting for transverse shear, cf. [6, 7] and the discussion in [17]. The fourth-order isotropic projectors are given as follows.

$$\mathcal{P}_1^{\mathfrak{S}} = \frac{1}{2}\underset{\sim}{\boldsymbol{P}} \otimes \underset{\sim}{\boldsymbol{P}}, \qquad \mathcal{P}_2^{\mathfrak{S}} = \mathcal{P}^{\mathrm{sym}} - \mathcal{P}_1^{\mathfrak{S}}. \tag{3.17}$$

Herein, $2\mathcal{P}^{\mathrm{sym}} = \boldsymbol{e}_\alpha \otimes \boldsymbol{e}_\beta \otimes (\boldsymbol{e}_\alpha \otimes \boldsymbol{e}_\beta + \boldsymbol{e}_\beta \otimes \boldsymbol{e}_\alpha)$ is the symmetric part of the fourth-order identity of the surface.

Finally, the equilibrium equations of the two-dimensional planar continuum are written as

$$\boldsymbol{\nabla} \cdot \underset{\sim}{\boldsymbol{N}} \qquad + \boldsymbol{s} = \boldsymbol{o}, \tag{3.18a}$$
$$\boldsymbol{\nabla} \cdot \boldsymbol{q} \qquad + p = 0, \tag{3.18b}$$
$$\boldsymbol{\nabla} \cdot \underset{\sim}{\boldsymbol{L}} - \boldsymbol{q} + \boldsymbol{m} \times \boldsymbol{n} = \boldsymbol{o}. \tag{3.18c}$$

where $\underset{\sim}{\boldsymbol{N}}, \underset{\sim}{\boldsymbol{L}}$ are the tensors of in-plane forces and out-of-plane moments, respectively, $\boldsymbol{q}$ is the vector of transverse shear forces, $\boldsymbol{m}$ is the surface moment vector (excluding drilling moments in line with the assumptions of a five degree of freedom continuum). The vector of surface forces $\boldsymbol{t}$ is split in two parts: $\boldsymbol{s}$ and $\boldsymbol{p}$, where the first one is the vector of tangent surface forces and the second one is the vector transverse surface loads.

$$\boldsymbol{t} = \boldsymbol{s} + \boldsymbol{p} \quad \text{with} \quad \boldsymbol{p} = p\boldsymbol{n} \quad \text{and} \quad \boldsymbol{s} = -s_\alpha \boldsymbol{e}_\alpha \tag{3.19}$$

It is worth mentioning that any general body forces are applied as surface loads to the planar structure ($\mathfrak{S}$):

$$\boldsymbol{t} := \int_{-\frac{H}{2}}^{+\frac{H}{2}} \rho \boldsymbol{b} \,\mathrm{d}X_3 = \boldsymbol{s} + \boldsymbol{p}. \tag{3.20}$$

where $\boldsymbol{t}$ is the general surface load (tangent and transverse) and ρ is the mass density of the body $\mathfrak{B}$. In a similar fashion, the moment of the body loads along directions 1 and 2 are applied as a surface moment vector (per unit area):

$$\boldsymbol{m} := -\int_{-\frac{H}{2}}^{+\frac{H}{2}} \rho X_3 \underset{\sim}{\boldsymbol{P}} \cdot \boldsymbol{b} \,\mathrm{d}X_3. \tag{3.21}$$

By excluding any mixed boundary conditions on the surface $\mathfrak{S}$, only independent Drichlet (on $\partial\mathfrak{S}_\mathrm{D}$) and Neumann boundary conditions (on $\partial\mathfrak{S}_\mathrm{N}$) are considered over the boundary of the domain $\partial\mathfrak{S}$:

$$\partial\mathfrak{S}_\mathrm{D} \cup \partial\mathfrak{S}_\mathrm{N} = \partial\mathfrak{S}, \qquad \partial\mathfrak{S}_\mathrm{D} \cap \partial\mathfrak{S}_\mathrm{N} = \varnothing. \tag{3.22}$$

where $\partial\mathfrak{S}_\mathrm{D}$ and $\partial\mathfrak{S}_\mathrm{N}$ are non-intersecting non-trivial subsets of the boundary $\partial\mathfrak{S}$ over which an outward normal direction vector $\boldsymbol{\upsilon}$ is defined, i.e., $\boldsymbol{\upsilon} \cdot \boldsymbol{n} = 0$. In addition, the arc-length variable ℓ parametrises the distance along the contour $\partial\mathfrak{S}$.

Consequently, the Drichlet boundary conditions consist of prescribed kinematic values for in-plane displacements $\boldsymbol{v}_\star(\ell)$, deflections $w_\star(\ell)$, and rotations $\boldsymbol{\varphi}_\star(\ell)$:

$$\text{on } \partial\mathfrak{S}_\text{D}: \qquad \boldsymbol{v} \overset{!}{=} \boldsymbol{v}_\star(\ell), \tag{3.23a}$$

$$w \overset{!}{=} w_\star(\ell), \tag{3.23b}$$

$$\boldsymbol{\varphi} \overset{!}{=} \boldsymbol{\varphi}_\star(\ell), \tag{3.23c}$$

In addition, the Neumann boundary conditions consists of prescribed static values (distributed per unit length) for in-plane forces $\boldsymbol{s}_\star(\ell)$, transverse forces $\boldsymbol{p}_\star(\ell)$, and in-plane moments $\boldsymbol{m}_\star(\ell)$:

$$\text{on } \partial\mathfrak{S}_\text{N}: \qquad \boldsymbol{v} \cdot \underset{\sim}{\boldsymbol{N}} \overset{!}{=} \boldsymbol{s}_\star(\ell), \tag{3.24a}$$

$$\boldsymbol{v} \cdot \boldsymbol{q} \overset{!}{=} p_\star(\ell), \tag{3.24b}$$

$$\boldsymbol{v} \cdot \underset{\sim}{\boldsymbol{L}} \overset{!}{=} \boldsymbol{m}_\star(\ell). \tag{3.24c}$$

3.3 Thermal Effects and Hygroscopic Impact

Environmental thermal cycles result in expansion/contraction of materials with non-zero coefficients of thermal expansion ($\alpha^\theta \neq 0$). A homogeneous temperature change in an unconstraint isotropic material results in a volumetric change, i.e., it nullifies the thermal curvature change tensor $\underset{\sim}{\boldsymbol{K}}^\theta = \underset{\sim}{\boldsymbol{0}}$ and thermal transverse shear vector $\boldsymbol{g}^\theta = \mathbf{o}$. Thus, the following Duhamel-Neumann form [18, 19, 20] can be written if only normal in-plane thermal strains are considered:

$$\underset{\sim}{\boldsymbol{N}} = \mathcal{A} : (\underset{\sim}{\boldsymbol{G}} - \underset{\sim}{\boldsymbol{G}}^\theta), \tag{3.25a}$$

$$\underset{\sim}{\boldsymbol{L}} = \mathcal{D} : \underset{\sim}{\boldsymbol{K}}, \tag{3.25b}$$

$$\boldsymbol{q} = \underset{\sim}{\boldsymbol{Z}} \cdot \boldsymbol{g}, \tag{3.25c}$$

where $\underset{\sim}{\boldsymbol{G}}^\theta$ is the tensor of thermal in-plane strains:

$$\underset{\sim}{\boldsymbol{G}}^\theta = \alpha^\theta \Delta\theta \underset{\sim}{\boldsymbol{P}}. \tag{3.26}$$

Polymeric materials can absorb moisture and undergo swelling. Such a volumetric change is similar to what happens during thermal expansion in the sense that both are considered to be dilatoric eigenstrains ($\boldsymbol{q}^\mu = \mathbf{o}$). By assuming a uniform through-thickness moisture absorption (Δm), an isotropic plane will experience no curvature change ($\underset{\sim}{K}^{\mu} = \mathbf{0}$) but only planar eigenstrains:

$$\underset{\sim}{N} = \mathcal{A} : (\underset{\sim}{G} - \underset{\sim}{G}^{\mu}), \tag{3.27a}$$

$$\underset{\sim}{L} = \mathcal{D} : \underset{\sim}{K}, \tag{3.27b}$$

$$\boldsymbol{q} = \underset{\sim}{Z} \cdot \boldsymbol{g}, \tag{3.27c}$$

where $\underset{\sim}{G}^{\mu}$ includes the in-plane moisture strains:

$$\underset{\sim}{G}^{\mu} = \alpha^{\mu} \Delta m \underset{\sim}{P}, \tag{3.28}$$

where α^{μ} is the coefficient of moisture expansion, and Δm is the change in the moisture mass (absorption/desorption) per unit mass of the plate.
It is noteworthy that the mechanical properties of the material are deemed independent of the hygrothermal effects. Moreover, this smeared representation does not include individual constituents (e.g., fibres within a matrix) and their mesoscopic structures. More importantly, the hygroscopic component does not entail any residual stresses due to violating the compatibility condition—hence, it is a stress-free eigenstrain, see [21, 22].

3.4 Residual Kinetics

Based on the concept of residual stresses, we also introduce residual kinetic quantities, i.e. in-plane forces, moments, and transverse shear forces. Such quantities may be present even in the absence of any external loading, thermal gradients etc. These quantities are often associated with plastic deformations, inhomogeneous micro-structural transformations or hygrothermal effects.

In the context of present approach, a general set of incompatible strains results in residual fields for in-plane forces $\underset{\sim}{N}^0$, moments $\underset{\sim}{L}^0$, and transverse shear forces $\boldsymbol{q}^0$. In analogy to the problems of three-dimensional continua, these quantities can only be measured by the available non-destructive and destructive methods where the deformations corresponding to the residual stresses are measured. In order not to lapse into details, we refer to [23] or [24] and only disclose the transfer of measured residual Cauchy stresses $\underset{\sim}{T}^0 = T^0_{ij} \boldsymbol{e}_i \otimes \boldsymbol{e}_j$ to the quantities introduced here. In this sense, the following residual forces and moments are introduced:

$$\underset{\sim}{N}^0 = \int_{-\frac{H}{2}}^{+\frac{H}{2}} \underset{\sim}{P} \cdot \underset{\sim}{T}^0 \cdot \underset{\sim}{P} \, \mathrm{d}X_3, \tag{3.29a}$$

$$\underset{\sim}{L}^0 = \int_{-\frac{H}{2}}^{+\frac{H}{2}} X_3 \underset{\sim}{P} \cdot \underset{\sim}{T}^0 \cdot \underset{\sim}{P} \, \mathrm{d}X_3, \tag{3.29b}$$

$$q^0 = \int_{-\frac{H}{2}}^{+\frac{H}{2}} \underset{\sim}{P} \cdot \underset{\sim}{T}^0 \cdot n \, \mathrm{d}X_3. \tag{3.29c}$$

The strain energy function now should be extended:

$$\begin{aligned} W(\underset{\sim}{G},\underset{\sim}{K},g) =& \frac{1}{2}\left(\underset{\sim}{G}:\mathcal{A}:\underset{\sim}{G}+\underset{\sim}{K}:\mathcal{D}:\underset{\sim}{K}+g\cdot\underset{\sim}{Z}\cdot g\right) \\ +& \ \underset{\sim}{N}^0:\underset{\sim}{G}+\underset{\sim}{L}^0:\underset{\sim}{K}+q^0\cdot g \end{aligned} \tag{3.30}$$

Considering the constitutive equations originally introduced, we can now extend this set as follows.

$$\underset{\sim}{N} = \mathcal{A}:\underset{\sim}{G}+\underset{\sim}{N}^0 \tag{3.31a}$$

$$\underset{\sim}{L} = \mathcal{D}:\underset{\sim}{K}+\underset{\sim}{L}^0 \tag{3.31b}$$

$$q = \underset{\sim}{Z}\cdot g + q^0 \tag{3.31c}$$

3.5 Conclusion

In present treatise, the direct approach was adopted to set up the field equations of a generalized plate theory under various influences. More specifically, thermal and hygroscopic effects along with residual quantities are considered. When all the quantities taken into account are combined, the Duhamel-Neumann analogy takes subsequent form.

$$\underset{\sim}{N} = \mathcal{A}:(\underset{\sim}{G}-\underset{\sim}{G}^{\theta}-\underset{\sim}{G}^{\mu})+\underset{\sim}{N}^0, \tag{3.32a}$$

$$\underset{\sim}{L} = \mathcal{D}:(\underset{\sim}{K})+\underset{\sim}{L}^0, \tag{3.32b}$$

$$q = \underset{\sim}{Z}\cdot(g)+q^0. \tag{3.32c}$$

Obviously, hygroscopic and thermal loads effects the in-plane force tensor solely, at least in present case. Contrary, residual quantities effect all three states. The field equations of Eq. (3.18) my then be rearranged with the reduced forces and moments of Eqs. (3.32) to set up the linear hygro-thermo-static equilibrium with residual quantities. A concise visual representation can be found in Fig. 3.3.

To be honest, present execution concerns the simplest case of a two-dimensional theory where the reference surface halves the plate thickness. In absence of such geometrical symmetry, quantities like thermal moments etc. also occur then. Future work should take these quantities into account. Moreover, we are here limited to purely static considerations. In continuation, fluctuations, at least of temperature and moisture, should find entrance in the direct approach.

However, potential applications of present ansatz can be found in the analysis of composite structures as presented for example in [25].

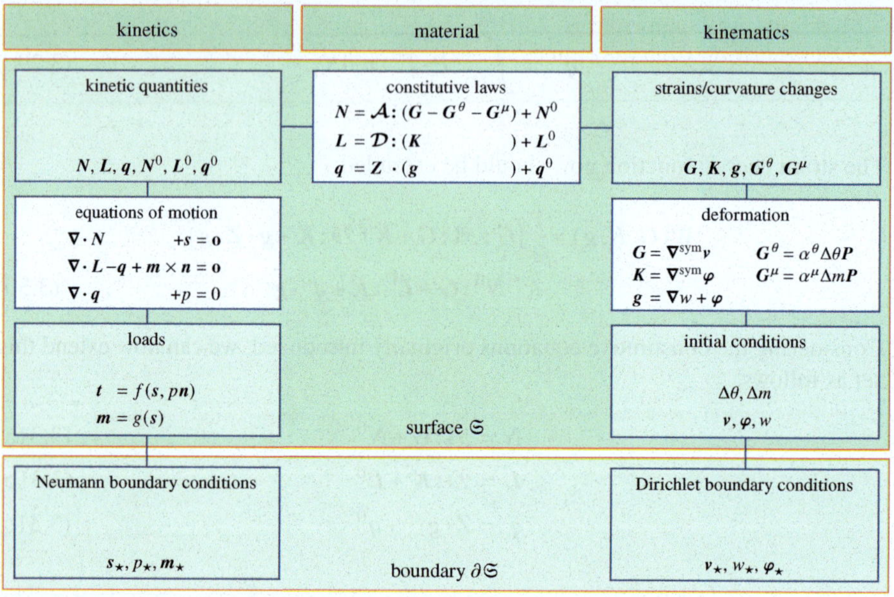

Fig. 3.3: Tonti diagram [26] for the composed plate problem of present treatise.

References

[1] Javanbakht Z, Aßmus M, Naumenko K, Öchsner A, Altenbach H (2019) On thermal strains and residual stresses in the linear theory of anti-sandwiches, ZAMM - Journal of Applied Mathematics and Mechanics / Zeitschrift für Angewandte Mathematik und Mechanik **99**(8):e201900062. DOI 10.1002/zamm.201900062

[2] Zhilin PA (1976) Mechanics of deformable directed surfaces, International Journal of Solids and Structures **12**(9):635 – 648. DOI 10.1016/0020-7683(76)90010-X

[3] Cosserat F, Cosserat E (1909) Théorie des corps déformables, A. Hermann et fils, Paris. URL http://jhir.library.jhu.edu/handle/1774.2/34209

[4] Ericksen JL, Truesdell C (1957) Exact theory of stress and strain in rods and shells. Archive for Rational Mechanics and Analysis **1**(1):295–323. DOI 10.1007/BF00298012

[5] Kirchhoff GR (1850) Über das Gleichgewicht und die Bewegung einer elastischen Scheibe, Crelles Journal für die reine und angewandte Mathematik **40**:51–88. DOI 10.1515/crll.1850.40.51

[6] Reissner E (1945) The effect of transverse shear deformation on the bending of elastic plates, Journal of Applied Mechanics **12**(2):A69–A77. DOI 10.1115/1.4009435

[7] Mindlin RD (1951) Influence of rotatory inertia and shear on flexural motions of isotropic, elastic plates, Journal of Applied Mechanics **18**(1):31–38. DOI 10.1115/1.4010217

[8] Aßmus M, Naumenko K, Altenbach H (2019) Subclasses of mechanical problems arising from the direct approach for homogeneous plates, In: Altenbach H, Chróścielewski J, Eremeyev V, Wiśniewski K (Eds) Recent Developments in the Theory of Shells, Advanced Structured Materials, vol 110, Springer, Singapore, pp 1–21. DOI 10.1007/978-3-030-17747-8_4
[9] Šilhavý M (1997) The Mechanics and Thermodynamics of Continuous Media. Springer, Berlin · Heidelberg
[10] Eremeyev VA, Cloud MJ, Lebedev LP (2018) Applications of Tensor Analysis in Continuum Mechanics. World Scientific, DOI 10.1142/10959
[11] Germain S (1821) Recherches sur la théorie des surfaces élastiques, Veuve Courtier, Paris. URL www.cambridge.org/9781108050371
[12] Lagrange JL (1828) Annales de Chimie 39:149,207
[13] Navier M (1823) Mémoire sur les lois de l'équilibre et du mouvement des corps solides élastiques, Bulletin de la Société Philomathique de Paris pp 177–181
[14] Poisson SD (1829) Sur l'équilibre et mouvement des corps élastiques, Mémoires de l'Académie des Sciences VIII
[15] Altenbach H, Eremeyev V (2017) Thin-walled structural elements: Classification, classical and advanced theories, new applications, In: Altenbach H, Eremeyev V (Eds) Shell-like Structures: Advanced Theories and Applications, CISM International Centre for Mechanical Sciences, vol 572, Springer International Publishing, Cham, pp 1–62. DOI 10.1007/978-3-319-42277-0_1
[16] Aßmus M, Eisenträger J, Altenbach H (2017) Projector representation of isotropic linear elastic material laws for directed surfaces, Zeitschrift für Angewandte Mathematik und Mechanik **97**:1–10. DOI 10.1002/zamm.201700122
[17] Altenbach H (2000) An alternative determination of transverse shear stiffnesses for sandwich and laminated plates, International Journal of Solids and Structures **37**(25):3503–3520. DOI 10.1016/S0020-7683(99)00057-8
[18] Neumann FE (1885) Vorlesungen über mathematische Physik: Vorlesungen über die Theorie der Elasticität der festen Körper, Vol 4, BG Teubner, Leipzig
[19] Duhamel JMC (1837) Second mémoire sur les phénomènes thermomécaniques, Journal de l'École polytechnique **15**(25):1–57
[20] Duhamel JMC (1838) Mémoire sur le calcul des actions moléculaires développées par les changements de température dans les corps solides, Mémoires présentés par Divers Savants à l'Academie Royale des sciences de l'Institut de France **5**:440–498
[21] Korsunsky AM (2008) Eigenstrain analysis of residual strains and stresses, The Journal of Strain Analysis for Engineering Design **44**(1):29–43. DOI 10.1243/03093247JSA423
[22] Korsunsky A (2017) A Teaching Essay on Residual Stresses and Eigenstrains, Butterworth-Heinemann, Amsterdam
[23] Lu J, for Experimental Mechanics (US) S (1996) Handbook of Measurement of Residual Stresses, Handbook of Measurement of Residual Stresses, Fairmont Press. URL https://books.google.de/books?id=h4h9QgAACAAJ
[24] Schajer GS (2013) Practical Residual Stress Measurement Methods, John Wiley & Sons, Ltd, Chichester. DOI 10.1002/9781118402832

[25] Aßmus M (2019) Structural Mechanics of Anti-Sandwiches. An Introduction. SpringerBriefs in Continuum Mechanics, Springer, Cham, DOI 10.1007/978-3-030-04354-4
[26] Tonti E (2013) The Mathematical Structure of Classical and Relativistic Physics, Birkhäuser.

Chapter 4
A Technique for Determining True Deformation Diagrams Under Dynamic Tension Using DIC

Artem V. Basalin, Anatoly M. Bragov, Aleksandr Yu. Konstantinov, and Andrey R. Filippov

Abstract This article describes a pneumodynamic experimental setup created in the laboratory of dynamic tests of the Research Institute of Mechanics of the Nizhny Novgorod State University and designed to study the dynamic behavior of structural materials at average strain rates of the order of 10-100 s^{-1}. Modern means of registration of parameters of dynamic processes are applied. The results of approbation are given on the example of dynamic tests of samples of sheet steel 09G2S. The technical and true strain diagrams in the static and dynamic range of strain rates are determined.

Key words: Strain rate · Experimental study · Plastic strain · Yield strength · DIC · Experiment · Strain rate

4.1 Introduction

Interest in the study of dynamic deformation and fracture of materials is due to the fact that a large number of important problems in assessing the strength of new critical structures requires specifying the strength and deformation characteristics of materials as functions of the strain rate. Such dependences are extremely important for obtaining reliable estimates of the stress-strain state and strength of structures experiencing dynamic loads as a result of technological accidents, terrorist attacks, falls, etc. To determine the characteristics of materials at high deformation rates, the Hopkinson split bar method [1] and its modifications are often used [2]-[10].

Artem V. Basalin · Anatoly M. Bragov · Aleksandr Yu. Konstantinov · Andrey R. Filippov
National Research Lobachevsky State University of Nizhny Novgorod, 23 Gagarin Avenue, Nizhny Novgorod 603022, Russian Federation,
e-mail: artem.basalin@mech.unn.ru, bragov@mech.unn.ru, konstantinov@mech.unn.ru, filippov@mech.unn.ru

H. Altenbach et al. (eds.), *Sixty Shades of Generalized Continua*,
Advanced Structured Materials 170, https://doi.org/10.1007/978-3-031-26186-2_4

The range of deformation rates from 10^0 s^{-1} to 10^2 s^{-1} is currently poorly understood, but this range of deformation rates is critically important for assessing the strength of structures during a fall and low-speed impact. There are a number of works devoted to the study of the behavior of structural materials in the range of average deformation rates. In [11], the influence of the deformation rate on the mechanical properties during stretching of the industrial aluminum alloy AA7081 was analyzed. Tests at medium strain rate were carried out on a hydropneumatic machine (HPM) [12]. The method of conducting the test is described in detail in [13]-[15]. The deformation rate obtained with this installation was about 30 s^{-1}.

To study the strength and deformation characteristics of sandstone, stress-strain relationships were obtained under uniaxial compression at strain rates of $10^{-2} \sim 55\,s^{-1}$ using MTS and the drop weight test in [16]. Groth et al. [17], Zhang and Zhao [18], as well as Ma et al. [2, 20] performed low-speed and high-speed impact tests of rocks in the deformation velocity ranges of $10^{-5} \sim 10^{-2}$ s^{-1} and $10^1 \sim 10^2$ s^{-1} using a hydraulic testing machine and an RSG installation with larger bars diameter. One of the difficulties of testing at moderate strain rates is the short duration of the load pulse. Thus, for a limited time of loading at a low speed, it is not possible to achieve significant deformations of the sample. This problem is partially solved in schemes with a prestressed loading elastic bar. The duration of the load in this case turns out to be significant. In [21, 22] the effects of strain rate, temperature, and size effects of three steels used in the nuclear industry were studied at the facility. In [23], the study of the behavior of steels Armox 500T and Armox 600T was carried out in the range of strain rates 100-1000 s^{-1}. A single bar tensile Hopkinson apparatus was used. The tensile force was generated with pendulum. Zabotkin et al. [24] showed the design concept of an instrumented falling weight impact (IFWI) tester and tested 316L stainless steel, titanium alloy, and alloy C22 under medium strain rates.

In [25] an investigation into the effect that medium strain rates have on the mechanical properties of cold drawn mild steel and 304L stainless steel was carried out using two 6 mm diameter by 14 mm long specimens machined from each grade of steel. The tests were performed using an Instron test machine under a quasi-static strain rate, and a high capacity instrumented falling weight axial impact (IFWI) test rig developed at the University of Wollongong was used to con-duct the medium strain rate tests.

In [26] high-speed tensile tests were performed on a Zwick/Roell HTM5020 high-speed testing system (Ulm, Germany) with a maximum test velocity of 20 m/s, as well as a maximum load of 50 kN. The mechanical properties and failure behavior of long glass fiber reinforced thermoplastic composites were investigated. The dynamic tensile tests were carried out at the strain rates of 1, 10, 100, 200, and 400 s^{-1}.

Determination of the true deformation diagrams of materials from static and dynamic stretching experiments is complicated by the process of localization of deformation (neck formation), which violates the uniformity of the VAT and leads to a local change in the distribution of deformations and the type of stress state, which makes it impossible to directly determine stresses and deformations in the sample based on integral displacements and forces, which are measured in the experiment. In a dynamic experiment, the localization of deformations also affects the history of changes in the deformation rate, which increases sharply in the neck region. These features of the interpretation of the results of the dynamic experiment are considered in [27, 28]. In this case, correction of the deformation curve beyond the point of localization of deformation is required [29].

Various methods have been proposed to evaluate effective von Mises stresses in a tensile experiment [30]-[38]. In [39], the DIC system was used to obtain deformation fields. It has been proved that this method is an effective method for registering the deformation of composites [40]-[44]. The article describes a pneumodynamic experimental setup created in the dynamic testing laboratory of the Nizhny Novgorod State University Mechanics Research Institute to study the dynamic behavior of structural materials at deformation rates of the order of 100 s^{-1}. The technique of obtaining true deformation diagrams using DIC is also considered.

4.2 Pneumatic Dynamic Installation for Testing Materials at a Deformation Rate of the Order of 10-100 s^{-1}

Currently, there is a need to develop and implement an experimental scheme and methodology that allows achieving the specified strain rates. This paper describes the unique pneumatic dynamic installation created for testing materials at a deformation rate of the order of $10 - 10^2$ s^{-1}. The scheme of installation and obtaining information used to determine the mechanical characteristics of materials and their dependences on the deformation rate is described.

4.2.1 Installation Scheme

The general scheme of the experimental installation is shown in Fig. 4.1. The sample dimensions are shown in the Fig. 4.2. The compressor 1 is used to create high pressure in the chamber 2. The pressure value is controlled using a pressure gauge 3. The pressure from the cylinder can be directed either to the lower or upper part of the pneumatic cylinder using a distributor 6. Depending on this, it is possible to ensure the movement of the rod either up or down. The stroke of the cylinder rod is 600 mm. The diameter of the hydraulic cylinder piston is 200 mm. The maximum pressure for which the distributor is designed is 10 bar. The magnitude of the applied force is controlled by the pressure in the cylinder chambers. The maximum force generated by the pneumatic cylinder is 31 kN. The upper part of sample 9 is rigidly fixed in the installation frame. The lower part is connected to a movable rod. The upper blade part of the sample (right part on the figure) is longer than the lower one for installation of the strain gauges.

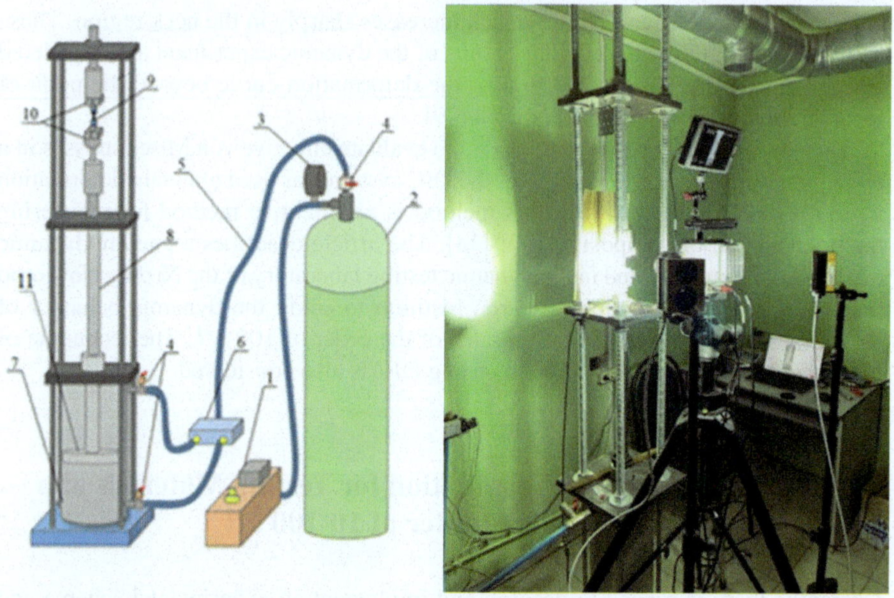

Fig. 4.1: Experimental installation: 1 - compressor, 2 – high pressure chamber, 3 - pressure gauge, 4 - ball valves, 5 - connecting hoses, 6 - solenoid valves, 7 - pneumatic cylinder, 8 - lower movable bar, 9 - sample, 10 - grips, 11 - the piston.

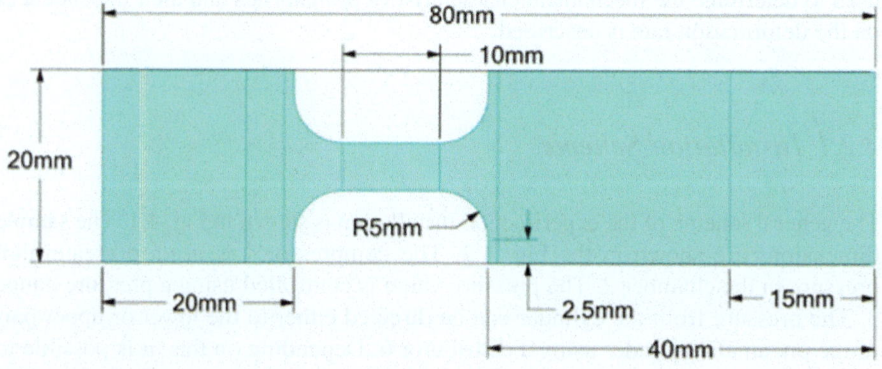

Fig. 4.2: Sample dimension.

4.2.2 Methods of Obtaining and Processing Information in the Experiment

In the laboratory of the Research Institute of Mechanics of the UNN has a licensed VIC-2D non-contact measurement of deformation fields by Correlated Solutions, which includes a high-speed Phantom VEO 710 L camera with high resolution for video recording of the deformation process of a sample during dynamic tests (compression, stretching, shear, three- or a four-point bend), a powerful LED illuminator, a computer with specialized DIC (digital image correlation or digital image correlation method) software installed, and a number of additional devices and devices: a tripod for the camera, stands for illuminators, a device for applying speckles to samples.

The preparatory stage consists in the preliminary application to the surface of the test sample of a contrasting finely dispersed spray, which is a collection of randomly located black dots on a white matte base, the so-called speckle structure, Fig. 4.3 (left). The object under examination is illuminated by a LED monochrome uniform illumination system, which illuminates the outer edges and the central part of the inspection surface with equal intensity, which significantly improves the quality of the measurement results. Next, the camera records images of the investigated object surface corresponding to different phases of the object deformation process. Changes in the position of points of an inhomogeneous surface structure are processed by special high-precision correlation algorithms. The location of points for further analysis of the deformation processes of the samples is shown to the right (Fig. 4.3). The deformation of the sample was observed in two cases.

The average deformation of the working part was determined by the displacements of the reference points P0 and P1 according to the formula:

$$\varepsilon_{\mathrm{avg}} = \ln\left(\frac{L_{\mathrm{P1-P0}}}{L^{0}_{\mathrm{P1-P0}}}\right) \tag{4.1}$$

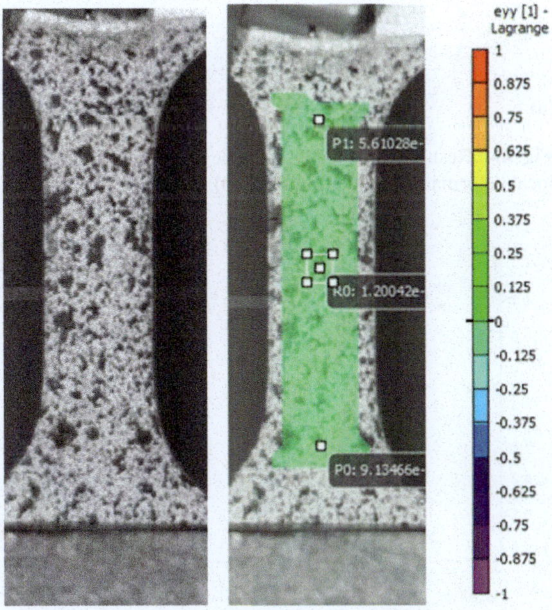

Fig. 4.3 Surface of the test sample: speckle structure (left), location of points for further analysis of the deformation processes (right).

where L_{P1-P0} is the distance between points P1 and P0 at the current time, L^0_{P1-P0} - distance between points P1 and P0 at time 0. The local deformation was determined by averaging the effective von Mises deformations calculated by the DIC method over a small region R0 in the zone of deformation localization.

The strain rates were determined by differentiating the time dependences of the strains. The loading rate was varied by changing the pressure in the working chamber of the installation. Tests were carried out with pressures of 2, 4 and 6 atm in a high pressure chamber. The results of the analysis of the processes of dynamic deformation of the samples are shown in the figure. Figure 4.4a) shows the time dependences of local deformations, Fig. 4.4b) shows the strain rates.

Figures 4.5 and 4.6 show the results of comparing the time histories of average and local strains (a) and strain rates (b) for two loading modes. It can be seen that up to a certain moment (localization time 15 ms), the average and local values are

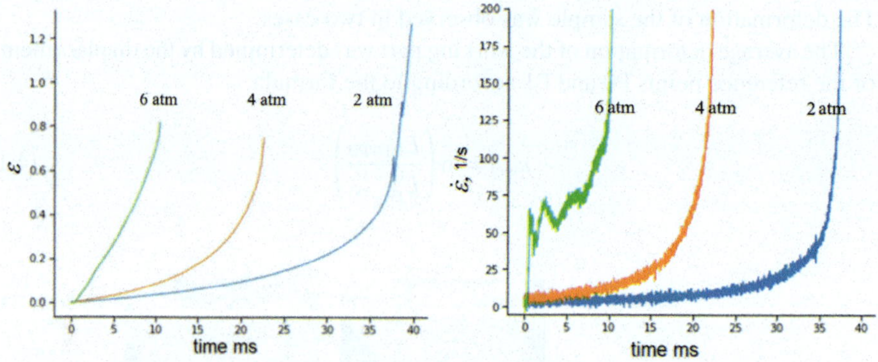

(a) (b)

Fig. 4.4: Results of the analysis of the processes of dynamic deformation: time dependences of local deformations (a), strain rates (b).

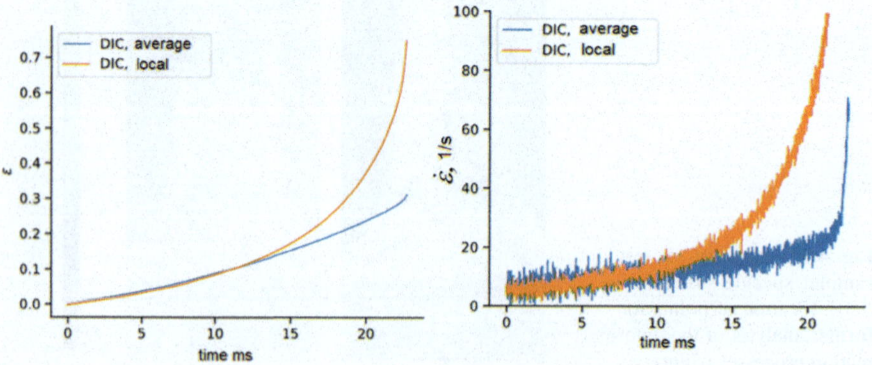

(a) (b)

Fig. 4.5: Time histories of average and local strains (a) and strain rates (b).

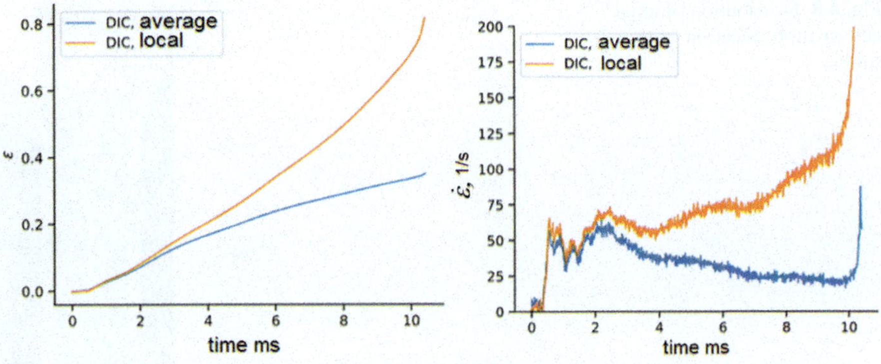

(a) (b)

Fig. 4.6: Time histories of average and local strains (a) and strain rates (b).

in good agreement. Up to this point in time, the values of stresses and strains are distributed uniformly along the length of the sample. Later, a neck is formed in the sample.

Figure 4.7 illustrates the strain fields (strains along the vertical axis) at different time points. One can see the localization of deformation (2 ms) in the middle part of the sample. The right side of Figs. 4.8 and 4.9 shows graphs characterizing the distribution of axial deformation of the sample at different points in time. The line on which these distributions were built is illustrated on the left side of Figs. 4.8 and 4.9. The coordinates of the points on the line are plotted along the horizontal axis, and the deformation values are plotted along the vertical axis. One can observe the process of deformation localization.

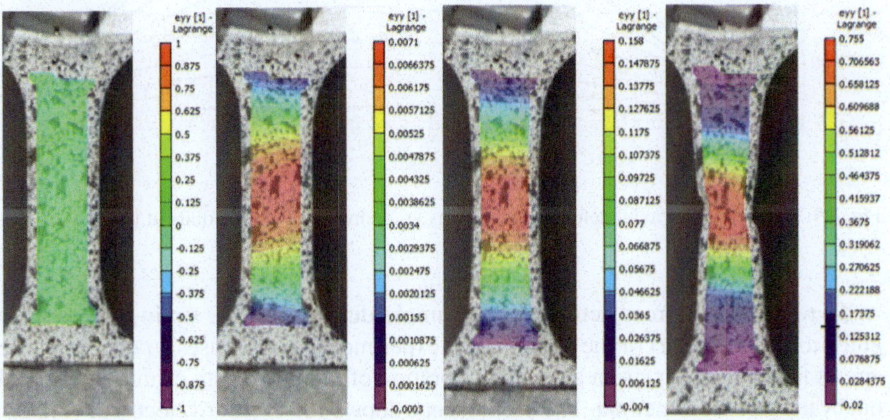

Fig. 4.7: Strain fields (strains along the vertical axis) at different times.

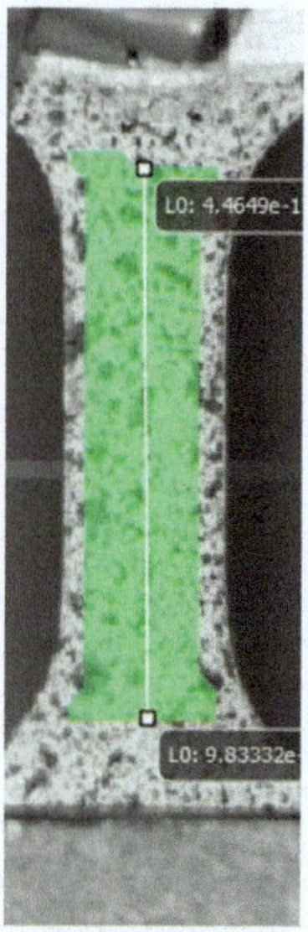

Fig. 4.8 Distribution of axial deformation: location of the strains.

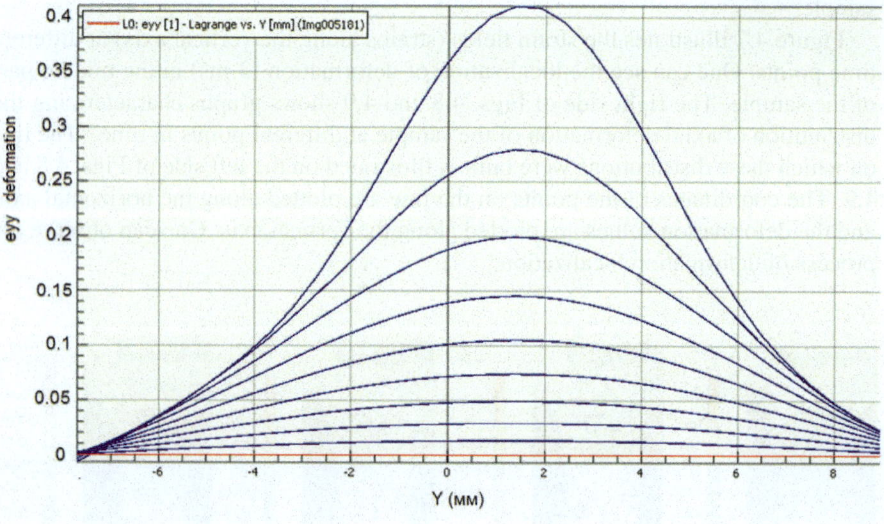

Fig. 4.9: Distribution of axial deformation: strains vs. coordinates of the points of the line.

To register the forces acting on the sample during loading, strain gauges were glued to the blade part of the sample. The experimental circuit uses a potentiometric circuit for connecting strain gauges. Calibration of the measuring channel is carried out by the electrical method. The oscilloscope allows you to record electrical voltages in the power supply circuit of sensors. Strain gauge deformation is associated with a change in its resistance by the main strain gauge dependence:

$$k\varepsilon = \frac{\Delta R}{R},$$

where k is the strain gauge sensitivity coefficient, ε is the strain gauge deformation, ΔR is the change in strain gauge resistance and R is the initial strain gauge resistance.

The calibration procedure is used to determine the conversion factor of the voltage registered by the oscilloscope into the strain gauge deformation. For this, a known resistance R_c is periodically switched on in the strain gauge power supply circuit using a relay. Thus, the change in the resistance of the strain gauge by the value of Rc is modeled. With the help of an oscilloscope, the corresponding change in the electrical voltage in the sensor supply circuit ΔU_c is recorded.

Since the voltage jump is proportional to the change in the resistance of the circuit, you can make the following proportion:

$$\frac{\Delta U_c}{R_c} = \frac{\Delta U}{\Delta R},$$

where ΔU_c is the voltage jump during the calibration procedure, R_c is the additional resistance, ΔU is the voltage jump during experiment and ΔR is the change of strain gauge resistance during experiment.

From the above formulas it follows:

$$\varepsilon = \frac{R_c}{k\Delta U_c R}\Delta U$$

Thus, the conversion coefficient (calibration coefficient) is equal to:

$$K = \frac{R_c}{k\Delta U_c R}.$$

It has the dimension of [1/V].

The signal recorded with the help of strain gauges glued on the blade part of the sample is shown in Fig. 4.10.

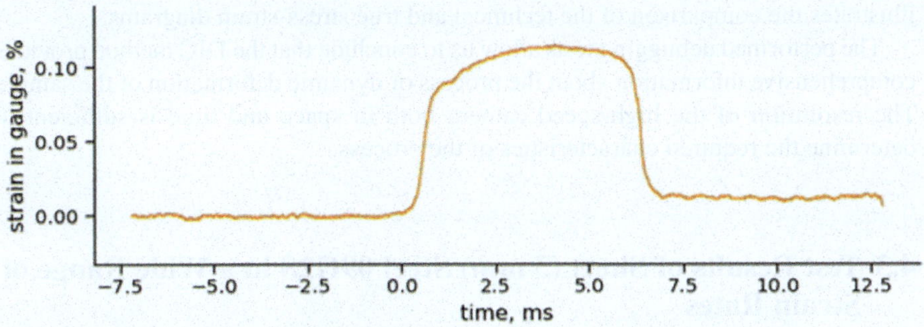

Fig. 4.10: Strains in gauge vs. time.

The force that acts on the sample during loading is calculated as follows:

$$F(t) = \varepsilon_e(t) E w_g h_g \tag{4.2}$$

where $\varepsilon_e(t)$ - is the elastic deformation of a certain section of the blade part of the sample, E - is the Young's modulus of the sample material, $A_g = w_g h_g$ - is the sectional area of the sample in the blade part at the location of the strain gauges, w_g and h_g are the width and thickness of the blade parts of the sample, respectively. The technical stress is determined by dividing the magnitude of the force by the initial cross-sectional area of the sample:

$$\sigma_{tech}(t) = \frac{F(t)}{w_0 h_0} \tag{4.3}$$

Excluding time as a parameter from the dependences $\varepsilon(t)$ and $\sigma(t)$, we obtain the strain curve $\sigma(\varepsilon)$ with the known law of change in the strain rate $\dot{\varepsilon}(\varepsilon)$ during loading of the sample. To determine the true stress at any given time, using high-speed recording data, the force acting on the sample is divided by the current area of the minimum section of the sample. Determining the width of the sample based on the results of video recording is carried out automatically using machine vision tools. The process of tracking the position of reference points is shown in Fig. 4.11. Figure 4.12 illustrates the time dependence of the width of the working part of the sample.

It is assumed that the thinning of the sample in the direction normal to the plane of the plate is proportional to the change in the width of the working part. Then the true voltage at each moment of time can be calculated by the formula:

$$\sigma_{true}(t) = \frac{F(t)}{w(t)h(t)} = \frac{F(t)}{w^2(t)\frac{h_0}{w_0}} \tag{4.4}$$

where $w(t)$ - is the width of the working part of the sample at time t, $h(t)$ - is the thickness of the working part of the sample at time t, w_0 and h_0 - are the initial values of the width and thickness of the working part of the sample. Figure 4.13 compares the time dependences of the technical and true stress in the specimen, and Fig. 4.14 illustrates the comparison of the technical and true stress-strain diagrams.

The performed debugging tests allow us to conclude that the DIC method provides comprehensive information about the process of dynamic deformation of the sample. The resolution of the high-speed camera both in space and time is sufficient to determine the required characteristics of the process.

4.3 Test Results of Sheet (3 mm) Steel 09G2S in a Wide Range of Strain Rates

Steel 09G2S belongs to the class of structural low-alloy steels. This material is used for the manufacture of various parts and elements of welded metal structures

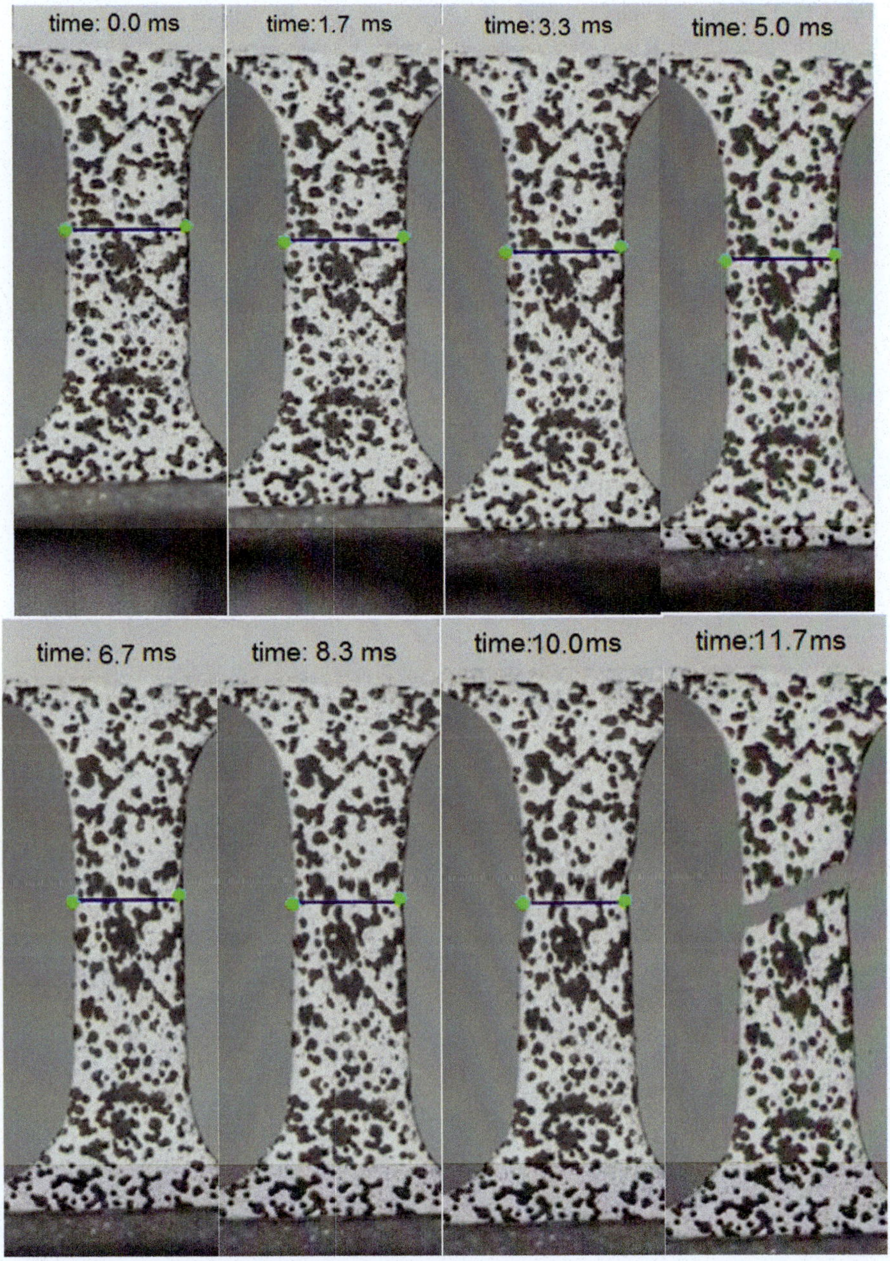

Fig. 4.11: Process of tracking the position of reference points.

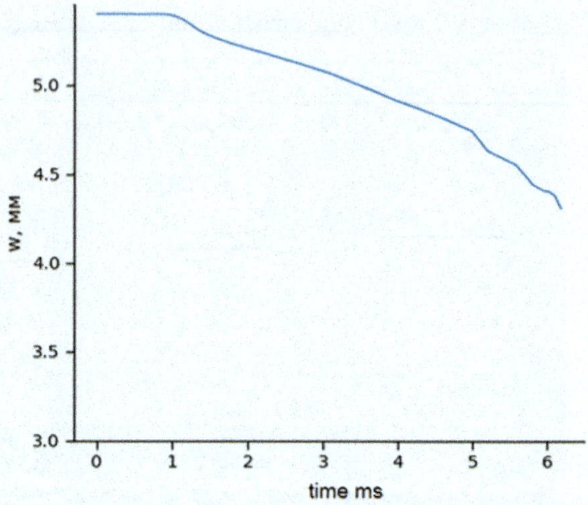

Fig. 4.12 Time of the width of the working part of the sample.

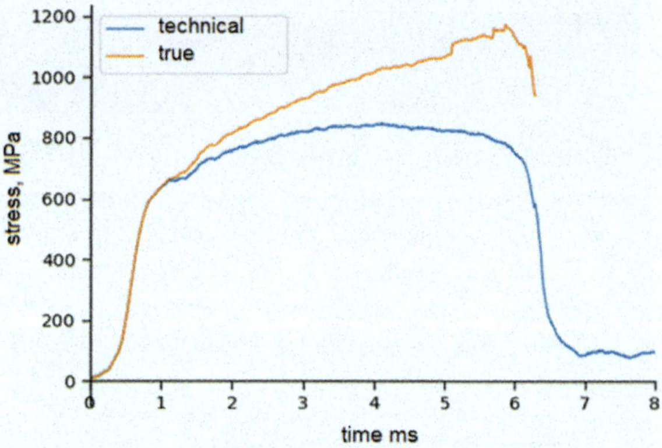

Fig. 4.13 Time vs. technical and true stress.

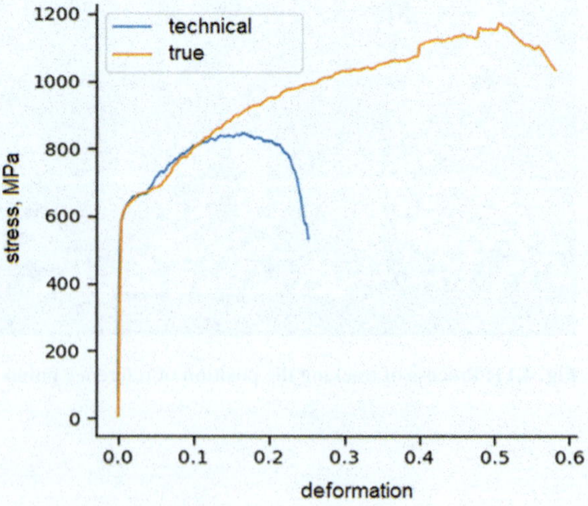

Fig. 4.14 Technical and true stress-strain diagrams.

operating at temperatures from -70 to 425° C under pressure. The high strength of the material and good indicators of deformation properties in a wide temperature range, the ability to change the properties of the alloy after heat treatment, a wide choice of different grades, a variety of thicknesses have led to the wide use of this alloy in many areas of production and engineering.

Building structures, pipes for transporting various liquids (water, oil, etc.) and gases, tanks for various purposes, steam boilers, oilfield equipment and various machine parts, including agricultural ones, are made from 09G2S steel. Of interest are the rate dependences of the mechanical characteristics of this alloy for predicting the behavior of structures made of this material under shock loads. In this section, for the first time, the results of studying the deformation and fracture of specimens from the 09G2S alloy in a wide range of deformation rates are presented.

The test specimens were made from a 3 mm thick sheet. Static tensile tests were carried out on the Z100 ZWICK-ROEL test facility. High-speed tests (with strain rates above 500 1/s) under compression and tension were carried out on a setup that implements the Kolsky method. The tests were carried out on high-strength steel measuring bars with a diameter of 20 mm. Tensile tests with intermediate strain rates (of the order of 100 1/s) were carried out on a pneumodynamic experimental setup created in the course of the dissertation research.

For static tests, samples with a working part width of 20 mm, a thickness of 3 mm, and a working part length of 120 mm were used. The tests were carried out until the destruction of the sample. The view of the sample after testing is shown in Fig. 4.15. Figure 4.16 illustrates the technical deformation curves determined during stat-ic tests of specimens made of steel 09G2S. It can be seen from the presented diagram that the deformation diagram has a yield point (A-B).

To build a true strain curve, the section of the diagram located to the right of the strain localization point (the point of maximum stress on the technical strain diagram) was discarded. For the point corresponding to the fracture moment of the sample, plastic deformation and average stress were determined as follows:

$$\varepsilon_\mathrm{f} = \ln\left(\frac{1}{1-\psi}\right), \tag{4.5}$$

$$\sigma_\mathrm{f} = \frac{\sigma_\mathrm{y}}{1-\psi} \tag{4.6}$$

where ψ is the relative constriction at rupture, σ_y is the conditional (technical) stress at which the rupture occurred.

The true deformation curves of steel 09G2S are shown in Fig. 4.17. The dotted

Fig. 4.15: Sample after testing.

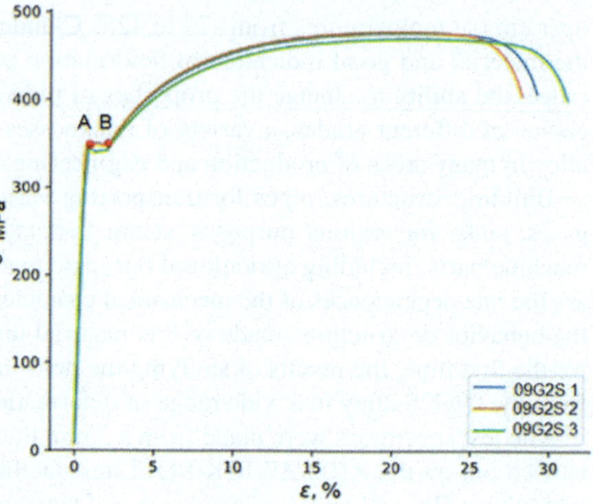

Fig. 4.16 Engineering stress-strain diagrams.

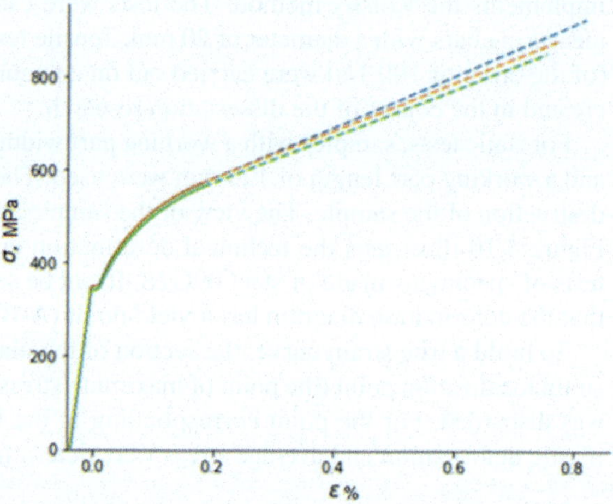

Fig. 4.17 True stress-strain diagrams.

line shows the section of the linear approximation of the diagram, connecting the point of the beginning of the localization of deformation and the point of rupture of the sample. High-speed tensile tests were carried out using a modification of the Kolsky direct tensile method.

To determine strain diagrams in tension at a strain rate of about 100 1/s, the previously described pneumodynamic setup was used. Samples-plates with a working part length of 10 mm, a working part width of 5.4 mm, and a thickness of 0.7 mm were tested. The test results are shown in Fig. 4.18. The dotted line shows the history of change in the strain rate of the samples during loading (right vertical axis). It can be seen that a deformation rate close to a constant value is realized. Figure 4.19 shows the results of static and dynamic tests.

The dependence of the initial tensile yield strength on the strain rate is shown in Fig. 4.20. The rate dependences of the ultimate fracture characteristics of steel 09G2S

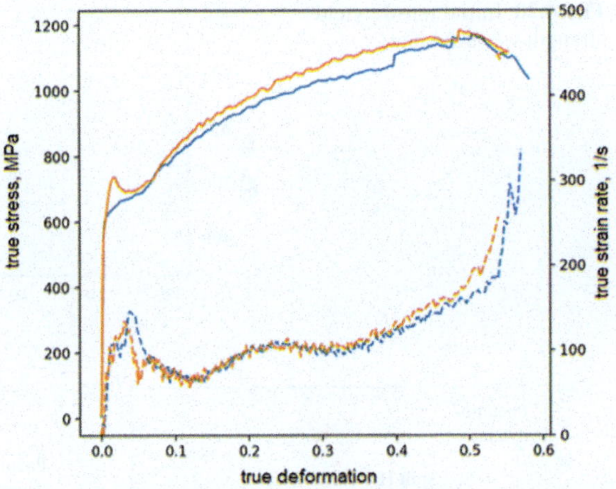

Fig. 4.18 True stress-strain and stress-strain rate diagrams.

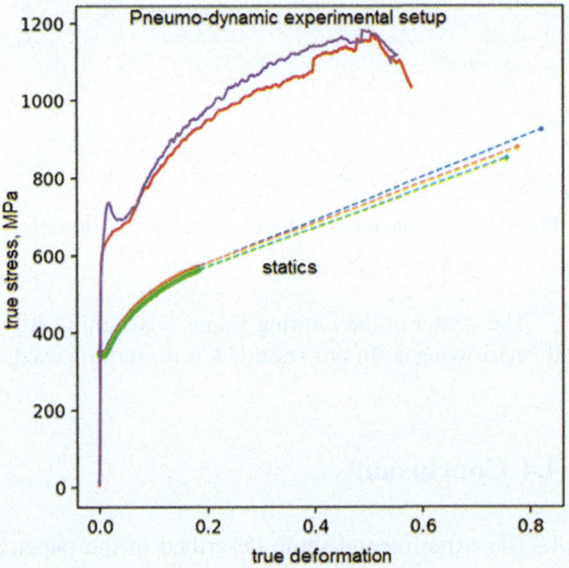

Fig. 4.19 True stress-strain diagrams.

are shown in Fig. 4.21 (ψ relative narrowing after rupture, δ relative elongation after rupture). Blue stars represent static test data, red round markers - data obtained on a pneumodynamic installation, orange squares - results of testing samples with a base of 8 mm on the split Hopkinson bar, green triangles - results of testing samples with a base of 5 mm on the split Hopkinson bar. It can be seen that the ultimate elongation after rupture (left side of the figure) strongly depends on the sample base. The limiting narrowing, which characterizes the plastic deformation of fracture under conditions of pronounced localization of deformation, is practically independent of the strain rate.

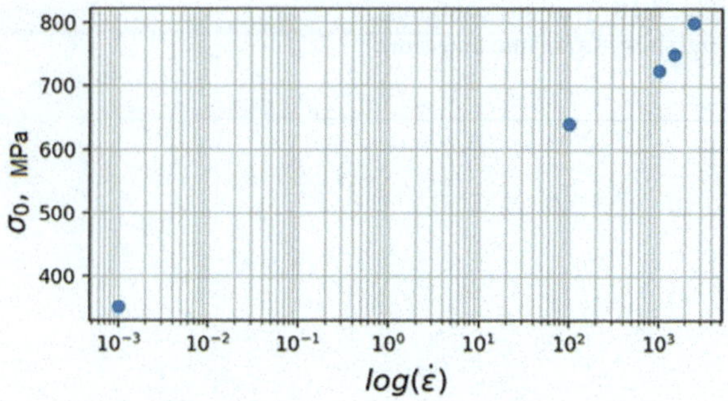

Fig. 4.20 Initial tensile yield strength vs. strain rate.

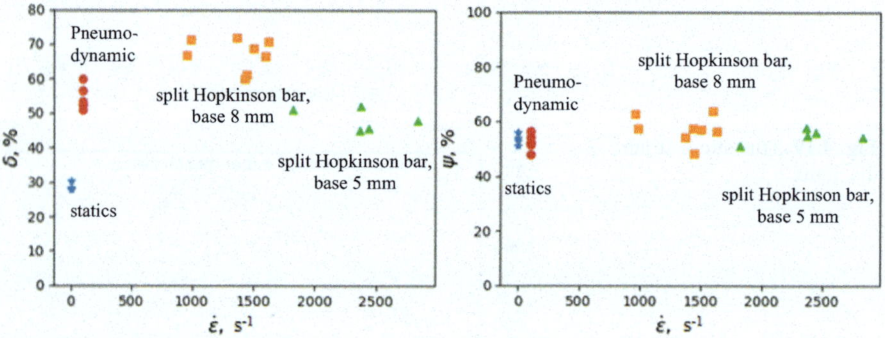

Fig. 4.21: Ultimate fracture characteristics of steel 09G2S.

The scatter of the limiting values δ is significantly higher for the limiting value of the narrowing ψ. In this regard, ψ is most often used in the strain-type failure criteria.

4.4 Conclusion

1. The experimental setup described in the paper makes it possible to study the processes of deformation and destruction of lamellar specimens at moderate strain rates.
2. To improve the accuracy of measuring the strains of the working part of the sample, high-speed video registration was used in combination with the DIC tool.
3. The DIC method was used to construct true strain diagrams taking into account the heterogeneity of strain fields in the working part of the specimen.
4. The results of testing specimens of steel 09G2S in a wide range of strain rates are presented, showing a significant effect of strain rate on the yield strength of the studied material. The limiting characteristics of plasticity in this case are practically independent of the strain rate.

Acknowledgements The work was supported by the Russian Science Foundation grant no. 21-19-00283.

References

[1] Hopkinson B (1914) A method of measuring the pressure produced in the detonation of high explosives or by the impact of bullets, Philosophical Transactions of the Royal Society of London, Series A Containing Papers of a Mathematical or Physical Character **213**:437–456. DOI 10.1098/rsta.1914.0010

[2] (Rusty) Gray GT III (2000) Classic Split-Hopkinson Pressure Bar Testing, In: Mechanical Testing and Evaluation, Vol 8, ASM Handbook (Kuhn H, Medlin D Eds), ASM Int, Materials Park OH, pp 462–476. DOI 10.31399/asm.hb.v08.a0003296

[3] Bragov AM, Lomunov AK (1984) Features of the construction of strain diagrams by the Kolsky method (in Russ), Applied Problems of Strength and Plasticity **28**:125-137.

[4] Caverzan A, Cadoni E, di Prisco M (2012) Tensile behaviour of high performance fibre-reinforced cementitious composites at high strain rates, International Journal of Impact Engineering **45**:28-38. DOI 10.1016/j.ijimpeng.2012.01.006

[5] Davies RM (1948) A critical study of the Hopkinson pressure bar, Philosophical Transactions of the Royal Society of London, Series A **240**:375–457. DOI 10.1098/rsta.1948.0001

[6] Eskandari H, Nemes JA (2000) Dynamic testing of composite laminates with a tensile split Hopkinson bar, Journal of Composite Materials **34**(4):260-273. DOI 10.1177/002199830003400401

[7] Gama BA, Lopatnikov SL, Gillespie JW Jr (2004) Hopkinson bar experimental technique: A critical review. Applied Mechanics Reviews **57**4:223-250. DOI 10.1115/1.1704626

[8] Jiang B, Zhang R (2006) Tensile properties in the through-thickness direction for a carbon fiber woven reinforced composite at impact loading rate, Journal de Physique IV **134**(1):1071–1075. DOI 10.1051/jp4:2006134164

[9] Kolsky H (1949) An investigation of the mechanical proper-ties of materials at very high rates of loading, Proceedings of the Physical Society, Section B, **62**(11):676–700. DOI 10.1088/0370-1301/62/11/302

[10] Lomunov AK (1987) Methodology for studying the processes of viscoplastic deformation and material properties based on the split Hopkinson bar (in Russ), PhD thesis, 1987.

[11] Cadoni E, Dotta M, Forni D, Kaufmann H (2016) Effects of strain rate on mechanical properties in tension of a commercial aluminium alloy used in armour applications, Procedia Structural Integrity **2**:986-993. DOI 10.1016/j.prostr.2016.06.126

[12] Cadoni E, Dotta M, Forni D, Bianchi S, Kaufmann H (2012) Strain rate effects on mechanical properties in tension of aluminium alloys used in armour applications, EPJ Web of Conferences **26**:05004. DOI 10.1051/epjconf/20122605004
[13] Cadoni E, Dotta M, Forni D, Spaetig P (2011) Strain-rate behavior in tension of the tempered martensitic reduced activation steel Eurofer97, Journal of Nuclear Materials **414**(3):360–366. DOI 10.1016/j.jnucmat.2011.05.002
[14] Cadoni E, Dotta M, Forni D, Tesio N (2011) Dynamic behaviour of reinforcing steel bars in tension, In: Applied Mechanics and Materials (Eds Cadoni E, di Prisco, M) **82**:86–91, Trans Tech Publications Ltd. DOI 10.4028/www.scientific.net/AMM.82.86
[15] Asprone D, Cadoni E, Prota A, Manfredi G (2009) Strain-rate sensitivity of a pultruded E-glass/polyester composite, Journal of Composites for Construction **13**(6):558–564. DOI 10.1061/(ASCE)CC.1943-5614.0000036
[16] Shi Y, Shi J, Luo Z, Liu H, Wang D, Shen H (2020) Experimental investigation on strength and deformation characteristics of red sandstone at strain rates of $10^{-2} \sim 55$ s^{-1}, Advances in Civil Engineering, Special Issue Advancements in the Analysis and Design of Protective Structures against Extreme Loadings 2020 **2020**:8882976. DOI 10.1155/2020/8882976
[17] Grote DL, Park SW, Zhou M (2001) Dynamic behavior of concrete at high strain rates and pressures: I. experimental characterization, International Journal of Impact Engineering **25**(9):869–886. DOI 10.1016/S0734-743X(01)00020-3
[18] Zhang QB, Zhao J. (2014) Quasi-static and dynamic fracture behaviour of rock materials: phenomena and mechanisms, International Journal of Fracture **189**(1):1–32, 2014. DOI 10.1007/s10704-014-9959-z
[19] Ma L, Li Z, Liu J, Duan L, Wu J (2019) Mechanical properties of coral concrete subjected to uniaxial dynamic compression, Construction and Building Materials **199**:244–255. DOI 10.1016/j.conbuildmat.2018.12.032
[20] Ma L, Wu J, Wang M, Dong L, Wei H (2020) Dynamic compressive properties of dry and saturated coral rocks at high strain rates, Engineering Geology **272**:105615. DOI 10.1016/j.enggeo.2020.105615
[21] Solomos G, Albertini C, Labibes K, Pizzinato V, Viaccoz B (2004) Strain rate effects of nuclear steels in room and higher temperatures, Nuclear Engineering and Design **229**(2-3):139-149. DOI 10.1016/j.nucengdes.2003.10.006
[22] Albertini C, Montagnani M (1977) Dynamic material properties of several steels for fast breeder reactor safety analysis. Final report, EUR 5787 EN, Joint Research Centre Ispra - Applied Mechanics Division.
[23] Nilson M (2003) Constitutive model for Armox 500T and Armox 600T at low and medium strain rates, Swedish defence research agency, Technical report.
[24] Zabotkin K, O'Toole B, Trabia M (2003) Identification of the dynamic tensile properties of metals under moderate strain rates. 16th ASCE Engineering Mechanics Conference, July 16-18, 2003 University of Washington, Seattle.
[25] Kong SY, Remennikov A, Uy B (2010) The effect of medium strain rates on the mechanical properties of high performance steels, 21st Australasian Conference on the Mechanics of Structures and Materials, Melbourne, Victoria, Australia

[26] Cui J, Wang S, Wang S, Li G, Wang P, Liang C (2019) The effects of strain rates on mechanical properties and failure behavior of long glass fiber reinforced thermoplastic composites, Polymers **2019**(11):2019. DOI 10.3390/polym11122019

[27] Mirone G (2013) The dynamic effect of necking in Hopkinson bar tension tests, Mechanics of Materials **58**:84–96. DOI 10.1016/j.mechmat.2012.11.006

[28] Mirone G, Coralloa D, Barbagallo R (2016) Interaction of strain rate and necking on the stress-strain response of uniaxial tension tests by Hopkinson bar, Procedia Structural Integrity **2**:974–985. DOI 10.1016/j.prostr.2016.06.125

[29] Bridgman P (1964) Studies in Large Plastic Flow and Fracture, with Special Emphasis on the Effects of Hydrostatic Pressure, Harvard University Press. DOI 10.4159/harvard.9780674731349

[30] Alves M, Jones N (1999) Influence of hydrostatic stress on failure of axisymmetric notched specimens, Journal of the Mechanics and Physics of Solids **47**:643–667. DOI 10.1016/S0022-5096(98)00060-X

[31] Dietrich L, Miastkowski J, Szczepinski W (1970) Nowość graniczna elementów konstrukcji (Limiting Capacity of the Construction Elements, in Polish), PWN, Warsaw.

[32] Gromada M, Mishuris G, Öchsner A (2011) Correction Formulae for the Stress Distribution in Round Tensile Specimens at Neck Presence, SpringerBriefs in Applied Sciences and Technology, Computational Mechanics, Springer, Berlin, Heidelberg. DOI 10.1007/978-3-642-22134-7

[33] La Rosa G, Mirone G, Risitano A (2003) Postnecking elasto-plastic characterization: degree of approximation in the Bridgman method and properties of the flow-stress/true-stress ratio, Metallurgical and Materials Transactions A **34**(3):615–624. DOI 10.1007/s11661-003-0096-y

[34] Ling Y (2004) Uniaxial true stress–strain after necking, AMP Journal of Technology **5**:37–48.

[35] Malinin NN, Rżysko J (1981) Mechanika materiałów (Mechanics of Materials, in Polish), PWN, Warsaw.

[36] Mirone G (2004) A new model for the elastoplastic characterization and the stress–strain determination on the necking section of a tensile specimen, International Journal of Solids and Structures **41**(13):3545–3564. DOI 10.1016/j.ijsolstr.2004.02.011

[37] Zhang ZL, Hauge M, Odegard J, Thaulow C (1999) Determining material true stress–strain curve from tensile specimens with rectangular cross section, International Journal of Solids and Structures **36**(23):2386–2405. DOI 10.1016/S0020-7683(98)00153-X

[38] Davidenkov NN, Spiridonova NN (1945) Analysis of the stress state in the neck of a stretched specimen (in Russ.), Zavodskaya Laboratoriya (6):583-593.

[39] Mehdikhani M, Aravand M, Sabuncuoglu B, Callens MG, Lomov SV, Gorbatikh L (2016) Full-field strain measurements at the micro-scale in fiber-reinforced composites using digital image correlation, Composite Structures **140**:192–201. DOI 10.1016/j.compstruct.2015.12.020

[40] Jones I, Iadicola ME (Eds) (2018) A Good Practices Guide for Digital Image Correlation, International Digital Image Correlation Society. DOI 10.32720/idics/gpg.ed1/print.format

[41] Li L, Sun L, Dai Z, Xiong Z, Huang B, Zhang Y (2019) Experimental investigation on mechanical properties and failure mechanisms of polymer composite-metal hybrid materials processed by direct injection-molding adhesion method, Journal of Materials Processing Technology **263**:385–395. DOI 10.1016/j.jmatprotec.2018.08.039

[42] Codolini A, Li QM, Wilkinson A (2018) Mechanical characterization of thin injection-moulded polypropylene specimens under large in-plane shear deformations, Polymer Testing **69**:485–489. DOI 10.1016/j.polymertesting.2018.06.010

[43] Röhrig C, Scheer T, Diebels S (2017) Mechanical characterization of a short fiber-reinforced polymer at roomtemperature: Experimental setups evaluated by an optical measurement system, Continuum Mechanics and Thermodynamics **29**:1093–1111. DOI 10.1007/s00161-017-0560-3

[44] McCormick N, Lord J (2010) Digital Image Correlation, Materialstoday **13**(12):52–54. DOI 10.1016/S1369-7021(10)70235-2

Chapter 5
Strain Gradient Elasticity and Dual Internal Variables

Arkadi Berezovski

Abstract The framework of dual internal variables approach is used to formulate a strain gradient elasticity model. The elimination of internal variables results in an equation of motion with no causality issues. Furthermore, the Aifantis-type strain gradient model is reproduced if the primary internal variable is proportional to the Laplacian of the strain gradient.

Key words: Strain gradient elasticity · Internal variables · Causality

5.1 Introduction

The strain gradient elasticity is a natural extension of classical theory that takes into account the internal length scale [1]. Classical linear elasticity theory is characterized by the Hooke law which in one-dimensional case takes the form of the stress-strain relation

$$\sigma = \rho c^2 \varepsilon, \tag{5.1}$$

where σ is the stress, ρ is the density, ε is the strain, c is the elastic wave speed. The corresponding equation of motion in terms of the displacement u is the well known wave equation

$$u_{tt} = c^2 u_{xx}. \tag{5.2}$$

Accordingly, the free energy density W is a quadratic function of the strain

$$W = \frac{\rho c^2}{2} \varepsilon^2. \tag{5.3}$$

Arkadi Berezovski
Department of Cybernetics, School of Science, Tallinn University of Technology, Ehitajate tee 5, 19086 Tallinn, Estonia, e-mail: arkadi.berezovski@cs.ioc.ee

H. Altenbach et al. (eds.), *Sixty Shades of Generalized Continua*,
Advanced Structured Materials 170, https://doi.org/10.1007/978-3-031-26186-2_5

Taking the strain gradient into account, the quadratic free energy is represented as

$$W = \frac{\rho c^2}{2}\varepsilon^2 + A\varepsilon\varepsilon_x + \frac{1}{2}B(\varepsilon_x)^2. \tag{5.4}$$

Then the stress is generalized, by definition,

$$\sigma = \frac{\partial W}{\partial \varepsilon} = \rho c^2 \varepsilon + A\varepsilon_x, \tag{5.5}$$

and the double stress is, correspondingly,

$$\mu = \frac{\partial W}{\partial \varepsilon_x} = A\varepsilon + B\varepsilon_x. \tag{5.6}$$

In terms of stresses, the equation of motion takes the form [2]

$$\rho u_{tt} = \sigma_x - \mu_{xx}, \tag{5.7}$$

which, in terms of displacement, is transformed to

$$\rho u_{tt} = \rho c^2 u_{xx} - B u_{xxxx}. \tag{5.8}$$

It is instructive to visualize dispersion properties of this equation. Considering harmonic wave with the frequency ω and wavenumber k

$$u(x,t) = \hat{u} e^{\mathrm{i}(kx-\omega t)}, \tag{5.9}$$

we arrive at the dispersion relation

$$\rho\omega^2 = \rho c^2 k^2 + Bk^4. \tag{5.10}$$

In terms of dimensionless frequency and wavenumber

$$\omega' = \frac{\omega}{\omega_0}, \quad k' = \frac{ck}{\omega_0}. \tag{5.11}$$

we have

$$\omega'^2 = k'^2 + B'k'^4, \tag{5.12}$$

where ω_0 is a characteristic frequency and

$$B' = \frac{B\omega_0^2}{\rho c^4}.$$

The corresponding dispersion curve is shown in Fig. 5.1. As can be seen, only acoustical branch of the dispersion curve is reproduced [3, 4, c.f.]. The dispersion curve with positive B is unrealistic [5, 6].

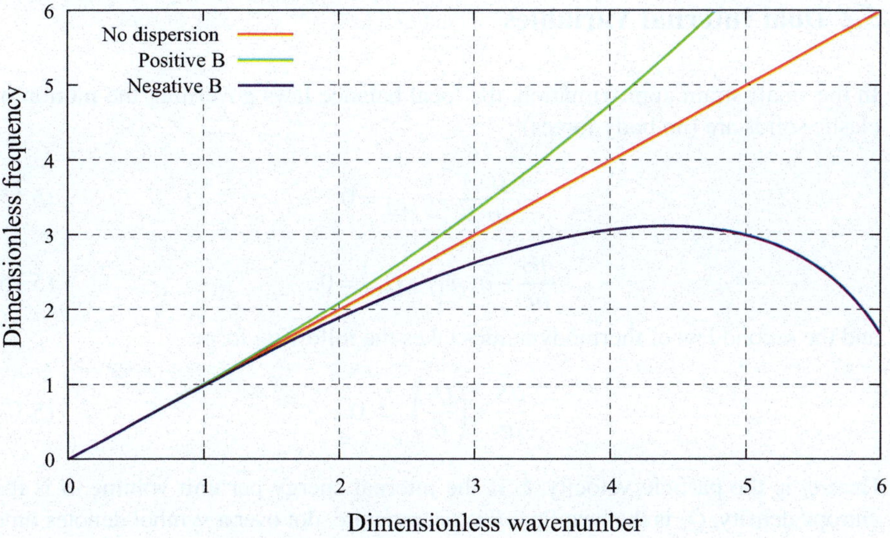

Fig. 5.1: Dispersion curves for $B' = 0.0256$.

The same situation holds true for a low frequency, long wavelength approximation of the micromorphic elasticity [7] with vanishing relative deformation [2], [8]-[13] based on the free energy density for linear isotropic elasticity in the form [2]

$$\begin{aligned} W =& \frac{1}{2}\lambda\varepsilon_{ii}\varepsilon_{jj} + \mu\varepsilon_{ij}\varepsilon_{ij} + a_1\varepsilon_{ik,i}\varepsilon_{jj,k} + a_2\varepsilon_{jj,i}\varepsilon_{kk,i} \\ &+ a_3\varepsilon_{ik,i}\varepsilon_{jk,j} + a_4\varepsilon_{jk,i}\varepsilon_{jk,i} + a_5\varepsilon_{jk,i}\varepsilon_{ij,k}, \end{aligned} \tag{5.13}$$

where ε_{ij} is the strain tensor, λ and μ are Lamé constants, a_i are additional constitutive parameters.

Another approach to strain gradient elasticity is represented by a Laplacian-based theory of gradient elasticity provided by Aifantis and coworkers [15]-[17]

$$\sigma_{ij} = \lambda\delta_{ij}\varepsilon_{kk} + 2\mu\varepsilon_{ij} - l^2(\lambda\delta_{ij}\varepsilon_{kk,mm} + 2\mu\varepsilon_{ij,mm}), \tag{5.14}$$

where σ_{ij} is the stress tensor and l is a length scale parameter. This idea, which has been extensively elaborated by Askes and coworkers [5, 6], [18]-[23] as well as by other researchers [24]-[29], is unable to reveal the optical branch of the dispersion curve due to the absence of higher order time derivatives in the equations of motion. This is related to the problem of causality [30] and the possibility of infinite wave propagation speed. The causality problem can be solved by either phenomenologically adding the higher order time derivatives [20, 30] or by using internal variables [31]-[35]. The goal of this paper is to show how dual internal variables can be used to model strain gradient elasticity while avoiding the causality problem. Numerous strain gradient applications [14, 22] necessitate this.

5.2 Dual Internal Variables

In the small-strain approximation, the local balance laws governing the motion of elastic solids are (no body forces)

$$\frac{\partial(\rho v_i)}{\partial t} - \sigma_{ij,j} = 0, \tag{5.15}$$

$$\frac{\partial \mathcal{E}}{\partial t} - \sigma_{ij}\dot{\varepsilon}_{ij} + Q_{i,i} = 0, \tag{5.16}$$

and the second law of thermodynamics takes the following form:

$$\frac{\partial S}{\partial t} + \left(\frac{Q_i}{\theta}\right)_{,i} \geq 0. \tag{5.17}$$

Here v_i is the particle velocity, $\mathcal{E}$ is the internal energy per unit volume, S is the entropy density, Q_i is the heat flux, θ is temperature, dot over a symbol denotes time derivative.

Applying the dual internal variable concept [36], we consider internal variables as second-order tensors. In this approach, the free energy per unit volume W is represented as a sufficiently regular function of strain, temperature, and two unspecified internal variables φ_{ij} and ψ_{ij}

$$W = \overline{W}(\varepsilon_{ij}, \theta, \varphi_{ij}, \psi_{ij}). \tag{5.18}$$

The stress and entropy are defined as usual

$$\sigma_{ij} = \frac{\partial \overline{W}}{\partial \varepsilon_{ij}}, \qquad S = -\frac{\partial \overline{W}}{\partial \theta}, \tag{5.19}$$

and two quantities conjugated to internal variables are introduced as follows:

$$A_{ij} := -\frac{\partial \overline{W}}{\partial \varphi_{ij}}, \qquad B_{ij} := -\frac{\partial \overline{W}}{\partial \psi_{ij}}. \tag{5.20}$$

In terms of the free energy

$$W = \mathcal{E} - \theta S,$$

entropy inequality (5.17) is transformed to

$$-\frac{\partial \overline{W}}{\partial t} - S\frac{\partial \theta}{\partial t} + \left(\frac{Q_i}{\theta}\right)\theta_{,i} + \sigma_{ij}\dot{\varepsilon}_{ij} \geq 0. \tag{5.21}$$

The time derivative of the free energy is determined by accepted functional dependence (5.18) and equations of state (5.19)

$$
\begin{aligned}
\frac{\partial \overline{W}}{\partial t} &= \frac{\partial \overline{W}}{\partial \varepsilon_{ij}}\frac{\partial \varepsilon_{ij}}{\partial t} + \frac{\partial \overline{W}}{\partial \theta}\frac{\partial \theta}{\partial t} + \frac{\partial \overline{W}}{\partial \varphi_{ij}}\frac{\partial \varphi_{ij}}{\partial t} + \frac{\partial \overline{W}}{\partial \psi_{ij}}\frac{\partial \psi_{ij}}{\partial t} \\
&= \sigma_{ij}\dot{\varepsilon}_{ij} - S\dot{\theta} - A_{ij}\dot{\varphi}_{ij} - B_{ij}\dot{\psi}_{ij}.
\end{aligned} \tag{5.22}
$$

Here dot over a symbol denotes time derivative. When we combine relationship (5.22) and dissipation inequality (5.21) we get

$$
-\frac{Q_i}{\theta}\theta_{,i} + A_{ij}\dot{\varphi}_{ij} + B_{ij}\dot{\psi}_{ij} \geq 0. \tag{5.23}
$$

Consequently, dissipation is governed by thermal and internal factors.

5.2.1 Evolution Equations

In the isothermal case, the dissipation inequality is expressed in terms of internal variables

$$
A_{ij}\dot{\varphi}_{ij} + B_{ij}\dot{\psi}_{ij} \geq 0. \tag{5.24}
$$

If the case is non-dissipative, the most straightforward solution of Eq. (5.24) is

$$
\dot{\varphi}_{ij} = B_{ij}, \tag{5.25}
$$

$$
\dot{\psi}_{ij} = -A_{ij}. \tag{5.26}
$$

In this case, the evolution of one internal variable is driven by another, indicating the duality of the internal variables.

5.2.2 Quadratic Free Energy

A quadratic function of the arguments is used to represent the free energy

$$
\overline{W} = \frac{1}{2}\lambda\varepsilon_{ii}\varepsilon_{jj} + \mu\varepsilon_{ij}\varepsilon_{ij} + D_1\varepsilon_{ij}\varphi_{ij} + \frac{D_2}{2}\varphi_{ij}\varphi_{ij} + \frac{D_3}{2}\psi_{ij}\psi_{ij}. \tag{5.27}
$$

The internal variable φ_{ij} is regarded as primary while the internal variable ψ_{ij} is considered auxiliary. Equations of state that correspond to them are

$$
\begin{aligned}
\sigma_{ij} &= \frac{\partial \overline{W}}{\partial \varepsilon_{ij}} = \lambda\delta_{ij}\varepsilon_{kk} + 2\mu\varepsilon_{ij} + D_1\varphi_{ij}, \\
A_{ij} &= -\frac{\partial \overline{W}}{\partial \varphi_{ij}} = -D_1\varepsilon_{ij} - D_2\varphi_{ij}, \quad B_{ij} = -\frac{\partial \overline{W}}{\partial \psi_{ij}} = -D_3\psi_{ij}.
\end{aligned} \tag{5.28}
$$

In terms of the displacement, the balance of linear momentum takes the form

$$\rho \ddot{u}_i = (\lambda+\mu)u_{j,ij} + \mu u_{i,jj} + D_1 \varphi_{ij,j}. \tag{5.29}$$

The evolution equation for the internal variable φ is determined by elimination the second internal variable using equations (5.25) and (5.26)

$$\ddot{\varphi}_{ij} = -D_3 \dot{\psi}_{ij} = D_3 A_{ij} = -D_3 D_1 \varepsilon_{ij} - D_3 D_2 \varphi_{ij}. \tag{5.30}$$

Furthermore, we can write a single equation of motion in terms of displacement differentiating equation of motion (5.29) twice with respect to time and applying evolution equation (5.30)

$$\begin{aligned} \rho \ddddot{u}_i - (\lambda+\mu)\ddot{u}_{j,ij} - \mu \ddot{u}_{i,jj} &= -D_3 D_1^2 \frac{1}{2}(u_{i,jj} + u_{j,ij}) \\ &\quad - D_3 D_2 (\rho \ddot{u}_i - (\lambda+\mu)u_{j,ij} - \mu u_{i,jj}). \end{aligned} \tag{5.31}$$

We expect that the internal variable φ_{ij} is a function of the strain gradient, i.e.,

$$\varphi_{ij} = \varphi(\varepsilon_{ij,k}).$$

That is, the second rank tensor φ_{ij} must be expressed in terms of the third rank tensor $\varepsilon_{ij,k}$. The obvious choice for such a relationship is the proportionality of internal variable φ_{ij} to the Laplacian of the strain gradient

$$\varphi_{ij} = k\varepsilon_{ij,kk} = k\Delta\varepsilon_{ij}, \tag{5.32}$$

where k is an appropriate constant. The application of Eq. (5.32) in the definition of the stress $(5.28)_1$ results in the Aifantis strain gradient model [5, 6, 15, 16, 17, 22]

$$\sigma_{ij} = \lambda\delta_{ij}\varepsilon_{kk} + 2\mu\varepsilon_{ij} + kD_1\Delta\varepsilon_{ij}. \tag{5.33}$$

However, equation of motion (5.31) is free from the causality problem.

5.3 Concluding Remarks

The use of dual internal variables allows to eliminate the causality problems in the corresponding dispersive wave equation in the context of strain gradient elasticity. The Aifantis model for the strain gradient elasticity is reproduced by proportionality of the internal variable to the Laplacian of the strain tensor. Since the third-order strain gradient tensor is not explicitly included in the presented model, it does not replace existing strain gradient models. The model can be combined with other strain gradient models that are currently available [9, 19, 22, 23, 26, 27], [37]-[41].

Acknowledgements The work was supported by the Estonian Research Council under Research Project RPG1227 and by the grant project with No. 22-00863K of the Czech Science Foundation (CSF) within institutional support RVO:61388998.

References

[1] Aifantis EC (2016) Internal length gradient (ILG) material mechanics across scales and disciplines, Advances in Applied Mechanics **49**:1-110. DOI 10.1016/bs.aams.2016.08.001

[2] Mindlin RD, Eshel NN (1968) On first strain-gradient theories in linear elasticity, International Journal of Solids and Structures **4**(1):109-124. DOI 10.1016/0020-7683(68)90036-X

[3] Lombardo M, Askes H (2012) Higher-order gradient continuum modelling of periodic lattice materials, Computational Materials Science **52**(1):204-208. DOI 10.1016/j.commatsci.2011.05.025

[4] Madeo A, Neff P, Ghiba I-D, Placidi L, Rosi G (2013) Wave propagation in relaxed micromorphic continua: modeling metamaterials with frequency band-gaps, Continuum Mechanics and Thermodynamics **27**(4-5):551-570. DOI 10.1007/s00161-013-0329-2

[5] Askes H, Suiker ASJ, Sluys LJ (2002) A classification of higher-order strain-gradient models – linear analysis, Archive of Applied Mechanics **72**(2-3):171-188. DOI 10.1007/s00419-002-0202-4

[6] Metrikine AV, Askes H (2002) One-dimensional dynamically consistent gradient elasticity models derived from a discrete microstructure, European Journal of Mechanics - A/Solids **21**(4):555-572. DOI 10.1016/S0997-7538(02)01218-4

[7] Mindlin RD (1964) Micro-structure in linear elasticity, Archive for Rational Mechanics and Analysis **16**(1):51-78. DOI 10.1007/BF00248490

[8] Fleck NA, Hutchinson JW (1997) Strain gradient plasticity, In: Advances in Applied Mechanics, Vol 33, pp 295-361, Elsevier. DOI 10.1016/S0065-2156(08)70388-0

[9] Lam DCC, Yang F, Chong ACM, Wang J, Tong P (2003) Experiments and theory in strain gradient elasticity, Journal of the Mechanics and Physics of Solids **51**(8):1477-1508. DOI 10.1016/S0022-5096(03)00053-X

[10] Zervos A (2008) Finite elements for elasticity with microstructure and gradient elasticity, International Journal for Numerical Methods in Engineering **73**(4):564-595. DOI 10.1002/nme.2093

[11] Dell'Isola F, Sciarra G, Vidoli S (2009) Generalized Hooke's law for isotropic second gradient materials, Proceedings of the Royal Society A: Mathematical, Physical and Engineering Sciences **465**(2107):2177-2196. DOI 10.1098/rspa.2008.0530

[12] Forest S, Sab K (2021) Stress gradient continuum theory, Mechanics Research Communications **40**:16-25. DOI 10.1016/j.mechrescom.2011.12.002

[13] Rosi G, Placidi L, Auffray N (2018) On the validity range of strain-gradient elasticity: A mixed static-dynamic identification procedure, European Journal of Mechanics - A/Solids **69**:179-191. DOI 10.1016/j.euromechsol.2017.12.005

[14] Aifantis EC (2020) A concise review of gradient models in mechanics and physics, Frontiers in Physics **7**:239. DOI 10.3389/fphy.2019.00239

[15] Aifantis EC (1992) On the role of gradients in the localization of deformation and fracture, International Journal of Engineering Science, **30**(10):1279-1299. DOI 10.1016/0020-7225(92)90141-3

[16] Ru CQ, Aifantis EC (1993) A simple approach to solve boundary-value problems in gradient elasticity, Acta Mechanica **101**(1-4):59-68. DOI 10.1007/BF01175597

[17] Altan BS, Aifantis EC (1997) On some aspects in the special theory of gradient elasticity, Journal of the Mechanical Behavior of Materials **8**(3):231-282. DOI 10.1515/JMBM.1997.8.3.231

[18] Askes H, Metrikine AV (2002) One-dimensional dynamically consistent gradient elasticity models derived from a discrete microstructure: Part 2: Static and dynamic response, European Journal of Mechanics - A/Solids **21**(4):573-588. DOI 10.1016/S0997-7538(02)01217-2

[19] Askes H, Aifantis EC (2006) Gradient elasticity theories in statics and dynamics - a unification of approaches, International Journal of Fracture **139**(2):297-304. DOI 10.1007/s10704-006-8375-4

[20] Askes H, Metrikine AV, Pichugin AV, Bennett T (2008) Four simplified gradient elasticity models for the simulation of dispersive wave propagation, Philosophical Magazine **88**(28-29):3415-3443. DOI 10.1080/14786430802524108

[21] Askes H, Gitman IM (2010) Review and critique of the stress gradient elasticity theories of Eringen and Aifantis, In: Maugin G, Metrikine A (Eds) Mechanics of Generalized Continua, Advances in Mechanics and Mathematics, Vol 21, pp 203-210, Springer, New York, NY. DOI 10.1007/978-1-4419-5695-8_21

[22] Askes H, Aifantis EC (2011) Gradient elasticity in statics and dynamics: An overview of formulations, length scale identification procedures, finite element implementations and new results, International Journal of Solids and Structures **48**(13):1962-1990. DOI 10.1016/j.ijsolstr.2011.03.006

[23] DeDomenico D, Askes H (2016) A new multi-scale dispersive gradient elasticity model with micro-inertia: Formulation and C^0-finite element implementation, International Journal for Numerical Methods in Engineering **108**(5):485-512. DOI 10.1002/nme.5222

[24] Polizzotto C (2014) Stress gradient versus strain gradient constitutive models within elasticity, International Journal of Solids and Structures, **51**(9):1809-1818. DOI 10.1016/j.ijsolstr.2014.01.021

[25] Broese C, Tsakmakis C, Beskos D (2016) Mindlin's micro-structural and gradient elasticity theories and their thermodynamics, Journal of Elasticity **125**(1):87-132. DOI 10.1007/s10659-016-9572-7

[26] Polizzotto C (2017) A hierarchy of simplified constitutive models within isotropic strain gradient elasticity, European Journal of Mechanics - A/Solids **61**:92-109. DOI 10.1016/j.euromechsol.2016.09.006

[27] Khakalo S, Niiranen J (2018) Form II of Mindlin's second strain gradient theory of elasticity with a simplification: For materials and structures from nano-to macro-scales, European Journal of Mechanics-A/Solids **71**:292-319. DOI 10.1016/j.euromechsol.2018.02.013

[28] Forest S (2020) Strain gradient elasticity from capillarity to the mechanics of nano-objects, In: Bertram A, Forest S (Eds) Mechanics of Strain Gradient Materials, CISM International Centre for Mechanical Sciences, Vol 600, pp 37-70, Springer, Cham. DOI 10.1007/978-3-030-43830-2_3

[29] Broese C, Papargyri-Beskou S, Tsakmakis C (2021) Comparison of Laplacian based explicit and implicit gradient elasticity with reference to one-dimensional problems in statics and dynamics, European Journal of Mechanics - A/Solids **88**:104270. DOI 10.1016/j.euromechsol.2021.104270

[30] Metrikine AV (2006) On causality of the gradient elasticity models, Journal of Sound and Vibration **297**(3-5):727-742. DOI 10.1016/j.jsv.2006.04.017

[31] Engelbrecht J, Berezovski A, Pastrone F, Braun M (2005) Waves in microstructured materials and dispersion, Philosophical Magazine **85**(33-35):4127-4141. DOI 10.1080/14786430500362769

[32] Berezovski A., Engelbrecht J, Maugin GA (2009) One-dimensional microstructure dynamics, In: Ganghoffer JF, Pastrone F (Eds) Mechanics of Microstructured Solids, Lecture Notes in Applied and Computational Mechanics, Vol 46, pp 21-28, Springer, Berlin, Heidelberg. DOI 10.1007/978-3-642-00911-2_3

[33] Berezovski A, Engelbrecht J, Berezovski M (2011) Waves in microstructured solids: a unified viewpoint of modeling, Acta Mechanica **220**(1-4):349-363. DOI 10.1007/s00707-011-0468-0

[34] Engelbrecht J, Berezovski A (2015) Reflections on mathematical models of deformation waves in elastic microstructured solids, Mathematics and Mechanics of Complex Systems **3**(1):43-82. DOI 10.2140/memocs.2015.3.43

[35] Berezovski A, Yildizdag ME, Scerrato D (2020) On the wave dispersion in microstructured solids, Continuum Mechanics and Thermodynamics **32**(3):569-588. DOI 10.1007/s00161-018-0683-1

[36] Ván P, Berezovski A, Engelbrecht J (2008) Internal variables and dynamic degrees of freedom, Journal of Non-Equilibrium Thermodynamics **33**(3):235-254. DOI 10.1515/JNETDY.2008.010

[37] Bacigalupo A, Gambarotta L (2014) Second-gradient homogenized model for wave propagation in heterogeneous periodic media, International Journal of Solids and Structures **51**(5):1052-1065. DOI 10.1016/j.ijsolstr.2013.12.001

[38] Zhou S, Li A, Wang B (2016) A reformulation of constitutive relations in the strain gradient elasticity theory for isotropic materials, International Journal of Solids and Structures **80**:28-37. DOI 10.1016/j.ijsolstr.2015.10.018

[39] Madeo A, Neff P, Barbagallo G, d'Agostino MV, Ghiba I-D (2017) A review on wave propagation modeling in band-gap metamaterials via enriched continuum models, In: dell'Isola F, Sofonea M, Steigmann D. (Eds) Mathematical Modelling in Solid Mechanics, Advanced Structured Materials, vol 69, pp 89-105, Springer, Singapore. DOI 10.1007/978-981-10-3764-1_6

[40] DeDomenico D, Askes H, Aifantis EC (2019) Gradient elasticity and dispersive wave propagation: Model motivation and length scale identification procedures in concrete and composite laminates, International Journal of Solids and Structures **158**:176-190. DOI 10.1016/j.ijsolstr.2018.09.007

[41] Fu G, Zhou S, Qi L (2020) On the strain gradient elasticity theory for isotropic materials, International Journal of Engineering Science **154**:103348. DOI /10.1016/j.ijengsci.2020.103348

Chapter 6
On the Coercivity of Strain Energy Functions in Generalized Models of 6-Parameter Shells

Mircea Bîrsan and Patrizio Neff

Abstract In this paper we consider geometrically nonlinear 6-parameter shell models. We establish some existence proofs by the direct methods of the calculus of variations. In contrast to more classical approaches, we also investigate models up to order h^5 in the shell thickness, where the form of the equations is determined by a dimensional descent from a three-dimensional Cosserat model.

Key words: Geometrically nonlinear shells · 6-parameter shells · Cosserat models · Strain energy density · Coercivity · Existence of minimizers

6.1 Introduction

The theory of shells is a vast subject. It is useful to distinguish three types of models: the shells of Kirchhoff–Love–Koiter type (with normality assumption, see e.g. [1]-[6]), the 5-parameter shells of Reissner–Mindlin type (allowing for transverse shear, see e.g. [7]-[9]), and the 6-parameter shells (in addition allowing for in-plane drill [10]-[12]). In the linearized setting, all these models are well-posed: one can prove existence and uniqueness of solutions (see, e.g., [13]-[16]).

However, in the geometrically nonlinear setting astounding differences appear. Indeed, as will be shown in a forthcoming paper, the Koiter and the Reissner–Mindlin

Mircea Bîrsan
Faculty of Mathematics, University Duisburg-Essen, Thea-Leymann Str. 9, 45127 Essen, Germany & Alexandru Ioan Cuza University of Iaşi, Department of Mathematics, Blvd. Carol I, no. 11, 700506 Iaşi, Romania,
e-mail: mircea.birsan@uni-due.de

Patrizio Neff
Lehrstuhl für Nichtlineare Analysis und Modellierung, Faculty of Mathematics, University Duisburg-Essen, Thea-Leymann Str. 9, 45127 Essen, Germany,
e-mail: patrizio.neff@uni-due.de

H. Altenbach et al. (eds.), *Sixty Shades of Generalized Continua*,
Advanced Structured Materials 170, https://doi.org/10.1007/978-3-031-26186-2_6

models (based on quadratic strain and curvature energies) are both not well-posed in the sense that global minimizers do not exist. This may come as a surprise since they are often used by engineers. On the contrary, the general 6-parameter (Cosserat, micropolar) shell model is well-posed. This has been shown for the first time in [17] for the flat shell problem (see also [18, 19]) and in general in [20], and here we will expand on our knowledge of that. Also from an engineering point of view, 6-parameter shells have certain advantages, e.g. the imposition of boundary conditions is transparent, and these shells can easily be coupled with beam elements. Therefore, we strongly advocate the use of 6-parameter shell models, although it means that a numerical code implementing such models must be able to handle the rotation map.

Outline of the paper. In Sect. 6.2 we present briefly the governing equations of 6-parameter shells, including the differential geometry of the reference midsurface, the strain measures and the general stress-strain relations. In Sect. 6.3 we introduce the isotropic Cosserat 6-parameter shell model of order $O(h^3)$ and show that the strain energy function is coercive under certain conditions on the constitutive coefficients. Then, we apply the direct methods of the calculus of variations to prove the existence of minimizers to the variational problem. In Sect. 6.4 we consider the higher order Cosserat 6-parameter shell model and show the coercivity of the areal strain energy density. Also, we use this coercivity property to prove existence results for the minimization problem associated to equilibrium of Cosserat (6-parameter) shells of order $O(h^5)$.

Summary of notations. We present first some useful notations which will be used throughout this paper. The Latin indices $i, j, k, \ldots$ range over the set $\{1,2,3\}$, while the Greek indices $\alpha, \beta, \gamma, \ldots$ range over the set $\{1,2\}$. The Einstein summation convention over repeated indices is used. A subscript comma preceding an index i (or α) designates partial differentiation with respect to the variable x_i (or x_α, respectively), e.g. $f_{,i} = \frac{\partial f}{\partial x_i}$. We denote by δ_i^j the Kronecker symbol and employ the direct tensor notation. Thus, $\otimes$ designates the dyadic product, while $\mathrm{axl}(\boldsymbol{W})$ stands for the axial vector of any skew-symmetric tensor $\boldsymbol{W}$. For any second order tensor $\boldsymbol{X}$, let $\mathrm{tr}(\boldsymbol{X})$ designate the trace, $\mathrm{sym}(\boldsymbol{X})$ the symmetric part, and $\mathrm{skew}(\boldsymbol{X})$ the skew-symmetric part of $\boldsymbol{X}$. The scalar product between any second order tensors $\boldsymbol{A}$ and $\boldsymbol{B}$ is denoted by $\langle \boldsymbol{A}, \boldsymbol{B} \rangle = \mathrm{tr}(\boldsymbol{A}^T \boldsymbol{B})$. For any vector $\boldsymbol{v}$ and second order tensor $\boldsymbol{A}$ we write also $\boldsymbol{v}\boldsymbol{A} = \boldsymbol{A}^T \boldsymbol{v}$.

6.2 General 6-Parameter Elastic Shells. Governing Equations

Let us present first the geometry and kinematics of general 6-parameter shells. Denote by $\mathcal{S}_c$ the deformed (current) configuration of the shell and by $\mathcal{S}_\xi$ its reference configuration. The midsurface of the reference configuration is denoted by ω_ξ. This surface is determined by the position vector $\boldsymbol{y}_0(x_1, x_2)$, where $\boldsymbol{y}_0 : \omega \subset \mathbb{R}^2 \to \omega_\xi \subset \mathbb{R}^3$ is a parametric representation. The curvilinear coordinates (x_1, x_2) are convected coordinates on the surface ω_ξ.

Concerning the differential geometry of the midsurface ω_ξ, we introduce the covariant base vectors $\boldsymbol{a}_\alpha$ and the contravariant base vectors $\boldsymbol{a}^\alpha$ by

$$\boldsymbol{a}_\alpha = \frac{\partial \boldsymbol{y}_0}{\partial x_\alpha}, \qquad \boldsymbol{a}^\alpha \cdot \boldsymbol{a}_\beta = \delta^\alpha_\beta \qquad (\alpha,\beta = 1,2). \tag{6.1}$$

Moreover, we introduce the vectors

$$\boldsymbol{a}_3 = \boldsymbol{a}^3 = \boldsymbol{n}_0, \qquad \text{where} \qquad \boldsymbol{n}_0 = \frac{\boldsymbol{a}_1 \times \boldsymbol{a}_2}{\|\boldsymbol{a}_1 \times \boldsymbol{a}_2\|} \tag{6.2}$$

is the unit normal vector to the surface ω_ξ. Let $\boldsymbol{a}$ and $\boldsymbol{b}$ be the first fundamental tensor and the second fundamental tensor of the midsuface ω_ξ, respectively, i.e.

$$\begin{aligned} \boldsymbol{a} &= \mathrm{Grad}_s\, \boldsymbol{y}_0 = \boldsymbol{a}_\alpha \otimes \boldsymbol{a}^\alpha = a_{\alpha\beta} \boldsymbol{a}^\alpha \otimes \boldsymbol{a}^\beta = a^{\alpha\beta} \boldsymbol{a}_\alpha \otimes \boldsymbol{a}_\beta, \\ \boldsymbol{b} &= -\mathrm{Grad}_s\, \boldsymbol{n}_0 = -\boldsymbol{n}_{0,\alpha} \otimes \boldsymbol{a}^\alpha = b_{\alpha\beta}\, \boldsymbol{a}^\alpha \otimes \boldsymbol{a}^\beta = b^\alpha_\beta\, \boldsymbol{a}_\alpha \otimes \boldsymbol{a}^\beta, \end{aligned} \tag{6.3}$$

where Grad_s is the surface gradient operator defined by $\mathrm{Grad}_s\, \boldsymbol{f} = \boldsymbol{f}_{,\alpha} \otimes \boldsymbol{a}^\alpha$ for any $\boldsymbol{f}$. Also, we employ the surface divergence operator given by $\mathrm{Div}_s\, \boldsymbol{T} = \boldsymbol{T}_{,\alpha}\, \boldsymbol{a}^\alpha$ for any second order tensor $\boldsymbol{T}$. The so-called alternator tensor $\boldsymbol{c}$ of the surface is

$$\boldsymbol{c} = \frac{1}{\sqrt{a}}\, \epsilon_{\alpha\beta}\, \boldsymbol{a}_\alpha \otimes \boldsymbol{a}_\beta = \sqrt{a}\, \epsilon_{\alpha\beta}\, \boldsymbol{a}^\alpha \otimes \boldsymbol{a}^\beta, \tag{6.4}$$

where

$$a = \det\big(a_{\alpha\beta}\big)_{2\times 2} > 0$$

and $\epsilon_{\alpha\beta}$ is the two-dimensional alternator

$$\epsilon_{12} = -\epsilon_{21} = 1,\ \epsilon_{11} = \epsilon_{22} = 0.$$

Let

$$H = \frac{1}{2}\mathrm{tr}\, \boldsymbol{b} = \frac{1}{2} b^\alpha_\alpha$$

be the mean curvature and

$$K = \det \boldsymbol{b} = \det\big(b^\alpha_\beta\big)_{2\times 2}$$

be the Gauß curvature of the surface ω_ξ. Then, the relation of Cayley-Hamilton type holds

$$\boldsymbol{b}^2 - 2H\boldsymbol{b} + K\boldsymbol{a} = \boldsymbol{0}. \tag{6.5}$$

The last relation is equivalent to $\boldsymbol{b}(2H\boldsymbol{a} - \boldsymbol{b}) = K\boldsymbol{a}$. Hence, we introduce the tensor

$$\boldsymbol{b}^* = 2H\boldsymbol{a} - \boldsymbol{b}, \tag{6.6}$$

which can be regarded as the cofactor of $\boldsymbol{b}$ in the tangent plane. We also denote by κ_1, κ_2 the principal curvatures of the reference midsurface and we assume as usual that $|\kappa_\alpha h| < 1$ $(\alpha = 1,2)$.

To describe its deformation, we refer the shell to a Cartesian coordinate frame $Ox_1x_2x_3$ with orthonormal base vectors $\{\boldsymbol{e}_1, \boldsymbol{e}_2, \boldsymbol{e}_3\}$. The reference configuration is characterized by the position vector $\boldsymbol{y}_0$ and the initial microrotation tensor $\boldsymbol{Q}_0$ as

$$\begin{aligned} &\boldsymbol{y}_0 : \omega \subset \mathbb{R}^2 \to \omega_\xi \subset \mathbb{R}^3, \qquad &&\boldsymbol{y}_0 = \boldsymbol{y}_0(x_1,x_2), \\ &\boldsymbol{Q}_0 : \omega \subset \mathbb{R}^2 \to \mathrm{SO}(3), \qquad &&\boldsymbol{Q}_0 = \boldsymbol{d}_i^0(x_1,x_2) \otimes \boldsymbol{e}_i\,. \end{aligned} \tag{6.7}$$

The parameter domain ω is a bounded open domain with Lipschitz boundary $\partial\omega$ in the Ox_1x_2 plane. The reference directors $\{\boldsymbol{d}_1^0, \boldsymbol{d}_2^0, \boldsymbol{d}_3^0\}$ are orthonormal and the third director $\boldsymbol{d}_3^0$ is chosen to coincide with the unit normal in the reference configuration, i.e. $\boldsymbol{d}_3^0 = \boldsymbol{n}_0$. The shell deformation is characterized by the deformation function $\boldsymbol{m}$ and the microrotation tensor $\boldsymbol{Q}_e$ given by

$$\begin{aligned} &\boldsymbol{m} : \omega \to \omega_c\,, \qquad &&\boldsymbol{m} = \boldsymbol{m}(x_1,x_2), \\ &\boldsymbol{Q}_e : \omega \to \mathrm{SO}(3), \qquad &&\boldsymbol{Q}_e = \boldsymbol{Q}_e(x_1,x_2) = \boldsymbol{d}_i \otimes \boldsymbol{d}_i^0\,. \end{aligned} \tag{6.8}$$

Here, ω_c is the deformed midsurface and $\{\boldsymbol{d}_1, \boldsymbol{d}_2, \boldsymbol{d}_3\}$ is the orthonormal triad of directors in the deformed configuration.

We introduce the strain measures of 6-parameter shells as follows [10, 11, 21]: the shell strain tensor is

$$\boldsymbol{E}^e = \boldsymbol{Q}_e^T \mathrm{Grad}_s \boldsymbol{m} - \boldsymbol{a} \tag{6.9}$$

and the shell bending-curvature tensor is

$$\boldsymbol{K}^e = \mathrm{axl}(\boldsymbol{Q}_e^T \boldsymbol{Q}_{e,\alpha}) \otimes \boldsymbol{a}^\alpha. \tag{6.10}$$

The local equilibrium equations for 6-parameter shells have the following form (see, e.g. [21, 22])

$$\mathrm{Div}_s \boldsymbol{N} + \boldsymbol{f} = \boldsymbol{0}, \qquad \mathrm{Div}_s \boldsymbol{M} + \mathrm{axl}\big(\boldsymbol{N}\boldsymbol{F}^T - \boldsymbol{F}\boldsymbol{N}^T\big) + \boldsymbol{l} = \boldsymbol{0}, \tag{6.11}$$

where $\boldsymbol{N}$ is the internal surface stress tensor and $\boldsymbol{M}$ the internal surface couple stress tensor (of the first Piola-Kirchhoff type). The tensor

$$\boldsymbol{F} = \mathrm{Grad}_s \boldsymbol{m} \;=\; \boldsymbol{m}_{,\alpha} \otimes \boldsymbol{a}^\alpha$$

is the shell deformation gradient, while the vectors $\boldsymbol{f}$ and $\boldsymbol{l}$ are the external body forces and body couples, respectively. We consider the following boundary conditions of mixed type prescribed on the boundary curve $\partial\omega_\xi$ [23, 24, 20]

$$\begin{aligned} &\boldsymbol{N}\boldsymbol{\nu}=\boldsymbol{N}^{*},\ \boldsymbol{M}\boldsymbol{\nu}=\boldsymbol{M}^{*} \ \text{ along } \partial\omega_f\,, \\ &\boldsymbol{m}=\boldsymbol{m}^{*},\quad \boldsymbol{Q}_e=\boldsymbol{Q}^{*} \quad \text{along } \partial\omega_d\,, \end{aligned} \tag{6.12}$$

where $\partial\omega_f\cup\partial\omega_d=\partial\omega_\xi$ is a disjoint partition of the boundary curve $\partial\omega_\xi$. Here, $\boldsymbol{N}^*$ and $\boldsymbol{M}^*$ are the external boundary force and couple vectors respectively, applied along the deformed boundary curve, but measured per unit length of $\partial\omega_f$. The vector $\boldsymbol{\nu}$ is the outer unit normal to the boundary curve $\partial\omega_\xi$, lying in the tangent plane.

Let the areal strain energy density for 6-parameter shells be given as a function of the strain measures in the form

$$\mathcal{W}_{\text{shell}}=\mathcal{W}_{\text{shell}}(\boldsymbol{E}^e,\boldsymbol{K}^e). \tag{6.13}$$

Under hyperelasticity assumptions, the stress and couple stress tensors are expressed by the following constitutive relations

$$\boldsymbol{Q}_e^T\boldsymbol{N}=\frac{\partial\mathcal{W}_{\text{shell}}}{\partial\boldsymbol{E}^e}\,,\qquad \boldsymbol{Q}_e^T\boldsymbol{M}=\frac{\partial\mathcal{W}_{\text{shell}}}{\partial\boldsymbol{K}^e}\,. \tag{6.14}$$

To obtain the explicit form of the stress-strain relations we need the specific expression of the strain energy function $\mathcal{W}_{\text{shell}}$. In [21], Eremeyev and Pietraszkiewicz have presented the general form of a quadratic energy density for isotropic shells, but the constitutive coefficients are not determined in terms of three-dimensional material constants. For instance, the following simplified expression of the energy density is proposed [21]

$$\begin{aligned} 2\widehat{\mathcal{W}}_{\text{shell}}(\boldsymbol{E}^e,\boldsymbol{K}^e)=&\;\alpha_1\big[\text{tr}(\boldsymbol{a}\boldsymbol{E}^e)\big]^2+\alpha_2\text{tr}\big[(\boldsymbol{a}\boldsymbol{E}^e)^2\big]+\alpha_3\|\boldsymbol{a}\boldsymbol{E}^e\|^2+\alpha_4\|\boldsymbol{n}_0\boldsymbol{E}^e\|^2\\ &+\beta_1\big[\text{tr}(\boldsymbol{a}\boldsymbol{K}^e)\big]^2+\beta_2\text{tr}\big[(\boldsymbol{a}\boldsymbol{K}^e)^2\big]+\beta_3\|\boldsymbol{a}\boldsymbol{K}^e\|^2+\beta_4\|\boldsymbol{n}_0\boldsymbol{K}^e\|^2, \end{aligned}$$

or equivalently, in view of the relation $\text{tr}(\boldsymbol{X}^2)=\|\text{sym}(\boldsymbol{X})\|^2-\|\text{skew}(\boldsymbol{X})\|^2$,

$$\begin{aligned} 2\widehat{\mathcal{W}}_{\text{shell}}(\boldsymbol{E}^e,\boldsymbol{K}^e)=&\;(\alpha_2+\alpha_3)\|\text{sym}(\boldsymbol{a}\boldsymbol{E}^e)\|^2+(\alpha_3-\alpha_2)\|\text{skew}(\boldsymbol{a}\boldsymbol{E}^e)\|^2\\ &+\alpha_1\big[\text{tr}(\boldsymbol{a}\boldsymbol{E}^e)\big]^2+\alpha_4\|\boldsymbol{n}_0\boldsymbol{E}^e\|^2+(\beta_2+\beta_3)\|\text{sym}(\boldsymbol{a}\boldsymbol{K}^e)\|^2\\ &+(\beta_3-\beta_2)\|\text{skew}(\boldsymbol{a}\boldsymbol{K}^e)\|^2+\beta_1\big[\text{tr}(\boldsymbol{a}\boldsymbol{K}^e)\big]^2+\beta_4\|\boldsymbol{n}_0\boldsymbol{K}^e\|^2, \end{aligned} \tag{6.15}$$

where α_k and β_k $(k=1,2,3,4)$ are constant constitutive coefficients.

Let us introduce the surface deviator operator dev_s defined in [25] by

$$\text{dev}_s\boldsymbol{T}=\boldsymbol{T}-\frac{1}{2}(\text{tr}\boldsymbol{T})\boldsymbol{a}$$

for any $\boldsymbol{T}$. Then, we can decompose any tensor of the type $\boldsymbol{X}=X_{i\alpha}\boldsymbol{a}^i\otimes\boldsymbol{a}^\alpha$ as a direct sum (orthogonal decomposition) in the form

$$\boldsymbol{X}=\text{dev}_s(\text{sym}\,\boldsymbol{X})+\text{skew}\,\boldsymbol{X}+\frac{1}{2}(\text{tr}\,\boldsymbol{X})\boldsymbol{a}. \tag{6.16}$$

Hence, it follows that

$$\|\mathrm{sym}\,\boldsymbol{X}\|^2 = \|\mathrm{dev}_s(\mathrm{sym}\,\boldsymbol{X})\|^2 + \frac{1}{2}(\mathrm{tr}\,\boldsymbol{X})^2 \tag{6.17}$$

and relation (6.15) can be written in the form

$$\begin{aligned} 2\,\widehat{W}_{\mathrm{shell}}(\boldsymbol{E}^e,\boldsymbol{K}^e) = {} & (\alpha_2+\alpha_3)\|\mathrm{dev}_s\mathrm{sym}(\boldsymbol{a}\boldsymbol{E}^e)\|^2 + (\alpha_3-\alpha_2)\|\mathrm{skew}(\boldsymbol{a}\boldsymbol{E}^e)\|^2 \\ & + \Big(\alpha_1 + \frac{\alpha_2+\alpha_3}{2}\Big)\big[\mathrm{tr}(\boldsymbol{a}\boldsymbol{E}^e)\big]^2 + \alpha_4\|\boldsymbol{n}_0\boldsymbol{E}^e\|^2 \\ & + (\beta_2+\beta_3)\|\mathrm{dev}_s\mathrm{sym}(\boldsymbol{a}\boldsymbol{K}^e)\|^2 + (\beta_3-\beta_2)\|\mathrm{skew}(\boldsymbol{a}\boldsymbol{K}^e)\|^2 \\ & + \Big(\beta_1 + \frac{\beta_2+\beta_3}{2}\Big)\big[\mathrm{tr}(\boldsymbol{a}\boldsymbol{K}^e)\big]^2 + \beta_4\|\boldsymbol{n}_0\boldsymbol{K}^e\|^2. \end{aligned} \tag{6.18}$$

From the last relation we see that the strain energy function $\widehat{W}_{\mathrm{shell}}(\boldsymbol{E}^e,\boldsymbol{K}^e)$ is coercive in terms of $\boldsymbol{E}^e$ and $\boldsymbol{K}^e$ provided that the coefficients verify the conditions

$$\begin{aligned} & 2\alpha_1+\alpha_2+\alpha_3>0, \qquad \alpha_2+\alpha_3>0, \qquad \alpha_3-\alpha_2>0, \qquad \alpha_4>0, \\ & 2\beta_1+\beta_2+\beta_3>0, \qquad \beta_2+\beta_3>0, \qquad \beta_3-\beta_2>0, \qquad \beta_4>0. \end{aligned} \tag{6.19}$$

Indeed, if we denote by

$$c_1 = \min\{2\alpha_1+\alpha_2+\alpha_3\,,\,\alpha_2+\alpha_3\,,\,\alpha_3-\alpha_2\,,\,\alpha_4\} > 0$$

and

$$c_2 = \min\{2\beta_1+\beta_2+\beta_3\,,\,\beta_2+\beta_3\,,\,\beta_3-\beta_2\,,\,\beta_4\} > 0,$$

then from (6.16) and (6.18) we obtain

$$\begin{aligned} 2\,\widehat{W}_{\mathrm{shell}}(\boldsymbol{E}^e,\boldsymbol{K}^e) \geq {} & c_1\Big(\|\mathrm{dev}_s\mathrm{sym}(\boldsymbol{a}\boldsymbol{E}^e)\|^2 + \|\mathrm{skew}(\boldsymbol{a}\boldsymbol{E}^e)\|^2 + \frac{1}{2}\big[\mathrm{tr}(\boldsymbol{a}\boldsymbol{E}^e)\big]^2 \\ & + \|\boldsymbol{n}_0\boldsymbol{E}^e\|^2\Big) + c_2\Big(\|\mathrm{dev}_s\mathrm{sym}(\boldsymbol{a}\boldsymbol{K}^e)\|^2 + \|\mathrm{skew}(\boldsymbol{a}\boldsymbol{K}^e)\|^2 \\ & + \frac{1}{2}\big[\mathrm{tr}(\boldsymbol{a}\boldsymbol{K}^e)\big]^2 + \|\boldsymbol{n}_0\boldsymbol{K}^e\|^2\Big) \\ \geq {} & c_1\Big(\|\boldsymbol{a}\boldsymbol{E}^e\|^2 + \|\boldsymbol{n}_0\boldsymbol{E}^e\|^2\Big) + c_2\Big(\|\boldsymbol{a}\boldsymbol{K}^e\|^2 + \|\boldsymbol{n}_0\boldsymbol{K}^e\|^2\Big) \\ = {} & c_1\|\boldsymbol{E}^e\|^2 + c_2\|\boldsymbol{K}^e\|^2, \end{aligned} \tag{6.20}$$

i.e., the strain energy function is coercive. Remark that the same conditions (6.19) have been imposed in [15] to establish existence results for linear 6-parameter shells, see also [26, 27].

The energy function $\widehat{W}_{\mathrm{shell}}$ has been employed to solve shell problems in [11, 28]. To this aim, the following values of the coefficients α_k and β_k have been chosen for isotropic shells made of a material with Poisson ratio ν and Young modulus E:

$$\begin{aligned} &\alpha_1 = C\,\nu, \qquad \alpha_2 = 0, \qquad \alpha_3 = C\,(1-\nu), \qquad \alpha_4 = \alpha_s\,C\,(1-\nu),\\ &\beta_1 = D\,\nu, \qquad \beta_2 = 0, \qquad \beta_3 = D\,(1-\nu), \qquad \beta_4 = \alpha_t\,D\,(1-\nu). \end{aligned} \tag{6.21}$$

Here, we denote by $C = E\,h/(1-\nu^2)$ the stretching (membrane) stiffness of the shell, $D = E\,h^3/12(1-\nu^2)$ is the bending stiffness, h is the thickness, and $\alpha_s = 5/6$, $\alpha_t = 7/10$ are two shear correction factors. Inserting (6.21) into (6.15) and using the relations

$$\begin{aligned} &C\,\frac{1+\nu}{2} = h\,\frac{\mu(3\lambda+2\mu)}{\lambda+2\mu}, \qquad C(1-\nu) = 2\mu h,\\ &D\,\frac{1+\nu}{2} = \frac{h^3}{12}\,\frac{\mu(3\lambda+2\mu)}{\lambda+2\mu}, \qquad D(1-\nu) = \frac{\mu h^3}{6}, \end{aligned} \tag{6.22}$$

where λ and μ are the Lamé constants of the isotropic elastic material, we see that the specific form of the strain energy density for 6-parameter shells commonly used in the literature is

$$\begin{aligned} \widehat{\mathcal{W}}_{\text{shell}}(\boldsymbol{E}^e,\boldsymbol{K}^e) = h\Big[\mu\,\|\boldsymbol{a}\boldsymbol{E}^e\|^2 + \frac{\lambda\mu}{\lambda+2\mu}\big[\text{tr}(\boldsymbol{a}\boldsymbol{E}^e)\big]^2 + \mu\,\alpha_s\,\|\boldsymbol{n}_0\boldsymbol{E}^e\|^2\Big]\\ + \frac{h^3}{12}\Big[\mu\,\|\boldsymbol{a}\boldsymbol{K}^e\|^2 + \frac{\lambda\mu}{\lambda+2\mu}\big[\text{tr}(\boldsymbol{a}\boldsymbol{K}^e)\big]^2 + \mu\,\alpha_t\,\|\boldsymbol{n}_0\boldsymbol{K}^e\|^2\Big]. \end{aligned} \tag{6.23}$$

In view of inequalities (6.19) and relations (6.21) and (6.22), we deduce that strain energy density (6.23) is coercive, provided that the Lamé constants satisfy the conditions

$$\mu > 0, \qquad 3\lambda + 2\mu > 0. \tag{6.24}$$

These conditions are usually assumed to hold for isotropic elastic materials.

In the next sections we consider two generalized models of 6-parameter shells made of Cosserat material, in which the constitutive relations are more elaborate. For these models we investigate the coercivity of the strain energy functions.

6.3 The Order h^3 Model of 6-Parameter Shells made of Cosserat Material

Starting from an isotropic three-dimensional Cosserat parent model, we have performed in [29, 30] a dimensional reduction and have obtained a generalized 6-parameter shell model of higher order, which has been investigated mathematically in [31]. Then, using a different method suggested by the classical shell theory [5, 6], we have derived in [22] a related 6-parameter Cosserat shell model of order $O(h^3)$, which will be analysed in details in this section.

Thus, the explicit form of the areal strain energy density $\mathcal{W}_{\text{shell}}$ has been obtained in [22] as a quadratic function of $(\boldsymbol{E}^e,\boldsymbol{K}^e)$, in which the coefficients are expressed in terms of the three-dimensional material constants and depend also on the curvature of the reference midsurface. Let λ, μ, and μ_c denote the Lamé constants and the Cosserat couple modulus, respectively, of the three-dimensional Cosserat material. We introduce the bilinear form $W_{\text{Coss}}(\cdot,\cdot)$ defined for any tensors $\boldsymbol{X} = X_{i\alpha}\boldsymbol{a}^i \otimes \boldsymbol{a}^\alpha$, $\boldsymbol{Y} = Y_{i\alpha}\boldsymbol{a}^i \otimes \boldsymbol{a}^\alpha$ by

$$\begin{aligned} W_{\text{Coss}}(\boldsymbol{X},\boldsymbol{Y}) = \mu\langle \text{sym}(\boldsymbol{a}\boldsymbol{X}), \text{sym}(\boldsymbol{a}\boldsymbol{Y})\rangle + \mu_c\langle \text{skew}(\boldsymbol{a}\boldsymbol{X}), \text{skew}(\boldsymbol{a}\boldsymbol{Y})\rangle \\ + \frac{\lambda\mu}{\lambda+2\mu}\text{tr}(\boldsymbol{a}\boldsymbol{X})\,\text{tr}(\boldsymbol{a}\boldsymbol{Y}) + \frac{2\mu\,\mu_c}{\mu+\mu_c}(\boldsymbol{n}_0\boldsymbol{X})\cdot(\boldsymbol{n}_0\boldsymbol{Y}), \end{aligned} \tag{6.25}$$

as well as the associated quadratic form

$$\begin{aligned} W_{\text{Coss}}(\boldsymbol{X}) = \mu\,\|\text{sym}(\boldsymbol{a}\boldsymbol{X})\|^2 + \mu_c\|\text{skew}(\boldsymbol{a}\boldsymbol{X})\|^2 + \frac{\lambda\mu}{\lambda+2\mu}\big[\text{tr}(\boldsymbol{a}\boldsymbol{X})\big]^2 \\ + \frac{2\mu\,\mu_c}{\mu+\mu_c}\|\boldsymbol{n}_0\boldsymbol{X}\|^2. \end{aligned} \tag{6.26}$$

Moreover, let $W_{\text{curv}}(\cdot,\cdot)$ be the bilinear form defined by

$$\begin{aligned} W_{\text{curv}}(\boldsymbol{X},\boldsymbol{Y}) = \mu L_c^2\Big[\, b_1\langle \text{sym}\boldsymbol{X}, \text{sym}\boldsymbol{Y}\rangle + b_2\langle \text{skew}\boldsymbol{X}, \text{skew}\boldsymbol{Y}\rangle \\ + \big(b_3 - \frac{b_1}{3}\big)(\text{tr}\boldsymbol{X})(\text{tr}\boldsymbol{Y})\Big] \end{aligned} \tag{6.27}$$

and the associated quadratic form

$$W_{\text{curv}}(\boldsymbol{X}) = \mu L_c^2\Big[\, b_1\,\|\,\text{sym}\,\boldsymbol{X}\|^2 + b_2\,\|\,\text{skew}\,\boldsymbol{X}\|^2 + \big(b_3 - \frac{b_1}{3}\big)(\text{tr}\,\boldsymbol{X})^2\Big], \tag{6.28}$$

where the coefficients b_1, b_2, b_3 are dimensionless constitutive coefficients and the parameter $L_c > 0$ introduces an internal length (characteristic for the Cosserat material, see details in [17]-[19], [32]).

With these notations, the explicit form of the shell strain energy density for the order h^3 model is given by (see [22, Eq. (68)])

$$\begin{aligned} \mathcal{W}^{(3)}_{\text{shell}}(\boldsymbol{E}^e,\boldsymbol{K}^e) = \Big(h - K\frac{h^3}{12}\Big)\Big[W_{\text{Coss}}(\boldsymbol{E}^e) + W_{\text{curv}}(\boldsymbol{K}^e)\Big] \\ + \frac{h^3}{12}\Big[W_{\text{Coss}}(\boldsymbol{E}^e\boldsymbol{b} + \boldsymbol{c}\boldsymbol{K}^e) - 2W_{\text{Coss}}(\boldsymbol{E}^e, \boldsymbol{c}\boldsymbol{K}^e\boldsymbol{b}^*) + W_{\text{curv}}(\boldsymbol{K}^e\boldsymbol{b})\Big], \end{aligned} \tag{6.29}$$

Notice that this strain energy function satisfies the invariance properties required by the local symmetry group of isotropic 6-parameter shells, which have been established by Eremeyev and Pietraszkiewicz in [21, Sect. 9].

6.3.1 Coercivity Results for the Model of Order $O(h^3)$

With a view toward proving the coercivity of the strain energy function, let us verify first that the quadratic forms $W_{\text{Coss}}(\boldsymbol{X})$ and $W_{\text{curv}}(\boldsymbol{X})$, which appear in (6.29), are coercive. Indeed, using (6.17), (6.26) and (6.28) we can write the equivalent forms

$$\begin{aligned} W_{\text{Coss}}(\boldsymbol{X}) &= \mu\,\|\text{dev}_{\text{s}}\text{sym}(\boldsymbol{a}\boldsymbol{X})\|^2 + \mu_c\|\text{skew}(\boldsymbol{a}\boldsymbol{X})\|^2 + \frac{\mu(3\lambda+2\mu)}{2(\lambda+2\mu)}\,(\text{tr}\boldsymbol{X})^2 \\ &\quad + \frac{2\mu\,\mu_c}{\mu+\mu_c}\,\|\boldsymbol{n}_0\boldsymbol{X}\|^2, \\ W_{\text{curv}}(\boldsymbol{X}) &= \mu L_c^2\Big(b_1\|\text{dev}_{\text{s}}\text{sym}(\boldsymbol{a}\boldsymbol{X})\|^2 + b_2\|\text{skew}(\boldsymbol{a}\boldsymbol{X})\|^2 + \big(b_3 + \frac{b_1}{6}\big)(\text{tr}\boldsymbol{X})^2 \\ &\quad + \frac{b_1+b_2}{2}\,\|\boldsymbol{n}_0\boldsymbol{X}\|^2\Big). \end{aligned} \tag{6.30}$$

Then, under the usual assumptions on material constants (6.24) from classical elasticity, together with the conditions $\mu_c > 0$ and $b_i > 0$ for the Cosserat material, we see that quadratic forms (6.30) are coercive, since it holds

$$W_{\text{Coss}}(\boldsymbol{X}) \geq \bar{C}_1\,\|\boldsymbol{X}\|^2 \qquad \text{and} \qquad W_{\text{curv}}(\boldsymbol{X}) \geq \bar{C}_2\,\|\boldsymbol{X}\|^2, \tag{6.31}$$

where the positive constants are

$$\bar{C}_1 = \min\Big\{\mu, \mu_c, \frac{\mu(3\lambda+2\mu)}{\lambda+2\mu}, \frac{2\mu\,\mu_c}{\mu+\mu_c}\Big\} > 0$$

and

$$C_2 = \mu L_c^2 \min\Big\{b_1, b_2, 2b_3 + \frac{b_1}{3}, \frac{b_1+b_2}{2}\Big\} > 0.$$

For the sake of brevity and for later convenience, let us denote by $\boldsymbol{\Phi}^e$ the mixed bending tensor

$$\boldsymbol{\Phi}^e = \boldsymbol{E}^e\boldsymbol{b} + c\boldsymbol{K}^e. \tag{6.32}$$

Then, using (6.5) and (6.32) we can decompose the strain energy density (6.29) as follows

$$\mathcal{W}^{(3)}_{\text{shell}}(\boldsymbol{E}^e, \boldsymbol{K}^e) = \mathcal{W}^{(3)}_{\text{memb,bend}}(\boldsymbol{E}^e, \boldsymbol{K}^e) + \mathcal{W}^{(3)}_{\text{bend,curv}}(\boldsymbol{K}^e), \tag{6.33}$$

where the membrane-bending part is given by

$$\begin{aligned} \mathcal{W}^{(3)}_{\text{memb,bend}}(\boldsymbol{E}^e, \boldsymbol{K}^e) &= \Big(h + K\frac{h^3}{12}\Big)W_{\text{Coss}}(\boldsymbol{E}^e) + \frac{h^3}{12}\,W_{\text{Coss}}(\boldsymbol{\Phi}^e) \\ &\quad - 2\frac{h^3}{12}\,W_{\text{Coss}}(\boldsymbol{E}^e, \boldsymbol{\Phi}^e\boldsymbol{b}^*). \end{aligned} \tag{6.34}$$

and the bending-curvature part is

$$\mathcal{W}^{(3)}_{\text{bend,curv}}(\boldsymbol{K}^e) = \Big(h - K\frac{h^3}{12}\Big)W_{\text{curv}}(\boldsymbol{K}^e) + \frac{h^3}{12}\,W_{\text{curv}}(\boldsymbol{K}^e\boldsymbol{b}). \tag{6.35}$$

Firstly, the bending-curvature part $\mathcal{W}^{(3)}_{\text{bend,curv}}$ of the energy density is obviously coercive, since for $|\kappa_\alpha h| < 1$ we have

$$|K|\,h^2 = |\kappa_1 h|\cdot|\kappa_2 h| < 1\,, \tag{6.36}$$

so we can write using $(6.31)_2$

$$\mathcal{W}^{(3)}_{\text{bend,curv}}(\boldsymbol{K}^e) \ge \Big(h - K\frac{h^3}{12}\Big)W_{\text{curv}}(\boldsymbol{K}^e) \ge \Big(h - \frac{h}{12}\Big)W_{\text{curv}}(\boldsymbol{K}^e) \ge \frac{11}{12}h\bar{C}_2\|\boldsymbol{K}^e\|^2. \tag{6.37}$$

Secondly, for the membrane-bending part of the energy we establish the following result.

Lemma 6.1. *Assume that the constitutive coefficients satisfy the conditions*

$$\mu > 0, \qquad 3\lambda + 2\mu > 0, \qquad \textit{and} \qquad \mu_c > 0. \tag{6.38}$$

Let κ be the maximum of the absolute value of the principal curvatures $|\kappa_\alpha|$ on the midsurface ω_ξ, i.e. $\kappa = \max_{\omega_\xi}\{|\kappa_1|,|\kappa_2|\}$. Assume that the product κh satisfies

$$\kappa h < \min\left\{\frac{1}{2}\,,\ \Big(\frac{47}{32}\cdot\frac{\min\{\lambda+2\mu\,,3\lambda+2\mu\}}{\lambda+\mu}\Big)^{1/2},\ \Big(\frac{47}{8}\cdot\frac{\min\{\mu,\mu_c\}}{\mu+\mu_c}\Big)^{1/2}\right\}. \tag{6.39}$$

Then, there exists a positive constant $C_1 > 0$ such that

$$\mathcal{W}^{(3)}_{\text{memb,bend}}(\boldsymbol{E}^e,\boldsymbol{K}^e) \ge C_1\,\|\boldsymbol{E}^e\|^2. \tag{6.40}$$

Proof. We notice first that the radicands in (6.39) are always positive, in view of the conditions (6.38).

In order to estimate the coupling term in energy function (6.34) we employ a similar technique as in the classical shell theory (see, e.g. [6]). Using the Cauchy-Schwarz inequality for the scalar product $W_{\text{Coss}}(\cdot,\cdot)$ we can write

$$\big|W_{\text{Coss}}(\boldsymbol{E}^e,\boldsymbol{\Phi}^e\boldsymbol{b}^*)\big| \le \sqrt{W_{\text{Coss}}(\boldsymbol{E}^e)}\,\sqrt{W_{\text{Coss}}(\boldsymbol{\Phi}^e\boldsymbol{b}^*)}\,. \tag{6.41}$$

To express the last term, we use the spectral representation of the tensor $\boldsymbol{b}$ in the form

$$\boldsymbol{b} = \kappa_1\,\boldsymbol{u}_1\otimes\boldsymbol{u}_1 + \kappa_2\,\boldsymbol{u}_2\otimes\boldsymbol{u}_2\,, \tag{6.42}$$

where $\boldsymbol{u}_1$ and $\boldsymbol{u}_2$ are the orthonormal principal vectors. Then, we have

$$\boldsymbol{a} = \boldsymbol{u}_1\otimes\boldsymbol{u}_1 + \boldsymbol{u}_2\otimes\boldsymbol{u}_2$$

and the cofactor $\boldsymbol{b}^*$ given by (6.6) can be written as

$$
\begin{aligned}
\boldsymbol{b}^* &= H\boldsymbol{a} + (H\boldsymbol{a} - \boldsymbol{b}) \\
&= H\boldsymbol{a} + \Big[\frac{\kappa_1+\kappa_2}{2}(\boldsymbol{u}_1\otimes\boldsymbol{u}_1+\boldsymbol{u}_2\otimes\boldsymbol{u}_2) - (\kappa_1\boldsymbol{u}_1\otimes\boldsymbol{u}_1+\kappa_2\boldsymbol{u}_2\otimes\boldsymbol{u}_2)\Big] \\
&= H\boldsymbol{a} + \frac{\kappa_1-\kappa_2}{2}(\boldsymbol{u}_2\otimes\boldsymbol{u}_2-\boldsymbol{u}_1\otimes\boldsymbol{u}_1).
\end{aligned}
\tag{6.43}
$$

Thus, we obtain

$$
\boldsymbol{\Phi}^e\boldsymbol{b}^* = H\boldsymbol{\Phi}^e + \frac{\kappa_1-\kappa_2}{2}\,\boldsymbol{\Phi}^e\boldsymbol{\delta} \tag{6.44}
$$

where

$$
\boldsymbol{\delta} := \boldsymbol{u}_2\otimes\boldsymbol{u}_2-\boldsymbol{u}_1\otimes\boldsymbol{u}_1 .
$$

We notice that

$$
\Big(\frac{\kappa_1-\kappa_2}{2}\Big)^2 = \Big(\frac{\kappa_1+\kappa_2}{2}\Big)^2 - \kappa_1\kappa_2 = H^2-K,
$$

so we can estimate

$$
\begin{aligned}
W_{\text{Coss}}(\boldsymbol{\Phi}^e\boldsymbol{b}^*) &= W_{\text{Coss}}\Big(H\boldsymbol{\Phi}^e + \frac{\kappa_1-\kappa_2}{2}\,\boldsymbol{\Phi}^e\boldsymbol{\delta}\Big) \\
&\le 2\big[H^2W_{\text{Coss}}(\boldsymbol{\Phi}^e) + (H^2-K)W_{\text{Coss}}(\boldsymbol{\Phi}^e\boldsymbol{\delta})\big].
\end{aligned}
\tag{6.45}
$$

With the notation

$$
\kappa = \max_{\omega_\xi}\{|\kappa_1|,|\kappa_2|\}
$$

we can write

$$
\begin{aligned}
H^2 &= \Big(\frac{\kappa_1+\kappa_2}{2}\Big)^2 \le \frac{\kappa_1^2+\kappa_2^2}{2} \le \kappa^2 \\
&\text{and} \\
H^2-K &= \Big(\frac{\kappa_1-\kappa_2}{2}\Big)^2 \le \frac{\kappa_1^2+\kappa_2^2}{2} \le \kappa^2 .
\end{aligned}
\tag{6.46}
$$

Hence, relation (6.45) yields

$$
W_{\text{Coss}}(\boldsymbol{\Phi}^e\boldsymbol{b}^*) \le 2\kappa^2\big[W_{\text{Coss}}(\boldsymbol{\Phi}^e) + W_{\text{Coss}}(\boldsymbol{\Phi}^e\boldsymbol{\delta})\big]. \tag{6.47}
$$

To estimate the sum in the last brackets we decompose

$$
\begin{aligned}
W_{\text{Coss}}(\boldsymbol{\Phi}^e) &= W_{\text{Coss}}(\boldsymbol{\Phi}^e(\boldsymbol{u}_1\otimes\boldsymbol{u}_1+\boldsymbol{u}_2\otimes\boldsymbol{u}_2)) \\
&= W_{\text{Coss}}(\boldsymbol{\Phi}^e\boldsymbol{u}_1\otimes\boldsymbol{u}_1) + W_{\text{Coss}}(\boldsymbol{\Phi}^e\boldsymbol{u}_2\otimes\boldsymbol{u}_2) \\
&\quad + 2\,W_{\text{Coss}}(\boldsymbol{\Phi}^e\boldsymbol{u}_1\otimes\boldsymbol{u}_1,\, \boldsymbol{\Phi}^e\boldsymbol{u}_2\otimes\boldsymbol{u}_2)
\end{aligned}
\tag{6.48}
$$

and

$$
\begin{aligned}
W_{\text{Coss}}\big(\boldsymbol{\Phi}^e\boldsymbol{\delta}\big) &= W_{\text{Coss}}\big(\boldsymbol{\Phi}^e(\boldsymbol{u}_2\otimes\boldsymbol{u}_2-\boldsymbol{u}_1\otimes\boldsymbol{u}_1)\big)\\
&= W_{\text{Coss}}\big(\boldsymbol{\Phi}^e\boldsymbol{u}_2\otimes\boldsymbol{u}_2\big)+W_{\text{Coss}}\big(\boldsymbol{\Phi}^e\boldsymbol{u}_1\otimes\boldsymbol{u}_1\big) \qquad (6.49)\\
&\quad -2\,W_{\text{Coss}}\big(\boldsymbol{\Phi}^e\boldsymbol{u}_1\otimes\boldsymbol{u}_1\,,\,\boldsymbol{\Phi}^e\boldsymbol{u}_2\otimes\boldsymbol{u}_2\big),
\end{aligned}
$$

so we get by addition

$$
W_{\text{Coss}}\big(\boldsymbol{\Phi}^e\big)+W_{\text{Coss}}\big(\boldsymbol{\Phi}^e\boldsymbol{\delta}\big) = 2\,W_{\text{Coss}}\big(\boldsymbol{\Phi}^e\boldsymbol{u}_1\otimes\boldsymbol{u}_1\big)+2\,W_{\text{Coss}}\big(\boldsymbol{\Phi}^e\boldsymbol{u}_2\otimes\boldsymbol{u}_2\big). \qquad (6.50)
$$

Then, inequality (6.47) becomes

$$
W_{\text{Coss}}\big(\boldsymbol{\Phi}^e\boldsymbol{b}^*\big) \le 4\kappa^2\big[W_{\text{Coss}}\big(\boldsymbol{\Phi}^e\boldsymbol{u}_1\otimes\boldsymbol{u}_1\big)+W_{\text{Coss}}\big(\boldsymbol{\Phi}^e\boldsymbol{u}_2\otimes\boldsymbol{u}_2\big)\big]. \qquad (6.51)
$$

Inserting this in (6.41), we obtain for membrane-bending energy (6.34) the inequality

$$
\begin{aligned}
\mathcal{W}^{(3)}_{\text{memb,bend}}(\boldsymbol{E}^e,\boldsymbol{K}^e) &\ge \Big(h+K\frac{h^3}{12}\Big)W_{\text{Coss}}\big(\boldsymbol{E}^e\big)+\frac{h^3}{12}\,W_{\text{Coss}}\big(\boldsymbol{\Phi}^e\big)\\
&\quad -4\kappa\frac{h^3}{12}\sqrt{W_{\text{Coss}}\big(\boldsymbol{E}^e\big)}\cdot\sqrt{W_{\text{Coss}}\big(\boldsymbol{\Phi}^e\boldsymbol{u}_1\otimes\boldsymbol{u}_1\big)+W_{\text{Coss}}\big(\boldsymbol{\Phi}^e\boldsymbol{u}_2\otimes\boldsymbol{u}_2\big)}\,.
\end{aligned}
\qquad (6.52)
$$

To show that the right-hand side is positive, let us decompose the tensor $\boldsymbol{\Phi}^e$ using the orthonormal basis $\{\boldsymbol{u}_1,\boldsymbol{u}_2,\boldsymbol{n}_0\}$: denoting its components by $\varphi_{i\alpha}$, we have

$$
\boldsymbol{\Phi}^e = \varphi_{\alpha\beta}\,\boldsymbol{u}_\alpha\otimes\boldsymbol{u}_\beta+\varphi_{3\beta}\,\boldsymbol{n}_0\otimes\boldsymbol{u}_\beta\,.
$$

Then,

$$
\boldsymbol{\Phi}^e\boldsymbol{u}_1\otimes\boldsymbol{u}_1 = \varphi_{\alpha1}\,\boldsymbol{u}_\alpha\otimes\boldsymbol{u}_1+\varphi_{31}\,\boldsymbol{n}_0\otimes\boldsymbol{u}_1
$$

and

$$
\boldsymbol{\Phi}^e\boldsymbol{u}_2\otimes\boldsymbol{u}_2 = \varphi_{\alpha2}\,\boldsymbol{u}_\alpha\otimes\boldsymbol{u}_2+\varphi_{32}\,\boldsymbol{n}_0\otimes\boldsymbol{u}_2\,,
$$

and using definitions (6.25), (6.26) we compute directly

$$
\begin{aligned}
W_{\text{Coss}}\big(\boldsymbol{\Phi}^e\boldsymbol{u}_1\otimes\boldsymbol{u}_1\big) &= \frac{2\mu(\lambda+\mu)}{\lambda+2\mu}\,\varphi_{11}^2+\frac{\mu+\mu_c}{2}\,\varphi_{21}^2+\frac{2\mu\mu_c}{\mu+\mu_c}\,\varphi_{31}^2\,,\\
W_{\text{Coss}}\big(\boldsymbol{\Phi}^e\boldsymbol{u}_2\otimes\boldsymbol{u}_2\big) &= \frac{2\mu(\lambda+\mu)}{\lambda+2\mu}\,\varphi_{22}^2+\frac{\mu+\mu_c}{2}\,\varphi_{12}^2+\frac{2\mu\mu_c}{\mu+\mu_c}\,\varphi_{32}^2\,, \qquad (6.53)\\
W_{\text{Coss}}\big(\boldsymbol{\Phi}^e\boldsymbol{u}_1\otimes\boldsymbol{u}_1\,,\,\boldsymbol{\Phi}^e\boldsymbol{u}_2\otimes\boldsymbol{u}_2\big) &= \frac{\lambda\mu}{\lambda+2\mu}\,\varphi_{11}\varphi_{22}+\frac{\mu-\mu_c}{2}\,\varphi_{12}\varphi_{21}\,.
\end{aligned}
$$

Substituting (6.53) in (6.48) we obtain

$$
\begin{aligned}
W_{\text{Coss}}\big(\boldsymbol{\Phi}^e\big) &= \frac{2\mu(\lambda+\mu)}{\lambda+2\mu}\,(\varphi_{11}^2+\varphi_{22}^2)+\frac{2\lambda\mu}{\lambda+2\mu}\,\varphi_{11}\varphi_{22}+\frac{\mu+\mu_c}{2}\,(\varphi_{12}^2+\varphi_{21}^2)\\
&\quad +(\mu-\mu_c)\,\varphi_{12}\varphi_{21}+\frac{2\mu\mu_c}{\mu+\mu_c}\,(\varphi_{31}^2+\varphi_{32}^2).
\end{aligned}
\qquad (6.54)
$$

Since

$$|\kappa_\alpha h| < \frac{1}{2}$$

we have

$$|K|\,h^2 = |\kappa_1 h| \cdot |\kappa_2 h| < \frac{1}{4} \qquad \text{and} \qquad |H|\,h \leq \frac{1}{2}\big(|\kappa_1 h| + |\kappa_2 h|\big) < \frac{1}{2}\,, \tag{6.55}$$

so we can write the inequality

$$h + K\frac{h^3}{12} > h - \frac{h}{48} = \frac{47}{48}h = \frac{\delta}{48}h + \frac{47-\delta}{48}h \tag{6.56}$$

for any $\delta > 0$. The choice of δ will be specified later in (6.61). Using relations (6.53), (6.54) and (6.56) we can transform inequality (6.52) as follows

$$\begin{aligned}\mathcal{W}^{(3)}_{\text{memb,bend}}(\boldsymbol{E}^e, \boldsymbol{K}^e) \geq \frac{\delta h}{48} W_{\text{Coss}}(\boldsymbol{E}^e) + h\Bigg[\frac{47-\delta}{48} W_{\text{Coss}}(\boldsymbol{E}^e) - \frac{h^2\kappa}{3}\sqrt{W_{\text{Coss}}(\boldsymbol{E}^e)} \\ \times\left(\frac{2\mu(\lambda+\mu)}{\lambda+2\mu}(\varphi_{11}^2+\varphi_{22}^2) + \frac{\mu+\mu_c}{2}(\varphi_{12}^2+\varphi_{21}^2) + \frac{2\mu\mu_c}{\mu+\mu_c}(\varphi_{31}^2+\varphi_{32}^2)\right)^{1/2} \\ + \frac{h^2}{12}\bigg(\frac{2\mu(\lambda+\mu)}{\lambda+2\mu}(\varphi_{11}^2+\varphi_{22}^2) + \frac{2\lambda\mu}{\lambda+2\mu}\varphi_{11}\varphi_{22} + \frac{\mu+\mu_c}{2}(\varphi_{12}^2+\varphi_{21}^2) \\ +(\mu-\mu_c)\varphi_{12}\varphi_{21} + \frac{2\mu\mu_c}{\mu+\mu_c}(\varphi_{31}^2+\varphi_{32}^2)\bigg)\Bigg].\end{aligned} \tag{6.57}$$

To prove that the expression in square brackets in (6.57) is positive, we regard it as a quadratic function in

$$\sqrt{W_{\text{Coss}}(\boldsymbol{E}^e)}$$

and show that its discriminant Δ is negative. Indeed, the discriminant is

$$\begin{aligned}\Delta = \frac{h^4\kappa^2}{9}\left(\frac{2\mu(\lambda+\mu)}{\lambda+2\mu}(\varphi_{11}^2+\varphi_{22}^2) + \frac{\mu+\mu_c}{2}(\varphi_{12}^2+\varphi_{21}^2) + \frac{2\mu\mu_c}{\mu+\mu_c}(\varphi_{31}^2+\varphi_{32}^2)\right) \\ -\frac{47-\delta}{12}\frac{h^2}{12}\bigg(\frac{2\mu(\lambda+\mu)}{\lambda+2\mu}(\varphi_{11}^2+\varphi_{22}^2) + \frac{2\lambda\mu}{\lambda+2\mu}\varphi_{11}\varphi_{22} + \frac{\mu+\mu_c}{2}(\varphi_{12}^2+\varphi_{21}^2) \\ +(\mu-\mu_c)\varphi_{12}\varphi_{21} + \frac{2\mu\mu_c}{\mu+\mu_c}(\varphi_{31}^2+\varphi_{32}^2)\bigg).\end{aligned}$$

Then, we can combine the terms and write

$$\frac{9}{h^2}\Delta = \left[\Big(\kappa^2h^2 - \frac{47-\delta}{16}\Big)\frac{2\mu(\lambda+\mu)}{\lambda+2\mu}\,(\varphi_{11}^2+\varphi_{22}^2) - \frac{47-\delta}{16}\,\frac{2\lambda\mu}{\lambda+2\mu}\,\varphi_{11}\varphi_{22}\right]$$
$$+\left[\Big(\kappa^2h^2 - \frac{47-\delta}{16}\Big)\frac{\mu+\mu_c}{2}\,(\varphi_{12}^2+\varphi_{21}^2) - \frac{47-\delta}{16}\,(\mu-\mu_c)\,\varphi_{12}\varphi_{21}\right]$$
$$+\Big(\kappa^2h^2 - \frac{47-\delta}{16}\Big)\frac{2\mu\mu_c}{\mu+\mu_c}\,(\varphi_{31}^2+\varphi_{32}^2),$$

or, equivalently,

$$\frac{9}{h^2}\Delta = \Big(\kappa^2h^2 - \frac{47-\delta}{16}\Big)\left[\frac{2\mu(\lambda+\mu)}{\lambda+2\mu}\Big(\varphi_{11}^2+\varphi_{22}^2 - \frac{47-\delta}{16\kappa^2h^2-(47-\delta)}\right.$$
$$\times\frac{\lambda}{\lambda+\mu}\,\varphi_{11}\varphi_{22}\Big) + \frac{\mu+\mu_c}{2}\Big(\varphi_{12}^2+\varphi_{21}^2 - \frac{47-\delta}{16\kappa^2h^2-(47-\delta)}\,\frac{\mu-\mu_c}{\mu+\mu_c}\,2\varphi_{12}\varphi_{21}\Big)$$
$$\left.+\,\frac{2\mu\mu_c}{\mu+\mu_c}\,(\varphi_{31}^2+\varphi_{32}^2)\right]. \tag{6.58}$$

Notice that it holds

$$\frac{\min\{\lambda+2\mu\,,3\lambda+2\mu\}}{2(\lambda+\mu)} = 1-\frac{|\lambda|}{2(\lambda+\mu)}$$
$$\text{and} \tag{6.59}$$
$$\frac{2\min\{\mu,\mu_c\}}{\mu+\mu_c} = 1-\frac{|\mu-\mu_c|}{\mu+\mu_c}.$$

Then, inequality (6.39) can be written equivalently

$$\kappa h < \min\left\{\frac{1}{2}\,,\;\frac{\sqrt{47}}{4}\Big(1-\frac{|\lambda|}{2(\lambda+\mu)}\Big)^{1/2},\;\frac{\sqrt{47}}{4}\Big(1-\frac{|\mu-\mu_c|}{\mu+\mu_c}\Big)^{1/2}\right\}. \tag{6.60}$$

From the last relation we deduce

$$47-16\kappa^2h^2\Big(1-\frac{|\lambda|}{2(\lambda+\mu)}\Big)^{-1} > 0 \qquad\text{and}\qquad 47-16\kappa^2h^2\Big(1-\frac{|\mu-\mu_c|}{\mu+\mu_c}\Big)^{-1} > 0.$$

Hence, we can choose the positive constant $\delta > 0$ such that

$$\delta < \min\left\{43\,,\,47-16\kappa^2h^2\Big(1-\frac{|\lambda|}{2(\lambda+\mu)}\Big)^{-1},\,47-16\kappa^2h^2\Big(1-\frac{|\mu-\mu_c|}{\mu+\mu_c}\Big)^{-1}\right\}. \tag{6.61}$$

From $\delta < 43$ and $\kappa h < \frac{1}{2}$ it follows that

$$\kappa^2h^2 - \frac{47-\delta}{16} < \kappa^2h^2 - \frac{4}{16} < 0. \tag{6.62}$$

Further, from (6.61) we also derive that

$$\left|\frac{47-\delta}{16\kappa^2h^2-(47-\delta)}\cdot\frac{\lambda}{2(\lambda+\mu)}\right|<1 \quad \text{and} \quad \left|\frac{47-\delta}{16\kappa^2h^2-(47-\delta)}\cdot\frac{\mu-\mu_c}{\mu+\mu_c}\right|<1. \tag{6.63}$$

Using inequalities (6.62) and (6.63) in (6.58) we deduce that

$$\begin{aligned}\frac{9}{h^2}\Delta < \left(\kappa^2h^2-\frac{47-\delta}{16}\right)\Bigg[&\frac{2\mu(\lambda+\mu)}{\lambda+2\mu}\big(|\varphi_{11}|-|\varphi_{22}|\big)^2+\frac{\mu+\mu_c}{2}\big(|\varphi_{12}|-|\varphi_{21}|\big)^2\\ &+\frac{2\mu\mu_c}{\mu+\mu_c}\big(\varphi_{31}^2+\varphi_{32}^2\big)\Bigg]\leq 0.\end{aligned} \tag{6.64}$$

Thus, the discriminant Δ is negative, so the expression in square brackets in (6.57) is positive. Hence, relations (6.57) and $(6.31)_1$ infer

$$\mathcal{W}^{(3)}_{\text{memb,bend}}(\boldsymbol{E}^e,\boldsymbol{K}^e)\geq\frac{\delta}{48}\,h\,W_{\text{Coss}}(\boldsymbol{E}^e)\geq C_1\,\|\boldsymbol{E}^e\|^2,$$

for some positive constant $C_1>0$. The proof is complete.

Let us present now another auxiliary result (a variant of Lemma 6.1), in which inequality (6.39) is replaced by an alternative assumption. Thus, we prove the following coercivity result which is complementary to the previous lemma.

Lemma 6.2. *Assume that the constitutive coefficients fulfil the conditions* (6.38) *and that the product* $\kappa h<1$ *is small enough such that*

$$\kappa h<\min\left\{\left(\frac{12\min\{\lambda+2\mu,3\lambda+2\mu\}}{8(\lambda+\mu)+\min\{\lambda+2\mu,3\lambda+2\mu\}}\right)^{1/2},\left(\frac{12\min\{\mu,\mu_c\}}{2(\mu+\mu_c)+\min\{\mu,\mu_c\}}\right)^{1/2}\right\}. \tag{6.65}$$

Then, the membrane-bending energy $\mathcal{W}^{(3)}_{\text{memb,bend}}(\boldsymbol{E}^e,\boldsymbol{K}^e)$ *satisfies the inequality* (6.40) *for some positive constant* C_1.

Proof. We begin the proof in the same way as in the case of Lemma 6.1 and establish relations (6.41)-(6.54). If we insert (6.53) and (6.54) into Eq. (6.52) and use the relation

$$1+K\frac{h^2}{12}\geq 1-\frac{\kappa^2h^2}{12}=\epsilon+\left(1-\frac{\kappa^2h^2}{12}-\epsilon\right)\qquad\text{for any }\epsilon>0, \tag{6.66}$$

then we obtain the inequality

$$
\begin{aligned}
\mathcal{W}^{(3)}_{\text{memb,bend}}(\boldsymbol{E}^e,\boldsymbol{K}^e) \geq\ & h\epsilon\, W_{\text{Coss}}(\boldsymbol{E}^e) + h\Bigg[\Big(1-\frac{\kappa^2h^2}{12}-\epsilon\Big)W_{\text{Coss}}(\boldsymbol{E}^e)\\
&-\frac{h^2\kappa}{3}\sqrt{W_{\text{Coss}}(\boldsymbol{E}^e)}\Bigg(\frac{2\mu(\lambda+\mu)}{\lambda+2\mu}(\varphi_{11}^2+\varphi_{22}^2)+\frac{\mu+\mu_c}{2}(\varphi_{12}^2+\varphi_{21}^2)\\
&+\frac{2\mu\mu_c}{\mu+\mu_c}(\varphi_{31}^2+\varphi_{32}^2)\Bigg)^{1/2}+\frac{h^2}{12}\Bigg(\frac{2\mu(\lambda+\mu)}{\lambda+2\mu}(\varphi_{11}^2+\varphi_{22}^2)+\frac{2\lambda\mu}{\lambda+2\mu}\varphi_{11}\varphi_{22}\\
&+\frac{\mu+\mu_c}{2}(\varphi_{12}^2+\varphi_{21}^2)+(\mu-\mu_c)\varphi_{12}\varphi_{21}+\frac{2\mu\mu_c}{\mu+\mu_c}(\varphi_{31}^2+\varphi_{32}^2)\Bigg)\Bigg].
\end{aligned}
\tag{6.67}
$$

We regard the expression in square brackets in (6.67) as a quadratic function in $\sqrt{W_{\text{Coss}}(\boldsymbol{E}^e)}$ and denote by D its discriminant. To show that the discriminant D is negative, we compute

$$
\begin{aligned}
D =\ & \frac{\kappa^2h^4}{9}\Bigg(\frac{2\mu(\lambda+\mu)}{\lambda+2\mu}(\varphi_{11}^2+\varphi_{22}^2)+\frac{\mu+\mu_c}{2}(\varphi_{12}^2+\varphi_{21}^2)+\frac{2\mu\mu_c}{\mu+\mu_c}(\varphi_{31}^2+\varphi_{32}^2)\Bigg)\\
&-\frac{h^2}{3}\Big(1-\frac{\kappa^2h^2}{12}-\epsilon\Big)\Bigg(\frac{2\mu(\lambda+\mu)}{\lambda+2\mu}(\varphi_{11}^2+\varphi_{22}^2)+\frac{2\lambda\mu}{\lambda+2\mu}\varphi_{11}\varphi_{22}\\
&\qquad+\frac{\mu+\mu_c}{2}(\varphi_{12}^2+\varphi_{21}^2)+(\mu-\mu_c)\varphi_{12}\varphi_{21}+\frac{2\mu\mu_c}{\mu+\mu_c}(\varphi_{31}^2+\varphi_{32}^2)\Bigg),
\end{aligned}
$$

or, equivalently,

$$
\begin{aligned}
\frac{9}{h^2}D =\ & \frac{2\mu(\lambda+\mu)}{\lambda+2\mu}\Bigg[\Big(\kappa^2h^2-(3-\frac{\kappa^2h^2}{4}-3\epsilon)\Big)(\varphi_{11}^2+\varphi_{22}^2)-(3-\frac{\kappa^2h^2}{4}-3\epsilon)\\
&\times\frac{\lambda}{\lambda+\mu}\varphi_{11}\varphi_{22}\Bigg]+\frac{\mu+\mu_c}{2}\Bigg[\Big(\kappa^2h^2-(3-\frac{\kappa^2h^2}{4}-3\epsilon)\Big)(\varphi_{12}^2+\varphi_{21}^2)\\
&-(3-\frac{\kappa^2h^2}{4}-3\epsilon)\frac{\mu-\mu_c}{\mu+\mu_c}2\varphi_{12}\varphi_{21}\Bigg]+\frac{2\mu\mu_c}{\mu+\mu_c}\Big(\kappa^2h^2-(3-\frac{\kappa^2h^2}{4}-3\epsilon)\Big)(\varphi_{31}^2+\varphi_{32}^2).
\end{aligned}
$$

Hence, we get

$$\frac{3}{h^2}D = -\left(1-\frac{5}{12}\kappa^2h^2-\epsilon\right)\left\{\frac{2\mu(\lambda+\mu)}{\lambda+2\mu}\left[(\varphi_{11}^2+\varphi_{22}^2)+\frac{1-\frac{1}{12}\kappa^2h^2-\epsilon}{1-\frac{5}{12}\kappa^2h^2-\epsilon}\right.\right.$$
$$\left.\times\frac{\lambda}{\lambda+\mu}\varphi_{11}\varphi_{22}\right]+\frac{\mu+\mu_c}{2}\left[(\varphi_{12}^2+\varphi_{21}^2)+\frac{1-\frac{1}{12}\kappa^2h^2-\epsilon}{1-\frac{5}{12}\kappa^2h^2-\epsilon}\cdot\frac{\mu-\mu_c}{\mu+\mu_c}2\varphi_{12}\varphi_{21}\right]$$
$$\left.+\frac{2\mu\mu_c}{\mu+\mu_c}(\varphi_{31}^2+\varphi_{32}^2)\right\}. \tag{6.68}$$

Now, we want to choose a positive number ϵ such that the following inequalities are fulfilled

$$1-\frac{5}{12}\kappa^2h^2-\epsilon>0 \quad \text{and} \quad \frac{1-\frac{1}{12}\kappa^2h^2-\epsilon}{1-\frac{5}{12}\kappa^2h^2-\epsilon}\cdot\frac{|\lambda|}{\lambda+\mu}<1$$
$$\text{and} \quad \frac{1-\frac{1}{12}\kappa^2h^2-\epsilon}{1-\frac{5}{12}\kappa^2h^2-\epsilon}\cdot\frac{|\mu-\mu_c|}{\mu+\mu_c}<1. \tag{6.69}$$

If we solve inequations (6.69) with respect to ϵ, then we find, respectively

$$\epsilon<1-\frac{5}{12}\kappa^2h^2 \quad \text{and} \quad \epsilon<1-\frac{\kappa^2h^2}{3}\left[\frac{1}{4}+\left(1-\frac{|\lambda|}{2(\lambda+\mu)}\right)^{-1}\right]$$
$$\text{and} \quad \epsilon<1-\frac{\kappa^2h^2}{3}\left[\frac{1}{4}+\left(1-\frac{|\mu-\mu_c|}{\mu+\mu_c}\right)^{-1}\right]. \tag{6.70}$$

Further, using relations (6.59) we can write

$$\frac{1}{4}+\left(1-\frac{|\lambda|}{2(\lambda+\mu)}\right)^{-1}=\frac{8(\lambda+\mu)+\min\{\lambda+2\mu,3\lambda+2\mu\}}{4\min\{\lambda+2\mu,3\lambda+2\mu\}} \quad \text{and}$$
$$\frac{1}{4}+\left(1-\frac{|\mu-\mu_c|}{\mu+\mu_c}\right)^{-1}=\frac{2(\mu+\mu_c)+\min\{\mu,\mu_c\}}{4\min\{\mu,\mu_c\}}. \tag{6.71}$$

Substituting (6.71) into (6.70) we find the equivalent form

$$\epsilon<1-\frac{5}{12}\kappa^2h^2 \quad \text{and} \quad \epsilon<1-\kappa^2h^2\cdot\frac{8(\lambda+\mu)+\min\{\lambda+2\mu,3\lambda+2\mu\}}{12\min\{\lambda+2\mu,3\lambda+2\mu\}}$$
$$\text{and} \quad \epsilon<1-\kappa^2h^2\cdot\frac{2(\mu+\mu_c)+\min\{\mu,\mu_c\}}{12\min\{\mu,\mu_c\}}. \tag{6.72}$$

By virtue of assumptions (6.65), we see that the right-hand sides of inequalities (6.72) are all positive, so there exists a constant $\epsilon>0$ having properties (6.72), i.e., ϵ satisfies inequalities (6.69). Then, using (6.69) in Eq. (6.68) we derive

$$\frac{3}{h^2}D < -\Big(1-\frac{5}{12}\kappa^2h^2-\epsilon\Big)\bigg[\frac{2\mu(\lambda+\mu)}{\lambda+2\mu}\big(|\varphi_{11}|-|\varphi_{22}|\big)^2+\frac{\mu+\mu_c}{2}\big(|\varphi_{12}|-|\varphi_{21}|\big)^2$$
$$+\frac{2\mu\mu_c}{\mu+\mu_c}\big(\varphi_{31}^2+\varphi_{32}^2\big)\bigg] \leq 0.$$

Thus, the discriminant D is negative and, hence, the expression in square brackets in relation (6.67) is always positive. Consequently,inequations (6.67) and $(6.31)_1$ imply that there exists a constant $C_1 > 0$ such that

$$\mathcal{W}^{(3)}_{\text{memb,bend}}(\boldsymbol{E}^e,\boldsymbol{K}^e) \geq h\epsilon W_{\text{Coss}}(\boldsymbol{E}^e) \geq C_1\,\|\boldsymbol{E}^e\|^2,$$

which completes the proof.

We are now able to prove the coercivity of the strain energy density $\mathcal{W}^{(3)}_{\text{shell}}(\boldsymbol{E}^e,\boldsymbol{K}^e)$. We obtain the following result as a consequence of the above lemmas.

Theorem 6.1. *Assume that the constitutive coefficients fulfil the inequalities*

$$\mu > 0, \qquad 3\lambda+2\mu > 0, \qquad \mu_c > 0, \qquad \textit{and} \qquad b_i > 0, \tag{6.73}$$

and that κh satisfies at least one of the conditions (6.39) *and* (6.65)*. Then, the areal strain energy density $\mathcal{W}^{(3)}_{\text{shell}}(\boldsymbol{E}^e,\boldsymbol{K}^e)$ given by* (6.33)-(6.35) *is coercive, i.e. there exist some positive constants $C_1 > 0$, $C_2 > 0$ such that*

$$\mathcal{W}^{(3)}_{\text{shell}}(\boldsymbol{E}^e,\boldsymbol{K}^e) \geq C_1\,\|\boldsymbol{E}^e\|^2+C_2\,\|\boldsymbol{K}^e\|^2. \tag{6.74}$$

Proof. In view of the hypotheses of the theorem, we can apply Lemma 6.1 or Lemma 6.2. In both case, inequality (6.40) holds true. Then, taking into account relations (6.37) and (6.40) we obtain

$$\mathcal{W}^{(3)}_{\text{shell}}(\boldsymbol{E}^e,\boldsymbol{K}^e) = \mathcal{W}^{(3)}_{\text{memb,bend}}(\boldsymbol{E}^e,\boldsymbol{K}^e)+\mathcal{W}^{(3)}_{\text{bend,curv}}(\boldsymbol{K}^e) \geq C_1\|\boldsymbol{E}^e\|^2+C_2\|\boldsymbol{K}^e\|^2$$

for some positive constants C_1, C_2. The proof is complete.

Remark 6.1. We mention that Theorem 6.1 is the main original result in this work. It improves a similar coercivity result established for a related Cosserat shell model of order $O(h^3)$. This related Cosserat shell model has been derived and investigated in [29, 30, 31] and the comparison with the present model has been presented in [22, Sect. 5.2]. More precisely, Theorem 6.1 improves the result in [31, Proposition 4.1] in two ways: Firstly, the conditions on constitutive coefficients given by (6.73) are less restrictive, since in [31, Proposition 4.1] the inequalities $\mu > 0$ and $2\lambda+\mu > 0$ are assumed, which are more restrictive. Secondly, the conditions on the product κh given by (6.39) and (6.65) are more convenient to check, since they involve only the constitutive coefficients. In contrast, the corresponding conditions (i) and (ii) listed in [31, Proposition 4.1] involve also the smallest and largest eigenvalues of the quadratic forms $W_{\text{curv}}(\boldsymbol{X})$, $W_{\text{shell}}(\boldsymbol{X})$ and, hence, they are more difficult to check.

Remark 6.2. Notice that hypothesis (6.65) in Lemma 6.2 can be written in the equivalent form

$$\kappa h < \min\left\{1,\, 2\sqrt{3}\left(\frac{2(\lambda+\mu)-|\lambda|}{10(\lambda+\mu)-|\lambda|}\right)^{1/2},\, 2\sqrt{3}\left(\frac{\mu+\mu_c-|\mu-\mu_c|}{5(\mu+\mu_c)-|\mu-\mu_c|}\right)^{1/2}\right\}. \tag{6.75}$$

Indeed, by a straightforward calculation we verify that

$$\begin{aligned} &\frac{1}{4}+\left(1-\frac{|\lambda|}{2(\lambda+\mu)}\right)^{-1} = \frac{1}{4}\cdot\frac{10(\lambda+\mu)-|\lambda|}{2(\lambda+\mu)-|\lambda|} \qquad \text{and} \\ &\frac{1}{4}+\left(1-\frac{|\mu-\mu_c|}{\mu+\mu_c}\right)^{-1} = \frac{1}{4}\cdot\frac{5(\mu+\mu_c)-|\mu-\mu_c|}{\mu+\mu_c-|\mu-\mu_c|}\,. \end{aligned} \tag{6.76}$$

Then, from relations (6.71) and (6.76) we deduce

$$\begin{aligned} &\frac{8(\lambda+\mu)+\min\{\lambda+2\mu,3\lambda+2\mu\}}{\min\{\lambda+2\mu,3\lambda+2\mu\}} = \frac{10(\lambda+\mu)-|\lambda|}{2(\lambda+\mu)-|\lambda|} \qquad \text{and} \\ &\frac{2(\mu+\mu_c)+\min\{\mu,\mu_c\}}{\min\{\mu,\mu_c\}} = \frac{5(\mu+\mu_c)-|\mu-\mu_c|}{\mu+\mu_c-|\mu-\mu_c|}\,. \end{aligned} \tag{6.77}$$

Hence, conditions (6.65) and (6.75) are indeed equivalent.

Remark 6.3. In the case $\lambda \geq 0$ the hypotheses of Theorem 6.1 (namely relations (6.39) and (6.65)) simplify. Indeed, for $\lambda \geq 0$ we get

$$\left(\frac{47}{32}\cdot\frac{\min\{\lambda+2\mu,3\lambda+2\mu\}}{\lambda+\mu}\right)^{1/2} = \left(\frac{47}{32}\cdot\frac{\lambda+2\mu}{\lambda+\mu}\right)^{1/2} > 1$$

and condition (6.39) in the statements of Lemma 6.1 and Theorem 6.1 reduces to the simpler form

$$\kappa h < \min\left\{\frac{1}{2},\left(\frac{47}{8}\cdot\frac{\min\{\mu,\mu_c\}}{\mu+\mu_c}\right)^{1/2}\right\}. \tag{6.78}$$

Also, if $\lambda \geq 0$ the inequalities (6.38) imply

$$\frac{12\min\{\lambda+2\mu,3\lambda+2\mu\}}{8(\lambda+\mu)+\min\{\lambda+2\mu,3\lambda+2\mu\}} = \frac{12(\lambda+2\mu)}{8(\lambda+\mu)+\lambda+2\mu} > 1.$$

Hence, in the case $\lambda \geq 0$ condition (6.65) can be reduced to

$$\kappa h < \min\left\{1,\left(\frac{12\min\{\mu,\mu_c\}}{2(\mu+\mu_c)+\min\{\mu,\mu_c\}}\right)^{1/2}\right\}. \tag{6.79}$$

Remark 6.4. In the case $\lambda < 0$ condition (6.39) from Lemma 6.1 takes the following form

$$\kappa h < \min\left\{ \frac{1}{2}, \left(\frac{47}{32} \cdot \frac{3\lambda+2\mu}{\lambda+\mu} \right)^{1/2}, \left(\frac{47}{8} \cdot \frac{\min\{\mu,\mu_c\}}{\mu+\mu_c} \right)^{1/2} \right\}, \tag{6.80}$$

while condition (6.65) from Lemma 6.2 reduces to

$$\kappa h < \min\left\{ 1, 2\sqrt{3}\left(\frac{3\lambda+2\mu}{11\lambda+10\mu} \right)^{1/2}, \left(\frac{12\min\{\mu,\mu_c\}}{2(\mu+\mu_c)+\min\{\mu,\mu_c\}} \right)^{1/2} \right\}. \tag{6.81}$$

6.3.2 *Existence of Minimizers*

Let us write the variational formulation for equilibrium of Cosserat shells. To this aim, we consider the usual Lebesgue and Sobolev spaces for vectors and tensors

$$\begin{aligned} L^p(\omega,\mathbb{R}^3) &= \{\boldsymbol{v} = v_i \boldsymbol{e}_i \,|\, v_i \in L^p(\omega)\}, \\ L^p(\omega,\mathbb{R}^{3\times 3}) &= \{\boldsymbol{T} = T_{ij} \boldsymbol{e}_i \otimes \boldsymbol{e}_j \,|\, T_{ij} \in L^p(\omega)\} \qquad (p \geq 1), \\ H^1(\omega,\mathbb{R}^3) &= \{\boldsymbol{v} = v_i \boldsymbol{e}_i \,|\, v_i \in H^1(\omega)\}, \\ H^1(\omega,\mathbb{R}^{3\times 3}) &= \{\boldsymbol{T} = T_{ij} \boldsymbol{e}_i \otimes \boldsymbol{e}_j \,|\, T_{ij} \in H^1(\omega)\}, \end{aligned} \tag{6.82}$$

We also introduce the subsets

$$L^p(\omega,\mathrm{SO}(3)) = \{\boldsymbol{Q} \in L^p(\omega,\mathbb{R}^{3\times 3}) \,|\, \boldsymbol{Q}(x_1,x_2) \in \mathrm{SO}(3) \text{ for a.e. } (x_1,x_2) \in \omega\}$$

by abuse of notation, with the induced strong topology of $L^p(\omega,\mathbb{R}^{3\times 3})$, as well as

$$H^1(\omega,\mathrm{SO}(3)) = \{\boldsymbol{Q} \in H^1(\omega,\mathbb{R}^{3\times 3}) \,|\, \boldsymbol{Q}(x_1,x_2) \in \mathrm{SO}(3) \text{ for a.e. } (x_1,x_2) \in \omega\}$$

with the induced strong and weak topologies of $H^1(\omega,\mathbb{R}^{3\times 3})$.

We assume that the boundary data in (6.12) satisfy the regularity $\boldsymbol{m}^* \in H^1(\omega,\mathbb{R}^3)$ and $\boldsymbol{Q}^* \in H^1(\omega,\mathrm{SO}(3))$, and we define the set of admissible pairs $(\boldsymbol{m},\boldsymbol{Q}_e)$ by

$$\mathcal{A} = \{(\boldsymbol{m},\boldsymbol{Q}_e) \in H^1(\omega,\mathbb{R}^3) \times H^1(\omega,\mathrm{SO}(3)) \,|\, \boldsymbol{m}\big|_{\partial\omega_d} = \boldsymbol{m}^*, \boldsymbol{Q}_e\big|_{\partial\omega_d} = \boldsymbol{Q}^* \}, \tag{6.83}$$

where the boundary conditions hold in the sense of traces. For boundary-value problem (6.11), (6.12) we assume the existence of the potential $\Lambda(\boldsymbol{m},\boldsymbol{Q}_e)$ of external surface loads $\boldsymbol{f}$, $\boldsymbol{l}$ and boundary loads $\boldsymbol{N}^*$, $\boldsymbol{M}^*$ (cf. [23]), such that the total energy functional can be written as

$$
\begin{aligned}
\mathcal{E}^{(3)}(\boldsymbol{m},\boldsymbol{Q}_e) &= \iint_{\omega_\xi} \mathcal{W}^{(3)}_{\text{shell}}(\boldsymbol{E}^e,\boldsymbol{K}^e)\,\mathrm{d}a - \Lambda(\boldsymbol{m},\boldsymbol{Q}_e) \\
&= \iint_{\omega} \mathcal{W}^{(3)}_{\text{shell}}(\boldsymbol{E}^e,\boldsymbol{K}^e)\,a(x_1,x_2)\,\mathrm{d}x_1\mathrm{d}x_2 - \Lambda(\boldsymbol{m},\boldsymbol{Q}_e),
\end{aligned}
\tag{6.84}
$$

where the tensors $\boldsymbol{E}^e,\boldsymbol{K}^e$ are given in terms of $\boldsymbol{m},\boldsymbol{Q}_e$ by Eqs. (6.9), (6.10). Here, the external loading potential has the form

$$
\Lambda(\boldsymbol{m},\boldsymbol{Q}_e) = \iint_{\omega_\xi} \boldsymbol{f}\cdot\boldsymbol{u}\,\mathrm{d}a + \Pi_{\omega_\xi}(\boldsymbol{Q}_e) + \int_{\partial\omega_f} \boldsymbol{N}^*\cdot\boldsymbol{u}\,\mathrm{d}s + \Pi_{\partial\omega_f}(\boldsymbol{Q}_e), \tag{6.85}
$$

where $\boldsymbol{u} := \boldsymbol{m} - \boldsymbol{y}_0$ is the displacement vector and we assume that $\boldsymbol{f} \in L^2(\omega,\mathbb{R}^3)$ and $\boldsymbol{N}^* \in L^2(\partial\omega_f,\mathbb{R}^3)$. The potential $\Pi_{\omega_\xi} : L^2(\omega,\mathrm{SO}(3)) \to \mathbb{R}$ of the external surface couples $\boldsymbol{l}$ and the potential $\Pi_{\partial\omega_f} : L^2(\partial\omega_f,\mathrm{SO}(3)) \to \mathbb{R}$ of the external boundary couples $\boldsymbol{M}^*$ are assumed to be continuous and bounded operators.

We can prove now the existence of minimizers for the shell model of order $O(h^3)$ following closely the initial idea presented in [17].

Theorem 6.2. *Consider the minimization problem for the equilibrium of Cosserat 6-parameter elastic shells:*

$$
\text{minimize} \quad \mathcal{E}^{(3)}(\boldsymbol{m},\boldsymbol{Q}_e) \quad \text{w.r.t.} \quad (\boldsymbol{m},\boldsymbol{Q}_e) \in \mathcal{A}, \tag{6.86}
$$

where the total energy functional $\mathcal{E}^{(3)}$ is given by (6.84) *and the admissible set $\mathcal{A}$ is defined by* (6.83)*. Assume that the constitutive coefficients satisfy inequalities* (6.73) *and that κh is small enough such that at least one of conditions* (6.39) *and* (6.65) *holds. Moreover, the external loads and boundary data are assumed to satisfy the regularity conditions*

$$
\boldsymbol{f} \in L^2(\omega,\mathbb{R}^3), \quad \boldsymbol{N}^* \in L^2(\partial\omega_f,\mathbb{R}^3), \quad \boldsymbol{m}^* \in H^1(\omega,\mathbb{R}^3), \quad \boldsymbol{Q}^* \in \boldsymbol{H}^1(\omega,SO(3)), \tag{6.87}
$$

while the reference configuration of the shell fulfils the regularity conditions

$$
\boldsymbol{y}_0 \in H^2(\omega,\mathbb{R}^3), \quad \boldsymbol{Q}_0 \in H^1(\omega,\mathrm{SO}(3)), \quad \boldsymbol{a}_\alpha \in L^\infty(\omega,\mathbb{R}^3), \quad a(x_1,x_2) \geq a_0 > 0, \tag{6.88}
$$

where a_0 is a positive constant. Then, minimization problem (6.86) *admits at least one minimizing solution pair $(\hat{\boldsymbol{m}},\hat{\boldsymbol{Q}}_e)$ in the admissible set $\mathcal{A}$.*

Proof. To prove this assertion we employ the general existence result established in [20] for 6-parameter shells. We can verify that the hypotheses of Theorem 6 from [20] are satisfied, i.e. that the strain energy density $\mathcal{W}^{(3)}_{\text{shell}}$ is a coercive and uniformly convex quadratic function of $(\boldsymbol{E}^e,\boldsymbol{K}^e)$.

Indeed, in view of Theorem 6.1 the function $\mathcal{W}^{(3)}_{\text{shell}}(\boldsymbol{E}^e,\boldsymbol{K}^e)$ is coercive. Since $\mathcal{W}^{(3)}_{\text{shell}}$ is a quadratic form in terms of $(\boldsymbol{E}^e,\boldsymbol{K}^e)$, which is also positive definite, we see that it is also convex. Thus, all the hypotheses of Theorem 6 from [20] are fulfilled. Applying this general result, we can derive the existence of a minimizing pair $(\hat{\boldsymbol{m}},\hat{\boldsymbol{Q}}_e)\in\mathcal{A}$. In what follows, we present only the main steps of the existence proof and refer to [20, Theorem 6] for further details.

We estimate first the external loading potential and show that

$$|\Lambda(\boldsymbol{m},\boldsymbol{Q}_e)|\leq c_1\left(\|\boldsymbol{m}\|_{H^1(\omega)}+1\right)\qquad\text{for any}\quad(\boldsymbol{m},\boldsymbol{Q}_e)\in\mathcal{A},\tag{6.89}$$

for some positive constant c_1. Then, using coercivity relation (6.74) we deduce that there exist some constants $c_2>0$ and c_3, c_4 such that

$$\mathcal{E}^{(3)}(\boldsymbol{m},\boldsymbol{Q}_e)\geq c_2\|\nabla\boldsymbol{m}\|^2_{L^2(\omega)}-c_3\|\boldsymbol{m}\|_{H^1(\omega)}-c_4.\tag{6.90}$$

Here, we denote by $\|\cdot\|_{L^2(\omega)}$ and $\|\cdot\|_{H^1(\omega)}$ the norms in the Lebesgue and Sobolev spaces, respectively. Using the Poincaré–inequality we deduce from (6.90) that

$$\mathcal{E}^{(3)}(\boldsymbol{m},\boldsymbol{Q}_e)\geq c_5\|\boldsymbol{m}-\boldsymbol{m}^*\|^2_{H^1(\omega)}-c_6\|\boldsymbol{m}-\boldsymbol{m}^*\|_{H^1(\omega)}+c_7\quad\text{for any }(\boldsymbol{m},\boldsymbol{Q}_e)\in\mathcal{A},\tag{6.91}$$

where $c_5>0$ and c_6, c_7 are some constants. Hence, the functional $\mathcal{E}^{(3)}(\boldsymbol{m},\boldsymbol{Q}_e)$ is bounded from below over $\mathcal{A}$. Therefore, there exists an infimizing sequence $(\boldsymbol{m}_n,\boldsymbol{Q}_n)_{n\in\mathbb{N}}$ such that

$$\lim_{n\to\infty}\mathcal{E}^{(3)}(\boldsymbol{m}_n,\boldsymbol{Q}_n)=\inf\left\{\mathcal{E}^{(3)}(\boldsymbol{m},\boldsymbol{Q}_e)\,\middle|\,(\boldsymbol{m},\boldsymbol{Q}_e)\in\mathcal{A}\right\}.\tag{6.92}$$

For this infimizing sequence we show that $(\boldsymbol{m}_n)$ is bounded in $H^1(\omega,\mathbb{R}^3)$ and $(\boldsymbol{Q}_n)$ is bounded in $H^1(\omega,\mathbb{R}^{3\times3})$. Then, there exist some subsequences (not relabeled) and the limit pair $(\hat{\boldsymbol{m}},\hat{\boldsymbol{Q}}_e)\in\mathcal{A}$ such that the following weak and strong convergences hold

$$\begin{aligned}&\boldsymbol{m}_n\rightharpoonup\hat{\boldsymbol{m}}\quad\text{in }H^1(\omega,\mathbb{R}^3)&&\text{and}&&\boldsymbol{m}_n\to\hat{\boldsymbol{m}}\quad\text{in }L^2(\omega,\mathbb{R}^3),\\&\boldsymbol{Q}_n\rightharpoonup\hat{\boldsymbol{Q}}_e\quad\text{in }H^1(\omega,\mathbb{R}^{3\times3})&&\text{and}&&\boldsymbol{Q}_n\to\hat{\boldsymbol{Q}}_e\quad\text{in }L^2(\omega,\mathbb{R}^{3\times3}).\end{aligned}\tag{6.93}$$

Since the pairs $(\boldsymbol{m}_n,\boldsymbol{Q}_n)$ and $(\hat{\boldsymbol{m}},\hat{\boldsymbol{Q}}_e)$ are elements of the admissible set $\mathcal{A}$, we can construct the corresponding shell strain measures $(\boldsymbol{E}^e_n,\boldsymbol{K}^e_n)$ and $(\hat{\boldsymbol{E}}^e,\hat{\boldsymbol{K}}^e)$, respectively, using definitions (6.9), (6.10). Then, we can extract some subsequences (not relabeled) such that we have the following weak convergences

$$\boldsymbol{E}^e_n\rightharpoonup\hat{\boldsymbol{E}}^e\quad\text{in }L^2(\omega,\mathbb{R}^{3\times3})\qquad\text{and}\qquad\boldsymbol{K}^e_n\rightharpoonup\hat{\boldsymbol{K}}^e\quad\text{in }L^2(\omega,\mathbb{R}^{3\times3}).\tag{6.94}$$

We use now the convexity of the energy density function $\mathcal{W}^{(3)}_{\text{shell}}$ and obtain

$$\iint_\omega\mathcal{W}^{(3)}_{\text{shell}}(\hat{\boldsymbol{E}}^e,\hat{\boldsymbol{K}}^e)\,a\,\mathrm{d}x_1\mathrm{d}x_2\leq\liminf_{n\to\infty}\iint_\omega\mathcal{W}^{(3)}_{\text{shell}}(\boldsymbol{E}^e_n,\boldsymbol{K}^e_n)\,a\,\mathrm{d}x_1\mathrm{d}x_2.\tag{6.95}$$

Finally, by virtue of (6.84) and (6.95) we get

$$\mathcal{E}^{(3)}(\hat{\boldsymbol{m}},\hat{\boldsymbol{Q}}_e) \leq \liminf_{n\to\infty} \mathcal{E}^{(3)}(\boldsymbol{m}_n,\boldsymbol{Q}_n), \tag{6.96}$$

so $(\hat{\boldsymbol{m}},\hat{\boldsymbol{Q}}_e)$ is a minimizing solution pair of minimization problem (6.86).

6.4 The Higher Order Model of Cosserat 6-Parameter Shells

The first Cosserat 6-parameter shell model of order $O(h^5)$ has been established in [29, 30, 31] by a dimensional descent from the three-dimensional nonlinear Cosserat elasticity. Then, using an alternative derivation procedure suggested by the classical shell theory [5, 6], we have derived in [33] a refined higher order Cosserat shell model, in which we have optimized some terms.

Let us investigate in this section the 6-parameter shell model of order $O(h^5)$ derived in [33] for shells made of Cosserat material. Thus, the following areal strain energy density has been obtained (see [33, Eq. (119)])

$$\begin{aligned}\mathcal{W}^{(5)}_{\text{shell}}(\boldsymbol{E}^e,\boldsymbol{K}^e) &= \Big(h - K\frac{h^3}{12}\Big)\Big[W_{\text{Coss}}(\boldsymbol{E}^e) + W_{\text{curv}}(\boldsymbol{K}^e)\Big] - \frac{h^3}{6} W_{\text{Coss}}(\boldsymbol{E}^e, \boldsymbol{c}\boldsymbol{K}^e\boldsymbol{b}^*)\\ &+\Big(\frac{h^3}{12} - K\frac{h^5}{80}\Big)\Big[W_{\text{Coss}}(\boldsymbol{E}^e\boldsymbol{b} + \boldsymbol{c}\boldsymbol{K}^e) + W_{\text{curv}}(\boldsymbol{K}^e\boldsymbol{b})\Big]\\ &+\frac{h^5}{80}\Big[W_{\text{Coss}}((\boldsymbol{E}^e\boldsymbol{b} + \boldsymbol{c}\boldsymbol{K}^e)\boldsymbol{b}) + W_{\text{curv}}(\boldsymbol{K}^e\boldsymbol{b}^2)\Big].\end{aligned} \tag{6.97}$$

If we employ relation (6.5) and notation (6.32), we can decompose the above energy density in the form

$$\mathcal{W}^{(5)}_{\text{shell}}(\boldsymbol{E}^e,\boldsymbol{K}^e) = \mathcal{W}^{(5)}_{\text{memb,bend}}(\boldsymbol{E}^e,\boldsymbol{K}^e) + \mathcal{W}^{(5)}_{\text{bend,curv}}(\boldsymbol{K}^e) \qquad \text{with} \tag{6.98}$$

$$\begin{aligned}\mathcal{W}^{(5)}_{\text{memb,bend}}(\boldsymbol{E}^e,\boldsymbol{K}^e) &= \Big(h + K\frac{h^3}{12}\Big)W_{\text{Coss}}(\boldsymbol{E}^e) - 2\frac{h^3}{12}W_{\text{Coss}}(\boldsymbol{E}^e,\boldsymbol{\Phi}^e\boldsymbol{b}^*)\\ &+\Big(\frac{h^3}{12} - K\frac{h^5}{80}\Big)W_{\text{Coss}}(\boldsymbol{\Phi}^e) + \frac{h^5}{80}W_{\text{Coss}}(\boldsymbol{\Phi}^e\boldsymbol{b}),\end{aligned} \tag{6.99}$$

$$\begin{aligned}\mathcal{W}^{(5)}_{\text{bend,curv}}(\boldsymbol{K}^e) &= \Big(h - K\frac{h^3}{12}\Big)W_{\text{curv}}(\boldsymbol{K}^e) + \Big(\frac{h^3}{12} - K\frac{h^5}{80}\Big)W_{\text{curv}}(\boldsymbol{K}^e\boldsymbol{b})\\ &+\frac{h^5}{80}W_{\text{curv}}(\boldsymbol{K}^e\boldsymbol{b}^2).\end{aligned} \tag{6.100}$$

Let us consider first the bending-curvature part $\mathcal{W}^{(5)}_{\text{bend,curv}}(\boldsymbol{K}^e)$ and prove its coercivity. Since $|\kappa_\alpha h| < 1$, we have $|Kh^2| < 1$ (cf. (6.36)) and, hence,

$$h - K\frac{h^3}{12} > h - \frac{h}{12} = \frac{11}{12}h \qquad \text{and} \qquad \frac{h^3}{12} - K\frac{h^5}{80} > \frac{h^3}{12} - \frac{h^3}{80} = \frac{17}{240}h^3. \tag{6.101}$$

Then, Eqs. (6.100), (6.101) and $(6.31)_2$ yield

$$
\begin{aligned}
\mathcal{W}^{(5)}_{\text{bend,curv}}(\boldsymbol{K}^e) &\geq \frac{11}{12}\,h\,W_{\text{curv}}\big(\boldsymbol{K}^e\big) + \frac{17}{240}\,h^3 W_{\text{curv}}\big(\boldsymbol{K}^e\boldsymbol{b}\big) + \frac{h^5}{80}\,W_{\text{curv}}\big(\boldsymbol{K}^e\boldsymbol{b}^2\big) \\
&\geq \frac{11}{12}\,h\,W_{\text{curv}}\big(\boldsymbol{K}^e\big) \;\geq\; C_3\,\|\boldsymbol{K}^e\|^2,
\end{aligned}
\tag{6.102}
$$

for some positive constant C_3. Thus, the energy function $\mathcal{W}^{(5)}_{\text{bend,curv}}(\boldsymbol{K}^e)$ is coercive.

We turn our attention now to membrane-bending part (6.99) and establish an auxiliary result.

Lemma 6.3. *Assume that the condition*

$$
|\kappa_\alpha h| < \frac{1}{2}
$$

holds and the constitutive coefficients satisfy inequalities (6.38). *Then, there exist some positive constants $C_4 > 0$, $C_5 > 0$ such that the membrane-bending energy density* (6.99) *satisfies inequality*

$$
\mathcal{W}^{(5)}_{\text{memb,bend}}(\boldsymbol{E}^e,\boldsymbol{K}^e) \geq C_4\,\|\boldsymbol{E}^e\|^2 + C_5\,\|\boldsymbol{\Phi}^e\|^2. \tag{6.103}
$$

Proof. We can put the membrane-bending energy density (6.99) in the form

$$
\begin{aligned}
\mathcal{W}^{(5)}_{\text{memb,bend}}(\boldsymbol{E}^e,\boldsymbol{K}^e) &= \Big(h + K\frac{h^3}{12}\Big)W_{\text{Coss}}\big(\boldsymbol{E}^e\big) - 2\frac{h^3}{12}\,W_{\text{Coss}}\big(\boldsymbol{E}^e,\boldsymbol{\Phi}^e\boldsymbol{b}^*\big) \\
&\quad + \Big(\frac{h^3}{12} + (4H^2 - K)\frac{h^5}{80}\Big)W_{\text{Coss}}\big(\boldsymbol{\Phi}^e\big) + \frac{h^5}{80}W_{\text{Coss}}\big(\boldsymbol{\Phi}^e\boldsymbol{b}^*\big) - 4H\frac{h^5}{80}W_{\text{Coss}}\big(\boldsymbol{\Phi}^e,\boldsymbol{\Phi}^e\boldsymbol{b}^*\big) \\
&= \Big(\frac{11}{36}\,h + K\frac{h^3}{12}\Big)W_{\text{Coss}}\big(\boldsymbol{E}^e\big) + h\Big[\frac{25}{36}\,W_{\text{Coss}}\big(\boldsymbol{E}^e\big) - 2\frac{h^2}{12}\,W_{\text{Coss}}\big(\boldsymbol{E}^e,\boldsymbol{\Phi}^e\boldsymbol{b}^*\big) \\
&\quad + \frac{4}{5}\,\frac{h^4}{80}\,W_{\text{Coss}}\big(\boldsymbol{\Phi}^e\boldsymbol{b}^*\big)\Big] + \Big(\frac{h^3}{12} - (16H^2 + K)\frac{h^5}{80}\Big)W_{\text{Coss}}\big(\boldsymbol{\Phi}^e\big) \\
&\quad + \frac{h^5}{80}\Big[20H^2 W_{\text{Coss}}\big(\boldsymbol{\Phi}^e\big) - 4H W_{\text{Coss}}\big(\boldsymbol{\Phi}^e,\,\boldsymbol{\Phi}^e\boldsymbol{b}^*\big) + \frac{1}{5}\,W_{\text{Coss}}\big(\boldsymbol{\Phi}^e\boldsymbol{b}^*\big)\Big] \\
&= \frac{h}{12}\Big(\frac{11}{3} + Kh^2\Big)W_{\text{Coss}}\big(\boldsymbol{E}^e\big) + h\,W_{\text{Coss}}\Big(\frac{5}{6}\,\boldsymbol{E}^e - \frac{h^2}{10}\,\boldsymbol{\Phi}^e\boldsymbol{b}^*\Big) \\
&\quad + \Big(\frac{h^3}{12} - H^2\frac{h^5}{5} - K\frac{h^5}{80}\Big)W_{\text{Coss}}\big(\boldsymbol{\Phi}^e\big) + \frac{h^5}{16}\,W_{\text{Coss}}\big(2H\boldsymbol{\Phi}^e - \frac{1}{5}\,\boldsymbol{\Phi}^e\boldsymbol{b}^*\big).
\end{aligned}
\tag{6.104}
$$

In view of

$$
|\kappa_\alpha h| < \frac{1}{2}
$$

we have

$$
|K|\,h^2 = |\kappa_1 h|\cdot|\kappa_2 h| < \frac{1}{4} \qquad \text{and} \qquad |H|\,h \leq \frac{1}{2}\big(|\kappa_1 h| + |\kappa_2 h|\big) < \frac{1}{2}, \tag{6.105}
$$

so it holds

$$\frac{11}{3} + Kh^2 > \frac{11}{3} - \frac{1}{4} = \frac{41}{12}$$
and
$$\frac{h^3}{12} - H^2\frac{h^5}{5} - K\frac{h^5}{80} > h^3\Big(\frac{1}{12} - \frac{1}{4}\cdot\frac{1}{5} - \frac{1}{4}\cdot\frac{1}{80}\Big) = \frac{29}{960}\,h^3. \tag{6.106}$$

Using inequalities (6.106) and $(6.31)_1$, from relation (6.104) we get

$$\begin{aligned} \mathcal{W}^{(5)}_{\text{memb,bend}}(\boldsymbol{E}^e,\boldsymbol{K}^e) &\geq \frac{41}{144}\,h\,W_{\text{Coss}}(\boldsymbol{E}^e) + \frac{29}{960}\,h^3 W_{\text{Coss}}(\boldsymbol{\Phi}^e) \\ &\geq C_4\,\|\boldsymbol{E}^e\|^2 + C_5\,\|\boldsymbol{\Phi}^e\|^2, \end{aligned} \tag{6.107}$$

for some positive constants C_4, C_5. The lemma is proved.

We are now able to prove the coercivity of the shell strain energy function for the higher order model.

Theorem 6.3. *Assume that*
$$|\kappa_\alpha h| < \frac{1}{2}$$
and the constitutive coefficients satisfy inequalities (6.73). *Then, the areal strain energy density* $\mathcal{W}^{(5)}_{\text{shell}}$ *given by* (6.98)-(6.100) *is coercive, i.e. there exist some positive constants* $C_3 > 0$, $C_4 > 0$ *such that*

$$\mathcal{W}^{(5)}_{\text{shell}}(\boldsymbol{E}^e,\boldsymbol{K}^e) \geq C_4\,\|\boldsymbol{E}^e\|^2 + C_3\,\|\boldsymbol{K}^e\|^2 . \tag{6.108}$$

Proof. By virtue of the relation (6.102) and the Lemma 6.3 we deduce that

$$\mathcal{W}^{(5)}_{\text{memb,bend}}(\boldsymbol{E}^e,\boldsymbol{K}^e) + \mathcal{W}^{(5)}_{\text{bend,curv}}(\boldsymbol{K}^e) \geq C_4\,\|\boldsymbol{E}^e\|^2 + C_5\,\|\boldsymbol{\Phi}^e\|^2 + C_3\,\|\boldsymbol{K}^e\|^2,$$

so coercivity inequality (6.108) holds true.

In a similar way as in Subsect. 6.3.2 we can prove the existence of minimizers for the Cosserat 6-parameter shell model of order $O(h^5)$. In this case, the total energy functional is expressed by

$$\mathcal{E}^{(5)}(\boldsymbol{m},\boldsymbol{Q}_e) = \iint_\omega \mathcal{W}^{(5)}_{\text{shell}}(\boldsymbol{E}^e,\boldsymbol{K}^e)\,a(x_1,x_2)\,\mathrm{d}x_1\mathrm{d}x_2 - \Lambda(\boldsymbol{m},\boldsymbol{Q}_e), \tag{6.109}$$

where the areal strain energy density $\mathcal{W}^{(5)}_{\text{shell}}$ is given by (6.98)-(6.100) and the external loading potential Λ has the form (6.85).

Theorem 6.4. *Consider the minimization problem for the equilibrium of Cosserat 6-parameter elastic shells:*

$$\text{minimize} \quad \mathcal{E}^{(5)}(\boldsymbol{m},\boldsymbol{Q}_e) \quad \text{w.r.t.} \quad (\boldsymbol{m},\boldsymbol{Q}_e) \in \mathcal{A}, \tag{6.110}$$

where the total energy functional $\mathcal{E}^{(5)}$ is given by (6.109) *and the admissible set $\mathcal{A}$ is defined by* (6.83). *Assume that $|\kappa_\alpha h| < \frac{1}{2}$ and the constitutive coefficients satisfy the inequalities* (6.73). *Also, the reference configuration of the shell is assumed to satisfy regularity conditions* (6.88), *while the external loads and boundary data fulfil regularity conditions* (6.87). *Then, the minimization problem* (6.110) *admits at least one minimizing solution pair $(\hat{\boldsymbol{m}}, \hat{\boldsymbol{Q}}_e)$ in the admissible set $\mathcal{A}$.*

Proof. We proceed similarly as in the proof of Theorem 6.2. Since all the hypotheses of Theorem 6.3 are satisfied, we deduce that the strain energy density $\mathcal{W}^{(5)}_{\text{shell}}(\boldsymbol{E}^e, \boldsymbol{K}^e)$ is coercive. Further, from relations (6.98)-(6.100) we see that this function is a quadratic form in $(\boldsymbol{E}^e, \boldsymbol{K}^e)$, which is positive definite. Then, $\mathcal{W}^{(5)}_{\text{shell}}(\boldsymbol{E}^e, \boldsymbol{K}^e)$ is also convex. We can apply the general existence theorem in [20], since all its hypotheses are fulfilled. From [20, Theorem 6] we obtain the existence of minimizers to our minimization problem (6.110) for the Cosserat 6-parameter shell model of order $O(h^5)$. We refer to [20] for the details of the proof.

Remark 6.5. We mention that Theorem 6.4 is similar to the existence result presented previously in [31, Theorem 3.3] for a related Cosserat shell model including terms up to order $O(h^5)$. This related Cosserat shell model has been investigated in [30, 31, 34] using the matrix formulation. A detailed comparison between the two Cosserat approaches to 6-parameter shells (which employ either matrix formulation or tensorial notation) can be found in [33, Sect. 5.3].

Acknowledgements This research has been funded by the Deutsche Forschungsgemeinschaft (DFG, German Research Foundation) – Project no. 415894848: BI 1965/2-1 (M. Bîrsan) and NE 902/8-1 (P. Neff).

References

[1] Love A (1944) A Treatise on the Mathematical Theory of Elasticity, Dover, New York.
[2] Novozhilov V (1959) The Theory of Thin Shells, Noordhoff, Groningen.
[3] Koiter W (1966) On the nonlinear theory of thin elastic shells, Proc K Ned Akad Wet B **69**:1-54.
[4] Ciarlet P (2000) Mathematical Elasticity, Vol. III: Theory of Shells, North-Holland, Amsterdam.
[5] Steigmann D (2013) Koiter's shell theory from the perspective of three-dimensional nonlinear elasticity, J Elasticity **111**:91-107.
[6] Steigmann D, Bîrsan M, Shirani M (2023) Lecture Notes on the Theory of Plates and Shells, Series Solid Mechanics and Its Applications, Springer, to appear.
[7] Reissner E (1974) Linear and nonlinear theory of shells, In: YC Fung, EE Sechler (Eds) Thin Shell Structures, pp 29-44, Prentice-Hall, Englewood Cliffs, New Jersey.

[8] Zhilin P (1976) Mechanics of deformable directed surfaces, Int J Solids Struct **12**:635-648.

[9] Altenbach H, Zhilin P (2004) The theory of simple elastic shells, In: R Kienzler, H Altenbach, I Ott (Eds) Theories of Plates and Shells. Critical Review and New Applications, Euromech Colloquium 444, Lecture Notes in Applied and Computational Mechanics, vol. 16, pp 1-12, Springer, Heidelberg

[10] Libai A, Simmonds J (1998) The Nonlinear Theory of Elastic Shells, 2nd edn, Cambridge University Press, Cambridge.

[11] Chróścielewski J, Makowski J, Pietraszkiewicz W (2004) Statics and Dynamics of Multifold Shells: Nonlinear Theory and Finite Element Method (in Polish), Wydawnictwo IPPT PAN, Warsaw.

[12] Eremeyev V, Zubov L (2008) Mechanics of Elastic Shells (in Russ), Nauka, Moscow.

[13] Ciarlet P (1998) Introduction to Linear Shell Theory, Gauthier-Villars, Paris.

[14] Bîrsan M, Altenbach H (2010) A mathematical study of the linear theory for orthotropic elastic simple shells, Math Methods Appl Sci **33**:1399-1413.

[15] Eremeyev V, Lebedev L (2011) Existence theorems in the linear theory of micropolar shells, Z Angew Math Mech **91**:468-476.

[16] Ghiba I, Bîrsan M, Neff P (2022) A linear isotropic Cosserat shell model including terms up to $O(h^5)$. Existence and uniqueness, (submitted), arXiv:2208.04574v1.

[17] Neff P (2004) A geometrically exact Cosserat-shell model including size effects, avoiding degeneracy in the thin shell limit. Part I: Formal dimensional reduction for elastic plates and existence of minimizers for positive Cosserat couple modulus, Cont Mech Thermodynamics **16**:577-628.

[18] Neff P (2007) A geometrically exact planar Cosserat shell-model with microstructure: Existence of minimizers for zero Cosserat couple modulus, Math Mod Meth Appl Sci **17**:363-392.

[19] Neff P, Bîrsan M, Osterbrink F (2015) Existence theorem for geometrically nonlinear Cosserat micropolar model under uniform convexity requirements, J Elasticity **121**:119-141.

[20] Bîrsan M, Neff P (2014) Existence of minimizers in the geometrically nonlinear 6-parameter resultant shell theory with drilling rotations, Math Mech Solids **19**:376-397.

[21] Eremeyev V, Pietraszkiewicz W (2006) Local symmetry group in the general theory of elastic shells, J Elasticity **85**:125-152.

[22] Bîrsan M (2020) Derivation of a refined six-parameter shell model: descent from the three-dimensional Cosserat elasticity using a method of classical shell theory, Math Mech Solids **25**:1318-1339.

[23] Eremeyev V, Pietraszkiewicz W (2004) The nonlinear theory of elastic shells with phase transitions, J Elasticity **74**:67-86.

[24] Pietraszkiewicz W (2011) Refined resultant thermomechanics of shells, Int J Engng Sci **49**:1112-1124.

[25] Bîrsan M, Neff P (2017) Analysis of the deformation of Cosserat elastic shells using the dislocation density tensor, In: F dell'Isola, M Sofonea, D Steigmann

(Eds) Mathematical Modelling in Solid Mechanics, Advanced Structured Materials, vol. 69, pp 13-30, Springer Nature, Singapore.
[26] Altenbach H, Eremeyev V (2009) On the linear theory of micropolar plates, Z Angew Math Mech **89**:242-256.
[27] Altenbach H, Eremeyev V (2013) Cosserat-type shells, In: H Altenbach, V Eremeyev (Eds) Generalized Continua - from the Theory to Engineering Applications, CISM Courses and Lectures, vol. 541, pp 131-178, Springer, Wien.
[28] Chróścielewski J, Pietraszkiewicz W, Witkowski W (2010) On shear correction factors in the non-linear theory of elastic shells, Int J Solids Struct **47**:3537-3545.
[29] Bîrsan M, Ghiba I, Martin R, Neff P (2019) Refined dimensional reduction for isotropic elastic Cosserat shells with initial curvature, Math Mech Solids **24**:4000-4019.
[30] Ghiba I, Bîrsan M, Lewintan P, Neff P (2020) The isotropic Cosserat shell model including terms up to $O(h^5)$. Part I: Derivation in matrix notation, J Elasticity **142**:201-262.
[31] Ghiba I, Bîrsan M, Lewintan P, Neff P (2020) The isotropic Cosserat shell model including terms up to $O(h^5)$. Part II: Existence of minimizers, J Elasticity **142**:263-290.
[32] Sander O, Neff P, Bîrsan M (2016) Numerical treatment of a geometrically nonlinear planar Cosserat shell model, Comput Mech **57**:817-841.
[33] Bîrsan M (2011) Alternative derivation of the higher-order constitutive model for six-parameter elastic shells, Z Angew Math Phys **72**:50.
[34] Ghiba I, Bîrsan M, Lewintan P, Neff P (2021) A constrained Cosserat-shell model including terms up to $O(h^5)$, J Elasticity **146**:83-141.

Chapter 7
Solving the Equations of Nonlinear Model of Crystalline Media with Complex Lattice and Some Structures of Plane Deformation

Anatolii N. Bulygin and Yuri V. Pavlov

Abstract A nonlinear model of deformation of solid state with a complex crystal lattice is considered. A general solution of dynamic equations is given for the macro-stress tensor and the macro-shift vector for the case of plane deformation of cubic symmetry media under simplifying assumptions. Solution for the micro-shifts field is reduced to the solution of sine-Gordon type equation and sine-Gordon with a variable amplitude equations. Examples of exact analytical solutions of micro-field equations are given and the pictures of their corresponding deformation structures are presented.

Key words: Nonlinear model · Crystal lattice · Sine-Gordon equation

7.1 Introduction

The nonlinear model of the deformation of crystalline media with a complex lattice has been proposed in [1, 2]. This model describes many physical and mechanical processes that are implemented in experiments (super-lattice formation, occurrence of defects, phase transitions such as martensitic, etc.), which the classical linear model does not describe. In classical models of crystalline media [3] displacements are assumed to be small and do not shift atoms beyond the crystal cell. This limitation does not allow to describe large deformations of the crystal lattice and cardinal rearrangement of the structure of the crystalline medium under the action of intense external influences. Many modern methods of obtaining new materials use intensive force influences. In this case, the initial crystal structure is radically rebuilt, phase transformations and the appearance of defects can occur. Analytical models capable

Yuri V. Pavlov
Institute for Problems in Mechanical Engineering of Russian Academy of Sciences, 61 Bol'shoy, V.O., Saint Petersburg 199178, Russian Federation,
e-mail: yuri.pavlov@mail.ru

H. Altenbach et al. (eds.), *Sixty Shades of Generalized Continua*,
Advanced Structured Materials 170, https://doi.org/10.1007/978-3-031-26186-2_7

of describing such phenomena become necessary for the development of the scientific foundations for the technology of obtaining new materials.

7.2 Nonlinear Model of Deformation of Crystal Media

In the nonlinear model, medium deformation is described by the macro-shift vector $\boldsymbol{U}(t,x,y,z)$ (acoustic mode) and micro-shift vector $\boldsymbol{u}(t,x,y,z)$ (optical mode). Consider the case of a crystal lattice consisting of two sublattices, with atoms with masses m_1 and m_2. The vector $\boldsymbol{U}(t,x,y,z)$ is defined as the displacement of the center of inertia of a pair of atoms of a elementary cell

$$\boldsymbol{U} = \frac{m_1\boldsymbol{r}_1 + m_2\boldsymbol{r}_2}{m_1+m_2}, \tag{7.1}$$

where $\boldsymbol{r}_1$ and $\boldsymbol{r}_2$ is the displacement of the atoms of the corresponding sublattices. The vector $\boldsymbol{u}(t,x,y,z) = \boldsymbol{r}_1 - \boldsymbol{r}_2$ takes into account the change in near order in the crystal lattice.

The equations of motion for acoustic and optical modes can be derived from the variational principle for the action

$$I = \int L(u_i, \dot{u}_i, u_{i,j}, \dot{U}_i, U_{i,j})\, dt\, dV, \tag{7.2}$$

where $i,j = 1,2,3$, comma in index denotes partial derivative in respect to the corresponding variable. The density of the Lagrangian can be represented as the difference between the density of the kinetic Q and potential D energy of deformation

$$I = Q - D, \tag{7.3}$$

where

$$Q = \frac{\rho}{2}\dot{U}_i\dot{U}_i + \frac{\mu_0}{2}\dot{u}_i\dot{u}_i, \tag{7.4}$$

summation for repeating indices is implied, ρ is the average mass density of the elementary cell atoms

$$\rho = \frac{m_1+m_2}{2V}, \tag{7.5}$$

V is the volume of elementary cell, μ_0 is the reduced atomic density

$$\mu_0 = \frac{m_1+m_2}{(m_1+m_2)V}. \tag{7.6}$$

The density D of the potential energy is chosen in the form

$$D = \frac{1}{2}\lambda_{ilmn}e_{il}e_{mn} + \frac{1}{2}k_{ilmn}\varepsilon_{il}\varepsilon_{mn} + \frac{1}{2}C_{ilmn}e_{il}\varepsilon_{mn} + R\Phi(u_s), \tag{7.7}$$

where λ_{ilmn}, k_{ilmn} are the coefficients of elasticity and microelasticity,, the tensor C_{ilmn} describes the interaction of acoustic and optical modes, e_{ij}, ε_{ij} are the strain and microstrain tensors:

$$e_{ij} = \frac{1}{2}\left(U_{i,j} + U_{j,i}\right), \quad \varepsilon_{ij} = \frac{1}{2}\left(u_{i,j} + u_{j,i}\right). \tag{7.8}$$

The first two terms in (7.7) describe the contribution to the energy density of elastic macro- and micro-deformations. The third term describes the interaction of macro- and micro-deformation fields. Its choice is discussed in detail in [4].

The last term in (7.7) corresponds to the energy of rigid shift of the sublattices. For reasons of invariance with respect to symmetry transformations, the argument in the last term is chosen as

$$u_s = \boldsymbol{B} \cdot \boldsymbol{u}, \tag{7.9}$$

where $\boldsymbol{B}$ is a inverse lattice vector [5]. For cubic symmetry crystals with cell size b one has

$$\boldsymbol{B} = \frac{2\pi}{b}(\boldsymbol{i} + \boldsymbol{j} + \boldsymbol{k}) \tag{7.10}$$

and

$$u_s = \frac{2\pi}{b}(u_x + u_y + u_z). \tag{7.11}$$

The function $\Phi(u_s)$ describes the interaction energy of sublattices. It should be even and periodic with respect to the periods of the crystal lattice. In the fundamental contribution [6] and in most subsequent works [7, 8] it is accepted

$$\Phi(u_s) = 1 - \cos u_s. \tag{7.12}$$

The multiplier

$$R = p - s_{ij} e_{ij}, \tag{7.13}$$

represents the activation energy of micro and macro shift bonds. The term p is half of the energy of activation of the rigid shift of sublattices, and s_{ij} is the mechanostriction tensor.

The equations of motion of the nonlinear model follow from the variational principle and have the form [4, 9]

$$\rho \frac{\partial^2 U_i}{\partial t^2} = \sigma_{ij,j}, \tag{7.14}$$

$$\mu_0 \frac{\partial^2 u_i}{\partial t^2} = \chi_{ij,j} - R \frac{\partial \Phi(u_s)}{\partial u_i}. \tag{7.15}$$

where σ_{ik}, χ_{ik} are the stress and micro-stress tensors:

$$\sigma_{ij} = \lambda_{ijmn} e_{mn} + C_{ijmn} \varepsilon_{mn} - s_{ij} \Phi(u_s), \tag{7.16}$$

$$\chi_{ij} = k_{ijmn} \varepsilon_{mn} + C_{ijmn} e_{mn}. \tag{7.17}$$

7.3 General Solution of Dynamic Equations of Plane Deformation of the Nonlinear Model

The deformed state will be referred to as plane, parallel to axis OZ, if

$$U_x = U_x(t,x,y), \quad U_y = U_y(t,x,y), \quad U_z = 0, \tag{7.18}$$

$$u_x = u_x(t,x,y), \quad u_y = u_y(t,x,y), \quad u_z = 0. \tag{7.19}$$

Consider the case of crystalline media of cubic symmetry and let's assume that the interaction of the acoustic and optical modes is carried out only due to the last nonlinear term in (7.7), i.e. we put $C_{ijmn} = 0$. In the case of cubic lattice symmetry media, only some components of material tensors λ_{ijmn}, k_{ijmn} are non-zero:

$$\lambda_{1111} = \lambda_{2222} = \lambda_{3333}, \quad \lambda_{1122} = \lambda_{1133}, \quad \lambda_{1212} = \lambda_{1313}. \tag{7.20}$$

$$k_{1111} = k_{2222} = k_{3333}, \quad k_{1122} = k_{1133}, \quad k_{1212} = k_{1313}. \tag{7.21}$$

For independent components the following notation is used [10]:

$$\lambda_{1122} = \lambda_{12} = \lambda, \quad \lambda_{1212} = \lambda_{44} = \mu, \quad \lambda_{1111} = \lambda_{11}, \tag{7.22}$$
$$k_{1122} = k_{12}, \quad k_{1212} = k_{44}, \quad k_{1111} = k_{11}. \tag{7.23}$$

In general case λ_{ijmn} and k_{ijmn} have the form of Huang tensor [11]

$$\begin{pmatrix} \lambda_{ijmn} \\ k_{ijmn} \end{pmatrix} = \begin{pmatrix} \lambda \\ k_{12} \end{pmatrix} \delta_{ij}\delta_{mn} + \begin{pmatrix} \mu \\ k_{44} \end{pmatrix} (\delta_{jn}\delta_{im} + \delta_{in}\delta_{jm})$$
$$+ \begin{pmatrix} \lambda_{11} - \lambda_{12} - 2\lambda_{44} \\ k_{11} - k_{12} - 2k_{44} \end{pmatrix} \delta_{ijmn}. \tag{7.24}$$

Here δ_{ij} is a unit tensor, and $\delta_{ijmn} = 1$, if all indices are the same and equal to zero in other cases. The last term in (7.24) describes the anisotropy of the medium. In solid state physics [12] the medium anisotropy factor (a_1, a_2) is introduced

$$\lambda_{11} - \lambda_{12} - 2\lambda_{44} = (\lambda_{11} - \lambda_{12})(1 - a_1), \quad k_{11} - k_{12} - 2k_{44} = (k_{11} - k_{12})(1 - a_2). \tag{7.25}$$

For media with weak anisotropy $a_1 \approx 1, a_2 \approx 1$ and cubic symmetry $s_{ij} = s\delta_{ij}$ material ratios (7.16), (7.17) in case of plane deformation and $C_{ijmn} = 0$ take the form

$$\begin{aligned} \sigma_{xx} &= 2\mu U_{x,x} + \lambda(U_{x,x} + U_{y,y}) - s\Phi(u_s), \\ \sigma_{yy} &= 2\mu U_{y,y} + \lambda(U_{x,x} + U_{y,y}) - s\Phi(u_s), \\ \sigma_{xy} &= \mu(U_{x,y} + U_{y,x}), \end{aligned} \tag{7.26}$$

$$\chi_{xx} = 2k_{44} u_{x,x} + k_{12}(u_{x,x} + u_{y,y}),$$

$$\chi_{yy} = 2k_{44}u_{y,y} + k_{12}(u_{x,x} + u_{y,y}), \qquad (7.27)$$
$$\chi_{xy} = k_{44}(u_{x,y} + u_{y,x}).$$

Stress tensor components $\sigma_{xx}, \sigma_{yy}, \sigma_{xy}$ must satisfy Beltrami–Michell conditions [13]. They are reduced to one equation for plane deformation in the model under consideration

$$\left(\Delta - \frac{1}{V_1^2}\frac{\partial^2}{\partial t^2}\right)(\sigma_{xx} + \sigma_{yy}) + \frac{2s\mu}{\lambda + 2\mu}\left(\Delta - \frac{1}{V_2^2}\frac{\partial^2}{\partial t^2}\right)\Phi(u_s) = 0, \qquad (7.28)$$

$$\Delta = \frac{\partial^2}{\partial x^2} + \frac{\partial^2}{\partial y^2}, \quad V_1 = \sqrt{\frac{\lambda + 2\mu}{\rho}}, \quad V_2 = \sqrt{\frac{\mu}{\rho}}. \qquad (7.29)$$

Here V_1 is longitudinal wave velocity and V_2 is velocity of shifts. The tensor σ_{ij} and the vector U_i are expressed through an arbitrary function $Q(t,x,y)$ [14]:

$$\sigma_{xx} = L_{11}(Q) = \left(\frac{\partial^2}{\partial y^2} - \frac{1}{2V_2^2}\frac{\partial^2}{\partial t^2}\right)Q,$$

$$\sigma_{yy} = L_{22}(Q) = \left(\frac{\partial^2}{\partial x^2} - \frac{1}{2V_2^2}\frac{\partial^2}{\partial t^2}\right)Q,$$

$$\sigma_{xx} + \sigma_{yy} = \theta = \left(\Delta - \frac{1}{V_2^2}\frac{\partial^2}{\partial t^2}\right)Q, \qquad (7.30)$$

$$\left(\Delta - \frac{1}{2V_2^2}\frac{\partial^2}{\partial t^2}\right)\left(\sigma_{xy} + \frac{\partial^2 Q}{\partial x \partial y}\right) = 0,$$

$$2\mu U_{x,x} = \sigma_{xx} - \frac{\lambda}{2(\lambda + \mu)}\theta + \frac{\mu}{\lambda + \mu}s\Phi(u_s),$$
$$2\mu U_{y,y} = \sigma_{yy} - \frac{\lambda}{2(\lambda + \mu)}\theta + \frac{\mu}{\lambda + \mu}s\Phi(u_s). \qquad (7.31)$$

If (7.30) is substituted in (7.28), then one has the equation for finding the function $Q(t,x,y)$

$$\left(\Delta - \frac{1}{V_1^2}\frac{\partial^2}{\partial t^2}\right)\left[\left(\Delta - \frac{1}{V_2^2}\frac{\partial^2}{\partial t^2}\right)Q + 2s\frac{V_2^2}{V_1^2}\Phi(u_s)\right] = 0. \qquad (7.32)$$

Equation (7.32) shows that $Q(t,x,y)$ is a dynamic analogue of the Airy function, which is introduced to solve static problems of classical plane deformation. The function $Q(t,x,y)$, unlike the Airy function, satisfies a non-homogeneous dynamic biharmonic equation. The function $\Phi(u_s)$ plays the role of volumetric sources of macro stresses and macro deformations. The general solution of equation (7.32) can be written as the sum of two terms

$$Q(t,x,y) = F(t,x,y) + Q_0(t,x,y). \tag{7.33}$$

The function $F(t,x,y)$ satisfies to the homogeneous biharmonic equation

$$\left(\Delta - \frac{1}{V_1^2}\frac{\partial^2}{\partial t^2}\right)\left(\Delta - \frac{1}{V_2^2}\frac{\partial^2}{\partial t^2}\right)F(t,x,y) = 0, \tag{7.34}$$

and $Q_0(t,x,y)$ is a solution of non-homogeneous equation

$$\left(\Delta - \frac{1}{V_1^2}\frac{\partial^2}{\partial t^2}\right)Q_0(t,x,y) + 2s\frac{V_2^2}{V_1^2}\Phi(u_s) = 0. \tag{7.35}$$

The obtained general solution of equations of the macro-field (7.14), (7.15) allows to set various dynamic problems and find exact analytical solutions based on the non-linear model.

7.4 Solving the Micro-Field Equations

Micro-field equations (7.15) can be written in component form if the micro-stress tensor χ_{ij} (7.27) is substituted into equation (7.15) and the form of the tensor k_{ijmn} is taken into account:

$$\begin{aligned}
\mu_0\frac{\partial^2 u_x}{\partial t^2} &= k_{44}\Delta u_x + (k_{12}+k_{44})(u_{x,xx}+u_{y,xy}) + (k_{11}-k_{12}-2k_{44})u_{x,xx} - \frac{R}{b}\sin u_s,\\
\mu_0\frac{\partial^2 u_y}{\partial t^2} &= k_{44}\Delta u_y + (k_{12}+k_{44})(u_{x,xy}+u_{y,yy}) + (k_{11}-k_{12}-2k_{44})u_{y,yy} - \frac{R}{b}\sin u_s.
\end{aligned} \tag{7.36}$$

Instead of a component (u_x, u_y), one enters u_s and $u_m = (u_x - u_y)/b$. Then the sum and the difference of Eqs. (7.36) are written as follows

$$\begin{aligned}
2\mu_0\frac{\partial^2 u_s}{\partial t^2} &= (k_{11}+k_{44})\Delta u_s + 2(k_{12}+k_{44})u_{s,xy} + (k_{11}-k_{44})(u_{m,xx}-u_{m,yy}) - \frac{4R}{b^2}\sin u_s,\\
2\mu_0\frac{\partial^2 u_m}{\partial t^2} &= (k_{11}+k_{44})\Delta u_m - 2(k_{12}+k_{44})u_{m,xy} + (k_{11}-k_{44})(u_{s,xx}-u_{s,yy}).
\end{aligned} \tag{7.37}$$

Equations (7.37) are separated if $k_{11} = k_{44}$. Let us take this condition and instead of variables (x,y) one enters

$$q_1 = L_1(x+y), \quad q_2 = L_2(x-y), \tag{7.38}$$

where

$$L_1 = \frac{1}{b}\sqrt{\frac{2p}{(k_{11}+k_{44})(1+k_0)}}, \quad L_2 = \frac{1}{b}\sqrt{\frac{2p}{(k_{11}+k_{44})(1-k_0)}}, \quad k_0 = \frac{k_{12}+k_{44}}{k_{11}+k_{44}}.$$

In terms of (q_1,q_2) Eqs. (7.37) for static solutions will take the form

$$\frac{\partial^2 u_s}{\partial q_1^2}+\frac{\partial^2 u_s}{\partial q_2^2}=\frac{R}{p}\sin u_s, \tag{7.39}$$

$$\omega^2\frac{\partial^2 u_m}{\partial q_1^2}+\frac{\partial^2 u_m}{\partial q_2^2}=0, \quad \omega=\sqrt{\frac{1-k_0}{1+k_0}}. \tag{7.40}$$

From relations (7.13), (7.26) one finds

$$\begin{aligned}\frac{R}{p}&=p_1+2p_2\cos u_s,\\ p_1&=1-2p_2-\frac{p_2}{s}(\sigma_{xx}+\sigma_{yy}),\\ p_2&=\frac{s^2}{2p(\lambda+\mu)}.\end{aligned} \tag{7.41}$$

Taking into account (7.41) Eq. (7.39) takes the form

$$\frac{\partial^2 u_s}{\partial q_1^2}+\frac{\partial^2 u_s}{\partial q_2^2}=p_1\sin u_s+p_2\sin 2u_s. \tag{7.42}$$

Equation (7.42) differs from the classical double sine-Gordon equation in that the amplitude p_1 is not a constant, but a function of (t,x,y,u_s). There are no analytical methods for solving such an equation in the literature. For this reason, we use assumptions that reduce (7.42) to equation that has exact analytical solutions. For the medium where $s^2 \ll 2p(\lambda+\mu)$ one can be accepted $p_2=0$ and $p_1=1$. Then Eq. (7.42) will become the classic sine-Gordon equation

$$\frac{\partial^2 u_s}{\partial q_1^2}+\frac{\partial^2 u_s}{\partial q_2^2}=\sin u_s. \tag{7.43}$$

If $s^2 \ll 2p(\lambda+\mu)$, but $s(\sigma_{xx}+\sigma_{yy})/2p(\lambda+\mu)$ is not negligible, then (7.42) takes the form of a sine-Gordon equation with variable amplitude

$$\frac{\partial^2 u_s}{\partial q_1^2}+\frac{\partial^2 u_s}{\partial q_2^2}=p(q_1,q_2)\sin u_s, \tag{7.44}$$

$$p(q_1,q_2)=1-s(\sigma_{xx}+\sigma_{yy})/2p(\lambda+\mu).$$

The sine-Gordon equation (7.43) has been studied in detail in the literature. For the sine-Gordon equation with variable amplitude, analytic solutions built only for the particular function type $p(q_1,q_2)$ [15]–[17].

Below are some examples of solutions and corresponding deformed crystal structures. In Fig. 7.1 is shown the medium which is deformed according to the following solution of Eq. (7.43):

$$u = 4\,\mathrm{arctg}(A\,\mathrm{dn}\,(\xi,\nu_1)\,\mathrm{dn}\,(\eta,\nu_2)) \tag{7.45}$$

where

$$\xi = \frac{q_1}{H}\mathrm{K}(\nu_1), \quad \eta = \frac{q_2}{B}\mathrm{K}(\nu_2),$$

$$\begin{aligned} H^2 &= \frac{1}{A^2}(1+A^2)\left[1+(1-\nu_1^2)A^2\right]\mathrm{K}^2(\nu_1), \\ B^2 &= \frac{1}{A^2}(1+A^2)\left[1+(1-\nu_2^2)A^2\right]\mathrm{K}^2(\nu_2) \end{aligned} \tag{7.46}$$

with

$$A^2 = \frac{1}{\sqrt{(1-\nu_1^2)(1-\nu_2^2)}}, \qquad 0 \le \nu_1, \nu_2 \le 1,$$

Here $\mathrm{K}(\nu)$ is complete elliptic integral of the first kind, $\mathrm{dn}\,(\eta,\nu)$, and further $\mathrm{cn}\,(\eta,\nu)$ are Jacobi elliptic functions. Images of the deformed medium are given in the following form. In figures at the left side shifts u_s are in direction vertical axis, and at the right side shifts u_s are represented as perpendicular to the plane (q_1, q_2).

The function $\mathrm{dn}\,(x,\nu)$ is double periodic. It has no zeroes and poles. For this reason solution (7.45) describes deformation when shift of atoms is less than the size of the cell. In this case the lattice will not be reorganized significantly (its topology does not change).

The function

$$u = 4\,\mathrm{arctg}\left(A\frac{\mathrm{cn}\,(\xi,\nu_1)}{\mathrm{cn}\,(\eta,\nu_2)}\right) \tag{7.47}$$

will be the solution of (7.43), if

$$H^2 = \frac{1}{A^2}(1+A^2)\left[(1+A^2)\nu_1^2 - A^2\right]\mathrm{K}^2(\nu_1),$$

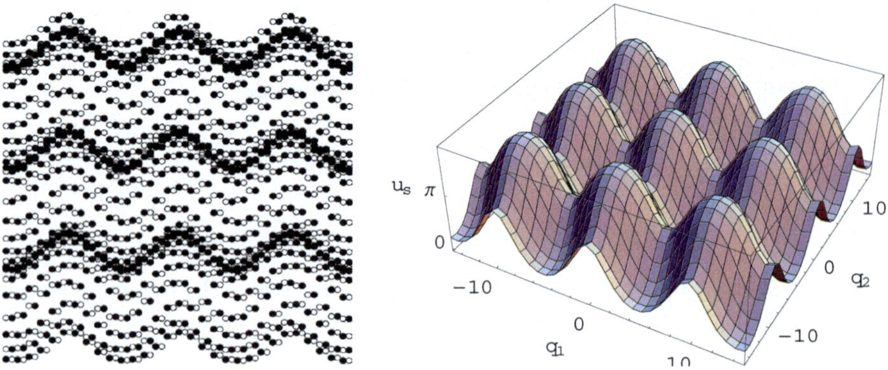

Fig. 7.1: Crystal deformation corresponding to the solution (7.45) for $\nu_1 = \nu_2 = 0.99$.

$$B^2 = \frac{\nu_2^2}{\nu_1^2}(1+A^2)\left[(1+A^2)\nu_1^2 - A^2\right] \mathrm{K}^2(\nu_2), \tag{7.48}$$

with

$$A^2 = \frac{\nu_1}{\nu_2}\sqrt{\frac{1-\nu_2^2}{1-\nu_1^2}}, \quad \nu_1^2+\nu_2^2 \geq 1, \quad 0 \leq \nu_1, \nu_2 \leq 1.$$

Figure 7.2 shows the medium that is deformed according to Eq. (7.47). The function cn (x, ν) is double periodic also, but it has both zeroes and poles. In this case shifts of atoms can exceed the cell size, and the crystal structure can significantly be reorganized. The medium is stratified, micropores appear everywhere.

Figure 7.3 represents medium which microshifts are described by the following solution of Eq. (7.43)

$$u = 4\,\mathrm{arctg}\left[\tan\psi \frac{\sinh(q_1\cos\psi)}{\sinh(q_2\sin\psi)}\right], \quad \psi = \mathrm{const} \tag{7.49}$$

The half-space $q_1 < 0$ contracts, and $q_1 > 0$ extends. As a result on the line of $q_2 = 0$, $q_1 > 0$ defect like the main crack arises.

Figure 7.4 shows medium which deformation is described by the equation

$$u = 4\,\mathrm{arctg}\left[\coth\psi \frac{\cos(q_1\sinh\psi)}{\sinh(q_2\cosh\psi)}\right], \quad \psi = \mathrm{const}. \tag{7.50}$$

In this case defects like micropores or ruptures are located along an axis (Oq_1). The medium is almost not deformed at far from the line.

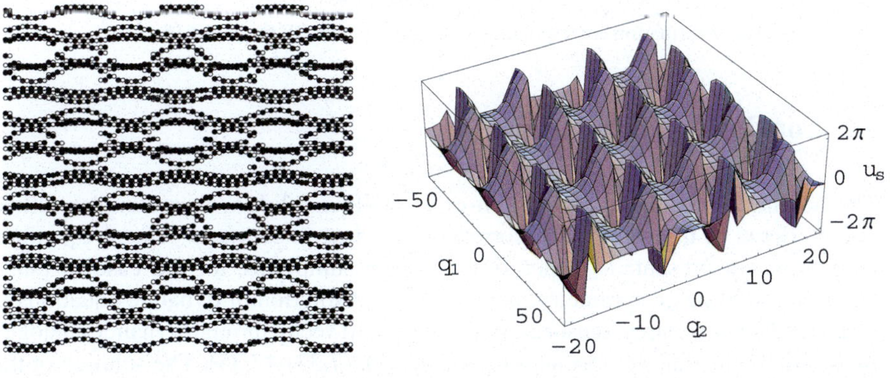

Fig. 7.2: Crystal deformation corresponding to the solution (7.47) for $\nu_1 = 0.5$, $\nu_2 = 0.9999$.

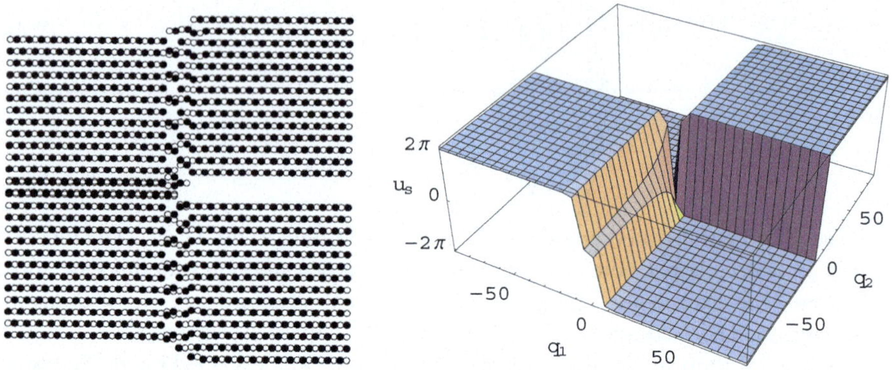

Fig. 7.3: Crystal deformation corresponding to the solution (7.49) for $\psi = 0.01$.

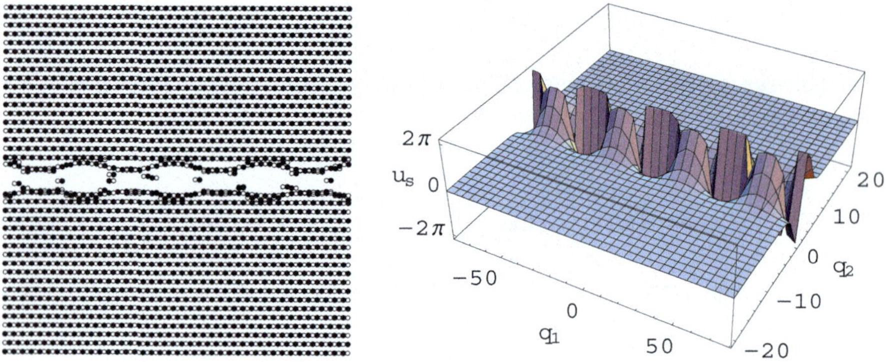

(a) (b)

Fig. 7.4: Crystal deformation corresponding to the solution (7.50) for $\psi = 0.15$.

7.5 Conclusion

The nonlinear model of deformation of crystalline solids with a complex lattice is considered in this work. The corresponding Lagrangian and general equations of motion are presented. Under simplifying assumptions, for the case of plane deformation of cubic symmetry media, a general solution of dynamic equations is given for the macro-stress tensor and the macro-shift vector through dynamic analogue of the Airy function, see formulas (7.30), (7.31), (7.33)–(7.35). The solution of the equations of the nonlinear model macro-stress tensor and the macro-shift vector is reduced to the solution to biharmonic equation (7.34) and nonhomogeneous wave equation (7.35), in which additional term depends on the field of micro-shifts.

Under simplifying assumptions, solutions are presented for the field of the micro-shifts in the static case. Depending on the magnitude of the parameters of the nonlinear model, equation for micro-fields (7.39) reduced to sine-Gordon equation (7.43), double sine-Gordon equation (7.42) and sine-Gordon type equation with variable amplitude (7.44). The field of micro-shifts in the static case determines the structure of the deformed crystal under external influences. In the work the examples of solutions of micro-field equations and corresponding deformation structures are presented, see (7.45), (7.47), (7.49) and (7.50). For example, various systems of micropores, the main cracks and rupture lines are formed under such deformations. The set of solutions of nonlinear equations of the sine-Gordon type is very rich, the properties of such solutions vary greatly. This allows us to argue that in the nonlinear model of deformation of crystalline media with a complex lattice, different structures appear naturally in the form of solutions of field equations corresponding to deformation of materials under strong external influences. This indicates the possibility of using the nonlinear model in the development of modern technologies for producing materials with a given structure.

References

[1] Aero EL (2000) Microscale deformations in a two-dimensional lattice: Structural transitions and bifurcations at critical shear, Physics of the Solid State **42**:1147-1153. DOI 10.1134/1.1131331
[2] Aero EL (2006) Micromechanics of a double continuum in a model of a medium with variable periodic structure, Journal of Engineering Mathematics **55**:81-95. DOI 10.1007/s10665-005-9012-3
[3] Born M, Huang K (1954) Dynamical Theory of Crystal Lattices, Clarendon, Oxford.
[4] Aero EL, Bulygin AN, Pavlov YuV (2018) Nonlinear deformation model of crystal media allowing martensite transformations: solution of static equations, Mechanics of Solids **53**:623-632. DOI 10.3103/s0025654418060043
[5] Shaskol'skaya MP (1984) Crystallography (in Russ.), Vysshaya Shkola, Moscow.
[6] Frenkel J, Kontorova T (1939) On the theory of plastic deformation and twinning, Acad. Sci. U.S.S.R. J. Phys. **1**:137-149.
[7] Braun OM, Kivshar YS (2004) The Frenkel-Kontorova Model. Concepts, Methods, and Applications, Springer, Berlin.
[8] Porubov AV, Aero EL, Maugin GA (2009) Two approaches to study essentially nonlinear and dispersive properties of the internal structure of materials, Phys. Rev. E **79**:046608. DOI 10.1103/PhysRevE.79.046608
[9] Aero EL, Bulygin AN, Pavlov YV (2019) Nonlinear model of deformation of crystalline media allowing for martensitic transformations: plane deformation, Mechanics of Solids **54**:797-806. DOI 10.3103/S0025654419050029
[10] Voigt W (1910) Lehrbuch der Kristalphysik, Leipzig.
[11] Leibfried G (1955) Gittertheorie der mechanischen und thermischen Eigenschaften der Kristalle, Handbuch der Physik, Band 7, Teil 2, Springer, Berlin.
[12] Kittel C (1956) Introduction to Solid State Physics, Wiley, New York.

[13] Muskhelishvili NI (1977) Some Basic Problems of the Mathematical Theory of Elasticity, Springer, Netherlands. DOI 10.1007/978-94-017-3034-1

[14] Bulygin AN, Pavlov YV (2022) Solution of dynamic equations of plane deformation for nonlinear model of complex crystal lattice, In: Polyanskiy VA, Belyaev AK (Eds) Mechanics and Control of Solids and Structures, Advanced Structured Materials, vol. 164, pp 115-136, Springer, Cham. DOI 10.1007/978-3-030-93076-9_6

[15] Aero EL, Bulygin AN, Pavlov YV (2015) Solutions of the sine-Gordon equation with a variable amplitude, Theor. Math. Phys. 184:961-972. DOI 10.1007/s11232-015-0309-8

[16] Aero EL, Bulygin AN, Pavlov YV (2018) Exact analytical solutions for nonautonomic nonlinear Klein-Fock-Gordon equation, In: dell'Isola F, Eremeyev, VA, Porubov, A (Eds) Advances in Mechanics of Microstructured Media and Structures, Advanced Structured Materials, vol. 87, pp 21-33, Springer, Cham. DOI 10.1007/978-3-319-73694-5_2

[17] Aero EL, Bulygin AN, Pavlov YV (2020) Some solutions of dynamic and static nonlinear nonautonomous Klein-Fock-Gordon equation, In: Altenbach H, Eremeyev V, Pavlov I, Porubov A (Eds) Nonlinear Wave Dynamics of Materials and Structures, Advanced Structured Materials, vol. 122, pp 107-120, Springer, Cham. DOI 10.1007/978-3-030-38708-2_7

Chapter 8
Modal Analysis of a Second-Gradient Annular Plate made of an Orthogonal Network of Logarithmic Spiral Fibers

Alessandro Ciallella, Francesco D'Annibale, Francesco dell'Isola, Dionisio Del Vescovo, and Ivan Giorgio

Abstract A dynamical study of small vibrations of an annular system made of an orthogonal network of fibers around the undeformed configuration is performed. Logarithmic spirals characterize the net of fibers connected by deformable cylinders. This particular arrangement is chosen because it produces a tough material with an economy of matter.

Key words: Modal analysis · Second gradient surfaces · Homogenized nets · Metamaterials

8.1 Introduction

In this contribution, we analyze the case of a bi-dimensional generalized continuum characterized by a microstructure [1]-[12], which bestows on it high performance, namely high strength with a lightweight. An orthogonal net of curved fibers, i.e., logarithmic spirals, constitutes this microstructure optimized to have one of the most efficient patterns for the structural viewpoint, as established in [13].

Among new manufacturing technologies, 3D printing allows us to build these kinds of microstructured mechanical systems, opening unimaginable possibilities until a

Alessandro Ciallella · Francesco D'Annibale · Francesco dell'Isola · Ivan Giorgio
Department of Civil, Construction-Architectural and Environmental Engineering (DICEAA) & International Research Center on the Mathematics and Mechanics of Complex Systems (M&MoCS), University of L'Aquila, Italy,
e-mail: alessandro.ciallella@univaq.it, francesco.dannibale@univaq.it, francesco.dellisola@univaq.it, ivan.giorgio@univaq.it

Dionisio Del Vescovo
Department of Mechanical and Aerospace Engineering, SAPIENZA Università di Roma, Rome & International Research Center on the Mathematics and Mechanics of Complex Systems (M&MoCS), University of L'Aquila, Italy,
e-mail: dionisio.delvescovo@uniroma1.it

H. Altenbach et al. (eds.), *Sixty Shades of Generalized Continua*,
Advanced Structured Materials 170, https://doi.org/10.1007/978-3-031-26186-2_8

few decades ago [14]-[16]. For this reason, many researchers initiated the set up of systematic methods for synthesizing new micro-structures and, therefore, conceiving novel materials with a macroscopic behavior that fulfills one or more required functionalities. This way of proceeding is typical of a new branch of generalized continua that focuses on designing the so-called metamaterials [17]-[19].

In this contribution, we focus our attention on the dynamic response of the considered system of fibers examining its small vibrations around the undeformed configuration. Locally this fibers network has many features in common with the pantographic sheet [20]. Concerning dynamic behavior, we recall the works by [21]-[26].

This chapter is organized as follows: in Sect. 8.2, we briefly recall the used model to describe the considered network of fibers. In Sect. 8.3, we provide the modal analysis of the system. Finally, in Sect. 8.4, we draw some conclusions and future developments.

8.2 The Model for a Fiber Net Arranged in Logarithmic Spirals

In the continuum mechanics framework, we develop a plate model based on an elastic second gradient theory [27]-[30] to study a mechanical system composed of a network of orthogonal fibers. The aim is to take into account the deformation modes of the fibers that, in a classical plate theory, are not considered, namely, geodesic bending and twisting [3], [31]-[35]. Specifically, the fibers are curved and shaped in the undeformed reference configuration according to logarithmic spirals. To build the net, the fibers are connected with cylindrical pivots, which allow them to have a relative rotation around their pivotal axis [36, 37, 38]. In the examined example, the network is arranged in an annular fashion on a plate (see Fig. 8.1).

According to [35], [39], we define the reference configuration of the elastic surface, $\mathcal{S}$, by the material points $\boldsymbol{X} = (X_1, X_2) \in \Omega \subset \mathbb{R}^2$ that are mapped in the location of the current configuration, $\boldsymbol{x} = (x_1, x_2, x_3) \in \mathbb{R}^3$, by the placement map χ with the standard notation:

$$\boldsymbol{x} = \chi(\boldsymbol{X}) = X_\alpha \boldsymbol{e}_\alpha + u_i(\boldsymbol{X})\boldsymbol{e}_i \tag{8.1}$$

where u_i are the components of the displacement field, the Greek indexes range from 1 to 2, the Latin ones from 1 to 3, being Einstein's convention adopted, and the orthonormal vectors $\{\boldsymbol{e}_i\}$ set as a basis that spans the associated vector space of the Euclidian affine space.

The logarithmic spirals that represent the middle lines of the fibers are parameterized by φ and ψ as follows:

$$\begin{cases} X_1 = R_0\, e^{(\varphi+\psi)} \cos(\varphi-\psi) \\ X_2 = R_0\, e^{(\varphi+\psi)} \sin(\varphi-\psi) \end{cases} \tag{8.2}$$

where R_0 is a constant which specifies the inner radius of the annular plate. In what follows, we adopt the unit-speed parametrization S_β ($\beta = 1,2$) to characterize the two fiber families in the reference configuration given in Eq. (8.2). Therefore, we express the two curvilinear abscissae using the relationships $\varphi = \hat{\varphi}(S_1)$ and $\psi = \hat{\psi}(S_2)$ [39].

The unit tangent vector fields to the fibers in the reference configuration become

$$\boldsymbol{D}_\beta(\boldsymbol{X}) = \frac{\mathrm{d}X_\alpha}{\mathrm{d}S_\beta}\boldsymbol{e}_\alpha \tag{8.3}$$

For the two families of fibers in the reference configuration, we have $\boldsymbol{D}_1 \cdot \boldsymbol{D}_2 = 0$; thus, the fibers are orthogonal. The tangent vectors in the current configuration can be evaluated as

$$\lambda_\beta(\boldsymbol{X})\,\boldsymbol{d}_\beta(\boldsymbol{X}) = \frac{\mathrm{d}\chi}{\mathrm{d}S_\beta}(\boldsymbol{X}) = \boldsymbol{F}(\boldsymbol{X})\,\boldsymbol{D}_\beta(\boldsymbol{X}) \tag{8.4}$$

where $\lambda_\beta = \|\boldsymbol{F}\,\boldsymbol{D}_\beta\|$, $\boldsymbol{F} = \nabla_X\,\chi$, and $\boldsymbol{d}_\beta$ represent the unit tangent vector fields in the current configuration. The unit vector normal to the surface $\mathcal{S}$ in the current configuration is

$$\boldsymbol{n}(\boldsymbol{X}) = \frac{\boldsymbol{d}_1 \times \boldsymbol{d}_2}{\|\boldsymbol{d}_1 \times \boldsymbol{d}_2\|} \tag{8.5}$$

To fully describe the kinematics of the fibers, we also introduce two rotation tensor fields, one for each fiber family representative of their cross-section orientation as

$$\boldsymbol{R}_\beta : \{\boldsymbol{e}_i\} \mapsto \{\boldsymbol{d}_\beta,\ \boldsymbol{n}\times\boldsymbol{d}_\beta,\ \boldsymbol{n}\} \tag{8.6}$$

and, accordingly, by treating the fibers as an infinite distribution of Kirchhoff's rods [40]-[42], to specify their current geometric features, we define the curvature tensor as follows

$$\boldsymbol{W}_\beta = \boldsymbol{R}_\beta^\top \frac{\mathrm{d}\boldsymbol{R}_\beta}{\mathrm{d}S_\beta} \tag{8.7}$$

The skew-symmetric tensor (8.7), indeed, allow us to define the twisting, out-of-plane, and geodesic curvature, i.e.,

$$\kappa_{T\beta} = \left(\boldsymbol{n}\times\boldsymbol{d}_\beta\right)\cdot\frac{\mathrm{d}\boldsymbol{n}}{\mathrm{d}S_\beta}, \quad \kappa_{n\beta} = \boldsymbol{n}\cdot\frac{\mathrm{d}\boldsymbol{d}_\beta}{\mathrm{d}S_\beta}, \quad \kappa_{g\beta} = -\left(\boldsymbol{n}\times\boldsymbol{d}_\beta\right)\cdot\frac{\mathrm{d}\boldsymbol{d}_\beta}{\mathrm{d}S_\beta} \tag{8.8}$$

respectively. The measures of deformation characterizing the elastic response of the surface $\mathcal{S}$ are: i) the *fiber stretching*, $\varepsilon_\beta = \lambda_\beta - 1$; ii) the *shear distortion angle*, $\gamma = \operatorname{asin}(\boldsymbol{d}_1 \cdot \boldsymbol{d}_2)$; iii) the *curvature change*, $\Delta\kappa_{T\beta} = \kappa_{T\beta}$, $\Delta\kappa_{n\beta} = \kappa_{n\beta}$, $\Delta\kappa_{g\beta} = \kappa_{g\beta} - \kappa^0_{g\beta}$ in which the superscript '0' refers to the reference configuration and, in particular, the reference geodesic curvature is

$$\kappa^0_{g\beta} = -\left(\boldsymbol{e}_3\times\boldsymbol{D}_\beta\right)\cdot\frac{\mathrm{d}\boldsymbol{D}_\beta}{\mathrm{d}S_\beta} \tag{8.9}$$

The deformation energy can be assumed as a quadratic form in the measures of deformation [39] as a first attempt to model the elastic behavior of the examined fiber net:

$$2U[\chi(\cdot)] = \int_{\Omega} \sum_{\beta=1}^{2} K_{e\beta} (\lambda_\beta - 1)^2 \, \mathrm{d}\Omega + \int_{\Omega} K_s \gamma^2 \, \mathrm{d}\Omega + \int_{\Omega} \sum_{\beta=1}^{2} K_{g\beta} (\kappa_{g\beta} - \kappa_{g\beta}^0)^2 \, \mathrm{d}\Omega + \int_{\Omega} \sum_{\beta=1}^{2} K_{n\beta} (\kappa_{n\beta})^2 \, \mathrm{d}\Omega + \int_{\Omega} \sum_{\beta=1}^{2} K_{T\beta} (\kappa_{T\beta})^2 \, \mathrm{d}\Omega \tag{8.10}$$

in which $K_{e\beta}$, K_s, $K_{g\beta}$, $K_{n\beta}$, and $K_{T\beta}$ are material stiffnesses related to stretching, shearing, geodesic and normal bending, and twisting, respectively. They are evaluated by the expressions:

$$K_e = \eta_e \frac{YA}{p}, \quad K_s = \eta_s \frac{GJ_p}{h_p p^2}, \quad K_T = \eta_T \frac{GJ_t}{p}, \quad K_g = \eta_g \frac{YJ_{fg}}{p}, \quad K_n = \eta_n \frac{YJ_{fn}}{p} \tag{8.11}$$

borrowed by the beam theory. Since the hypotheses behind these expressions are not adequately satisfied, we introduce corrective coefficients, η_i, to overtake this issue. Moreover, A is the area of the fibers cross-section, $J_t = 0.196 h^3 b$ and $J_{fn} = h^3 b/12, J_{fg} = b^3 h/12$ are the torsional, the out-of-plane and in-plane bending second moment of area of the fiber cross-sections, respectively; $J_p = \pi r_p^4/2$ is the torsional second moment of area of the pivot. Because the two families of fibers are equal, so the related stiffnesses are. The variable pitch between two adjacent fibers is denoted by p.

To define the kinetic energy of the plate, we introduce the skew tensor

$$\boldsymbol{V}_\beta = \boldsymbol{R}_\beta^\top \dot{\boldsymbol{R}}_\beta \tag{8.12}$$

where the dot denotes differentiation with respect to the time. The significant components of the tensor (8.12) are the angular velocity fields of the cross-section of the fibers, in detail

$$\omega_{\beta 1} = (\boldsymbol{n} \times \boldsymbol{d}_\beta) \cdot \dot{\boldsymbol{n}}; \quad \omega_{\beta 2} = \boldsymbol{n} \cdot \dot{\boldsymbol{d}}_\beta; \quad \omega_{\beta 3} = -(\boldsymbol{n} \times \boldsymbol{d}_\beta) \cdot \dot{\boldsymbol{d}}_\beta \tag{8.13}$$

Therefore, the kinetic energy of the equivalent continuum annular plate can be evaluated by

$$2T[\chi(\cdot), \dot{\chi}(\cdot)] = \int_{\Omega} \varrho_s (\dot{u}_1^2 + \dot{u}_2^2 + \dot{u}_3^2) \, \mathrm{d}\Omega + \int_{\Omega} \sum_{\beta=1}^{2} \left[J_{\beta 1} (\omega_{\beta 1})^2 + J_{\beta 2} (\omega_{\beta 2})^2 + J_{\beta 3} (\omega_{\beta 3})^2 \right] \mathrm{d}\Omega \tag{8.14}$$

where ϱ_s is the apparent mass density per unit area, J_i (the same for the two families of fibers) are related to the inertia of the cross-section of fibers and can be evaluated by the expression $\varrho I_i/p$, being ϱ the apparent mass density per unit volume, I_i the second moments of area with respect to the directions $\boldsymbol{D}_1$, $\boldsymbol{D}_2$, $\boldsymbol{e}_3$.

The governing equation of the annular plate, therefore, can be deduced by the stationary-action principle, using the Lagrangian function $\mathscr{L} = U - T$.

8.3 Modal Analysis

In this section, we evaluate the vibration modes of an annular plate made of a logarithmic spiral network of fibers using the second-gradient bi-dimensional model, linearized around the undeformed configuration.

The modal shapes are evaluated using the finite element software COMSOL Multiphysics, which allows the straightforward implementation of the Lagrangian function, $\mathscr{L}$, based on Eq. (8.10) and (8.14). The finite element discretization employs the Argyris elements since the model is two-dimensional and is characterized by a second-gradient elastic theory. In this context, it is paramount to use interpolating functions in the Sobolev space H^2. The same problem might be solved more efficiently with an *ad hoc* homemade code via the isogeometric formulation [43]-[48], discrete Hencky models based on the geometry of the microstructure of the fibers, as done in [49]-[54], or alternatively, as recently proposed, based on swarm robotics [55], [56].

Specifically, a representative sample of an annular plate is investigated. The fibers are assumed to have a rectangular cross-section of size $b \times h$, being $b = 1.6$ mm the base and $h = 1$ mm the height. They are linked together by cylindrical pivots having radius $r_p = 0.45$ mm and height $h_p = 1$ mm. The radius of the inner circle is set to be $R_0 = 50$ mm and the outer radius is $R_e = 260$ mm. The pitch p is geometrically evaluated from a 3D CAD model as shown in Fig. 8.1 by considering the positions of the barycenters of the pivots. In particular, this parameter is characterized by a linear dependence on the annular plate radius according to the relationship:

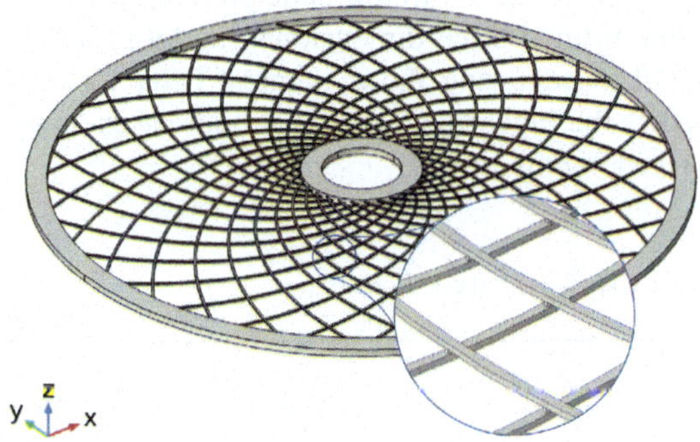

Fig. 8.1 A net prototype feasible with 3D printing.

$$p = m\sqrt{2\left({X_1}^2 + {X_2}^2\right)} + q \tag{8.15}$$

in which $q = -4\times 10^{-2}$ mm and $m = 0.1085$.

The coefficients η_i conceived to correct the material stiffnesses are estimated by a fitting procedure made with the aim of reproducing the same modal shapes and natural frequencies of a more refined three-dimensional elastic model (see, e.g., [57], [58] for a similar approach). The obtained values are listed in Table 8.1. We assumed the material of the annular plate to be polyamide, homogeneous, and isotropic. It is characterized by Young's modulus $Y = 1600$ MPa and Poisson's ratio $\nu = 0.4$.

The apparent mass density per unit area ϱ_S, clearly not uniform, is also estimated using the 3D CAD model. The idea is to divide the annular plate into a large number of parts within cylindrical annuli, evaluate the mass of each piece, and divide it by the equivalent area of the underlying surface. The result of such a procedure is reported in Fig. 8.2 as a function of the radius because of the symmetry of the geometry.

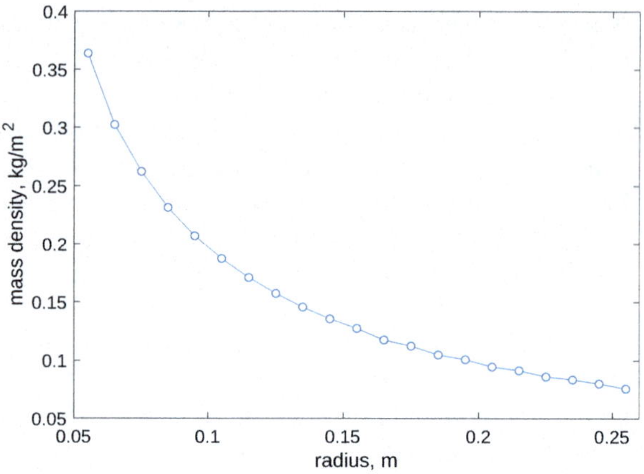

Fig. 8.2 Apparent mass density per unit area vs. radius of the annular plate.

The first twelve modal shapes associated with their natural frequencies are determined and shown in Figs. 8.3–8.5. The out-of-plane displacement is displayed by color in the figures. In the same pictures, we also exhibit the same results obtained with a refined three-dimensional model to assess the correctness of the proposed model in capturing the dynamical behavior of the considered system. Table 8.2 reports the eigenfrequencies for both cases analyzed and compares them with the corresponding errors.

Table 8.1: The fitted coefficients η.

η_e	η_s	η_g	η_n	η_T
1.0	1.0	0.8	1.9	0.8

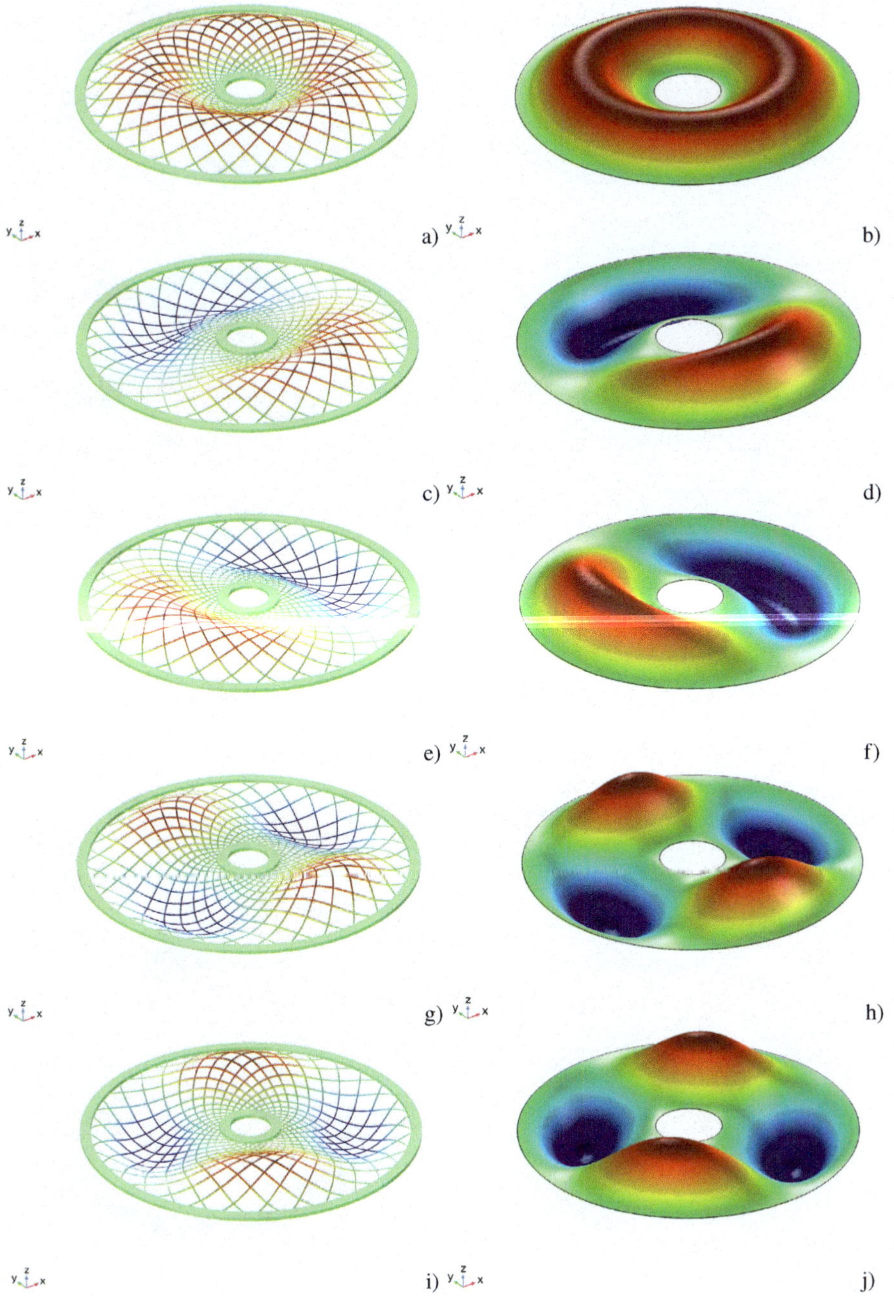

Fig. 8.3: Modal shapes from the first to the fifth, comparison between a 3D model (a, c, e, g, i) and the reduced 2D model (b, d, f, h, j)

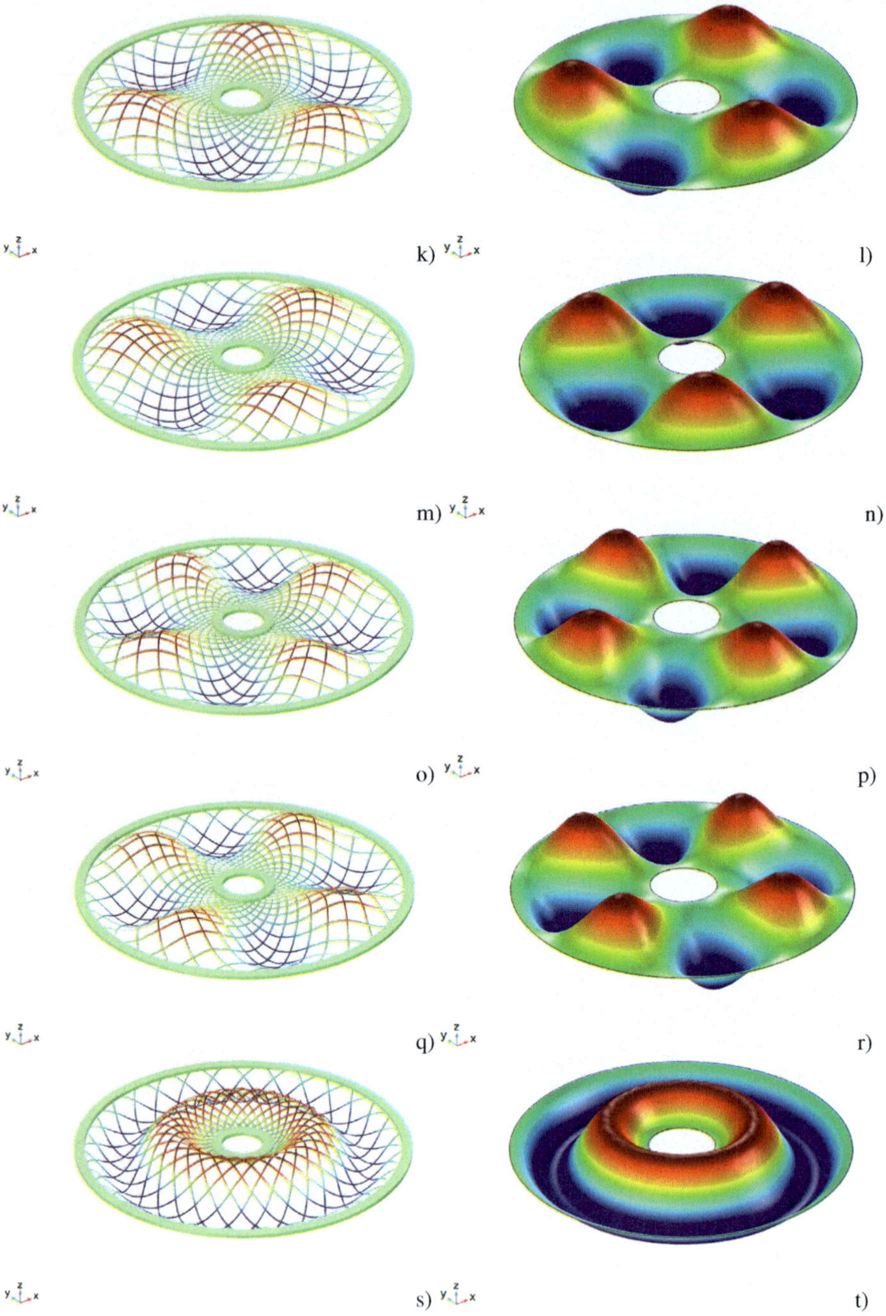

Fig. 8.4: Modal shapes from the sixth to the tenth, comparison between a 3D model (k, m, o, q, s) and the reduced 2D model (l, n, p, r, t).

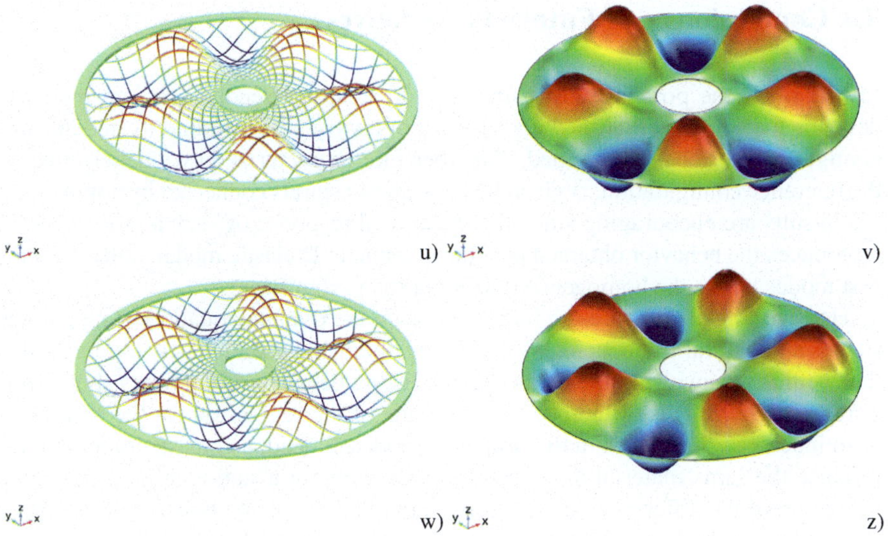

Fig. 8.5: Modal shapes from the eleventh to the twelfth, comparison between a 3D model (u, w) and the reduced 2D model (v, z).

These results show an excellent agreement for the modal shapes. The natural frequencies are rather close for the two models considered for the first seven modes; the others are still close, but a more considerable discrepancy is detected for the remaining modes. These more significant errors, however, can be easily explained by comparing the length of the unit cell of the network to be homogenized with the wavelength of the modal shape. By increasing the frequency, these two quantities become of a compatible magnitude; therefore, the continuum approach is less suitable for accurately describing the annular plate behavior. Besides, the fact that the qualitative behavior is well captured, after all confirms some results obtained for wide-knit pantographic structures [59], even though in a static regime.

Table 8.2: Natural eigenfrequencies, rad/s.

Mode #	3D model	2D reduced model	difference %
1	167.86	162.37	-3.27
2, 3	180.86	174.13	-3.72
4, 5	221.52	210.42	-5.01
6, 7	272.79	271.41	-0.51
8, 9	322.73	353.95	9.67
10	382.55	436.73	14.16
11, 12	385.38	453.67	17.70

8.4 Conclusions and Future Perspectives

In this work, the possibility of using a bi-dimensional second gradient model to describe the small oscillations of a network of curved fibers around its undeformed configuration has been examined. The fibers are characterized by a pattern that, in the reference configuration, is given by logarithmic spirals connected by a weak joint. The results are encouraging since the dynamical response of such a model closely reproduces the behavior obtained by a more refined 3D elastic model, at least for the first modes where the homogenized formulation is significant.

However, if we compare the fitting based on the dynamic quantities, i.e., modal shapes and natural frequencies, with the one obtained by static tests, we notice that the coefficients involving the flexural and twisting behavior of the fibers show some differences [39]. This difference in the results of the regression for the stiffnesses is an index of some issues in the proposed model. Indeed, a consistent model must produce the same material parameters in a dynamic or a static regime. Analyzing the nature of the difference, namely related to the flexural and twisting deformation of the fibers and the current shape of the fibers in the considered modal shapes (see Fig. 8.6), it seems reasonable that the coupling between the twisting of a fiber and the bending of the connected fiber plays a considerable role in the behavior of such a system. This observation leads us to believe that the diagonal constitutive behavior given by Eq. (8.10) is too simplistic in this case. Therefore, we plan to examine, both in static and dynamic regimes, coupling constitutive behaviors between the flexural and twisting energy contributions of fibers as future developments to see if employing more comprehensive energy could overcome the detected discrepancy.

Acknowledgements The authors express their gratitude to Victor Eremeyev for the stimulating discussions and his friendly criticism.

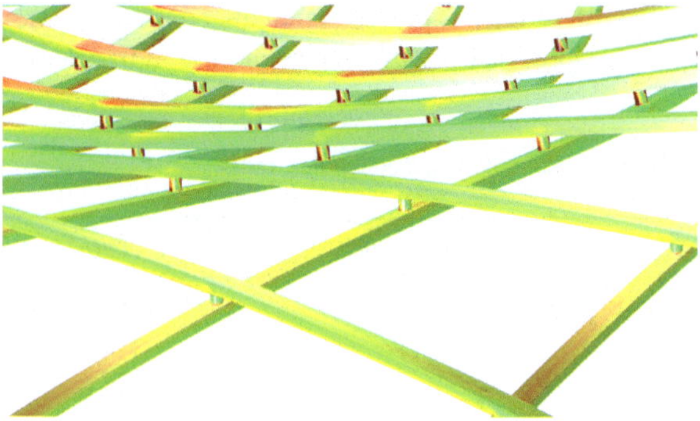

Fig. 8.6 Zoom of a modal shape where a flexion-torsion coupling appears.

References

[1] Eremeyev VA, Pietraszkiewicz W (2016) Material symmetry group and constitutive equations of micropolar anisotropic elastic solids, Mathematics and Mechanics of Solids **21**(2):210-221.

[2] Eremeyev VA (2018) On the material symmetry group for micromorphic media with applications to granular materials, Mechanics Research Communications **94**:8-12.

[3] Eremeyev VA, dell'Isola F, Boutin C, Steigmann D (2018) Linear pantographic sheets: existence and uniqueness of weak solutions, Journal of Elasticity **132**(2):175-196.

[4] Altenbach H, Eremeyev VA, Naumenko K (2015) On the use of the first order shear deformation plate theory for the analysis of three-layer plates with thin soft core layer, ZAMM-Zeitschrift für Angewandte Mathematik und Mechanik **95**(10):1004-1011.

[5] Altenbach H, Eremeyev VA (2011) On the shell theory on the nanoscale with surface stresses, International Journal of Engineering Science **49**(12):1294-1301.

[6] Altenbach J, Altenbach H, Eremeyev VA (2010) On generalized Cosserat-type theories of plates and shells: a short review and bibliography, Archive of Applied Mechanics **80**(1):73-92.

[7] Altenbach H, Eremeyev VA (2008) Analysis of the viscoelastic behavior of plates made of functionally graded materials, ZAMM-Zeitschrift für Angewandte Mathematik und Mechanik **88**(5):332-341.

[8] Altenbach H, Eremeyev VA (2008) Direct approach-based analysis of plates composed of functionally graded materials, Archive of Applied Mechanics **78**(10):775-794.

[9] Solyaev Y (2022) Effective length scale parameters of the fiber-reinforced composites, Lobachevskii Journal of Mathematics **43**(7):1993-2002.

[10] La Valle G, Massoumi S (2022) A new deformation measure for micropolar plates subjected to in-plane loads, Continuum Mechanics and Thermodynamics **34**(1):243-257.

[11] Karathanasopoulos N, Reda H, Ganghoffer JF (2017) Designing two-dimensional metamaterials of controlled static and dynamic properties, Computational Materials Science **138**:323-332.

[12] De Cicco S, De Angelis F (2020) A plane strain problem in the theory of elastic materials with voids, Mathematics and Mechanics of Solids **25**(1):46-59.

[13] Michell AGM (1904) LVIII The limits of economy of material in frame-structures, The London, Edinburgh, and Dublin Philosophical Magazine and Journal of Science **8**(47):589-597.

[14] Abali BE, Barchiesi E (2021) Additive manufacturing introduced substructure and computational determination of metamaterials parameters by means of the asymptotic homogenization, Continuum Mechanics and Thermodynamics **33**(4):993-1009.

[15] Stilz M, Plappert D, Gutmann F, Hiermaier S (2022) A 3D extension of pantographic geometries to obtain metamaterial with semi-auxetic properties, Mathematics and Mechanics of Solids **27**(4):673-686.

[16] Yang H, Abali BE, Timofeev D, Müller WH (2020) Determination of metamaterial parameters by means of a homogenization approach based on asymptotic analysis, Continuum Mechanics and Thermodynamics **32**(5):1251-1270.

[17] Barchiesi E, dell'Isola F, Seppecher P, Turco E (2022) A beam model for duoskelion structures derived by asymptotic homogenization and its application to axial loading problems, European Journal of Mechanics-A/Solids **98**:104,848.

[18] Vangelatos Z, Gu GX, Grigoropoulos CP (2019) Architected metamaterials with tailored 3D buckling mechanisms at the microscale, Extreme Mechanics Letters **33**:100,580.

[19] Barchiesi E, Spagnuolo M, Placidi L (2019) Mechanical metamaterials: a state of the art, Mathematics and Mechanics of Solids **24**(1):212-234.

[20] dell'Isola F, Seppecher P, Alibert JJ, et al (2019) Pantographic metamaterials: an example of mathematically driven design and of its technological challenges, Continuum Mechanics and Thermodynamics **31**(4):851-884.

[21] Laudato M, Manzari L (2020) Linear dynamics of 2D pantographic metamaterials: numerical and experimental study, In: Developments and novel approaches in biomechanics and metamaterials, Springer, Cham, pp 353-375.

[22] Laudato M, Barchiesi E (2019) Non-linear dynamics of pantographic fabrics: modelling and numerical study, In: Wave dynamics, mechanics and physics of microstructured metamaterials, Springer, Cham, pp 241-254.

[23] Laudato M, Manzari L, Barchiesi E, Di Cosmo F, Göransson P (2018) First experimental observation of the dynamical behavior of a pantographic metamaterial, Mechanics Research Communications **94**:125-127.

[24] Engelbrecht J, Berezovski A, Pastrone F, Braun M (2005) Waves in microstructured materials and dispersion, Philosophical Magazine **85**(33-35):4127-4141.

[25] Eugster SR (2022) Numerical analysis of nonlinear wave propagation in a pantographic sheet, Mathematics&Mechanics of Complex Systems **9**:293-310.

[26] Turco E, Barchiesi E, dell'Isola F (2022) A numerical investigation on impulse-induced nonlinear longitudinal waves in pantographic beams, Mathematics and Mechanics of Solids **27**(1):22-48.

[27] Fedele R (2022) Piola's approach to the equilibrium problem for bodies with second gradient energies. Part I: First gradient theory and differential geometry, Continuum Mechanics and Thermodynamics **34**(2):445-474.

[28] Fedele R (2022) Third-gradient continua: nonstandard equilibrium equations and selection of work conjugate variables, Mathematics and Mechanics of Solids **27**(10):2046-2072.

[29] Rezaei N, Barchiesi E, Timofeev D, Tran CA, Misra A, Placidi L (2022) Solution of a paradox related to the rigid bar pull-out problem in standard elasticity, Mechanics Research Communications **126**:104,015

[30] Eremeyev VA, Lurie SA, Solyaev YO, dell'Isola F (2020) On the well posedness of static boundary value problem within the linear dilatational strain gradient elasticity, Zeitschrift für angewandte Mathematik und Physik **71**(6):1-16.

[31] Steigmann DJ, dell'Isola F (2015) Mechanical response of fabric sheets to three-dimensional bending, twisting, and stretching, Acta Mechanica Sinica **31**(3):373-382.

[32] Giorgio I, Rizzi NL, Turco E (2017) Continuum modelling of pantographic sheets for out-of-plane bifurcation and vibrational analysis, Proceedings of the Royal Society A: Mathematical, Physical and Engineering Sciences **473**(2207):20170,636.

[33] Shirani M, Luo C, Steigmann DJ (2019) Cosserat elasticity of lattice shells with kinematically independent flexure and twist, Continuum Mechanics and Thermodynamics **31**(4):1087-1097.

[34] Giorgio I, Ciallella A, Scerrato D (2020) A study about the impact of the topological arrangement of fibers on fiber-reinforced composites: some guidelines aiming at the development of new ultra-stiff and ultra-soft metamaterials, International Journal of Solids and Structures **203**:73-83.

[35] Giorgio I (2021) Lattice shells composed of two families of curved Kirchhoff rods: an archetypal example, topology optimization of a cycloidal metamaterial, Continuum Mechanics and Thermodynamics **33**(4):1063-1082.

[36] Spagnuolo M, Cazzani AM (2021) Contact interactions in complex fibrous metamaterials, Continuum Mechanics and Thermodynamics **33**(4):1873-1889.

[37] Spagnuolo M, Peyre P, Dupuy C (2019) Phenomenological aspects of quasi-perfect pivots in metallic pantographic structures, Mechanics Research Communications **101**:103,415.

[38] Ciallella A, Pasquali D, D'Annibale F, Giorgio I (2022) Shear rupture mechanism and dissipation phenomena in bias-extension test of pantographic sheets: Numerical modeling and experiments, Mathematics and Mechanics of Solids **27**(10):2170-2188.

[39] Ciallella A, D'Annibale F, Del Vescovo D, Giorgio I (2022) Deformation patterns in a second-gradient lattice annular plate composed of "spira mirabilis" fibers, Continuum Mechanics and Thermodynamics DOI 10.1007/s00161-022-01169-6.

[40] Jamun Kumar N, Dhas B, Srinivasa AR, Reddy JN, Roy D (2022) A novel four-field mixed FE approximation for Kirchhoff rods using Cartan's moving frames, Computer Methods in Applied Mechanics and Engineering **402**:115,094.

[41] Harsch J, Capobianco G, Eugster SR (2021) Finite element formulations for constrained spatial nonlinear beam theories, Mathematics and Mechanics of Solids **26**(12):1838-1863.

[42] Greco L (2020) An iso-parametric G^1-conforming finite element for the nonlinear analysis of Kirchhoff rod. Part I: the 2D case, Continuum Mechanics and Thermodynamics **32**:1473–1496.

[43] Schulte J, Dittmann M, Eugster SR, Hesch S, Reinicke T, dell'Isola F, Hesch C (2020) Isogeometric analysis of fiber reinforced composites using Kirchhoff-Love shell elements, Computer Methods in Applied Mechanics and Engineering **362**:112,845.

[44] Maurin F, Greco F, Desmet W (2019) Isogeometric analysis for nonlinear planar pantographic lattice: discrete and continuum models, Continuum Mechanics and Thermodynamics **31**(4):1051-1064.

[45] Greco L, Cuomo M, Contrafatto L (2018) A reconstructed local B formulation for isogeometric Kirchhoff-Love shells, Computer Methods in Applied Mechanics and Engineering **332**:462-487.

[46] Greco L, Cuomo M, Castello D, Scrofani A (2022) An updated lagrangian Bézier finite element formulation for the analysis of slender beams, Mathematics and Mechanics of Solids **27**(10):2110-2138.

[47] Greco L, Scrofani A, Cuomo M (2021) A non-linear symmetric G1-conforming Bézier finite element formulation for the analysis of Kirchhoff beam assemblies. Computer Methods in Applied Mechanics and Engineering **387**:114,176.

[48] Cazzani A, Malagù M, Turco E (2016) Isogeometric analysis of plane-curved beams, Mathematics and Mechanics of Solids **21**(5):562-577.

[49] Turco E, Barchiesi E, Ciallella A, dell'Isola F (2022) Nonlinear waves in pantographic beams induced by transverse impulses, Wave Motion **115**:103064.

[50] Turco E (2022) Modeling of three-dimensional beam nonlinear vibrations generalizing Hencky's ideas, Mathematics and Mechanics of Solids **27**(10):1950-1973.

[51] Turco E (2019) Numerically driven tuning of equilibrium paths for pantographic beams, Continuum Mechanics and Thermodynamics **31**(6):1941-1960.

[52] Barchiesi E, dell'Isola F, Bersani AM, Turco E (2021) Equilibria determination of elastic articulated duoskelion beams in 2D via a Riks-type algorithm., International Journal of Non-Linear Mechanics **128**:1-24.

[53] Eremeyev VA, Turco E (2020) Enriched buckling for beam-lattice metamaterials, Mechanics Research Communications **103**:103,458.

[54] Yildizdag ME, Placidi L, Turco E (2022) Modeling and numerical investigation of damage behavior in pantographic layers using a hemivariational formulation adapted for a Hencky-type discrete model, Continuum Mechanics and Thermodynamics DOI 10.1007/s00161-022-01154-z

[55] dell'Erba R (2019) Swarm robotics and complex behaviour of continuum material, Continuum Mechanics and Thermodynamics **31**(4):989-1014.

[56] dell'Erba R (2018) Position-based dynamic of a particle system: a configurable algorithm to describe complex behaviour of continuum material starting from swarm robotics, Continuum Mechanics and Thermodynamics **30**(5):1069-1090.

[57] Rosi G, Placidi L, Auffray N (2018) On the validity range of strain-gradient elasticity: a mixed static-dynamic identification procedure, European Journal of Mechanics-A/Solids **69**:179-191.

[58] Fantuzzi N, Tornabene F, Bacciocchi M, Dimitri R (2017) Free vibration analysis of arbitrarily shaped functionally graded carbon nanotube-reinforced plates, Composites Part B: Engineering **115**:384-408.

[59] Spagnuolo M, Yildizdag ME, Andreaus U, Cazzani AM (2020) Are higher-gradient models also capable of predicting mechanical behavior in the case of wide-knit pantographic structures? Mathematics and Mechanics of Solids DOI 10.1177/1081286520937339

Chapter 9
Non-Linear Simplest Reduced Kelvin's Medium in the Vicinity of the Spherical Stress State: Waves and Instabilities

Mikhail A. Drepin and Elena F. Grekova

Abstract We obtain constitutive equations of a simplest elastic nonlinear reduced Kelvin's medium, a medium consisting of axisymmetric particles performing free fast rotation about their axes, as well as finite translations and rotations. Constitutive relations are chosen in such a way that they give linear theory for infinitesimal displacements. We consider small perturbations near nonlinear equilibrium for the spherical stress state and investigate wave propagation along the axis of a body point and orthogonal to it. The medium appears to be a smart acoustic metamaterial in the domain of stability. Its properties can be tuned by means of the initial stress. The change of the metamaterial type is possible under sufficiently large extension. There are various mechanisms of the material stability loss, under shear perturbations as in the classical medium, under rotational perturbations as in the Cosserat medium, and also under hydrostatic perturbations.

9.1 Introduction and Notation

The authors are happy to dedicate this work to Prof. Victor Eremeyev on the occasion of his 60th anniversary, with their sincere best wishes. One of the authors, Elena Grekova, was fortunate to meet Victor in 2000. Ever since she has been appreciating the rigour and audacity of his scientific works. She is grateful to Victor for fruitful, kind and humorous discussions, and friendship.

Many works in last decades are dedicated to various Cosserat-type media. Among seminal contributions one cannot avoid to mention the book by Cosserat brothers [3] and the article on the full 3D Cosserat theory developed by A. Kafadar and C.A. Eringen [18]. The reader will enjoy also recent books on this subject, for instance, [1, 6].

Mikhail A. Drepin · Elena F. Grekova
Institute for Problems in Mechanical Engineering of the Russian Academy of Sciences, 199178, Bolshoy pr. V.O., 61, St. Petersburg, Russia, e-mail: mihail.drepin@gmail.com, elgreco@pdmi.ras.ru

H. Altenbach et al. (eds.), *Sixty Shades of Generalized Continua*,
Advanced Structured Materials 170, https://doi.org/10.1007/978-3-031-26186-2_9

Micropolar science started dealing with Lagrange formulation and infinitesimal rigid body as a body point. Since that time, many generalisations have been made, not only micromorphic continua [7], but also spatial approach has been developed and destruction of particle is introduced into the theory [24]. Complex particles possessing rotational degrees of freedom, sometimes multicomponent, are used to model thermal and electromagnetic phenomena [16, 17] and granular media [8].

A special kind of Cosserat media are gyrocontinua, where particles possess finite intrinsic dynamic spin.

The topic of this work was inspired in 1990s by P.A. Zhilin who suggested to consider Kelvin's medium, a medium which consists of axisymmetric body points, freely and fastly rotating about their axes, and performing nonlinear translational and rotational motion. Pavel Andreyevich Zhilin gave to this medium the name of Kelvin since the great scientist desired to introduce it in science but did not express his ideas in equations. Zhilin's idea was that the medium of this kind could describe some elastomagnetic phenomena. Later on, the term "Kelvin's medium" has been included into the dictionary of modern non-classical continuum mechanics [20]. The reason for this is that indeed, there was established an exact analogy between equations of Kelvin's medium and elastic ferromagnetic continua under some restrictions [9]. The method to obtain constitutive equations for Kelvin's medium firstly was suggested for 2D Cosserat media (shells) in [2]. It is a medium with specific symmetry group which influences the choice of possible strain measures and the constitutive equations (for various types of Cosserat and micromorphic continua see discussion at [4, 5, 21]).

Apart from specific for Kelvin's medium restriction, given by axial symmetry of its particles, we want to impose another requirement on the strain energy of the medium, namely, to consider reduced Kelvin's medium, which does not react to the gradient of microrotation. If the elastic energy of a continuum with an ordinary inertial term depends on special generalised vector co-ordinates, but does not depend on their gradient, we will call such a continuum reduced. The special co-ordinate may have rotational or translational nature, for instance, mass-in-mass model is also a metamaterial [22]. Under certain symmetry conditions, such a medium will be an acoustic metamaterial [10]. However, in the latter work the inertia term differs from the case of gyro-media, in particular, from the case of Kelvin's medium, due to the absence of finite kinetic moment of gyroscopic origin.

Acoustic metamaterials are media with anomalous wave properties. Single negative acoustic metamaterials have band gaps, and double negative acoustic metamaterials have a decreasing part of the dispersion curve. They are used to control wave propagation [14, 19].

The simplest linear reduced Kelvin's medium is a single negative acoustic metamaterial with respect to the shear-rotation waves, which are polarised in direction of proper rotation (it has a band gap) [11, 12]. The nonlinear stress state in the reduced and complete Cosserat medium without gyroscopic inertial term changes the effective elastic constants [15] and affects the wave propagation and stability [13], creating new mechanisms of the stability loss. We will consider the effects of the preliminary stress on the dynamic properties of Kelvin's medium.

The constitutive relations of the Kelvin medium, which is a special case of the Cosserat medium, will have a special form and the dynamic term in them will differ due to the symmetry of the body points.

We denote the position vector of the center of mass of a body-point by $\boldsymbol{R}$ in the actual configuration and by $\boldsymbol{r}$ in the reference configuration. Let us introduce partial derivatives with respect to Lagrangian co-ordinates q^i:

$$\boldsymbol{r}_i = \frac{\partial \boldsymbol{r}}{\partial q^i} \equiv \partial_i \boldsymbol{r}, \quad \boldsymbol{R}_i = \frac{\partial \boldsymbol{R}}{\partial q^i} \equiv \partial_i \boldsymbol{R}, \tag{9.1}$$

We choose q^i in such a way that $\boldsymbol{r}_i$ form an orthonormal base. We consider two gradient operators in the reference and actual configurations:

$$\overset{\circ}{\nabla} = \frac{\partial}{\partial \boldsymbol{r}} = \boldsymbol{r}^i \partial_i, \quad \nabla = \frac{\partial}{\partial \boldsymbol{R}} = \boldsymbol{R}^i \partial_i, \tag{9.2}$$

where $\boldsymbol{r}^i, \boldsymbol{R}^i$ define mutual bases for $\boldsymbol{r}_i, \boldsymbol{R}_i$. Here and below the Einstein summation rule is used for repeated indices.

Let $\boldsymbol{m}_0$ be a unit vector of the axis of a body-point in the reference configuration. We can represent the rotation tensor of a body-point (so called micro-rotation) as

$$\boldsymbol{P} = \boldsymbol{P}_\perp \cdot \boldsymbol{P}_\varphi(\varphi \boldsymbol{m}_0), \tag{9.3}$$

where

$$\boldsymbol{P}_\varphi = (1 - \cos\varphi)\boldsymbol{m}_0\boldsymbol{m}_0 + \cos\varphi \boldsymbol{E} + \sin\varphi \boldsymbol{m}_0 \times \boldsymbol{E} \tag{9.4}$$

describes the proper rotation of a body-point about its initial axis $\boldsymbol{m}_0$, and $\boldsymbol{P}_\perp$ the rotation of this axis.

The angular velocity of a body-point equals

$$\omega = \omega_\perp + \dot{\varphi}\boldsymbol{P}_\perp \cdot \boldsymbol{m}_0 = \omega_\perp + \dot{\varphi}\boldsymbol{m}_0, \tag{9.5}$$

where $\dot{\boldsymbol{P}}_\perp = \omega_\perp \times \boldsymbol{P}_\perp$, and $\boldsymbol{m} = \boldsymbol{P} \cdot \boldsymbol{m}_0 = \boldsymbol{P}_\perp \cdot \boldsymbol{m}_0$ is an axis of a body point in the actual configuration.

Strains in the Cosserat continuum can be described by the Cosserat deformation tensor $\boldsymbol{A}$ and transposed wryness tensor $\boldsymbol{K}$:

$$\boldsymbol{A} = \overset{\circ}{\nabla}\boldsymbol{R} \cdot \boldsymbol{P}, \quad \boldsymbol{K} = \boldsymbol{r}^i \otimes \boldsymbol{\Phi}_i \cdot \boldsymbol{P}, \tag{9.6}$$

where $\boldsymbol{\Phi}_i$ is the solution of the Poisson equation with respect to spatial co-ordinates:

$$\partial_i \boldsymbol{P} = \boldsymbol{\Phi}_i \times \boldsymbol{P}, \tag{9.7}$$

Denote by ρ the density of the medium. The mass density of the tensor of inertia of a point-body is equal to

$$\boldsymbol{I} = \boldsymbol{P} \cdot \boldsymbol{I}_0 \cdot \boldsymbol{P}^\top = \boldsymbol{P}_\perp \cdot \boldsymbol{I}_0 \cdot \boldsymbol{P}_\perp, \tag{9.8}$$

where $\boldsymbol{I}_0 = I_\perp(\boldsymbol{E} - \boldsymbol{m}_0\boldsymbol{m}_0) + I\boldsymbol{m}_0\boldsymbol{m}_0$ is the mass density of the tensor of inertia of an axisymmetric body point in the reference configuration.

We consider rapidly rotating particles with small moments of inertia $I_\perp, I$, so that the projection of the angular momentum on the axis of symmetry $\rho I\dot{\varphi}$ (corresponding to the proper rotation component of the intrinsic spin) is finite. This results in the analogy between equations of Kelvin's medium and equations of elastic ferromagnetic insulators in the state of magnetic saturation, approximation of quasimagnetostatics [9].

Denote by $\boldsymbol{\tau}$ the Cauchy force stress tensor, and $\boldsymbol{T} = \overset{\circ}{\nabla}\boldsymbol{R}^{-\top} \cdot \boldsymbol{\tau} \det \overset{\circ}{\nabla}\boldsymbol{R}$ – the first Piola-Kirchhoff stress tensor. Note that the stress tensor $\boldsymbol{\tau}$, generally speaking, is not symmetric. Apart from the force stress tensor, in Cosserat media there exist a couple stress tensor $\boldsymbol{\mu}$. In this work it will be equal to zero as in other reduced Cosserat-type media.

9.2 Basic Equations for the Isotropic Elastic Reduced Kelvin's Medium in the Vicinity of a Preliminary Stress State

In this section we will write down constitutive equations of the simplest reduced non-linear Kelvin's medium that give a linear theory in the vicinity of the natural state. We will consider zero external loads. The strain energy of the simplest reduced Kelvin's medium does not depend on the gradient of the micro-rotation, therefore it is a function only of the Cosserat deformation tensor $\boldsymbol{A}$ and does not depend on $\boldsymbol{K}$. However, it must have a specific type to provide the condition that no stresses are induced in the medium by means of the proper rotation, i.e. the strain energy U does not depend on $\varphi, \nabla\varphi$. We use Piola–Kirchoff stress tensor for the sake of convenience. Afterwards we linearize these equations in the vicinity of a spherical stress state.

9.2.1 Dynamic Laws of a Nonlinear Reduced Kelvin Medium

The balance of energy in Zhilin's form for the Cosserat medium [9] can be written as

$$\rho\dot{U} = \boldsymbol{\tau}^\top \cdot\cdot \dot{\boldsymbol{A}} + \boldsymbol{\mu}^\top \cdot\cdot \dot{\boldsymbol{K}}. \tag{9.9}$$

Since we consider a **reduced** medium, the deformation energy U does not depend on $\boldsymbol{K}$ (no power is produced on $\nabla\boldsymbol{\omega}$). This yields in the absence of the couple stress tensor: $\boldsymbol{\mu} = \boldsymbol{0}$.

Therefore the dynamic relations for the Kelvin's medium of general type [9] are simplified. Taking into account the absence of mass density of external forces, we write the balance of forces in local form:

$$\nabla \cdot \boldsymbol{\tau} = \rho\ddot{\boldsymbol{u}}, \tag{9.10}$$

and, in terms of the first Piola–Kirchoff stress tensor,

$$\overset{\circ}{\nabla} \cdot \boldsymbol{T} = \rho_0 \ddot{\boldsymbol{u}}. \tag{9.11}$$

In the absence of external moments, since the couple stress tensor equals zero, the balance of moments in the local form looks as

$$\boldsymbol{\tau}_\times = (\boldsymbol{I} \cdot \boldsymbol{\omega})^{\cdot} \approx \boldsymbol{\omega} \times \rho I \dot{\varphi} \boldsymbol{m}. \tag{9.12}$$

and, in terms of the first Piola–Kirchoff stress tensor,

$$[\overset{\circ}{\nabla} \boldsymbol{R}^\top \cdot \boldsymbol{T}]_\times = \boldsymbol{\omega} \times \rho_0 I \dot{\varphi} \boldsymbol{m}. \tag{9.13}$$

The inertial term $(\boldsymbol{I} \cdot \boldsymbol{\omega})^{\cdot}$ equals the gyroscopic term $\boldsymbol{\omega} \times \rho I \dot{\varphi} \boldsymbol{m}$ up to the second order terms, since we consider the case when $I_\perp, I$ are infinitesimal, $\dot{\varphi}$ is large, and $I\dot{\varphi}$ is finite [9].

9.2.2 Constitutive Relations for a Nonlinear Reduced Kelvin Medium

We will derive the constitutive relations in the same way as [9], as well as [18], [13], from the law of balance of energy. As we have stated in the previous section, for an elastic reduced nonlinear Cosserat medium in case of adiabatic or isothermic processes it takes form

$$\rho \dot{U} = \boldsymbol{\tau}_*^\top \cdot\cdot \dot{\boldsymbol{A}}, \tag{9.14}$$

where $\boldsymbol{\tau}_*$ is an energetic stress tensor,

$$\boldsymbol{\tau}_* = \overset{\circ}{\nabla} \boldsymbol{R}^{-\top} \cdot \boldsymbol{\tau} \cdot \boldsymbol{P}, \tag{9.15}$$

which, as well as Cosserat deformation tensor $\boldsymbol{A}$ does not change under rigid rotation. This allows us to consider the strain energy U as a function of $\boldsymbol{A}$ and obtain the constitutive equation from (9.14):

$$\boldsymbol{\tau}_* = \rho \frac{\partial U}{\partial \boldsymbol{A}}, \tag{9.16}$$

which yields in the following expressions for $\boldsymbol{\tau}$ and $\boldsymbol{T}$:

$$\boldsymbol{\tau} = \rho \overset{\circ}{\nabla} \boldsymbol{R}^\top \cdot \frac{\partial U}{\partial \boldsymbol{A}} \cdot \boldsymbol{P}^\top, \tag{9.17}$$

$$\boldsymbol{T} = \rho_0 \frac{\partial U}{\partial \boldsymbol{A}} \cdot \boldsymbol{P}^\top. \tag{9.18}$$

Couple stress tensor $\boldsymbol{\mu} = \mathbf{0}$. Note that the strain energy does not depend on $\nabla\varphi$ since U does not depend on $\boldsymbol{K}$.

The elastic energy of Kelvin's medium must not depend on φ. We have to use as arguments of U only those functions of $\boldsymbol{A}$ that do not depend on φ, altogether containing 6 independent scalar measures of deformations.

We can also consider any complete set of strain measures for full Kelvin's medium (see [9]) and exclude from it those depending on $\boldsymbol{K}$. For instance, we can consider the strain energy for the full Kelvin's medium as a function of strain measures

$$\boldsymbol{G} = \boldsymbol{G}^\top = (\boldsymbol{A}\cdot\boldsymbol{A}^\top - \boldsymbol{E})/2, \tag{9.19}$$

$$\boldsymbol{F} = \boldsymbol{K}\cdot\tilde{\boldsymbol{a}}\cdot\boldsymbol{A}^\top, \tag{9.20}$$

where $\tilde{\boldsymbol{a}} = \boldsymbol{E} - \boldsymbol{m}_0\boldsymbol{m}_0$, and

$$\boldsymbol{\gamma} = \boldsymbol{A}\cdot\boldsymbol{m}_0. \tag{9.21}$$

To consider a reduced Kelvin's medium, we have to exclude $\boldsymbol{F}$ from this set since it depends on $\boldsymbol{K}$. The strain energy U will be a function of $\boldsymbol{G},\boldsymbol{\gamma}$.

Instead of $\boldsymbol{G},\boldsymbol{\gamma}$ we may use strain measures $\boldsymbol{G},\tilde{\boldsymbol{\gamma}}$, where $\tilde{\boldsymbol{\gamma}}$ is a linear combination of $\boldsymbol{\gamma}$ and $\boldsymbol{G}$:

$$\tilde{\boldsymbol{\gamma}} = \boldsymbol{\gamma} - \boldsymbol{m}_0 - \boldsymbol{G}\cdot\boldsymbol{m}_0 = \boldsymbol{A}\cdot\boldsymbol{m}_0 - \boldsymbol{m}_0 - \boldsymbol{G}\cdot\boldsymbol{m}_0. \tag{9.22}$$

We do this to deal with infinitesimal (linear) strain tensors in the vicinity of the natural state. Indeed, linearization near the natural state gives:

$$\boldsymbol{G} \approx (\nabla\boldsymbol{u})^S, \quad \tilde{\boldsymbol{\gamma}} \approx (\boldsymbol{\theta} - \nabla\times\boldsymbol{u}/2)\times\boldsymbol{m}_0, \tag{9.23}$$

which corresponds to infinitesimal deformations in the linear theory.

This is not the only way to specify a system of deformation measures, but we will focus on this option.

To obtain constitutive relations for stress tensors in terms of the chosen strain measures, we write the increment of U in two ways:

$$\begin{cases} dU(\boldsymbol{A}) = \dfrac{\partial U}{\partial \boldsymbol{A}}\cdot\cdot\, d\boldsymbol{A}^\top \\ dU(\boldsymbol{G},\tilde{\boldsymbol{\gamma}}) = \dfrac{\partial U}{\partial \boldsymbol{G}}\cdot\cdot\, d\boldsymbol{G}^\top + \dfrac{\partial U}{\partial \tilde{\boldsymbol{\gamma}}}\cdot d\tilde{\boldsymbol{\gamma}} \end{cases} \tag{9.24}$$

We express $d\boldsymbol{G}$ and $d\tilde{\boldsymbol{\gamma}}$ via $d\boldsymbol{A}$:

$$d\boldsymbol{G} = d\boldsymbol{G}^\top = \frac{1}{2}(d\boldsymbol{A}\cdot\boldsymbol{A}^\top + \boldsymbol{A}\cdot d\boldsymbol{A}^\top), \tag{9.25}$$

$$d\tilde{\boldsymbol{\gamma}} = d\boldsymbol{A}\cdot\boldsymbol{m}_0 - d\boldsymbol{G}\cdot\boldsymbol{m}_0 = d\boldsymbol{A}\cdot\boldsymbol{m}_0 - d\boldsymbol{A}\cdot(\boldsymbol{m}_0\cdot\boldsymbol{A})/2 - \boldsymbol{A}\cdot(\boldsymbol{m}_0\cdot d\boldsymbol{A})/2, \tag{9.26}$$

and, substituting these expressions to (9.24), due to the symmetry of $\boldsymbol{G}$ we obtain

$$\frac{\partial U}{\partial \boldsymbol{A}} = \frac{\partial U}{\partial \boldsymbol{G}}\cdot\boldsymbol{A} - \left(\frac{\partial U}{\partial \tilde{\boldsymbol{\gamma}}}\boldsymbol{m}_0\right)^S\cdot\boldsymbol{A} + \frac{\partial U}{\partial \tilde{\boldsymbol{\gamma}}}\boldsymbol{m}_0. \tag{9.27}$$

This expression, the constitutive equation (9.15) and definition of $\boldsymbol{A} = \overset{\circ}{\nabla}\boldsymbol{R} \cdot \boldsymbol{P}$ give us the constitutive relations for a nonlinear reduced Kelvin medium in the following form:

$$\boldsymbol{\tau} = \rho \overset{\circ}{\nabla}\boldsymbol{R}^{\top} \cdot \left(\left(\frac{\partial U}{\partial \boldsymbol{G}} - \left(\frac{\partial U}{\partial \tilde{\boldsymbol{\gamma}}} \boldsymbol{m}_0 \right)^S \right) \cdot \overset{\circ}{\nabla}\boldsymbol{R} + \frac{\partial U}{\partial \tilde{\boldsymbol{\gamma}}} \boldsymbol{m} \right), \tag{9.28}$$

$$\boldsymbol{T} = \rho_0 \left(\left(\frac{\partial U}{\partial \boldsymbol{G}} - \left(\frac{\partial U}{\partial \tilde{\boldsymbol{\gamma}}} \boldsymbol{m}_0 \right)^S \right) \cdot \overset{\circ}{\nabla}\boldsymbol{R} + \frac{\partial U}{\partial \tilde{\boldsymbol{\gamma}}} \boldsymbol{m} \right). \tag{9.29}$$

9.2.3 *The Simplest Nonlinear Reduced Kelvin Medium. Strain Energy.*

We will consider a physically linear but geometrically nonlinear medium. We define the strain energy U as the quadratic form of the nonlinear strain tensors $\boldsymbol{G} = \boldsymbol{G}^{\top}, \tilde{\boldsymbol{\gamma}}$:

$$\rho_0 U = \frac{1}{2} \boldsymbol{G} \cdot\cdot \boldsymbol{C} \cdot\cdot \boldsymbol{G} + 2\alpha \tilde{\boldsymbol{\gamma}}^2, \tag{9.30}$$

where $\boldsymbol{C}$ is a fourth rank constant tensor, and α is a scalar.

We consider the **simplest case** of a Kelvin medium, i.e. we suppose that the anisotropy of the medium exists only due to the fact that rotations about $\boldsymbol{m}$ do not create stresses. In this case, the coupling in the energy between the tensor $\boldsymbol{G}$ and $\tilde{\boldsymbol{\gamma}}$ is zero, which is already reflected in the formula above, and the elasticity tensor ${}^4\boldsymbol{C}$ is isotropic and has the form:

$${}^4\boldsymbol{C} - \lambda \boldsymbol{E}\boldsymbol{E} + 2\mu (\boldsymbol{e}_m \boldsymbol{e}_n)^S (\boldsymbol{e}^m \boldsymbol{e}^n)^S, \tag{9.31}$$

where $\boldsymbol{e}_n$ is an orthonormal basis.

Substituting the energy, expressed in terms of the constants λ, μ, α, into constitutive relations (9.28),(9.29) allows us to relate stresses and strains in the material:

$$\tau = \frac{\rho}{\rho_0} \overset{\circ}{\nabla}\mathbb{R}^{\top} \cdot ((\lambda \mathrm{tr} \boldsymbol{G} \boldsymbol{E} + 2\mu \boldsymbol{G} - 4\alpha (\tilde{\boldsymbol{\gamma}} \boldsymbol{m}_0)^S) \cdot \overset{\circ}{\nabla}\boldsymbol{R} + 4\alpha \tilde{\boldsymbol{\gamma}} \boldsymbol{m}), \tag{9.32}$$

$$\boldsymbol{T} = ((\lambda \mathrm{tr} \boldsymbol{G} \boldsymbol{E} + 2\mu \boldsymbol{G} - 4\alpha (\tilde{\boldsymbol{\gamma}} \boldsymbol{m}_0)^S) \cdot \overset{\circ}{\nabla}\boldsymbol{R} + 4\alpha \tilde{\boldsymbol{\gamma}} \boldsymbol{m}). \tag{9.33}$$

9.2.4 Medium in the Vicinity of a Homogeneous Nonlinear Spherical Deformation State

Let us consider a homogeneous spherical strain state $\boldsymbol{R}_s = A\boldsymbol{r}$. The medium expands if $A > 1$ and contracts if $A < 1$. We denote by subindex "s" the magnitudes related to this non-linear spherically deformed state.

The position vector $\boldsymbol{R}_s$, the proper rotation-independent part of the rotation tensor $\boldsymbol{P}_{\perp s}$, Cosserat deformation tensor $\boldsymbol{A}_s$, and the vector of the axis of the body point $\boldsymbol{m}_s$, with a uniform spherical deformation, take the form:

$$\boldsymbol{R}_s = A\boldsymbol{r}, \quad \boldsymbol{P}_{\perp s} = \boldsymbol{E}, \quad \boldsymbol{A}_s = A\boldsymbol{E}, \quad \boldsymbol{m} = \boldsymbol{m}_0, \tag{9.34}$$

and therefore strain measures $\boldsymbol{G}, \tilde{\boldsymbol{\gamma}}$ in this state equal

$$\boldsymbol{G}_s = (A^2 - 1)\boldsymbol{E}/2, \qquad \tilde{\boldsymbol{\gamma}}_s = -(A-1)^2\boldsymbol{m}_0/2. \tag{9.35}$$

We find stress tensors in the state of spherical deformation via constitutive equations (9.32),(9.33):

$$\boldsymbol{T}_s = T\boldsymbol{E} + T_1\boldsymbol{m}_0\boldsymbol{m}_0, \tag{9.36}$$

where

$$T = A(A^2 - 1)(\frac{3}{2}\lambda + \mu), \quad T_1 = 2\alpha(A-1)^3 \tag{9.37}$$

$$\boldsymbol{\tau}_s = \boldsymbol{T}_s/A^2. \tag{9.38}$$

The gradient operator with respect to the co-ordinates in the preliminary strain state is:

$$\overset{s}{\nabla} = \frac{\partial}{\partial \boldsymbol{R}_s} = A^{-1}\overset{\circ}{\nabla}. \tag{9.39}$$

We consider small perturbations of this preliminary strain state:

$$\boldsymbol{R} = A\boldsymbol{r} + \boldsymbol{u}, \qquad \boldsymbol{P}_\perp = \boldsymbol{E} + \boldsymbol{\theta} \times \boldsymbol{E}, \tag{9.40}$$

where $\boldsymbol{\theta}$ is an infinitesimal rotation vector, $\boldsymbol{\theta} \cdot \boldsymbol{m}_0 = 0$.

Figure 9.1 illustrates an idea of the model of the linear reduced Kelvin's medium in the vicinity of the preliminary spherical strain state:

Let us linearize the strain measures:

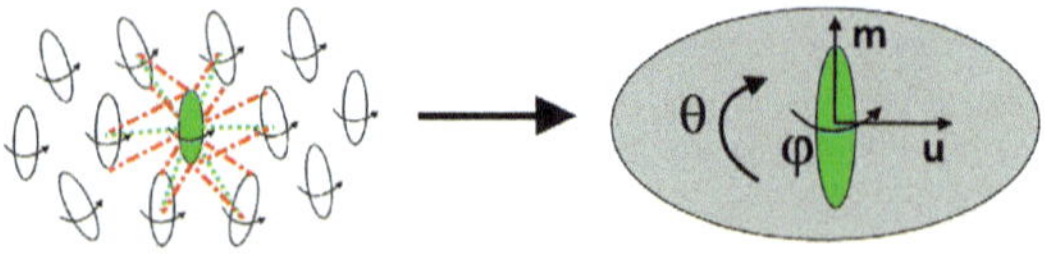

Fig. 9.1 Reduced Kelvin medium.

$$\boldsymbol{A} = \overset{\circ}{\nabla}\mathbb{R} \cdot \boldsymbol{P} \approx A(\boldsymbol{E} + \overset{s}{\nabla}\boldsymbol{u} + \boldsymbol{\theta} \times \boldsymbol{E}) \cdot \boldsymbol{P}_\varphi(\varphi \boldsymbol{m}_0), \tag{9.41}$$

$$\boldsymbol{G} \approx \frac{1}{2}(A^2 - 1)\boldsymbol{E} + A^2(\overset{s}{\nabla}\boldsymbol{u})^S, \tag{9.42}$$

$$\tilde{\boldsymbol{\gamma}} \approx -\frac{1}{2}(A-1)^2\boldsymbol{m}_0 + A(\overset{s}{\nabla}\boldsymbol{u} + \boldsymbol{\theta} \times \boldsymbol{E}) \cdot \boldsymbol{m}_0 - A^2(\overset{s}{\nabla}\boldsymbol{u})^S \cdot \boldsymbol{m}_0. \tag{9.43}$$

We denote by $[\,]_n, n \in \mathbb{N}$ the term of n-th order in the expansion of a certain magnitude in $\boldsymbol{u}, \boldsymbol{\theta}$. Expanding (9.33), we obtain the linear increment of $\boldsymbol{T}$:

$$\begin{aligned}
[\boldsymbol{T}]_1 = \boldsymbol{C}_1 \cdot \overset{s}{\nabla}\boldsymbol{u} + A^3(\lambda \overset{s}{\nabla} \cdot \boldsymbol{u}\boldsymbol{E} + 2\mu \overset{s}{\nabla}\boldsymbol{u}^S)& \\
+ 2\alpha A^2 \boldsymbol{E} \times (\boldsymbol{m}_0 \times (\overset{s}{\nabla}\boldsymbol{u} - A\overset{s}{\nabla}\boldsymbol{u}^S) \cdot \boldsymbol{m}_0) + 4\alpha(1-A)A(\overset{s}{\nabla}\boldsymbol{u} - A\overset{s}{\nabla}\boldsymbol{u}^S) \cdot \boldsymbol{m}_0\boldsymbol{m}_0& \\
+ \boldsymbol{C}_2 \times \boldsymbol{\theta} + 4\alpha(1-A)A\boldsymbol{\theta} \times \boldsymbol{m}_0\boldsymbol{m}_0,&
\end{aligned} \tag{9.44}$$

where

$$\boldsymbol{C}_1 = \boldsymbol{T}_s - 4\alpha A \tilde{\boldsymbol{\gamma}}_s \boldsymbol{m}_0 = T\boldsymbol{E} + 4\alpha(A-1)^2(A - 1/2)\boldsymbol{m}_0\boldsymbol{m}_0, \tag{9.45}$$

$$\boldsymbol{C}_2 = 2\alpha(A^2\boldsymbol{E} - 2\tilde{\boldsymbol{\gamma}}_s\boldsymbol{m}_0) = 2\alpha(A^2\boldsymbol{E} + (A-1)^2\boldsymbol{m}_0\boldsymbol{m}_0). \tag{9.46}$$

Then balance of forces (9.11) in the linear approximation gives:

$$\begin{aligned}
A\overset{s}{\nabla} \cdot (\boldsymbol{C}_1 \cdot \overset{s}{\nabla}\boldsymbol{u}) + A^4((\lambda + 2\mu)\overset{s}{\nabla}\overset{s}{\nabla} \cdot \boldsymbol{u} - \mu \overset{s}{\nabla} \times (\overset{s}{\nabla} \times \boldsymbol{u}))& \\
+ 2\alpha A^3 \overset{s}{\nabla} \times (\boldsymbol{m}_0 \times (\overset{s}{\nabla}\boldsymbol{u} - A\overset{s}{\nabla}\boldsymbol{u}^S) \cdot \boldsymbol{m}_0) + 4\alpha(1-A)A(\overset{s}{\Delta}\boldsymbol{u} - A\overset{s}{\nabla} \cdot (\overset{s}{\nabla}\boldsymbol{u}^S)) \cdot \boldsymbol{m}_0\boldsymbol{m}_0& \\
+ A\overset{s}{\nabla} \cdot (\boldsymbol{C}_2 \times \boldsymbol{\theta}) + 4\alpha(A-1)A\overset{s}{\nabla} \cdot (\boldsymbol{m}_0 \times \boldsymbol{\theta})\boldsymbol{m}_0 = \rho_0 \ddot{\boldsymbol{u}}.&
\end{aligned} \tag{9.47}$$

Let us also find the linear increment of the vectorial invariant of the stress tensor $\boldsymbol{\tau}$:

$$[\boldsymbol{\tau}_\times]_1 = -2\frac{\rho}{\rho_0}\alpha A(\boldsymbol{m}_0 \cdot ((3A-1)(A-1)\overset{s}{\nabla}\boldsymbol{u}^S + (A^2+1)\overset{s}{\nabla}\boldsymbol{u}^A) \times \boldsymbol{m}_0 + (A^2+1)\boldsymbol{\theta})$$

Then the balance of moments in the linear approximation is as follows:

$$-\boldsymbol{m}_0 \cdot (\omega_s \overset{s}{\nabla}\boldsymbol{u}^S + \omega_0 \overset{s}{\nabla}\boldsymbol{u}^A) \times \boldsymbol{m}_0 = \omega_0 \boldsymbol{\theta} + \dot{\boldsymbol{\theta}} \times \boldsymbol{m}_0, \tag{9.48}$$

where

$$\omega_0 = \frac{2\alpha A(A^2+1)}{\rho_0 I \dot{\varphi}} > 0, \qquad \omega_s = \frac{2\alpha A(3A-1)(A-1)}{\rho_0 I \dot{\varphi}}. \tag{9.49}$$

Equations (9.47),(9.48) are dynamic equations in displacements for the simplest reduced Kelvin's medium in the vicinity of the preliminary spherical strain state. We will use them to obtain the dispersion relations of the medium and investigate its stability.

9.3 Dispersion Relations of the Simplest Elastic Reduced Kelvin Medium in the Vicinity of Preliminary Spherical Strain State for Special Directions of Perturbation Propagation. Waves and Stability

Dispersion relations for the simplest reduced linear gyrocontinuum and Kelvin's medium in the vicinity of natural state were derived in [12, 11]. It was shown that both plane and polarised waves are present in these media, and that these media behave as acoustic metamaterials with respect to some type of waves. For instance, under zero external loads simplest reduced linear Kelvin's medium has a band gap for the wave propagating rather in the direction of proper rotation (if the wave vector's projection on $\boldsymbol{m}_0$ is positive, considering $\dot{\varphi} > 0$). External follower torque changes dispersion curve and, probably, may even change the type of acoustic metamaterial, though the last question needs to be analysed in more details than it is done in [11].

The question of criteria of stability is a sophisticated issue that is rather controversial in case of acoustic metamaterials. Indeed, usually when they discuss stability of the material in a certain state, they superimpose small perturbations on this finite deformation and assume that the medium is stable if and only if these small perturbations remain small, see, for instance, the classical work [23]. If linearised equations give exponential solution instead of harmonic waves, the material is supposed to be unstable. This happens, for instance, under large homogeneous hydrostatic compression in physically linear classical elastic medium (shear wave disappears and converts into exponential solution, e.g. the medium behaves as a liquid). This corresponds to the case $\omega^2(k) < 0$ for $k \in \mathbb{R}$, where ω is the wave frequency, and k is the wave number. If we draw the dispersion curve, we will see a band gap for low frequencies.

However, single negative acoustic metamaterials possess band gaps, which are called "working zones" of the metamaterials. They are used in practical purposes exactly in these zones, though, formally speaking, these zones also correspond to the instabilities, since $\omega^2(k) < 0$ in these zones and there are two solutions, one exponentially growing and one exponentially decreasing. In practice, we do not see unstable solutions. This is probably due to the fact that band gaps in metamaterials are related to the existence of "zero elastic modulus", i.e. to the absence of the reaction on a certain strain, a certain gradient of a generalised co-ordinate. In practice this, probably, leads to localisation phenomena and not to instabilities. For instance, Green functions for the isotropic reduced Cosserat medium for harmonic point loads, when the frequency of harmonic loads belongs to the band gap, give us decaying solutions if we use the principle of limit absorption [10], or, which is the same, if we introduce an infinitesimal dissipation in the problem. Thus we should somehow distinguish between the "true", "practical" instability (when the material is destroyed) and the "formal" instability, which corresponds to the working zones of metamaterial and, possibly, to some localised solutions.

This would be a very hard task to look for Green functions for harmonic point loads in all cases when we obtain a band gap $\omega^2(k) < 0$. We may suggest another definition to distinguish between instability "in practice" and working zones of the metamaterial, though it is not at all clear if it is an equivalent scheme.

Suppose that we have obtained a domain $\omega^2(k) < 0$ in our dispersion relation for a medium indifferent to a certain gradient of a special generalised co-ordinate but reacting to this special generalised co-ordinate itself. Let us introduce into the model an infinitesimal positive elastic modulus providing reaction to the strain which was absent in the acoustic metamaterial. If domain $\omega^2(k) < 0$ converts into the domain of the harmonic wave solution $\omega^2(k) > 0$, we assume that it was a working zone of the metamaterial, stable in practice. Otherwise, it is a domain of "true" instability.

9.3.1 Harmonic Waves

We look for a time-harmonic solution $\boldsymbol{u} = \boldsymbol{U}e^{i\omega t}, \boldsymbol{\theta} = \boldsymbol{\Theta}e^{i\omega t}$ of equations in displacements (9.47),(9.48). It will be convenient to express $\boldsymbol{\theta}$ via $\boldsymbol{u}$ from balance of moment (9.48), provided that $\omega \neq \omega_0$:

$$\theta = (\frac{\omega^2}{\omega_0^2} - 1)^{-1}(\boldsymbol{E} + i\frac{\omega}{\omega_0}\boldsymbol{m}_0 \times \boldsymbol{E}) \cdot (\boldsymbol{m}_0 \cdot (\frac{\omega_s}{\omega_0}\overset{S}{\nabla}\boldsymbol{u}^S + \overset{S}{\nabla}\boldsymbol{u}^A) \times \boldsymbol{m}_0). \tag{9.50}$$

Substituting this expression into the balance of forces (9.47) and looking for a solution in the form $\boldsymbol{u} = \boldsymbol{u}_0 \exp(i(\boldsymbol{k} \cdot \boldsymbol{R}_s - \omega t))$ would allow us to formulate a reduced spectral problem for an arbitrary direction of wave propagation.

To avoid cumbersome calculations, we will perform this scheme in two particular cases: when waves propagate along the proper rotation and in the orthogonal direction. In both cases two types of waves are possible, longitudinal and transverse ones. In the case of arbitrary direction of wave propagation the waves will be mixed, but we will not consider here this case.

9.3.2 Propagation of Harmonic Waves Along the Body Point Axis ($\hat{\boldsymbol{k}} = \pm \boldsymbol{m}_0$)

Given the direction of $\hat{\boldsymbol{k}}$, we can rewrite equation (9.50) for $\boldsymbol{\theta}$ in the following form:

$$\theta = \pm\frac{ik}{2}(\frac{\omega_s}{\omega_0} + 1)\frac{\omega_0^2}{\omega^2 - \omega_0^2}(-\boldsymbol{m}_0 \times \boldsymbol{u} + i\frac{\omega}{\omega_0}(\boldsymbol{E} - \boldsymbol{m}_0\boldsymbol{m}_0) \cdot \boldsymbol{u}). \tag{9.51}$$

When we substitute this expression into (9.47), after some simplifications we obtain the following form for the balance of force in displacements, which represents now a reduced spectral problem:

$$\boldsymbol{m}_0\boldsymbol{m}_0 \cdot \boldsymbol{u}(\rho_0\omega^2 - \rho_0 k^2 c_p^2)$$

$$+(\boldsymbol{E}-\boldsymbol{m}_0\boldsymbol{m}_0)\cdot\boldsymbol{u}\left(\rho_0\omega^2-k^2\left(\mu_{\text{eff}}+\alpha_{\text{eff}}\left(1+\frac{\omega_0^2}{\omega^2-\omega_0^2}\right)\right)\right.$$
$$\left.-ik^2\alpha_{\text{eff}}\frac{\omega\omega_0}{\omega^2-\omega_0^2}\boldsymbol{m}_0\times\boldsymbol{u}=\boldsymbol{0}. \quad (9.52)$$

In this case, we obtain effective constants that vary depending on the preliminary strain:

$$c_p^2=\frac{A}{\rho_0}(\lambda A(5A^2-3)/2+\mu A(3A^2-1)+2\alpha(A-1)^2(2A+1)), \quad (9.53)$$

$$\alpha_{\text{eff}}=\alpha A(1+\frac{\omega_s}{\omega_0})(2A^2-2A+1)=2\alpha A\frac{(A^2+(A-1)^2)(2A^2-2A+1)}{A^2+1}>0 \quad (9.54)$$

$$\mu_{\text{eff}}=3\lambda A^2(A^2-1)/2+2\mu A^2(A^2-1/2)$$
$$+\alpha A(4(A-1)^2(A-1/2)+A^3)-\alpha_{\text{eff}}=$$
$$3\lambda A^2(A^2-1)/2+2\mu A^2(A^2-1/2)+\alpha A(4(A-1)^2(A-1/2)+A^3)-$$
$$2\alpha A\frac{(A^2+(A-1)^2)^2}{A^2+1} \quad (9.55)$$

We can show that the term proportional to α in μ_{eff} changes its sign only once, at $A\approx 0.83$, and also vanishes at $A=1$. It is negative at strong compression and positive or zero otherwise.

First two terms in μ_{eff} give together a negative value at

$$A^2<\frac{2+3\lambda/\mu}{4+4\lambda/\mu}\in(0,5;0.75). \quad (9.56)$$

Therefore at $A<0.5$ we have $\mu_{\text{eff}}<0$, at $A>0.83$ (approximately) we have $\mu_{\text{eff}}>0$, between these values it depends on the relation of the parameters.

In what follows at this section dedicated to the perturbations propagating along the axis of body points, we will see that preliminary strain may lead to instabilities of the material. These are "true" instabilities, since they take place also at low frequencies and therefore cannot be eliminated by introducing an infinitesimal reaction to the gradient of micro-rotation.

9.3.2.1 Non-Dispersive Longitudinal Waves Propagating Along the Body Point Axis ($\boldsymbol{u}||\boldsymbol{m}_0,\boldsymbol{u}||\hat{\boldsymbol{k}}$).

If $\boldsymbol{u}_0=u_0\boldsymbol{m}_0=u_0\hat{\boldsymbol{k}}$, only the first term in reduced spectral problem (9.52) differs from zero, and in case $c_p^2>0$ we obtain a classical dispersion relation for longitudinal waves:

$$\omega=c_pk \quad (9.57)$$

Depending on the elastic constants λ, μ, α, function $c_p^2(A)$ may be monotone or not. Stability with respect to this kind of perturbation increases as α increases. In the domain of stability of the material, this is a non-dispersive wave.

We show two numerical examples. In Fig. 9.2 we see a propagating harmonic wave, whose velocity first decreases as compression increases (slight compression), then it increases (medium compression), and then

In Fig. 9.3 the minimum point of function $c_p^2(A)$ is below 0. This means that in the interval for A [0.23,0.72], our medium loses stability under longitudinal perturbation, and c_p^2 is not a square of the wave velocity in this domain (the wave converts into an exponentially growing function). This is a new mechanism of the loss of stability for the medium (provided that other kind of perturbations that could cause instability, are absent).

In the stability domain longitudinal waves propagating along axis of the body point are similar to the linear case (non-dispersive). At $A = 1$ they coincide with it. If $c_p^2 < 0$, the medium is unstable with respect to the longitudinal perturbations. Note that $c_p^2(\alpha)$ increases at $\alpha > 0$ (which is the case under consideration). At $\alpha \to 0$ we have

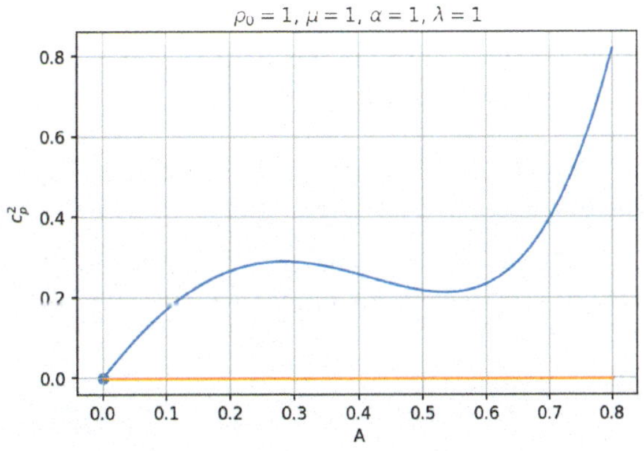

Fig. 9.2 Plot c_p^2 for the case $c_{p_{min}}^2 > 0$, longitudinal perturbations, $\hat{\boldsymbol{k}} = \boldsymbol{m}_0$.

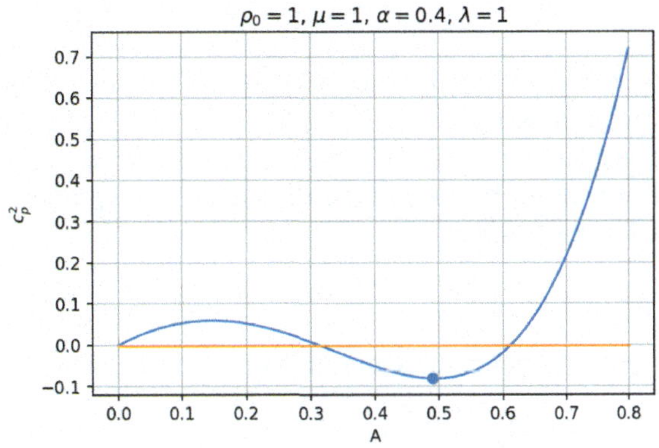

Fig. 9.3 Plot c_p^2 for $c_{p_{min}}^2 <$ 0, longitudinal perturbations, $\hat{\boldsymbol{k}} = \boldsymbol{m}_0$.

$$\frac{\rho_0}{A^2} c_p^2 \to A^2(5\lambda/2+3\mu)-(3\lambda/2+\mu). \tag{9.58}$$

For instability of the material we have to require necessary but not sufficient condition

$$A^2 < \frac{3\lambda+2\mu}{5\lambda+6\mu} \tag{9.59}$$

At small α we have a domain of instability somewhat below this treshold. When we compress more, the term with α prevails at small A and we find again a longitudinal wave instead of increasing perturbations.

At large λ and small μ, α we can find instability with respect to longitudinal perturbations, for instance, at A somewhat below 0.75. We have an example of such instability at $\lambda = 12\mu$, $\alpha = \mu$, $A = 0.75$.

This is a new mechanism of the loss of stability, which we do not observe in classical or non-gyroscopic Cosserat media with convex energy, where the stability is lost only due to the shear or rotational perturbations.

However, it may happen that the stability of the medium will be lost at these values of compression due to other kind of perturbations, for instance, propagating in other directions.

9.3.2.2 Polarised Shear-Rotation Wave Propagating Along the Body Point Axis ($\boldsymbol{u}\cdot\boldsymbol{m}_0 = 0, \boldsymbol{u}||\hat{\boldsymbol{k}}$).

If $\boldsymbol{u}\cdot\boldsymbol{m}_0 = 0$, the first term in reduced spectral problem (9.52) vanishes, and the dispersion relation takes the form:

$$\boldsymbol{u}(\rho_0\omega^2 - k^2(\mu_{\text{eff}} + \alpha_{\text{eff}}(1+\frac{\omega_0^2}{\omega^2-\omega_0^2}))) - ik^2\alpha_{\text{eff}}\frac{\omega\omega_0}{\omega^2-\omega_0^2}\boldsymbol{m}_0\times\boldsymbol{u} = \boldsymbol{0}. \tag{9.60}$$

Let us denote

$$\omega_1^2 = \omega_0^2\frac{\mu_{\text{eff}}}{\mu_{\text{eff}}+\alpha_{\text{eff}}}, \tag{9.61}$$

$$c_{s\alpha}^2 = \frac{\mu_{\text{eff}}+\alpha_{\text{eff}}}{\rho_0}. \tag{9.62}$$

In case when $\mu_{\text{eff}} > 0$, we have $\omega_1^2 > 0$, and we obtain the same reduced spectral problem as for the case linearised in the vicinity of the natural state, which was considered in [12]. There exist two waves with circular polarisation: $\boldsymbol{u} = \boldsymbol{u}_0(\boldsymbol{e}_1 \pm i\boldsymbol{e}_2)$, For waves polarised in the direction of the proper rotation, the dispersion relation takes the form:

$$k^2 = \frac{\omega^2}{c_{s\alpha}^2}\frac{\omega+\omega_0}{\omega+\omega_1^2/\omega_0}, \tag{9.63}$$

For waves polarised against the proper rotation, the dispersion relation takes the form:

$$k^2 = \frac{\omega^2}{c_{s\alpha}^2} \frac{\omega - \omega_0}{\omega - \omega_1^2/\omega_0}. \tag{9.64}$$

At low frequencies the dispersion relations for both polarised waves converge, i.e. both waves travel almost at the same velocity, close to $\sqrt{\mu_{\text{eff}}/\rho_0}$.

If $\mu_{\text{eff}} < 0$, at low frequencies we have an unstable material with respect to shear–rotational waves: the contribution of terms proportional to α_{eff} vanishes at small ω, and the material becomes "liquified", as in the case of classical medium or non-gyroscopic Cosserat media, i.e. it is unstable with respect to quasistatic shear–rotational perturbations. As it was mentioned above, it is always so at $A < 1/2$ independently on the parameters. At $A > 0.83$ (approximately) we always have wave propagation for this type of perturbations.

In Fig. 9.4 the domain of wave propagation corresponds to the branch $\omega_1^2 > 0$ which is situated to the right of the red dotted line. In this case $\mu_{\text{eff}} > 0$. We see that strong compression yields in the material instability.

By means of preliminary strain we can influence the position of the band gap and its width (Fig. 9.5). In this case, for all three cases (slight compression, undeformed medium, and tension) the material is stable. The wave, polarised contrary to the direction of the proper rotation, is strongly dispersive and has a band gap that would disappear if we introduce an infinitesimal resistance to $\nabla\theta$. The wave, polarised in the direction of proper rotation, is slightly dispersive (Fig. 9.6) and has no band gap.

9.3.3 Propagation of Harmonic Waves in the Direction Orthogonal to the Body Point Axis ($\hat{\boldsymbol{k}} \perp \boldsymbol{m}_0$)

Let us introduce a vector $\boldsymbol{l} = \hat{\boldsymbol{k}} \times \boldsymbol{m}_0$, which forms together with $\hat{\boldsymbol{k}}, \boldsymbol{m}_0$ an orthonormal basis in three-dimensional space.

We have to recalculate expression (9.50) for $\boldsymbol{\theta}$ via $\boldsymbol{u}$ and substitute it into the balance of force (9.47).

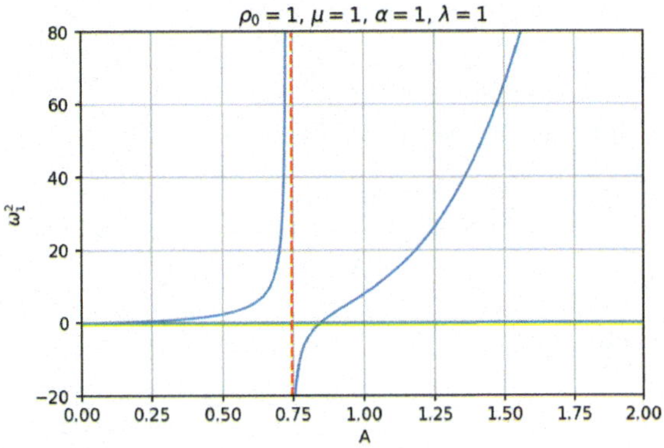

Fig. 9.4 Function $\omega_1^2(A)$ for shear–rotational perturbations, $\hat{\boldsymbol{k}} = \boldsymbol{m}_0$. Domain of wave propagation (stability of the material): $\omega_1^2 > 0$, right branch.

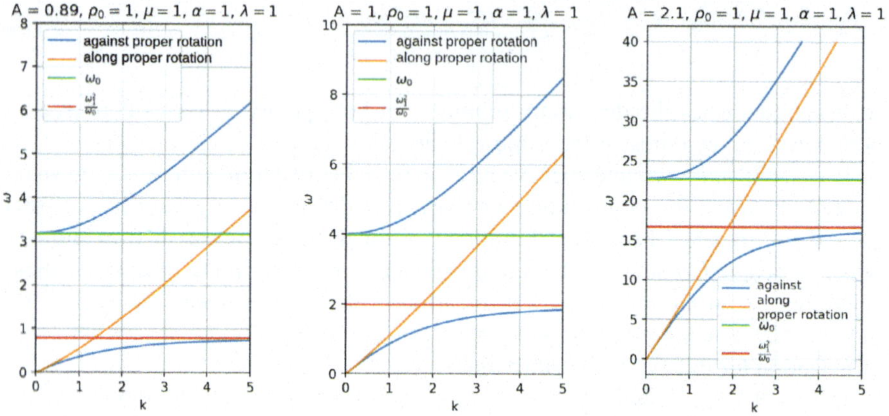

Fig. 9.5: Dispersion curves for polarised shear–rotational waves propagating along the body point axis, from left to right: low compression area, natural state, tension area (stable domains). The wave polarised in the direction of proper rotation is less dispersive and does not have a band gap. The wave polarised contrary to the proper rotation has a band gap and is strongly dispersive. The band gap can be tuned by preliminary strain.

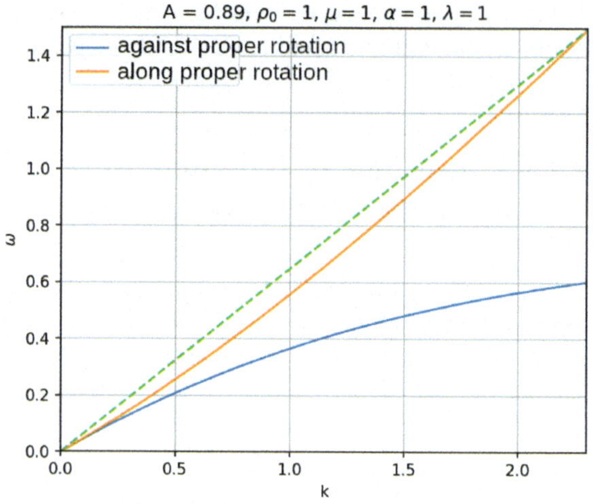

Fig. 9.6 Small dispersion of shear–rotational waves propagating along the body point axis.

It is easy to see that in this case

$$\boldsymbol{\theta} = \frac{ik}{2}\frac{\frac{\omega_s}{\omega_0}-1}{\frac{\omega^2}{\omega_0^2}-1}(\boldsymbol{l}+i\frac{\omega}{\omega_0}\hat{\boldsymbol{k}})\boldsymbol{m}_0\cdot\boldsymbol{u} = ik\Theta(\boldsymbol{l}+i\frac{\omega}{\omega_0}\hat{\boldsymbol{k}})\boldsymbol{m}_0\cdot\boldsymbol{u}, \tag{9.65}$$

where

$$\Theta = \frac{1}{2}\frac{\frac{\omega_s}{\omega_0}-1}{\frac{\omega^2}{\omega_0^2}-1}. \tag{9.66}$$

For elastic tensors $\boldsymbol{C}_1, \boldsymbol{C}_2$, defined by eq. (9.45), (9.46), we will use the notation:

$$\boldsymbol{C}_1 = \boldsymbol{C}_{11}\boldsymbol{E} + \boldsymbol{C}_{12}\boldsymbol{m}_0\boldsymbol{m}_0, \tag{9.67}$$

$$\boldsymbol{C}_2 = \boldsymbol{C}_{21}\boldsymbol{E} + \boldsymbol{C}_{22}\boldsymbol{m}_0\boldsymbol{m}_0, \tag{9.68}$$

where

$$C_{11} = A(A^2-1)(\frac{3}{2}\lambda+\mu), \quad C_{12} = 4\alpha(A-1)^2(A-1/2) \tag{9.69}$$

$$C_{21} = 2\alpha A^2, \qquad C_{22} = 2\alpha(A-1)^2. \tag{9.70}$$

After substituting the expression for $\boldsymbol{\theta}$ into balance of force (9.47) we obtain a new reduced spectral problem for waves propagating in the direction orthogonal to $\boldsymbol{m}_0$ and then we write the dispersion relations. We will see that in this case $(\hat{\boldsymbol{k}} \perp \boldsymbol{m}_0)$ free harmonic waves turn out to be plane, as in non-gyroscopic media, in contrast to waves propagating along the spin, which are polarised.

9.3.3.1 Non-Dispersive Longitudinal Waves Propagating in the Direction Orthogonal to the Body Point Axis $(\boldsymbol{u} || \hat{\boldsymbol{k}}, \hat{\boldsymbol{k}} \cdot \boldsymbol{m}_0 = 0)$

In this case from (9.50) we obtain $\boldsymbol{\theta} = \boldsymbol{0}$. The medium behaves as a classical one. In balance of force (9.47) all terms, proportional to α, disappear, and only classical terms (part of the first line) remain. The wave is plane. We obtain the following dispersion relation:

$$k^2 = \frac{\rho_0\omega^2}{AC_{11} + A^4(\lambda+2\mu)} \tag{9.71}$$

This is a non-dispersive wave, the same as compression wave in the classical medium. It is easy to see that this wave converts into a growing solution under sufficiently large compression, namely, if and only if

$$A^2 < \frac{3\lambda+2\mu}{5\lambda+6\mu}. \tag{9.72}$$

This is illustrated by Fig. 9.7.

We see that it is the same threshold for instability as in the case of waves propagating along the axis of the body point if we take $\alpha = 0$ in the latter case. Thus we see that longitudinal perturbation propagating orthogonal to the body point axis leads to instability before it takes place for these perturbations propagating along the body point axis.

9.3.3.2 Shear-Rotational Waves Propagating in the Direction Orthogonal to the Body Point axis $(\boldsymbol{u} \cdot \hat{\boldsymbol{k}} = 0, \hat{\boldsymbol{k}} \cdot \boldsymbol{m}_0 = 0)$

The vector $\boldsymbol{u}$ has no $\hat{\boldsymbol{k}}$ component, so we can write:

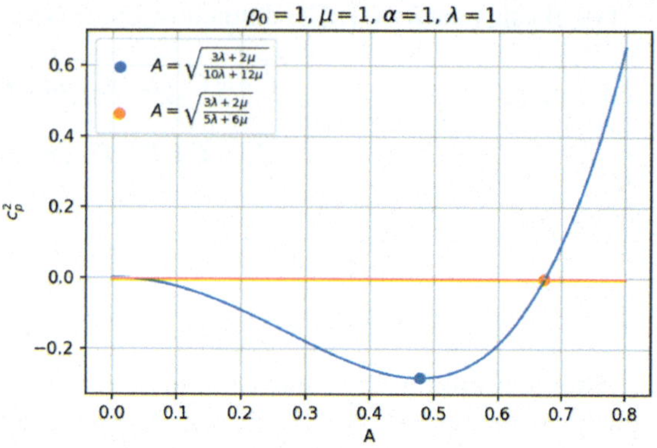

Fig. 9.7 Graph of $c_p^2(A)$ for longitudinal perturbations, $\hat{\boldsymbol{k}} \cdot \boldsymbol{m}_0 = 0$. Domain of wave propagation: $A^2 > \dfrac{3\lambda+2\mu}{5\lambda+6\mu}$.

$$\boldsymbol{u} = (u_m \boldsymbol{m}_0 + u_l \boldsymbol{l}) e^{i(\boldsymbol{k}\cdot\boldsymbol{R}_s - \omega t)} \tag{9.73}$$

In this case $\hat{\boldsymbol{k}} \times (\hat{\boldsymbol{k}} \times \boldsymbol{u}) = -\boldsymbol{u}$.

Balance of force (9.47) and expression (9.65) give us the reduced spectral problem for this case:

$$\begin{aligned} k^2 A(C_{11}\boldsymbol{u} + A^3\mu\boldsymbol{u} + \alpha A^2(2-A)\boldsymbol{m}_0\boldsymbol{m}_0\cdot\boldsymbol{u} \\ + \alpha(2-2A)(2-A)\boldsymbol{m}_0\boldsymbol{m}_0\cdot\boldsymbol{u} \\ - \Theta C_{21}\boldsymbol{m}_0\boldsymbol{m}_0\cdot\boldsymbol{u} + 4\Theta\alpha(A-1)\boldsymbol{m}_0\boldsymbol{m}_0\cdot\boldsymbol{u}) = \rho_0\omega^2\boldsymbol{u} \end{aligned} \tag{9.74}$$

We see now that two different solutions exist: $\boldsymbol{u} = u_l\boldsymbol{l}$ and $\boldsymbol{u} = u_m\boldsymbol{m}_0$, again they correspond to plane waves.

Non-Dispersive Shear Wave, $\boldsymbol{u}\cdot\boldsymbol{m}_0 = \hat{\boldsymbol{k}}\cdot\boldsymbol{m}_0 = \boldsymbol{u}\cdot\hat{\boldsymbol{k}} = 0, \boldsymbol{\theta} = \boldsymbol{0}$

In the first case, $\boldsymbol{u} = u_l\boldsymbol{l}$, wave displacement is orthogonal to the axis of a body-point and to the direction of wave propagation. Note that in this case $\boldsymbol{\theta} = \boldsymbol{0}$. The wave has the same character as the shear wave in a classical elastic medium:

$$\omega^2 = c_s^2 k^2 \tag{9.75}$$

where

$$c_s^2 = \frac{AC_{11} + A^4\mu}{\rho_0} = \frac{A^2(A^2-1)(\frac{3}{2}\lambda+\mu) + A^4\mu}{\rho_0} \tag{9.76}$$

We see that this perturbation may lead to instability at large compression, if

$$A^2 < \frac{3\lambda+2\mu}{3\lambda+4\mu}. \tag{9.77}$$

Note that this condition is stronger than the condition for the instability of the shear–rotational waves propagating along the spin. Indeed, if this condition is reached, then $A < 0.83$, and μ_{eff} in case $\hat{\boldsymbol{k}} = \boldsymbol{m}_0$ is less than μ_{eff} in this section (a negative term with α has to be added). In other words, quasistatic shear perturbations along the body point axis need less preliminary compression to cause instability of the material.

Dispersive Shear-Rotational Wave, $\boldsymbol{u}||\boldsymbol{m}_0, \hat{\boldsymbol{k}} \cdot \boldsymbol{m}_0 = \boldsymbol{u} \cdot \hat{\boldsymbol{k}} = 0$

Now consider another type of the transverse wave: $\boldsymbol{u} = u_m \boldsymbol{m}_0$. The reduced spectral problem admits this solution if we satisfy the dispersion relation

$$k^2 \left(\mu_{\text{eff}} + \alpha_{\text{eff}} \left(1 + \frac{\omega_0^2}{\omega^2 - \omega_0^2} \right) \right) = \rho_0 \omega^2, \tag{9.78}$$

where

$$\alpha_{\text{eff}} = \alpha A(A^2 - 2A + 2)(1 - \frac{\omega_s}{\omega_0}) = 2\alpha A^2 (A^2 - 2A + 2) \frac{2 - A}{A^2 + 1} \tag{9.79}$$

$$\mu_{\text{eff}} = 3\lambda A^2 (A^2 - 1)/2 + 2\mu A^2 (A^2 - 1/2) + \alpha A(A^2 - 2A + 2)(2 - A) - \alpha_{\text{eff}} =$$
$$3\lambda A^2 (A^2 - 1)/2 + 2\mu A^2 (A^2 - 1/2) + \alpha(2 - A)A((A - 1)^2 + 1) \frac{(A - 1)^2}{A^2 + 1} \tag{9.80}$$

Note that $\mu_{\text{eff}}, \alpha_{\text{eff}}$ here take values different from the section devoted to the wave propagating along the axis of the body.

We introduce the following notation:

$$\omega_1^2 = \omega_0^2 \frac{\mu_{\text{eff}}}{\mu_{\text{eff}} + \alpha_{\text{eff}}}, \tag{9.81}$$

$$c_{s\alpha}^2 = \frac{\mu_{\text{eff}} + \alpha_{\text{eff}}}{\rho_0}. \tag{9.82}$$

The dispersion relation can be rewritten as

$$k^2 = \frac{\omega^2}{c_{s\alpha}^2} \frac{\omega^2 - \omega_0^2}{\omega^2 - \omega_1^2}, \tag{9.83}$$

This dispersion relation is similar to the dispersion relation for the non-gyroscopic linear elastic reduced Cosserat medium. It has a band gap between ω_1 and ω_0. This band gap would disappear if we introduce an infinitesimal positive elastic reaction to $\nabla\boldsymbol{\theta}$, thus it is a working zone of an acoustic metamaterial (with respect to this particular kind of perturbations).

However, there are some differences with the reduced non-gyroscopic Cosserat medium. For instance, ω_0 is defined by Eq. (9.49), it is always positive, and is not related to α_{eff}. Therefore $\boldsymbol{\alpha_{\text{eff}}}$ **has not necessarily to be positive to provide the stability** of the material, in contrast to the reduced Cosserat medium. However, $\boldsymbol{\mu_{\text{eff}}}$ **and** $\boldsymbol{\mu_{\text{eff}}+\alpha_{\text{eff}}}$ **must be positive in a stable material**, otherwise quasistatic or high-frequency shear perturbations will grow, and this cannot be helped by introducing medium's reaction to $\nabla\boldsymbol{\theta}$.

Note that $\alpha_{\text{eff}} > 0$ for $A < 2$ and $\alpha_{\text{eff}} < 0$ for $A > 2$.

Stability at Low Frequencies: Properties of μ_{eff}

The first two terms in μ_{eff}, depending on λ, μ, give together a positive value if

$$A^2 > \frac{3\lambda+2\mu}{3\lambda+4\mu} \in (0,5;0,75), \tag{9.84}$$

and negative otherwise (for strong compression). This is the same situation as in the case of $\hat{\boldsymbol{k}} = \boldsymbol{m}_0$. However, the term depending on α, has a different behaviour in these two cases. In case $\hat{\boldsymbol{k}} = \boldsymbol{m}_0$ it was negative at $A < 0.83$, zero at $A = 1$ and positive otherwise. If $\hat{\boldsymbol{k}} \cdot \boldsymbol{m}_0 = 0$, it is positive at $A < 2$ and negative at $A > 2$. This means that at large α in case $\hat{\boldsymbol{k}} = \boldsymbol{m}_0$ the medium could become unstable at $A < 0.83$ and become stable at $A > 0.83$, and in case $\hat{\boldsymbol{k}} \cdot \boldsymbol{m}_0 = 0$ the medium can become unstable at large tension ($A > 2$), in contrast to the previous case.

Change of Metamaterial Type at Tension $A > 2$

When $\alpha_{\text{eff}} = 0$ (at $A = 2$), the band gap disappears, and when α_{eff} moves to the negative domain, it appears again, but characteristic frequencies ω_0 and ω_1^2/ω_0 interchange their order. Decreasing parts of dispersion curves appear, and the medium not only continues to be a single negative acoustic metamaterial in the domain of band gap, but also has two other working zones where it is a double negative acoustic metamaterial. (A similar effect is observed in the linear simplest Kelvin's medium under negative external follower torque [12].) This takes place in the domain of parameters where $\mu_{\text{eff}} > 0, \mu_{\text{eff}} + \alpha_{\text{eff}} > 0$, but $\alpha_{\text{eff}} < 0$ ($A > 2$ but is not too large). When $\mu_{\text{eff}} + \alpha_{\text{eff}}$ becomes negative, the medium loses its stability, as it may happen also for the nonlinear non-gyroscopic Cosserat media [13]. An illustration is given in Figs. 9.8-9.10.

Possible Instability at Large Tension

At least for $\alpha > 3\lambda/2 + 2\mu$ at very large A we have $\mu_{\text{eff}} + \alpha_{\text{eff}} < 0$. We have not investigated if this condition is necessary to reach instability at some tension, but it is sufficient.

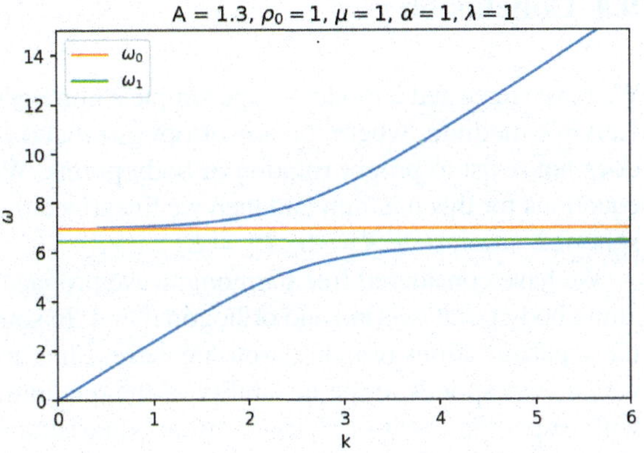

Fig. 9.8 Dispersion curves for shear–rotational waves, propagating in the direction orthogonal to the body point axis $(A < 2)$.

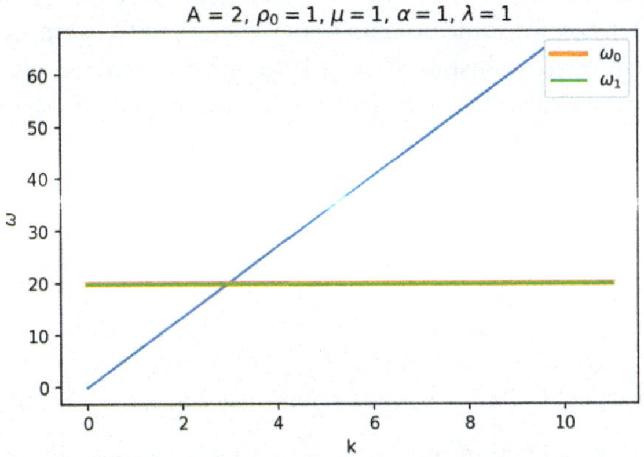

Fig. 9.9 Dispersion curves for shear waves propagating in the direction orthogonal to the body point axis $(A = 2)$.

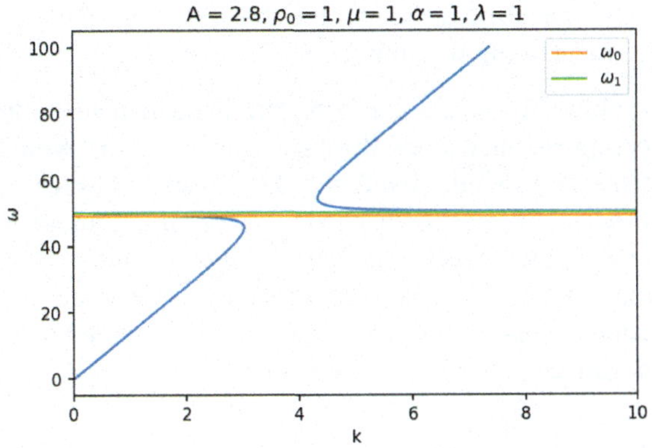

Fig. 9.10 Dispersion curves for shear waves propagating in the direction orthogonal to the body point axis ($A > 2$ but not too large to cause instabilities).

9.4 Conclusion

We have suggested a model for the simplest homogeneous non-linear elastic reduced Kelvin's medium, where the anisotropy is related only to the fact that the medium does not resist to proper rotation of body points. We have obtained the constitutive equations for this medium and then we linearized them in the vicinity of preliminary spherical strain state $\boldsymbol{R} = A\boldsymbol{r}$, $\boldsymbol{P}_{\perp} = \boldsymbol{E}$.

We have considered free harmonic wave propagation in two particular directions: along body point axis $\boldsymbol{m}_0$ and orthogonal to it. In some domains of preliminary strain these perturbations remain harmonic waves, and in others they grow exponentially, which corresponds to the instability of the medium. We have observed instabilities with respect to shear or shear-rotational quasistatic perturbations under large compression (as it occurs in classical elasticity and non-gyroscopic Cosserat medium), they always exist at $A < 0.5$ independently on the medium parameters. Under large tension for some domain of parameters (if Cosserat couple modulus α is large enough) we observe instabilities at large tension with respect to rotational–shear wave, as it occurs also for non-gyroscopic nonlinear full and reduced elastic Cosserat media under some conditions.

We also have discovered another possible mechanism of the loss of stability, not present in any previously studied Cosserat-type or classical media, namely, with respect to longitudinal perturbations under large compression (for the wave propagating orthogonal to the axis of the point body the critical compression is less than for the wave propagating along the axis of the point body).

In the domain of stability the medium behaves as an acoustic metamaterial with respect to various shear–rotational waves:

1. propagating along the axis of the body point and polarised against the proper rotation
2. propagating orthogonal to the axis of the body point, with translational displacement along this axis.

Preliminary strain can tune the band gap in both cases (in numerical examples compression widens the band gap). For the waves of the second type at critical tension $A = 2$ the band gap disappears, and then appears again together with two zones of decreasing dispersion curves above and below it, and increases from zero as tension increases, while the condition of material stability is satisfied. Thus tension may convert a single negative acoustic metamaterial into a combination of single and double negative ones. Similar effect is observed in the linear case under the action of negative external follower torque.

Acknowledgements This work is carried out within the state assignment of Ministry of Science and Higher Education of the Russian Federation in IPME RAS.

References

[1] Altenbach H, Eremeyev VA (2013) Cosserat media, In: Generalized Continua from the Theory to Engineering Applications, Springer, pp 65-130.

[2] Altenbach H, Zhilin P (1988) A general theory of elastic simple shells, Uspekhi Mekhaniki **11**(4):107–148.

[3] Cosserat E, Cosserat F (1909) Théorie des corps déformables, Librairie Scientifique A. Hermann et Fils.

[4] Eremeyev V (2018) On the material symmetry group for micromorphic media with applications to granular materials, Mechanics Research Communications **94**:8–12.

[5] Eremeyev V, Pietraszkiewicz W (2012) Material symmetry group of the non-linear polar-elastic continuum, International Journal of Solids and Structures **49**(14):1993–2005.

[6] Eremeyev VA, Lebedev LP, Altenbach H (2012) Foundations of micropolar mechanics, Springer Science & Business Media.

[7] Eringen AC (2012) Microcontinuum Field Theories: I. Foundations and Solids, Springer Science & Business Media.

[8] Giorgio I, Hild F, Gerami E, Dell'Isola F, Misra A (2022) Experimental verification of 2D Cosserat chirality with stretch-micro-rotation coupling in orthotropic metamaterials with granular motif, Mechanics Research Communications **126**:104020.

[9] Grekova E, Zhilin P (2001) Basic equations of Kelvin's medium and analogy with ferromagnets, Journal of Elasticity and the Physical Science of Solids **64**(1):29-70.

[10] Grekova E, Kulesh M, Herman G (2009) Waves in linear elastic media with microrotations, part 2: Isotropic reduced cosserat model, Bulletin of the Seismological Society of America **99**(2B):1423-1428.

[11] Grekova EF (2018) Harmonic waves in the simplest reduced kelvin's and gyrostatic media under an external body follower torque. In: 2018 Days on Diffraction (DD), IEEE, pp 142-148.

[12] Grekova EF (2018) Simplest linear homogeneous reduced gyrocontinuum as an acoustic metamaterial. In: Generalized Models and Non-classical Approaches in Complex Materials 1, Springer, pp 375-386.

[13] Grekova EF (2019) Nonlinear isotropic elastic reduced and full cosserat media: waves and instabilities. Continuum Mechanics and Thermodynamics 31(6):1805–1824

[14] Guenneau S, Movchan A, Pétursson G, Ramakrishna SA (2007) Acoustic metamaterials for sound focusing and confinement, New Journal of physics **9**(11):399.

[15] Ieşan D (2008) A theory of prestressed thermoelastic cosserat continua. ZAMM-Journal of Applied Mathematics and Mechanics/Zeitschrift für Angewandte Mathematik und Mechanik: Applied Mathematics and Mechanics **88**(4):306-319.

[16] Ivanova EA (2020) Thermal effects by means of two-component cosserat continuum, Encyclopedia of Continuum Mechanics pp 2455-2466.
[17] Ivanova EA (2022) A new approach to modeling of thermal and electrical conductivities by means of the cosserat continuum. Continuum Mechanics and Thermodynamics **34**(5):1313-1342.
[18] Kafadar C, Eringen AC (1971) Micropolar media—i the classical theory. International Journal of Engineering Science **9**(3):271-305.
[19] Kaina N, Lemoult F, Fink M, Lerosey G (2015) Negative refractive index and acoustic superlens from multiple scattering in single negative metamaterials, Nature **525**(7567):77-81.
[20] Maugin GA (2017) Non-classical Continuum Mechanics, Springer.
[21] Pietraszkiewicz W, Eremeyev V (2009) On natural strain measures of the non-linear micropolar continuum. International Journal of Solids and Structures **46**(3-4):774-787.
[22] Porubov A, Antonov I (2021) On control of harmonic waves in an acoustic metamaterial, Mechanics Research Communications **116**:103,745.
[23] Toupin R, Bernstein B (1961) Sound waves in deformed perfectly elastic materials. acoustoelastic effect, The Journal of the Acoustical Society of America **33**(2):216-225.
[24] Vilchevskaya EN, Müller WH, Eremeyev VA (2022) Extended micropolar approach within the framework of 3M theories and variations thereof, Continuum Mechanics and Thermodynamics **34**(2):533-554.

Chapter 10
On the Spectrum of Relaxation Times of Coupled Diffusion and Rheological Processes in Media with Microstructure

Dmitrii S. Dudin and Ilya E. Keller

Abstract The effective diffusion and viscous linear coefficients is studied for the coupled set of equations with incorporating the processes of interdiffusion in a binary alloy, viscoelastic deformations, and microstructure changes. The technique developed in works of G.B. Stephenson, Z. Suo, and L. Brassart is used, but to study abnormally fast processes of chemical composition changing in metal alloys under intensive plastic deformations, the presence of microstructure was consistently regarded in the model. The microstructure is described by a scalar internal variable being the concentration of grain boundaries, and by the intensity of deviatoric inelastic strains associated with dislocation density of the crystal lattice. The model, preferred by specialists in atomic diffusion and based on the experimental method for determining the Kirkendall – Darken interdiffusion coefficients, is adopted as the equations of interdiffusion in a medium experiencing elastic and viscous deformations. The model problem of free axial deformation of a rod, constrained on the sides, with torsion is considered. For this problem, effective diffusion and viscous linear coefficients are determined for perturbating a homogeneous stationary solution. The equations for perturbations are reduced to an eigenvalue problem, which is addressed through computer algebra. In the long-wave and short-wave approximations, the relaxation of perturbations is characterized by a spectrum of effective diffusion coefficients and viscosity. The sensitivity of diffusion coefficients to thermodynamic and kinetic coefficients associated with microstructure is established. One of the effective diffusion coefficients can drastically increase, when those parameters are changed, and it can explain the abnormally fast diffusion of the alloy components.

Key words: Interdiffusion · Metal alloys · Viscoelastic deformations · Microstructure evolution · Coupled processes · High pressure torsion · Effective coefficients · Fast diffusion

---

Dmitrii S. Dudin · Ilya E. Keller
Institute of Continuous Media Mechanics of the Ural Branch of Russian Academy of Science, Academician Korolev St., 1, Perm 614013, Russian Federation,
e-mail: dudin.d@icmm.ru, kie@icmm.ru

H. Altenbach et al. (eds.), *Sixty Shades of Generalized Continua*,
Advanced Structured Materials 170, https://doi.org/10.1007/978-3-031-26186-2_10

10.1 Introduction

The work is devoted to study the possibilities of describing the coupled processes of viscoelastic deformations and interdiffusion in metal alloys by an effective model with consisting of connected diffusion and viscous elements and having a certain spectrum of characteristic times. It is some phenomenological description of the quite complex physical process to understand some anomalies of mechanodiffusion processes during mechanical alloying and intensive plastic deformations of solid and powder metals. In particular, some observed processes of decomposition, stratification of metallic solid solutions may have a rate many orders of magnitude higher than the rate of concentration diffusion [1]. Therefore, it is necessary to find out to what extent the coupled diffusion and rheological processes can explain these anomalies. In addition, the processes are usually accompanied by the evolution of microstructure in the crystal lattice of metal, and the effective kinetic coefficients are very structurally sensitive. Therefore, the connectivity of diffusion processes with microstructural ones should also be considered in this study. The work continues the researches started in [2, 3, 4], and the methodology of these works is taken as a basis. Elastic and viscous deformations are regarded within the shear compression scheme, and the evolution and migration of microstructure is described by a scalar variable. The connectivity of processes is considered both by thermodynamic coefficients and by the kinematics of diffusion processes in a deformable body. The work is based on the theory [4], in which, to superimpose diffusion processes on deformation ones, the velocity of material markers, relative to which the laws of diffusion are determined, is identified with the convective velocity. Another approaches to study coupled mechanodiffusion processes were considered in our work [5], and we afford to give some of them [6]-[11].

10.2 Deformations

It is assumed that at the reference time instant $t = 0$ the position of the material element of the continuum is given by the radius vector $\boldsymbol{R}$, and the volume, occupied by the material element, is denoted by dV_0. For the current time respective values $\boldsymbol{r}$, dV are introduced along with the displacement $\boldsymbol{u} = \boldsymbol{r} - \boldsymbol{R}$ and the total volumetric strain $\varepsilon_V = (dV - dV_0)/dV_0$. The total strain of the material element is assumed to be small and described by a tensor

$$\boldsymbol{\varepsilon} = \frac{1}{2}(\nabla \boldsymbol{u} + \boldsymbol{u}\nabla), \tag{10.1}$$

where ∇ is the gradient operator in the current configuration. The total strain consists of elastic $\boldsymbol{\varepsilon}^e$, viscous $\boldsymbol{\varepsilon}^v$, and diffusive $\boldsymbol{\varepsilon}^d$ components

$$\boldsymbol{\varepsilon} = \boldsymbol{\varepsilon}^e + \boldsymbol{\varepsilon}^v + \boldsymbol{\varepsilon}^d,$$

which are decomposed into spherical and deviatoric parts

$$\boldsymbol{\varepsilon} = \frac{1}{3}\varepsilon_V \boldsymbol{I} + \boldsymbol{e},\ \boldsymbol{\varepsilon}^e = \frac{1}{3}\varepsilon_V^e \boldsymbol{I} + \boldsymbol{e}^e,\ \boldsymbol{\varepsilon}^d = \frac{1}{3}\varepsilon_V^d \boldsymbol{I},\ \varepsilon^v = \boldsymbol{e}^v,$$
$$\mathrm{sp}(\boldsymbol{e}) = \mathrm{sp}(\boldsymbol{e}^e) = \mathrm{sp}(\boldsymbol{e}^v) = 0. \tag{10.2}$$

The kinematics of the continuum, concurrently undergoing deformations of rheological and diffusive nature, was considered by a number of authors [2, 3, 4, 11, 12, 13, 14]. They regarded both elastic and diffusive, viscous and diffusive, or all three components. The diffusion component of deformation is caused by a change in the chemical composition of the material element. In our work [5] it is noted that consideration of deformations of diffusive and rheological nature, regardless of deformation magnitude, requires an additional hypothesis which is related to the method used for measuring diffusion coefficients (Boltzmann method relative to the Matano plane or Darken – Kirkendall method relative to the inert markers). A number of similar hypotheses were formulated in [5] and nonequivalence of the corresponding coupled models of diffusion and deformation processes was shown. In this paper, as in [4], the diffusion frame of reference is associated with the inert markers, but elastic deformations are also considered along with viscous and diffusive ones.

Here a binary system, consisting of A and B components, is considered. The material element contains dN_A and dN_B moles of the corresponding components and it is characterized by the molar concentrations $C_A = dN_A/dV_0$ and $C_B = dN_B/dV_0$ in the reference configuration, and $c_A = dN_A/dV$, $c_B = dN_B/dV$ in the current configuration. For the material element, the molecular incompressibility condition is assumed. It means that the material element is continuously filled with the molar volumes of components V_A and V_B

$$dV = V_A dN_A + V_B dN_B + \varepsilon_V^e dV_0,$$

and it is written in terms of the swelling ratio $\Omega = dV/dV_0$

$$\Omega = V_A(\xi)\, C_A + V_B(\xi)\, C_B + \varepsilon_V^e, \tag{10.3}$$

where $\xi = C_B/(C_A + C_B)$ is the composition variable. Supposing $\Omega = \Omega\left(C_A, C_B, \varepsilon_V^e\right)$ it is obtained

$$\dot{\Omega} = \frac{\partial \Omega}{\partial C_A}\dot{C}_A + \frac{\partial \Omega}{\partial C_B}\dot{C}_B + \frac{\partial \Omega}{\partial \varepsilon_V^e}\dot{\varepsilon}_V^e, \tag{10.4}$$

where

$$\dot{a} = \frac{da}{dt} = \frac{\partial a}{\partial t} + \boldsymbol{v} \cdot \nabla a$$

is the material derivative of some variable a and $\boldsymbol{v} = \dot{\boldsymbol{u}}$ is the velocity of the material element. On the other hand, the material derivative of (10.3) is

$$\dot{\Omega} = \dot{V}_A C_A + \dot{V}_B C_B + V_A \dot{C}_A + V_B \dot{C}_B + \dot{\varepsilon}_V^e \tag{10.5}$$

Direct comparison of (10.4) with (10.5) allows for deducing the compatibility conditions

$$V_A = \frac{\partial \Omega}{\partial C_A}, \quad V_B = \frac{\partial \Omega}{\partial C_B}, \quad \dot{V}_A C_A + \dot{V}_B C_B = 0, \quad \frac{\partial \Omega}{\partial \varepsilon_V^e} = 1. \tag{10.6}$$

Because of weak dependence of molar volumes on the composition variable for solid deformable bodies [15, 16] it is assumed that $V_A = \text{const}$, $V_B = \text{const}$. Therefore, condition $(10.6)_3$ is fulfilled automatically.

In the current configuration, explicit expressions for the volumetric strains ε_V^e and ε_V^d are useful to write. To do that, the swelling ratio Ω is represented as

$$\Omega = 1 + \varepsilon_V,$$

and using decomposition (10.2) of volumetric strains $\varepsilon_V = \varepsilon_V^d + \varepsilon_V^e$, expressions (10.3) and the definition for the molar concentrations in the current configuration $c_A = dN_A/dV$, $c_B = dN_B/dV$ is deduced

$$\varepsilon_V^e = (1+\varepsilon_V)(1 - V_A c_A - V_B c_B), \ \varepsilon_V^d = V_A C_A + V_B C_B - 1. \tag{10.7}$$

According to the second equation in (10.7), the volume strain due to diffusion ε_V^d is a homogeneous function of the first order for C_A and C_B. Definitions (10.7) generalize the molecular incompressibility condition $c_A V_A + c_B V_B = 1$ in the case of elastic strains.

10.3 Balance Equations

The system of coupled equations of diffusion and rheological processes with changes in microstructure includes the balance equations for variables describing the local chemical composition of a binary alloy, the momentum and microstructure density, and equations of state. As variables characterizing the chemical composition of a two-component mixture, the molar concentrations c_k, $k = A, B$ are taken and the diffusion fluxes relative to the material velocity $\boldsymbol{v}$ are introduced

$$\boldsymbol{j}_k = c_k (\boldsymbol{v}_k - \boldsymbol{v}),$$

where $\boldsymbol{v}_k$, $k = A, B$ are the velocities of the components in the current configuration. These diffusion fluxes are not coupled ($\boldsymbol{j}_A + \boldsymbol{j}_B \neq 0$), since the material velocity is not identified with the average velocity of multicomponent medium. For the same reason the balance equations for molar concentrations written without the volume production of components induced by chemical reactions

$$\frac{dc_k}{dt} + c_k \nabla \cdot \boldsymbol{v} = -\nabla \cdot \boldsymbol{j}_k, \ k = A, B \tag{10.8}$$

are independent. Similarly, the diffusion flux of microstructure is written relative to the material velocity

$$j_H = h(v_H - v),$$

where h is the volume concentration of microstructure (the area of grain boundaries per unit volume of the material element in the current configuration), v_H is the migration rate of microstructure. The variable h satisfies the balance equation

$$\frac{dh}{dt} + h\nabla \cdot v = -\nabla \cdot j_H + p_H, \tag{10.9}$$

which includes the volume production of microstructure p_H. It is assumed that mechanical processes are much faster than diffusion ones, therefore the equilibrium equation is used

$$\nabla \cdot \sigma = 0, \tag{10.10}$$

where σ is the Cauchy stress tensor.

During the derivation of the thermodynamic inequality, Eqs. (10.8) and (10.9), written in terms of the reference configuration, will be needed

$$\frac{dC_k}{dt} = -\overset{0}{\nabla} \cdot J_k, \quad \frac{dH}{dt} = -\overset{0}{\nabla} \cdot J_H + P_H, \quad \overset{0}{\nabla} = \frac{\partial}{\partial R}, \quad k = A, B, \tag{10.11}$$

where $H = \Omega h$ is the microstructure concentration in the reference configuration, $P_H = \Omega p_H$ is the volumetric production of microstructure, and for the diffusion fluxes J_k and microstructure flux J_H in the reference configuration the connections $\overset{0}{\nabla} \cdot J_k = \Omega \nabla \cdot j_k$, $k = A, B$ and $\overset{0}{\nabla} \cdot J_H = \Omega \nabla \cdot j_H$ are valid [17].

10.4 The Helmholtz Free Energy

The Helmholtz free energy is key in the construction of constitutive relations, which will be derived in Sect. 10.6, and it affects the nature of the connectivity of the processes in question. In addition to the energy of grain boundaries, it is desirable to consider the dislocation energy, which is assumed to be proportional to the intensity of deviatoric viscous strains. It is supposed that for an isotropic material the density of free energy, assigned per unit volume of the reference configuration, is represented as

$$\psi = \psi\left(C_A, C_B, H, \varepsilon_V^e, I_2^e, I_2^v\right), \tag{10.12}$$

where H is the microstructure concentration, defined per unit volume of the reference configuration, $I_2^e = e^e : e^e$ is the second invariant of elastic deviatoric strain, $I_2^v = e^v : e^v$ is the second invariant of viscous deviatoric strain. The material derivative of (10.12) allows for getting

$$\dot{\psi} = \sum_{k=A,B} F_k \dot{C}_k + F_H \dot{H} + F_V^e \dot{\varepsilon}_V^e + F_I^e \dot{I}_2^e + F_\upsilon \dot{I}_2^\upsilon,$$

$$F_k = \frac{\partial \psi}{\partial C_k},\ k = A, B,\ F_H = \frac{\partial \psi}{\partial H},\ F_V^e = \frac{\partial \psi}{\partial \varepsilon_V^e},\ F_I^e = \frac{\partial \psi}{\partial I_2^e},\ F_\upsilon = \frac{\partial \psi}{\partial I_2^\upsilon} \qquad (10.13)$$

where $F_k,\ k = A, B$ are the partial mixing energies, F_H is the microstructure energy, F_V^e and F_I^e are the constituents of elastic deformation energies, F_υ is the dislocation energy. In the study of linearized problems, the following expansion of the function ψ is accepted

$$\psi \approx f_A \frac{(\Delta C_A)^2}{2} + f_B \frac{(\Delta C_B)^2}{2} + f_H \frac{(\Delta H)^2}{2} + K \frac{(\Delta \varepsilon_V^e)^2}{2} + f_\upsilon \frac{(\Delta I_2^\upsilon)^2}{2} + G I_2^e$$
$$+ f_{AH} \Delta C_A \Delta H + f_{BH} \Delta C_B \Delta H + f_{H\upsilon} \Delta H \Delta I_2^\upsilon + L(C_A, C_B, H, \varepsilon_V^e, I_2^\upsilon), \quad (10.14)$$

up to quadratic terms near the perturbed state $C_A \equiv C_A^0$, $C_B \equiv C_B^0$, $H \equiv H_0$, $\varepsilon_V^e \equiv \varepsilon_V^{e0}$, $I_2^e \equiv I_2^{e0} = 0$ and $I_2^\upsilon \equiv I_2^{\upsilon 0}$. Expansion (10.14) contains the most significant thermodynamic coefficients describing the coupled processes. In the derivation of (10.14) Eq. (10.12) is used, and the following designations are introduced: L is the linear function of its arguments, $\Delta(\cdot) = (\cdot) - (\cdot)_0$ is the increment of a variable, K is the bulk modulus, and G is the shear modulus. The quadratic terms considered significantly affect the type of physical relations discussed in Sect. 10.6. Components of the Helmholtz free energy (10.13), given (10.14), have the form

$$\begin{aligned} F_A &= f_A \Delta C_A + f_{AH} \Delta H + F_A^0,\ F_B = f_B \Delta C_B + f_{BH} \Delta H + F_B^0, \\ F_H &= f_H \Delta H + f_{AH} \Delta C_A + f_{BH} \Delta C_B + f_{H\upsilon} \Delta I_2^\upsilon + F_H^0, \\ F_V^e &= K \Delta \varepsilon_V^e + F_e^0,\ F_I^e = G,\ F_\upsilon = f_\upsilon \Delta I_2^\upsilon + f_{H\upsilon} \Delta H + G_\upsilon^0, \end{aligned} \qquad (10.15)$$

All coefficients, situated on the right side of expressions (10.16), are functions of the variables at the perturbed state $f = f\left(C_A^0, C_B^0, H_0, \varepsilon_V^{e0}, I_2^{\upsilon 0}\right)$.

10.5 Thermodynamic Inequality

To derive a restriction on the form of constitutive relations the theory of classical thermodynamics of irreversible processes is employed. Quasi-static processes occurring under isothermal conditions are considered. For them the second law of thermodynamics is written as a requirement that the power supplying the system does not exceed the rate of change of free energy

$$\int_{V_0} \dot{\psi} dV_0 + \int_{S_0} (\mu_A \boldsymbol{J}_A + \mu_B \boldsymbol{J}_B + \mu_H \boldsymbol{J}_H) \cdot \boldsymbol{N} dS_0 - \int_{V_0} \boldsymbol{\sigma} : \dot{\boldsymbol{\varepsilon}} \Omega dV_0 \leq 0, \qquad (10.16)$$

where $\mu_k,\ k=A,B$ are the chemical potentials, $\boldsymbol{J}_k,\ k=A,B$ are the diffusion fluxes at the reference configuration, $\boldsymbol{\sigma}$ is the Cauchy stress tensor, $S_0=\partial V_0$ and V_0 are the surface and volume of body at the reference configuration. Using the divergence theorem and Eqs. (10.11) it can be obtained that

$$\int\limits_{V_0}\left(\Omega\boldsymbol{\sigma}:\dot{\boldsymbol{\varepsilon}}-\dot{\psi}+\sum_{k=A,B}\left(\mu_k\dot{C}_k-\boldsymbol{J}_k\cdot\overset{0}{\nabla}\mu_k\right)+\mu_H\dot{H}\right.$$
$$\left.-\boldsymbol{J}_H\cdot\overset{0}{\nabla}\mu_H-P_H\mu_H\right)dV_0\geq 0. \qquad (10.17)$$

Then the local form of thermodynamic inequality (10.16) is deduced

$$\Omega\boldsymbol{\sigma}:\dot{\boldsymbol{\varepsilon}}+\sum_{k=A,B}\left((\mu_k-F_k)\dot{C}_k-\boldsymbol{J}_k\cdot\overset{0}{\nabla}\mu_k\right)+(\mu_H-F_H)\dot{H}-\boldsymbol{J}_H\cdot\overset{0}{\nabla}\mu_H$$
$$-\mu_H P_H-F^e_V\dot{\varepsilon}^e_V-F^e_I\dot{I}^e_2-F_v\dot{I}^v_2\geq 0. \qquad (10.18)$$

In (10.18) expressions (10.13) and (10.14) are used. Applying to (10.18) a decomposition $\boldsymbol{\sigma}=\sigma_m\boldsymbol{I}+\boldsymbol{s}$, its consequence $\boldsymbol{\sigma}:\dot{\boldsymbol{\varepsilon}}=\sigma_m\dot{\varepsilon}_V+\boldsymbol{s}:\dot{\boldsymbol{e}}$, Eq. (10.2), and definitions $I^v_2=\boldsymbol{e}^v:\boldsymbol{e}^v$, $I^e_2=\boldsymbol{e}^e:\boldsymbol{e}^e$, inequality (10.18) takes the form:

$$\sum_{k=A,B}\left(\left(\frac{\mu_k-F_k}{V_k}+\sigma_m\right)\frac{V_k\dot{C}_k}{\Omega}-\frac{\boldsymbol{J}_k\cdot\overset{0}{\nabla}\mu_k}{\Omega}\right)+(\mu_H-F_H)\frac{\dot{H}}{\Omega}-\frac{\boldsymbol{J}_H\cdot\overset{0}{\nabla}\mu_H}{\Omega}$$
$$-\mu_H\frac{P_H}{\Omega}+\left(\boldsymbol{s}-\frac{2F_v\boldsymbol{e}^v}{\Omega}\right):\dot{\boldsymbol{e}}^v+\left(\boldsymbol{s}-\frac{2F^e_I\boldsymbol{e}^e}{\Omega}\right):\dot{\boldsymbol{e}}^e+\left(\sigma_m-\frac{F^e_V}{\Omega}\right)\dot{\varepsilon}^e_V\geq 0. \qquad (10.19)$$

To write the constitutive equations in the actual configuration, the relations between the state variables, specified per unit volume in the actual and reference configurations, are used: the molar concentrations $c_k=C_k/\Omega$, $k=A,B$; the true volumetric penetration rates $i_k=V_k\dot{C}_k/\Omega=V_k\dot{c}_k+c_kV_k\nabla\cdot\boldsymbol{v}$, $k=A,B$ [4]; the diffusion fluxes $\boldsymbol{J}_k\cdot\overset{0}{\nabla}\mu_k=\Omega\boldsymbol{j}_k\cdot\nabla\mu_k$, $k=A,B$; the true rate of microstructure change $i_H=\dot{H}/\Omega=\dot{h}+h\nabla\cdot\boldsymbol{v}$; the microstructure flux $\boldsymbol{J}_H\cdot\overset{0}{\nabla}\mu_H=\Omega\boldsymbol{j}_H\cdot\nabla\mu_H$; the microstructure production $p_H=P_H/\Omega$. Finally, thermodynamic inequality (10.19) takes the form:

$$\sum_{k=A,B}\left(\left(\frac{\mu_k-F_k}{V_k}+\sigma_m\right)i_k-\boldsymbol{j}_k\cdot\nabla\mu_k\right)+(\mu_H-F_H)i_H-\boldsymbol{j}_H\cdot\nabla\mu_H$$
$$-\mu_H p_H+\left(\boldsymbol{s}-\frac{2F_v\boldsymbol{e}^v}{\Omega}\right):\dot{\boldsymbol{e}}^v+\left(\boldsymbol{s}-\frac{2F^e_I\boldsymbol{e}^e}{\Omega}\right):\dot{\boldsymbol{e}}^e+\left(\sigma_m-\frac{F^e_V}{\Omega}\right)\dot{\varepsilon}^e_V\geq 0. \qquad (10.20)$$

The equality sign in (10.20) corresponds to the course of reversible elastic deformation processes. Microstructure changes, viscous deformations, and diffusive mass transfer are dissipative processes and they correspond to a strict inequality in (10.20).

10.6 Constitutive Equations

Thermodynamic inequality (10.20) is split into reversible and irreversible parts

$$\sum_{k=A,B}\left(\left(\frac{\mu_k - F_k}{V_k}+\sigma_m\right)i_k\right)+(\mu_H-F_H)i_H+\left(s-\frac{2F_I^e\boldsymbol{e}^e}{\Omega}\right):\dot{\boldsymbol{e}}^e$$
$$+\left(\sigma_m-\frac{F_V^e}{\Omega}\right)\dot{\varepsilon}_V^e=0,$$
$$-\sum_{k=A,B}\boldsymbol{j}_k\cdot\nabla\mu_k-\boldsymbol{j}_H\cdot\nabla\mu_H-\mu_H p_H+\left(s-\frac{2F_v\boldsymbol{e}^v}{\Omega}\right):\dot{\boldsymbol{e}}^v\geq 0. \qquad (10.21)$$

In [4] the terms with $i_k,\ k=A,B$ are assumed dissipative; here $i_k,\ k=A,B$ and i_H are assumed as conservative because of their weak effect on the relaxation times presented in Sect. 10.8. Since the velocities $i_k,\ k=A,B,\ i_H,\ \dot{\boldsymbol{e}}^e$ and $\dot{\varepsilon}_V^e$ are independent, the solution of the first equation of system (10.21) is given by the relations

$$\mu_k=F_k-\sigma_m V_k,\ k=A,B,\quad \mu_H=F_H,\quad s=\frac{2F_I^e\boldsymbol{e}^e}{\Omega},\quad \sigma_m=\frac{F_V^e}{\Omega}. \qquad (10.22)$$

Applying the molecular incompressibility condition leads to appearing the term σ_m in expressions for chemical potentials μ_k [4]. The last two equations in (10.22), given F_V^e, F_I^e from (10.16) and the smallness of deformations $\Omega\approx 1$, are reduced to Hooke's law

$$s=2G\boldsymbol{e}^e,\ \sigma_m=K\varepsilon_V^e. \qquad (10.23)$$

Particular solution of inequality (10.21) without cross terms is written in the form of quasi-linear relations

$$\boldsymbol{j}_k=-c_k M_k\nabla\mu_k,\ \boldsymbol{j}_H=-hM_H\nabla\mu_H,\ p_H=-\beta_H\mu_H,\ s=\frac{2F_v\boldsymbol{e}^v}{\Omega}+2\eta\dot{\boldsymbol{e}}^v,$$
$$M_k>0,\ M_H>0,\ \beta_H>0,\ \eta>0,\ k=A,B \qquad (10.24)$$

Using expressions for potentials μ_k, μ_H from (10.22), components of the Helmholtz free energy (10.16), and the definition of variables in the current configuration $C_A=\Omega c_A,\ C_B=\Omega c_B,\ H=\Omega h$ with $\Omega\approx 1$, the following expressions are obtained

$$\begin{aligned} j_k &= -c_k M_k \left(f_k \nabla c_k + f_{kH} \nabla h - V_k \nabla \sigma_m\right), \; k = A, B, \\ j_H &= -h M_H \left(f_H \nabla h + f_{AH} \nabla c_A + f_{BH} \nabla c_B + f_{Hv} \nabla I_2^v\right), \\ p_H &= -\beta_H \left(f_H \Delta h + f_{AH} \Delta c_A + f_{BH} \Delta c_B + f_{Hv} \Delta I_2^v + F_H^0\right), \\ s &= 2\eta \dot{e}^v + 2\left(f_v \Delta I_2^v + f_{Hv} \Delta H + G_v^0\right) e^v \end{aligned} \tag{10.25}$$

To describe the effect of abnormally fast diffusion of components under intensive plastic deformations, in equations for diffusion fluxes (10.25) the gradients of microstructural variable and average stresses are provided, which are capable of both accelerating and decelerating diffusion process. The nonzero flux of microstructural variable j_H due to the connectivity of the equations also affect the diffusion fluxes. Constitutive relations (10.23), (10.25), approximations (10.1), (10.7), and balance equations (10.8) — (10.10) are the system of coupled equations of diffusion and deformation spatial problem with microstructural changes in an isotropic media.

10.7 Model Problem and its Equations

The influence of rheological processes and microstructure evolution on diffusion under the formulated coupled set of equations will be investigated using the simplest one-dimensional problem, which is expressed by the following assumptions:

- Components are able to diffuse only along one coordinate

$$c_A = c_A(x,t), \; c_B = c_B(x,t) \tag{10.26}$$

- All components of the total strain tensor are equal to zero except for

$$\varepsilon_{xx} = \varepsilon(x,t) \neq 0, \; \varepsilon_{yz} \equiv \gamma_0 \neq 0 \tag{10.27}$$

- All components of the stress tensor are zero except for

$$\sigma_{yy} = \sigma_{zz} = \sigma(x,t) \neq 0, \; \sigma_{yz} = \tau(x,t) \neq 0 \tag{10.28}$$

- Microstructural changes occur only along the considered coordinate

$$h = h(x,t) \tag{10.29}$$

The examined type of stress and strain fields satisfies by default the equilibrium equation and the deformation compatibility conditions. Applying hypotheses (10.26) — (10.29) to model equations (10.1), (10.7), (10.23) and (10.25) — (10.10) gives a nonlinear system of one-dimensional partial differential equations having a homogeneous stationary solution

$$\sigma_m(x,t) \equiv \sigma_m^0, \; \tau(x,t) \equiv \tau_0, \; h(x,t) \equiv h_0, \; c_A(x,t) \equiv c_A^0, \; c_B(x,t) \equiv c_B^0. \tag{10.30}$$

In the vicinity of solution (10.30) the behavior of small perturbations is investigated in Sect. 10.8.

To obtain the set of equations, the certain treatment is conducted. First, in Eqs. (10.24) the hypothesis $\eta = 0$ is accepted, which is justified by the fact that the shear viscosity does not affect the diffusion processes in the perturbed state [13].

Then the expressions for ε_V and σ_m are deduced. For that it is derived $\boldsymbol{e}^e = \boldsymbol{s}/(2G)$, $\boldsymbol{e}^v = \boldsymbol{s}/(2F_v)$, therefore the total deviatoric strain is defined by $\boldsymbol{e} = \boldsymbol{e}^e + \boldsymbol{e}^v = (1/G + 1/F_v)\boldsymbol{s}/2 = \boldsymbol{s}/(2G_{ev})$. The components of deviatoric tensors are determined $s_{xx} = -2\sigma/3 = -\sigma_m$, $e_{xx} = 2\varepsilon_V/3$, which, given the constitutive relation $s_{xx} = 2G_{ev}e_{xx}$, provide $\sigma_m = -4G_{ev}\varepsilon_V/3$ (other diagonal components give the same expression). Because of $\varepsilon_V = \varepsilon_{xx}$ and $\varepsilon_{xx} \ll 1$, G_{ev} can be considered as constant in linear approximation of σ_m, thus the relation $\sigma_m = -4G_0\varepsilon_V/3$ with $1/G_0 = 1/G + 1/G_v^0$ is valid. Eventually, the system of three equations is solved: $\sigma_m = -4G_0\varepsilon_V/3, \sigma_m = K\varepsilon_V^e$, $\varepsilon_V^e = (1+\varepsilon_V)(1 - c_AV_A - c_BV_B) \approx \varepsilon_V^{e0}(1+\varepsilon_V) - V_A\Delta c_A - V_B\Delta c_B$ at $1 - V_Ac_A^0 - V_Bc_B^0 = \varepsilon_V^{e0}/(1+\varepsilon_V^{e0}) \approx \varepsilon_V^{e0}$, $\varepsilon_V^{e0} \ll 1$. The solution is the following:

$$\varepsilon_V = \frac{V_A\Delta c_A + V_B\Delta c_B - \varepsilon_V^{e0}}{k_G + \varepsilon_V^{e0}}, \; \sigma_m = -\frac{4G_0}{3}\varepsilon_V, \; k_G = \frac{4G_0}{3K}, \; G_0 = \frac{GG_v^0}{G + G_v^0}. \tag{10.31}$$

The above variables and coefficients with the upper index 0 correspond to the perturbed state and they coincide with the designations in Sect. 10.4.

A relation for perturbation of the second invariant of viscous strains is also required. For that purpose, I_2^v is expanded in series up to linear terms by the components of viscous deviatoric strains $e_{xx}^v = \varepsilon_v$, $e_{yy}^v = \varepsilon_{v\perp}$, $e_{yz}^p = \gamma_v/2$ (the rest equal to zero) that gives $\Delta I_2^v = 2\varepsilon_v^0\Delta\varepsilon_v + 2\varepsilon_{v\perp}^0\Delta\varepsilon_{v\perp} + \gamma_v^0\Delta\gamma_v$. Further, the relation $\gamma_v^0 \gg \varepsilon_v^0, \varepsilon_{v\perp}^0$ is taken, which provides $\Delta I_2^v = \gamma_v^0\Delta\gamma_v$. The expression for $\Delta\gamma_v$ follows from $\gamma_v = \gamma_0 - \gamma_e = \gamma_0 - \tau/G$, where γ_0 is the shear strain in the perturbed state, and it gives $\Delta\gamma_v = -\Delta\tau/G$ with $\Delta\tau = \Delta G_{ev}\gamma_0 = \tau_0\Delta G/G_0 = \tau_0 G/(2G_v^0(G + G_v^0))\,(f_{Hv}\Delta h + f_v\Delta I_2^v)$. Solving the equation $\Delta I_2^v = \gamma_v^0\Delta\gamma_v(\Delta I_2^v)$ relative to ΔI_2^v with $\Delta\gamma_v = -\Delta\tau/G$ and $\gamma_v^0 = \tau_0/F_v^0 = \gamma_0 - \tau_0/G$ it is finally received

$$\Delta I_2^v = -\frac{f_{Hv}}{f_v + \frac{2(G_v^0 + G)}{\gamma_0^2}\left(1 + \frac{G_v^0}{G}\right)^2}\Delta h, \; \tau_0 = G_0\gamma_0. \tag{10.32}$$

From (10.25) expressions for the diffusion fluxes and microstructure flux are defined. They are written in a linearized form

$$\begin{aligned} j_k &= -D_k\left(\frac{\partial c_k}{\partial x} + \frac{f_{kH}}{f_k}\frac{\partial h}{\partial x} - \frac{V_k}{f_k}\frac{\partial \sigma_m}{\partial x}\right), \; k = A, B, \\ j_H &= -D_H\left(\frac{\partial h}{\partial x} + \frac{f_{AH}}{f_H}\frac{\partial c_A}{\partial x} + \frac{f_{BH}}{f_H}\frac{\partial c_B}{\partial x} + \frac{f_{Hv}}{f_H}\frac{\partial I_2^v}{\partial x}\right), \end{aligned} \tag{10.33}$$

where $D_k \equiv c_k^0 M_k f_k$ is the diffusion coefficients of components k, $D_H \equiv h_0 M_H f_H$ is the coefficient determining microstructure mobility.

Linearized balance equations of components (10.8) and microstructure (10.9)

$$\frac{\partial c_k}{\partial t}+c_k^0\frac{\partial \varepsilon_V}{\partial t}=-\frac{\partial j_k}{\partial x},\ \frac{\partial h}{\partial t}+h_0\frac{\partial \varepsilon_V}{\partial t}=-\frac{\partial j_H}{\partial x}+p_H,\ k=A,B,$$
$$p_H=-\beta_H\left(f_H\Delta h+f_{AH}\Delta c_A+f_{BH}\Delta c_B+f_{Hv}\Delta I_2^v+F_H^0\right) \tag{10.34}$$

along with (10.31) — (10.33) are the system of linear coupled differential equations of the model problem.

10.8 Diffusion Coefficients in Coupled System

For a set of equations (10.31) — (10.34) linearized in the vicinity of a homogeneous stationary solution (10.30) the perturbed solutions are considered

$$c_k(x,t)=c_k^0+\hat{c}_k\exp\left(-\frac{t}{\tau}\right)\sin\left(\frac{2\pi x}{\lambda}\right),\ k=A,B$$
$$h(x,t)=h_0+\hat{h}\exp\left(-\frac{t}{\tau}\right)\sin\left(\frac{2\pi x}{\lambda}\right) \tag{10.35}$$

where $\hat{c}_A,\ \hat{c}_B,\ \hat{h}\in R, |\hat{c}_A|,\ |\hat{c}_B|,\ |\hat{h}|\ll 1, \tau$ is the relaxation time, λ is the perturbation wavelength.

Solution (10.35) must satisfy the linearized system of equations, which is equivalent to solving an eigenvalue problem with a third-order eigenvalue equation for τ. Positive solutions of the equation $\tau=\tau_k(\lambda),\ k=1,2,3$ and the corresponding eigenvectors $\hat{\boldsymbol{u}}_k=\left(\hat{c}_{Ak},\hat{c}_{Bk},\hat{h}_k\right),\ k=1,2,3$ are determined. If for a particular branch $\tau=\tau_{k0}(\lambda)$ the eigenvector has the form $\hat{\boldsymbol{u}}_{k0}=\left(0,0,\hat{h}_{k0}\right)$, then relaxation processes are controlled by microstructure changes. Otherwise, at $\hat{\boldsymbol{u}}_{k0}=(\hat{c}_{Ak0},\hat{c}_{Bk0},0)$ the relaxation is determined by diffusive mass transfer and stresses, inextricably associated with changes in c_A and c_B (10.31). Finally, if $\hat{\boldsymbol{u}}_{k0}=\left(\hat{c}_{Ak0},\hat{c}_{Bk0},\hat{h}_{k0}\right)$ is carried out, all the processes in question affect the relaxation. This method was used earlier in [2, 4].

In Fig. 10.1 dependences $\tau=\tau_k(\lambda),\ k=1,2,3$ are shown. The limiting cases $\lambda\to 0$ and $\lambda\to\infty$ correspond to viscous or diffusive relaxation processes:

- The viscous (horizontal) asymptotic is determined by the constant relaxation time $\tau\equiv\tau_v$. In this limit, the processes are described by an equation of the type

$$\frac{\partial \boldsymbol{u}}{\partial t}=-\frac{\boldsymbol{u}}{\tau_v},$$

  with $\boldsymbol{u}=\hat{\boldsymbol{u}}\exp\left(-t/\tau_v\right)\sin\left(2\pi x/\lambda\right)$

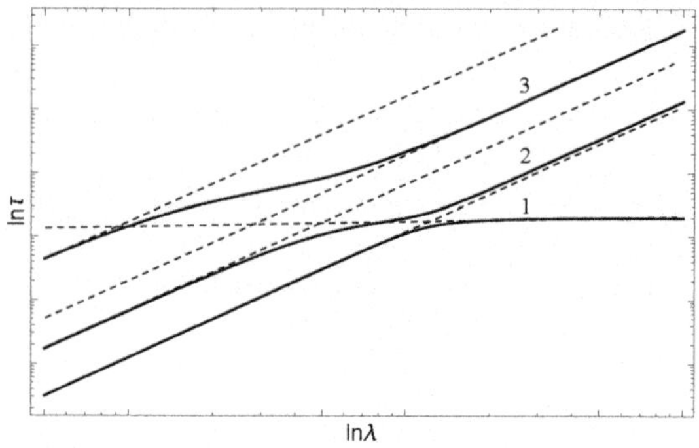

Fig. 10.1 Branches of the relaxation times and their asymptotics

- The diffusion (oblique) asymptotics are determined by the interdiffusion coefficient D at $\tau = \lambda^2/4\pi^2 D$. In this limit, the relaxation process is described by an equation of the type
$$\frac{\partial \boldsymbol{u}}{\partial t} = D\frac{\partial^2 \boldsymbol{u}}{\partial x^2}$$
with $\boldsymbol{u} = \hat{\boldsymbol{u}} \exp\left(-4\pi^2 t/\lambda^2\right) \sin\left(2\pi x/\lambda\right)$. The equation written for concentrations is called Fick's second law.

In the limit at $\lambda \to 0$ a cubic equation for D_0 is obtained. The exact solution of it is cumbersome and difficult to analyze, and for this reason an approximate solution is used. According to Vieta's theorem, solutions of the cubic equation $aD_0^3 + bD_0^2 + cD_0 + d = 0$ satisfy the system

$$D_1^0 + D_2^0 + D_3^0 = -\frac{b}{a},\ D_1^0 D_2^0 + D_1^0 D_3^0 + D_2^0 D_3^0 = \frac{c}{a},\ D_1^0 D_2^0 D_3^0 = -\frac{d}{a} \tag{10.36}$$

Suppose that $D_1^0 \gg D_2^0 \gg D_3^0$, then the system (10.36) has the solution

$$D_1^0 \approx -\frac{b}{a},\ D_2^0 \approx -\frac{c}{b},\ D_3^0 \approx -\frac{d}{c} \tag{10.37}$$

Approximate solution (10.37) has an acceptable accuracy (not exceeding 11% for any coefficient) already with the ratio $D_1^0 = 10D_2^0 = 100D_3^0$. For D_k^0, $k = 1,2,3$ the eigenvectors have the form $\hat{\boldsymbol{u}}_k^0 = \left(\hat{c}_{Ak}^0, \hat{c}_{Bk}^0, \hat{h}_k^0\right)$.

For the maximum of three diffusion coefficients, there is the following expression

$$\begin{aligned}
D_1^0 &= e_A D_A + e_B D_B + e_H D_H,\\
e_k &= \frac{1}{k_G}\left(1 + k_G - \phi_k^0 - h_0 V_k \frac{f_{kH}}{f_k} + K k_G \frac{V_k^2}{f_k}\right),\ k = A, B,\\
e_H &= \frac{1+k_G}{k_G}\left(1 - \frac{f_{Hv}^2}{f_H\left(f_v + \frac{2(G+G_v^0)}{\gamma_0^2}\left(1+\frac{G_v^0}{G}\right)\right)}\right),
\end{aligned} \tag{10.38}$$

where the coefficients e_k, $k = A, B$, e_H are dimensionless, and $\phi_k^0 = c_k^0 V_k$, $k = A, B$ are the volume fractions of components k at the perturbed state. The coefficient D_1^0 in (10.38) meets a structural scheme with a sequential connection of three diffusion elements and the diffusion rate is limited by the fastest component. For example, assume that $D_A \gg D_B \gg D_H$, then $D_1^0 = e_A D_A$ can be derived. In the absence of stress gradients and microstructure, the Darken interdiffusion coefficient $D_1^0 = D_A + D_B$, explained the Kirkendall experiment, is occured [18, 19]. The value of D_2^0 is defined as

$$
\begin{aligned}
D_2^0 &= \frac{e_{AH} D_A D_H + e_{BH} D_B D_H + e_{AB} D_A D_B}{e_A D_A + e_B D_B + e_H D_H}, \\
e_{AB} &= 1 + \frac{\varepsilon_V^{e0}}{k_G} + K\left(\frac{V_A^2}{f_A} + \frac{V_B^2}{f_B}\right) - \frac{h_0}{k_G}\left(\frac{f_{AH} V_A}{f_A} + \frac{f_{BH} V_B}{f_B}\right), \\
e_{kH} &= (c_0 - c_k^0) V_k \frac{f_{AH} f_{BH}}{k_G f_k f_H} + \frac{e_H}{1 + k_G}\left(1 + k_G - \phi_k^0 + K k_G \frac{V_k^2}{f_k}\right) \\
&\quad - \frac{f_{kH}^2}{f_k f_H k_G}(1 + k_G - \phi_k^0), \; k = A, B.
\end{aligned}
\tag{10.39}
$$

Similarly, suppose $D_A \gg D_B \gg D_H$ in (10.39) that gives $D_2^0 = e_{AB} D_B$. Thus, the diffusion rate is determined by the slow flux of component B. In the case of concentration diffusion, there is the Nazarov – Gurov interdiffusion coefficient $D_2^0 = D_A D_B / (D_A + D_B)$ [20]. The third diffusion coefficient has the following structure

$$
\begin{aligned}
D_3^0 &= \frac{e_{AB}^H D_A D_B D_H}{e_{AH} D_A D_H + e_{BH} D_B D_H + e_{AB} D_A D_B}, \\
e_{AB}^H &= \frac{k_G e_H}{1 + k_G}\left(1 + \frac{c_V^{e0}}{k_G} + K\left(\frac{V_A^2}{f_A} + \frac{V_B^2}{f_B}\right)\right) - \frac{K (f_{AH} V_B - f_{BH} V_A)^2}{f_A f_B f_H} \\
&\quad - \frac{k_G + \varepsilon_V^{e0}}{k_G f_H}\left(\frac{f_{AH}^2}{f_A} + \frac{f_{BH}^2}{f_B}\right).
\end{aligned}
\tag{10.40}
$$

The coefficient D_3^0 in (10.40) meets a structural scheme with a parallel connection of diffusion elements and for the case $D_A \gg D_B \gg D_H$ the relation $D_3^0 = e_{AB}^H D_H / e^{AB}$ is fair. Thereby, the relaxation time is determined by the microstructure flux. The coefficients in (10.38) — (10.40) correspond to small relaxation times because of $\lim_{\lambda \to 0} \tau = \lim_{\lambda \to 0} \lambda^2 / 4\pi^2 D = 0$, and they are able to describe the fast interdiffusion processes caused by the microstructural features and (or) the dynamics of microstructure development.

At $\lambda \to \infty$, the quadratic equation $aD^2 + bD + c = 0$ and the linear equation for τ_3^∞ are obtained. The approximate solution of the quadratic equation is constructed similar to (10.37) under the assumption $D_1^\infty \gg D_2^\infty$. On diffusion asymptotes the eigenvectors have the form $\hat{\boldsymbol{u}}_k^\infty = \left(\hat{c}_{Ak}^\infty, \hat{c}_{Bk}^\infty, \hat{h}_k^\infty\right)$, $k = 1, 2$. The first coefficient D_1^∞ is defined as follows

$$D_1^\infty = e_A^\infty D_A + e_B^\infty D_B,$$
$$e_k^\infty = (c_0 - c_k^0)\frac{f_{AH} f_{BH}}{k_G f_H e_H f_k} + \frac{1}{1+k_G}\left(1 + k_G - \phi_k^0 + k_G K \frac{V_k^2}{f_k}\right) - \frac{1+k_G-\phi_k^0}{k_G f_H e_H}\frac{f_{kH}^2}{f_k},\ k = A, B \tag{10.41}$$

and it meets a structural scheme with a sequential connection of two diffusion elements. In the absence of stresses, microstructure, and shear deformations, the coefficient D_1^∞ in (10.41) determines the Darken interdiffusion. The second coefficient D_2^∞ corresponds to a structural scheme with a parallel connection of diffusion elements

$$D_2^\infty = \frac{e_{AB}^\infty D_A D_B}{e_A^\infty D_A + e_B^\infty D_B},$$
$$e_{AB}^\infty = \frac{K V_A^2 V_B^2}{f_A f_B f_H e_H}\left(\frac{f_{BH}}{V_B} - \frac{f_{AH}}{V_A}\right)^2 + \frac{k_G + \varepsilon_V^{e0}}{k_G f_H e_H}\sum_{k=A,B}\frac{f_{kH}^2}{f_k} - \frac{1}{1+k_G}\left(k_G + \varepsilon_V^{e0} + k_G K \textstyle\sum_{k=A,B}\frac{V_k^2}{f_k}\right) \tag{10.42}$$

and it is reduced to the Nazarov – Gurov interdiffusion coefficient at concentration diffusion. The coefficients in (10.41) — (10.42) can describe rather slow diffusion processes due to their conformity to large relaxation times $\lim_{\lambda\to\infty}\tau = \lim_{\lambda\to\infty}\lambda^2/4\pi^2 D \to \infty$. The viscous relaxation time for the third branch τ_v^∞ has the form

$$\tau_3^\infty = \frac{1+k_G}{k_G \beta_H f_H e_H}. \tag{10.43}$$

The asymptote meets the eigenvector $\hat{\boldsymbol{u}}_3^\infty = \left(0, 0, \hat{h}_3^\infty\right)$, therefore it characterizes the microstructure relaxation in the absence of its mobility and diffusion of components.

In Fig. 10.2 the influence of thermodynamic f_{AH}, f_{BH}, e_H and kinetic D_H coefficients on the interdiffusion coefficients, determined for the asymptotic cases $\lambda \to 0$ and $\lambda \to \infty$, is shown for a coupled system. For this purpose, the dimensionless functions $\tilde{D}_k^{0,\infty}(e_H) = D_k^{0,\infty}(e_H)/D_k^{0,\infty}(1)$, $\tilde{D}_k^0(D_H) = D_k^0(D_H)/D_k^0(0)$, and $\tilde{D}_k^{0,\infty}(f_{kH}) = D_k^{0,\infty}(f_{kH})/D_k^{0,\infty}(0)$, $k = A, B$ are considered. They are constructed in the coordinates of their argument, when the remaining coefficients being parameters of function are fixed. Depending on the parameter values, the qualitative graphs, shown by solid or dotted line, are different. The accepted parameters and range of arguments provide a small change in the value of the functions, which allows for placing dependences for different diffusion coefficients on the same figure. Using other parameter values and ranges of argument can dramatically increase the values of functions (by several orders of magnitude), which indicates the coefficients f_{AH}, f_{BH}, e_H and D_H significantly affect the interdiffusion coefficients during the course of coupled physical and mechanical processes. It was found that:

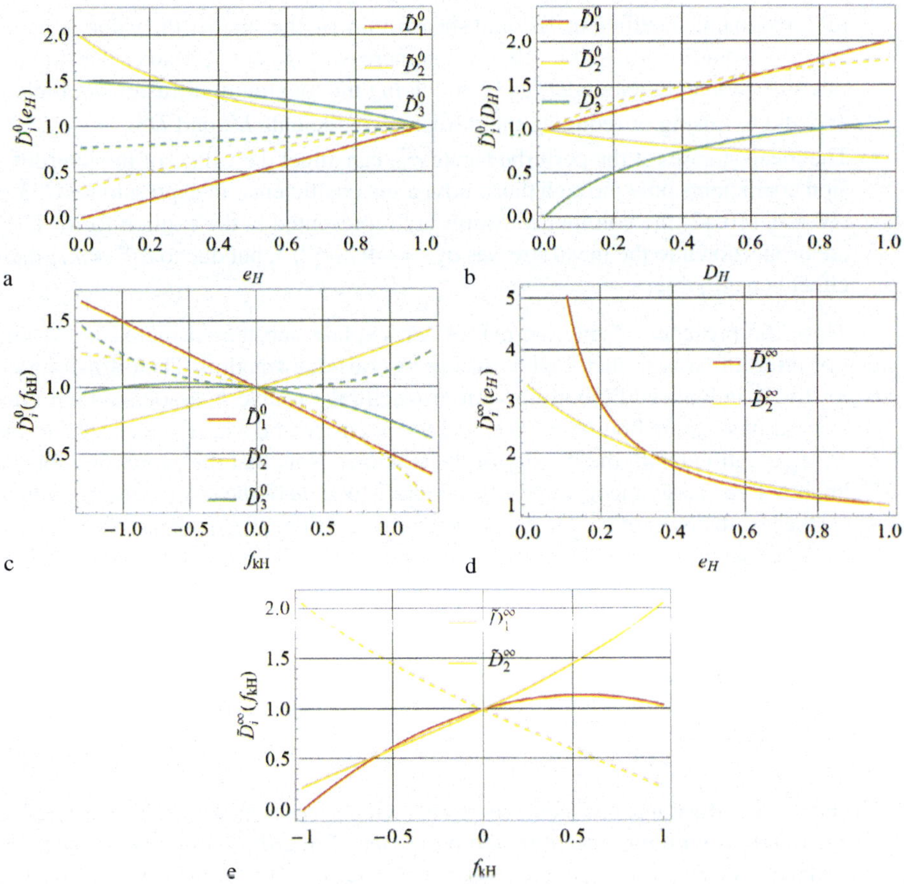

Fig. 10.2: Qualitative dependences of the dimensionless diffusion coefficients.

- The migration of grain boundaries can significantly accelerate the interdiffusion at $\lambda \ll 1$, which may correspond to a fine-grained structure. According to Fig. 10.2b with increasing the microstructure mobility coefficient D_H, the interdiffusion coefficients at small perturbation wavelengths D_1^0, D_3^0 increase, and D_2^0 can both increase and decrease. At large perturbation wavelengths, the diffusion coefficients D_1^∞ and D_2^∞ do not depend on D_H, which is obvious from (10.41) and (10.42).
- For any perturbation wavelengths, the rate of interdiffusion can both significantly increase and decrease depending on the connectivity of the diffusion fluxes and microstructure flux, expressed in coefficients f_{AH}, f_{BH}. In Fig. 10.2c,e the qualitative dependence of the interdiffusion coefficients on the coefficients f_{AH}, f_{BH} is shown.
- The coupling of viscous deformations and microstructure accelerates interdiffusion at large scales $\lambda \gg 1$ and it can both decelerate and accelerate diffusion at small scales $\lambda \ll 1$. According to equation $(10.38)_3$, the greater the value of

thermodynamic coefficient f_{Hv} and shear strain γ_0, the smaller the value of e_H is. Thereby from Fig. 10.2a,d it follows that increasing f_{Hv} or γ_0 leads to raising the interdiffusion coefficients D_1^∞, D_2^∞, reducing the maximum diffusion coefficient D_1^0, and to raising or reducing the diffusion coefficients D_2^0 and D_3^0.
- The mean stresses at the perturbed state σ_m^0 can affect the values of the interdiffusion coefficients only through thermodynamic coefficients. In expressions (10.39), (10.40), (10.42) the volumetric strains are represented in the explicit form. They are proportional to the mean stresses $\sigma_m^0 = -4G_0\varepsilon_V^{e0}/3$, but due to $\varepsilon_V^{e0} \ll k_G$ their effect is negligible.

Thus, the presence of microstructure can explain substantial increasing in the rate of interdiffusion in metal alloy during viscous deformation. The diffusion rate is positively influenced by both the microstructural variable h, associated with the concentration of grain boundary area, and the intensity of deviatoric inelastic strains I_2^v, being identified with the density of dislocations. Note that the expression for the Helmholtz free energy incorporates the quadratic degree of the last parameter, which meets the fourth power of $\boldsymbol{e}^v$. The thermodynamic coefficient f_{Hv}, shear strain γ_0, and kinetic coefficient D_H significantly contribute to the effective diffusion coefficient (10.38). It also emphasizes the connectivity (generally speaking, not obvious) of diffusion properties and microstructure.

10.9 Conclusion

The effective diffusion and viscous linear coefficients has been studied for the coupled set of equations with incorporating the processes of interdiffusion in a binary alloy, viscoelastic deformations, and microstructure changes. The technique developed in works [2, 3, 4] was used, but to study abnormally fast processes of chemical composition changing in metal alloys under intensive plastic deformations, the presence of microstructure was consistently regarded in the model. The microstructure is described by a scalar internal variable being the concentration of grain boundaries, and by the intensity of deviatoric inelastic strains associated with dislocations of the crystal lattice. Both variables are added to the Helmholtz free energy and the latter is to the second power, which turned out to be necessary. The model problem in question [2, 4], for which effective diffusion and viscous linear coefficients are determined, is also generalized to come close to the loading scheme in Bridgman anvils [1]. To be more precise, the scheme corresponds to the free axial deformation of the rod, constrained on the sides, with torsion. Although it would be more correct to consider a torsion compression scheme, however, to compare with the already existing results of simpler diffusion models in a deformable body, that scheme was adopted. In the future, a torsion compression scheme should be considered. Finally, the model, preferred by specialists in atomic diffusion and based on the experimental method for determining the Kirkendall – Darken interdiffusion coefficients, was adopted as the equations of interdiffusion in a medium experiencing elastic and viscous deformations (see review [5]).

The considered set of coupled equations, linearized in the vicinity of homogeneous stationary solution, was reduced to an eigenvalue problem. For the limiting cases of the perturbation wavelengths, the solution is characterized by a spectrum of effective diffusion coefficients and viscosity. Each of the effective diffusion coefficients depends on the thermodynamic and kinetic coefficients and they can drastically differ from the concentration diffusion coefficients of alloy components. The effective interdiffusion coefficients in the simplest cases meet structural schemes with sequential or parallel connection of diffusion elements.

It was defined that the values of diffusion fluxes are very sensitive to the coefficients associated with microstructural contributions to free energy and the migration of microstructure. Within linear analysis and without estimating the limits of variation of these coefficients, it is difficult to estimate how much the rate of effective diffusion of alloy components may differ from the concentration diffusion coefficients. Qualitative results suggest that regarding effective (and not concentration) diffusion coefficients in a coupled system is necessery to assess the speed of processes and explain their abnormally fast speed.

The analysis of solutions of the eigenvalue problem allows (in limiting cases) for studying not just the behavior of eigenvalues (effective kinetic coefficients) from the parameters (primarily microstructural) of the model. To drastically contribute to the diffusion fluxes, the structure of the eigenvectors corresponding to the "fast" diffusion coefficients obliges the concentration gradients of microstructure and elastic stresses have large values.

Some non-obvious results connecting the diffusion rate with the microstructure in an inelastic deformable crystalline body can be used as some point of view along with the physics of metals.

References

[1] Straumal BB, Baretzky B, Mazilkin AA, Phillipp F, Kogtenkova OA, Volkov MN, Valiev RZ (2004) Formation of nanograined structure and decomposition of supersaturated solid solution during high pressure torsion of Al-Zn and Al-Mg alloys, Acta Mater. **52**(15):4469-4478. DOI 10.1016/j.actamat.2004.06.006

[2] Stephenson GB (1988) Deformation during interdiffusion, Acta Metallurg. **36**(10):2663-2683. DOI 10.1016/0001-6160(88)90114-9

[3] Suo Z (2004) A continuum theory that couples creep and self-diffusion, J. Appl. Mech. **71**(5):646-651. DOI 10.1115/1.1781176

[4] Brassart L, Liu Q, Suo Z (2018) Mixing by shear, dilation, swap, and diffusion, J. Mech. Phys. Solids **112**:253-272. DOI 10.1016/j.jmps.2017.12.008

[5] Dudin DS, Keller IE (2022) Review of approaches to formulating coupled equations of interdiffusion in viscoelastic media, Chem. Phys. Mesoscopy **24**(3):296-311. DOI 10.15350/17270529.2022.3.24

[6] Knyazeva AG (2020) Application of irreversible thermodynamics to diffusion in solids with internal surfaces. J. Non-Equil. Thermodyn. **45**(4):401-417. DOI 10.1515/jnet-2020-0021
[7] Anand L (2011) A thermo-mechanically-coupled theory accounting for hydrogen diffusion and large elastic-viscoplastic deformations of metals, J. Solids Struct. **48**(6):962-971. DOI 10.1016/j.ijsolstr.2010.11.029
[8] Aouadi M, El Dhaba AR, Ghaleb AF (2018) Stability aspects in strain gradient theory of thermoelasticity with mass diffusion, J. Appl. Math. Mech. **98**:1794–1812, DOI 10.1002/zamm.201800043
[9] Frolova KP, Vilchevskaya EN, Polyanskiy VA, Yakovlev YA (2021) Modeling the skin effect associated with hydrogen accumulation by means of the micropolar continuum, Continuum Mech. Thermodyn. **33**:697–711. DOI 10.1007/s00161-020-00948-3
[10] Fischer FD, Svoboda J (2015) Stress, deformation and diffusion interactions in solids – a simulation study, J. Mech. Phys. Solids **78**:427-442. DOI 10.1016/j.jmps.2015.02.018
[11] Dantzig JA, Boettinger WJ, Warren JA, McFadden GB, Coriell SR, Sekerka RF (2006) Numerical modeling of diffusion-induced deformation, Metallurg. Mat. Trans. **37**:2701–2714. DOI 10.1007/BF02586104
[12] Dudin DS, Keller IE (2021) On the spectrum of relaxation times in coupled diffusion and rheological processes in metal alloys, In: dell'Isola F, Igumnov L (Eds) Dynamics, Strength of Materials and Durability in Multiscale Mechanics, Advanced Structured Materials, vol. 137, pp 41-55, Springer, Cham. DOI 10.1007/978-3-030-53755-5_3
[13] Dudin DS, Keller IE (2021) On description of fast diffusion in a coupled multicomponent system with microstructure within the framework of the thermodynamics of irreversible processes, In: Altenbach H, Eremeyev VA, Igumnov LA (Eds) Multiscale Solid Mechanics, Advanced Structured Materials, vol. 141, pp 81-95, Springer, Cham. DOI 10.1007/978-3-030-54928-2_8
[14] Brassart L, Suo Z (2013) Reactive flow in solids, J. Mech. Phys. Solids **61**(1):61-77. DOI 10.1016/j.jmps.2012.09.007
[15] Paul A, Laurina T, Vuorinen V, Divinscki SV (2014) Thermodynamics, Diffusion and the Kirkendall Effect in Solids, Springer, Heidelberg.
[16] Mehrer H (2007) Diffusion in Solids, Springer, Heidelberg.
[17] Hong W, Zhao X, Zhou J, Suo Z (2008) A theory of coupled diffusion and large deformation in polymer gels, J Mech Phys Solids **56**(5):1779-1793. DOI 10.1016/j.jmps.2007.11.010
[18] Kirkendall EO (1942) Diffusion of zinc in alpha brass, Trans Metallurg Soc AIME **147**:104-110.
[19] Darken LS (1948) Diffusion, mobility and their interrelation through free energy in binary metallic systems, Trans AIME **175**:184-201.
[20] Nazarov AV, Gurov KP (1974) The kinetic theory of interdiffusion in binary system. Concentration of vacancies during mutual diffusion, Phys Met Metallurg **37**(3):41-47.

Chapter 11
Representative Volume Element Size and Length Scale Identification in Generalised Magneto-Elasticity

Sinan Eraslan, Inna M. Gitman, Mingxiu Xu, Harm Askes, and René de Borst

Abstract A generalised magneto-elastic continuum model with gradients of strain, magnetic field, and piezo-magnetic coupling terms has been discussed in this study. Characteristic length scale parameters, accompanying the higher order components in the model, have been identified in terms of representative volume element (RVE) sizes, in order to introduce the microstructural information to material properties on the macro-level. Following this qualitative definition, a quantitative determination, and the analysis of RVE sizes in different physical phenomena, a RVE and, hence, length-scale enriched generalised gradient magneto-elastic continuum model has been employed in a benchmark numerical example demonstrating the removal of singularities from magnetic and mechanical fields.

Key words: Magnetorheological elastomers · RVE size · Length scale · Generalised magneto-elastic continuum

Sinan Eraslan
The University of Sheffield, Department of Mechanical Engineering, Sheffield, United Kingdom,
e-mail: seraslan1@sheffield.ac.uk

Inna M. Gitman · Harm Askes
University of Twente, Faculty of Engineering Technology, Enschede, The Netherlands,
e-mail: i.m.gitman@utwente.nl, h.askes@utwente.nl

Mingxiu Xu
University of Science and Technology Beijing, Department of Applied Mechanics, Beijing, P.R. China,
e-mail: xumx@ustb.edu.cn

René de Borst
The University of Sheffield, Department of Civil and Structural Engineering, Sheffield, United Kingdom,
e-mail: r.deborst@sheffield.ac.uk

H. Altenbach et al. (eds.), *Sixty Shades of Generalized Continua*,
Advanced Structured Materials 170, https://doi.org/10.1007/978-3-031-26186-2_11

11.1 Introduction

Smart composite materials have been the topic of many research investigations in the last few decades due to being responsive to external stimulus such as temperature, pH, electric or magnetic field [1, 2]. One type of these responsive materials, known as magnetorheological elastomers (MREs), consist of magnetic particles embedded in a silicone-based elastomer, and present a coupling between magnetism and elasticity via the magnetostriction effect [1]-[4]. Magnetostriction is a phenomenon in which a magnetic field leads to a change in shape in ferromagnetic materials. In MREs, particles will exhibit magnetostrictive behaviour when a magnetic field is applied. Thus, forces will be exerted to the polymer matrix and the composite material will deform [3, 5]. This coupling phenomena proposes various potential applications in many engineering fields, including actuators, sensors, vibration isolation and control, sensing of ultrasonic waves, micro beams/plates in micro-electro-mechanical systems (MEMS) and dialysis membranes in biomedical applications [1]-[3], [6]-[8]. Various magnetic materials such as Terfenol-D, cobalt ferrite, certain earth metals and iron alloys can be used as magnetic filler with several alternatives for the elastic matrix such as natural rubber, silicone rubber, vinyl rubber or polyurethane [2, 4, 9, 10].

MREs can be categorised as macroscopically isotropic or anisotropic depending on the particle distribution in the matrix. Polymer and particles are mixed mechanically with some additives, and then the mixture is cured without or with magnetic field to obtain isotropic or anisotropic MREs, respectively [2, 4, 9]. Various reviews have been conducted on the fundamentals of magnetostriction, manufacturing/modelling of MREs, and their applications. Magnetic responsive polymer composites (MRPCs) have been reviewed by Thévenot et al. [2] and Filipcsei et al. [1] by presenting different types of MRPCs, detailed fabrication methods including used products, preparation steps and chemical processes. Progress after 2000 and the current state of those materials have been evaluated by Elhajjar et al. [3] with some examples of different compositions (such as galfenol alloys, cobalt ferrites, Terfenol-D alloys and carbonyl iron). Ekreem et al. [5] explained how magnetostriction occurs and presented measurement procedures of this phenomenon. In this overview, the advantages and disadvantages of the procedures were discussed to conclude the most common and sensitive methods.

Researchers have proposed various microscopic and macroscopic material models of MREs and ferrogels since their introduction in the late 1990s. A microscopic model has been presented by Wood and Camp [10] to study the relationship between physical properties of the ferrogels and features of the microstructure using Monte Carlo computer simulations. These authors demonstrated that it is feasible to control the elastic modulus of the ferrogel by applying a uniform magnetic field to the samples at the gel-formation stage. Another microscopically motivated approach, derived from the continuum formulation of a magneto-mechanical boundary value problem, has been developed by Kalina et al. [11] to study the influence of mechanical preloads on the deformation behaviour of MREs. Attaran et al. [12] presented a macroscopic model to capture the mechanical deformation of ferrogels. They proposed this macroscopic model as a simplified form of a previously developed continuum model, and they demonstrated a good agreement between numerically predicted and experimental results. Further, Raikher and Stolbov [13] have followed a continuum approach for MREs by considering them as a homogeneous and isotropic elastic medium. Besides limitations and drawbacks, the model can present practical and experimentally acceptable results.

It is known that microscopic properties affect the macroscopic behaviour of heterogeneous materials. Therefore, introducing microstructural information in the macroscopic continuum results in a significant improvement in accurately describing and predicting the response of an MRE. Accordingly, a multi-scale approach can be used in the analysis of these materials as various scales have to be studied simultaneously [14]. So called analytical homogenisation is a multi-scale approach in which the material is described as a heterogeneous medium on the micro-level, by considering each component's configuration and constitutive properties explicitly; simultaneously, material is described as homogeneous on the macro-level, and the constants in constitutive relations of the macrostructure appear in the form of effective properties (see, among others [15, 16]). The Representative Volume Element (RVE) approach is typically employed to model the material on the micro-scale. Here, RVE is the smallest part of a material which is large enough to be constitutively valid [15]. The unit cell (representative unit cell, or, indeed an RVE) is considered as input parameter in analytical homogenisation approaches.

On the other hand, several authors have proposed nonlocal continuum theories to include the information from the micro-scale in the macroscopic continuum via additional material constants such as characteristic length scale parameters [8, 16, 17]. It was already shown in [15, 16] that there is a link between the RVE size on the micro-scale and additional characteristic length scale parameters on the macro-scale. The authors have extended the theory to magneto-elasticity, and derived macro-level magneto-elastic constitutive relations enriched with higher-order terms accompanied by characteristic length scale parameters. They also proposed a statistical method to determine the RVE sizes for different phenomena in an MRE material [18]. However, they have not used the determined RVE sizes, and hence identified length scale parameters, in the macro-level magneto-elastic constitutive relations to show the effects of the additional parameters on static magneto-elastic behavior.

The aim of this study is to present the influence of identified length scale parameters in removing mechanical and magnetic singularities by using the determined RVE sizes as model parameters in generalised magneto-elasticy, thereby bridging the gap between RVE-based approaches and generalised/nonlocal continuum approaches. In Sect. 11.2, the macro-level magneto-elastic constitutive formulation with identified length scale parameters and determined RVE sizes will be discussed. In Sect. 11.3, numerical results will be presented for a two-dimensional in-plane problem to remove singularities in mechanical and magnetic fields. Finally, some closing remarks are presented in Sect. 11.4.

11.2 Formulation

Magnetostrictive materials typically exhibit non-linear material behaviour. However, mechanical pre-stresses and bias magnetic fields are generally applied in most applications. These bias conditions allow to use linear piezomagnetic equations to describe the material behaviour. In a static magnetic field (curl-free), the constitutive equations of a linear piezomagnetic medium in classical continuum theory can be given as [8], [18]-[21]

$$\sigma_{ij} = C_{ijkl}\varepsilon_{kl} - Q_{nij}H_n, \qquad B_i = Q_{ikl}\varepsilon_{kl} + \mu_{in}H_n \tag{11.1}$$

with the kinematic relationships

$$\varepsilon_{ij} = \frac{1}{2}(u_{i,j} + u_{j,i}) \text{ and } H_i = -\varphi_{,i} \tag{11.2}$$

where $\boldsymbol{\sigma}$ and $\boldsymbol{B}$ are the stress tensor and the magnetic induction vector, $\boldsymbol{C}$ and $\boldsymbol{Q}$ are the stiffness tensor and the piezomagnetic coupling tensor, $\boldsymbol{\mu}$ is the magnetic permeability, $\boldsymbol{\varepsilon}$ and $\boldsymbol{H}$ are the strain tensor and the magnetic field vector, and $\boldsymbol{u}$ and φ are the displacement field vector and the scalar magnetic potential.

11.2.1 Homogenisation and Macroscopic Characteristic Length Scale Parameters

As discussed above, micro and macro scales are linked to each other by model parameters in a multi-scale framework. The MRE material is considered as heterogeneous to reflect the real structure on the micro-level, but it is modelled as a piezomagnetic material with homogeneous effective material properties and model parameters on the macro-level. The concept of an RVE is used to quantify the micro scale behaviour and, hence, the size of the RVE becomes a model parameter in the multi-scale approach. This approach combines the advantage of computational efficiency with an accurate description of the material behaviour on the macro-level.

In the multi-scale analysis, the macroscopic stress and induction in the MRE can be defined as the volume average of the microscopic counterparts in the RVE:

$$\begin{aligned}
\sigma_{ij}^{\mathrm{M}} &= \frac{1}{V_{\mathrm{RVE}}}\int_{V_{\mathrm{RVE}}} \sigma_{ij}^{\mathrm{m}}\mathrm{d}V = \frac{1}{V_{\mathrm{RVE}_1}}\int_{V_{\mathrm{RVE}_1}} C_{ijkl}^{\mathrm{m}}\varepsilon_{kl}^{\mathrm{m}}\mathrm{d}V - \frac{1}{V_{\mathrm{RVE}_2}}\int_{V_{\mathrm{RVE}_2}} Q_{nij}^{\mathrm{m}}H_n^{\mathrm{m}}\mathrm{d}V \\
B_i^{\mathrm{M}} &= \frac{1}{V_{\mathrm{RVE}}}\int_{V_{\mathrm{RVE}}} B_i^{\mathrm{m}}\mathrm{d}V = \frac{1}{V_{\mathrm{RVE}_3}}\int_{V_{\mathrm{RVE}_3}} Q_{ikl}^{\mathrm{m}}\varepsilon_{kl}^{\mathrm{m}}\mathrm{d}V + \frac{1}{V_{\mathrm{RVE}_4}}\int_{V_{\mathrm{RVE}_4}} \mu_{in}^{\mathrm{m}}H_n^{\mathrm{m}}\mathrm{d}V
\end{aligned} \tag{11.3}$$

It can initially be assumed that an MRE has different RVEs as given in Eq. (11.3), with V_{RVE_1} and V_{RVE_4} representing purely mechanical and purely magnetic phenomena, while V_{RVE_2} and V_{RVE_3} correspond to coupling phenomena. The superscripts m and M denote the micro and the macro level, respectively.

Next, by using a second-order homogenisation scheme and Eq. (11.3), macroscopic constitutive and governing equations can be derived. These equations include gradients in terms of the RVE sizes to introduce the micro-level information into the macro-level. The associated local coordinate systems are assumed to have their origin at the centre of the RVE. Before applying second-order homogenisation, linearisations of spatially dependent stiffness, strain, coupling, permeability and magnetic field can be presented as

$$\begin{aligned} C^{\mathrm{m}}_{ijkl} = C^{\mathrm{M}}_{ijkl} + C^{\mathrm{M}}_{ijkl,o}\delta x_o, \quad \varepsilon^{\mathrm{m}}_{kl} = \varepsilon^{\mathrm{M}}_{kl} + \varepsilon^{\mathrm{M}}_{kl,p}\delta x_p, \quad Q^{\mathrm{m}}_{nij} = Q^{\mathrm{M}}_{nij} + Q^{\mathrm{M}}_{nij,o}\delta x_o \\ H^{\mathrm{m}}_n = H^{\mathrm{M}}_n + H^{\mathrm{M}}_{n,p}\delta x_p, \quad \mu^{\mathrm{m}}_{in} = \mu^{\mathrm{M}}_{in} + \mu^{\mathrm{M}}_{in,o}\delta x_o \end{aligned} \tag{11.4}$$

The constitutive relation for the macroscopic stress and magnetic induction (Eq. (11.3)) can be rewritten as

$$\begin{aligned} \sigma^{\mathrm{M}}_{ij} =& \frac{1}{V_{\mathrm{RVE}_1}} \int_{V_{\mathrm{RVE}_1}} \left(C^{\mathrm{M}}_{ijkl}\varepsilon^{\mathrm{M}}_{kl} + C^{\mathrm{M}}_{ijkl}\varepsilon^{\mathrm{M}}_{kl,p}\delta x_p + C^{\mathrm{M}}_{ijkl,o}\varepsilon^{\mathrm{M}}_{kl}\delta x_o + C^{\mathrm{M}}_{ijkl,o}\varepsilon^{\mathrm{M}}_{kl,p}\delta x_o\delta x_p \right) \mathrm{d}V \\ &- \frac{1}{V_{\mathrm{RVE}_2}} \int_{V_{\mathrm{RVE}_2}} \left(Q^{\mathrm{M}}_{nij}H^{\mathrm{M}}_n + Q^{\mathrm{M}}_{nij}H^{\mathrm{M}}_{n,p}\delta x_p + Q^{\mathrm{M}}_{nij,o}H^{\mathrm{M}}_n\delta x_o + Q^{\mathrm{M}}_{nij,o}H^{\mathrm{M}}_{n,p}\delta x_o\delta x_p \right) \mathrm{d}V \\ B^{\mathrm{M}}_i =& \frac{1}{V_{\mathrm{RVE}_3}} \int_{V_{\mathrm{RVE}_3}} \left(Q^{\mathrm{M}}_{ikl}\varepsilon^{\mathrm{M}}_{kl} + Q^{\mathrm{M}}_{ikl}\varepsilon^{\mathrm{M}}_{kl,p}\delta x_p + Q^{\mathrm{M}}_{ikl,o}\varepsilon^{\mathrm{M}}_{kl}\delta x_o + Q^{\mathrm{M}}_{ikl,o}\varepsilon^{\mathrm{M}}_{kl,p}\delta x_o\delta x_p \right) \mathrm{d}V \\ &+ \frac{1}{V_{\mathrm{RVE}_4}} \int_{V_{\mathrm{RVE}_4}} \left(\mu^{\mathrm{M}}{}_{in}H^{\mathrm{M}}_n + \mu^{\mathrm{M}}_{in}H^{\mathrm{M}}_{n,p}\delta x_p + \mu^{\mathrm{M}}_{in,o}H^{\mathrm{M}}_n\delta x_o + \mu^{\mathrm{M}}_{in,o}H^{\mathrm{M}}_{n,p}\delta x_o\delta x_p \right) \mathrm{d}V \end{aligned} \tag{11.5}$$

In Eq. (11.5), $C^{\mathrm{M}}_{ijkl}, \varepsilon^{\mathrm{M}}_{kl}, Q^{\mathrm{M}}_{nij}, Q^{\mathrm{M}}_{ikl}, \mu^{\mathrm{M}}_{in}$ and H^{M}_n can be taken out of the integral since they are the values at the centre of the RVEs. Assuming a square RVE with its centre acting as origin of a Cartesian coordinate system, the linear terms of δx cancel as they consist of odd functions integrated over a symmetric domain. The quadratic terms are integrated by parts as follows

$$\begin{aligned} \int_{V_{\mathrm{RVE}_1}} C^{\mathrm{M}}_{ijkl,o}\varepsilon^{\mathrm{M}}_{kl,p}\delta x_o\delta x_p \mathrm{d}V =& \int_S C^{\mathrm{M}}_{ijkl}\varepsilon^{\mathrm{M}}_{kl,p}n_o\delta x_o\delta x_p \mathrm{d}S - \int_{V_{\mathrm{RVE}_1}} \Big(C^{\mathrm{M}}_{ijkl}\varepsilon^{\mathrm{M}}_{kl,op}\delta x_o\delta x_p \\ &+ C^{\mathrm{M}}_{ijkl}\varepsilon^{\mathrm{M}}_{kl,p}\delta x_{o,o}\delta x_p + C^{\mathrm{M}}_{ijkl}\varepsilon^{\mathrm{M}}_{kl,p}\delta x_o\delta x_{p,o} \Big) \mathrm{d}V \\ \int_{V_{\mathrm{RVE}_2}} Q^{\mathrm{M}}_{nij,o}H^{\mathrm{M}}_{n,p}\delta x_o\delta x_p \mathrm{d}V =& \int_S Q^{\mathrm{M}}_{nij}H^{\mathrm{M}}_{n,p}n_o\delta x_o\delta x_p \mathrm{d}S - \int_{V_{\mathrm{RVE}_2}} \Big(Q^{\mathrm{M}}_{nij}H^{\mathrm{M}}_{n,op}\delta x_o\delta x_p \\ &+ Q^{\mathrm{M}}_{nij}H^{\mathrm{M}}_{n,p}\delta x_{o,o}\delta x_p + Q^{\mathrm{M}}_{nij}H^{\mathrm{M}}_{n,p}\delta x_o\delta x_{p,o} \Big) \mathrm{d}V \\ \int_{V_{\mathrm{RVE}_3}} Q^{\mathrm{M}}_{ikl,o}\varepsilon^{\mathrm{M}}_{kl,p}\delta x_o\delta x_p \mathrm{d}V =& \int_S Q^{\mathrm{M}}_{ikl}\varepsilon^{\mathrm{M}}_{kl,p}n_o\delta x_o\delta x_p \mathrm{d}S - \int_{V_{\mathrm{RVE}_3}} \Big(Q^{\mathrm{M}}_{ikl}\varepsilon^{\mathrm{M}}_{kl,op}\delta x_o\delta x_p \\ &+ Q^{\mathrm{M}}_{ikl}\varepsilon^{\mathrm{M}}_{kl,p}\delta x_{o,o}\delta x_p + Q^{\mathrm{M}}_{ikl}\varepsilon^{\mathrm{M}}_{kl,p}\delta x_o\delta x_{p,o} \Big) \mathrm{d}V \end{aligned}$$

$$\int_{V_{\mathrm{RVE}_4}} \mu^{\mathrm{M}}_{in,o} H^{\mathrm{M}}_{n,p} \delta x_o \delta x_p \mathrm{d}V = \int_S \mu^{\mathrm{M}}_{in} H^{\mathrm{M}}_{n,p} n_o \delta x_o \delta x_p \mathrm{d}S - \int_{V_{\mathrm{RVE}_4}} \Big(\mu^{\mathrm{M}}_{in} H^{\mathrm{M}}_{n,op} \delta x_o \delta x_p + \mu^{\mathrm{M}}_{in} H^{\mathrm{M}}_{n,p} \delta x_{o,o} \delta x_p + \mu^{\mathrm{M}}_{in} H^{\mathrm{M}}_{n,p} \delta x_o \delta x_{p,o} \Big) \mathrm{d}V \tag{11.6}$$

Assuming periodic boundary conditions, the boundary integrals vanish and the last two terms in each of Eq. (11.6) vanish since they consist of odd functions. Furthermore, the integrals with $\delta x_o \delta x_p$ can be evaluated as

$$\int_{V_{\mathrm{RVE_i}}} \delta x_o \delta x_p \mathrm{d}V = \int_{-\frac{L_i}{2}}^{\frac{L_i}{2}} \int_{-\frac{L_i}{2}}^{\frac{L_i}{2}} \int_{-\frac{L_i}{2}}^{\frac{L_i}{2}} \delta x_o \delta x_p \mathrm{d}x_1 \mathrm{d}x_2 \mathrm{d}x_3 = \frac{1}{12} L_i^5 \delta_{op} \quad (i = 1,2,3,4) \tag{11.7}$$

where δ_{op} is the Kronecker delta, $V_{\mathrm{RVE}_i} = L_i^3$ and L_i is the size of the ith RVE.

The piezomagnetic macroscopic constitutive equations with gradients of strain, magnetic field, and magneto-mechanical coupling terms can thus be expressed as

$$\begin{aligned} \sigma^{\mathrm{M}}_{ij} &= C^{\mathrm{M}}_{ijkl} \left(\varepsilon^{\mathrm{M}}_{kl} - \frac{L_1^2}{12} \varepsilon^{\mathrm{M}}_{kl,pp} \right) - Q^{\mathrm{M}}_{nij} \left(H^{\mathrm{M}}_n - \frac{L_2^2}{12} H^{\mathrm{M}}_{n,pp} \right) \\ B^{\mathrm{M}}_i &= Q^{\mathrm{M}}_{ikl} \left(\varepsilon^{\mathrm{M}}_{kl} - \frac{L_3^2}{12} \varepsilon^{\mathrm{M}}_{kl,pp} \right) + \mu^{\mathrm{M}}_{in} \left(H^{\mathrm{M}}_n - \frac{L_4^2}{12} H^{\mathrm{M}}_{n,pp} \right) \end{aligned} \tag{11.8}$$

As seen in Eq. (11.8), new gradient terms, accompanied with multipliers L_i, appear in addition to the material coefficients of the macroscopic constitutive equations. It can be seen that Eq. (11.8) follows the structure of the gradient enriched piezomagnetic model proposed by Xu et. al [8], and written in terms of phenomenological parameters ℓ_i:

$$\begin{aligned} \sigma_{ij} &= C_{ijkl} \left(\varepsilon_{kl} - \ell_1^2 \varepsilon_{kl,mm} \right) - q_{ijk} \left(H_k - \ell_2^2 H_{k,mm} \right) \\ B_i &= q_{ijk} \left(\varepsilon_{jk} - \ell_3^2 \varepsilon_{jk,mm} \right) + \mu_{ij} \left(H_j - \ell_4^2 H_{j,mm} \right) \end{aligned} \tag{11.9}$$

Comparing Eqs. (11.8) and (11.9), it can be seen that the link between phenomenological parameters, representing internal characteristic length scale parameters ℓ_i, and RVE sizes L_i can be established as follows: $\ell_i^2 = L_i^2/12$.

11.2.2 Determination of RVE Sizes and Identification of Characteristic Length Scale Parameters

In [18], Eqs. (11.3-11.8) were derived to motivate the determination of the RVE sizes. The detailed methodology to determine the RVE size for magneto-elastic material has been presented in the previous study of the authors [18], however the main steps and results of this analysis will be summarised here. The methodology consists of a statistical numerical analysis of a heterogeneous material, undergoing mechanical and magnetic loading via prescribed values of the displacement $\boldsymbol{u}$ and the magnetic potential φ (see Fig. 11.1a). Boundary value problems are solved for multiple realisations of unit cells (see Fig. 11.1b). Resulting values of stresses and magnetic induction are then averaged over a unit cell. A statistical analysis is then performed and the coefficient of variation is compared with a desired accuracy (97% in our case). If the desired accuracy is reached, the current unit cell can be considered representative (i.e. the RVE is found); otherwise the unit cell size is increased (see Fig. 11.1c) and the procedure is repeated for the new larger unit cell size.

As mentioned above, the MRE material, used in this study, consists of circular magnetic inclusions (where a uniform distribution of 100–300 μm in diameter is assumed) and a non-magnetisable polymer matrix; the material parameters of the constituents are presented in Table 11.1. Two hundred different realisations of each unit cell size (ranging from 0.5 to 2.5 mm) have been analysed and the coefficient of

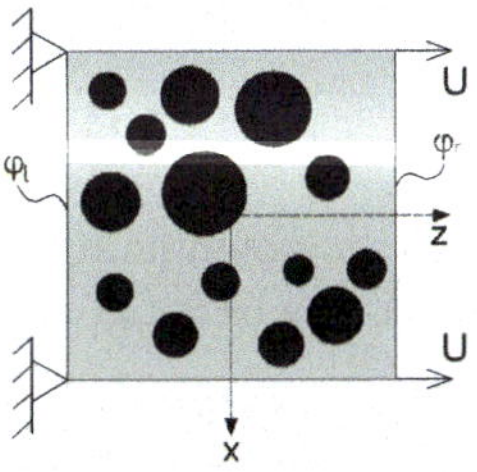

(a) Boundary value problem of a unit cell on micro level

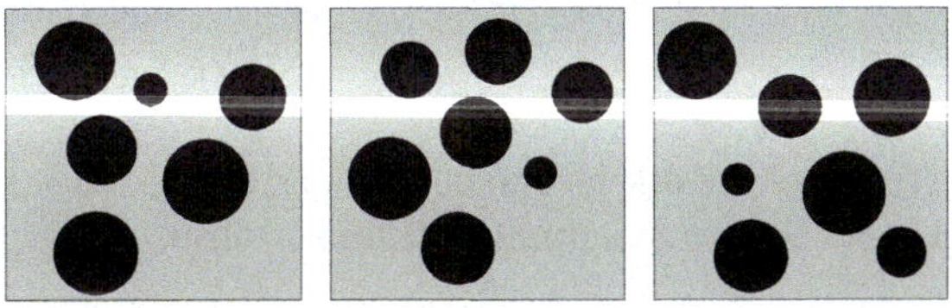

(b) Different realisations of the unit cell (size 1x1 mm^2).

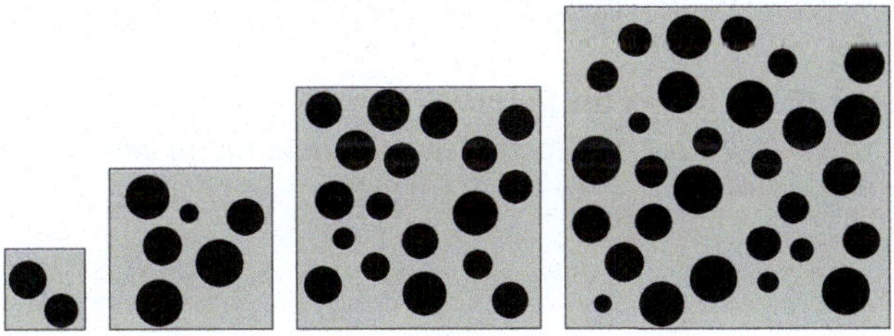

(c) Different sizes of unit cells (from left to right: 0.5x0.5 mm^2, 1x1 mm^2, 1.5x1.5 mm^2, 2x2 mm^2).

Fig. 11.1: Determination of RVE size [18].

Table 11.1: Material properties of magnetic inclusions and polymer matrix.

	C_{11}	C_{13}	C_{33}	C_{55}	Q_{31}	Q_{33}	Q_{15}	μ_{11}	μ_{33}
Terfenol-D [22]	35	23	46	4	-32.5	195	68.75	8.9	1.8
Polymer [23]	7.8	4.7	7.8	1.6	0	0	0	μ_0	μ_0

C_{ij} in GPa, Q_{ij} in N/Am, μ_{ij} in 10^{-6}N/A^2. ($\mu_0 = 4\pi 10^{-6}$ N/A^2)

variation, which is a statistical measure of relative variability, has been calculated for each unit cell size. These coefficients have been compared to the value of 0.03 (i.e. requiring 97% accuracy), and, hence, lower bounds of the RVE sizes (L_i) have been defined (see Fig. 11.2).

For an MRE material, the authors have demonstrated that lower bounds of the RVE sizes can be defined by following the proposed statistical analysis. As can be seen from Fig. 11.2, for the analysed MRE material, the difference between the coupling RVE sizes L_2 and L_3 is negligible in line with thermodynamic consistency requirements, while RVE sizes L_1 and L_4 are clearly smaller and different. According to numerical results, it can be seen that the largest determined RVE size for L_2 and L_3 also covers the lower bound condition for L_1 and L_4. Therefore, it can be concluded that it is sufficient to use only these largest sizes for practical purposes and set $L_1 = L_2 = L_3 = L_4 = \max(L_i) \equiv L$ to introduce the magneto-elastic information from micro-level.

Eventually, the field equations of the problem on the macro scale can be obtained by combining the kinematic relations, balance equations and constitutive equations:

$$\varepsilon_{ij}^{\mathrm{M}} = \frac{1}{2}(u_{i,j}^{\mathrm{M}} + u_{j,i}^{\mathrm{M}}) \quad \text{and } H_i^{\mathrm{M}} = -\varphi_{,i}^{\mathrm{M}} \tag{11.10}$$

$$\sigma_{ij,j}^{\mathrm{M}} = 0 \quad \text{and } B_{i,i}^{\mathrm{M}} = 0 \tag{11.11}$$

where u_i^{M} is the displacement field and φ^{M} is the scalar magnetic potential on the macro-level. The governing equations in terms of the primary unknowns $\boldsymbol{u}$ and φ are

$$\begin{aligned} C_{ijkl}^{\mathrm{M}}\left(u_{k,jl}^{\mathrm{M}} - \frac{L^2}{12}u_{k,jlpp}^{\mathrm{M}}\right) + Q_{nij}^{\mathrm{M}}\left(\varphi_{,jn}^{\mathrm{M}} - \frac{L^2}{12}\varphi_{,jnpp}^{\mathrm{M}}\right) = 0 \\ Q_{ikl}^{\mathrm{M}}\left(u_{k,il}^{\mathrm{M}} - \frac{L^2}{12}u_{k,ilpp}^{\mathrm{M}}\right) - \mu_{in}^{\mathrm{M}}\left(\varphi_{,in}^{\mathrm{M}} - \frac{L^2}{12}\varphi_{,inpp}^{\mathrm{M}}\right) = 0 \end{aligned} \tag{11.12}$$

with L chosen as the largest of the four RVE sizes as discussed above. In the previous study, the authors have not addressed the application of macro-level governing

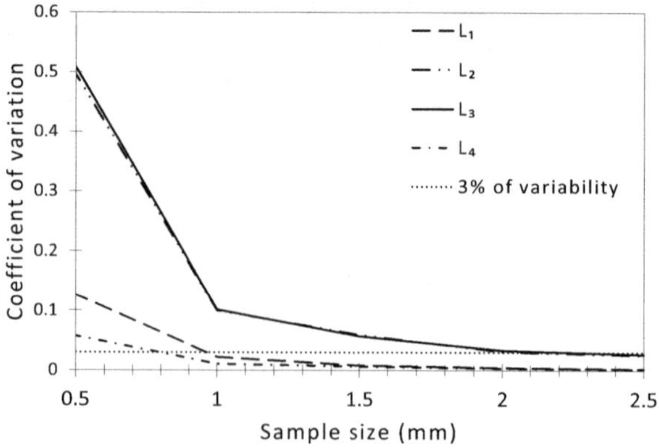

Fig. 11.2 Convergence of the results [18]

equations with identified length scale parameters in terms of RVE size. Here, by using the finite element implementation given in [8], the determined RVE sizes can be used instead of predefined phenomenological length scale parameters in a generalised magnetoelasticity example.

11.3 Numerical Results and Discussion

In this section, the standard benchmark numerical example, demonstrating the removal of singularities in mechanical and magnetic fields, will be analysed to demonstrate the performance of the model. The detailed finite element implementation of gradient enriched piezomagnetic continuum model for in-plane problems has been presented in [8]. For this, the authors extended the Ru-Aifantis approach to gradient magneto-elasticity by considering only one phenomenological length scale parameter in their model (Eq. (11.9)) (see [8] for the details of FEM formulation and in particular the $\boldsymbol{\varepsilon}\&\boldsymbol{H}$-RA approach described therein).

In the numerical example, a homogeneous MRE plate was modelled on the macro-scale with the effective properties (Table 11.2), and external loadings were applied as shown in Fig. 11.3. The plate is polarized along the x-direction with a uniform load $\boldsymbol{q}$ = 10 MPa and magnetic field $\boldsymbol{H}_o$ = 100 A/m. As discussed, it was assumed that $L_1 = L_2 = L_3 = L_4 \equiv L$ in Eq. (11.12), and L is the size of the RVE determined as 2.3 mm (see Fig. 11.2). The crack in the model has been considered as an inclusion with vacuum permeability and zero values for elastic constants, piezo-magnetic constants and RVE size L. Figure 11.4 presents the distributions of $\boldsymbol{\varepsilon}$ and

Fig. 11.3 MRE Plate with a crack (Units: mm and thickness = 5 mm)

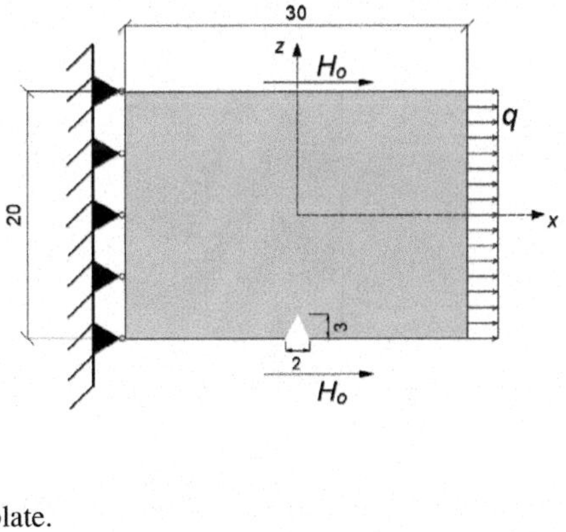

Table 11.2: Effective material properties of MRE plate.

	C_{11}	C_{13}	C_{33}	C_{55}	Q_{31}	Q_{33}	Q_{15}	μ_{11}	μ_{33}
MRE Plate [8]	31.1	15.2	35.6	13.6	156.8	108.3	-60.9	5.4	5.4

C_{ij} in GPa, Q_{ij} in N/Am, μ_{ij} in 10^{-6}N/A^2

H components in the x-direction around the crack tip. To investigate the effectiveness

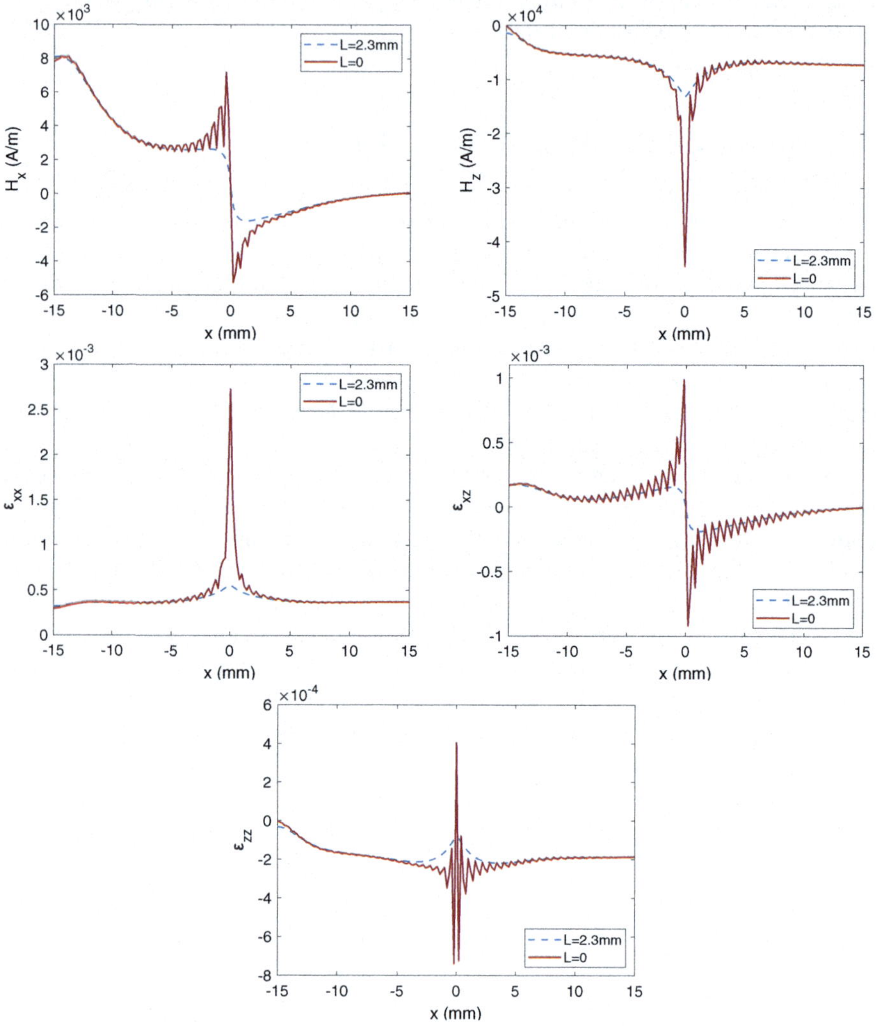

Fig. 11.4: $\boldsymbol{\varepsilon}$ and $\boldsymbol{H}$ distributions along x-axis based on $\boldsymbol{\varepsilon}$&$\boldsymbol{H}$-RA approach.

of the gradient formulation, two different cases have been considered: one with length scale $\ell^2 = L_{\text{RVE}}^2/12$ with the RVE size 2.3 mm that considers the gradient dependence, and the other with the length scale $\ell_i = 0$ that represents classical magneto-elasticity in which the microstructural information is absent. It can be seen (Fig. 11.4) that using the gradient enhanced, RVE-based magneto-elasticity formulation can effectively remove the singularities of all $\boldsymbol{\varepsilon}$ and $\boldsymbol{H}$ components, while singularities appear at the crack tip for the classical formulation.

11.4 Conclusions

In this study, an in-plane problem has been addressed by using the previously developed piezo-magnetic continuum model that includes gradients of strain, magnetic field and piezo-magnetic coupling accompanied by characteristic length scale parameter found in terms of RVE size. With this generalised magneto-elastic continuum model, a solution scheme is formulated and implemented based on the finite element method and the Ru-Aifantis theorem adopted from the study by Xu et al. [8]. In the continuum model, representative volume elements (RVEs) have been defined and included to introduce the microstructural information into the macroscopic behaviour of the material. By using the determined model parameter RVE size L, it was observed that the singularities can be removed in mechanical and magnetic fields for a homogenised MRE plate on the macro scale.

Acknowledgements This work was supported by International Graduate Education Scholarship (YLSY) programme funded by The Ministry of National Education of Türkiye.

References

[1] Filipcsei G, Csetneki I, Szilágyi A, Zrínyi M (2007) Magnetic Field-Responsive Smart Polymer Composites, In: Oligomers - Polymer Composites - Molecular Imprinting. Advances in Polymer Science, Vol 206, pp 137-189, Springer, Berlin, Heidelberg.

[2] Thévenot J, Oliveira H, Sandre O, Lecommandoux S (2013) Magnetic responsive polymer composite materials. Chem. Soc. Rev. **42**:7099-7116.

[3] Elhajjar R, Law CT, Pegoretti A (2018) Magnetostrictive polymer composites: Recent advances in materials, structures and properties, Progress in Materials Science **97**:204-229.

[4] Liu T, Xu Y (2019) Magnetorheological elastomers: Materials and applications, In: Xufeng Dong, editor, Smart and Functional Soft Materials, chapter 4. IntechOpen, Rijeka.

[5] Ekreem NB, Olabi AG, Prescott T, Rafferty A, Hashmi MSJ (2007) An overview of magnetostriction, its use and methods to measure these properties, Journal of Materials Processing Technology **191**(1):96-101.

[6] Davis LC (1999) Model of magnetorheological elastomers, Journal of Applied Physics **85**(6):3348-3351.

[7] Liu TY, Chan TY, Wang KS, Tsou HM (2015) Influence of magnetic nanoparticle arrangement in ferrogels for tunable biomolecule diffusion, RSC Adv. **5**:90098-90102.

[8] Xu M, Gitman IM, Askes H (2019) A gradient-enriched continuum model for magneto-elastic coupling: Formulation, finite element implementation and in-plane problems, Computers and Structures **212**:275-288.

[9] Bastola AK, Hoang VT, Li L (2017) A novel hybrid magnetorheological elastomer developed by 3d printing, Materials Design **114**:391-397.
[10] Wood DS, Camp PJ (2011) Modeling the properties of ferrogels in uniform magnetic fields, Phys. Rev. E **83**:011402.
[11] Kalina KA, Metsch P, Kästner M (2016) Microscale modeling and simulation of magnetorheological elastomers at finite strains: A study on the influence of mechanical preloads, International Journal of Solids and Structures **102-103**:286-296.
[12] Attaran A, Brummund J, Wallmersperger T (2017) Modeling and finite element simulation of the magneto-mechanical behavior of ferrogels, Journal of Magnetism and Magnetic Materials **431**:188-191.
[13] Raikher YL, Stolbov OV (2008) Numerical modeling of large field-induced strains in ferroelastic bodies: a continuum approach, Journal of Physics: Condensed Matter **20**(20):204126.
[14] Gitman IM, Askes H, Sluys LJ (2004) Representative volume size as a macroscopic length scale parameter. Fracture Mechanics for Concrete and Concrete Structures, 1:483-491.
[15] Gitman IM (2006) Representative volumes and multi-scale modelling of quasi-brittle materials, Doctoral thesis, Technische Universiteit Delft.
[16] Gitman IM, Askes H, Aifantis EC (2007) Gradient elasticity with internal length and internal inertia based on the homogenisation of a representative volume element, Journal of the Mechanical Behavior of Materials **18**(1):1-16.
[17] Ke L, Wang Y, Yang J, Kitipornchai S (2014) Free vibration of size-dependent magneto-electro-elastic nanoplates based on the nonlocal theory, Acta Mechanica Sinica **30**:516-525.
[18] Eraslan S, Gitman IM, Askes H, de Borst R (2022) Determination of representative volume element size for a magnetorheological elastomer, Computational Materials Science **203**:111070.
[19] Lan M, Wei P (2014) Band gap of piezoelectric/piezomagnetic phononic crystal with graded interlayer, Acta Mechanica **225**:1779–1794.
[20] Mane H (2019) Mathematical Modeling and Numerical Simulation of Magnetoelastic Coupling, Doctoral thesis, Technische Universität Kaiserslautern.
[21] Pang Y, LiuJ, Wang Y, Fang D (2008) Wave propagation in piezoelectric/piezomagnetic layered periodic composites, Acta Mechanica Solida Sinica **21**:483-490.
[22] Claeyssen F, Lhermet N, Barillot F, Le Letty R (2006) Giant dynamic strains in magnetostrictive actuators and transducers, ISAGMM 2006 CONF.
[23] Wang YZ, Li FM, Huang WH, Jiang X, Wang YS, Kishimoto K (2008) Wave band gaps in two-dimensional piezoelectric/piezomagnetic phononic crystals, International Journal of Solids and Structures **45**:4203-4210.

Chapter 12
Rayleigh Waves in the Cosserat Half-Space (Reduced Model) and Half-Space of Damaged Material

Vladimir Erofeev, Artem Antonov, Anna Leonteva, and Alexey Malkhanov

Abstract The peculiarities of propagation of Rayleigh surface waves along the free boundary of the half-space of the Cosserat medium (reduced model), as well as along the free boundary of the half-space made of damaged material, are studied. It is shown that, in contrast to the classical Rayleigh surface wave, a wave propagating along the boundary of the Cosserat half-space has dispersion. In the plane "phase velocity – frequency" for such waves, there are two dispersion branches: lower ("acoustic") and upper ("optical"). As the frequency increases, the phase velocity of the wave related to the lower dispersion branch decreases. The phase velocity of the wave belonging to the upper dispersion branch increases with increasing frequency. The phase velocity of the surface wave in the entire frequency range exceeds the phase velocity of the bulk shear wave. The stresses and displacements arising in the zone of propagation of a surface wave are calculated.

For an isotropic elastic half-space with damage of its material, a self-consistent problem is formulated that includes the dynamic equation of elasticity theory and the kinetic equation of damage accumulation in the material of the medium. This system of equations with boundary conditions expressing the absence of stresses at the boundary of the half-space reduces to a complex dispersion equation. A surface wave propagating along the boundary of a damaged half-space decays in the direction of propagation, and low-frequency perturbations have frequency-dependent dissipation and dispersion. It is shown that the dispersion has an anomalous character. It has been established that with a decrease in the value of the damage factor, in the region of high frequencies, the value of the phase velocity increases, and the group velocity decreases. At very low frequencies, both velocities increase as the damage factor decreases.

Vladimir Erofeev · Artem Antonov · Anna Leonteva · Alexey Malkhanov
Mechanical Engineering Research Institute of Russian Academy of Sciences, 85 Belinskogo str., Nizhny Novgorod 603024, Russian Federation,
e-mail: erof.vi@yandex.ru, artem.antonov@autorambler.ru, aleonav@mail.ru, alexey.malkhanov@gmail.com

H. Altenbach et al. (eds.), *Sixty Shades of Generalized Continua*,
Advanced Structured Materials 170, https://doi.org/10.1007/978-3-031-26186-2_12

Key words: Cosserat half-space · Damaged material · Rayleigh waves · Dispersion · Attenuation

12.1 Introduction

In 1885, Lord Rayleigh (John William Strett) theoretically demostrated that along the plane boundary of a solid elastic half-space with a vacuum or a sufficiently rarefied medium (for example, with air), waves can propagate, the amplitude of which rapidly decreases with depth [1]. These waves, called Rayleigh surface waves, depending on the frequency range, have different applied directions.

Almost immediately it became obvious that Rayleigh waves in the low frequency range (1–100 Hz) are the main type of waves observed during earthquakes. Therefore, they have been studied in detail in seismology for nearly 140 years [2].

The main regularities of the propagation of Rayleigh waves are as follows: the absence of dispersion, i.e. the speed of the wave does not depend on its frequency and is constant for each material; this speed reaches 0.87–0.96 of the speed of the bulk shear wave; the displacement vector has longitudinal and transverse components, while the transverse component always exceeds the longitudinal one.

In the cycle of works by Krylov [3]-[9], awarded by the Acoustics Institute of Great Britain the Rayleigh Medal (often called the Nobel Prize in Acoustics) in 2000, dedicated to the study of elastic earth vibrations generated by trains and motor vehicles, a very high level of earth vibrations generated by high-speed railways was theoretically predicted. These vibrations are generated by high-speed trains moving at a speed higher than the speed of Rayleigh surface waves in the ground. Krylov's theory was experimentally confirmed in 1997–1998 (with his direct participation) on a new high-speed line in Sweden (Gothenburg-Malmo), where on some sections of the route the Rayleigh wave speed was only 45 m/s, and the train speed of 160 km/h was enough to observe the effect. The discovered effect began to be called "ground vibration shock" (by analogy with the well-known sonic boom from a supersonic aircraft), and the generation sources began to be called "trans-Rayleigh trains" [10].

It should be noted that the existence of critical speeds of movement of loads along rail guides, when exceeded, bending waves are generated in the guides, was discussed in the first half of the 1980s in the works of Vesnitsky and his school [11], as well as in the works of G.G. Denisov, V.V. Novikov and E.K. Kugusheva [12]. However, the calculated critical speeds showed the practical unattainability of the effect of generation of bending waves by the vehicle in the guides. It turned out to be easier for the load to overcome the speed of the Rayleigh wave in the soil located under the guide rail, and the guide itself, along with the system of sleepers and ballast, acted as an intermediary between the source of wave generation and the medium in which these waves arose.

At present, the problems of the stability of the movement of high-speed objects along rail guides, the problems of generating bending and bending-torsional waves in rail guides are recognized as relevant and their results serve as methodological and computational support in setting up experiments on high-speed acceleration of the payload on rocket tracks [13]-[15].

Since the 1950s, Rayleigh waves in the ultrasonic range (frequencies of the order of 10^6 Hz) have been widely used. The state of the surface layer of a sample can be controlled by their assistance (detection of surface and near-surface defects in samples made of metal, glass, plastic and other materials – ultrasonic surface defectoscopy). The influence of the properties of the surface layer of the sample on the velocity and damping of Rayleigh waves makes it possible to use the latter to determine the residual stresses of the surface layer of the metal, the thermal and mechanical properties of the surface layer of the sample [16].

Since the early 1970s, Rayleigh waves with frequencies of 10^7–10^{10} Hz have been widely used in miniature solid-state information processing devices (ultrasonic delay lines, bandpass filters, signal couplers, phase shifters, etc.). At the junction of ultra- and hypersonic acoustics, on the one hand, and solid-state electronics, on the other hand, a branch of knowledge called acoustoelectronics has emerged. Such an important industry as microelectronics based on SAW (i.e., on surface acoustic waves) is based on the principles of acoustoelectronics [17].

Along with the classical continuum model, generalized continuum models are widely used in the mechanics of a deformable solid body [18]-[39]. The most well-known generalized (nonclassical) continuums include the micropolar Cosserat medium and the gradient elastic medium. When working with a micropolar medium, one should distinguish between the general model of a continuum proposed by the brothers Eugène and François Cosserat in 1909, each point of which has three translational and three rotational degrees of freedom [40], its particular case – a model of a micropolar medium with constrained rotation of particles [41], and a reduced model of a micropolar medium, proposed by L. Schwartz, D. Johnson and S. Feng in 1984 [42]-[45]. The model of a gradient elastic medium also appeared at the beginning of the 20th century and is associated with the names of Leroux (1911, 1913) [46, 47] and Jaramillo (1929) [48]. A number of authors have studied Rayleigh surface waves within the framework of the Cosserat continuum [49] and gradient elastic medium [50] models. Below we study the main regularities of Rayleigh wave propagation along the free boundary of the half-space of the Cosserat medium (reduced model), as well as along the free boundary of the half-space made of damaged material (another version of the generalized continuum model).

12.2 Rayleigh Waves in the Cosserat Half-space (Reduced Model)

Let us consider the problem of propagation of elastic surface wave in a reduced dynamic model of the Cosserat medium, which take place in the intermediate state between the classical dynamics theory and the actual model of the Cosserat medium, which has an asymmetry of the stress tensor and the presence of load moments, are considered. In contrast to the latter, in the simplified model, the three of six elasticity constants are supposed to be equal to zero and, as a result, there is no couple-stress tensor.

The equations of the dynamics of the Cosserat continuum have the form [41]:

$$\begin{aligned} &\rho\frac{\partial^2 \boldsymbol{u}}{\partial t^2} - (\lambda+2\mu)\,\mathrm{grad}\,\mathrm{div}\,\boldsymbol{u} + (\mu+\alpha)\,\mathrm{rot}\,\mathrm{rot}\,\boldsymbol{u} - 2\alpha\,\mathrm{rot}\,\boldsymbol{\theta} = 0, \\ &I\frac{\partial^2 \boldsymbol{\theta}}{\partial t^2} - (\beta+2\gamma)\,\mathrm{grad}\,\mathrm{div}\,\boldsymbol{\theta} + (\gamma+\varepsilon)\,\mathrm{rot}\,\mathrm{rot}\,\boldsymbol{\theta} - 2\alpha\,\mathrm{rot}\,\boldsymbol{\theta} + 4\alpha\boldsymbol{\theta} = 0. \end{aligned} \tag{12.1}$$

Here $\boldsymbol{u}$ – displacements vector; $\boldsymbol{\theta}$ – rotation vector; ρ – density of the medium, I – a constant characterizing the inertial properties of macrovolume, equal to the product of the moment of inertia of a particle of a substance around any axis passing through its center of gravity by the number of particles per unit volume; λ, μ – Lame constants, $\alpha, \beta, \gamma, \varepsilon$ – new elastic constants of micropolar material satisfying the constraints [41]:

$$\alpha \geq 0,\ \gamma+\varepsilon \geq 0,\ 3\beta+2\gamma \geq 0,\ -(\gamma+\varepsilon) \geq \gamma-\varepsilon \geq (\gamma+\varepsilon). \tag{12.2}$$

In [49] the following dependence was found between these elastic constants:

$$\mu\,(2\gamma+\beta) = (\alpha+\mu)\,(\gamma+\varepsilon). \tag{12.3}$$

Along with the general case, a simplified version of the micropolar medium (Cosserat pseudo-continuum) is also considered, in which a rigid dependence of the rotation vector on the displacement rotor, which coincides with the relations of the classical theory of elasticity, is assumed ($\boldsymbol{\theta} = \frac{1}{2}\mathrm{rot}\,\boldsymbol{u}$ – cramped rotation), but at the same time, the moment stresses and asymmetry of the stress tensor are preserved. In such a medium, the symmetric part of the stress tensor depends on the symmetric strain tensor in the same way as in the classical theory of elasticity.

The equations of dynamics of the Cosserat pseudo-continuum have the form [41]:

$$\rho\ddot{\boldsymbol{u}} - (\lambda+\mu)\,\mathrm{grad}\,\mathrm{div}\,\boldsymbol{u} - \mu\Delta\boldsymbol{u} - \frac{1}{4}(\gamma+\varepsilon)\,\mathrm{rot}\,\mathrm{rot}\,\Delta\boldsymbol{u} + \frac{1}{4}\mathrm{rot}\,\mathrm{rot}\,\ddot{\boldsymbol{u}} = 0. \tag{12.4}$$

In addition to this particular case, another special case of the Cosserat medium model is known – the Cosserat reduced medium model, for which three of the six elastic constants, specifically $\beta, \gamma, \varepsilon$, are equal to zero and, as a consequence, there is no moment stress tensor. This model was first proposed in [42] for the description of bulk materials.

The following system of equations of dynamics is used to describe the reduced Cosserat medium [42]:

$$\begin{aligned} &\rho\frac{\partial^2 \boldsymbol{u}}{\partial t^2} - (\lambda+2\mu)\,\mathrm{grad}\,\mathrm{div}\,\boldsymbol{u} + (\mu+\alpha)\,\mathrm{rot}\,\mathrm{rot}\,\boldsymbol{u} - 2\alpha\,\mathrm{rot}\,\boldsymbol{\theta} = 0, \\ &I\frac{\partial^2 \boldsymbol{\theta}}{\partial t^2} - 2\alpha\,\mathrm{rot}\,\boldsymbol{\theta} + 4\alpha\boldsymbol{\theta} = 0, \end{aligned} \tag{12.5}$$

which reduces to a single vector equation in displacements

$$(\lambda+2\mu)\nabla(\nabla\cdot\boldsymbol{u})-\mu\nabla*(\nabla*\boldsymbol{u})-J\frac{\partial^2}{\partial t^2}\nabla*(\nabla*\boldsymbol{u})=\rho\frac{\partial^2\boldsymbol{u}}{\partial t^2}. \tag{12.6}$$

where the following designation ∇ is introduced in order to shorten the notation ($\nabla\cdot\boldsymbol{u}\equiv\mathrm{div}\boldsymbol{u}$, $\nabla*\boldsymbol{u}\equiv\mathrm{rot}\boldsymbol{u}$).

Let us introduce scalar φ and vector $\boldsymbol{\psi}$ potentials thus displacements vector $\boldsymbol{u}$ can be written in the following form:

$$\boldsymbol{u}=\nabla\varphi+\nabla*\boldsymbol{\psi}. \tag{12.7}$$

In the case of considering a plane problem, the vector potential will have only one component different from zero. We denote this component by ψ.

Let us substitute displacement vector (12.7) into (12.6), we obtain:

$$\nabla\left[(\lambda+2\mu)\Delta\varphi-\rho\frac{\partial^2\varphi}{\partial t^2}\right]+\nabla*\left[\mu\Delta\psi+J\Delta\frac{\partial^2\psi}{\partial t^2}-\rho\frac{\partial^2\psi}{\partial t^2}\right]=0. \tag{12.8}$$

Equation (12.8) is satisfied if each term of the expressions in square brackets is zero:

$$\begin{aligned}\Delta\varphi-\frac{1}{c_1^2}\frac{\partial^2\varphi}{\partial t^2}&=0,\\ \Delta\psi+G\Delta\frac{\partial^2\psi}{\partial t^2}-\frac{1}{c_2^2}\frac{\partial^2\psi}{\partial t^2}&=0.\end{aligned} \tag{12.9}$$

The following notations are introduced here: $c_1^2=\frac{(\lambda+2\mu)}{\rho}$, $c_2^2=\frac{\mu}{\rho}$, where c_1 – velocity of longitudinal wave, and c_2 - shear wave velocity in a classical medium, $G=J/\mu$.

We will seek a solution to the equations in the form of harmonic waves which propagate in the direction of the axis x and have a heterogeneous structure in z:

$$\varphi=\Phi(z)\,e^{-i(\omega t-kx)},\qquad \psi=\Psi(z)\,e^{-i(\omega t-kx)}. \tag{12.10}$$

Substituting formulas (12.10) into (12.9), we obtain from the latter ordinary differential equations:

$$\frac{\partial^2\Phi}{\partial z^2}-\nu_1^2\Phi=0,\qquad \frac{\partial^2\Psi}{\partial z^2}-\nu_2^2\Psi=0, \tag{12.11}$$

where

$$\nu_1=\left(k^2-\frac{\omega^2}{c_1^2}\right)^{\frac{1}{2}},\qquad \nu_2=\left(k^2-\frac{\omega^2}{c_2^2(1-G\omega^2)}\right)^{\frac{1}{2}}.$$

From the solutions of Eqs. (12.11) we select only those that correspond to a decrease in the wave amplitudes with depth:

$$\Phi=Ae^{-\nu_1 z},\qquad \Psi=Be^{-\nu_2 z}. \tag{12.12}$$

The postulate adopted here on the extinction of waves with depth entails the assertion that ν_α ($\alpha=1,2$) should be real and positive values, i.e. $\nu_\alpha>0$.

The final solution of equations (12.9) will have the form:

$$\varphi = Ae^{-\nu_1 z - i(\omega t - kx)}, \qquad \psi = Be^{-\nu_2 z - i(\omega t - kx)}. \tag{12.13}$$

Assuming that the boundary $z = 0$ is free of stresses, we have three conditions:

$$\sigma_{zz}|_{z=0} = 0, \; \sigma_{zx}|_{z=0} = 0, \; \sigma_{zy}|_{z=0} = 0. \tag{12.14}$$

In terms of displacements, these stress components are expressed as follows:

$$\sigma_{zx} = \mu\left(\frac{\partial u}{\partial z} + \frac{\partial w}{\partial x}\right) + J\left(\frac{\partial^3 w}{\partial x \partial t^2} - \frac{\partial^3 u}{\partial z \partial t^2}\right), \quad \sigma_{zz} = \lambda\left(\frac{\partial u}{\partial x} + \frac{\partial w}{\partial z}\right) + 2\mu\frac{\partial w}{\partial z}, \tag{12.15}$$

and the relation of displacements and potentials (see (12.7)) is given by the expressions:

$$u = \frac{\partial \varphi}{\partial x} - \frac{\partial \psi}{\partial z}, \qquad w = \frac{\partial \varphi}{\partial z} + \frac{\partial \psi}{\partial x}, \tag{12.16}$$

stresses can be expressed with the help of potentials in the following form:

$$\begin{gathered} \sigma_{zz} = \lambda\left(\frac{\partial^2 \varphi}{\partial x^2} + \frac{\partial^2 \varphi}{\partial z^2}\right) + 2\mu\left(\frac{\partial^2 \varphi}{\partial z^2} + \frac{\partial^2 \psi}{\partial x \partial z}\right), \\ \sigma_{zx} = \mu\left(2\frac{\partial^2 \varphi}{\partial x \partial z} + \frac{\partial^2 \psi}{\partial x^2} - \frac{\partial^2 \psi}{\partial z^2}\right) - J\left(\frac{\partial^4 \psi}{\partial x^2 \partial t^2} + \frac{\partial^4 \psi}{\partial z^2 \partial t^2}\right). \end{gathered} \tag{12.17}$$

Substituting expressions (12.13) into (12.17) and using boundary conditions (12.14), we obtain the following system of algebraic equations:

$$\begin{gathered} \left[\lambda\left(\nu_1^2 - k^2\right) + 2\mu\nu_1^2\right] A - 2\mu i k \nu_2 B = 0, \\ 2\mu i k \nu_1 A + \left(\mu k^2 + \mu\nu_2^2 - J\nu_2^2\omega^2 + J\omega^2 k^2\right) B = 0. \end{gathered} \tag{12.18}$$

From the compatibility condition of this system we obtain the relation

$$\left[\lambda\left(\nu_1^2 - k^2\right) + 2\mu\nu_1^2\right]\left[\mu\left(k^2 + \nu_2^2\right) + J\omega^2\left(k^2 - \nu_2^2\right)\right] - 4\mu\nu_1\nu_2 k^2 = 0, \tag{12.19}$$

which by introducing notations

$$\xi = \frac{c_2^2}{c_1^2} < 1, \eta = \frac{c^2}{c_2^2}$$

and account of dependences

$$\frac{c_1^2}{c_2^2} = \frac{\lambda + 2\mu}{\mu}, \; \frac{\lambda}{\mu} = \frac{c_1^2}{c_2^2} - 2, \; \nu_1^2 = k^2 - \frac{\omega^2}{c_1^2}, \; \nu_2^2 = k^2 - \frac{\omega^2}{c_2^2\left(1 - G\omega^2\right)},$$

can be transformed to the following form:

$$\eta\left[\eta^3-8\eta^2+\left(24-16\frac{\xi}{1-\frac{J}{\mu}\omega^2}\right)\eta-16\left(2-\frac{1}{1-\frac{J}{\mu}\omega^2}-\xi\right)\right]=0. \tag{12.20}$$

Note that (12.20) is a dispersion equation for calculating the phase velocity (c) of Rayleigh surface wave.

Figure 12.1 presents the dependence of the square of the phase velocity of the surface wave $\eta = c_R^2$ on frequency ω. The curves are given in dimensionless form: squared surface wave velocity c_R^2 is divided on squared shear velocity c_2^2.

The graph shows that here, in contrast to the classical case, the Rayleigh surface wave has dispersion. In the plane "phase velocity - frequency" there are two dispersion branches: the lower ("acoustic") and the upper ("optical"). As the frequency increases, the phase velocity of the wave related to the lower dispersion branch also decreases at infinity the square of the velocity of the surface wave $\eta \to 0.7$. The phase velocity of the wave related to the upper dispersion branch increases with increasing frequency. For frequencies $\omega > 9$ this growth becomes unlimited. Therefore, the upper dispersion branch describes wave processes in the frequency range $0 < \omega < 9$, further, the process ceases to be wave.

A plane shear wave is described by the second of equations (12.9):

$$\frac{\partial^2\psi}{\partial x^2}+G\frac{\partial^4\psi}{\partial x^2\partial t^2}-\frac{1}{c_2^2}\frac{\partial^2\psi}{\partial t^2}=0.$$

The solution to this equation we will look in the form:

$$\psi = Be^{i(\omega t-kx)},$$

which leads to dispersion equation

$$-k^2+Gk^2\omega^2+\frac{1}{c_2^2}\omega^2=0. \tag{12.21}$$

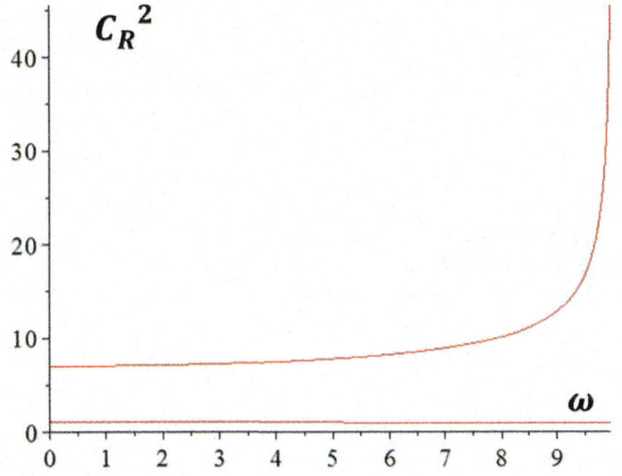

Fig. 12.1 The dependence of squared phase velocity of the surface wave on frequency

From (12.21) we determine the relationship between the frequency and the wave number of the shear wave:

$$\omega^2 = \frac{k^2 c_2^2}{1 + \frac{Gk^2}{c_2^2}}$$

and the square of the phase velocity of this wave:

$$v_{\phi\tau}^2 = \frac{\omega^2}{k^2} = \frac{c_2^2}{1 + \frac{Gk^2}{c_2^2}}.$$

Figure 12.2 presents two dependences: the square of the surface wave velocity c_R^2 (red line) and the square of the phase velocity of the shear wave v_ϕ^2 (blue line). The curves are given in dimensionless form: the square of the surface wave velocity is divided on the square of the shear wave velocity of the classical medium c_2^2, and the square of the phase velocity of the shear wave $v_{\phi\tau}^2$ is divided on c_2^2 as well. The graph shows that the phase velocity of the surface wave in the entire frequency range exceeds the phase velocity of the shear wave, which originates in unity and monotonically decreases to zero when $c_R^2 \to 0.7$, with the condition $\omega \to \infty$.

Relations (12.13) and (12.16) allow us to calculate the displacements:

$$u = (ikAe^{-\nu_1 z} + \nu_2 Be^{-\nu_2 z})\, e^{i(kx-\omega t)}, \qquad w = (-A\nu_1 e^{-\nu_1 z} + Bike^{-\nu_2 z})\, e^{i(kx-\omega t)}. \tag{12.22}$$

The constant A can be expressed through B, with the help of the second equation from (12.18):

$$B = -\frac{2i\nu_1 k\mu}{\mu k^2 + \mu\nu_2^2 - J\nu_2^2\omega^2 + Jk^2\omega^2} A,$$

then components u and w take the following form:

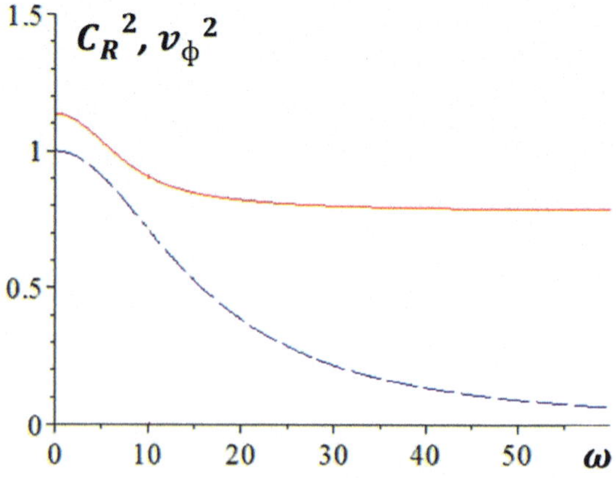

Fig. 12.2 Dependences of the squared surface wave velocity c_R^2 (red line) and the square of the phase velocity of the shear wave v_ϕ^2 (blue line) on frequency

$$u = \left(ikAe^{-\nu_1 z} - \frac{2i\nu_1 k\mu}{\mu k^2 + \mu\nu_2^2 - J\nu_2^2\omega^2 + Jk^2\omega^2} A\nu_2 e^{-\nu_2 z} \right) e^{i(kx-\omega t)},$$
$$w = \left(-A\nu_1 e^{-\nu_1 z} + \frac{2\nu_1 k^2\mu}{\mu k^2 + \mu\nu_2^2 - J\nu_2^2\omega^2 + Jk^2\omega^2} Ae^{-\nu_2 z} \right) e^{i(kx-\omega t)}. \tag{12.23}$$

Taking the real part of Eqs. (12.23), we obtain the final formulas for the displacements:

$$u = -Ak \left(e^{-\nu_1 z} - \frac{2\nu_1 \nu_2 \mu}{\mu k^2 + \mu\nu_2^2 - J\nu_2^2\omega^2 + Jk^2\omega^2} e^{-\nu_2 z} \right) \sin(kx - \omega t),$$
$$w = -A\nu_1 \left(e^{-\nu_1 z} - \frac{2k^2\mu}{\mu k^2 + \mu\nu_2^2 - J\nu_2^2\omega^2 + Jk^2\omega^2} e^{-\nu_2 z} \right) \cos(kx - \omega t). \tag{12.24}$$

Figure 12.3 shows the depth dependences of the displacement amplitudes u and w of the Rayleigh wave on depth in Cosserat medium and classic continuum. The curves are shown in the dimensionless form: the amplitudes of displacements are related to the amplitude of normal displacement on the surface $w|_{z=0} = 0$. Depth is shown in fractions of a wavelength.

Calculations by formulas (12.24) show that the amplitude of the displacement normal to the surface increased by about 25% compared with the classical case. The amplitude of movement parallel to the surface also increased by 90% and changed sign at depth $z = 0.18\lambda$. The graph shows that the offset parallel to the surface can exceed the transverse component in a thin near-surface layer. The trajectories of particles motion during the passage of a surface wave, as in the classical case, are ellipses.

Displacements (12.24) allow us to calculate the components of the stress tensor:

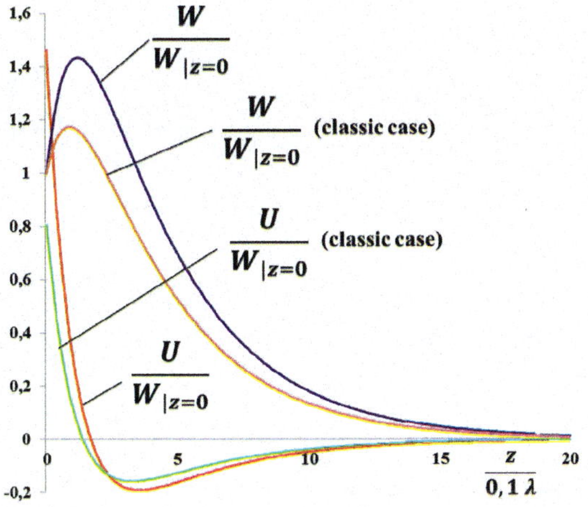

Fig. 12.3 Dependence of displacement amplitudes u and w in a Rayleigh wave on depth: amplitude of normal displacement to the surface in the simplified Cosserat model (blue line), amplitude of normal displacement to the surface in the classical model (burgundy line), amplitude of parallel displacement to the surface in the simplified Cosserat model (red line), amplitude of parallel movement to the surface in the classical model (green line).

$$\sigma_{xx} = A\left[\left(\lambda\left(\nu_1^2-k^2\right)+2\mu\nu_1^2\right)\mathrm{e}^{-\nu_1 y}+\frac{4\nu_1\nu_2k^2\mu^2}{\mu k^2+\mu\nu_2^2-J\nu_2^2\omega^2+Jk^2\omega^2}\mathrm{e}^{-\nu_2 y}\right]\cos(kx-\omega t),$$
$$\sigma_{xz} = 2A\mu\nu_1 k\left[e^{-\nu_1 y}-e^{-\nu_2 y}\right]\sin(kx-\omega t),$$
$$\sigma_{zz} = A\left[\left(\lambda\left(\nu_1^2-k^2\right)+2\mu\nu_1^2\right)e^{-\nu_1 y}-\frac{4\nu_1\nu_2k^2\mu^2}{\mu k^2+\mu\nu_2^2-J\nu_2^2\omega^2+Jk^2\omega^2}e^{-\nu_2 y}\right]\cos(kx-\omega t),$$
$$\sigma_{zx} = 2A\mu\nu_1 k\left[e^{-\nu_1 y}-\frac{\mu k^2+\mu\nu_2^2+J\nu_2^2\omega^2-Jk^2\omega^2}{\mu k^2+\mu\nu_2^2-J\nu_2^2\omega^2+Jk^2\omega^2}e^{-\nu_2 y}\right]\sin(kx-\omega t).$$

Figure 12.4 presents stress changes $\sigma_{xx},\sigma_{xz},\sigma_{zz},\sigma_{zx}$ in a Rayleigh wave from the depths of half-space. The curves are presented in a dimensionless form: the stress amplitudes are related to the amplitude of the normal stress on the surface $\sigma_{xx}|_{z=0}$. The graphs also show that σ_{xx} changes sign, whereas $\sigma_{xz},\sigma_{zz},\sigma_{zx}$ reach a maximum at approximately $z = 1.5$ and then exponentially decreases with depth. It is also seen that the stress tensor in this case is asymmetric: the stress σ_{xz} reaches a greater value than σ_{zx}. The normal stresses in this case decrease with the depth of half-space in the same way as it does with normal stresses for the classical Rayleigh wave.

12.3 Rayleigh Waves in the Half-space of Damaged Material

Currently, the mechanics of damaged media is being intensively developed, which studies both the stress-strain state of media and the accumulation of damage in their materials. Damage is usually understood as a reduction in the elastic response of the body due to a reduction in the effective area that transfers internal forces from one part of the body to another part of it, which, in turn, is caused by the appearance

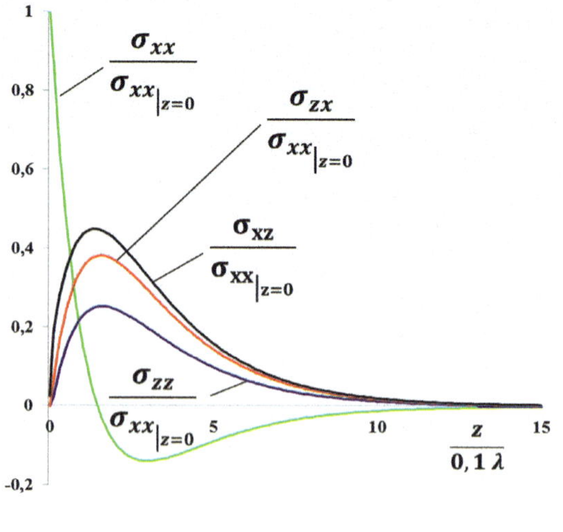

Fig. 12.4 Dependences of stress amplitudes in a surface wave on depth

and development of a scattered field of microdefects (microcracks – in elasticity, dislocations – in plasticity, micropores – in creep, surface microcracks in fatigue) [51, 52, 53].

The continuum damage mechanics has been intensively developed, starting from the fundamental works of Kachanov, generalized in [54], and Rabotnov, summarized in [55]. In traditional calculations, a measure of damage in the process of deformation development can be expressed through the scalar damage parameter $\psi(x,t)$, which characterizes the relative density of microdefects uniformly scattered per unit volume. This parameter is equal to zero when there is no damage, and is close to one at the moment of destruction.

As a rule, in the mechanics of a deformable solid, the problems of dynamics are considered separately from the problems of damage accumulation. When developing such methods, it is customary to postulate in advance that the elastic wave velocity is a given function of damage, and then experimentally determine the proportionality coefficients. The phase velocity of the wave and its attenuation are usually considered to be power functions of frequency and linear damage functions. With undoubted advantages (simplicity), this approach has a number of disadvantages, like any approach that is not based on mathematical models of processes and systems.

Let us consider a self-consistent problem with damage to the material, which includes the dynamic equation of the theory of elasticity and the kinetic equation of damage accumulation in the material of the medium. We assume that the damage is uniformly distributed in the material of the medium. Let us study the propagation of a surface wave along the free boundary of the damaged half-space. We assume that all processes are homogeneous along the third axis.

In [56, 57, 58] the problem to be self-consistent is considered, which includes, in addition to the damage development equation, the dynamic equation of elasticity theory:

$$\begin{aligned} \rho\frac{\partial^2 \boldsymbol{u}}{\partial t^2} &= \left(K+\frac{1}{3}G\right)\operatorname{grad}\operatorname{div}\boldsymbol{u} + G\Delta\boldsymbol{u} - \beta_1 \operatorname{grad}\psi, \\ \frac{\partial\psi}{\partial t} &+ \alpha\psi - \beta_2 \operatorname{div}\boldsymbol{u} = 0. \end{aligned} \tag{12.25}$$

Here $\boldsymbol{u}$ – displacement vector, K – modulus of dilatation, G – shear modulus, ρ – density of the material, t – time, $\alpha = \frac{1}{\tau_*}$, β_1, β_2 – constant parameters characterizing the damage of the material and the relationship of cyclic processes with the processes of damage accumulation ($\alpha > 0$); τ_* – relaxation time.

When considering the problem of the propagation of a Rayleigh surface wave in an isotropic elastic half-space in the presence of damage to its material, we restrict ourselves to the two-dimensional case when all processes are homogeneous along the axis x_2. System of Eqs. (12.25) then becomes two-dimensional and acquires the following form:

$$\rho\frac{\partial^2 u_1}{\partial t^2} - (\lambda+2\mu)\frac{\partial^2 u_1}{\partial x_1^2} - \mu\frac{\partial^2 u_1}{\partial x_3^2} - (\lambda+\mu)\frac{\partial^2 u_3}{\partial x_1 \partial x_3} = -\beta_1\frac{\partial\psi}{\partial x_1}, \tag{12.26}$$

$$\rho \frac{\partial^2 u_3}{\partial t^2} - (\lambda + 2\mu) \frac{\partial^2 u_3}{\partial x_3^2} - \mu \frac{\partial^2 u_3}{\partial x_1^2} - (\lambda + \mu) \frac{\partial^2 u_1}{\partial x_1 \partial x_3} = -\beta_1 \frac{\partial \psi}{\partial x_3}, \tag{12.27}$$

$$\frac{\partial \psi}{\partial t} + \alpha \psi = \beta_2 (\lambda + 2\mu) \left(\frac{\partial u_1}{\partial x_1} + \frac{\partial u_3}{\partial x_3} \right), \tag{12.28}$$

where $u_1(x_1, x_3, t)$, $u_3(x_1, x_3, t)$ – displacement vector components along the axes x_1 and x_3 respectively; λ, μ – Lamé's constants.

System of Eqs. (12.26)–(12.27) must be supplemented with boundary conditions expressing the absence of stresses on the boundary of the half-space:

$$\left[\frac{\partial u_1}{\partial x_3} + \frac{\partial u_3}{\partial x_1} \right] \Bigg|_{x_3=0} = 0, \tag{12.29}$$

$$\left[\frac{\partial u_3}{\partial x_3} + \left(1 - 2\frac{c_\tau^2}{c_l^2} \right) \frac{\partial u_1}{\partial x_1} \right] \Bigg|_{x_3=0} = 0, \tag{12.30}$$

where $c_l = \sqrt{\frac{\lambda+2\mu}{\rho}}$ and $c_\tau = \sqrt{\frac{\mu}{\rho}}$ – propagation velocities of dilation and shear waves in an infinite medium.

Eliminating the damage function from system of Eqs. (12.26)–(12.27), we obtain two equations for longitudinal and transverse displacements:

$$\begin{aligned} &\frac{\partial^2 u_1}{\partial t^2} - c_l^2 \left(1 - \frac{\beta_1 \beta_2}{\alpha} \right) \frac{\partial^2 u_1}{\partial x_1^2} - c_\tau^2 \frac{\partial^2 u_1}{\partial x_3^2} - \left(c_m^2 - c_l^2 \frac{\beta_1 \beta_2}{\alpha} \right) \frac{\partial^2 u_3}{\partial x_1 \partial x_3} + \\ &+ \frac{1}{\alpha} \frac{\partial^3 u_1}{\partial t^3} - \frac{c_l^2}{\alpha} \frac{\partial^3 u_1}{\partial x_1^2 \partial t} - \frac{c_\tau^2}{\alpha} \frac{\partial^3 u_1}{\partial x_3^2 \partial t} - \frac{c_m^2}{\alpha} \frac{\partial^3 u_3}{\partial x_1 \partial x_3 \partial t} = 0, \end{aligned} \tag{12.31}$$

$$\begin{aligned} &\frac{\partial^2 u_3}{\partial t^2} - c_l^2 \left(1 - \frac{\beta_1 \beta_2}{\alpha} \right) \frac{\partial^2 u_3}{\partial x_3^2} - c_\tau^2 \frac{\partial^2 u_3}{\partial x_1^2} - \left(c_m^2 - c_l^2 \frac{\beta_1 \beta_2}{\alpha} \right) \frac{\partial^2 u_1}{\partial x_1 \partial x_3} + \\ &+ \frac{1}{\alpha} \frac{\partial^3 u_3}{\partial t^3} - \frac{c_l^2}{\alpha} \frac{\partial^3 u_3}{\partial x_3^2 \partial t} - \frac{c_\tau^2}{\alpha} \frac{\partial^3 u_3}{\partial x_1^2 \partial t} - \frac{c_m^2}{\alpha} \frac{\partial^3 u_1}{\partial x_1 \partial x_3 \partial t} = 0, \end{aligned} \tag{12.32}$$

where $c_m = \sqrt{\frac{\lambda+\mu}{\rho}}$ $(c_\tau < c_m < c_l)$, note that $c_m^2 = c_l^2 - c_\tau^2$.

Consider a perturbation that propagates along the boundary $x_3 = 0$ and attenuates in the direction of the axis x_3. The components of the displacement vector takes the following form:

$$u_1 = \frac{\partial \varphi}{\partial x_1} - \frac{\partial \theta}{\partial x_3}, \qquad u_3 = \frac{\partial \varphi}{\partial x_3} + \frac{\partial \theta}{\partial x_1}, \tag{12.33}$$

where φ – scalar potential (rot grad $\varphi = 0$), θ – nonzero component of the vector potential $\boldsymbol{\theta} = \{0, \theta, 0\}$ (div rot $\boldsymbol{\theta} = 0$). The expressions for the potentials are as follows:

$$\varphi = A_1 e^{-q_1 x_3 + i(kx_1 - \omega t)}, \qquad \theta = A_2 e^{-q_2 x_3 + i(kx_1 - \omega t)}, \tag{12.34}$$

where ω – frequency, k – wave number, q_1, q_2 – positive coefficients characterizing the decay of perturbations deep into the half-space, A_1, A_2 – arbitrary constants.

From system (12.29)–(12.32), taking into account (12.33)–(12.34), we find the dispersion equation

$$-c_l^2\left(k^2-q_1^2\right)\left(k^2+q_2^2\right)+2c_\tau^2k^2\left(k^2+q_2^2-2q_1q_2\right)=0, \qquad (12.35)$$

in which the coefficients q_1, q_2 are defined with the following relations

$$q_1^2=k^2-\frac{(i\omega-\alpha)\,\omega^2}{(i\omega-\alpha+\beta_1\beta_2)\,c_l^2}, \qquad q_2^2=k^2-\frac{\omega^2}{c_\tau^2}. \qquad (12.36)$$

From relations (12.35), (12.36) it follows that the wave number k is a complex valued $k=k_1+ik_2$. Substituting this expression into the complex dispersion equation and separating the real and imaginary parts, we obtain a system of two nonlinear algebraic equations.

In the limiting case, when there is no damage in the material, the system of algebraic equations is reduced to the dispersion equation for the Rayleigh wave, and the frequency dependence of the wave number is given by the expression

$$\omega^6-8k_1^2\omega^4+8k_1^4\left(3-2a_1^2\right)\omega^2-16k_1^6\left(1-a_1^2\right)=0 \qquad (12.37)$$

where $a_1=\frac{c_\tau}{c_l}$. Dispersion equation (12.37) is written in dimensionless variables $\overline{\omega}, \overline{k}$ ($\omega=\alpha\overline{\omega}$, $k=\frac{\alpha}{c_\tau}\overline{k}$). The dimensionless propagation velocity of the Rayleigh wave is defined as $c_R=\frac{\omega}{k_1}$. It is known that the classical Rayleigh wave propagates along the free boundary of the half-space without attenuation and does not have dispersion, its velocity is a constant value for each material.

It can be seen from (12.35) and (12.36) that in the presence of damage in the material, the Rayleigh wave attenuates during propagation. Frequency dependences of the real and imaginary parts of the wavenumber for a change in the damage parameter and a fixed parameter a_1 are shown in the Figs. 12.5 and 12.6. Parameter a_1 can be expressed through Poisson coefficient (ν):

$$a_1=\sqrt{\frac{1-2\nu}{2-2\nu}}$$

and its value varies within

$$0\le a_1\le\frac{\sqrt{2}}{2}.$$

Dimensionless parameter

$$a_2=\frac{\beta_1\beta_2}{\alpha}$$

characterizes the damage of the material, the sign of this parameter depends on the signs of the initial parameters β_1, β_2, more often $a_2<0$. The dispersion curves have fundamentally different forms for positive and negative values of this parameter.

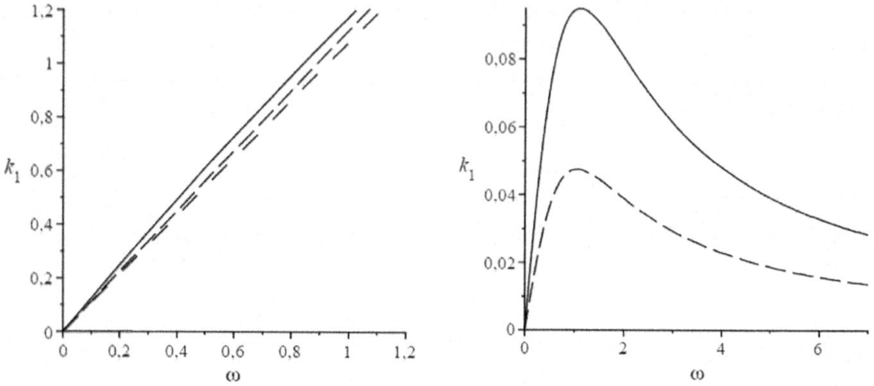

Fig. 12.5: Frequency dependences of the real part of the wave number $k_1(\omega)$ for different values of the parameter a_2: $a_2^{(1)} = -1$ (solid line), $a_2^{(2)}$ (dashed line), $a_2^{(3)} = 0$ (dashed line with spaces), $a_2^{(1)} < a_2^{(2)} < a_2^{(3)}$.

When $a_2 \neq 0$ the dispersion curves of the real and imaginary parts of the frequency have two branches, and the positive values k_1 correspond to positive values k_2, and for negative values k_1 – negative values k_2. Therefore, it is sufficient to consider these dependences in the first quarter. Figure 12.5 shows the dependences $k_1(\omega)$ for different values of the parameter a_2, here are depicted two dependences

$$f_i = k_1\left(\omega, \quad a_2 = a_2^{(i)}\right) - k_1(\omega, a_2 = 0), \qquad i = 1, 2,$$

which clearly show the existing deviations of the curve under consideration from the dispersion line corresponding to the classical Rayleigh wave.

From Figs. 12.6 and 12.7 it can be seen that the dependence $k_2(\omega)$, has a horizontal asymptote at $\omega \to +\infty$, and the damping factor $\gamma = \frac{k_2}{k_1}$ tends to zero at large values ω. When $a_2 = 0$ the graph of the dependence of the real part of the wave number on frequency is a straight line, while the imaginary part is absent.

The dependences of phase

$$v_{ph} = \frac{\omega}{k_1}$$

and group

$$v_{gr} = \frac{d\omega}{dk_1}$$

velocities on frequency are shown in Fig. 12.8. It can be seen that for a fixed non-zero value of the parameter a_2 the group velocity curve is located above the phase velocity curve; dispersion is evident at low frequencies. With a decrease in the absolute value of the damage factor, the amplitudes of both curves decrease, while the value of the phase velocity in the entire frequency range increases, and the value of the group velocity increases at lower frequencies and decreases at higher frequencies; the

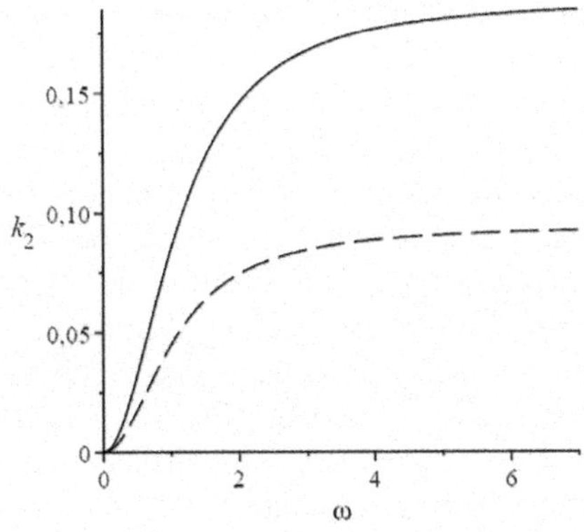

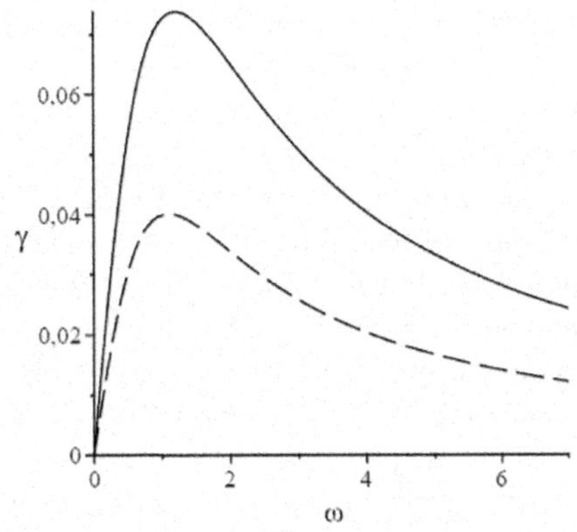

Fig. 12.6: Frequency dependences of the imaginary part of the wave number $k_2(\omega)$ for different values of the parameter a_2: $a_2^{(1)} = -1$ (solid line), $a_2^{(2)}$ (dashed line), $a_2^{(1)} < a_2^{(2)} < 0$.

Fig. 12.7: Attenuation coefficient dependences $\gamma(\omega)$ for different values of the parameter a_2: $a_2^{(1)} = -1$ (solid line), $a_2^{(2)}$ (dashed line), $a_2^{(1)} < a_2^{(2)} < 0$.

Fig. 12.8 Dependences of phase $v_{ph}(\omega)$ (1) and group velocities $v_{gr}(\omega)$ (2) velocities at different values of the parameter a_2: $a_2^{(1)} = -1$ (solid line), $a_2^{(2)}$ (dashed line), $a_2^{(3)} = 0$ (dashed line with spaces), $a_2^{(1)} < a_2^{(2)} < a_2^{(3)}$

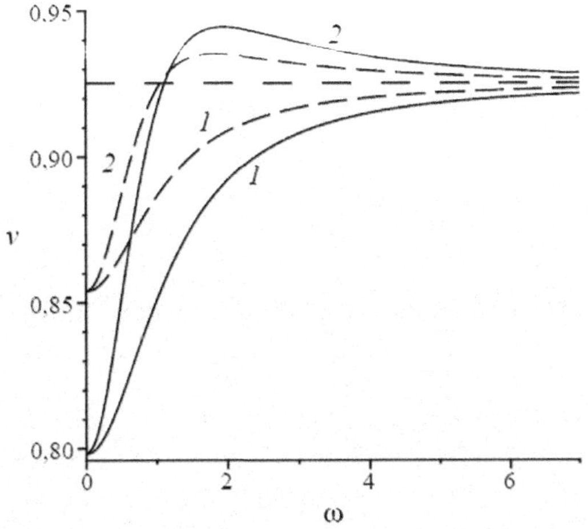

maximum of the group velocity function drops and shifts towards low frequencies. When $a_2 = 0$ curves of phase and group velocities coincide $v_{ph} = v_{gr} = \text{const}$, where the constant is given by equation (12.37) and depends only on Poisson coefficient.

12.4 Conclusion

We consider a simplified (reduced) dynamic model of a Cosserat medium, which occupies an intermediate position between the classical dynamic theory of elasticity and the proper Cosserat medium model, which has asymmetry in the stress tensor and the presence of moment stresses. In contrast to the latter, in the simplified model, three of the six elastic constants are zero and, as a result, there is no moment stress tensor.

In the two-dimensional formulation for the model of a reduced medium, the problem of the propagation of an elastic surface wave along the half-space boundary was solved. The solution of the equations was described as the sum of the scalar and vector potentials, and only one component of the vector potential is nonzero. It is shown that such a wave, in contrast to the classical surface Rayleigh wave, has a dispersion. The phase velocity of the surface wave in the entire frequency range exceeds the phase velocity of the bulk shear wave. The stresses and displacements arising in the zone of propagation of the surface wave are calculated.

In the presence of damage in the medium, surface waves attenuate in the process of propagation along the boundary of the half-space and have dispersion. The presence of damage contributes to the appearance of insignificant dispersion in the low-frequency range. The smaller the value of the damage coefficient, the less its manifestation. The dispersion is anomalous. In the absence of damage in the medium, the surface wave propagates without dispersion and attenuation.

Acknowledgements The work was supported by Russian Science Foundation (project 20-19-00613).

References

[1] Lord Rayleigh (1985) On waves propagated along the plane surface of an elastic solid, Proceedings of the London Mathematical Society **S1-17**(1):4-11. DOI 10.1112/plms/s1-17.1.4
[2] Aki K, Richards PG (2002) Quantitative Seismology: Theory and Methods, 2nd ed, New York, Freeman, University Science Books.
[3] Krylov VV (1994) On the theory of railway-induced ground vibrations, Journal de Physique IV Proceedings **4**(C5):769-772. DOI 10.1051/jp4:19945167
[4] Krylov VV, Ferguson CC (1994) Calculation of low-frequency ground vibrations from railway trains, Applied Acoustics **42**:199-213. DOI 10.1016/0003-682X(94)90109-0
[5] Krylov VV (1995) Generation of ground vibrations by superfast trains, Applied Acoustics. **44**:49-164. DOI 10.1016/0003-682X(95)91370-I
[6] Krylov VV (1996) Vibrational impact of high-speed trains. I. Effect of track dynamics, Journal of the Acoustical Society of America **101**(6):3121-3134. DOI 10.1121/1.417123
[7] Krylov VV (1997) Spectra of low-frequency ground vibrations generated by high-speed trains on layered ground, Journal of Low Frequency Noise, Vibration and Active Control **16**(4):257-270. DOI 10.1177/026309239701600404

[8] Krylov VV (1998) Effect of track properties on ground vibrations generated by high-speed trains. Acustica united with Acta Acustica **84**(1):78-90.

[9] Krylov VV (1999) Ground vibration boom from high-speed trains, Journal of Low Frequency Noise, Vibration and Active Control **18**(4):207-218. DOI 10.1177/026309239901800405

[10] Krylov VV (Ed) (2001) Noise and Vibration from High Speed Trains, London, Thomas Telford Publishing.

[11] Vesnitsky AI (2001) Waves in Systems with Moving Boundaries and Loadings (in Russ), Moscow, Fizmatlit.

[12] Denisov GG, Novikov VV, Kugusheva EK (1985) On the problem of the stability of one-dimensional unbounded elastic systems, Journal Applied Mathematics and Mechanics **49**(4):533-537. DOI 10.1016/0021-8928(85)90065-6

[13] Butova SV, Gerasimov SI, Erofeev VI, Kamchatnyi VG (2015) Stability of high-speed objects moving along a rocket track guide, J Mach Manuf Reliab **44**:1-5. DOI 10.3103/S1052618815010021

[14] Gerasimov SI, Erofeev VI (2016) Calculation of flexural-and-torsional vibrations of a rocket track rail, J Mach Manuf Reliab **45**:211-213. DOI 10.3103/S1052618816030055

[15] Gerasimov SI, Erofeev VI, Kamchatnyi VG, Odzerikho IA (2018) The sliding contact condition in stability analysis of stage motion for a rocket sled track facility, J Mach Manuf Reliab **47**:221-226. DOI 10.3103/S105261881803007X

[16] Ermolov IN, Lange YV (2004) Ultrasound Control (in Russ), vol 3 of Kluev VV (Ed) Nondestructive Testing: a Guide in 7 Volumes, Moscow, Mashinostroenie.

[17] Gulyaev YV (2005) Acoustoelectronics (historical review), Physics – Uspekhi **48**:847-855. DOI 10.1070/PU2005v048n08ABEH002840

[18] Maugin GA, Metrikine AV (Eds) (2010) Mechanics of Generalized Continua: on Hundred Years after the Cosserats, Advanced in Mathematics and Mechanics, vol 21, Springer, Berlin. DOI 10.1007/978-1-4419-5695-8

[19] Altenbach H, Maugin GA, Erofeev V (Eds) Mechanics of Generalized Continua, Advanced Structured Matherials, vol 7, Springer, Berlin-Heidelberg. DOI 10.1007/978-3-642-19219-7

[20] Altenbach H, Forest S, Krivtsov A (Eds) (2013) Generalized Continua as Models with Multi-Scale Effects or Under Multi-Field Actions, Advanced Structured Matherials, vol 22, Springer, Berlin-Heidelberg. DOI 10.1007/978-3-642-36394-8

[21] Altenbach H, Eremeyev VA (Eds) (2013) Generalized Continua – from the Theory to Engineering Applications, CISM International Centre for Mechanical Sciences, vol 541,Springer, Wien. DOI 10.1007/978-3-7091-1371-4

[22] Bagdoev AG, Erofeyev VI, Shekoyan AV (2016) Wave Dynamics of Generalized Continua, Advanced Structured Materials, vol 24, Springer, Berlin-Heidelberg. DOI 10.1007/978-3-642-37267-4

[23] Altenbach H, Forest S (Eds) (2016) Generalized Continua as Models for Classical and Advanced Materials, Advanced Structured Matherials vol 42, Springer, Berlin-Heidelberg. DOI 10.1007/978-3-319-31721-2

[24] Maugin GA. (2017) Non-Classical Continuum Mechanics, Advanced Structured Matherials, vol 51, Springer, Singapore. DOI 10.1007/978-981-10-2434-4

[25] dell'Isola F, Eremeyev VA, Porubov A (2018) Advanced in Mechanics of Microstructured Media and Structures, Advanced Structured Matherials, vol 87, Springer, Cham. DOI 10.1007/978-3-319-73694-5

[26] Altenbach H, Pouget J, Rousseau M, Collet B, Michelitsch T (2018) Generalized Models and Non-Classical Approaches in Complex Materials 2, Advanced Structured Matherials, vol 89, Springer, Cham. https://doi.org/10.1007/978-3-319-72440-9

[27] Altenbach H, Pouget J, Rousseau M, Collet B, Michelitsch T (2018) Generalized Models and Non-Classical Approaches in Complex Materials 2, Advanced Structured Matherials, vol 90, Springer, Cham. DOI 10.1007/978-3-319-77504-3

[28] Erofeev V, Porubov A, Sargsyan S (2018) Materials Physics and Mechanics **35** (Nonlinear Wave Dynamics of Generalized Continua, Special issue dedicated to the memory EL Aero and GA Maugin).

[29] Altenbach H, Belyaev A, Eremeyev V (Eds) (2019) Dynamical Processes in Generalized Continua and Structures, Advanced Structured Matherials, vol 103, Springer, Cham. DOI 10.1007/978-3-030-11665-1

[30] Sumbatyan MA (Ed) (2019) Wave Dynamics, Mechanics and Physics of Microstructured Metamaterials, Advanced Structured Matherials, vol 109, Springer, Cham. DOI 10.1007/978-3-030-17470-5

[31] Altenbach H, Müller WH, Abali BE (Eds) (2019) Higher Gradient Matherials and Related Generalized Continua, Advanced Structured Matherials, vol 120, Springer, Cham, DOI 10.1007/978-3-030-30406-5

[32] Erofeev VI, Pavlov IS (2019) Structural Modeling of Metamaterials, Advanced Structured Matherials, vol 144, Springer, Cham. DOI 10.1007/978-3-030-60330-4

[33] Lur'ye SA (2020) On the paradox of anomalous relative bending stiffness of ultrathin beams in the gradient theory of elasticity, Mech Solids **55**(3):340-347. DOI 10.3103/S0025654420030085

[34] Eremeev V.A, Lebedev LP (2020) On solvability of boundary value problems for elastic micropolar shells with rigid inclusions, Mech Solids **55**:852-856. DOI 10.3103/S0025654420050052

[35] Radaev YN (2020) Factorization of the main hyperbolic differential operator of the micropolar elasticity theory, Mech Solids **55**:776-783. DOI 10.3103/S0025654420060126

[36] Tarlakovskii DV, Nguyen VL (2021) Estimation of taking into account the moment properties of the medium on the example of a nonstationary axisymmetric problem, Mech Solids **56**:1013-1019. DOI 10.3103/S0025654421060182

[37] Murashkin EV, Radaev YN (2021) Generalization of the algebraic Hamilton – Cayley theory, Mech Solids **56**:996-1003. DOI 10.3103/S0025654421060145

[38] Vasil'ev VV, Lurie SA, Salov VA (2021) New solution to the problem of a crack in an orthotropic plate under tension, Mech Solids **56**:902-910. DOI 10.3103/S0025654421060200

[39] Murashkin EV, Radaev YN (2022) On theory of oriented rensor elements of area for a micropolar continuum immersed in an external plane space, Mech Solids **57**:205-213. DOI 10.3103/S0025654422020108
[40] Cosserat E, Cosserat F (1909) Theorie des Corps Deformables, Paris: Librairie Scientifique A Hermann et Fils. DOI 10.1038/081067a0
[41] Nowacki W (1986) Theory of Asymmetric Elasticity, Pergamon Press.
[42] Schwartz LM, Jonson DL, Feng S (1984) Vibrational models in granular materials, Physical Review Letters **52**(10):831-834. DOI doi.org/10.1103/PhysRevLett.52.831
[43] Grekova EF (2016) Plane waves in the linear elastic reduced Cosserat medium with a finite axially symmetric coupling between volumetric and rotational strains, Math Mech Solids **21**(1):73-93. DOI 10.1177/1081286515577042
[44] Grekova EF (2018) Waves in elastic reduced Cosserat medium with anisotropy in the term coupling rotational and translational strains or in the dynamic term, In: dell'Isola F, Eremeyev VA, Porubov A (Eds) Advances in Mechanics of Microstructured Media and Structures, Advanced Structured Materials, vol. 87, pp 143-156, Springer, Cham. DOI 10.1007/978-3-319-73694-5_9
[45] Grekova EF (2019) Reduced enhanced elastic continua as acoustic metamaterials, In: Altenbach H, Belyaev A, Eremeyev VA, Krivtsov A, Porubov AV (Eds) Dynamical Processes in Generalized Continua and Structures, Advanced Structured Materials, vol 103, pp 253-268, Springer, Cham. DOI 10.1007/978-3-030-11665-1_14
[46] Le Roux J (1911) Étude géométrique de la torsion et de la flexion dans la déformation infinitésimale d'un milieu continu, Annales scientifiques de l'ÉNS, 3e série **28**:523-579. DOI 10.24033/asens.643
[47] Le Roux J (1913) Recherches sur la géométrie des déformations finies, Annales scientifiques de l'ÉNS, 3e série **30**:193-245. DOI 10.24033/asens.659
[48] Jaramillo TJ (1929) A Generalization of the Energy Function of Elasticity Theory, PhD Dissertation, Chicago, University of Chicago.
[49] Erofeyev VI (2003) Wave Processes in Solids with Microstructure, World Scientific Publishing, New Jersey et al.
[50] Antonov AM, Erofeev VI (2018) Rayleigh wave on the boundary of gradient-elastic semi-space (in Russ), Herald of the Bauman Moscow State Tech Univ, Nat Sci (4):59-72. DOI 10.18698/1812-3368-2018-4-59-72
[51] Maugin GA (1992) The Thermomechanics of Plasticity and Fracture, Cambridge University Press.
[52] Lemaitre J (1992) A Course on Damage Mechanics, Springer, Berlin.
[53] Krajcinovic D (1996) Damage Mechanics, Elsevier Science, Amsterdam.
[54] Kachanov LM (1986) Introduction to Continuum Damage Mechanics, Springer, New York.
[55] Rabotnov YN (1969) Creep Problems in Structural Members, North-Holland Series in Applied Mathematics and Mechanics, North-Holland Publishing Company, Amsterdam.

[56] Erofeev VI, Nikitina EA (2010) The self-consistent dynamic problem of estimating the damage of a material by an acoustic method, Acoustical Physics **56**(4):584-587. DOI 10.1134/S106377101004024X
[57] Erofeev VI, Nikitina EA, Sharabanova AV (2010) Wave propagation in damaged materials using a new generalized continuum, In: Maugin GA, Metrikine AV (Eds) Mechanics of Generalized Continua. One Hundred Years After the Cosserats, Advances in Mechanics and Mathematics, vol 21, Springer, Heidelberg, pp 143–148. DOI doi10.1007/978-1-4419-5695-8_15
[58] Stulov A, Erofeev V (2016) Frequency-dependent attenuation and phase velocity dispersion of an acoustic wave propagating in the media with damages, In: Altenbach H, Forest S (Eds) Generalized Continua as Models for Classical and Advanced Materials, Advances Structured Materials, vol 42, Springer, Cham, pp 413–423. DOI 10.1007/978-3-319-31721-2_19

Chapter 13
Validation of a Hemi-Variational Block-Based Approach to the Modelling of Common In-plane Failures in Masonry Structures

José Manuel Torres Espino, Jaime Heman Espinoza Sandoval, Chuong Anthony Tran, Roberto Fedele, Emilio Turco, Francesco dell'Isola, Luca Placidi, Anil Misra, Francisco James León Trujillo, and Emilio Barchiesi

José Manuel Torres Espino · Jaime Heman Espinoza Sandoval · Francisco James León Trujillo
Universidad de Lima, Instituto de Investigación Científica, Santiago de Surco, Peru,
e-mail: 20181896@aloe.ulima.edu.pe, jhespino@ulima.edu.pe, fleon@ulima.edu.pe

Chuong Anthony Tran · Francesco dell'Isola
Università degli Studi dell'Aquila, Dipartimento di Ingegneria Civile, Edile-Architettura, Ambientale (DICEAA) & Università degli Studi dell'Aquila, International Research Center for the Mathematics and Mechanics of Complex Systems (M&MoCS), L'Aquila, Italy,
e-mail: tcanth@outlook.com, francesco.dellisola@univaq.it

Roberto Fedele
Politecnico di Milano, Dipartimento di Ingegneria Civile e Ambientale (DICA), Milan & Università degli Studi dell'Aquila, International Research Center for the Mathematics and Mechanics of Complex Systems (M&MoCS), L'Aquila, Italy,
e-mail: roberto.fedele@polimi.it

Emilio Turco
Università degli Studi di Sassari, Dipartimento di Architettura, design e urbanistica (DADU), Alghero & Università degli Studi dell'Aquila, International Research Center for the Mathematics and Mechanics of Complex Systems (M&MoCS), L'Aquila, Italy,
e-mail: eturco@uniss.it

Luca Placidi
Uninettuno International Telematic University, Faculty of Engineering, Rome & Università degli Studi dell'Aquila, International Research Center for the Mathematics and Mechanics of Complex Systems (M&MoCS), L'Aquila, Italy,
e-mail: luca.placidi@uninettunouniversity.net

Anil Misra
Civil, Environmental and Architectural Engineering Department, University of Kansas, Lawrence, USA & Università degli Studi dell'Aquila, International Research Center for the Mathematics and Mechanics of Complex Systems (M&MoCS), L'Aquila, Italy,
e-mail: amisra@ku.edu

Emilio Barchiesi
Università degli Studi di Sassari, Dipartimento di Architettura, design e urbanistica (DADU), Alghero, Università degli Studi dell'Aquila, International Research Center for the Mathematics and Mechanics of Complex Systems (M&MoCS), L'Aquila, Italy & Universidad de Lima, Instituto de Investigación Científica, Santiago de Surco, Peru,
e-mail: ebarchiesi@uniss.it

H. Altenbach et al. (eds.), *Sixty Shades of Generalized Continua*,
Advanced Structured Materials 170, https://doi.org/10.1007/978-3-031-26186-2_13

Abstract Numerical simulations of several planar failure modes of masonry structures are presented, based on the model and solving code from a recent hemivariational block-based model inspired from granular micromechanics (Tran et al. [33]). The numerical tests include a comparison with literature results for a constant shearing load, a parametric study of the influence of mortar thickness, and the simulation of common failures such as bending, shear sliding, and rocking. The results show that the studied model is able to capture common failure modes. Conclusions and insights towards the extension of the studied model are then given.

Key words: Masonry structures · Block-Based model · Masonry failure · Variational model · Hemivariational model · numerical simulations

13.1 Introduction

Masonry or infill walls, formed by brick units joined together by mortar, has been used since ancient times and is still common today, including in reinforced concrete buildings. Generally, masonry elements are used to close or delimit certain spaces, although in some cases they help to support the loads to which a structure is attached, fulfilling the function of a bearing wall. When analyzing structures and evaluating their performance in response to a seismic event, it is very common not to take into account the fill walls, since they are frequently categorized as non-structural elements, assuming that they do not contribute to the rigidity of the structure, and therefore become considered as secondary elements. However, post-quake assessments show that the failure or detachment of the infill walls are responsible for large losses of human life or require large investments for repair [1].

Therefore, it is necessary to provide a more in-depth investigation into the behavior of masonry, which is a very complex issue to address, given that its behavior involves a number of factors that present a great variability, such as: the properties of mortar, the properties of bricks or masonry elements, the quality of labor, the thickness of the mortar layer, the type of bricks, etc. [2].

There were several attempts to incorporate infill walls in the analysis of a structure, such as the contribution of [3], which is still used in certain analyses. According to this proposal, the masonry enclosed between two adjacent columns could be replaced by connecting rod-type elements working under axial load. This approach can be considered as a macro-modeling of masonry walls.

However, thanks to computational advances, it is feasible to resort to micro-modeling, where it is possible to replace the bricks with homogeneous units joined by a layer of mortar with homogeneous properties. In this way, the origin of the faults and the monitoring of these faults in a wall could be detected in a much more precise way, using numerical methods such as finite elements [4, 5] or other ones. More specifically, mixed finite element methods have also been used for the elastoplastic [6] and multiscale [7] analysis of historical masonry structures.

There are several forms of micro-modeling that have been developed, as well as various approaches to model the way masonry units interact with mortar. One such modeling approach lies in using variational principles. Those principles have had a fundamental historical role in deriving mechanical models; we refer the interested reader to [8]-[10] for general historical information about those approaches, [11] for a synthetic presentation and interpretation, and [12, 13] for specific insights regarding the development generalized theories derived from variational principles. One of the advantages of such principles is their self-consistency: they lead to mathematically well-posed problems, as presented in [14, 15]. Variational principles can be declined into different specific versions, depending on how refined one wants their model to be. The simplest example would be the principle of minimum potential energy [16]. It can be generalized into other minimization principles, such as least action principle [17]-[19]. As presented by [20], least action principles can themselves be generalized into virtual work or power principles [21]-[23] or, in the case of variational inequalities, maximum-energy release principles [14, 24].

One way to take advantage of the modeling capabilities of variational principles while accounting for a precise geometrical description of the studied system is to base such principles on discrete kinematics. Indeed, the resulting models allow a refined description of the specific characteristics of the system to be modeled [25]-[28]. This makes them most useful when complex behaviours are arising from complex structures [29]-[32]. The specific example of discrete description which is used in the present work is the "block-based" approach to masonry modeling.

The present work gathers precisely the contributions of Tran et al. [33], who propose the micro-modeling of masonry with a discrete approach, where the base element is an "augmented brick" made of a brick (which is assumed rigid) and a layer of surrounding mortar. Such augmented masonry units kinematically have the possibility of planar translations and rotations in response to an external action that can act on the wall. In this approach, given the rigidity of masonry units, it can be considered that there are certain energy barriers that impose fault behaviors only in the mortar.

In fact, many recent studies on the numerical modeling of masonry have been based on the consideration of rigid blocks that interact with each other through their contact surfaces [34]-[36]. These surfaces, when interacting, tend to compress the mortar at the same time that a cutting effect is generated that tries to cause the sliding of one unit over another, so the role of friction is vital. The development of these approaches is based on variational calculation and Heyman's limit analysis.

The interest of this work has focused on expanding the proposal of Tran et al. [33], which is a numerical modeling of masonry walls based on micro-modeling. Indeed, the model proposed in their work was adapted from continuous granular micromechanics models. Such models have been studied in many recent research efforts [37]-[41], in order to model precisely the grain interactions and their effect on the global behavior [42]-[44]. More specifically, the model studied by Tran et al. is inspired from a recent hemivariational approach to granular micromechanics [24]-[45].

In the present work, we will study how such a model is able to predict several types of faults which have been reported as common in the evaluation of walls. In this way, different masonry walls will be subjected to a progressive monotonous load, evidencing different types of failures, as well as validate the model with other studies.

13.2 Mathematical Formulation

The present mathematical formulation is based on the model reported in [33], where more details can be seen for this approach. First, a masonry structure is considered as being formed by rigid bricks with a layer of mortar around. As seen in Fig. 13.1, each of these mortar brick blocks that make up the structure will be regular and identical, and will be called augmented bricks. Since the (internal) bricks are rigid, only the mortar will suffer deformations and damage.

13.2.1 Vertex Springs and Stiffnesses

The elastic behavior of the mortar will be modeled using two springs at the extremities of the contact surface between every pair of neighboring augmented bricks. The elongation of the spring, applying elementary operations of linear algebra, can be calculated as:

$$\underline{\Delta r_v} = \left[\underset{\sim}{R}_{\theta'} - \underset{\sim}{I}\right] \cdot \underline{C'S'} - \left[\underset{\sim}{R}_{\theta} - \underset{\sim}{I}\right] \cdot \underline{CS} + u_\eta \underline{\eta} + u_\tau \underline{\tau} \tag{13.1}$$

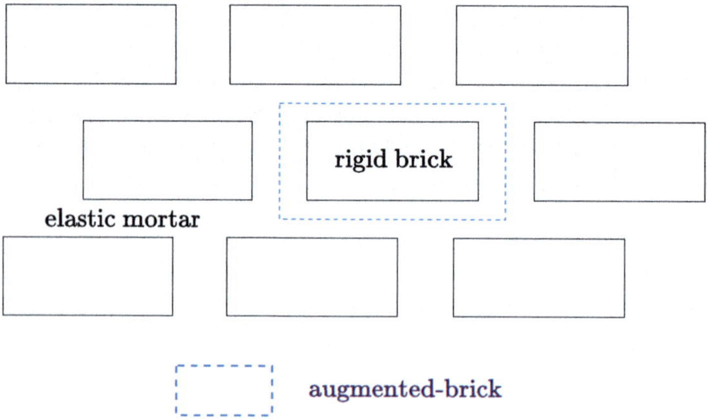

Fig. 13.1: An augmented-brick is defined as a rigid brick together with its surrounding layer of elastic mortar [33].

Where $\underline{\Delta r_\nu}$ represents the elongation of a vertex spring ν, and $\underset{\sim}{R}_\theta$, $\underset{\sim}{R}_{\theta'}$ are rotation matrices. The rest of the parameters correspond to geometric values that can be observed in Figure 13.2. This equation indicates that if we know the relative rotations and the relative displacements between a pair of augmented bricks, we can know the elongation of the two vertex springs that join them.

The stiffnesses $k^{\eta,t}_{\nu,d}$, $k^{\eta,c}_{\nu,d}$ and $k^{\tau}_{\nu,d}$ are now defined. The indices η, t, c, ν, τ, d mean: normal, tension, compression, vertex spring, tangential and damage respectively. In addition, damages $D^{\eta,t}_{\nu}$, $D^{\eta,c}_{\nu}$ and D^{τ}_{ν} are introduced as belonging to $[0,1]$. They are the normal (tension and compression) and tangential damages, respectively, associated with each stiffness. f_S is a surface factor that takes into account the contact surface area between the augmented bricks. It can be expressed as half the ratio of

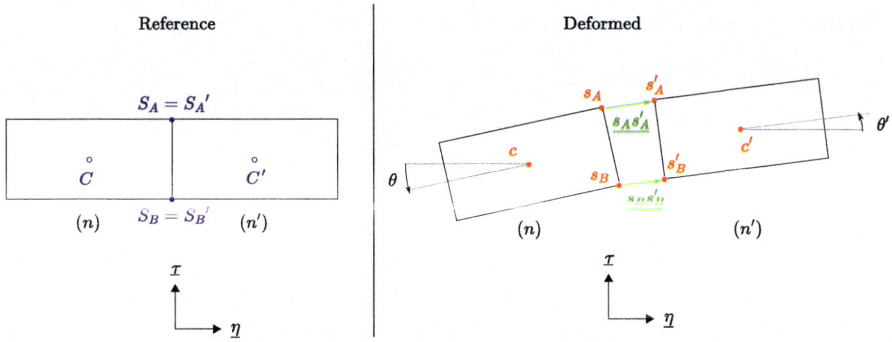

(a) Vertex springs for horizontal neighbours.

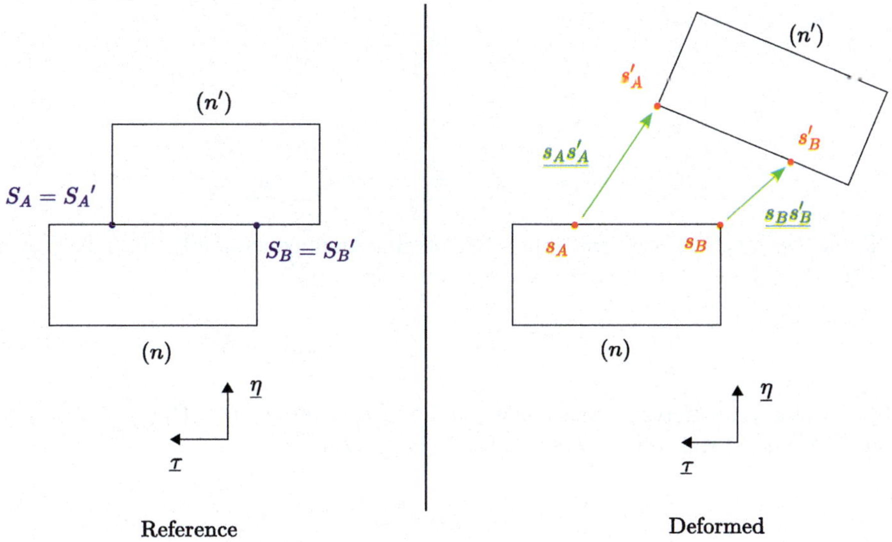

(b) Vertex springs for vertical neighbours.

Fig. 13.2: Vertex springs for a pair of augmented-bricks (n, n') [33].

the length of the shared segment and the total length of the associated brick.

$$\begin{aligned} k^{\eta,t}_{\nu,d} &= fs\, k^{\eta,t}_{\nu}\left(1 - D^{\eta,t}_{\nu}\right), && \text{(normal, tension)} \\ k^{\eta,c}_{\nu,d} &= fs\, k^{\eta,c}_{\nu}\left(1 - D^{\eta,c}_{\nu}\right), && \text{(normal, compression)} \\ k^{\tau}_{\nu,d} &= fs\, k^{\tau}_{\nu}\left(1 - D^{\tau}_{\nu}\right), && \text{(tangential)} \end{aligned} \tag{13.2}$$

13.2.2 Deformation Energy and Impenetrability Potential

Because the internal bricks are rigid, an energy barrier will be used to prevent them from overlapping. For this purpose, the elastic deformation energy associated with the relative displacement of a vertex spring is defined as:

$$^{(n,n')}\mathcal{U}^{\nu} = \frac{C_{\nu}}{\underline{\Delta r_{\nu}} \cdot \underline{\eta} - \sigma} + \frac{1}{2} k^{\tau}_{\nu,d}\left(\underline{\Delta r_{\nu}} \cdot \underline{\tau}\right)^2 \tag{13.3}$$

The compression component of this potential energy referred to a fixed pair of augmented bricks (n,n') can have two forms. It shall be quadratic until the normal relative displacement $\underline{\Delta r_{\nu}} \cdot \underline{\eta}$ reaches a limit u^{*}_{η} (defined as negative by being in compression). Beyond that limit, the normal potential energy becomes an inverse function, which has its asymptote equal to σ, whose absolute value is equal to half the thickness of the mortar. So, as normal relative displacement tends to σ, potential energy tends to infinity. In this way, the normal potential energy, depending on its relative displacement, becomes:

$$\begin{cases} \mathcal{U}^{\nu}_{\eta,\text{wall}} = \dfrac{C_{\nu}}{\left(\underline{\Delta r_{\nu}} \cdot \underline{\eta} - \sigma\right)} + \mathcal{E}_0 & \text{if } s < \underline{\Delta r_{\nu}} \cdot \underline{\eta} < u^{*}_{\eta} \\ \mathcal{U}^{\nu}_{\eta,\text{quad}} = \dfrac{1}{2} k^{\eta}_{\nu,d}\left(\underline{\Delta r_{\nu}} \cdot \underline{\eta}\right)^2 & \text{if } u^{*}_{\eta} < \underline{\Delta r_{\nu}} \cdot \underline{\eta} \end{cases} \tag{13.4}$$

C_{ν} and $\mathcal{E}_0$ are parameters that adjust the potential for impenetrability. It can be shown that, when the values of these new parameters are:

$$u^{*}_{\eta} = \frac{\sigma}{3}, \quad C_{\nu} = \frac{4}{27} k^{\eta}_{\nu,d} |\sigma|^3, \quad \mathcal{E}_0 = -\frac{1}{6} k^{\eta}_{\nu} \sigma^2,$$

the resulting function becomes twice continuously differentiable. This will allow the calculation of local rigidities detailed in later sections.

13.2.3 Damage Laws

The dissipated energy can be defined for each vertex spring as a function of the damage introduced in the stiffness of the Eq. (13.2). These functions are described by Timofeev et al. [45] where the formulation can be seen in more detail. The main idea of this dissipated energy is that it takes into account the anisotropic behavior that exists in granular materials, where it is known that there is a greater resistance in compression than in tension and/or shear (tangential). As described in [45], when applying the considered hemivariational principle, the dissipated energy leads to the following damage thresholds, in the normal (tension and compression) and tangential directions:

$$\begin{cases} D_\nu^{\eta,t} = 1 - \exp\left(-\dfrac{\underline{\Delta r_\nu} \cdot \underline{\eta}}{B_\nu^{\eta,t}}\right) \\ D_\nu^{\eta,c} = \dfrac{2}{\pi}\arctan\left(-\dfrac{\underline{\Delta r_\nu} \cdot \underline{\eta}}{B_\nu^{\eta,c}}\right) \\ D_\nu^{\tau} = 1 - \exp\left(-\dfrac{\left|\underline{\Delta r_\nu} \cdot \underline{\tau}\right|}{B_\nu^{\tau}}\right) \end{cases} \tag{13.5}$$

where:

$$B_\nu^\tau = \begin{cases} B_{\nu 0}^\tau, & \text{if } \underline{\Delta r_\nu} \cdot \underline{\eta} > 0 \\ B_{\nu 0}^\tau - \alpha_2 \underline{\Delta r_\nu} \cdot \underline{\eta}, & \text{if } \dfrac{1-\alpha_1}{\alpha_2} B_{\nu 0}^\tau \le \underline{\Delta r_\nu} \cdot \underline{\eta} \le 0 \\ \alpha_1 B_{\nu 0}^\tau, & \text{if } \underline{\Delta r_\nu} \cdot \underline{\eta} < \dfrac{1-\alpha_1}{\alpha_2} B_{\nu 0}^\tau \end{cases} \tag{13.6}$$

The displayed values of B_ν^τ allow damage to continue to increase or to be maintained. The values of $B_{\nu 0}^\tau$, α_1 and α_2 model the asymmetry between tension and compression. It is also important to understand the behavior of these functions in the evolution of damage. It can be observed graphically that the tension damage function $D_\nu^{\eta,t}$ converges to 1 faster than the compression damage function $D_\nu^{\eta,c}$, when the spring elongation takes large values (positive in tension, negative in compression).

13.2.4 Principle of Minimum Potential Energy

Although the base model presented by Timofeev et al. [45] considered a hemivariational principle, in the present work we will follow the same presentation as Tran et al. [33], who take the damage equations as damage laws independently from the elastic equations. Thus, for a fixed damage state of the masonry system, the elastic equations can be derived from a usual principle of minimum potential energy. Such a separation is legitimate since it is used in the solving algorithm on which is based the program used for the numerical simulations presented in the Results section.

The total potential energy is thus defined as the sum of all deformation energies for every augmented brick pairs (n,n') having v_A and v_B as pairs of vertex springs.

$$\mathcal{E} = \sum_{(n,n')} \left({}^{(n,n')}\mathcal{U}^{v_A} + {}^{(n,n')}\mathcal{U}^{v_B} \right) \tag{13.7}$$

Thus, the stationarity condition for the total potential energy shall be given by equalizing the gradient of the total energy to the null vector: $\nabla\mathcal{E} = \underline{0}$.

13.2.5 Numerical Model

For the numerical model used by Tran et al. [33] a system of $N = n_x \times n_y$ bricks arranged in a regular pattern is considered, as seen in Figure 13.3. Each augmented brick is l_x wide and l_y high. The mortar layer between two bricks is l_m thick. A ratio $\alpha \in (0,1)$ is defined to represent the horizontal space or overlap between two adjacent rows.

The numerical variables for the calculation of iteration i are: U^i the list of all kinematic variables (displacements and rotations), D^i the list of all damages, K^i the stiffness matrix and b^i the loads. Those loads will be calculated from imposed displacements, instead of external forces.

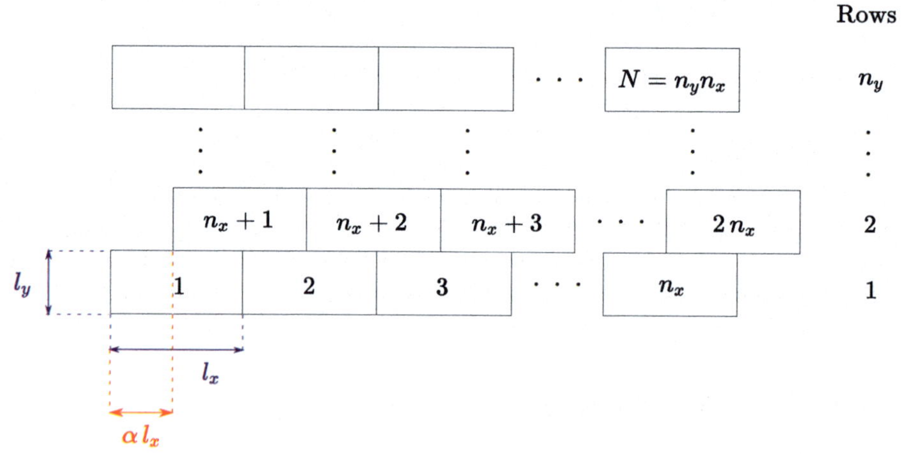

Fig. 13.3: Numerical masonry model considered in this work [33].

13.2.6 Stiffness Matrix

The local stiffnesses for a fixed pair of augmented bricks (n,n') and a fixed pair of vertex springs ν are calculated from the second order gradient of the deformation energy ${}^{(n,n')}\mathcal{U}^{\nu}$. Then, each of these local stiffnesses are assembled into a global stiffness matrix.

$$^{(n,n')}K_{\nu}(u_{\eta},u_{\tau},\theta,\theta') := \nabla^{2}\left[{}^{(n,n')}\mathcal{U}^{\nu}(u_{\eta},u_{\tau},\theta,\theta')\right] \tag{13.8}$$

13.2.7 Algorithm

The algorithm for performing numerical calculations is now described. For an iteration n that has already been calculated, the global displacements and rotations U^{n}, the damages D^{n}, the stiffness matrix K^{n} and the load matrix b^{n}, are known.

For the next step $(n+1)$, the displacements U^{n+1} are unknown, so the increment of the displacement δU^{n+1} will be calculated. It is defined such that:

$$U^{(n+1)} = U^{(n)} + \delta U^{(n+1)} \tag{13.9}$$

In addition, the loads shall be held as follows:

$$b^{(n+1)} = b^{(n)} + \delta b^{(n+1)} \tag{13.10}$$

By elementary matrix operations and by relating the stiffness matrix to displacements and loads, the displacement increase is obtained:

$$\delta U^{(n+1)} = \left(\left[K^{(n)}\right]^{-1} - \left[K^{(n-1)}\right]^{-1}\right)\cdot b^{(n)} + \left[K^{(n)}\right]^{-1}\cdot \delta b^{(n+1)} \tag{13.11}$$

In this way, with equation (11) we can calculate the increment of the displacement δU^{n+1} and with it, perform the calculation of the displacement U^{n+1}, which then allows to update the damage D^{n+1} and the stiffness matrix K^{n+1}.

13.3 Results

In this section, four case studies will be described with the aim of expanding the descriptive results that can be obtained with the model described by Tran et al. [33]. The first corresponds to a comparative result of a simple wall. The second assesses the influence of mortar on masonry behavior, while the last two aim to visualize and explain some common in-plane failure phenomena.

13.3.1 Comparative Result

In order to numerically compare the model formulated by Tran et al. [33], it is intended to evaluate a simple masonry wall subjected to a lateral monotone force on its superior side. Preciado [46] evaluated this scenario by the finite element method in order to validate its proposed model. The study model can be seen graphically in Fig. 13.4.

The geometric parameters of this model are detailed below. The width l_x and height l_y correspond to 0.31 meters and 0.20 meters, respectively. In addition, the mortar thickness l_m is equal to 0.01 meters. It should be taken into account that both the width and height correspond to dimensions of the augmented bricks. The elastic rigidities and damage parameters used in the model can be seen in Table 13.1. In this way, the results were obtained in a Force-Displacement graph of the top of the masonry wall. The result shows a similar behavior with the one evaluated in [46],

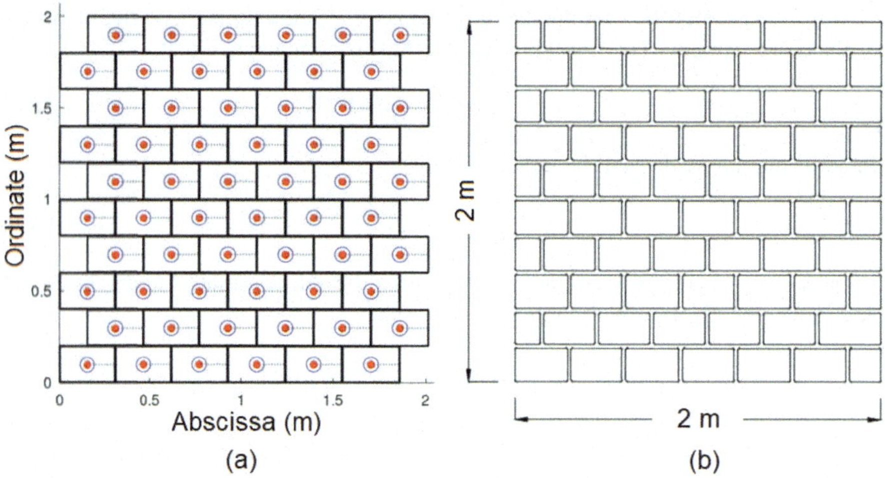

Fig. 13.4: Masonry: (a) Masonry model of the present work. (b) Masonry wall studied in [46].

Table 13.1: Elastic stiffnesses and damage parameters..

Parameter	Value	Unit
$k_\nu^{\eta,t}$	2.05×10^8	N/m
$k_\nu^{\eta,c}$	2.05×10^8	N/m
k_ν^{τ}	2.271×10^8	N/m
$B_\nu^{\eta,t}$	6×10^{-5}	m
$B_\nu^{\eta,c}$	6×10^{-5}	m
$B_{\nu0}^{\tau}$	6×10^{-5}	m
α_1	20	–

where a maximum load of 47 kN is observed with a displacement of approximately 0.8 mm, and a maximum displacement of 1.75 mm, which are graphically compared in Fig. 13.5. As Lourenço indicated in his doctoral thesis [4], the normal and tangential rigidities are directly related to Young's module and the shear modulus, respectively. Where it is also known that the shear modulus is smaller than the Young module in masonry, which is also reflected in the values of the normal and tangential stiffnesses indicated in Table 13.1. On the other hand, it is valid to mention the presence of friction, which is characteristic in masonry, due to the α_1 parameter.

In addition, by analyzing Fig. 13.6 of the evolution of damage in masonry, the behavior of the Force-Displacement curve can be explained. As iterations progress, damage in the normal (compression) and tangential direction increases until the maximum force is reached. Then, damage in the normal direction remains constant, while tangential damage continues to increase until it reaches its maximum, leading to softening.

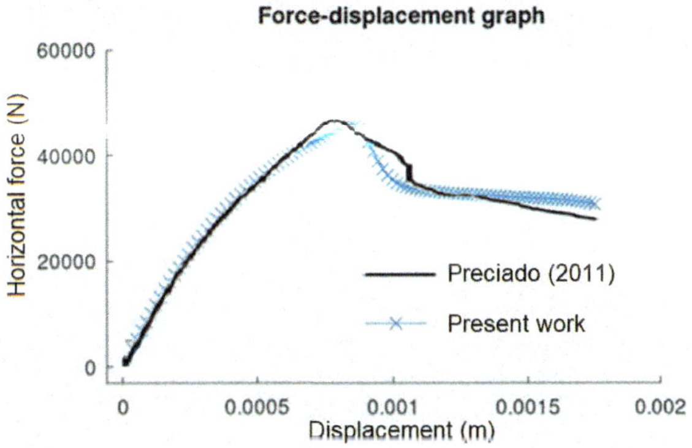

Fig. 13.5 Comparison of force-displacement graphs.

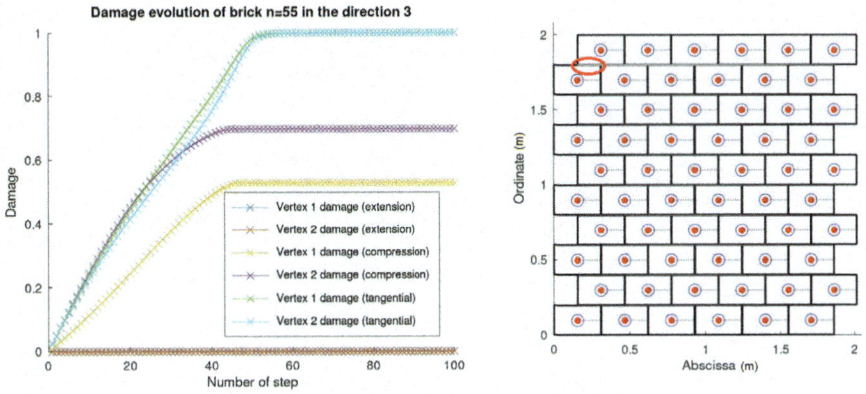

Fig. 13.6: Evolution of damage with respect to iterations in a specific part of the masonry.

13.3.2 Influence of Mortar Thickness on Masonry Performance

In order to evaluate the influence of the mortar thickness on the masonry, three scenarios will be evaluated, where each will have different mortar thicknesses, but the rest of parameters will remain constant (Fig. 13.7). These thicknesses correspond to 10, 20 and 30 mm (Table 13.2). In addition, the elastic rigidities and the parameters constituting the damage correspond to the values of Table 13.1. For the three cases, a standard brick geometry corresponding to 215 x 65 mm in width and height respectively shall be used, where a horizontal monotone force shall be applied to the upper left corner.

When evaluating these scenarios, Force-Displacement curves were obtained for each case. As can be seen in Figure 13.8, as the mortar thickness is reduced, the maximum horizontal force that the masonry supports is reduced. It is common to observe in some studies as in Caldeira et al. [47], where it is reported that a high mortar thickness may be susceptible to cause a weakening in the structural behavior

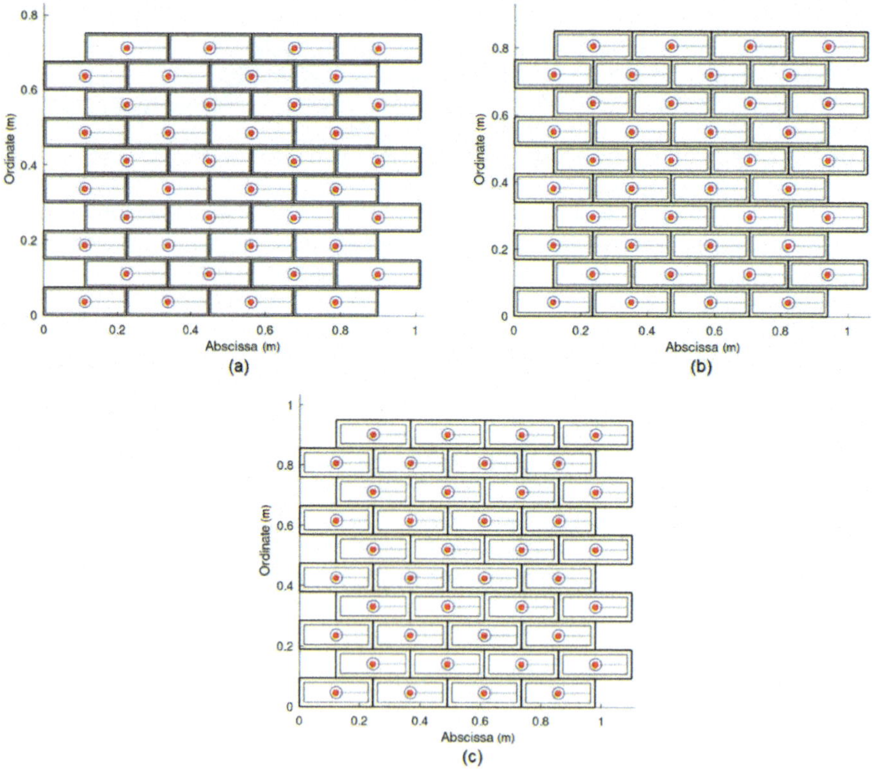

Fig. 13.7: Masonry with different mortar thicknesses. (a) 10 mm (b) 20 mm (c) 30 mm.

Table 13.2: Geometric parameters corresponding to the augmented bricks.

Thickness (l_m)	**Width** (l_x)	**Height** (l_y)
10 mm	0.225 m	0.075 m
20 mm	0.235 m	0.085 m
30 mm	0.245 m	0.095 m

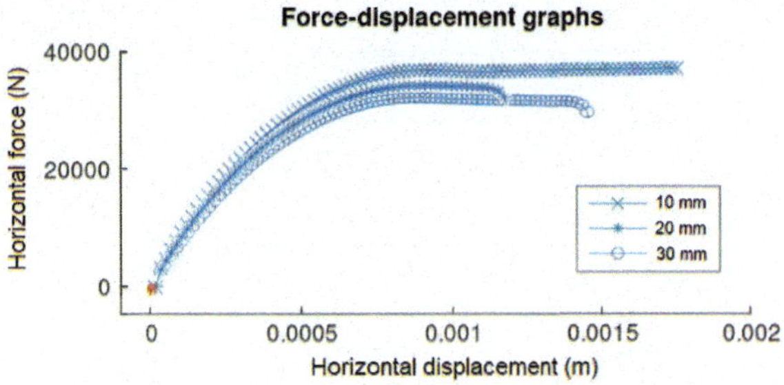

Fig. 13.8 Force-Displacement graphics according to mortar thickness.

of masonry in general. In the model, this could be reflected in the decrease of the maximum force in the Force-Displacement curves.

13.3.3 Bending and Shear Sliding

With the aim of studying the bending and shear sliding phenomena in masonry walls reported in several studies as in [48]-[50], a scenario will be evaluated where the characteristic length of damage *B* will be modified. In the same way, the reference values of Table 13.1 will be taken for a first case and, for the second case, the value of B that will be the same in the normal and tangential direction, will be reduced. The graphical result is shown in Fig. 13.9, where, in the case of the reference value B, the wall tends to produce a bending, whereas, in the case of the reduced value B, the wall tends to slide by shear. It is then said that, the wall will fail by shear as long as the value of the characteristic length of the damage B is reduced, which has a direct relation with the properties of the mortar. Figure 13.10 shows in greater detail the movement of the bricks for each type of wall failure evaluated.

This is corroborated in some studies [51], where it is mentioned that this phenomenon of sliding by shear is produced when poor or weak mortar mixtures are present and there are low axial forces on the top of the wall.

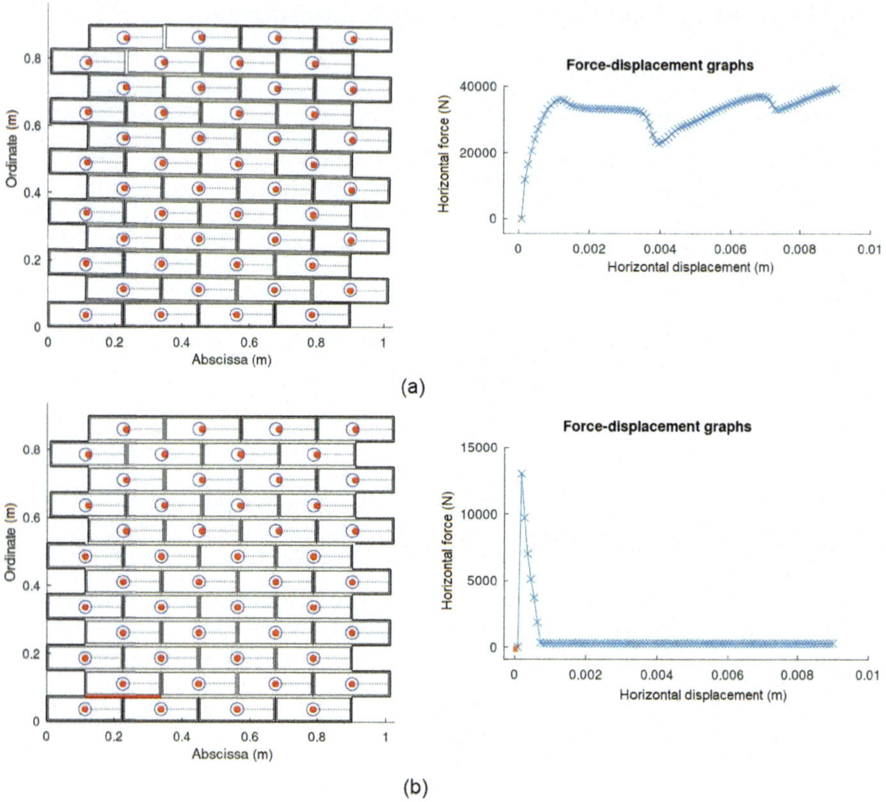

Fig. 13.9: Masonry and its respective Force-Displacement curves. (a) Bending $B = 6 \times 10^{-5}$ m (b) Shear sliding $B = 6 \times 10^{-6}$ m.

Fig. 13.10 Row number five (from the top): (a) bending, (b) shear sliding.

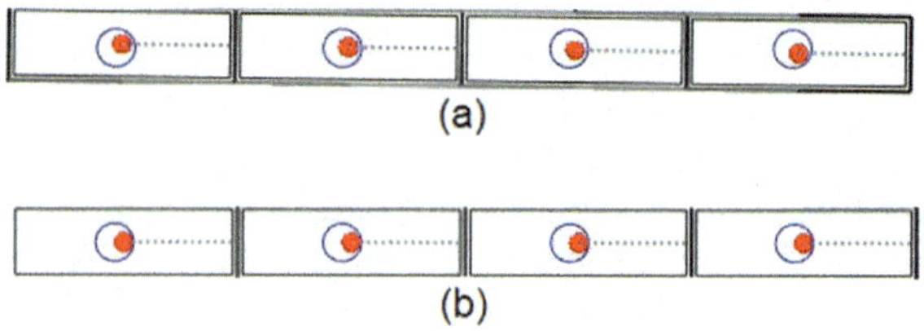

13.3.4 Rocking

The rocking phenomenon has also been reported in numerous studies evaluating in-plane failures of simple masonry walls [48]-[50]. For the present work, it will be sought to simulate the same behavior, depending on the parameters used previously and the displacements imposed on the wall, which are calculated as forces.

The parameters correspond to those of Table 13.1 and Table 13.2 for a masonry of 10 mm thick mortar. The imposed displacements (u_x, u_y) correspond to (0.02 m,

0.02 m). In addition, the damage characteristic length values in Table 13.3 were used. The results in Fig. 13.11 show that due to the vertical and horizontal displacements imposed, the masonry tends to balance with respect to the lower right, where in addition a failure in the base is observed. It is important to note that this also occurs due to the embedding of the base with respect to the wall.

13.4 Conclusions and Future Challenges

The study that has been developed is an extension of what was done by Tran et al. [33], who worked on a structure of nine bricks, while the present work has been developed on structures with a greater number of bricks. Thus, it has been possible to monitor each of the augmented bricks, allowing to detect various types of shear, bending and rocking failures.

The results obtained show an acceptable approximation by the proposal of Tran et al. [33] of previous investigations made in this regard, as is the case of Preciado [46]. From this, it has been possible to detect some characteristic properties in masonry as the case of friction and a tangential stiffness less than the normal stiffness.

The thickness of the mortar layer influences the structural behavior of a wall, although this behavior will depend on whether the wall is subjected to compression or tension. In the case of a wall subjected to compression, as is the case with the load-bearing walls, it has been observed that, the smaller the thickness of the mortar

Table 13.3: Characteristic lengths of damage.

Parameter	Value	Unit
$B_{\nu}^{\eta,t}$	3×10^{-5}	m
$B_{\nu}^{\eta,c}$	3×10^{-5}	m
$B_{\nu0}^{\tau}$	3×10^{-5}	m

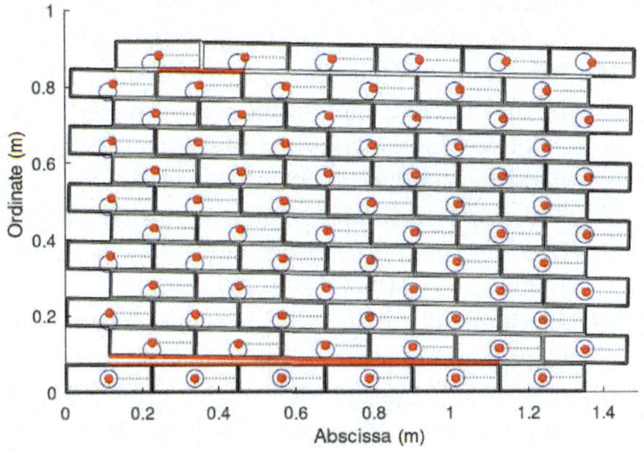

Fig. 13.11 Rocking in masonry wall.

layer, the lower the damage. This has been observed in the present work, for the case of the evaluation of the influence of the mortar thickness on the walls.

While it is true that anti-seismic structures require a certain ductility that allows a controlled deformation, and thereby reduce seismic energy, a sharp reduction in the rigidity of the mortar could cause its resistance to shearing to decrease, causing a potential displacement of one row or set of rows from others, as evidenced by the results of shear sliding.

The model of Tran et al. [33] assumes that a brick has an infinite rigidity, so that the only element that has the possibility of failure is the mortar. This hypothesis should be reviewed and the rigidity of a brick and its possible fracture incorporated into the model. Therefore, the proposed model does not explain the diagonal failure that a wall may present, due to a failure by shearing of the bricks. In this respect, it would be desirable for the model consider the failure lines of the bricks and the evolution of these during the iterations.

References

[1] Derakhshan H, Griffith MC. Ingha, JM (2016) Out-of-plane seismic response of vertically spanning urm walls connected to flexible diaphragms, Earthquake Engineering & Structural Dynamics **45**(4):563-580.

[2] Quinteros RD (2014) Modelización del comportamiento de estructuras de mampostería mediante la teoría de homogeneización, Ph.D. thesis, Tesis doctoral, Facultad de Ingeniería, Universidad Nacional de Salta.

[3] Polyakov S (1960) On the interaction between masonry filler walls and enclosing frame when loaded in the plane of the wall, Translations in Earthquake Engineering **2**(3):36-42.

[4] Lourenço P (1996) Computational strategies for masonry structures [ph. d. thesis], Civil Engineering and Geosciences. Eindhoven, Netherland, Delft University.

[5] Xu C, Xiangli C, Bin L (2012) Modeling of influence of heterogeneity on mechanical performance of unreinforced masonry shear walls, Construction and Building Materials **26**(1):90-95.

[6] Bilotta A, Turco E (2017) Elastoplastic analysis of pressure-sensitive materials by an effective three-dimensional mixed finite element, ZAMM-Journal of Applied Mathematics and Mechanics/Zeitschrift für Angewandte Mathematik und Mechanik **97**(4):382-396.

[7] Tedesco F, Bilotta A, Turco E (2017) Multiscale 3d mixed fem analysis of historical masonry constructions, European Journal of Environmental and Civil Engineering **21**(7-8):772-797.

[8] Berdichevsky VL (2009) Variational principles. In: Variational Principles of Continuum Mechanics, pp 3-44. Springer.

[9] dell'Isola F, Eugster SR, Spagnuolo M, Barchiesi E (2021) Evaluation of Scientific Sources in Mechanics: Heiberg's Prolegomena to the Works of Archimedes

and Hellinger's Encyclopedia Article on Continuum Mechanics, Advanced Structured Materials, vol. 152, Springer Nature.

[10] dell'Isola F, Della Corte A (2020) History of the Principle of Virtual Work. In: Encyclopedia of Continuum Mechanics, pp 1190–1197, Springer.

[11] dell'Isola F, Placidi L (2011) Variational principles are a powerful tool also for formulating field theories, In: Variational models and methods in solid and fluid mechanics, pp 1–15. Springer.

[12] dell'Isola F, Andreaus U, Placidi L (2015) At the origins and in the vanguard of peridynamics, non-local and higher-gradient continuum mechanics: an underestimated and still topical contribution of Gabrio Piola, Mathematics and Mechanics of Solids **20**(8):887-928.

[13] dell'Isola F, Corte AD, Giorgio I (2017) Higher-gradient continua: The legacy of piola, mindlin, sedov and toupin and some future research perspectives, Mathematics and Mechanics of Solids **22**(4):852-872.

[14] Barchiesi E, Hamila N: Maximum mechano-damage power release-based phase-field modeling of mass diffusion in damaging deformable solids, Zeitschrift für angewandte Mathematik und Physik **73**(1):1-21.

[15] Maugin GA (2013) The principle of virtual power: from eliminating metaphysical forces to providing an efficient modelling tool, Continuum Mechanics and Thermodynamics **25**(2):127-146.

[16] Tran CA, Gołaszewski M, Barchiesi E (2020) Symmetric-in-plane compression of polyamide pantographic fabrics–modelling, experiments and numerical exploration, Symmetry **2020**(12):693.

[17] Abali BE, Müller WH, dell'Isola F (2017) Theory and computation of higher gradient elasticity theories based on action principles, Archive of Applied Mechanics **87**(9):1495-1510.

[18] Auffray N, dell'Isola F, Eremeyev VA, Madeo A, Rosi G (2015) Analytical continuum mechanics à la Hamilton–Piola least action principle for second gradient continua and capillary fluids, Mathematics and Mechanics of Solids **20**(4):375-417.

[19] Giorgio I (2021) Lattice shells composed of two families of curved Kirchhoff rods: an archetypal example, topology optimization of a cycloidal metamaterial, Continuum Mechanics and Thermodynamics **33**(4):1063-1082.

[20] dell'Isola F, Seppecher P, Placidi L, Barchiesi E, Misra A (2020) Least action and virtual work principles for the formulation of generalized continuum models, Discrete and Continuum Models for Complex Metamaterials, pp 327-394 (2020)

[21] Germain P (1973) The method of virtual power in continuum mechanics. Part 2: Microstructure, SIAM Journal on Applied Mathematics **25**(3):556-575.

[22] Germain P (2020) The method of virtual power in the mechanics of continuous media, I: Second-gradient theory, Mathematics and Mechanics of Complex Systems **8**(2):153-190.

[23] Maugin G (1980) The method of virtual power in continuum mechanics: application to coupled fields, Acta Mechanica **35**(1):1-70.

[24] Placidi L, Barchiesi E, dell'Isola F, Maksimov V, Misra A, Rezaei N, Scrofani A, Timofeev D (2020) On a hemi-variational formulation for a 2D elasto-plastic-

damage strain gradient solid with granular microstructure, Mathematics in Engineering **5**(1):1-24.
[25] Giorgio I (2020) A discrete formulation of Kirchhoff rods in large-motion dynamics, Mathematics and Mechanics of Solids **25**(5):1081-1100.
[26] Turco E, Barchiesi E, Giorgio I, dell'Isola F (2020) A Lagrangian Hencky-type non-linear model suitable for metamaterials design of shearable and extensible slender deformable bodies alternative to Timoshenko theory, International Journal of Non-Linear Mechanics **123**:103481.
[27] Alibert JJ, Seppecher P, dell'Isola F (2003) Truss modular beams with deformation energy depending on higher displacement gradients, Mathematics and Mechanics of Solids **8**(1):51-73.
[28] Baroudi D, Giorgio I, Battista A, Turco E, Igumnov LA (2019) Nonlinear dynamics of uniformly loaded elastica: Experimental and numerical evidence of motion around curled stable equilibrium configurations, ZAMM-Journal of Applied Mathematics and Mechanics/Zeitschrift für Angewandte Mathematik und Mechanik **99**:e201800121
[29] Turco E, Barchiesi E (2019) Equilibrium paths of Hencky pantographic beams in a three-point bending problem, Mathematics and Mechanics of Complex Systems **7**(4):287–310.
[30] Misra A, Placidi L, Turco E (2020) Variational methods for discrete models of granular materials, Encyclopedia of Continuum Mechanics, pp 2621-2634.
[31] Turco E, Misra A, Pawlikowski M, dell'Isola F, Hild F (2018) Enhanced Piola–Hencky discrete models for pantographic sheets with pivots without deformation energy: numerics and experiments, International Journal of Solids and Structures **147**:94-109.
[32] dell'Isola F, Giorgio I, Pawlikowski M, Rizzi NL (2016) Large deformations of planar extensible beams and pantographic lattices: heuristic homogenization, experimental and numerical examples of equilibrium, Proceedings of the Royal Society A: Mathematical, Physical and Engineering Sciences **472**(2185):20150790.
[33] Tran CA, Barchiesi E, Placidi L, León Trujillo FJ (2021) A block-based variational elasto-damage model for masonry analysis inspired from granular micromechanics: Preliminary study, Mechanics Research Communications **118**:103802.
[34] Gagliardo R, Portioli F, Cascini L, Landolfo R, Lourenço P (2021) A rigid block model with no-tension elastic contacts for displacement-based assessment of historic masonry structures subjected to settlements, Engineering Structures **229**:111609.
[35] Pepe M, Sangirardi M, Reccia E, Pingaro M, Trovalusci P, De Felice G (2020) Discrete and continuous approaches for the failure analysis of masonry structures subjected to settlements, Frontiers in Built Environment **6**:43.
[36] Portioli FP, Godio M, Calderini C, Lourenço PB (2021) A variational rigid-block modeling approach to nonlinear elastic and kinematic analysis of failure mechanisms in historic masonry structures subjected to lateral loads, Earthquake Engineering & Structural Dynamics **50**(12):3332-3354.

[37] Placidi L, Barchiesi E, Misra A, Timofeev D (2021) Micromechanics-based elasto-plastic-damage energy formulation for strain gradient solids with granular microstructure, Continuum Mechanics and Thermodynamics **33**:2213–2241.

[38] Placidi L, Misra A, Barchiesi E (2019) Simulation results for damage with evolving microstructure and growing strain gradient moduli, Continuum Mechanics and Thermodynamics **31**(4):1143-1163.

[39] Misra A, Singh V: Nonlinear granular micromechanics model for multi-axial rate-dependent behavior, International Journal of Solids and Structures **51**(13):2272-2282.

[40] Barchiesi E, Misra A, Placidi L, Turco E (2021) Granular micromechanics-based identification of isotropic strain gradient parameters for elastic geometrically nonlinear deformations, ZAMM-Journal of Applied Mathematics and Mechanics/Zeitschrift für Angewandte Mathematik und Mechanik **101**:e202100059.

[41] Maksimov V. Barchiesi E, Misra A, Placidi L, Timofeev D (2021) Two-dimensional analysis of size effects in strain-gradient granular solids with damage-induced anisotropy evolution, Journal of Engineering Mechanics **147**(11):04021098-1.

[42] Nejadsadeghi N, Misra A (2020) Extended granular micromechanics approach: a micromorphic theory of degree n, Mathematics and Mechanics of Solids **25**(2):407-429.

[43] Giorgio I, dell'Isola F, Misra A (2020) Chirality in 2D Cosserat media related to stretch-micro-rotation coupling with links to granular micromechanics, International Journal of Solids and Structures **202**:28-38.

[44] Misra A, Poorsolhjouy P (2017) Grain-and macro-scale kinematics for granular micromechanics based small deformation micromorphic continuum model, Mechanics Research Communications **81**:1-6.

[45] Timofeev D, Barchiesi E, Misra A, Placidi L (2021) Hemivariational continuum approach for granular solids with damage-induced anisotropy evolution. Mathematics and Mechanics of Solids **26**(5):738-770.

[46] Preciado Quiroz A (2011) Seismic vulnerability reduction of historical masonry towers by external prestressing devices, Ph.D. thesis, Dissertation, Braunschweig, Technische Universität Braunschweig.

[47] Caldeira FE, Nalon GH, de Oliveira DS, Pedroti LG, Ribeiro JCL, Ferreira FA, de Carvalho JMF (2020) Influence of joint thickness and strength of mortars on the compressive behavior of prisms made of normal and high-strength concrete blocks, Construction and Building Materials **234**:117419.

[48] D'Altri AM, Sarhosis V, Milani G, Rots J, Cattari S, Lagomarsino S, Sacco E, Tralli A, Castellazzi G, de Miranda S (2020) Modeling strategies for the computational analysis of unreinforced masonry structures: review and classification, Archives of Computational Methods in Engineering **27**:1153–1185

[49] Dong F, Wang H., Jiang F, Xing Q, Yu J: In-plane shear behavior of masonry panels strengthened with ultra-high ductile concrete (HHDC), Engineering Structures **252**:113609.

[50] Mezrea PE, Ispir M, Balci IA, Bal IE, Ilki A (2021) Diagonal tensile tests on historical brick masonry wallets strengthened with fabric reinforced cementitious mortar, Structures **33**:935-946.
[51] Celano T, Argiento LU, Ceroni F, Casapulla C (2021) Literature review of the in-plane behavior of masonry walls: Theoretical vs. experimental results, Materials **14**(11):3063.

Chapter 14
Size Effects in Cosserat Crystal Plasticity

Samuel Forest and Flavien Ghiglione

Abstract Some Cosserat elastoplasticity models for single crystals are reviewed in the present chapter. Their size-dependent response is evaluated in the case of a simple one-dimensional shear test involving one single slip system and vanishing microrotation prescribed at the boundaries of a material strip of width $2L$. The inhomogeneous distribution of slip in the channel mimics the piling-up of dislocations against the boundaries. The free energy density function depends on the elastic strain and Cosserat curvature tensors. Two types of potentials are examined with respect to the curvature tensor, namely a quadratic function of its norm, on the one hand, and the norm itself, on the second hand. The first model is very often used but turns out to be non-physical since, according to physical metallurgy, the stored energy is proportional to the dislocation density (here the density of geometrically dislocations) rather than its square. The scaling laws predicted by these models are shown to be L^{-2} or L^{-1}, respectively. The latter scaling is reminiscent of Orowan's law of yielding [24]. The chapter ends with the combination of both quadratic and rank one contributions in a unified formulation applied to grain boundary modelling.

Key words: Crystal plasticity · Cosserat elasticity · Cosserat plasticity · Micropolar media · Dislocation density tensor · Dislocation pile-up · Phase field model · Grain boundary

14.1 Introduction

Cosserat modelling has been recognized as a good candidate for continuum modelling of plasticity in crystals due to the relation between inhomogeneous dislocation distributions and lattice curvature [1, 2]. Alternative generalized continuum theories

Samuel Forest · Flavien Ghiglione
Mines Paris PSL University, Centre des Matériaux, CNRS UMR 7633, France,
e-mail: samuel.forest@ensmp.fr, flavien.ghiglione@minesparis.psl.eu

H. Altenbach et al. (eds.), *Sixty Shades of Generalized Continua*,
Advanced Structured Materials 170, https://doi.org/10.1007/978-3-031-26186-2_14

to the Cosserat approach are gradient or micromorphic plasticity, as discussed in [3, 4].

Cosserat crystal plasticity constitutive equations involve elastic contributions, a generalized Schmid law and size–dependent hardening induced by the curvature development, as exposed in the formulations presented in the references [5]-[9]. A critical ingredient of the model is the choice of the dependence of the Helmholtz free energy potential on the curvature tensor. Quadratic dependence has been classically used, especially in strain gradient plasticity models [10, 11]. The quadratic choice was questioned in several contributions like [12, 13], because it predicts unrealistic scaling laws. Alternative potentials include energy functions proportional to the norm of the curvature or dislocation density tensor or its logarithm [14, 15]. Both rank one and logarithmic potentials are non-differentiable at zero curvature which leads to difficulties in implementing these models and requires the consideration of discontinuities in analytical solutions.

To evaluate the various available Cosserat models of elastoplasticity, a simple shear test of a material strip with constrained boundaries is considered, for which analytical solutions can be derived. The microrotation is clamped at the two boundaries, $\theta(-L) = \theta(L) = 0$ and an overall glide amount $\bar{\gamma}$ is applied to the strip. In the following, analytical solutions are provided for quadratic, rank one and combined potentials, considering pure elasticity first, and then crystal plasticity. One single slip system is considered for simplicity and the usual Schmid law is extended to the Cosserat case. In particular the scaling laws with respect to the system size are derived and solutions are compared to existing ones in strain gradient plasticity models involving the dislocation density tensor [15, 16].

In the last section, it is shown that the combined quadratic and rank one potential can be incorporated in a phase field model to describe grain boundary formation in a crystal following [17].

14.2 Problem Setting

14.2.1 Field Equations

Cosserat material points are characterized by their displacement, $\underline{\boldsymbol{u}}$, and microrotation degrees of freedom [18, 19]. The latter microrotation is represented by the axial vector $\underline{\boldsymbol{\theta}}$. Within the small strain, small rotation and small curvature assumption, the Cosserat deformation measures are the relative deformation and curvature tensors defined as

$$\underset{\sim}{\boldsymbol{e}} = \underline{\boldsymbol{u}} \otimes \boldsymbol{\nabla} + \underset{\approx}{\boldsymbol{\epsilon}} \cdot \underline{\boldsymbol{\theta}}, \quad e_{ij} = u_{i,j} + \epsilon_{ijk}\theta_k \tag{14.1}$$

$$\underset{\sim}{\boldsymbol{\kappa}} = \underline{\boldsymbol{\theta}} \otimes \boldsymbol{\nabla}, \quad \kappa_{ij} = \theta_{i,j} \tag{14.2}$$

in an orthonormal basis.

The conjugate stress quantities to the previous deformation measures are the simple stress tensor, $\underset{\sim}{\sigma}$, and the couple stress tensor, $\underset{\sim}{m}$, which fulfil the static balance laws for momentum and moment of momentum

$$\underset{\sim}{\sigma} \cdot \nabla = 0, \quad \sigma_{ij,j} = 0 \tag{14.3}$$

$$\underset{\sim}{m} \cdot \nabla - \underset{\approx}{\epsilon} : \underset{\sim}{\sigma} = 0, \quad m_{ij,j} - \epsilon_{ijk}\sigma_{jk} = 0 \tag{14.4}$$

in the absence of body forces and couples. These field equations are accompanied by appropriate Neumann boundary conditions in the form

$$\underline{t} = \underset{\sim}{\sigma}.\underline{n}, \quad t_i = \sigma_{ij}n_j, \quad \underline{m} = \underset{\sim}{m}.\underline{n}, \quad m_i = m_{ij}n_j \tag{14.5}$$

where $\underline{t}$ and $\underline{m}$ are surface traction and couple stress vectors.

14.2.2 Constitutive Equations

The Cosserat deformation tensor is decomposed into elastic and plastic contributions

$$\underset{\sim}{e} = \underset{\sim}{e}^e + \underset{\sim}{e}^p \tag{14.6}$$

In the present work, no such decomposition is introduced for curvature, for the sake of simplicity following [20, 21].

The Helmholtz free energy density function $\psi(\underset{\sim}{e}^e, \underset{\sim}{\kappa})$ is here assumed to depend on elastic deformation and total curvature. Stress–deformation relations are derived from this free energy potential in the form

$$\underset{\sim}{\sigma} = \frac{\partial \psi}{\partial \underset{\sim}{e}^e}, \quad \underset{\sim}{m} = \frac{\partial \psi}{\partial \underset{\sim}{\kappa}} \tag{14.7}$$

Special forms of the free energy function will be considered in the proposed analysis, including a quadratic potential and more general power law potentials.

Crystal plasticity is based on Schmid's yield function, written here for a single slip system

$$f(\underset{\sim}{\sigma}, \underline{\ell}, \underline{n}) = |\tau| - \tau_c, \quad \text{with} \quad \tau = \underset{\sim}{\sigma} : (\underline{\ell} \otimes \underline{n}) \tag{14.8}$$

where τ is the resolved shear stress and τ_c the critical resolved shear stress. The plastic slip system is characterized by the slip direction $\underline{\ell}$ and the normal to the slip plane $\underline{n}$. Normality is a property of the crystal plasticity flow rule

$$\dot{\underset{\sim}{e}}^p = \dot{\gamma}\frac{\partial f}{\partial \underset{\sim}{\sigma}} = \dot{\gamma}\,\underline{\ell} \otimes \underline{n}\ (\operatorname{sign} \tau) \tag{14.9}$$

where $\dot{\gamma}$ is the plastic multiplier.

The Schmid law is generalized here for Cosserat media accounting for generally non-symmetric stress tensor [3]-[5], [9]. A special case limiting the Schmid law to the symmetric part of the stress tensor will also be considered in the following analysis.

14.2.3 Studied Boundary Value Problem

The one-dimensional simple shear problem of Fig. 14.1 is considered successively for several choices of the free energy potential and Schmid law. It involves a single slip system $(\underline{\boldsymbol{\ell}},\underline{\boldsymbol{n}})$ respectively parallel to the two first vectors of the orthonormal basis $(\underline{\boldsymbol{e}}_1,\underline{\boldsymbol{e}}_2,\underline{\boldsymbol{e}}_3)$. The origin is located at the centre of the strip of width $2L$ and height H. The length H is regarded as infinite so that the solution is invariant along the y−direction.

The unknown displacement and microrotation fields takes the simple form:

$$\underline{\boldsymbol{u}} = \bar{\gamma} y \underline{\boldsymbol{e}}_1 + u(x)\underline{\boldsymbol{e}}_2, \quad \underline{\boldsymbol{\theta}} = \theta(x)\underline{\boldsymbol{e}}_3 \tag{14.10}$$

where the mean shear amount $\bar{\gamma}$ is prescribed to the strip. The total deformation and curvature tensors follow

$$\underset{\sim}{\boldsymbol{e}} = (\bar{\gamma}+\theta)\underline{\boldsymbol{e}}_1 \otimes \underline{\boldsymbol{e}}_2 + (u'-\theta)\underline{\boldsymbol{e}}_2 \otimes e_1, \quad \underset{\sim}{\boldsymbol{\kappa}} = \theta' \underline{\boldsymbol{e}}_3 \otimes \underline{\boldsymbol{e}}_1 \tag{14.11}$$

where $u(x)$ and $\theta(x)$ are the main unknowns of the problem. The notation u' is set for the derivative of $u(x)$ with respect to x.

The following boundary conditions are enforced:

$$u(0) = 0, \quad u(-L) = u(L), \quad \theta(-L) = \theta(L) = 0 \tag{14.12}$$

The first condition sets the rigid body translation and the second conditions correspond to periodicity requirement of the fluctuation $u(x)$. The micro-clamping conditions for the Cosserat rotation mimic the piling-up of dislocations at the left and boundaries.

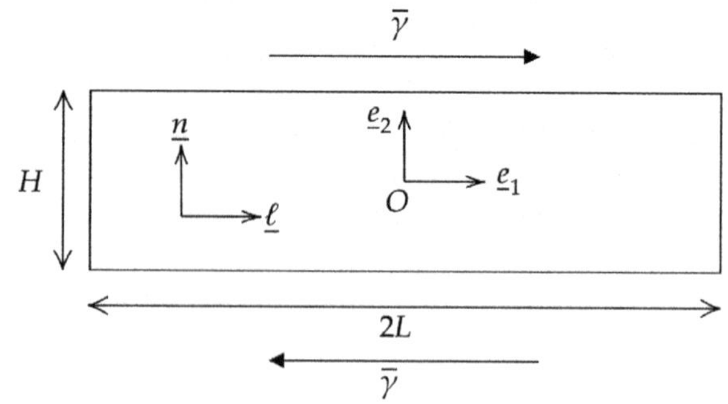

Fig. 14.1 Geometry of the simple shear boundary value problem in crystal plasticity involving a single slip system.

14.3 Cosserat Elastoplasticity Based on a Quadratic Potential

In this section, the Helmholtz free energy potential is assumed to be a quadratic function of the deformation measures

$$\psi(\underset{\sim}{e}^e, \underset{\sim}{\kappa}) = \frac{1}{2}\underset{\sim}{e}^e : \underset{\approx}{\Lambda} : \underset{\sim}{e}^e + \frac{1}{2}\underset{\sim}{\kappa} : \underset{\approx}{C} : \underset{\sim}{\kappa} \tag{14.13}$$

where $\underset{\approx}{\Lambda}$ and $\underset{\approx}{C}$ are the fourth order tensors of Cosserat elastic moduli. Point symmetry was assumed, thus excluding coupling terms between elastic deformation and curvature. The Cosserat elastic laws follow from Eq. (14.7)

$$\underset{\sim}{\sigma} = \underset{\approx}{\Lambda} : \underset{\sim}{e}^e, \quad \underset{\sim}{m} = \underset{\approx}{C} : \underset{\sim}{\kappa} \tag{14.14}$$

14.3.1 Simple Glide in Isotropic Elasticity

The boundary value problem of Sect. 14.2.3 is first solved in the case of linear isotropic elasticity, i.e. in the absence of plasticity. The isotropic elasticity laws read

$$\underset{\sim}{\sigma} = \lambda \operatorname{trace}(\underset{\sim}{e})\underset{\sim}{1} + 2\mu \operatorname{sym}(\underset{\sim}{e}) + 2\mu_c \operatorname{skew}(\underset{\sim}{e}), \tag{14.15}$$

$$\underset{\sim}{m} = \alpha \operatorname{trace}(\underset{\sim}{\kappa})\underset{\sim}{1} + 2\beta \operatorname{sym}(\underset{\sim}{\kappa}) + 2\beta_2 \operatorname{skew}(\underset{\sim}{\kappa}) \tag{14.16}$$

where $\lambda, \mu, \mu_c, \alpha, \beta, \beta_2$ are the six Cosserat elastic moduli required in the isotropic case. In the examples given in the present work, the parameter α is not active and we will assume $\beta_2 = \beta$ for simplicity [22].

In the particular case of simple shear, the non-vanishing components of the stress tensors are

$$\sigma_{12} = (\mu+\mu_c)\bar{\gamma} + (\mu-\mu_c)u' + 2\mu_c\theta, \tag{14.17}$$

$$\sigma_{21} = (\mu-\mu_c)\bar{\gamma} + (\mu+\mu_c)u' - 2\mu_c\theta, \tag{14.18}$$

$$m_{31} = 2\beta\theta' \tag{14.19}$$

The force stress balance tells that $\sigma_{12,2} = \sigma_{21,1} = 0$ so that σ_{21} is uniform. This provides a first differential equation

$$(\mu+\mu_c)u'' - 2\mu_c\theta' = 0 \quad \Longrightarrow \quad u'' = \frac{2\mu_c}{\mu+\mu_c}\theta' \tag{14.20}$$

On the other hand, the couple stress balance equation yields

$$m_{31,1} - (\sigma_{12} - \sigma_{21}) = 0 \tag{14.21}$$

$$\beta\theta'' - \mu_c(\bar{\gamma} - u' + 2\theta) = 0 \tag{14.22}$$

The combination of the previous equations leads to the following differential equation for θ

$$\theta''' = \omega^2 \theta', \quad \text{with} \quad \omega = \sqrt{\frac{2\mu\mu_c}{\beta(\mu+\mu_c)}} \tag{14.23}$$

from which the microrotation function is deduced, after consideration of the boundary conditions $(14.12)_3$,

$$\theta(x) = a(\cosh(\omega x) - \cosh(\omega L)) \tag{14.24}$$

The displacement function then follows from Eq. (14.20):

$$u(x) = \frac{2\mu_c}{\mu+\mu_c}\frac{a}{\omega}\sinh(\omega x) + bx + c \tag{14.25}$$

The boundary conditions $(14.12)_{1,2}$ are used to determine the constants b and c:

$$u(x) = \frac{2\mu_c}{\mu+\mu_c}\frac{a}{\omega}(\sinh(\omega x) - \frac{x}{L}\sinh(\omega L)) \tag{14.26}$$

The integration constant a is finally determined by inserting the found functions in Eq. (14.22):

$$a = \frac{\bar{\gamma}}{2(\cosh(\omega L) - \frac{\mu_c}{\mu+\mu_c}\frac{\sinh(\omega L)}{\omega L})} \tag{14.27}$$

Numerical simulations are carried out using the finite element code Z-Set [23]. More details regarding the implementation of the Cosserat element are given in reference [6]. The parameters given in Table 14.1 are used throughout this work, unless explicitly specified otherwise. Comparisons of the analytical and FEM solutions for the displacement and microrotation fields are plotted in Fig. 14.2, which shows a perfect agreement between the solutions.

Table 14.1: Parameters used for computing analytical and numerical solutions.

Parameter	value
2L [mm]	20
H [mm]	1
E [MPa]	70000
ν	0.3
μ_c [MPa]	10000
α [N]	0
β [N]	26923.8

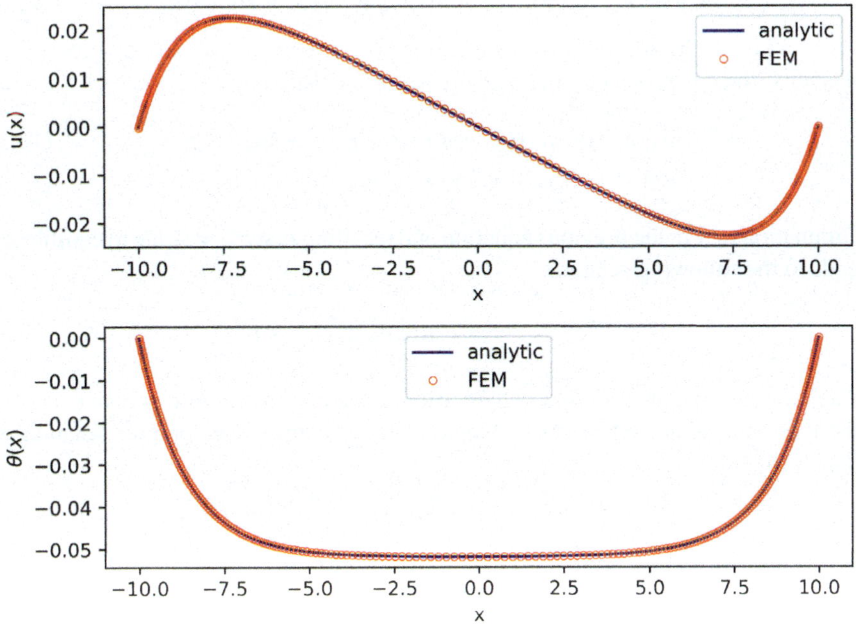

Fig. 14.2: Cosserat elasticity, displacement and microrotation fields. Finite Element simulations are compared to the analytical solution, based on the material parameters of Table 14.1.

14.3.2 Crystal Plasticity Based on the Full Stress Tensor

The resolved shear stress is computed as

$$\tau = \underset{\sim}{\sigma} : (\underline{\ell} \otimes \underline{n}) = \underset{\sim}{\sigma} : (\underline{e}_1 \otimes \underline{e}_2) = \sigma_{12} \tag{14.28}$$

In the plastic regime, assuming positive shear, $\tau = \tau_c$, the critical resolved shear stress is taken as a constant parameter (no hardening). Assuming that this plasticity threshold is reached in the whole specimen, the stress components σ_{12} and σ_{21} are therefore uniform during further straining. Space derivation of Eq. (14.21) implies that

$$m''_{31} = 0 \quad \Longrightarrow \quad \theta''' = 0 \tag{14.29}$$

It follows that the microrotation distribution in the strip is parabolic:

$$\theta(x) = a(x^2 - L^2) \tag{14.30}$$

with the integration constant a. The plastic and elastic deformation tensors take the form

$$\underset{\sim}{e}^p = \gamma \underline{\ell} \otimes \underline{n} = \gamma \underline{e}_1 \otimes \underline{e}_2 \tag{14.31}$$

$$\underset{\sim}{e}^e = \underset{\sim}{e} - \underset{\sim}{e}^p = (\bar{\gamma} + \theta - \gamma)\underline{e}_1 \otimes \underline{e}_2 + (u' - \theta)\underline{e}_2 \otimes \underline{e}_1 \tag{14.32}$$

where $\gamma(x)$ is the plastic slip distribution to be determined. In the presence of plasticity, the equations (14.17) and (14.18) are replaced by

$$\sigma_{12} = (\mu + \mu_c)(\bar{\gamma} - \gamma) + (\mu - \mu_c)u' + 2\mu_c\theta = \tau_c, \tag{14.33}$$

$$\sigma_{21} = (\mu - \mu_c)(\bar{\gamma} - \gamma) + (\mu + \mu_c)u' - 2\mu_c\theta \tag{14.34}$$

Elimination of γ in the previous equation and recalling that $\sigma'_{21} = 0$ due to equilibrium, leads to the following relation

$$u'' = \theta' \quad \Longrightarrow \quad u(x) = a\frac{x}{3}(x^2 - L^2) \tag{14.35}$$

where the displacement boundary conditions have been taken into account. The slip distribution is obtained from (14.33) and the found expressions for microrotation and displacement:

$$\gamma(x) = \bar{\gamma} - \frac{\tau_c}{\mu + \mu_c} + a\left(x^2 - \frac{\mu + 5\mu_c}{\mu + \mu_c}\frac{L^2}{3}\right) \tag{14.36}$$

The coefficient a is determined after insertion of $u(x)$ and $\theta(x)$ in the moment of momentum equation Eq. (14.21) which can be worked out as

$$(\mu + \mu_c)\beta\theta'' - \tau_c\mu_c + 2\mu\mu_c(u' - \theta) = 0 \tag{14.37}$$

Finally

$$a = \frac{\tau_c}{2\beta(1 + \mu/\mu_c) + 4\mu L^2/3} = \frac{\tau_c/2\mu}{\beta(1/\mu + 1/\mu_c) + 2L^2/3} \tag{14.38}$$

To highlight the size effect induced by the Cosserat model, the stress component σ_{21} is computed as a function of the parameter β and structural length L:

$$\sigma_{21} = \frac{\mu - \mu_c}{\mu + \mu_c}\tau_c + \frac{4\mu\mu_c}{\mu + \mu_c}(u' - \theta) \tag{14.39}$$

$$= \frac{\mu - \mu_c}{\mu + \mu_c}\tau_c + \frac{8}{3}\frac{\mu\mu_c}{\mu + \mu_c}aL^2 \tag{14.40}$$

$$= \tau_c\left(\frac{\mu - \mu_c}{\mu + \mu_c} + \frac{4\mu\mu_c}{\mu + \mu_c}\frac{L^2}{3\beta(1 + \mu/\mu_c) + 2\mu L^2}\right) \tag{14.41}$$

The limit case $\mu_c \to \infty$ can be more easily interpreted:

$$\lim_{\mu_c \to \infty} \sigma_{21}/\tau_c = -1 + \frac{4\mu L^2}{3\beta + 2\mu L^2} = \tau_c\frac{1 - 3\beta/2\mu L^2}{1 + 3\beta/2\mu L^2} \tag{14.42}$$

Where $\beta \to 0$, the classical limit $\sigma_{21} = \sigma_{12} = \tau_c$ with a symmetric stress tensor, is retrieved. For vanishing system size $L \to 0$, $\sigma_{21} = -\sigma_{12} = -\tau_c$ and the stress tensor is skew-symmetric.

14.3.3 Schmid Law Limited to the Symmetric Part of the Stress Tensor

A variant of the previous analysis is the consideration of a modification of the generalized Schmid law (14.8):

$$f(\underset{\sim}{\sigma},\underline{\ell},\underline{n}) = |\tau| - \tau_c, \quad \text{with} \quad \text{sym}(\underset{\sim}{\sigma}) : (\underline{\ell} \otimes \underline{n}) \tag{14.43}$$

According to this variant, the resolved shear stress is computed from the projection of the symmetric part of the stress tensor instead of the full stress tensor.

The analysis of the one-dimensional shear layer problem is modified as follows.

$$\tau = (\sigma_{12} + \sigma_{21})/2 = \tau_c \tag{14.44}$$

assuming positive shear loading. This yield condition combined with balance of momentum implies that both σ_{12} and σ_{21} are uniform as in the previous section. It follows that the microrotation profile (14.30) is unchanged in the analysis. This holds true for Eqs. (14.31) and (14.32).

A difference arises in the evaluation of the resolved shear stress

$$\sigma_{12} + \sigma_{21} = 2\mu(\bar{\gamma} - \gamma) + 2\mu u' = 2\tau_c \tag{14.45}$$

but the same relation $u'' = \theta'$ is finally obtained, implying the same displacement profile (14.35). The constant a is determined from Eq. (14.37) which is modified here as

$$\beta\theta'' - \frac{\mu_c}{\mu}\tau_c + 2\mu_c(u' - \theta) = 0 \tag{14.46}$$

and finally

$$a = \frac{\tau_c/2\mu}{\beta/\mu_c + 2L^2/3} \tag{14.47}$$

$$\gamma(x) = \bar{\gamma} - \frac{\tau_c}{\mu} + a(x^2 - L^2/3) \tag{14.48}$$

which are slightly different from (14.38) and (14.36).

The stress components are then computed as

$$\sigma_{12} = \tau_c(1 + \frac{\mu_c}{\mu}) - \frac{4}{3}\mu_c a L^2 \tag{14.49}$$

$$\sigma_{21} = \tau_c(1 - \frac{\mu_c}{\mu}) + \frac{4}{3}\mu_c a L^2 \tag{14.50}$$

The first equation yields the following scaling with the size L of the system

$$\sigma_{12}/\tau_c = 1 + \frac{\beta/\mu}{\beta/\mu_c + 2L^2/3} \tag{14.51}$$

In the limit case,

$$\lim_{\mu_c \to \infty} \sigma_{12}/\tau_c = 1 + \frac{3\beta}{2\mu L^2} \tag{14.52}$$

making the $1/L^2$ scaling clearly visible. This scaling is rather questionable according to physical metallurgy which rather predicts $1/L$ (Orowan) or $1/\sqrt{L}$ (Hall-Petch) scaling laws in plasticity [13, 24].

14.3.4 Comparison with the Curl H^p *Model*

According to the theory developed in [15], the displacement gradient is split into elastic and plastic contributions:

$$\underset{\sim}{\boldsymbol{H}} = \operatorname{grad} \underline{\boldsymbol{u}} = \underset{\sim}{\boldsymbol{H}}^e + \underset{\sim}{\boldsymbol{H}}^p \tag{14.53}$$

The free energy density of the Curl H^p model is taken as a quadratic form

$$\psi(\underset{\sim}{\boldsymbol{\varepsilon}}^e, \operatorname{curl} \underset{\sim}{\boldsymbol{H}}^p) = \frac{1}{2} \underset{\sim}{\boldsymbol{\varepsilon}}^e : \underset{\approx}{\boldsymbol{\Lambda}} : \underset{\sim}{\boldsymbol{\varepsilon}}^e + A \|\operatorname{curl} \underset{\sim}{\boldsymbol{H}}^p\|^2 \tag{14.54}$$

where the elastic strain $\underset{\sim}{\boldsymbol{\varepsilon}}^e$ is the symmetric part of $\underset{\sim}{\boldsymbol{H}}^e$ and A the higher order modulus. The curl H^p model involves a size-dependent back-stress

$$x = -A(\operatorname{curl} \operatorname{curl} H^p) : (\underline{\boldsymbol{\ell}} \otimes \underline{\boldsymbol{n}}) \tag{14.55}$$

The Schmid law is generalized into

$$f(\underset{\sim}{\boldsymbol{\sigma}}, x, \underline{\boldsymbol{\ell}}, \underline{\boldsymbol{n}}) = |\tau - x| - \tau_c \tag{14.56}$$

and

$$\dot{\underset{\sim}{\boldsymbol{H}}}{}^p = \dot{\gamma} \underline{\boldsymbol{\ell}} \otimes \underline{\boldsymbol{n}} \operatorname{sign}(\tau - x) \quad \text{with} \quad \tau = \underset{\sim}{\boldsymbol{\sigma}} : (\underline{\boldsymbol{\ell}} \otimes \underline{\boldsymbol{n}}) \tag{14.57}$$

When applied to the studied boundary value problem, the Curl H^p model predicts the following:

$$\underset{\sim}{\boldsymbol{H}} = \bar{\gamma} \underline{\boldsymbol{e}}_1 \otimes \underline{\boldsymbol{e}}_2 + u' \underline{\boldsymbol{e}}_2 \otimes \underline{\boldsymbol{e}}_1, \quad \underset{\sim}{\boldsymbol{H}}^p = \gamma \underline{\boldsymbol{e}}_1 \otimes \underline{\boldsymbol{e}}_2 \tag{14.58}$$

$$\operatorname{curl} \underset{\sim}{\boldsymbol{H}}^p = -\gamma' \underline{\boldsymbol{e}}_1 \otimes \underline{\boldsymbol{e}}_3, \quad \operatorname{curl} \operatorname{curl} H^p = -\gamma'' \underline{\boldsymbol{e}}_1 \otimes \underline{\boldsymbol{e}}_2, \quad x = -A\gamma'' \tag{14.59}$$

The single non-vanishing component of the stress tensor is

$$\sigma_{12} = \sigma_{21} = \mu(\bar{\gamma} - \gamma + u') \tag{14.60}$$

The Schmid law then stipulates that

$$\sigma_{12} + A\gamma'' = \tau_c \tag{14.61}$$

Equilibrium $\sigma_{21,1} = 0$ requires that the shear stress is uniform. The previous equations then implies that $\gamma''' = 0$ so that the slip distribution is parabolic as in the Cosserat case. Boundary conditions must be chosen to represent the fact that no plasticity occurs at the boundaries:

$$\gamma(\pm L) = 0 \quad \Longrightarrow \quad \gamma(x) = a(L^2 - x_2) \tag{14.62}$$

Note that these conditions are only approximately equivalent to those chosen for the Cosserat medium, namely vanishing lattice rotation $\theta(\pm L) = 0$.

The stress is then related to coefficient a by

$$\sigma_{12} = \tau_c - Aa \tag{14.63}$$

The displacement is derived from Eq. (14.60):

$$\begin{aligned} u' &= \frac{\sigma_{12}}{\mu} - \bar{\gamma} + \gamma \\ u &= (\frac{\tau_c - Aa}{\mu} - \bar{\gamma} + aL^2)x - a\frac{x^3}{3} + Cste \end{aligned} \tag{14.64}$$

The constant vanishes since $u(0) = 0$. The condition $u(-L) = u(L)$ is used to determine the remaining constant a:

$$a = \frac{3}{2L^2}\left(\frac{\tau_c}{\mu} - \bar{\gamma}\right) \tag{14.65}$$

and finally, the shear stress value

$$\sigma_{12} = \tau_c - \frac{3A}{2L^2}\left(\frac{\tau_c}{\mu} - \bar{\gamma}\right) \tag{14.66}$$

The $1/L^2$ scaling is clearly observed and is in agreement with the Cosserat solutions [16]. Quadratic potentials in Cosserat or strain gradient plasticity therefore suffer from the same limitations compared to common knowledge in mechanical metallurgy.

14.4 Rank One Energy Potential

In this section, a non-quadratic free energy potential is adopted

$$\psi(\underset{\sim}{e}^e, \underset{\sim}{\kappa}) = \frac{1}{2}\underset{\sim}{e}^e : \underset{\approx}{\Lambda} : \underset{\sim}{e}^e + A\|\underset{\sim}{\kappa}\|, \quad \text{with} \quad \|\underset{\sim}{\kappa}\| = \sqrt{\underset{\sim}{\kappa} : \underset{\sim}{\kappa}} \tag{14.67}$$

involving the norm of the curvature tensor. The simple and couple stress tensors are then given by

$$\underset{\sim}{\sigma} = \frac{\partial \psi}{\partial \underset{\sim}{e}^e} = \underset{\approx}{\Lambda} : \underset{\sim}{e}^e \tag{14.68}$$

$$\underset{\sim}{m} = \frac{\partial \psi}{\partial \underset{\sim}{\kappa}} = A \frac{\underset{\sim}{\kappa}}{||\underset{\sim}{\kappa}||} \tag{14.69}$$

It is apparent that the couple stress tensor is singular in the case of vanishing curvature.

In the one-dimensional shear problem studied here, the single non-vanishing component of the couple stress tensor is

$$m_{31} = A \frac{\theta'}{|\theta'|} \tag{14.70}$$

14.4.1 Elasticity Solution

In the elastic case, $\underset{\sim}{e}^e \equiv \underset{\sim}{e}$. The microrotation profile is searched for in the following form:

$$\theta(x) = \bar{\theta}(H(x+L) - H(x-L)) \tag{14.71}$$

$$\theta'(x) = \bar{\theta}(\delta(x+L) - \delta(x-L)) \tag{14.72}$$

which means that the microrotation takes the unknown uniform value $\bar{\theta}$ in $]-L:L]$ and is indeterminate at the ends of the interval. The expression involves the Heaviside function and the Dirac distribution such that

$$H'(x-a) = \delta(x-a) \tag{14.73}$$

The uniform rotation field corresponds to the classical solution in the absence of curvature. The difference in the Cosserat case is that curvature energy is now concentrated at the boundaries.

It follows that the couple stress component is indeterminate due to the vanishing of θ'. The total work balance equation is therefore used for a suitable treatment of the distribution functions:

$$\int_V \underset{\sim}{\sigma} : (\underline{u} \otimes \nabla + \underset{\approx}{\epsilon} \cdot \underline{\theta}) + \underset{\sim}{m} : \underset{\sim}{\kappa}\, dV = \int_{\partial V} \underline{t} \cdot \underline{u} + \underline{m} \cdot \underline{\theta}\, dS \tag{14.74}$$

The volume V is the infinite ribbon $[-L:L] \times \mathbb{R}$ and invariance along y is assumed. The application of the divergence theorem and the Neumann condition $\underline{t} = \underset{\sim}{\sigma} \cdot \underline{n}$ are used to eliminate the $\underline{u} \otimes \nabla$ and $\underline{t} \cdot \underline{u}$ terms Eq. (14.74), while the last term vanishes due to the vanishing microrotation boundary conditions. It remains

$$\int_V \underset{\sim}{\sigma} : \underset{\approx}{\epsilon} \cdot \underline{\theta} + \underset{\sim}{m} : (\underline{\theta} \otimes \nabla)\, dV = 0 \tag{14.75}$$

Keeping only the non-vanishing components, the latter becomes

$$\int_{-L}^{L} (\sigma_{12} - \sigma_{21})\theta + m_{31}\theta' \, dx = \int_{-L}^{L} (\sigma_{12} - \sigma_{21})\theta + A|\theta'| \, dx = 0 \tag{14.76}$$

where $m_{31}\theta' = A\theta'^2/|\theta'|$ has been used. After considering Eq. (14.72), the last integral is evaluated as

$$\int_{-L}^{L} A|\theta'|dx = A|\bar{\theta}| \int |(\delta(x+L) - \delta(x-L))|dx = 2A|\bar{\theta}| \tag{14.77}$$

The first contribution in Eq. (14.76) is then evaluated as

$$\int_{-L}^{L} (\sigma_{12} - \sigma_{21})\theta \, ds = \int_{-L}^{L} 2\mu_c(\bar{\gamma} - u' + 2\theta)\theta \, ds = 4\mu_c\bar{\theta}L(\bar{\gamma} + 2\bar{\theta}) \tag{14.78}$$

taking the periodicity of u into account. Finally, the combination of the two found relations provides the value

$$\bar{\theta} = -\frac{\bar{\gamma}}{2} - \frac{A}{4\mu_c L}\mathrm{sign}\bar{\theta} = -\frac{\bar{\gamma}}{2} + \frac{A}{4\mu_c L}\mathrm{sign}\bar{\gamma} \tag{14.79}$$

where $\bar{\theta}$ and $\bar{\gamma}$ have opposite signs for sufficiently high values of μ_c. The limit case $\mu_c \longrightarrow +\infty$ constrains $\bar{\theta}$ to coincide with the material rotation $-\bar{\gamma}/2$.

Finally, the displacement field is obtained from the shear stress component

$$\sigma_{21} = (\mu + \mu_c)u' + (\mu - \mu_c)\bar{\gamma} - 2\mu_c\theta \tag{14.80}$$

According to equilibrium, this component is uniform from which we deduce that $u'' = 0$ so that $u(x)$ is linear.

The couple stress component m_{31} is indeterminate in the interval $]-L:L[$ where θ' vanishes. Let us assume that it is uniform in this interval. This, combined with the balance of moment of momentum equation, implies that the skew symmetric part of the stress tensor also vanishes. Finally,

$$u = (\bar{\gamma} + 2\theta)x = \frac{A}{2\mu_c L}x \, \mathrm{sign}\,\bar{\gamma}, \quad \forall x \in]-L:L[\tag{14.81}$$

This fluctuation vanishes for $\mu_c \longrightarrow \infty$ or vanishing length scale.

14.4.2 Crystal Plasticity

In the crystal plasticity case, the shear stress component σ_{12} is equal to the critical resolved shear stress value. Eq. (14.76) is still valid and now provides the relation

$$(\tau_c - \sigma_{21})\bar{\theta}2L + 2A|\bar{\theta}| = 0 \tag{14.82}$$

Equilibrium still implies that σ_{21} is uniform. Its value follows

$$\sigma_{21} = \tau_c + \frac{A}{L}\text{sign}\bar{\theta} \tag{14.83}$$

The scaling in $1/L$ is clearly visible and is distinct from the $1/L^2$ scaling law found for the quadratic potential.

On the other hand, the elasticity law tells us that

$$\sigma_{12} - \sigma_{21} = 2\mu_c(\bar{\gamma} - \gamma(x) - u'(x)) + 4\mu_c\theta \tag{14.84}$$

Averaging this relation over the interval $[-L : L]$ and assuming periodicity of displacement give

$$\tau_c - \sigma_{21} = 2\mu_c < \bar{\gamma} - \gamma > +4\mu_c\bar{\theta} \tag{14.85}$$

The average $< \bar{\gamma} - \gamma >$ is deduced from Eq. (14.33) as

$$< \bar{\gamma} - \gamma >= (\tau_c - 2\mu_c\bar{\theta})/(\mu + \mu_c) \tag{14.86}$$

and finally

$$\frac{2\mu_c}{\mu + \mu_c}(\tau_c + 2\mu\bar{\theta}) = -\frac{A}{L}\text{sign}\bar{\theta} \tag{14.87}$$

from which the constant $\bar{\theta}$ is derived:

$$\bar{\theta} = -\frac{\tau_c}{2\mu} - \frac{A}{L}\frac{\mu + \mu_c}{4\mu\mu_c}\text{sign}\bar{\theta} \tag{14.88}$$

14.4.3 Comparison with the Curl H^p Model

A rank one potential was also considered in [15, 25]:

$$\psi(\underset{\sim}{\varepsilon}^e, \text{curl}\,\underset{\sim}{\boldsymbol{H}}^p) = \frac{1}{2}\underset{\sim}{\varepsilon}^e : \underset{\approx}{\boldsymbol{\Lambda}} : \underset{\sim}{\varepsilon}^e + A\|\text{curl}\,\underset{\sim}{\boldsymbol{H}}^p\| \tag{14.89}$$

involving the norm of the dislocation density $\underset{\sim}{\textbf{curl}}\,H^p$. The symmetric stress tensor and the higher order stress are then computed as

$$\underset{\sim}{\sigma} = \underset{\approx}{\boldsymbol{\Lambda}} : \underset{\sim}{\varepsilon}^e, \quad \underset{\sim}{\boldsymbol{m}} = A\frac{\text{curl}\,\underset{\sim}{\boldsymbol{H}}^p}{\|\text{curl}\,\underset{\sim}{\boldsymbol{H}}^p\|} \tag{14.90}$$

In the studied problem,

$$\text{curl}\,\underset{\sim}{\boldsymbol{H}}^p = -\gamma'\,\underline{\boldsymbol{e}}_1 \otimes \underline{\boldsymbol{e}}_3, \quad \underset{\sim}{\boldsymbol{m}} = -A\text{sign}\,\gamma' \tag{14.91}$$

It was shown in [15, 25] that the total work balance reduces to the integral

$$\int_V \underset{\sim}{\boldsymbol{s}} : \underset{\sim}{\boldsymbol{H}}^p + \underset{\sim}{\boldsymbol{m}} : \text{curl}\,H^p\,dV = 0 \tag{14.92}$$

where the involved generalized stress tensors fulfil the balance law, $\underset{\sim}{s} + \operatorname{curl} \underset{\sim}{m} = 0$. For the studied problem, this amounts to

$$\int_{-L}^{L} s_{12}\gamma + A|\gamma'|\, dx = 0 \tag{14.93}$$

According to the Schmid law,

$$\sigma_{12} + s_{12} = \tau_c \tag{14.94}$$

where both σ_{12} and s_{12} are uniform.
The plastic slip distribution is uniform in $]-L:L[$:

$$\gamma(x) = \bar{\gamma}(H(x+L) - H(x-L)) \tag{14.95}$$

$$\gamma'(x) = \bar{\gamma}(\delta(x+L) - \delta(x-L)) \tag{14.96}$$

Finally, the previous integral is calculated as

$$(\tau_c - \sigma_{12})2L\bar{\gamma} + 2A\bar{\gamma} = 0 \tag{14.97}$$

so that the stress value

$$\sigma_{12} = \tau_c + \frac{A}{L} \tag{14.98}$$

exhibits a $1/L$ size dependence, in the same way as the previous Cosserat model.

14.5 Combined Potential

In this section, a free energy involving both quadratic and norm potentials is considered

$$\psi(\underset{\sim}{e}^e, \underset{\sim}{\kappa}) = \frac{1}{2}\underset{\sim}{e}^e : \underset{\approx}{\Lambda} : \underset{\sim}{e}^e + \frac{1}{2}\underset{\sim}{\kappa} : \underset{\approx}{C} : \underset{\sim}{\kappa} + A\|\underset{\sim}{\kappa}\|, \quad \text{with} \quad \|\underset{\sim}{\kappa}\| = \sqrt{\underset{\sim}{\kappa} : \underset{\sim}{\kappa}} \tag{14.99}$$

The simple and couple stress tensors are then given by

$$\underset{\sim}{\sigma} = \frac{\partial \psi}{\partial \underset{\sim}{e}^e} = \underset{\approx}{\Lambda} : \underset{\sim}{e}^e \tag{14.100}$$

$$\underset{\sim}{m} = \frac{\partial \psi}{\partial \underset{\sim}{\kappa}} = \underset{\approx}{C} : \underset{\sim}{\kappa} + A\frac{\underset{\sim}{\kappa}}{\|\underset{\sim}{\kappa}\|} \tag{14.101}$$

In the problem considered in this work, the only non-vanishing component of the couple-stress tensor is

$$m_{31} = 2\beta\theta' + A\frac{\theta'}{|\theta'|} \tag{14.102}$$

More precisely,

$$m_{31}(x) = 2\beta\theta - A, \quad \text{if} \quad x < 0 \tag{14.103}$$

$$m_{31}(x) = 2\beta\theta + A, \quad \text{if} \quad x > 0 \tag{14.104}$$

considering the situation where $\theta(x)$ is decreasing (resp. increasing) on $[-L,0]$ (resp. $[0,L]$). In that case, the curvature is indeterminate only at $x = 0$. The field equations (14.21) to (14.23) from section 14.3.1 are valid in the present context. The solution has cosh form but there is a jump of sign θ' at zero. The micro-rotation field is therefore assumed to be of the following form:

$$\theta(x) = a^+\cosh(\omega x) + b^+\sinh(\omega x) + c^+, \quad \text{if} \quad x > 0 \tag{14.105}$$

$$= a^-\cosh(\omega x) + b^-\sinh(\omega x) + c^-, \quad \text{if} \quad x < 0 \tag{14.106}$$

The microtation field is an even function, $\theta(x) = \theta(-x)$, which results in

$$a^+ = a^- = a, \quad b^+ = -b^- = b, \quad c^+ = c^- \tag{14.107}$$

The boundary conditions, $\theta(-L) = \theta(L) = 0$, give

$$c = -(a\cosh(\omega L) + b\sinh(\omega L)) \tag{14.108}$$

The integration constant b can be found by ensuring the continuity of the couple stress at $x = 0$, $m_{31}(0^-) = m_{31}(0^+)$, due to the unhindered transmission of the couple stress vector,

$$2\beta\omega b + A = -2\beta\omega b - A \quad \Longrightarrow \quad b = \frac{-A}{2\beta\omega} \tag{14.109}$$

According to Eqs. (14.103) and (14.104), the continuity of m_{31} results in a jump of θ' at $x = 0$. Finally, the last integration constant a is obtained after integrating the equilibrium equation for the couple-stress

$$\int_{-L}^{L} m_{31,1}dx = \int_{-L}^{L} \sigma_{12} - \sigma_{21}dx \tag{14.110}$$

The couple stress $m_{31}(x)$ being an odd function, the left hand-side of the above equation is evaluated as

$$\int_{-L}^{L} m_{31,1}dx = 2\lim_{\epsilon\to 0}\int_{\epsilon}^{L} m_{31,1}^{+}dx \tag{14.111}$$

$$\int_{-L}^{L} m_{31,1}dx = 4\beta\omega\,(a\sinh(\omega L) + b\cosh(\omega L)) + 2A \tag{14.112}$$

whereas the second integral to be evaluated gives

$$\int_{-L}^{L} \sigma_{12} - \sigma_{21}dx = \int_{-L}^{L} 2\mu_c\,(\bar{\gamma} - u' + 2\theta)\,dx \tag{14.113}$$

$$\int_{-L}^{L} \sigma_{12} - \sigma_{21}dx = 4\mu_c\left[\bar{\gamma}L + 2\left(\frac{a}{\omega}\sinh(\omega L) + \frac{b}{\omega}(\cosh(\omega L) - 1) + cL\right)\right] \tag{14.114}$$

where the periodicity of u and θ has been accounted for. Eq. (14.110) now becomes

$$
\begin{aligned}
&4\beta\omega\,(a\sinh(\omega L) + b\cosh(\omega L)) + 2A \\
&\quad = 4\mu_c\left[\bar{\gamma}L + 2\left(\frac{a}{\omega}\sinh(\omega L) + \frac{b}{\omega}(\cosh(\omega L) - 1) + cL\right)\right]
\end{aligned}
\tag{14.115}
$$

After replacing the constants b and c by their found values one gets

$$
a = \frac{2\mu\mu_c L\bar{\gamma} + A(L\omega(\mu+\mu_c)\sinh(\omega L) + \mu_c(1-\cosh(\omega L))}{4\mu\mu_c L\cosh(\omega L) - 2\mu_c\beta\omega\sinh(\omega L)} \tag{14.116}
$$

The microrotation field is continuous at $x = 0$ but its first derivative experiences a jump at this location.

14.6 Application to Grain Boundary Behaviour

The Cosserat model can be combined with a phase field approach to model grain boundary migration in polycrystals [26]. It originates from Kobayashi, Warren and Carter's model [17, 29] which relies on two phase field variables, the crystal orientation θ and the order parameter, called cristallinity parameter, ϕ. The KWC model is limited here to the 2D case, for which the orientation pseudo-vector reduces to the scalar θ and curvature vector $\nabla\theta$, for the sake of brevity. After showing the connections between the KWC and Cosserat models, we report on an analytical solution for a sharp grain boundary and smooth order parameter profile, after [29, 30]. The KWC model is interpreted as a Cosserat-phase field model of grain boundaries [26].

14.6.1 Cosserat-Phase Field Model of Grain Boundaries

The Helmholtz free energy function of the KWC model is

$$
\psi(\phi,\nabla\phi,\nabla\theta) = f(\phi) + \frac{\alpha^2}{2}||\nabla\phi||^2 + sg(\phi)||\nabla\theta|| + \frac{\varepsilon^2}{2}h(\phi)||\nabla\theta||^2 \tag{14.117}
$$

using the original notations of the authors. The potential (14.117) involves the two functions

$$
f(\phi) = \frac{w^2}{2}(1-\phi)^2, \quad g(\phi) = h(\phi) = \phi^2 \tag{14.118}
$$

and the material parameters $w, \alpha, s, \varepsilon^2$. The situation $\phi = 1$ corresponds to the perfect crystal whereas ϕ is minimal inside the grain boundary as a sign of disorder.

It is apparent that the potential combines a quadratic part for the curvature $\nabla\theta$ and rank one part of the potential. The non-differentiability of the potential at $\nabla\theta = 0$ is the source of mathematical difficulties, already seen in the Sect. 14.4 dedicated to the similar rank-one potential. A mathematical discussion can be found in [31].

The field equations can be derived from the method of virtual power by introducing the following power densities of generalized internal and contact forces:

$$p^{(i)} = a\dot{\phi} + \underline{\boldsymbol{b}} \cdot \boldsymbol{\nabla}\dot{\phi} + 2\overset{\times}{\sigma}\dot{\theta} + \underline{\boldsymbol{m}} \cdot \boldsymbol{\nabla}\dot{\theta} \tag{14.119}$$

$$p^{(c)} = a^c\dot{\phi} + m\dot{\theta} \tag{14.120}$$

The balance laws are derived in the form

$$a = \operatorname{div}\underline{\boldsymbol{b}}, \quad 2\overset{\times}{\sigma} = \operatorname{div}\underline{\boldsymbol{m}} \tag{14.121}$$

The second equation can be interpreted as part of the moment of momentum balance equation for a Cosserat medium, with $\underline{\boldsymbol{m}}$ having the physical dimension of couple stresses. They are complemented by the Neumann conditions

$$a^c = \underline{\boldsymbol{b}} \cdot \underline{\boldsymbol{n}}, \quad m = \underline{\boldsymbol{m}} \cdot \underline{\boldsymbol{n}} \tag{14.122}$$

The Clausius-Duhem inequality can be exploited to propose the following state laws:

$$\underline{\boldsymbol{b}} = \frac{\partial \psi}{\partial \boldsymbol{\nabla}\phi}, \quad \underline{\boldsymbol{m}} = \frac{\partial \psi}{\partial \boldsymbol{\nabla}\theta}, \quad a = \frac{\partial \psi}{\partial \phi} + a^{dis} \tag{14.123}$$

The introduction of the dissipative contribution a^{dis} was proposed by Fried and Gurtin [27, 28]. The residual dissipation rate then is

$$a^{dis}\dot{\phi} + 2\overset{\times}{\sigma}\dot{\theta} \geq 0 \tag{14.124}$$

The following dissipation potential is introduced

$$\Omega(a^{dis}, \overset{\times}{\sigma}) = \frac{1}{2M}(a^{dis})^2 + \frac{\overset{\times}{\sigma}^2}{\eta} \tag{14.125}$$

from which the evolution equations are derived

$$\dot{\phi} = \frac{\partial \Omega}{\partial a^{dis}} = a^{dis}/\beta = (a - \frac{\partial \psi}{\partial \phi})/\beta \tag{14.126}$$

$$\dot{\theta} = \frac{\partial \Omega}{\partial \overset{\times}{\sigma}} = 2\overset{\times}{\sigma}/\eta = (\operatorname{div}\underline{\boldsymbol{m}})/\eta = (\operatorname{div}\frac{\partial \psi}{\partial \boldsymbol{\nabla}\theta})/\eta \tag{14.127}$$

After substitution of the state laws derived from the free energy potential (14.117), the evolution equations are explicited as

$$\beta\dot{\phi} = \alpha^2\Delta\phi + w^2(1-\phi) - 2s\phi||\boldsymbol{\nabla}\theta|| - \varepsilon^2\phi||\theta||^2 \tag{14.128}$$

$$2\eta = \operatorname{div}\left(s\phi^2\frac{\boldsymbol{\nabla}\theta}{||\boldsymbol{\nabla}\theta||} + \varepsilon^2\phi^2\boldsymbol{\nabla}\phi\right) \tag{14.129}$$

The fields $\phi = 1, \nabla\theta = 0$ are stationary solutions of the previous evolution equations. It corresponds to bulk crystal without lattice rotation and perfectly ordered crystal $\phi = 1$.

14.6.2 Analytical Solution of a Single Flat Grain Boundary

Analytical solutions of grain boundary formation with the KWC model is not an easy task. Such a solution can be worked out in the case $\varepsilon = 0$, assuming a sharp interface with respect to orientation and smooth with respect to the order parameter, see [29, 30]. The problem is 1D along x with an infinite flat grain boundary perpendicular to the -axis. The orientation is assumed to be piece-wise uniform with a jump at $x = 0$ in the 1D system. The orientation profile is taken of the form

$$[[\theta]] = \theta^+ - \theta^-, \quad \nabla\theta(x) = [[\theta]]\delta(x) \tag{14.130}$$

We then search for the order parameter field $\phi(x)$, solution of Eq. (14.128):

$$\alpha^2\phi'' + w^2(1-\phi) - 2s\phi|\theta'| = 0 \tag{14.131}$$

The second equation (14.129) becomes

$$\mathrm{div}\left(s\phi^2\frac{\theta'}{|\theta'|}\right) = 0 \tag{14.132}$$

It is identically fulfilled almost everywhere.

For $x \neq 0$, Eq. (14.131) becomes

$$\alpha^2\phi'' + w^2(1-\phi) = 0 \tag{14.133}$$

which provides the solution

$$\phi(x) = C_1^+\exp(-wx/\alpha) + C_2^+\exp(wx/\alpha) + 1, \quad \forall x > 0 \tag{14.134}$$

$$\phi(x) = C_1^-\exp(-wx/\alpha) + C_2^-\exp(wx/\alpha) + 1, \quad \forall x < 0 \tag{14.135}$$

Full cristallinity is enforced at $x = \pm\infty$, $\phi(\pm\infty) = 1$ so that $C_2^+ = C_1^- = 0$. Continuity of ϕ is assumed at $x = 0$, $\phi(0^+) = \phi(0^-) = \phi_{min}$ so that $C_1^+ = C_2^-$ and finally

$$\phi(x) = 1 - C\exp(-w|x|/\alpha), \quad C = 1 - \phi_{min} \tag{14.136}$$

$$\phi'(x) = C\frac{w}{\alpha}\exp(-w|x|/\alpha)\mathrm{sign}\,(x) \tag{14.137}$$

Determination of ϕ_{min}

For that purpose, the field equation (14.131) is integrated over the interval $[-\epsilon:\epsilon]$:

$$\int_{-\epsilon}^{\epsilon} \alpha^2\phi'' dx + \int_{-\epsilon}^{\epsilon} w^2(1-\phi)dx - 2s\int_{-\epsilon}^{\epsilon} \phi|\theta'| = 0 \tag{14.138}$$

Letting $\epsilon \to 0$, the second term vanishes due to continuity of the integrand and it remains

$$\alpha^2[[\phi']] = 2s\phi(0)[[\theta]] \tag{14.139}$$

and finally

$$\alpha^2(\phi'(0^+) - \phi'(0^-)) = 2\alpha^2(1-\phi_{min})\frac{w}{\alpha} = 2s\phi_{min}[[\theta]] \tag{14.140}$$

The last equation provides the value of ϕ_{min}:

$$\phi_{min} = \frac{w\alpha}{w\alpha + s[[\theta]]} = \left(1 + \frac{s}{w\alpha}[[\theta]]\right)^{-1} \tag{14.141}$$

This expression displays two limit cases: vanishing orientation jump gives $\phi_{min}([[\theta]] = 0) = 1$, as it should be, and

$$\lim_{[[\theta]]\to\infty} \phi_{min} = 0 \tag{14.142}$$

which means that lower values of the order parameter are reached for higher misorientation between the grains.

14.6.3 Grain Boundary Energy

The energy of the diffuse grain boundary is computed from the following integral of the energy density over the entire domain

$$\gamma_{GB} = \int_{-\infty}^{+\infty} \frac{w^2}{2}(1-\phi)^2 + \frac{\alpha^2}{2}\phi'^2 + s\phi^2|\theta'|\, dx \tag{14.143}$$

The first two integrals are computed as follows

$$\begin{aligned}\int_{-\infty}^{+\infty} (1-\phi)^2\, dx &= C^2\int_{-\infty}^{+\infty} \exp(-\frac{2w}{\alpha}|x|)\, dx \\ &= \int_{-\infty}^{0} \exp(\frac{2w}{\alpha}x)\, dx + \int_{0}^{+\infty} \exp(-\frac{2w}{\alpha}x)\, dx \\ &= C^2\frac{\alpha}{w} \end{aligned}\tag{14.144}$$

$$\int_{-\infty}^{+\infty} \phi'^2\, dx = C^2\frac{w}{\alpha} \tag{14.145}$$

Finally

$$\begin{aligned}\gamma_{GB} &= C^2\alpha w + s[[\theta]]\phi_{min}^2 \\ &= (1-\phi_{min})^2\alpha w + s[[\theta]]\phi_{min}^2 \\ &= \left(\frac{1}{s[[\theta]]} + \frac{1}{w\alpha}\right)^{-1}\end{aligned} \tag{14.146}$$

This energy vanishes when $[[\theta]]$, as expected. This formula and the choice of the diffuse interface width enable calibration of the parameters s, w, α. It is complemented by proper fitting to the Read-Shockley energy profile for low angle grain boundaries.

After proper non-dimensionalising of the model with respect to ω and introduction of time and length scales [29], the profiles of phase-fields ϕ and θ are given in Fig. 14.3 for arbitrarily chosen parameters $w = 1, \alpha = 0.3, s = 2$. A dimensionless grain boundary energy-misorientation curve is given in Fig. 14.4.

A calibration of the phase-field parameters can also be carried out. Whereas parameter ϵ is a choice allowing to change the interfaces width, parameters w, α, s are given by an optimisation procedure aiming to fit the grain boundary energy-misorientation curve of the model to the Read-Shockley part of the one given by real data. For more details on the calibration procedure the reader is referred to Appendix B of reference [26].

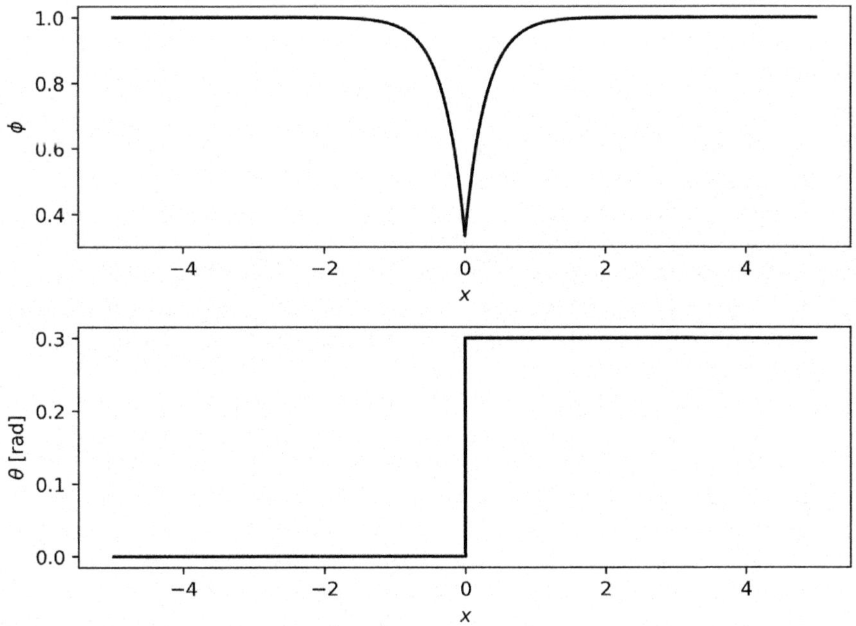

Fig. 14.3: Profiles of ϕ and θ in the sharp interface case.

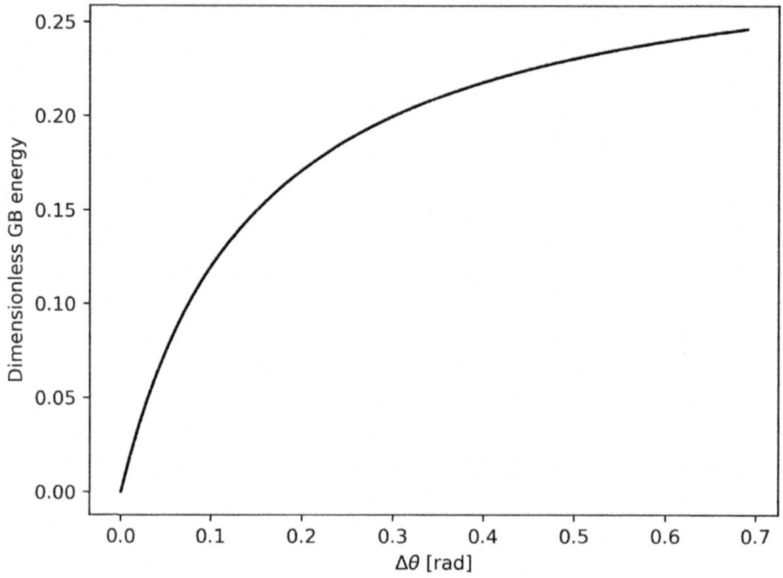

Fig. 14.4: Dimensionless grain boundary energy-misorientation curve.

References

[1] McClintock F, André P, Schwerdt K, Stoeckly R (1958) Interface couples in crystals, Nature **4636**:652–653.

[2] Kröner E (1963) On the physical reality of torque stresses in continuum mechanics, Int J Engng Sci **1**:261–278.

[3] Mayeur JR, McDowell DL, Forest S (2018) Micropolar crystal plasticity, In: GZ Voyiadjis (Ed) Handbook of Nonlocal Continuum Mechanics for Materials and Structures, pp 1–47, Springer International Publishing. DOI 10.1007/978-3-319-22977-5_48-1

[4] Forest S, Mayeur JR, McDowell DL (2018) Micromorphic crystal plasticity, In: GZ Voyiadjis (Ed) Handbook of Nonlocal Continuum Mechanics for Materials and Structures, pp 1–44, Springer International Publishing. DOI 10.1007/978-3-319-22977-5_49-1

[5] Forest S, Cailletaud G, Sievert R (1997) A Cosserat theory for elastoviscoplastic single crystals at finite deformation, Archives of Mechanics **49**(4):705–736.

[6] Forest S, Barbe F, Cailletaud G (2000) Cosserat modelling of size effects in the mechanical behaviour of polycrystals and multiphase materials, International Journal of Solids and Structures **37**:7105–7126.

[7] Mayeur J, McDowell D, Bammann D (2011) Dislocation–based micropolar single crystal plasticity: Comparison of multi– and single criterion theories, Journal of the Mechanics and Physics of Solids **59**:398–422.
[8] Mayeur J, McDowell D (2014) A comparison of Gurtin type and micropolar theories of generalized single crystal plasticity, International Journal of Plasticity **57**:29–51. DOI http://dx.doi.org/10.1016/j.ijplas.2014.01.010
[9] Rys M, Stupkiewicz S, Petryk H (2022) Micropolar regularization of crystal plasticity with the gradient-enhanced incremental hardening law, International Journal of Plasticity **156**:103355. DOI 10.1016/j.ijplas.2022.103355
[10] Gurtin M (2003) On a framework for small–deformation viscoplasticity: free energy, microforces, strain gradients, International Journal of Plasticity **19**:47–90.
[11] Gurtin M, Anand L (2009) Thermodynamics applied to gradient theories involving the accumulated plastic strain: The theories of Aifantis and Fleck & Hutchinson and their generalization, Journal of the Mechanics and Physics of Solids **57**:405–421.
[12] Ohno N, Okumura D (2007) Higher–order stress and grain size effects due to self–energy of geometrically necessary dislocations, Journal of the Mechanics and Physics of Solids **55**:1879–1898.
[13] Forest S (2013) Questioning size effects as predicted by strain gradient plasticity, Journal of the Mechanical Behavior of Materials **22**:101–110.
[14] Forest S, Guéninchault N (2013) Inspection of free energy functions in gradient crystal plasticity, Acta Mechanica Sinica **29**:763–772.
[15] Wulfinghoff S, Forest S, Böhlke T (2015) Strain gradient plasticity modeling of the cyclic behavior of laminate microstructures, Journal of the Mechanics and Physics of Solids **79**:1–20. DOI 10.1016/j.jmps.2015.02.008
[16] Cordero NM, Gaubert A, Forest S, Busso E, Gallerneau F, Kruch S (2010) Size effects in generalised continuum crystal plasticity for two–phase laminates, Journal of the Mechanics and Physics of Solids **58**:1963–1994.
[17] Kobayashi R, Warren JA, Carter WC (2000) A continuum model of grain boundaries, Physica D **140**(1-2):141–150.
[18] Nowacki W (1986) Theory of Asymmetric Elasticity, Pergamon.
[19] Eringen A (1999) Microcontinuum Field Theories, Springer, New York (1999)
[20] Ghiglione F, Forest S (2022) On the torsion of isotropic elastoplastic Cosserat circular cylinders, Journal of Micromechanics and Molecular Physics **6**:1–14. DOI 10.1142/S2424913021420078
[21] Russo R, Girot Mata FA, Forest S (2020) Thermomechanics of cosserat medium: Modeling adiabatic shear bands in metals, Continuum Mechanics and Thermodynamics. DOI 10.1007/s00161-020-00930-z
[22] deBorst R (1993) A generalization of J_2-flow theory for polar continua, Computer Methods in Applied Mechanics and Engineering **103**: 347–362. DOI 10.1016/0045-7825(93)90127-J
[23] Besson J, Foerch R (1997) Large scale object–oriented finite element code design, Computer Methods in Mechanical Engineering **142**:165–187.
[24] Hirth J, Lothe J (1982) Theory of Dislocations, Wiley Intersciences.

[25] Mesarovic S, Forest S, Zbib H (Eds) (2019) Mesoscale Models. From Micro-Physics to Macro-Interpretation, Springer, CISM International Centre for Mechanical Sciences, Vol 587. DOI 10.1007/978-3-319-94186-8

[26] Ask A, Forest S, Appolaire B, Ammar K, Salman OU (2018) A Cosserat crystal plasticity and phase field theory for grain boundary migration, Journal of the Mechanics and Physics of Solids **115**:167-194. DOI https://doi.org/10.1016/j.jmps.2018.03.006

[27] Fried E, Gurtin M (1993) Continuum theory of thermally induced phase transitions based on an order parameter, Physica D **68**:326–343.

[28] Gurtin M (1996) Generalized Ginzburg–Landau and Cahn–Hilliard equations based on a microforce balance, Physica D **92**:178–192.

[29] Warren JA, Kobayashi R, Lobkovsky AE, Carter W (2003) Extending phase field models of solidification to polycrystalline materials, Acta Materialia **51**:6035–6058.

[30] Lobkovsky AE, Warren JA (2001) Sharp interface limit of a phase-field model of crystal grains, Physical Review E **63**:051605.

[31] Giga MH, Giga Y (1998) Evolving graphs by singular weighted curvature, Arch Rational Mech Anal **141**:117–198. DOI 10.1007/s002050050075

Chapter 15
On the Influence of Poisson's Ratio on Phase Transformations Limiting Surfaces

Alexander B. Freidin and Leah L. Sharipova

Abstract The influence of the sign and value of Poisson's ratios of the phases on the limiting surfaces of stress-induced phase transformations is studied. Relationships defining the limiting surfaces in the case of auxetic phases are derived and the difference with the case of positive Poisson's ratio is highlighted. Limiting surfaces for phase transformations in the cases if one or both phases are auxetic materials are compared with the limiting surfaces constructed for 'normal' phases.

Key words: Stress-induced phase transformations · Limiting phase transformation surfaces · Metamaterials · Negative Poisson ratio · Auxetics

15.1 Introduction

Poisson's ratio of a material is defined for a material undergoing tension as the ratio of the lateral contractile strain to the longitudinal tensile strain, i.e., it characterizes, for example, how much a cylindrical specimen becomes thinner or thicker if it is stretched. Most of materials have positive Poisson's ratio. Materials and structures with negative Poisson's ratio (NPR) exhibit a counter-intuitive behaviour. Under uniaxial compression (tension), these materials and structures contract (expand) transversely. Such materials and structures are termed auxetics.

The possibility of negative Poisson's ratio was pointed out by Love [1], who reported $\nu = -1/7$ for some directions in an anisotropic crystal of iron pyrites. In 1985, Kolpakov [2] and Almgren [3] described a material with NPR theoretically. Lakes [4] was apparently the first who produced in 1987 a re-entrant foam structure

Alexander B. Freidin · Leah L. Sharipova
Institute for Problems in Mechanical Engineering RAS, Bolshoy pr., 61, V.O., St. Petersburg, 199178, Russian Federation,
e-mail: alexander.freidin@gmail.com, sleah07@gmail.com

H. Altenbach et al. (eds.), *Sixty Shades of Generalized Continua*,
Advanced Structured Materials 170, https://doi.org/10.1007/978-3-031-26186-2_15

which exhibited NPR. Four years latter, materials with NPR were named 'auxetics' or 'auxetic materials' by Evans et al. [5].

Auxetics possess better mechanical properties if to compare to conventional materials, such as enhanced indentation resistance [6], bending stiffness and shear resistance [7], high dissipated energy per unit volume under compressive cyclic loading [8]-[10] (see also reviews [11]-[17] and monographs [18]-[22]).

Among the fields of application of auxetics, one can mention medical applications [23]-[26], use in materials for sports equipment [27, 28], in the development of new textile fabrics [29, 30] Also the interest to such materials is motivated by intensive developing new technologies of their creation, particulary on the base of 3D printing [31]-[33].

Nowadays, metamaterials, including auxetics, are also the subject of many theoretical studies, see, e.g., monograph [34], papers [35]-[41] and reference therein. In the present paper we will focus on phase transformations associated with auxetics.

Phase transformations from normal material to auxetic are observed in auxetic foams [42]. The foam specimens were transformed from a state with conventional Poisson's ratio to auxetic, returned to conventional and once again to auxetic under multiple mechanical and thermal loading. The phase transition from the normal to the auxetic state was also revealed using molecular dynamics modeling [43] and in an experiment on carbon honeycombs [44].

Earlier we have developed a procedure for the construction of limiting phase transformations surfaces in a strain space for stress-induced phase transformations [45]-[47]. The phase transformation cannot start until the limiting surface is achieved on a given straining path, analogously to the yield surface in plasticity theory. There are two limiting surfaces in the case of a phase transformation: for direct and reverse transformations, and stress-strain diagrams on the transformation pathes were constructed. During these studies, the problems of optimal composite microstructures were also solved: given average strain a minimizing microstructure was found and corresponding energy was derived (see also [48]-[50]). But at that time only the case of positive Poisson's ratios of the phases was implied in derivations.

In the present paper we study in detail how the formulae defining the transformation surfaces and the shapes of the surfaces are changed depending on the sign and value of Poisson's ratios of the phases. Various combinations of the signs of Poisson's ratios of the phases are considered.

The paper is arranged as follows. In Sect. 15.2 we recall the notions of phase equilibrium and phase transition zones and derive the relationships for the construction of phase transition zones in the case of negative Poisson's ratio, highlighting the difference with the case of positive Poisson's ratio.

In Sect. 15.3 we summarize the results on optimal laminates, minimizing microstructures obtained in [45]-[47] and discuss how to construct the phase transformations limiting surfaces in the case of arbitrary Poisson's ratio, including the negative ratio. In Sect. 15.4 phase transformation limiting surfaces for phase transformations 'auxetic $\leftrightarrow$ normal material' and 'auxetic $\leftrightarrow$ auxetic' are constructed and compared with the limiting surface for the case of phases with positive Poisson's ratio. Section 15.5 contains conclusions. In Appendix the formulae related to the construction of the limiting surfaces are derived.

15.2 Phase Equilibrium and Phase Transition Zones for Phases with Positive and Negative Poisson's Ratios

The notion of phase transition zones (PTZ) appeared as a result of the analysis of the conditions across an equilibrium phase boundary (an interface) in an elastic solid [51]-[53]. The PTZ is formed in a strain space by all deformations that can exist at the equilibrium interfaces in a given material, whatever the loading conditions.

The usefulness of the PTZ construction can be motivated as follows

- PTZ is defined exclusively by the material properties (elasticity moduli of the phases, transformation strain) and the energy parameter defined by the temperature.
- Each point of the PTZ boundary corresponds to some piecewise homogeneous two - phase configuration with a plane interface. The external boundaries of the PTZ can be considered as the limiting surfaces of the nucleation of layers of a new phase. The PTZ represents locally all possible equilibrium interfaces.
- The PTZ construction may be related with the interface stability analysis. For elastic solids, it was shown that instability of the interface was not observed if deformations at the interface coincided with deformations at the external boundaries of the PTZ[54]-[56]. Then it was proved that belonging the deformations at the interface to the external boundaries of the PTZ is a necessary stability condition [57, 58].
- Even if a new phase appears in a more complex form than simple layers, in order to construct the phase transformations limiting surfaces, it is necessary to find the maximum of the quadratic form, which is examined when constructing the PTZ [45]-[47].

In the case of small strains, the construction of the PTZ was considered for the general case of isotropic phases but implemented for positive Poisson's coefficients of phases (see, e.g., [52, 53], [59]-[61]. The result with positive Poisson's ratio were used also in finding energy minimizing microstructures and construction of phase transformations limiting surfaces [45]-[47]. In the present paper we focus on the case of negative Poisson's ratio. For the sake of completeness, we repeat briefly some general derivations and results.

In the case of small strains a problem on equilibrium two-phase configurations of elastic bodies is reduced to finding the interface Γ and corresponding displacement field $\boldsymbol{u}(\boldsymbol{x})$, which is smooth enough at material points $\boldsymbol{x} \notin \Gamma$, continuous across Γ

$$[[\boldsymbol{u}]] = 0, \quad \boldsymbol{x} \in \Gamma \tag{15.1}$$

and satisfies boundary conditions and equilibrium conditions

$$\boldsymbol{x} \notin \Gamma : \quad \nabla \cdot \boldsymbol{\sigma} = 0, \tag{15.2}$$

$$\boldsymbol{x} \in \Gamma: \quad [[\boldsymbol{\sigma}]] \cdot \boldsymbol{n} = 0, \quad [[f]] - \langle \boldsymbol{\sigma} \rangle : [[\boldsymbol{\varepsilon}]] = 0, \tag{15.3}$$

where $\boldsymbol{n}$ is the unit normal to the boundary, $\boldsymbol{\sigma}$ and $\boldsymbol{\varepsilon} = (\nabla \boldsymbol{u} + \nabla \boldsymbol{u}^T)/2$ are the stress and strain tensors, square and angle brackets denote the jump across the interface and arithmetic mean: $[[a]] = a_+ - a_-$, $\langle a \rangle = 1/2(a_+ + a_-)$, subscripts "–" and "+" refer the values to a material being in the "–" and "+" phase states, respectively.

Thermodynamic condition $(15.3)_2$ follows from the conditions on the equilibrium phase boundary in a nonlinear elastic material, see [62] and reference therein. Note that from displacement and traction continuity it follows that

$$\langle \boldsymbol{\sigma} \rangle : [[\boldsymbol{\varepsilon}]] = \boldsymbol{\sigma}_+ : [[\boldsymbol{\varepsilon}]] = \boldsymbol{\sigma}_- : [[\boldsymbol{\varepsilon}]].$$

The volume density of the Helmholtz free energy is taken in a form

$$f(\boldsymbol{\varepsilon}) = \min_{-,+} \left\{ f^-(\boldsymbol{\varepsilon}), f^+(\boldsymbol{\varepsilon}) \right\} \tag{15.4}$$

where the energies of the phases are

$$f_\pm(\boldsymbol{\varepsilon}) = f_\pm^0 + w_\pm(\boldsymbol{\varepsilon}), \quad w_\pm(\boldsymbol{\varepsilon}) = \frac{1}{2}(\boldsymbol{\varepsilon} - \boldsymbol{\varepsilon}_\pm^*) : \boldsymbol{C}_\pm : (\boldsymbol{\varepsilon} - \boldsymbol{\varepsilon}_\pm^*), \tag{15.5}$$

$w_\pm(\boldsymbol{\varepsilon})$ are the strain energies, $\boldsymbol{C}_\pm$ are the elasticity tensors, $f_\pm^0$ are the chemical energies of the phases defined by the temperature, $\boldsymbol{\varepsilon}_\pm^*$ are the strains in stress-free states; if one of the strains $\boldsymbol{\varepsilon}_\pm^*$ is zero then another one is the transformation strain $\boldsymbol{\varepsilon}^{tr}$.

Constitutive equations take the form

$$\boldsymbol{\sigma}_\pm(\boldsymbol{\varepsilon}) = \boldsymbol{C}_\pm : \left(\boldsymbol{\varepsilon} - \boldsymbol{\varepsilon}_\pm^*\right), \quad \boldsymbol{\varepsilon} \in \mathcal{E}_\pm \tag{15.6}$$

where domains of definition of phases "–" and "+"

$$\mathcal{E}_- = \{\boldsymbol{\varepsilon} : \ \psi(\boldsymbol{\varepsilon}) > 0\}, \ \mathcal{E}_+ = \{\boldsymbol{\varepsilon} : \ \psi(\boldsymbol{\varepsilon}) < 0\}, \\ \psi(\boldsymbol{\varepsilon}) = f^+(\boldsymbol{\varepsilon}) - f^-(\boldsymbol{\varepsilon}). \tag{15.7}$$

It follows from $(15.3)_1$, (15.6) and (15.1) that [63]

$$[[\boldsymbol{\varepsilon}]] = -\boldsymbol{K}_\mp(\boldsymbol{n}) : \boldsymbol{q}_\pm, \quad \boldsymbol{q}_\pm = [[\boldsymbol{C}]] : \boldsymbol{\varepsilon}_\pm - [[\boldsymbol{C} : \boldsymbol{\varepsilon}^*]], \\ \boldsymbol{K}_\pm(\boldsymbol{n}) = \{\boldsymbol{n} \otimes \boldsymbol{G}_\pm \otimes \boldsymbol{n}\}^s, \quad \boldsymbol{G}_\pm = (\boldsymbol{n} \cdot \boldsymbol{C}_\pm \cdot \boldsymbol{n})^{-1}, \tag{15.8}$$

s means the symmetrization: $K_{ijkl} = n_{(i}G_{j)(k}n_{l)}$.

Substituting (15.5), (15.6) into $(15.3)_2$ and using (15.8) leads to the thermodynamic equilibrium condition in the forms [64]

$$2\gamma + [[\boldsymbol{\varepsilon}^* : \boldsymbol{C} : \boldsymbol{\varepsilon}^*]] + \boldsymbol{\varepsilon}_\pm : [[\boldsymbol{C}]] : \boldsymbol{\varepsilon}_\pm - 2\boldsymbol{\varepsilon}_\pm : [[\boldsymbol{C} : \boldsymbol{\varepsilon}^*]] \pm \boldsymbol{q}_\pm : \boldsymbol{K}_\mp(\boldsymbol{n}) : \boldsymbol{q}_\pm = 0 \tag{15.9}$$

where upper and lower signs and indices '+" and "–" correspond to each other, $\gamma = [[f_0]]$ acts as temperature in absence of thermal stresses.

Any of two equations (15.9) determines one-parametric family of unit normals depending on γ and strains on one side ("+" or "−") of the interface. Those $\boldsymbol{\varepsilon}_-$ and $\boldsymbol{\varepsilon}_+$ at which the equation is solvable for the unit normal form a phase transition zone in strain space that consists of two subzones $\mathcal{E}_-$ and $\mathcal{E}_+$ [52, 53, 59, 60].

If the inverse tensor $[[\boldsymbol{C}]]^{-1}$ exists, then (15.9) can be rewritten in terms of $\boldsymbol{q}$ as

$$\varphi(\boldsymbol{q}_\pm) \pm \mathcal{K}_\mp(\boldsymbol{q}_\pm, \boldsymbol{n}) = 0, \tag{15.10}$$

$$\varphi(\boldsymbol{q}_\pm) = 2\gamma_* + \boldsymbol{q}_\pm : [[\boldsymbol{C}]]^{-1} : \boldsymbol{q}_\pm, \quad \mathcal{K}_\mp(\boldsymbol{q}_\pm, \boldsymbol{n}) = \boldsymbol{q}_\pm : \boldsymbol{K}_\mp(\boldsymbol{n}) : \boldsymbol{q}_\pm,$$

$$\gamma_* = \gamma + \frac{1}{2}[[\boldsymbol{\varepsilon}^*]] : [[\boldsymbol{B}]]^{-1} : [[\boldsymbol{\varepsilon}^*]], \quad \boldsymbol{B}_\pm = \boldsymbol{C}_\pm^{-1}$$

Tensors $\boldsymbol{q}$ for which Eqs. (15.10) can be solved for the unit normal $\boldsymbol{n}$ form the phase transition zone in q-space that consists of two subzones $Q_\pm$. The subzones $Q_\pm$ and corresponding subzones $\mathcal{E}_\pm$ are defined by inequalities

$$\mathcal{K}_\mp^{\min}(\boldsymbol{q}_\pm) \le \mp\varphi(\boldsymbol{q}_\pm) \le \mathcal{K}_\mp^{\max}(\boldsymbol{q}_\pm), \tag{15.11}$$

$$\mathcal{K}_\mp^{\max}(\boldsymbol{q}_\pm) = \max_{\boldsymbol{n}} \mathcal{K}_\mp(\boldsymbol{q}, \boldsymbol{n}), \quad \mathcal{K}_\mp^{\min}(\boldsymbol{q}_\pm) = \min_{\boldsymbol{n}} \mathcal{K}_\mp(\boldsymbol{q}_\pm, \boldsymbol{n}),$$

The normals

$$\boldsymbol{n}_{\text{ex}}^\pm(\boldsymbol{q}) = \arg\max_{\boldsymbol{n}} \mathcal{K}_\mp(\boldsymbol{q}, \boldsymbol{n}), \quad \boldsymbol{n}_{\text{in}}^\pm(\boldsymbol{q}) = \arg\min_{\boldsymbol{n}} \mathcal{K}_\mp(\boldsymbol{q}, \boldsymbol{n}) \tag{15.12}$$

to the interface correspond to external and internal boundaries of the subzones $Q_\pm$ and $\mathcal{E}_\pm$ [52, 53].

The external boundaries of the subzones are given by the equations

$$\varphi(\boldsymbol{q}_+) + \mathcal{K}_-^{\max}(\boldsymbol{q}_+) = 0, \quad \mathcal{K}_+^{\max}(\boldsymbol{q}_-) - \varphi(\boldsymbol{q}_-) = 0. \tag{15.13}$$

The internal boundaries are given by the equations

$$\varphi(\boldsymbol{q}_+) + \mathcal{K}_-^{\min}(\boldsymbol{q}_+) = 0, \quad \mathcal{K}_+^{\min}(\boldsymbol{q}_-) - \varphi(\boldsymbol{q}_-) = 0. \tag{15.14}$$

The surface dividing the domains of definition of the phases, i.e. corresponding to (15.8), passes between the internal boundaries (15.14). Note that in relationships (15.10)–(15.14) tensors $\boldsymbol{\varepsilon}_\pm^*$ related to the transformation strain are hidden in $\boldsymbol{q}$ and parameter γ_*.

Further we consider isotropic phases with the elasticity tensors

$$\boldsymbol{C}_\pm = \lambda_\pm \boldsymbol{E} \otimes \boldsymbol{E} + 2\mu_\pm \boldsymbol{I}, \tag{15.15}$$

where $\lambda_\pm$ and $\mu_\pm$ are Lamé's coefficients, $\boldsymbol{E}$ and $\boldsymbol{I}$ are the second and forth rank unit tensors, in Cartesian coordinates $E_{ij} = \delta_{ij}$,

$$I_{ijkl} = \frac{1}{2}(\delta_{ik}\delta_{jl} + \delta_{il}\delta_{jk}),$$

δ_{ij} is Kroneker's delta. Then

$$\boldsymbol{K}_{-}(\boldsymbol{n}) = \frac{1}{\mu_{-}}\left((\boldsymbol{n}\otimes\boldsymbol{E}\otimes\boldsymbol{n})^{s} - a_{-}\boldsymbol{n}\otimes\boldsymbol{n}\otimes\boldsymbol{n}\otimes\boldsymbol{n}\right) \tag{15.16}$$

$$\mathcal{K}_{-}(\boldsymbol{q}_{+},\boldsymbol{n}) = \frac{1}{\mu_{-}}F(N_1,N_2), \tag{15.17}$$

where $F(N_1,N_2) = N_2 - aN_1^2$, $a = (2(1-\nu_{-}))^{-1}$, ν_{-} is Poisson's ratio, $N_1 = \boldsymbol{n}\cdot\boldsymbol{q}_{+}\cdot\boldsymbol{n}$ and $N_2 = \boldsymbol{n}\cdot\boldsymbol{q}_{+}^2\cdot\boldsymbol{n}$ are the orientation invariants.

Representation (15.17) admits the following geometrical interpretation of the equation (15.10) and inequalities (15.11) (Fig. 15.1). Given $\boldsymbol{q}_{+}$, (15.10) takes the form of a relationship between orientation invariants

$$N_2 = aN_1^2 - \mu_{-}\varphi(\boldsymbol{q}_{+}) \tag{15.18}$$

which defines a parabola on N_1N_2-plane. Note that $0 < a \leq 1$ if $\nu \leq 0.5$.

Since a couple of orientation invariants N_1, N_2 has to determine the unit normal $\boldsymbol{n}$, there are restrictions on N_1, N_2. In the basis of eigenvectors of $\boldsymbol{q}_{+} = \sum q_i \boldsymbol{e}_i \otimes \boldsymbol{e}_i$ the normal is determined by the system of linear equations for $p_i = n_i^2$ ($n_i = \boldsymbol{n}\cdot\boldsymbol{e}_i$, i=1,2,3)

$$\sum p_i = 1, \quad \sum q_i p_i = N_1, \quad \sum q_i^2 p_i = N_2 \tag{15.19}$$

Since the solution of system (15.19) is to be non–negative, the admissible values domain $\mathcal{D}$ for orientation invariants N_1, N_2 is a triangle with vertexes (q_i, q_i^2) $(i = 1,2,3)$ (see Fig. 15.1). The vertexes correspond to $\boldsymbol{n} = \boldsymbol{e}_i$, points of ij-side

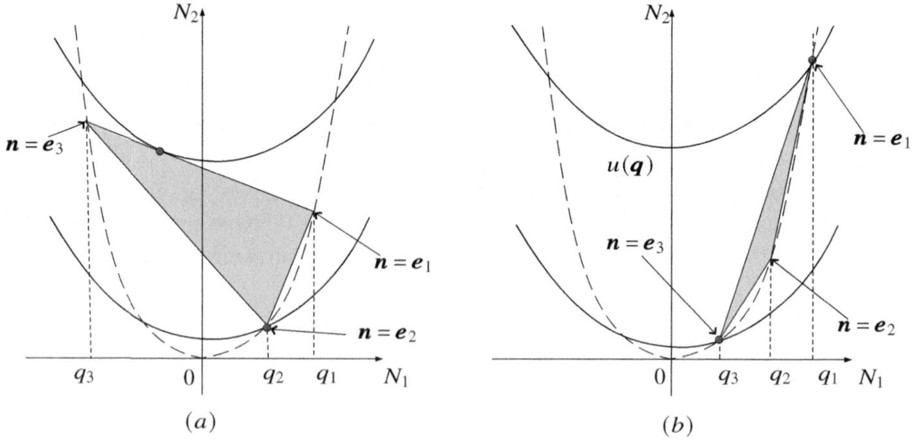

Fig. 15.1: Domains (triangles) of admissible values of orientation invariants and limiting curves of the one-parametric family of solution: (a) - parabola touches the side corresponding to the maximal and minimal eigenvalues of the tensor $\boldsymbol{q}$ (b) - parabola (15.18) passes through vertexes of the triangle.

of the triangle correspond to normal $\boldsymbol{n}$ lying in the ij-principal plane of $\boldsymbol{q}_+$. If q_i $(i = 1,2,3)$ change, then vertexes of $\mathcal{D}$ move along the skeleton curve $N_2 = N_1^2$.

On the other hand, the parabola (15.18) moves along N_2-axis if $\boldsymbol{q}_+$ changes. The line of intersection of the parabola and $\mathcal{D}$ gives one parameter family of solutions. The subzone Q_+ is formed by all $\boldsymbol{q}_+$ such that this intersection exists.

One can see that, depending on the eigenvalues of $\boldsymbol{q}_+$, the maximum value corresponds to the point where the parabola touches the upper side of the triangle, corresponding to maximal and minimal eigenvalues (Fig. 15.1(a), the normal lies in the corresponding eigenplane in tis case), or passes via the vertex corresponding to the eiginvalue, having the maximal absolute value (Fig. 15.1(b), the normal coinsides with the corresponding eigenvector).

The minimal value correspond to passing the parabola via the vertex corresponding to the eiginvalue with the minimal absolute value, the normal coinsides with the corresponding eigenvector.

The above considerations can be formalized as follows (see also Appendix):

1. Assume that the tensor $\boldsymbol{q}_+$ has different eigenvalues, and $q_{\min}$, $q_{\max}$ and q_{mid}, and $\boldsymbol{e}_{\max}$, $\boldsymbol{e}_{\min}$ and $\boldsymbol{e}_{\mathrm{mid}}$ are the minimal, maximal and intermediate eigenvalues and corresponding eigenvectors of $\boldsymbol{q}_+$; $n_{\min}$, $n_{\max}$ and n_{mid} are components of the normal $\boldsymbol{n}$ in the basis of eigenvectors of $\boldsymbol{q}_+$, $|q|_{\max}$ and $|q|_{\min}$ are the maximal and minimal absolute values of the eigenvalues, $\boldsymbol{e}_{|q|\max}$ and $\boldsymbol{e}_{|q|\min}$ are the corresponding eigenvectors.

   If $\nu_- > 0$ and

$$q_{\min}q_{\max} < 0 \quad \text{or} \quad \begin{cases} q_{\min}q_{\max} > 0 \\ (1-\nu_-)|q|_{\min} < \nu_-|q|_{max}. \end{cases} \tag{15.20}$$

   or $\nu_- < 0$ and

$$\begin{cases} q_{\min}q_{\max} < 0 \\ (1-\nu_-)|q|_{\min} + \nu_-|q|_{\max} > 0. \end{cases} \tag{15.21}$$

   then the normal $\boldsymbol{n}_{\mathrm{ex}}^+$ to the intertface corresponding to the external PTZ boundary lies in the plane of maximal and minimal eigenvalues of $\boldsymbol{q}_+$: $n_{\mathrm{mid}} = 0$,

$$n_{\max}^2 = \frac{(1-\nu_-)q_{\max} - \nu_- q_{\min}}{q_{\max} - q_{\min}}, \quad n_{\min}^2 = \frac{\nu_- q_{\max} - (1-\nu_-)q_{\min}}{q_{\max} - q_{\min}}, \tag{15.22}$$

   Then for both $\nu_- > 0$ and $\nu_- < 0$

$$\mathcal{K}_-^{\max} = \frac{1-\nu_-}{2\mu_-}(q_{\max}^2 + q_{\min}^2) - \frac{\nu_-}{\mu_-} q_{\max} q_{\min}. \tag{15.23}$$

   If conditions (15.20) for $\nu_- > 0$ or (15.21) for $\nu_- < 0$ are not satisfied then

$$\boldsymbol{n}_{ex}^+ = \boldsymbol{e}_{|q|\max}, \quad \mathcal{K}_-^{\max} = \frac{1-2\nu_-}{2\mu_-(1-\nu_-)}|q|_{\max}^2. \tag{15.24}$$

We emphasize that if $\nu_- < 0$ then only the case $q_{\min}q_{\max} < 0$ with additional restrictions allows normal (15.22) lying in the eigenplane of $\boldsymbol{q}_+$
The normal to the interface corresponding to the internal PTZ boundary and $\mathcal{K}_-^{\min}$ are given by formulae:

$$\boldsymbol{n}_{\text{in}}^+ = \boldsymbol{e}_{|q|\min}, \quad \mathcal{K}_-^{\min} = \frac{1-2\nu_-}{2\mu_-(1-\nu_-)}|q|^2_{\max}. \tag{15.25}$$

2. Assume that the tensor $\boldsymbol{q}_+$ is axisymmetric:

$$\boldsymbol{q}_+ = q(\boldsymbol{E} - \boldsymbol{k}\otimes\boldsymbol{k}) + q_k\boldsymbol{k}\otimes\boldsymbol{k}.$$

Then the domain $\mathcal{D}$ degenerates into the straight segment with the ends on the skeleton parabola.
If $\nu_- > 0$ and

$$q_k q < 0 \quad \text{or} \quad \begin{cases} q_k q > 0 \\ (1-\nu_-)|q|_{\min} - \nu_-|q|_{\max} < 0. \end{cases} \tag{15.26}$$

or $\nu_- < 0$ and

$$\begin{cases} q_k q < 0 \\ (1-\nu_-)|q|_{\min} + \nu_-|q|_{\max} > 0 \end{cases} \tag{15.27}$$

$$|q|_{\min} = \min\{|q_k|, |q|\}, \quad |q|_{\max} = \max\{|q_k|, |q|\},$$

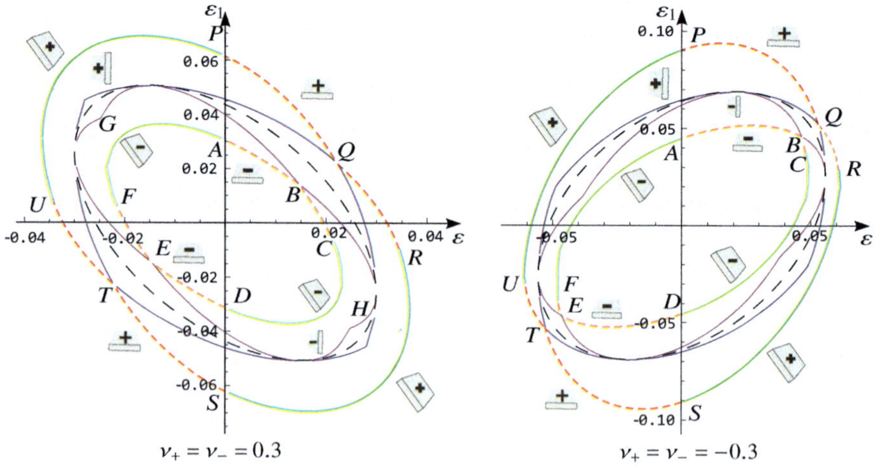

Fig. 15.2: The cross sections of the phase transition zones by the plane $\varepsilon_1 = \varepsilon_2 = \varepsilon$ for a normal materials and for auxetics. Initial undeformed state is phase "+" ($\vartheta^{tr} = 0$).

Table 15.1: Parameters.

Fig.	μ_+	μ_-	ν_+	ν_-	K_+	K_-	λ_+	λ_-	ϑ^{tr}	γ
Fig. 15.4	55	15	-0.3	0.3	16	32.5	-20	22	-0.06	-0.05
Fig. 15.5	55	15	0.3	-0.3	119	4	82	-5	-0.06	-0.05
Fig. 15.2,15.7	30	15	·	·	·	·	·	·	0	-0.05

then the maximum of $\mathcal{K}_-$ is reached at any normal with the square of the projection onto the axe $\boldsymbol{k}$ equal to

$$n_k^2 = \frac{(1-\nu_-)q_k - \nu_- q}{q_k - q}. \tag{15.28}$$

Then

$$\mathcal{K}_-^{\max} = \frac{1-\nu_-}{2\mu_-}(q_k^2 + q^2) - \frac{\nu_-}{\mu_-} q_k q, \tag{15.29}$$

If conditions (15.26), (15.27) are not satisfied then the maximum of $\mathcal{K}_-$

$$\mathcal{K}_-^{max} = \frac{1-2\nu_-}{2\mu_-(1-\nu_-)}|q|^2_{\max} \tag{15.30}$$

is achieved at

$$\boldsymbol{n}^* = \boldsymbol{k} \quad \text{if} \quad |q_k| > |q|,$$
$$\boldsymbol{n}^* \perp \boldsymbol{k} \quad \text{if} \quad |q_k| < |q|.$$

3. Assume that the tensor $\boldsymbol{q}_+$ is spherical: $\boldsymbol{q}_+ = q\boldsymbol{E}$. Then $\mathcal{K}_-$ does not depend on $\boldsymbol{n}$,

$$\boldsymbol{q}_+ : \boldsymbol{K}_-(\boldsymbol{n}) : \boldsymbol{q}_+ = cq^2, \quad c = \frac{1-2\nu_-}{2\mu_-(1-\nu_-)} = \left(k_- + \frac{4}{3}\mu_-\right)^{-1}, \tag{15.31}$$

k_- is the bulk modulus.

The examples of the PTZ cross section by plane $\varepsilon_2 = \varepsilon_3$, constructed by formulae (15.13), (15.14) for the cases $\nu_+ = \nu_- > 0$ and $\nu_+ = \nu_- < 0$ are shown in Fig. 15.2. We take the phase "+" as an initial phase and assume that the transformation strain is spherical. Then $f_+^0 < f_-^0$, $\boldsymbol{\varepsilon}_+^* = 0$, $\boldsymbol{\varepsilon}_-^* = \boldsymbol{\varepsilon}^{tr} = (\vartheta^{tr}/3)\boldsymbol{E}$, i.e. $\gamma < 0$. The parameters are given in Table 15.1. The units are not shown, but Lamé constants and corresponding bulk moduli can be considered as given in GPa. The choice of parameter values was made according to reasons of better visualization of the variety of graphs here and further and consistency with the small strains approach.

In the figures, lines $ABCDEFA$ (direct phase transformation) and $PQRSTUPQ$ (reverse phase transformation) are the external boundaries of the subzones $\mathcal{E}_+$ and $\mathcal{E}_-$, respectively. Lines $BGEHB$ and $QKTLQ$ are the internal PTZ boundaries of the subzones $\mathcal{E}_+$ and $\mathcal{E}_-$.

Dashed black lines, dividing the domains of definition of phases "+" and "−", are determined by the equation $\psi(\boldsymbol{\varepsilon}) = 0$ (see (15.7)).

The parts ABC, FED, PQR, UTS of the external boundaries of the PTZ correspond to the normal, orientation of which coincides with the eigenvector of the tensor $\boldsymbol{q}$ (the interface is perpendicular to an eigenvector of $\boldsymbol{q}$). The parts AF, DC, UP and SR correspond to the normal lying the principal plane of the tensor $\boldsymbol{q}$ (the interface is oriented similar to shear band with the angle depending on the eigenvalues of $\boldsymbol{q}$).

In terms of minimizing laminates (see the next section), ABC and FED are the parts of the limiting surface corresponding to the nucleation of direct first-rank laminates; AF and DC are the parts of the limiting surface corresponding to the nucleation of inclined first-rank laminates.

Note that the external boundary $PQRSTUPQ$ of the subzone $\mathcal{E}_-$ is non-convex. The phase transformations limiting surface construction leads to the convexication with the use of microstructure which are not reduced to simple layers.

One can see that the sign of the Poisson's ratio does not change the expressions of $\mathcal{K}_-^{max}$ but changes the conditions for switching the orientation of the normal. The orientation of the PTZ with respect to strain axes is also changed.

15.3 Optimal Laminates and Phase Transformations Limiting Surfaces

Two phase states, corresponding to the energy minimization may not be limited to simple layers. By the Gibbs principle, at given temperature and average strains, average Helmholtz free energy of a two-phase microstructure

$$F = \int_{\Omega} \left((1 - H(\boldsymbol{x})) f_-(\boldsymbol{\varepsilon}) + H(\boldsymbol{x}) f_+(\boldsymbol{\varepsilon}) \right) d\Omega, \tag{15.32}$$

in the case of equilibrium, is to be minimal with respect to the new phase volume fraction and the geometry parameters of the microstructure. Here subscripts "−" and "+" indicate values related to the phases "−" and "+", $\Omega = \Omega_- \bigcup \Omega_+$ is a unit cell divided into subdomains Ω_- and Ω_+ occupied by the phases "−" and "+", $H(\boldsymbol{x})$ is the characteristic function of the subdomain Ω_+:

$$H(\boldsymbol{x}) = \begin{cases} 1, & \boldsymbol{x} \in \Omega_+, \\ 0, & \boldsymbol{x} \in \Omega_-. \end{cases}$$

We take the free energy density in a form (15.4), (15.5), and the constitutive equations in the form (15.6). Consider further the case when the shear module μ_+ of the phase "+" is greater then the shear module of the phase "−", i.e. $\mu_+ > \mu_-$.

It was proved [45] that the infimum of energy (15.32), (15.5) can be constructed with the use of laminates of the first, second and third rank (Fig. 15.3). The formulae for the construction of the phase transformations limiting surface were derived and then example were done for the case of positive Poisson's ratios [45]–[47]. The construction is based, in particular, on the use of extreme properties of the quadratic form $\mathcal{K}_-(\boldsymbol{q}_+,\boldsymbol{n})$ described in Sect. 15.2.

The first-rank laminate consists of alternating planar layers occupied by homogeneous phases "–" and "+". Then the second-rank laminate consists of alternating layers of the phase "–" and layers which are themselves first-rank laminates. It is characterized by two normals to the layers and sublayers of phase "–" and by an additional parameter related with the volume fraction of the pure layers of the phase "–". The third-rank laminates consists of the layers of the phase "–" and layers which are second-rank laminates. They are characterized by three normals and two additional parameters related with the volume fractions of pure layers of phase "–" and the volume fraction of phase "–" in the included second rank laminates. Note that in the case of the second-rank and third-rank laminates, the phase "+" with greater shear module occupies the inner domains surrounded by the phase "–". The characteristic sizes of the laminates decrease at every step of the microstructure construction and the strains in such layered material tend to piece-wise constant fields, see, e.g., [65].

Five cases ('regimes') were distinguished.

1. Direct first-rank laminates.
   The eigenvalues of $\boldsymbol{q}_+$ are different and do not satisfy conditions (15.20), (15.21). The normal and quadratic forma are defined by Eq. (15.24)
2. Inclined first-rank laminates.
   The eigenvalues of $\boldsymbol{q}_+$ are different and satisfy one the conditons conditons (15.20), (15.21) which depend on the sign of the Poisson ration. The normal and quadratic form are defined by Eq. (15.22), (15.23).
3. Skew second-rank laminates.
   Tensor $\boldsymbol{q}_+$ is axisymmetric. The eigenvalues satisfy conditon (15.26) or (15.27). The quadratic form is defined by Eq. (15.29).

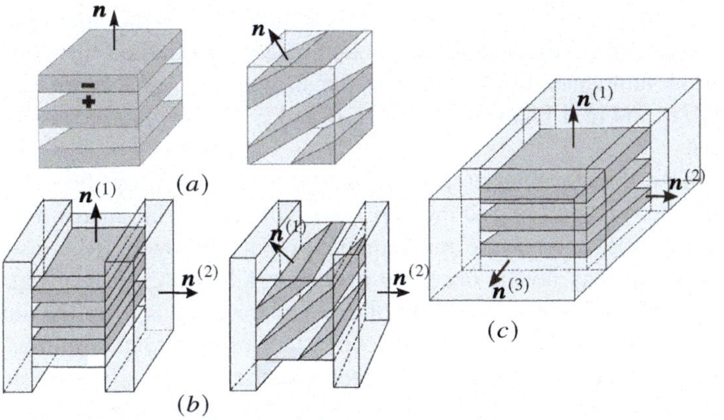

Fig. 15.3: Laminates for optimal microstructures: (a) – direct and inclined first-rank laminates, (b) – direct and skew second-rank laminates, (c) – third-rank laminate.

4. Direct second-rank laminates.
   Tensor $\boldsymbol{q}_+$ is axisymmetric. The eigenvalues do not satisfy conditons (15.26) and (15.27). The quadratic form is defined by Eq. (15.30).
5. Third-rank laminates. Tensor $\boldsymbol{q}_+$ is spherical. The quadratic form is defined by Eq. (15.31).

The external average strain can be related with the tensor $\boldsymbol{q}_+$. Then the limiting surface can be constructed in strain space. Basic relationships and detais of the procedure can be found in [45]-[47]. The influence of Poisson's ratio is highlighted in the next section.

15.4 Results

The cross sections of phase transformation limiting surfaces and surfaces of the appearance of different laminates by the plane $\varepsilon_2 = \varepsilon_3$ are shown in Figs. 15.4, 15.5, 15.7. The parameters taken in calculations are given in Table 15.1.

Materials with positive and negative Poisson's ratios are referred as normal materials and auxetics, respectively. Fig. 15.4 demonstrates the cross-sections for the direct and reverse phase transformations between the auxetic and normal material. Fig. 15.5 represents the cross-sections for the phase transformations if the initial phase state is a normal material which transforms into the auxetic.

Figure 15.6 represents the sketches of the dependencies of the free energy on volume strains: Fig. 15.6(a) corresponds to Fig. 15.4; Fig. 15.6(b) corresponds to Fig. 15.5. Note that in the first case the jump of the elasticity tensor is not sign-definite: the bulk module increases while the shear module decreases due to the phase transformation from the auxetic to the normal material. This leads to unclosed limiting surfaces, as it was mentioned in [45, 59].

Various lines correspond to various types of laminates. Red and orange lines in Figs. 15.4 and 15.5 correspond to the appearance of the direct first-rank laminates, green dashed lines – to skew first-rank laminates, blue dashed lines – to the direct second-rank laminates, black lines – to the skew second-rank laminates, segments AB and $A'B'$ in Fig. 15.4 and segments CD and $C'D'$ in Fig. 15.5 – to the third-rank laminates.

The phase transformations limiting surfaces are given by the combination of the surfaces, corresponding to the appearance of various laminates. Recall that the laminate micrtostructures are mathematical objects allowing to construct energy minimizers and may not be observed in a material. In the case of Fig. 15.4 lines *FOG* represent the limiting surfaces of the transformations from the auxetic to the normal material. Lines *SC'B'A'T* and *SABCT* correspond to the limiting surfaces of the transformations from the normal material to the auxetic.

In the case of Fig. 15.5 the limiting surface of the direct transformation from the normal material to the auxetic is presented by line *PQRS*. The reverse transformation from the auxetic to normal material is presented by line *BCDEB'C'D'E'B*. Dashed black lines divides areas of definition of phases "+" and "–".

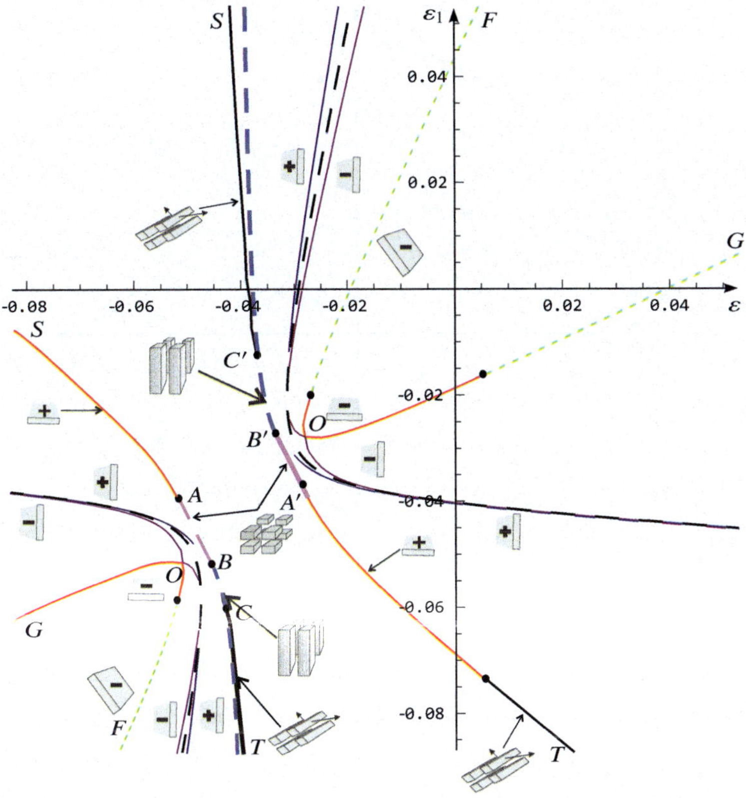

Fig. 15.4: Limiting surface for the phase transformation from the auxetic (phase "+") to normal material (phase "−").

If the energy parameter γ, defined by the temperature, decreases then the limiting surfaces expand and, therefore, the direct phase transformation starts later while the reverse transformation — earlier.

Phase transformations limiting surfaces for various values of Poisson's ratios of the phases in the case of phase transformations between normal phases with equal positive and zero Poisson' ratios and between auxetics with equal negative Poisson's ratios are shown in Fig. 15.7. One can observe that the orientation of the limiting surface with respect to strain axes and its shape are different for positive and negative Poisson' ratios. In the case of positive Poisson's ratio the cross-section is elongated along the axis $\epsilon_1 = -\epsilon_2 = -\epsilon_3$. If $\nu_+ = \nu_- \rightarrow 0.5$ then the limiting surface stretches in direction of this axe $\epsilon_1 = \epsilon_2 = \epsilon_3$ and shrinks in the perrpendicular direction. In the case of negative positive Poisson's ratios the cross-section is elongated along the axis $\epsilon_1 = \epsilon_2 = \epsilon_3$. If $\nu_+ = \nu_- \rightarrow -1$ then the limiting surface stretches in the direction of the axe $\epsilon_1 = \epsilon_2 = \epsilon_3$ and shrinks in the perpendicular direction.

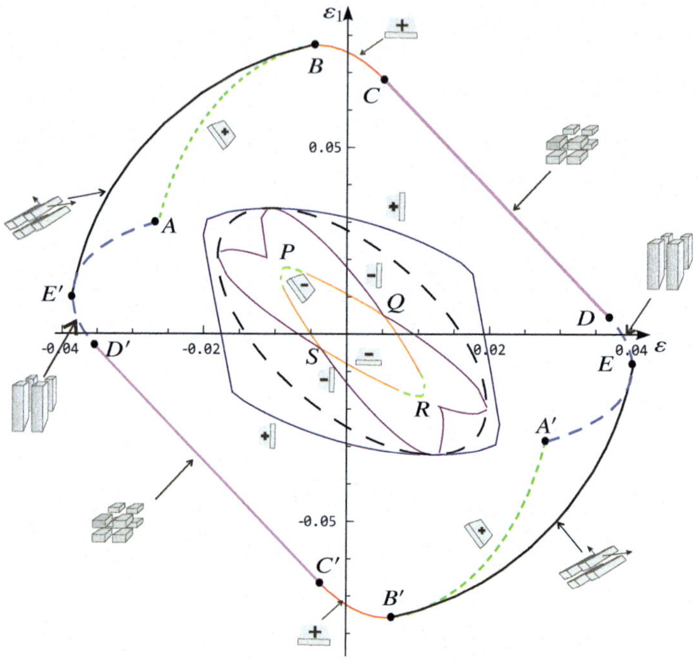

Fig. 15.5: Limiting surfaces for the phase transformation from the normal material (phase "+") to auxetic (phase "–").

15.5 Conclusions

We developed a procedure for the construction of phase transformation limiting surfaces of stress-induced phase transformation – a phase diagram in a strain space – for the case of auxetic phases. Adopting the procedures developed earlier for the case of positive Poisson's ratio, and implementing the approach based on the energy minimizer construction we found minimizing microstructures and derived formulae

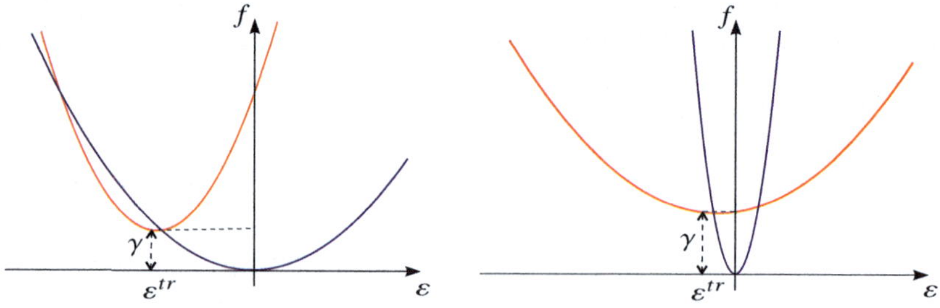

Fig. 15.6: Free energy in the cases corresponding to: (a) – Fig. 15.4; (b) – Fig. 15.5.

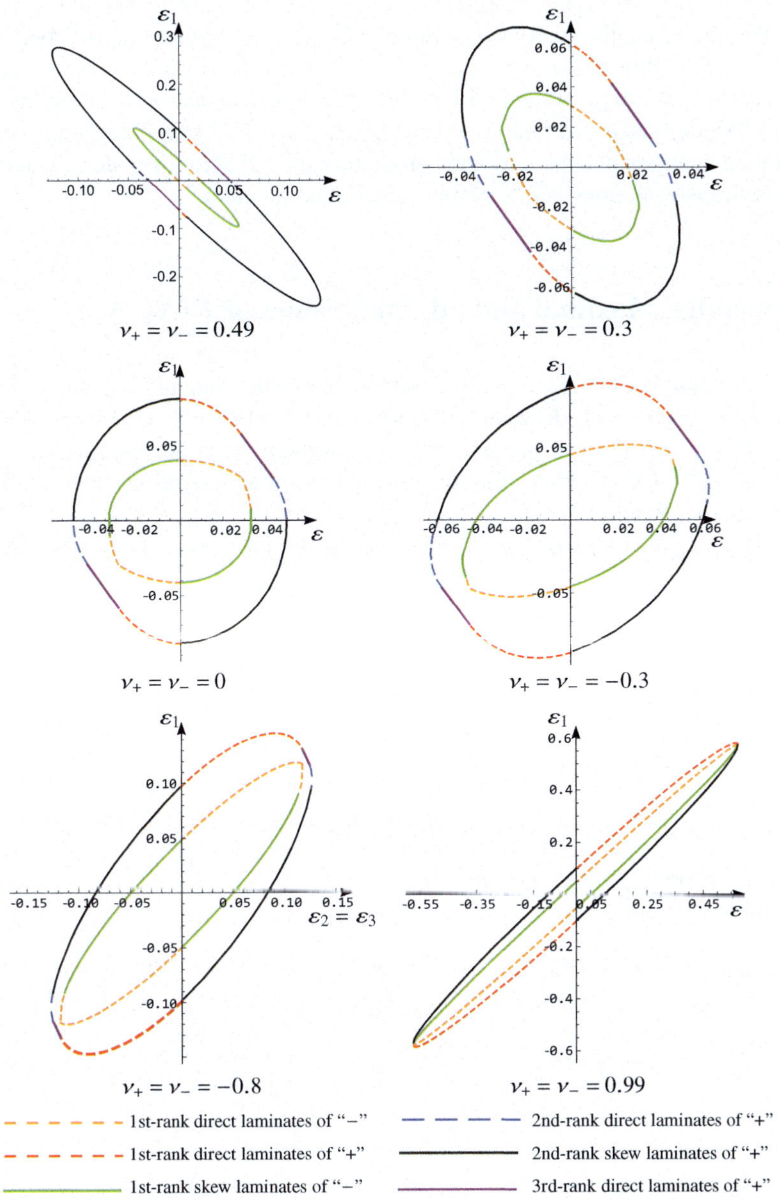

Fig. 15.7: Limiting surfaces for direct and reverse phase transformations in normal materials and auxetics. Initial state is phase "+".

for limiting surfaces construction for phase transformations 'auxetic ↔ normal material' and 'auxetic ↔ auxetic'.

We studied qualitatively and quantitatively how the sign and value of Poisson's ratio affect the conditions for switching the regimes corresponding to various laminates and change the shape of the limiting surface, and its orientation with respect to strain axes. We demonstrated that the construction of the PTZ and phase transformations limiting surfaces may be a tool for predicting the behavior of materials undergoing phase transformations and selecting material parameters.

Appendix. Maximal and minimal values of $F(N_1,N_2)$

This is a simple exercise to find the normals corresponding to maximal and minimal values of $F(N_1,N_2)$, depending on relationships between eigenvalues q_1,q_2,q_3. Results for $\nu > 0$ were presented earlier (see e.g. [53, 66, 60] or recent papers [45, 46, 47]). Nevertheless, for convenience, we give derivations here, highlighting the cases $\nu > 0$ and $\nu < 0$.

Let $\boldsymbol{q} = \sum q_k \boldsymbol{e}_k \boldsymbol{e}_k$, $n_k = \boldsymbol{n}\cdot\boldsymbol{e}_k$, $p_k = n_k^2$. To find the normal $\boldsymbol{n}^{ex}$ maximizing the function

$$F(N_1,N_2) = N_2 - aN_1^2, \quad a = \frac{1}{2(1-\nu)} \tag{A-1}$$

$$N_1 = \Sigma q_k p_k, \quad N_2 = \sum q_k^2 p_k, \quad \sum p_k = 1, \quad 0 \le p_k \le 1 \tag{A-2}$$

we take the Lagrange function

$$\Phi = F(N_1,N_2) - \lambda \Sigma p_k \tag{A-3}$$

where λ is the Lagrange multiplier.

From the conditions

$$\frac{\partial \Phi}{\partial p_k} = q_k^2 - 2aN_1 q_k - \lambda = 0, \quad k = 1,2,3 \tag{A-4}$$

it follows

$$q_1^2 - 2aN_1 q_1 = q_2^2 - 2aN_1 q_2 = q_3^2 - 2aN_1 q_3 = \lambda. \tag{A-5}$$

Then

$$q_1^2 - q_2^2 = 2aN_1(q_1-q_2),\ q_1^2 - q_3^2 = 2aN_1(q_1-q_3),\ q_2^2 - q_3^2 = 2aN_1(q_2-q_3). \tag{A-6}$$

If q_i are different, i.e. $q_1 \neq q_2$, $q_1 \neq q_2$, $q_2 \neq q_3$, then, by (A-6),

$$q_1 + q_2 = q_1 + q_3 = q_2 + q_3 = 2aN_1. \tag{A-7}$$

Eq. (A-7) can be satisfied only if $q_1 = q_2 = q_3$, which is the contradiction.

Thus, if q_i are different then (A-5) cannot be fulfilled, maximum and minimum of $F(N_1,N_2)$ are reached at the boundary of the domain $\sum p_k = 1$, $0 \le p_k \le 1$, i.e. if

$p_1 = 0$ or $p_2 = 0$ or $p_3 = 0$. This corresponds to the analysis presented graphically in Section 2, see Fig 15.1. That is why let us order the eigenvalues of $\boldsymbol{q}$ such that $q_1 > q_2 > q_3$ and consider the normals such that $n_2 = 0$. Then

$$N_1 = (q_1 - q_3)p_1 + q_3, \quad N_2 = (q_1^2 - q_3^2)p_1 + q_3^2 \tag{A-8}$$

Then F becomes the function of p_1, and from the condition

$$\frac{dF}{dp_1} = 0 \tag{A-9}$$

it follows that the the extrema is achieved if

$$q_1 + q_3 - 2a\{(q_1 - q_3)p_1 + q_3\} = 0 \tag{A-10}$$

This corresponds to

$$N_1 = \frac{q_1 + q_3}{2a}, \quad N_2 = \frac{(q_1 + q_3)^2}{2a} - q_1 q_3 \tag{A-11}$$

$$F = \frac{(q_1 + q_3)^2}{4a} - q_1 q_3 = \frac{1-\nu}{2}(q_1^2 + q^2) - \nu q_1 q_3 \tag{A-12}$$

Since

$$\frac{d^2F}{dp_1^2} = -2a(q_1 - q_3)^2 < 0,$$

the extrema, if it is achieved, indeed, is the minimum.

From (A-10) it follows that

$$p_1 = \frac{(1-\nu)q_1 - \nu q_3}{q_1 - q_3}. \tag{A-13}$$

Formula (A-13) defines the normal only if

$$0 \leq \frac{(1-\nu)q_1 - \nu q_3}{q_1 - q_3} \leq 1 \tag{A-14}$$

Since $q_1 > q_3$, conditions (A-14) are equivalent to the inequalities

$$0 < (1-\nu)q_1 - \nu q_3 < q_1 - q_3 \tag{A-15}$$

which can be rewritten as

$$A = (1-\nu)q_1 - \nu q_3 > 0 = q_1 - \nu(q_1 + q_3), \tag{A-16}$$

$$B = \nu q_1 - (1-\nu)q_3 > 0 = \nu(q_1 + q_3) - q_3. \tag{A-17}$$

Since $A - B = (1-2\nu)(q_1 + q_3)$, and $-1 < \nu < 0.5$, the following implications are valid

$$q_1 + q_3 > 0 \quad \Longrightarrow \quad A > B, \quad \text{(A-16) follows from (A-17)},$$
$$q_1 + q_3 < 0 \quad \Longrightarrow \quad A < B, \quad \text{(A-17) follows from (A-16)}.$$

Then the normal can be defined by (A-13) in two cases for which

$$q_1 + q_3 > 0 \quad \text{and} \quad \nu(q_1 + q_3) > q_3, \tag{A-18}$$

or

$$q_1 + q_3 < 0 \quad \text{and} \quad \nu(q_1 + q_3) < q_1. \tag{A-19}$$

First consider $\nu > 0$.

If $q_1 q_3 < 0$, then $q_1 > 0$, $q_3 < 0$ and one of the cases (A-18) or (A-19) is met.

If $q_1 q_3 > 0$ and $q_1 > q_3 > 0$, then only the case (A-18) can take place for which

$$\nu q_1 - (1-\nu) q_3 > 0. \tag{A-20}$$

If $q_1 q_3 > 0$ and $q_3 < q_1 < 0$, then only the case (A-19) can take place for which

$$-\nu q_3 - (1-\nu) q_1 > 0 \tag{A-21}$$

that is equivalent to

$$\nu |q_3| - (1-\nu)|q_1| > 0. \tag{A-22}$$

Finally obtain (15.26): the normal is defined at $\nu > 0$ by (A-15) if

$$q_1 q_3 < 0 \tag{A-23}$$
$$\text{or} \quad \{q_1 q_3 > 0 \quad \text{and} \quad \nu |q|_{\max} - (1-\nu)|q|_{\min} > 0\} \tag{A-24}$$
$$|q|_{\max} = \max\{|q_1|, |q_3|\}, \quad |q|_{\min} = \min\{|q_1|, |q_3|\} \tag{A-25}$$

Consider now $\nu < 0$. From both (A-18) and (A-19) it follows that $q_1 > 0$, $q_3 < 0$, i.e. $q_1 q_3 < 0$. Then from (A-18) it follows that

$$\nu q_1 - (1-\nu) q_3 > 0,$$

and from (A-19) it follows that

$$-\nu q_3 + (1-\nu) q_1 > 0.$$

Both inequalities can be rewritten in a unified form (15.27):

$$q_1 q_3 < 0, \quad \nu |q|_{\max} + (1-\nu)|q|_{\min} > 0 \tag{A-26}$$

The case of the axisymmetric $\boldsymbol{q}$ can be considered analogously.

Acknowledgements Authors acknowledge the financial support of the Russian Science Foundation (Grant No. 19-19-00552–П)

References

[1] Love AEH (1994) A Treatise on the Mathematical Theory of Elasticity, 4th ed, New York, Dover.

[2] Kolpakov AG (1985) Determination of the average characterisitics of elastic frameworks, PMM USSR **49**(6):739–745.

[3] Almgren RF (1985) An isotropic three-dimensional structure with Poisson's ratio= −1, J. Elasticity **15**:427–430.

[4] Lakes R (1987) Foam structures with a negative Poisson's ratio, Science **235**:1038–40.

[5] Evans KE, Nkansah MA, Hutchinson IJ, Rogers SC (1991) Molecular network design, Nature **353**:124–124.

[6] Lakes RS, Elms K (1993) Indentability of conventional and negative Poisson's ratio foams, Journal of Composite Materials **27**:1193–1202.

[7] Lakes RS (1993) Design considerations for negative Poisson's ratio materials, Journal of Mechanical Design **115**:696–700.

[8] Bezazi A, Scarpa F (2007) Mechanical behaviour of conventional and negative Poisson's ratio thermoplastic polyurethane foams under compressive cyclic loading, International Journal of Fatigue **29**(5):922–930.

[9] Bezazi A, Scarpa F (2009) Tensile fatigue of conventional and negative Poisson's ratio open cell PU foams, International Journal of Fatigue **31**(3):488–494.

[10] Scarpa F, Tomlinson G (2000) Theoretical characteristics of the vibration of sandwich plates with in-plane negative Poisson's ratio values, Journal of Sound and Vibration **230**:45–67.

[11] Greaves GN, Greer AL, Lakes RS, Rouxel T (2011) Poisson's ratio and modern materials, Nature Mater. **10**(11):823–37.

[12] Saxena KK, Das R, Calius E.P. (2016) Three decades of auxetics research – materials with negative Poisson's ratio: a review, Adv. Eng. Mater. **18**:1847–1870.

[13] Saadatmand S, Rasoulian A, Ashjari M (2017) Auxetic materials materials with negative poisson's ratio, Material Sci & Eng Int J. **1**(2):62–64.

[14] Lakes RS (2017) Negative-Poisson's-Ratio Materials: Auxetic Solids, Annu. Rev. Mater. Res. **47**:1.1–1.19.

[15] Ren X, Das R, Tran P, Ngo TD, Xie YM (2018) Auxetic metamaterials and structures: a review, Smart Mater. Struct. **27**:023001.

[16] Yang W, Li ZM, Shi W, Xie B-H, Yang M-B (2004) Review on auxetic materials. Journal of Materials Science **39**:3269–3279.

[17] Mir M, Ali MA, Sami J, Ansari U (2014) Review of Mechanics and Applications of Auxetic Structures, Advances in Materials Science and Engineering **2014**:753496.

[18] Lim T-C (2015) Auxetic Materials and Structures, in: Engineering Materials, Springer, Singapore.

[19] Hu H, Zhang M, Liu Y (2019) Auxetic Textiles. Woodhead, Duxford

[20] Lim T-C (2020) Mechanics of Metamaterials with Negative Parameters, in Engineering Materials, Springer, Singapore.

[21] Cui TJ, Smith DR, Liu R (Eds.) (2010) Metamaterials, Springer, New Yourk.
[22] Lakes R (2020) Composites and Metamaterials, World Scientific, New Jersey.
[23] Ko J, Bhullar SK, Mohtaram NK, Willerth SM, Jun MBG (2014) Using mathematical modeling to control topographical properties of poly (ε-caprolactone) melt electrospun scaffolds, Journal of Micromechanics and Microengineering **24**(6):1–13.
[24] Ali M, Busfield JC, Rehman I (2014) Auxetic oesophageal stents: structure and mechanical properties, Journal of Materials Science: Materials in Medicine **25**(2):527–553.
[25] Miller W, Ren Z, Smith CW, Evans KE (2012) A negative Poisson's ratio carbon fibre composite using a negative Poisson's ratio yarn reinforcement, Composites Science and Technology **72**(7):761–766.
[26] Evans KE (1991) Auxetic polymers: a new range of materials, Endeavour **15**(4):170–174.
[27] Sanamia M, Ravirala N, Alderson K, Alderson A (2014) Auxetic materials for sports applications, Procedia Engineering **72**:453–458.
[28] Duncan O, Shepherd T, Moroney C, Foster L, Venkatraman PD, Winwood K, Allen T, Alderson A (2018) Review of Auxetic Materials for Sports Applications: Expanding Options in Comfort and Protection, Applied Sciences **8**(6):941.
[29] Sloan MR, Wright JR, Evans KE (2011) The helical auxetic yarn - a novel structure for composites and textiles; geometry, manufacture and mechanical properties, Mech. Mater. **43**(9):476–486.
[30] Wang Z, Hu H (2014) Auxetic wrap-knitted spacer fabrics, Physica Status Solidi **251**(2):281–288.
[31] Lei M, Hong W, Zhao Z, Hamel C, Chen M, Lu H, and Qi HJ (2019) 3D Printing of auxetic metamaterials with digitally reprogrammable shape, ACS Appl. Mater. Interfaces **11**(25):22768–22776.
[32] Alomaraha A, Masood SH, Sbarski I, Faisal B, Gao Z, Ruan D (2020) Compressive properties of 3D printed auxetic structures: experimental and numerical studies, Virtual and Physical Prototyping **15**(1):1-21.
[33] Choudhry NK, Panda B, Kumar S (2021) In-plane energy absorption characteristics of a modified re-entrant auxetic structure fabricated via 3D printing, Composites Part B: Engineering **228**:109437.
[34] dell'Isola F, Steigmann DJ (2020) Discrete and Continuum Models for Complex Metamaterials, Cambridge University Press, Cambridge.
[35] Girchenko AA, Eremyev VA, Altenbach H (2012) Interaction of a helical shell with a nonlinear viscous fluid, International Journal of Engineering Science **61**:53–58.
[36] Eremeyev VA, dell'Isola F, Boutin C, Steigmann D (2018) Linear pantographic sheets: existence and uniqueness of weak solutions, J. Elasticity **132**:175–196.
[37] dell'Isola F, Seppecher P, Alibert JJ, Lekszycki T, Grygoruk R, Pawlikowski M, Steigmann D, Giorgio I, Andreaus U, Turco E, Gołaszewski M, Rizzi N, Boutin C, Eremeyev VA, Misra A, Placidi L, Barchiesi E, Greco L, Cuomo M, Cazzani A, Corte AD, Battista A, Scerrato D, Zurba IE, Rahali Y, Ganghoffer JF, Müller W, Ganzosch G, Spagnuolo M, Pfaff A, Barcz K, Hoschke K, Neggers

J, Hild F (2019) Pantographic metamaterials: an example of mathematically driven design and of its technological challenges, Continuum Mechanics and Thermodynamics **31**(4):851-884.

[38] Eremeyev VA, Turco E (2020) Enriched buckling for beam-lattice metamaterials, Mechanics Research Communications **103**:103458.

[39] Turco E, Barchiesi E, dell'Isola F (2022) A numerical investigation on impulse-induced nonlinear longitudinal waves in pantographic beams, Mathematics and Mechanics of Solids **27**(1):22–48.

[40] Stilz M, dell'Isola F, Giorgio I, Eremeyev VA, Ganzenmuller G, Hiermaier S (2022) Continuum models for pantographic blocks with second gradient energies which are incomplete, Mechanics Research Communications **125**:103988.

[41] Misra A, Hild F, Eremeyev VA (2023) Design of metamaterials: Preface, Mechanics Research Communications **127**:104036.

[42] Bianchi M, Scarpa F, Smith CW (2010) Shape memory behaviour in auxetic foams: Mechanical properties, Acta Materialia **58**:858–865.

[43] Wang W, He C, Xie L, Peng Q (2019) The temperature-sensitive anisotropic negative Poisson's ratio of carbon honeycomb, Nanomaterials **9**(4):487.

[44] Li Y, Wang S, Yang B (2021) Auxetic carbon honeycomb: strain-tunable phase transitions and novel negative Poisson's ratio, ACS Omega **6**:14896–14902.

[45] Antimonov MA, Cherkaev A, Freidin AB (2016) Phase transformations surfaces and exact energy lower bounds, Int. J. Engineering Sci. **98**:153–182.

[46] Freidin A, Sharipova L (2019) Two-phase equilibrium microstructures against optimal composite microstructures, Arch. Appl. Mech. **89**(3):561–580.

[47] Freidin A, Sharipova L, Cherkaev A (2021) On equilibrium two-phase microstructures at plane strain, Acta Mech. **232**:2005–2021.

[48] Cherkaev AV (2000) Variational Methods for Structural Optimization. New York, Springer-Verlag.

[49] Milton GW (2004) The Theory of Composites. Cambridge Monogr. Appl. Comput. Math. **6** Cambridge University Press, Cambridge.

[50] Chenchiah IV, Bhattacharya K (2008) The relaxation of two-well energies with possibility unequal moduli, Archive for Rational Mechanics and Analysis **187**(3):409–479.

[51] Freidin AB, Chiskis AM (1994) Regions of phase transitions in nonlinear-elastic isotropic materials. Part 1: Basic relations, Mech. Solid, **29**(4):91–109. Translated from Izvestia RAN, Mekhanika Tverdogo Tela **4**:91–109.

[52] Freidin AB (1999) Small strains approach to the theory on solid-solid phase transformations under the process of deformation, Studies on Elasticity and Plasticity (St. Petersburg State University) **18**:266–290 (in Russian).

[53] Morozov NF, Freidin AB (1998) Zones of phase transition zones and phase transformations in elastic bodies under various stress states, Proceedings of the Steklov Mathematical Institute **223**:219–232.

[54] Eremeev VA, Freidin AB, Sharipova LL (2003) Nonuniqueness and stability in problems of equilibrium of elastic two-phase bodies, Doklady Physics **48**:359–363.

[55] Yeremeyev VA, Freidin AB, Sharipova LL (2007) The stability of the equilibrium of two-phase elastic solids, Journal of Applied Mathematics and Mechanics **71**:61–84.
[56] Fu YB, Freidin AB (2004) Characterization and stability of two-phase piecewise-homogeneous deformation, Proc Roy Soc London Ser A **460**:3065–94.
[57] Grabovsky Y, Truskinovsky L (2011) Roughening instability of broken extremals, Archive for Rational Mechanics and Analysis **200**:183–202.
[58] Grabovsky Y, Truskinovsky L (2013) Marginal material stability, Journal of Nonlinear Science **23**:891–969.
[59] Freidin AB, Sharipova LL (2003) Equilibrium two-phase deformations and phase transitions zones within framework of small strains, Nonlinear problems of continuum mech. Izvestia Vuzov. North-Caucasus region. Special issue 291–299 (in Russian).
[60] Freidin A, Sharipova L (2006) On a model of heterogenous deformation of elastic bodies by the mechanism of multiple appearance of new phase layers, Meccanica **41**(3):321–339.
[61] Freidin AB, Vilchevskay EN, Sharipova LL (2002) Two-phase deformations within the framework of phase transition zones, Theor. Apll. Mech. **28–29**:149–172.
[62] Grinfeld M (1991) Thermodynamic Methods in the Theory of Heterogeneous Systems. Longman Sc & Tech.
[63] Kunin IA (1983) Elastic Media with Microstructure II. Springer-Verlag, Berlin, New York, etc.
[64] Kublanov LB, Freidin AB (1988) Solid phase seeds in a deformable material, Journal of Applied Mathematics and Mechanics (PMM USSR) **52**:382–389 (translated from Prikladnaia Matematica i Mekhanika (1988) **52**:493–501.
[65] Briane M (1994) Correctors for the homogenization of a laminate, Adv. Math. Sci. Appl. **2**:357–379.
[66] Freidin AB (2007) On new phase inclusions in elastic solids, Z Angew Math Mech (ZAMM) **81**(2):102–116.

Chapter 16
Application of Nonlocal Fick's Law Within Micropolar Approach

Ksenia Frolova, Nikolay Bessonov, and Elena Vilchevskaya

Abstract The paper is concerned with modeling of stress-induced diffusion within the framework of the micropolar approach. The specificity of the model accounting for stress-strain state of the host material by means of additional thermodynamic driving force in the form of gradient of chemical potential is discussed. The considered model represents a generalization of the known corresponding model developed within the framework of the classical approach. The proposed generalization allows one to account for size effects while modeling the diffusion process. The main focus is on modeling of the skin effect manifested near the perturbation surface. It is shown that accounting for couple stress interactions between material particles results in a significant excess of diffusing impurity in the vicinity of the border. An axially symmetric problem is solved and investigated in detail. Various types of boundary condition on the outer border are used to demonstrate the general specificity of the developed model. Qualitative and quantitative effect of non-classical material parameters is discussed.

Key words: Stress-induced diffusion · Chemical potential · Micropolar continua · Size effect · Skin effect

16.1 Introduction

The paper models stress-induced diffusion within the framework of the micropolar approach to account for size effects. Problem of diffusion in solids has applications in various fields of science such as geology, metallurgical engineering, etc. The accumulation of harmful impurities in metals can worsen material properties and

Ksenia Frolova · Nikolay Bessonov · Elena Vilchevskaya
Institute for Problems in Mechanical Engineering RAS, Bolshoy pr., 61, V.O., St. Petersburg, 199178, Russian Federation,
e-mail: fkp@ipme.ru, bnm@ipme.ru, ven@ipme.ru

H. Altenbach et al. (eds.), *Sixty Shades of Generalized Continua*,
Advanced Structured Materials 170, https://doi.org/10.1007/978-3-031-26186-2_16

even lead to its damage. In particular, a widespread problem of material science is hydrogen embrittlement of metal alloys [1, 2]. It is known that mechanical loading, as well as the presence of inhomogeneities of different nature can affect mass transport process in a host material. In particular, a local increase of impurity can take place [3]-[5]. The question arises of accounting for the influence of stress-strain state and microstructure of solid on diffusion of impurity.

The diffusion process depends on both the diffusion characteristics of impurity itself and on the state of the host material lattice. In general, diffusion can be modeled from the standpoint of rational mechanics or within the framework of thermodynamics. In the first case modeling of the diffusion process is based on the principles of continuum mechanics which is implementation of conversation laws for multi-component media supplemented by appropriate constitutive equations [6, 7]. The diffusion flux can be expressed in terms of the velocity of impurity. It depends on various fields, in particular, on the concentration gradient, as well as stresses and strains in the host material. The solution of the corresponding nonlinear kinematic equation is a challenging task, so simplifying assumptions on interaction between the impurity and the main structure are usually made [7]-[9]. Within a more widely used thermodynamic approach, the diffusion flux is expressed directly as a gradient of chemical potential. In the most simple case of the chemical potential of ideal gas, the constitutive equation for the diffusion flux reduces to the phenomenological Fick's first law relating the diffusion flux to the concentration gradient. Fick's law can be considered as a local one in that it relates the local flux to the local concentration gradient [10]. Li et al. [11] and Larche and Cahn [12, 13] considered a more complex expression for the chemical potential and accounted for the influence of the trace of the stress tensor caused by the eigenstrains due to accumulation of impurity. Larche and Cahn [12, 13] referred to the amended Fick's law as to nonlocal one, since the chemical potential is expressed in terms of the local stress, which depends on the solution of a boundary-value problem. Wu [10] studied the mutual influence of self-stresses and another forms of internal stresses and took chemical potential of the isotropic host material to be proportional to the trace of the Eshelby energy-momentum tensor introduced in [14] within classical elasticity.

So that, the introduction of the chemical potential of the system "impurity-solid host material" allows one to describe stress-induced diffusion. Obviously, such an approach can be extended to description of diffusion process in media modeled by microstructure-informed continua to account for size effects like the presence of inhomogeneities or skin effects. It is known that gradient theory accounts for the microstructure by means of accounting for deformation gradient in the expression for the elastic strain energy [15]-[18], micropolar theory accounts for traction at micro-scale by means of introduction couple stress interactions between material particles [19]-[21]. Such an accounting for internal degrees of freedom leads to necessity to introduce a more general expression for the Eshelby energy-momentum tensor, which depends on the stress-strain state of material. Within the frame of the present research we focus on modeling of stress-induced diffusion within micropolar approach. Expressions for the Eshelby tensors of non-linear and linear micropolar continua were obtained respectively in [22, 23]. In [24] we developed a micropolar model of stress-induced diffusion accounting for the chemical potential in a form of the trace of the Eshelby energy-momentum tensor. The specificity of this model was investigated on the base of the boundary-value problem with Dirichlet condition on the outer border. The solution of the problem allowed to describe skin effect that is accumulation of impurity in the vicinity of the border. The present research discusses the specificity of the model in a broader sense and solves the problems with different types of the boundary condition to demonstrate that skin effect is caused by the presence of inhomogeneities in material, but not by specific conditions of saturation.

16.2 Diffusion in Media Modeled by Micropolar Continuum

The paper is concerned with modeling the diffusion process in a solid taking into account its microstructure influence. To this end, we generalize a known model of stress-induced diffusion usually used within the framework of the classical theory to describe mass transport process in a solid modeled by micropolar continuum. So that, the present research is multidisciplinary in that sense it makes an attempt to combine ideas of two theories to account for the influence of stress-strain state of the host material on diffusion process in the vicinity of some kind of perturbation area that is inhomogeneity, outer border, etc. The paper accounts for the influence of inner stresses and strains on diffusion process by means of chemical potential of host material as suggested in [10, 25]. The basic ideas of micropolar theory can be found in the reviews [19]-[21].

For simplicity we restrict ourselves to modeling diffusion in physically isotropic and geometrically linear solid host material. In the following we first provide the main formulas of diffusion problem and then the main formulas of micropolar elasticity.

The concentration of impurity c can be obtained from the mass balance that is

$$\frac{\mathrm{d}}{\mathrm{dt}}\int_V c\mathrm{dV} = -\int_{\mathrm{S}} \boldsymbol{n}\cdot\boldsymbol{J}\mathrm{dS}, \tag{16.1}$$

where d/dt is the total derivative with respect to time, $\boldsymbol{n}$ is the outer normal to surface S surrounding the elementary domain V, $\boldsymbol{J}$ is the diffusion flux.

The mass balance in local form reduces to

$$\frac{\mathrm{dc}}{\mathrm{dt}} = -\nabla\cdot\boldsymbol{J}. \tag{16.2}$$

In accordance with linear nonequilibrium thermodynamics, diffusion flux is related to the gradient of chemical potential μ of the system "impurity-host material" as

$$\boldsymbol{J} = -Mc\nabla\mu, \tag{16.3}$$

where M is the impurity mobility ($M = D/RT$, here D is the diffusion coefficient, R is the universal gas constant, T is the absolute temperature).

The chemical potential μ is a sum of the chemical potentials of the diffusing impurity and the host material. Equating the chemical potential of the impurity to the chemical potential of the ideal gas, one can rewrite Eq. (16.3) in the following form:

$$\boldsymbol{J} = -D\left(\nabla c + c\frac{1}{RT}\nabla\mu_S\right), \tag{16.4}$$

where μ_S represents the chemical potential of solid host material.

The first term in Eq. (16.4) represents the diffusion flux due to the concentration gradient, whereas the second term represent the diffusion flux due to the additional thermodynamic driving force $\boldsymbol{F} = \nabla\mu_S/RT$. The chemical potential of a solid is as follows [10]:

$$\mu_S = M_S\left(f_S^0 + \frac{1}{\rho_S}\mathrm{tr}\boldsymbol{\Sigma}\right), \tag{16.5}$$

where M_S is the molar mass of solid host material, f_S^0 is the free energy density of host material in a stress-free state, ρ_S is the density of host material, $\boldsymbol{\Sigma}$ is the Eshelby energy-momentum tensor, "tr" denotes trace of a second-order tensor. So, the additional thermodynamic force reads

$$\boldsymbol{F} = \frac{M_S}{RT\rho_S}\nabla\left(\mathrm{tr}\boldsymbol{\Sigma}\right). \tag{16.6}$$

The Eshelby energy-momentum tensor $\boldsymbol{\Sigma}$ depends on the stress-strain state of the host material. A material particle of micropolar continuum represents an infinitesimal rigid body. Its kinematic is described by independent displacement vector $\boldsymbol{u}$ and microrotation vector $\boldsymbol{\theta}$. Two strain measures, stretch tensor $\boldsymbol{\varepsilon}$ and wryness tensor $\boldsymbol{\kappa}$ can be expressed in terms of displacements and microrotations as follows [20]:

$$\boldsymbol{\varepsilon} = \nabla\boldsymbol{u} + \boldsymbol{I}\times\boldsymbol{\theta}, \qquad \boldsymbol{\kappa} = \nabla\boldsymbol{\theta}. \tag{16.7}$$

where $\boldsymbol{I}$ is the unit second-order tensor, operation $\times$ denotes vector product. Vector product of second-order tensor $\boldsymbol{A} = \boldsymbol{a}_k\boldsymbol{b}_k$ and vector $\boldsymbol{c}$ is given by $\boldsymbol{A}\times\boldsymbol{c} = \boldsymbol{a}_k\left(\boldsymbol{b}_k\times\boldsymbol{c}\right)$, hereafter the Einstein rule of summation over the repeated index is used.

The total strain tensors can be decomposed into elastic parts and eigenstrains due to accumulation of impurity, i.e. strains that would exist in material in the absence of constraint. Accumulation of gas is usually assumed to lead to dilatation only, so the total wryness tensor coincides with the elastic one given by the second equation (16.7), whereas the stretch tensor represents a sum of elastic stretch tensor $\boldsymbol{\varepsilon}^{el}$ and eigenstrains $\boldsymbol{\varepsilon}^* = \alpha c\boldsymbol{I}$ (α is the linear expansion coefficient) that is

$$\boldsymbol{\varepsilon} = \boldsymbol{\varepsilon}^{el} + \boldsymbol{\varepsilon}^*. \tag{16.8}$$

The strain energy density can be written as [26]

$$W = \frac{1}{2}\boldsymbol{\varepsilon}^{el}\cdot\cdot{}^4\boldsymbol{C}\cdot\cdot\boldsymbol{\varepsilon}^{el} + \boldsymbol{\varepsilon}^{el}\cdot\cdot{}^4\boldsymbol{B}\cdot\cdot\boldsymbol{\kappa} + \frac{1}{2}\boldsymbol{\kappa}\cdot\cdot{}^4\boldsymbol{D}\cdot\cdot\boldsymbol{\kappa}, \tag{16.9}$$

where ${}^4\boldsymbol{C}$, ${}^4\boldsymbol{B}$, ${}^4\boldsymbol{D}$ are tensors of elastic moduli.

The constitutive relations for the stress tensor $\boldsymbol{T}$ and couple stress tensor $\boldsymbol{M}$ read

$$\boldsymbol{T} = \frac{\partial W}{\partial \boldsymbol{\varepsilon}^{el}} = {}^4\boldsymbol{C} : \boldsymbol{\varepsilon}^{el} + {}^4\boldsymbol{B} : \boldsymbol{\kappa}, \qquad \boldsymbol{M} = \frac{\partial W}{\partial \boldsymbol{\kappa}} = {}^4\boldsymbol{B} : \boldsymbol{\varepsilon}^{el} + {}^4\boldsymbol{D} : \boldsymbol{\kappa}. \tag{16.10}$$

For an isotropic material ${}^4\boldsymbol{B} = 0$, and tensors of elastic moduli ${}^4\boldsymbol{C}$, ${}^4\boldsymbol{D}$ can be represented as linear combinations of isotropic fourth-order tensors ${}^4\boldsymbol{I}_1 = \boldsymbol{II}$, ${}^4\boldsymbol{I}_2 = \boldsymbol{e}_k \boldsymbol{I} \boldsymbol{e}_k$ and ${}^4\boldsymbol{I}_3 = \boldsymbol{e}_k \boldsymbol{e}_m \boldsymbol{e}_k \boldsymbol{e}_m$ as

$$\begin{aligned} {}^4\boldsymbol{C} &= \lambda {}^4\boldsymbol{I}_1 + 2G {}^4\boldsymbol{E} + \frac{\kappa}{2}\left({}^4\boldsymbol{I}_3 - {}^4\boldsymbol{I}_2\right), \\ {}^4\boldsymbol{D} &= \beta_1^4 \boldsymbol{I}_1 + (\beta_2 + \beta_3)\, {}^4\boldsymbol{E} + \frac{\beta_3 - \beta_2}{2}\left({}^4\boldsymbol{I}_3 - {}^4\boldsymbol{I}_2\right), \end{aligned} \tag{16.11}$$

where ${}^4\boldsymbol{E} = 1/2\left({}^4\boldsymbol{I}_2 + {}^4\boldsymbol{I}_3\right)$, and λ, G are Lamé coefficients, $\kappa, \beta_1, \beta_2, \beta_3$ are micropolar elastic moduli.

The micropolar elastic moduli can be expressed in terms of engineering constants, namely, coupling number N, polar ratio ψ, characteristic length for torsion l_t and characteristic length for bending l_b in accordance with [27, 28] as follows

$$\begin{aligned} N &= \sqrt{\frac{\kappa}{2G+\kappa}}, \quad N \in [0,1], \qquad \psi = \frac{\beta_2+\beta_3}{\beta_1+\beta_2+\beta_3}, \quad \psi \in [0, 1.5], \\ l_t &= \sqrt{\frac{\beta_2+\beta_3}{2G}}, \qquad l_b = \sqrt{\frac{\beta_3}{4G}}. \end{aligned} \tag{16.12}$$

The steady-state of material is determined by the balances of linear and angular momentum that in the absence of body forces and couples are as follows:

$$\int_S \boldsymbol{n} \cdot \boldsymbol{T} \mathrm{dS} = \boldsymbol{0}, \qquad \int_S (\boldsymbol{n} \cdot \boldsymbol{M} + \boldsymbol{r} \times (\boldsymbol{n} \cdot \boldsymbol{T}))\, \mathrm{dS} = \boldsymbol{0}, \tag{16.13}$$

where $\boldsymbol{r}$ is the position vector of a material particle.

Equations (16.13) in local form reduce to

$$\nabla \cdot \boldsymbol{T} = \boldsymbol{0}, \qquad \nabla \cdot \boldsymbol{M} + \boldsymbol{T}_\times = \boldsymbol{0}, \tag{16.14}$$

where $\boldsymbol{A}_\times = (\boldsymbol{a}_k \boldsymbol{b}_k)_\times = \boldsymbol{a}_k \times \boldsymbol{b}_k$ is the vectorial invariant. Its invariant representation is given by $\boldsymbol{A}_\times = -\boldsymbol{A} \times\cdot \boldsymbol{I}$, where the mixed product of two second-order tensors is introduced as $\boldsymbol{A} \times\cdot \boldsymbol{B} = \boldsymbol{a}_k \boldsymbol{b}_k \times\cdot \boldsymbol{c}_m \boldsymbol{d}_m = (\boldsymbol{b}_k \times \boldsymbol{c}_m)(\boldsymbol{a}_k \cdot \boldsymbol{d}_m)$.

Within the micropolar linear elasticity, the Eshelby energy-momentum tensor appearing in Eq. (16.5) is defined as [23]

$$\boldsymbol{\Sigma} = W\boldsymbol{I} - (\nabla \boldsymbol{u} - \boldsymbol{\varepsilon}^*) \cdot \boldsymbol{T}^T - \nabla \boldsymbol{\theta} \cdot \boldsymbol{M}^T, \tag{16.15}$$

where the upper index T denotes a transposed second-order tensor ($\boldsymbol{A}^T = (\boldsymbol{a}_k \boldsymbol{b}_k)^T = \boldsymbol{b}_k \boldsymbol{a}_k$).

Substituting Eqs. (16.15), $(16.7)_1$–(16.10) in Eq. (16.6) and considering isotropic material, we obtain the following expression for the additional thermodynamic driving force:

$$\boldsymbol{F} = \frac{M_S}{RT\rho_S}\nabla\left(\frac{1}{2}(\nabla\boldsymbol{u}-\alpha c\boldsymbol{I})\cdot\cdot{}^4\boldsymbol{C}\cdot\cdot(\nabla\boldsymbol{u}-\alpha c\boldsymbol{I})+\right. \tag{16.16}$$

$$\left.+2(\nabla\boldsymbol{u}-\alpha c\boldsymbol{I})\cdot\cdot{}^4\boldsymbol{C}\cdot\cdot(\boldsymbol{I}\times\boldsymbol{\theta})+\frac{3}{2}(\boldsymbol{I}\times\boldsymbol{\theta})\cdot\cdot{}^4\boldsymbol{C}\cdot\cdot(\boldsymbol{I}\times\boldsymbol{\theta})+\frac{1}{2}\nabla\boldsymbol{\theta}\cdot\cdot{}^4\boldsymbol{D}\cdot\cdot\nabla\boldsymbol{\theta}\right).$$

The last three terms in Eq. (16.16) appear due to the presence of independent microrotations of material particles and couple stress interactions between them. Their presence increases the value of the thermodynamic driving force. These terms vanish within the framework of the classical elasticity accounting for only translation degrees of freedom of material particles. The value of displacement's gradient, in turn, depends on stress-strain state caused by both accumulation of impurity and couple stress interactions between material particles. As a result, accounting for couple stress interactions between material particles results in increasing the first term in Eq. (16.16). So that, modeling of material particles by means of infinitesimal rigid bodies results in increasing the thermodynamic driving force. The question arises on the influence of such an increase on diffusion process.

The paper focuses on description of skin effects, so that in the following we seek for solution of initial boundary-value problem.

16.3 Axially Symmetric Problem

We restrict ourselves to consideration of an axially symmetric problem and model non-stationary stress-induced diffusion in a long cylinder of radius r_0. Cylindrical coordinate system (r, φ, z), where r, z denote respectively radial and axial coordinates and φ denotes azimuthal angle, is introduced, $\boldsymbol{e}_r$, $\boldsymbol{e}_\varphi$, $\boldsymbol{e}_z$ denote the basic vectors. Initial concentration inside cylinder is assumed to be zero.

To demonstrate the specificity of the influence of the additional thermodynamic driving force introduced in the previous section on the concentration profile regardless of the boundary conditions (BCs), we consider three possible types of BC on the lateral surface of the cylinder. In the first case a constant concentration is prescribed, that is

$$c|_{r=r_0} = c_0. \tag{16.17}$$

In the second case the diffusion flux is prescribed, i.e

$$D\left.\frac{\partial c}{\partial r}\right|_{r=r_0} = J_0. \tag{16.18}$$

In the third case the diffusion flux is supposed to be proportional to the concentration difference, so

$$D\frac{\partial c}{\partial r}\bigg|_{r=r_0} = k\,(c - c_0)\,, \tag{16.19}$$

where k is the surface mass transfer coefficient, c_0 is a constant concentration maintained in a working chamber.

The lateral surface of the cylinder is supposed to be traction free, whereas a non-classical static boundary condition is used, so that

$$\boldsymbol{e}_r \cdot \boldsymbol{T}|_{r=r_0} = \boldsymbol{0}, \tag{16.20}$$

$$\boldsymbol{e}_r \cdot \boldsymbol{M}|_{r=r_0} = M_0 \boldsymbol{e}_\varphi. \tag{16.21}$$

Note that setting of microrotations on the border instead of external couples does not affect the solution qualitatively, see [29] for details.

The cylinder is not loaded axially, hence

$$\int_S \boldsymbol{e}_z \cdot \boldsymbol{T} \mathrm{d}S\bigg|_{z=0,z=L} = \boldsymbol{0}, \qquad \int_S \boldsymbol{e}_z \cdot \boldsymbol{M} \mathrm{d}S\bigg|_{z=0,z=L} = \boldsymbol{0}, \tag{16.22}$$

where L is the length of the cylinder.

Assuming all cross-sections behave similarly, we seek a solution with the following ansatz:

$$c = c\,(r)\,, \tag{16.23}$$

$$\boldsymbol{u} = u_r\,(r)\,\boldsymbol{e}_r + U_z\,(r,z)\,\boldsymbol{e}_z, \tag{16.24}$$

$$\varepsilon_{zz} = \frac{\partial U_z}{\partial z} = \text{const}, \tag{16.25}$$

$$\boldsymbol{\theta} = \theta_r\,(r)\,\boldsymbol{e}_r + \theta_z\,(r)\,\boldsymbol{e}_z. \tag{16.26}$$

From Eq. (16.25) it follows that

$$U_z\,(r,z) = u_z\,(r) + \widetilde{u}_z\,(z)\,. \tag{16.27}$$

At the center of the cylinder the following boundary conditions are prescribed:

$$\frac{\partial c}{\partial r}\bigg|_{r=0} = 0, \tag{16.28}$$

$$u_r|_{r=0} < \infty, \qquad u_z|_{r=0} = 0, \qquad \theta_r|_{r=0} = \theta_\varphi\big|_{r=0} = 0. \tag{16.29}$$

System of equations for concentration c, axial stretch strains $\varepsilon_{zz} = \partial U_z/\partial z$, radial displacement u_r, axial displacement u_z and microrotation θ_φ takes form

$$r\dot{c} = [rD\,(c' + cF)]'\,, \quad \text{where} \tag{16.30}$$

$$F = \frac{M_S}{RT\rho_S}\left[\frac{\lambda}{2}\left(\varepsilon_{zz} + u_r' + \frac{u_r}{r}\right)^2 + G\left(\left(\frac{u_r}{r}\right)^2 + (u_r')^2 + (\varepsilon_{zz})^2\right) - \right. \tag{16.31}$$

$$-2\,(3\lambda + 2G)\,\alpha\left(\varepsilon_{zz} + u_r' + \frac{u_r}{r}\right)c + \frac{9}{2}\,(3\lambda + 2G)\,\alpha^2 c^2 +$$

$$+G\left(\frac{3}{2N^2}\left(u_z'\right)^2+2l_b^2\left[\left(\theta_\varphi'\right)^2+\left(\frac{\theta_\varphi}{r}\right)^2\right]-2\left(l_t^2-2l_b^2\right)\frac{\theta_\varphi}{r}\theta_\varphi'\right)\Bigg]',$$

$$\int_0^{r_0}\left[(\lambda+2G)\,\varepsilon_{zz}+\lambda\left(u_r'+\frac{u_r}{r}\right)-(3\lambda+2G)\,\alpha c\right]rdr=0, \tag{16.32}$$

$$\left(ru_r'\right)'-\frac{u_r}{r}-\frac{(3\lambda+2G)}{(\lambda+2G)}\alpha\left((rc)'-c\right)=0, \tag{16.33}$$

$$(ru_z')'+2N^2\left(r\theta_\varphi\right)'=0, \tag{16.34}$$

$$2l_b^2\left(r\theta_\varphi'\right)'-\left(\frac{2}{r}l_b^2+r\frac{2N^2}{1-N^2}\right)\theta_\varphi-r\frac{N^2}{1-N^2}u_z'=0, \tag{16.35}$$

where $\dot{c}$ is the partial derivative of concentration with respect to time, $'$ denotes derivation with respect to radial coordinate. Note that equation for θ_r has a trivial solution and is not provided.

Let us analyze the expression for the additional thermodynamic driving force F given by Eq. (16.31). In fact, it consists of two parts. The first part depends on the solution of the elasticity problem due to the mass transport, whereas the second one depends on the solution of the micropolar boundary-value problem. Solutions of these two problems are independent, since Eqs. (16.34)–(16.35) can be solved independently on the general system of equations and provide behavior of a non-saturated solid, which is not changed with time. In [29] we demonstrated that an approximate solution of the micropolar boundary-value problem can be found analytically. In accordance with [29], this solution decreases exponentially with distance from the border as

$$u_z\approx-\frac{M_0}{2G}\exp\left(-\frac{N}{l_b}\left(r_0-r\right)\right),\qquad \theta_\varphi\approx\frac{M_0}{4GNl_b}\exp\left(-\frac{N}{l_b}\left(r_0-r\right)\right). \tag{16.36}$$

The microrotation is a rapidly decreasing function, so that terms with $\theta_\varphi\theta_\varphi'/r$ and $\left(\theta_\varphi/r\right)^2$ in Eq. (16.31) can be neglected. Accounting for the couple stress interactions between material particles increases the value of the thermodynamic driving force in the vicinity of the border. The value of this driving force depends on the distributed couple stress M_0 and non-classical material parameters l_b, N. Increasing of M_0 results in increasing the absolute values of F, but does not change significantly the area of function's decay, which is mainly determined by the ratio N/l_b. Increasing of the value N/l_b, in turn, results both in increasing the absolute value of the thermodynamic driving force and in decreasing the area of function's decay. As a result, an increase of the ratio of the coupling number and characteristic length for bending leads to a more dramatic change in the concentration profile in comparison with the result obtained under consideration of a smaller value. Such a change can be caused both by increasing the coupling number and by decreasing the characteristic length for bending. At the same time, decreasing of l_b plays a more significant role, since in fact F is determined by terms proportional to $N\exp\left(-N\left(r_0-r\right)/l_b\right)/l_b^3$.

We seek for numerical solution of the partially coupled diffusion-elasticity problem. To this end, a numerical scheme is developed. The main specificity of the scheme is discussed in the Appendix. In [29] we demonstrated that an approximate analytical solution to the micropolar boundary-value problem is in good agreement with the corresponding numerical solution, so that the above discussion on behavior of the additional thermodynamic driving force must be true for both for analytical and numerical solutions.

16.4 Results and Discussion

We investigate the specificity of the developed model by an example of diffusion of hydrogen in a standard cylindrical steel specimen of radius 4 mm and length 40 mm. Experiment on the hydrogen saturation of such a specimen and the subsequent estimation of the inner concentration were carried out in [5, 30]. The tested specimen was incubated at room temperature for 96 hours in a deaerated solution based on distilled water with 5 wt% NaCl and 0.5 wt% CH_3COOH. The gaseous hydrogen sulfide was purged through the solution by bubbling. A constant concentration of 2500 mg/l of hydrogen sulfide was maintained in the working chamber.

Material parameters of steel and characteristics of impurity are given in Table 16.1, see [24] for details. We consider three values of the distributed couple stress, namely, $M_0 = -150\,\text{kPa}\cdot\text{m}$, $M_0 = -100\,\text{kPa}\cdot\text{m}$ and $M_0 = -50\,\text{kPa}\cdot\text{m}$.

We first assume that a constant concentration c_0 is prescribed on the border. According to [5], 2500 mg/l of hydrogen sulfide maintained on the boundary corresponds to concentration of about 50 ppm w.r.t. mass. Figure 16.1 compares the results obtained under consideration classical Fick's law ($\mu_S = 0$), nonlocal Fick's

Table 16.1: Material parameters of steel and characteristics of impurity.

Parameter	Value
Steel	
Molar mass, M_S	$0.056\,\text{kg/mol}$
Density, ρ_S	$7800\,\text{kg/m}^3$
Lamé coefficients, λ and G	120 GPa and 80 GPa
Characteristic length for bending, l_b	$7\,\mu\text{m}$, $10\,\mu\text{m}$, $13\,\mu\text{m}$
Coupling number, N	0.7, 0.9
Impurity	
Molar mass, M_H	$0.002\,\text{kg/mol}$
Linear expansion coefficient, α	$1.5\cdot 10^{-6}$
Diffusion coefficient, D	$5\cdot 10^{-16}\,\text{m}^2/\text{s}$

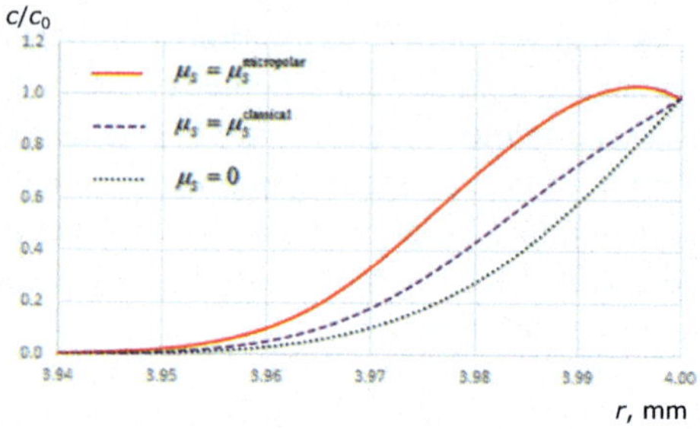

Fig. 16.1 Concentration profiles obtained at different constitutive equations for the diffusion flux

law accounting for the chemical potential of solid host material calculated within the classical elasticity ($\mu_S = \mu_S^{\text{classical}}$) and nonlocal Fick's law accounting for the chemical potential of solid host material calculated within the micropolar elasticity ($\mu_S = \mu_S^{\text{micropolar}}$ at $l_b = 13\,\mu\text{m}$, $N = 0.9$, $M_0 = -100\,\text{kPa}\cdot\text{m}$). It is seen that in the case of nonlocal diffusion laws the impurity tends to accumulate in the vicinity of the border more intensive than in the case of local law. Accounting for the couple stress interactions between material particles results in the largest local increase in hydrogen concentration. The main specificity of the model with $\mu_S = \mu_S^{\text{micropolar}}$ is that starting at some point in time concentration in the vicinity of the border becomes bigger than the one prescribed on the border ($c/c_0 > 1$).

As discussed in Sect. 16.2, the expression of the diffusion flux in terms of a sum of chemical potentials of components of the system "impurity-host material" results in consideration of thermodynamic driving forces of different nature. The first one is due to the concentration difference, whereas the second one is due to the stress-strain state of the host material. Obviously, these forces can be unidirectional or multidirectional. The chemical potential of solid matrix, in turn, depends on the stress-strain state caused both by the self-stresses due to accumulation of hydrogen and by the presence of the distributed couples on the lateral surface. Accumulation of impurity results in increasing the chemical potential within the hydrogenated area, and accounting for the couple stress interactions results in increasing the chemical potential in the vicinity of the border. So that, the chemical potential of solid takes its maximum directly on the border and rapidly decreases with distance. As a result, diffusion flux due to the gradient of the chemical potential of solid host material is always directed from the outer areas into the inner areas of material.

Figure 16.2 presents concentration profiles at different time of saturation the metal specimen with hydrogen. Solution of the problem obtained within the micropolar approach is considered. One can see a fast accumulation of impurity near the border, which does not have time to diffuse into deeper areas of the material before a new portion of impurity comes. Moreover, despite we consider an open system and take into account that impurity in principle can be extracted from the cylinder, the resulting diffusion flux is directed from the border inside cylinder, but not from the cylinder to the working chamber even after concentration exceeds the one prescribed on the surface. So that, presence of couple stress interactions in material can be considered as a reason of the appearance of a source term in the diffusion equation.

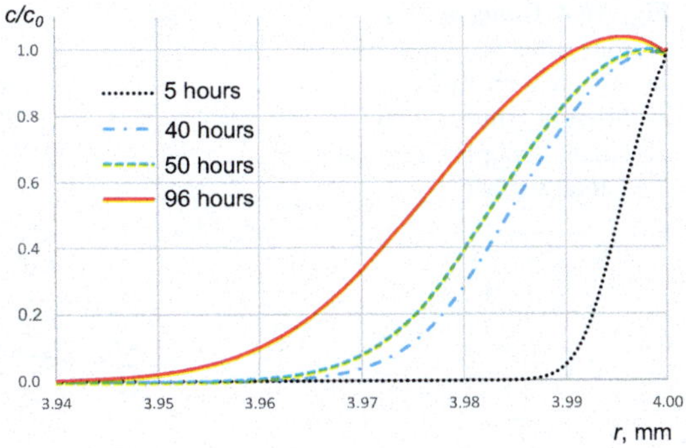

Fig. 16.2 Concentration profiles at different time of saturation with hydrogen

Figure 16.3 shows the influence of non-classical material parameters on the concentration profiles after 96 hours of saturation with hydrogen. It is seen that as discussed in the previous section, increasing of the coupling number and decreasing of the material length for bending results in a more dramatic change of the concentration profile in the vicinity of the border.

Figure 16.4 compares the concentration profiles at two values of the distributed couple stress on the lateral surface at $l_b = 13\,\mu\text{m}$, $N = 0.9$. Increasing of M_0 also results in a more dramatic change of the concentration profile in the vicinity of the border.

Instead of the Dirichlet BC given by Eq. (16.17) one can use the Robin BC given by Eq. (16.19) at an infinite surface mass transfer coefficient ($k \to \infty$). In this case concentration on the border tends to instantly take on value c_0 maintained in a working chamber. However, real diffusion processes have a finite velocity, so that $k < \infty$ and $c|_{r=r_0} - c_0 \neq 0$. Figure 16.5 presents the concentration profiles at two values of k at different time ($l_b = 13\,\mu\text{m}$, $N = 0.9$, $M_0 = -100\,\text{kPa}\cdot\text{m}$). Again it is seen that impurity accumulates rapidly in the vicinity of the border resulting in an excess of hydrogen. Decreasing of the mass transfer coefficients results in smaller values of concentration, but does not change the result qualitatively.

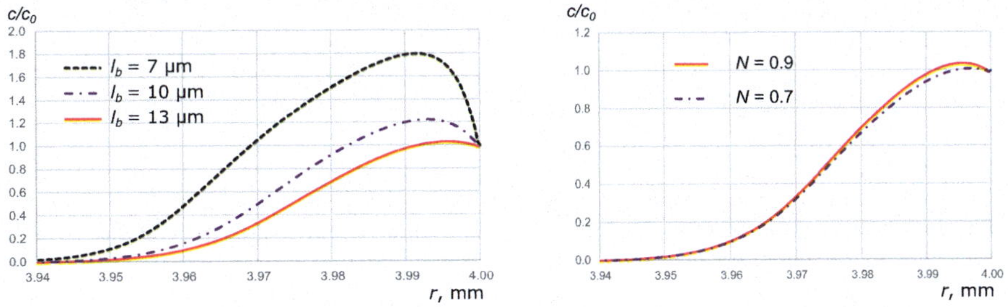

Fig. 16.3: Concentration profiles after 96 hours of saturation with hydrogen at fixed $N = 0.9$ (left) and fixed $l_b = 13\,\mu\text{m}$ (right).

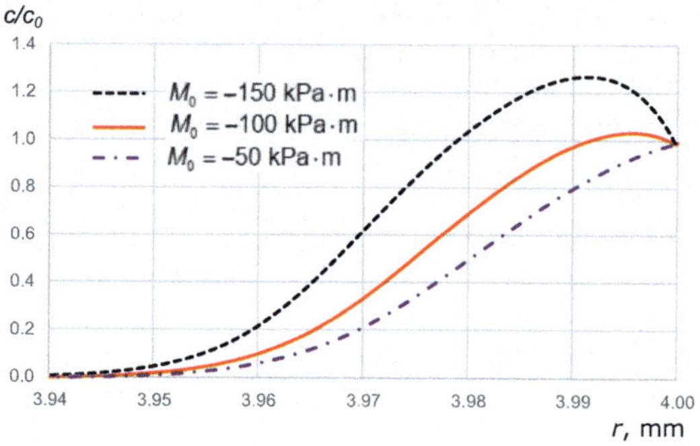

Fig. 16.4 Concentration profiles after 96 hours of saturation with hydrogen at different values of the distributed couple stress on the lateral surface

Let us finally prescribe a constant nonzero flux in a working chamber that can be described by the Neumann BC given by Eq. (16.18), $J_0 = 5 \cdot 10^{-10}\,\mathrm{m/s}$. Comparison of the results presented at Figs. 16.2–16.6 allows to conclude that the presence of a "hump" on the concentration profile is a general specificity of the model of stress-induced diffusion accounting for the chemical potential calculated within the micropolar theory. The presence of a relatively large distributed couple stress on the border results in a significantly non-uniform stress-strain state of material and leads to a local increase of concentration in the vicinity of the outer surface. In principle, such a dramatic change in chemical potential of the solid could be observed even in the case of modeling the host material by classical continuum. However, the appearance of such a big eigenstrains seems to be non-physical. At the same time, couple stress interactions between material particles indeed may play a significant role in the vicinity of the border. Note that values of M_0 used in the paper correspond to microrotation of surface elements less than a few degrees that correlates with physical concepts [31].

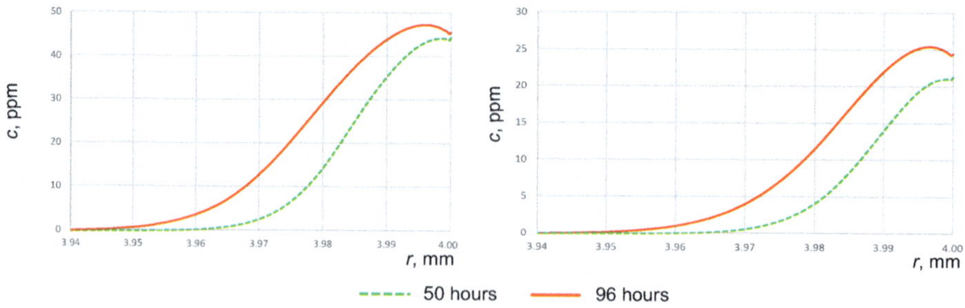

Fig. 16.5: Concentration profiles at different time of saturation with hydrogen in the case of the Robin BC on the lateral surface of the cylinder: $k = 5 \cdot 10^{-10}\,\mathrm{m/s}$ (left) and $k = 5 \cdot 10^{-11}\,\mathrm{m/s}$ (right).

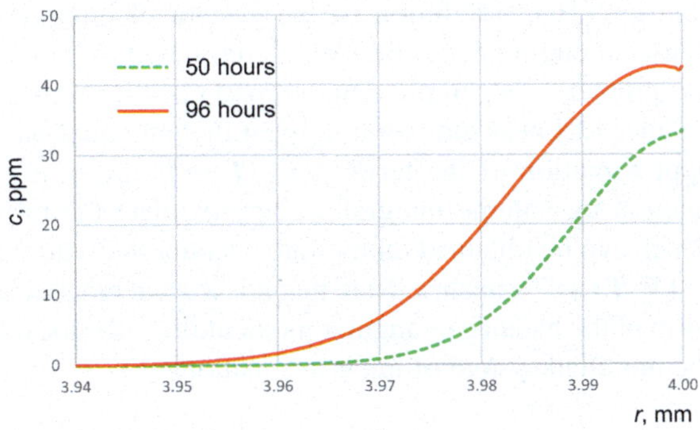

Fig. 16.6 Concentration profiles at different time of saturation with hydrogen in the case of the Neumann BC on the lateral surface of the cylinder

16.5 Conclusions

The paper proposed a model of stress-induced diffusion due to the gradient of chemical potential of solid host material modeled by means of the microstructure-informed micropolar continuum. The model is a generalization of the corresponding model developed in literature within classical theory. Consideration of general expression for the Eshelby energy-momentum tensor allows one to account for couple stress interaction between material particles that plays a role in the vicinity of the area of disturbance. As a result, the model can be implemented to describe size effects due to the presence of inhomogeneities. The paper focused on investigation of skin effect that is an excess of concentration in the vicinity of the saturated border. It was shown that application of nonlocal Fick's law results in a fast saturation of the boundary layer of material. In the case of relatively big values of couple stresses fluxes caused by the concentration gradient and by the solid's chemical potential gradient become multi-directional at some time instant. As a result, the concentration in the vicinity of the border may exceed on the border. Such a situation leads to a significant local increase of impurity concentration, which can worsen material properties and even lead to its damage and so that must be taken into account.

Appendix: Some Remarks on Numerical Approximation

Numerical methods can be both conservative and non-conservative that means they ensure conservation of physical characteristics that are mass, linear and angular momentum, or do not [32]. Within the research we tried our best to follow the principle of "conservativity". In the following we briefly provide the main stages of our numerical approximation.

To obtain a conservative numerical approximation, it is preferred to deal with integral equations describing the physical laws, but not with differential equations providing corresponding local balances. So that it is preferred to deal with mass balance given by Eq. (16.1), and balances of linear and angular momentum given by Eqs. (16.13).Solution of the system of integral equations in the given form is a challenging task due to the complexity of solving Eq. (16.13). To overcome the problem, we simplify the second equation (16.13) through accounting for the balance of linear momentum given by first equation. So that, instead of numerical approximation of the term "$-\nabla\cdot(\boldsymbol{T}\times\boldsymbol{r})$" appearing after replacement of surface integral by volume integral we approximate the term "$-\boldsymbol{T}\times\cdot\nabla\boldsymbol{r}=-\boldsymbol{T}\times\cdot\boldsymbol{I}=\boldsymbol{T}_\times$", which can be obtained taking into account that after parenthesis expansion the term "$-(\nabla\cdot\boldsymbol{T})\times\boldsymbol{r}$" vanishes. Note that it is a usual practice when passing from the integral form of the balance of angular momentum to the local form. As a result, we deal with the modified system of integral equations

$$\int_V \dot{c}\mathrm{dV} = -\int_V \nabla\cdot\boldsymbol{J}\mathrm{dV}, \quad \int_V \nabla\cdot\boldsymbol{T}\mathrm{dV}=\boldsymbol{0}, \quad \int_V (\nabla\cdot\boldsymbol{M}+\boldsymbol{T}_\times)\,\mathrm{dV}=\boldsymbol{0}.$$

In fact, Eqs. (16.1), the first equation of (16.13) remain unchanged, and the second one is modified. This makes the form of integral equation corresponding to balance of angular momentum to be non-conservative, whereas forms of the mass balance and balance of linear momentum are conservative.

Numerical approximation of divergence of tensor function $\boldsymbol{\Phi}$ is also made on the base of a known definition in terms of integrals written for an elementary domain V surrounded by surface S:

$$\nabla\cdot\boldsymbol{\Phi} = \lim_{V\to 0}\frac{\int_S \boldsymbol{n}\cdot\boldsymbol{\Phi}\mathrm{dS}}{V}.$$

In the case of cylindrical coordinate system divergence of an arbitrary tensor function $\boldsymbol{\Phi}$ is defined as

$$\nabla\cdot\boldsymbol{\Phi} = \frac{1}{r}\boldsymbol{e}_r\cdot\frac{\partial}{\partial r}(r\boldsymbol{\Phi}) + \frac{1}{r}\frac{\partial}{\partial\varphi}\left(\boldsymbol{e}_\varphi\cdot\boldsymbol{\Phi}\right) + \boldsymbol{e}_z\cdot\frac{\partial\boldsymbol{\Phi}}{\partial z}. \tag{16.37}$$

Note that application of Eq. (16.37) without removing the brackets makes a sense only in the case of numerical solution to obtain a conservative approximation. In the case of analytical solution, brackets can be removed as usually done.

Solution of the problem considered in the paper depends only on the radial coordinate. So that, a one-dimensional mesh is introduced to obtain a numerical solution. Concentration, displacements and microrotations are specified in mesh nodes, whereas fluxes, deformations and stresses are specified in mesh cells. Differential equations (16.30)–(16.35) were obtained under application of Eq. (16.37) written for ∇-operator in cylindrical coordinates. These equations are further approximated by finite differences to define the stress–strain state in the inner nodes and cells. Solution in the end nodes, in turn, can be found both by "fixing" boundary conditions or "naturally" from integral conversation laws. The first way is used when kinematic boundary conditions are specified within elastic problem (Eqs. (16.17), (16.29)) and when concentration is prescribed on the border (Eq. (16.17)). The second way is used when static boundary conditions are prescribed within the the framework of elasticity problem (Eq. (16.20), (16.21)) and when diffusion fluxes are given within diffusivity problem (Eqs. (16.18), (16.19), (16.28)).

References

[1] Koyama M, Akiyama E, Lee Y-K, Raabe D, Tsuzaki K (2017) Overview of hydrogen embrittlement in high-Mn steels, Int J Hydrog Energy **42**(17):12706-12723.

[2] Wasim M, Djukic MB (2020) Hydrogen embrittlement of low carbon structural steel at macro-, micro-and nano-levels, Int J Hydrogen Energy **45**(3):2145-2156.

[3] Wu TI, Wu JC (2008) Effects of cathodic charging and subsequent solution treating parameters on the hydrogen redistribution and surface hardening of Ti–6Al–4V alloy, J Alloys Compd **466**(1-2):153-159.

[4] Martinsson Å, Sandström R (2012) Hydrogen depth profile in phosphorus-doped, oxygen-free copper after cathodic charging, J Mater Sci **47**(19):6768-6776.

[5] Polyanskiy VA, Belyaev AK, Alekseeva EL, Polyanskiy AM, Tretyakov DA, Yakovlev YA (2019) Phenomenon of skin effect in metals due to hydrogen absorption, Continuum Mech Thermodyn **31**(6):1961-1975.

[6] Aifantis EC (1980) On the problem of diffusion in solids, Acta Mech **37**(3):265-296.

[7] Indeitsev D, Mochalova Y (2014) Mechanics of multi-component media with exchange of mass and non-classical supplies, In: Irschik H, Belyaev AK (Eds) Dynamics of Mechanical Systems with Variable Mass, pp 165-194, Springer, Vienna.

[8] Indeitsev DA, Mochalova YA (2017) On the problem of diffusion in materials under vibrations, In: Altenbach H, Goldstein R, Murashkin E (Eds) Mechanics for Materials and Technologies, pp 183-193, Springer, Cham.

[9] Belyaev AK, Polyanskiy VA, Yakovlev YA (2012) Stresses in a pipeline affected by hydrogen, Acta Mech **223**(8):1611-1619.

[10] Wu CH (2001) The role of Eshelby stress in composition-generated and stress-assisted diffusion, J Mech Phys Solids **49**(8):1771-1794

[11] Li JCM, Oriani RA, Darken LS (1966) The thermodynamics of stressed solids, Z Phys Chem **49**(3-5):271-290.

[12] Larché F, Cahn JW (1973) A linear theory of thermochemical equilibrium of solids under stress, Acta Metall **21**(8):1051-1063.

[13] Larchté, F.C., Cahn, J.L.: The effect of self-stress on diffusion in solids. Acta Metall. **30**(10), 1835–1845 (1982)

[14] Eshelby JD (1975) The elastic energy–momentum tensor, J Elast **5**(3-4):321-335.

[15] Mindlin RD (1965) Second gradient of strain and surface-tension in linear elasticity, Int J Solids Struct **1**(4):417-438.

[16] Mindlin RD, Eshel NN: On first strain-gradient theories in linear elasticity, Int J Solids Struct **4**(1):109-124.
[17] Toupin R (196(2) Elastic materials with couple-stresses, Arch Ration Mech Anal **11**(1):385-414.
[18] Aifantis EC (2003 Update on a class of gradient theories, Mech Mater **35**(3-6), 259-280.
[19] Altenbach, H., Eremeyev, VA (2012) Generalized Continua - from the Theory to Engineering Applications, Springer Science & Business Media.
[20] Eremeyev VA, Lebedev LP, Altenbach H (2012) Foundations of micropolar mechanics. Springer Science & Business Media (2012)
[21] Maugin GA, Metrikine AV (Eds): Mechanics of Generalized Continua, Springer, New York (2010)
[22] Eremeev VA, Zubov LM Phase-equilibrium conditions in nonlinear-elastic media with microstructurel Doklady Akad Nauk Minerologia USSR **322**(6):1052-1056.
[23] Lazar M, Kirchner HOK (2007) The Eshelby stress tensor, angular momentum tensor and scaling flux in micropolar elasticity, Int J Solids Struct **44**(14-15):4613-4620.
[24] Frolova KP, Vilchevskaya EN, Bessonov NM (2022) On modeling of stress-induced diffusion within micropolar and classical approaches, ZAMM - Z für Angew Math Mech **102**(6):e202100505 (2022)
[25] Cui Z, Gao F, Qu J (2012) A finite deformation stress-dependent chemical potential and its applications to lithium ion batteries, J Mech Phys Solids **60**(7):1280-1295.
[26] Nowacki W (1974) The linear theory of micropolar elasticity In: Nowacki W, Olszak W (Eeds) Micropolar Elasticity, International Centre for Mechanical Sciences, vol 151, pp 1-43, Springer, Vienna.
[27] Eringen AC (1999) Theory of micropolar elasticity. Microcontinuum field theories, Springer, New York, NY.
[28] Lakes R (2001) Elastic and viscoelastic behavior of chiral materials, Int J Mech Sci **43**(7):1579-1589.
[29] Frolova K, Vilchevskaya E, Bessonov N, Müller W, Polyanskiy V, Yakovlev Y (2022) Application of micropolar theory to the description of the skin effect due to hydrogen saturation, Math Mech Solids **27**(6):1092-1110.
[30] Alekseeva EL, Belyaev AK, Zegzhda AS, Polyanskiy AM, Polyanskiy VA, Frolova KP, Yakovlev YA (2018) Boundary layer influence on the distribution of hydrogen concentrations during hydrogen-induced cracking test of steels, Diagnostics, Resource and Mechanics of Materials and Structures **3**:43-57.
[31] Merson E, Kudrya AV, Trachenko VA, Merson D, Danilov V, Vinogradov A (2016) The Use of Confocal Laser Scanning Microscopy for the 3D Quantitative Characterization of Fracture Surfaces and Cleavage Facets, Procedia Struct Integr **2**:533-540.
[32] Samarskii AA (2001) The Theory of Difference Schemes, vol. 240, CRC Press, Boca Raton.

Chapter 17
Geometrically Nonlinear Cosserat Elasticity with Chiral Effects Based upon Granular Micromechanics

Ivan Giorgio, Anil Misra, and Luca Placidi

Abstract In this short contribution, we exemplify how the Cosserat kinematics, introduced into classical continuum mechanics at the turn of the XXth century, arises when we consider the granular microstructure inherent in all matter. The discussed discrete-continuum identification, that follows Piola's *ansatz*, is utilized to develop kinematic measures in the framework of finite deformations. These kinematic measures are then utilized to express the internal deformation (strain) energy in terms of both grain-scale and continuum kinematic measures. As a result, we have identified the macroscopic elastic constants of geometrically nonlinear Cosserat continuum model in terms of the grain scale parameters.

Key words: Cosserat media · Granular micromechanics · Microstructured solids · Generalized continua · Finite deformations

---

Ivan Giorgio
Department of Civil, Construction-Architectural and Environmental Engineering (DICEAA)& International Research Center on the Mathematics and Mechanics of Complex Systems (M&MoCS), University of L'Aquila, Italy,
e-mail: ivan.giorgio@univaq.it

Anil Misra
Civil, Environmental and Architectural Engineering Department, The University of Kansas, 1530 W. 15th Street, Lawrence, KS 66045-7609 & International Research Center on the Mathematics and Mechanics of Complex Systems (M&MoCS), University of L'Aquila, Italy,
e-mail: amisra@ku.edu

Luca Placidi
Faculty of Engineering, International Telematic University Uninettuno, Rome & International Research Center on the Mathematics and Mechanics of Complex Systems (M&MoCS), University of L'Aquila, Italy,
e-mail: luca.placidi@uninettunouniversity.net

H. Altenbach et al. (eds.), *Sixty Shades of Generalized Continua*,
Advanced Structured Materials 170, https://doi.org/10.1007/978-3-031-26186-2_17

17.1 Introduction

All materials are characterized by hierarchies of microstructures, which at some spatial scale, can be described as composed of discrete sub-bodies or structural elements [1]-[8]. The relative geometrical arrangements and inter-connectivity (interphases or interfaces) of these discrete elements (grains) forms the microstructure of the matter irrespective of the spatial scale [9]-[15]. Furthermore, the geometrical/mechanical attributes of the interconnections informs the micromechanics of the grain motions [16]-[19]. The symbiotic effects produced through interaction of the microstructure and micromechanics can be termed as micro-mechano-morphology [20]-[25]. The mathematical description of the material deformation by accounting for the micro-mechano-morphological effects requires introduction of additional kinematical descriptors in continuum models [26]-[32]. At the turn of the XXth century, the Cosserat brothers developed the continuum model of deformable bodies by introducing the rotational kinematic degree-of-freedom associated with every continuum material point [33]-[35]. The Cosserat model and its refinements/modifications, known as the micro-polar continuum theory, have come to play an essential role in explaining certain size-dependent phenomena exhibited by the so-called "micro-structured" solids [36]-[43], including materials with granular microstructures [35]. In this sense, the Cosserat or the micro-polar models are considered an important step in the endeavor to capture in greater details the effect of the microstructure in continuum models of material deformation [20].

In this paper, we derive a geometrically nonlinear Cosserat continuum theory on the basis of granular micromechanics by linking the grain-scale deformation to the continuum kinematic measures [44]-[47]. In this regard, we first employ Piola's *ansatz* to make the discrete-continuum identification [26],[48]-[53] such that the objective relative motion between grains can be related to the continuum strain measures. Moreover, we can then define the elastic strain energy at the grain-scale utilizing the objective relative motion between grains, thus laying the foundation for the micro-macro identification of the elastic constants for a hemitropic Cosserat media exhibiting chirality.

17.2 Discrete and Continuous Models for Granular Systems

17.2.1 Identification via Piola's Ansatz

In the discrete model, the reference configuration of the considered set of n grains is given by their position

$$\{X_1, X_2, \ldots, X_n\} \in \left(E^2\right)^n,$$

where E^2 is the Euclidean two dimensional space and reference angular arrangement,

$$\{\mathfrak{R}_1 = I, \mathfrak{R}_2 = I, \ldots, \mathfrak{R}_n = I\} \in \left(\mathrm{Orth}^+\right)^n,$$

where Orth^+ is the space of orthogonal tensor with positive determinant (thus, the space of rotation). For the sake of simplicity, the reference angular arrangement is assumed to be constant in space and equal to the identity tensor I.

They displace, respectively, with the following displacement functions

$$u_1(X_1) = \chi_1(X_1) - X_1, u_2(X_2) = \chi_2(X_2) - X_2, \ldots, u_n(X_n) = \chi_n(X_n) - X_n,$$

and rotate as

$$\{R_1, R_2, \ldots, R_n\} \in \left(\mathrm{Orth}^+\right)^n,$$

where $\chi_i(X_i)$ is the placement of the i-th grain eventually depending on time t and where Orth^+ is the space of orthogonal tensor with positive determinant (thus, the space of rotation).

In the continuum model, we have a continuous body $\mathfrak{B}$ which, in the reference configuration, is constituted by infinite particles having position X, i.e. $X \in \mathcal{B}$. Each particle has displacement $u(X) = \chi(X) - X$, where $\chi(X)$ is the placement function of the continuous body $\mathfrak{B}$ and rotation $R(X)$.

In the continuum-discrete models identification, the following Piola's ansatz will be assumed

$$\begin{array}{llll} u(X_1) = u_1(X_1), & u(X_2) = u_2(X_2), & \ldots & u(X_n) = u_n(X_n), \\ R(X_1) = R_1(X_1), & R(X_2) = R_2(X_2), & \ldots & R(X_n) = R_n(X_n), \end{array}$$

which means that the displacements and rotation of the n grains correspond to the displacement $u(X)$ and rotation $R(X)$ of the continuous body $\mathfrak{B}$ evaluated at the points X_i with $i = 1, \ldots, n$ where the grains are located in the reference configuration.

17.2.2 Relative Intergranular Displacement and Related Continuum Deformation Measures

Let us assume that the distance between the particles at X_n and X_p is L and the unit vector $\hat{c}$ is defined as follows,

$$X_n - X_p = \hat{c}L. \tag{17.1}$$

In the reference configuration, therefore, the vector attached to the position X_n and pointing to the position X_p is $\hat{c}L$ and is given in (17.1).

In the current configuration, the positions of the two particles at X_n and X_p are, respectively, $\chi(X_n)$ and $\chi(X_p)$. The vector in (17.1) in the current figuration yields

$$\chi(X_n) - \chi(X_p). \tag{17.2}$$

The difference between the vectors in (17.1) and (17.2) is called the relative displacement δ_{np} of the two grains n and p,

$$\delta^{np} = \chi(X_n) - \chi(X_p) - (X_n - X_p) = u_n(X_n) - u_p(X_p). \tag{17.3}$$

In order to define an objective relative displacement (i.e. a relative displacement that is a measure of the contribution of the $n-p$ pair to the deformation of the granular assembly), we consider both the deformation gradient

$$F = \mathcal{F}(X_p) = \nabla\chi, \tag{17.4}$$

as the gradient of the placement function χ and the rotation

$$R = \mathcal{R}(X_p). \tag{17.5}$$

Thus, we define two objective relative displacements, i.e. the objective macro relative displacement u^{np} and the objective micro-macro relative displacement γ^{np} as follows,

$$u^{np} = \mathcal{F}^T(X_p)\left[\chi(X_n) - \chi(X_p)\right] - (X_n - X_p), \tag{17.6}$$

$$\gamma^{np} = \mathcal{R}^T(X_p)\left[\chi(X_n) - \chi(X_p)\right] - (X_n - X_p). \tag{17.7}$$

In the current configuration, the rotations of the two particles at X_n and X_p are, respectively, $\mathcal{R}(X_n)$ and $\mathcal{R}(X_p)$. Thus, we define an objective relative tensor,

$$m^{np} = \mathcal{R}^T(X_p)\mathcal{R}(X_n) - I \tag{17.8}$$

that is called the intergranular micro deformation. Let us now assume that the two grains n and p are neighboring grains. Let us restrict the present model to the case they place and rotate similarly in the present configuration, and therefore the following Taylor's series expansions are possible and yield

$$\chi(X_n) \cong \chi(X_p) + (\nabla\chi)_{X_p}(X_n - X_p) = \chi(X_p) + \mathcal{F}(X_p)(X_n - X_p). \tag{17.9}$$

$$R(X_n) \cong \mathcal{R}(X_p) + (\nabla\mathcal{R})_{X_p}(X_n - X_p). \tag{17.10}$$

It is notable that this simplification may loose applicability for materials that possess strong variations and discontinuities of stiffnesses or have highly disordered geometries, which could be deliberately designed as in pantographic metamaterials [11, 21, 49], [54]-[68] or in granular (meta)materials [69]. In these cases, Taylor expansion is not representative and additional kinematic descriptors may be introduced to accurately describe the response as in Cosserat or micromorphic media [20, 36].

The Green–Saint-Venant tensor G,

$$G = \frac{1}{2}\left(F^T F - I\right), \tag{17.11}$$

is defined in terms of the deformation gradient F as well as a relative micro-macro Green–Saint-Venant tensor Υ

$$\Upsilon = \frac{1}{2}\left(R^T F - I\right). \tag{17.12}$$

The last four equations (17.9-17.11-17.10-17.12) involve third order tensors and it is more convenient to show them in index notation

$$\chi_i^n = \chi_i^p + F_{ij}^p \hat{c}_j L, \qquad G_{ij}^p = \frac{1}{2}\left(F_{ai}^p F_{aj}^p - \delta_{ij}\right),$$
$$R_{aj}^n = R_{aj}^p + R_{aj,h}^p \hat{c}_h L, \qquad \Upsilon_{ij}^p = \frac{1}{2}\left(R_{ai}^p F_{aj}^p - \delta_{ij}\right), \tag{17.13}$$

where superscripts n or p denote that the functions are evaluated, respectively, at the points X_n or X_p. Thus, the objective relative displacements in (17.6) and (17.7) and tensor (17.8) are in index form,

$$u_i^{np} = F_{ai}^p (\chi_a^n - \chi_a^p) - (X_i^n - X_i^p),$$
$$\gamma_i^{np} = R_{ai}^p (\chi_a^n - \chi_a^p) - (X_i^n - X_i^p),$$
$$m_{ij}^{np} = R_{ai}^p R_{aj}^n - \delta_{ij},$$

that, from (17.1) and (17.13)$_1$ yields

$$u_i^{np} = F_{ai}^p \left(F_{aj}^p \hat{c}_j L\right) - \hat{c}_i L = F_{ai}^p F_{aj}^p \hat{c}_j L - \delta_{ij}\hat{c}_j L = 2G_{ij}^p \hat{c}_j L, \tag{17.14}$$

$$\gamma_i^{np} = R_{ai}^p \left(F_{aj}^p \hat{c}_j L\right) - \hat{c}_i L = \left[R_{ai}^p F_{aj}^p - \delta_{ij}\right] \hat{c}_j L = 2\Upsilon_{ij}^p \hat{c}_j L, \tag{17.15}$$

$$m_{ij}^{np} = R_{ai}^p R_{aj}^n - \delta_{ij} = R_{ai}^p \left(R_{aj}^p + R_{aj,h}^p \hat{c}_h L\right) - \delta_{ij} = R_{ai}^p R_{aj,h}^p \hat{c}_h L, \tag{17.16}$$

where δ_{ij} is the Kronecker symbol. In absolute notation (17.14), (17.15) and (17.16) are

$$u^{np} = 2LG\hat{c}, \qquad \gamma^{np} = 2L\Upsilon\hat{c}, \qquad m^{np} = LR^T(\nabla R)\hat{c}, \tag{17.17}$$

where all the superscripts p have been omitted because all the quantities are locally evaluated at $X = X_p$.

17.2.3 On the Objective (Macro and Micro-macro) Displacement Vectors

The projections of the objective relative displacements on the unit vector $\hat{c}$, defined in (17.1), is the so called normal displacement u_η, and normal micro relative displacement that are defined,

$$u_\eta = \frac{1}{2} u^{np} \cdot \hat{c}, \qquad \gamma_\eta = \frac{1}{2}\gamma^{np} \cdot \hat{c}. \tag{17.18}$$

Insertion of (17.14) and (17.15) into (17.18) provides

$$u_\eta = LG_{ij}\hat{c}_i\hat{c}_j, \qquad \gamma_\eta = L\Upsilon_{ij}\hat{c}_i\hat{c}_j. \tag{17.19}$$

Their squares are,

$$u_\eta^2 = L^2 \hat{c}_i \hat{c}_j \hat{c}_a \hat{c}_b G_{ij} G_{ab}, \qquad \gamma_\eta^2 = L^2 \hat{c}_i \hat{c}_j \hat{c}_a \hat{c}_b \Upsilon_{ij} \Upsilon_{ab}. \tag{17.20}$$

The objective relative displacement vectors u^{np} and γ^{np} projected orthogonally to the unit vector $\hat{c}$, defined in (17.1), are the so called tangent displacements u_τ and γ_τ that are vectors and defined as follows,

$$u_\tau = u^{np} - (u^{np} \cdot \hat{c})\, \hat{c}, \qquad \gamma_\tau = \gamma^{np} - (\gamma^{np} \cdot \hat{c})\, \hat{c}. \tag{17.21}$$

The only objective quantities derived from u_τ, γ_τ and $\hat{c}$ are the squares of both u_τ and γ_τ and their scalar product, i.e.

$$u_\tau^2 = [u^{np} - (u^{np} \cdot \hat{c})\, \hat{c}] \cdot [u^{np} - (u^{np} \cdot \hat{c})\, \hat{c}] = u^{np} \cdot u^{np} - (u^{np} \cdot \hat{c})^2, \tag{17.22}$$

$$\gamma_\tau^2 = [\gamma^{np} - (\gamma^{np} \cdot \hat{c})\, \hat{c}] \cdot [\gamma^{np} - (\gamma^{np} \cdot \hat{c})\, \hat{c}] = \gamma^{np} \cdot \gamma^{np} - (\gamma^{np} \cdot \hat{c})^2, \tag{17.23}$$

$$\begin{aligned} u_\tau \cdot \gamma_\tau &= [u^{np} - (u^{np} \cdot \hat{c})\, \hat{c}] \cdot [\gamma^{np} - (\gamma^{np} \cdot \hat{c})\, \hat{c}] \\ &= u^{np} \cdot \gamma^{np} - (u^{np} \cdot \hat{c})\, (\gamma^{np} \cdot \hat{c}), \end{aligned} \tag{17.24}$$

or, in index notation and taking (17.20) into account,

$$u_\tau^2 = (2G_{ij}\hat{c}_j L)\,(2G_{ik}\hat{c}_k L) - 4u_\eta^2, \tag{17.25}$$

$$\gamma_\tau^2 = (2\Upsilon_{ij}\hat{c}_j L)\,(2\Upsilon_{ik}\hat{c}_k L) - 4\gamma_\eta^2, \tag{17.26}$$

$$u_\tau \cdot \gamma_\tau = (2G_{ij}\hat{c}_j L)\,(2\Upsilon_{ik}\hat{c}_k L) - 4u_\eta \gamma_\eta, \tag{17.27}$$

or in a more compact form,

$$u_\tau^2 = 4L^2 G_{ij} G_{ab} \left(\delta_{ia}\hat{c}_j \hat{c}_b - \hat{c}_i \hat{c}_j \hat{c}_a \hat{c}_b\right), \tag{17.28}$$

$$\gamma_\tau^2 = 4L^2 \Upsilon_{ij} \Upsilon_{ab} \left(\delta_{ia}\hat{c}_j \hat{c}_b - \hat{c}_i \hat{c}_j \hat{c}_a \hat{c}_b\right), \tag{17.29}$$

$$u_\tau \cdot \gamma_\tau = 4L^2 G_{ij} \Upsilon_{ab} \left(\delta_{ia}\hat{c}_j \hat{c}_b - \hat{c}_i \hat{c}_j \hat{c}_a \hat{c}_b\right). \tag{17.30}$$

17.2.4 On the Objective Tensor

In the 2D case, the rotation tensor R is assumed to be a rotation of an angle φ around the third axis orthogonal to the 2D plane. Besides, the unit vector $\hat{c}$ is parameterized by the anticlockwise angle θ with respect to the horizontal axis within the 2D plane and the displacement vector u has only two components, i.e. $u_3 = 0$. Finally, the deformation gradient F is an in-plane tensor. In formulas we have,

$$R = \begin{pmatrix} \cos\varphi & \sin\varphi & 0 \\ -\sin\varphi & \cos\varphi & 0 \\ 0 & 0 & 1 \end{pmatrix}, \qquad \hat{c} = \begin{pmatrix} \cos\theta \\ \sin\theta \\ 0 \end{pmatrix},$$

$$u = \begin{pmatrix} u_1 \\ u_2 \\ 0 \end{pmatrix}, \quad F = \begin{pmatrix} 1+u_{1,1} & u_{1,2} & 0 \\ u_{2,1} & 1+u_{2,2} & 0 \\ 0 & 0 & 0 \end{pmatrix}. \tag{17.31}$$

The gradient of the rotation tensor is a third order tensor but its projection along the unit vector $\hat{c}$ is a second order tensor with the following components

$$[(\nabla R)\,\hat{c}]_{ij} = (\nabla R)_{ijh}\,\hat{c}_h = R_{ij,h}\hat{c}_h. \tag{17.32}$$

Thus, evaluating component by component the second order tensor defined in (17.32) we have,

$$\begin{aligned}
[(\nabla R)\,\hat{c}]_{11} &= R_{11,1}\hat{c}_1 + R_{11,2}\hat{c}_2 = -\sin\varphi\,(\varphi_{,1})\cos\theta - \sin\varphi\,(\varphi_{,2})\sin\theta,\\
[(\nabla R)\,\hat{c}]_{12} &= R_{12,1}\hat{c}_1 + R_{12,2}\hat{c}_2 = \cos\varphi\,(\varphi_{,1})\cos\theta + \cos\varphi\,(\varphi_{,2})\sin\theta,\\
[(\nabla R)\,\hat{c}]_{21} &= R_{21,1}\hat{c}_1 + R_{21,2}\hat{c}_2 = -\cos\varphi\,(\varphi_{,1})\cos\theta - \cos\varphi\,(\varphi_{,2})\sin\theta,\\
[(\nabla R)\,\hat{c}]_{22} &= R_{22,1}\hat{c}_1 + R_{22,2}\hat{c}_2 = -\sin\varphi\,(\varphi_{,1})\cos\theta - \sin\varphi\,(\varphi_{,2})\sin\theta,
\end{aligned}$$

or in a matrix compact form,

$$(\nabla R)\,\hat{c} = \begin{pmatrix} -\sin\varphi\,(\varphi_{,1}\cos\theta + \varphi_{,2}\sin\theta) & \cos\varphi\,(\varphi_{,1}\cos\theta + \varphi_{,2}\sin\theta) & 0 \\ -\cos\varphi\,(\varphi_{,1}\cos\theta + \varphi_{,2}\sin\theta) & -\sin\varphi\,(\varphi_{,1}\cos\theta + \varphi_{,2}\sin\theta) & 0 \\ 0 & 0 & 0 \end{pmatrix},$$

that can be reduced as follows,

$$(\nabla R)\,\hat{c} = \nabla\varphi\cdot\hat{c}\begin{pmatrix} -\sin\varphi & \cos\varphi & 0 \\ -\cos\varphi & -\sin\varphi & 0 \\ 0 & 0 & 0 \end{pmatrix}.$$

Keeping this in mind, the objective relative tensor in (17.16) is easily represented from $(17.31)_1$ as follows,

$$\begin{aligned}
m^{np} = LR^T\,(\nabla R)\,\hat{c} &= L\nabla\varphi\cdot\hat{c}\begin{pmatrix} \cos\varphi & -\sin\varphi & 0 \\ \sin\varphi & \cos\varphi & 0 \\ 0 & 0 & 1 \end{pmatrix}\begin{pmatrix} -\sin\varphi & \cos\varphi & 0 \\ -\cos\varphi & -\sin\varphi & 0 \\ 0 & 0 & 0 \end{pmatrix}\\
&= \kappa\begin{pmatrix} 0 & 1 & 0 \\ -1 & 0 & 0 \\ 0 & 0 & 1 \end{pmatrix},
\end{aligned} \tag{17.33}$$

where,

$$\kappa = L\nabla\varphi\cdot\hat{c}. \tag{17.34}$$

The representation (17.33) of the objective relative tensor in terms of a single scalar is worth noting.

17.2.5 The Objective Scalar Deformation Measures

Finally, we can list the scalar quantities of the relative deformation measures of the pairs (17.6), (17.7) and (17.8),

$$u_\eta, \gamma_\eta, u_\tau^2, \gamma_\tau^2, \kappa, u_\tau \cdot \gamma_\tau. \tag{17.35}$$

from (17.18), (17.28), (17.29), (17.30) and (17.34).

17.3 Elastic Energy Function

The elastic energy function U^{np} for a given couple of particles, say the couple $n-p$ considered in Section 17.2.2, is assumed to be, in the discrete and (because of the Piola's ansatz) in the continuum models, a quadratic form of normal and tangent parts components of the objective relative displacements listed in (17.35),

$$\begin{aligned} U^{np} &= \frac{1}{2}k_\eta\left(1-D_\eta\right)u_\eta^2 + \frac{1}{2}k_{\gamma\eta}\left(1-D_{\gamma\eta}\right)\gamma_\eta^2 + \frac{1}{2}k_\tau\left(1-D_\tau\right)u_\tau^2 + \\ &+ \frac{1}{2}k_{\gamma\tau}\left(1-D_{\gamma\tau}\right)\gamma_\tau^2 + \frac{1}{2}k_\kappa\left(1-D_\kappa\right)\kappa^2 + \\ &+ k_{u\tau\gamma\tau}u_\tau\cdot\gamma_\tau + k_{u\eta\gamma\eta}u_\eta\gamma_\eta + k_{u\eta\kappa}u_\eta\kappa + k_{\gamma\eta\kappa}\gamma_\eta\kappa, \end{aligned} \tag{17.36}$$

where the coefficients of this quadratic form are the stiffness of the grain pair elastic interaction, that can be divided into the coefficients of the quadratic terms,

$$k_\eta, k_{\gamma\eta}, k_\tau, k_{\gamma\tau}, k_\kappa, \tag{17.37}$$

and that of the coupling terms,

$$k_{u\tau\gamma\tau}, k_{u\eta\gamma\eta}, k_{u\eta\kappa}, k_{\gamma\eta\kappa}. \tag{17.38}$$

For the sake of simplicity, we have assumed that damage affects only the stiffness listed in (17.37) with the following, and respectively, damage variables

$$D_\eta, D_{\gamma\eta}, D_\tau, D_{\gamma\tau}, D_\kappa.$$

The total elastic energy density, for a given position X, is the some of that of every pairs or, in the continuum approximation, the integral over all the orientation

$$
\begin{aligned}
U = & \int_{S^1} \frac{1}{2} k_\eta \left(1 - D_\eta\right) \left(L^2 \hat{c}_i \hat{c}_j \hat{c}_a \hat{c}_b G_{ij} G_{ab}\right) \\
& + \int_{S^1} \frac{1}{2} k_{\gamma\eta} \left(1 - D_{\gamma\eta}\right) L^2 \hat{c}_i \hat{c}_j \hat{c}_a \hat{c}_b \Upsilon_{ij} \Upsilon_{ab} \\
& + \int_{S^1} \frac{1}{2} k_\tau \left(1 - D_\tau\right) \left(4L^2 G_{ij} G_{ab} \left(\delta_{ia} \hat{c}_j \hat{c}_b - \hat{c}_i \hat{c}_j \hat{c}_a \hat{c}_b\right)\right) \\
& + \int_{S^1} \frac{1}{2} k_{\gamma\tau} \left(1 - D_{\gamma\tau}\right) 4L^2 \Upsilon_{ij} \Upsilon_{ab} \left(\delta_{ia} \hat{c}_j \hat{c}_b - \hat{c}_i \hat{c}_j \hat{c}_a \hat{c}_b\right) \\
& + \int_{S^1} \frac{1}{2} k_\kappa \left(1 - D_\kappa\right) L^2 \varphi_{,i} \varphi_{,j} \hat{c}_i \hat{c}_j \\
& + \int_{S^1} k_{u\tau\gamma\tau} 4L^2 G_{ij} \Upsilon_{ab} \left(\delta_{ia} \hat{c}_j \hat{c}_b - \hat{c}_i \hat{c}_j \hat{c}_a \hat{c}_b\right) \\
& + \int_{S^1} k_{u\eta\gamma\eta} 2L \hat{c}_i \hat{c}_j G_{ij} 2L \Upsilon_{ab} \hat{c}_a \hat{c}_b \\
& + \int_{S^1} k_{u\eta\kappa} 2L \hat{c}_i \hat{c}_j \hat{c}_h G_{ij} \varphi_{,h} \\
& + \int_{S^1} k_{\gamma\eta\kappa} L \Upsilon_{ij} \hat{c}_i \hat{c}_j \varphi_{,h} \hat{c}_h
\end{aligned} \tag{17.39}
$$

where insertion of (17.18), (17.28), (17.29), (17.30) and (17.34) have been used and the integral over the unit circle is intended as follows,

$$
\int_{S^1} (\bullet) = \int_0^{2\pi} (\bullet)\, d\theta.
$$

Grouping the terms of (17.39), we have

$$
\begin{aligned}
U = & \int_{S^1} \frac{1}{2} L^2 \left[k_\eta \left(1 - D_\eta\right) \left(\hat{c}_i \hat{c}_j \hat{c}_a \hat{c}_b\right)\right] G_{ij} G_{ab} \\
& + \int_{S^1} \frac{1}{2} L^2 \left[k_\tau \left(1 - D_\tau\right) \left(\left(\delta_{ia} \hat{c}_j \hat{c}_b + \delta_{ib} \hat{c}_j \hat{c}_a + \delta_{ja} \hat{c}_i \hat{c}_b + \delta_{jb} \hat{c}_i \hat{c}_a\right)\right.\right. \\
& \left.\left. - 4 \hat{c}_i \hat{c}_j \hat{c}_a \hat{c}_b\right)\right] G_{ij} G_{ab} \\
& + \int_{S^1} \frac{1}{2} L^2 \left[k_{\gamma\eta} \left(1 - D_{\gamma\eta}\right) \hat{c}_i \hat{c}_j \hat{c}_a \hat{c}_b\right. \\
& \left. + k_{\gamma\tau} \left(1 - D_{\gamma\tau}\right) 4 \left(\delta_{ia} \hat{c}_j \hat{c}_b - \hat{c}_i \hat{c}_j \hat{c}_a \hat{c}_b\right)\right] \Upsilon_{ij} \Upsilon_{ab}
\end{aligned}
$$

$$+\int_{S^1} \frac{1}{2} L^2 k_\kappa (1-D_\kappa) \hat{c}_i \hat{c}_j \varphi_{,i} \varphi_{,j}$$
$$+\int_{S^1} L^2 4 \left[k_{u\tau\gamma\tau} \left(\frac{1}{2} (\delta_{ia} \hat{c}_j \hat{c}_b + \delta_{ja} \hat{c}_i \hat{c}_b) - \hat{c}_i \hat{c}_j \hat{c}_a \hat{c}_b \right) + k_{u\eta\gamma\eta} \hat{c}_i \hat{c}_j \hat{c}_a \hat{c}_b \right] G_{ij} \Upsilon_{ab}$$
$$+\int_{S^1} k_{u\eta\kappa} 2L \hat{c}_i \hat{c}_j \hat{c}_h G_{ij} \varphi_{,h}$$
$$+\int_{S^1} k_{\gamma\eta\kappa} L \hat{c}_i \hat{c}_a \hat{c}_b \Upsilon_{ab} \varphi_{,i}$$

or, in a compact form, it yields

$$U = \frac{1}{2} \mathbb{C}_{ijab} G_{ij} G_{ab} + \frac{1}{2} \mathbb{C}^\gamma{}_{ijab} \Upsilon_{ij} \Upsilon_{ab} + \frac{1}{2} \mathbb{K}_{ij} \varphi_{,i} \varphi_{,j}$$
$$+ \mathbb{K}^{u\gamma}_{ijab} G_{ij} \Upsilon_{ab} + \mathbb{K}^{u\kappa}_{ijh} G_{ij} \varphi_{,h} + \mathbb{K}^{\kappa\gamma}_{iab} \varphi_{,i} \Upsilon_{ab}, \tag{17.40}$$

where the elastic stiffness $\mathbb{C}$, $\mathbb{C}^\gamma$, $\mathbb{K}$, $\mathbb{K}^{u\gamma}$, $\mathbb{K}^{u\kappa}$, and $\mathbb{K}^{\kappa\gamma}$ are identified in (17.40) as follows, with the symmetrization induced by the symmetry of the strain tensor G,

$$\mathbb{C}_{ijab} = L^2 \int_{S^1} k_\eta (1-D_\eta) (\hat{c}_i \hat{c}_j \hat{c}_a \hat{c}_b)$$
$$+4L^2 \int_{S^1} k_\tau (1-D_\tau) \left(\frac{1}{4} (\delta_{ia} \hat{c}_j \hat{c}_b + \delta_{ib} \hat{c}_j \hat{c}_a + \delta_{ja} \hat{c}_i \hat{c}_b + \delta_{jb} \hat{c}_i \hat{c}_a) - \hat{c}_i \hat{c}_j \hat{c}_a \hat{c}_b \right) \tag{17.41}$$

$$\mathbb{C}^\gamma{}_{ijab} = L^2 \int_{S^1} k_{\gamma\eta} (1-D_{\gamma\eta}) (\hat{c}_i \hat{c}_j \hat{c}_a \hat{c}_b)$$
$$+4L^2 \int_{S^1} k_{\gamma\tau} (1-D_{\gamma\tau}) (\delta_{ia} \hat{c}_j \hat{c}_b - \hat{c}_i \hat{c}_j \hat{c}_a \hat{c}_b) \tag{17.42}$$

$$\mathbb{K}_{ij} = L^2 \int_{S^1} k_\kappa (1-D_\kappa) \hat{c}_i \hat{c}_j \tag{17.43}$$

$$\mathbb{K}^{u\gamma}_{ijab} = 4L^2 \int_{S^1} (k_{u\eta\gamma\eta} \hat{c}_i \hat{c}_j \hat{c}_a \hat{c}_b)$$
$$+4L^2 \int_{S^1} \left\{ k_{u\tau\gamma\tau} \left[\frac{1}{2} (\delta_{ia} \hat{c}_j \hat{c}_b + \delta_{ja} \hat{c}_i \hat{c}_b) - \hat{c}_i \hat{c}_j \hat{c}_a \hat{c}_b \right] \right\} \tag{17.44}$$

$$\mathbb{K}^{u\kappa}_{ijh} = 2L \int_{S^1} (k_{u\eta\kappa} \hat{c}_i \hat{c}_j \hat{c}_h) \tag{17.45}$$

$$\mathbb{K}^{\kappa\gamma}_{iab} = L \int_{S^1} \left(k_{\gamma\eta\kappa} \hat{c}_i \hat{c}_a \hat{c}_b \right) \tag{17.46}$$

17.4 Identification of the Undamaged Isotropic Case

17.4.1 Characterization of the Undamaged Isotropic Case

In the non-damaged isotropic case, we have both

$$D_\eta = D_\tau = D_{\gamma\eta} = D_{\gamma\tau} = D_\kappa = 0,$$

and constant distribution of stiffnesses with respect to the orientation

$$k_\eta = \widetilde{k}_\eta(\theta) = \frac{\bar{k}_\eta}{2\pi}, \qquad k_\tau = \widetilde{k}_\tau(\theta) = \frac{\bar{k}_\tau}{2\pi},$$
$$k_{\gamma\eta} = \widetilde{k}_{\gamma\eta}(\theta) = \frac{\bar{k}_{\gamma\eta}}{2\pi}, \qquad k_{\gamma\tau} = \widetilde{k}_{\gamma\tau}(\theta) = \frac{\bar{k}_{\gamma\tau}}{2\pi},$$
$$k_\kappa = \widetilde{k}_\kappa(\theta) = \frac{\bar{k}_\kappa}{2\pi},$$
$$k_{u\eta\gamma\eta} = \widetilde{k}_{u\eta\gamma\eta}(\theta) = \frac{\bar{k}_{u\eta\gamma\eta}}{2\pi}, \qquad k_{u\tau\gamma\tau} = \widetilde{k}_{u\tau\gamma\tau}(\theta) = \frac{\bar{k}_{u\tau\gamma\tau}}{2\pi},$$
$$k_{u\eta\kappa} = \widetilde{k}_{u\eta\kappa}(\theta) = \frac{\bar{k}_{u\eta\kappa}}{2\pi}, \qquad k_{\gamma\eta\kappa} = \widetilde{k}_{\gamma\eta\kappa}(\theta) = \frac{\bar{k}_{\gamma\eta\kappa}}{2\pi}$$

where k_η, k_τ $\bar{k}_{\gamma\eta}$, $\bar{k}_{\gamma\tau}$ $\bar{k}_\kappa$, $\bar{k}_{u\eta\gamma\eta}$, $\bar{k}_{u\tau\gamma\tau}$, $\bar{k}_{u\eta\kappa}$ and $\bar{k}_{\gamma\eta\kappa}$ are the integrated stiffness over the set of possible orientations, that are defined in the general anisotropic case as follows,

$$\bar{k}_\eta = \int_0^{2\pi} \widetilde{k}_\eta(\theta)\,d\theta, \qquad \bar{k}_\tau = \int_0^{2\pi} \widetilde{k}_\tau(\theta)\,d\theta,$$
$$\bar{k}_{\gamma\eta} = \int_0^{2\pi} \widetilde{k}_{\gamma\eta}(\theta)\,d\theta, \qquad \bar{k}_{\gamma\tau} = \int_0^{2\pi} \widetilde{k}_{\gamma\tau}(\theta)\,d\theta,$$
$$\bar{k}_\kappa = \int_0^{2\pi} \widetilde{k}_\kappa(\theta)\,d\theta,$$
$$\bar{k}_{u\eta\gamma\eta} = \int_0^{2\pi} \widetilde{k}_{u\eta\gamma\eta}(\theta)\,d\theta, \quad \bar{k}_{u\tau\gamma\tau} = \int_0^{2\pi} \widetilde{k}_{u\tau\gamma\tau}(\theta)\,d\theta,$$

$$\bar{k}_{u\eta\kappa} = \int_0^{2\pi} \widetilde{k}_{u\eta\kappa}(\theta)\, d\theta, \quad \bar{k}_{\gamma\eta\kappa} = \int_0^{2\pi} \widetilde{k}_{\gamma\eta\kappa}(\theta)\, d\theta.$$

With these hypotheses, stiffness tensors (17.41), (17.42), (17.43), (17.44), and (17.46) reduce to the following simplified form

$$\begin{aligned}\mathbb{C}_{ijab} &= 2L^2\frac{\bar{k}_\eta}{\pi}\int_{S^1}(\hat{c}_i\hat{c}_j\hat{c}_a\hat{c}_b)\\ &+ 2L^2\frac{\bar{k}_\tau}{\pi}\int_{S^1}\left[\frac{1}{4}(\delta_{ia}\hat{c}_j\hat{c}_b+\delta_{ib}\hat{c}_j\hat{c}_a+\delta_{ja}\hat{c}_i\hat{c}_b+\delta_{jb}\hat{c}_i\hat{c}_a)-\hat{c}_i\hat{c}_j\hat{c}_a\hat{c}_b\right]\end{aligned} \tag{17.47}$$

$$\mathbb{C}^{\gamma}{}_{ijab} = 2L^2\frac{\bar{k}_{\gamma\eta}}{\pi}\int_{S^1}(\hat{c}_i\hat{c}_j\hat{c}_a\hat{c}_b) + 2L^2\frac{\bar{k}_{\gamma\tau}}{\pi}\int_{S^1}(\delta_{ia}\hat{c}_j\hat{c}_b-\hat{c}_i\hat{c}_j\hat{c}_a\hat{c}_b) \tag{17.48}$$

$$\mathbb{K}_{ij} = \frac{\bar{k}_\kappa}{2\pi}L^2\int_{S^1}(\hat{c}_i\hat{c}_j) \tag{17.49}$$

$$\begin{aligned}\mathbb{K}^{u\gamma}_{ijab} &= 2L^2\frac{\bar{k}_{u\eta\gamma\eta}}{\pi}\int_{S^1}(\hat{c}_i\hat{c}_j\hat{c}_a\hat{c}_b)\\ &+ 2L^2\frac{\bar{k}_{u\tau\gamma\tau}}{\pi}\int_{S^1}\left[\left(\frac{1}{2}(\delta_{ia}\hat{c}_j\hat{c}_b+\delta_{ja}\hat{c}_i\hat{c}_b)-\hat{c}_i\hat{c}_j\hat{c}_a\hat{c}_b\right)\right]\end{aligned} \tag{17.50}$$

$$\mathbb{K}^{u\kappa}_{ijh} = 0 \tag{17.51}$$

$$\mathbb{K}^{\kappa\gamma}_{iab} = 0. \tag{17.52}$$

It is worth noting that the coupling tensor

$$\mathbb{K}^{u\gamma}_{ij[ab]} = L^2\frac{\bar{k}_{u\tau\gamma\tau}}{\pi}\int_{S^1}\left[\frac{1}{2}(\delta_{ia}\hat{c}_j\hat{c}_b+\delta_{ja}\hat{c}_i\hat{c}_b)-\frac{1}{2}(\delta_{ib}\hat{c}_j\hat{c}_a+\delta_{jb}\hat{c}_i\hat{c}_a)\right]$$

is null for the isotropic case,

$$\mathbb{K}^{u\gamma}_{ij[ab]} = L^2\frac{\bar{k}_{u\tau\gamma\tau}}{2\pi}\left[\pi\delta_{ia}\delta_{jb}+\pi\delta_{ib}\delta_{ja}-\pi\delta_{ib}\delta_{ja}-\pi\delta_{ia}\delta_{jb}\right]=0.$$

17.4.2 Macroscopic Isotropic Stiffness Matrices

The standard isotropic representation of 2nd and 4th order stiffness tensors, for the 2D case, is given by following expressions (17.53), (17.55), (17.57), and (17.59),

$$\mathbb{K}_{ij} = \kappa\delta_{ij} \tag{17.53}$$

$$\kappa = \mathbb{K}_{11} \tag{17.54}$$

$$\mathbb{C}_{ijab} = \mu\delta_{ia}\delta_{jb} + \mu\delta_{ib}\delta_{ja} + \lambda\delta_{ij}\delta_{ab} \tag{17.55}$$

$$\mathbb{C}_{1212} = \mu, \qquad \mathbb{C}_{1122} = \lambda, \tag{17.56}$$

$$\mathbb{C}^{\gamma}_{ijab} = \mu^{\gamma}_{1}\delta_{ia}\delta_{jb} + \mu^{\gamma}_{2}\delta_{ib}\delta_{ja} + \lambda^{\gamma}\delta_{ij}\delta_{ab} \tag{17.57}$$

$$\mathbb{C}^{\gamma}_{1212} = \mu^{\gamma}_{1}, \quad \mathbb{C}^{\gamma}_{1221} = \mu^{\gamma}_{2}, \quad \mathbb{C}^{\gamma}_{1122} = \lambda^{\gamma}, \tag{17.58}$$

$$\mathbb{K}^{u\gamma}_{ijab} = \mu^{u\gamma}_{1}\delta_{ia}\delta_{jb} + \mu^{u\gamma}_{2}\delta_{ib}\delta_{ja} + \lambda^{u\gamma}\delta_{ij}\delta_{ab} \tag{17.59}$$

$$\mathbb{K}^{u\gamma}_{1212} = \mu^{u\gamma}_{1}, \quad \mathbb{K}^{u\gamma}_{1221} = \mu^{u\gamma}_{2}, \quad \mathbb{K}^{u\gamma}_{1122} = \lambda^{u\gamma} \tag{17.60}$$

where identifications of Lamè coefficients in terms of the stiffness matrix components have also been expressed in (17.54), (17.56), (17.58) and (17.60).

17.4.3 Identification of the Macroscopic Isotropic Stiffness Matrices

In the 2D case, a standard representation of the unit vector $\hat{c}$ is

$$\hat{c}_1 = \cos\theta, \ \hat{c}_2 = \sin\theta, \tag{17.61}$$

where θ is the anti-clockwise angle from the first unit vector $\hat{e}_1$ of the frame of reference to $\hat{c}$. Besides, trivial analytical results are as follows

$$\int_0^{2\pi} \left(\sin^2\theta\right) d\theta = \int_0^{2\pi} \left(\cos^2\theta\right) d\theta = \pi, \tag{17.62}$$

$$\int_0^{2\pi} \left(\sin^2\theta\cos^2\theta\right) d\theta = \frac{1}{4}\pi, \tag{17.63}$$

$$\int_0^{2\pi} \sin^4\theta d\theta = \int_0^{2\pi} \cos^4\theta d\theta = \frac{3}{4}\pi. \tag{17.64}$$

From (17.54), (17.49) and (17.62)

$$\kappa = \mathbb{K}_{11} = \frac{\bar{k}_{\kappa}}{2\pi}L^2 \int_0^{2\pi} \left(\cos^2\theta\right) = \frac{\bar{k}_{\kappa}}{2}L^2. \tag{17.65}$$

From (17.56)$_1$, (17.47) and (17.63)

$$\mu = \mathbb{C}_{1212} = 2L^2 \frac{\bar{k}_\eta}{\pi} \int_0^{2\pi} \left(\sin^2\theta \cos^2\theta \right) + 2L^2 \frac{\bar{k}_\tau}{\pi} \int_0^{2\pi} \left[\frac{1}{4} - \sin^2\theta \cos^2\theta \right] = \frac{1}{2} L^2 \left(\bar{k}_\eta + \bar{k}_\tau \right). \tag{17.66}$$

From (17.56)$_2$, (17.47) and (17.63)

$$\lambda = \mathbb{C}_{1122} = 2L^2 \frac{\bar{k}_\eta}{\pi} \int_0^{2\pi} \left(\sin^2\theta \cos^2\theta \right) + 2L^2 \frac{\bar{k}_\tau}{\pi} \int_0^{2\pi} \left[-\sin^2\theta \cos^2\theta \right] = \frac{1}{2} L^2 \left(\bar{k}_\eta - \bar{k}_\tau \right). \tag{17.67}$$

From (17.58)$_1$, (17.57), (17.64) and (17.63)

$$\mu_1^\gamma = \mathbb{C}^\gamma_{1212} = \mathbb{C}^\gamma_{2121} = 2L^2 \frac{\bar{k}_{\gamma\eta}}{\pi} \int_0^{2\pi} \left(\sin^2\theta \cos^2\theta \right) + 2L^2 \frac{\bar{k}_{\gamma\tau}}{\pi} \int_0^{2\pi} \left[\sin^2\theta - \sin^2\theta \cos^2\theta \right] = \frac{1}{2} L^2 \left(\bar{k}_{\gamma\eta} + 3\bar{k}_{\gamma\tau} \right). \tag{17.68}$$

From (17.58)$_2$, (17.57) and (17.63)

$$\mu_2^\gamma = \mathbb{C}^\gamma_{1221} = 2L^2 \frac{\bar{k}_{\gamma\eta}}{\pi} \int_0^{2\pi} \left(\sin^2\theta \cos^2\theta \right) + 2L^2 \frac{\bar{k}_{\gamma\tau}}{\pi} \int_0^{2\pi} \left[-\sin^2\theta \cos^2\theta \right] = \frac{1}{2} L^2 \left(\bar{k}_{\gamma\eta} - \bar{k}_{\gamma\tau} \right). \tag{17.69}$$

From (17.58)$_3$, (17.57) and (17.63)

$$\lambda^\gamma = \mathbb{C}^\gamma_{1122} = 2L^2 \frac{\bar{k}_{\gamma\eta}}{\pi} \int_0^{2\pi} \left(\sin^2\theta \cos^2\theta \right)$$

$$+2L^2 \frac{\bar{k}_{\gamma\tau}}{\pi} \int_0^{2\pi} \left[-\sin^2\theta \cos^2\theta\right] = \frac{1}{2} L^2 \left(\bar{k}_{\gamma\eta} - \bar{k}_{\gamma\tau}\right). \quad (17.70)$$

From $(17.60)_1$, (17.50), (17.64) and (17.63)

$$\mu_1^{u\gamma} = \mathbb{K}_{1212}^{u\gamma} = 2L^2 \frac{\bar{k}_{u\eta\gamma\eta}}{\pi} \int_0^{2\pi} \left(\sin^2\theta \cos^2\theta\right) +$$

$$+2L^2 \frac{\bar{k}_{u\tau\gamma\tau}}{\pi} \int_0^{2\pi} \left[\sin^2\theta - \sin^2\theta \cos^2\theta\right] = \frac{1}{2} L^2 \left(\bar{k}_{u\eta\gamma\eta} + 3\bar{k}_{u\tau\gamma\tau}\right). \quad (17.71)$$

From $(17.60)_2$, (17.50), (17.64) and (17.63)

$$\mu_2^{u\gamma} = \mathbb{K}_{1221}^{u\gamma} = 2L^2 \frac{\bar{k}_{u\eta\gamma\eta}}{\pi} \int_0^{2\pi} \left(\sin^2\theta \cos^2\theta\right) +$$

$$+2L^2 \frac{\bar{k}_{u\tau\gamma}}{\pi} \int_0^{2\pi} \left[-\sin^2\theta \cos^2\theta\right] = \frac{1}{2} L^2 \left(\bar{k}_{u\eta\gamma\eta} - \bar{k}_{u\tau\gamma\tau}\right). \quad (17.72)$$

From $(17.60)_2$, (17.50), (17.64) and (17.63)

$$\lambda^{u\gamma} = \mathbb{K}_{1122}^{u\gamma} = 2L^2 \frac{\bar{k}_{u\eta\gamma\eta}}{\pi} \int_0^{2\pi} \left(\sin^2\theta \cos^2\theta\right) +$$

$$2L^2 \frac{\bar{k}_{u\tau\gamma\tau}}{\pi} \int_0^{2\pi} \left[-\sin^2\theta \cos^2\theta\right] = \frac{1}{2} L^2 \left(\bar{k}_{u\eta\gamma\eta} - \bar{k}_{u\tau\gamma\tau}\right). \quad (17.73)$$

17.5 Conclusion

The primary result of the presented work is the deduction of a set of elastic material constants for a isotropic Cosserat media derived by exploiting micro-macro identification of granular micro-mechano-morphology. To this end, we have utilized Piola's *ansatz* to make the identification of a discrete granular system with a continuum body to develop relationships between grain-scale and continuum kinematic measures in the framework of finite deformations. The internal deformation (strain) energy is then expressed in terms of these kinematic measures leading to the mentioned Cosserat model.

References

[1] Chen J, Ouyang L, Rulis P, Misra A, Ching WY (2005) Complex nonlinear deformation of nanometer intergranular glassy films in β-Si_3N_4, Physical Review Letters **95**(25):256,103

[2] Eremeyev VA (2018) On the material symmetry group for micromorphic media with applications to granular materials, Mechanics Research Communications **94**:8-12.

[3] Jia H, Misra A, Poorsolhjouy P, Liu C (2017) Optimal structural topology of materials with micro-scale tension-compression asymmetry simulated using granular micromechanics, Materials & Design **115**:422-432.

[4] Misra A, Jiang H (1997) Measured kinematic fields in the biaxial shear of granular materials, Computers and Geotechnics **20**(3-4):267-285.

[5] Misra A, Poorsolhjouy P (2015) Granular micromechanics model for damage and plasticity of cementitious materials based upon thermomechanics, Mathematics and Mechanics of Solids **25**(10):1778-1803.

[6] Misra A, Singh V (2015) Thermomechanics-based nonlinear rate-dependent coupled damage-plasticity granular micromechanics model, Continuum Mechanics and Thermodynamics **27**(4-5):787-817.

[7] Turco E, dell'Isola F, Misra A (2019) A nonlinear lagrangian particle model for grains assemblies including grain relative rotations, International Journal for Numerical and Analytical Methods in Geomechanics **43**(5):1051-1079.

[8] Massoumi S, La Valle G (2022) Static analysis of 2d micropolar model for describing granular media by considering relative rotations, Mechanics Research Communications **119**:103,812.

[9] Spagnuolo M, Franciosi P, dell'Isola F (2020) A Green operator-based elastic modeling for two-phase pantographic-inspired bi-continuous materials, International Journal of Solids and Structures **188**:282-308.

[10] Berezovski A, Yildizdag ME, Scerrato D (2020) On the wave dispersion in microstructured solids, Continuum Mechanics and Thermodynamics **32**(3):569-588.

[11] dell'Isola F, Madeo A, Seppecher P (2009) Boundary conditions at fluid-permeable interfaces in porous media: A variational approach, International Journal of Solids and Structures **46**(17):3150-3164.

[12] Valmalle M, Vintache A, Smaniotto B, Gutmann F, Spagnuolo M, Ciallella A, Hild F (2022) Local-global DVC analyses confirm theoretical predictions for deformation and damage onset in torsion of pantographic metamaterial, Mechanics of Materials **172**:194379.

[13] Casalotti A, D'Annibale F, Rosi G (2020) Multi-scale design of an architected composite structure with optimized graded properties, Composite Structures **252**:112,608.

[14] Dell'Erba R (2022) Rules governing swarm robot in continuum mechanics, Mathematics and Mechanics of Solids **27**(10):1930-1949.

[15] Amar M, Andreucci D, Bisegna P, Gianni R (2010) Homogenization limit and asymptotic decay for electrical conduction in biological tissues in the high

radiofrequency range, Communications on Pure & Applied Analysis **9**(5):1131-1160.

[16] dell'Isola F, Madeo A, Seppecher P (2016) Cauchy tetrahedron argument applied to higher contact interactions, Archive for Rational Mechanics and Analysis **219**(3):1305-1341.

[17] Ivanova EA (2018) On the use of the continuum mechanics method for describing interactions in discrete systems with rotational degrees of freedom, Journal of Elasticity **133**(2):155-199.

[18] Seppecher P, Alibert JJ, dell'Isola F (2011) Linear elastic trusses leading to continua with exotic mechanical interactions, In: Journal of Physics: Conference Series, IOP Publishing, vol. 319.

[19] Dimitri R, Rinaldi M, Trullo M, Tornabene F (2022) Theoretical and computational investigation of the fracturing behavior of anisotropic geomaterials, Continuum Mechanics and Thermodynamics. DOI 10.1007/s00161-022-01141-4

[20] Nejadsadeghi N, Misra A (2020) Extended granular micromechanics approach: a micromorphic theory of degree n, Mathematics and Mechanics of Solids **25**(2):407-429.

[21] Abali B, Müller W, dell'Isola F (2017) Theory and computation of higher gradient elasticity theories based on action principles, Archive of Applied Mechanics **87**(9):1495–1510.

[22] Barchiesi E, Eugster SR, Dell'isola F, Hild F (2020) Large in-plane elastic deformations of bi-pantographic fabrics: asymptotic homogenization and experimental validation, Mathematics and Mechanics of Solids **25**(3):739-767.

[23] Barchiesi E, dell'Isola F, Hild F, Seppecher P (2020) Two-dimensional continua capable of large elastic extension in two independent directions: asymptotic homogenization, numerical simulations and experimental evidence, Mechanics Research Communications **103**:103466.

[24] La Valle G, Ciallella A, Falsone G (2022) The effect of local random defects on the response of pantographic sheets, Mathematics and Mechanics of Solids **27**(10):2147-2169.

[25] Stilz M, Plappert D, Gutmann F, Hiermaier S (2022) A 3d extension of pantographic geometries to obtain metamaterial with semi-auxetic properties, Mathematics and Mechanics of Solids **27**(4):673-686.

[26] dell'Isola F, Andreaus U, Placidi L (2015) At the origins and in the vanguard of peridynamics, non-local and higher-gradient continuum mechanics: An underestimated and still topical contribution of Gabrio Piola, Mathematics and Mechanics of Solids **20**(8):887-928.

[27] Turco E (2018) Discrete is it enough? The revival of Piola–Hencky keynotes to analyze three-dimensional Elastica, Continuum Mechanics and Thermodynamics **30**(5):1039-1057.

[28] Auffray N, dell'Isola F, Eremeyev V, Madeo A, Rosi G (2015) Analytical continuum mechanics à la Hamilton-Piola least action principle for second gradient continua and capillary fluids, Mathematics and Mechanics of Solids **20**(4):375-417.

[29] Eremeyev VA, Lazar M (2022) Strong ellipticity within the Toupin–Mindlin first strain gradient elasticity theory, Mechanics Research Communications **124**:103,944.
[30] Aydin G, Sarar BC, Yildizdag ME, Abali BE (2022) Investigating infill density and pattern effects in additive manufacturing by characterizing metamaterials along the strain-gradient theory, Mathematics and Mechanics of Solids **27**(10):2002-2016.
[31] Barchiesi E, dell'Isola F, Bersani AM, Turco E (2021) Equilibria determination of elastic articulated duoskelion beams in 2D via a Riks-type algorithm, International Journal of Non-Linear Mechanics **128**:103,628.
[32] Eremeyev VA, Ganghoffer JF, Konopińska-Zmysłowska V, Uglov NS (2020) Flexoelectricity and apparent piezoelectricity of a pantographic micro-bar, International Journal of Engineering Science **149**:103,213.
[33] Cosserat E, Cosserat F (1909) Théorie des corps déformables, Hermann et Fils (reprint 2009), Paris.
[34] Shirani M, Luo C, Steigmann DJ (2019) Cosserat elasticity of lattice shells with kinematically independent flexure and twist, Continuum Mechanics and Thermodynamics **31**(4):1087-1097.
[35] Vardoulakis I (2019) Cosserat Continuum Mechanics With Applications to Granular Media, Lecture Notes in Applied and Computational Mechanics (Ed by P Wriggers, P Eberhard) , Vol 87, Springer, Cham.
[36] Giorgio I, De Angelo M, Turco E, Misra A (2019) A Biot–Cosserat two-dimensional elastic nonlinear model for a micromorphic medium, Continuum Mechanics and Thermodynamics 32(5):1357–1369.
[37] Shirani M, Steigmann DJ (2020) A Cosserat model of elastic solids reinforced by a family of curved and twisted fibers, Symmetry **12**(7):1133.
[38] Altenbach J, Altenbach H, Eremeyev VA (2010) On generalized Cosserat-type theories of plates and shells: a short review and bibliography, Archive of Applied Mechanics **80**(1):73-92.
[39] de Felice G, Rizzi N (1999) Macroscopic modelling of Cosserat media. Trends and applications of Mathematics to Mechanics, Monographs and Surveys in Pure and Applied Mathematics **106**:58-65.
[40] La Valle G, Massoumi S (2022) A new deformation measure for micropolar plates subjected to in-plane loads, Continuum Mechanics and Thermodynamics **34**(1):243-257.
[41] La Valle G (2022) A new deformation measure for the nonlinear micropolar continuum. Zeitschrift für angewandte Mathematik und Physik **73**(2):1–26
[42] Fedele R (2022) Deformation-induced coupling of the generalized external actions in third-gradient materials, Zeitschrift für angewandte Mathematik und Physik **73**(5):1-29.
[43] Eugster S, dell'Isola F, Steigmann D (2019) Continuum theory for mechanical metamaterials with a cubic lattice substructure, Mathematics and Mechanics of Complex Systems **7**(1):75-98.
[44] Cazzani A, Stochino F, Turco E (2016) An analytical assessment of finite element and isogeometric analysis of the whole spectrum of Timoshenko beams, ZAMM-

Journal of Applied Mathematics and Mechanics/Zeitschrift für Angewandte Mathematik und Mechanik **96**(10):1220-1244.
[45] dell'Isola F, Seppecher P, Alibert JJ, Lekszycki T, Grygoruk R, Pawlikowski M, Steigmann D, Giorgio I, Andreaus U, Turco E, Golaszewski M, Rizzi N, Boutin C, Eremeyev VA, Misra A, Placidi L, Barchiesi E, Greco L, Cuomo M, Cazzani A, Corte AD, Battista A, Scerrato D, Zhurba-Eremeeva I, Rahali Y, Ganghoffer JF, Mueller W, Ganzosch G, Spagnuolo M, Pfaff A, Barcz K, Hoschke K, Neggers J, Hild F (2018) Pantographic metamaterials: an example of mathematically driven design and of its technological challenges, Continuum Mechanics and Thermodynamics **31**(4):851-884.
[46] Spagnuolo M, Yildizdag ME, Andreaus U, Cazzani AM (2021) Are higher-gradient models also capable of predicting mechanical behavior in the case of wide-knit pantographic structures? Mathematics and Mechanics of Solids **26**(1):18-29.
[47] Turco E, dell'Isola F, Cazzani A, Rizzi NL (2016) Hencky-type discrete model for pantographic structures: numerical comparison with second gradient continuum models, Zeitschrift für angewandte Mathematik und Physik **67**(4):1-28.
[48] dell'Isola F, Maier G, Perego U, Andreaus U, Esposito R, Forest S (2014) The complete works of Gabrio Piola: Volume I Commented English Translation - English and Italian Edition, Advanced Structured Materials (Ed by A Öchsner, LFM da Silva, H Altenbach) Vol 38, Springer, Cham.
[49] dell'Isola F, Giorgio I, Pawlikowski M, Rizzi N (2016) Large deformations of planar extensible beams and pantographic lattices: heuristic homogenization, experimental and numerical examples of equilibrium, Proc R Soc A **472**(2185):20150790.
[50] Alibert JJ, Seppecher P, dell'Isola F (2003) Truss modular beams with deformation energy depending on higher displacement gradients, Mathematics and Mechanics of Solids **8**(1):51-73.
[51] Turco E, Misra A, Pawlikowski M, dell'Isola F, Hild F (2018) Enhanced Piola–Hencky discrete models for pantographic sheets with pivots without deformation energy: numerics and experiments, International Journal of Solids and Structures **147**:94-109.
[52] dell'Isola F, Eugster SR, Fedele R, Seppecher P (2022) Second-gradient continua: From Lagrangian to Eulerian and back, Mathematics and Mechanics of Solids **27**(12):2715-2750.
[53] Eugster SR, dell'Isola F, Fedele R, Seppecher P (2022) Piola transformations in second-gradient continua, Mechanics Research Communications **120**:103,836
[54] Eugster SR (2022) Numerical analysis of nonlinear wave propagation in a pantographic sheet, Mathematics and Mechanics of Complex Systems **9**(3):293-310.
[55] Andreaus U, dell'Isola F, Giorgio I, Placidi L, Lekszycki T, Rizzi N (2016) Numerical simulations of classical problems in two-dimensional (non) linear second gradient elasticity, International Journal of Engineering Science **108**:34-50.

[56] Cuomo M, dell'Isola F, Greco L, Rizzi N (2016) First versus second gradient energies for planar sheets with two families of inextensible fibres: Investigation on deformation boundary layers, discontinuities and geometrical instabilities, Composites Part B: Engineering **115**:423-448.
[57] Cuomo M, dell'Isola F, Greco L (2016) Simplified analysis of a generalized bias test for fabrics with two families of inextensible fibres, Zeitschrift für angewandte Mathematik und Physik **67**(3):1-23.
[58] dell'Isola F, Lekszycki T, Pawlikowski M, Grygoruk R, Greco L (2015) Designing a light fabric metamaterial being highly macroscopically tough under directional extension: first experimental evidence, Zeitschrift für angewandte Mathematik und Physik **66**:3473-3498.
[59] Ciallella A, Giorgio I, Eugster SR, Rizzi NL, dell'Isola F (2022) Generalized beam model for the analysis of wave propagation with a symmetric pattern of deformation in planar pantographic sheets, Wave Motion **113**:102,986.
[60] Eremeyev VA, dell'Isola F, Boutin C, Steigmann D (2017) Linear pantographic sheets: existence and uniqueness of weak solutions, Journal of Elasticity **132**(2):175-196.
[61] Ganzosch G, dell'Isola F, Turco e, Lekszycki T, Müller W (2016) Shearing tests applied to pantographic structures, Acta Polytechnica CTU Proceedings **7**:1-6.
[62] Turco E, Barchiesi E, Ciallella A, dell'Isola F (2022) Nonlinear waves in pantographic beams induced by transverse impulses, Wave Motion **115**:103064.
[63] Ciallella A, Steigmann DJ (2022) Unusual deformation patterns in a second-gradient cylindrical lattice shell: Numerical experiments, Mathematics and Mechanics of Solids DOI 10.1177/10812865221101
[64] Sciarra G, dell'Isola F, Coussy O (2007) Second gradient poromechanics, International Journal of Solids and Structures **44**(20):6607-6629.
[65] Spagnuolo M, Barcz K, Pfaff A, dell'Isola F, Franciosi P (2017) Qualitative pivot damage analysis in aluminum printed pantographic sheets: numerics and experiments, Mechanics Research Communications 83:47-52.
[66] Steigmann D, dell'Isola F (2015) Mechanical response of fabric sheets to three-dimensional bending, twisting, and stretching, Acta Mechanica Sinica **31**(3):373-382.
[67] Turco E, dell'Isola F, Rizzi N, Grygoruk R, Müller W, Liebold C (2016) Fiber rupture in sheared planar pantographic sheets: Numerical and experimental evidence, Mechanics Research Communications **76**:86-90.
[68] De Angelo M, Spagnuolo M, D'Annibale F, Pfaff A, Hoschke K, Misra A, Dupuy C, Peyre P, Dirrenberger J, Pawlikowski M (2019) The macroscopic behavior of pantographic sheets depends mainly on their microstructure: experimental evidence and qualitative analysis of damage in metallic specimens, Continuum Mechanics and Thermodynamics **31**(4):1181-1203.
[69] Misra A, Nejadsadeghi N, De Angelo M, Placidi L (2020) Chiral metamaterial predicted by granular micromechanics: verified with 1d example synthesized using additive manufacturing, Continuum Mechanics and Thermodynamics **32**(5):1497-1513.

Chapter 18
Study of the Dynamic Properties of Reinforced Concrete Under High-Speed Compression

Mikhail E. Gonov, Vladimir V. Balandin, Anatoly M. Bragov, and Aleksandr Yu. Konstantinov

Abstract An experimental study of the dynamic properties of fine-grained concrete reinforced with fine-meshed meshes under dynamic uniaxial compression relative to the original fine-grained concrete has been carried out. Reinforced samples were made by pouring concrete into molds with pre-installed mesh frames. Dynamic tests have been carried out. Dynamic compression tests were carried out using the Kolsky method at strain rates from 30 to 600 s^{-1}. The paper presents the composition of the test material, test parameters, as well as a comparative analysis of the data obtained. The introduction of reinforcing meshes into the original fine-grained concrete increased the dynamic strength of the material. The dependences obtained demonstrate that the maximum breaking stresses achieved in the experiments increase linearly with the growth of the strain rate, as do the corresponding limiting strains. The time before the onset of fracture decreases with increasing strain rate according to a power law.

Key words: Dynamic strength · Stress · Strain · Strain rate · Concrete · Reinforced concrete · Dynamic tests · Kolsky method · Experiment

18.1 Introduction

The relevance of the study is due to the fact that in parallel with the growth of the technogenicity of our civilization, the number of emergency situations is growing, exposing the structures of buildings and structures for civil, industrial or military purposes to high-speed dynamic effects in the form of shocks and explosions. In

---

Mikhail E. Gonov · Vladimir V. Balandin · Anatoly M. Bragov · Aleksandr Yu. Konstantinov
National Research Lobachevsky State University of Nizhny Novgorod, 23 Gagarin Avenue, Nizhny Novgorod 603022, Russian Federation,
e-mail: gonov_mikhail@mech.unn.ru, bragov@mech.unn.ru, rustydog2007@yandex.ru, constantinov.al@yandex.ru

H. Altenbach et al. (eds.), *Sixty Shades of Generalized Continua*,
Advanced Structured Materials 170, https://doi.org/10.1007/978-3-031-26186-2_18

addition to man-made disasters, the increasing frequency of natural disasters, as well as the increased terrorist threat in the form of various explosive and combustible substances, are coming to the fore. The main danger of such phenomena is the loss of human lives, both with the loss of the bearing capacity of buildings and structures, and with the occurrence of many injuries from fragments of fragile materials formed during their dynamic destruction. Since concrete is currently one of the most common building materials, a key method to increase its strength is the addition of finely dispersed reinforcement: fiber reinforcement and fine mesh.

The practical applicability of the obtained results is related to the field of three-dimensional computer modeling, in which the behavior of real materials and structures under the action of normal or extreme (emergency) loads is re-placed by their mathematical model, which requires various experimental data obtained in the articles, a lot of parameters and constants. Ordinary concrete has relatively low tensile strength and low ductility, so it is prone to cracking. The formation of cracks leads to the early onset of the destruction of concrete or rein-forced concrete. It has been proven that fine-mesh reinforcement increases the tensile strength and plasticity of concrete [1]-[5]. We note right away that there are much fewer studies of the dynamic properties of fine-grained concretes rein-forced with meshes with a small cell size (up to 10 mm) than fine-grained concretes reinforced with fiber.

In [6] a new environmentally friendly building material is presented - high-performance geopolymer concrete. Both numerical and experimental studies have been carried out on a new type of structural plates reinforced with steel wire mesh for protection against explosions near the ground surface. To compare the results, conventional concrete slabs reinforced with steel reinforcement and subjected to explosive loading were also investigated. An experimental study was carried out to study the mechanism of slab damage. It was found that a geopolymer concrete slab reinforced with steel wire mesh showed less damage when exposed to 50 kg of trinitrotoluene (TNT) at distances of 3 m, 5 m and 7 m compared to a concrete slab. Numerical analysis was then carried out to further study the dynamic characteristics of the slab. The combination of steel wire mesh reinforcement with geopolymer concrete can help improve blast resistance, resulting in a promising and environmentally friendly structural protective material.

It was noted in [7] that the use of high-strength fiber-reinforced concrete in practical construction is very expensive due to the high content of steel fibers (> 2.5% by volume). The study proposes two column designs (2000 × 168 × 168 mm). Hollow core components and steel wire mesh reinforced components were cast from high-strength concrete containing 1.5% steel fiber by volume, and the impact resistance of both types of structures was also studied. The test pieces included two hollow core square and round hollow core columns and two steel wire mesh reinforced columns with 6 and 10 layers of wire reinforcement. The impact scenario was simulated using a 411 kg jackhammer freely falling from a height of 1.25 m to the middle of the span of the test specimen. The results showed that all specimens retained flexural strength with minimal damage. The developed numerical model recorded the impact force, structural deformation and damage with sufficient accuracy. Using the tested model, the distribution of dynamic shear and moment, the level of damage to the columns after the impact were evaluated. The influence of the shape and ratio of hollow sections, the level of axial load and the coefficient of longitudinal reinforcement for columns with a hollow core, as well as the influence of layers of steel wire mesh for columns reinforced with steel wire mesh, was investigated. Compared to other hollow columns at impact velocities from 4.95 m/s to 6.64 m/s, round hollow columns with a 15% cavity ratio proved to be the most effective in terms of cost and impact resistance. A column reinforced with steel wire mesh, reinforced in the entire section, has better impact resistance than its column, which had wire mesh reinforcement only in the tension zone.

In [8] experimental and numerical studies of the impact resistance of targets made of reactive powder concrete (RPC) reinforced with 44-layer steel wire meshes are presented. Steel projectiles with an average mass of 330 g and striking speeds from 550 m/s to 800 m/s were fired at RPC cylindrical targets with a diameter of 750 mm and a thickness of 700 mm. The results showed that the helical distribution of the steel wire mesh reinforcement in the RPC, perpendicular to projectile penetration, can effectively reflect and refract the stress wave generated by projectile impact, resulting in better impact resistance relative to projectile penetration. Steel wire mesh reinforcement in RPC targets effectively improves impact resistance and reduces crater diameter compared to the previous target study. The results of the experiment show that with an in-crease in the impact velocity of the projectile, the diameter of the crater increases.

In [9], an increase in electromagnetic protection and impact resistance of reinforced concrete walls of protective structures was considered. Electromagnetic interference shielding has received attention in the design of radiation reduction processes to protect people or equipment. Several studies have examined the use of hybrid amplification techniques for shielding effectiveness or examined the relationship between damaged areas and shielding effectiveness. In the work, a rein-forced concrete (RC) wall was prepared, subjected to various impact loads, containing steel fiber, metal mesh - mesh reinforcement and hybrid reinforcement. Compared to the 300 mm thick RC specimens and the same type reinforced specimens, the hybrid reinforced specimens had about 25.9 times and 2.4 times higher shielding efficiency, respectively. Samples reinforced with metal mesh showed a free area ratio of less than 150% and a shielding efficiency of more than 30 dB. Hybrid metal mesh and optional metal mesh have reached more than 40db. The hybrid reinforced specimens have demonstrated improved impact resistance and high shielding performance in damaged conditions.

In [10], a comparative evaluation of steel wire mesh, steel fiber and high-strength concrete slabs reinforced with polyethylene fiber was carried out during explosion tests. Due to the relatively low tensile strength, concrete tends to tensile failure and crack under external loads. To improve the tensile strength and ductility of the concrete material, the study considered possible solutions, including fiber reinforcement and steel mesh reinforcement. Ultra high molecular weight polyethylene fiber and steel wire meshes were mixed at different volume fractions in a concrete matrix. Static testing of reinforced concrete slab materials, including uniaxial compression and bending tests, showed that the addition of steel fiber resulted in a better increase in strength, while ultra high molecular weight polyethylene fiber provided better material ductility. Concrete samples with hybrid steel fiber reinforcement and steel mesh showed high strength and ductility. Field blast tests are designed to study the behavior of blasts at close range. Explosion tests show different damage profiles. The results showed the benefits of replacing the steel fiber with ultra-high molecular weight polyethylene or steel wire mesh.

18.2 Test Method

High-speed tests under conditions of a one-dimensional compressive stress state were carried out on an experimental setup SHPB-60 [11]-[13], the scheme of which is shown in Fig. 18.1. The experimental setup according to the Kolsky method includes a system of measuring rods, between which a test sample is installed, a gas gun for accelerating a cylindrical impactor, strain gauges, a velocity meter, recording and computing equipment with a software package. Measuring rods, both loading and supporting, as well as a cylindrical striker, are made of D16T duralumin alloy. The striker length in all experiments was 300 mm. The amplitude of the loading wave and, accordingly, the value of the realized strain rate were varied by changing the impactor velocity. The loading modes were chosen in such a way that a gradual increase in the strain rate was observed. In one of the rods, after the impact of the impactor, a one-dimensional elastic compression wave is excited, which propagates along the rods at the speed of sound. Upon reaching the sample, this wave splits due to the difference in the acoustic stiffness of the materials of the rod and the sample, as well

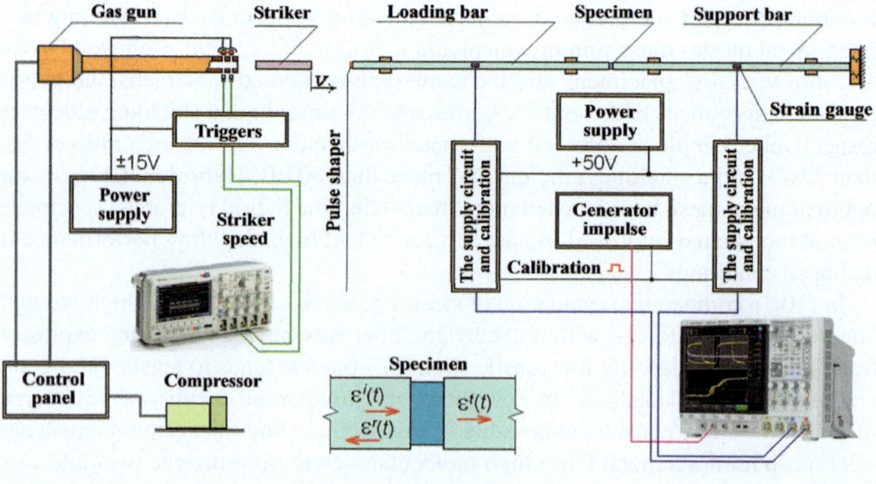

Fig. 18.1: Scheme of the installation for testing under compression in the condition of a uniaxial stress state.

as the areas of their cross sections: part of it is reflected back by a tension wave, and part passes through the sample into the second rod by a compression wave. In this case, the sample undergoes elastic-plastic deformation or destruction, while the rods are deformed elastically. By registering elastic strain pulses in measuring rods with strain gauges according to the formulas proposed for the first time by Kolsky, it is possible to determine the stresses, strains, and strain rates in the sample as a function of time.

To fulfill the main premise of the Kolsky method - a uniaxial stress state in the sample, the main recommendations were taken into account, namely: the ratio of the length of the sample to the diameter within 0.3 - 1.0; treatment of the ends of the measuring rods before the start of each experiment with Litol lubricant to reduce the effect of friction forces during the radial expansion of the sample.

18.3 Characteristics of the Tested Materials

This section presents the results of tests of reinforced fine-grained concrete of class B25, made by pouring concrete into molds in which mesh frames were previously installed (Fig. 18.2). Reinforcing fine-mesh mesh is made of light O/N wire in accordance with GOST 3282-74 from steel grade 1PS, mesh size 10x10 mm, wire

Fig. 18.2: Reinforcing mesh frame and poured concrete blank.

thickness 1.4 mm. The composition of fine-grained concrete is presented in Table 18.1.

After keeping the manufactured reinforced beams until the standard strength was reached, samples were made from it. Using a Hilti DD130 drilling rig, cores with a diameter of ~54 mm were drilled from the beam, which were then cut into individual reinforced samples ~30 mm thick using a Cedima CTS-57-G stone-cutting machine (Fig. 18.3).

18.4 Results of Dynamic Tests for Uniaxial Compression

High-speed tests under conditions of one-dimensional stress state in compression were carried out on the experimental setup SHPB-60, on samples with a diameter of 53 mm and a length of 30 mm. Reinforced concrete was tested at four different speed modes, 2-4 shots for each mode. The strain rates were in the range from 30 to 600 s^{-1}. Test parameters are shown in Table 18.2.

Below is a set of dependences of stress change and strain rate for each mode in the form of averaged dependences with a confidence interval. The initial sections of the resulting deformation diagrams are linear, and the fall-ing branches of the diagrams and examination of the samples indicate that the samples were destroyed during the experiment, which is confirmed by the photo-graphs of the destroyed samples in Fig. 18.4.

Figure 18.5 shows the averaged diagrams of the deformation of reinforced concrete with the chronology of the change in the strain rate. In Fig. 18.5, the straight line shows the dependence on stress, the dotted line shows the dependence on strain rate.

In the above diagrams, in the stress–strain axes $\sigma(\varepsilon)$, in the initial section of loading, one can distinguish a section where the growth of stress and strain occurs according to a law close to linear, and with further deformation, when the limiting stress values are reached, the concrete is intensively destroyed (in modes $2, 3, 4$), which is accompanied by a decrease in stresses and an increase in deformations. For each of the resulting diagrams, characteristic points with the maximum achieved stresses ($\sigma_{\max}$) were identified, after which the destruction of the samples began. For these points, the corresponding values of ultimate strains ($\varepsilon_{\max}$) and time to failure ($\tau_{\max}$)

Table 18.1: The composition of fine-grained concrete class B25.

No.	Material	Consumption per 1 m^3 (kg)
1	Cement (grade D500)	480
2	Crushed stone (module 1-3 mm)	1250
3	Medium sand	390
4	Superplasticizer	2.4
5	Water	205

Fig. 18.3: Manufacture of reinforced concrete specimens.

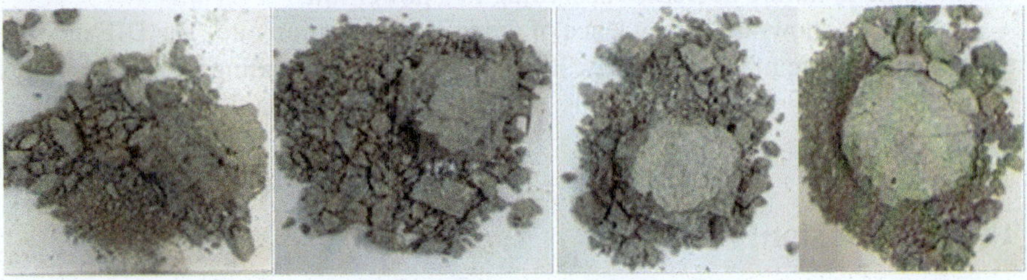

Specimen No. 3 Specimen No. 5 Specimen No. 7 Specimen No. 12

Fig. 18.4: The nature of the destruction of the specimens.

Table 18.2: Test parameters for reinforced concrete under dynamic compression.

Mode	Experiment Code	Loading parameters		Sample parameters			Experiment results				
		Pressure, bar	Striker speed, m/s	D, mm	$S_{c.s.}$, mm^2	L, mm	Ultimate stress, MPa	Ultimate strain, %	Strain rate, 1/s	Lifetime, ms	DIF
1	c669-01	0,5	7,14	53,7	2265	29,75	52,01	0,6	36,4	99	2,1
	c669-02	0,5	4,3	53,7	2265	29,2	23,00	0,5	43,5	108	0,9
2	c669-03	1	14,5	53,4	2240	30,4	98,57	1,9	165,1	98	3,9
	c669-04	1	15,5	53,4	2240	30,16	99,94	2,0	176,7	100	4,0
3	c669-05	2,5	19	52,73	2184	30,69	124,15	2,5	219,1	108	5,0
	c669-06	2,5	20,4	53,5	2248	30,3	105,41	2,5	319,5	65	4,2
	c669-07	2,5	22,2	53,4	2240	30,13	113,10	3,1	313,9	90	4,5
	c669-08	2,5	22,7	53,5	2248	31	103,06	2,7	329,0	82	4,1
4	c669-09	3	26,5	53,2	2223	30,85	124,52	3,4	390,9	92	5,0
	c669-10	3	28,4	53,8	2273	30,9	101,73	4,5	579,5	90	4,1
	c669-11	3	29	53,4	2240	31	100,54	3,5	595,1	78	4,0
	c669-12	3	25,5	54,2	2307	30,3	120,60	3,1	374,5	80	4,8

DIF - Dynamic increase factor

are determined. The strain rates ($\dot{\varepsilon}$) were taken to be maximum before the samples began to fail, since they change during deformation. Figure 18.6 shows the speed dependences of the main characteristics of reinforced concrete. The dependences obtained demonstrate that with an increase in the strain rate, the maximum stresses increase, the corresponding limiting strains also in-crease (according to a linear law), and the time to failure decreases according to a power law.

Figure 18.7 shows the dependence of the DIF on the strain rate. The DIF index in the obtained range of strain rates ~ 30–600 s^{-1} varies depending on the strain rate from 2.9 to 5.

18.5 Conclusion

An experimental study of the dynamic properties of fine-grained concrete rein-forced with steel fine-mesh mesh under uniaxial compression has been carried out. Diagrams of dynamic deformation at strain rates from 30 to 600 s^{-1} are obtained, the results of which determine the characteristics of strength and deformability, and determine the dependences on various strain rates.

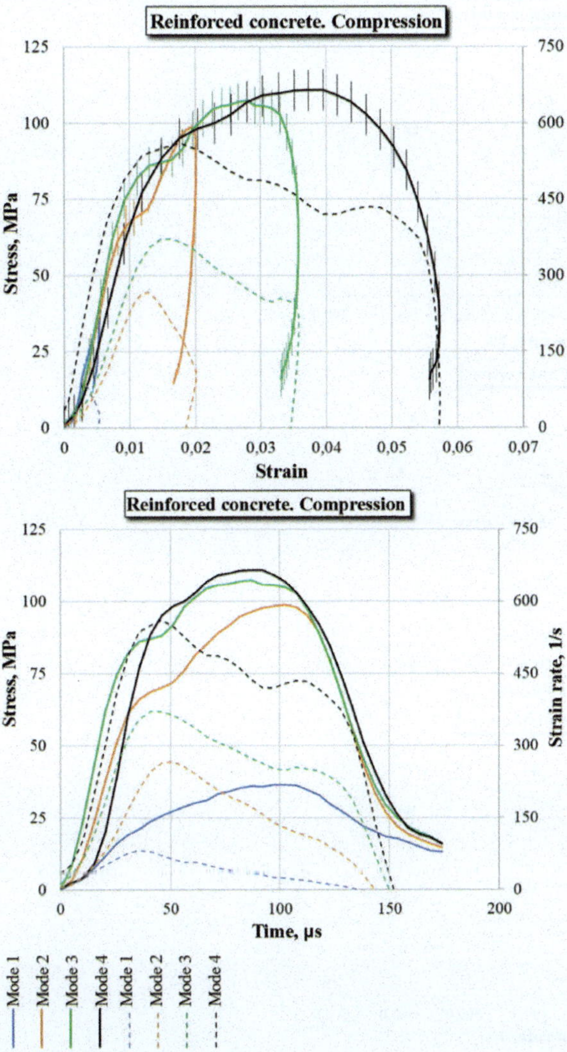

Fig. 18.5: Averaged diagrams $\sigma(\varepsilon)$ and $\dot{\varepsilon}(\varepsilon)$, $\sigma(t)$ and $\dot{\varepsilon}(t)$ under compression for all modes.

The introduction of a reinforcing fine-mesh mesh into the original fine-grained concrete increased the dynamic strength of the material at strain rates from 200 to 400 s^{-1} by an average of 45% relative to the previously tested fine-grained concrete [14]. The DIF values reached an average value of 4.0.

Acknowledgements The work was supported by the Russian Science Foundation grant No. 22-19-00138.

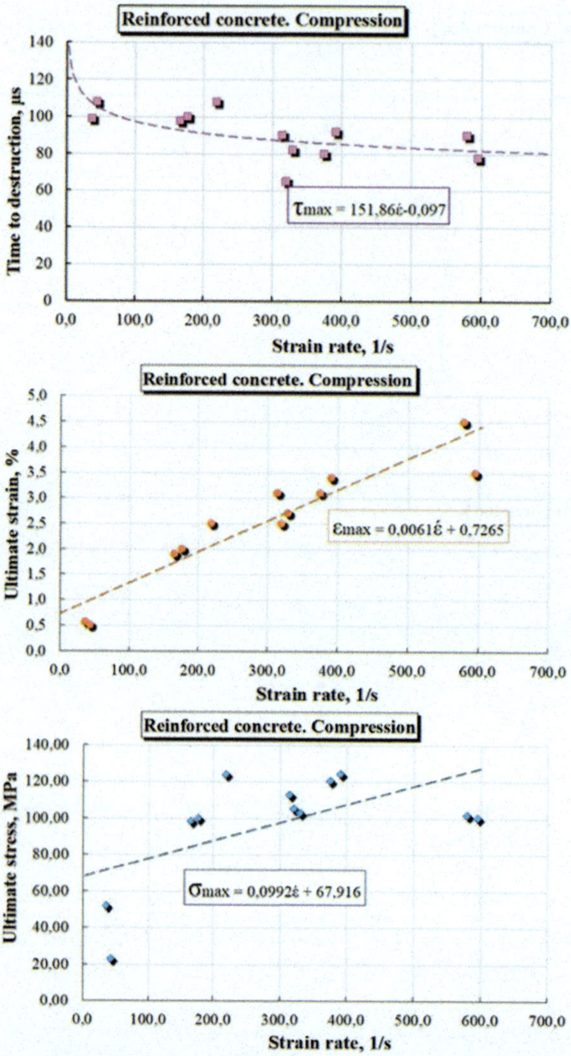

Fig. 18.6: Influence of strain rate on the mechanical properties of reinforced concrete under dynamic compression.

References

[1] Yu R, van Beers L, Spiesz P, Brouwers HJH (2016) Impact resistance of a sustainable Ultra-High Performance Fibre Reinforced Concrete (UHPFRC) under pendulum impact loadings, Construction and Building Materials **107**:203–215. DOI 10.1016/j.conbuildmat.2015.12.157

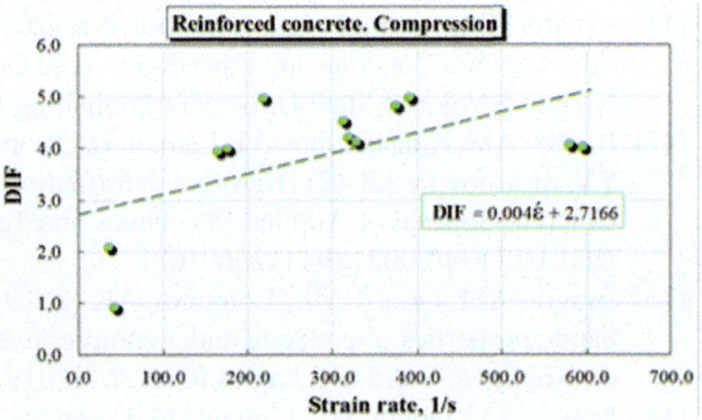

Fig. 18.7 Dependence of DIF on strain rate

[2] Wang D, Shi C, Wu Z, Xiao J, Huang Z, Fang Z. (2015) A review on ultra high performance concrete: Part II. Hydration, microstructure and properties, Construction and Building Materials **96**:368–377. DOI 10.1016/j.conbuildmat.2015.08.095

[3] Li PP, Yu QL (2019) Responses and post-impact properties of ultra-high performance fibre reinforced concrete under pendulum impact, Composite Structures **208** (2019) 806–815. DOI 10.1016/j.compstruct.2018.10.071

[4] Li PP, Brouwers HJH, Yu Q (2020) Influence of key design parameters of ultra-high performance fibre reinforced concrete on in-service bullet resistance, International Journal of Impact Engineering **136**:103434. DOI 10.1016/j.ijimpeng.2019.103434

[5] Khosravani MR, Weinberg K (2018) A review on split Hopkinson bar experiments on the dynamic characterisation of concrete, Construction and Building Materials **190**:1264–1283. DOI 10.1016/j.conbuildmat.2018.09.187

[6] Meng Q, Wu C, Su Y, Li J, Liu J, Pang J (2019) A study of steel wire mesh reinforced high performance geopolymer concrete slabs under blast loading, Journal of Cleaner Production **210**:1150-1163. DOI 10.1016/j.jclepro.2018.11.083

[7] Wei J, Li J, Wu C (2020) Behaviour of hollow-core and steel wire mesh reinforced ultra-high performance concrete columns under lateral impact loading, International Journal of Impact Engineering **146**:103726. DOI 10.1016/j.ijimpeng.2020.103726

[8] Liu J., Wu C., Li J., Su Y., Shao R., Liu Z, Chen G (2017) Experimental and numerical study of reactive powder concrete reinforced with steel wire mesh against projectile penetration, International Journal of Impact Engineering **109**:131-149. DOI 10.1016/j.ijimpeng.2017.06.006

[9] Yuan T-F, Choi J-S, Hong S-H, Yoon Y-S (2021) Enhancing the electromagnetic shielding and impact resistance of a reinforced concrete wall for protective structures, Cement and Concrete Composites **122**:104148. DOI 10.1016/j.cemconcomp.2021.104148

[10] Li J, Wu C, Liu Z-X (2018) Comparative evaluation of steel wire mesh, steel fibre and high performance polyethylene fibre reinforced concrete slabs in blast tests, Thin-Walled Structures **126**:117-126. DOI 10.1016/j.tws.2017.05.023

[11] Bragov AM, Igumnov LA, Lomunov AK (2015) High-Speed Deformation of Fine-Grained Concrete and Fiber-Reinforced Concrete (in Russ.), N. Novgorod: Nizhny Novgorod State University Publishing House
[12] Bragov AM, Konstantinov AY, Lamzin DA, Lomunov AK, Karihaloo BL, Petrov YV, Smirnov IV (2012) High-rate deformation and fracture of fiber reinforced concrete, Journal of Applied Mechanics and Technical Physics **53**(6):926–933. DOI 10.1134/S0021894412060168
[13] Bragov AM, Gonov ME, Lomunov AK (2019) Experimental study of the dynamic properties of concrete under compressive load, International Conference on Nonlinear Solid Mechanics ICoNSoM2019 June 16-19, 2019, pp 403-412.
[14] Bragov AM, Gonov ME, Lamzin DA, Lomunov AK, Modin IA (2021) Response of fine-grained fiber-reinforced concretes under dynamic compression, Materials Physics and Mechanics **47**(6):962-967. DOI 10.18149/MPM.4762021_14

Chapter 19
Multistability of Convective Flows in a Porous Enclosure

Vasily Govorukhin, Mezhlum Sumbatyan, and Vyacheslav Tsybulin

Abstract We discuss multistability for the problem of convection in a porous medium. Nontrivial phenomena of extreme multistability arise in the mathematical model of the Darcy convection with a linear temperature profile on the boundary. It manifests in the appearance of one-parameter families of steady states. An illustrative example concerns the anisotropic convection problem. We review here some issues of transition from extreme to standard multistability. Then we describe numerical and algorithmic aspects of the extreme multistability and provide necessary references.

Key words: Convection · Porous medium · Darcy-Boussinesq equations · Anisotropy · Multistability · Cosymmetry

19.1 Introduction

Investigation of the convective motions of fluids in porous media is an important subject of interest to many researchers [1]-[4]. Geophysics and some branches of industry require new models, to describe various observed phenomena [5, 6]. Nonlinearity of the corresponding mathematical models typically leads to the multistability – a phenomenon of coexistence of different solutions.

Often, multistability is a consequence of some symmetry group in a system. Existence of a one–parameter family of solutions (extreme multistability) may be caused by a continuous symmetry group (translational or rotational invariance). Another reason for this phenomena is cosymmetry, which was introduced by Yudovich, to explain the existence of one–parameter family of steady states in the problem with no continuous symmetry [7]-[9]. Such an extreme multistability was found in the

Vasily Govorukhin · Mezhlum Sumbatyan · Vyacheslav Tsybulin
Vorovich Institute of Mathematics, Mechanics and Computer Science, Southern Federal University, 344090, Milchakova Street 8a, Rostov-on-Don, Russian Federation,
e-mail: vngovoruhin@sfedu.ru, masumbatyan@sfedu.ru, vgcibulin@sfedu.ru

H. Altenbach et al. (eds.), *Sixty Shades of Generalized Continua*,
Advanced Structured Materials 170, https://doi.org/10.1007/978-3-031-26186-2_19

planar Darcy convection problem [7], [10]-[12]. The cosymmetry is a vector field which is orthogonal to a given vector field in a Hilbert space H. For example, cosymmetry of a differential equation $\dot{u} = \Phi(u)$, $u \in R^n$ is a nontrivial vector field $L(u)$, such that $(\Phi(u), L(u)) = 0$ at each point of R^n. In [7] there are formulated the conditions for existence of the family of equilibria in the system with cosymmetry. Assume u_0 to be a solution of $\Phi(u_0) = 0$, and $J = \nabla\Phi$ is the Jacobian matrix. Let the zero be a simple eigenvalue of the operator $J(u_0)$, and ϕ_0 is the basis vector in $\ker J(u_0)$. Then the given differential equation has the one–parameter family of solutions: $u_\alpha = u_0 + \alpha\phi_0 + O(\alpha^2)$, for a sufficiently small θ. This family of the equilibria cannot be a result of any symmetry group [8]. Moreover, the stability spectrum of the equilibria u_α depends on parameter α. The cosymmetry differs from the symmetry case, where the spectrum is identical for all equilibria. The nontrivial cosymmetry indicates that the system has a hidden free parameter, i.e. it is underdetermined.

An analytic study of the one-parameter cosymmetric family of steady-state regimes is possible only for those values of parameters which are close to a bifurcation in its appearance [7, 8]. For other situations only numerical approaches can be used as a rule. Some special numerical methods, to study cosymmetric dynamical systems, were developed in [13, 14] and continued in [14]-[25].

Yudovich proposed a selective function for studying the situation when the cosymmetry does not takes place and the continuous families of steady states break down or disappear [9]. Two scenarios are possible for a family of equilibria: the disinteg,ration of the family to a finite number of equilibria (ordinary multistability) or the appearance of slow periodic motions. The destruction of the family of steady states was investigated in [10, 19, 26, 27]. There is observed a memory effect about broken cosymmetry: relaxation oscillations and slow dynamics in a vicinity of the disappeared family of equilibria [19, 28].

Let us present a short survey of some results in this field, where authors' papers play a key role. We outline the numerical approaches for investigation both cosymmetric and close to cosymmetric problems. This permits the study of scenarios of bifurcations for some families [29, 30]. A spectral-difference method was derived for the convection in the rectangle [20]. The convection of a multi-component fluid is studied in [31]. A discretization in cylindrical coordinates is applied in [25, 32, 33]. A three–dimensional problem is studied in [24, 34, 35]. An anisotropy in a gravitational convection for rectangular enclosure is analyzed in [15, 26, 36, 37]. A selection of states under extreme multistability is considered in [17, 19, 38].

We consider the gravitational convection in a porous medium saturated by the heat-conducting fluid. The enclosure is heated from below. We are interested in an appearance of convective flows when the quiescent state (mechanical equilibrium) loses its stability. The resulting system is cosymmetric under some additional conditions for the parameters of the problem. It means that the continuous family of stationary regimes branches off from the quiescent state. After discretization with preserving the cosymmetry we obtain a system of ordinary differential equations. To compute the extreme multistability, a special continuation procedure is developed.

19.2 Mathematical Formulation of the Problem

Let us consider the problem of heating a rectangular container

$$\Omega = [0,a] \times [0,b],$$

see Fig. 19.1. Let us set the impermeability conditions on the boundary $\partial\Omega$, as well as a linear temperature profile. The gravity acts in the direction opposite to coordinate y:

$$\begin{aligned} u(0,y,t) &= u(a,y,t) = 0, \\ v(x,0,t) &= v(x,b,t) = 0, \\ T_*(y) &= T_0 + A - \frac{y}{b}A, \end{aligned}$$

where u, v are the horizontal and vertical velocities respectively, T_0 and $T_0 + A$ correspond to the temperature at the upper ($y = b$) and lower ($y = 0$) boundaries. The following system describes anisotropic convection in the porous medium:

$$0 = \frac{\partial p}{\partial x} + \mu(M_{11}u + M_{12}v), \tag{19.1}$$

$$0 = \frac{\partial p}{\partial y} + \mu(M_{21}u + M_{22}v) - \rho_0\beta(T - \hat{T})g, \tag{19.2}$$

$$0 = \frac{\partial u}{\partial x} + \frac{\partial v}{\partial y}, \tag{19.3}$$

$$\begin{aligned} (\rho c_p)_m \frac{\partial T}{\partial t} &= -(\rho c_p)_f \left(u\frac{\partial T}{\partial x} + v\frac{\partial T}{\partial y}\right) + \frac{\partial}{\partial x}\left(\Lambda_{11}\frac{\partial T}{\partial x} + \Lambda_{12}\frac{\partial T}{\partial y}\right) \\ &+ \frac{\partial}{\partial y}\left(\Lambda_{21}\frac{\partial T}{\partial x} + \Lambda_{22}\frac{\partial T}{\partial y}\right). \end{aligned} \tag{19.4}$$

Here t is time, x, y are the Cartesian coordinates, p is the pressure, T is the temperature, μ is the fluid viscosity, ρ is the density, g is the gravity acceleration, Λ_{ij} and M_{ij} are the components of tensors of the temperature conduction and the inverse permeability,

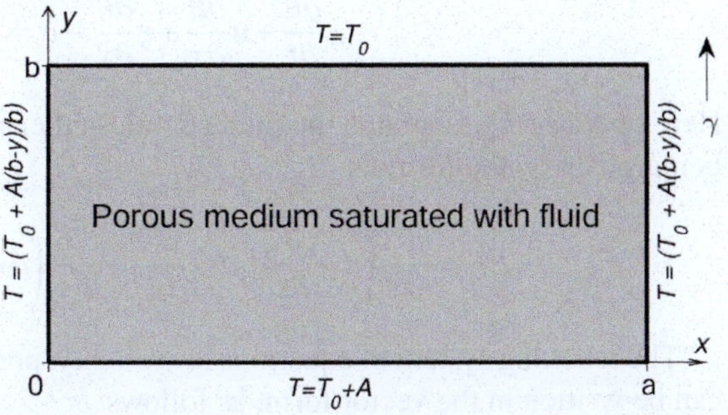

Fig. 19.1 Flow region

correspondingly. Then, ρ_0 is the density at the reference temperature T_0, the parameter $(\rho c_p)_m$ characterizes the product of the density by the specific heat at constant pressure averaged for the fluid and the porous medium, and $(\rho c_p)_f$ is an analogous characteristics for the fluid; β is the thermal expansion coefficient.

We introduce the deviation of the temperature θ as

$$T(x,y,t) = T_*(y) + \theta(x,y,t)$$

and dimensionless variables and parameters, as follows:

$$t = t_*\tilde{t}, \quad x = h_*\tilde{x}, \quad y = h_*\tilde{y}, \quad u = v_*\tilde{u}, \quad v = v_*\tilde{v}, \quad \theta = \tau_*\tilde{\theta},$$

$$\hat{P} = P_*\tilde{P}, \quad \hat{P} = p - \int_0^y \rho_0\beta g(T_* - \hat{T})dy, \quad M_{ij} = M_{22}\mu_{ij}, \quad d_{ij} = \frac{1}{\Lambda_{22}}\Lambda_{ij}.$$

The physical characteristics of length h_*, velocity v_*, pressure P_*, time t_*, and temperature τ_* are, respectively:

$$h_* = b, \quad v_* = \frac{\Lambda_{22}}{(\rho c_p)_f b}, \quad P_* = \frac{\mu\Lambda_{22}M_{22}}{(\rho c_p)_f} \quad \tau_* = At_* = \frac{b^2(\rho c_p)_m}{\Lambda_{22}} \tag{19.5}$$

The Rayleigh number λ is defined by the formula

$$\lambda = \frac{(\rho c_p)_f A b \rho_0 \beta g}{M_{22}\mu\Lambda_{22}}. \tag{19.6}$$

As a result, for dimensionless quantities (the "tilde" sign above the variables is omitted), the following system is obtained:

$$\begin{aligned} \frac{\partial p}{\partial x} + \mu_{11}u + \mu_{12}v &= 0, \\ \frac{\partial p}{\partial y} + \mu_{21}u + \mu_{22}v - \lambda\theta &= 0. \end{aligned} \tag{19.7}$$

$$\begin{aligned} \frac{\partial u}{\partial x} + \frac{\partial v}{\partial y} &= 0, \\ \frac{\partial \theta}{\partial t} + u\frac{\partial \theta}{\partial x} + v\frac{\partial \theta}{\partial y} - v &= L_D\theta, \end{aligned} \tag{19.8}$$

Here operator L_D contains the dimensionless thermal conductivity coefficients d_{ij}, being given by the formula

$$L_D = \frac{\partial}{\partial x}\left(d_{11}\frac{\partial}{\partial x} + d_{12}\frac{\partial}{\partial y}\right) + \frac{\partial}{\partial y}\left(d_{21}\frac{\partial}{\partial x} + d_{22}\frac{\partial}{\partial y}\right). \tag{19.9}$$

The resulting system of equations for velocity, pressure, and temperature deviation can be written in the vector form, as follows:

$$\nabla p + \boldsymbol{M} \cdot \boldsymbol{V} + \lambda\theta\boldsymbol{\gamma} = 0, \tag{19.10}$$

with

$$\nabla \cdot \boldsymbol{V} = 0$$

and

$$\dot{\theta} = L_D\theta - \boldsymbol{V} \cdot \boldsymbol{\gamma} - (\boldsymbol{V} \cdot \nabla)\theta. \tag{19.11}$$

Here dot-sign denotes differentiation with respect to time t, $\boldsymbol{M}$ represents the tensor of dimensionless inverse permeability coefficients μ_{ij}, $\boldsymbol{\gamma} = (0, -1)$ is a unit vector which determines the direction of the gravity.

These equations are supplemented with the boundary conditions of impermeability and homogeneity of the temperature deviation at the boundary $\partial\Omega$

$$\begin{aligned} u(0,y,t) &= u(a,y,t) = 0, \\ v(x,0,t) &= v(x,b,t) = 0, \end{aligned} \tag{19.12}$$

$$\theta(x,0,t) = \theta(0,y,t) = \theta(a,y,t) = \theta(x,b,t) = 0. \tag{19.13}$$

To analyze the two-dimensional problem, it is convenient to introduce the stream function ψ by $u = \psi_y, v = -\psi_x$, to come to a system for two unknowns functions: stream function ψ and temperature θ (temperature deviation). By applying differentiation to first equation (19.7) with respect to y, and to second one – with respect to x, and then taking their difference, one obtains the following equation:

$$0 = \mu_{11}\psi_{yy} - (\mu_{12} + \mu_{21})\psi_{xy} + \mu_{22}\psi_{xx} + \lambda\theta_x, \tag{19.14}$$

Besides, Eq. (8) is reduced to:

$$\dot{\theta} = L_D\theta - \psi_x - J(\psi, \theta), \quad J = \theta_x\psi_y - \theta_y\psi_x \tag{19.15}$$

The boundary conditions follow from (19.12):

$$\psi(x,0,t) = \psi(0,y,t) = \psi(a,y,t) = \psi(x,b,t) = 0. \tag{19.16}$$

The system for the isotropic porous medium follows from equations (19.14)-(19.16), (19.9) with

$$\mu_{ii} = d_{ii} = 1, (i = 1,2)$$

and

$$\mu_{ij} = d_{ij} = 0, (i \neq j).$$

It is shown in [7] that in this case the problem has a cosymmetry, and there may arise a family of stationary convective motions.

The following properties of the Jacobian $J(\psi, \theta)$ hold

$$\int_\Omega J(\psi,\theta)\psi\, dxdy = 0, \quad \int_\Omega J(\psi,\theta)\theta\, dxdy = 0. \tag{19.17}$$

To discretize the equations (19.14)-(19.15) properly, we must preserve the discrete analogues of the integral identities (19.17). The numerical schemes constructed appear to be conservative.

The system (19.14)-(19.15) is invariant under transformation

$$R_{xy} : \{x, y, \psi, \theta\} \mapsto \{a - x, b - y, \psi, -\theta\}.$$

If $\mu_{12} = \mu_{21} = d_{12} = d_{21} = 0$, then there is also an invariance

$$R_x : \{x, y, \psi, \theta\} \mapsto \{a - x, y, -\psi, \theta\}.$$

To analyze convective motions, system (19.9), (19.14)-(19.16) must be supplemented with the initial temperature distribution

$$\theta(x, y, 0) = \theta_0(x, y). \tag{19.18}$$

The initial condition for the stream function is not required, and the corresponding distribution for $t = 0$ is found from problem (19.14), (19.16) applied to the given function $\theta_0(x, y)$. The resulting initial-boundary problem describes the non–stationary gravitational anisotropic convection of a heat-conducting fluid in a porous rectangle heated from below.

The formulated problem demonstrates nontrivial effects of thermal and mass transfer. In the isotropic case, the onset of the convection is accompanied with an extreme multistability. Such phenomena make clear that there appears a one-parameter continuum family of the steady convective regimes. It means that there exist a hidden parameter in the system. The reason for this is the cosymmetry, the conditions for its existence are established in [7].

In the anisotropic case, the cosymmetry is the vector function

$$L = (\theta, -\psi)$$

and this exists if the following conditions are fulfilled

$$\mu_{11} = d_{22}, \quad \mu_{22} = d_{11}, \quad \mu_{21} = -d_{21}, \quad \mu_{12} = -d_{12}, \tag{19.19}$$

see [36]. In primitive variables the cosymmetry is given by the vector $(\theta_y, -\theta_x, 0, \psi)$.

If the conditions for parameters (19.19) are violated, then vector-function

$$L = (d_{22}\theta, -\mu_{11}\psi)$$

is not a cosymmetry. As a result, only the finite number of convective regimes (steady or time-dependent) exist, instead of the one-parametric family. To analyze this case, Yudovich developed the selective function techniques [9]. In the considered problem, the selective function has the form

$$S(s) = \int_{\Omega} \psi(s)\theta(s)(d_{22}\mu_{22} - \mu_{11}d_{11})dxdy. \tag{19.20}$$

Here s is a hidden parameter, $\psi(s)$ and $\theta(s)$ are the members of a family. Thus, for the analysis of destruction, one needs to calculate solutions for any s.

19.3 Numerical Methods and Extreme Multistability

Numerical analysis of cosymmetric and near-cosymmetric systems results in specific computational problems. First, the numerical method should preserve the fundamental properties of the problem. This includes preserving cosymmetry through discretization of the partial differential equations. Violation of the cosymmetry by discretization can lead to a wrong behavior, such as destruction of the family of steady states. Another challenging problem arises, due to an inability to apply standard methods to study cosymmetric problems because of its strong degeneration [39, 40].

Based on the Galerkin method, the study of the continuum family of stationary solutions in isotropic convection problems in porous media has been started in [16, 17]. To analyze dynamics in a wide range of parameters, some special methods for calculation of convective steady states and continuation along a hidden parameter are developed in [18, 41, 42]. Subsequently, finite-difference and spectral-difference approaches preserving cosymmetry were derived in [21]-[23].

With all above mentioned variants of discretization, one obtains a system of ordinary differential equations in the form

$$\dot{\Theta} = \Phi(\Theta) + \delta K(\Theta) \tag{19.21}$$

The right-hand side of this equation contains both cosymmetric (Φ) and noncosymmetric (δK) parts. Vector $\Theta(t)$ is a discrete approximation of $\theta(x,y,t)$. Cosymmetry $\hat{L}$ for Φ follows from the approximation of L. So, the case $\delta = 0$ corresponds to a cosymmetric finite-dimensional problem. Further, we define a discrete version of the selective function as $\hat{S} = (K, \hat{L})$.

19.3.1 Spectral Global Galerkin Method

Let us briefly describe the Galerkin method for analysis of problem (19.14)–(19.15), see [17, 42]. Functions ψ and θ are approximated by the series

$$\psi = \sum_{i=1}^{n}\sum_{j=1}^{n} \psi_{i,j}(t)\phi_{i,j}(x,y), \quad \theta = \sum_{i=1}^{n}\sum_{j=1}^{n} \theta_{i,j}(t)\phi_{i,j}(x,y), \tag{19.22}$$

with

$$\phi_{i,j}(x,y) = \frac{2}{\sqrt{ab}} \sin\left(i\frac{\pi x}{a}\right) \sin\left(j\frac{\pi y}{b}\right)$$

Substitution of (19.22) to (19.14)–(19.15) and projection operations lead to a system of ODEs of the order $N = n^2$ in form (19.21). Here

$$\Theta = (\theta_{11}, \theta_{12}, \ldots, \theta_{1n}, \theta_{21}, \ldots, \theta_{nn})$$

Similarly, one obtains an approximating system for the selective function (19.20). It is easy to prove that approximation (19.22) preserves the cosymmetry of the isotropic problem.

19.3.2 Cosymmetry Preserving Finite-Difference Approximations

The approximation of the problem in primitive variables (19.9)-(19.11) is carried out based on the staggered grids. The implementation for the isotropic case is presented in [22]. Firstly, we introduce the mesh on coordinates x and y:

$$0 = x_0 < x_1 < \ldots < x_n < x_{n+1} = a, 0 = y_0 < y_1 < \ldots < y_m < y_{m+1} = b.$$

The nodes (x_i, y_j) form the main grid on which the temperature θ_i^j is determined:

$$\omega_\theta = \{ (x_i, y_j), \quad i = 0, \cdots, n+1, \;\; j = 0, \cdots, m+1 \}.$$

We place auxiliary nodes in the middle of the intervals formed by neighboring nodes of the main grid

$$x_{i+1/2} = \frac{1}{2}(x_i + x_{i+1}), \quad y_{j+1/2} = \frac{1}{2}(y_j + y_{j+1}).$$

As a result, three additional grids are obtained, offset relative to the main one:

$$\begin{aligned}
\omega_u &= \{ (x_i, y_{j+1/2}) \; , \; i = 0, \cdots, n+1, \;\; j = 0, \cdots, m \}, \\
\omega_v &= \{ (x_{i+1/2}, y_j) \; , \; i = 0, \cdots, n, \;\; j = 0, \cdots, m+1 \}, \\
\omega_p &= \{ (x_{i+1/2}, y_{j+1/2}) \; , \; i = 0, \cdots, n, \;\; j = 0, \cdots, m \}.
\end{aligned}$$

The horizontal and vertical components of the velocity vector are calculated at grids ω_u and ω_v, and the pressure at node ω_p is calculated. The following notation is used for grid variables:

$$\begin{aligned}
u_i^{j+1/2} &= u(x_i, y_{j+1/2}), \quad v_{i+1/2}^j = u(x_{i+1/2}, y_j), \\
p_{i+1/2}^{j+1/2} &= u(x_{i+1/2}, y_{j+1/2}), \quad \theta_i^j = u(x_i, y_j).
\end{aligned}$$

To approximate (19.9)-(19.11), discrete analogues of differential operators are used. First-order difference operators are introduced on the two-point stencil

$$(\delta_x f)_{i+1/2}^{j+1/2} = \frac{f_{i+1}^{j+1/2} - f_i^{j+1/2}}{x_{i+1} - x_i}, \tag{19.23}$$

$$(\delta_y f)_{i+1/2}^{j+1/2} = \frac{f_{i+1/2}^{j+1} - f_{i+1/2}^{j}}{y_{j+1} - y_j}. \tag{19.24}$$

and operators for calculating the weighted average on the interval

$$(\delta_0^x f)_{i+1/2}^{j+1/2} = \frac{(x_{i+1} - x_{i+1/2}) f_{i+1}^{j+1/2} + (x_{i+1/2} - x_i) f_i^{j+1/2}}{x_{i+1} - x_i}, \tag{19.25}$$

$$(\delta_0^y f)_{i+1/2}^{j+1/2} = \frac{(y_{j+1} - y_{j+1/2}) f_{i+1/2}^{j+1} + (y_{j+1/2} - y_j) f_{i+1/2}^{j}}{y_{j+1} - y_j}. \tag{19.26}$$

Formulas (19.23)-(19.26) are valid for integer and half-integer values of indices i and j. With the help of the introduced operators, the difference relations on the three-point templates are determined:

$$(D_x f)_i^j = (\delta_x \delta_0^x f)_i^j, \quad (D_y f)_i^j = \left(\delta_y \delta_0^y f\right)_i^j.$$

Using the introduced operators, differential operator L_D (19.9) is approximated as follows:

$$L_D^h = \delta_x d_{11} \delta_x + d_x d_{12} d_y + d_y d_{21} d_x + \delta_y d_{22} \delta_y, \tag{19.27}$$

since

$$d_x d_y = D_x D_y$$

and the operators d_x, d_y and D_x, D_y are commutative. It should be noted that for the calculation of terms with d_{11} and d_{22}, respective operators on three-dot templates are used, and the use of operators $d_x d_x$ and $d_y d_y$ leads to the loss of cosymmetry for the grid analogue of the problem.

Next, first-order difference derivatives and operators for calculating the average on a rectangular template are introduced:

$$d_0 = \delta_0^x \delta_0^y \equiv \delta_0^y \delta_0^x, \quad d_x = \delta_x \delta_0^y \equiv \delta_0^y \delta_x, \quad d_y = \delta_y \delta_0^x \equiv \delta_0^x \delta_y. \tag{19.28}$$

Due to the non-uniformity of the grid, the operators on the rectangles are different for half-integer and integer index values, see [23].

To approximate the Jacobian J, a linear combination

$$(1-\alpha) J_D + \alpha J_d$$

of two discrete analogs of J is used

$$J_D = D_x(\theta D_y \psi) - D_y(\theta D_x \psi), \quad J_d = d_x(d_0 \theta d_y \psi) - d_y(d_0 \theta d_x \psi). \tag{19.29}$$

The value of the parameter $\alpha = 1/3$ makes it possible to satisfy the cosymmetry.

To approximate the problem (19.14)–(19.15), we use the approach [21, 23]. There are introduced the uniform grids

$$x_i = ih, i = 0, 1, \ldots, n+1, h = a/(n+1), y_j = jg, j = 0, 1, \ldots, m+1, g = b/(m+1),$$

with respective values of the stream functions ψ_{ji} and the temperatures θ_{ji} at gridpoints (x_i, y_j).

The resulting system of ordinary differential equations can be written as:

$$\dot{\Theta} = A\Theta + B\Psi - F(\Theta, \Psi), \quad \Psi = \lambda A^{-1} B\Theta \tag{19.30}$$

where the block-three-diagonal matrix A corresponds to the approximation of the Laplacian, and matrix B corresponds to the differential operator of the first order. The nonlinear vector-function $F(\Theta, \Psi)$ comes from a finite-difference approximation of the Jacobian. The problem at hand can also be reduced to (19.21).

19.3.3 Continuation on the Hidden Parameter Method

The continuation method is based on the cosymmetric implicit function theorem, see [43]. The first version of this method is presented in [14]. The one-parameter family of steady state regimes corresponds to the curve of equilibria $\hat{\Theta}(s)$ (here s is a hidden parameter) of approximating system of ordinary differential equations (19.21). This equilibrium curve may be continued with respect to parameter s from the point $\hat{\Theta}_0$, solving the following Cauchy problem

$$\frac{d\hat{\Theta}(s)}{ds} = \phi(s), \qquad \hat{\Theta}(0) = \hat{\Theta}_0 \tag{19.31}$$

Here $\phi(s)$ is a vector from the kernel of the linearized operator $\Phi'\hat{\Theta}(s)$.

Let us briefly describe the algorithm step by step, as follows:

1. Find any point $\hat{\Theta}_0$ on the curve by the modified Newton method.
2. Calculate the kernel of Jacobi matrix $\Phi'\hat{\Theta}_0$ (SVD method), check whether this is not degenerated (otherwise all next calculations are impossible) and choose the direction of the continuation along the curve.
3. Do one step of the Runge-Kutta method for the problem (19.31), where respective right-hand-side is an eigenvector corresponding to a trivial eigenvalue of $\Phi'\hat{\Theta}(s)$.
4. Accuracy checking. If the accuracy of calculations is not satisfactory, correct the solution.
5. Calculation of the selection function $S(s)$, stability analysis, verification of the exit conditions, then go to step 3 to find next point on the equilibria curve.

The described algorithm was applied to analyze a number of problems of filtration convection with various functions $K(\Theta)$.

19.4 Multistability

When conditions (19.19) are satisfied, the extreme multistability occurs simultaneously with convection. This results in the formation of a family of stationary states, see Fig. 19.2. Firstly, all stationary regimes are stable, but differ by the flow structure and heat distribution. At the same time, selection mechanisms, i.e. realizations of certain convective states, are very complicted, see [42]. As the Rayleigh parameter increases, the family of stationary flows becomes more complex: the structure of streamlines, the stability spectrum, etc. A further increase in heating leads to the appearance of instability on the family, see the right side of Fig. 19.2. Unstable states can form arcs, see [16]-[18], [20].

In the presence of heat sources inside and at the boundary, the fluid penetrates through the boundaries, with other influences, that leads to violation of the cosymmetry. In such cases the extreme multistability is lost [9, 19, 38]. Here, different scenarios for the transformation of convective movements are possible. The most typical is a multistability, which is expressed in the coexistence of a finite set of convective regimes. The selective function technique [9] allows one to determine the number of remaining regimes. An example of the destruction of the family is the inclusion of internal heat sources [19]. In Fig. 19.2 the signs on the central curves mark the regimes which are implemented for various options, if placing heat sources inside the region. With so doing, bold signs correspond to stable states.

The number of regimes which remain, when the cosymmetry is broken, can be different. It depends on the type of perturbation and the Rayleigh number. Figure 19.3 shows the results of exploring selective function (19.20) when the conditions (19.19) are violated. The analysis shows that there remain eight different convective regimes in this case. Thus, the presented examples show the transition from extreme to ordinary multistability.

Acknowledgements This research was carried out in the Laboratory of Computational Mechanics of Southern Federal University, with a financial support of the Government of the Russian Federation (Contract no. 075-15-2019-1928).

Besides, the authors are deeply grateful to Professor Victor Eremeyev for many years of scientific friendship: positive examples, good advice, kind help, as well as for creating an atmosphere of goodwill.

References

[1] Huang P-W, Wellmann F (2021) An Explanation to the Nusselt–Rayleigh discrepancy in naturally convected porous media, Transport in Porous Media **137**(1):195-214 . DOI 10.1007/s11242-021-01556-8

[2] Nield DA, Bejan A (2017) Convection in Porous Media, New York, Springer. DOI 10.1007/978-3-319-49562-0

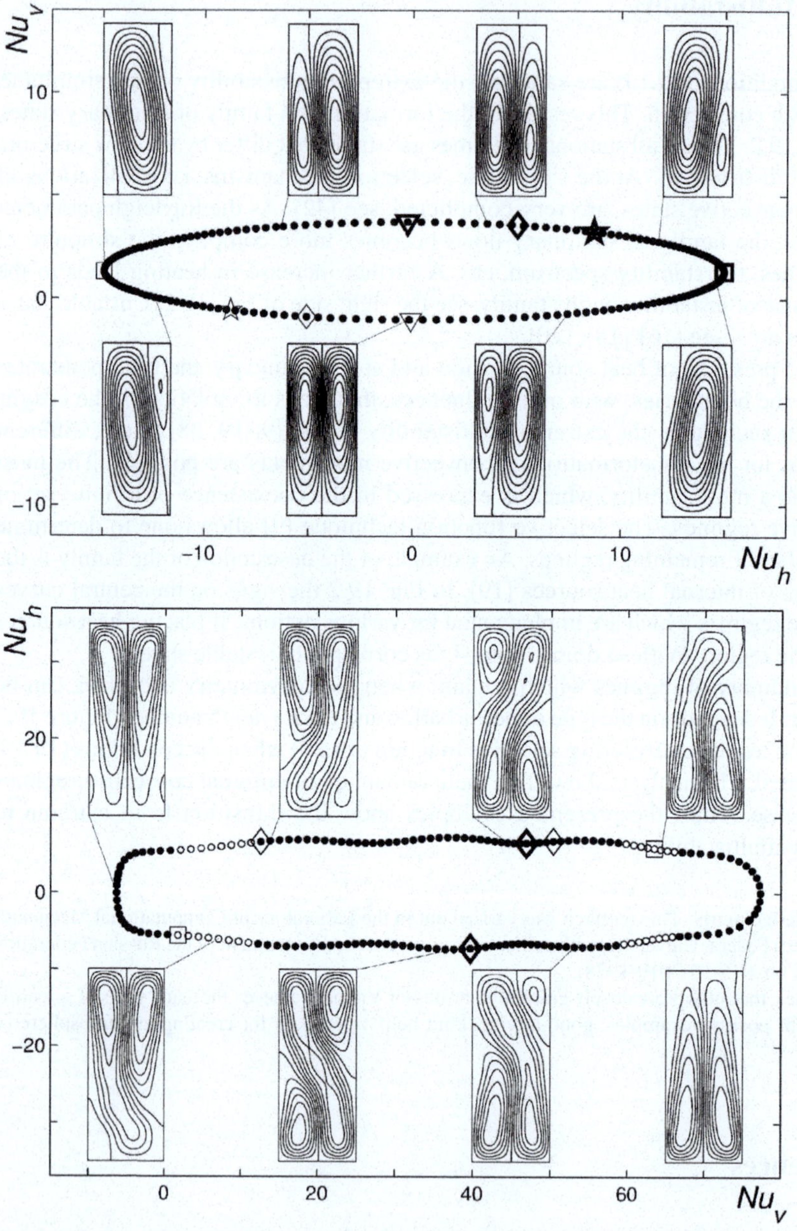

Fig. 19.2: Families of steady states and streamlines of convective flows: at top – after branching off from a quiescent state, at bottom – close to loss of stability on the family. Filled circle marks stable state, and unstable one is depicted by empty circle. .

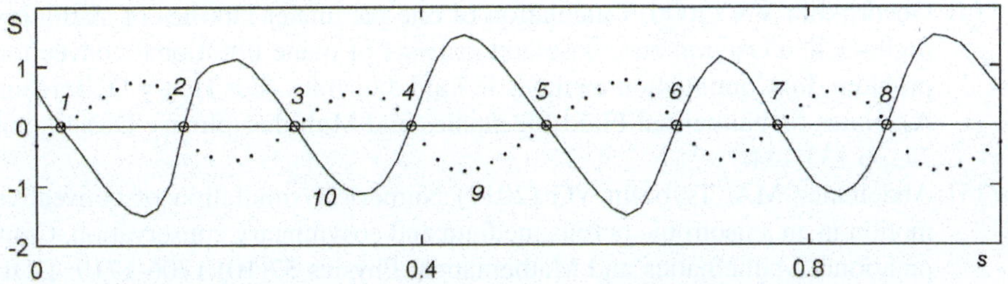

Fig. 19.3: Selective function S: $\mu_{11} = d_{22} = 1.2$ and $\mu_{22} = 1.0$; $\lambda = 270$, (1–8) are the zeros of the selective function, $d_{11} = 0.95$ and 1.1 (curves 9 and 10, respectively). .

[3] Soboleva EB (2019) A method for numerical simulation of Haline convective flows in porous media as applied to geology, Computational Mathematics and Mathematical Physics **59**(11):1893-1903. DOI 10.1134/S0965542519110101

[4] Storesletten L, Rees DAS (2019) Onset of convection in an inclined anisotropic porous layer with internal heat generation, Fluids **4**(2):75. DOI 10.3390/fluids4020075

[5] Maryshev B, Lyubimova T, Lyubimov D (2013) Two-dimensional thermal convection in porous enclosure subjected to the horizontal seepage and gravity modulation, Physics of Fluids **25**(8):084105. DOI 10.1063/1.4817375

[6] Rionero S (2012) Absence of subcritical instabilities and global nonlinear stability for porous ternary diffusive-convective fluid mixtures, Physics of Fluids **24**(10):104101. DOI 10.1063/1.4757858

[7] Yudovich VI (1991) Cosymmetry, degeneration of solutions of operator equations, and onset of a filtration convection, Mathematical Notes of the Academy of Sciences of the USSR **49**(5):540-545. DOI 10.1007/BF01142654

[8] Yudovich VI (1995) Secondary cycle of equilibria in a system with cosymmetry, its creation by bifurcation and impossibility of symmetric treatment of it, Chaos **5**(2):402-411. DOI 10.1063/1.166110

[9] Yudovich VI (2004) Bifurcations under perturbations violating cosymmetry, Doklady Physics **49**(9):522-526. DOI 10.1134/1.1810578

[10] Bratsun DA, Lyubimov DV, Roux B (1995) Co-symmetry breakdown in problems of thermal convection in porous medium, Physica D: Nonlinear Phenomena **82**(4):398-417. DOI 10.1016/0167-2789(95)00045-6

[11] Glukhov AF, Putin GF (1999) Experimental study of convective structures in a liquid-saturated porous medium near the threshold of instability of mechanical equilibrium (in Russ.), Hydrodynamics **12**:104-120.

[12] Lyubimov DV (1975) Convective motions in a porous medium heated from below, Journal of Applied Mechanics and Technical Physics **16**:257-261. DOI 10.1007/BF00858924

[13] Govorukhin VN (1996) Computer experiments with cosymmetric models, ZAMM Zeitschrift fur Angewandte Mathematik und Mechanik **76**(Suppl. 4):559-562.

[14] Govorukhin VN (2000) Calculation of one-parameter families of stationary regimes in a cosymmetric case and analysis of plane filtrational convection problem, In: Continuation methods in fluid dynamics (Eds Henry D, Bergeon A), Notes on Numerical Fluid Mechanics and Multidisciplinary Design, vol. 74, pp 133-144.
[15] Abdelhafez MA, Tsybulin VG (2017) Numerical simulation of convective motion in an anisotropic porous medium and cosymmetry conservation, Computational Mathematics and Mathematical Physics **57**(10):1706-1719. DOI 10.1134/S0965542517100025
[16] Govorukhin VN (1998) Numerical simulation of the loss of stability for secondary steady regimes in the Darcy plane-convection problem, Doklady Physics **43**(12):806-808.
[17] Govorukhin VN (1999) Analysis of families of secondary steady-state regimes in the problem of plane flow through a porous medium in a rectangular vessel, Fluid Dynamics **34**(5):652-659.
[18] Govorukhin VN, Shevchenko IV (2003) Numerical investigation of the second transition in the problem of plane convective flow through a porous medium, Fluid Dynamics **38**(5):760-771. DOI 10.1023/B:FLUI.0000007838.46669.1a
[19] Govorukhin VN (2014) On the action of internal heat sources on convective motion in a porous medium heated from below, J. Applied Mechanics and Technical Phys. **55**(2):225-233. DOI 10.1134/S0021894414020059
[20] Kantur OY, Tsybulin VG (2002) A spectral-difference method for computing convective fluid motions in a porous medium and cosymmetry preservation, Comput. Mathematics and Mathematical Phys. **42**(6):878-888.
[21] Karasözen B, Tsybulin VG (1999) Finite-difference approximations and cosymmetry conservation in filtration convection problem, Physics Letters, Section A **262**:321-329. DOI 10.1016/S0375-9601(99)00599-X
[22] Karasözen B, Tsybulin VG (2005) Cosymmetry preserving finite-difference methods for convection equations in a porous medium, Applied Numerical Mathematics **55**(1):69-82. DOI 10.1016/j.apnum.2004.10.008
[23] Karasözen B, Tsybulin VG (2005) Mimetic discretization of two-dimensional darcy convection, Computer Physics Communications **167**(3):203-213. DOI 10.1016/j.cpc.2004.12.012
[24] Karasözen B, Nemtsev AD, Tsybulin VG (2012) Staggered grids for three-dimensional convection of a multicomponent fluid in a porous medium, Computers and Mathematics with Applications **64**(6):1740-1751. DOI 10.1016/j.camwa.2012.02.007
[25] Karasözen B, Trofimova AV, Tsybulin VG (2012) Natural convection in porous annular domains: Mimetic scheme and family of steady states, J. Comput. Physics **231**(7):2995-3005. DOI 10.1016/j.jcp.2012.01.004
[26] Abdelhafez MA, Tsybulin VG (2018) Anisotropic problem of Darcy convection: family of steady flows and its disintegration during the destruction of cosymmetry, Fluid Dynamics **53**(6)738-749. DOI 10.1134/S0015462818060125

[27] Tsybulin VG, Karasözen B (2008) Destruction of the family of steady states in the planar problem of Darcy convection, Physics Letters A **372**(35):5639-5643. DOI 10.1016/j.physleta.2008.07.006
[28] Govorukhin VN, Tsybulin VG (2021) Multistability, scattering and selection of equilibria in the mechanical system with constraint, Communications in Nonlinear Science and Numerical Simulation **95**:105602. DOI 10.1016/j.cnsns.2020.105602
[29] Govorukhin VN, Shevchenko IV (2013) Selection of steady regimes of a one-parameter family in the problem of plane convective flow through a porous medium, Fluid Dynamics **48**(4):523-532. DOI 10.1134/S001546281304011X
[30] Karasözen B, Tsybulin VG (2004) Cosymmetric families of steady states in Darcy convection and their collision, Physics Letters, Section A: General, Atomic and Solid State Physics **323**(1-2):67-76. DOI 10.1016/j.physleta.2004.01.053
[31] Kantur OYu, Tsibulin VG (2004) Numerical investigation of the plane problem of convection of a multicomponent fluid in a porous medium, Fluid Dynamics **39**(3):464-473. DOI 10.1023/B:FLUI.0000038565.09347.ac
[32] Trofimova AV, Tsibulin BG (2011) Convective motions in a porous ring sector, J. Applied Mechanics and Technical Phys. **52**(3):427-435. DOI 10.1134/S0021894411030138
[33] Trofimova AV, Tsibulin VG (2014) Filtration convection in an annular domain and branching of a family of steady-state regimes, Fluid Dynamics **49**(4):481-490. DOI 10.1134/S0015462814040085
[34] Karasözen B, Nemtsev AD, Tsybulin VG (2008) Staggered grids discretization in three-dimensional Darcy convection, Computer Physics Commun. **178**(12):885-893. DOI 10.1016/j.cpc.2008.02.004
[35] Nemtsev AD, Tsibulin VG (2007) Numerical investigation of the first transition in the three-dimensional problem of convective flow in a porous medium, Fluid Dynamics **42**(4):637-643. DOI 10.1007/s10697-007-0066-y
[36] Abdelhafez MA, Tsybulin VG (2017) Anisotropy effect on the convection of a heat-conducting fluid in a porous medium and cosymmetry of the Darcy problem, Fluid Dynamics **52**(1):49-57. DOI 10.1134/S0015462817010057
[37] Abdelhafez MA, Tsybulin VG (2018) Modeling of anisotropic convection for the binary fluid in porous medium, Computer Research and Modeling **10**(6):801-816. DOI 10.20537/2076-7633-2018-10-6-801-816
[38] Tsybulin VG, Karasözen B, Ergench T (2006) Selection of steady states in planar Darcy convection, Physics Letters, Section A **356**(3):189-194. DOI 10.1016/j.physleta.2006.03.043
[39] Eremeyev VA, Lebedev LP, Altenbach H (2013) Foundations of Micropolar Mechanics, Springer, Heidelberg.
[40] Sumbatyan MA, Scalia A (2005) Equations of Mathematical Diffraction Theory, CRC Press, Boca Raton, Florida.
[41] Govorukhin VN, Shevchenko IV (2006) Scenarios of the onset of unsteady regimes in the problem of plane convective flow through a porous medium, Fluid Dynamics **41**(6):967-975. DOI 10.1007/s10697-006-0111-2

[42] Govorukhin VN, Shevchenko IV (2017) Multiple equilibria, bifurcations and selection scenarios in cosymmetric problem of thermal convection in porous medium, Physica D: Nonlinear Phenomena, **361**:42-58. DOI 10.1016/j.physd.2017.08.012

[43] Yudovich VI (1996) Implicit function theorem for cosymmetric equations, Mathematical Notes **60**(2):235-238. DOI 10.1007/bf02305191

Chapter 20
Hydrogen Transport in Framework of Linear Non-Equilibrium Thermodynamics Approach

Polina M. Grigoreva and Vladimir A. Polyanskiy

Abstract The paper considers the problem of the hydrogen diffusion through the surface of the semi-infinite semi-circular crack in the metal. To describe the experimentally observed boundary layer with high hydrogen concentrations, the diffusion equation is written within the framework of linear non-equilibrium thermodynamics approach. This mass balance equation takes into account the mutual influence of the stress-strain state of the body on the absorption of gas into it from the first principles. In this paper, linear elastic stresses and strains are considered, and the leak term in the diffusion equation obtained using the concept of the damageability tensor is also taken into account. The boundary value problem is solved numerically for equations with and without considering the stock term. The results obtained are compared with experiments.

Key words: Hydrogen diffusion · Skin-effect · Plane strain-state of solid

20.1 Introduction

There are a number of phenomena that have a pronounced surface nature and, at the same time, directly affect the most important mechanical properties of the entire bulk material. The most well-known and well-researched phenomenon is the special effect of the metal surface on the yield strength and various stages of deformation hardening [1]. Back in the early 20th century, it was noticed that residual stresses arising from plastic deformation of flat metal samples completely disappear during chemical or electrochemical etching of the metal surface layer [2, 3].

Polina Grigoreva · Vladimir Polyanskiy
Institute for Problems in Mechanical Engineering RAS, Bolshoy pr., 61, V.O., St. Petersburg 199178, Russian Federation,
e-mail: polina.grigoreva239@gmail.com, vapol@mail.ru

H. Altenbach et al. (eds.), *Sixty Shades of Generalized Continua*,
Advanced Structured Materials 170, https://doi.org/10.1007/978-3-031-26186-2_20

Numerous studies show that the surface effect during plastic deformation has a depth of about 10-50 microns and is observed in both polycrystalline and monocrystalline metal samples. Mechanical or chemical removal of a layer of such thickness from the surface of massive samples completely removes residual stresses, hardening that occurred as a result of plastic deformation of samples and often such removal is equivalent to annealing.

Hydrogen embrittlement is rarely associated with the surface phenomenon [4]. Usually, the surface and volume sorption of hydrogen [5] is considered, but in case of surface sorption the surface is considered as a metal layer, the thickness of which is about 10 nm (dozens of monolayers of atoms). In a number of works, it is proved that on the surface of the metal [6] molecular hydrogen dissociates into atoms and is transported into the metal in atomic form. Experimental investigations [7]-[10], however, show that molecular hydrogen diffuses into metals, and forms a thin near-surface layer there, in which its concentration is tens or even hundreds of times higher than in the whole metal. Recent experimental work [11]-[13] suggests the strongest influence of this thin layer (its thickness is about one micrometer) on the elastic properties, brittleness and viscosity of the metal. Thus, the modeling of a thin boundary layer resulting from the diffusion of hydrogen from the environment into the metal and the description of the mutual influence between the diffusion process and the mechanical properties of the metal is of considerable theoretical and practical interest.

There is no certain opinion in the scientific literature on how to take into account the influence of mechanical stresses on diffusion. It is believed that with volumetric hydrogen sorption, diffusion trapping mechanisms work (see [14, 15] and many others, for example [16]-[20]). In [21], the redistribution of a substance diffusing in a solid in the region of tensile stresses is justified (Gorsky's law for hydrogen). The basic equations of hydrogen transport include Gorsky's Law as an addition of chemical potential to the diffusion equations [22]. Later, models based on the thermodynamics of irreversible processes and representing a generalization of Fourier-type equations, according to which irreversible processes can be described by linear differential equations with constant coefficients, became widespread. Within the framework of this approach, the diffusion rate is proportional to the thermodynamic forces, which can be expressed in terms of gradients of the corresponding potentials, depending, in particular, on the stress-strain state (see, for example, [22]-[24]). These additional thermodynamic forces are usually associated with a change in the volume or stiffness of a solid material or with a change in gas concentration. In a number of works, the thermodynamic force, taking into account the influence of stresses, is recorded through the chemical potentials of materials, using the Eshelby tensor as the chemical potential of a deformable solid [16], [25]-[27]. However, such models are rarely used due to some difficulties associated with the definition of model constants. In addition, experiments show that hydrogen diffusion is multichannel and it cannot be considered only as volumetric or trap modes, but it is necessary to take into account several diffusion channels [8], [28]-[30]. It should also be noted that most trapping and volume diffusion models have been verified for small gradients of hydrogen concentration in the material. This does not allow taking into account the observed boundary layer and requires significant modification of these models.

For a more accurate description of the experimentally observed near-surface layer with a high concentration of hydrogen, which is relatively stable in time and does not expand into the metal due to diffusion, we also considered introducing a leak term. Instead of the classical McNabb model [14] we use an alternative model of two–channel diffusion, which separately takes into account the flow through inhomogeneities inside the metal [31]. The resulting equations were solved for a one-dimensional problem [32], and the models were investigated for their applicability to describe the occurrence of the skin effect. However, the flat problem of hydrogen distribution at the crack boundaries is also of great importance for practice. This problem has already been considered within the framework of the McNabb and Oriani approach [33, 34], however, the mutual influence of the stress-strain state of the metal on hydrogen absorption was taken into account phenomenologically, namely, by specifying the distribution of traps from the amount of plastic deformation, while plastic deformation was assumed to depend on the concentration of hydrogen. We have shown that even in the case of a one-dimensional problem, taking into account the linear elastic stress state of the metal can lead to significant gradients in the concentration of hydrogen inside the sample, so a more detailed study of this problem is of considerable interest. Moreover, in the work of Sofronis [34], the threshold local concentrations of hydrogen at which the influence of the HELP mechanism of hydrogen embrittlement begins is indicated as 10^3 ppm range for different metals, which is critical value for most metals used nowadays [35].

In this paper we consider a plane boundary value problem with a circular crack in an infinite plane, which experiences the action of stresses, and through the surface of which hydrogen diffuses. We believe that the deformations and stresses that occur in the system (including those induced by hydrogen) do not go beyond the limits considered by the linear theory of elasticity. The hydrogen diffusion is described within the framework of linear non-equilibrium thermodynamics, and as additional diffusion channels we introduce stock term using the damageability tensor.

20.2 Problem Statement and Governing Equations

In our previous works, we considered the diffusion of hydrogen into the metal within the framework of the approach of linear non-equilibrium thermodynamics, which assumes that the diffusion fluxes are proportional to the gradient of the chemical potential. Within the framework of this approach, in the case of an ideal gas and a solid, the chemical potentials of gas μ_g and linear elastic solid μ_s look like following:

$$\mu_g = \mu_0(T) + RT \ln \frac{c}{c_*}, \tag{20.1}$$

where T is temperature of the system (we assume that it is constant for both solid and gas components and we do not have any heat fluxes), R is the gas constant, c is the concentration of the gas and c_* is a reference concentration which we assume as the concentration on the boundary:

$$\mu_s = \frac{M}{\rho}\left(3\eta(T) + \frac{3}{2}\boldsymbol{\varepsilon}_e \cdot\cdot \boldsymbol{C} \cdot\cdot \boldsymbol{\varepsilon}_e - \boldsymbol{\sigma} \cdot\cdot \boldsymbol{\varepsilon}_e\right), \tag{20.2}$$

where $\boldsymbol{\varepsilon}_e = \boldsymbol{\varepsilon} - \boldsymbol{\varepsilon}_*$ is elastic strain tensor, $\boldsymbol{\sigma}$ is the Cauchy stress tensor, $\boldsymbol{C}$ is the stiffness tensor, $\boldsymbol{\varepsilon}_* = \alpha c \boldsymbol{I}$ is a tensor of hydrogen induced strains, with α as a proportionality coefficient and $\boldsymbol{I}$ as a unit tensor. Assuming the solid material is isotropic and behaves according to the Hooke's law, we finally obtain the diffusion equation:

$$\frac{\partial c}{\partial t} = \nabla \cdot (D_{\text{eff}} \nabla c + \boldsymbol{V} c), \tag{20.3}$$

where effective diffusion coefficient is equal to

$$D_{\text{eff}} = D_0\left(1 - \frac{cM}{RT\rho}\frac{E}{1-2\nu}\alpha \operatorname{tr}\boldsymbol{\varepsilon}\right), \tag{20.4}$$

and proportionality coefficient of the additional term of diffusion equation is

$$\boldsymbol{V} = \frac{D_0 M E}{RT\rho}\frac{\nabla(\operatorname{tr}\boldsymbol{\varepsilon})}{1-2\nu}\left(\frac{\nu}{1+\nu}\operatorname{tr}\boldsymbol{\varepsilon} - \alpha c\right), \tag{20.5}$$

All above M, ρ are molar mass and density of the solid; D_0 – temperature-only dependent diffusion coefficient; E and ν are Young's modulus and Poisson's coefficients of the solid. This diffusion equation takes into account linear elastic stress-strain state of the solid body and its mutual influence on diffusion of the gas into it. To enhance this equation, we introduce the stock term.

McNabb [14] suggested dividing the diffusing hydrogen into two parts, the one which diffuses through the lattice of the metal, and the other which diffuses through the trapping modes. Considering the right part of the diffusion equation (20.3) works for the hydrogen that diffuses through the lattice, and assuming that the concentration in the trap modes is the product of their concentration and occupancy, McNabb [14] suggested additional equation for the occupancy of the trap modes.

We recall that trapping modes are in fact the inhomogeneities of the metal: dislocations, nanopores, microvoids and microcracks, etc. [36]-[38]. Thus, we can assume that the flux of the trapped hydrogen is proportional to the damageability of the material [31]. Not to limit ourselves to the one dimension or to the fully isotropic problem we use the damageability tensor [39, 40] and rewrite the diffusion equation (20.3) as follows:

$$\frac{\partial c}{\partial t} = \nabla \cdot [(\boldsymbol{I} - \boldsymbol{D}) \cdot (D_{\text{eff}} \nabla c + \boldsymbol{V} c)]\,. \tag{20.6}$$

Here $\boldsymbol{D}$ is a damageability tensor; the major advantage of its usage is that its components can be measured separately from any diffusion problem (unlike concentration and occupancy in McNabb and Oriani models).

To solve the diffusion problem we need the solution of the elasticity equation. We use the mass balance and apply Hooke's law since we assume that there are no additional mechanical fields in solid:

$$\nabla \cdot \left[\frac{\nu}{1-2\nu} (\mathrm{tr}\boldsymbol{\varepsilon} - 3\alpha c) \boldsymbol{I} + (\boldsymbol{\varepsilon} - \alpha c \boldsymbol{I}) \right] = 0. \quad (20.7)$$

Both the Eq. (20.7) and any of the diffusion equations (20.3) or (20.6) cannot be solved separately, thus, we have to solve a coupled problem.

20.3 Plane Boundary Value Problem

We consider an infinite body with a semi-infinite crack of an initial semi-circular opening r_0 (see Fig. 20.1). We apply asymptotic boundary conditions at the circular distance $R = 10000 r_0$ from the crack tip, which allows us consider our problem in a small strain approach and not contradict with the elasticity equations. We assume that at the time instant there is no hydrogen in the solid.

The boundary value problem (for both the elasticity and the diffusion equations) is solved numerically using an explicit finite difference scheme that corresponds to the finite volume method. In the scheme, instead of differential equations of local balances, we considered discrete variants of the corresponding integral balance equations. Displacements and concentrations were set in the grid nodes, deformations,

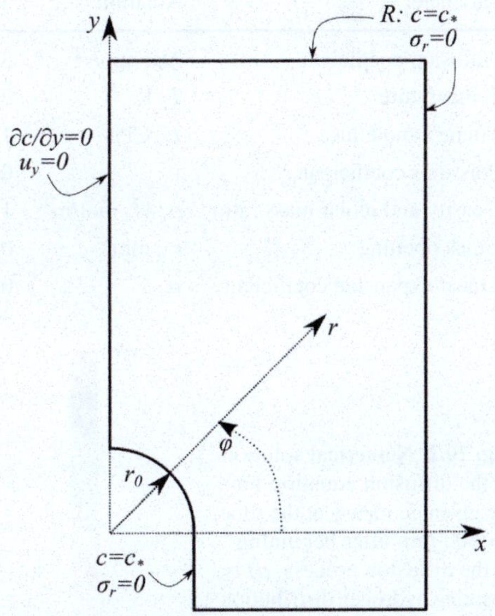

Fig. 20.1 Geometry of the infinite plane with semi-infinite opening of the radius r_0. All the boundary conditions are shown

stresses and diffusion coefficients were set in the grid cells. The scheme itself is verified and investigated for convergence.

Experimental data for calculations are taken from [41] for steel T24. The parameter corresponding to the expansion due to hydrogen diffusion is chosen so that all deformations of the solid remain in the range of the linear elastic region. The selected parameters are shown in the Table 20.1. Values of the damageability tensor required for calculating the radial and angle component depending on the coordinate are taken from [39].

The results for the numerical solution are shown on the Figs. 20.2-20.4. We can see that using full chemical potential even in case of linear elasticity leads to the formation of so-called skin-effect. Taking hydrogen trapping into account, we obtain the boundary layer that goes deeper into the solid material in the crack tip. It could have happened due to the higher values of the strains along the curvature of the crack, and the diffusion flux becomes more susceptible to the changes in the strains value. This corresponds with observed experiments which show that hydrogen embrittlement makes crack more unstable comparing to crack behavior in hydrogen-free environment. Highly uneven hydrogen distribution along the entire surface of the crack may also be the cause of cracking across the direction of crack growth observed experimentally (see, e.g. [20, 42]).

From the Fig. 20.3 one can see that even absence of the stock term produces high gradients of the concentration at the surface of the crack. However, linear elastic

Table 20.1: Experimental data.

Parameter	Notation	Value
Diffusion coefficient	D_0, mm/s^2	$3.5{\cdot}10^{-9}$
Temperature	T, K	293
Young's modulus	E, GPa	182
Poisson's coefficient	ν	0.295
Density and molar mass ratio	ρ/M, mol/m^3	$1.45{\cdot}10^{4}$
Crack opening	r_0, mm	0.1
Linear expansion coefficient	α	0.03

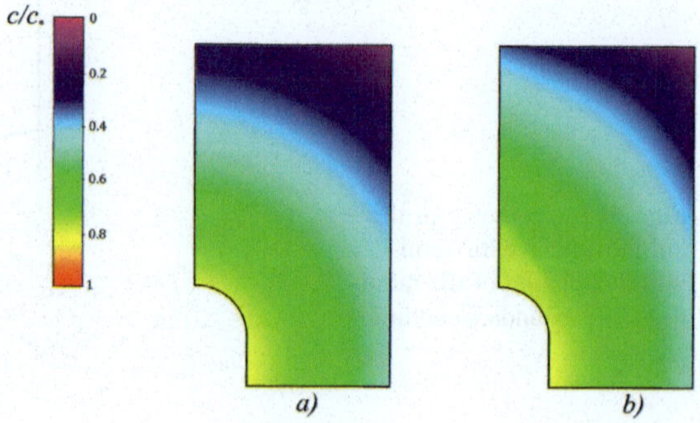

Fig. 20.2 Numerical solution of the diffusion equation for the given geometry at the time t = 600 min after beginning of the diffusion process; *a)* is resulting hydrogen distribution for the diffusion equation (20.3); *b)* – for the diffusion with stock term (20.6)

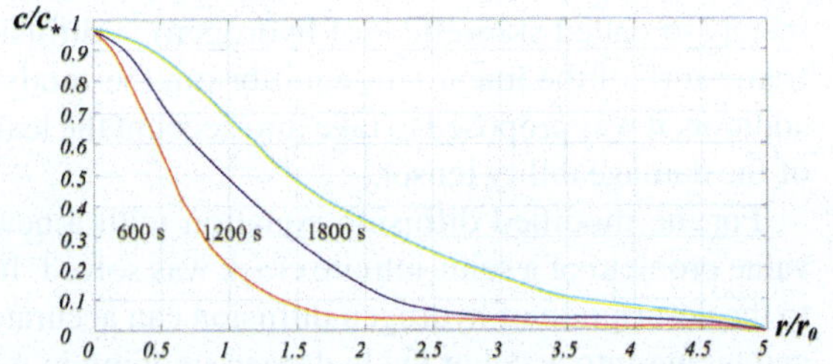

Fig. 20.3 Dependence of hydrogen distribution on the coordinate at various periods of time, no leak term is considered

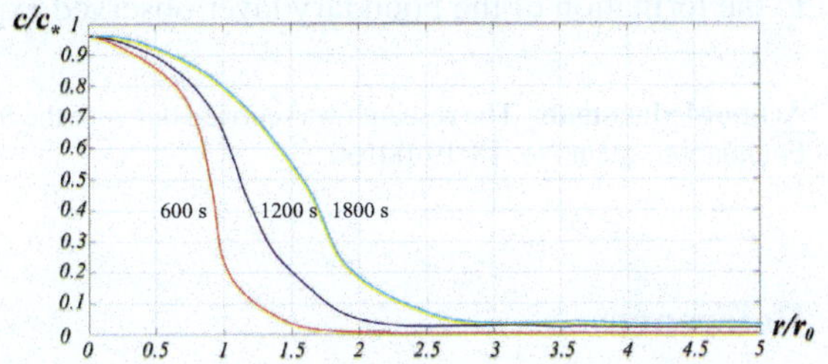

Fig. 20.4 Dependence of hydrogen distribution on the coordinate at various periods of time, leak term using damageability tensor is considered

stress field cannot hold this concentration gradient, and the boundary layer begins to blur over time, as if the diffusion proceeded without impact of stress-strain state at all. Since we do not observe this blurring experimentally, we can conclude that, in general, the mass balance equation we have obtained can serve as a basis for describing the diffusion of hydrogen in steel, but additional modifications of this equation are necessary.

Taking into account the leak term significantly changes the picture of the distribution of hydrogen (see Fig. 20.4). The concentration gradients that appear are retained over time, and the boundary layer does not blur but moves deeper into the metal, maintaining high gradients of hydrogen concentration. It should be noted that the results obtained were not obtained using the McNabb trapping model, which is phenomenological and requires refinement of material parameters from experiment to experiment, but using the concept of the damageability tensor, which values can be taken from an independent experiment. We can also assume that the results obtained indicate that models for metal micro-damages accounting can significantly improve the description of hydrogen diffusion into metals without the need to use the contradictory McNabb model.

20.4 Conclusion

In this paper we consider the diffusion of hydrogen into a solid body in a stress-strain state and the interaction between deformations inside the body and the transport of hydrogen from the external environment surrounding the solid. For a more accurate description of the experimentally observed boundary layer with a high hydrogen concentration, which is relatively stable in time and does not propagate into the metal (so called skin-effect of hydrogen), a diffusion equation obtained within the framework of the linear non-equilibrium thermodynamics approach is proposed. In addition, it was proposed to take into account the leak term obtained using the concept of the damageability tensor.

For the modified diffusion equation with a new leak term, the plane boundary value problem of a semi-infinite crack was solved. It was found that such an approach to the description of hydrogen diffusion can accurately describe experimental results, and taking into account the hydrogen trapping in metal inhomogeneities contributes to the formation of the boundary layer observed experimentally.

Acknowledgements The research was carried out with the financial support of the Russian Science Foundation, grant No. 18-19-00160.

References

[1] Alekhin VP (1983) Physics of Strength and Plasticity of Surface Layers of Materials (in Russ), Nauka, Moscow.
[2] Wassermann G (1935) Quenching Stresses, Mitt. KW Inst. Eisenforschung **17**:167-174.
[3] Pearson WB (1958) A Handbook of Lattice Spacings and Structures of Metals and Alloys: International Series of Monographs on Metal Physics and Physical Metallurgy, Vol 1, Pergamon Press.
[4] Kolachev B (1989) Hydrogen Embrittlement of Metals (in Russ.), Itogi Nauki Tekh, Metalloved Term Obrab **23**:3-46.
[5] Nechaev YS (2008) Metallic materials for the hydrogen energy industry and main gas pipelines: complex physical problems of aging, embrittlement, and failure. Physics-Uspekhi **51(7)**:681.
[6] Hirth JP (1980) Effects of hydrogen on the properties of iron and steel, Metallurgical Transactions A **11(6)**:861-890.
[7] Frolova K, Vilchevskaya E, Polyanskiy V, Alekseeva E (2019) Modelling of a hydrogen saturated layer within the micropolar approach, In: New Achievements in Continuum Mechanics and Thermodynamics, Advanced Structured Materials, Vol 108, pp 117–128, Springer, Cham.
[8] Lopez-Suarez A, Valencia CE, Lopez-Patino J, Vargas MC, Fuentes BE (2015) Improvement of titanium hydrogenation by low energy ion irradiation, Int J Hydrog Energy **40(11)**:4194–4199.
[9] Wu T-I, Wu J-Ch (2008) Effects of cathodic charging and subsequent solution treating parameters on the hydrogen redistribution and surface hardening of Ti–6Al–4V alloy, J Alloys Compd **466**(1–2):153–159.
[10] Martinsson A, Sandstrom R (2012) Hydrogen depth profile in phosphorus-doped, oxygen-free copper after cathodic charging, J Mater Sci **47**(19):6768-6776.

[11] Arseniev DG, Belyaev A, Polyanskiy AM, Polyanskiy VA, Yakovlev YA (2019) Benchmark study of measurements of hydrogen diffusion in metals, In: Dynamical Processes in Generalized Continua and Structures, Advanced Structured Materials, Vol 103, Springer, Cham, pp 37-61.

[12] Polyanskiy VA., Belyaev AK, Alekseeva EL, Polyanskiy AM, Tretyakov DA, Yakovlev YA (2019) Phenomenon of skin effect in metals due to hydrogen absorption, Continuum Mechanics and Thermodynamics, **31**(6):1961-1975.

[13] Andronov DY, Arseniev DG, Polyanskiy AM, Polyanskiy VA, Yakovlev YA (2017) Application of multichannel diffusion model to analysis of hydrogen measurements in solid, Int J Hydrog Energy **42**(1)(2017):699-710.

[14] McNabb A, Foster PK (1963) A new analysis of diffusion of hydrogen in iron and ferritic steels, Trans Metallurg Soc AIME **227**(3);618-627.

[15] Oriani RA (1970) The diffusion and trapping of hydrogen in steel, Acta Metall **18**(1):147-157.

[16] Toribio J, Kharin V, Lorenzo M, Vergara D (2011) Role of drawing-induced residual stresses and strains in the hydrogen embrittlement susceptibility of prestressing steels, Corr Sci **53**(10):3346-3355.

[17] Oudriss A, Creus J, Bouhattate J, Conforto E, Berziou C, Savall C, Feaugas X (2012) Grain size and grain-boundary effects on diffusion and trapping of hydrogen in pure nickel, Acta Mater **60**(19):6814-6828.

[18] Takahashi J, Kawakami K, Kobayashi Y (2018) Origin of hydrogen trapping site in vanadium carbide precipitation strengthening steel, Acta Mater **153**(July):193-204.

[19] Depover T, Van den Eeckhout E, Verbeken K (2018) Hydrogen induced mechanical degradation in tungsten alloyed steels, Mater Charact **136**(February):84-93.

[20] Zhang Z, Moore KL, McMahon G, Morana R, Preuss M (2019) On the role of precipitates in hydrogen trapping and hydrogen embrittlement of a nickel-based superalloy, Corr Sci **146**(January):58–69.

[21] Gorsky WS (1935) Theorie der elastischen Nachwirkung in ungeordneten Mischkristallen (elastische Nachwirkung zweiter Art), Physik Z Sowjetunion **8**:457-471.

[22] Birnbaum HK, Sofronis P. (1994) Hydrogen-enhanced localized plasticity — a mechanism for hydrogen-related fracture, Mat Sc Engng: A. **176**(1-2):191-202.

[23] Stashchuk M, Dorosh M (2017) Analytical evaluation of hydrogen induced stress in metal, Int J Hydrog Energy **42**(9):6394-6400.

[24] Drexler A, Bergmann C, Manke G, Kokotin V, Mraczek K, Pohl M, Ecker W (2021) On the local evaluation of the hydrogen susceptibility of cold-formed and heat treated advanced high strength steel (AHSS) sheets, Mat Sc Engng: A **800**:140276.

[25] Larché FC, Cahn JL (1982) The effect of self-stress on diffusion in solids, Acta Metallurg **30**(10):1835-1845.

[26] Wu CH (2001) The role of Eshelby stress in composition-generated and stress-assisted diffusion, J Mech Phys Solids **49**(8):1771-1794.

[27] Knyazeva AG (2003) Cross effects in solid media with diffusion, J Appl Mech Techn Phys **44**(3):373-384.

[28] Hadam U, Zakroczymski T (2009) Absorption of hydrogen in tensile strained iron and high-carbon steel studied by electrochemical permeation and desorption techniques, Int J Hydrog Energy **34**(5):2449–2459.

[29] Turnbull A (2015) Perspectives on hydrogen uptake, diffusion and trapping, Int J Hydrog Energy **40**(47):16961–16970.

[30] Liu Q, Venezuela J, Zhang M, Zhou Q, Atrens A (2016) Hydrogen trapping in some advanced high strength steels, Corr Sc **111**(October):770–785.

[31] Grigoreva PM (2022) The influence of the trapping model choice on the hydrogen diffusion into metals from the external environment, J Phys Math St Petersburg State Polytechnic University **15**(3).

[32] Grigoreva PM, Vilchevskaya EN, Polyanskiy VA (2021) Influence of linear elastic stresses on hydrogen diffusion into metals, In: Advances in Hydrogen Embrittlement Study (Ed. by VA Polyanskiy and AK Belyaev), pp 143–157, Springer, Cham.

[33] Taha A, Sofronis P (2001) A micromechanics approach to the study of hydrogen transport and embrittlement, Engng Fract Mech **68**(6):803–837.

[34] Sofronis P, Lufrano J (1999) Interaction of local elastoplasticity with hydrogen: embrittlement effects, Mat Sc Engng: A **260**(1–2):41–47.

[35] Nie Y, Kimura Y, Inoue T, Yin F, Akiyama E, Tsuzaki K (2012) Hydrogen embrittlement of a 1500-MPa tensile strength level steel with an ultrafine elongated grain structure, Metallurgical and Materials Trans A **43**(5):1670-1687.

[36] Hu J, Liu J, Lozano-Perez S, Grovenor CR, Christensen M, Wolf W, Mader EV (2019) Hydrogen pickup during oxidation in aqueous environments: The role of nano-pores and nano-pipes in zirconium oxide films, Acta Materialia **180**:105–115.

[37] Pressouyre GM, Bernstein IM (1981) An example of the effect of hydrogen trapping on hydrogen embrittlement, Metallurg Trans A. **12**(5):835–844.

[38] Chen L, Xiong X, Tao X, Su Y, Qiao L (2020) Effect of dislocation cell walls on hydrogen adsorption, hydrogen trapping and hydrogen embrittlement resistance, Corr Sc **166**:108428.

[39] Semenov AS, Polyanskiy VA, Shtukin LV, Tretyakov DA (2018) Effect of surface layer damage on acoustic anisotropy, J Appl Mech Techn Phys **59**(6)1136–1134.

[40] Belyayev AK, Polyanskiy VA, Tretyakov DA (2020) Estimating of mechanical stresses, plastic deformations and damage by means of acoustic anisotropy (in Russ), PNRPU Mechanics Bulletin **4**:130–151.

[41] Rhode M, Mente T, Steppan E, Steger J, Kannengiesser T (2018) Hydrogen trapping in T24 Cr-Mo-V steel weld joints—microstructure effect vs experimental influence on activation energy for diffusion, Welding in the World **62**(2):277–287.

[42] Drexler A, Bergmann C, Manke G, Kokotin V, Mraczek K, Pohl M, Ecker W (2021) On the local evaluation of the hydrogen susceptibility of cold-formed and heat treated advanced high strength steel (AHSS) sheets, Mat Sc Engng: A **800**:140276.

Chapter 21
Classical and Non-Classical Models of Changes in the Young Modulus of Geomaterials Under Alternating Loads

Mikhail A. Guzev, Evgenii P. Riabokon, Mikhail S. Turbakov,
Vladimir V. Poplygin, Evgenii V. Kozhevnikov, and Evgenii A. Gladkikh

Abstract The paper proposes the classical and the non-classical models of changes in the Young's modulus of a geomaterial under the effect of an alternating load. The classical model is based on the Newton's equation and the Jaeger's idealized "mass-on-spring" model. The non-classical model is based on the gradient theory, which takes into account the internal heterogeneous structure (scale) of a geomaterial.

Key words: Mathematical model · Young's modulus · Heterogeneous Structure · Geomaterial

21.1 Introduction

In geotechnical engineering, in order to build reliable underground structures (mines, wells, etc.) and develop the reservoirs, geomaterials (rocks) are the main object of research. Particular attention is paid to the study and description of the behavior of the elastic mechanical characteristics of rocks under various loading conditions. In a quasi-static situation, the behavior of the elastic characteristics of geomaterials is well known, methods and research approaches are well developed. Under conditions of dynamic loading, researchers note nonlinear effects, in particular, the dispersion of the Young modulus, in which the Young modulus increases with increasing

---

Mikhail A. Guzev
Institute for Applied Mathematics of the Far Eastern Branch of the Russian Academy of Sciences, Vladivostok & Perm National Research Polytechnic University, Perm, Russian Federation,
e-mail: guzev@iam.dvo.ru

Evgenii P. Riabokon · Mikhail S. Turbakov · Vladimir V. Poplygin · Evgenii V. Kozhevnikov · Evgenii A. Gladkikh
Perm National Research Polytechnic University, Perm, Russian Federation,
e-mail: riabokon.evgenii@gmail.com, msturbakov@gmail.com, poplygin@bk.ru,
kozhevnikov_evg@mail.ru, gladkih.ea@mail.ru

H. Altenbach et al. (eds.), *Sixty Shades of Generalized Continua*,
Advanced Structured Materials 170, https://doi.org/10.1007/978-3-031-26186-2_21

frequency of the applied load [1, 2]. For the moment, the issue of describing the dynamic mechanical characteristics is still of particular interest. For example, using the ultrasonic pulse method in [3], the experimental dependence of the dynamic Young modulus on the static one was obtained. In [4], the authors obtained the dispersion of the Young modulus with an increase in the frequency of elastic vibrations. Using the Resonance Frequency Meter test apparatus, the correlation of the dynamic Young modulus of elasticity from the uniaxial compressive strength and tensile strength was obtained in [5].

An analysis of the existing theoretical and experimental works on the study of the dynamics of elastic characteristics indicates the widespread occurrence of such a phenomenon as the dispersion of the Young modulus. Some researchers suggest that the reason for the growth of the Young modulus are inertial effects in the rock sample (for example, [6]). Other researchers believe that due to the high heterogeneity of the geomaterial under the action of dynamic loads, the apparent increase in the Young modulus of the rock is associated with the closing, opening, displacement of inhomogeneities (pores, cracks, cavities) and dislocations in its structure [1].

The above grounds for the proposed approaches are objective, but the authors do not offer mathematical models that would describe the phenomenon of dispersion of the Young modulus of rocks under alternating loading. In this regard, the purpose of this work is to propose the classical and the non-classical models of the dispersion of the Young modulus, which are based on mechanical equations and physically reasonable relationships. In the classical model, the approach [7] is used, according to which the strain of the sample is set by the displacement of the particle in the field of the elastic force. This corresponds to the classical model of an elastic continuum. In the non-classical model, it is assumed that the geomaterial has an internal structure, which is the cause of the dispersion of the dynamic Young modulus. From the point of view of continuum mechanics, this means that the rheological ratio of the material changes. Non-classical continuum models [8]-[10] are actively used to model the behavior of materials with different structures. Among non-classical models, gradient models should be singled out [1]-[15]. The gradient model was first proposed in [16]. Its modern application allowed to obtain solutions that refine the description of macroscopic bodies in which the stresses and strains of the classical theory of elasticity have singularities. For example, variants of gradient models developed in [17, 18] allowed to construct the nondegenerate solutions for describing materials with a structure. Gradient models are also used when considering nonclassical effects of rock behavior [19, 20]. It was shown in [21] that, in the linear approximation, the non-Euclidean and gradient models lead to the same result when describing the phenomenon of zonal disintegration in a material.

This paper is organized as follows. Section 21.2 gives a brief description and results of the experimental study and shows the dispersion of the Young modulus depending on the frequency and amplitude of the alternating load. Section 21.3 formulates the classical model and selects its parameters for one of the samples. Section 21.4 is devoted to the non-classical model, followed by a conclusion.

21.2 Experimental Studies

To study the dependence of Young's modulus on a frequency and an amplitude of loading of rock samples, laboratory studies were performed on an experimental setup. When conducting experiments on a rock sample with a length l = 0.0156 m and a diameter d = 0.0075 m a static preload was applied F_{st} = 700 N and a dynamic load F_{dyn}, whose amplitude A varied from 0 to 250 N, and the loading frequency ω was chosen from 15 to 40 Hz (Fig. 21.1b). The displacement of the sample surfaces was recorded by sensors such as optical for horizontal displacement and eddy current probe for vertical displacement (Fig. 21.1a). A detailed description of the experiment is presented in [22].

The minimum l_{min} and maximum l_{max} sample lengths include static (determined by preload) and dynamic (determined by alternating load) components. Sample lengths l_{min} and l_{max} at vertical displacement of its upper end surface u=$u(t)$ correspond to the peak values of the dynamic load F_{dyn} spaced apart in time equal to half the oscillation period $T/2$, where T=2 π/ω (Fig. 21.2). The value l_0 corresponds to the length of the sample in the preloaded state.

The behavior of the Young modulus under the action of a variable load F_{dyn} was analyzed under the condition that the compressive stress σ and the relative

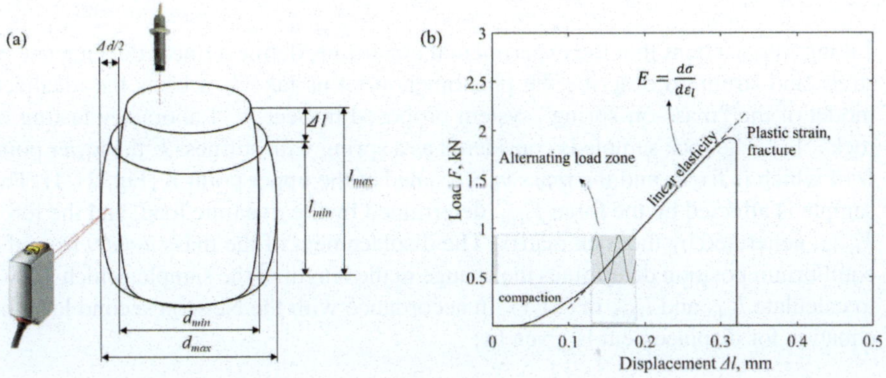

Fig. 21.1: Experimental conditions: (a) sample deformation under alternating load (schematically); (b) stress state of the sample at the moment of loading.

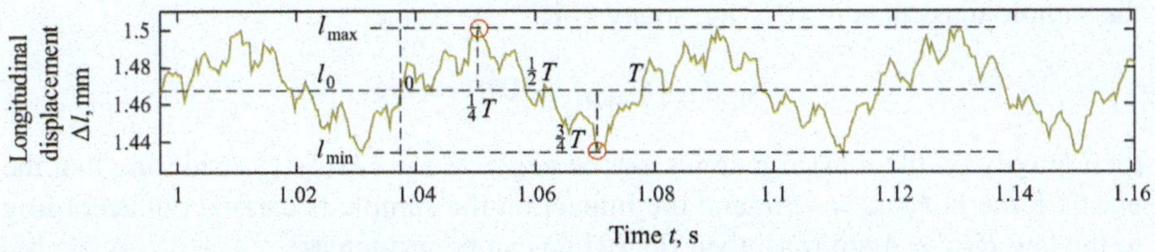

Fig. 21.2: Longitudinal elongation and contraction of the sample under the alternating load at F_{st}= 700 N, A = 250 N, ω = 25 Hz.

longitudinal deformation of the sample ε are connected by a linear relationship, which corresponds to the highlighted area in Fig. 21.1b. Then the dynamic component of the Young modulus E_{dyn} can be introduced by analogy with the static Young modulus and is calculated based on the relation

$$E_{dyn} = \frac{\sigma}{|\varepsilon|}, \quad \sigma = \frac{2A}{S}, \quad |\varepsilon| = \frac{\Delta l}{l}, \quad |\Delta l| = |u_{max} - u_{min}|, \tag{21.1}$$

where S is the cross-sectional area of the sample.

Under the alternating loading of nine rock samples at 30 combinations of amplitudes and frequencies, the nonlinear nature of the Young modulus was obtained. A preliminary analysis of the averaged experimental data processed on the basis of relation (21.1) indicates the dependence of E_{dyn} on the amplitude A (Fig. 21.3a) and the frequency ω (Fig. 21.3b) of the dynamic load, i.e. $E_{dyn} = E_{dyn}(\omega, A)$. For a mathematical description of the behavior of the dynamic Young modulus it is required to develop physically justified relationships based on the laws of motion and state. Sections 21.3 and 21.4 comprise constructed classical and nonclassical models.

21.3 Classical Model of Changes in the Young Modulus Under an Alternating Load

Taking into account that the experimental conditions define a linear dependence of stress and strain, to describe the phenomenon let us take as a basis the idealized model of the "mass-on-spring" system proposed in Sect. 6 "Laboratory testing of rocks" [7]. The rock sample is considered as a spring with stiffness k, the lower point O of which is fixed, and the mass m is located at the upper point A (Fig. 21.4). The sample is affected by the force F_{dyn}, determined by the dynamic load, and the force F_{resist}, generated by the rock matrix. The displacement of the mass u=$u(t)$ from the equilibrium position determines the change in the length of the sample, which allows to calculate l_{min} and l_{max} in (21.1). In accordance with the Newton second law, the equation for displacement is given by:

$$m\ddot{u} = F_{dyn} + F_{resist} - mg, \tag{21.2}$$

in which the last term takes into account the action of gravity. In order of magnitude, the sample mass is $m \sim 10^{-3}$ kg, so $mg \sim 10^{-2}$ N. Since

$$|F_{dyn}| \sim |F_{resist}| \sim 10^2\text{N} \gg mg,$$

then gravity on the mass dynamics can be neglected in (21.2). Considering that the elastic force is $F_{resist} = -ku$, and the impact on the sample is carried out according to the law $F_{dyn} = A \sin(\omega t)$, then Eq. (21.2) can be written as

$$m\ddot{u} = A\sin(\omega t) - ku. \tag{21.3}$$

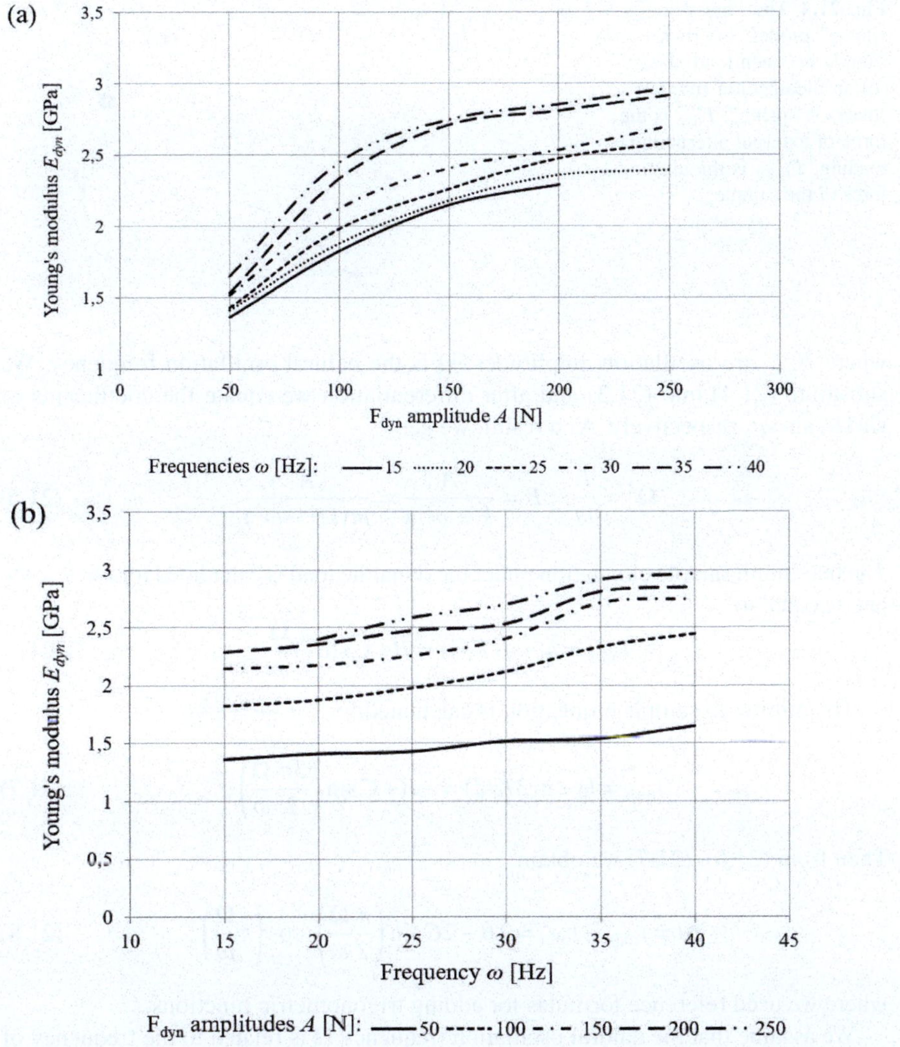

Fig. 21.3: Experimental data on the change in the dynamic Young modulus when alternating: (a) the frequency of the alternating load for different amplitudes; (b) the amplitude of the alternating load for different frequencies.

Thus, the description of the movement of the upper end of a rock sample is mathematically reduced to the study of the dynamics of the mass m on a spring.

It is well known [23] that the general solution of equation of motion (21.3) consists of the general solution of the homogeneous equation (free oscillation) plus the particular solution of the inhomogeneous equation:

$$u(t) = C \sin \Omega t + B \sin \omega t, \tag{21.4}$$

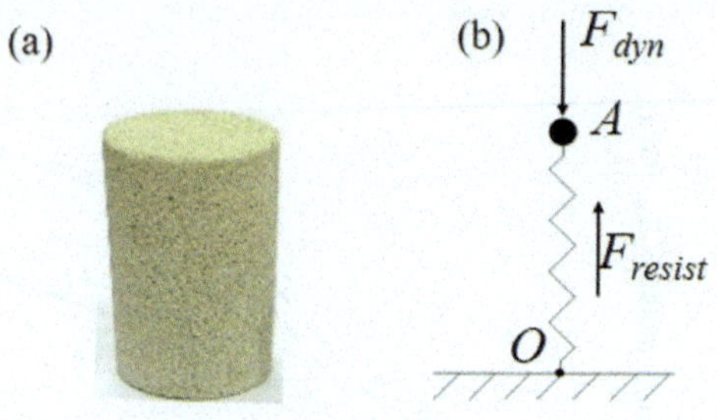

Fig. 21.4 The "mass-on-spring" model: (a) rock sample between load plates; (b) an idealization (model) "mass-on-spring"; F_{dyn} is the force of external alternating loading; F_{resist} is the elastic force of the sample

where B, C are oscillation amplitudes; Ω is the natural oscillation frequency. We substitute (21.4) into (21.3) and after differentiation we equate the coefficients of $\sin\Omega t$, $\sin\omega t$, respectively. As a result, we get

$$\Omega^2 = \frac{k}{m}, \quad B = \frac{A}{k-\omega^2 m} = \frac{A}{m(\Omega^2-\omega^2)}. \tag{21.5}$$

The maximum sample elongation under the dynamic load is calculated for $t = \frac{T}{4} = \frac{\pi}{2\omega}$ and is equal to

$$l_{\max} = l_0 + u(T/4) = B + C\sin\frac{\pi}{2}\frac{\Omega}{\omega}. \tag{21.6}$$

The minimum sample elongation is calculated for $t = \frac{3T}{4} = \frac{3\pi}{2\omega}$:

$$l_{\min} = l_0 + u(3T/2) = -B + C\sin\left(\frac{3\pi}{2}\frac{\Omega}{\omega}\right). \tag{21.7}$$

Then from (21.6), (21.7) we obtain

$$\Delta l = l_{\max} - l_{\min} = 2B - 2C\sin\left(\frac{\pi}{2}\frac{\Omega}{\omega}\right)\cos\left(\pi\frac{\Omega}{\omega}\right), \tag{21.8}$$

where we used reference formulas for adding trigonometric functions.

We assume that the natural oscillation frequency Ω is related to the frequency of the perturbing oscillations ω by the relation $\Omega = \alpha\omega$, where the coefficient $\alpha > 0$. Since the coefficient B is determined from (21.5), then (21.8) can be represented as

$$|\Delta l| = \left|\frac{2A}{m[\alpha^2-1]\omega^2} - 2C\sin\frac{\pi\alpha}{2}\cos\pi\alpha\right|. \tag{21.9}$$

From here and from (21.1) follows the formula for the dynamic component of Young modulus:

$$E_{\text{dyn}} = \frac{A}{\frac{S}{l}\left|\frac{A}{m[\alpha^2-1]\omega^2} - C\sin\frac{\pi\alpha}{2}\cos\pi\alpha\right|}. \tag{21.10}$$

Knowing the characteristics of the rock image (length l, diameter d, mass m), as well as choosing the parameters C and α using formula (21.10), one can calculate the dynamic Young modulus E_{dyn}.

Let us choose model parameters (21.10) for one of nine samples. To do this, we substitute the characteristics of the sample (length l, diameter d, mass m) into model (21.10). An analysis of the experimental data gives the value of the parameter C equal to 5×10^{-4} m, and the value of the parameter α varies from 1420 to 4300. The result of substituting the characteristics and choosing the parameters of model (21.10) is an approximation of the Young modulus depending on a frequency and an amplitude of the load (Fig. 21.5). The selection model parameters C and α for experimental data averaged over 9 samples can be found in [24].

21.4 Non-classical Model of Change in the Young Modulus Under the Alternating Load

21.4.1 Formulation of the Non-classical Model

As can be seen from Eq. (21.1), in a one-dimensional situation, the classical Hooke law determines the relationship between stress σ and strain ε:

$$\sigma = E_{st}\varepsilon, \quad \varepsilon = \frac{\partial U}{\partial x}, \tag{21.11}$$

where E_{st} is the Young modulus; U is the displacement of sample particles. Application of the variational principle [25] allows to modify the classical model. The idea is that the bulk energy density of the material is assumed to be a function of the strain ε and the strain gradient ε_x. Then the use of the variational principle leads to a new equation of state of the material:

$$\sigma = E_{st}\varepsilon + \gamma\varepsilon_{xx}. \tag{21.12}$$

Comparison of (21.12) with (21.11) shows that the stress contains an additional contribution associated with the inhomogeneous distribution of deformation in a geomaterial. The phenomenological parameter γ is the reaction of the rock to an alternating load, takes into account the internal heterogeneous structure of a geomaterial and is determined experimentally. In the above relations for determining the Young modulus, the loading conditions are quasi-static. Since in this paper we study the influence of the frequency of the applied load, we will call the corresponding Young modulus dynamic and denote it E_{dyn}. According to the conditions of the experiment, the dynamic component of the Young modulus at the end of the sample is defined as

$$E_{dyn} = \frac{|\Delta\sigma|}{|\varepsilon|}, \tag{21.13}$$

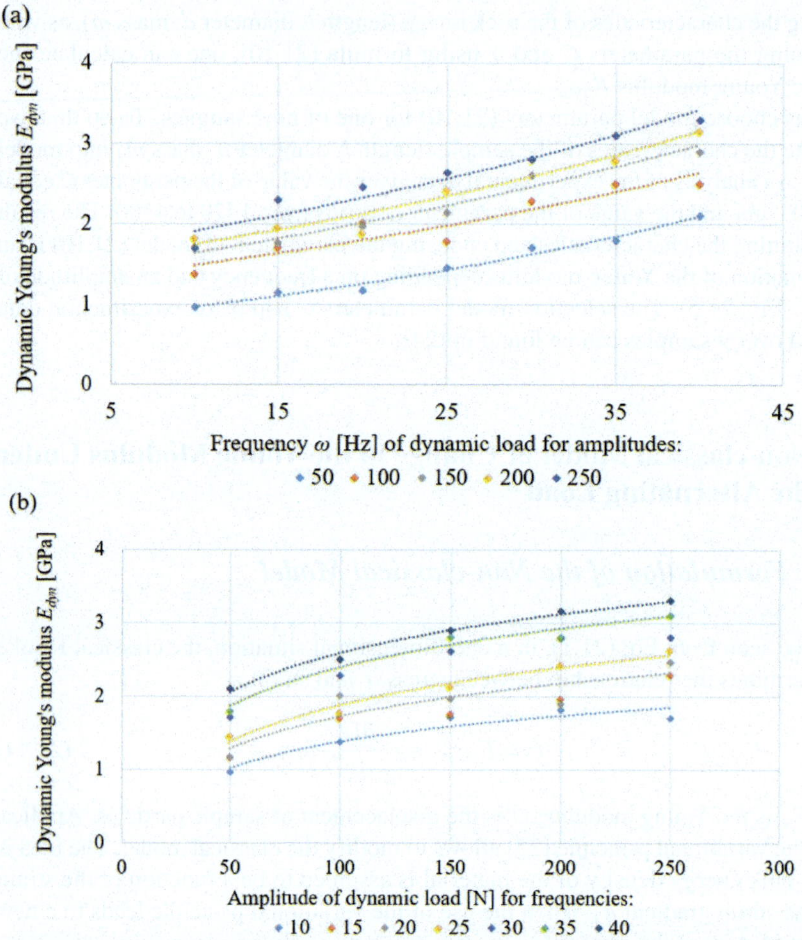

Fig. 21.5: Approximation of experimental data using model (21.10) for one of nine samples: (a) dependence of E_{dyn} on frequency at different load amplitudes; (b) dependence of E_{dyn} on the load amplitude at different load frequencies.

$$E_{\text{dyn}} = \frac{|E_{\text{st}}\varepsilon + \gamma\varepsilon_{xx}|}{|\varepsilon|} = |E_{\text{st}} + \frac{\gamma}{\varepsilon}\varepsilon_{xx}|.$$

Given that

$$E_{\text{st}} > |\frac{\gamma\varepsilon_{xx}}{\varepsilon}|,$$

then at the end of the sample the introduced additive will be positive and, taking into account the inhomogeneity of the distribution of grains

$$E_{\text{dyn}} = E_{\text{st}} + \frac{\gamma\varepsilon_{xx}}{\varepsilon}\bigg|_{x=l_0}. \tag{21.14}$$

Let us write the momentum conservation law for a continuous medium

$$\rho_0 \frac{\partial^2 U}{\partial t^2} = \frac{\partial \sigma}{\partial x}. \tag{21.15}$$

Substituting (21.12) into (21.15), we get:

$$\rho_0 U_{tt} = E_{\text{st}} \varepsilon_x + \gamma \varepsilon_{xxx}. \tag{21.16}$$

Differentiating relation (21.16) with respect to x gives the equation:

$$\rho_0 \varepsilon_{tt} = E_{\text{st}} \varepsilon_{xx} + \gamma \varepsilon_{xxxx}. \tag{21.17}$$

In the framework of the experiment, the boundary conditions for equation (21.17) have the form

$$\sigma|_{x=l_0} = E_{\text{st}} \varepsilon + \gamma \epsilon_{xx}|_{x=l_0} = \frac{A}{S} \sin \omega t, \quad \sigma|_{x=0} = 0. \tag{21.18}$$

21.4.2 Building the Solution

Now let us construct a solution to Eq. (21.17) using the Fourier method. We represent ε as:

$$\varepsilon = \alpha \sin \omega t \sin \lambda x + \sum_{k=1}^{\infty} \varepsilon_k \sin \Omega_k t \sin \lambda_k x, \tag{21.19}$$

where λ, Ω_k and λ_k are unknown. Substitution (21.19) into (21.17) leads to the relation between the parameters:

$$\omega^2 = \lambda^2 \frac{E_{\text{st}}}{\rho_0} - \gamma \frac{\lambda^4}{\rho_0}; \quad \Omega_k^2 = \lambda_k^2 \frac{E_{\text{st}}}{\rho_0} - \gamma \frac{\lambda_k^4}{\rho_0}. \tag{21.20}$$

Since $E_{\text{st}}/\rho_0 = \upsilon_{\text{st}}^2$, where υ_{st} is the velocity of the elastic wave in the material, then (21.20) will be written as

$$\gamma \frac{\lambda^4}{\rho_0} - \lambda^2 \upsilon_{\text{st}}^2 + \omega^2 = 0; \quad \Omega_k^2 = \lambda_k^2 \upsilon_{\text{st}}^2 - \gamma \frac{\lambda_k^4}{\rho_0}. \tag{21.21}$$

The first equation allows to find λ^2:

$$\lambda^2 = \frac{\upsilon_{\text{st}}^2 - \sqrt{\upsilon_{\text{st}}^4 - \frac{4\gamma\omega^2}{\rho_0}}}{\frac{2\gamma}{\rho_0}}. \tag{21.22}$$

If

$$v_{\text{st}}^4 \gg \frac{4\gamma\omega^2}{\rho_0},$$

then from (21.22) we get

$$\lambda^2 \cong \frac{\omega^2}{v_{\text{st}}^2}. \tag{21.23}$$

This shows that λ has the meaning of the wavelength propagating in the material under external action.

The second equation (21.21) for a given λ_k determines Ω_k. For the classical model $\gamma \to 0$, then

$$\Omega_k^2 \cong \lambda_k^2 v_{\text{st}}^2.$$

Thus, we get a set of frequencies Ω_k corresponding to the wavelengths of λ_k. To determine λ_k, we use boundary conditions (21.18). The second of them is satisfied automatically for the solution (21.19). Substituting (21.19) into first relation (21.18), we obtain

$$\sigma|_{l_0} = (E_{\text{st}} - \lambda^2\gamma)\alpha \sin\omega t \sin\lambda l_0 + \sum_{k=1}^{\infty} \varepsilon_k (E_{\text{st}} - \lambda_k^2\gamma) \sin\Omega_k \sin\lambda_k l_0 = \frac{A}{S} \sin\omega t.$$

The fulfillment of this condition leads to restrictions

$$\alpha(E_{\text{st}} - \lambda^2\gamma) \sin\lambda l_0 = \frac{A}{S}, \quad \sin\lambda_k l_0 = 0. \tag{21.24}$$

The first equation allows to find the amplitude λ. From the second relation we determine λ_k:

$$\lambda_k l_0 = \pi k,$$

where k is a natural number.

Substituting (21.19) into (21.14) gives

$$E_{\text{dyn}} = E_{\text{st}} - \gamma\lambda^2. \tag{21.25}$$

When condition (21.23) is satisfied, we obtain a quadratic dependence of the Young modulus on frequency:

$$E_{\text{dyn}} = E_{\text{st}} - \gamma\frac{\omega^2}{v_{\text{st}}^2}.$$

Note that over the dimension

$$[\omega] = 1/[t], \quad [v_{\text{st}}] = [l]/[t],$$

then the dimension is

$$[E_{\text{st}}] = [\gamma]/[l]^2.$$

Thus, the value of the internal scale of the rock and the unit of measurement are determined from the experiment. Considering that λ_k^2 is determined from relation (21.22), we obtain a more accurate approximation of E_{dyn}:

$$E_{\text{dyn}} = E_{\text{st}} + \rho_0 \frac{-\upsilon_{st}^2 + \sqrt{\upsilon_{st}^4 - (4\gamma\omega^2)/\rho_0}}{2}. \tag{21.26}$$

Let us choose the phenomenological parameter γ of model (21.26) for one of the nine samples. To do this, let us substitute the value of the Young modulus of the geomaterial E_{st} into the model under quasi-static loading of the sample up to a level of 700 N. Let us take the sound wave propagation velocity in the sample $\upsilon_{\text{st}} = 2500$ m/s. An analysis of the experimental data gives the value of the γ parameter varying in the range from 2×10^{12} N to 4.6×10^{12} N.

The result of substituting the characteristics and choosing the model parameter (21.26) for the same sample is an approximation of the Young modulus depending on the frequency and amplitude of the load. Approximation by models (21.26) and (21.10) gives the same result (Fig. 21.6).

21.5 Conclusion

In this paper, two models of changes in the Young modulus of the geomaterial under the alternating loads are constructed: the classical model and the non-classical model. The adjustment and selection of the phenomenological parameters of the models was carried out. Both models reflect the nonlinear nature of the Young modulus under the action of an alternating load on the sample. According to the results of the work, the following can be noted:

1. The classical model includes such sample parameters as mass, length, diameter. The model also contains the C parameter, which takes into account the experimental conditions, as well as the α parameter, which reflects the ratio of the natural oscillation frequency of the geomaterial at the moment of loading to the frequency of the applied load. It can be said that the classical model corresponds to the continuum model.
2. The non-classical model is based on the gradient theory. The model contains the phenomenological parameter γ, which takes into account the internal heterogeneous structure (scale) of the geomaterial. In addition to the additive, the model takes into account the velocity of sound wave propagation in the geomaterial, the density of the geomaterial, and the frequency of the external applied load.
3. From the point of view of ease of use, due to fewer parameters and assuming a constant density and speed of sound wave propagation in the rock under loading in different modes (different combinations of frequencies and load amplitudes), the non-classical model seems to be more utilitarian. At the same time, these assumptions do not have a significant effect on the result of the approximation of experimental data. The classical model can also be used for approximation, but taking into account the substitution of the values for more parameters.

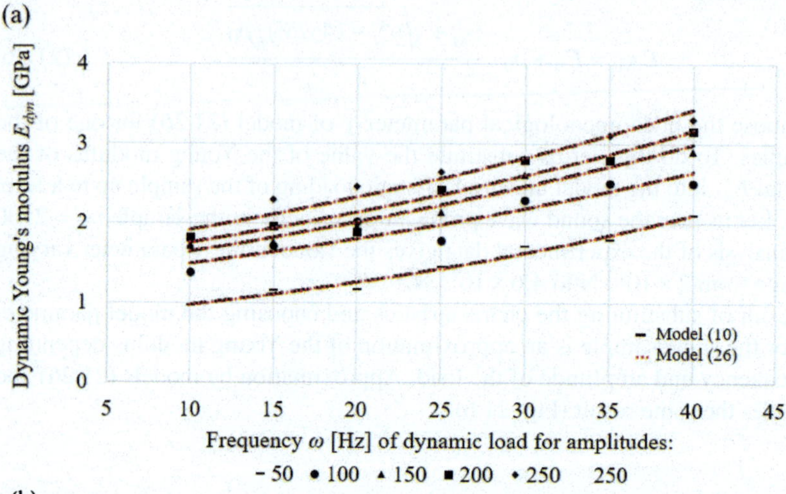

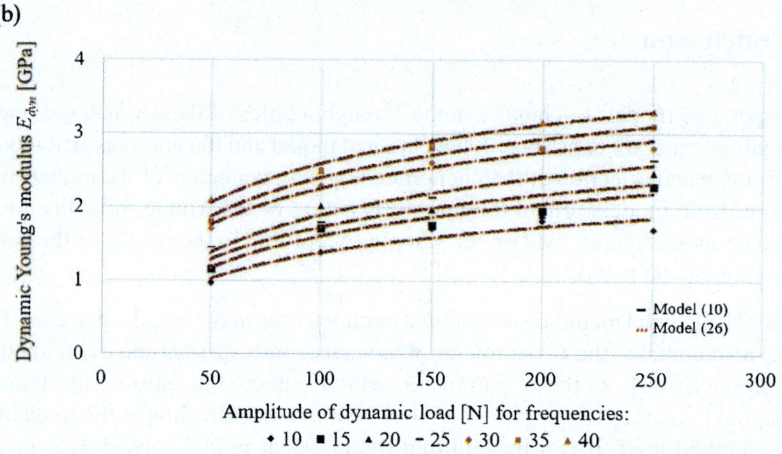

Fig. 21.6: Approximation of experimental data using models (21.26) and (21.10) for one of nine samples: (a) dependence of E_{dyn} on frequency at different load amplitudes; (b) dependence of E_{dyn} on the load amplitude at different load frequencies.

4. Based on different approaches, the obtained models give the same approximation of the experimental data and qualitatively reflect the dispersion of the Young modulus under the alternating loading of the geomaterial.

Acknowledgements The research was supported by a grant from the Russian Science Foundation (Project No. 22-19-00447, https://rscf.ru/project/22-19-00447/).

References

[1] Tutuncu AN, Podio AL, Gregory AR, Sharma MM (1998) Nonlinear viscoelastic behavior of sedimentary rocks, Part I: Effect of frequency and strain amplitude, Geophysics **63**:184-194. DOI 10.1190/1.1444311

[2] Bauer A, Bhuiyan MH, Fjaer E, Holt RM, Lozovyi S, Pohl M, Szewczyk D. (2016) Frequency-dependent wave velocities in sediments and sedimentary rocks: Laboratory measurements and evidences, The Leading Edge **35**:474-560. DOI 10.1190/tle35060490.1

[3] Kotsanis D, Nomikos P, Rozos D (2021) Comparison of static and dynamic Young's modulus of prasinites, Mater. Proc. **2021**(5):54. DOI 10.3390/materproc2021005054

[4] Borgomano JVM, Gallagher A, Sun C, Fortin J (2020) An apparatus to measure elastic dispersion and attenuation using hydrostatic- and axial-stress oscillations under undrained conditions, Review of Scientific Instruments **91**(3):034502. DOI 10.1063/1.5136329

[5] Waqas U, Ahmed MF (2020) Prediction modeling for the estimation of dynamic elastic Young's modulus of thermally treated sedimentary rocks using linear–nonlinear regression analysis, regularization, and ANFIS, Rock Mech. Rock Eng. **53**:5411-5428. DOI 10.1007/s00603-020-02219-8

[6] Zhang QB, Zhao J (2014) A review of dynamic experimental techniques and mechanical behaviour of rock materials. Rock Mech. Rock Eng. **47**:1411-1478. DOI 10.1007/s00603-013-0463-y

[7] Jaeger JC, Cook NGW, Zimmerman RW (2007) Laboratory testing of rocks. In: Fundamentals of Rock Mechanics, 4th ed., Blackwell Publishing, Malden, MA, USA, pp 145-167.

[8] Forest S, Sievert R (2003) Elastoviscoplastic constitutive frameworks for generalized continua, Acta Mechanica **160**(1-2):71-111, DOI 10.1007/s00707-002-0975-0

[9] Smolin IY (2005) On the application of the Cosserat model to the description of plastic deformation at the mesoscale, Fizicheskaya Mezomekhanika (3):49-62.

[10] Bazant ZP, Jirasek M (2002) Nonlocal integral formulations of plasticity and damage: Survey of progress, Journal of Engineering Mechanics **128**(11):1119-1149. DOI 10.1061/(ASCE)0733-9399(2002)128:11(1119)

[11] Lurie SA, Kalamkarov AL, Solyaev YO, Volkov AV (2021) Dilatation gradient elasticity theory, European Journal of Mechanics, A/Solids **88**:104258. DOI 10.1016/j.euromechsol.2021.104258

[12] Placidi L, Misra A, Barchiesi E (2019) Simulation results for damage with evolving microstructure and growing strain gradient moduli, Continuum Mechanics and Thermodynamics **31**(4):1143-1163. DOI 10.1007/s00161-018-0693-z

[13] Zhang G, Zheng C, Mi C, Gao X-L (2021) A microstructure-dependent Kirchhoff plate model based on a reformulated strain gradient elasticity theory, Mechanics of Advanced Materials and Structures **29**(17):2521-2530. DOI 10.1080/15376494.2020.1870054

[14] Nazarenko L, Glüge R, Altenbach H (2021) Positive definiteness in coupled strain gradient elasticity, Continuum Mechanics and Thermodynamics **33**(3):713-725. DOI 10.1007/s00161-020-00949-2
[15] Sidhardh S, Ray MC (2018) Element-free Galerkin model of nano-beams considering strain gradient elasticity, Acta Mechanica **229**(7):2765-2786. DOI 10.1007/s00707-018-2139-x
[16] Mindlin RD (1964) Micro-structure in linear elasticity, Archive for Rational Mechanics and Analysis **16**(1):51-78. DOI 10.1007/BF00248490
[17] Aifantis EC (2012) A note on gradient elasticity and nonsingular crack fields, Journal of the Mechanical Behavior of Materials **20**(4-5):103-105. DOI 10.1515/jmbm-2012-0002
[18] Parisis K, Konstantopoulos I, Aifantis EC (2018) Nonsingular solutions of GradEla models for dislocations: An extension to fractional GradEla, Journal of Micromechanics and Molecular Physics **3**(3-4):1840013. DOI 10.1142/S2424913018400131
[19] Guzev MA Non-classical solutions of a continuum model for rock descriptions, Journal of Rock Mechanics and Geotechnical Engineering **6**(3):180-185. DOI 10.1016/j.jrmge.2014.03.001
[20] Qi CZ, Li KR, Bai JP, Chanyshev AI, Liu P. (2017) Strain gradient model of zonal disintegration of rock mass near deep-level tunnels, Journal of Mining Science **53**:21-33. DOI 10.1134/S1062739117011808
[21] Lavrikov SV, Revuzhenko AF (2019) Mathematical modeling of deformation of self-stress rock mass surrounding a tunnel, Im: Wu W (ed) Desiderata Geotechnica, Springer Series in Geomechanics and Geoengineering pp 79-85. DOI 10.1007/978-3-030-14987-1_9
[22] Guzev M, Kozhevnikov, E, Turbakov M, Riabokon E, Poplygin V (2020) Experimental studies of the influence of dynamic loading on the elastic properties of sandstone, Energies **2020**(13):6195. DOI 10.3390/en13236195
[23] Goldstein H, Poole Jr C, Safko J (2002) Classical Mechanics, 3rd ed, Addison Wesley, San Francisco.
[24] Guzev M, Riabokon E, Turbakov M, Kozhevnikov E, Poplygin V (2020) Modelling of the dynamic Young's modulus of a sedimentary rock subjected to nonstationary loading, Energies **2020**(13):6461. DOI 10.3390/en13236461
[25] Lomakin EV, Lurie SA, Rabinskiy LN, Solyaev YO (2019) On the refined stress analysis in the applied elasticity problems accounting of gradient effects, Doklady Physics **489**(6):585-591. DOI 10.1134/S1028335819120103

Chapter 22
Two Approaches to Modeling Viscoelastic Cosserat Continua

Elena A. Ivanova

Abstract We focus our attention on the differential equations relating the gradients of translational and angular velocities to the corresponding strain tensors. We show that these equations can be transformed to the integral form, and they have the structure of the balance equations. We consider the possibility to add the source terms in the balance equations for strains. We show that a special choice of these source terms allows us to obtain a model of viscoelastic continuum even if the constitutive equations for the stress tensor and the moment stress tensor are the same as in the case of the elastic continuum. We compare our approach to modeling viscoelastic materials with the approach developed by Zhilin. We show that these two approaches do not contradict each other, but rather complement each other.

Key words: Viscoelasticity · Constitutive equations · Strain balance equations

22.1 Introduction

Theories of the viscoelastic Cosserat continuum are developed since the middle of the 20th century by many authors, see, e.g., [1]-[14].In the present study, we focus on three questions. The first one concerns the description of kinematics of the continuum. The second one concerns the definition of strain tensors and different forms of the differential equations relating the strain tensors to the velocity vector and the angular velocity vector. The third question concerns the derivation of the constitutive equations. Certainly, all three questions are interrelated.

Elena A. Ivanova
Peter the Great St.Petersburg Polytechnic University, Dept. of Theoretical Mechanics and Mathematical Physics, Polytechnicheskaya, 29, 195251, Saint-Petersburg & Institute for Problems in Mechanical Engineering of Russian Academy of Sciences, Bolshoy pr. V.O., 61, 199178, Saint-Petersburg, Russian Federation,
e-mail: elenaivanova239@gmail.com

H. Altenbach et al. (eds.), *Sixty Shades of Generalized Continua*,
Advanced Structured Materials 170, https://doi.org/10.1007/978-3-031-26186-2_22

It is well known that, in continuum mechanics, there are two methods of description of continuum motion: the material description [15]-[18] and the spatial description [19]-[21]. For a long time, the spatial description was applied only in fluid and gas dynamics [19]-[23], where the elastic part of the stress tensor is completely determined by pressure. The constitutive equations for pressure are written through the density of mass, and therefore there is no need to introduce the strain tensors in fluid and gas dynamics. When modeling dynamic deformation of solids, the material description is usually used. However, in the case of inelastic behaviour of solids, it can be problematic to correctly use the material description. That is why, in solid dynamics, along with the material description, the spatial description is also used, see, e.g., [24]-[33]. In a number of commercial software packages, such as Abaqus and LS Dyna, modeling of dynamic processes in solids is implemented not only using the finite element method, but also using the finite volume method, which is based on the spatial description. In recent times, the spatial description has been further developed and now it is applied not only in modeling of classical solids, but also in modeling of the Cosserat continua [34]-[39] and the Cosserat continua with microstructure [40]-[46].

In the material description, we can introduce several strain tensors. All of them have the clear geometrical meaning since they characterize the change in volume and shape of an elementary material volume fixed in the reference configuration. From the mathematical point of view, strain tensors used in the spatial description are determined by the same expressions as the corresponding strain tensors used in the material description. However, there are two important differences. Firstly, in the spatial description, we consider only current configuration, and hence, we introduce only one position vector and only one gradient operator. That is why, not all strain tensors used in the material description have their counterparts among the strain tensors used in the spatial description. Secondly, in the spatial description, the displacement vector cannot be defined by the simple algebraic relation since the reference position vector is not introduced at all. In the spatial description, the displacement vector is defined as a solution of the partial differential equation relating it to the velocity vector [37]. The displacement vector defined in such a way is a formal characteristic, which does not correspond to the real displacement of some particle of matter in most of cases. In the spatial description, the rotation tensor is also a formal characteristic and it does not correspond to the real rotation of some particle of matter. As a result, the strain tensors defined through the gradient of the displacement vector and the gradient of the rotation tensor lose the clear geometric meaning that they have in the material description. That is why, in the spatial description, the definitions of the strain tensors by the equations that relate them to the velocity gradient and the angular velocity gradient are of interest. These equations are known [34]-[39], and they have the similar structure in the spatial description and in the material description. However, in the material description the use of these equations is not expedient, whereas in the spatial description these equations become very useful.

We consider two approaches to modeling viscoelastic Cosserat continua: Zhilin's approach and our approach. Zhilin's approach is based on transformation of the energy balance equation to a special form called the reduced energy balance equation.

This form of the energy balance is obtained by separation of the stress tensor and the moment stress tensor into elastic and dissipative components and introduction of quantities characterizing the physical processes associated with neglected degrees of freedom, e.g., temperature and entropy. Zhilin's approach allows one to obtain the model of viscoelastic Cosserat continuum similar to the Kelvin–Voigt model. Detailed descriptions of Zhilin's approach and its modifications can be found in [34]-[49]. A summary of the main ideas of this approach is given below. Our approach to modeling viscoelastic continua is based on the idea to add source terms to the equations for the strain tensors. Assuming that the internal energy of the continuum depends on the strain tensors and does not depend on their time derivatives, we use the standard method of continuum mechanics to derive the constitutive equations. In doing so, due to the special choice of the source terms in the equations for the strain tensors, we arrive at the model of viscoelastic Cosserat continuum, which is similar to Maxwell's model. We emphasize that we construct our model of viscoelastic continuum without any assumptions concerning the dependence of constitutive equations on the time derivatives of the strain tensors. The presented research develops ideas of our previous studies [31]-[50].

The paper is organized as follows. We starts with basic equation of the Cosserat continuum. After that, we discuss Zhilin's approach to modeling viscoelastic materials. Next, we study the differential equations relating the strain tensors to the velocity gradient and the angular velocity gradient, aiming to transform these equations to the integral form, to be exact, to the form of integral balance equations for the strain tensors. We discuss the idea to add the source terms to the strain balance equations and to use the obtained equations for modeling viscoelastic materials. Finally, we compare our approach with Zhilin's approach.

22.2 Kinematics and Balance Equations

In the spatial description, all equations are formulated for a part of matter that occupies a fixed control volume at a given time point. Since the particles of matter move, at each time point, some particles enter the control volume and some other particles leave the control volume. Thus, different particles of matter occupy the considered control volume at different time points. The position vector $\boldsymbol{r}$, determining the control volume position in the reference frame, does not depend on time and has nothing to do with the motion of matter. Since $d\boldsymbol{r}/dt = 0$ and $\nabla\boldsymbol{r} = \boldsymbol{E}$, where $\boldsymbol{E}$ is the unit tensor, the relation between the velocity vector $\boldsymbol{v}(\boldsymbol{r},t)$ and the material time derivative of the position vector,

$$\frac{\delta \boldsymbol{r}}{\delta t} = \frac{d\boldsymbol{r}}{dt} + \boldsymbol{v}\cdot\nabla\boldsymbol{r} = \boldsymbol{v}, \tag{22.1}$$

is an identity, and therefore, it cannot be used as a definition of the velocity vector, see [51]. Now, we formulate the kinematic relations for the Cosserat continuum in the spatial description. They are

$$\frac{\delta \boldsymbol{u}}{\delta t} = \boldsymbol{v}, \qquad \frac{\delta \boldsymbol{P}}{\delta t} = \boldsymbol{\omega} \times \boldsymbol{P}, \tag{22.2}$$

where $\boldsymbol{u}(\boldsymbol{r},t)$ is the displacement vector, $\boldsymbol{P}(\boldsymbol{r},t)$ is the rotation tensor, $\boldsymbol{\omega}(\boldsymbol{r},t)$ is the angular velocity vector.

It is important to note that the first equation in (22.2) is the definition of the displacement vector $\boldsymbol{u}$ rather than the velocity vector $\boldsymbol{v}$. This means that the displacement vector $\boldsymbol{u}$ is defined as a solution of the differential equation, see [37]. The second equation in (22.2) was firstly introduced in [34, 35], see also [36]. The form of this equation coincides with the kinematic relation used in the material description. The only difference is that the material time derivative is used instead of the total time derivative [51, 52]. The second equation in (22.2) can be interpreted in two ways. On the one hand, we can consider this equation as a definition of the angular velocity vector. On the other hand, we can consider it as a definition of the rotation tensor field. Indeed, the rotation tensor, as well as the displacement vector, is not an instantaneous characteristic. It depends on the reference configuration, which is not determined in the spatial description. That is why, in the spatial description, the angular velocity field plays an independent role, similar to the role of the velocity field. In this case, the rotation tensor is, in fact, a secondary and auxiliary characteristic of rotational motion, similar to that the displacement vector is a secondary and auxiliary characteristic of translational motion.

Now, we formulate four balance equations: the mass balance equation, the first law of the Eulerian dynamics, the second law of the Eulerian dynamics, and the energy balance equation. The mass balance equation is

$$\frac{\delta \rho}{\delta t} + \rho \nabla \cdot \boldsymbol{v} = 0, \tag{22.3}$$

where ρ is the density of mass. The first law of the Eulerian dynamics (the linear momentum balance equation) and the second law of the Eulerian dynamics (the angular momentum balance equation) are

$$\nabla \cdot \boldsymbol{\tau} + \rho \boldsymbol{f} = \rho \frac{\delta \boldsymbol{v}}{\delta t}, \qquad \nabla \cdot \boldsymbol{\mu} + \boldsymbol{\tau}_{\times} + \rho \boldsymbol{L} = \rho J \frac{\delta \boldsymbol{\omega}}{\delta t}, \tag{22.4}$$

where $\boldsymbol{\tau}$ is the stress tensor, $\boldsymbol{f}$ is the external force per unit mass, $\boldsymbol{\mu}$ is moment stress tensor, $\boldsymbol{L}$ is the external moment per unit mass, J is the moment of inertia per unit mass, $(\ )_{\times}$ denotes the vector invariant of a tensor that is defined for an arbitrary dyad as $(\boldsymbol{ab})_{\times} = \boldsymbol{a} \times \boldsymbol{b}$. For simplicity sake, we consider an isotropic continuum. That is why, the inertia tensor per unit mass $\boldsymbol{J}$ is assumed to be spherical one: $\boldsymbol{J} = J\boldsymbol{E}$. The local form of the energy balance equation is written as

$$\rho \frac{\delta U}{\delta t} = \boldsymbol{\tau}^T \cdot\cdot (\nabla \boldsymbol{v} + \boldsymbol{E} \times \boldsymbol{\omega}) + \boldsymbol{\mu}^T \cdot\cdot \nabla \boldsymbol{\omega} - \nabla \cdot \boldsymbol{h} + \rho q, \tag{22.5}$$

where U is the specific internal energy, $\boldsymbol{h}$ is the heat flux vector, q is the energy supply per unit mass per unit time, the double contraction of arbitrary dyads is defined as $\boldsymbol{ab}\cdot\cdot\boldsymbol{cd} = (\boldsymbol{b}\cdot\boldsymbol{c})(\boldsymbol{a}\cdot\boldsymbol{d})$, the curl product of a second rank tensor and a vector is defined as follows: if $\boldsymbol{A} = \boldsymbol{ab}$ then $\boldsymbol{A}\times\boldsymbol{c} = \boldsymbol{ab}\times\boldsymbol{c} = \boldsymbol{a}(\boldsymbol{b}\times\boldsymbol{c})$.

22.3 Differential Equations Relating the Strain Tensors to the Velocity Vector and the Angular Velocity Vector

In the spatial description, the stretch tensor $\boldsymbol{g}(\boldsymbol{r},t)$ and the wryness tensor $\boldsymbol{\Theta}(\boldsymbol{r},t)$ are introduced in the same way as it is done in the material description:

$$\boldsymbol{g} = \boldsymbol{E} - \nabla\boldsymbol{u}, \qquad \nabla\boldsymbol{P} = \boldsymbol{\Theta}\times\boldsymbol{P}. \tag{22.6}$$

It is easy to show that from the first equation in (22.2) and the first equation in (22.6) follows the relation between the stretch tensor and the velocity gradient:

$$\frac{\delta\boldsymbol{g}}{\delta t} = -(\nabla\boldsymbol{v})\cdot\boldsymbol{g}. \tag{22.7}$$

In the spatial description and in the material description, relation (22.7) has the same form. The difference is in the meaning of the time derivatives. The derivation of Eq. (22.7) in the case of the material description is well known and it can be found in many books. The derivation of this equation in the case of the spatial description is less common; it can be found, e.g., in [36].

The relation analogues to (22.7) can be obtained for the wryness tensor. As shown in [36], from the second equation in (22.2) and the second equation in (22.6) follows the differential equation for the wryness tensor, which has the form

$$\frac{\delta\boldsymbol{\Theta}}{\delta t} = \nabla\boldsymbol{\omega} - \boldsymbol{\Theta}\times\boldsymbol{\omega} - (\nabla\boldsymbol{v})\cdot\boldsymbol{\Theta}. \tag{22.8}$$

We note that Eq. (22.8) has a more complex structure than the analogous equation in the material description. The difference is not only in the meaning of the time derivatives, but also in the fact that, in the spatial description, Eq. (22.8) depends on the velocity of translational motion, whereas in the material description, the analogous equation does not depend on it.

22.4 The Reduced Energy Balance Equation and the Heat Conduction Equation: Zhilin's Method

In accordance with Zhilin's method [34]-[36], [40], [47]-[49], we represent the stress tensor $\boldsymbol{\tau}$ and the moment stress tensor $\boldsymbol{\mu}$ as

$$\boldsymbol{\tau} = \boldsymbol{\tau}_e + \boldsymbol{\tau}_d, \qquad \boldsymbol{\mu} = \boldsymbol{\mu}_e + \boldsymbol{\mu}_d, \tag{22.9}$$

where $\boldsymbol{\tau}_e$ and $\boldsymbol{\mu}_e$ are the elastic components of the stress tensor and the moment stress tensor, respectively; $\boldsymbol{\tau}_d$ and $\boldsymbol{\mu}_d$ are their dissipative components. In view of Eq. (22.9), the energy balance equation (22.5) takes the form

$$\rho \frac{\delta U}{\delta t} = \boldsymbol{\tau}_e^T \cdot\cdot (\nabla \boldsymbol{v} + \boldsymbol{E} \times \boldsymbol{\omega}) + \boldsymbol{\mu}_e^T \cdot\cdot \nabla \boldsymbol{\omega} + \\ + \boldsymbol{\tau}_d^T \cdot\cdot (\nabla \boldsymbol{v} + \boldsymbol{E} \times \boldsymbol{\omega}) + \boldsymbol{\mu}_d^T \cdot\cdot \nabla \boldsymbol{\omega} - \nabla \cdot \boldsymbol{h} + \rho q. \tag{22.10}$$

In accordance with Zhilin's method, the absolute temperature T and the specific entropy $\mathcal{H}$ are introduced as follows, see [34]-[36], [40], [47]-[49]:

$$\rho T \frac{\delta \mathcal{H}}{\delta t} = \boldsymbol{\tau}_d^T \cdot\cdot (\nabla \boldsymbol{v} + \boldsymbol{E} \times \boldsymbol{\omega}) + \boldsymbol{\mu}_d^T \cdot\cdot \nabla \boldsymbol{\omega} - \nabla \cdot \boldsymbol{h} + \rho q. \tag{22.11}$$

Equation (22.11) is the heat conduction equation. In view of Eqs. (22.7), (22.8), (22.11), the energy balance equation (22.10) can be transformed to the form

$$\rho \frac{\delta U}{\delta t} = \boldsymbol{\tau}_*^T \cdot\cdot \frac{\delta \boldsymbol{g}_*}{\delta t} + \boldsymbol{\mu}_*^T \cdot\cdot \frac{\delta \boldsymbol{\Theta}_*}{\delta t} + \rho T \frac{\delta \mathcal{H}}{\delta t}, \tag{22.12}$$

where the energy stress tensor $\boldsymbol{\tau}_*$ and the energy moment stress tensor $\boldsymbol{\mu}_*$ are determined as

$$\boldsymbol{\tau}_* = \boldsymbol{g}^T \cdot \boldsymbol{\tau}_e \cdot \boldsymbol{P}, \qquad \boldsymbol{\mu}_* = \boldsymbol{g}^T \cdot \boldsymbol{\mu}_e \cdot \boldsymbol{P}, \tag{22.13}$$

and also the energy stretch tensor $\boldsymbol{g}_*$ and the energy wryness tensor $\boldsymbol{\Theta}_*$ are calculated by the formulas

$$\boldsymbol{g}_* = \boldsymbol{g}^{-1} \cdot \boldsymbol{P}, \qquad \boldsymbol{\Theta}_* = \boldsymbol{g}^{-1} \cdot \boldsymbol{\Theta} \cdot \boldsymbol{P}. \tag{22.14}$$

Equation (22.12) is called the reduced energy balance equation. It allows to obtain the Cauchy–Green relations

$$\boldsymbol{\tau}_* = \rho \frac{\partial U(\boldsymbol{g}_*, \boldsymbol{\Theta}_*, \mathcal{H})}{\partial \boldsymbol{g}_*}, \qquad \boldsymbol{\mu}_* = \rho \frac{\partial U(\boldsymbol{g}_*, \boldsymbol{\Theta}_*, \mathcal{H})}{\partial \boldsymbol{\Theta}_*}, \qquad T = \frac{\partial U(\boldsymbol{g}_*, \boldsymbol{\Theta}_*, \mathcal{H})}{\partial \mathcal{H}}. \tag{22.15}$$

Specifying the function $U(\boldsymbol{g}_*, \boldsymbol{\Theta}_*, \mathcal{H})$ and taking into account Eq. (22.13), we can obtain the constitutive equations for the elastic component of the stress tensor $\boldsymbol{\tau}_e$, the elastic component of the moment stress tensor $\boldsymbol{\mu}_e$, and temperature. In order to close the obtained system of equations we should supplement it with the constitutive equations for the dissipative component of the stress tensor $\boldsymbol{\tau}_d$, the dissipative component of the moment stress tensor $\boldsymbol{\mu}_d$, and the heat flux vector $\boldsymbol{h}$. These constitutive equations can be, e.g.,

$$\boldsymbol{\tau}_d = \boldsymbol{B}_1 \cdot\cdot (\nabla \boldsymbol{v} + \boldsymbol{E} \times \boldsymbol{\omega}), \qquad \boldsymbol{\mu}_d = \boldsymbol{B}_2 \cdot\cdot (\nabla \boldsymbol{\omega}), \qquad \boldsymbol{h} = -\lambda \nabla T, \tag{22.16}$$

where λ is the thermal conductivity, $\boldsymbol{B}_1$ and $\boldsymbol{B}_2$ are the fourth rank tensors characterizinf viscous properties of a material. These tensors can be constants or some functions. Here we briefly outline Zhilin's method that allows us to describe some type of viscoelastic materials. A more detailed discussion of this method can be found in [34]-[36], [47]-[49].

22.5 Integral Equations Relating the Strain Tensors to the Velocity Vector and the Angular Velocity Vector

The system of the basic equations describing the deformation processes in solids consists of the equations of motion, the constitutive equations and the equations for the strain tensors. All parameters of a material are contained in the constitutive equations. The equations of motion and the equations for the strain tensors have in common that they have the same form for all materials. However, in contrast to the equations of motion, which can be written in both differential and integral forms, the equations for the strain tensors are always written in differential form. At the same time, it is not difficult to show that the equations for the strain tensors (22.7), (22.8) can be rewritten in integral form.

Let us start with Eq. (22.7). As shown in [32], this equation can be transformed to the form

$$\frac{d\boldsymbol{g}}{dt} = -\nabla(\boldsymbol{v}\cdot\boldsymbol{g}). \tag{22.17}$$

It is evident that Eq. (22.17) can be rewritten as

$$\frac{d\boldsymbol{g}}{dt} = -\nabla\cdot\boldsymbol{J}_g, \tag{22.18}$$

where the third rank tensor $\boldsymbol{J}_g$ has the form

$$\boldsymbol{J}_g = \boldsymbol{E}\,\boldsymbol{v}\cdot\boldsymbol{g}. \tag{22.19}$$

Integrating Eq. (22.18) over the volume fixed in the reference frame and making use the divergence theorem, we obtain

$$\frac{d}{dt}\int\limits_{\mathcal{V}} \boldsymbol{g}\,d\mathcal{V} = -\int\limits_{S} \boldsymbol{n}\cdot\boldsymbol{J}_g\,dS, \tag{22.20}$$

where vector $\boldsymbol{n}$ is the unit outer normal to the surface S that bounds the control volume $\mathcal{V}$. It is easy to see that Eq. (22.20) has the form of the balance equation for the stretch tensor, where tensor $\boldsymbol{J}_g$ plays the role of the stretch tensor flux.

Now, we consider Eq. (22.8). As shown in [50], this equation can be transformed to the form

$$\frac{d\boldsymbol{\Theta}}{dt} = \nabla(\boldsymbol{\omega}-\boldsymbol{v}\cdot\boldsymbol{\Theta}) - \boldsymbol{\Theta}\times(\boldsymbol{\omega}-\boldsymbol{v}\cdot\boldsymbol{\Theta}). \tag{22.21}$$

Equation (22.21) can be rewritten as

$$\frac{d\boldsymbol{\Theta}}{dt} = -\nabla \cdot \boldsymbol{J}_\Theta + \boldsymbol{\Upsilon}^*_\Theta, \tag{22.22}$$

where the third rank tensor $\boldsymbol{J}_\Theta$ and the second rank tensor $\boldsymbol{\Upsilon}^*_\Theta$ have the form

$$\boldsymbol{J}_\Theta = \boldsymbol{E}\,(\boldsymbol{v}\cdot\boldsymbol{\Theta} - \boldsymbol{\omega}), \qquad \boldsymbol{\Upsilon}^*_\Theta = \boldsymbol{\Theta}\times(\boldsymbol{v}\cdot\boldsymbol{\Theta} - \boldsymbol{\omega}). \tag{22.23}$$

Integrating Eq. (22.22) over the volume fixed in the reference frame and using the divergence theorem, we arrive at the following equation:

$$\frac{d}{dt}\int_{\mathcal{V}} \boldsymbol{\Theta}\, d\mathcal{V} = -\int_{S} \boldsymbol{n}\cdot\boldsymbol{J}_\Theta\, dS + \int_{\mathcal{V}} \boldsymbol{\Upsilon}^*_\Theta\, d\mathcal{V}. \tag{22.24}$$

It is evident that Eq. (22.24) has the form of the balance equation, where tensor $\boldsymbol{J}_\Theta$ plays the role of the wryness tensor flux and tensor $\boldsymbol{\Upsilon}^*_\Theta$ plays the role of the wryness tensor production inside the considered control volume.

Comparing the wryness tensor balance equation (22.24) with the stretch tensor balance equation (22.20), we infer that there is an important difference between these equations: Eq. (22.24) contains the term that has the meaning of the production inside the control volume, whereas Eq. (22.20) does not contain an analogous term.

22.6 Source Terms in the Strain Balance Equations

Since we have transformed the equations for the strain tensors to the form of the balance equations, it seems logical to consider the possibility to add source terms to these equations. Adding a source term $\boldsymbol{\Upsilon}_g$ to the stretch tensor balance equation (22.20) yields

$$\frac{d}{dt}\int_{\mathcal{V}} \boldsymbol{g}\, d\mathcal{V} = -\int_{S} \boldsymbol{n}\cdot\boldsymbol{J}_g\, dS + \int_{\mathcal{V}} \boldsymbol{\Upsilon}_g\, d\mathcal{V}. \tag{22.25}$$

The local form of Eq. (22.25) is

$$\frac{\delta \boldsymbol{g}}{\delta t} = -(\nabla\boldsymbol{v})\cdot\boldsymbol{g} + \boldsymbol{v}\cdot\nabla\boldsymbol{g} - (\nabla\boldsymbol{g}^T)\cdot\boldsymbol{v} + \boldsymbol{\Upsilon}_g. \tag{22.26}$$

Equation (22.26) is a generalization of Eq. (22.7). It is important to note that $\boldsymbol{v}\cdot\nabla\boldsymbol{g} - (\nabla\boldsymbol{g}^T)\cdot\boldsymbol{v} = 0$ if we assume that the first equation in (22.6) is valid. However, when we add the source term $\boldsymbol{\Upsilon}_g$ to the stretch tensor balance equation (22.20), we obtain Eq. (22.26), which does not follow from the system of the first equation in (22.2) and the first equation in (22.6). Therefore, if we use Eq. (22.26), we should reject either the first equation in (22.2) or the first equation in (22.6). A detailed discussion of this issue can be found in [32]. Here, we reject the first equation in (22.6). In this case, firstly, $\boldsymbol{v}\cdot\nabla\boldsymbol{g} - (\nabla\boldsymbol{g}^T)\cdot\boldsymbol{v} \neq 0$, and secondly, Eq. (22.26) should be considered as a definition of the stretch tensor.

Although wryness tensor balance equation (22.24) contains the term $\boldsymbol{\Upsilon}^*_\Theta$, which has the meaning of the wryness tensor production inside the control volume, we can add an additional source term $\boldsymbol{\Upsilon}_\Theta$, just as we have added the source term to the stretch tensor balance equation. As a result, we have

$$\frac{d}{dt}\int\limits_{\mathcal{V}} \boldsymbol{\Theta}\, d\mathcal{V} = -\int\limits_{S} \boldsymbol{n}\cdot\boldsymbol{J}_\Theta\, dS + \int\limits_{\mathcal{V}} (\boldsymbol{\Upsilon}^*_\Theta + \boldsymbol{\Upsilon}_\Theta)\, d\mathcal{V}. \tag{22.27}$$

The local form of Eq. (22.27) is

$$\frac{\delta\boldsymbol{\Theta}}{\delta t} = \nabla\boldsymbol{\omega} - \boldsymbol{\Theta}\times\boldsymbol{\omega} - (\nabla\boldsymbol{v})\cdot\boldsymbol{\Theta} + \boldsymbol{v}\cdot\nabla\boldsymbol{\Theta} - (\nabla\boldsymbol{\Theta}^T)\cdot\boldsymbol{v} + \boldsymbol{\Theta}\times(\boldsymbol{v}\cdot\boldsymbol{\Theta}) + \boldsymbol{\Upsilon}_\Theta. \tag{22.28}$$

Equation (22.28) is a generalization of Eq. (22.8). We note that we can prove that $\boldsymbol{v}\cdot\nabla\boldsymbol{\Theta} - (\nabla\boldsymbol{\Theta}^T)\cdot\boldsymbol{v} + \boldsymbol{\Theta}\times(\boldsymbol{v}\cdot\boldsymbol{\Theta}) = 0$ and remove these terms from Eq. (22.28) only if we assume that the second equation in (22.6) is valid. However, Eq. (22.28) does not follow from the system of the second equation in (22.2) and the second equation in (22.6). Hence, if we use Eq. (22.28), we should reject either the second equation in (22.2) or the second equation in (22.6). Here, we reject the second equation in (22.6) and consider Eq. (22.28) as a definition of the wryness tensor.

22.7 The Reduced Energy Balance Equation and the Heat Conduction Equation: A new Method

Now, we apply Zhilin's method, but with two modifications. The first modification consists in that we do not divide the stress tensor and the moment stress tensor into elastic and dissipative components. The second one consists in that we use the modified definitions of the strain tensors. Namely, we use Eq. (22.26) instead of Eq. (22.7) for the stretch tensor, and Eq. (22.28) instead of Eq. (22.8) for the wryness tensor.

In view of Eqs. (22.26), (22.28), energy balance equation (22.5) takes the form

$$\begin{aligned}\rho\frac{\delta U}{\delta t} &= \boldsymbol{\tau}^T_*\cdot\cdot\frac{\delta\boldsymbol{g}_*}{\delta t} + \boldsymbol{\mu}^T_*\cdot\cdot\frac{\delta\boldsymbol{\Theta}_*}{\delta t} - \nabla\cdot\boldsymbol{h} + \rho q + \\ &+ \left(\boldsymbol{\tau}^T + \boldsymbol{\Theta}\cdot\boldsymbol{\mu}^T\right)\cdot\cdot\left[\left(\boldsymbol{v}\cdot\nabla\boldsymbol{g} - (\nabla\boldsymbol{g}^T)\cdot\boldsymbol{v} + \boldsymbol{\Upsilon}_g\right)\cdot\boldsymbol{g}^{-1}\right] + \\ &+ \boldsymbol{\mu}^T\cdot\cdot\left[(\nabla\boldsymbol{\Theta}^T)\cdot\boldsymbol{v} - \boldsymbol{v}\cdot\nabla\boldsymbol{\Theta} - \boldsymbol{\Theta}\times(\boldsymbol{v}\cdot\boldsymbol{\Theta}) - \boldsymbol{\Upsilon}_\Theta\right],\end{aligned} \tag{22.29}$$

where the energy stress tensor $\boldsymbol{\tau}_*$ and the energy moment stress tensor $\boldsymbol{\mu}_*$ are calculated by the formulas

$$\boldsymbol{\tau}_* = \boldsymbol{g}^T\cdot\boldsymbol{\tau}\cdot\boldsymbol{P}, \qquad \boldsymbol{\mu}_* = \boldsymbol{g}^T\cdot\boldsymbol{\mu}\cdot\boldsymbol{P}, \tag{22.30}$$

and also the energy stretch tensor $\boldsymbol{g}_*$ and the energy wryness tensor $\boldsymbol{\Theta}_*$ are determined by Eq. (22.14).

In accordance with Zhilin's method, the absolute temperature T and the specific entropy $\mathcal{H}$ are introduced as

$$\rho T\frac{\delta\mathcal{H}}{\delta t}=\left(\boldsymbol{\tau}^T+\boldsymbol{\Theta}\cdot\boldsymbol{\mu}^T\right)\cdot\cdot\left[\left(\boldsymbol{v}\cdot\nabla\boldsymbol{g}-(\nabla\boldsymbol{g}^T)\cdot\boldsymbol{v}+\boldsymbol{\Upsilon}_g\right)\cdot\boldsymbol{g}^{-1}\right]+ \\ +\boldsymbol{\mu}^T\cdot\cdot\left[(\nabla\boldsymbol{\Theta}^T)\cdot\boldsymbol{v}-\boldsymbol{v}\cdot\nabla\boldsymbol{\Theta}-\boldsymbol{\Theta}\times(\boldsymbol{v}\cdot\boldsymbol{\Theta})-\boldsymbol{\Upsilon}_\Theta\right]-\nabla\cdot\boldsymbol{h}+\rho q. \tag{22.31}$$

Equation (22.31) is the heat conduction equation. In view of Eq. (22.31), the energy balance equation (22.29) can be transformed to the form of the reduced energy balance equation (22.12). This equation allows us to obtain the Cauchy–Green relations, which have the form of Eq. (22.15). Specifying the internal energy $U\big(\boldsymbol{g}_*,\,\boldsymbol{\Theta}_*,\,\mathcal{H}\big)$ and taking into account Eq. (22.30), we can obtain the constitutive equations for the stress tensor $\boldsymbol{\tau}$, the moment stress tensor $\boldsymbol{\mu}$, and temperature.

In order to close the system of equations we should supplement it with the constitutive equations for the source term $\boldsymbol{\Upsilon}_g$ in stretch tensor balance equation (22.26), source term $\boldsymbol{\Upsilon}_\Theta$ in wryness tensor balance equation (22.28), and the heat flux vector $\boldsymbol{h}$. These constitutive equations can be, e.g.,

$$\left(\boldsymbol{\Upsilon}_g+\boldsymbol{v}\cdot\nabla\boldsymbol{g}-(\nabla\boldsymbol{g}^T)\cdot\boldsymbol{v}\right)\cdot\boldsymbol{g}^{-1}=\boldsymbol{D}_1\cdot\cdot\left(\boldsymbol{\tau}+\boldsymbol{\mu}\cdot\boldsymbol{\Theta}^T\right), \\ \boldsymbol{\Upsilon}_\Theta-(\nabla\boldsymbol{\Theta}^T)\cdot\boldsymbol{v}+\boldsymbol{v}\cdot\nabla\boldsymbol{\Theta}+\boldsymbol{\Theta}\times(\boldsymbol{v}\cdot\boldsymbol{\Theta})=-\boldsymbol{D}_2\cdot\cdot\boldsymbol{\mu},\quad \boldsymbol{h}=-\lambda\nabla T, \tag{22.32}$$

where $\boldsymbol{D}_1$ and $\boldsymbol{D}_2$ are the fourth rank tensors characterizing viscous properties of a material. In order to show that constitutive equations (22.32) gives us a model of continuum with viscous properties, we eliminate from these equations the source terms $\boldsymbol{\Upsilon}_g$ and $\boldsymbol{\Upsilon}_\Theta$ by means Eqs. (22.26), (22.28). After that, we transform the obtained equations taking into account Eqs. (22.2), (22.14). As a result, we arrive at the following equations:

$$-\frac{\delta\boldsymbol{g}_*}{\delta t}+\boldsymbol{g}^{-1}\cdot(\nabla\boldsymbol{v}+\boldsymbol{E}\times\boldsymbol{\omega})\cdot\boldsymbol{P}=\boldsymbol{g}^{-1}\cdot\left[\boldsymbol{D}_1\cdot\cdot(\boldsymbol{\tau}+\boldsymbol{\mu}\cdot\boldsymbol{\Theta}^T)\right]\cdot\boldsymbol{P}, \\ -\frac{\delta\boldsymbol{\Theta}_*}{\delta t}+\boldsymbol{g}^{-1}\cdot(\nabla\boldsymbol{\omega})\cdot\boldsymbol{P}=\boldsymbol{g}^{-1}\cdot\left[\boldsymbol{D}_2\cdot\cdot\boldsymbol{\mu}+\boldsymbol{D}_1\cdot\cdot(\boldsymbol{\tau}\cdot\boldsymbol{\Theta}+\boldsymbol{\mu}\cdot\boldsymbol{\Theta}^T\cdot\boldsymbol{\Theta})\right]\cdot\boldsymbol{P}. \tag{22.33}$$

If we calculate the material time derivatives of the Cauchy–Green relations (22.15) and eliminate from the obtained equations the material time derivatives of the energy stretch tensor $\boldsymbol{g}_*$ and the energy wryness tensor $\boldsymbol{\Theta}_*$ by means of Eq. (22.33), we arrive at the constitutive equations that contain the stress tensor and its material time derivative, the moment stress tensor and its material time derivative, and also tensors $\nabla\boldsymbol{v}+\boldsymbol{E}\times\boldsymbol{\omega}$ and $\nabla\boldsymbol{\omega}$. Such constitutive equations are, in fact, the constitutive equations of a viscoelastic material.

22.8 A Comparison of two Approaches

Now, we compare the constitutive equations for the stress tensor and the moment stress tensor obtained by using Zhilin's method and by using our method. For simplicity sake, we consider the linear theories and ignore thermal expansion. We start with Zhilin's approach. In the linear approximation, the constitutive equations take the form

$$\begin{aligned}\boldsymbol{\tau} &= \boldsymbol{C}_1 \cdot\cdot\, \boldsymbol{\varepsilon} + \boldsymbol{B}_1 \cdot\cdot \frac{d\boldsymbol{\varepsilon}}{dt},\\ \boldsymbol{\mu} &= \boldsymbol{C}_2 \cdot\cdot\, \boldsymbol{\kappa} + \boldsymbol{B}_2 \cdot\cdot \frac{d\boldsymbol{\kappa}}{dt},\end{aligned} \tag{22.34}$$

where $\boldsymbol{C}_1$ and $\boldsymbol{C}_1$ are the stiffness tensors, $\boldsymbol{\varepsilon}$ and $\boldsymbol{\kappa}$ are the strain tensors, which are related to the velocity vector and the angular velocity vector as

$$\begin{aligned}\frac{d\boldsymbol{\varepsilon}}{dt} &= \nabla \boldsymbol{v} + \boldsymbol{E} \times \boldsymbol{\omega},\\ \frac{d\boldsymbol{\kappa}}{dt} &= \nabla \boldsymbol{\omega}.\end{aligned} \tag{22.35}$$

We see that Zhilin's approach leads to the Kelvin–Voigt model of viscoelastic material.

Now, we turn to the proposed approach. First of all, we note that relations (22.35) are not valid in our theory. They should be modified as

$$\begin{aligned}\frac{d\boldsymbol{\varepsilon}}{dt} &= \nabla \boldsymbol{v} + \boldsymbol{E} \times \boldsymbol{\omega} + \boldsymbol{\Upsilon}_{\varepsilon},\\ \frac{d\boldsymbol{\kappa}}{dt} &= \nabla \boldsymbol{\omega} + \boldsymbol{\Upsilon}_{\kappa},\end{aligned} \tag{22.36}$$

where $\boldsymbol{\Upsilon}_{\varepsilon}$ and $\boldsymbol{\Upsilon}_{\kappa}$ are the source terms. In the linear approximation, the constitutive equations are written as

$$\begin{aligned}\frac{d\boldsymbol{\tau}}{dt} + (\boldsymbol{G}_1 \cdot\cdot\, \boldsymbol{D}_1) \cdot\cdot\, \boldsymbol{\tau} &= \boldsymbol{G}_1 \cdot\cdot\, (\nabla \boldsymbol{v} + \boldsymbol{E} \times \boldsymbol{\omega}),\\ \frac{d\boldsymbol{\mu}}{dt} + (\boldsymbol{G}_2 \cdot\cdot\, \boldsymbol{D}_2) \cdot\cdot\, \boldsymbol{\mu} &= \boldsymbol{G}_2 \cdot\cdot\, (\nabla \boldsymbol{\omega}),\end{aligned} \tag{22.37}$$

where the fourth rank tensors $\boldsymbol{G}_1$ and $\boldsymbol{G}_1$ have the meaning of the stiffness tensors. It is easy to see that the proposed approach leads to Maxwell's model of viscoelastic material.

Thus, Zhilin's approach and the proposed approach lead to different models of viscoelastic material. These two approaches do not contradict each other, but rather complement each other. Moreover, combining these two approaches opens up the possibility to obtain more complex models of viscoelastic materials.

22.9 Discussion

Almost all the results presented above are obtained by rigorous mathematical transformations. The only thing that goes beyond pure mathematics is the idea of adding source terms to the balance equations for the strain tensors. Certainly, if we consider deformations as purely geometric characteristics, the idea of adding the source terms may seem very strange and even absurd. However, we pay attention to the fact that there is not only the geometric definition of the strain tensors (22.6), but also the energy definition, which was given by Zhilin, see [34]-[36]. According to the energy definition, strain tensors are those tensors on which the stress tensor and the moment stress tensor perform work. In other words, the strain tensors are defined as quantities conjugate to the stress tensor and the moment stress tensor in the energy balance equation. The energy definition of strain tensors can be successfully used both in the case of the material description, when strain tensors have a clear geometric meaning, and in the case of the spatial description, when strain tensors are rather geometric and kinematic characteristics than just geometric ones. Certainly, adding the source terms to the balance equations for the strain tensors, we change their kinematic and geometric meaning. However, the energy definition does not change. This means that adding the source terms to the balance equations for strain tensors does not lead to fundamental changes in the energy balance equation. By adding the source terms to the balance equations for strain tensors, we expand our views on the meaning of those quantities on which the specific internal energy can depend.

Now, we take a look at how the new approach based on the modified balance equations for the strain tensors is applied to solve some physical problems. In [14, 50], we consider the problem of modeling of thermal and electrical conductivities by means of mechanical models. In both papers, we develop linear theories of the Cosserat continuum possessing only rotational degrees of freedom. In both papers, we associate the main variables characterizing the stress–strain state of the continuum with quantities characterizing electrodynamic and thermodynamic processes. The difference between the models developed in [14] and [50] is as follows. In [14], we use the classical equation relating the wryness tensor to the angular velocity vector and the constitutive equation for the moment stress tensor similar to the constitutive equation that is used in Maxwell's model of viscoelastic material. In [50], the differential equation relating the wryness tensor to the angular velocity vector contains a source term and the constitutive equation for the moment stress tensor is the same as for the elastic continuum. Thanks to the special choice of the constitutive equation for the source term, we obtain the model of continuum that possesses some properties of Maxwell's model of viscoelastic material. In fact, the model developed in [50] is similar to the model considered in the present paper with only difference that in [50] we deal with the Cosserat continuum of a special type. It is important to note that the model constructed in [50] describes all physical processes that are described by using the model constructed in [14], but at the same time, it gives some additional results that cannot be obtained in the framework of the model presented in [14]. In particular, the model developed in [50] allows us to arrive at Maxwell's equations for conductors without modifying constitutive equations. It describes the conversion of electrical energy into thermal energy due to Joule heat. It also allows us to obtain the entropy balance equation.

We are deeply convinced that the proposed approach can be successfully used in solving many other problems from various fields of physics and mechanics.

References

[1] Maugin GA (1974) Acceleration waves in simple and linear viscoelastic micropolar materials, International Journal Engineering Science **12**:143-157.
[2] Marinov P (1976) On the thermomechanical behaviour of the viscoelastic Cosserat continuum, Acta Mechanica **25**:63-77.
[3] Zubov LM, Eremeev VA (1996) Equations for a viscoelastic micropolar fluid, Doklady Physics **41**:598-601.
[4] Sansour C (1998) A theory of the elasto-viscoplastic Cosserat continuum, Archives of Mechanics **50**(3):577-597.
[5] Sansour C (1998) A unified concept of elastic-viscoplastic Cosserat and micromorphic continua, Journal de Physique IV France, **08**(PR8):Pr8-341-Pr8-348.
[6] Yeremeyev VA, Zubov LM (1999) The theory of elastic and viscoelastic micropolar liquids, Journal of Applied Mathematics and Mechanics **63**:755-767.
[7] Neff P (2006) A finite-strain elastic-plastic Cosserat theory for polycrystals with grain rotations, International Journal of Engineering Science **44**(8-9):574-594.
[8] Ieşan D (2007) A Theory of thermoviscoelastic composites modelled as interacting Cosserat continua, Journal of Thermal Stresses **30**(12):1269-1289.
[9] Sedaghatizadeh N, Atefi G, Fardad AA, Barari A, Soleimani S, Khani S (2011) Analysis of blood flow through a viscoelastic artery using the Cosserat continuum with the large-amplitude oscillatory shear deformation model, Journal of the Mechanical Behavior of Biomedical Materials **4**(7):1123-1131.
[10] Erofeev VI, Malkhanov AO (2017) Macromechanical modelling of elastic and visco-elastic Cosserat continuum, Zeitschrift für Angewandte Mathematik und Mechanik **97**:1072-1077.
[11] Surana KS, Joy AD, Reddy JN (2018) Ordered rate constitutive theories for non-classical thermoviscoelastic fluids incorporating internal and Cosserat rotation rates, International Journal of Applied Mechanics **10**(2):1850012.
[12] Grekova EF, Porubov AV, dell'Isola F (2020) Reduced linear constrained elastic and viscoelastic homogeneous Cosserat media as acoustic metamaterials, Symmetry **12**(4)521.
[13] Pierro E, Carbone G (2021) A new technique for the characterization of viscoelastic materials: Theory, experiments and comparison with DMA, Journal of Sound and Vibration **515**:116462.

[14] Ivanova EA (2022) Modeling of thermal and electrical conductivities by means of a viscoelastic Cosserat continuum, Continuum Mech Thermodyn **34**:555–586.
[15] Eringen C (1980) Mechanics of Continua, Robert E. Krieger Publishing Company, Huntington, New York.
[16] Wilmanski K (1998) Thermomechanics of continua, Springer-Verlag, Berlin, Heidelberg, New York.
[17] Malvern E (1969) Introduction to the Mechanics of a Continuous Medium, Prentice-Hall, Inc, Englewood Cliffs, New Jersey.
[18] Truesdell, C (1972) A First Course in Rational Continuum Mechanics, The Johns Hopkins University, Baltimore, Maryland.
[19] Batchelor G (1970) An Introduction to Fluid Dynamics, Cambridge University Press, Cambridge.
[20] Loicyanskii LG (1987) Mekhanika Zhidkosti i Gaza (Mechanics of Fluids, in Russ), Nauka, Moscow.
[21] Daily J, Harleman D (1966) Fluid Dynamics, Addison-Wesley, Massachusetts, USA.
[22] Birkhoff G (1960) Hydrodynamics. A Study in Logic, Fact and Similitude, Princeton University Press, Princeton, New Jersey.
[23] Prandtl L, Tietjens O (1929) Hydro- und Aeromechanik, Springer, Berlin.
[24] Miller GH, Colella P (2002) A conservative three-dimensional Eulerian method for coupled fluid-solid shock capturing, Journal of Computational Physics **183**(1):26-82.
[25] Benson, DJ, Okazawa S (2004) Contact in a multimaterial Eulerian finite element fomulation, Computer Methods in Applied Mechanics and Engineering **193**:4277-4298.
[26] Al-Athel KS, Gadala MS (2011) Eulerian volume of solid (VOS) approach in solid mechanics and metal forming, Computer Methods in Applied Mechanics and Engineering **200**(25-28):2145-2159.
[27] Kamrin K, Rycroft CH, Nave JC (2012) Reference map technique for finite-strain elasticity and fluid-solid interaction, Journal of the Mechanics and Physics of Solids **60**(11):1952-1969.
[28] Schoch S, Nordin-Bates K, Nikiforakis N (2013) An Eulerian algorithm for coupled simulations of elastoplastic-solids and condensed-phase explosives, Journal of Computational Physics **252**:163-194.
[29] Ortega AL, Lombardini M, Pullin DI, Meiron DI (2014) Numerical simulation of elastic-plastic solid mechanics using an Eulerian stretch tensor approach and HLLD Riemann solver, Journal of Computational Physics **257**:414–441.
[30] Valkov B, Rycroft CH, Kamrin K (2015) Eulerian method for multiphase interactions of soft solid Bodies in fluids, Journal of Applied Mechanics **82**(4):041011.
[31] Ivanova EA, Matias DV, Stepanov MD (2019) Employment of Eulerian, Lagrangian, and arbitrary Lagrangian–Eulerian decription for crack opening problem, Materials Physics and Mechanics **42**:470-483.

[32] Ivanova EA, Jatar Montaño LE (2021) A new approach to solving the solid mechanics problems with matter supply, Continuum Mechanics and Thermodynamics **33**:1829-1855.

[33] Ivanova EA, Vilchevskaya EN (2022) A note on dependence of the inertia tensor on the strain measures, Continuum Mechanics and Thermodynamics. DOI 10.1007/s00161-022-01174-9

[34] Zhilin PA (2006) Advanced Problems in Mechanics, Vol 1, Institute for Problems in Mechanical Engineering, St. Petersburg (in Russ).

[35] Zhilin PA (2006) Advanced Problems in Mechanics, Vol 2, Institute for Problems in Mechanical Engineering, St. Petersburg.

[36] Zhilin PA (2012) Rational Continuum Mechanics, Polytechnic University Publishing House, St Petersburg (in Russ).

[37] Ivanova EA, Vilchevskaya EN (2016) Micropolar continuum in spatial description, Continuum Mechanics and Thermodynamics **28**(6):1759-1780.

[38] Ivanova EA (2019) On a micropolar continuum approach to some problems of thermo- and electrodynamics, Acta Mechanica **230**:1685-1715.

[39] Ivanova EA (2021) Modeling of electrodynamic processes by means of mechanical analogies, Zeitschrift für Angewandte Mathematik und Mechanik **101**(4):e202000076.

[40] Altenbach H, Naumenko K, Zhilin PA (2003) A micro-polar theory for binary media with application to phase-transitional flow of fiber suspensions, Continuum Mechanics and Thermodynamics **15**(6):539-570.

[41] Ivanova EA (2010) Derivation of theory of thermoviscoelasticity by means of two-component medium, Acta Mechanica **215**:261-286.

[42] Ivanova E (2011) On one model of generalised continuum and its thermodynamical interpretation, In: Altenbach H, Maugin GA, Erofeev V (Eds) Mechanics of Generalized Continua, Advanced Structured Materials, Vol 7, pp 151-174, Springer, Berlin

[43] Ivanova EA (2014) Description of mechanism of thermal conduction and internal damping by means of two-component Cosserat continuum, Acta Mechanica **225**:757-795.

[44] Ivanova EA (2015) A new model of a micropolar continuum and some electromagnetic analogies, Acta Mechanica **226**:697-721

[45] Brazgina OV, Ivanova EA, Vilchevskaya EN (2016) Saturated porous continua in the frame of hybrid description, Continuum Mechanics and Thermodynamics **28**(5):1553-1581.

[46] Ivanova EA (2017) Description of nonlinear thermal effects by means of a two-component Cosserat continuum, Acta Mechanica **228**:2299-2346.

[47] Ivanova E, Vilchevskaya E (2013) Description of Thermal and Micro-Structural Processes in Generalized Continua: Zhilin's Method and its Modifications, In: Altenbach H, Forest S, Krivtsov A (Eds) Generalized Continua as Models for Materials. Advanced Structured Materials, Vol 22, pp 179-197, Springer, Berlin, Heidelberg.

[48] Ivanova EA, Vilchevskaya EN (2020) Zhilin's Method and Its Modifications, In: Altenbach H, Öchsner A (eds) Encyclopedia of Continuum Mechanics, pp 2795-2804, Springer, Berlin.

[49] Ivanova EA, Vilchevskaya EN (2020) Truesdell's and Zhilin's approaches: derivation of constitutive equations, In: Altenbach H, Öchsner A (Eds) Encyclopedia of Continuum Mechanics, pp 2574-2584, Springer, Berlin.

[50] Ivanova EA (2022) A new approach to modeling of thermal and electrical conductivities by means of the Cosserat continuum, Continuum Mechanics and Thermodynamics **34**:1313-1342.

[51] Ivanova EA, Vilchevskaya EN, Müller WH (2016) Time derivatives in material and spatial description - what are the differences and why do they concern us?, In: Naumenko K, Aßmus M (eds.) Advanced Methods of Continuum Mechanics for Materials and Structures, Advanced Structured Materials, Vol 60, pp 3-28, Springer, Berlin.

[52] Ivanova EA, Vilchevskaya EN, Müller WH (2017) A study of objective time derivatives in material and spatial description, In: Altenbach H, Goldstein R, Murashkin E (Eds) Mechanics for Materials and Technologies. Advanced Structured Materials, Vol 46, pp 195-229, Springer, Cham.

Chapter 23
Porous Media Models Based on Generalized State Equations with Simple Examples

Anna Knyazeva and Nelli Nazarenko

Abstract The paper presents a variant of construction of flow models in porous media using thermodynamic equations of state in differential form. The hypothesis of pressure separation into elastic and viscous constituents is accepted. The first part follows from Gibbs thermodynamics relations, the second part is a consequence of Onsager's theory. The consistent application of general relations to different particular situations leads to coupled models including different cross-effects and their corresponding physically well understood parameters, as well as different generalizations of Darcy's law. We present examples of stationary models of flow of a non-ideal non-viscous gas; a one-component viscous fluid; a simple non-ideal fluid in a material with a double porous space structure; a non-viscous binary mixture with consideration of barodiffusion and pressure change due to composition change; a binary mixture in a material with double porosity.

Basic Designations

a - visibility
b_l - generalized thermodynamical forces corresponding to them
c_p,c_γ - isobaric and isochoric heat capacities
C_k - mass concentrations of species
f - free Helmholtz energy
g - free Gibbs energy
g_k - chemical potentials of species
G - gas (liquid) consumption
$\boldsymbol{J}_T$ - heat flux
$\boldsymbol{J}_k$ - diffusion fluxes

Anna Knyazeva · Nelli Nazarenko
Institute of Strength Physics and Materials Science of Siberian Branch Russian Academy of Sciences, 2/4, pr. Akademicheskii, Tomsk, 634055, Russian Federation,
e-mail: anna-knyazeva@mail.ru, nnelli@ispms.ru

H. Altenbach et al. (eds.), *Sixty Shades of Generalized Continua*,
Advanced Structured Materials 170, https://doi.org/10.1007/978-3-031-26186-2_23

k - permeability
k_1, k_2 - relative permeabilities
K_T - isothermal bulk module
m - porosity
m_1, m_2 - porosities of attributable to pore systems 1 and 2
$\boldsymbol{n}$ - normal to the section
p - pressure
p^e - elastic part of the pressure
p^V - viscous part of the pressure
s - entropy density
s_k - partial entropies
t - time
T - temperature
u - internal energy
V - volume
$\boldsymbol{v}$ - convective velocity vector
$\boldsymbol{w}$ - filtration velocity vector
V_x - component of velocity vector
x, y, z - spatial coordinates
$\boldsymbol{X}_T$, $\boldsymbol{X}_k$ - thermodynamical forces
α - some characteristic of capillary-porous medium or medium with double porosity, characterizing the exchange between pore systems
$\beta = 1/K_T$ - isothermal compressibility coefficient
$\gamma = \rho^{-1}$ - specific volume
η_1, η_2- saturations
μ - viscosity
ρ - density
σ_{ij} - stress tensor components

23.1 Introduction

Many natural objects and materials, fabrics of plant and animal origin are porous and extremely diverse in their structure [1, 2]. Mass transfer processes in them include jointly going phenomena of diffusion and filtration. Despite the huge number of publications in this field summarized in monographic and educational literature [1], [3]-[6], a number of phenomena accompanying flows in porous media remain poorly investigated. Among them is barodiffusion or diffusion under the action of pressure gradient. The mechanism of mass transfer under the action of pressure gradient differs from transfer by convective flow of liquid. In nonisothermal conditions mass transfer under action of temperature gradient or thermodiffusion also appears and also there are features in convective flow. These and other phenomena appear explicitly in the models when using the state equations in differential form [7, 8].

In the present work the possibilities of classical thermodynamics of irreversible processes for description of nonlinear processes of mass transfer in porous body including both diffusive and convective phenomena are illustrated. As a result, the model of a porous medium takes into account finite compressibility and concentration expansion related both to diffusion and to an additional mechanism of pressure change in the pore space volume; the phenomenon of components transfer under the action of pressure gradient.

23.2 Definitions and General Relationships

Filtration theory studies the motion of gases and liquids in porous media and is based on the notion that the porous medium and the fluid filling it are a continuum. Let us define the basic parameters of a porous medium, which will be used further. So, medium porosity m and visibility a are determined by formulae

$$m = \frac{V_p}{V} \text{ and } a = \frac{S_p}{S}$$

and are equivalent characteristics if the medium is isotropic. Here V_p is the volume occupied by the pores, V is the total volume including the pore volume and the volume occupied by the solid skeleton; S_p is the area occupied by the pores in an arbitrary section, the total area of which is S.

Determine the proper velocity of the gas (liquid) **v** and filtration velocity **w**

$$\boldsymbol{v} = \frac{G}{S_p}\boldsymbol{n} \text{ and } \boldsymbol{w} = \frac{G}{S}\boldsymbol{n},$$

where G is the gas (liquid) consumption, **n** is the normal to the section. Therefore,

$$\boldsymbol{w} = \frac{S_p}{S}\boldsymbol{v} = a\boldsymbol{v}. \tag{23.1}$$

The basic relation of filtration theory establishes the relationship between the filtration velocity vector and the pressure field, which causes filtration motion [5, 9]. Experimentally established for slow motions, the classical filtration law can be represented as

$$\boldsymbol{w} = -\frac{k}{\mu}\nabla p \tag{23.2}$$

where k is the permeability, μ is the dynamic viscosity. For the Cartesian coordinate system (23.2) can be written as follows

$$w_x = -\frac{k}{\mu}\frac{dp}{dx},\ w_y = -\frac{k}{\mu}\frac{dp}{dy},\ w_z = -\frac{k}{\mu}\frac{dp}{dz}. \tag{23.3}$$

Including the gravitational force directed along the axis 0Z results in additional term in the last of relations (23.3)

$$w_z = -\frac{k}{\mu}\left(\frac{dp}{dz} + \rho g\right) \tag{23.4}$$

where g is the acceleration of gravity. There are generalizations of this law for complex media and fast flows.

In an anisotropic medium, the permeability in (23.2) becomes a tensor characteristic

$$\boldsymbol{k} = \begin{Vmatrix} k_{xx} & k_{xy} & k_{xz} \\ k_{yx} & k_{yy} & k_{yz} \\ k_{zx} & k_{zy} & k_{zz} \end{Vmatrix}.$$

At high velocities [10], inertial effects become more and more significant. In this case, the matrix resistance can be represented as a superposition of the viscous and inertial components

$$-\frac{dp}{dz} = \alpha\mu w + \beta\rho w^2, \tag{23.5}$$

where $w = G\rho^{-1}$, α and β are experimentally determined viscosity and inertial resistance coefficients of the porous material. Their ratio has length dimensionality and can be used as a characteristic size of permeable structure. In more general case, two-term filtration law (Forchheimer's law) can be represented as [11]

$$-\nabla p = \frac{k}{\mu}\boldsymbol{w} + \frac{\gamma\rho}{\sqrt{k}}|w|\boldsymbol{w}$$

where γ as well as permeability, some function of porosity. In general, the equation of motion of a fluid in a porous medium can be written as

$$\rho\frac{\partial \boldsymbol{w}a^{-1}}{\partial t} + \rho\frac{\boldsymbol{w}}{a}\nabla\left(\frac{\boldsymbol{w}}{a}\right) = \nabla p + \rho(\boldsymbol{F}_1 + \boldsymbol{F}_2), \tag{23.6}$$

where

$$\boldsymbol{F}_1 = -\frac{1}{\rho}\nabla g z$$

is the force of gravity, $\boldsymbol{F}_2$ force of internal friction, which depends on the filtration velocity. For the so-called creeping flow, when the second term in (23.6) can be neglected, we have

$$\rho\frac{\partial \boldsymbol{w}a^{-1}}{\partial t} = \nabla p + \rho(\boldsymbol{F}_1 + \boldsymbol{F}_2).$$

In the limit of stationary filtration modes we get from above

$$\nabla p = -\rho(\boldsymbol{F}_1 + \boldsymbol{F}_2). \tag{23.7}$$

From (23.7) Darcy's law directly follows. Darcy's law is actually an analogue of the Fourier and Fick transport laws and follows from the Onsager principle of thermodynamics of irreversible processes as applied to the force of interfacial interaction between solid matrix and fluid [12, 13].

In addition to the equation of motion, filtration theory models require the continuity equation, which includes the possible inflow of mass from the porous skeleton

$$\frac{\partial}{\partial t} m\rho = -\nabla \cdot \rho \boldsymbol{w} + \sigma. \tag{23.8}$$

If the volume source for mass is absent and porosity is constant, this equation takes usual form

$$\frac{\partial}{\partial t} m\rho = -\nabla \cdot \rho a \boldsymbol{v} \; or \; \frac{\partial}{\partial t}\rho = -\nabla \cdot \rho \boldsymbol{v}. \tag{23.9}$$

When describing the motion in a porous medium of two immiscible fluids [5, 9, 11, 14] the concept of pore space saturation by each phase is introduced. Saturation is defined as the relative part of the medium's pore volume occupied by a given phase

$$\eta_i = \frac{V_{pi}}{V_p}, \; \eta_1 + \eta_2 = 1. \tag{23.10}$$

The movement of each phase is characterized by the filtration velocity vector of this phase w_i. If we denote $\eta_1 = \eta$, then $\eta_2 = 1 - \eta$. For each phase, generalized Darcy's law is considered to be true in the form:

$$w_i = -\frac{k_i^*}{\mu_i} \nabla p_i \tag{23.11}$$

Introducing relative phase permeabilities, $k_i = k_i^*/k$, which are a function of saturation,

$$w_i = -k \frac{k_i^*(\eta)}{\mu_i} \nabla p_i \tag{23.12}$$

The pressures in the phases in the general case do not coincide due to the action of surface tension

$$p_2 - p_1 = p_c(\eta), \tag{23.13}$$

where p_c is the capillary pressure, a function of saturation and pore space structure. In fact, it is assumed that each phase moves through its own pore space system. The function $p_c(\eta)$ is determined in different theories using various ideas.

The mass conservation equations in this case for each phase are

$$\frac{\partial}{\partial t} m\rho_1 \eta = -\nabla \cdot \rho_1 \boldsymbol{w}_1 + \sigma_1, \tag{23.14}$$

$$\frac{\partial}{\partial t} m\rho_2 (1-\eta) = -\nabla \cdot \rho_2 \boldsymbol{w}_2 + \sigma_2. \tag{23.15}$$

Phase densities are functions of pressures in the phases, and the possible change in porosity is usually assumed to depend on the mean pressure

$$\overline{p} = p_1\eta + p_2(1-\eta).$$

One variant of filtration models in a medium with double porosity or in a fractured-porous medium includes equations similar to (23.11). However, since we are talking about the same fluid, the Darcy law equations for pores of two types or pores and cracks include the same viscosity μ:

$$\boldsymbol{w}_1 = -\frac{k_1}{\mu_i}\boldsymbol{\nabla} p_1,\ \boldsymbol{w}_2 = -\frac{k_2}{\mu_i}\boldsymbol{\nabla} p_2, \tag{23.16}$$

The continuity equations are also similar to previous ones

$$\frac{\partial m_1\rho_1}{\partial t} + \boldsymbol{\nabla}\cdot\rho_1\boldsymbol{w}_1 + q_1 = 0,\ \frac{\partial m_2\rho_2}{\partial t} + \boldsymbol{\nabla}\cdot\rho_2\boldsymbol{w}_2 + q_2 = 0, \tag{23.17}$$

but include mass transfer between media (pore systems) in addition to possible inflow of liquid from the solid skeleton:

$$q_k = \pm\rho_0\alpha(p_2 - p_1) + \sigma_k. \tag{23.18}$$

In (23.16) - (23.18) we assume: ρ_k is density of fluid in media 1 and 2, m_1 and m_2 are fractions belonging to pores of different kinds in the total pore system; ρ_0 is the initial porosity of fluid, α is a characteristic of capillary-porous medium or medium with double porosity, characterizing the intensity of fluid exchange between media. Obviously, the value α is proportional to the specific surface area of capillaries or pores.

If the fluid consists of several components, diffusion equations should be added to above equations. For a simple (single-phase) multicomponent fluid in a porous space we have

$$\left(\frac{\partial m\rho C_i}{\partial t} + \boldsymbol{\nabla}\cdot\rho\boldsymbol{w}C_i\right) = -\boldsymbol{\nabla}\cdot\boldsymbol{J}_i + \varphi_i, \tag{23.19}$$

where

$$C_i = \frac{\rho_i}{\rho},\ \sum_{i=1}^{n} C_i = 1,\ \sum_{i=1}^{n}\boldsymbol{J}_i = 0,\ \varphi_i = C_i\sigma_k. \tag{23.20}$$

For a porous medium with two pore systems, it is true

$$\left(\frac{\partial m_k\rho_k C_{i,k}}{\partial t} + \boldsymbol{\nabla}\cdot\rho_k\boldsymbol{w}_k C_{i,k}\right) = -\boldsymbol{\nabla}\cdot\boldsymbol{J}_{i,k} + \varphi_{i,k}, \tag{23.21}$$

where

$$C_i = \frac{\rho_{ik}}{\rho_k},\ \sum_{i=1}^{n} C_{i,k} = 1,\ \sum_{i=1}^{n}\boldsymbol{J}_{i,k} = 0,\ k = 1,2, \tag{23.22}$$

$\varphi_{i,k}$ describes the exchange of components between media, $C_{i,k}$ is the concentration of the i - component in the medium k; $\boldsymbol{J}_{i,k}$ are diffusion fluxes of components in media k.

The energy equation for fluid with the reactions can be presented in the form

$$T\rho\frac{ds}{dt} = -\nabla\cdot\boldsymbol{J_T} + H, \tag{23.23}$$

where s is the entropy density, $\boldsymbol{J}_T$ is the heat flux; the term H includes dissipation dueto viscous fluid flow; sources and sinks associated with chemical reactions and heat exchange with the solid framework.

The equations written out contain the total pressure (or total pressures in the phases or pore systems) including "elastic" and "viscous" parts [7]

$$p = p^e + p^V \tag{23.24}$$

To construct the diffusion-filtration model applicable to different situations it is necessary to determine the values heat and substance fluxes $\boldsymbol{J}_T, \boldsymbol{J}_k$, viscous pressure p^V and sources terms σ_k. In this paper we will use the thermodynamics of irreversible processes.

23.3 Thermodynamical Relations

Next, it is necessary to recall briefly the main points of Gibbs thermodynamics and Onsager's theory [15, 16]. For the construction of generalized equations of state we will start from the Gibbs equation, which can be written down in different forms corresponding to different observational conditions. Thus, if the process under study proceeds at constancy of strains (volume) and temperature, it is convenient to use the Helmholtz free energy f as the basic potential:

$$df = -sdT + \sigma^e_{ij}\rho^{-1}d\varepsilon_{ij} + \sum_{i=1}^{n} g_k dC_k - \sum_{(l)} b_l dZ_l, \tag{23.25}$$

and if the process occurs under conditions of constant stress (pressure) and temperature, then the Gibbs energy is more suitable for establishing the defining relations:

$$df = -sdT - \varepsilon_{ij}\rho^{-1}d\sigma^e_{ij} + \sum_{i=1}^{n} g_k dC_k - \sum_{(l)} b_l dZ_l, \tag{23.26}$$

where $g = u - \rho^{-1}\sigma_{ij}\varepsilon_{ij} - Ts$; $f = u - Ts$; u is the internal energy; T is the temperature; σ_{ij} are stress tensor components; ε_{ij} are strain tensor components; $\rho$is the density; g_k are chemical potentials of species; C_k are their mass concentrations; n are species number; Z_l are parameters, reflecting the action of physical fields of different nature; B_l are generalized thermodynamical forces corresponding to them.

The form of Gibbs equations used here corresponds to the approximation of small strains. All quantities included in (23.25), (23.26) depend on spatial coordinates and time. The index "e" indicates that the Gibbs equations (in general, all thermodynamic relations) include the part of the stress tensor linearly the increments of connected with strains and other thermodynamic state variables [17]. All kinds of energies in (23.25), (23.26), as well as chemical potentials are measured in J/kg.

For a gas or liquid phase in the absence of external fields, Eqs. (23.25), (23.26) are simplified (we assume here that the stress tensor can be considered spherical). Then Gibbs equations take the form

$$df = -sdT + p^e d\gamma + \sum_{i=1}^{n} g_k dC_k, \tag{23.27}$$

$$dg = -sdT + \gamma dp^e + \sum_{i=1}^{n} g_k dC_k, \tag{23.28}$$

where $p^e = -\frac{1}{3}\sigma^e_{k,k}$ is the equivalent pressure, $\mathrm{N/m^2 = J/m^3}$; $\gamma = 1/\rho$ is the specific volume, $\mathrm{m^3/kg}$.

Because in (23.27) $f = f(T, \gamma, C_k)$, so and s, p, g_k are the functions of the same state variables. Hence,

$$ds = \left(\frac{\partial s}{\partial T}\right)_{\gamma,C_k} dT + \left(\frac{\partial s}{\partial \gamma}\right)_{T,C_k} d\gamma + \sum_{k=1}^{n}\left(\frac{\partial s}{\partial C_k}\right)_{\gamma,C_j,j\neq k} dC_k, \tag{23.29}$$

$$dp = \left(\frac{\partial p}{\partial T}\right)_{\gamma,C_k} dT + \left(\frac{\partial p^e}{\partial \gamma}\right)_{T,C_k} d\gamma + \sum_{k=1}^{n}\left(\frac{\partial p^e}{\partial C_k}\right)_{\gamma,C_j,j\neq k} dC_k, \tag{23.30}$$

$$dg_k = \left(\frac{\partial g_k}{\partial T}\right)_{\gamma,C_k} dT + \left(\frac{\partial g_k}{\partial \gamma}\right)_{T,C_k} d\gamma + \sum_{k=1}^{n}\left(\frac{\partial g_k}{\partial C_k}\right)_{\gamma,C_j,j\neq i} dC_i. \tag{23.31}$$

The coefficients near the differentials in (23.29)-(23.31) form a symmetric matrix

$$\begin{bmatrix}
\left(\frac{\partial s}{\partial T}\right)_{\gamma,C_k} & \left(\frac{\partial s}{\partial \gamma}\right)_{T,C_k} & \left(\frac{\partial s}{\partial C_1}\right)_{T,C_k k\neq 1} & \cdots & \left(\frac{\partial s}{\partial C_j}\right)_{T,C_k,k\neq j} & \cdots & \left(\frac{\partial s}{\partial C_n}\right)_{T,C_k,k\neq n} \\
\left(\frac{\partial p^e}{\partial T}\right)_{\gamma,C_k} & \left(\frac{\partial p^e}{\partial \gamma}\right)_{T,C_k} & \left(\frac{\partial p^e}{\partial C_1}\right)_{T,C_k k\neq 1} & \cdots & \left(\frac{\partial p^e}{\partial C_j}\right)_{T,C_k,k\neq j} & \cdots & \left(\frac{\partial p^e}{\partial C_n}\right)_{T,C_k,k\neq n} \\
\left(\frac{\partial g_1}{\partial T}\right)_{\gamma,C_k} & \left(\frac{\partial g_1}{\partial \gamma}\right)_{T,C_k} & \left(\frac{\partial g_1}{\partial C_1}\right)_{T,C_k k\neq 1} & \cdots & \left(\frac{\partial g_1}{\partial C_j}\right)_{T,C_k,k\neq j} & \cdots & \left(\frac{\partial g_1}{\partial C_n}\right)_{T,C_k,k\neq n} \\
\cdots & \cdots & \cdots & & \cdots\cdots & & \cdots \\
\left(\frac{\partial g_n}{\partial T}\right)_{\gamma,C_k} & \left(\frac{\partial g_n}{\partial \gamma}\right)_{T,C_k} & \left(\frac{\partial g_n}{\partial C_1}\right)_{T,C_k k\neq 1} & \cdots & \left(\frac{\partial g_n}{\partial C_j}\right)_{T,C_k,k\neq j} & \cdots & \left(\frac{\partial g_n}{\partial C_n}\right)_{T,C_k,k\neq n}
\end{bmatrix}$$

By definition, the diagonal coefficients of the coefficient matrix in (23.29)-(23.31) are

$$\left(\frac{\partial s}{\partial T}\right)_{\gamma,C_k} = -\frac{\partial}{\partial T}\left(\frac{\partial f}{\partial T}\right)_{\gamma,C_k} = -\left(\frac{\partial^2 f}{\partial T^2}\right)_{\gamma,C_k} = \frac{c_\gamma}{T};$$

$$\left(\frac{\partial p^e}{\partial \gamma}\right)_{T,C_k} = -\left(\frac{\partial^2 f}{\partial \gamma^2}\right)_{T,C_k} = -\frac{1}{\beta_T \gamma} = -\frac{K_T}{\gamma};$$

$$\left(\frac{\partial g_k}{\partial C_k}\right)_{\gamma,C_j,j\neq k} = -\left(\frac{\partial^2 f}{\partial C_l}\right)_{\gamma,C_j,j\neq k},$$

where c_γ is the isochoric neat capacity; K_T is the isothermal bulk modulus; β_T is the isothermal compressibility coefficient. Symmetry means, for example:

$$\left(\frac{\partial g_k}{\partial C_i}\right)_{\gamma,C_j,j\neq i} = \beta_i^k = \frac{\partial}{\partial C_i}\left(\frac{\partial f}{\partial C_k}\right) = \frac{\partial}{\partial C_k}\left(\frac{\partial f}{\partial C_i}\right) = \left(\frac{\partial g_i}{\partial C_k}\right)_{\gamma,C_j,j\neq k} = \beta_k^i;$$

$$p_k = \left(\frac{\partial p^e}{\partial C_k}\right)_\gamma = -\left[\frac{\partial}{\partial C_k}\left(\frac{\partial f}{\partial \gamma}\right)_{C_j}\right]_\gamma = -\left[\frac{\partial}{\partial \gamma}\left(\frac{\partial f}{\partial C_k}\right)_\gamma\right]_{C_j} = -\left(\frac{\partial p^e}{\partial \gamma}\right)_{C_j}.$$

Derivative

$$\left(\frac{\partial p^e}{\partial T}\right)_{\gamma,C_k} = -\frac{\partial}{\partial T}\left(\frac{\partial f}{\partial \gamma}\right) = -\frac{\partial}{\partial \gamma}\left(\frac{\partial f}{\partial T}\right) = \left(\frac{\partial s}{\partial \gamma}\right) = \gamma_T$$

determines an isochoric coefficient of thermal increase of pressure, and there is a linc

$$\gamma_T = \alpha_T K_T,$$

where α_T is the thermal expansion coefficient. The equalities

$$\left(\frac{\partial s}{\partial C_k}\right)_{\gamma,C_k,j\neq k} = -\frac{\partial}{\partial C_k}\left(\frac{\partial f}{\partial T}\right) = -\frac{\partial}{\partial T}\left(\frac{\partial f}{\partial C_k}\right) = -\left(\frac{\partial g_k}{\partial T}\right) = -s_k$$

suggest that the change in entropy with a change in composition is described by the same coefficients as the change in chemical potentials with temperature.

As a result, we find from (23.29)-(23.31):

$$ds = \frac{c_\gamma}{T}dT + \gamma_T d\gamma + \sum_{k=1}^{n} s_k dC_k; \tag{23.32}$$

$$dp^e = \gamma_T dT - \frac{K_T}{\gamma}d\gamma + \sum_{k=1}^{n} p_k dc_k; \tag{23.33}$$

$$dg_k = s_k dT - p_k d\gamma + \sum_{k=1}^{n} p_k dc_k. \tag{23.34}$$

This is the system of state equations in differential form. When writing it down, it is not necessary to know the specific type of thermodynamic potential f. However, if the coefficients in (23.32)-(23.34) corresponding to different physical processes are known from experiment or found from independent assumptions as functions of state variables $T,\ \gamma,\ C_k$, we can find both the potential and calculate the coefficients for other conditions. In any case, all physical parameters having thermodynamic definitions are considered to be experimentally determined.

If the basic potential is Gibbs energy, which corresponds to the experimental conditions when temperature and pressure can be controlled, the coefficients in the system of state equations will change. For example, the isobaric heat capacity

$$\left(\frac{\partial s}{\partial T}\right)_{p,C_k} = -\frac{\partial}{\partial T}\left(\frac{\partial f}{\partial T}\right)_{\gamma,C_k} = -\left(\frac{\partial^2 f}{\partial T^2}\right)_{\gamma,C_k} = \frac{c_p}{T}$$

and concentration expansion coefficients

$$\alpha_k = \frac{1}{\gamma}\left(\frac{\partial \gamma}{\partial C_k}\right)_{p,T} = \frac{1}{\gamma}\left[\frac{\partial}{\partial C_k}\left(\frac{\partial g}{\partial p^e}\right)_{C_j}\right]_{p,T} = \frac{1}{\gamma}\left[\frac{\partial}{\partial p^e}\left(\frac{\partial g}{\partial C_k}\right)_{C_j}\right]_{p,T} = \frac{1}{\gamma}\left(\frac{\partial g_k}{\partial p^e}\right)_{C_j,T}.$$

The value

$$p_k = \left(\frac{\partial p}{\partial C_k}\right)_{C_j,T,\gamma} = -\left(\frac{\partial g_k}{\partial \gamma}\right)_{C_j,T},$$

included in Eqs. (23.29) and (23.30), is called in physics the isochoric coefficient of thermal pressure increase. In principle. It can also be called the partial pressure of the component k . Similarly to γ_T we have a relationship between the coefficients

$$p_k = \alpha_k K_T.$$

Since only $n-1$ concentrations are independent of n concentrations, we can reduce the number of coefficients in (23.28)-(23.30) by expressing or through the remaining concentrations. In this case the definitions of some coefficients will change, which is more convenient to analyze in specific special cases.

Depending on the problem under study, different types of concentrations, e.g., mass, volume, atomic, and different reference systems are convenient to use [15, 18]. However, different types of concentrations are uniquely related to each other, so that all transformations eventually lead to equivalent ratios. For example, when taking into account chemical transformations in the source terms in (23.20) and (23.21), there will appear additional terms (w_i or w_{ik}) reflecting accumulation or losses of components in reactions:

$$\varphi_i = -C_i\sigma_k + \omega_i,$$

$$\varphi_{i,k} = \mp\rho_0\alpha C_{i,k}(p_2 - p_1) - C_{i,k}\sigma_k + \omega_{i,k}.$$

Examples of models with chemical reactions are not considered in this paper.

Further we will need another group of thermodynamic relations. In general case, to find the expressions for $\boldsymbol{J}_T,\ \boldsymbol{J}_k,\ p^V$ we could use the Onsager theory or experimental laws. The first way was described in details in thermodynamical literature. Recall the linear relations for vector fluxes, which will be used further in particular examples:

$$\boldsymbol{J}_T = \sum_{k=1}^{n} L_{Tk}\boldsymbol{X}_k + L_{TT}\boldsymbol{X}_T; \tag{23.35}$$

$$\boldsymbol{J}_k = \sum_{k=1}^{n} L_{kj}\boldsymbol{X}_j + L_{kT}\boldsymbol{X}_T, \tag{23.36}$$

where

$$\boldsymbol{X}_k = -\frac{\nabla T}{T}; \quad \boldsymbol{X}_k = -\nabla\left(\frac{g_k}{T}\right)$$

are thermodynamical forces; $L_{kj} = L_{jk}$; $L_{kT} = L_{Tk}$ are phenomenological coefficients.

In the absence of chemical reactions for the viscous part of the pressure, Onsager's theory gives

$$p^V = -\kappa_V \nabla \cdot \boldsymbol{v}, \tag{23.37}$$

where κ_V is the volume viscosity, in general case different from μ, which connects with the pore wall friction. An overview of model building possibilities for liquid and gaseous media with chemical reactions using nonequilibrium state equations is presented in [19].

Further, let us take into account that Eqs. (23.2), (23.5)-(23.7) and other equations of mechanics include total pressure, and thermodynamic relations like (23.27), (23.28), - only p^e.

Equation (23.33) can be written as

$$\nabla p^e = \gamma_T \nabla T - \frac{K_T}{\gamma}\nabla\gamma + \sum_{k=1}^{n} p_k \nabla C_k, \tag{23.38}$$

and also in the form

$$\dot{p}^e = \gamma_T \dot{T} - \frac{K_T}{\gamma}\dot{\gamma}\sum_{k=1}^{n} p_k \dot{C}_k, \tag{23.39}$$

where point under symbol indicates the time derivative.

Taking this into account will lead to some modification of the equations written out above. For example, instead of (23.6) we obtain

$$\frac{\mathrm{d}\boldsymbol{w}a^{-1}}{\mathrm{d}t} = -\nabla\left(\kappa_V \nabla \cdot \boldsymbol{v}\right) + \gamma_T \nabla T - \frac{K_T}{\gamma} d\gamma + \sum_{k=1}^{n} p_k \nabla C_k + \rho\left(\boldsymbol{F}_1 + \boldsymbol{F}_2\right). \tag{23.40}$$

All formulas can easily be generalized to the general case when (23.25), (23.26) contains a stress tensor of a general form (i.e., not spherical), and all kinds of viscous stresses are assumed in the equations of continua mechanics.

Below we consider only the simplest models to illustrate the influence of "new" factors appearing due to interrelation between phenomena of various physical natures.

23.4 Examples of Particular Problems

Let us consider examples showing that the use of thermodynamic equations of state in differential form leads to coupled equations involving different cross-effects. All models presented below can be generalized to the case of variable properties (depending on temperature and composition).

23.4.1 Compressible Nonviscous Gas

Let us consider the simplest situation when a non-viscous one-component gas, whose temperature is constant, flows through a porous flat wall. The porosity is constant, the medium is isotropic, and there is no mass inflow from the solid skeleton solid frame. For the stationary regime from (23.9) and (23.2) we have

$$\nabla \cdot \rho \boldsymbol{v} = 0 \quad \text{and} \quad \boldsymbol{v} = \frac{k}{\mu m} \nabla p^e. \tag{23.41}$$

The first relationship gives

$$\frac{\nabla \rho}{\rho} = -\frac{\nabla \boldsymbol{v}}{\boldsymbol{v}}. \tag{23.42}$$

Using relation (23.33), from which and from the thermodynamic definitions for these conditions equalities

$$\nabla p^e = -\frac{K_T}{\gamma} d\gamma \equiv \beta_T \frac{\mathrm{d}\rho}{\rho}; \quad \nabla p^e = \beta_T \frac{\nabla \rho}{\rho}, \tag{23.43}$$

follow, we arrive at the equation for pressure in the form (index "e" is omitted)

$$\nabla \cdot \nabla p + \beta_T (\nabla)^2 = 0. \tag{23.44}$$

For a flat layer we have a simple problem

$$\frac{\mathrm{d}^2 p}{\mathrm{d}x^2} + \beta_T \left(\frac{\mathrm{d}p}{\mathrm{d}x} \right)^2 = 0; \quad x = 0: \quad p = p_A; \quad x = L: \quad p = p_B, \tag{23.45}$$

the solution of which has a form

$$p = {\beta_T}^{-1} \left[\ln (x + A_1) A_2 \right]$$

$$V_x = \frac{k}{\mu m} \frac{1}{\beta_T} \frac{1}{x + A_1} = \frac{k}{\mu m} \frac{1}{\beta_T} \frac{1 - \exp[-(p_A - p_B)\beta_T]}{L + x(1 - \exp[-(p_A - p_B)\beta_T])},$$

where

$$A_1 = \frac{\beta_T L}{\exp(\beta_T (p_B - p_A)) - 1}; \quad A_2 = \frac{\exp(\beta_T p_B) - \exp(\beta_T p_A)}{\beta_T L}.$$

V_x is the velocity in the direction x. If the pressure gradient is small or there is an incompressible gas, neglecting the second term, we come to the usual Laplace equation for pressure, as in the classical models. In this case we have

$$p = p_A + (p_B - p_A)\frac{x}{L};$$

$$V = \frac{k}{\mu m}\frac{1}{\beta_T}\frac{p_A - p_B}{L_1}.$$

When a compressible fluid flows in a medium with double porosity, we have relations (23.16)

$$\boldsymbol{v}_1 = \frac{k_1}{\mu m_1}\nabla p_1, \quad \boldsymbol{v}_2 = \frac{k_2}{\mu m_2}\nabla p_2$$

and (again, for the stationary mode)

$$\nabla \cdot \rho_1 m_1 \boldsymbol{v}_1 + \rho_0 \alpha (p_2 - p_1) = 0;$$

$$\nabla \cdot \rho_2 m_2 \boldsymbol{v}_2 + \rho_0 \alpha (p_2 - p_1) = 0.$$

The state equation for each pore system takes the form

$$\nabla p_1 = \beta_T \frac{\nabla \rho_1}{\rho_1}; \quad \nabla p_2 = \beta_T \frac{\nabla \rho_2}{\rho_2}.$$

Since the same fluid flows in each pore system, we can assume, in first approximation, that its compressibility coefficient, as well as its viscosity, are the same.

For one dimensional situation

$$\frac{\mathrm{d}p_1}{\mathrm{d}x} = \frac{\beta_T}{\rho_1}\frac{d\rho_1}{dx}; \quad \frac{\mathrm{d}p_2}{\mathrm{d}x} = \frac{\beta_T}{\rho_2}\frac{d\rho_2}{dx}.$$

Consequently, we find the relationship of velocities with densities

$$V_1 = -\frac{k_1}{\mu m_1}\frac{\mathrm{d}}{\mathrm{d}x}p_1 = -\frac{\beta_T}{\rho_1}\frac{k_1}{\mu m_1}\frac{\mathrm{d}\rho_1}{\mathrm{d}x}, \quad V_2 = -\frac{k_2}{\mu m_2}\frac{\mathrm{d}}{\mathrm{d}x}p_2 = -\frac{\beta_T}{\rho_2}\frac{k_2}{\mu m_2}\frac{\mathrm{d}\rho_2}{\mathrm{d}x}.$$

We substitute the velocities in the continuity equations and, provided the properties are constant, we obtain:

$$-\frac{k_1\beta_T}{\mu}\frac{\mathrm{d}^2\rho_1}{\mathrm{d}x^2} + \rho_0\rho\alpha(p_2 - p_1) = 0;$$

$$-\frac{k_2\beta_T}{\mu}\frac{\mathrm{d}^2\rho_2}{\mathrm{d}x^2} + \rho_0\rho\alpha(p_2 - p_1) = 0.$$

Then we differentiate these equations by coordinate and replace again the pressure derivatives by the density derivatives

$$-\frac{k_1}{\mu}\frac{d^3\rho_1}{dx^3}+\rho\alpha\left(\frac{1}{\rho_2}\frac{d\rho_2}{dx}-\frac{1}{\rho_1}\frac{d\rho_1}{dx}\right)=0;\quad -\frac{k_2}{\mu}\frac{d^3\rho_2}{dx^3}+\rho\alpha\left(\frac{1}{\rho_2}\frac{d\rho_2}{dx}-\frac{1}{\rho_1}\frac{d\rho_1}{dx}\right)=0. \tag{23.46}$$

These two equations can be reduced to a single equation either for ρ_1, or for ρ_2, but in the quasi-stationary approximation, the compressibility coefficient has no effect on density. At the same time, the pressures and velocities of fluids in different pore systems depend on the compressibility coefficient. The unsteady equations cannot be separated.

23.4.2 Compressible Fluid with Bulk Viscosity

For a non-ideal one-component compressible fluid with volumetric viscosity, the equation of state does not change and has the form (23.43). However, taking into account (23.24), let us represent Darcy's law as

$$\boldsymbol{v}=-\frac{k}{\mu m}\boldsymbol{\nabla}p=-\frac{k}{\mu m}\left[\boldsymbol{\nabla}p^e+\boldsymbol{\nabla}p^V\right]. \tag{23.47}$$

Consequently, from (23.47), (23.43) and (23.37) we find

$$\boldsymbol{v}=-\frac{k}{\mu m}\left[\beta_T\frac{\boldsymbol{\nabla}\rho}{\rho}-\boldsymbol{\nabla}\left(\kappa_V\boldsymbol{\nabla}\cdot\boldsymbol{v}\right)\right]$$

or with (23.43)

$$\boldsymbol{v}\cdot\boldsymbol{v}=\frac{k}{\mu m}\left[\beta_T\boldsymbol{\nabla}\cdot\boldsymbol{v}+\boldsymbol{v}\boldsymbol{\nabla}\left(\kappa_V\boldsymbol{\nabla}\cdot\boldsymbol{v}\right)\right]. \tag{23.48}$$

This equation can be seen as an equation for finding velocity. The one-dimensional equation takes the form:

$$\frac{k}{\mu m}\left[\beta_T\frac{dV_x}{dx}+\frac{dV_x}{dx}\cdot\frac{d}{dx}\left(\kappa_V\frac{dV_x}{dx}\right)\right]-(V_x)^2=0.$$

Using (23.47), it is easy to go to the nonlinear equation for the total pressure:

$$\left[-\beta_T\frac{d^2p}{dx^2}+\frac{k\kappa_V}{\mu m}\frac{d^2p}{dx^2}\frac{d^3p}{dx^3}\right]-\left(\frac{dp}{dx}\right)^2=0.$$

23.4.3 Non-ideal Gas Under Non-isothermal Conditions

For a non-ideal non-viscous one-component gas under nonisothermal conditions, we require two equations of state following from (23.32) and (23.33). Considering the relation between the coefficients $\gamma_T=\alpha_T K_T$, we shall write

$$ds = \frac{c_\gamma}{T} dT + \alpha_T K_T d\gamma; \tag{23.49}$$

$$dp^e = \alpha_T K_T dT + \frac{K_T}{\rho} d\rho. \tag{23.50}$$

Equation (23.49) is equivalent to the following

$$\frac{\mathrm{d}s}{\mathrm{d}t} = \frac{c_\gamma}{T}\frac{\mathrm{d}T}{\mathrm{d}t} - \alpha_T K_T \frac{\mathrm{d}\gamma}{\mathrm{d}t}.$$

Then energy equation (23.23) in the absence of dissipation and other volumetric heat sources and sinks (except for heat exchange with the solid frame) will look like

$$c_\gamma \rho \frac{\mathrm{d}T}{\mathrm{d}t} = -\nabla \cdot \boldsymbol{J}_T - \alpha_T K_T T \frac{\mathrm{d}\gamma}{\mathrm{d}t} + \alpha (T - T_s).$$

From (23.51) and Darcy's law under the condition $p = p^e$ we find

$$\boldsymbol{v} = -\frac{k}{\mu m} dp^e = -\frac{k}{\mu m}\left[\alpha_T K_T T + \frac{K_T}{\rho}\nabla\rho\right].$$

The continuity equation remains unchanged.

In the particular case of a flat layer in the steady-state regime of heat transfer and flow, and provided that the heat flux satisfies the classical Fourier law, we arrive to the equations

$$c_\gamma \rho V_x \frac{\mathrm{d}T}{\mathrm{d}x} = \lambda_T \frac{\mathrm{d}T^2}{\mathrm{d}x^2} - \alpha_T K_T T V_x \frac{\mathrm{d}\gamma}{\mathrm{d}x} + \alpha (T - T_s); \tag{23.51}$$

$$V_x = -\frac{k}{\mu m}\left[\alpha_T K_T T + \frac{K_T}{\rho}\frac{\mathrm{d}\rho}{\mathrm{d}x}\right]; \tag{23.52}$$

$$\frac{\mathrm{d}}{\mathrm{d}x}(\rho V_x) = 0. \tag{23.53}$$

The second equation directly connects the flow velocity with changes in temperature and density. In addition, the heat conduction equation for the solid frame is required, which we present in the simplest form

$$\lambda_{Ts}\frac{\mathrm{d}^2 T_s}{\mathrm{d}x^2} - \alpha (T - T_s) = 0,$$

where λ_T, λ_{Ts} are the corresponding thermal conductivity coefficients. In this form, the equations are convenient when the filtration flow is caused mainly by a temperature difference. In principle, the system of equations can be reduced to the solution of two equations for pressure and for temperature. However, it is not possible to separate the equations.

23.4.4 Binary Non-viscous Imperfect Mixture

Consider a binary mixture under isothermal conditions. In this case

$$C_1 + C_2 = 1. \tag{23.54}$$

From Eqs. (23.32) and (23.34) we find

$$dp^e = -\rho\beta_T^{-1}d\gamma + p_1 dC_1 + p_2 dC_2;$$

$$dg_1 = -p_1 d\gamma + \beta_1^1 dC_1 + \beta_2^1 dC_2;$$

$$dg_2 = -p_2 d\gamma + \beta_1^2 dC_1 + \beta_2^2 dC_2.$$

Taking into account (23.54) and the ratio $p_k = \alpha_k K_T = \alpha_k \beta_T^{-1}$, we obtain

$$\begin{aligned} dp^e &= -\rho\beta_T^{-1}d\gamma + (\alpha_1 - \alpha_2)\beta_T^{-1}dC_1; \\ dg_1 &= -\alpha_1\beta_T^{-1}d\gamma + \left(\beta_1^1 - \beta_2^1\right)dC_1; \\ dg_2 &= -\alpha_2\beta_T^{-1}d\gamma + \left(\beta_1^2 - \beta_2^2\right)dC_1. \end{aligned} \tag{23.55}$$

Taking into account $\gamma = \rho^{-1}$ and after introducing the notation $\alpha_C = \alpha_1 - \alpha_2$, the first relation of the system (23.55) will take the form

$$dp^e = \beta_T^{-1}\left(\rho^{-1}d\rho + \alpha_C dC_1\right).$$

or

$$\nabla p^e = \frac{1}{\beta_T}\left(\frac{\nabla\rho}{\rho} + \alpha_C dC_1\right). \tag{23.56}$$

We obtain equivalent relations if we use Gibbs energy as the basic thermodynamic potential $g = f + p\gamma$. Then the Gibbs equation takes the form

$$dg = \gamma\nabla p^e + \sum_{k=1}^{n} g_k dC_k,$$

and the corresponding nonequilibrium equations of state take the form of

$$dg_k = \alpha_k\rho^{-1}dp^e + \sum_{k=1}^{n}\beta_i^k dC_k,$$

$$d\gamma = -\beta_T\rho^{-1}dp^e + \sum_{k=1}^{n}\alpha_k\rho^{-1}dC_k.$$

Obviously, here all physical values are now determined via partial derivatives of the Gibbs energy:

$$\beta_T = -\gamma^{-1}\left(\frac{\partial \gamma}{\partial p^e}\right)_{C_k} = -\gamma^{-1}\left(\frac{\partial^2 g}{\partial p^{e2}}\right)_{C_k},$$

$$\alpha_k = \frac{1}{\gamma}\left(\frac{\partial \gamma}{\partial C_k}\right)_p = \frac{1}{\gamma}\left[\frac{\partial}{\partial C_k}\left(\frac{\partial g}{\partial p^e}\right)_{C_j}\right]_p = \frac{1}{\gamma}\left(\frac{\partial g_k}{\partial p^e}\right)_{C_j},$$

$$\beta_i^k = \left(\frac{\partial g_k}{\partial C_i}\right)_\gamma = \left[\frac{\partial}{\partial C_i}\left(\frac{\partial g}{\partial C_k}\right)_{C_j}\right]_{C_k,\gamma} = \left(\frac{\partial g_i}{\partial C_k}\right)_p.$$

Consequently, for a binary mixture we find

$$d\gamma = -\beta_T \rho^{-1} dp^e + \alpha_1 \rho^{-1} dC_1 + \alpha_2 \rho^{-1} dC_2;$$

$$dg_1 = \alpha_1 \rho^{-1} dp^e + \left(\beta_1^1 - \beta_2^1\right) dC_1;$$

$$dg_2 = \alpha_2 \rho^{-1} dp^e + \left(\beta_2^2 - \beta_1^2\right) dC_2.$$

or

$$\nabla\gamma = -\beta_T \rho^{-1} \nabla p^e + (\alpha_1 - \alpha_2)\rho^{-1} dC_1;$$

$$\nabla g_1 = \alpha_1 \rho^{-1} \nabla p^e + \left(\beta_1^1 - \beta_2^1\right) \nabla C_1; \tag{23.57}$$

$$\nabla g_2 = \alpha_2 \rho^{-1} dp^e - \left(\beta_2^2 - \beta_1^2\right) dC_2.$$

It is easy to show that the first relation of system (23.57) at replacement of the specific volume by density takes the form (23.56).

Further, in order to write down the expression for the fluxes of components, it is necessary to redefine for this case the expressions for the generalized thermodynamic forces. For a binary non-viscous mixture under constant temperature:

$$\sigma_s = -\left[\boldsymbol{J}_1 \frac{(\nabla g_1)_T}{T} + \boldsymbol{J}_2 \frac{(\nabla g_2)_T}{T}\right]. \tag{23.58}$$

In this case the Gibbs-Duhem relations are valid entropy production σ_s connects with diffusion:

$$C_1 (\nabla g_1)_{T,p} = C_2 (\nabla g_2)_{T,p} = 0,$$

and the balance of diffusion fluxes

$$\boldsymbol{J}_1 + \boldsymbol{J}_2 = 0$$

suggests that to describe redistribution of concentrations we need not only one concentration, but also one flux. Hence, from (23.58) we find

$$\sigma_s = \left[\frac{\boldsymbol{J}_1}{T}\left((\nabla g_1)_{T,p} + \left(\frac{\partial g_1}{\partial p}\right)_{T,C_1} \nabla p^e\right) - \frac{\boldsymbol{J}_1}{T}\left((\nabla g_2)_{T,p} + \left(\frac{\partial g_2}{\partial p}_{T,C_2}\right)\nabla p^e\right)\right] =$$

$$= \boldsymbol{J}_1 \frac{1}{T}\left[\left(1+\frac{C_1}{C_2}\right)(\boldsymbol{\nabla} g_1)_{p,T} + \left(\left(\frac{\partial g_1}{\partial p}\right)_{T,C_1} - \left(\frac{\partial g_2}{\partial p}\right)_{T,C_2}\right)\boldsymbol{\nabla} p^e\right].$$

Hence, the generalized force causing the flux should be written as

$$\boldsymbol{X}_1 = \frac{1}{T}\left[\left(1+\frac{C_1}{C_2}\right)(\boldsymbol{\nabla} g_1)_{p,T} + \left(\left(\frac{\partial g_1}{\partial p}\right)_{T,C_1} - \left(\frac{\partial g_2}{\partial p}\right)_{T,C_2}\right)\boldsymbol{\nabla} p^e\right].$$

It includes both the chemical potential gradient of the diffusate and the pressure gradient. Thus

$$\boldsymbol{J}_1 = L_{11}\boldsymbol{H}_1.$$

Taking into account the definition of the coefficients of concentration expansion and dependence of chemical potential g_1 on the composition,

$$(\boldsymbol{\nabla} g_1)_{p,T} = TRM^{-1}(\boldsymbol{\nabla}\ln C_1), \varphi(C_1)$$

as well as the determination of the phenomenological coefficient L_{11} through the diffusion coefficient [20] ,

$$L_{11} = \frac{D\rho M C_1}{R},$$

we come to the following relation for diffusion flux:

$$\boldsymbol{J}_1 = -D\rho\left(1+\frac{C_1}{C_2}\right)\left(1+C_1\frac{\partial\varphi}{\partial C_1}\right)\boldsymbol{\nabla} C_1 - \frac{DC_1M}{RT}\alpha_C\boldsymbol{\nabla} p^e, \tag{23.59}$$

$\alpha_C = \alpha_1 - \alpha_2$ - is difference between concentration expansion coefficients of components.

When $\boldsymbol{\nabla} p^e = 0$, we naturally get Fick's diffusion law. Omitting indexes 1 and "e" we write:

$$\boldsymbol{J} = -D\rho(C)\boldsymbol{\nabla} C - \frac{DC_1M}{RT}\alpha_C\boldsymbol{\nabla} p. \tag{23.60}$$

Substituting (23.60) into balance equation (23.19) and omitting the source terms, we obtain the diffusion equation like

$$\left(\frac{\partial m\rho C}{\partial t} + \boldsymbol{\nabla}\cdot\rho a\boldsymbol{v}C\right) = \boldsymbol{\nabla}\cdot[\rho Df(C)\boldsymbol{\nabla} C] + \boldsymbol{\nabla}\cdot[BC\boldsymbol{\nabla} p^e], \tag{23.61}$$

where

$$B = \frac{DM\alpha_C}{RT} - \tag{23.62}$$

is the transfer coefficient under the stress action (similar to pressure diffusion coefficient); M for the molar mass of a diffusing component. For small concentration of the component as compared to the solvent concentration C_2 we may assume that $f(C) = 1$. Therefore, in the case under study we have the continuity equation, Darcy's law, state equations and diffusion equation.

For a steady flow we have a set of equations

$$\boldsymbol{v} = -\frac{k}{\mu a}\boldsymbol{\nabla} p, \quad \boldsymbol{\nabla}\cdot\rho\boldsymbol{v}a = 0,$$

$$\boldsymbol{\nabla}\rho = \frac{1}{\beta_T}\left(\frac{\boldsymbol{\nabla}\rho}{\rho} + \alpha_C\boldsymbol{\nabla}C\right); \tag{23.63}$$

$$\boldsymbol{\nabla}\cdot\rho a\boldsymbol{v}C_k \equiv \rho a\boldsymbol{v}\boldsymbol{\nabla}C_k \equiv \boldsymbol{\nabla}\cdot[\rho D f(C)\boldsymbol{\nabla}C] + \boldsymbol{\nabla}\cdot[BC\boldsymbol{\nabla}p^e].$$

Similarly to the previous one we pass to the one-dimensional problem, which under the condition of constancy of diffusion coefficient $D = \text{const}$ and $f(C) = 1$ will include equations

$$\frac{\mathrm{d}^2 p}{\mathrm{d}x^2} + \left[\beta_T\frac{\mathrm{d}p}{\mathrm{d}x} - \alpha_C\frac{\mathrm{d}C}{\mathrm{d}x}\right]\frac{\mathrm{d}p}{\mathrm{d}x} = 0, \tag{23.64}$$

$$D\frac{\mathrm{d}^2 C}{\mathrm{d}x^2} + \frac{D}{\rho}\frac{\mathrm{d}\rho}{\mathrm{d}x}\frac{\mathrm{d}C}{\mathrm{d}x} + \frac{B}{\rho}\frac{\mathrm{d}}{\mathrm{d}x}\left(C\frac{\mathrm{d}p}{\mathrm{d}x}\right) + \frac{k}{\mu}\frac{\mathrm{d}p}{\mathrm{d}x}\frac{\mathrm{d}C}{\mathrm{d}x}. \tag{23.65}$$

Using (23.56), we find the derivative of density

$$\frac{1}{\rho}\frac{\mathrm{d}\rho}{\mathrm{d}x} = \beta_T\frac{\mathrm{d}p}{\mathrm{d}x} - \alpha_C\frac{\mathrm{d}C}{\mathrm{d}x}.$$

Consequently, the diffusion equation is modified to the form

$$D\frac{\mathrm{d}^2 C}{\mathrm{d}x^2} + \left[\left(\beta_T D + \frac{k}{\mu}\right)\frac{\mathrm{d}p}{\mathrm{d}x} - \alpha_C D\frac{\mathrm{d}C}{\mathrm{d}x}\right]\frac{\mathrm{d}C}{\mathrm{d}x} + \frac{B}{\rho}\frac{\mathrm{d}}{\mathrm{d}x}\left(C\frac{\mathrm{d}p}{\mathrm{d}x}\right) = 0. \tag{23.66}$$

In the particular case of membranes [21], the redistribution of solute concentration occurs precisely due to the difference in solute concentrations to the left and right of the membrane. The boundary conditions of the first kind correspond to the condition of good solute mass exchange in the solution and in the membrane layer. At the joint action of different transport mechanisms it is required to set conditions both for the pressure in the media (solutions) on different sides of the membrane or from the porous layer separating different media and for the concentration. For the velocity in this case, it is not required to set conditions.

As a result, the problem for a flat layer will include Eqs. (23.26) and (23.30), as well as the boundary conditions

$$x = 0: \quad C = C_{10}; \quad x = L: \quad C = C_{20};$$

$$x = 0: \quad p = p_1; \quad x = L: \quad p = p_2.$$

The density and velocity can be found from the remaining relations:

$$V_x = -\frac{k}{\mu m}\frac{\mathrm{d}p}{\mathrm{d}x}$$

and

$$\frac{\mathrm{d}\ln\rho}{\mathrm{d}x} = -\frac{\mathrm{d}\ln V_x}{\mathrm{d}x} = \frac{\mathrm{d}\ln V_x^{-1}}{\mathrm{d}x},$$

or

$$\rho V_x = \text{const} = G.$$

The latter relation is a necessary condition for the stationarity of the process.

The study of this particular problem is carried out in [21] after passing to dimensionless variables

$$\overline{p} = p/p_1, \quad u = V/V_*, \quad \xi = x/L, \quad \overline{\rho} = \rho/\rho_*, \tag{23.67}$$

where

$$V_* = \frac{k}{\mu m}\frac{p_1}{L}, \quad \frac{G}{V_*\rho_*} = 1.$$

One dimensional problem includes the equation

$$\frac{d^2\overline{p}}{d\xi^2} + \left[\overline{\beta}_T \frac{d\overline{p}}{d\xi} - \alpha_C \frac{dC}{d\xi}\right]\frac{d\overline{p}}{d\xi} = 0, \tag{23.68}$$

$$\delta\frac{d^2C}{d\xi^2} - \frac{d\overline{p}}{d\xi}\omega\frac{d}{d\xi}\cdot\left[C\frac{d\overline{p}}{d\xi}\right] + \left[\left(\overline{\beta}_T\delta + 1\right)\frac{d\overline{p}}{d\xi} - \alpha_C\delta\frac{dC}{d\xi}\right]\frac{dC}{d\xi} = 0; \tag{23.69}$$

$$\overline{\rho} = \frac{1}{u} = -\frac{1}{d\overline{p}/d\xi}, \quad u = -\frac{d\overline{p}}{d\xi}. \tag{23.70}$$

with boundary conditions

$$\xi = 0: \quad C = C_{10}; \quad \xi = 1: \quad C = C_{20} < 1;$$

$$\xi = 0: \quad \overline{p} = 1; \quad \xi = 1: \quad \overline{p} = \overline{p}_2,$$

where $\overline{p} = p_2/p_1$ characterizes the pressure gradient, contains the physically clear parameters:

$$\overline{\beta}_T = p_1\beta_T-$$

is compressibility factor:

$$\omega = \frac{1}{\rho}\frac{\mu a L}{k p_1}\frac{D M \alpha_C p_1}{R T a L} = \frac{M\alpha_C p_1}{R T \rho_*}\frac{D}{V_* a L}$$

is the coupling coefficient (this parameter is responsible for barodiffusion).

Parameter

$$\delta = \frac{D}{V_* a L}$$

is the relation of diffusion and filtration rates; the inverse value is actually Pecle's number for this problem,

$$Pe = \frac{V_* a L}{D}.$$

An example of the distribution of velocity and diffusant concentration for convective (left) and diffusive (right) flow modes is given in Fig. 23.1. We see that the change of cocentration expansion coefficient can leads to the velocity change dramatically. In

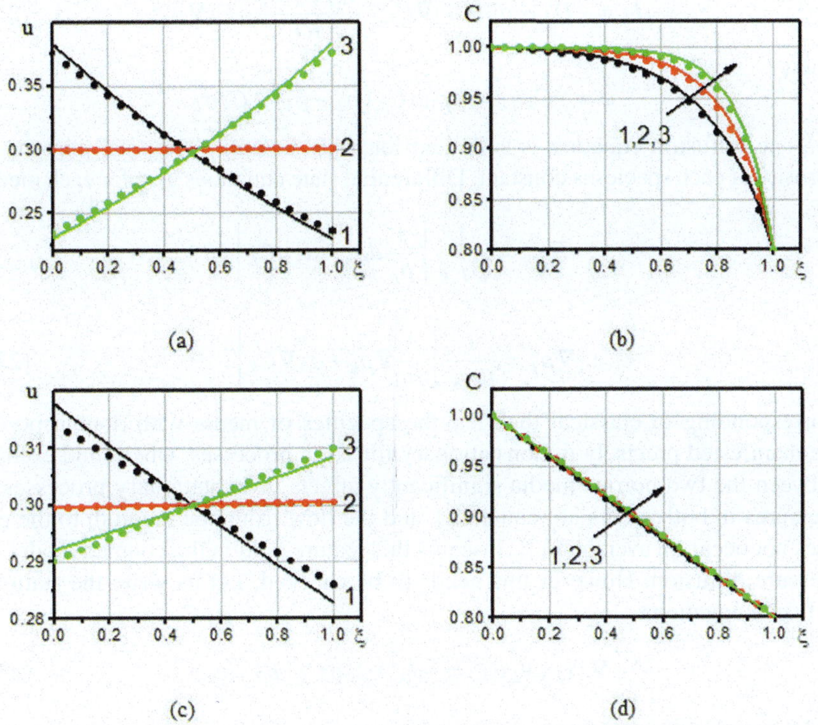

Fig. 23.1: Velocity and concentration in flat layer during fluid filtration α_C= 1) 0.57; 2) 0.0; 3) 0.3; $\overline{\beta}_T = 0.01$; $\overline{p}_2 = 0.7$; (a,b) $\delta = 0.05$; (c,d) $\delta = 100$; symbols: $\omega = 0$; solid lines: $\omega = 0.3$.

diffusion mode, when $Pe = 1/\delta \ll 1$,the concentration varies nonlinearly (Fig. 23.1 b); in convective mode, we have linear concentration distribution. Other examples are presented in [21] applicable to biophysical problems.

23.4.5 Diffusion and Filtration in Media with Double Porosity

Suppose that the nonviscous fluid consists of two components, i.e., similarly to (23.54) for each medium we have

$$C_{1,k} + C_{2,k} = 1. \tag{23.71}$$

Then for each medium we need one diffusion equation, in which we omit the numbers of components, and equations for diffusion fluxes

$$\left(\frac{\partial m_k \rho_k C_k}{\partial t} + \nabla \cdot \rho_k V_k C_k \right) = -\nabla \cdot \boldsymbol{J}_k + \varphi_k, \tag{23.72}$$

$$\boldsymbol{J}_k = -D_k \rho_k f_k(C_k) \nabla C_k - \frac{D_k C_k M_k}{RT} \alpha C_k \nabla p_k^e, \tag{23.73}$$

where

$$\varphi = -C_k q_k = \mp \rho_0 \alpha C_k (p_2 - p_1) - C_k \sigma_k,$$

i.e., one diffusion equation is sufficient for each medium. The total porosity and porosity of each species is constant. Differential state equations give for each medium

$$\nabla p_k^e = \frac{1}{\beta_{T,k}} \left[\frac{1}{\rho_k} d\rho_k + \alpha_k dC_k \right]$$

or

$$\nabla p_k^e = \frac{1}{\beta_{T,k}} \left[\frac{1}{\rho_k} \nabla \rho_k + \alpha_k \nabla C_k \right]. \tag{23.74}$$

Corresponding to classical theories, the specifics of media with double porosity are manifested precisely in nonstationary filtration processes, when fluid exchange between the two porous media significantly affects it. In stationary processes, the pressures in both media are equalized, and the flow from one medium to the other does not occur on average [22]. It seems this feature hold in the coupled model with pressure diffusion. However this needs to be checked. Let us write the stationary diffusion equations

$$\nabla \cdot (\rho_k V_k C_k) = -\boldsymbol{J}_k \mp \rho_0 \alpha C_k (p_2 - p_1) \tag{23.75}$$

and the continuity equations

$$\nabla \cdot \rho_k V_k \pm \rho_0 \alpha (p_2 - p_1) = 0. \tag{23.76}$$

The rest of the equations will not change.

Firstly we analyze the particular case of ideal gas. In this case $\beta_{T,k} \sim p_k^{-1}$, there is no volumetric viscosity $p_k^e = p_k$, the pressure caused by a change in concentration of the diffusate can be neglected. It follows from (23.74)

$$\frac{\nabla p_k}{p_k} = \frac{A}{\rho_k} \nabla \rho_k, \quad p_k = A \rho_k, \tag{23.77}$$

where A - is some constant.

With help of the of Darcy's law, from (23.76) we obtain

$$-\nabla \cdot \left(\frac{p_k}{A} \frac{k_k}{\mu} \nabla p_k \right) \pm \rho_0 \alpha (p_2 - p_1) = 0.$$

Believing, that viscosity is proportional to pressure, $\mu = \mu_0 p$, rewrite this equation in form

$$-\nabla \cdot \left(\frac{1}{A} \frac{k_k}{\mu_0} \nabla p_k \right) \pm \rho_0 \alpha (p_2 - p_1) = 0.$$

In the one-dimensional situation for a flat layer we have

$$-\frac{\mathrm{d}}{\mathrm{d}x}\left(\kappa_1\frac{\mathrm{d}}{\mathrm{d}x}p_1\right)+\rho_0\alpha\left(p_2-p_1\right)=0$$

and

$$-\frac{\mathrm{d}}{\mathrm{d}x}\left(\kappa_2\frac{\mathrm{d}}{\mathrm{d}x}p_2\right)+\rho_0\alpha\left(p_2-p_1\right)=0,$$

where

$$\kappa_k=\frac{1}{A}\frac{k_k}{\mu_0}.$$

From the first equation we find

$$p_2=\frac{\kappa_1}{\rho_0\alpha}\frac{\mathrm{d}^2p_1}{\mathrm{d}x^2}+p_1.$$

Substituting it in the second, we arrive at the fourth-order equation

$$\frac{\kappa_1\kappa_2}{\rho_0\alpha}\frac{\mathrm{d}^4p_2}{\mathrm{d}x^4}+\left(\kappa_1+\kappa_2\right)\frac{\mathrm{d}^2p_2}{\mathrm{d}x^2}=0. \tag{23.78}$$

We can obtain the same equation for p_1. That is, the pressures in the media satisfy the same equations. However, since the permeabilities of the media are different, the filtration rates in them will also be different. This will lead to differences in the concentration distribution not only because of different convective velocities, but also because of different diffusion coefficients and different transfer coefficients under the pressure gradient. However, for the same in different porous spaces.

State equations (23.77) can be written as follows

$$\frac{\rho_1}{p_1}=\frac{\rho_2}{p_2}=\frac{\rho_0}{p_0}.$$

Then, using the relations for fluxes, we obtain the diffusion equations coupled with pressure even in stationary conditions

$$\nabla\cdot\left(\frac{\rho_0}{p_0}p_1V_1C_1\right)=\nabla\cdot\left[D_1\frac{\rho_0}{p_0}p_1\nabla C_1+\frac{D_1C_1M}{RT}\alpha_1\nabla p_1\right]-\rho_0\alpha C_1\left(p_2-p_1\right),$$

$$\nabla\cdot\left(\frac{\rho_0}{p_0}p_2V_2C_2\right)=\nabla\cdot\left[D_2\frac{\rho_0}{p_0}p_2\nabla C_2+\frac{D_2C_2M}{RT}\alpha_2\nabla p_2\right]-\rho_0\alpha C_1\left(p_2-p_1\right).$$

Although the pressures in the media are the same, the concentration distributions in different pore systems may be different. Note, if $\mu=\text{const}$, equations turn out to be more complex.

23.4.6 Viscous Two-component Fluid in Media with Double Porosity

If fluid is viscous, Eqs. (23.75), (23.76) and the Darcy law include the total pressure. However, state equations and fluxes include only part of pressure p_k^e. The set of equations for each medium is presented in Table 23.1. Let us make transformations similar to Subsect. 23.4.2, for example, for medium 1 and take into account that for the same fluid in different porous spaces we can take $\beta_{T1} = \beta_{T2} = \beta_T$; $\alpha_{C1} = \alpha_{C2} = \alpha_C$. Taking into account the equation of state and the expression for the viscous pressure part, the Darcy equation and the continuity equation take the form

$$V_1 = -\frac{k_1}{\mu m_1 \beta_T}\left(\frac{\nabla\rho_1}{\rho_1} + \alpha_C \nabla C_1\right) + \frac{k_1 \kappa_V}{\mu m_1}\nabla(\nabla\cdot V_1);$$

$$\nabla(\nabla\cdot\rho_1 V_1) + \frac{\rho_0\alpha}{m_1}\times$$

$$\times\left[\frac{1}{\beta_T}\left(\frac{\nabla\rho_2}{\rho_2} - \frac{\nabla\rho_1}{\rho_1} + \alpha_C(\nabla C_2 - \nabla C_1)\right) - \kappa_V \nabla(\nabla\cdot V_2 - \nabla\cdot V_1)\right] = 0.$$

Table 23.1: Set of stationary equations.

1	2
$\nabla\cdot\rho_1 V_1 + \frac{\rho_0\alpha}{m_1}(p_2 - p_1) + \frac{\sigma_1}{m_1} = 0$	$\nabla\cdot\rho_2 V_2 + \frac{\rho_0\alpha}{m_2}(p_2 - p_1) + \frac{\sigma_2}{m_2} = 0$
$V_1 = -\frac{k_1}{\mu m_1}\left(p_1^e + p_1^V\right)$	$V_2 = -\frac{k_2}{\mu m_2}\left(p_2^e + p_2^V\right)$
$p_1^V = -\kappa_{V1}\nabla\cdot V_1$	$p_2^V = -\kappa_{V2}\nabla\cdot V_2$
$p_1 = p_1^e + p_1^V$	$p_2 = p_2^e + p_2^V$
$\nabla\cdot(\rho_1 V_1 C_1) = -\nabla\cdot \boldsymbol{J}_1 - \rho_0\alpha C_1(p_2 - p_1)$	$\nabla\cdot(\rho_2 V_2 C_2) = -\nabla\cdot \boldsymbol{J}_2 - \rho_0\alpha C_2(p_2 - p_1)$
$\boldsymbol{J}_1 = -D_1\rho_1 f(C_1)\nabla C_1 - \frac{D_1 C_1 M_1}{RT}\alpha_{C1}\nabla p_1^e$	$\boldsymbol{J}_2 = -D_2\rho_2 f(C_2)\nabla C_2 - \frac{D_2 C_2 M_2}{RT}\alpha_{C2}\nabla p_2^e$
$p_1^e = \frac{1}{\beta_{T_1}}\left(\frac{\nabla\rho_1}{\rho_1} + \alpha_{C_1}\nabla C_1\right)$	$p_2^e = \frac{1}{\beta_{T_2}}\left(\frac{\nabla\rho_2}{\rho_2} + \alpha_{C_2}\nabla C_2\right)$

The diffusion coefficient in the same liquid in different porous spaces in the first approximation also can be considered the same. Then, taking into account the equation of state, we obtain the stationary diffusion equation in the form:

$$\nabla \cdot (\rho_1 V_1 C_1) + \rho_0 \alpha C_1 (p_2 - p_1) =$$
$$= \left[D\rho_1 f(C_1) \nabla C_1 + \frac{DC_1 M_1}{RT} \alpha_C \frac{1}{\beta_T} \left(\frac{\nabla \rho_1}{\rho_1} + \alpha_C \nabla C_1 \right) \right].$$

I.e. we have a system of three equations for the three unknowns ρ_1, V_1, C_1, coupled with similar equations for the second medium. In principle, it is possible to go to the equations with respect to ρ_1, P_1, C_1. It is possible to separate the equations only with additional simplifying assumptions.

23.4.7 Nonviscous Two-component Fluid in Porous media with Nano- and Micro-pores

Let us consider interested situation. In one medium there is only diffusion (index 1), permeability is small (for example, in nanopores), and in the second medium (index 2) there is mainly filtration (in micron-sized pores). Volumetric viscosity in this particular case is also neglected. Then we come to the equations

$$-\nabla \cdot \boldsymbol{J}_1 - \rho_0 \alpha C_1 (p_2 - p_1) = 0,$$
$$\nabla p_1 = \frac{1}{\beta_T} \left[\frac{1}{\rho_1} + \alpha_1 \nabla C_1 \right] \approx \frac{1}{\beta_{T,1}(p)} \alpha_1 \nabla C_1, \qquad (23.79)$$
$$\boldsymbol{J}_1 = -D_1 \rho_1 \nabla C_1 - \frac{D_1 C_1 M}{RT} \alpha_1 \nabla p_1 = -D_1 \rho_1 \nabla C_1 - \frac{D_1 C_1 M}{RT} \frac{\alpha_1^2}{\beta_{T,1}(p)} \nabla C_1$$

for the first medium. From the first and third equations (23.79) we find

$$\nabla \cdot \left[D_1 \rho_1 \nabla C_1 + \frac{D_1 C_1 M}{RT} \frac{\alpha_1^2}{\beta_{T,1}(p)} \nabla C_1 \right] - \rho_0 \alpha C_1 (p_2 - p_1) = 0. \qquad (23.80)$$

For the second medium we get

$$\nabla \cdot \rho_2 V_2 - \rho_0 \alpha (p_2 - p_1) = 0;$$
$$\nabla \cdot (\rho_2 V_2 C_2) = \rho_0 \alpha C_2 (p_2 - p_1);$$
$$\nabla p_2 = \frac{1}{\beta_{T,2}(p)} \frac{1}{\rho_2} \nabla \rho_2;$$
$$V_2 = -\frac{k_2}{\mu(p)} \nabla p_2 = -\frac{k_2}{\mu(p)} \frac{1}{\beta_{T,2}(p)} \frac{1}{\rho_2} \nabla \rho_2;$$

It follows from the first two equations that there is no concentration gradient in the second medium. From the first and fourth equations we have

$$-\nabla\cdot\left(\frac{k_2}{\mu(p)}\frac{1}{\beta_{T,2}(p)}\nabla\rho_2\right)-\rho_0\alpha(p_2-p_1)=0$$

or

$$-\nabla\cdot\rho_2\frac{k_2}{\mu(p)}\nabla\rho_2-\rho_0\alpha(p_2-p_1)=0.$$

We assume again that compressibility is inversely proportional to pressure

$$\beta_{T,k}\sim\beta_{T,0}p_k^{-1}.$$

Then for the first medium we have

$$\frac{\mathrm{d}}{\mathrm{d}x}p_1=\frac{p_1}{\beta_{T,0}}\alpha_1\frac{\mathrm{d}C_1}{\mathrm{d}x}.$$

or

$$\ln p_1=\ln A_1+\frac{\alpha_1}{\beta_{T,0}}C_1,$$

or

$$p_1=p_0\exp\left(\frac{\alpha_1}{\beta_{T,0}}(C_1-C_0)\right)\approx p_0\left[1+\frac{\alpha_1}{\beta_{T,0}}(C_1-C_0)\right]$$

if $\frac{\alpha_1}{\beta_{T,0}}\ll 1$. The diffusion equation is obtained in the form

$$\frac{\mathrm{d}}{\mathrm{d}x}\left[D_1\rho_1\left(1+\frac{C_1M}{RT\rho_1}\frac{\alpha_1^2p_1}{\beta_{T,0}}\right)\frac{\mathrm{d}C_1}{\mathrm{d}x}\right]-\rho_0\alpha C_1\,(p_2-p_1)=0.$$

For the second medium we get

$$\frac{\mathrm{d}}{\mathrm{d}x}p_2=\frac{p_2}{\beta_{T,0}}\frac{1}{\rho_2}\beta_{T,0}\rho_2,$$

from where we find

$$\frac{\rho_2}{p_2}=\frac{\rho_0}{p_0}$$

or

$$\frac{\mathrm{d}}{\mathrm{d}x}\frac{\rho_0}{p_0}p_2\frac{k_2}{\mu(p)}\frac{\mathrm{d}}{\mathrm{d}x}p_2+\rho_0\alpha\,(p_2-p_1)=0.$$

If the viscosity is proportional to pressure, then

$$\frac{\mathrm{d}}{\mathrm{d}x}\left(\frac{\rho_0}{p_0}\frac{k_2}{\mu_0}\frac{\mathrm{d}}{\mathrm{d}x}p_2\right)+\rho_0\alpha\,(p_2-p_1)=0$$

or

$$\frac{\rho_0}{p_0}\frac{k_2}{\mu_0}\frac{\mathrm{d}^2p_2}{\mathrm{d}x^2}+\rho_0\alpha\,(p_2-p_1)=0.$$

Thus, to find the concentration distribution in nanopores, two coupled equations are required

$$D_1\rho_0\frac{\mathrm{d}^2C_1}{\mathrm{d}x^2}+\frac{D_1M}{RT}\frac{\alpha_1}{\beta_{T,0}}p_0\frac{\mathrm{d}}{\mathrm{d}x}\left[C_1\left(1+\frac{\alpha_1}{\beta_{T,0}}(C_1-C_0)\right)\frac{\mathrm{d}}{\mathrm{d}x}C_1\right]-$$
$$-\rho_0\alpha C_1\left(p_2-p_0\left[1+\frac{\alpha_1}{\beta_{T,0}}(C_1-C_0)\right]\right)=0$$

and

$$\frac{\rho_0}{p_0}\frac{k_2}{\mu_0}\frac{\mathrm{d}^2p_2}{\mathrm{d}x^2}+\rho_0\alpha\left(p_2-p_0\left[1+\frac{\alpha_1}{\beta_{T,0}}(C_1-C_0)\right]\right)=0.$$

In fact, mass transfer from one pore system to another leads to an increased role of pressure diffusion.

23.5 Conclusion

Presented here general equations and described particular situation speak that the approach to the model construction based on state equations in differential form leads to coupled models. The splitting of equations is possible only in the simplest cases. All particular problems contain physically clear parameters, which are determined in thermodynamics and can be found experimentally.

Acknowledgements The work was performed according to the Government research assignment for ISPMS SB RAS, project FWRW-2021-0007.

References

[1] Barenblatt GI, Entov VM, Ryzhik VM (1972) Theory of Non-stationary Filtration of Liquid and Gas (in Russ), Nedra, Moscow.
[2] Vafai K (2005) Handbook of Porous Media, 2nd ed., Taylor Francis Group, Boca Raton.
[3] Bear J (2018) Modeling Phenomena of Flow and Transport in Porous Media, Springer, New York.
[4] Nield DA, Bejan A (2013) Convection in Porous Media, Springer, New York, Heidelberg, Dordrecht, London.
[5] Vafai K (2015) Handbook of Porous Media, CRC Press, Boca Raton, New York, Singapore.
[6] Muskat M (1982) The Flow of Homogeneous Fluids Through Porous Media, International Human Resources Development Corp., Boston.

[7] Knyazeva AG (1997) Introduction into Locally Equilibrium Thermodynamics of Physicochemical Transformations in Deformable Media (in Russ), TSU Press, Tomsk.
[8] Knyazeva AG (2016) Thermodynamics with additional parameters for polycrystals, Int. J. Nanomechanics Sci. Technol. **7**(1):1-25.
[9] Barenblatt GI, Entov VM, Ryzhik VM (1984) Movement of Liquids and Gases in Natural Reservoirs (in Russ), Nedra, Moscow.
[10] Polyaev VM, Maiorov VA, Vasiliev LL (1988) Hydrodynamics and Heat Transfer in Porous Structural Elements of Aircraft (in Russ), Mashinostroenie, Moscow.
[11] Basniev KS, Kochina PYa, Maksimov VM (1993) Underground Hydromechanics (in Russ), Nedra, Moscow.
[12] Maksimov VM (2019) Generalized law of multiphase filtration and new effects of surface phenomena at two-phase flows in a porous medium, Georesources **21**(1):86-91.
[13] Kolesnichenko AV, Maksimov VM (2001) The generalized Darcy law of filtration as inquest of Stefan-Maxwell relations for heterogeneous medium, Matem. Mod. **13**(1):3-25.
[14] Szymkiewicz A (2013) Modelling Water Flow in Unsaturated Porous Media, Springer, Heidelberg, New York, Dordrecht, London.
[15] De Grot SR, Mazur P (1984(Non-equilibrium Thermodynamics, Dover Publications Inc.
[16] Diarmati, I. (1974) Non-equilibrium Thermodynamics (in Russ), Mir, Moscow.
[17] Kirillin VA, Sychev AV, Sheindlin AE (1983) Technical Thermodynamics. Energoatomizdat, Moscow.
[18] Haase R (1990) Thermodynamics of Irreversible Processes,Dover, New York.
[19] Knyazeva AG. (2006) Thermodynamic model of a viscous thermally conductive gas and its application to the description of combustion processes, Mathematical Modeling of Systems and Processes (in Russ). **14**:92-108.
[20] Kondepury D, Prigogine I (2002) Modern Thermodynamics, John Wiley & Sons, Chichester, New York, Weinheim, Brisbane, Toronto, Singapore.
[21] Knyazeva AG, Nazarenko NN (2022) Coupled model of a biological fluid filtration through a flat layer with due account for barodiffusion. Transp. Porous Med. **141**:331-358.
[22] Barenblatt GI, Zheltov YP, Kochina IN (1960) Basic concepts in the theory of seepage of homogeneous liquids in fractured rocks(in Russ), Prik. Matem. i Mekh. **24**(5):852-864.

Chapter 24
Ball Indentation of Perforated Circular Hyperelastic Membranes

Alexey M. Kolesnikov

Abstract This paper deals with the problem of ball indentation of a thin flat circular highly elastic membrane with a hole in the centre. The aim of this paper is to investigate the effect of friction on the passage of the ball through the hole. The problem is investigated within the framework of the theory of non-linear elastic membranes. The Gent model is chosen as the material model. The friction is described by Coulomb's law. The dependences between the indentation force and indenter displacement are calculated for different friction coefficients, ball indenter radius and hole radius. The ultimate force and displacement of the indenter, after which the ball passes completely through the hole in the membrane, are determined.

Key words: Membrane · Nonlinear · Elastic · Indentation · Contact · Coulomb's friction

24.1 Introduction

Many engineering, nanophysics and biomechanics problems involve the deflection of thin elastic membranes with a fixed boundary, e.g. pressure sensors, valves and actuators, biological tissues. Membrane deflection is used to determine mechanical properties of [1]-[3] materials, including biological tissues [4, 5] and graphene membranes [6, 7]. For rubber-like, elastomeric, biological and graphene membranes, large deflections and movements are acceptable. These materials are also characterised by a non-linear elastic behaviour.

The problems of indentation and puncture of highly elastic membranes at large deflections and deformations have been investigated by various authors. In [8, 9] the

Alexey M. Kolesnikov
Southern federal university, Milchakova St. 8a, Rostov-na-Donu, 344090 & Don state technical university, Gagarin square 1, Rostov-na-Donu, 344000, Russian Federation.
e-mail: Alexey.M.Kolesnikov@gmail.com

H. Altenbach et al. (eds.), *Sixty Shades of Generalized Continua*,
Advanced Structured Materials 170, https://doi.org/10.1007/978-3-031-26186-2_24

axisymmetric deformation of a circular highly elastic membrane under the action of a spherical and parabolic solid body is considered. In [10, 11] the piercing of a highly elastic membrane with a cylindrical indenter with a spherical tip is investigated. In [12, 13] the indentation of a curved highly elastic shell and membrane with a cylindrical indenter with a spherical tip was investigated. The papers [8], [10]-[12] study axisymmetric cases and neglects friction in the contact area.

Research on indentation of a circular highly elastic membrane with a friction is presented in [14]-[17]. In [14, 17] experimental results of ball indentation of rubber and silicone membranes are presented and the indentation process taking into account friction in the contact area is numerical simulated. In [14] the asymmetric case is considered in addition to the axisymmetric case. In [15, 16] the indenter is a cylinder and experimental data and numerical simulation results are presented.

Thus, indentation of circular hyperelastic membranes is an actively studied and demanded problem in various fields. The problem of indentation of a perforated highly elastic membrane at large deformations is one of the less studied problems. In [10, 11] the problem of indentation and puncture an initially circular membrane is considered. The problem of indentation of a highly elastic membrane with a hole is considered in [18]. In [18] a cylindrical indenter with a spherical tip is considered. The bifurcation of the solution during the indenter movement is investigated, as well as the effect of membrane pre-tensioning and the existence of an elastic substrate on the indentation process. Friction in the indenter and membrane contact area is neglected. At the same time the influence of friction in indentation problems may be significant, as shown in [14]-[17].

This paper investigates the effect of friction on the indentation of a rigid ball indenter in a circular membrane with a hole in the centre. The objective is to determine the effect of friction on the passage of a ball indenter through a hole in the elastic membrane and to determine the limit value of the indentation force and the movement of the indenter after which the ball will pass completely through the hole. The indentation process is modelled as quasi-static. In the analysis, the theory of non-linear elastic membranes is used. The friction is modelled by Coulomb's law [14]. The statics problem is reduced to a boundary value problem for the system of ordinary differential equations with a parameter (unknown boundary of a contact region). The derived equations are valid for arbitrary nonlinear-elastic incompressible material. In this work we consider the Gent model of material. The dependences between the indenting force and indenter displacement are constructed numerically. The limiting values of the indentation force and indenter displacement, at which the membrane-indenter system is in equilibrium, are determined.

24.2 Mathematical Model

24.2.1 Axisymmetric Problem of Non-linear Elastic Membranes

We assume that the thickness of the membrane with respect to the radius is small and large tensile forces occur in the membrane during deformation, so we can neglect the bending stiffness and bending stresses and use the theory of non-linear elastic membranes [19, 20, 21, 22] among others. In the undeformed state, the position of the points of the membrane middle surface is given in the cylindrical coordinate system as follows

$$\{r \in [r_A, r_B], \phi \in [0, 2\pi], z = 0\}.$$

For an axisymmetric deformation, the position of the points on the middle surface of the deformed membrane can be written in the same cylindrical coordinate system as $\{R(r), \phi, Z(r)\}$.

The principal stretches λ_α and curvature κ_β of the deformed membrane surface are expressed as

$$\lambda_1 = \sqrt{R'^2 + Z'^2}, \quad \lambda_2 = \frac{R}{r}, \quad \left(R' = \frac{dR}{dr}\right),$$
$$\kappa_1 = \frac{\psi'}{\lambda_1}, \quad \kappa_2 = \frac{\sin\psi}{r\lambda_2}, \quad \tan\psi = \frac{Z'}{R'}. \tag{24.1}$$

Here $\psi(r)$ is the angle of the tangent to the surface of the deformed membrane in the RZ plane.

The equilibrium equations for a non-linear elastic membrane under axisymmetric deformation are

$$\frac{dT_1}{dr} + \frac{T_1 - T_2}{R}\frac{dR}{dr} + \lambda_1 q_1 = 0,$$
$$T_1\kappa_1 + T_2\kappa_2 + q = 0. \tag{24.2}$$

Here $q_1 = q_1(r)$ and $q = q(r)$ are tangential and normal surface loads, T_1 and T_2 are the stress resultants in the membrane. For an elastic membrane composed from a highly elastic incompressible material, the defining relations can be written using the specific strain potential energy function $W(\lambda_1, \lambda_2)$ as follows

$$T_1 = \frac{hW_1}{\lambda_2}, \quad T_2 = \frac{hW_2}{\lambda_1}. \tag{24.3}$$

Hereinafter the following symbols are used

$$W_k = \frac{\partial W}{\partial \lambda_k}, \quad W_{mn} = \frac{\partial^2 W}{\partial \lambda_m \partial \lambda_n}.$$

24.2.2 Ball Indentation of Perforated Circular Hyperelastic Membranes

The deformed membrane can be divided into two parts: the part AC that is in contact with a solid ball of radius R_0 and the part CB that is free from surface loads, as shown in Fig. 24.1.

24.2.2.1 Contact Area

In the contact area, the membrane is tight against the ball indenter, then the surface of the membrane can be described as follows

$$R(\psi) = R_0 \sin\psi, \quad Z(\psi) = R_0(1-\cos\psi), \quad \psi \in [\psi_A, \psi_C]. \tag{24.4}$$

Here the angle ψ is the angular coordinate, which for a ball indenter is the same as the angle of the tangent (24.1).

For the tangential component of the surface load we assume that

$$|q_1| = f|q|.$$

That is, we assume that Coulomb's law is satisfied in the limiting case.

Equilibrium equations (24.2) are reduced to a second order differential equation for the function $\psi(r)$ and an expression for the normal pressure $q(r)$:

$$\begin{gathered}\frac{d^2\psi}{dr^2}W_{11}R_0 r + \frac{d\psi}{dr}(W_{12}R_0\cos\psi - W_1 r f) - W_{12}\frac{R_0}{r}\sin\psi + \\ +W_1 - W_2(\cos\psi + f\sin\psi) = 0, \\ q = -\frac{h}{R_0^2}\left[\frac{r}{\sin\psi}W_1 + W_2\left(\frac{d\psi}{dr}\right)^{-1}\right].\end{gathered} \tag{24.5}$$

24.2.2.2 Out of Contact Area

Equilibrium equations (24.2) of the out of contact part of the membrane can be represented as [16]

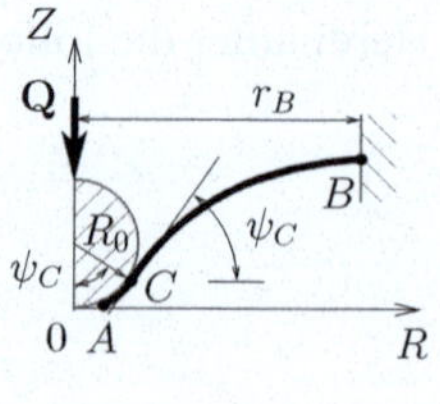

Fig. 24.1 Section of deformed membrane in the plane RZ.

$$W_{11}\frac{d\lambda_1}{dr}+\frac{1}{r}\Big[(\lambda_1\cos\psi-\lambda_2)W_{12}-W_2\cos\psi+W_1\Big]=0,$$
$$W_1\frac{d\psi}{dr}+W_2\frac{\sin\psi}{r}=0,\quad \frac{d\lambda_2}{dr}=\frac{1}{r}\Big(\lambda_1\cos\psi-\lambda_2\Big), \tag{24.6}$$
$$\frac{dR}{dr}=\lambda_1\cos\psi,\quad \frac{dZ}{dr}=\lambda_1\sin\psi.$$

24.2.2.3 Boundary Conditions

The following boundary conditions are imposed on the solutions of the differential equations in and out of the contact region. At point A the edge of the membrane is free

$$T_1(r_A)=0. \tag{24.7}$$

The other edge of the membrane (point B) is fixed

$$R(r_B)=r_B,\quad \lambda_2(r_B)=1. \tag{24.8}$$

At the point C the solutions on the two sections AC and CB are bound by the continuity and smoothness condition

$$R^{CB}(r_C)=R^{AC}(r_C),\quad Z^{CB}(r_C)=Z^{AC}(r_C),\quad \psi^{CB}(r_C)=\psi^{AC}(r_C),$$
$$\lambda_1^{CB}(r_C)=\lambda_1^{AC}(r_C),\quad \lambda_2^{CB}(r_C)=\lambda_2^{AC}(r_C). \tag{24.9}$$

For a system consisting of a ball and part of a membrane AC, the equation of equilibrium forces along the axis Z can be written as

$$Q-2\pi R_0(\sin\psi(r_C))^2T_1(r_C). \tag{24.10}$$

This condition is necessary to determine the unknown coordinate r_C from a given force Q. Or it can be used to determine the force Q for a given coordinate r_C. Thus, the static problem of a nonlinear-elastic membrane is reduced to a boundary value problem with parameter r_C for the ordinary differential equations, which is solved by the shoot method.

24.3 Numerical Results

In the numerical simulation we will use Gent's strain energy density function [23]

$$W=-\frac{\mu J_m}{2}\ln\left(1-\frac{I_1-3}{J_m}\right). \tag{24.11}$$

The following values of material parameters $\mu = 1$ (MPa), $J_m = 50$ were used in the numerical analysis. Note that with the dimensionless parameters adopted below, the parameter μ has no effect on the results.

In the experiment, the measured quantities are the indentation force Q and the displacement Δ of the indenter ball. In a given coordinate system, the indenter displacement can be calculated as

$$\Delta = Z_B - R_0 (1 - \cos\psi_{A0}), \quad \psi_{A0} = \arcsin\frac{r_A}{R_0}$$

We use the following geometrical parameters $r_B = 25$ (mm), $h = 0.5$ (mm) for the calculation. The parameters r_A and R_0 were varied. Let us introduce the dimensionless indentation force, the relative indenter displacement and the dimensionless coordinates

$$\tilde{Q} = \frac{Q}{2\pi R_0 \mu h}, \quad \tilde{\Delta} = \frac{\Delta}{R_0}, \quad \tilde{R} = \frac{R}{r_B}, \quad \tilde{Z} = \frac{Z}{r_B}.$$

Figure 24.2 shows the dependence between the dimensionless indentation force $\tilde{Q}$ and the relative indenter displacement $\tilde{\Delta}$ for different friction coefficients f, indenter ball raduis R_0 and membrane hole radius r_A. The curves in Fig. 24.2 correspond to friction coefficients of 0, 0.1, 0.2, ..., 1.1, starting from the bottom curve.

All of the curves shown have common features. At the start of the indentation process, friction has almost no effect. A small change in force leads to large indenter displacements. This is due to the neglect of bending stiffness of the thin elastic plate. A further increase in the indenter displacement is accompanied by an increase in the indentation force. The friction force begins to have a significant effect. The indentation force $\tilde{Q}$ has a maximum, which is obviously greater the greater the friction coefficient. If we set the force in the experiment, then an excess of this maximum means that the ball has passed through the hole.

After that an increase in indenter displacement is accompanied by a fall in the indenting force. In the experiment this is possible if we set the displacement of the indenter. The indenter displacement also has a maximum. The value of which is higher for larger friction coefficients. If this is exceeded, the ball passes through the hole in the experiment. Once the maximum displacement is reached, the curves in the figures continue with a decrease in both force and displacement. In Fig. 24.2 the two gray curves connecting the points (squares and triangles) correspond to the maximum indentation force and maximum displacement of indenter, respectively.

The dependence of the maximum indentation force and maximum indenter displacement on the friction coefficient is shown in Fig. 24.3 and 24.4 at various geometric parameters shown in Table 24.1. As we can se from Fig. 24.3a at $R_0/r_A \leq 2$ the dependence $\tilde{Q} - f$ is close to linear and weakly dependent on the radii of the ball-indenter and hole in the membrane. The dotted curves in Fig. 24.3 coincide. At $R_0/r_A = 3$ the dependence $\tilde{Q} - f$ becomes substantially nonlinear and the influence of the indenter ball and hole sizes is significant (see Fig. 24.3a).

The maximum indenter displacement increases slowly for all considered cases, except one case: $r_A = 2.5$, $R_0 = 3.75$ (mm) (the black solid curve in Fig. 24.4a).

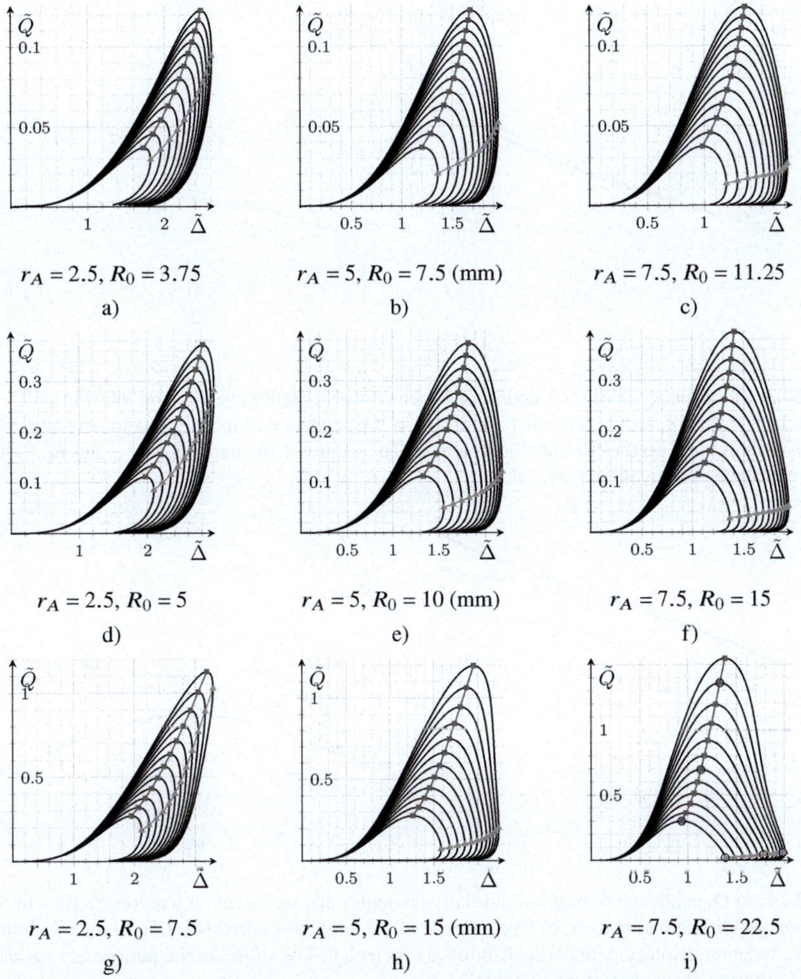

Fig. 24.2: Dependence between ball-indentor displacement Δ^* and indentation force Q^*. The friction coefficient $f = 0, 0.1, 0.2, \ldots, 1.1$, starting from the bottom curve.

Table 24.1: The parameter values r_A and R_0 for the curves in Fig. 24.3 and 24.4.

$\frac{R_0}{r_A}$ \ r_A	2.5	5	7.5
1.5	....	....	....
2	- - - - - - - -	- - - - - - - -	- - - - - - - -
3	————————	————————	————————

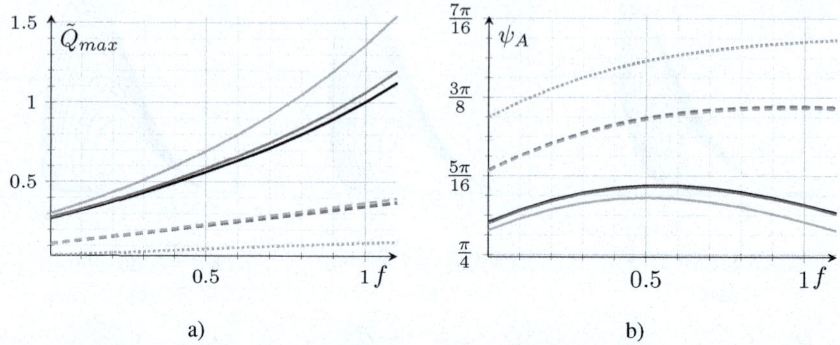

Fig. 24.3: a) Dependence of the maximum dimensionless indentation force on the friction coefficient f for different geometric parameters. b) Dependence of the angle ψ_A corresponding to the maximum force on the friction coefficient f. The values of the parameters r_A and R_0 for the different curves are given in Table 24.1.

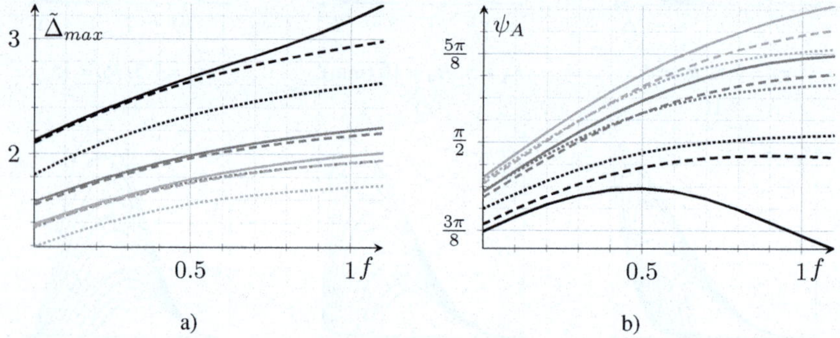

Fig. 24.4: a) Dependence of maximum relative indenter displacement on friction coefficient f for different geometric parameters. b) Dependence of the angle ψ_A corresponding to the maximum relative indenter displacement on the friction coefficient f. The values of the parameters r_A and R_0 for the different curves are given in Table 24.1.

Let us consider the relationship between ψ_A and the friction coefficient f for two special cases: at maximum indentation force (Fig. 24.3b), and at maximum indenter displacment (Fig. 24.4b). The angle ψ_A is the angle of inclination of the tangent to the membrane (24.1). For a ball indenter, it is also the angular coordinate (24.4), which determines the position of the edge of the mambrane hole on the ball indenter. This parameter describes how much of the ball has already passed through the hole. For example, $\psi_A = \pi/2$ means that one half of the ball is on one side of the membrane and the other half is on the other side.

As we can see in Fig. 24.3b, the maximum indentation force $\tilde{Q}$ is reached before the ball-indenter penetrates half of the hole ($\psi_A < \pi/2$). For $R_0/r_A = 1.5$ the angle ψ_A increases as the friction coefficient f increases. And for $R_0/r_a = 2, 3$, the dependence

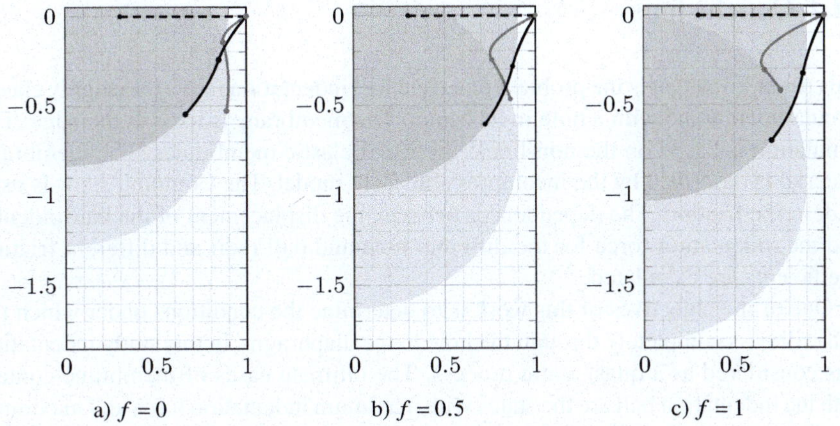

Fig. 24.5: Section of the deformed membrane in dimensionless coordinates at $r_A = 7.5$, $R_0 = 22.5$ (mm). The black curve is the shape of the membrane at maximum compressive force Q, the gray curve is the shape of the membrane at maximum displacement of indenter Δ, the black dashed line is the undeformed state of the membrane.

of ψ_A on f is not monotonic. This is particularly strong at $R_0/r_a = 3$ (the solid curves in Fig. 24.3b).

The angle ψ_A corresponding to the maximum relative indenter displacement $\tilde{\Delta}$ (while the indenter is in contact with the membrane) is shown in Fig. 24.4b. Without friction and with small friction coefficients, it is less than $\pi/2$, i.e. in the state of equilibrium corresponding to the maximum movement of the indenter ball, the indenter ball passes through the hole in the membrane by less than half, as shown in Fig. 24.5a ($f = 0$) by the gray line. Increase friction coefficient lead to increase the angle ψ_A increases. With a small hole radius of $r_A = 2.5$ (mm), the angle ψ_A does not exceed $\pi/2$. At $r_A = 5$ and $r_A = 7.5$ (mm), the angle ψ_A becomes greater than $\pi/2$ at $f \approx 0.15 - 0.3$ depending on the hole radius. That is, until the limit state is reached, the ball-indenter passes through the hole in the membrane by more than half, as shown in Fig. 24.5b,c ($f = 0.5$, 1, respectively).

Figure 24.5 shows the shapes of the membranes at $r_A = 7.5$, $R_0 = 22.5$ (mm) in the dimensionless coordinates $\{\tilde{R}, \tilde{Z}\}$ for the different f shown in the figure. The black curve corresponds to the deformed membrane at the maximum force Q, the grey curve corresponds to the deformed membrane at the maximum displacement of the ball-indenter Δ. The black dashed line shows the initial state of the membrane. The states corresponding to these shapes are marked in Fig. 24.2i by circles.

24.4 Conclusions

This paper investigates the problem of a rigid ball indentation into a thin highly elastic circular membrane with a hole in its center. The membrane is fixed at the edge. The simulation is based on the non-linear theory of elastic membranes. The membrane material is described by the incompressible Gent model. The Coulomb's law is used to describe friction. The dependences between the displacement of the ball-indenter and the indentation force for the different hole and ball radii and different friction coefficients are calculated.

One of the objectives of this work is to determine the conditions under which the ball will pass completely through the hole in the diaphragm. In this study indentation was considered as a quasi-static process. The ultimate cases of membrane contact with the indentation ball are the states with maximum indentation force and maximum displacement of the indenter.

If the control action is an indentation force, then we can assume that the ball will completely pass through the hole in the membrane when the indentation force exceeds the maximum value one. There is no equilibrium state in which the membrane is in contact with the indenter for force more than the maximum. Note that if the ratio of the indenter ball radius to the hole radius is less than or equal to two, then the size of the hole has almost no effect on the value of the ultimate force. If the control action is an indenter displacement, once the displacement exceeds the maximum value, it will not contact the indenter.

In the indentation process, the parameter for penetration of the ball-indenter through the hole in the membrane can be the angular coordinate ψ_A. It describes a location of the edge of the hole on the ball. For example, if $\psi_A = \pi/2$, then half of the ball has passed through the hole and the other half has not. For the maximum indentation force the angle ψ_A is less than $\pi/2$. It is mean that in the limiting state the ball passes the hole by less than half. Different cases are possible for the maximum displacement, which are depended on the sizes of the indenter and the hole and the friction coefficient. In the ultimate condition the ball can pass through the hole by either less than half or more than half. Thus, we note that the results obtained show the extreme importance of considering friction in the problems of indentation of elastic membranes.

Acknowledgements This research was supported by the Government of Russian Federation grant No. 14.Z50.31.0046.

References

[1] Liu KK, Ju BF (2001) A novel technique for mechanical characterization of thin elastomeric membrane, Journal of Physics D: Applied Physics **34**(15):L91-L94.
[2] Ju B, Ju Y, Saka M, Liu K, Wan K (2005) A systematic method for characterizing the elastic properties and adhesion of a thin polymer membrane, International

Journal of Mechanical Sciences **47**(3):319-332.

[3] Garnica-Palafox I, Álvarez-Camacho M, Sánchez-Arévalo F (2019) Macro-and micromechanical responses of an elastomeric membrane undergoing biaxial tension by indentation, Journal of Materials Science **54**(22):14255-14274.

[4] Aernouts J, Aerts J, Dirckx J (2012) Mechanical properties of human tympanic membrane in the quasi-static regime from in situ point indentation measurements, Hearing research **290**(1-2):45-54.

[5] Wei X, Kysar J (2012) Experimental validation of multiscale modeling of indentation of suspended circular graphene membranes, International Journal of Solids and Structures **49**(22):3201-3209.

[6] Chandler T, Vella D (2020) Indentation of suspended two-dimensional solids: The signatures of geometrical and material nonlinearity, Journal of the Mechanics and Physics of Solids **144**:104109.

[7] Höller R, Libisch F, Hellmich C (2022) A membrane theory for circular graphene sheets, based on a hyperelastic material model for large deformations, Mechanics of Advanced Materials and Structures **29**(5):651-661.

[8] Feng W (1986) Indentation of a plane membrane with a rigid paraboloid, International Journal of Non-Linear Mechanics **22**(3):377-383.

[9] Yang W, Hsu K (1971) Indentation of a circular membrane, Journal of Applied Mechanics **38**(1):227-230.

[10] Steigmann D (2005) Puncturing a thin elastic sheet, International Journal of Non-Linear Mechanics **40**(2-3):255-270.

[11] Nadler B, Steigmann D (2006) Modeling the indentation, penetration and cavitation of elastic membranes, Journal of the Mechanics and Physics of Solids **54**(10):2005-2029.

[12] Pearce S, King J, Holdsworth M (2011) Axisymmetric indentation of curved elastic membranes by a convex rigid indenter, International Journal of Nonlinear Mechanics **46**(9):1128-1138.

[13] Pearce S, King J, Steinbrecher T, Leubner-Metzger G, Everitt N, Holdsworth M (2018) Finite indentation of highly curved elastic shells, Proceedings of the Royal Society A: Mathematical, Physical and Engineering Sciences **474**(2209):20170482.

[14] Selvadurai A (2006) Deflections of a rubber membrane, Journal of the Mechanics and Physics of Solids **54**(6):1093-1119.

[15] Liu J, Chen Z, Liang X, Huang X, Mao G, Hong W, Yu H, Qu S (2018) Puncture mechanics of soft elastomeric membrane with large deformation by rigid cylindrical indenter, Journal of the Mechanics and Physics of Solids **112**:458-471.

[16] Kolesnikov A, Shatvorov N (2022) Indentation of a circular hyperelastic membrane by a rigid cylinder, International Journal of Non-Linear Mechanics **138**:103836.

[17] Liu J, Zhong D, Yin T, Chen Z, Liu B, Wang P, Qu S, Kang G (2022) Indentation of elastomeric membranes by sphere-tipped indenters: snap-through instability, shrinkage, and puncture Journal of the Mechanics and Physics of Solids **167**:04973.

[18] Jia F, Amar M (2020) Scaling laws and snap-through events in indentation of perforated membranes, Journal of the Mechanics and Physics of Solids **135**:103797.
[19] Green A, Adkins J (1960) Large Elastic Deformation and Non-linear Continuum Mechanics, Clarendon Press, Oxford.
[20] Naghdi P (1972) The theory of shells and plates, In: Truesdell C (ed), Mechanics of Solids II, Handbuch der Physik (Herausgeber Siegfried Flügge), vol VIa/2, pp 425-640, Springer-Verlag, Berlin.
[21] Zubov L (1982) The Methods of Nonlinear Elastisity in the Shell Theory (in Russ.), Rostov Univ. Press, Rostov on Don.
[22] Libai A, Simmonds J (1988) The Nonlinear Theory of Elastic Shells, Academic Press, San Diego.
[23] Gent AN (1996) A new constitutive relation for rubber, Rubber Chemistry and Technology **69**(1):59-61.

Chapter 25
Integrated Asymptotic Approach to the Structural Mechanics Models

Alexander G. Kolpakov and Sergey I. Rakin

Abstract Methods for calculating multi-element structures are discussed, in which the calculation of the frame and the local stress-strain state in the joints is considered as stages of one procedure. The problem is discussed from the point of view of the development of integrated engineering methodology and development of prospective engineering courses that combines the existing approaches with the recent progress in the analysis of thin-walled multi-element structures.

The recent researches demonstrate that the Structural Mechanics model is the phenomenon of asymptotic nature (namely, the leading term of asymptotic of the elasticity theory problem in a system of connected thin domains). The asymptotic analysis of the elasticity theory problem also provides us with the method of computation of local stress-strain state in the joints. The advantage of this approach is that it allows integrating all the stages of the calculation of a thin-walled structure into a step-by-step procedure suitable for computer implementation.

Key words: Multi-element structure · Thin-walled structure · Asymptotic analysis · Structural mechanics model · Joints

25.1 Motivation of the Research

Structure is a unity of structural elements and connecting units. The work and strength of structure are determined both the structural elements and connecting units. It would be reasonable to expect that the method of computation of structure would treat the

Alexander G. Kolpakov
SysAn (System Analysis in Engineering), A. Nevskogo str., 12a, 34, Novosibirsk, 630075, Russian Federation, e-mail: algk@ngs.ru

Sergey I. Rakin
Siberian Transport University, Dusi Koval'chuk Str., 191, Novosibirsk, 630049, Russian Federation, e-mail: rakinsi@ngs.ru

H. Altenbach et al. (eds.), *Sixty Shades of Generalized Continua*,
Advanced Structured Materials 170, https://doi.org/10.1007/978-3-031-26186-2_25

structural elements and connecting units in the frameworks of a uniform (integrated) procedure. In the contrary, it was happen that the computation of structural framework and connecting units are separated between Structural Mechanics and discipline called "Machine design", "Details of machines", see, e.g., [1, 2]. The names of the modern similar disciplines sounds as "Joints design", "Analysis of joints", etc. [3].

Model of single beam was developed in 1750s by Euler and Bernoulli. "The Birth of the Question" (the origin of Structural Mechanics) in [4] is associated with L. Euler "De pressione ponderis in planum cui incumbit" [5]. However, Timoshenko wrote in [6] that in 1844 there existed no theory for computation of frameworks in the modern sense. In [4], the development of the general methods for calculating elastic systems is associated with A. Clebsch "Theorie der Elasticität fester Körper" [7]. J.C. Maxwell in "On the Calculation of the Equilibrium and Stiffness of Frames" [8] formulated the definition of frame:

Definition 25.1 (Frame). Frame is a system of lines connecting a number of points.

This idea was expanded to other types of multi-element structures.

In the second part of XIX, the progress in the Structural Mechanics was fast, and the Structural Mechanics became soon the basis element of engineering education. Nevertheless, for the long time the foundation of the Structural Mechanics remains half-empiric, half-exact. Mention the problem of foundation of the conjugated conditions in connecting point for the beams. Solution to the problem, in the form of the condition of "infinite rigidity" of connecting unit, was justified quite recently [9, 10]. The condition is asymptotic in nature: it is mathematically proved that the connecting node behaves like a body of infinite rigidity, although it is usually made of the same elastic material as the connected beams. In reality this is impossible, but quite correct asymptotically.

25.2 Modern State of Structural Mechanics

In any structure, two types of parts are clearly distinguishable: the main parts of the structural elements and the connecting units. We conduct our presentation in relation to frames. The reader will be able to modify our presentation in relation to other types of structures.

25.2.1 Deformation of the "Main" Part of Frame

We call the "main part" the part of a certain structural element (rod, beam, plate, membrane, etc.) placed on a far distance from connections, joints and boundaries, so that the stress-strain state in this part is not subjected to influence of boundary layers and similar phenomena. Models and methods for computation of stress/strain state in the main parts of uniform structural elements (beams of constant cross-section, plates of constant thickness) made of homogeneous materials are based on the hypothesis about behavior of thin-walled structural elements proposed by Euler, Saint-Venant, Kirchhoff-Love as well as other prominent contributors.

Subsequent progress in the analysis of deformation in the "main parts" of structural elements was

- exact solution to some problems in the frameworks of the elasticity theory (for example, exact solution of the elasticity problem for the bending cylindrical beam [6]);
- development of the theories of the dimension reduction for rods, beams, plates, membranes of periodic structures [11]-[13].

At present, various methods for calculating deformation in the "main parts" of structural elements have been developed. In this paper, we do not touch the problems of the structural stability, structural dynamics and non-linear problems.

25.2.2 Deformation of Connecting Units of Frame

Until the second half of 20th circle, analysis of connecting units belongs to the discipline called "Machine design" and it was based on simplified or partial solutions. This becomes immediately clear, if to take into account that analysis of connecting units usually requires solution to the elasticity theory problem that is usually impossible without application of computer and proper numerical software. After the appearance of computers, numerical methods and computer programs, the computer methods began to intensively apply to analysis of joints [14]. The discipline called "Machine design" began to be complemented or replaced by disciplines as "Joints design", "Analysis of joints" and similar. The last disciplines often were oriented to solution of the specific engineering problems by using the numerical methods and computer software, but not the theoretical analysis of joint as a constitutive part of structures.

In 1980s-1990s, mathematicians demonstrated interest to the problem of connecting units. The significant contributions into analysis of joined elastic bodies by the mathematical methods were done in [9, 10], [15]-[17].

Note 25.1. The behavior of the end's parts of connecting structural elements are often constructed with the of boundary layer method. An attempt to expand the boundary layer approach to the connecting units, usually is low productive. The boundary layer approach seems to be too restricted to analyze connecting units. For example, it is difficult to imagine how, say, the stress-strain state in a bolted joint may be constructed in the frameworks of the boundary layer technique.

25.2.3 "Infinite Rigidity" of Connecting Units

The problem of deformation of connecting units is a two-fold problem. It is the principal problem in Analysis of Joints - computation of local stress-strain state in joints. It is also the principal problem in Structural Mechanics because it determines the conjugated conditions in the joint.

The models of Structural Mechanics are based on the following assumptions:

i) the connecting units are in equilibrium (sums of forces and moments are zero);
ii) the ends of the structural elements arriving to the same (αth) connecting unit has the same displacements $\boldsymbol{u}_\alpha$ and rotations $\boldsymbol{\omega}_\alpha$, $\alpha = 1, \ldots, N$ (N is the total number of the joints).

Condition i) is usually accepted with no doubt. Condition ii) may be interpreted in such a way that connecting units in structural mechanics behave themselves as the "rigid joints", i.e. the ends of the elements arriving at the connecting unit have no relative displacements and rotations. Apparently, this should be surprising from a common point of view. After all, the connecting nodes are made of materials with elastic characteristics similar to the elastic characteristics of the structural elements being connected, that is, the connecting units are no more rigid than the structural elements. It is also known that thin connecting units exhibit "flexibility" [18].

An explanation of the "rigid connection" phenomenon was recently given in [9, 10] for a planar system of connected beams. It demonstrates that the contradiction is resolved by the acceptance of the fact that Structural Mechanics is the asymptotic model, more precisely, the leading term of the asymptotic in the problem of elasticity theory from a thin region of complex geometry, The contributions [9, 10] marked the beginning of the completion of the justification of the Structural Mechanics model.

Note 25.2. The connecting unit behaves themselves as "rigid joint" if it is made of materials non very soft as compared with the materials of structural elements and not very thin as compared with the thickness of structural elements ("not degenerated" joint [19]). Else, for "degenerated" joints, there appear models with "compliant" conductivity/elasticity conditions. This is known both from mathematical analysis of the problem, see, e.g. [18] and as well as from engineering practice, see, e.g. [20].

Note 25.3. When an element of undamaged beam is degenerated (destroyed), it passes through the following stages: *undamaged element – partially destroyed element – mostly destroyed element – completely destroyed element.* If one tries to detect the level of destruction by measuring the macroscopic displacement of the structure, one cannot fix the transition from the stage "undamaged element" to "partially destroyed element" because at the stage "partially destroyed element" an analog of "none degenerated" joint arises but no "degenerated" joint arises. As a result, this transition is not identifying at the macroscopic level [19]. The first stage, which can be fixed by the macroscopic-scale measurements, is the transition to the "mostly destroyed element" stage, which corresponds to "degenerated" joint.

25.2.4 Structural Mechanics Achievements and Limitations

Classical Structural Mechanics model involves no information (and needs no information) about the individual properties (local structure, geometry, local material characteristics. etc.) of connecting units. The model has the form of the linear algebraic system or system of linear ordinary differential equations with proper boundary conditions. This is complete mathematical model (is solvable and has unique solution under proper boundary conditions). It is surprising that a model of such outstanding practical value is so abstract in fact.

The development of Structural Mechanics model made possible computation of complex structures (bridges, buildings, ships, etc.) without solution to the elasticity theory problem (that was especially important in the before computer era because it is practically impossible to solve the elasticity problems to typical joints without numerical methods). The limitation of Structural Mechanics model is that it allows computing displacements of structure and stress/strain state in the "main" parts of the beams. Structural Mechanics model provide us with no information about local stress/strain states in the connecting units. Since stress-strain state in the connecting unit and its vicinity has general form, it cannot be analyzed in details in the framework of simplified models.

Practical needs required methods for calculating connecting units. Especially because the connecting units are critical elements of structures - their strength often determined the strength of the entire structure. As a result, there were developed other theories, different from the Structural Mechanics. The methods of the developed theories were also different from the approaches of Structural Mechanics. As a result, the calculation of an integral structure becomes "torn" between several theories.

25.3 Integrated Approach to the Computation of Thin-walled Structures

The situation described above stimulates searches of the approach to the construction of the integrated method for computation of structures. Since the behavior of the structural elements far from the connecting units is described well in the frameworks of the Structural Mechanics, now, the attention of the researches is concentrated on the analysis of connecting units an integrated approaches to the analysis of structures, see, e.g., [15]-[19], [21, 22] (the list is not completed) and references in these papers.

25.3.1 Displacements in the "Main" Part of the Beams

The structural mechanics model allows computing the stress/strain state in the "main" parts of the beam. Denote $\boldsymbol{u}_{\alpha\beta}(\boldsymbol{x})$ the displacements in the "main" part of the beam

$B_{\alpha\beta}$ with the ends in the αth and the βth nodes. If forces applied to the structure are known and displacements $\boldsymbol{u}_\alpha$ and rotations $\boldsymbol{\omega}_\alpha$ are already determined, one can compute the asymptotic of the displacements $\boldsymbol{u}_{\alpha\beta}(\boldsymbol{x})$ in the "main" part of the beam.

For the sake of simplicity, we considered 2-D elastic problem. In this case, displacements $\boldsymbol{u}_{\alpha\beta}(\boldsymbol{x})$ are written as

$$\boldsymbol{u}_{\alpha\beta}(\boldsymbol{x}) \approx \boldsymbol{U}_{\alpha\beta}(\boldsymbol{x}) \tag{25.1}$$

where

$$\begin{aligned} \boldsymbol{U}_{\alpha\beta}(\boldsymbol{x}) = u_{\alpha\beta}(s_{\alpha\beta})\boldsymbol{e}_1 + w_{\alpha\beta}(s_{\alpha\beta})\boldsymbol{e}_2 - t_{\alpha\beta}w'_{\alpha\beta}(s_{\alpha\beta})\boldsymbol{e}_1 \\ + \varepsilon\boldsymbol{N}^0_{\alpha\beta}(\boldsymbol{x}/\varepsilon)u'_{\alpha\beta}(s_{\alpha\beta}) + \varepsilon\boldsymbol{N}^2_{\alpha\beta}(\boldsymbol{x}/\varepsilon)w''_{\alpha\beta}(s_{\alpha\beta}) \end{aligned} \tag{25.2}$$

where $s_{\alpha\beta}$ is the coordinate along the axis of the beam $B_{\alpha\beta}$; $t_{\alpha\beta}$ is the coordinate along the normal to the beam's axis, and functions $u_{\alpha\beta}$ are axial (tangential) displacement of the beam considered as 1-D structure, $w_{\alpha\beta}$ are normal deflection(s). These are computed as functions of displacements $\boldsymbol{u}_\alpha$ and rotations $\boldsymbol{\omega}_\alpha$ of the nodes and mass forces (if applied) in known way.

Functions $\boldsymbol{N}^0_{\alpha\beta}(\boldsymbol{y})$, $\boldsymbol{N}^1_{\alpha\beta}(\boldsymbol{y})$, and $\boldsymbol{N}^2_{\alpha\beta}(\boldsymbol{y})$ in (25.2) are solutions to the periodicity cell problems for beam (for details, see [13]). For 2-D beam of constant thickness made of homogeneous isotropic material [23]

$$\boldsymbol{N}^0_{\alpha\beta}(\boldsymbol{y}) = -\nu t_{\alpha\beta}\boldsymbol{e}_2, \quad \boldsymbol{N}^1_{\alpha\beta}(\boldsymbol{y}) = 0, \quad \boldsymbol{N}^2_{\alpha\beta}(\boldsymbol{y}) = t_{\alpha\beta}\boldsymbol{e}_1 - \nu(t^2_{\alpha\beta}/2 - t_{\alpha\beta}s_{\alpha\beta})\boldsymbol{e}_2$$

if the middle line of the beam coincides with the coordinate axis $t_{\alpha\beta} = 0$. Exact solutions similar to (25.2) are known also for 3-D beam.

Note 25.4. The representation similar (25.2) takes place to wide variety of structural elements (rods, beams, plates, etc.) [11]-[13] (the list is not complete). For some structural elements, representation of the form (25.2) coincides with the kinematics hypothesis accepted in classical theories, for some – not (an example is the beam bending below).

Note 25.5. Note as a curiosity that the local stress/strain states in the beam, predicted by the classical theory and the asymptotic theory not coincide one with another, but the stiffness of the beam computed with the classical theory and the asymptotic theories are the same.

25.3.2 "Rigid" Displacements of the Connecting Unit

The displacement of the connecting unit J_α as "infinitely rigid" object is are given by the formula

$$\boldsymbol{u}_\alpha + \boldsymbol{\omega}_\alpha \times \boldsymbol{x} \tag{25.3}$$

Formula (25.3) describes displacements of a rigid body and it is far from an adequate description of the local stress-strain states in the connecting unit.

Construct a set of the admissible displacements defined in the entire area occupied by the beam, which are consistent with displacements (25.2) in the main parts of the beams and with displacements (25.3) of the end of the beam. It should be taken into account that displacements (25.3) cannot be implemented in a connecting unit made of an elastic material(s) with mechanical properties similar to the material of the structural elements.

25.3.3 Conjugation of the Displacements in the Main Part of a Beam with the "Rigid Body" Displacements of the Connecting Unit

Let us construct an auxiliary set of admissible displacements, which would match displacements (25.2) and (25.3). For this, we consider single beam $B_{1\alpha}$. We assume that is $B_{1\alpha}$ 2-D or 3-D domain. Both functions (25.2) and (25.3) are defined in the domain $B_{1\alpha}$. The functions (25.2) and (25.3) will pass one to another when $\boldsymbol{x}$ goes from one side of S to another those side, if we introduce function

$$\boldsymbol{V}_{\alpha\beta} = \boldsymbol{U}_{\alpha\beta}(\boldsymbol{x})\eta(s_{\alpha\beta}) + (\boldsymbol{u}_{\alpha} + \boldsymbol{\omega}_{\alpha} \times \boldsymbol{x})(1 - \eta(s_{\alpha\beta})) \tag{25.4}$$

where conjugation function and it takes value 1 to the left of and 0 to the right of S (it is displayed in Fig. 25.1 (left)). The function (25.4) belongs $H_2^1(B_{1\alpha} \cup J_\beta)$ if $u_{1\beta}(s_{1\beta}) \in C^1$ and $w_{1\beta}(s_{1\beta}) \in C^2$. We assume that the last inclusion takes place.

On another end of the beam $B_{1\beta}$, we can construct the similar conjugation of the beam B_{1V} with the connecting unit J_β. As a result, function (25.4) is defined in the single-beam domain $J_1 \cup B_{1\beta} \cup J_\beta$, Fig. 25.2. We can construct functions of the form (25.4) for all single-beam domains $J_\alpha \cup B_{\alpha\beta} \cup J_\beta$. Denote the function constructed for the single-beam domains by $\boldsymbol{V}_{\alpha\beta}$.

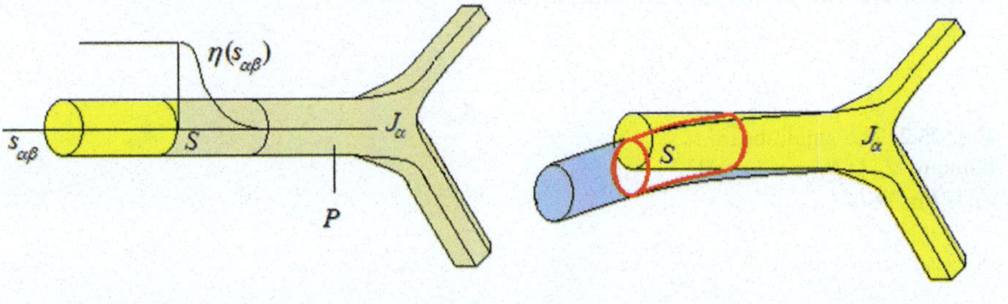

Fig. 25.1: Undeformed fragment of structure and function η - left, deformation corresponding to conjugation with function η (bold) – right.

25.3.4 Assembling the Single-beam Domain Functions into the Function in 2-D Frame

We construct the admissible displacements in the domain occupied the whole 2-D frame from the single-beam domain displacements. Any single-beam domain has the structure "connecting unit – beam - connecting unit". Thus, we have to identify the displacements in the domain J_β belonging to the domains $J_\alpha \cup B_{\alpha\beta} \cup J_\beta$ and $J_\beta \cup B_{\beta\gamma} \cup J_\gamma$, Fig.25.2.

By construction, the functions $\boldsymbol{V}_{\alpha\beta}$ and $\boldsymbol{V}_{\beta\gamma}$ coincide in the connecting unit J_β. In the result of the identification of the functions in the same-named connecting units, we assemble the function belonging to H_2^1 in whole frame and

$$\boldsymbol{V}(\boldsymbol{x}) = \boldsymbol{V}_{\alpha\beta}(\boldsymbol{x}) \text{ in } J_\alpha \cup B_{\alpha\beta} \cup J_\beta \tag{25.5}$$

is well-defined in the domain $Q = \bigcup_{\alpha\beta}(J_\alpha \cup B_{\alpha\beta} \cup J_\beta)$ occupied by the whole frame, and functions of the form (25.5) belong to the functional class $H_2^1(Q)$.

The function $\boldsymbol{V}(\boldsymbol{x})$ coincides with (25.1) in the main parts of the beams and coincides with (25.3) in the connecting units. Denote the class of functions of the form (25.5) by V_{Struct}.

Note 25.6. We call the main part of the beam the part placed sufficiently far from the ends of the beam, where the influence of the ends in the form of local perturbations (usually treated as boundary layers) may be neglected. For a uniform beam made of homogeneous isotropic material, stress-strain state in that main part has form (25.1). The end's conditions influence functions $u_{\alpha\beta}(s_{\alpha\beta})$ and $w_{\alpha\beta}(s_{\alpha\beta})$ in (25.1), but give no contribution in the form of fast changing terms.

Functions of the form (25.5) are suitable to describe the deformation of the main part of the beam. But they are not suitable to describe the deformation of the connecting unit and in the vicinity of the connecting unit.

Note 25.7. For the structure under consideration, we cannot determinately indicate where the connecting unit ends and beam starts. This is natural situation. It brings no problems in the present consideration.

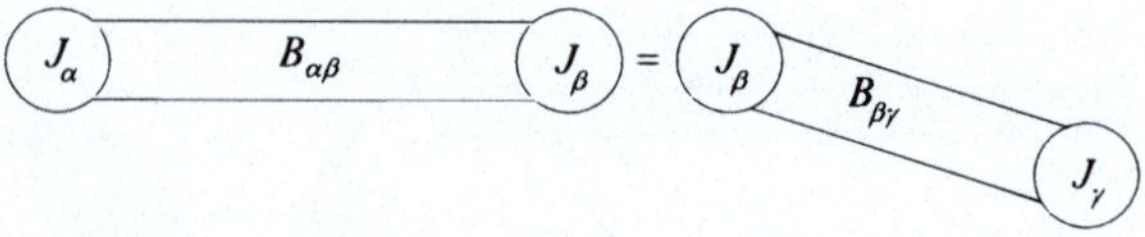

Fig. 25.2 The single-beam domains $J_\alpha \cup B_{\alpha\beta} \cup J_\beta$ and $J_\beta \cup B_{\beta\gamma} \cup J_\gamma$

25.3.5 Supplement of the Functions (25.5) with the Local Perturbations

The insufficient the function $\boldsymbol{V}(\boldsymbol{x})$ of the form (25.5) is that the functions $\boldsymbol{V}(\boldsymbol{x})$ (25.5) describes the connecting unit as "physically rigid" (not deformable) body. In addition, the conjugation between the "beam-type" solution and connecting unit must be given by a proper solution of the elasticity theory problem, which, it is expected, differs from conjugation (25.4).

To construct the suitable set of admissible displacements, we use the functions $\boldsymbol{V}(\boldsymbol{x})$ of the form (25.5) as the "skeleton" of the constructed functional set. We supplement the functions of the form $\boldsymbol{V}(\boldsymbol{x})$ by the sets of functions localized near the connecting units and "diverse enough" to describe the elastic behavior of the connecting units.

Consider the connecting unit J_1 and the set S_1 of beams connected to J_1. Introduce the set of functions

$$L_1 = \{\boldsymbol{w}(\boldsymbol{y}) \in H_2^1(Q) : \boldsymbol{w}(\boldsymbol{y}) \to 0 \text{ as } \mathrm{dist}(\boldsymbol{y}, J_1) \to 0 \text{ along } B_{1\alpha}, \alpha \in S_1\}.$$

In this case, the admissible displacements belong to the class H_2^1 in the connecting unit and in its vicinity. This class is, no doubt, diverse enough to describe the elastic behavior.

In the similar way, we introduce the set of functions $L_2, \ldots, L_N$ for all the remaining connecting units and the beams connected to the connecting units. Introduce the set of functions $L = L_1 \cup \ldots \cup L_N$. The set L corresponds to the idea of the "local perturbation" technique developed first for the thermoconductivity problem [19] and then expanded to some elastic problems [24, 21]. The advantage of the local perturbations $\boldsymbol{w}_\alpha(\boldsymbol{w}/\varepsilon)$ is that the set L

- contains boundary layer type components, which vanish in the beam far from the joint;
- contains the set of admissible displacements , which is sufficient to describe stress/strain state of general form inside the connecting unit;
- in the "main parts" of beams, contains the set of admissible displacements corresponding to the asymptotic form of solution, known for the "main parts" of beams.

The set of admissible displacements is constructed in the form

$$H_{\mathrm{Struct}} = \{\boldsymbol{V}(\boldsymbol{x}) + \boldsymbol{w}(\boldsymbol{x}) : \boldsymbol{V} \in V_{\mathrm{Struct}}, \boldsymbol{w} \in L\}. \tag{25.6}$$

We did not use above the small thickness of the elements of the structure. Now, we remind that the beams are thin (with characteristic thickness $\varepsilon \ll 1$, that is formalized as $\varepsilon \to 0$) and specify the functions in (25.1)-(25.6). Namely, we account that perturbations are small in value and the conjugation with function changes those value from 1 to 0 at the interval, which has the characteristic length ε. After that, we write the elasticity theory principle of minimum of potential energy

$$E(\boldsymbol{u}) = \frac{1}{2}\int_Q a_{ijkl}u_{i,j}u_{k,l}d\boldsymbol{x} \rightarrow \min. \tag{25.7}$$

Substituting (25.6) into (25.7), we arrive at the minimization problem

$$\begin{aligned} &E([u_{\alpha\beta}(s_{\alpha\beta})\boldsymbol{e}_1 - \varepsilon v t_{\alpha\beta}u'_{\alpha\beta}(s_{\alpha\beta})\boldsymbol{e}_2 + w''_{\alpha\beta}(s_{\alpha\beta})t_{\alpha\beta}\boldsymbol{e}_1 \\ &-\varepsilon v(t^2_{\alpha\beta}/2 - t_{\alpha\beta}s_{\alpha\beta})w''_{\alpha\beta}(s_{\alpha\beta})\boldsymbol{e}_2]\eta(s_{\alpha\beta}) + (\boldsymbol{u}_\alpha + \boldsymbol{\omega}_\alpha \times \boldsymbol{x})(1-\eta(s_{\alpha\beta})) \\ &+\varepsilon\boldsymbol{w}(\boldsymbol{x}/\varepsilon)) \rightarrow \min. \end{aligned} \tag{25.8}$$

with respect to the functions $u_{\alpha\beta}$ and $w_{\alpha\beta}$, vectors $\boldsymbol{u}_\alpha$ and $\boldsymbol{\omega}_\alpha$; and function $\boldsymbol{w}(\boldsymbol{x}/\varepsilon)$).

25.3.5.1 Erasing of the Individual Properties of Joints in Structural Mechanics Model

Rewrite function (25.4), which conjugates the function of the type "beam bending" with the function of the type "displacement of rigid bogy", in the form

$$\boldsymbol{V}_{\alpha\beta} = \boldsymbol{U}_{\alpha\beta}(\boldsymbol{x})\eta(s_{\alpha\beta}) + \boldsymbol{g}(\boldsymbol{x})(1-\eta(s_{\alpha\beta})), \tag{25.9}$$

where

$$\begin{aligned} \boldsymbol{U}_{\alpha\beta}(\boldsymbol{x}) &= u_{\alpha\beta}(s_{\alpha\beta})\boldsymbol{e}_1 - v t_{\alpha\beta}u'_{\alpha\beta}(s_{\alpha\beta})\boldsymbol{e}_2 + w_{\alpha\beta}(s_{\alpha\beta})\boldsymbol{e}_2 \\ &+ w''_{\alpha\beta}(s_{\alpha\beta})t_{\alpha\beta}\boldsymbol{e}_1 - v(t^2_{\alpha\beta}/2 - t_{\alpha\beta}s_{\alpha\beta})w''_{\alpha\beta}(s_{\alpha\beta})\boldsymbol{e}_2, \\ \boldsymbol{g}(\boldsymbol{x}) &= \boldsymbol{\omega}_\alpha + \boldsymbol{\omega}_\alpha \times \boldsymbol{x}. \end{aligned} \tag{25.10}$$

Admissible displacement belonging to H_{Struct} (25.6) has the form

$$\boldsymbol{V}(\boldsymbol{x}) + \varepsilon\boldsymbol{w}(\boldsymbol{x}/\varepsilon) = \boldsymbol{U}_{\alpha\beta}(\boldsymbol{x})\eta(s_{\alpha\beta}/\varepsilon) + \boldsymbol{g}(\boldsymbol{x})(1-\eta(s_{\alpha\beta}/\varepsilon)) + \varepsilon\boldsymbol{w}(\boldsymbol{x}/\varepsilon). \tag{25.11}$$

The function $\eta(s_{\alpha\beta}/\varepsilon) = 1$ for $s_{\alpha\beta} > S\varepsilon$. Correspondingly, $1-\eta(s_{\alpha\beta}/\varepsilon) = 0$ for $s_{\alpha\beta} > S\varepsilon$. For $0 < s_{\alpha\beta} < S\varepsilon$, groping terms in (25.11) , we form the function

$$\begin{aligned} \varepsilon\boldsymbol{W}(\boldsymbol{x}/\varepsilon) &= -\boldsymbol{U}_{\alpha\beta}(\boldsymbol{x})(1-\eta(s_{\alpha\beta}/\varepsilon)) + \boldsymbol{g}(\boldsymbol{x})(1-\eta(s_{\alpha\beta}/\varepsilon)) + \varepsilon\boldsymbol{w}(\boldsymbol{x}/\varepsilon) \\ &= (-\boldsymbol{U}_{\alpha\beta}(\boldsymbol{x}) + \boldsymbol{g}(\boldsymbol{x}))(1-\eta(s_{\alpha\beta}/\varepsilon)) + \varepsilon\boldsymbol{w}(\boldsymbol{x}/\varepsilon). \end{aligned} \tag{25.12}$$

In (25.12), we use that $|-\boldsymbol{U}_{\alpha\beta}(\boldsymbol{x}) + \boldsymbol{g}(\boldsymbol{x})| < C\varepsilon$, see Fig. 25.3. By using (25.12), we can rewrite (25.11) as

$$\boldsymbol{V}(\boldsymbol{x}) + \varepsilon\boldsymbol{w}(\boldsymbol{x}/\varepsilon) = U_{\alpha\beta}(\boldsymbol{x}) + \varepsilon\boldsymbol{w}(\boldsymbol{x}/\varepsilon) \tag{25.13}$$

This transformation separates the admissible displacement into the sum of the function of the type "beam bending" only and the function of the type "local perturbation" only, because in (25.12) the summand

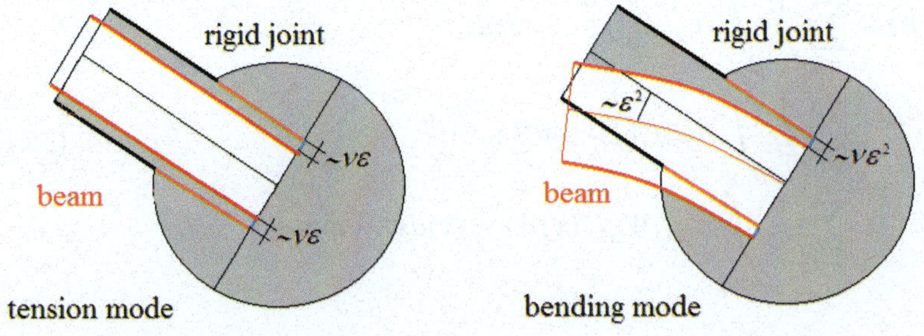

Fig. 25.3: Difference between displacements the "rigid" joint and connected beam.

$$-U_{\alpha\beta}(\boldsymbol{x})(1-\eta(s_{\alpha\beta}/\varepsilon))+\boldsymbol{g}(\boldsymbol{x})(1-\eta(s_{\alpha\beta}/\varepsilon))=0 \text{ for } s_{\alpha\beta}>S\varepsilon \qquad (25.14)$$

and has the form of local perturbation, too. Thus, (25.13) is again

$$H_{\text{Struct}}=\{\boldsymbol{V}(\boldsymbol{x})+\varepsilon\boldsymbol{w}(\boldsymbol{x}/\varepsilon):\boldsymbol{w}\in L\}. \qquad (25.15)$$

We arrive at the minimization problem

$$E(\boldsymbol{u})=\frac{1}{2\varepsilon^T}\int\limits_Q a_{ijkl}u_{i,j}u_{k,l}d\boldsymbol{x}\to\min,\boldsymbol{u}\in H_{\text{Struct}}. \qquad (25.16)$$

Here ε^T is the order of the stiffness if the structural elements in the dependence on those thickness: $T=1$ for frame (the structural elements are rods) and $T=3$ for truss (the structural elements are beams). Solution to (25.16) has the form

$$\boldsymbol{u}(\boldsymbol{x})=\boldsymbol{U}_{\alpha\beta}(\boldsymbol{x})+\varepsilon\boldsymbol{W}(\boldsymbol{x}/\varepsilon) \qquad (25.17)$$

in the beam $B_{\alpha\beta}(\boldsymbol{x})$, where $\boldsymbol{U}_{\alpha\beta}(\boldsymbol{x})$ is given by (25.2) and $\boldsymbol{W}(\boldsymbol{y})$ ($\boldsymbol{y}=\boldsymbol{x}/\varepsilon$) decays fast.

Consider a function $f(\boldsymbol{x},\boldsymbol{x}/\varepsilon)$. For the function $f(\boldsymbol{x},\boldsymbol{x}/\varepsilon)$ takes place the following two-scale differentiating rule [25]

$$\frac{\partial}{\partial x_i}f(\boldsymbol{x},\boldsymbol{x}/\varepsilon)=f_{,ix}(\boldsymbol{x},\boldsymbol{x}/\varepsilon)+f_{,iy}(\boldsymbol{x},\boldsymbol{x}/\varepsilon).$$

By using two-scale differentiating rule, we obtain from (25.17)

$$
\begin{aligned}
E(\boldsymbol{u}) &= \sum_{\alpha\beta} \frac{1}{2\varepsilon} \int\limits_{B_{\alpha\beta}} a_{ijkl} u'^2_{\alpha\beta}(s_{\alpha\beta}) ds dt \\
&+ \sum_{\alpha\beta} \frac{1}{2\varepsilon} \int\limits_{B_{\alpha\beta}} a_{ijkl} w''^2_{\alpha\beta}(s_{\alpha\beta}) t^2_{\alpha\beta} ds dt \\
&+ \sum_{\alpha\beta} \frac{1}{2\varepsilon} \int\limits_{B_{\alpha\beta}} a_{ijkl} W_{i,jy}(\boldsymbol{x}) W_{k,ly}(\boldsymbol{x}) d\boldsymbol{x} \to \min, \boldsymbol{u} \in H_{\mathrm{Struct}}.
\end{aligned}
\tag{25.18}
$$

25.3.5.2 "Erasing" of Individual Properties of Joint, $T = 1$ - the Frame Model

If $\boldsymbol{W}(\boldsymbol{x})$ integrable in $B_{\alpha\beta}$, then changing the variables $\boldsymbol{y} = \boldsymbol{x}/\varepsilon$, we arrive at

$$
\frac{1}{2\varepsilon} \int\limits_{B_{\alpha\beta}} a_{ijkl} W_{i,jy} W_{k,ly} d\boldsymbol{x} = \frac{1}{2\varepsilon} \varepsilon^d \int\limits_{B_{\alpha\beta}} a_{ijkl} W_{i,jy} W_{k,ly} d\boldsymbol{y} \to 0 \text{ as } \varepsilon \to 0, \tag{25.19}
$$

$d = 2, 3$ is dimension of the problem under consideration. In addition,

$$
\frac{1}{2\varepsilon} \int\limits_{B_{\alpha\beta}} a_{ijkl} w''^2_{\alpha\beta}(s_{\alpha\beta}) t^2_{\alpha\beta} ds dt \, \varepsilon^2 \to 0
$$

as $\varepsilon \to 0$ and (25.18) takes the form

$$
\begin{aligned}
E(\boldsymbol{u}) &= \sum_{\alpha\beta} \frac{1}{2\varepsilon} \int\limits_{B_{\alpha\beta}} a_{ijkl} u'^2_{\alpha\beta}(s_{\alpha\beta}) ds dt \\
&= \sum_{\alpha\beta} \frac{D_{\alpha\beta}}{2} \int\limits_{B_{\alpha\beta}} u'^2_{\alpha\beta}(s_{\alpha\beta}) ds dt \to \min, u_{\alpha\beta} \in V_{\mathrm{Struct}}.
\end{aligned}
\tag{25.20}
$$

The problem (25.20) is frame model, $D_{\alpha\beta}$ is the tensile stiffness of the corresponding rod. If solution to (25.20) is non-zero, it determines the leading term of energy. If solution to (25.20) is zero, we arrive at truss.

25.3.5.3 "Erasing" of Individual Properties of Joint, $T = 3$ - the Truss Model

In connecting fragments of beams

$$
|-U_{\alpha\beta}(\boldsymbol{x}) + \boldsymbol{g}(\boldsymbol{x})| < C\varepsilon,
$$

see Fig. 25.3, then perturbation has the form $\varepsilon^2 \boldsymbol{W}(\boldsymbol{x}/\varepsilon)$ and

$$\frac{1}{2\varepsilon^{-3}}\int\limits_{B_{\alpha\beta}} a_{ijkl}W_{i,jy}(\boldsymbol{x})W_{k,ly}(\boldsymbol{x})d\boldsymbol{x} = \frac{1}{2\varepsilon^{-3}}\varepsilon^4\int\limits_{B_{\alpha\beta}} a_{ijkl}W_{i,jy}(\boldsymbol{x})W_{k,ly}(\boldsymbol{x})d\boldsymbol{y} \to 0 \text{ as } \varepsilon \to 0. \tag{25.21}$$

25.4 Local Stress-strain State in the Connecting Units

It is suitable to formulate the problem of computation of local stress-strain state in the connecting unit and its vicinity as a problem with respect to the function

$$\boldsymbol{V}_{\alpha\beta}(\boldsymbol{x}/\varepsilon) = u_{\alpha\beta}(s_{\alpha\beta})\boldsymbol{e}_1 - \nu t_{\alpha\beta}u'_{\alpha\beta}(s_{\alpha\beta})\boldsymbol{e}_2 + w_{\alpha\beta}(s_{\alpha\beta})\boldsymbol{e}_2 + w''_{\alpha\beta}(s_{\alpha\beta})t_{\alpha\beta}\boldsymbol{e}_1 - \nu(t^2_{\alpha\beta}/2 - t_{\alpha\beta}s_{\alpha\beta})w''_{\alpha\beta}(s_{\alpha\beta})\boldsymbol{e}_2 + \varepsilon\boldsymbol{W}(\boldsymbol{x}/\varepsilon) \tag{25.22}$$

instead $\varepsilon w(\boldsymbol{x}/\varepsilon)$. The set of admissible displacements corresponding to (25.22) and (25.6) is $H^1_2(B_{\alpha\beta} \cup J_\alpha)$ with the condition

$$\boldsymbol{V}_{\alpha\beta} \to u_{\alpha\beta}(+0)\boldsymbol{e}_1 - \varepsilon\nu t_{\alpha\beta}u'_{\alpha\beta}(+0)\boldsymbol{e}_2 + w_{\alpha\beta}(+0)\boldsymbol{e}_2 + w''_{\alpha\beta}(+0)t_{\alpha\beta}\boldsymbol{e}_1 - \varepsilon\nu t^2_{\alpha\beta}/2 - w''_{\alpha\beta}(+0)\boldsymbol{e}_2 \tag{25.23}$$

in the beam $B_{\alpha\beta}$ as $\varepsilon \to 0$. Thus, (25.23) describes both the local stress-strain state in the connecting unit and the boundary layers in the beams connecting to in the connecting unit.

Note 25.8. The set of functions $H^1_2(B_{\alpha\beta} \cup J_\alpha)$ with the condition (25.23) is exactly the "local perturbation" functions as they were introduced in [19]. This set does not involve the auxiliary function $\eta(s_{\alpha\beta})$. Condition (25.23) resembles conditions used in [17].

The derivatives

$$\frac{\partial}{\partial x_k}\varepsilon\boldsymbol{W}_{\alpha\beta}(\boldsymbol{x}/\varepsilon) = \frac{\partial}{\partial y_k}\boldsymbol{W}_{\alpha\beta}(\boldsymbol{y}), \quad \boldsymbol{y} = \boldsymbol{x}/\varepsilon \tag{25.24}$$

are not small and have the order of the derivatives of solution in the main parts of the beams.

Let us formulate the following *problem of computation of the local perturbations*:

Problem 25.1 (Problem U). Determine displacements $\boldsymbol{U}_{\alpha\beta}(\boldsymbol{x}/\varepsilon)$ from solution to the elasticity theory problem with the boundary conditions (25.23).

25.5 Representative Fragment of Joint

The idea of the representative fragment of the connecting unit below shows a similarity with the consideration in [10]. Denote by $\Gamma_{\alpha\beta}$ the cross-section of the beam $B_{\alpha\beta}$ near the α-th connecting unit, see Fig. 25.4. The difference from [10] is that in the representative fragment, the lengths of adjacent fragments of the beam are taken in such way that the stress-strain states on the cross-sections $\Gamma_{\alpha\beta}$ of the adjacent fragments become close to the stress-strain state in the main parts of the corresponding beams.

By virtue of (25.23), one can change (25.23) at the cross-sections placed at proper distance from the joint, by the condition

$$\begin{aligned} V_{\alpha\beta}(y) = u_{\alpha\beta}(+0)\boldsymbol{e}_1 - \varepsilon v t_{\alpha\beta} u'_{\alpha\beta}(+0)\boldsymbol{e}_2 + w_{\alpha\beta}(+0)\boldsymbol{e}_2 \\ + w''_{\alpha\beta}(+0) t_{\alpha\beta}\boldsymbol{e}_1 - \varepsilon v t^2_{\alpha\beta}/2 - w''_{\alpha\beta}(+0)\boldsymbol{e}_2 \text{ on } \Gamma_{\alpha\beta}. \end{aligned} \tag{25.25}$$

Taking into account computations of the length of penetration of the boundary layers from [19, 24], we estimate the proper distance as $3\mathrm{div}\,5\varepsilon$. The connecting unit with the adjacent fragments of beams of proper length (estimated as) is called the *representative fragment of joint.*

For the representative fragment of joint, we formulate the following problem:

Problem 25.2 (Problem RU). Determine displacements from solution to the elasticity theory problem with the boundary conditions (25.25), where functions are solution to the structural mechanics problem for the framework.

Note 25.9. Solution of the **Problem U** may be represented as a linear combination with the coefficients, but only under condition that satisfy the proper solvability conditions.

Due to (25.23), solutions of the **Problem U** and **Problem RU**, in particular, local strains and stresses, are close one to another. This closeness justifies the term "representative fragment of joint". The solution of **Problem RU** defines the local stress-strain state in both the connecting unit and in the adjacent fragments of the

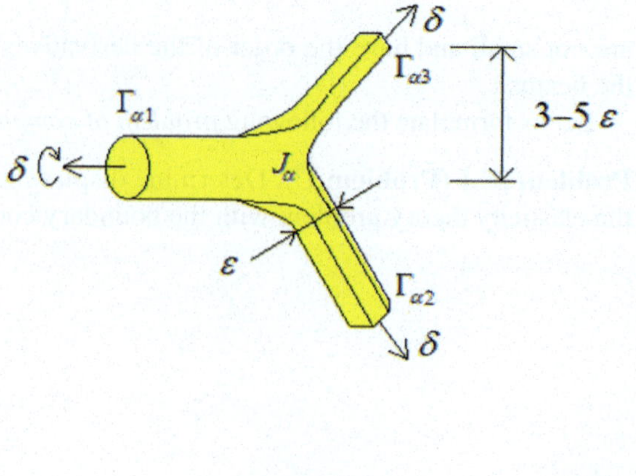

Fig. 25.4 Beams near the connecting unit

structural elements. The local stress-strain state in the adjacent fragments of the structural elements is usually treated as boundary layers.

25.6 Integrated Procedure for Computation of Framework

Summarizing the results obtained above, we arrive at the integrated procedure for computation of framework. The solution steps are the following:

1st: solve classical Structural Mechanics problem (with proper mass forces and boundary conditions), determine displacements and rotations of the connecting node;

2nd: solve the elasticity theory **Problem RU** (with zero mass forces and zero normal stresses on the lateral surface) for the representative element of connecting unit and determine local stress/strain state in the elements of the connecting unit and fragments of adjacent.

At the first step of solutions, one deal with the problem of Structural Mechanics, at the second step, one deals with the problem usually considered in the Details of Machines, Design of Joints and similar disciplines. In the proposed approach, the first and the second steps form the steps of a unit procedure. The thing, one has to do is the data transmission in accordance with the formula (25.25) from the problem of Structural Mechanics – 1st step, to the elasticity theory problem – 2nd step.

Note 25.10. It is necessary to be sure that the Saint-Venant principle takes place for the structure under consideration (usually if takes place) and the length of decay of the local perturbation in beams (in accordance with our computations [19] it is usually 3-5 thickness of the beam) to take the surfaces in good positions. The "good positions" for the surfaces $\Gamma_{\alpha\beta}$. In (25.25), $\Gamma_{\alpha\beta}$ are placed sufficiently far from the connecting unit (it is the necessary condition of the validity of the proposed approach) but not very far from the connecting unit (in order to minimize the domain where the elasticity problem will be solved numerically). The length of decay of the local perturbation is the best candidate for that.

25.7 Actual Problems

Among numerous problems arising, we mention

- conjugated conditions for various types of structural elements (beams, plates, membranes) and various types of connecting elements (pins, seams);
- connecting elements with degree(s) of freedom – hinges with one or more degree(s) of freedom;
- "rigidity/rot rigidity" of connecting units as asymptotic phenomenon. The justification of the "rigidity of joints" was given in [9, 10] for planar rod/beam system. It seems important to investigate the problem for the thin-walled systems.

25.8 Conclusions

The current research on the asymptotic analysis of thin-walled multi-element elastic structures brings new understanding of the classical theories of Structural Mechanics, Machine Design, Joint Design and similar theories. The recent researches indicate that the mentioned theories are, in fact, the terms of the asymptotic of the elasticity problem in thin multi-domains. On the basis of the asymptotic methods, it is possible to develop an approach to the thin-walled multi-element elastic structures, which integrates separated earlier theories of structural mechanics, machine design, joint design and similar theories.

Pin and bolt connections are characterized by the presence of contacts and initial stresses. The contact problem arises even the joint has no degrees of freedom. Joints of non-zero degrees of freedom forms another very large class of the joints with unilateral constraints. Problems in elasticity theory with contacts, in general form, are rigorously treated in the framework of elasticity theory with unilateral constraints [26] founded in 20th century in 60s. The periodic structures with initial stresses were considered in [23]. To the best knowledge of the authors, nowadays there exist no researches in the field of the asymptotic analysis of the thin-walled multi-structures with joints of non-zero degree of freedom similar to the research carried out for the thin-walled multi-structures with the "rigid joints". The last problem is also not completely solved yet.

References

[1] Bhandari VB (2019) Machine Design Data Book, 2nd ed, McGraw-Hill, New York.

[2] Lingaiah K (2003) Machine Design Databook, 2nd ed, McGraw-Hill, New York.

[3] Kulak GL, Fisher JW, Struik JHA (1987) Guide to Design Criteria for Bolted and Riveted Joints, 2nd ed, John Wiley and Sons, New York.

[4] Benvenuto E (1991) An Introduction to the History of Structural Mechanics: Part II: Vaulted Structures and Elastic Systems, Springer, Berlin.

[5] Euler L (1773) De pressione ponderis in planum cui incumbit, Novi Commentarii Academiae Scientiarum Imperialis Petropolitanae, vol. 18.

[6] Timoshenko SP (1983) History of Strength of Materials, Dover, New York.

[7] Clebsch A (1862) Theorie der Elasticität fester Körper, Leipzig.

[8] Maxwell JC (1864) On the calculation of the equilibrium and stiffness of frames, The London, Edinburgh, and Dublin Philosophical Magazine and Journal of Science **27**(182):294-299. DOI 10.1080/14786446408643668

[9] Zhikov VV (2002) Homogenization of elasticity problems on singular structures, Izvestiya: Mathematics **66**(4):299-365. DOI 10.1070/IM2002v066n02ABEH000380

[10] Zhikov VV, Pastukhova SE (2003) Homogenization for elasticity problems on periodic networks of critical thickness, Sbornik Mathematics **194**(5):697-732. DOI 10.1070/SM2003v194n05ABEH000735

[11] Caillerie D (1984) Thin elastic and periodic plates, Mathematical Methods in the Applied Sciences **6**:159-191. DOI 10.1002/mma.1670060112

[12] Kohn RV, Vogelius M (1984) New model for thin plates with rapidly varying thickness, International Journal of Solids and Structures **20**(4):333-350. DOI 10.1016/0020-7683(84)90044-1

[13] Trabucho L, Viano JM (1996) Mathematical Modeling of Rods, In: Handbook of Numerical Analysis, PG Ciarlet, JL Lions (Eds), Vol 4, pp 487-974 Elsevier, Amsterdam. DOI 10.1016/S1570-8659(96)80006-8

[14] Makerle J (2003) Finite element analysis of fastening and joining: A bibliography (1900-2002), International Journal of Pressure Vessels and Piping, **80**(4):253-271. DOI 10.1016/S0308-0161(03)00030-9

[15] Ciarlet PG (1991) Plates and Junctions in Elastic Multi-Structures: An Asymptotic Analysis, Springer, Berlin.

[16] Le Dret H (1991) Problèmes Variationnels dans les Multi-Domaines, Modélisation des Jonctions et Applications, Masson, Paris.

[17] Titeux I, Sanchez-Palencia E (2000) Junction of thin plates, European Journal of Mechanics - A/Solids **19**(3):377-400. DOI 10.1016/S0997-7538(00)00175-3

[18] Cabib E, Freddi L, Morassi A, Percivale D (2001) Thin notched beams, Journal of Elasticity and the Physical Science of Solids **64**:157-178. DOI 10.1023/A:1015210302849

[19] Gaudiello A, Kolpakov AG (2011) Influence of non degenerated joint on the global and local behavior of joined rods, International Journal of Engineering Science **49**(3):295-309. DOI 10.1016/j.ijengsci.2010.11.002

[20] Ivanyi M, Baniotopoulos CC (Eds) (2000)Semi-Rigid Connections in Structural Steelwork, CISM International Centre for Mechanical Sciences, Vol 419. DOI 10.1007/978-3-7091-2478-9

[21] Kolpakov AG, Andrianov IV, Markert B (2015) Asymptotic decomposition in the problem of joined elastic plates, ZAMM - Journal of Applied Mathematics and Mechanics / Zeitschrift für Angewandte Mathematik und Mechanik **95**(12), 1268-1281. DOI 10.1002/zamm.201400006

[22] Kolpakov AG, Andrianov IV, Rakin SI, Rogerson GA (2018) An asymptotic strategy to couple homogenized elastic structures, International Journal of Engineering Science **131**:26-39. DOI 10.1016/j.ijengsci.2018.04.006

[23] Kolpakov AG (1994) The stiffnesses of elastic cylindrical beams, Journal of Applied Mathematics and Mechanics **58**(2):293-301. DOI 10.1016/0021-8928(94)90058-2

[24] Kolpakov AG, Andrianov IV (2014) Asymptotic decomposition in the problem of joined elastic beams, ZAMM - Journal of Applied Mathematics and Mechanics / Zeitschrift für Angewandte Mathematik und Mechanik **94**(13):818-836. DOI 10.1002/zamm.201200278

[25] Sanchez-Palencia E (1972) Non-Homogeneous Media and Vibration Theory, Springer, Berlin.

[26] Fichera G (1972) Existence Theorems in Elasticity, Springer, New York.

Chapter 26
The Homogenized Delamination Criterion for Fiber-reinforced Plate

Alexander G. Kolpakov, Sergey I. Rakin, and Igor V. Andrianov

Abstract We consider a layered composite plate formed from the layers of unidirectional fibers - prepregs. While the prepregs are unidirectional materials, they are stacked with different orientation. In such composite, we distinguish the local stress-strain states in prepregs and at the prepregs interfaces. We investigate both the local stress-strain states and construct the corresponding homogenized strength criteria. Since the local stress-strain state at the prepregs interfaces has the form of boundary layer, we arrive at the notion of the strength criterion of the interface surface. The strength of a composite in a whole is determined by the strength of prepregs and the interfaces. In this paper, we consider in details the in-plane deformation of the fiber-reinforced plate, however, the methodology presented can be applied to many elastic structures containing internal interfaces.

Key words: Prepreg technology · Interface · Boundary layer · Homogenized strength criteria · Delamination

---

Alexander G. Kolpakov
SysAn (System Analysis in Engineering), A. Nevskogo str., 12a, 34, Novosibirsk, 630075, Russian Federation,
e-mail: algk@ngs.ru

Sergey I. Rakin
Siberian Transport University, Dusi Koval'chuk Str., 191, Novosibirsk, 630049, Russia Federation,
e-mail: rakinsi@ngs.ru

Igor V. Andrianov
Chair and Institute of General Mechanics, RWTH Aachen University, Eilfschornsteinstraße 18, Aachen, D-52062, Germany,
e-mail: igor.andrianov@gmail.com

H. Altenbach et al. (eds.), *Sixty Shades of Generalized Continua*,
Advanced Structured Materials 170, https://doi.org/10.1007/978-3-031-26186-2_26

26.1 Introduction

Many fiber-reinforced thin-walled structures are produced by using the prepreg technology based on the stacking layers of unidirectional fibers – prepregs, under alternating angles. For detailed description of the prepreg technology and layered composite thin-walled structures see [1]-[5]. The thickness of prepreg varies from 50 μm to 200 μm. For most industrial carbon-, glass- and polymer fibers, the fiber diameter lies within 10-50 μm [3]-[6]. The number of the fibers across the prepreg thickness may be from few ones to several tens of the fiber radius.

In this paper, we discuss the basic problem in the strength of laminated fiber-reinforced plates and the methodology of the failure analysis of laminated fiber-reinforced plates with the two-scale method and the corresponding computation methods. For this reason, we discuss the simplest reinforcement scheme – the orthogonal stacking of the fiber layers (prepregs). The problems arising are illustrated in Fig. 26.1(*a*), which displays the basic modes of fracture of the laminated composite:

A. delamination;
B. fibers fracture;
C. matrix fracture.

We assume that the prepregs have periodic structure. The periodic structure are characterized by a periodicity cell(s) (εY^- and εY^+ in Fig. 26.1(*b*)), hereinafter ε is the characteristic size of the microstructure. The prepregs are the same, thus the

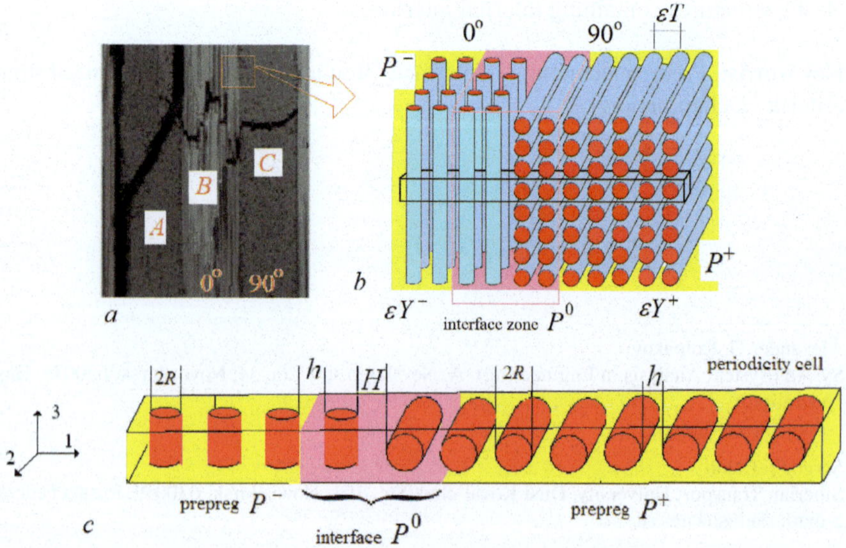

Fig. 26.1: Three types of destruction of the prepreg composite – (a); the scheme of composite (interface zone colored in pink) – (b); the periodicity cell of prepreg – (c).

periodicity cells εY^- and εY^+ are the same, but have different orientations. We denote: εR – the radius of the fibers, εh – the distance between the fibers in the main part of the prepregs, εH – the distance between the fibers in the adjacent prepregs, see Fig. 26.1(*b*).

The interface zone between the adjacent prepregs has complex structure, see Fig. 26.1(*b*), (*c*). This is the zone between two prepregs of periodic structure, which are turned on 90° one with respect to another. The stress-strain states in the zones of periodic structure may be computed by using the periodic functions of complex variables [7]-[9] or the homogenization theory [10]-[13]. It is evident, that the periodic solutions on the left and on the right from the interface zone do not coincide one with another. Thus, there exists a specific solution in the interface zone P^0. In this paper, we are concentrated on the failure in the interface zone between the prepregs. This kind of failure is usually associated with the so-called delamination, see Fig. 26.1(*a*).

26.2 Boundary Layer

The delamination, from the mathematical point of view, is associated with the boundary layer, which appears on the interface of 0°- and 90°-"thick" prepregs. The boundary layer near the interface of 0°- and 90°-prepregs is an asymptotic phenomena and lead to the notion of the interface strength. To the best knowledge of the authors, such boundary layer was not investigated earlier. Mention some theoretical problems [14]-[23] related to the boundary layers closed to the problem considered in this paper.

Between the prepregs P^- and P^+, there is some transition (interface) zone P^0 that differs from the connected structure (Fig. 26.1). The local perturbation method [21]-[23] makes it possible to describe both the local strain-stares state in the prepregs P^- and P^+, and in the interface zone P^0. The authors treat this as the main advantage of the proposed method.

Consider the problem of elasticity theory in an inhomogeneous periodic medium occupying the domain $P^- \cup P^0 \cup P^+$ (we assume that the characteristic size of this domain is 1). The variational statement of the problem has the form [24]

$$\int a_{ijkl}(\boldsymbol{x}/\varepsilon)u_{i,j}v_{k,l}y d\boldsymbol{x} = \int \boldsymbol{q}(\boldsymbol{x})\boldsymbol{v}(\boldsymbol{x})d\boldsymbol{x} \text{ for any } \boldsymbol{v} \in V, \tag{26.1}$$

where a_{ijkl} are the local elastic constants – the elastic constants of the constitutive elements of the composite, $\boldsymbol{u}$ – the vector of the displacements, $\boldsymbol{v}$ – the test function, V – the set of admissible displacements, $\boldsymbol{q}$ – the vector of the body forces. Equation (26.1) is the same for two- or three-dimensional problems.

While the fibers in the prepregs are distributed periodically with the periodicity cells εY^- and εY^+, the pair of the adjacent prepregs has another periodicity cell $P = P^- \cup P^0 \cup P^+$, see Fig. 26.1(*b*). The periodicity cell of composite is displayed in Fig. 26.1(*b*), (*c*).

At the distance from the interface zone, solution to the elasticity theory problem in the periodicity cell of composite tends to the periodic solutions with the periodicity cell εY^- in the left part of P and with the periodicity cell εY^- in the right part of P. Since "a priori" we do not know the length of the perturbation zone, the periodicity cell of composite must initially be taken "infinitely long".

Let us take, without loss of generality, the case of a macroscopically rectilinear boundary coinciding with the axis $x_1 = 0$. At the microscopic level, the boundary has a periodic geometry with a period εT, Fig. 26.1. Following [21]-[23], we look for a solution to problem (26.1) in the form

$$\boldsymbol{u} = \boldsymbol{u}_0(\boldsymbol{x}) + \varepsilon \boldsymbol{u}_{\mathrm{loc}}(\boldsymbol{x}/\varepsilon) + \varepsilon \boldsymbol{u}_{\mathrm{p}}(\boldsymbol{x}/\varepsilon). \tag{26.2}$$

Here $\varepsilon \boldsymbol{u}_{\mathrm{p}}(\boldsymbol{x}/\varepsilon)$ is the periodic corrector of the homogenization theory [19, 20], i.e., a function periodic with the periodicity cells Y^- and Y^+ to the left and to the right of the interface region. The function $\varepsilon \boldsymbol{u}_{\mathrm{p}}(\boldsymbol{x}/\varepsilon)$ is used to construct the local stress-strain state in the "main parts" of connected prepregs. The function $\varepsilon \boldsymbol{u}_{\mathrm{loc}}(\boldsymbol{x}/\varepsilon)$ is the local perturbation decaying as $|x_1| \to \infty$.

Trial function $\boldsymbol{v}$ is taken in the form similar to (26.2)

$$\boldsymbol{v} = \boldsymbol{v}_0(\boldsymbol{x}) + \varepsilon \boldsymbol{v}_{\mathrm{loc}}(\boldsymbol{x}/\varepsilon) + \varepsilon \boldsymbol{v}_{\mathrm{p}}(\boldsymbol{x}/\varepsilon). \tag{26.3}$$

Substituting (26.3) into (26.1), we arrive at the following variational problem

$$\int a_{ijkl}(\boldsymbol{x}/\varepsilon)(u_{\mathrm{loc}\,k,ly} + u_{\mathrm{p}\,k,ly} + u_{0\,k,lx})(v_{\mathrm{p}\,k,ly} + v_{1\,k,lx})d\boldsymbol{x} = 0 \tag{26.4}$$

$$\text{for any } \boldsymbol{v}_1 \in V_1, \boldsymbol{v}_{\mathrm{p}} \in V_{\mathrm{p}},$$

where functions $\boldsymbol{v}_{\mathrm{p}}$ are periodic with periodicity cells Y^- and Y^+ to the left and to the right of the interface zone and $\boldsymbol{v}_1$ decays as $|x_1| \to \infty$.

In the interface zone, we have no reasonable way to separate the functions $\boldsymbol{u}_{\mathrm{p}}$ and $\boldsymbol{u}_{\mathrm{loc}}$, and we consider the sum $\boldsymbol{u}_{\mathrm{p}} + \boldsymbol{u}_{\mathrm{loc}}$. The sum $\boldsymbol{u}_{\mathrm{p}} + \boldsymbol{u}_{\mathrm{loc}}$ is just the local perturbation function [21].

Equation (26.4) leads to the following boundary value problem

$$\begin{cases} (a_{ijkl}(\mathbf{y})((\boldsymbol{u}_{\mathrm{loc}}(\mathbf{y}) + \boldsymbol{u}_{\mathrm{p}}(\mathbf{y}))_{k,ly} + a_{ij\alpha\beta}(\mathbf{y})\boldsymbol{u}_{0\,\alpha,\beta x}(\boldsymbol{x}))_{,jy} = 0 \text{ in } P, \\ \boldsymbol{u}_{\mathrm{loc}}(\mathbf{y}) + \boldsymbol{u}_{\mathrm{p}}(\mathbf{y}) \text{ periodic in } y_2 \text{ and } y_3 \text{ with period } T, \\ \boldsymbol{u}_{\mathrm{loc}}(\mathbf{y}) + \boldsymbol{u}_{\mathrm{p}}(\mathbf{y}) \text{ tends to a function periodic with the periodicity} \\ \quad \text{cell } Y^- \text{ as } y_1 \to -\infty \text{ and the periodicity cell } Y^+ \text{ as } y_1 \to \infty. \end{cases} \tag{26.5}$$

Boundary value problem (26.5) is a problem in the "fast" variables $\mathbf{y} = \boldsymbol{x}/\varepsilon$. The functions of the "slow" variables are parameters in (26.5). It is suitable to use the function $\boldsymbol{v}_{\mathrm{p}} + \boldsymbol{v}_{\mathrm{loc}}$ as unknown in (26.5).

Boundary value problem (26.5) is not solvable for any $\boldsymbol{u}_{0\alpha,\beta,x}(\boldsymbol{x})$. To obtain the solvability condition for (26.5), we integrate Eq. (26.5) by parts and take into account the boundary conditions in (26.5). As a result, we arrive at the equation

$$A^{-}_{i1\alpha\beta}e^{-}_{\alpha\beta} = A^{+}_{i1\alpha\beta}e^{+}_{\alpha\beta}. \tag{26.6}$$

In (26.6), $e^{-}_{\alpha\beta}$ and $e^{+}_{\alpha\beta}$ are the strains corresponding to the displacements in the "main part" of the prepregs on the left and on the right from the interface; $A^{-}_{ij\alpha\beta}$ and $A^{+}_{ij\alpha\beta}$ are the homogenized elastic constants [10]-[14] of the prepregs placed on the left and on the right from the joint. Equation (26.6) is the necessary and sufficient condition of the solvability of the periodicity cell problem (26.5).

The appearance of solvability conditions due homogenization of boundary value problem with a wavy free boundary was noted earlier [15]. Solvability condition (26.6) differs from the condition obtained in [15]. The most essential difference is that solvability condition from [15] directly accounts the geometry of the boundary, while (26.6) involves only solutions of the limit problem and does not use the properties of the boundary. The presence of solvability condition (26.6) differs problem (26.5) from the periodicity cell problem of the homogenization theory.

Representation (26.2) is similar to the ansatz of the homogenization theory [10]-[14], but there are two basic differences:

1. In the homogenization theory, the corrector $\boldsymbol{u}_1(\boldsymbol{y})$ is a periodic function, while the local perturbation tends to periodic functions with different periodicity cells in one direction and periodic in the remaining directions.
2. In the homogenization theory $\boldsymbol{u}_0(\boldsymbol{y})$ is a C^1-function. In the case under consideration, some first derivatives of $\boldsymbol{u}_0(\boldsymbol{y})$ are discontinuous.

As noted above, the perturbation $\boldsymbol{u}_1(\boldsymbol{y})$ itself is not of practical interest and we need the full stress-strain state corresponding to the complete solution $\boldsymbol{u}_1(\boldsymbol{y})+\boldsymbol{u}_\mathrm{p}(\boldsymbol{y})$. This stress-strain state that determines the strength of the connection. It is important to solve problem (26.4) in a finite area containing the interface zone. For this aim, we consider the domain $P = P^{-} \cup P^{0} \cup P^{+}$, which contains the interface zone P^0 and fragments P^- and P^+ of adjoining prepregs, see Fig. 26.1(*c*). Let us take the left and the right boundaries of the region P placed "far enough" from the interface zone P^0, just so that the perturbation decays outside P^0. The meaning of the term "far enough" is discussed below.

Estimates of the speed of decaying of edge effects have been theoretically studied in various papers, see, for example, [16]. However, they do not give numerical values for the size of the attenuation zones of edge effects. Our numerical experiments demonstrate that the decay distance of the edge effects can be estimated as 1-2 thicknesses of the periodicity cell. Note that for connected rods, the decay distance of the edge effects is estimated as 3-5 periodicity cell thicknesses [21]-[23].

26.3 Numerical Computation Results

In the previous sections, we have presented arguments confirms the existence of the local perturbation of the stress-strain state between the adjacent prepregs. Now, we verify the predicted local perturbation numerically. For this, we solve numerically the cell problem (26.5) for the periodicity cell displayed in Fig. 26.1(*c*). In our computations Young's modulus and Poisson's ratio, strength limit of the fibers and matrix are: $E_f = 170$ GPa, $\nu_f = 0.35$, $\sigma_f^* = 1$ GPa and $E_b = 20$ GPa, $\nu_b = 0.4$, $\sigma_b^* = 0.2$ MPa, correspondingly. These values correspond to the carbon/epoxy composite [1]-[3]. Periodicity cell sizes are $1 \times 1 \times 2L$; L is a distance where the local perturbation vanishes. Our computational experiments show that $L \approx 2\text{div}\,5$. In the "fast" variables, the fiber radius is R, $R < 1$. In the computations below, $R = 0.4$ (highly filled composite) and $R = 0.3$ (medium filled composite). The distance between the fibers in the prepregs is $h = 2(1 - R)$. The thickness of the interface zone is H. In the computations below, H varies from 0.2 to 0.01.

26.3.1 Extension Along Oy- or Oy-axis

Figure 26.2 displays von Mises stress in the whole periodicity cell (the fibers and the matrix). Figure 26.3 displays von Mises stress in the matrix. The comparison of Fig. 26.2 and Fig. 26.3 illustrates the idea of multicomponent effect in the homogenization [7]. It is seen that the stress-strain state in the fibers is uniform and in the matrix is strongly nonuniform. The local stress-strain state is periodic on the left and on the right from the interface zone, but the periodic functions are different. In the interface zone, one periodic function turns out into another periodic function. The stress-strain state in the interface zone significantly differs from the stress-strain state on the left and on the right from the interface.

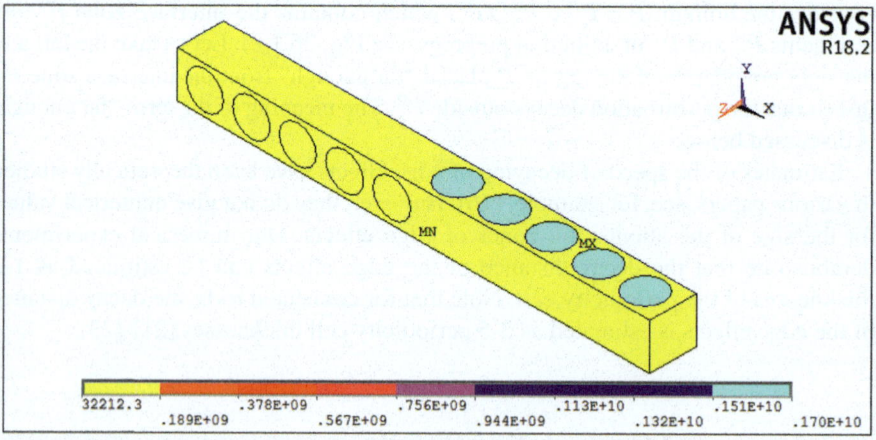

Fig. 26.2: The von Mises stress in the whole periodicity cell (fibers and matrix) for $R = 0.4$, $H = 0.1$.

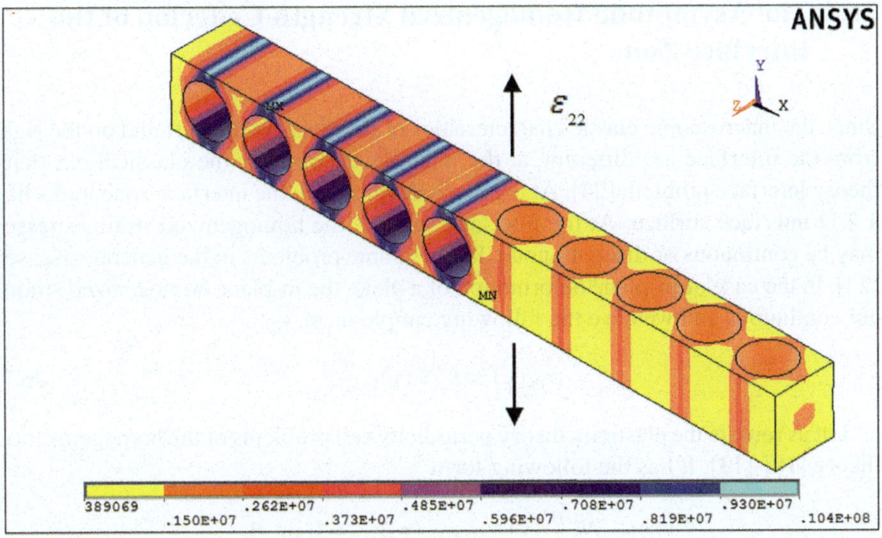

Fig. 26.3: The von Mises stress in the matrix for $R = 0.4$, $H = 0.1$.

26.3.2 Shift in Ox_2x_3-plane

Figure 26.3 displays the local von Mises stress in the matrix for the fiber radius $R = 0.3$ and the interface distance $H = 0.02$. It is displayed the interface zone between the orthogonal fibers of adjacent prepregs. The maximum von Mises stress occurs in this interface zone. It is the phenomenon similar to one discovered in [25]. The phenomenon results from the mutual rotation of the orthogonal fibers at the microlevel when the composite is subjected to the macroscopic shift. This phenomenon is parameter sensitive, first, it depends of the R and H. For $R = 0.4$ and $H = 0.1$, the maximum von Mises stress occurs at in the main parts of the prepregs. For $R = 0.3$ and $H = 0.02$, the maximum von Mises stress occurs interface zone between the adjacent prepregs. Second, it depends on the elastic characteristics of the fibers and the matrix.

When a bulk composite is considered, one can also take into account the local stresses corresponding to the following macroscopic stresses: extension along Ox-axis and shift in Oxy- or Oxz-plane. Stresses in the directions across the plate are very small in the thin plates satisfying the Kirchhoff-Love hypothesis [24]. We consider the thin plates satisfying the Kirchhoff-Love hypothesis and neglect the local stresses corresponding to these zero macroscopic stresses.

26.4 The Asymptotic Homogenized Strength Criterion of the Interface Zone

Since the macroscopic elastic characteristics of the layers on the left and on the right from the interface are different, at the macrolevel, we have the classical elasticity theory interface problem [24]. At the macroscopic level, the interface zone looks like a 2-D interface surface. At the interface surface, the homogenized strains/stresses may be continuous or discontinuous. It takes some problems in the general case, see [23]. In the case of in-plane deformation of a plate, the in-plane homogenized strains are continuous and we have the following simple form

$$\sigma_{ij}(\mathbf{y}) = K_{ij}^{\alpha\beta}(\mathbf{y}) e_{\alpha\beta}. \tag{26.7}$$

Let us rewrite the elasticity theory periodicity cell problems of the homogenization theory [10]-[13]. It has the following form

$$\begin{cases} (a_{ijkl} N_{k,ly}^{\alpha\beta}(\mathbf{y}) + a_{ij\alpha\beta}(\mathbf{y}))_{,jy} = 0 \text{ in } P, \\ \mathbf{N}^{\alpha\beta}(\mathbf{y}) \text{ periodic with period } T. \end{cases} \tag{26.8}$$

For in-plane deformation, the indices α and β take values 2 or 3.

Problem (26.8) describes the local stress-strain state in the periodicity cell of prepreg far from the interface zone. In particular, the local stress σ_{ij} in the prepreg far from the interface zone is

$$\sigma_{ij}(\mathbf{y}) = k_{ij}^{\alpha\beta}(\mathbf{y}) e_{\alpha\beta}, \tag{26.9}$$

where

$$k_{ij}^{\alpha\beta}(\mathbf{y}) = a_{ijkl} N_{k,ly}^{\alpha\beta}(\mathbf{y}) + a_{ij\alpha\beta}(\mathbf{y}) \tag{26.10}$$

is the local stress concentration tensor [26] and $e_{\alpha\beta}$ are the overall strains.

Formula (26.7) looks exactly like (26.9), but (26.7) follows not from the homogenization theory, but from the perturbation method [21, 22]. The coincidence of representations (26.9) and (26.7) is the specific occasions for the in-plane deformation of plate. In the general case, the representation for the local stresses in the interface zone differs from (26.7). The functions $K_{ij}^{\alpha\beta}(\mathbf{y})$ and $k_{ij}^{\alpha\beta}(\mathbf{y})$ are determined from problems (26.5) and (26.9), correspondingly, which are essentially different. The Figs. 26.3 and 26.4 illustrate the difference. The von Mises stress corresponding to the function $K_{ij}^{\alpha\beta}(\mathbf{y})$ is displayed in Fig. 26.4. The von Mises stress corresponding to the function $k_{ij}^{\alpha\beta}(\mathbf{y})$ is displayed in Fig. 26.3 on the left and on the right from the interface zone. Note that

$$K_{ij}^{\alpha\beta}(\mathbf{y}) \to k_{ij}^{\alpha\beta}(\mathbf{y}) \text{ as } |y_3| \to \infty. \tag{26.11}$$

Let the strength criterion of the fiber has the form $f_{\rm F}(e_{ij}) \le \sigma^*_{\rm F}$ and the strength criterion of the binder has the form $f_{\rm B}(e_{ij}) \le \sigma^*_B$. Introduce the following functions:

$$f(\boldsymbol{x},\sigma_{ij}) = \begin{cases} f_{\rm F}(e_{ij}) \\ f_{\rm B}(\sigma_{ij}) \end{cases} \qquad \sigma^*(\boldsymbol{x}) = \begin{cases} \sigma^*_{\rm F} \text{ if } \boldsymbol{x} \text{ in fiber,} \\ \sigma^*_{\rm B} \text{ if } \boldsymbol{x} \text{ in matrix.} \end{cases} \tag{26.12}$$

In this notations, we can write the strength criterion in the point $\boldsymbol{x}$ in the form

$$f(\boldsymbol{x},\sigma_{ij}) \le \sigma^*(\boldsymbol{x}). \tag{26.13}$$

Substituting σ_{ij} following (26.7), we obtain $f(\mathbf{y}, k^{\alpha\beta}_{ij}(\mathbf{y}) e_{\alpha\beta}) \le \sigma^*(\mathbf{y})$. Then the criterion for the absence of fracture at all points of the whole periodicity cell $P = P^- \cup P^0 \cup P^+$ (it includes the interface zone and the adjacent prepregs) can be written in the form [25]

$$F_{\rm composite}(e_{\alpha\beta}) = \max\nolimits_{\mathbf{y}\in P} \frac{f(\mathbf{y}, K^{\alpha\beta}_{ij}(\mathbf{y}) e_{\alpha\beta})}{\sigma^*(\mathbf{y})} \le 1. \tag{26.14}$$

The failure of elements of periodicity cell starts with $F_{\rm composite}(e_{\alpha\beta}) = 1$ and occurs at the point(s) $\mathbf{y}_0$ at which

$$f(\mathbf{y}_0, K^{\alpha\beta}_{ij}(\mathbf{y}_0) e_{\alpha\beta}) = \sigma^*(\mathbf{y}_0). \tag{26.15}$$

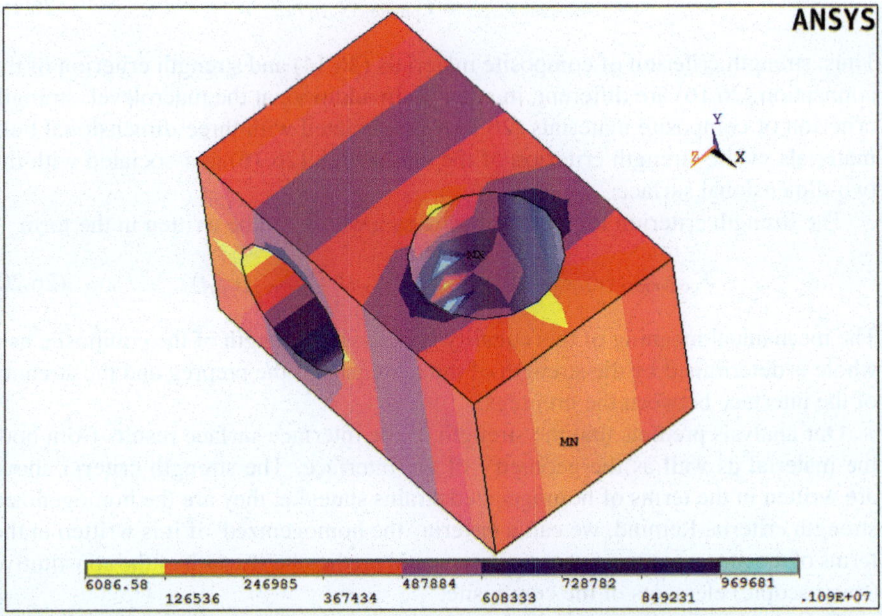

Fig. 26.4: The local von Mises stress in the matrix in the interface zone; $R = 0.3$, $H = 0.02$.

Since we are interesting in the strength criterion of the interface, we have to change in (26.14) the whole periodicity cell for the interface zone P^0. As result, we arrive at the homogenized interface zone strength criterion

$$F_{\text{delam}}(e_{\alpha\beta}) = \max_{\mathbf{y}\in P^0} \frac{f(\mathbf{y}, K_{ij}^{\alpha\beta}(\mathbf{y})e_{\alpha\beta})}{\sigma^*(\mathbf{y})} \leq 1. \tag{26.16}$$

We also can write the homogenized prepreg strength criterion:

$$F_{\text{prepreg}}(e_{\alpha\beta}) = \max_{\mathbf{y}\in P^\pm} \frac{f(\mathbf{y}, k_{ij}^{\alpha\beta}(\mathbf{y})e_{\alpha\beta})}{\sigma^*(\mathbf{y})} \leq 1. \tag{26.17}$$

Taking into account the periodicity structure of the main part of prepregs, we can compute maximum in (26.17) over the periodicity cells Y^- and Y^+, and write (26.17) in the following form:

$$F_{\text{prepreg}}(e_{\alpha\beta}) = \max_{\mathbf{y}\in Y^\pm} \frac{f(\mathbf{y}, k_{ij}^{\alpha\beta}(\mathbf{y})e_{\alpha\beta})}{\sigma^*(\mathbf{y})} \leq 1. \tag{26.18}$$

Strength criterion (26.18) corresponds to the homogenized strength criterion for composite of periodic structure discussed in [11]. Since the local stress-strain states in the periodicity cells $P^\pm$ and P^0 in the interface zone are significantly different (Fig.26.3 and Fig.26.4 illustrate this difference), then

$$F_{\text{delam}}(e_{\alpha\beta}) \neq F_{\text{prepreg}}(e_{\alpha\beta}). \tag{26.19}$$

Thus, strength criterion of composite materials (26.14) and strength criterion of the connection (26.16) are different, in general. In addition, at the macrolevel, strength criterion of composite materials (26.11) is associated with three-dimensional bulk materials, while strength criterion of the connection (26.16) is associated with the two-dimensional surface.

The strength criterion for the composite as a whole can be written in the form

$$F_{\text{composite}}(e_{\alpha\beta}) = \max(F_{\text{delam}}(e_{\alpha\beta}), F_{\text{prepreg}}(e_{\alpha\beta})). \tag{26.20}$$

The mechanical meaning of this equality is clear: the strength of the composite as a whole is determined by the strength of the main part of the prepreg and the strength of the interface between the prepregs.

Our analysis predicts that the strength of the interface surface results from both the material as well as the geometry of the interface. The strength criteria above are written in the terms of homogenized strains state, i.e. they are the homogenized strength criteria. Remind, we call a criteria "the homogenized" if it is written in the terms of the macroscopic stress-strain state and predict the strength of the constitutive (microscopic) elements of the composite.

The macroscopic strength criterion of the interface surface can be constructed in the following steps:

- Step 1: Compute $K_{ij}^{\alpha\beta}(\mathbf{y})$.
- Step 2: Form linear combinations (26.7) and select from them the combinations for which

$$f(\mathbf{y}, K_{ij}^{\alpha\beta}(\mathbf{y})e_{\alpha\beta}) = \sigma^*(\mathbf{y}) \tag{26.21}$$

or

$$f(\mathbf{y}, K_{ij}^{\alpha\beta}(\mathbf{y})e_{\alpha\beta}) < \sigma^*(\mathbf{y}) \tag{26.22}$$

if construct the safety zone.

Step 1 may be realized with FEM software, step 2 requires small additional programming to process the ANSYS data files and make computations indicated in (26.16) and (26.21). The coefficients of linear combinations (26.7) satisfying equality (26.16) form the macroscopic failure surface of the interface - the homogenized delamination criterion.

26.5 Constructing the Delaminating Strength Criterion

If the plate is subjected to in-plane deformation, the strength criterion is the function of $e_{\alpha\beta}$, $\alpha\beta = 2,3$. Since $e_{23} = e_{32}$, the safety zone is a three-dimensional objects in the coordinate system $Oe_{yy}e_{zz}e_{yz}$, see Fig. 26.5. In our computations, we use the von Mises strength criterion at the microlevel both for the fibers and the binder. Then $f(\sigma_{ij}) = \sigma_{\text{vM}}$. The von Mises stress corresponding to (26.7) is the function of the "fast coordinate" and the overall strains: $\sigma_{\text{vM}} = \sigma_{\text{vM}}(\mathbf{y}, e_{\alpha\beta})$.

As noted above, the strength limit of the fibers is greater than the strength limit of the binder. For this reason, we present numerically computed safety zone for the matrix: $S_{\text{B}} = \{(e_{yy}, e_{zz}, e_{yz}) : \sigma_{\text{vM}} \leq \sigma_{\text{B}}^*\}$. In Fig. 26.5, the half of the 3-D safety zone is displayed, namely, the zone $\{S_{\text{B}} : e_{yz} \geq 0\}$.

In Fig. 26.6, projections of the 3-D safety zone on the coordinate planes $Oe_{yy}e_{zz}$, $Oe_{yy}e_{yz}$ and $Oe_{zz}e_{yz}$ are displayed. Our numerical computations demonstrate that the stress concentration tensors are significantly different for $R = 0.4$ and $R = 0.3$.

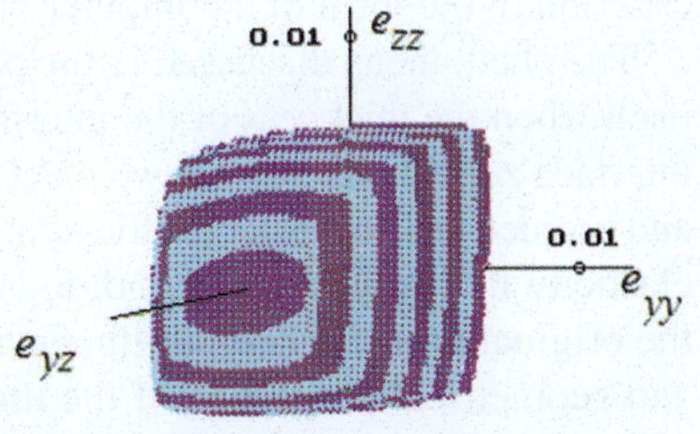

Fig. 26.5 Three-dimensional safety zone for the interface in the coordinate system e_{yy}, e_{zz}, e_{yz}

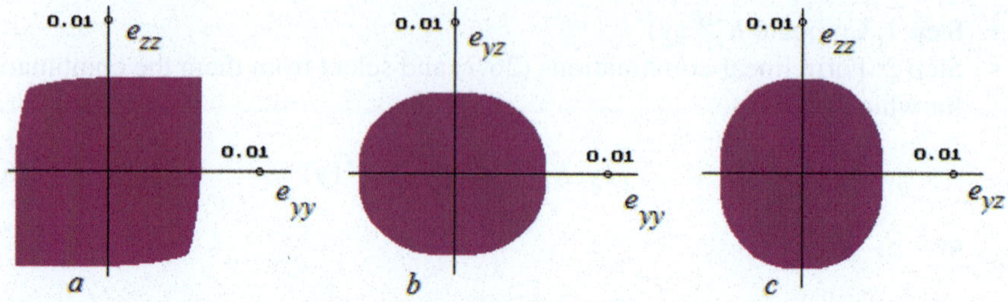

Fig. 26.6: Projections of the 3-D safety zone on the coordinate planes and 3-D reconstruction of the safety zone for the interface for $R = 0.3$, $H = 0.1$.

From the mechanical point of view, it means that a universal delamination criterion for the prepreg composites does not exist.

26.6 Conclusions

By using the theoretical approach developed in the papers [21, 22], we demonstrate that an interface is an independent object for fracture mechanics. In the composites produced using the prepreg technology, the interface strength criterion differs from the homogenized strength criteria of the connected prepregs. This difference results from the boundary layer which occurs in the zone between the prepregs. The local stress-strain state in the boundary layer, which defines the interface strength criterion, determines not only the strength properties of the connected prepregs but also elastic characteristics of the prepregs as well as the micro-geometry of the interface zone.

The interface strength criterion cannot be written in the terms of the "intrinsic" interface deformations (deformations of the interface surface considered as 2-D surface) only. It involves all components of the stress/strain tensors. The most compact form of the interface strength criterion is written in terms of the macroscopic "intrinsic" strains (these are the strains of the interface as 2-D surface) and the macroscopic "extrinsic" stresses. In some special cases the number of the "intrinsic" parameters may be reduced. In the general case, it is impossible to obtain the interface strength criterion in an explicit analytical form, but it is possible to construct this criterion in the form of a computer program.

The phenomena discussed in the paper have an asymptotic nature, thus, manifests itself when the thickness of the interface is small. In the limit, as the thickness of the interface zone tends to zero, we meet the effect of the losing of the individual elastic and geometrical properties of the joint [22, 27]. It means, that we arrive at the classical elasticity theory interface conditions formulated on a two-dimensional surface (if the original problem is three-dimensional). At the same time, the individual elastic and geometrical properties of the three-dimensional interface zone, as well as the strength properties, influence the strength properties of the interface. As a result, we arrive at the notion of the strength of the two-dimensional interface which differs from the strength of the main parts of the connected prepregs. In the reality, the thickness of the joint is small but non-zero, and the asymptotic solution provides us with approximations. The macroscopic strength criterion may be used in practice as a suitable approximation. Additional analysis is required to estimate the accuracy of these approximations.

The general principles presented in this paper may be applied to the various linear coupling problems (for example, molecular-level models or solid-to-molecular models, see references in [23], [27]-[31]). Naturally, the specific calculations may be very different, following the physical nature of the problems. In our illustrative examples, we used von Mises strength criterion both for the fibers and the binder. Our computer program may use strength criteria of the form (26.13) with arbitrary functions – it is sufficient to develop the proper computer codes for the functions in (26.12). In particular, one may develop the codes corresponding to the experimental strength criteria for the binder and the fiber.

References

[1] Lengsfeld H, Wolff-Fabris F, Krämer J, Lacalle J, Altstädt V (2021) Composite Technology: Prepregs and Monolithic Part Fabrication Technologies, Munich, Hanser Publ.

[2] ISO 8604:1988, Plastics - Prepregs - Definitions of terms and symbols for designations (2018), ANSI.

[3] Chawla KK (2012) Composite Materials, Springer, New York.

[4] Mazumdar SK (2002) Composites Manufacturing: Materials, Product, and Process Engineering, CRC Press, Boca Raton, FL.

[5] Agarwal BD, Broutman LJ, Chandrashekhara K (2017) Analysis and Performance of Fiber Composites, 4th ed, Wiley, Hoboken, NJ.

[6] Jawaid M, Thariq M (Eds) (2018) Sustainable Composites for Aerospace Applications, Woodhead.

[7] Mityushev V, Rogozin SV (2000) Constructive Methods for Linear and Nonlinear Boundary Value Problems of Analytic Function. Theory and Applications, CRC, Boca Raton, FL.

[8] Gluzman S, Mityushev V, Nawalaniec W (2018) Computational Analysis of Structured Media, Academic Press, Amsterdam.

[9] Drygas P, Gluzman S, Mityushev V, Nawalaniec W (2020) Applied Analysis of Composite Media: Analytical and Computational Results for Materials Scientists and Engineers, Elsevier, Amsterdam.

[10] Zhikov VV (2002) Homogenization of elasticity problems on singular structures, Izv. Math. **66**(3):299-365.

[11] Bakhvalov NS, Panasenko GP (1989) Averaging Processes in Periodic Media. Mathematical Problems in Mechanics of Composite Materials, Kluwer, Dordrecht.

[12] Sanchez-Palencia E (1980) Non-Homogeneous Media and Vibration Theory. Springer, Berlin, New York.

[13] Bensoussan A, Lions J-L, Papanicolaou G (1978) Asymptotic Analysis for Periodic Structures, North Holland, Amsterdam.

[14] Panasenko GP (1991) Multicomponent homogenization for processes in essentially nonhomogeneous structures, Sb. Math. **69**(1):143-153.

[15] Sanchez-Palencia E (1987) Boundary layers and edge effects in composites, In: Homogenization Techniques for Composite Materials (Eds Sanchez-Palencia E, Zaoui A), Springer, Berlin, pp 122-193.

[16] Oleinik OA, Shamaev AS, Yosifian GA (1992) Mathematical Problems in Elasticity and Homogenization, North-Holland, Amsterdam.

[17] Allaire G, Amar M (1999) Boundary layer tails in periodic homogenization, ESAIM Control Optim. Calc. Var. **4**:209-243.

[18] Mishuris G, Öchsner A (2005) Edge effects connected with thin interfaces in composite materials, Comp. Struct. 68:409-417.

[19] Panasenko GP(1981) Higher order asymptotic solutions of problems on the contact of periodic structures, Math. USSR-Sbornik **38**(8):465-494.

[20] Panasenko GP (2005) Multi-Scale Modeling for Structures and Composites, Springer, Berlin.

[21] Gaudiello A, Kolpakov AG (2011) Influence of non degenerated joint on the global and local behavior of joined rods, Int. J. Eng. Sci. **49**:295-309.

[22] Kolpakov AG, Andrianov IV (2014) Asymptotic decomposition in the problem of joined elastic beams, ZAMM **94**(23):818-836.

[23] Kolpakov AG, Andrianov IV, Rakin SI, Rogerson GA (2018) An asymptotic strategy to couple homogenized elastic structures, Int. J. Eng. Sci. **131**:26-39.

[24] Love AEH (2011) A Treatise on the Mathematical Theory of Elasticity, Dover Publications, Mineola, New York.

[25] Kalamkarov AL, Kolpakov AG (1997) Analysis, Design and Optimization of Composite Structures, Wiley, Chichester.

[26] Buryachenko VA (2007) Micromechanics of Heterogeneous Materials, Springer, New York.

[27] Kolpakov AA, Kolpakov AG (2009) Capacity and Transport in Contrast Composite Structures: Asymptotic Analysis and Applications, CRC Press, Boca Raton, FL.

[28] Xiao SP, Belytschko T (2004) A bridging domain method for coupling continua with molecular dynamics, Comp. Meth. Appl. Mech. Eng. **193**:1645-1669.

[29] Parks ML, Bache PB, Lehoucq RB(2008) Connecting atomistic-to-continuum coupling and domain decomposition, Multisc. Model Simul. **7**:362-380.

[30] Pfaller S, Possart G, Steinmann P, Rahimi M, Müller-Plathe F, Böhm MC (2012) A comparison of staggered solution schemes for coupled particle-continuum systems modeled with the Arlequin method Comp. Mech. **49**:565-579.

[31] Annin BD, Kolpakov AG, Rakin SI (2022) Interface strength criterion for elastic bodies, Continuum Mechanics and Thermodynamics. Online first https://doi.org/10.1007/s00161-022-01146-z

Chapter 27
Lightly Loaded Hydrodynamic Thrust Bearing Lubricated by a Non-Newtonian Fluid

Ilya I. Kudish, Sergei S. Volkov, and Andrey S. Vasiliev

Abstract A lubrication problem for a lightly loaded thrust bearing in a plane formulation is considered. The bearing is lubricated by a non-Newtonian fluid described by the Giesekus rheological model. The fluids described by the Giesekus model correctly determine the behavior of real lubricants at low and high shear stresses. The Giesekus rheology takes into account not only the rheology of the base stock/solvent but also the rheology of a polymeric additive dissolved in the base stock. The case of the viscosities of the base stock and polymeric additive as well as the relaxation time of the polymeric additives are functions of contact pressure. The problem is solved using the perturbation method with respect to the small mobility parameter α describing the degree to which the additive polymeric molecules are lined up with the lubricant flow. It was assumed that the value of the parameter α is significantly higher than the ratio of the characteristic gap between the surfaces and contact length. All problem parameters such as contact pressure, stress and lubricant viscosity components etc. are represented by asymptotic series in $\alpha \ll 1$ with retained only two first terms. The solutions are obtained in analytical form. Some numerical results are presented.

Key words: Fluid with non-Newtonian nonlinear Giesekus model · Hydrodynamic lubrication problem for a thrust bearing · Perturbation analysis · Lubricant viscosity and relaxation time dependent on pressure

---

Ilya I. Kudish
ASME Fellow, ILRIMA Consulting LLC, Millersburg, MI 49759, USA,
e-mail: ilyakudish@gmail.com

Sergei S. Volkov · Andrey S. Vasiliev
Laboratory of Functionally Graded and Composite Materials, Research and Education Center "Materials", Don State Technical University, 1 Gagrina sq., 344001, Rostov-on-Don, Russian Federation,
e-mail: fenix_rsu@mail.ru, andre.vasiliev@gmail.com

H. Altenbach et al. (eds.), *Sixty Shades of Generalized Continua*,
Advanced Structured Materials 170, https://doi.org/10.1007/978-3-031-26186-2_27

27.1 Introduction

Proper lubrication of machine components such as bearings and gears leads to reduction of friction and energy losses and increased component durability. For example, the usage of proper lubricants with low viscosity in contemporary engines of internal combustion provides reduction of energy losses and, therefore, fuel consumption. Over the years most of the theoretical studies were concerned with lubrication of machine elements by fluids with Newtonian rheology [1] - [10]. However, for already several decades almost all lubricants are formulated as a base stock with dissolved in it polymeric additives. Due to the presence of polymeric additives these lubricants behave as non-Newtonian fluids for which the tensor stress and strain component are no longer proportional as it is for fluids with Newtonian rheology. For example, one of the simplest rheological relationships among generalized Newtonian fluids [11] is the Ree-Eyring one [13] - [15]. An extensive review and analysis of different fluid rheologies is provided in [16]. The other classes of popular rheological models of non-Newtonian fluids are described by various Maxwell, Jeffreys, Oldroyd fluids A and B, etc. [17]. These models are used for taking into account such an important parameters as lubricant relaxation time as well as other structural parameters of non-Newtonian fluids. Some studies of this kind of fluids were presented in [18] - [23].

The main defect of these kind of models is their inability to adequately describe fluid rheological behavior for low and high fluid shear stresses when usually lubricant viscosity approaches to two different limiting values. The rheological fluid model that is free of this defect is the Giesekus model [17, 24, 25]. Specifically, besides introducing the fluid relaxation time this model provides for relatively high fluid viscosity at low fluid shear stresses and relatively low fluid viscosity at relatively high fluid shear stresses. This model is nonlinear and, therefore, it is much harder to analyze lubrication problems involving lubricants with such a rheology. A relatively simple case of a Giesekus fluid flow between two parallel flat surfaces is considered in [26, 27]. In the paper on lubrication of a plain thrust bearing lubricated by a fluid with the Giesekus rheology [28] the analysis is performed with the help of a perturbation method. However, the obtained solution does not consider the fluid convective terms and the presence of the solvent (base stock). Some other applications of perturbation techniques to steady and non-steady problems can be founded in [29, 30].

The present paper is the continuation of the series of authors' studies of hydrodynamic and elastohydrodynamic lubrication problems for Newtonian [31] - [34] and non-Newtonian lubricants [11] including a study of a lubrication problem for a line contact of a moving cylinder and a plane separated by a Giesekus fluid [35]. In this paper, the Giesekus rheology is used for consistent modeling of lubrication process of one rigid surface moving over another rigid surface being at rest. The surfaces are separated by an incompressible fluid described by a nonlinear non-Newtonian Giesekus rheology. A general case when the solvent and polymeric additive viscosities and the lubricant relaxation time are functions of pressure. The problem is analyzed using a regular perturbation method. The problem solution is obtained in a simple analytical form. The approximate solution of the problem considers the fluid convective terms, polymer relaxation time and viscosity as well as the presence of the lubricant solvent (base stock). The analytical expressions for contact pressure distribution, contact energy loss, friction force etc. were obtained.

The paper is organized as following. In the first section, the formulation of the hydrodynamic lubrication problem for a line contact is presented. In the second section, the proper simplification of the rheology equations and the equations of the motion pertinent to the case of steady lubricant flow in a narrow long channel is described. The third sections are dedicated to obtaining the components of lubricant velocity, derivation of Reynolds equations of different orders, and their analytical solution, respectively. Some specific examples of the obtained analytical solution and their analysis are presented in the last section. In particular, some examples of pressure distributions.

27.2 Formulation of the Lubrication Problem

Let us consider a steady plane problem for a lubricated contact modeling a two-dimensional hydrodynamic thrust bearing (see Fig. 27.1) lubricated by an incompressible non-Newtonian Giesekus fluid [17] with constant viscosity μ and relaxation time λ_1. The coordinate system is introduced in such a way that the $x-$axis is directed along the surface of the rigid runner moving with the linear velocity $\overline{u}_1$ while the $z-$axis is perpendicular to it and directed upward. The $y-$axis is directed in the solids. The fixed rigid pad (the linear velocity of which is $\overline{u}_2 = 0$) and the runner are completely separated by the lubrication film. The components of the lubricant velocity are represented by functions $u(x, y, z)$, $v(x, y, z) = 0$, and $w(x, y, z)$. The problem parameters are independent of the coordinate y. The equations of the motion of such a fluid are described by the solvent and additive rheology equations as follows [11, 12]

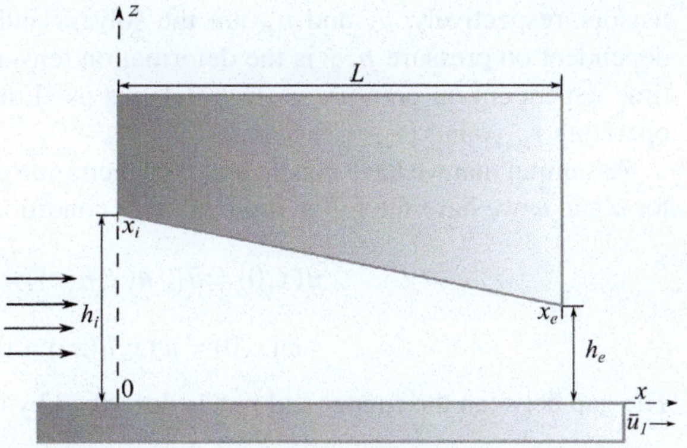

Fig. 27.1 General view of a lubricated contact

$$
\begin{aligned}
u\frac{\partial u}{\partial x}+w\frac{\partial u}{\partial z} &= \frac{1}{\rho}\left(\frac{\partial p_{xx}}{\partial x}+\frac{\partial p_{zx}}{\partial z}\right),\\
\frac{\partial p_{xy}}{\partial x}+\frac{\partial p_{zy}}{\partial z} &= 0,\\
u\frac{\partial w}{\partial x}+w\frac{\partial w}{\partial z} &= \frac{1}{\rho}\left(\frac{\partial p_{zx}}{\partial x}+\frac{\partial p_{zz}}{\partial z}\right).
\end{aligned}
\tag{27.1}
$$

In addition to that for an incompressible fluid with the fluid density $\rho(x,z) = constant$ we have the continuity equation

$$\frac{\partial u}{\partial x}+\frac{\partial w}{\partial z}=0. \tag{27.2}$$

In this case the stress tensor components are as follows

$$
\begin{aligned}
p_{xx} &= -p+\tau_{xx},\ p_{xy}=\tau_{xy}=0,\ \ p_{zx}=\tau_{zx},\\
p_{zz} &= -p+\tau_{zz},\ \ p_{zy}=\tau_{zy}=0,
\end{aligned}
\tag{27.3}
$$

where p is the pressure and τ_{xx}, τ_{xy}, τ_{zx}, τ_{zz}, and τ_{zy} are additional stress components acting in the corresponding directions. These tensor components satisfy the Giesekus fluid model which is a nonlinear model and takes into account the degree to which the additive polymeric molecules are aligned with the lubricant flow which is characterized by the mobility parameter α, $0 \le \alpha \le 1$. The rheological equations are as follows [17]

$$
\begin{aligned}
&\tau=\tau_s+\tau_p,\ \mu=\mu_s+\mu_p, \tau_s=\mu_s\dot{\gamma},\\
&\tau_p+\lambda_1\tau_{p(1)}-\alpha\frac{\lambda_1}{\mu_p}\{\tau_p\cdot\tau_p\}=\mu_p\dot{\gamma},\\
&\mu_s(x)=\mu_{s0}e^{Q_1 p(x,z)},\quad \mu_p(x)=\mu_{p0}e^{Q_2 p(x,z)},\quad \lambda_1(x)=\lambda_{10}e^{Q_3 p(x,z)}
\end{aligned}
\tag{27.4}
$$

where τ is the full stress tensor while τ_s and τ_p are the solvent and polymer stress tensors, respectively, μ_s and μ_p are the solvent and polymer dynamic viscosities dependent on pressure p, $\dot{\gamma}$ is the deformation tensor [17], and λ_1 is the relaxation time dependent on pressure p. In (27.4) we used the the definitions of the tensor operators $\tau_{p(1)}$ and $\{\tau_p\cdot\tau_p\}$ from [17].

Assuming that we have no slip and no penetration conditions on the solid surfaces for u and w we have the following boundary conditions on the lubricated surfaces

$$u(x,0)=\overline{u_1},\ u(x,h(x))=0, \tag{27.5}$$

$$w(x,0)=w(x,h(x))=0. \tag{27.6}$$

The gap between the runner and pad is described by the function

$$h(x)=h_i+mx,\ m=\frac{h_e-h_i}{L}<0,\ m=m_0\epsilon,\ m_0=O(1),\ \epsilon\ll 1, \tag{27.7}$$

where L is the actual length of the bearing, h_i and h_e are the gaps between the runner and pad at the inlet and exit from the contact, respectively, and $\epsilon = h_i/L \ll 1$ is a small parameter.

As outside of the bearing the lubricant pressure is atmospheric which is much lower the pressure in the lubricated contact for pressure p we have the following boundary conditions

$$p(0,z) = p(L,z) = 0. \tag{27.8}$$

It is assumed here that the inlet point in the lubricated contact is located at $x = 0$ while the exit point from the lubricated contact is located at $x = L$.

Our goal is to determine such components of the solution as contact pressure $p(x,z)$, the components of the tensor $\tau(x,z)$ in the fluid and its velocity components $u(x,z)$ and $w(x,z)$. We will assume that

$$\lambda_{10} = \lambda_{00}\epsilon,\ \lambda(x) = \lambda_{00}e^{Q_3 p(x)},\ \lambda = O(1),\ \epsilon \ll 1. \tag{27.9}$$

Here λ are nonnegative constants. Also, we will assume that

$$Re_0 = \frac{\rho \overline{u_1} h_i}{\mu_*} = O(1),\ \epsilon \ll 1, \tag{27.10}$$

were Re_0 is the effective local Reynolds number in the lubricant flow and μ_* is the ambient lubricant viscosity.

27.3 Asymptotic Analysis of the Rheological and Motion Equations

Let us introduce the following dimensionless variables

$$\begin{gathered}
\lambda_1' = \frac{\overline{u_1}}{L}\lambda_1,\ x' = \frac{x}{L},\ \{z', h_i', h_e'\} = \frac{1}{h_i}\{z, h_i, h_e\}, \\
p' = p\frac{h_i^2}{\mu_* \overline{u}_1 L},\ Q_i' = Q_i\frac{\mu_* \overline{u}_1 L}{h_i^2},\ i = 1,2,3, \\
u' = \frac{u}{\overline{u}_1},\ w' = \frac{w}{U_z},\ \{\mu', \mu_s', \mu_p'\} = \frac{1}{\mu_*}\{\mu, \mu_s, \mu_p\}, \\
\{\tau_{xx}', \tau_{sxx}', \tau_{pxx}', \tau_{zx}', \tau_{szx}', \tau_{pzx}', \tau_{zz}', \tau_{szz}', \tau_{pzz}'\} \\
= \frac{h_i}{\mu_* \overline{u}_1}\{\tau_{xx}, \tau_{sxx}, \tau_{pxx}, \tau_{zx}, \tau_{szx}, \tau_{pzx}, \tau_{zz}, \tau_{szz}, \tau_{pzz}\},
\end{gathered} \tag{27.11}$$

where U_z is the characteristic velocity of the lubricating fluid in the direction of the $z-$axis.

By introducing the above scaling in continuity equation (27.2) we obtain

$$\frac{\partial u}{\partial x}+\frac{U_z}{U_x}\frac{L_x}{L_z}\frac{\partial w}{\partial z}=0. \tag{27.12}$$

To retain both terms in Eq. (27.12) we assume that

$$\frac{U_z}{U_x}\frac{L_x}{L_z}=1. \tag{27.13}$$

Then Eq. (27.12) becomes

$$\frac{\partial u}{\partial x}+\frac{\partial w}{\partial z}=0. \tag{27.14}$$

By introducing scaling (27.11) in Eqs. (27.1), (27.3), (27.4) we obtain

$$\begin{gathered}\epsilon\frac{\rho U_x L_z}{\mu_*}\{u\frac{\partial u}{\partial x}+w\frac{\partial u}{\partial z}\}=-\frac{\partial p}{\partial x}+\epsilon\frac{\partial \tau_{xx}}{\partial x}+\frac{\partial \tau_{zx}}{\partial z},\\ \epsilon^3\frac{\rho U_x L_z}{\mu_*}\{u\frac{\partial w}{\partial x}+w\frac{\partial w}{\partial z}\}=\epsilon^2\frac{\partial \tau_{zx}}{\partial x}-\frac{\partial p}{\partial z}+\epsilon\frac{\partial \tau_{zz}}{\partial z},\end{gathered} \tag{27.15}$$

$$\begin{gathered}\tau_{xx}=\tau_{sxx}+\tau_{pxx},\ \tau_{zx}=\tau_{szx}+\tau_{pzx},\ \tau_{zz}=\tau_{szz}+\tau_{pzz},\\ \tau_{sxx}=\epsilon 2\mu_s\frac{\partial u}{\partial x},\ \tau_{szx}=\mu_s(\epsilon^2\frac{\partial w}{\partial x}+\frac{\partial u}{\partial z}),\ \tau_{szz}=\epsilon 2\mu_s\frac{\partial w}{\partial z}=-\epsilon 2\mu_s\frac{\partial u}{\partial x},\end{gathered} \tag{27.16}$$

$$\begin{aligned}&\tau_{pxx}+\lambda\epsilon[u\frac{\partial \tau_{pxx}}{\partial x}+w\frac{\partial \tau_{pxx}}{\partial z}-\frac{2}{\epsilon}\frac{\partial u}{\partial z}\tau_{pzx}-2\frac{\partial u}{\partial x}\tau_{pxx}]\\ &\quad-\alpha\frac{\lambda}{\mu_p}(\tau_{pxx}^2+\tau_{pzx}^2)=2\epsilon\mu_p\frac{\partial u}{\partial x},\\ &\tau_{pzx}+\lambda\epsilon[u\frac{\partial \tau_{pzx}}{\partial x}+w\frac{\partial \tau_{pzx}}{\partial z}-\frac{1}{\epsilon}\frac{\partial u}{\partial z}\tau_{pzz}-\frac{\partial u}{\partial x}\tau_{pzx}-\frac{\partial w}{\partial z}\tau_{pzx}-\epsilon\frac{\partial w}{\partial x}\tau_{pxx}]\\ &\quad-\alpha\frac{\lambda}{\mu_p}(\tau_{pxx}+\tau_{pzz})\tau_{pzx}=\mu_p(\frac{\partial u}{\partial z}+\epsilon^2\frac{\partial w}{\partial x}),\\ &\tau_{pzz}+\lambda\epsilon[u\frac{\partial \tau_{pzz}}{\partial x}+w\frac{\partial \tau_{pzz}}{\partial z}-2\frac{\partial w}{\partial z}\tau_{pzz}-2\epsilon\frac{\partial w}{\partial x}\tau_{pzx}]\\ &\quad-\alpha\frac{\lambda}{\mu_p}(\tau_{pzx}^2+\tau_{pzz}^2)=2\epsilon\mu_p\frac{\partial w}{\partial z}.\end{aligned} \tag{27.17}$$

For simplicity in the further analysis the primes at the dimensionless variables are dropped. Then the dimensionless $h_i=1$ and the problem solution is searched within the interval $0\le x\le 1$.

Due to nonlinearity and complexity of the problem it is impossible to develop any analytical solutions except the perturbation ones which will be used in this analysis. We will search the problem solution in the form of the following series in α

$$\begin{gathered}\{\tau_{sxx}(x,z),\tau_{sxz}(x,z),\tau_{szz}(x,z)\}=\{\tau_{sxx0}(x,z),\tau_{sxz0}(x,z),\\ \tau_{szz0}(x,z)\}+\alpha\{\tau_{sxx1}(x,z),\tau_{sxz1}(x,z),\tau_{szz1}(x,z)\}+\ldots,\end{gathered} \tag{27.18}$$

$$\begin{aligned} p(x,z) &= p_0(x,z) + \alpha p_1(x,z) + \ldots, \\ u(x,z) &= u_0(x,z) + \alpha u_1(x,z) + \ldots, \\ w(x,z) &= w_0(x,z) + \alpha w_1(x,z) + \ldots, \end{aligned} \tag{27.19}$$

where $p_0(x,z)$, $u_0(x,z)$, $w_0(x,z)$, $\tau_{sxx0}(x,z)$, $\tau_{sxz0}(x,z)$, $\tau_{szz0}(x,z)$, $\tau_{pxx0}(x,z)$, $\tau_{pxz0}(x,z), \tau_{pzz0}(x,z), p_1(x,z), u_1(x,z), w_1(x,z), \tau_{sxx1}(x,z), \tau_{sxz1}(x,z), \tau_{szz1}(x,z)$, $\tau_{pxx1}(x,z)$, $\tau_{pxz1}(x,z)$, $\tau_{pzz1}(x,z)$ are the unknown main and first-order approximations of the corresponding functions while the gap $h(x)$ between the runner and pad and functions $h_0(x_0)$ and $h_1(x_0)$ are described by the equations (obtained from (27.7), using (27.11))

$$\begin{aligned} h(x) &= h_0(x) + \epsilon h_1(x), \\ h_0(x) &= 1 + m_0 x, \ h_1(x) = 0. \end{aligned} \tag{27.20}$$

Now, let us simplify equations (27.14)-(27.17) while keeping the terms proportional to α and neglecting all the terms proportional to ϵ^m, $m \geq 1$, we have silently assumed that

$$\alpha \gg \epsilon, \ \epsilon \ll 1. \tag{27.21}$$

This is due to the fact that we dropped from the equations (and, therefore, from our approximate solution) all terms proportional to ϵ, $\epsilon^2, \ldots$ If $\alpha \sim \epsilon$ or $\alpha \ll \epsilon$ then before taking into account the terms proportional to α we would have to retain a certain number of terms which are much greater and of the same order as α with respect to $\epsilon \ll 1$. The case of $\alpha \sim \epsilon \ll 1$ is more complex and will be considered in the future in a separate publication.

Therefore, we obtain the equations

$$\frac{\partial p_i}{\partial x} - \frac{\partial \tau_{zxi}}{\partial z} = 0, \ \frac{\partial p_i}{\partial z} = 0, \ i = 0, 1, \tag{27.22}$$

$$\begin{aligned} &\tau_{xx} = \tau_{sxx} + \tau_{pxx}, \ \tau_{zx} = \tau_{szx} + \tau_{pzx}, \ \tau_{zz} = \tau_{szz} + \tau_{pzz}, \\ &\tau_{sxxi} = 0, \ \tau_{szzi} = 0, \ i = 0, 1 \\ &\tau_{szx0} = \mu_{s0} e^{Q_1 p_0} \frac{\partial u_0}{\partial z}, \ \tau_{szx1} = \mu_{s0} e^{Q_1 p_0} [\frac{\partial u_1}{\partial z} + Q_1 p_1 \frac{\partial u_0}{\partial z_0}], \end{aligned} \tag{27.23}$$

$$\begin{aligned} &\tau_{pxx0} = 2\lambda_{00}\mu_{p0} e^{(Q_2+Q_3)p_0} (\frac{\partial u_0}{\partial z})^2, \ \tau_{pzx0} = \mu_{p0} e^{Q_2 p_0} \frac{\partial u_0}{\partial z_0}, \ \tau_{pzz0} = 0, \\ &\tau_{pxx1} = \lambda_{00}\mu_{p0} e^{(Q_2+Q_3)p_0} \{10(\lambda_{00} e^{Q_3 p_0})^2 (\frac{\partial u_0}{\partial z})^4 + (\frac{\partial u_0}{\partial z})^2 \\ &\quad +4 \frac{\partial u_0}{\partial z} \frac{\partial u_1}{\partial z} + 2(Q_2 + Q_3) p_1 (\frac{\partial u_1}{\partial z})^2\}, \\ &\tau_{pzx1} = \mu_{p0} e^{Q_2 p_0} \{3(\lambda_{00} e^{Q_3 p_0})^2 (\frac{\partial u_0}{\partial z})^3 + \frac{\partial u_1}{\partial z} + Q_2 p_1 \frac{\partial u_1}{\partial z}\}, \\ &\tau_{pzz1} = \lambda_{00}\mu_{p0} e^{(Q_2+Q_3)p_0} (\frac{\partial u_0}{\partial z})^2. \end{aligned} \tag{27.24}$$

$$\frac{\partial u_i}{\partial x}+\frac{\partial w_i}{\partial z}=0,\ i=0,1. \tag{27.25}$$

As it will be shown below functions $p_0(x,z)$ and $p_1(x,z)$ are independent of z (i.e. $p_0(x,z)=p_0(x)$ and $p_1(x,z)=p_1(x)$) .

It is important to realize that the boundary conditions on the terms of the expansions of u, w, and p in $\alpha \ll 1$ are

$$u_0(x,0)=1,\ u_0(x,h_0)=0,\ u_1(x,0)=0,\ u_1(x,h_0)=0, \tag{27.26}$$

$$w_0(x,0)=0,\ w_0(x,h_0)=0,\ w_1(x,0)=0,\ w_1(x,h_0)=0. \tag{27.27}$$

Also, from (27.8) we have

$$p_0(0)=p_1(0)=0,\ p_0(1)=p_1(1)=0. \tag{27.28}$$

Solving Eq. (27.22) with boundary conditions (27.26)-(27.28) and taking into account (27.25), we obtain

$$u_0(x,z)=1-\frac{z}{h_0}+(z^2-zh_0)\frac{1}{2\mu_0}\frac{dp_0}{dx}, \tag{27.29}$$

$$\begin{aligned} w_0(x,z)=&\frac{z^3}{6\mu_0}\Big(\frac{1}{\mu_0}\frac{\partial p_0}{\partial x}\frac{\partial \mu_0}{\partial x}-\frac{\partial^2 p_0}{\partial x^2}\Big)-\frac{z^2}{2h_0^2}\frac{\partial h_0}{\partial x}\\ &+\frac{z^2}{4\mu_0}\Big(h_0\frac{\partial^2 p_0}{\partial x^2}+\frac{\partial p_0}{\partial x}\frac{\partial h_0}{\partial x}-\frac{h_0}{\mu_0}\frac{\partial p_0}{\partial x}\frac{\partial \mu_0}{\partial x}\Big), \end{aligned} \tag{27.30}$$

$$\begin{aligned} u_1(x,z)=&-\frac{(z-h_0)z}{4\mu_0^4h_0^2}\Big(3\lambda_0^2\mu_p^*h_0^2\Big(z^2-zh_0+\frac{h_0^2}{2}\Big)\Big(\frac{\partial p_0}{\partial x}\Big)^3-2\mu_0^3h_0^2\frac{\partial p_1}{\partial x}\\ &-12\lambda_0^2\mu_p^*\mu_0h_0\Big(z-\frac{h_0}{2}\Big)\Big(\frac{\partial p_0}{\partial x}\Big)^2+2\mu_0^2(9\lambda_0^2\mu_p^*+p_1\mu_1h_0^2)\frac{\partial p_0}{\partial x}\Big), \end{aligned} \tag{27.31}$$

where

$$\begin{aligned} &\mu_1(x)=Q_1\mu_{s0}e^{Q_1p_0}+Q_2\mu_{p0}e^{Q_2p_0},\mu_0(x)=\mu_{s0}e^{Q_1p_0}+\mu_{p0}e^{Q_2p_0},\\ &\mu_s^*(x)=\mu_{s0}e^{Q_1p_0},\mu_p^*(x)=\mu_{p0}e^{Q_2p_0},\lambda_0(x)=\lambda_{00}e^{Q_3p_0} \end{aligned} \tag{27.32}$$

Using expansions (27.18)-(27.20) in Eqs. (27.1)-(27.6), and (27.8) taking into account (27.9) and (27.10), and equating terms of the same order of α these equations can be simplified and reduced to solution of the problems for Reynolds equations of order zero

$$\frac{d}{dx}\int\limits_0^{h_0}u_0(x,z)dz=0, \tag{27.33}$$

of order one

$$\frac{d}{dx}\int\limits_0^{h_0}u_1(x,z)dz=0, \tag{27.34}$$

Keeping in mind that $h_0(x) = 1 + m_0 x$ the above mentioned problem for the Reynolds equation of order zero has the form

$$\frac{d}{dx}\{\frac{h_0^3}{6\mu_0}\frac{dp_0}{dx} - h_0\} = 0,\ h_0(x) = 1 + m_0 x,\ p_0(0) = p_0(1) = 0, \tag{27.35}$$

and

$$\frac{d}{dx}\{\frac{h_0}{\mu_0}(h_0^2\frac{dp_1}{dx} - \frac{h_0\mu_1 p_1 + 9\lambda_0^2\mu_p^*}{\mu_0}\frac{dp_0}{dx} - \frac{9\lambda_0^2\mu_p^* h_0^4}{20\mu_0^3}(\frac{dp_0}{dx})^3)\} = 0, \tag{27.36}$$

$$p_1(0) = p_1(1) = 0,$$

In case of $Q_1 = Q_2 = Q_3$, Eq. (27.35) has an analytical solution in the form

$$p_0(x) = \frac{-1}{Q_1}\ln[\frac{\{m_0^3 + 2m_0^2 - 6Q_1\mu_{00}m_0\}x^2 + \{2m_0^2 + (4 + 6Q_1\mu_{00})m_0\}x + m_0 + 2}{(m_0 + 2)(m_0 x + 1)^2}],$$
$$\mu_{00} = \mu_{s0} + \mu_{p0}, \tag{27.37}$$

Substituting (27.35) in (27.36) we obtain the following equations

$$\frac{d}{dx}\{(1 + m_0 x)[-2 - m_0^3 x^2 - 2(x^2 + x)m_0^2 + \{-1 + 6Q_1\mu_{00}x^2 - 2(2 + 3Q_1\mu_{00})x\}m_0]\frac{dp_1}{dx}$$
$$+6m_0 Q_1\mu_{00}(m_0 x + 2x - 1)p_1 + g(x)\} = 0,$$
$$p_1(0) = p_1(1) = 0, \tag{27.38}$$

where

$$g(x) = \frac{18m_0(m_0 x + 2x - 1)\lambda_0^2\mu_{p0}[10 + 7m_0^4 x^2 + (28x^2 - 4x)m_0^3}{5[-2 - m_0^3 x^2 - 2(x^2 + x)m_0^2 + (-1 + 6Q_1\mu_{00}x^2 - 2(2 + 3Q_1\mu_{00})x)m_0]^2}$$
$$+ \frac{(28x^2 + 2x + 7)m_0^2 + 10(2x + 1)m_0]}{5[-2 - m_0^3 x^2 - 2(x^2 + x)m_0^2 + (-1 + 6Q_1\mu_{00}x^2 - 2(2 + 3Q_1\mu_{00})x)m_0]^2}, \tag{27.39}$$

The solution of Eq. (27.38) has the form

$$p_1(x) = \frac{\mu_{00}[(x(m_0+1)^2(xm_0+2)\int\limits_0^1 \frac{\frac{dg(z)}{dz}}{(m_0 z+1)^2}dz}{m_0[m_0(-2m_0^2-4m_0+12Q_1\mu_{00})x^2-4(m_0^2+(2+3\mu_{00}Q_1)m_0)x-2(2+m_0)]}$$
$$-\frac{(m_0+2)(xm_0+1)^2\int\limits_0^x \frac{\frac{dg(z)}{dz}}{(m_0 z+1)^2}dz]}{m_0[m_0(-2m_0^2-4m_0+12Q_1\mu_{00})x^2-4(m_0^2+(2+3\mu_{00}Q_1)m_0)x-2(2+m_0)]}$$
$$+\frac{\mu_{00}[(m_0x^2-m_0+2x-2)g(0)}{m_0[m_0(-2m_0^2-4m_0+12Q_1\mu_{00})x^2-4(m_0^2+(2+3\mu_{00}Q_1)m_0)x-2(2+m_0)]}$$
$$-\frac{(m_0x^2+2x)g(1)+(2+m_0)g(x)]}{m_0[m_0(-2m_0^2-4m_0+12Q_1\mu_{00})x^2-4(m_0^2+(2+3\mu_{00}Q_1)m_0)x-2(2+m_0)]}. \quad (27.40)$$

Now, let us consider the energy losses in the contact E and the friction force F_{fr} determined by the formulas,in dimensional terms

$$E = \int\limits_0^L dx \int\limits_0^h \tau_{xz}(x,z)\frac{\partial u(x,z)}{\partial z}dz, \quad F_{\mathrm{fr}} = \int\limits_0^L \tau_{xz}(x,0)dx. \quad (27.41)$$

If we scale E and F_{fr} as follows $E = \mu_* \overline{u}_1^2 L E'$ and $F_{\mathrm{fr}} = \mu_* \overline{u}_1 L F'_{\mathrm{fr}}$ (further primes are dropped) then the dimensionless energy loss E and friction force F_{fr} are

$$E = \int\limits_0^1 dx \int\limits_0^{h_0} \tau_{xz0}(x,z)\frac{\partial u_0(x,z)}{\partial z}dz + \alpha[\int\limits_0^1 dx \int\limits_0^{h_0} \tau_{xz0}(x,z)\frac{\partial u_1(x,z)}{\partial z}dz$$
$$+\int\limits_0^1 dx \int\limits_0^{h_0} \tau_{xz1}(x,z)\frac{\partial u_0(x,z)}{\partial z}dz] + \ldots, \quad (27.42)$$
$$F_{\mathrm{fr}} = \int\limits_0^1 \tau_{xz0}(x,0)dx + \alpha \int\limits_0^1 \tau_{xz1}(x,0)dx + \ldots.$$

27.4 Examples of Some Specific Lubrication Problem Solutions and Discussion

Now, let us consider some results which can be extracted from the obtained approximate solution. We will assume that always $\mu_{00} = 1$. We will take as the basic set the following values: $\epsilon = 0.01$, $\mu_{p0} = 0.5$, $\lambda_{00} = 0.5$, $\alpha = 0.1$, and $m_0 = -0.1$.

Plots of functions $p_0(x)$ and $p_1(x)$ as functions of x for different values of parameter Q_1 are presented in Figs. 27.2a) and 27.2b) obtained for the above stated parameters. It is clear that functions $p_0(x)$ and $p_1(x)$ increase with the increase in Q_1. Moreover, $p_1(x)$ grows faster than $p_0(x)$ for increasing Q_1. Figure 27.3

shows the behavior of $p(x)$ obtained for $\alpha = 0.1$ according to formula (27.37) and using the data from Figs. 27.2a) and 27.2b). The case of $Q_1 = 0$ corresponds to the case of constant solvent and additive viscosities and lubricant relaxation time. As it is clear from Fig. 27.3 the pressure maximum is slightly shifted toward the exit from the contact which is typical for such kind of bearings in these conditions [41]. This shift of the maximum pressure increases with the increase of $|m_0|$. Figure 27.4 shows plots of $p_1(x)$ obtained for the presented above basic set of data and different values of parameter λ_{00}. Obviously, the values of $p_1(x)$ increase with the increase of λ_{00}. Because $p_0(x)$ is independent of λ_{00} the whole contribution related to λ_{00} comes from $p_1(x)$. Figure 27.5 presents plots of $p_1(x)$ obtained for the basic set of data and different values of μ_{p0}. Similar to Fig. 27.4 the values of $p_1(x)$ increase as the viscosity of the polymeric additive μ_{p0} increases. Therefore, we can conclude that the variations in pressure distribution $p(x)$ related to the variations in the ambient polymeric additive viscosity μ_{p0}, ambient lubricant relaxation time λ_{00}, piezo-coefficient of viscosity Q_1 are determined by the term $\alpha p_1(x)$. The plots in Fig. 27.6a) and 27.6b) demonstrate the energy loss E as a function of Q_1 for different

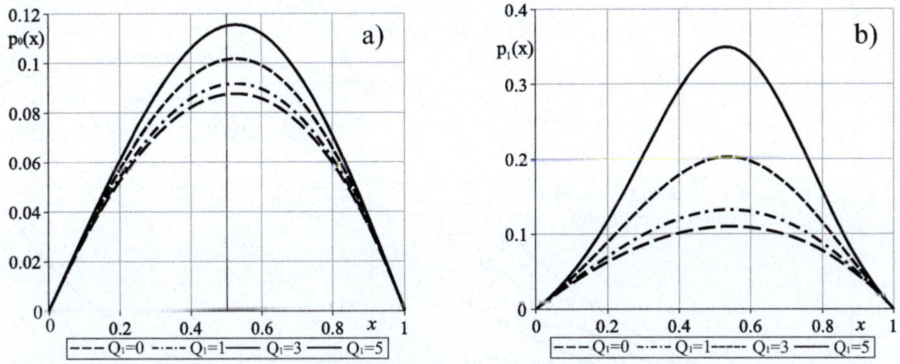

Fig. 27.2: Plots of $p_0(x)$ and $p_1(x)$ obtained for the basic set of data and different values of parameter Q_1 which determines the dependence of the solution and polymeric additive viscosities and lubricant relaxation time on pressure $p(x)$..

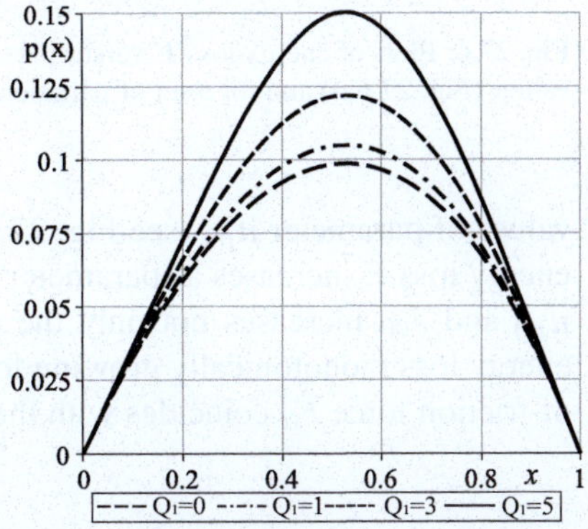

Fig. 27.3 Plots of pressure $p(x)$ obtained for the basic set of data and different values of parameter Q_1.

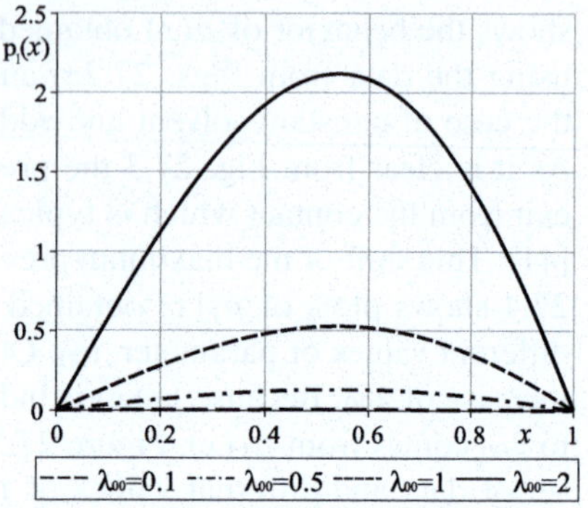

Fig. 27.4 Plots of $p_1(x)$ obtained for the basic set of data and different values of parameter λ_{00}.

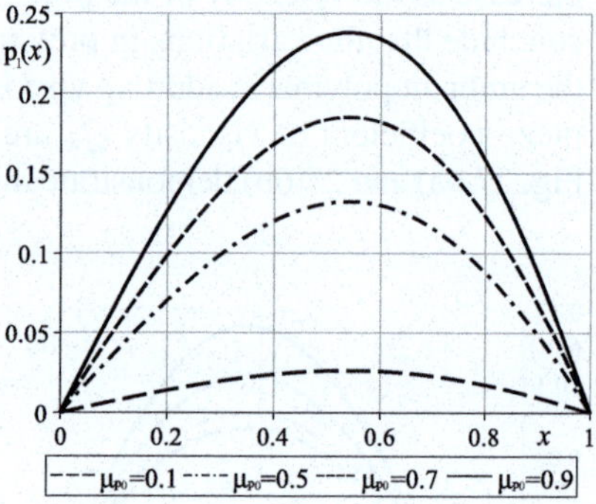

Fig. 27.5 Plots of $p_1(x)$ obtained for the basic set of data and different values of parameter μ_{p0}.

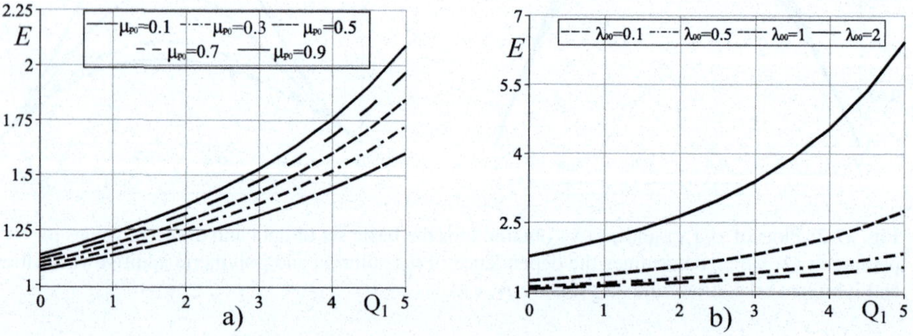

Fig. 27.6: Plots of energy loss E versus pressure coefficient Q_1 obtained for a set of parameter μ_{p0} values (Fig. 27.6a)) and for a set of parameter λ_{00} values (Fig. 27.6b))..

values of parameter μ_{p0} (see Fig. 27.6a)) and parameter λ_{00} (see Fig. 27.6b)). The energy loss E increases as parameter Q_1 grows. The increase in the ambient values μ_{p0} and λ_{00} increases not only the energy loss E for $Q_1 = 0$ but also keeps the energy loss monotonically growing for increasing values of μ_{p0} and λ_{00}. The value of friction force F_{fr} coincides with the value of energy loss E.

27.5 Closure

A new perturbation solution is obtained for a hydrodynamically lubricated thrust bearing. The lubricant rheology is modeled according to the non-Newtonian Giesekus model with the base oil and polymeric additive viscosities as well as the lubricant relaxation time dependent on pressure. The rheology equations of the lubricant were simplified using the scale analysis assuming that the size of the lubricant layer across it is much smaller than along it. That allowed for derivation of the Reynolds equations of the zero and first orders. For the specific case of $Q_1 = Q_2 = Q_3$ an analytical solution is obtained and analyzed for different set of problem input parameters. It is determined that the energy losses E grow with the growth of of the ambient polymeric additive viscosity and lubricant relaxation time as well as the viscosity pressure coefficient Q_1.

Acknowledgements The study was supported by the Russian Science Foundation grant No. 19-19-00444, https://rscf.ru/project/19-19-00444/.

References

[1] Chapkov AD, Bair S, Cann P, Lubrecht AA (2007) Film thickness in point contacts under generalized Newtonian EHL conditions: Numerical and experimental analysis, Tribol Intern **40**:1474–1478. DOI 10.1016/j.triboint.2007.01.002
[2] Yang P, Cui J, Jin ZM, Dowson D (2005) Transient elastohydrodynamic analysis of elliptical contacts. Part 2: Thermal and Newtonian lubricant solution, Proc Inst Mech Eng, Part J: J Engng Tribol **219**:187–200. DOI 10.1243/135065005X9826
[3] Li B, Sun J, Zhu S, Fu Y, Zhao X, Wang H, Teng Q, Ren Y, Li Y, Zhu G (2019) Thermohydrodynamic lubrication analysis of misaligned journal bearing considering the axial movement of journal, Tribol Intern, **135**:397–407. DOI 10.1016/j.triboint.2019.03.031
[4] Wang Z, Hu Y, Wang W, Wang H (2009) Numerical analysis of point contact EHL on coated substrates. B: ASME/STLE 2009 Intern Joint Tribol Conference, pp 471–473, ASMEDC, Memphis, Tennessee, USA. DOI 10.1115/ijtc2009-15038
[5] Su J, Song H-X, Ke L-L (2020) Elastohydrodynamic lubrication line contact based on surface elasticity theory, J Appl Mech **87**(8):081004. DOI 10.1115/1.4047088
[6] Charles P, Elfassi M, Lubrecht AA (2010) Double-Newtonian rheology in a model piston-ring cylinder-wall contact, Tribol Intern **43**:1902–1907. DOI 10.1016/j.triboint.2009.11.010
[7] Katyal P, Kumar P. (2014 On the role of second Newtonian viscosity in EHL point contacts using double Newtonian shear-thinning model, Tribol Intern **71**:140–148. DOI 10.1016/j.triboint.2013.11.009

[8] D'Agostino V, Petrone V, Senatore A (2013) Effects of the lubricant piezo-viscous properties on EHL line and point contact problems, Tribol Lett **49**:385–396. DOI 10.1007/s11249-012-0079-5

[9] Kazama T (2002) A comparative Newtonian and thermal EHL analysis using physical lubricant properties, Tribol Series **40**:435–446. DOI 10.1016/S0167-8922(02)80048-7

[10] Liu Y, Wang QJ, Bair S, Vergne P (2007) A quantitative solution for the full shear-thinning EHL point contact problem including traction, ASME/STLE 2007 Intern Joint Tribol Conference, Parts A and B, pp 1031–1032, ASMEDC, San Diego, CA, USA. DOI 10.1115/IJTC2007-44100

[11] Kudish II (2013) Elastohydrodynamic Lubrication for Line and Point Contacts. Asymptotic and Numerical Approaches, Chapman & Hall/CRC Press.

[12] Kudish II, Covitch MJ (2010) Introduction to Modeling and Analytical Methods in Tribology, Chapman & Hall/CRC.

[13] Ree T, Eyring H (1955) Theory of non-Newtonian Flow. I. Solid plastic system, J Appl Phys. **26**:793–800. DOI 10.1063/1.1722098

[14] Cui J, Yang P, Jin ZM, Dowson D (2007) Transient elastohydrodynamic analysis of elliptical contacts. Part 3: Non-Newtonian lubricant solution under isothermal and thermal conditions, Proc Inst Mech Eng, Part J: J Engng Tribol **221**:63–73. DOI 10.1243/13506501JET165

[15] Su J, Song H, Ke L, Aizikovich SM (2021) The size-dependent elastohydrodynamic lubrication contact of a coated half-plane with non-Newtonian fluid, Appl. Math. Mech. (Engl. ed.) **42**:915–930. DOI 10.1007/s10483-021-2744-7

[16] Cherizol R, Sain M, Tjong J (2015) Review of non-Newtonian mathematical models for rheological characteristics of viscoelastic composites, Green and Sustainable Chemistry **5**:6–14. DOI 10.4236/gsc.2015.51002

[17] Bird RB, Curtis CF, Armstrong RC, Hassager O (1987) Dynamics of Polymeric Liquids, Vol. 1: Fluid Mech, 2nd ed, John Wiley & Sons, New York.

[18] Tichy JA (1996) Non-Newtonian lubrication with the convected Maxwell model, J Tribol **118**:344–348. DOI 10.1115/1.2831307

[19] Huang P, Li Z, Meng Y, Wen S (2002) Study on thin film lubrication with second-order fluid, J Tribol **124**:547–552. DOI 10.1115/1.1467636

[20] Akyildiz FT, Bellout H (2004) Viscoelastic lubrication with Phan-Thein-Tanner fluid (PTT), J Tribol **126**:288–291. DOI 10.1115/1.1651536

[21] Gwynllyw DR, Phillips TN (2008) The influence of Oldroyd-B and PTT lubricants on moving journal bearing systems, J Non-Newtonian Fluid Mech **150**:196–210. DOI 10.1016/j.jnnfm.2007.10.014

[22] Covitch MJ, Trickett KJ (2015) How polymers behave as viscosity index improvers in lubricating oils. ACES **5**:134–151. DOI 10.4236/aces.2015.52015

[23] Sawyer WG, Tichy JA (1998) Non-Newtonian lubrication with the second-order fluid, J Tribol **120**:622–628. DOI 10.1115/1.2834596

[24] Giesekus H (1982) A simple constitutive equation for polymer fluids based on the concept of deformation-dependent tensorial mobility, J Non-Newtonian Fluid Mech **11**:69–109. DOI 10.1016/0377-0257(82)85016-7

[25] Giesekus H (1982) A unified approach to a variety of constitutive models for polymer fluids based on the concept of configuration-dependent molecular mobility, Rheol Acta **21**:366–375. DOI 10.1007/BF01534296

[26] Yoo JY, Choi HC (1989) On the steady simple shear flows of the one-mode Giesekus fluid, Rheol Acta **28**:13–24. DOI 10.1007/BF01354764

[27] Raisi A, Mirzazadeh M, Dehnavi AS, Rashidi F (2008) An approximate solution for the Couette–Poiseuille flow of the Giesekus model between parallel plates, Rheol Acta **47**:75–80. DOI 10.1007/s00397-007-0212-9

[28] Abbaspur A, Norouzi M, Akbarzadeh P, Vaziri SA (2021) Analysis of nonlinear viscoelastic lubrication using Giesekus constitutive equation, Proc Inst Mech Eng, Part J: J Engng Tribol **235**:1124–1138. DOI 10.1177/1350650120944280

[29] Van-Dyke M (1964) Perturbation Methods in Fluid Mech, New York, Academic Press.

[30] Kaplunov JD, Nolde EV, Shorr BF (2005) A perturbation approach for evaluating natural frequencies of moderately thick elliptic plates, J Sound Vibr **281**:905–919. DOI 10.1016/j.jsv.2004.02.046

[31] Kudish II, Volkov SS, Aizikovich SM, Ke L (2020) One simple case of lubricated line contact for double-layered elastic solids, Probl Strength Plasticity **84**:5-14. DOI 10.32326/1814-9146-2022-84-1-15-24

[32] Kudish II, Volkov SS, Vasiliev AS, Aizikovich SM (2021) Characterization of the behavior of different contacts with double coating, Math Mech Compl Sys **9**:179–202. DOI 10.2140/memocs.2021.9.179

[33] Kudish II, Pashkovski E, Volkov SS, Vasiliev AS, Aizikovich SM (2020) Heavily loaded line EHL contacts with thin adsorbed soft layers, Math Mech Solids **25**:1011–1037. DOI 10.1177/1081286519898878

[34] Kudish II, Volkov SS, Vasiliev AS, Aizikovich SM (2018) Lubricated point heavily loaded contacts of functionally graded materials. Part 2. Lubricated contacts, Math Mech Solids **23**:1081–1103. DOI 10.1177/1081286517704690

[35] Kudish I, Pashkovski E, Patterson R (2022) Line contact lubricated by a fluid described by non-Newtonian Giesekus model. IMA J Appl Math. DOI 10.1093/imamat/hxac019

[36] Liu T-J, Wang Y-S, Zhang C (2008) Axisymmetric frictionless contact of functionally graded materials, Archive of Appl Mech **78**:267–282. DOI 10.1007/s00419-007-0160-y

[37] Zelentsov VB, Mitrin BI, Lubyagin IA (2017) Wear resistance of coating materials under the frictional heating conditions, J Frict Wear **38**:265–271. DOI 10.3103/S1068366617040158

[38] Vasiliev A, Volkov S, Aizikovich SM (2018) Approximated analytical solution of contact problem on indentation of elastic half-space with coating reinforced with inhomogeneous interlayer, Materials Phys Mech **35**:175–180. DOI 10.18720/MPM.3512018_20

[39] Ke L-L, Wang Y-S (2006) Two-dimensional contact mechanics of functionally graded materials with arbitrary spatial variations of material properties, Intern J Solids Struct **43**:5779–5798. DOI 10.1016/j.ijsolstr.2005.06.081

[40] Su J, Ke L-L, Wang Y-S, Xiang Y (2017) Axisymmetric torsional fretting contact between a spherical punch and an FGPM coating, Appl Math Model. **52**:576–589. DOI 10.1016/j.apm.2017.08.010
[41] Orlov P. (1977) Fundamentals of Machine Design (in Russ) Mashinostroenie, Moscow.

Chapter 28
The Idea of Using Adhesive Bonds in Shaping of Cold-formed Thin-walled Beam-columns

Marcin Kujawa, Antonio Cazzani, Lukasz Smakosz, Violetta Konopińska-Zmysłowska, Karol Winkelmann, Faizullah Jan, and Czesław Szymczak

Abstract The chapter concerns the idea of using adhesive bonds in the shaping of cold-formed thin-walled beam-columns. The idea of adhesive-bonded connections is not new but is unlikely to have been used in the design and shaping of thin-walled beam-columns in civil engineering up to date.

Metal bar structures, including (due to economic reasons) thin-walled structures, are the most common and effective means to build engineering superstructures/load-bearing structures. Over the years, classical techniques of connection between individual structural elements, such as bolting or welding, have mainly been used. However, the classical types of connections have their disadvantages, eliminable by using less conventional solutions, e.g. the adhesive-bonded connections. The use of structural adhesives in engineering, particularly in metal construction design, unfortunately still raises questions about their durability and load-bearing capacity. An important issue in the case of bonded structures is the effect of creep on the behavior of such joints. These issues undoubtedly require further research in the context of bonded joints.

The topic still needs fundamental research of empirical and theoretical nature that will contribute to the acquisition of new knowledge on the mechanical phenomena occurring in adhesively bonded metal structures. The authors also expect that in the long run, the idea will be widely used in modern-day civil engineering.

Key words: Thin-walled structures · Adhesive bonds

Marcin Kujawa · Lukasz Smakosz · Violetta Konopińska-Zmysłowska · Karol Winkelmann · Faizullah Jan · Czesław Szymczak
Gdańsk University of Technology, G. Narutowicza St. 11/12, 80-233 Gdańsk, Poland,
e-mail: mark@pg.edu.pl, luksmako@pg.edu.pl, viokonop@pg.edu.pl, karolwin@pg.edu.pl, faizullah.jan@pg.edu.pl, szymcze@pg.edu.pl

Antonio Cazzani
University of Cagliari, Marengo St. 2, 09123 Cagliari, Italy,
e-mail: antonio.cazzani@unica.it

H. Altenbach et al. (eds.), *Sixty Shades of Generalized Continua*,
Advanced Structured Materials 170, https://doi.org/10.1007/978-3-031-26186-2_28

28.1 Introduction

Thin-walled cold-formed beam-columns [1] are among the most commonly used in civil engineering. However, such structural members, in addition to their many advantages, have their disadvantages. One of them may be the inability to shape cross-sections in any way. The aforementioned problem stems from the lack of physical capacity to manufacture certain cross-section geometries. For example, it is impossible to bend a bisymmetric cross-section from a single sheet of steel. Hence, cold-formed thin-walled members usually have cross-sections that are monosymmetric or have no axis of symmetry at all. Such a situation, unfortunately, usually forces additional stress on such structural members related to their torsion. In the case of thin-walled open sections, this poses a serious problem. This happens in the case of constrained torsion where the standard normal stresses present in the member can increase even up to 30%, compared to the baseline stress, without taking torsion into account. Among other reasons, it is advisable in structural design to use structural beam-column members with bisymmetric cross-sections.

In bar metal structures, the norm at present is to use bolted or welded connections. However, this type of structural connection has its drawbacks. The main disadvantages include uneven stress distribution or lack of weldability. The drawbacks could be eliminated by using other less conventional solutions, e.g. the adhesive-bonded connections. The use of structural adhesives in engineering, particularly in metal construction design, unfortunately still raises questions about their durability and load-bearing capacity.

For the above reasons, the idea of forming thin-walled beam-columns by permanently bonding two cold-formed steel channels by their external web surfaces with adhesive in such a way seems reasonable (Fig. 28.1). In this way, it would be possible to shape I-beam-columns with web surfaces not only completely flat but also embossed along their entire length (Fig. 28.2). Web embossing can be in the form of longitudinal ribs of different geometries, with congruent or convex walls. In the case of convex ribs, the inner spaces of the ribs may be hollow or, for example, filled with plastic foam [2]. Of course, many questions arise from the idea.

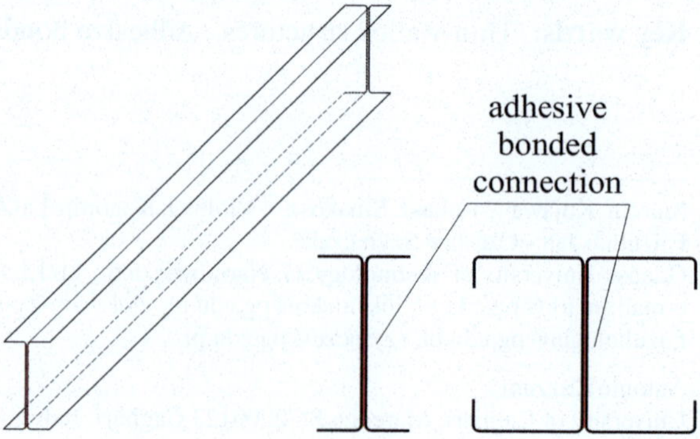

Fig. 28.1 Schematic diagram of the member

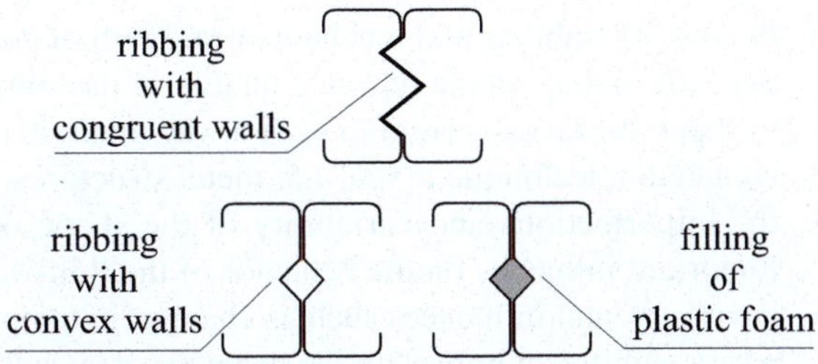

Fig. 28.2 Concept diagrams of web ribbing

The first and primary problem is how to fabricate these thin-walled beam-column members. Furthermore, it is also necessary to answer a number of key questions. What influence does an adhesive connection have on the load-bearing capacity of the member? How will the adhesive connection behave under different loading conditions, e.g., in tension or shear of the members? How will the mechanical strength of the adhesive-bonded area be affected by the layer protecting the material from corrosion, and does it have any effect at all? What effects are imposed on the stability and mechanical strength of the elements by the occurrence of imperfections of different types? What effect will web ribbing have on the member's resistance?

The second problem concerns the construction of correct numerical models. Proper modeling of loss of stability and failure phenomena will allow for further advanced studies with the use of these refined numerical models and will answer a few important questions, e.g.: what effect does the change in the strength parameters of the adhesive have on the bearing capacity of the members? How does the thickness of the adhesive coating itself affect the behavior of the sample? How to efficiently model adhesive layers using FEM? What effect does a patchy adhesive distribution on the contact surfaces have on the member's load-carrying capacity?

Thirdly, several questions about environmental influences on the behavior of the analyzed thin-walled members should be addressed. In this context, one has to try to answer the questions: how are adhesive and adhesive-bonded members affected by basic weather conditions such as changes in temperature or humidity? What effect does the aging of an adhesive layer have on the load-carrying capacity of the I-section member? Also, in the general sense, some remarks should be made on the questions related to failure analysis, e.g.: why does failure occur? How does failure occur? How to avoid recurrence?

The optimization [3, 4] of the cross-section shape is another important element of theoretical-numerical research, especially the shape of the contact surface at the connection of channel webs. During these analyses, the response has to be sought to crucial questions in the matter, e.g.: how to optimally shape the geometry of the adhesive-layered surface? What is the minimum bonding area? How do the adhesive parameters affect the load-bearing capacity of the member? The problems of how geometric imperfections or the quality of the adhesive coating affect the behavior of members should ideally be formulated in terms of reliability-based optimization.

To sum up, the main research hypotheses can be formulated as follows:

- the loss of stability and mechanical strength of composite thin-walled I-columns depends mainly on the type and quality of the adhesive material, but it is possible to shape the cross-section in such a way that it is rational to apply this innovative connection technique in real-life metal structures;
- the imperfections and variability of the shape of the I-member walls have an important influence on the behavior of the thin-walled member;
- environmental influences such as changes in temperature and humidity also have a very significant impact on the behavior of such thin-walled composite members;
- the effect of adhesive creep is also an important consideration in the design of structural components such as glued thin-walled beam-columns.

28.2 State of the Art

28.2.1 Static, Dynamics, and Stability Analysis of Thin-walled Members

Metal thin-walled member structures are the most common and essential form of constructing engineering structures. The main source of knowledge on the mechanics of thin-walled elastic beams in terms of classical theoretical foundations is given by Vlasov [5]. Vlasov described the general theory of statics, stability, and vibration of thin-walled beams of an open and closed cross-section. According to the original assumptions on the kinematics of the beam profile, Vlasov theory, unfortunately, does not account for the effects of cross-sectional deformation and any local in-plane wall deformations. Therefore, over time, some new methods were proposed as alternative concepts that develop the model proposed by Vlasov. They usually combine the theory of thin-walled beams with the theory of plates. From the beginning the theoretical, numerical, and experimental analysis of the problem was related to the analysis of global, local, and distortional stability in linear and nonlinear ranges. In this regard, special attention should be given to the method based on the classical Vlasov thin-walled beam theory where we treat the member's cross-section like a folded plate which is divided into longitudinal beams which are connected monolithically [6]. The folded-plate theory gave rise to one of the most frequently used numerical methods - the Finite Strip Method (FSM) [7]. The FSM is a particular case of the well-known and widely used Finite Element Method (FEM) [8]. The FSM was originally formulated by Cheung [9]. On the other hand, Hancock began the application of FSM in structural analysis of members, such as hot-rolled or cold-formed steel sections [10, 11, 12]. Based on the preliminary papers by Hancock, a team led by Schafer has developed the numerical software under the acronym CUFSM (the Constrained and Unconstrained Finite Strip Method) with the Direct Strength Method (DSM) of design of cold-formed sections [13, 14, 15, 16]. A different yet similar approach to the theoretical and numerical analysis of thin-walled members is the Generalized Beam Theory (GBT). The GBT method is originally proposed by Schardt [17] and further developed by Renton [18] and others [19, 20, 21, 22, 23, 24, 25, 26]. The GBT is very similar to the FSM, in terms of its principles and general concept. In the case of complex nonlinear problems [27] in the loss of stability issues or problems describing the successive phases of failure of structural systems, the FEM is still the most used one [28]. Together with advanced theoretical and numerical research, experimental studies have been conducted [29, 30, 31, 32, 33]. Unfortunately, most of the papers are focused on the classical connection methods and none of them are devoted to the problem of stability and failure analysis of thin-walled structures created by bonding cold-formed members with adhesives.

28.2.2 Influence of Imperfections

In the analysis of thin-walled members, crucial attention should be paid to the influence of random variation of structural (non-uniform distribution of the material's mechanical properties), technological (pre-stresses), and geometrical (any deviation of the geometry of the structure from the assumed ideal shape) imperfections of members on their loss of stability [28] and failure. It should be emphasized that initial imperfections are inevitable and are present at all stages of the fabrication of the structure. Not considering initial imperfections of structure in the analysis can lead to significant errors in the evaluation of structural behavior. Currently, in the design standard of metal structures in the case of single members resistance analysis, imperfections are considered in the form of predefined parameters concerning shapes and scales (EN 1993-1-1:2005 2005). The variability of imperfection parameters can be analyzed in deterministic, probabilistic, and statistical terms. For the deterministic approach, the algebraic method also known as the worst-case method is most commonly used [34]. Imperfection occurrence is a random phenomenon and can be analyzed as such [35]. It is usually assumed that the imperfection variation corresponds to a normal distribution [36]. In cases where the probability distributions cannot be assumed to be normal, advanced statistical methods are used. Most often in these cases, the Monte Carlo simulation method [37] is used in the analysis, although other methods are also used e.g.: the central moment method [38], a hybrid method that combines both aforementioned techniques (the approach leads to a significant reduction in the number of iterations), or numerical convolution method [39]. The experimental imperfection measurement methods and procedures are very well developed [40, 41]. They operate mostly on reverse engineering of structural modeling, such methods are also well known and widely used [42, 43].

28.2.3 Failure Analysis with Attention to Creep in Adhesive Bonded Metal Members/Structures

Adhesive technology is increasingly used in engineering structures and can be used to effectively join thin sheets of materials [44, 45]. It should be noted that the bonded joints may behave better than classical mechanical fastening due to the relatively smooth stress distribution, as compared to the highly localized stress concentration in mechanically fastened joints [46]. Moreover, adhesive bonding may be especially advisable when assembling thin-walled structures, as it works well in other similar fields of industry. Unfortunately, adhesive bonding is not very widespread in steel structures but rather in other areas of civil engineering, e.g.: reinforcing concrete structures by bonding composite materials, to bond wood or glass elements. Bonding is also used to reinforce damaged metal structures, but still very rarely due to the creep phenomenon [47]. The modeling of adhesive bonds is quite well described in the literature, both in terms of analytical and numerical analysis [48, 49, 50]. The simplest geometric configuration of bonded structures is probably the single-lap joint, which is widely used as a standard specimen for finding adhesive properties. In the case of single-lap joints, basic analytical models describing the stress states in an adhesive joint are as follows: the average shear stress model, the shear lag model, the adhesive-beam model, the model based on two-dimensional elastic theory, and the model based on load update and bending moment factor. There are also more complex models, such as models with the effects of transverse shear and geometric nonlinearity, coupled formulation with adhesive shear deformation effect but no overlap nonlinearity, coupled formulation with the shear and nonlinear effects but no peeling effect, or models with the fully coupled nonlinear formulation. For other joints, several models are used, for example, the symmetric linear model based on the Euler beam theory, the nonlinear model based on the Timoshenko beam theory, and others. The failure criteria models based on continuum or fracture mechanics and adhesively bonded structures are developed and widely used. With the help of these models, the damage both in the glued parts, in the adhesive, and on the contact surface may be described. With the development of computational units, numerical methods began to be used and developed in structural analysis. Some basic methods are widely used, i.e.: the Finite Difference Method (FDM), the Boundary Element Method (BEM), the Finite Volume Method (FVM), the Finite Strip Method (FSM), and the Finite Element Method (FEM) [8]. Nowadays, the FEM is the most commonly used method, particularly in the case of advanced analysis of nonlinear structures in a material or geometric sense. An increase in the sophistication of the FE analysis (FEA) has been accompanied by a shift from in-house software codes to powerful, user-friendly commercial software such as Abaqus (www.3ds.com). The most important feature of the present-day commercial software is the possibility to implement personal modifications to the software, i.e. by utilizing proprietary codes and scripts assembled in Fortran or Python programming languages. An important area of the current development of numerical methods covers the damage and failure modeling [51, 52, 53]. It should be noted that only a nonlinear finite element analysis allows for predicting structural failure. Nonlinear complex failure models are now standard features of commercial FEM software. Damage mechanics implemented in the FEA are usually based on the stress intensity factor or energy approaches. Failure in the adhesive layer is known as the cohesive failure, while failure in the region of the adhesive interface is known as the interfacial failure. Of course, nowadays both types of failures may be incorporated into the FE model even in standard commercial software. Failure analysis differs from the stress analysis approach because a suitable failure criterion is provided as part of the material definition in the FEM. For numerical damage modeling, we typically use either the failure initiation model or the crack propagation model [54]. In failure initiation models a state variable is used to indicate a failed material, but no change in material stiffness is considered. In contrast, in models with crack propagation, the elements satisfying the failure criteria are considered to have no stiffness and can usually be removed from the analysis. Furthermore, contrary to the stress analysis, the elements surpassing a failure criterion remain a part of the analysis and they contribute to the overall stiffness of the material. Naturally, the damage criteria can be based not only on the stress state but also on the strain state. In summary, the crack propagation model assumes that a crack exists in the material and grows when the selected fracture criterion is satisfied, in contrast to the failure initiation model, where the model does not need the existence of a pre-crack to determine a solution. Other techniques are also known, that are similar to the failure initiation model but can also predict damage evolution in a material without assuming that a pre-crack exists. In recent years, cohesive zone models have been increasingly used to simulate crack initiation, propagation, and failure [55, 56]. The cohesive zone model allows multiple cracks, and the direction of crack propagation need not be known, but cohesive zone elements need to be placed at all possible crack paths [57, 58]. The main problem with using cohesive zone models in numerical analysis is the proper selection of a traction-separation constitutive law to predict failure initiation, tracking damage, and failure. This is due to the difficulties in obtaining the proper experimental shape of the traction-separation law. Several propositions of shapes for the traction-separation law (e.g. bilinear, exponential, and trapezoidal) can be found in the literature. The shape of the traction-separation law is dependent mainly on the specimen geometry and material behavior. In addition, the traction-separation law also affects the numerical performance and convergence during analysis. Therefore, its proper selection is crucial in the numerical approach. Usually, a trapezoidal curve is considered more suitable for ductile materials, and a bilinear curve is habitually used for brittle and composite materials. Several methods for calibration of the traction-separation laws are known, but in general, the problem is not fully clarified. One method of finding the cohesive zone model parameters is the experimental-numerical approach [59]. Some other researchers have successfully used inverse methods to determine a unique set of cohesive law parameters [60, 61]. Furthermore, as one expects, adhesively bonded joints are usually subjected to several environmental factors (e.g. temperature, moisture, ultraviolet light) which affect their performance. That issue is also indicated in the literature [62, 63, 64]. Behavior and failure analysis of members made only of steel, with metal plasticity models, are well known and very well described [65, 66, 67, 68, 69, 70].

In bonded structures, the phenomenon of creep/cold flow [71, 72, 73] is observed in the form of quite large and increasing deformations over time. The problem, of course, is related mainly to the adhesive layer. In addition, with operating loads usually of long duration, we quite often have to deal with varying temperatures, promoting the formation of noticeable deformations. The successive gradual damage generated by creep results in the destruction of the entire structure by loss of adhesive cohesion. Therefore, in the analysis of bonded joints, the phenomenon of creep must certainly not be forgotten. Creep occurs in stages, and the durability of the member can be determined by the changes. Usually, in phase one, when first exposed to high temperatures or loads, deformation occurs almost immediately. The material goes through this phase quite quickly. In the second phase, creep enters a steady state. Then the primary creep rate decreases. The material remains in the second phase longer and in this state, the changes tend to be linear. In the third stage, the changes are already microstructural, at which point grain shifts and internal cracks occur. To sum up, in one sentence, we have to remember that proper creep resistance is one of the most important factors in the production of a member with plastic parts.

28.3 The Rationale for Addressing the Research Problem

Metal bar structures, including thin-walled structures, due to economic reasons are the most common and essential form of constructing engineering structures. Over the years, classical techniques of connection between individual structural elements, such as bolting, or welding have been mainly used. However, the classical types of connections have their disadvantages (residual stresses for welds as well as weakening of the cross-section for bolts and screws) [74], which are particularly relevant for thin-walled structures that are easy to eliminate by using other less conventional solutions, such as adhesively bonded connections [75]. The use of structural adhesive connections in engineering, particularly in metal structures, unfortunately still raises questions about durability and load-bearing capacity [76, 77]. Although it is less common, several research projects demonstrate its benefits even in civil engineering [78]. The majority of experimental, numerical, and theoretical studies are small-sample with single-lap adhesive joints [79, 80, 81, 44, 82]. Large or medium-scale metal column laboratory studies are still mainly focused on structures using classical techniques of connection. Undoubtedly, further research in this matter is needed to improve the innovative connection technique of adhesive bonding and to apply it to metal structures, especially thin-walled structures. Furthermore, the issue of environmental factors on the thin-walled composite members also still needs to be discussed more.

28.4 Conclusions

Adhesive bonding in joining steel structures is not a new idea. Nevertheless, to the authors' best knowledge nowadays there is no research dedicated to the process of fabrication of new cross-section shapes using adhesive bonding of cold-formed members. In recent years, plastics for adhesive bonding, are used progressively and successfully in various areas of mechanical or even civil engineering. Composite materials as well as composite structures are currently among the fastest developing fields of engineering science [83, 84] but are still very rare in civil engineering of metal structures. According to the authors' opinion, the proposed innovative approach to the shaping of new cross-sections from adhesive-bonded thin-walled members is undoubtedly an idea that is worthy of attention.

References

[1] Altenbach H, Eremeyev VA (Eds) (2017) Shell-like Structures. Advanced Theories and Applications, CISM International Centre for Mechanical Sciences, vol. 572, Springer, Cham.

[2] Altenbach H, Eremeyev VA (2011) Thin-walled structures made of foams, In: Cellular and Porous Materials in Structures and Processes (Eds H Altenbach, A Öchsner), CISM International Centre for Mechanical Sciences, vol. 521, pp 167-242, Springer, Vienna.

[3] Sarker RA, Newton CS (2008) Optimization Modelling. A Practical Approach, CRC Press.

[4] Lebedev LP, Cloud MJ, Eremeyev VA (2012) Advanced Engineering Analysis, World Scientific.

[5] Vlasov VZ (1961) Thin-walled Elastic Beams, Office of Technical Services, US Department of Commerce, Washington DC.

[6] Bar-Yoseph P, Hersckovitz I (1989) Analysis of folded plate structures. Thin-Walled Structures **7**(2):139-158.

[7] Lazzari JA, Batista EM (2021) Finite strip method computer application for buckling analysis of thin-walled structures with arbitrary cross-sections, REM - International Engineering Journal **74**(3):337-344.

[8] Zienkiewicz OC, Taylor RL, Fox DD (2014) The Finite Element Method for Solid & Structural Mechanics, Elsevier.

[9] Cheung YK, Tham LG (2020) Finite Strip Method, CRC Press,.

[10] Hancock GJ (1978) Local, distortional, and lateral buckling of I-beams, ASCE Journal of the Structural Division **104**(11):1787-1798.

[11] Hancock GJ, Kwon YB, Stefan Bernard E (1994) Strength design curves for thin-walled sections undergoing distortional buckling, Journal of Constructional Steel Research **31**(2):169-186.

[12] Hancock GJ, Pham CH (2022) Analysis and Design of Plated Structures, chapter Finite Strip Methods for stability analysis of thin-walled members with applications to the Direct Strength Method of design, Elsevier.
[13] Schafer BW, Peköz T (1998) Computational modeling of cold-formed steel: characterizing geometric imperfections and residual stresses Journal of Constructional Steel Research **47**(3):193-210.
[14] Schafer BW (2008) Review: the Direct Strength Method of cold-formed steel member design, Journal of Constructional Steel Research **64**(7-8):766-778.
[15] Kumar MVA, Kalyanaraman V (2010) Evaluation of Direct Strength Method for CFS compression members without stiffeners, Journal of Structural Engineering 1**36**(7):879-885.
[16] Nguyen VV, Hancock GJ, Pham CH (2020) Consistent and simplified Direct Strength Method for design of cold-formed steel structural members under localized loading, Journal of Structural Engineering **146**(6):04020090.
[17] Schardt R (1994) Generalized Beam Theory - an adequate method for coupled stability problems, Thin-Walled Structures **19**(2-4):161-180.
[18] Renton JD Generalized Beam Theory applied to shear stiffness, International Journal of Solids and Structures **27**(15):1955-1967.
[19] Silvestre N, Camotim (2003) Nonlinear Generalized Beam Theory for cold-formed steel members, International Journal of Structural Stability and Dynamics **3**(3):461-490.
[20] Silvestre N, Camotim (2013) Shear deformable Generalized Beam Theory for the analysis of thin-walled composite members, Journal of Engineering Mechanics **139**(8):1010-1024.
[21] Ranzi G, Luongo A (2011) A new approach for thin-walled member analysis in the framework of GBT, Thin-Walled Structures **49**(11):1404-1414.
[22] Taig G, Ranzi G (2014) Generalised Beam Theory (GBT) for stiffened sections, International Journal of Steel Structures **14**(2):381-397.
[23] Piccardo G, Ranzi G, Luongo A (2014) A complete dynamic approach to the Generalized Beam Theory cross-section analysis including extension and shear modes, Mathematics and Mechanics of Solids **19**(8):900-924.
[24] Piccardo G, Ranzi G, Luongo A (2014) A direct approach for the evaluation of the conventional modes within the GBT formulation, Thin-Walled Structures **74**:133-145.
[25] Piccardo G, Ferrarotti A, Luongo A (2017) Nonlinear Generalized Beam Theory for open thin-walled members, Mathematics and Mechanics of Solids **22**(10):1907-1921.
[26] Davies JM (2000) Recent research advances in cold-formed steel structures, Journal of Constructional Steel Research **55**(1-3):267-288.
[27] Doyle JF (2001) Nonlinear Analysis of Thin-walled Structures, Statics, Dynamics, and Stability, Springer.
[28] Kujawa M (2020) Selected local stability problems of channel section flanges made of aluminium alloys, Continuum Mechanics and Thermodynamics **32**(3):647-664.

[29] Stone TA, LaBoube RA (2005) Behavior of cold-formed steel built-up I-sections, Thin-Walled Structures **43**(12):1805-1817.

[30] Zhang JH, Young B (2012) Compression tests of cold-formed steel I-shaped open sections with edge and web stiffeners, Thin-Walled Structures **52**:1-11.

[31] Craveiro HD, Rodrigues JPC, Laím L (2016) Buckling resistance of axially loaded cold-formed steel columns, Thin-Walled Structures **106**:358-375.

[32] Zhang JH, Young B (2018) Experimental investigation of cold-formed steel built-up closed section columns with web stiffeners, Journal of Constructional Steel Research **147**:380-392.

[33] Fratamico DC, Torabian S, Zhao X, Rasmussen KJR, Schafer BW (2018) Experiments on the global buckling and collapse of built-up cold-formed steel columns, Journal of Constructional Steel Research **144**:65-80.

[34] Greenwood WH, Chase KW (1988) Worst case tolerance analysis with nonlinear problems, Journal of Engineering for Industry **110**(3):232-235.

[35] Ostwald PF, Huang J (1977) A method for optimal tolerance selection, Journal of Engineering for Industry **99**(3):558-565.

[36] Mansoor EM (1963) The application of probability to tolerances used in engineering designs, Proceedings of the Institution of Mechanical Engineers **178**(1):29-39.

[37] Zio E (2013) The Monte Carlo Simulation Method for System Reliability and Risk Analysis, Springer.

[38] Grubbström RW, Tang O (2006) The moments and central moments of a compound distribution, European Journal of Operational Research **170**(1):106-119.

[39] Varghese P, Braswell RN, Wang B, Zhang C (1996) Statistical tolerance analysis using FRPDF and numerical convolution, Computer-Aided Design **28**(9):723-732.

[40] McAnallen LE, Padilla-Llano DA, Zhao X, Moen CD, Schafer BW, Eatherton MR (2014) Initial geometric imperfection measurement and characterization of cold-formed steel C-section structural members with 3D non-contact measurement techniques, In Proceedings of the Annual Stability Conference Structural Stability Research Council, Toronto, Canada.

[41] Georgieva I, Schueremans L, de Roeck G, Pyl L (2011) Experimental investigation of the deformation of built-up members of cold-formed steel profiles, Applied Mechanics and Materials **70**:416-421.

[42] Kacmarcik J, Zaimovic-Uzunovic N, Lemes S (2020) New Technologies, Development and Application III, chapter Reverse engineering using 3D scanning and FEM analysis, Springer.

[43] Raja V, Fernandes KJ (Eds) (2008) Reverse Engineering, an Industrial Perspective, Springer.

[44] Adams RD, Comyn J, Wake WC (1997) Structural Adhesive Joints in Engineering, Springer, 2nd ed.

[45] Tong L, Steven GP (1999) Analysis and Design of Structural Bonded Joints, Springer.

[46] Pasternak H, Schwarzlos A, Schimmack N (2004) The application of adhesives to connect steel members, Journal of Constructional Steel Research **60**(3-5):649-658.

[47] Pasternak H, Meinz J (2006) Investigation of adhesive reinforcements in steel building structures. Versuche zu geklebten Verstärkungen im Stahlbau, Bauingenieur **81**:212-217.

[48] da Silva LFM, Öchsner A (Eds) (2008) Modeling of Adhesively Bonded Joints, Springer.

[49] da Silva LFM, Campilho RDSG (2012) Advances in Numerical Modelling of Adhesive Joints, Springer.

[50] Tserpes K, Barroso-Caro A, Carraro PA, Beber VC, Floros I, Gamon W, Kozłowski M, Santandrea F, Shahverdi M, Skejić D, Bedon C, Rajčić V (2021) A review on failure theories and simulation models for adhesive joints, The Journal of Adhesion **98**(12):1855-1915.

[51] Otegui JL (2014) Failure Analysis, Fundamentals and Applications in Mechanical Components, Springer.

[52] Altenbach H, Eremeyev VA, Igumnov LA (Eds) (2021) Multiscale Solid Mechanics. Strength, Durability, and Dynamics Advanced Structured Materials, vol. 141, Springer.

[53] Contro R, Poggi C, Cazzani A (1988) Numerical analysis of fire effects on beam structures, Engineering Computations **5**(1):53-58.

[54] Cui W, Huang X, Wang F (2014) Basic Concepts of Fracture Mechanics, Springer.

[55] Park K, Paulino GH (2011) Cohesive zone models: a critical review of traction-separation relationships across fracture surfaces, Applied Mechanics Reviews **64**(6):060802.

[56] Wciślik K, Pała T (2021) Selected aspects of cohesive zone modeling in fracture mechanics, Metals **11**(2):302.

[57] Campilho RDSG, Banea MD, Neto JABP, da Silva LFM (2013) Modelling adhesive joints with cohesive zone models: effect of the cohesive law shape of the adhesive layer, International Journal of Adhesion and Adhesives **44**:48-56.

[58] Demiral M, Kadioglu F (2018) Failure behaviour of the adhesive layer and angle ply composite adherends in single lap joints: a numerical study, International Journal of Adhesion and Adhesives **87**:181-190.

[59] Rudawska A, Stančeková D, Cubonova N, Vitenko T, Müller M, Valášek P (2017) Adhesive properties and adhesive joints strength of graphite/epoxy composites, Journal of Physics: Conference Series **842**(1):012073.

[60] Azevedo JCS, Campilho RDSG, da Silva LFM, Faneco TMS, Lopes RM. Cohesive law estimation of adhesive joints in mode II condition, Theoretical and Applied Fracture Mechanics **80**:143-154.

[61] Desai CK, Basu S, Parameswaran V (2016) Determination of traction separation law for interfacial failure in adhesive joints at different loading rates, The Journal of Adhesion **92**(10):819-839.

[62] Korta J, Mlyniec A, Uhl T (2015) Experimental and numerical study on the effect of humidity-temperature cycling on structural multi-material adhesive joints, Composites Part B: Engineering **79**:621-630.

[63] Zhang J, Cheng X, Guo X, Bao J, Huang W (2020) Effect of environment conditions on adhesive properties and material selection in composite bonded joints, International Journal of Adhesion and Adhesives **96**:102302.

[64] Brito CBG, Sales RCM, Donadon MV Effects of temperature and moisture on the fracture behaviour of composite adhesive joints, International Journal of Adhesion and Adhesives **100**:102607.

[65] Symonds PS (1967) Survey of methods of analysis for plastic deformation of structures under dynamic loading, Division of Engineering, Brown University.

[66] Marciniak Z, Kuczyński K (1967) Limit strains in the processes of stretch-forming sheet metal, International Journal of Mechanical Sciences **9**(9):609-612.

[67] Lindholm US, Bessey RL A survey of rate dependent strength properties of metals, Technical report, Defense Technical Information Center.

[68] Eleiche ASM (1972) A literature survey of the combined effects of strain rate and elevated temperature on the mechanical properties of metals, Technical report, Defense Technical Information Center.

[69] Anduquia-Restrepo J, Narváez-Tovar C, Rodríguez-Baracaldo R (2018) Computational and numerical analysis of ductile damage evolution under load-unload tensile test in dual phase steel Journal of Mechanical Engineering **64**(5):339-348.

[70] Sharifimehr S, Fatemi A (2019) Fatigue analysis of ductile and brittle behaving steels under variable amplitude multiaxial loading, Fatigue & Fracture of Engineering Materials & Structures 42(8):1722-1742.

[71] Altenbach H, Eisenträger J (2019) Encyclopedia of Continuum Mechanics, chapter Introduction to creep mechanics, Springer.

[72] Betten J (2008) Creep Mechanics, Springer, 3rd ed.

[73] Lemaitre J, Desmorat R (2005) Engineering Damage Mechanics. Ductile, Creep, Fatigue and Brittle Failures, Springer.

[74] Boracchini A (2018) Design and Analysis of Connections in Steel Structures, Fundamentals and Examples, Wiley.

[75] da Silva LFM, Öchsner A, Adams RD (Eds) (2018) Handbook of Adhesion Technology, Springer.

[76] Ciupack Y, Pasternak H, Schiel M, Ince E (2014) Adhesive bonded joints in steel structures, Steel Construction **7**(3):178-182.

[77] Ciupack Y, Pasternak H, Mette C, Stammen E, Dilger K (2017) Adhesive bonding in steel construction - challenge and innovation, Procedia Engineering **172**:186-193.

[78] Piekarczyk M, Grec R (2012) Application of adhesive bonding in steel and aluminium structures, Archives of Civil Engineering **58**(3):309-329.

[79] Goland M, Reissner E (1944) The stresses in cemented joints, American Society of Mechanical Engineers **11**(1):17-27.

[80] Hart-Smith LJ (1973) Adhesive-bonded single-lap joints, Technical report, NASA.

[81] Adams RD, Wake WC (1984) Structural Adhesive Joints in Engineering, Elsevier Applied Science Publishers.
[82] Cheng S, Chen D, Shi Y (1991) Analysis of adhesive-bonded joints with nonidentical adherends, Journal of Engineering Mechanics **117**(3):605-623.
[83] Gay D, Hoa H, Tsai T (2003) Composite Materials, Design and Application, CRC Press.
[84] Morozov E, Vasiliev V (2018) Advanced Mechanics of Composite Materials and Structures, Elsevier.

Chapter 29
Dissipative Mechano-Electro-Magnetism Simulations in Electronic Components

Yiming Liu, Wolfgang H. Müller, and Bilen Emek Abali

Abstract Especially at high frequencies, as they occur in 5G or relevant applications, electromagnetism in electronic components plays a significant role in system behavior. Moreover, dissipative effects are triggered by mechanical deformation. Indeed, dissipation is an energy loss and thus the industry tries to minimize this effect by optimizing the design. We examine this justification by using multiphysics simulations by means of the Finite Element Method (FEM) and solve deformation and dissipative dielectric material equations for a simple electronic component in a monolithic manner.

Key words: Electromagnetism · Mechanics · Computation · Finite element method

29.1 Introduction

Higher frequencies are required in today's electronic components. An accurate modeling is necessary in a computational study for predicting the system response and for optimizing the design of electronic devices. We basically know that electronic components are dissipative, for example they get heated up during use. There are various phenomena producing heat. Electric current as well as viscoelastic deformation may be reasons for heat production. Especially at higher frequencies, such irreversible (dissipative) effects start to become significant. Mostly the quality of

Yiming Liu · Wolfgang H. Müller
Technische Universität Berlin, Chair of Continuum Mechanics and Constitutive Theory, Institute of Mechanics, Berlin, Germany,
e-mail: liuym0507@icloud.com, wolfgang.h.mueller@tu-berlin.de

Bilen Emek Abali
Uppsala University, Division of Applied Mechanics, Department of Materials Science and Engineering, Uppsala, Sweden,
e-mail: bilenemek@abali.org

H. Altenbach et al. (eds.), *Sixty Shades of Generalized Continua*,
Advanced Structured Materials 170, https://doi.org/10.1007/978-3-031-26186-2_29

an electronic component is measured by how low dissipation is occurring. Since dissipation decreases efficiency, even small amendments for increasing efficiency has a great value. Therefore, we are interested to understand the phenomena and model them as accurately as possible. Reduced order models are lacking an accurate estimation of the system behavior [1]. The main reason is oversimplifying the system down to a constant (in space) electric field within the component. Geometry plays a decisive role and a heterogeneous distribution of the electric field results in a different system response than predicted by 1-D models that are often used in electrical engineering. We propose to use continuum mechanics based models for simulating electromagnetism. In this way, a multiphysics simulation may be exploited for understanding the details of an engineering system as well as to make design changes.

In electronics a passive component, such as a capacitor, is very often used by the hundreds on a motherboard. This component is a dielectric material between two parallel plates such that the capacitor works for storing energy. This storage is performed for deploying electric current in large quantities, because by contrast to batteries, capacitors are not lagging when used as a supply. A dielectric material isolating two conductors is shown in Fig. 29.1. Analytic results exist in a capacitor problem for calculating the electric field. Edge or fringing effects are considered to improve the accuracy even in very early studies as in [2, 3]. Of course, there are several assumptions leading to numerical and analytical approaches [4, 5]. Remarkably, the interface between an isolator and conductor is challenging to model since a surface charge may arise. Such a model incorporating surface charge and their numerical implementations are difficult to find in the literature [6, 7]. Governing equations are normally given in the bulk and an interface condition is not implemented in general. We emphasize that numerical stability problems may emerge by skipping interface conditions. Hence, in this work, we develop the weak form of the governing equations and discuss how one may implement such additional terms. As a multiphysics

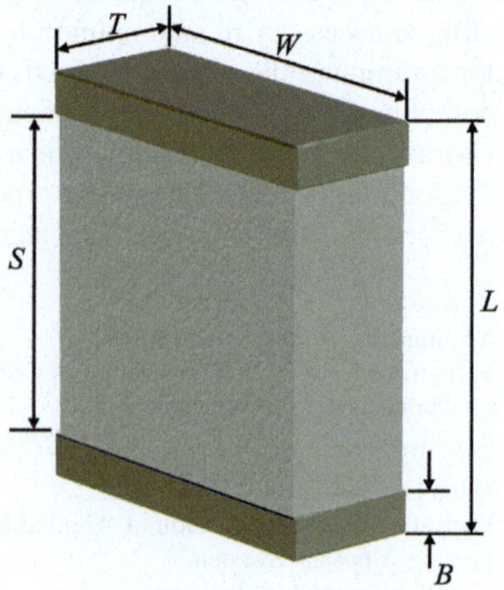

Fig. 29.1 Industrial example of the capacitor C2220 with the geometric sizes: $L = 5.7$ mm, $W = 5.0$ mm, $T = 2$ mm, $B = 0.6$ mm, $S = 4.5$ mm. Two metal conductors on top and bottom (deep brown) are isolated by a dielectric material in the middle (gray).

simulation, we compute the displacement and electromagnetic potentials (leading to electromagnetic fields) by solving coupled partial differential equations. Furthermore, we discuss material modeling for dissipative systems and motivate these constitutive equations based on thermodynamics.

29.2 Governing Equations

We use standard continuum mechanics notation with $\boldsymbol{x}$, denoting space positions, and t, being the time. In electromagnetism, fields are measured in a laboratory frame. However, in mechanics, a material frame is used where coordinates denote material particles. One rather simpler way of combining these intrinsically different variables relies on using the small deformation assumption and skip a distinction between current and reference configuration [8, 9]. Otherwise, there are challenges in theory and implementation [10]. All quantities are expressed in Cartesian coordinates. Over repeated indices, $i, j, k \ldots$ we understand an EINSTEIN summation convention from 1 to 3 in 3-D space. By using a comma notation, we denote a (partial) spatial derivative.

For displacement and electromagnetic fields, we need to have governing equations. Electric charge and mass are fundamental properties. We model a separation of the electric charge in bound and free charges. Free charges move macroscopic distances, their motion is the reason of electric current. Bound charges move interatomic distances and their motion is understood as electric and magnetic polarization. We separate the MAXWELL equations into two sets. The first set of equations reads

$$\frac{\partial B_i}{\partial x_i} = 0 \,, \quad \frac{\partial B_i}{\partial t} + \epsilon_{ijk}\frac{\partial E_k}{\partial x_j} = 0 \,, \tag{29.1}$$

with the magnetic flux, $\boldsymbol{B}$ in T (tesla), and the electric field, $\boldsymbol{E}$ in V/m (volt per meter); they are measurable quantities. These MAXWELL equations are solved by using

$$E_i = -\frac{\partial \phi}{\partial x_i} - \frac{\partial A_i}{\partial t} \,, \quad B_i = \epsilon_{ijk}\frac{\partial A_k}{\partial x_j} \,, \tag{29.2}$$

where we have introduced the electric potentials, ϕ in V, magnetic potential, A_i in T m, as unknowns to be solved in space and time. We represent 6 differential equations in Eq. (29.1) by 4 solution functions, $\phi, \boldsymbol{A}$. Hence we are free to choose two gauge conditions. The choice of gauge conditions fails to alter the electromagnetic fields, $\boldsymbol{E}, \boldsymbol{B}$. We propose the LORENZ's gauge:

$$\frac{\partial \phi}{\partial t} = -c^2 \frac{\partial A_i}{\partial x_i} \,, \tag{29.3}$$

where the speed of light in vacuum, c, is also a universal constant. This gauge condition resolves some of the numerical stability problems at higher frequencies.

The second set of MAXWELL equations reads:

$$\rho z = \frac{\partial D_i}{\partial x_i} \,, \quad -\frac{\partial D_i}{\partial t} + \epsilon_{ijk}\frac{\partial H_k}{\partial x_j} = J_i \,, \tag{29.4}$$

where the specific charge, z in C/kg (coulomb per kilogram), is involved in massive particles with a given mass density, ρ in kg/m^3, creating the divergence of the charge potential, $\boldsymbol{D}$ in C/m^2. The current potential, $\boldsymbol{H}$ in A/m (ampere per meter), and the charge potential, $\boldsymbol{D}$, need to be defined by constitutive equations. The electric current, $\boldsymbol{J}$ in A/m^2, involves a current because of free and bound charge motions. We use the MAXWELL–LORENTZ aether relations:

$$D_i = \varepsilon_0 E_i \,, \quad H_i = \frac{1}{\mu_0} B_i \,, \tag{29.5}$$

where $\varepsilon_0 = 8.85 \cdot 10^{-12}$ A s/(V m) and $\mu_0 = 12.6 \cdot 10^{-7}$ V s/(A m) are universal constants with the relation

$$\varepsilon_0 \mu_0 = \frac{1}{c^2} \,. \tag{29.6}$$

Electric polarization is the deviation of charges from their equilibrium positions. These bound charges move such that the center of negative charges and the center of positive charges fail to overlap. By convention the electric polarization is directed from the negative charge center to the positive charge center, which is along the opposite direction of (positive) charge potential, $\boldsymbol{D}$. Therefore, we have

$$D_i = \mathfrak{D}_i - P_i \,, \tag{29.7}$$

where $\mathfrak{D}_i$ denotes a charge potential due to free charges and P_i due to bound charges (electric polarization). In analogy, we comprehend the magnetic polarization (magnetization), $\boldsymbol{M}$, as a consequence of bound charges doing a circular/spiral motion in interatomistic distances. Also a circular motion of polarized particles create $\boldsymbol{P} \times \boldsymbol{v}$ that is indistinguishable from the magnetization, $\boldsymbol{M}$, hence, we have

$$\mathscr{M}_i = M_i + \epsilon_{ijk} P_j v_k \,. \tag{29.8}$$

Therefore, the total current potential reads

$$H_i = \mathfrak{H}_i + \mathscr{M}_i \,. \tag{29.9}$$

This distinction is for isolators (or even in some cases for semi-conductors). Indeed, a moving electric charge creates an electric current, $\boldsymbol{J}^{\text{fr.}}$. The freely moving electric current is much greater than the displacement current occurring due to the electric polarization. Practically, an electric polarization occurs only in an insulator. Now by inserting Eqs. (29.7) and (29.9) into Eq. (29.4) we obtain

$$
\begin{aligned}
-\frac{\partial \mathfrak{D}_i}{\partial t}+\frac{\partial P_i}{\partial t}+\epsilon_{ijk}\frac{\partial \mathfrak{H}_k}{\partial x_j}+\epsilon_{ijk}\frac{\partial \mathcal{M}_k}{\partial x_j} &= J_i \ , \\
-\frac{\partial \mathfrak{D}_i}{\partial t}+\epsilon_{ijk}\frac{\partial \mathfrak{H}_k}{\partial x_j} &= J_i^{\text{fr.}} \ ,
\end{aligned} \tag{29.10}
$$

with

$$
J_i = J_i^{\text{fr.}} + \frac{\partial P_i}{\partial t} + \epsilon_{ijk}\frac{\partial \mathcal{M}_k}{\partial x_j} \ . \tag{29.11}
$$

In other words, the total electric current, $\boldsymbol{J}$, is due to bound charges moving.

We refer to [11, Chap. 3.5] and use balance of momentum for calculating displacement, $\boldsymbol{u}$, with a given mass density, ρ_0, and volumetric supply term such as gravitational specific force, $\boldsymbol{f}$,

$$
\rho_0 \ddot{u}_i - \sigma_{ji,j} - \rho_0 f_i = \mathcal{F}_i \ , \tag{29.12}
$$

where stress $\boldsymbol{\sigma}$ and electromagnetic force density, $\boldsymbol{\mathcal{F}}$, need to be defined under further assumptions. We use the model known as MINKOWSKI momentum and obtain

$$
\rho_0 \ddot{u}_i + \frac{\partial \mathcal{G}_i}{\partial t} - \sigma_{ji,j}^{\text{tot.}} - \rho_0 f_i = 0 \ , \quad \sigma_{ji}^{\text{tot.}} = \sigma_{ji} + m_{ji} \ , \tag{29.13}
$$

with the following electromagnetic momentum and electromagnetic stress:

$$
\mathcal{G}_i = \epsilon_{ijk}\mathfrak{D}_j B_k \ , \quad m_{ji} = -\frac{1}{2}\delta_{ij}(H_k B_k + D_k E_k) + H_i B_j + D_j E_i \ , \tag{29.14}
$$

respectively.

29.3 Constitutive Equations

In order to close the system of equations, we need to define constitutive equations by using a model based on thermodynamics [12]. For a thermodynamics analysis, we follow [13] and decompose constitutive equations into reversible and dissipative (irreversible) terms

$$
\sigma_{ij} = {}^{\text{r}}\sigma_{ij} + {}^{\text{d}}\sigma_{ij} \ , \quad P_i = {}^{\text{r}}P_i + {}^{\text{d}}P_i \ , \quad \mathcal{M}_i = {}^{\text{r}}\mathcal{M}_i + {}^{\text{d}}\mathcal{M}_i \ . \tag{29.15}
$$

Then the First Law of Thermodynamics (balance of internal energy) results in the following reversible terms:

$$
\begin{aligned}
{}^{\text{r}}\sigma_{ij} &= C_{ijkl}\varepsilon_{kl} - \tilde{T}_{kij}E_k + \tilde{S}_{kji}B_k \ , \\
{}^{\text{r}}P_i &= \tilde{T}_{ijk}\varepsilon_{jk} + \varepsilon_0 \chi_{ij}^{\text{el.}} E_j + \tilde{R}_{ji}B_j \ , \\
{}^{\text{r}}\mathcal{M}_i &= \tilde{S}_{ijk}\varepsilon_{jk} + \tilde{R}_{ij}E_j + \mu_0^{-1}\chi_{ij}^{\text{mag.}} B_j \ .
\end{aligned} \tag{29.16}
$$

We assume that coupling effects, such as piezoelectric, $\tilde{\boldsymbol{T}} = 0$, piezomagnetic, $\tilde{\boldsymbol{S}}$, and electromagnetic coupling effects, $\tilde{\boldsymbol{R}} = 0$, vanish. Then for the reversible stress we use HOOKE's law with the stiffness tensor, $\boldsymbol{C}$, electric and magnetic susceptibilities, $\boldsymbol{\chi}^{\text{el.}}$, $\boldsymbol{\chi}^{\text{mag.}}$. Hence, we have

$$ {}^{\mathrm{r}}\sigma_{ij} = C_{ijkl}\varepsilon_{kl} \ , \quad {}^{\mathrm{r}}P_i = \varepsilon_0 \chi_{ij}^{\text{el.}} E_j \ , \quad {}^{\mathrm{r}}\mathcal{M}_i = \mu_0^{-1} \chi_{ij}^{\text{mag.}} B_j \ . \tag{29.17} $$

By using the balance of entropy and an assertion that the entropy production is positive, also known as the Second Law of Thermodynamics, we obtain a positive term in form of multiplications

$$ \left({}^{\mathrm{d}}\sigma_{ji} - P_j E_i + \mathcal{M}_i B_j \right) \dot{\varepsilon}_{ij} + \mathcal{E}_i \mathcal{J}_i^{\text{fr.}} + {}^{\mathrm{d}}P_i \dot{E}_i + B_i {}^{\mathrm{d}}\dot{\mathcal{M}}_i \geq 0 \tag{29.18} $$

For simplicity, we neglect coupling terms and use the CURIE principle such that same rank tensors depend on each other. Furthermore, we analyze a polarize material such that the free current vanishes. Also we assume that the dissipative magnetic polarization is negligible. Under these simplifications, we obtain

$$ {}^{\mathrm{d}}\sigma_{ji} = P_j E_i - \mathcal{M}_i B_j \ , \quad {}^{\mathrm{d}}P_i = K_{ij} \dot{E}_j \ . \tag{29.19} $$

Measurements for the $\boldsymbol{K}$ term are usually employed by decomposing dielectric properties in real and imaginary terms. An analogy is known from viscoelastic and stress rate dependent response in mechanics [14]. We demonstrate this fact by using polarization measurement for time varying (harmonic) electric fields, written by the EULER equation:

$$ \boldsymbol{E} = \boldsymbol{E}_{\text{amp}} \exp(\mathrm{i}\,\omega t), \tag{29.20} $$

with the complex susceptibility:

$$ \boldsymbol{\chi} = \boldsymbol{\chi}' + \mathrm{i}\,\boldsymbol{\chi}'' \tag{29.21} $$

we obtain

$$ P_i = \varepsilon_0 \left(\chi'_{ij} + \mathrm{i}\,\chi''_{ij} \right) E_j \ . \tag{29.22} $$

When we use the rate by taking the derivative, $\dot{\boldsymbol{E}} = \mathrm{i}\,\omega \boldsymbol{E}_{\text{amp}} \exp(\mathrm{i}\,\omega t) = \mathrm{i}\,\omega \boldsymbol{E}$, we immediately obtain

$$ P_i = \varepsilon_0 \left(\chi'_{ij} E_j + \frac{1}{\omega} \chi''_{ij} \dot{E}_j \right) . \tag{29.23} $$

In the aforementioned notation, we see the analogy $K_{ij} = \varepsilon_0 \chi''_{ij} / \omega$.

29.4 Generating Weak Forms

We aim for computing the electric potential, ϕ, the magnetic potential, $\boldsymbol{A}$, and the displacement, $\boldsymbol{u}$. We use the finite element method in space discretization and the finite difference method in time discretization. We skip a complete discussion of the balance equations on singular surfaces, yet we need to fulfill them between surfaces shared by finite elements. Especially an interface between two materials is of interest, where functions may undergo a discontinuity across a surface, called a singular surface without mass (non-material). There is a jump of the function's value across the singular surface, but we model electromagnetic potentials, ϕ, $\boldsymbol{A}$, and displacement, $\boldsymbol{u}$, as continuous functions. Charge and current potentials or equivalently electromagnetic polarizations or electric current or stress are related to their gradients. Between two different materials, with different material parameters for polarizations, current, stress a jump may occur across this interface. The balance equations on singular surfaces take the simple form

$$\boldsymbol{n}\cdot[\![\mathfrak{D}]\!]=q_{\mathrm{s}}^{\mathrm{fr.}}\;,\quad \boldsymbol{n}\times[\![\mathfrak{H}]\!]=j_{\mathrm{s}}^{\mathrm{fr.}}\;,\quad \boldsymbol{n}\cdot[\![\sigma^{\mathrm{tot.}}]\!]=0\;, \tag{29.24}$$

where squared brackets indicate a jump with emerging surface (free) charges, $q_{\mathrm{s}}^{\mathrm{fr.}}$, and surface (free) currents, $j_{\mathrm{s}}^{\mathrm{fr.}}$. The jump value is simply calculated by evaluating on the boundary at one side and subtracting from the evaluated value at the other side. We identify both sides of the interface by plus and minus with the plane normal, $\boldsymbol{n}^{+}$, pointing against the plane normal $\boldsymbol{n}^{-}$. In other words, $\boldsymbol{n}^{+}=-\boldsymbol{n}^{-}$. Then the balance equations read

$$\begin{aligned}
n_i[\![\mathfrak{D}_i]\!]=n_i(\mathfrak{D}_i^{+}-\mathfrak{D}_i^{-})=n_i^{+}\mathfrak{D}_i^{+}+n_i^{-}\mathfrak{D}_i^{-}=&\quad q_{\mathrm{s}}^{\mathrm{fr.}}\;,\\
\epsilon_{ijk}n_j[\![\mathfrak{H}_k]\!]=\epsilon_{ijk}n_j(\mathfrak{H}_k^{+}-\mathfrak{H}_k^{-})=\epsilon_{ijk}(n_j^{+}\mathfrak{H}_k^{+}+n_j^{-}\mathfrak{H}_k^{-})=&\quad j_{\mathrm{s}}^{\mathrm{fr.}}\;,\\
n_i[\![\sigma_{ij}]\!]=t_j^{+}-t_j^{-}=0\;,n_i[\![m_{ij}]\!]=&\quad 0\;,
\end{aligned} \tag{29.25}$$

by choosing $\boldsymbol{n}=\boldsymbol{n}^{+}$, where this choice is arbitrary. No mechanical stress jump is known as *actio* = *reactio* where the traction vector $\boldsymbol{t}$ is continuous across the singular surface without cracks or fracture. No electromagnetic stress jump is a simple consequence of MAXWELL–LORENTZ aether relations.

For the electric potential, ϕ, we exploit the balance of electric charge and insert in the Maxwell equation Eq. (29.4)$_1$, $\rho z=D_{i,i}$, and the total current from Eq. (29.11), so we get

$$\frac{\partial D_{i,i}}{\partial t}+\left(J_i^{\mathrm{fr.}}+\frac{\partial P_i}{\partial t}+\epsilon_{ijk}\mathscr{M}_{k,j}\right)_{,i}=0\;. \tag{29.26}$$

Within a domain without interfaces, for the same material where $\boldsymbol{P}$ is continuous, we interchange the order of space and time derivative,

$$\frac{\partial\mathfrak{D}_{i,i}}{\partial t}+\left(J_i^{\mathrm{fr.}}+\epsilon_{ijk}\mathscr{M}_{k,j}\right)_{,i}=0\;. \tag{29.27}$$

For the time discretization, by using a EULER backward scheme for an arbitrary field, $(\cdot)$,

$$\frac{\partial(\cdot)}{\partial t}=\frac{(\cdot)-(\cdot)^0}{\mathrm{d}t}\;, \tag{29.28}$$

we obtain a stable method and use for the sake of simplicity a constant time step, $\mathrm{d}t$. By multiplying with the test function, $\delta\phi$, and integrating over one finite element and then integrating by parts for the gradient terms, we acquire a variational form. Regarding the methodology we emphasize that we select one single finite element as the integration domain, since we guarantee the continuity of charge and current potentials within this domain. No interface is allowed within an element, which is also the usual procedure in meshing for the FEM. The weak form in the unit of energy (by multiplying the form with dt) reads

$$\begin{aligned} \mathrm{F}_\phi =& \int_{\Omega^*} \big(-(\mathfrak{D}_i - \mathfrak{D}_i^0)\delta\phi_{,i} - \mathrm{d}t J_i^{\mathrm{fr.}} \delta\phi_{,i} - \mathrm{d}t \epsilon_{ijk} \mathscr{M}_{k,j} \delta\phi_{,i}\big)\,\mathrm{d}v + \\ &+ \int_{\partial\Omega^*} n_i(\mathfrak{D}_i - \mathfrak{D}_i^0 + \mathrm{d}t J_i^{\mathrm{fr.}} + \mathrm{d}t \epsilon_{ijk}\mathscr{M}_{k,j})\delta\phi\,\mathrm{d}a\,. \end{aligned} \tag{29.29}$$

In the case of a structure with several materials, $\Omega = \sum \Omega^*$, we assemble simply by adding over all finite elements. Since an interface is a surface between two elements, $\partial\Omega^*$, a summation over all elements leads to the jump brackets for an area integral over interfaces, $\partial\Omega^I$, to be used for fulfilling the governing equations on singular surfaces,

$$\begin{aligned} \mathrm{F}_\phi =& \int_{\Omega} \big(-(\mathfrak{D}_i - \mathfrak{D}_i^0)\delta\phi_{,i} - \mathrm{d}t J_i^{\mathrm{fr.}} \delta\phi_{,i} - \mathrm{d}t \epsilon_{ijk} \mathscr{M}_{k,j} \delta\phi_{,i}\big)\,\mathrm{d}v + \\ &+ \int_{\partial\Omega^I} n_i [\![\mathfrak{D}_i - \mathfrak{D}_i^0 + \mathrm{d}t J_i^{\mathrm{fr.}} + \mathrm{d}t \epsilon_{ijk}\mathscr{M}_{k,j}]\!] \delta\phi\,\mathrm{d}a + \\ &+ \int_{\partial\Omega} n_i(\mathfrak{D}_i - \mathfrak{D}_i^0 + \mathrm{d}t J_i^{\mathrm{fr.}} + \mathrm{d}t \epsilon_{ijk}\mathscr{M}_{k,j})\delta\phi\,\mathrm{d}a\,. \end{aligned} \tag{29.30}$$

We stress that the whole structure has a boundary as well, $\partial\Omega$, where (boundary) conditions need to be defined. After a straight-forward rewriting

$$\begin{aligned} \epsilon_{ijk} n_j [\![H_k - \mathscr{M}_k]\!] &= j_{\mathrm{s}}^{\mathrm{fr.}}\,, \\ \epsilon_{ijk} n_j [\![H_k]\!] - j_{\mathrm{s}}^{\mathrm{fr.}} &= \epsilon_{ijk} n_j [\![\mathscr{M}_k]\!]\,, \\ \epsilon_{ijk} n_j \frac{1}{\mu_0} \epsilon_{klm} [\![A_{m,l}]\!] - j_{\mathrm{s}}^{\mathrm{fr.}} &= \epsilon_{ijk} n_j [\![\mathscr{M}_k]\!]\,, \end{aligned} \tag{29.31}$$

and using the identity $\epsilon_{ijk}\epsilon_{klm} = \delta_{il}\delta_{jm} - \delta_{im}\delta_{jl}$, we obtain

$$\epsilon_{ijk} n_j [\![\mathscr{M}_k]\!] = n_j \frac{1}{\mu_0} [\![A_{j,i} - A_{i,j}]\!] - j_{\mathrm{s}}^{\mathrm{fr.}}\,. \tag{29.32}$$

By inserting the latter into the weak form, the following results for computing the electric potential:

$$
\begin{aligned}
\mathrm{F}_\phi = & \int_\Omega \big(-(\mathfrak{D}_i - \mathfrak{D}_i^0)\delta\phi_{,i} - \mathrm{d}t J_i^{\mathrm{fr.}} \delta\phi_{,i} - \mathrm{d}t \epsilon_{ijk} \mathscr{M}_{k,j} \delta\phi_{,i} \big) \mathrm{d}v + \\
& + \int_{\partial\Omega^I} \Big((q_\mathrm{s}^{\mathrm{fr.}} - {}^0q_\mathrm{s}^{\mathrm{fr.}})\delta\phi + n_i \mathrm{d}t [\![J_i^{\mathrm{fr.}}]\!] \delta\phi + n_i \mathrm{d}t \epsilon_{ijk} [\![\mathscr{M}_{k,j}]\!]) \delta\phi \Big) \mathrm{d}a + \\
& + \int_{\partial\Omega} n_i (\mathfrak{D}_i - \mathfrak{D}_i^0 + \mathrm{d}t J_i^{\mathrm{fr.}} + \mathrm{d}t \epsilon_{ijk} \mathscr{M}_{k,j}) \delta\phi \, \mathrm{d}a \,.
\end{aligned}
\tag{29.33}
$$

For parts of the boundary where we employ DIRICHLET boundary conditions on $\partial\Omega$ and describe ϕ values, we set $\delta\phi = 0$.

For the magnetic potential, $\boldsymbol{A}$, we use MAXWELL's Eq. (29.4)$_2$, and also apply the LORENZ gauge in Eq. (29.3). After a straight-forward manipulation, we obtain

$$
\varepsilon_0 \frac{\partial^2 A_i}{\partial t^2} - \frac{1}{\mu_0} \frac{\partial^2 A_i}{\partial x_j \partial x_j} = J_i^{\mathrm{fr.}} + \frac{\partial P_i}{\partial t} + \epsilon_{ijk} \mathscr{M}_{k,j} \,.
\tag{29.34}
$$

We emphasize that the LORENZ gauge is of importance to change the type of differential equation so that it is more suitable for FEM. Therefore, we use standard elements without numerical difficulties. After discretizing in time, we generate the weak form by multiplying with the test function, $\delta\boldsymbol{A}$, and lowering the differentiability by integrating by parts and acquire

$$
\begin{aligned}
\mathrm{F}_{\boldsymbol{A}} = & \int_\Omega \Big(\varepsilon_0 \frac{A_i - 2A_i^0 + A_i^{00}}{\mathrm{d}t\mathrm{d}t} \delta A_i + \frac{1}{\mu_0} A_{i,j} \delta A_{i,j} - J_i^{\mathrm{fr.}} \delta A_i - \frac{P_i - P_i^0}{\mathrm{d}t} \delta A_i + \\
& + \epsilon_{ijk} \mathscr{M}_k \delta A_{i,j} \Big) \mathrm{d}v - \int_{\partial\Omega^I} \Big(\frac{1}{\mu_0} [\![A_{i,j}]\!] + \epsilon_{ijk} [\![\mathscr{M}_k]\!] \Big) \delta A_i n_j \, \mathrm{d}a - \\
& - \int_{\partial\Omega} \Big(\frac{1}{\mu_0} A_{i,j} + \epsilon_{ijk} \mathscr{M}_k \Big) \delta A_i n_j \, \mathrm{d}a \,.
\end{aligned}
\tag{29.35}
$$

Therefore, the weak form for computing the magnetic potential reads:

$$
\begin{aligned}
\mathrm{F}_{\boldsymbol{A}} = & \int_\Omega \Big(\varepsilon_0 \frac{A_i - 2A_i^0 + A_i^{00}}{\mathrm{d}t\mathrm{d}t} \delta A_i + \frac{1}{\mu_0} A_{i,j} \delta A_{i,j} - J_i^{\mathrm{fr.}} \delta A_i - \frac{P_i - P_i^0}{\mathrm{d}t} \delta A_i + \\
& + \epsilon_{ijk} \mathscr{M}_k \delta A_{i,j} \Big) \mathrm{d}v - \int_{\partial\Omega^I} \Big(n_j \frac{1}{\mu_0} [\![A_{j,i}]\!] - j_\mathrm{s}^{\mathrm{fr.}} \Big) \delta A_i \, \mathrm{d}a - \\
& - \int_{\partial\Omega} \Big(\frac{1}{\mu_0} A_{i,j} + \epsilon_{ijk} \mathscr{M}_k \Big) \delta A_i n_j \, \mathrm{d}a \,.
\end{aligned}
\tag{29.36}
$$

From the balance of linear momentum, by stressing that across the interface the jump conditions vanish, we acquire the following weak form in the unit of energy:

$$
\begin{aligned}
\mathrm{F}_{\boldsymbol{u}} = & \int_\Omega \Big(\rho_0 \frac{u_i - 2u_i^0 + u_i^{00}}{\mathrm{d}t\mathrm{d}t} \delta u_i + \frac{\epsilon_{ijk}(\mathfrak{D}_j B_k - \mathfrak{D}_j^0 B_k^0)}{\mathrm{d}t} \delta u_i - \sigma_{ji,j}^{\mathrm{tot.}} \delta u_i - \\
& - \rho_0 f_i \delta u_i \Big) \mathrm{d}v - \int_{\partial\Omega} (\hat{t}_i - N_j \sigma_{ji}) \delta u_i \, \mathrm{d}a \,.
\end{aligned}
\tag{29.37}
$$

The traction vector, $\hat{t}_i = N_j \hat{\sigma}^{\text{tot.}}_{ji}$, is given on the boundaries. When the boundary is a free surface, no traction occurs, $\hat{\boldsymbol{t}} = 0$. Alternatively there is a load measured in Pa and stated by the given traction vector, $\hat{\boldsymbol{t}}$, with its direction. This loading via the traction on the surface is incorporating the mechanical and electromagnetic stress together. Therefore, under the same traction vector different combinations might occur where a high mechanical stress is compensated by an electromagnetic stress, or otherwise. The electromagnetic stress is generated by the electric and magnetic potentials computed by their corresponding weak forms. By using the weak form:

$$\text{Form} = \text{F}_\phi + \text{F}_{\boldsymbol{A}} + \text{F}_{\boldsymbol{u}},$$

we solve $\{\phi, \boldsymbol{A}, \boldsymbol{u}\}$ in all nodes of the domain.

29.5 Implementation and Results

Weak forms are a set of nonlinear equations. We solve them by using FEniCS packages[1]. It is possible to use standard FEM elements. They are of order q and span $\mathcal{P}_q$ on Ω^{E}. Shape functions are piecewise (elementwise) continuous over the whole domain with a compact support on the element such that nodal values have influence only locally that ensures a monotonic convergence [15]. This representation is adequate for approximation in Hilbertian Sobolev space $\mathcal{H}^1$. Linear Lagrange elements are used for $\{\phi, \boldsymbol{A}, \boldsymbol{u}\}$, with an interpolation polynomial of first order as shape functions for each element, Ω^{E}, in scalar and vector spaces and calculated on each node 7 values,

$$\mathcal{V} = \left\{ \{\phi, \boldsymbol{A}, \boldsymbol{u}\} \in \mathcal{H}^1_7(\Omega) : \{\phi, A_1, A_2, A_3, u_1, u_2, u_3\}\Big|_{\Omega^{\text{E}}} \in \mathcal{P}_1(\Omega^{\text{E}}) \,, \forall \Omega^{\text{E}} \in \mathcal{T} \right\}. \tag{29.38}$$

We emphasize that we solve the electromagnetic potentials instead of electromagnetic fields and employ the Lorenz gauge. As a consequence, no numerical problems arise even in the case of using standard elements and the accuracy is adequate as verified in [16].

We construct the C2220 capacitor in Fig. 29.1. The geometry consists of two different parts, dielectric core and metal caps. For generating the geometry and meshing, i.e., for preprocessing, we use Salome and obtain the model in Fig. 29.2. Two metal caps are modeled as copper and the dielectric between two metal plates is made of a technical ceramics. We stress that we use the imaginary parts of susceptibility measurement in an experiment denoting its dissipative character. We estimate the material parameters of technical ceramics as follows:

$$\begin{aligned} \rho_0 = 7500\,\text{kg/m}^3 \,,\ \nu = 0.31 \,,\ \chi'_{\text{el.}} = 3400 \,,\ \chi''_{\text{el.}} = 64 \,, \\ S_{11} = 15.6 \cdot 10^{-12}\,\text{m}^2/\text{N} \,,\ S_{33} = 20 \cdot 10^{-12}\,\text{m}^2/\text{N} \,. \end{aligned} \tag{29.39}$$

[1] The FEniCS computing platform, https://fenicsproject.org/

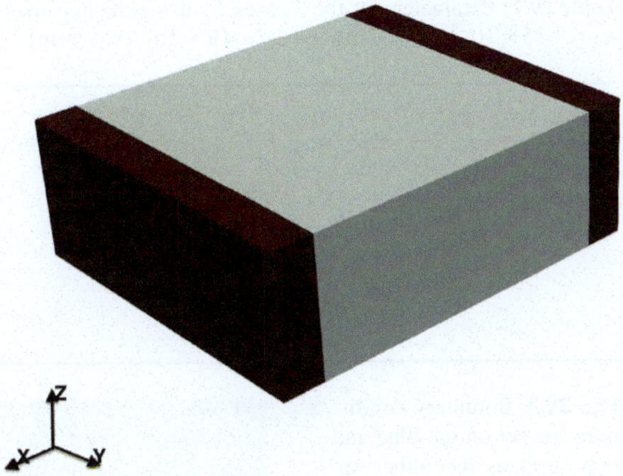

Fig. 29.2 The geometry is a capacitor with two copper plates (dark brown) on both sides of the PZT-5H core (light gray).

The compliance matrix, S_{IJ} , in the VOIGT notation reads:

$$S_{IJ} = \begin{pmatrix} S_{11} & -\nu S_{11} & -\nu S_{11} & 0 & 0 & 0 \\ -\nu S_{11} & S_{11} & -\nu S_{11} & 0 & 0 & 0 \\ -\nu S_{11} & -\nu S_{11} & S_{33} & 0 & 0 & 0 \\ 0 & 0 & 0 & (1+\nu)S_{11} & 0 & 0 \\ 0 & 0 & 0 & 0 & (1+\nu)S_{11} & 0 \\ 0 & 0 & 0 & 0 & 0 & (1+\nu)S_{11} \end{pmatrix} . \tag{29.40}$$

The inverse of the compliance matrix is the stiffness matrix in VOIGT notation reads:

$$(S_{JI})^{-1} = C_{IJ} = \begin{pmatrix} C_{1111} & C_{1122} & C_{1133} & C_{1123} & C_{1113} & C_{1112} \\ C_{2211} & C_{2222} & C_{2233} & C_{2223} & C_{2213} & C_{2212} \\ C_{3311} & C_{3322} & C_{3333} & C_{3323} & C_{3313} & C_{3312} \\ C_{2311} & C_{2322} & C_{2333} & C_{2323} & C_{2313} & C_{2312} \\ C_{1311} & C_{1322} & C_{1333} & C_{1323} & C_{1313} & C_{1312} \\ C_{1211} & C_{1222} & C_{1233} & C_{1223} & C_{1213} & C_{1212} \end{pmatrix} . \tag{29.41}$$

Choosing $\omega = 10\,\text{Hz}$, $1\,\text{kHz}$, and $500\,\text{kHz}$, we calculate the 3 cases as shown in Table 29.1 in order to demonstrate the effect of dissipation under different frequencies. Referring to Fig. 29.3, we choose the time-varying electric potential boundary condition $\phi = 100\sin(\omega t)$ in V on the blue surface and 0 V (grounding) on gray surface; these are indeed both metal caps acting as electrodes. For the displacement, we only limit the displacement $\boldsymbol{u} = (0,0,0)$ on the gray surface and the others are free for the sake of a simpler visualization. In a technical setting, both caps may be soldered on a board. The dielectric core is sandwiched between two caps and we simulate one side of the electrode clamped and grounded. We apply a periodically varying voltage

Table 29.1: Parameters for the 3 cases, by using the universal constants, $\varepsilon_0 = 8.85 \times 10^{-12}$ A s/(V m) and $\mu_0 = 4\pi \times 10^{-7}$ V s/(A m).

$\chi'_{\text{el.}}$	$\chi''_{\text{el.}}$	ω
3200	64	10 Hz 1 kHz 100 kHz 200 kHz 500 kHz

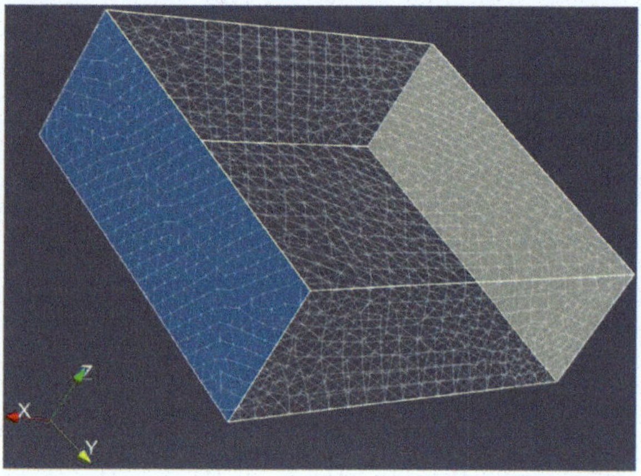

Fig. 29.3 Boundary conditions are set on the blue and grey surfaces. The other surfaces are shown as the light mesh without claiming any boundary conditions. The highlighted white lines are the outlines of the capacitor.

to the other electrode. Electric field creates an electric polarization. Electromagnetic fields and polarization generates an electromagnetic momentum and stress causing that the dielectric core is deformed. Such a deformation is small in size but we emphasize the presence of the imaginary part in Eq. (29.21). Hence, a dissipative term in electric polarization exists and is related to the frequency ω in typical measurements. We use a given frequency, ω, as an input such that the model is fairly simple yet showing a significant effect of the dissipation. By using the same susceptibility but simulating for different frequencies, displacement is correspondingly different. Due to the inherent frequency of the system, the frequency of the external excitation (the periodically varying voltage) affects the amplitude (maximum displacement) of the system response (vibration characteristics).

First, we present the displacement responses at 10 Hz, 1 kH, and 100 kHz in Fig. 29.4. Since 10 Hz is a low frequency, the inertia of the system has insignificant effect on the response, yet the phase shift of the displacement results is more obvious compared to 1 kHz, which is due to the dissipation term. For a simpler analysis, we assume that $\chi''_{\text{el.}}$ is constant yet we stress that for many materials this imaginary part of susceptibility is a function of frequency as well. By simplifying the system response with a constant susceptibility, the dissipative effect of $\chi''_{\text{el.}}$ looks to be more

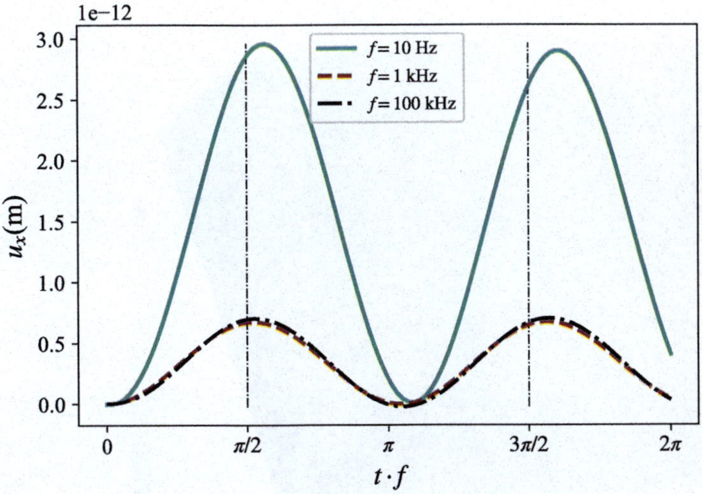

Fig. 29.4: Displacement curve in the x-direction at the midpoint of the blue surface in one circle. The teal solid line represents the result at 10 Hz. The maroon dashed line represents the result at 1 kHz. The black dashdotted line represents the result at 100 kHz.

significant at a low frequency than at a high frequency given by Eq. (29.23) and we stress that inertia plays an important role such that we solve the transient problem in time domain. The deformation results at 10 Hz and 1 kHz are shown in Fig. 29.5.

Second, we also compare the results at 100 kHz, 200 kHz, and 500 kHz shown in Fig. 29.6. When the frequency reaches this level, the increase in phase shift with increasing frequency is mainly due to system inertia. Also the geometry, and thus natural frequency of the system, plays a role that is already incorporated in a transient solution in the time domain.

29.6 Conclusion

Electromagnetism and mechanics are inherently coupled by two different phenomena. One well-known relation is at the material level where mechanical stress is caused by electric field called a piezoelectric effect. Another less known effect is the electromagnetic force, even its definition is part of the so-called ABRAHAM–MINKOWSKI debate or controversy. For a short review we refer to [17]. Herein we use MINKOWSKI definition and, motivated by thermodynamics, propose a dissipative electric polarization. Its direct consequence at the deformation is justified by the electromagnetic momentum and stress, results are presented by means of a numerical simulation with the finite element method. Such a dissipative characteristic in the deformation is well-known in electronics and often related to a quality factor. Of

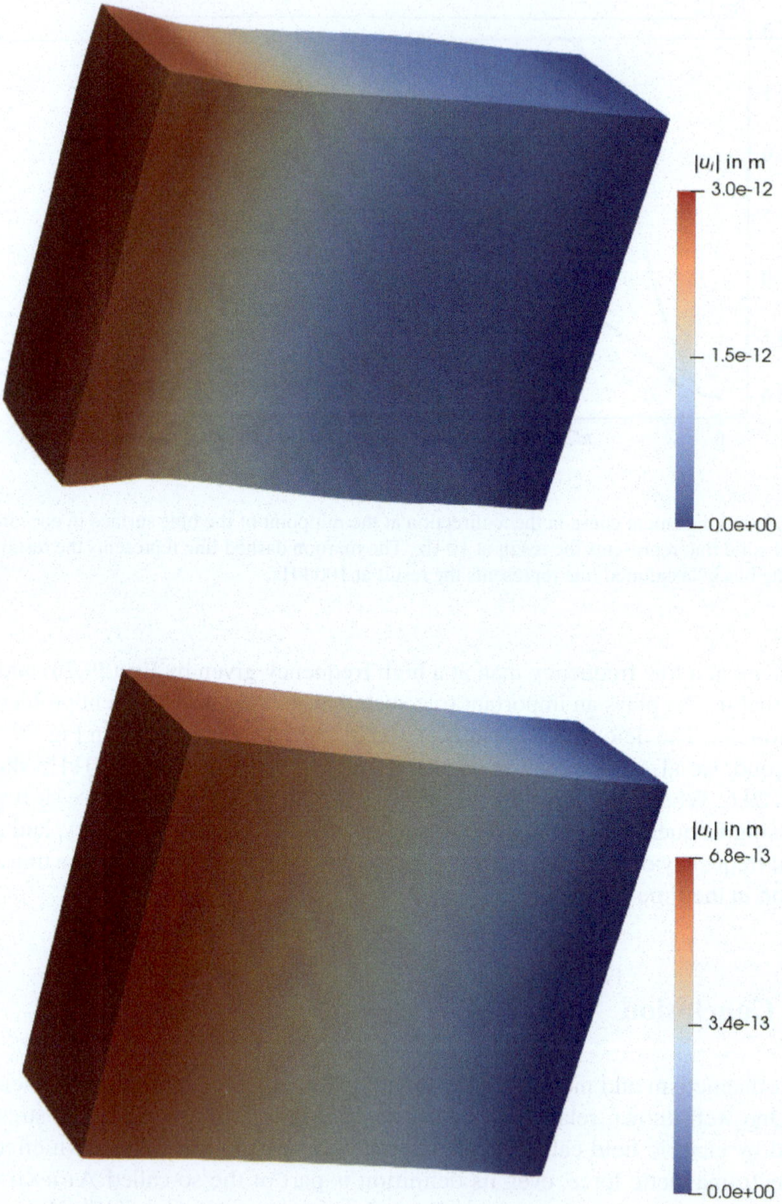

Fig. 29.5: The deformation of the capacitor at 10 Hz (top) and 1 kHz (bottom) at the first quarter of the first cycle with a scale factor of 1×10^8. Colors indicate the norm of displacement in m.

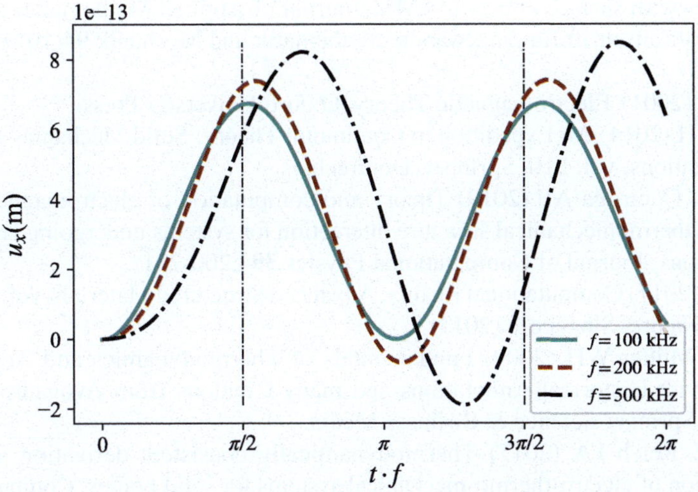

Fig. 29.6: The teal solid line represents the result at 100 kHz. The maroon dashed line represents the result at 200 kHz. The black dashdotted line represents the result at 500 kHz.

course, less dissipation is more favorable such that research is based on amending material properties. Yet we consider such an analysis also useful for understanding which electromagnetic force model is more adequate.

References

[1] Abali BE, Zohdi TI (2018) On the accuracy of reduced-order integrated circuit simulators for computing the heat production on electronic components, Journal of Computational Electronics **17**(2), 625-636.

[2] Kirchhoff GR (1879) Zur Theorie des Condensators, Monatsberichte der Königlich Preußischen Akademie der Wissenschaften zu Berlin, pp 48-50.

[3] Hutson V (1963) The circular plate condenser at small separations, Mathematical Proceedings of the Cambridge Philosophical Society, **59**(1):211-224.

[4] Wintle H, Kurylowicz S (1985) Edge Corrections for Strip and Disc Capacitors, IEEE Transactions on Instrumentation and Measurement, **IM-34**(1):41-47.

[5] Milovanović GV, Joksimović D. (2013) Properties of Boubaker polynomials and an application to Love's integral equation, Applied Mathematics and Computation **224**:74-87

[6] Nasedkin AV, Eremeyev VA (2013) Spectral properties of piezoelectric bodies with surface effects, In: Altenbach H, Morozov N (Eds) Surface Effects in Solid Mechanics, Advanced Structured Materials, Vol 30, Springer, Berlin, Heidelberg, pp 105-121.

[7] Nasedkin AV, Eremeyev VA (2014) Harmonic vibrations of nanosized piezoelectric bodies with surface effects, ZAMM-Journal of Applied Mathematics and Mechanics/Zeitschrift für Angewandte Mathematik und Mechanik **94**(10):878-892.
[8] Kovetz A (2001) Electromagnetic Theory, Oxford University Press.
[9] Müller WH (2014) An Expedition to Continuum Theory, Solid Mechanics and Its Applications, vol. 210, Springer, Dordrecht.
[10] Abali BE, Queiruga AF (2019) Theory and computation of electromagnetic fields and thermomechanical structure interaction for systems undergoing large deformations, Journal of Computational Physics **394**:200-231.
[11] Abali BE (2017) Computational Reality, Advanced Structured Materials, vol. 55, Springer Nature, Singapore (2017)
[12] Müller I, Müller WH (2009) Fundamentals of Thermodynamics and Applications: with Historical Annotations and many Citations from Avogadro to Zermelo, Springer Science & Business Media.
[13] Abali BE, Reich FA (2017) Thermodynamically consistent derivation and computation of electro-thermo-mechanical systems for solid bodies, Computer Methods in Applied Mechanics and Engineering **319**:567-595.
[14] Abali BE, Wu CC, Müller WH (2016) An energy-based method to determine material constants in nonlinear rheology with applications, Continuum Mechanics and Thermodynamics **28**(5):1221-1246.
[15] Zohdi TI (2018) Finite Element Primer for Beginners, SpringerBriefs in Applied Sciences and Technology, Springer, Cham.
[16] Abali BE, Reich FA (2020) Verification of deforming polarized structure computation by using a closed-form solution, Continuum Mechanics and Thermodynamics **32**(3):693-708.
[17] Reich FA, Rickert W, Müller WH (2018) An investigation into electromagnetic force models: differences in global and local effects demonstrated by selected problems, Continuum Mechanics and Thermodynamics **30**(2):233-266.

Chapter 30
On Possible Reduction of Gradient Theories of Elasticity

Sergey A. Lurie, Petr A. Belov, and Yury O. Solyaev

Abstract Mindlin–Tupin gradient theories of elasticity are considered, and a procedure is proposed to simplify general models with five moduli of elasticity to a gradient two-parameter model. It is shown that the simplified models are related to gradient models of the vector type, for which the quadratic form of the potential energy density is a convolution of vectors, and the number of physical additional parameters corresponds to the number of scaling characteristics in the system of resolving equations. Such models, called "vector" models, provide a classic view of static boundary conditions. For vector models, despite the fourth order of the resolving equations, there are no boundary conditions on the surface edges.

Key words: Gradient theories · Gradient elastic modulus tensor · Basis tensors · Divergent components · Vector gradient models

30.1 Introduction

Currently, there is an active development of gradient theories and their ever wider application in various applied problems, including the problems of estimating the effective properties of nanocomposites, nanofluids, the strength of bodies, the scale-dependent properties of ultrathin plates, rods, weakened by cracks, etc. At the same time, fundamental issues of constructing of these theories [1]- [7], and the problem of correct formulation of equilibrium equations and boundary conditions [8]- [14], as well as the problem of constructing applied models taking into account the features of gradient elasticity theories, for example, the requirement of symmetry conditions [15]-[18], etc. are discussed.

Sergey A. Lurie · Petr A. Belov · Yury O. Solyaev
Institute of Applied Mechanics RAS, Moscow, Russian Federation,
e-mail: salurie@mail.ru, belovpa@yandex.ru, yos@iam.ras.ru

H. Altenbach et al. (eds.), *Sixty Shades of Generalized Continua*,
Advanced Structured Materials 170, https://doi.org/10.1007/978-3-031-26186-2_30

In applications, there are used the simplified versions of the gradient theory of elasticity, containing a smaller number of additional parameters, the determination of which requires the use of specific experimental approaches [19]-[21] or molecular dynamics modeling [22]-[24]. Usually, simplified models are used, containing two parameters [16] instead of five [2, 3] or a single additional scale parameter [4]-[6], [10]-[12]. It was made an attempt to classify simplified models of the gradient theory of elasticity [10, 15, 25]. The issue of building simplified gradient elasticity models is very relevant, because it is directly related to applications, the physical interpretation of gradient elastic moduli, and the problem of identifying additional parameters.

In this paper, we discuss a new procedure for constructing the gradient theory of elasticity, which leads to significant simplifications of gradient theories and makes it possible to write static boundary conditions and equilibrium equations in terms of the same generalized stress components. Such stresses are proposed to be called semiclassical, because the equilibrium equations represent the divergence from these stresses, and the boundary conditions represent their convolution with the unit normal vector to the surface of the body. The class of gradient models considered in the paper is attractive from a practical point of view, since they eliminate the problem of boundary conditions on surface edges, which is a source of significant problems in theoretical and numerical modeling within the framework of gradient elasticity theory.

30.2 Variational Formulation of Gradient Theories

Gradient theories of elasticity are characterized by the fact that the arguments of the density of potential energy U are not only the components of the distortion tensors $d_{ij} = R_{i,j}$ (in the general case), but also their derivatives – the components of the tensors of the third rank $\kappa_{ijk} = R_{i,jk}$. It is assumed that the work A performed on the displacement vector R_i by external volumetric P_i^V and surface P_i^F forces, given respectively in the volume of the body V and on the surface of the body F, has a classical form. Note that it is possible to formally expand the expressions for the work of external force factors due to the work of moment force factors on normal derivatives of the displacement vector. However, such an extension is usually not given due to the fact that in reality it is difficult to substantiate and implement the assignment of moment force factors.

For linear gradient elasticity theories, with physical properties determined by fourth C_{ijlm} and sixth rank C_{ijklmn} tensor, the Lagrangian L can be written as:

$$
\begin{aligned}
&L = A - U, \\
&A = \int\limits_V P_i^V R_i \, dV + \int\limits_F P_i^F R_i \, dF, \\
&U = \int\limits_V U^V \, dV = \frac{1}{2} \int\limits_V (C_{ijlm} R_{i,j} R_{l,m} + C_{ijklmn} R_{i,jk} R_{l,mn}) \, dV.
\end{aligned}
\tag{30.1}
$$

The force model is introduced by determining the Cauchy stress tensor σ_{ij} and the couple stress tensor σ_{ijk} in the volume of the body. Therefore, the expression for the strain energy density can be written as:

$$U = \int_V U^V \, dV = \frac{1}{2}\int_V (\sigma_{lm} R_{l,m} + \sigma_{lmn} R_{l,mn})\, dV. \tag{30.2}$$

During the solving in displacements for linear elasticity, the defining relations are completely determined by the given quadratic form of distortions (or deformations) for the potential energy density. The physical (defining) equations of the material model are established by Green's formulas:

$$\sigma_{ij} = \frac{\partial U^V}{\partial R_{i,j}} = C_{ijlm} R_{l,m}, \qquad \sigma_{ijk} = \frac{\partial U^V}{\partial R_{i,jk}} = C_{ijklmn} R_{l,mn}. \tag{30.3}$$

The mathematical formulation of gradient elasticity is completely formulated using the Lagrange principle (30.1)–(30.3). The variational equation, which is a consequence of the stationarity condition for the Lagrange functional, has the following form:

$$\begin{aligned}
\delta L &= \int_V P_i^V \delta R_i \, dV + \int_F P_i^F \delta R_i \, dF - \int_V (C_{ijlm} R_{l,m} \delta R_{i,j} + C_{ijklmn} R_{l,mn} \delta R_{i,jk})\, dV \\
&= \int_V P_i^V \delta R_i \, dV + \int_F P_i^F \delta R_i \, dF - \int_V (C_{ijlm} R_{l,m} - C_{ijklmn} R_{l,mnk}) \delta R_{i,j} \, dV \\
&- \int_F C_{ijklmn} R_{l,mn} \delta R_{i,j} n_k \, dF = \int_V (C_{ijlm} R_{l,mj} - C_{ijklmn} R_{l,mnkj} + P_i^V) \delta R_i \, dV \\
&+ \int_F [P_i^F - (C_{ijlm} R_{l,m} - C_{ijklmn} R_{l,mnk}) n_j] \delta R_i \, dF \\
&- \int_F C_{ijklmn} n_k R_{l,mn} \delta R_{i,r} \delta^*_{rj} \, dF - \int_F C_{ijklmn} n_k R_{l,mn} \delta R_{i,r} n_r n_j \, dF \\
&= \int_V (C_{ijlm} R_{l,mj} - C_{ijklmn} R_{l,mnkj} + P_i^V) \delta R_i \, dV \\
&+ \int_F [P_i^F - (C_{ijlm} R_{l,m} - C_{ijklmn} R_{l,mnk}) n_j \\
&+ (C_{ijklmn} n_k R_{l,mn})_{,r} \delta^*_{rj} + 2H C_{ijklmn} R_{l,mn} n_k n_j] \delta R_i \, dF \\
&- \int_F C_{ijklmn} n_j n_k R_{l,mn} \delta(R_{i,r} n_r) \, dF \sum \oint C_{ijklmn} \upsilon_j n_k R_{l,mn} \delta R_i ds = 0,
\end{aligned} \tag{30.4}$$

where $\upsilon_i = n_j s_k e_{ijk}$, $\boldsymbol{\nu} = \{\nu_i\}$, υ_i is the unit vector of the co-normal to the edge, belongs to the tangent plane to the smooth surface bounded by the contour of the edges, s_k is the unit vector of the tangent to the edge, e_{ijk} is the Levi-Civita tensor, $H = -(1/2)(n_i)_{,j}\delta^*_{ij}$, $\delta^*_{ij} = \delta_{ij} - n_i n_j$. In variational equation (30.4), obtained using integration by parts, the contour integral is taken over all edges S_k of closed, piecewise-smooth surface of the body F with the normal $\boldsymbol{n} = \{n_j\}$.

In general, for non-linear anisotropic elasticity, the variational equation can be written in terms of stresses directly from the principle of possible displacements by integrating by parts:

$$\begin{aligned}
\delta L = \delta A - \int\limits_V (\sigma_{ij} - \sigma_{ijk,k})\delta R_{i,j}\, dV - \int\limits_F \sigma_{ijk} n_k \delta R_{i,j}\, dF \\
= \int\limits_V (\sigma_{ij,j} - \sigma_{ijk,jk} + P_i^V)\delta R_i\, dV \\
+ \int\limits_F [P_i^F - (\sigma_{ij} - \sigma_{ijk,k})n_j + (\sigma_{ijk} n_k)_{,j} - (\sigma_{ijk} n_k)_{,p} n_p n_j + 2H\sigma_{ijk} n_k n_j]\delta R_i\, dF \\
- \int\limits_F (\sigma_{ijk} n_k) n_j \delta(R_{i,p} n_p)\, dF - \sum_k \oint\limits_{S_k} (\sigma_{ijk} n_k)\upsilon_j \delta R_i ds = 0,
\end{aligned} \tag{30.5}$$

where $H = (1/2)\nabla_{Sj} n_j$, $\nabla_{Sj}(\cdot) = (\delta_{jk} - n_j n_k)(\cdot)_{,k}$. Equality (30.5) shows that in gradient elasticity the equilibrium equations (Euler) are written in divergent form with respect to the stresses $\tau_{ij} = \sigma_{ij} - \sigma_{ijk}$, which we will call semiclassical. At the same time, the generalized equations of Hooke's law (30.3) and the boundary conditions are generally written in terms of total stresses $\sigma_{ij}$and couple stresses σ_{ijk}. The static boundary conditions in stresses and in displacements, respectively, are determined by factors for variations in the displacement vector in surface integrals (30.5) and (30.4).

30.3 Structure of Sixth Rank Tensors

Let us consider the tensor of gradient properties of the considered body, focusing on the structure of the basis tensors. In the general case, the tensor C_{ijklmn} has the structure of a linear combination of sixth-rank basis tensors composed of tensor products of three Kronecker tensors with all possible index permutations

$$\begin{aligned}
C_{ijklmn} = a_1\delta_{ij}\delta_{kl}\delta_{mn} + a_2\delta_{ij}\delta_{km}\delta_{ln} + a_3\delta_{ij}\delta_{kn}\delta_{lm} + a_4\delta_{ik}\delta_{jl}\delta_{mn} \\
+ a_5\delta_{ik}\delta_{jm}\delta_{ln} + a_6\delta_{ik}\delta_{jn}\delta_{lm} + a_7\delta_{il}\delta_{jk}\delta_{mn} + a_8\delta_{il}\delta_{km}\delta_{jn} \\
+ a_9\delta_{il}\delta_{kn}\delta_{jm} + a_{10}\delta_{im}\delta_{kl}\delta_{jn} + a_{11}\delta_{im}\delta_{kj}\delta_{ln} + a_{12}\delta_{im}\delta_{kn}\delta_{lj} \\
+ a_{13}\delta_{in}\delta_{kl}\delta_{mj} + a_{14}\delta_{in}\delta_{km}\delta_{lj} + a_{15}\delta_{in}\delta_{kj}\delta_{lm}.
\end{aligned} \tag{30.6}$$

In the general case, for an isotropic body, the number of basis tensors and corresponding modules in expansion (30.6) is fifteen [1]-[3]. The condition for the existence of a potential energy density in the gradient theory or, in other words, the requirement for the reversibility of deformation processes leads to a reduction in the number of modules to eleven

$$\begin{aligned}C_{ijklmn} &= a_1(\delta_{ij}\delta_{kl}\delta_{mn}+\delta_{lm}\delta_{ni}\delta_{jk})+a_2(\delta_{ij}\delta_{km}\delta_{ln}+\delta_{lm}\delta_{nj}\delta_{ik})\\&+a_3\delta_{ij}\delta_{kn}\delta_{lm}+a_4(\delta_{ik}\delta_{jl}\delta_{mn}+\delta_{ln}\delta_{mi}\delta_{jk})\\&+a_5\delta_{ik}\delta_{jm}\delta_{ln}+a_7\delta_{il}\delta_{jk}\delta_{mn}+a_8\delta_{il}\delta_{km}\delta_{jn}+a_9\delta_{il}\delta_{kn}\delta_{jm}\\&+a_{10}(\delta_{im}\delta_{kl}\delta_{jn}+\delta_{lj}\delta_{ni}\delta_{mk})+a_{12}\delta_{im}\delta_{kn}\delta_{lj}+a_{13}\delta_{in}\delta_{kl}\delta_{mj}.\end{aligned} \tag{30.7}$$

Let us take into account that when we determine the moment stress tensor

$$\sigma_{ijk}=\frac{\partial U^V}{\partial R_{i,jk}}=C_{ijklmn}R_{l,mn},$$

the tensor of the gradient elastic moduli $C_{ijklmn}$is included in the convolution with the curvature tensor $R_{l,mn}$, which is symmetric in differentiation indices. The symmetry of the second derivatives of the displacement vector with respect to the order of differentiation is a necessary and sufficient condition for the existence of continuous distortions. Therefore, the tensor C_{ijklmn} in (30.7) must additionally have the property of symmetry when the last indices in each of the triples (differentiation indices) are permuted. Accounting for this additional symmetry property $C_{ijklmn}=C_{ijklnm}=C_{ikjlmn}$ in (30.7) leads to additional connections between modules and reduces the number of independent basis tensors to five. As a result, the sixth rank elastic modulus tensor C_{ijklmn} is determined by five physical constants a_1, a_2, a_7, a_8, a_{11} :

$$\begin{aligned}C_{ijklmn} &= a_1(\delta_{ij}\delta_{kl}\delta_{mn}+\delta_{lm}\delta_{ni}\delta_{jk}+\delta_{ik}\delta_{jl}\delta_{mn}+\delta_{ln}\delta_{mi}\delta_{jk})\\&+a_2(\delta_{ij}\delta_{km}\delta_{ln}+\delta_{lm}\delta_{nj}\delta_{ik}+\delta_{ij}\delta_{kn}\delta_{lm}+\delta_{ik}\delta_{jm}\delta_{ln})\\&+a_7\delta_{il}\delta_{jk}\delta_{mn}+a_8(\delta_{il}\delta_{km}\delta_{jn}+\delta_{il}\delta_{kn}\delta_{jm})\\&+a_{10}(\delta_{im}\delta_{kl}\delta_{jn}+\delta_{lj}\delta_{ni}\delta_{mk}+\delta_{im}\delta_{kn}\delta_{lj}+\delta_{in}\delta_{kl}\delta_{mj}).\end{aligned} \tag{30.8}$$

We note a feature of the structure of the five basis tensors of the sixth rank, which, as will be shown below, is also related to the feature of the physical relationships of gradient models as a whole. The basis tensors in (30.8) have structures of two types. The first three basis tensors in (30.8) will be referred to as the basis tensors of the first type:

$$\begin{aligned}&(\delta_{ij}\delta_{kl}\delta_{mn}+\delta_{lm}\delta_{ni}\delta_{jk}+\delta_{ik}\delta_{jl}\delta_{mn}+\delta_{ln}\delta_{mi}\delta_{jk}),\\&(\delta_{ij}\delta_{km}\delta_{ln}+\delta_{lm}\delta_{nj}\delta_{ik}+\delta_{ij}\delta_{kn}\delta_{lm}+\delta_{ik}\delta_{jm}\delta_{ln}),\\&\delta_{il}\delta_{jk}\delta_{mn}.\end{aligned} \tag{30.9}$$

They are formed in such a way that in each term in the basis tensors (30.9) there are two Kronecker tensors with two indices, respectively, from the first and second triples of indices C_{ijklmn}, and the third Kronecker tensor contains one index from the first and second triples of indices.

To the second type of basis tensors we refer the remaining two basis tensors (fourth and fifth) in (30.8)

$$(\delta_{il}\delta_{km}\delta_{jn}+\delta_{il}\delta_{kn}\delta_{jm}),\ (\delta_{im}\delta_{kl}\delta_{jn}+\delta_{lj}\delta_{ni}\delta_{mk}+\delta_{im}\delta_{kn}\delta_{lj}+\delta_{in}\delta_{kl}\delta_{mj}) \quad (30.10)$$

They have another structure. In them, in each of the terms, the Kronecker tensors have one index from each triple.

Let us show that the structure of basis tensors of the first type (30.9) is reduced to the form of convolutions with respect to one index (with respect to the index $a$in the cases under consideration) of two tensors of the fourth rank. To do this, we represent each Kronecker tensor which has indices from different triples as a convolution of two Kronecker tensors by one index, for example: $\delta_{kl}=\delta_{ka}\delta_{la}$. Let us transform the structures of each basic tensor of the first type (30.9) separately. We have, respectively, the following expressions for the first basis tensor in (30.9):

$$\begin{aligned}&(\delta_{ij}\delta_{kl}\delta_{mn}+\delta_{lm}\delta_{ni}\delta_{jk}+\delta_{ik}\delta_{jl}\delta_{mn}+\delta_{ln}\delta_{mi}\delta_{jk})\\&\quad=(\delta_{ij}\delta_{ka}+\delta_{ik}\delta_{ja}-\delta_{jk}\delta_{ia}2/3)\delta_{mn}\delta_{la}\\&\quad+(\delta_{lm}\delta_{na}+\delta_{ln}\delta_{ma}-\delta_{mn}\delta_{la}2/3)\delta_{jk}\delta_{ia}+(\delta_{jk}\delta_{ia})(\delta_{mn}\delta_{la})4/3,\end{aligned} \quad (30.11)$$

for the second basis tensor in (30.9):

$$\begin{aligned}&(\delta_{ij}\delta_{km}\delta_{ln}+\delta_{lm}\delta_{nj}\delta_{ik}+\delta_{ij}\delta_{kn}\delta_{lm}+\delta_{ik}\delta_{jm}\delta_{ln})\\&\quad=(\delta_{ij}\delta_{ka}+\delta_{ik}\delta_{ja}-\delta_{jk}\delta_{ia}2/3)(\delta_{lm}\delta_{na}+\delta_{ln}\delta_{ma}-\delta_{mn}\delta_{la}2/3)\\&\quad+(\delta_{lm}\delta_{na}+\delta_{ln}\delta_{ma}-\delta_{mn}\delta_{la}2/3)\delta_{jk}\delta_{ia}2/3\\&\quad+(\delta_{ij}\delta_{ka}+\delta_{ik}\delta_{ja}-\delta_{jk}\delta_{ia}2/3)\delta_{mn}\delta_{la}2/3+\delta_{jk}\delta_{ia}\delta_{mn}\delta_{la}4/9\end{aligned} \quad (30.12)$$

for the third basis tensor in (30.9):

$$\delta_{il}\delta_{jk}\delta_{mn}=(\delta_{jk}\delta_{ia})(\delta_{la}\delta_{mn}). \quad (30.13)$$

Let us substitute the obtained relations for basis tensors (30.11)–(30.13) into (30.8) and write down the gradient part of the volume potential energy density:

$$\begin{aligned}&C_{ijklmn}R_{i,jk}R_{l,mn}=C_1R_{i,ia}R_{j,ja}+C_2R_{i,ia}(\Delta R_a-R_{j,ja}) \quad (30.14)\\&\quad+C_3(\Delta R_a-R_{i,ia})(\Delta R_a-R_{j,ja})+a_8R_{i,jk}R_{i,jk}+2a_{10}R_{i,jk}(R_{k,ij}+R_{j,ki}).\end{aligned}$$

Thus, it is shown that the basis tensors of the first type generate a special kind of potential energy density, which is a convolution of two vectors $R_{i,ia}$, $(\Delta R_a-R_{j,ja})$. Note that in this case the vector field formed by the second derivatives of displacements and determining the potential energy density is represented as a potential and vortex components. The first vector $R_{i,ia}$, whose rotor is equal to zero, determines the potential field, the second vector $(\Delta R_a-R_{j,ja})$ characterizes the vortex field, since its divergence is zero.

As a result, we can introduce into consideration three other gradient elastic moduli C_1, C_2, C_3, which are linear combinations of the original three gradient moduli in

(30.8) for the basis tensors of the first type:

$$C_1 = (4a_2 + 4a_1 + a_7), \qquad C_2 = (4a_1 + 2a_7), \qquad C_3 = a_7. \tag{30.15}$$

The inverse relations are also satisfied

$$a_1 = C_2/4 - C_3/2, \qquad a_2 = (C_1 - C_2 + C_3)/4, \qquad a_7 = C_3. \tag{30.16}$$

Taking into account (30.15), (30.16), we obtain the following form of the potential energy density formed by the basis tensors of the first type:

$$\begin{aligned} U^V &= (4a_2 + 4a_1 + a_7)R_{i,ia}R_{j,ja} + a_7(\Delta R_a - R_{i,ia})(\Delta R_a - R_{j,ja}) \\ &+ (4a_1 + 2a_7)R_{i,ia}(\Delta R_a - R_{j,ja}) \\ &= C_1 R_{i,ia}R_{j,ja} + C_2 R_{i,ia}(\Delta R_a - R_{j,ja}) + C_3(\Delta R_a - R_{i,ia})(\Delta R_a - R_{j,ja}). \end{aligned} \tag{30.17}$$

There are other options for writing a quadratic form of the first type with vectors $R_{i,ia}$, ΔR_a or with vectors ΔR_a, $(\Delta R_a - R_{j,ja})$. However, it seems that quadratic form (30.17) is the most optimal, since the first and third terms are positive definite quadratic forms, and the term at a constant C_2 is completely converted into the potential energy of the surface.

30.4 Special Structure of Gradient Part of Potential Energy. Hypothesis on Absence of Divergent Terms

Let us write the gradient part of the potential energy, taking into account (30.14)–(30.17):

$$\begin{aligned} U &= \frac{1}{2}\int_V C_{ijklmn} R_{i,jk} R_{l,mn}\, dV \\ &= \frac{1}{2}\int_V [C_1 R_{i,ia}R_{j,ja} + C_2 R_{i,ia}(\Delta R_a - R_{j,ja}) + C_3(\Delta R_a - R_{i,ia})(\Delta R_a - R_{j,ja}) \\ &+ a_8 R_{i,jk}R_{i,jk} + 2a_{10} R_{i,jk}(R_{k,ij} + R_{j,ki})]\, dV. \end{aligned} \tag{30.18}$$

In expression for potential energy (30.18), the tensors of the first type form a quadratic form of two vectors $(\Delta R_a - R_{i,ia})$ and ΔR_a. If we consider a particular gradient model, assuming the modules a_8 and a_{10} equal to zero, then the gradient part of the potential energy (30.18) can be reduced to the quadratic canonical form of two vectors, which makes it possible to attribute such a particular model to the class of applied "vector" gradient elasticity theories [26]. Let us present additional considerations that allow us to consider always general model (30.18) as a vector one.

The last two terms in (30.18) correspond to the fourth and fifth basis vectors (30.10) and have a structure of the second type. Let us consider them separately and transform the potential energy determined by the modules a_8 and a_{10} so that the corresponding volumetric potential energy density is also written in terms of the vectors $(\Delta R_a - R_{i,ia})$ and $R_{i,ia}$. Using the procedure of integration by parts, we obtain the following equality for the term with modulus a_8:

$$\int_V R_{i,jk} R_{i,jk}\, dV$$
$$= \int_V [(\Delta R_i - R_{j,ji})(\Delta R_i - R_{k,ki}) + 2R_{j,ji}(\Delta R_i - R_{k,ki}) + R_{j,ji} R_{k,ki}]\, dV$$
$$+ \int_F (R_{i,jk} R_{i,j} - \Delta R_i R_{i,k}) n_k\, dF. \tag{30.19}$$

Similarly, for the strain energy with modulus a_{10}, we obtain the following expression:

$$\int_V (R_{i,jk} R_{k,ij} + R_{i,jk} R_{j,ki})\, dV = \int_V [R_{i,ij}(\Delta R_j - R_{k,kj}) + 2R_{i,ij} R_{k,kj}]\, dV$$
$$+ \int_F (R_{i,j} R_{k,ij} + R_{i,j} R_{j,ki} - R_{i,k} R_{j,ji} - R_{k,j} \Delta R_j) n_k\, dF. \tag{30.20}$$

As a result, taking into account (30.19), (30.20), the gradient part of the potential energy (30.18) takes the form:

$$U = \frac{1}{2} \int_V [(C_1 + a_8 + 4a_{10}) R_{i,ia} R_{j,ja} + (C_2 + 2a_8 + 4a_{10}) R_{i,ia}(\Delta R_a - R_{j,ja})$$
$$+ (C_3 + a_8)(\Delta R_a - R_{i,ia})(\Delta R_a - R_{j,ja})]\, dV + \frac{1}{2} \int_F a_8 (R_{i,jk} R_{i,j} - \Delta R_i R_{i,k}) n_k$$
$$+ 2a_{10}(R_{i,j} R_{k,ij} + R_{i,j} R_{j,ki} - R_{i,k} R_{j,ji} - R_{k,j} \Delta R_j) n_k\, dF. \tag{30.21}$$

Relation (30.21) defines the gradient part of the potential energy as the sum of the integral over the volume of the quadratic form of two vectors $(\Delta R_a - R_{i,ia})$ and ΔR_a, and an additional part – the integral over the surface of the body, which can be interpreted as an integral of some surface potential energy density.

Let us formulate an additional assumption about the potential energy density on the surface arising in (30.21). We can see that the potential energy density on the surface of the body determines the adhesive activity of the surface. However, the surface density in (30.21) is determined not by the adhesion moduli, but by the bulk moduli of elasticity. In other words, if we follow the obtained result, then this potential energy of "adhesion" cannot be controlled by any physical and chemical influences on the surface of the body (because it is determined by volume modules), which does not correspond to experimental data.

In this regard, we believe that the assumption is fair: if, during the constructing the potential energy in the canonical form with respect to convolutions of vectors $(\Delta R_a - R_{i,ia})$ and $R_{i,ia}$ in the volume of the body, there are terms for which there can be distinguished the divergent terms, leading to the appearance of a surface energy density proportional to the elastic moduli that determine the physical properties medium in volume, then these physical characteristics should be taken equal to zero. The assumption made above requires that in the potential energy density defined in volume, the terms corresponding to the bilinear forms of tensor objects (derivatives of displacements of various orders) with scalar elastic moduli be taken equal to zero. Following this assumption, in the considered gradient model with potential energy (30.21) one should accept:

$$a_8 = 0, \qquad a_{10} = 0. \tag{30.22}$$

Similar considerations apply to the bilinear term in (30.17) with the modulus of elasticity C_2. Therefore, it should be accepted that

$$C_2 = 0. \tag{30.23}$$

Indeed, this bilinear term is completely "divergent" and, taking into account the property $(\Delta R_a - R_{j,ja})_{,a} = 0$, can be represented by the potential energy of "specific adhesion":

$$\begin{aligned}
&\int\limits_V [C_2 R_{i,ia}(\Delta R_a - R_{j,ja})]\, dV \\
&= \int\limits_F [C_2 R_{i,i} n_a(\Delta R_a - R_{j,ja})]\, dF - \int\limits_V [C_2 R_{i,i}(\Delta R_a - R_{j,ja})_{,a}]\, dV \\
&= \int\limits_F [C_2 R_{i,i} n_a(\Delta R_a - R_{j,ja})]\, dF.
\end{aligned} \tag{30.24}$$

Note that the potential energy of "specific adhesion" in (30.24) can be represented by the canonical non-positive definite quadratic form of two scalars on the surface of the body $R_{i,i}$ and $n_a(\Delta R_a - R_{i,ia})$ for any value of the modulus C_2 :

$$\begin{aligned}
&\int\limits_V [C_2 R_{i,ia}(\Delta R_a - R_{j,ja})]\, dV = \int\limits_F [C_2 R_{i,i} n_a(\Delta R_a - R_{j,ja})]\, dF \\
&= \frac{1}{2} \int\limits_F \{C_2[R_{i,i} + n_a(\Delta R_a - R_{i,ia})][R_{j,j} + n_b(\Delta R_b - R_{j,jb})] \\
&- C_2[R_{i,i} - n_a(\Delta R_a - R_{i,ia})][R_{j,j} - n_b(\Delta R_b - R_{j,jb})]\}\, dF.
\end{aligned}$$

This can serve as an additional argument for the validity of the general assumption made above. As a result, the gradient part of potential energy (30.21) acquires the following canonical form:

$$U = \int_V [C_1 R_{i,ia} R_{j,ja} + C_3(\Delta R_a - R_{i,ia})(\Delta R_a - R_{j,ja})]\, dV. \tag{30.25}$$

Remark 30.1. It is known that the expression for the potential energy density completely determines both the physical model of the medium and the mathematical model, i.e. system of resolving equations and boundary conditions. However, the question of constructing a physical model and setting a boundary value problem according to the type of potential energy density seems to be more complicated and is not unambiguous at all. The problem is related to the restoration of the sixth rank moduli tensor for gradient elasticity, which defines the physical model. This problem is solved ambiguously and is directly related to the symmetry properties in gradient elasticity.

For example, to restore correctly the sixth rank elastic modulus tensor for a model with potential energy density (30.14), with full preservation of the initial assumptions about potentiality and symmetry properties, one should express the moduli a_k using (30.16) through the elastic moduli C_1, C_2, C_3, to introduce new basis tensors :

$$\begin{aligned}
\delta^1_{ijklmn} &= (\delta_{ij}\delta_{km}\delta_{ln} + \delta_{lm}\delta_{nj}\delta_{ik} + \delta_{ij}\delta_{kn}\delta_{lm} + \delta_{ik}\delta_{jm}\delta_{ln})/4,\\
\delta^2_{ijklmn} &= (\delta_{ij}\delta_{kl}\delta_{mn} + \delta_{lm}\delta_{ni}\delta_{jk} + \delta_{ik}\delta_{jl}\delta_{mn} + \delta_{ln}\delta_{mi}\delta_{jk})/4 -\\
&\quad - (\delta_{ij}\delta_{km}\delta_{ln} + \delta_{lm}\delta_{nj}\delta_{ik} + \delta_{ij}\delta_{kn}\delta_{lm} + \delta_{ik}\delta_{jm}\delta_{ln})/4,\\
\delta^3_{ijklmn} &= -(\delta_{ij}\delta_{kl}\delta_{mn} + \delta_{lm}\delta_{ni}\delta_{jk} + \delta_{ik}\delta_{jl}\delta_{mn} + \delta_{ln}\delta_{mi}\delta_{jk})/2 +\\
&\quad + (\delta_{ij}\delta_{km}\delta_{ln} + \delta_{lm}\delta_{nj}\delta_{ik} + \delta_{ij}\delta_{kn}\delta_{lm} + \delta_{ik}\delta_{jm}\delta_{ln})/4 + \delta_{il}\delta_{jk}\delta_{mn},\\
\delta^4_{ijklmn} &= (\delta_{il}\delta_{km}\delta_{jn} + \delta_{il}\delta_{kn}\delta_{jm}),\\
\delta^5_{ijklmn} &= (\delta_{im}\delta_{kl}\delta_{jn} + \delta_{lj}\delta_{ni}\delta_{mk} + \delta_{im}\delta_{kn}\delta_{lj} + \delta_{in}\delta_{kl}\delta_{mj}).
\end{aligned}$$

As a result, taking into account the transition to a new tensor basis, the general form of the six rank elasticity modulus tensor is restored:

$$C_{ijklmn} = C_1\delta^1_{ijklmn} + C_2\delta^2_{ijklmn} + C_3\delta^3_{ijklmn} + a_8\delta^4_{ijklmn} + a_{10}\delta^5_{ijklmn}.$$

Only this way of restoring the sixth rank gradient moduli tensor is correct. Otherwise, one can lose the properties of the potentiality of modules and the property of symmetry when we determine the moments.

30.5 Features of Variational Formulation of Vector-Type Gradient Models

Let us write the Lagrangian of the "vector" gradient theory of elasticity, taking into account (30.25):

$$L = A - U,$$

$$
\begin{aligned}
A &= \int_V P_i^V R_i \, dV + \int_F P_i^F R_i \, dF, \\
U &= \frac{1}{2} \int_V [(2\mu/3+\lambda) R_{i,ia} R_{j,ja} \\
&+ 2\mu (R_{i,j}/2 + R_{j,i}/2 - R_{k,k}\delta_{ij}/3)(R_{i,j}/2 + R_{j,i}/2 - R_{r,r}\delta_{ij}/3) \\
&+ C_1 R_{i,ia} R_{j,ja} + C_3 (\Delta R_a - R_{i,ia})(\Delta R_a - R_{j,ja})] \, dV.
\end{aligned}
\tag{30.26}
$$

and obtain a variational equation corresponding to the stationarity condition for functional (30.26):

$$
\begin{aligned}
\delta L &= \int_V \{[(2\mu+\lambda) R_{j,ji} + \mu(\Delta R_i - R_{j,ji}) \\
&+ C_1 \Delta R_{j,ji} - C_3 \Delta(\Delta R_i - R_{j,ji}) + P_i^V] \delta R_i\} \, dV \\
&+ \int_F \{P_i^F - [(2\frac{\mu}{3}+\lambda) R_{k,k}\delta_{ij} + 2\mu(\frac{R_{i,j}}{2} + \frac{R_{j,i}}{2} - R_{k,k}\frac{\delta_{ij}}{3}) \\
&+ C_1 \Delta R_{k,k}\delta_{ij} - C_3 \Delta(R_{i,j} - R_{j,i})] n_j] \delta R_i\} \, dF \\
&- \int_F [C_1 R_{k,ka} n_a \, \delta R_{j,j} + C_3 (\Delta R_a - R_{i,ia}) \delta(R_{a,j} - R_{j,a}) n_j] \, dF = 0.
\end{aligned}
\tag{30.27}
$$

Variational formulation (30.27) gives a complete mathematical formulation for the theory under consideration. From (30.27) the equilibrium equations follow:

$$
(2\mu+\lambda) R_{j,ji} + \mu(\Delta R_i - R_{j,ji}) - C_1 \Delta R_{j,ji} - C_3 \Delta(\Delta R_i - R_{j,ji}) + P_i^V = 0 \tag{30.28}
$$

and three pairs of alternative boundary conditions:

$$
\int_F \{P_i^F - [(2\mu/3+\lambda) R_{k,k}\delta_{ij} + 2\mu\gamma_{ij} - C_1 \Delta R_{k,k}\delta_{ij} - C_3 \Delta(R_{i,j} - R_{j,i})] n_j\} \delta R_i \, dF = 0. \tag{30.29}
$$

Boundary conditions (30.29) will be called "classical" boundary conditions, because the force vector on the surface of the body, following (30.29), is determined by the same expression

$$
[(2\mu/3+\lambda) R_{k,k}\delta_{ij} + 2\mu\gamma_{ij} - C_1 \Delta R_{k,k}\delta_{ij} - C_3 \Delta(R_{i,j} - R_{j,i})],
$$

the divergence of which gives the equilibrium equation (30.28). The last term on the right side of (30.27)

$$
\int_F [C_1 R_{k,ka} n_a \delta R_{j,j} + C_3 (\Delta R_a - R_{i,ia}) \delta(R_{a,j} - R_{j,a}) n_j] \, dF = 0 \tag{30.30}
$$

defines non-classical boundary conditions that are associated with the work on the surface of the body of specific moment force factors on variations in volume change and on variations in the tangential components of the rotation tensor. Consider separately the expressions in (30.30). Boundary dilatation conditions (30.30) define one pair of nonclassical alternative boundary conditions

$$\int_F C_1 R_{j,ja} n_a \delta R_{i,i}\, dF = 0, \tag{30.31}$$

which can be written in the form

$$C_1 R_{j,ja} n_a = p \quad \text{or} \quad R_{i,i} = \theta_0 \quad \text{for} \quad x_i \in F, \tag{30.32}$$

p and θ_0 under the conditions are considered to be given functions on the surface of the body.

Boundary conditions of the moment type on combinations of angles of rotation $(R_{a,j} - R_{j,a})n_j$ determine only two pairs of nonclassical alternative boundary conditions:

$$\int_F C_3 (R_{a,j} - R_{j,a})_{,j} \delta(R_{a,i} - R_{i,a}) n_i\, dF = 0 \tag{30.33}$$

Indeed, boundary conditions (30.33) do not contain variations of the third component of the rotation tensor:

$$(R_{a,i} - R_{i,a}) n_i n_a \equiv 0,$$

since the tensor $R_{a,i} - R_{i,a})$ is antisymmetric, and the tensor $n_i n_a$ is symmetric when the indices are interchanged. Thus, conditions (30.33) give, in addition to (30.32), two more pairs of alternative nonclassical boundary conditions:

$$C_3 (R_{a,j} - R_{j,a})_{,j} = m_a \quad \text{or} \quad (R_{a,i} - R_{i,a}) n_i = \omega_a, \tag{30.34}$$

where

$$\omega_s = (R_{a,i} - R_{i,a}) n_i s_a, \quad \omega_v = (R_{a,i} - R_{i,a}) n_i v_a, \quad \omega_n = (R_{a,i} - R_{i,a}) n_i n_a \equiv 0,$$

and $m_a = m_b \delta^*_{ab}$, $\omega_a = \omega_{v(s)}$ – are the given functions on the surface of the body.

As a result, expressions (30.29), (30.31) and (30.33) give six pairs of alternative boundary conditions at each non-singular point of the body surface. In this case, the variational formulation of the mathematical model is completely closed and correct; it is determined by resolving equation (30.28) and boundary conditions (30.29), (30.32), (30.34).

Another traditional version of the mathematical formulation can be obtained from the surface part of the variational equation by separating the derivatives of the displacements of the normal component in the variations and integrating by parts up to the variation of the displacements on the surface. In this case, the classical static boundary conditions are distorted due to moment force factors (30.29) and, in addition to non-classical boundary conditions, with variations in the normal derivatives of displacements, additional boundary conditions arise on the surface edges (if they take place). Such a formulation seems to be more complicated, because it is not possible to single out an explicitly auxiliary static boundary value problem analogous to classical elasticity. Moreover, a boundary value problem arises on edges, for which the mathematical problems of the existence of solutions are not completely solved and which, as is known, lead to significant numerical problems in solving applied problems.

30.6 System of Governing Equilibrium Equations

Let us consider governing equilibrium equation (30.28) and, after some transformations, rewrite these equations, separating the Lamé operator L_{ij}, in the resolving operator, in the following form

$$L_{ij}H_{jk}(R_k)+P_i^V=0, \tag{30.35}$$

where

$$\begin{gathered} L_{ij}(\dots)=\mu\Delta(\dots)\delta_{ij}+(\mu+\lambda)(\dots)_{,ij}, \\ H_{jk}(\dots)=[(\dots)\delta_{jk}-a^2\Delta(\dots)\delta_{jk}-b^2(\dots)_{,ij}], \\ a^2=\frac{C_3}{\mu}, \qquad b^2=\frac{C_1}{2\mu+\lambda}-\frac{C_3}{\mu}. \end{gathered}$$

In Eq. (30.35) $L_{ij}(\dots)$ is the Lamé operator, $H_{ij}(\dots)$ is the generalized Helmholtz operator, in which a^2 and b^2 are scale parameters of the length square order. Equation (30.35) corresponds to the general theorem on the representation of the governing equation of the general gradient theory of elasticity of the second order as the product of the Lamé operator (classical elasticity operator) and the Helmholtz operator with two scale parameters. The governing equation of the considered variant of gradient elasticity can also be written in terms of the vector potential ϕ_i

$$\begin{gathered} R_i=(\mu+\lambda)\phi_{j,ji}-(2\mu+\lambda)\Delta\phi_i-(C_1-C_3)\Delta\phi_{j,ji}+C_1\Delta\Delta\phi_i, \\ \Delta(\dots)\Delta(\dots)\left[(\dots)-\frac{C_3}{\mu}\Delta(\dots)\right]\left[(\dots)-\frac{C_1}{2\mu+\lambda}\Delta(\dots)\right]\phi_i-\frac{P_i^V}{\mu(2\mu+\lambda)}=0 \end{gathered} \tag{30.36}$$

The solution of equation (30.36) can be represented as

$$\begin{gathered} \phi_i=\Psi_i+\phi_i^\theta+\phi_i^\omega+\Psi_i^*, \quad \Delta\Delta\Psi_i=0, \quad \phi_i^\theta-l_\theta^2\Delta\phi_i^\theta=0, \quad \phi_i^\omega-l_\omega^2\Delta\phi_i^\omega=0, \\ \Delta\Delta[(\dots)-l_\theta^2\Delta(\dots)][(\dots)-l_\omega^2\Delta(\dots)]\Psi_i^*-\frac{P_i^V}{\mu(2\mu+\lambda)}=0, \\ C_1=(2\mu+\lambda)l_\theta^2, \quad C_3=\mu l_\omega^2. \end{gathered} \tag{30.37}$$

In (30.37), the potential Ψ_i^* is a particular solution of the inhomogeneous equation, the biharmonic potential Ψ_i corresponds to the general representation of the classical elasticity solution in terms of the biharmonic potential, the vector potentials ϕ_i^θ, ϕ_i^ω are the solutions of the Helmholtz equations, with the corresponding scale parameters l_θ^2, l_ω^2.

To find out the physical meaning of vector potentials ϕ_i^θ, ϕ_i^ω, we substitute them one by one into the representation for displacements (30.36) and take into account their properties as Helmholtz potentials (30.37). We get

$$\begin{aligned} R_i^\theta &= (\mu+\lambda)\phi^\theta_{j,ji} - (2\mu+\lambda)\Delta\phi_i^\theta - (C_1 - C_3)\Delta\phi^\theta_{j,ji} + C_1\Delta\Delta\phi_i^\theta \\ &= -(\mu+\lambda)\phi^\theta_{j,ji} + C_3\Delta\phi^\theta_{j,ji} = \left[-(\mu+\lambda) + (2\mu+\lambda)\frac{C_3}{C_1}\right]\phi^\theta_{k,ki}, \end{aligned} \quad (30.38)$$

$$\begin{aligned} R_i^\omega &= -\mu\phi^\omega_{j,ji} - (2\mu+\lambda)(\Delta\phi_i^\omega - \phi^\omega_{j,ji}) + C_1\Delta(\Delta\phi_i^\omega - \phi^\omega_{j,ji}) + C_3\Delta\phi^\omega_{j,ji} \\ &= -(2\mu+\lambda)(\Delta\phi_i^\omega - \phi^\omega_{j,ji}) + C_1\frac{\mu}{C_3}(\Delta\phi_i^\omega - \phi^\omega_{j,ji} \\ &= \left[\mu\frac{C_1}{C_3} - (2\mu+\lambda)\right](\Delta\phi_i^\omega - \phi^\omega_{j,ji}). \end{aligned} \quad (30.39)$$

Equality (30.38) shows that the part of the displacement vector, which is determined by the potential vector ϕ_i^θ, is a potential field, since it is the gradient of the scalar $\phi^\theta_{k,k}$. It follows from (30.39) that the component of the displacement vector, determined by the vector potential ϕ_i^ω, is a vortex field, since its divergence is identically equal to zero

$$(\Delta\phi_i^\omega - \phi^\omega_{j,ji})_{,i} \equiv 0.$$

In the general case, equation (30.36) for displacements is defined through the potential and vortex displacement fields, each of which has its own characteristic length: the potential field has the characteristic length l_θ, and the vortex field has the characteristic length l_ω. Thus, the "vector" gradient theory is determined with sufficient generality by two nonclassical parameters that determine multiscale effects: first one for potential displacement fields, and second one for vortex ones. We note that the separation of boundary value problems with the sequential solution of two separate problems of the second order cannot be carried out in the general case. It is easy to verify if we write the classical stresses through the auxiliary potentials $\Psi_i, \phi_i^\theta, \phi_i^\omega, \Psi_i^*$

$$\tau_{ij} = \tau_{ij}^\Psi + \tau_{ij}^\theta + \tau_{ij}^\omega + \tau_{ij}^{\Psi^*},$$

where

$$\begin{aligned} \tau_{ij}^\Psi &= \lambda\delta_{ij}\left[-\mu\Delta\Psi_{k,k}\right] \\ &\quad + \mu\left[2(\mu+\lambda)\Psi_{k,kij} - (2\mu+\lambda)\Delta(\Psi_{i,j} + \Psi_{j,i}) - 2(C_1 - C_3)\Delta\Psi_{k,kij}\right], \\ \tau_{ij}^\theta &= A\left[\lambda\delta_{ij}\Delta\phi^\theta_{k,k} + 2\mu\phi^\theta_{k,kij} - C_1\Delta\Delta\phi^\theta_{k,k}\delta_{ij}\right], \\ \tau_{ij}^\omega &= \mu B(\Delta\phi^\omega_{i,j} + \Delta\phi^\omega_{j,i} - 2\phi^\omega_{k,kij}) - C_3 B\Delta\Delta(\phi^\omega_{i,j} - \phi^\omega_{j,i}), \\ A &= \left[-(\mu+\lambda) + (2\mu+\lambda)\frac{C_3}{C_1}\right], \qquad B = \left[\mu\frac{C_1}{C_3} - (2\mu+\lambda)\right]. \end{aligned}$$

Note, that the separation of a closed problem into classical stresses in applied problems can be useful, and can significantly simplify the problem as a whole. So, for example, for the case $\lambda = 0$ there is a complete division of the problem in the case of static conditions, when the classical stress is found from a separate boundary value problem, the displacement field U_i of which is determined explicitly only by the potentials Ψ_i and Ψ_i^*. This displacement field can be called classical, since

$$L_{ij}(U_j) + P_i^V = 0.$$

To determine the entire solution of gradient elasticity, it is required to solve a boundary value problem for total displacements for the corresponding boundary value problem with the second order operator

$$H_{jk}(R_k) = U_j.$$

Let us consider again the governing equation in the form (30.35) and rewrite it using the characteristic length parameters l_θ^2, l_ω^2. We get

$$\begin{cases} \tau_{ij,j} + P_i^V = L_{ij} H_{jk}(R_k), \quad a^2 = l_\omega^2, \quad b^2 = l_\theta^2 - l_\omega^2, \\ L_{ij}(\ldots) = \mu\Delta(\ldots)\delta_{ij} + (\mu + \lambda)(\ldots)_{,ij}, \\ H_{jk}(\ldots) = [(\ldots)\delta_{jk} - l_\omega^2 \Delta(\ldots)\delta_{jk} - (l_\theta^2 - l_\omega^2)(\ldots)_{,ij}]. \end{cases}$$

In particular, if

$$b^2 = l_\theta^2 - l_\omega^2 = 0,$$

then the equilibrium equation of the particular model of gradient elasticity under consideration coincides with the equation of the well-known model proposed by Aifantis [4, 5] and built on the assumption of a special particular structure of the tensor of gradient elasticity moduli

$$C_{ijklmn} = C_{ijlm}\delta_{kn} l^2.$$

It should be noted that despite the fact that the equilibrium equation of the recorded particular case of the vector model coincides with the equilibrium equation of the applied one-parameter Aifantis model [4, 5], the variational formulations of the compared models will be different.

30.7 On Uniqueness of Reduced Vector-Type Models

We believe that the gradient elasticity model is completely determined by setting the density of the gradient potential energy. In the general case of isotropic elasticity, the potential energy density is represented as an expansion in fifteen basis vectors (30.6) with gradient elastic moduli $a_1 \operatorname{div} a_{15}$. Among them, the basic tensors of the first type include the second, third, fifth, sixth, seventh and eleventh basic tensors, which are reduced immediately after the convolution of vectors to quadratic canonical forms for $R_{i,im}R_{k,km}$, $R_{i,jj}R_{i,kk}$, and bilinear form $R_{i,jj}R_{l,il}$ (for the eleventh basic tensor). The remaining basis tensors are basis tensors of the second type and, after integration by parts, are reduced to only canonical quadratic forms with $R_{i,im}R_{k,km}$, $R_{i,jj}R_{i,kk}$, up to the corresponding integrals over the surface associated with the separation of the divergent form. All the corresponding surface integrals are grouped into the following independent groups of terms

$$(a_1+a_4)(a_1+a_4)\int\limits_F (R_{i,i}R_{k,jj}-R_{i,i}R_{j,jk})n_k\,dF,$$

$$(a_8+a_9)\int\limits_F (R_{i,j}R_{i,jk}-R_{i,k}R_{i,jj})n_k\,dF,$$

$$(a_{10}+a_{12}+a_{13}+a_{14})\int\limits_F (R_{i,j}R_{k,ij}-R_{k,i}R_{j,ji})n_k\,dF,$$

$$a_{15}\int\limits_F (R_{i,k}R_{j,ji}-R_{k,i}R_{j,ji})n_k\,dF.$$

Note that the bilinear form with $R_{i,jj}R_{l,il}$ for the eleventh basis vector

$$a_{11}\int\limits_V \delta_{im}\delta_{kj}\delta_{ln}R_{i,jk}R_{l,mn}\,dV = a_{11}\int\limits_V R_{i,jj}R_{l,il}\,dV$$

can be reduced to the canonical form relatively to $R_{i,im}R_{k,km}$ and $R_{i,jj}R_{i,kk}$ again, up to a divergent form, i.e. up to the corresponding surface integral.

Following the proposed criterion for the absence of divergent terms in the volume potential energy density, i.e. the absence of surface terms with moduli defined in the volume of the body, we must accept that the following relations hold for correct general models

$$\begin{aligned}(a_1+a_4)&=0,\\ (a_8+a_9)&=0,\\ (a_{10}+a_{12}+a_{13}+a_{14})&=0,\\ a_{15}&=0,\\ a_{11}&=0.\end{aligned}\tag{30.40}$$

As a result, the potential energy density in general terms is represented as

$$2U = [(a_1+a_4)+a_2+a_3+a_5+a_6+a_{10}+a_{12}+a_{13}+a_{14}+a_{15}]\int\limits_V R_{i,ik}R_{j,jk}\,dV$$

$$+\ [a_7+(a_8+a_9)]\int\limits_V R_{i,jj}R_{i,kk}\,dV + a_{11}\int\limits_V R_{i,jj}R_{l,il}\,dV.$$

Taking into account equalities (30.40), we finally find the following two-parameter vector model

$$\begin{aligned}2U &= (a_2+a_3+a_5+a_6)\int\limits_V R_{i,ik}R_{j,jk}\,dV + a_7\int\limits_V R_{i,jj}R_{i,kk}\,dV\\ &= \int\limits_V [D_1 R_{i,ik}R_{j,jk} + D_2 R_{i,jj}R_{i,kk}]\,dV.\end{aligned} \tag{30.41}$$

If the elastic moduli of the general gradient model are given, then the two-parameter model is written using (30.41). The criterion for the absence of divergent terms in the expression for the potential energy density necessarily requires the fulfillment of relations (30.40) for the initial elastic moduli. Consequently, for all models from the subspace of models (30.40), two-parameter model (30.41) is constructed in a unique way. However, the converse assertion does not hold. Generally speaking, there is no uniqueness in restoring the gradient model in the general Mindlin form through tensors of the sixth rank. Indeed, model (30.41) is valid for all gradient elasticity models with moduli tensor (30.6), in which $a_{15} = 0$, $a_{11} = 0$, and the remaining thirteen moduli are related by potentiality relations and three additional relations

$$\begin{aligned}(a_1+a_4) &= 0,\\ (a_8+a_9) &= 0,\\ (a_{10}+a_{12}+a_{13}+a_{14}) &= 0.\end{aligned}$$

Note 30.1. Earlier it was noted that the procedure for simplifying gradient theories founded on a physical hypothesis, which states that the physical moduli of elasticity that characterize the properties of the gradient model in the body volume cannot determine the properties of the body surface. There is also direct proving that the presence of a divergent term in the transformation of a fragment of the volume density of the potential energy gradient to a quadratic form containing $R_{i,im}R_{k,km}$ and $R_{i,jj}R_{i,kk}$ necessarily leads to the equality of the constant to zero for the considered fragment of the volume integral.

30.8 Conclusions

The proposed method for simplifying and reducing the gradient theory of Mindlin 1 to the gradient theory of elasticity of the vector type has the following advantages:

- The number of non-classical parameters has been reasonably reduced from five, as in the Mindlin I theory, to two.
- Classical static boundary conditions do not contain gradient corrections, which means that the definitions of classical stresses in the volume and on the surface of the body do not contradict each other.

- Non-classical boundary conditions do not lead to the need for additional taking by parts on the surface. This simplifies the solution of applied problems, because it does not require the definition of meniscus forces on the contours of the edges of a piecewise-smooth surface that bounds an elastic body and does not introduce momentary corrections into the formulation of "classical" static boundary conditions.

Acknowledgements This work was supported by the Russian Science Foundation under the grant 20-41-04404 issued to the Institute of Applied Mechanics of Russian Academy of Sciences.

The authors consider it their duty to express their sincere gratitude to Professor Victor Eremeyev and Professor Holm Altenbach for discussion and friendly criticism.

References

[1] Toupin RA (1962) Elastic materials with couple-stresses, Archive for Rational Mechanics and Analysis **11**:385-414. DOI 10.1007/BF00253945

[2] Mindlin RD (1964) Micro-structure in linear elasticity, Archive for Rational Mechanics and Analysis **16**(1):51-78. DOI 10.1007/BF00248490

[3] Mindlin RD, Eshel NN (1968) On first strain-gradient theories in linear elasticity, International Journal of Solids and Structures **4**(1):109-124. DOI 10.1016/0020-7683(68)90036-X

[4] Ru CQ, Aifantis EC (1993) A simple approach to solve boundary-value problems in gradient elasticity, Acta Mechanica **101**(1-4):59-68. DOI 10.1007/BF01175597

[5] Gao XL, Park SK (2007) Variational formulation of a simplified strain gradient elasticity theory and its application to a pressurized thick-walled cylinder problem, International Journal of Solids and Structures **44**(22-23):7486-7499. DOI 10.1016/j.ijsolstr.2007.04.022

[6] Askes H, Aifantis EC (2011) Gradient elasticity in statics and dynamics: an overview of formulations, length scale identification procedures, finite element implementations and new results, International Journal of Solids and Structures **48**(13):1962-1990. DOI 10.1016/j.ijsolstr.2011.03.006

[7] Polizzotto C (2014) Stress gradient versus strain gradient constitutive models within elasticity, International Journal of Solids and Structures **51**(9):809-1818. DOI 10.1016/j.ijsolstr.2014.01.021

[8] Polizzotto C (2015) A unifying variational framework for stress gradient and strain gradient elasticity theories, European Journal of Mechanics - A/Solids **49**:430-440.

[9] Polizzotto C (2016) A note on the higher order strain and stress tensors within deformation gradient elasticity theories. Physical interpretations and comparisons, International Journal of Solids and Structures **90**:116-121. DOI 10.1016/j.ijsolstr.2016.04.001

[10] Belov PA, Lurie SA (2014) Mathematical theory of defectness media. Gradient theories. Formulations, hierarchy. Analysis. Applications, Palmarium Academic Publishing.
[11] Forest S, Aifantis EC (2010) Some links between recent gradieForest S. and Aifantisnt thermo-elasto-plasticity theories and the thermomechanics of generalized continua, International Journal of Solids and Structures **47**(25-26):3367-3376. DOI 10.1016/j.ijsolstr.2010.07.009
[12] Aifantis EC (2016) Internal length gradient (ILG) material mechanics across scales & disciplines, In: SPA Bordas, and DS Balint (Eds) Advances in Applied Mechanics Vol 49, pp 1-110, Elsevier. DOI 10.1016/bs.aams.2016.08.001
[13] Abali BE, Müller WH, and Dell'Isola F (2017) Theory and computation of higher gradient elasticity theories based on action principles, Archive of Applied Mechanics **87**(9)1495-1510. DOI 10.1007/s00419-017-1266-5
[14] Papanicolopulos SA, Gulib F, Marinelli A (2019) A novel efficient mixed formulation for strain gradient models, Int International Journal for Numerical Methods in Engineering **117**:926–937. DOI 10.1002/nme.5985
[15] Polizzotto C (2014) Stress gradient versus strain gradient constitutive models within elasticity, International Journal of Solids and Structures **51**(9):1809-1818. DOI 10.1016/j.ijsolstr.2014.01.021
[16] Gusev AA, Lurie SA (2017) Symmetry conditions in strain gradient elasticity, Mathematics and Mechanics of Solids **22**(4)683-691. DOI 10.1177/10812865156069
[17] Vasiliev VV, Lurie SA (2016) On correct nonlocal generalized theories of elasticity, Physical Mesomechanics **19**(3):269-281. DOI 10.1134/S102995991603005X
[18] Lurie S, Belov P, Lykosova E (2021) Specifics of symmetry conditions in gradient elasticity theories, Materials Physics and Mechanics **47**(6):905-920. DOI 10.18149/MPM.4762021_10
[19] Vasiliev V, Lurie S, Salov V (2019) Estimation of the strength of plates with cracks based on the maximum stress criterion in a scale-dependent generalized theory of elasticity, Physical Mesomechanics **226**:456-462. DOI 10.1134/S102995991906002X
[20] Lam DCC, Yang F, Chong ACM, Wang J, Tong P (2003) Experiments and theory in strain gradient elasticity, Journal of the Mechanics and Physics of Solids **51**:1477-1508. DOI 10.1016/S0022-5096(03)00053-X
[21] Liebold C, Müller WH (2018) Measuring material coefficients of higher gradient elasticity by using AFM techniques and raman-spectroscopy, In: Generalized Continua as Models for Materials with Multi-scale Effects or Under Multi-field Actions (Eds H Altenbach, S Forest, A Krivtsov), Advanced Structured Materials, Vol 22, pp 255-271. DOI 10.1007/978-3-642-36394-8_14
[22] Gusev AA, Lurie SA (2010) Strain-gradient elasticity for bridging continuum and atomistic estimatesof stiffness of binary Lennard-Jones crystals, Advanced Engineering Materials **12**(6)529-533. DOI 10.1002/adem.201000004
[23] Marangantia R, Sharma P (2007) A novel atomistic approach to determine strain-gradient elasticity constants: Tabulation and comparison for various

metals, semiconductors, silica, polymers and the (Ir) relevance for nanotechnologies, Journal of the Mechanics and Physics of Solids **55**:1823-1852. DOI 10.1016/j.jmps.2007.02.011

[24] Lurie SA, Solyaev YO (2017) Identification of gradient elasticity parameters based on interatomic interaction potentials accounting for modified Lorentz-Berthelot rules, Physical Mesomechanics **20**(4)392-398. DOI 10.1134/S1029959917040038

[25] Polizzotto C (2017) A hierarchy of simplified constitutive models within isotropic strain gradient elasticity, European Journal of Mechanics - A/Solids **61**:92-109. DOI 10.1016/j.euromechsol.2016.09.006

[26] Lurie SA, Belov PA, Solyaev YO, Aifantis EC (2017) On one class of applied gradient models with simplified boundary problems, Materials Physics and Mechanics **32**:353-369.

Chapter 31
Dynamics of a Rectangular Rigid Body on a Movable Base

Vladimir S. Metrikin, Leonid A. Igumnov, and Elena I. Komarova

Abstract On the basis of general theory of dynamical systems (DS), vibrations of a rigid rectangular body on a horizontally vibrating with acceleration $\ddot{U}(t)$ rigid sup-porting plane are considered. It is assumed that the friction force between the rigid body and the surface is so great that there is no sliding of the base of the solid body along the plane, but oscillations occur relative to the corner points of the body support. The loss of energy occurs due to impact interactions of bodies with the coefficient of "recovery" of the angular velocity. For an arbitrary set of geometrical dimensions of a rigid rectangular body and acceleration parameters $\ddot{U}(t)$ of the supporting plane, the equations of point mappings of Poincaré surfaces are presented by numerical-analytical methods. Based on them, bifurcation diagrams were obtained, which made it possible to establish both the regions of existence in the multidimensional space of DS parameters of symmetric and asymmetric stable periodic motion modes of a rigid body and the scenario for the occurrence of chaotic motions.

Key words: Mathematical model · Point mapping · Stability · Bifurcation diagrams · Chaos

31.1 Introduction

A rigid rectangular block, the dynamics of which is considered in this paper, is either a model of ancient structures (arches, buildings) or individual objects of railway transport (cars), or a model of plane oscillations of the earth's crust during earthquakes, etc. An important feature of these models is that they do not have a foundation. The

Vladimir S. Metrikin · Leonid A. Igumnov · Elena I. Komarova
National Research Lobachevsky State University of Nizhny Novgorod, 23 Gagarin Avenue, Nizhny Novgorod 603022, Russian Federation,
e-mail: v.s.metrikin@mail.ru, igumnov@mech.unn.ru, komarova@mech.unn.ru

H. Altenbach et al. (eds.), *Sixty Shades of Generalized Continua*,
Advanced Structured Materials 170, https://doi.org/10.1007/978-3-031-26186-2_31

problem of stabilization of such and similar structures is currently quite relevant in seismically active areas [1]-[5]. Such systems are also interesting to study because, in practice, structures that are very unstable at first glance (such as individual stone pillars or elevated water tanks) sometimes survive earthquakes better than seemingly more stable structures. In [1] Fig. 1 shows the electrical equipment at the Sylmar Converter Station, destroyed during the earthquake in 1971 in San Fernando and Fig. 2 presents a train that fell during the 1906 Point Reyes earthquake in San Francisco. Studies on the rocking of a solid block on a horizontal base were considered by Housner [1]. In this study, the platform accelerations were presented as rectangular or sinusoidal time functions acting on only one period. Expressions were obtained for the minimum values of the acceleration amplitudes at which the rectangular block falls. However, such Hausner impulses are not physically realizable, so the solutions obtained for minimum accelerations are only of theoretical interest. Since this seminal work by Housner [1], a large amount of research has been presented to solve the complex dynamics of one of the simplest man-made structures, a free-standing rectilinear block [6]-[25]. The swinging response of free-standing rectangular blocks under a harmonic steady load was studied in detail by Spence and Koch [6], who identified "dangerous" and "safe" regions in phase space and developed analytical methods to determine the periodic and subharmonic regimes of the system. In [7], on the basis of qualitative methods controlled by the DS, "dangerous" and "safe" areas in the phase space for an arbitrary load are given. Their study was continued by Hogan [8], who further explained the mathematical structure of the problem by using the concepts of orbital stability and Poincaré secant surfaces. This study takes a deep look at the response of swings of free-standing blocks to cosine and sine acceleration impulses. These two trigonometric pulses are physically realizable and in some cases resemble the motion components recorded near the source of strong earthquakes [10]. In this paper, based on the general theory of dynamical systems (DS), oscillations of a rigid rectangular body on a horizontally vibrating with acceleration $\ddot{U}(t)$ rigid reference plane are considered. It is assumed that friction force between the rigid body and the supporting surface is so great that there is no sliding of the base of the solid body along the plane but oscillations occur relative to the corner points of the body support. The loss of energy occurs due to impact interactions of bodies with the coefficient of "recovery" of the angular velocity. For an arbitrary set of geometrical dimensions of a solid rectangular body and acceleration parameters $\ddot{U}(t)$ of the supporting plane, the equations of point mappings of Poincare surfaces are presented by numerical-qualitative methods. Based on them, bifurcation diagrams were obtained, which made it possible to establish the regions of existence in the multidimensional space of DS parameters of symmetric and asymmetric stable periodic modes of motion of a rigid body, as well as to establish the scenario for the occurrence of chaotic motions.

31.2 Mathematical Model

We consider vibrations of a rectangular rigid block standing on a flat horizontal surface (Fig. 31.1). It is assumed that the surface vibrates strictly horizontally with an acceleration $\ddot{u}(t)$. The frictional force between the surface and the block is assumed to be large enough to prevent the block from slipping. Under these assumptions, the block can only perform rotational - oscillatory movements relative to two corner points O and O' in the lower side of the block. The vibrations of the block can be unambiguously characterized by the angle θ between the surface and block underside. The transition from rotation of one point to another occurs by means of an impact, which is considered instantaneous. The energy lost during the impact is taken into account by introducing the angular velocity recovery factor $0 < r \leq 1$, so that $\dot{\theta}^{+} = r\dot{\theta}^{-}$, where $\dot{\theta}^{-}$ is the angular velocity of the block before the impact, $\dot{\theta}^{+}$ is the angular velocity of the block after the impact.

The block motion equations in the time intervals between impacts can be writ-ten in the form of momentum balance. In a frame of reference moving with the vibrating surface, only one force acts on the block, namely gravity. Therefore, the equations of motion in this coordinate system can be written as follows

$$\begin{aligned} J\frac{d^2\theta}{dt^2} &= -mgR\sin(\alpha-\theta), \qquad \theta > 0, \\ J\frac{d^2\theta}{dt^2} &= +mgR\sin(\alpha+\theta), \qquad \theta < 0 \end{aligned} \tag{31.1}$$

equating the angular force of inertia of the block to the moment of gravity. In these equations, $\alpha = \theta_0$ is the angle between the side of the block and the diagonal and $2R$ is the diagonal length. The moment of inertia J of the block relative to the points O and O' is expressed as follows

$$J = \frac{4m}{3}R^2.$$

In order to switch to a fixed coordinate system, it is necessary to take into ac-count the transfer force of inertia, which arises due to the uneven surface movement. This

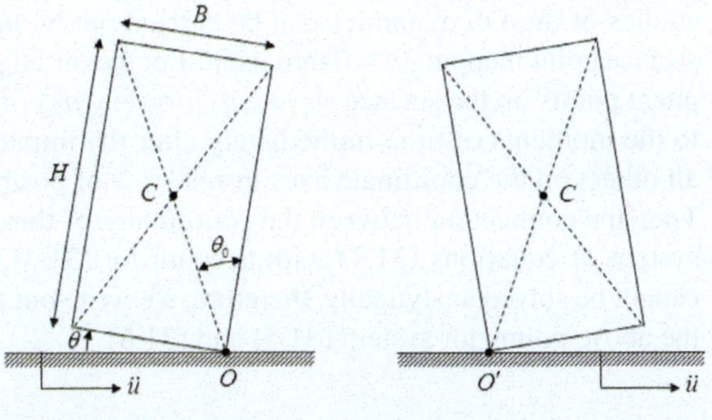

Fig. 31.1 Scheme of the considered system.

force acting on the mass center of block C is directed opposite to the surface acceleration s and is equal in magnitude $m\ddot{u}$. Taking into account the impact interactions of the block the equations can be rewritten as follows

$$
\begin{aligned}
&J\frac{d^2\theta}{dt^2} = -mgR\sin(\alpha-\theta) - mR\frac{d^2u}{dt^2}\cos(\alpha-\theta), \qquad \theta > 0,\\
&J\frac{d^2\theta}{dt^2} = +mgR\sin(\alpha+\theta) - mR\frac{d^2u}{dt^2}\cos(\alpha+\theta), \qquad \theta < 0,\\
&\dot{\theta}^+ = r\dot{\theta}^-, \qquad \theta = 0.
\end{aligned}
\tag{31.2}
$$

The first equation of system (31.2) describes the oscillatory motion of the block in the vicinity of the point O. The second equation of system (31.2) describes the oscillatory motion of the block in the vicinity of the point o'. The third equation of this system describes the impact on the block.

Introducing dimensionless time $\tau = t\sqrt{3g/2\sqrt{B^2+H^2}}$ and a new variable $x = \theta$, equations (31.2) can be rewritten as:

$$
\begin{cases}
\dot{x} = y,\\
\dot{y} = \begin{cases} \sin(x+\alpha) - u(\tau)\cos(x+\alpha), & -\pi/2 < x < 0,\\ \sin(x-\alpha) - u(\tau)\cos(x-\alpha), & 0 < x < \pi/2, \end{cases}
\end{cases}
\tag{31.3}
$$

$$
y^+ = ry^-, \qquad x = 0, \tag{31.4}
$$

where $u(\tau)$ is a dimensionless arbitrary piecewise continuous function y^-, y^+ are pre-shock and post-shock block velocities, respectively, $0 < r < 1$ a constant coefficient characterizing the energy loss upon impact, $0 < \alpha < \pi/2$, and dots over variables denote differentiation with respect to dimensionless time τ.

31.3 Construction of a Point Mapping of the Poincaré Surface

The form of Eqs. (31.3) and (31.4), as well as the behavior of the phase trajectories of the system described above, made it possible to establish that numerical-analytical studies of the DS dynamics can be carried out by investigating the properties of a surface point mapping $x = 0$ into itself. Let the initial point $A_0(\tau_0, \mu_0) \subset S$ and subsequent points on the surface $A_1(\tau_1, \mu_1)$, $A_2(\tau_2, \mu_2)$ of the point mapping correspond to the moments of time immediately after the impact (Fig. 31.2, in this figure and all others on the coordinate axes instead x, $\dot{x}$ of postponed θ, $\dot{\theta}$, and $\mu_i = \dot{x}(\tau_i) \in S$). Then the connection between the coordinates of these points is obtained by solving system of equations (31.3) with account for (31.4). It is obvious that this system cannot be solved analytically, therefore, we write out the coordinates dependences of the above points for system (31.5) and (31.6)

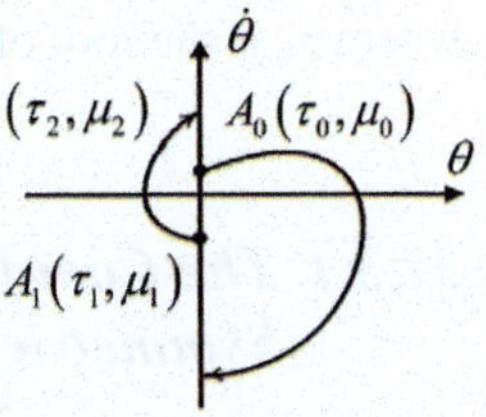

Fig. 31.2 Qualitative view of phase trajectories in the plane $\theta - \dot{\theta}$.

$$\begin{cases} \dot{x} = y, \\ \dot{y} = \begin{cases} x + \alpha - u(\tau), & -\pi/2 < x < 0, \\ x - \alpha - u(\tau), & 0 < x < \pi/2, \end{cases} \end{cases} \tag{31.5}$$

$$y^+ = ry^-, \qquad x = 0, \tag{31.6}$$

as:

$$\begin{aligned} 0 &= \mu_0 \sinh(\tau_1 - \tau_0) - a\cosh(\tau_1 - \tau_0) - \int_{\tau_0}^{\tau_1} u(z)\sinh(\tau_1 - z)dz + \alpha, \\ \mu_1 &= r\left[\mu_0 \cosh(\tau_1 - \tau_0) - a\sinh(\tau_1 - \tau_0) - \int_{\tau_0}^{\tau_1} u(z)\cosh(\tau_1 - z)dz\right], \end{aligned} \tag{31.7}$$

$$\begin{aligned} 0 &= \mu_1 \sinh(\tau_2 - \tau_1) + a\cosh(\tau_2 - \tau_1) - \int_{\tau_1}^{\tau_2} u(z)\sinh(\tau_2 - z)dz - \alpha, \\ \mu_2 &= r\left[\mu_1 \cosh(\tau_2 - \tau_1) + a\sinh(\tau_2 - \tau_1) - \int_{\tau_1}^{\tau_2} u(z)\cosh(\tau_2 - z)dz\right]. \end{aligned} \tag{31.8}$$

It is easy to check that if the following formal transformations are made in the systems of equations (31.7) and (31.8)

$$\tau_0 = -\tau_0^*, \qquad \tau_1 = \tau_1, \qquad \tau_2 = -\tau_2^*$$

then the mapping of the initial point $A_0(\tau_0, \mu_0) \subset S$ of the impact interaction surface S to a point $A_2(\tau_2^*, \mu_2)$ on the same surface is a doubly applied "half-period" mapping defined by the formulas

$$\begin{aligned} 0 &= \mu_0 \sinh(\tau_1 + \tau_0^*) + a\cosh(\tau_1 + \tau_0^*) - \int_{-\tau_0^*}^{\tau_1} u(z)\sinh(\tau_1 - z)dz - \alpha, \\ \mu_1 &= r\left[-\mu_0 \cosh(\tau_1 + \tau_0^*) + a\sinh(\tau_1 + \tau_0^*) + \int_{-\tau_0^*}^{\tau_1} u(z)\cosh(\tau_1 - z)dz\right], \end{aligned} \tag{31.9}$$

where τ_1 is the root of τ_0^* the first equation of system (31.9) closest to in magnitude.

31.3.1 The Coordinates of the Fixed Point Corresponding to the Symmetrical Periodic Motion. Sustainability

The coordinates of the starting point $A_0(\tau_0,\mu_0)$ of the display corresponding to the symmetrical two-impact periodic motions of the block (Fig. 31.3), are found from the system (31.9) under additional conditions

$$\mu_1 = -\mu_0 = -\mu_0^*, \qquad \tau_1 + \tau_0^* = \pi n,\ (n = 1,2,3,\ldots). \tag{31.10}$$

When (31.10) is satisfied, we obtain the coordinates of the fixed point τ_0^*, μ_0^* in the form

$$\begin{gathered} \mu_0^* = \left[\frac{r[a\sinh(\pi n) + I_1(\tau_0^*)]}{r+\cosh(\pi n)}\right], \\ \mu_0^* \sinh\sinh(\pi n) + \alpha(1-\cosh(\pi n)) + I_2(\tau_0^*) = 0, \end{gathered} \tag{31.11}$$

where the following notation is introduced

$$\begin{gathered} I_1(\tau_0^*) = \int_0^{\pi n} u(x+\tau_0^*)\cosh(x)dx, \\ I_2(\tau_0^*) = \int_0^{\pi n} u(x+\tau_0^*)\sinh(\pi n - x)dx. \end{gathered} \tag{31.12}$$

The second equation of the system (31.11) determines the initial time τ_0^*, and the first - the initial speed of the symmetric two-impact periodic motion of the block.

The stability of a fixed point (stability in the small) and, consequently, of the corresponding symmetric periodic motion, is determined by the roots value of the characteristic equation. To this end, we pass from the point mapping Eqs. (31.9) to the equations in variations in spatial coordinates μ_0, μ_1 and phases τ_0^*, τ_1 in the vicinity of the fixed point found and, assuming, we obtain $\delta\mu_1 = -z\delta\mu_0$, $\delta\tau_1 = -z\delta\tau_0^*$ the characteristic equation in the form

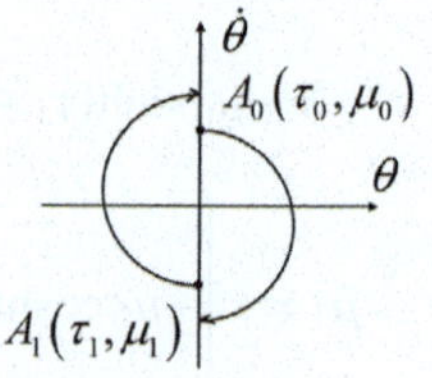

Fig. 31.3 Qualitative view of phase trajectories for symmetric periodic motion.

$$\chi_{2.1}(z) = a_{1.1}z^2 + a_{2.1}z + a_{3.1} = 0,$$

where its coefficients are defined as follows

$$\begin{aligned} a_{1.1} &= [I_3(\tau_0^*) + \alpha \sinh(\pi n) - \mu_0^* \cosh(\pi n)], \\ a_{2.1} &= [r(I_3(\tau_0^*)\cosh(\pi n) - I_2(\tau_0^*)\sinh(\pi n) - u(\tau_0^* + \pi n)\sinh(\pi n) - \mu_0^*) - \\ &\quad -\alpha \cosh(\pi n) + \mu_0^* \cosh(\pi n) - u(\tau_0^*)\sinh(\pi n)], \\ a_{3.1} &= r\mu_0^*, \\ I_3(\tau_0^*) &= \int_0^{\pi n} u(x + \tau_0^*)\cosh(x - \pi n)dx \end{aligned} \tag{31.13}$$

Symmetric periodic motion will be stable if all the roots $z_i, i = 1,2$ of the characteristic equation are located inside a circle of unit radius, that is, strict inequalities will be satisfied $|z_i| < 1, i = 1,2$ [26]. The boundaries of the stability region of the considered periodic motion mode in the parameter space are given by the equations N_+, N_-, N_φ, which are obtained by substitution into the characteristic equation $z = +1, z = -1, z = \exp(\pm i\varphi), 0 \le \varphi \le \pi$, respectively [26].

31.3.2 Equations for Determining the Coordinates of a Fixed Point Corresponding to Asymmetric Periodic Motions. Sustainability

To study the existence of stable asymmetric two-impact periodic motions, it is necessary to use point mappings (31.7) and (31.8). The equations for the existence of asymmetric two-impact periodic motions corresponding to fixed points of a doubly applied "half-period" point mapping, after introducing the asymmetry parameter $\varepsilon = ((\tau_1 - \tau_0)/\pi n) - 1$ and periodicity conditions

$$\mu_0 = \mu_2 = \mu_0^*, \qquad \tau_2 = \tau_0 + 2\pi n = \tau_0^* + 2\pi n$$

into consideration, are determined from a system of equations of the form

$$\begin{aligned} &[\alpha(\cosh(\beta) - 1) + I_{11}(\beta)]\sinh(T - \beta) = \\ &= r\{\alpha[1 + \sinh(T - \beta) - \cosh(T - \beta)] - I_{14}(\tau_1, \beta)\}\sinh(\beta), \\ &\{\alpha[1 - \cosh(T - \beta)] + I_{13}(\tau_1, \beta)\}\sinh(\beta) = r[\alpha(1 - \cosh(\beta)) + I_{12}(\beta)]\sinh(T - \beta), \\ &\sinh(\beta)\mu_0^* = \alpha(\cosh(\beta) - 1) + I_{11}(\beta), \\ &\sinh(\beta)\mu_1^* = -\alpha(\cosh(\beta) - 1) + I_{12}(\beta), \end{aligned} \tag{31.14}$$

where

$$\beta = (\varepsilon+1)\pi n,\ I_{11}(\beta) = \int_0^{\beta} u(\beta-x)\sinh(\beta-x)dx,\ I_{12}(\beta) = -\int_0^{\beta} u(\beta-x)\sinh(x)dx,$$

$$I_{13}(\tau_1,\beta) = \int_0^{T-\beta} u(\tau_1+x)\cosh(T-\beta-x)dx,\ I_{14}(\tau_1,\beta) = \int_0^{T-\beta} u(\tau_1+x)\sinh(x)dx.$$

The first two equations of system (31.14) determine the asymmetry parameter ε and the initial phase τ_0 of the periodic mode for given values of the system parameters, the last two equations determine the initial speed of the block after impacts μ_0^*, μ_1^*.

The characteristic equation, the roots of which determine the small-scale stability of two-impact asymmetric periodic motions is obtained from (31.7) and (31.8) after their linearization in a neighborhood of the fixed points found in (31.14) and has the form $\chi_{2.2}(z) = a_{1.2}z^2 + a_{2.2}z + a_{3.2} = 0$, where

$$\begin{gathered} a_{1.2} = (I_{11} + \alpha\sinh(x) - \mu_0\cosh(x))(\mu_1\cosh(y) + \alpha\sinh(y) - I_{13}), \\ a_{2.2} = \sum_{k=1}^{4} A_k J_{1k} + A_5, \\ a_{3.2} = -\mu_0\mu_1 r^2, \\ x = \tau_1 - \tau_0, \qquad y = \tau_2 - \tau_1. \end{gathered} \tag{31.15}$$

Recall that all the coefficients in the characteristic equation are calculated for the parameters values and phase variables corresponding to the obtained coordinates of fixed points.

The paper considers two types of periodic impact on the base:

$$\ddot{U}(\tau) = A\sin(\omega\tau), \tag{31.16}$$

$$\ddot{U}(\tau) = F\begin{cases} 1,\ 0 \le \tau < \gamma, \\ \dfrac{-\gamma\omega}{2\pi-\gamma\omega},\ \gamma \le \tau < \dfrac{2\pi}{\omega}. \end{cases} \tag{31.17}$$

31.4 Numerical Results for $\ddot{U}(\tau) = A\sin(\omega\tau)$

Substituting (31.16) into (31.11), (31.12) and calculating the integrals $I_1(\tau_0^*), I_2(\tau_0^*)$, we obtain a nonlinear system of equations for determining the coordinates of fixed points corresponding to symmetrical periodic modes of motion of the block in the form

$$\begin{cases} B_1 e^{\pi/\omega} + B_2 e^{-\pi/\omega} + \theta_0 + A\sin\left(\omega\left(\tau_0 + \dfrac{\pi}{\omega}\right)\right) = 0, \\ R\left(B_1 e^{\pi/\omega} - B_2 e^{-\pi/\omega} + A\omega\cos\left(\omega\left(\tau_0 + \dfrac{\pi}{\omega}\right)\right)\right) = -\mu_0, \end{cases} \tag{31.18}$$

$$B_1 = \frac{\mu_0 - \theta_0}{2} - \frac{A}{2}(\sin(\omega\tau_0) + \omega\cos(\omega\tau_0)),\ A = \frac{a}{g(\omega^2+1)},$$

$$B_2 = \frac{-\mu_0 - \theta_0}{2} - \frac{A}{2}(\sin(\omega\tau_0) - \omega\cos(\omega\tau_0)).$$

Resolving system (31.18) with respect to $\sin(\omega\tau_0)$ and $\cos(\omega\tau_0)$ and when the inequalities $-2rA \le \mu_0(1-r)\tanh(\pi/\omega) \le 2rA$ are satisfied, we derive the coordinates of fixed points $(\tau_0, \mu_0)_i$, $i = 1,2$ that correspond to symmetric periodic motions in the form

$$\begin{gathered} \sin(\omega\tau_0) = \frac{(r-1)\tanh(\pi/\omega)\mu_0}{2rA},\ \mathrm{ch}_{\pm} = \cosh(\pi/\omega) \pm 1, \\ \left[(r-1)^2\mathrm{ch}_- A^2\omega^2 + (1+r)^2\mathrm{ch}_+\right]\mu_0^2 - \\ -4(1-r)rA\theta_0\sinh(\pi/\omega)\mu_0 + 4r^2A^2\left[\theta_0^2\mathrm{ch}_- - \omega^2\mathrm{ch}_+\right] = 0. \end{gathered} \tag{31.19}$$

The domains of existence and stability of various (symmetric, asymmetric and chaotic) block motion modes for the first type of impact (31.16) were plotted using relations (31.11)-(31.15), (31.19) and the developed Microsoft Visual Studio 2008 in C++. The program allows you to find the coordinates of fixed points of mapping corresponding to periodic motions with a different number of impacts per period in the phase space of a dynamic system, to investigate their stability, to build bifurcation diagrams for any parameter of the entire set included in the description of the mathematical model under study. This made it possible to trace the main rearrangements of the motion modes, including the period doubling process, which leads to chaotic motions of the block [26].

Figure 31.4 presents a bifurcation diagram in terms of the amplitude A of the external impact the values of which are plotted along the abscissa axis, and the values of the block velocity μ at the moments of impacts at $\theta = 0$ are plotted along the ordinate axis. For parameter values $A \in [10; 12{,}02]$, as can be seen from Fig. 31.4, there exists a symmetrical periodic motion of the block with two impacts, the shape of which is shown in Fig. 31.5. For parameter values A within the range of values B, marked in Fig. 31.4, there correspond periodic movements of the block with a doubled period (four impacts per period), the shape of which is shown in Fig. 31.6 For parameter values A in the range of values C, marked in Fig. 31.4, the periodic motion modes of the block with a doubled period and the number of impacts per period can be estimated, the form of which is shown in Fig. 31.7. For the parameter values A within the range of D values presented in Fig. 31.4, there correspond chaotic motion modes, the shape of which is given in Fig. 31.8. Similar bifurcation diagrams are shown for the parameters θ_0 (Fig. 31.9a), ω (Fig. 31.9b) and R (Fig. 31.9c).

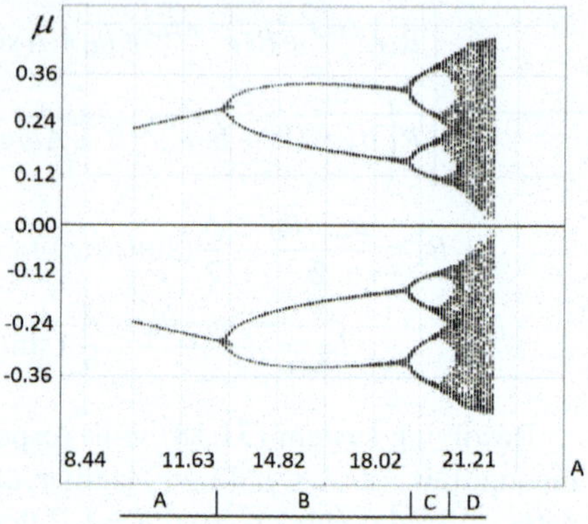

Fig. 31.4 Bifurcation diagram for the parameter $R = 0.4$, $\omega = 3$, $\theta_0 = 0.463648$

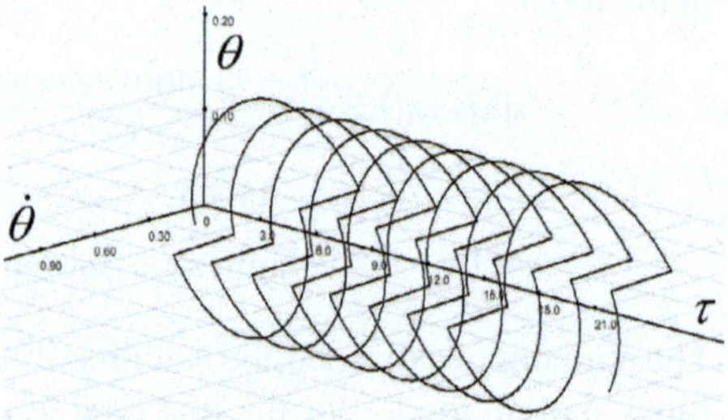

Fig. 31.5 View of the phase trajectories of the symmetrical periodic motion of the block

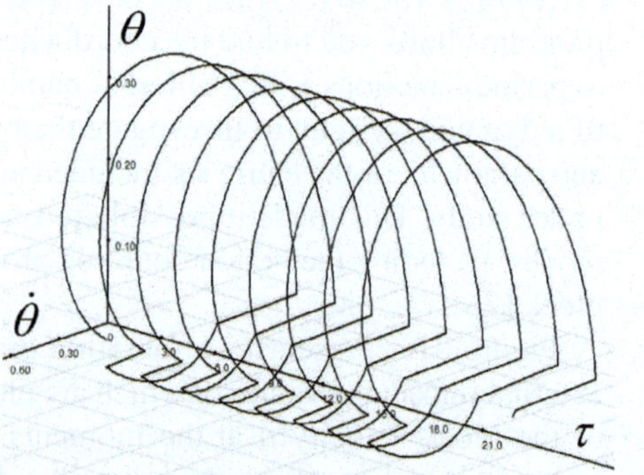

Fig. 31.6 View of the phase trajectories of the asymmetric periodic motion of the block

31.5 Numerical Results

Numerical results for the second type of platform acceleration

$$\ddot{U}_2(\tau) = \begin{cases} A, & 0 \le \tau < \dfrac{2\pi\gamma_H}{\omega} \\ \dfrac{-A\gamma_H}{1-\gamma_H}, & \dfrac{2\pi\gamma_H}{\omega} \le \tau < \dfrac{2\pi}{\omega} \\ \displaystyle\int_0^{2\pi/\omega} \ddot{U}_2(\tau)d\tau = 0, & 0 \le \gamma_H < 1 \end{cases}$$

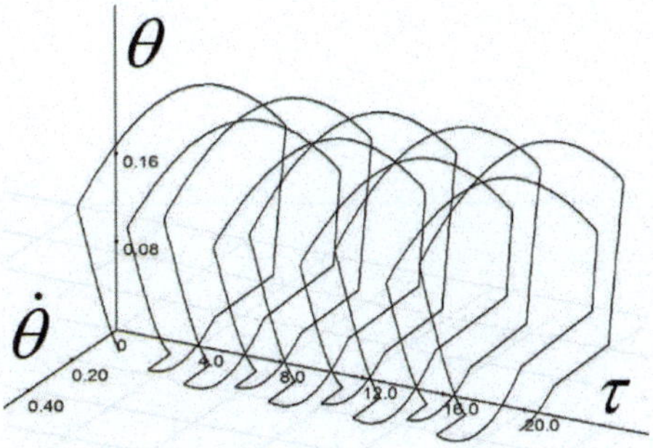

Fig. 31.7 Phase portrait of chaotic block motions

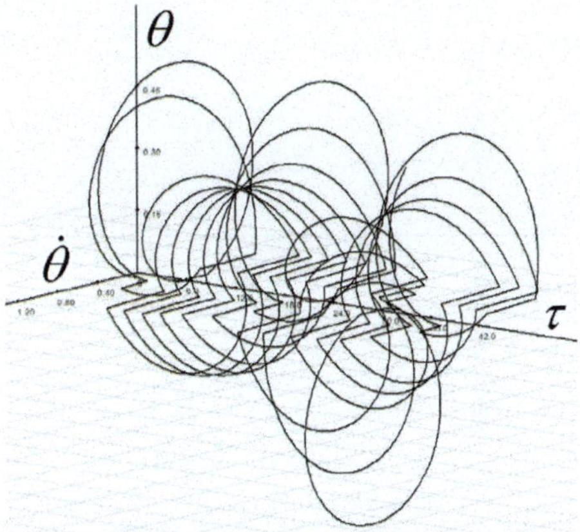

Fig. 31.8 Phase portrait of chaotic block motions

where $\gamma_H = \frac{\gamma\omega}{2\pi}$ (in the future, omit the index of γ_H). A graphical view of the second type of time-periodic acceleration is shown in the Fig. 31.10 Such a choice of $\ddot{U}_2(\tau)$ makes it possible, for different values of the off-duty parameter γ, to describe the dynamics of the block more close to real conditions under seismic impacts on objects.

The Fig. 31.11 shows the bifurcation diagram for the geometric parameter of the block θ_0 at $\omega = 3$, $R = 0.4$, $A = 13$, $\gamma = 0.3$.)

It can be seen from Fig. 31.12 that for θ_0 values belonging to region A, marked in Fig. 31.11, there are periodic stable asymmetric movements of the block with alternate oscillations around one and the other corner points with two hits of the block on the platform. We call such periodic motions one-turn.

The area of values for the parameter θ_0, belonging to area B (Fig. 31.11), corresponding to the asymmetric periodic two-turn motion of the block with four hits of the latter on the platform (Fig. 31.13).

The range of θ_0 parameter values belonging to the area C (Fig. 31.11) corresponds to the chaotic movements of the block (Fig. 31.14).

The Fig. 31.15 shows the bifurcation diagram for the acceleration amplitude of the platform A for the value of the remaining parameters equal to $\theta_0 = 0.48$, $\omega = 3$, $R = 0.4$, $\gamma = 0.3$.

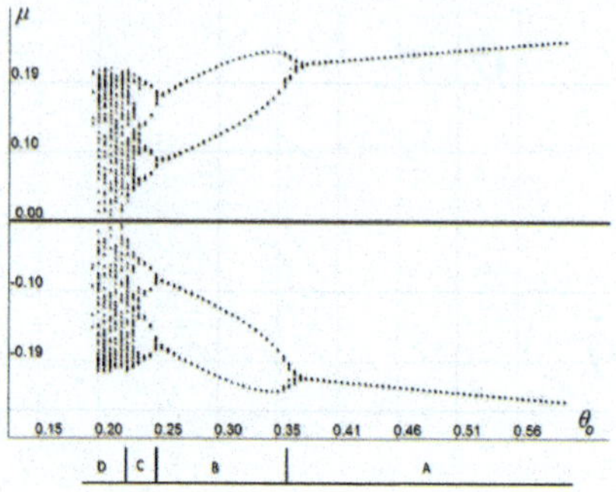

(a) $\omega = 3$, $R = 0.4$, $A = 10$, ω..

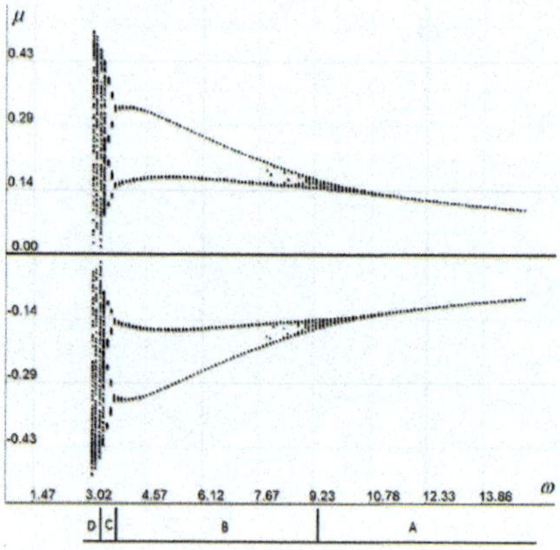

(b) $\theta_0 = 0.48$, $R = 0.4$, $A = 25$..

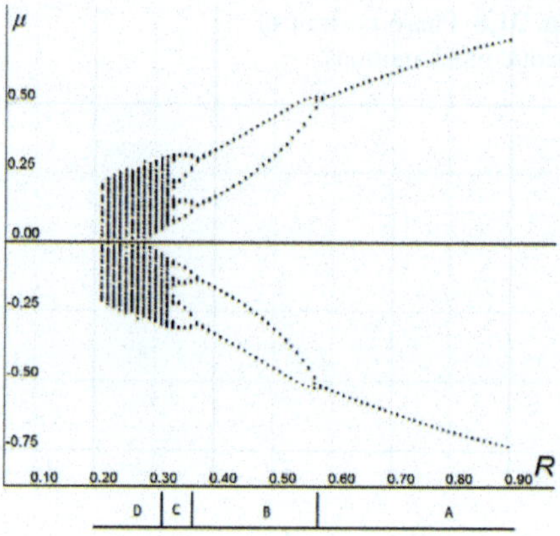

(c) $\theta_0 = 0.463648$, $\omega = 3$, $A = 18$..

Fig. 31.9: Bifurcation diagrams.

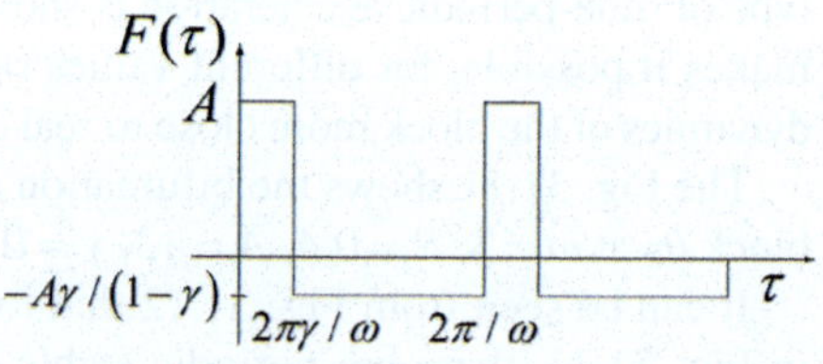

Fig. 31.10 Graphical view of platform acceleration.

From Fig. 31.15 and the numerical calculations carried out, it follows that for the amplitude values belonging to the first region A (Fig. 31.15), there correspond periodic single-turn asymmetric movements of the block with two hits of the latter on the platform. Area B (Fig. 31.15) corresponds to asymmetric two-turn periodic movements of the block with four hits of the block on the platform. Area C (Fig. 31.15) corresponds to the chaotic movement of the block.

The Fig. 31.16 shows the bifurcation diagram for the parameter ω for the values of the remaining parameters $\theta_0 = 0.48$, $R = 0.4$, $A = 17$, $\gamma = 0.3$.

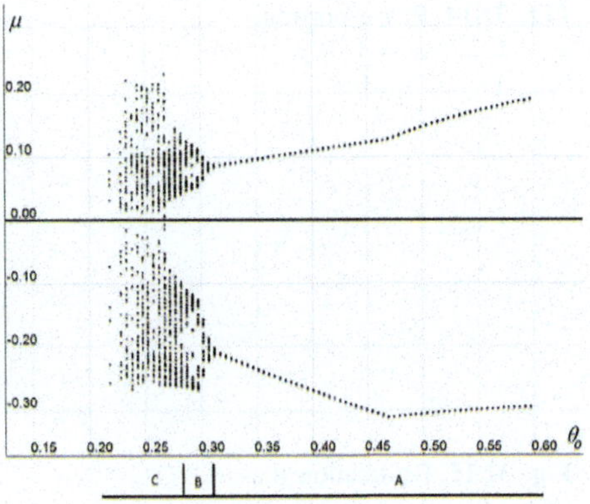

Fig. 31.11 Bifurcation diagram $at = 3$, $R = 0.4$, $A = 13$, $\gamma = 0.3$.

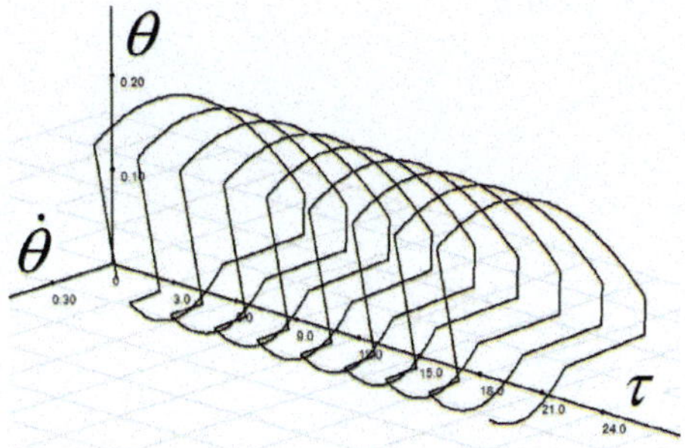

Fig. 31.12 Phase portrait. Parameters: $\theta_0 = 0.4$, $\omega = 3$, $R = 0.4$, $A = 13$, $\gamma = 0.3$.

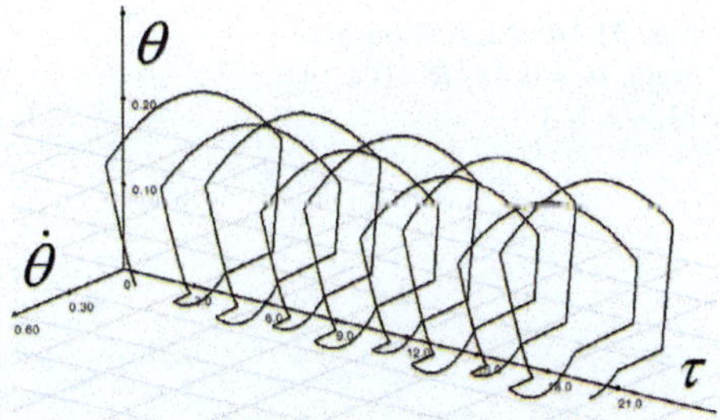

Fig. 31.13 Phase portrait. Parameters: $\theta_0 = 0.29$, $\omega = 3$, $R = 0.4$, $A = 13$, $\gamma = 0.3$.

From this diagram and the numerical analysis carried out, it is established that the area A (Fig. 31.16) corresponds to periodic single-turn asymmetric movements of the block around each of the corner points of the block with two hits of the block on the platform. Area B (Fig. 31.16) corresponds to periodic two-turn block movements with four hits of the block on the platform. Area C (Fig. 31.16) corresponds to the chaotic movement of the block.

The Fig. 31.17 shows the jufurcation diagram for the parameter R with the values of the remaining parameters equal to $\theta_0 = 0.48$, $\omega = 3$, $A = 23$, $\gamma = 0.3$

From Fig. 31.17 and the numerical analysis carried out, it was found that the range of R values belonging to A (Fig. 31.17) correspond to periodic one-turn asymmetric movements of the block with two hits of the block on the platform. Area B (Fig.

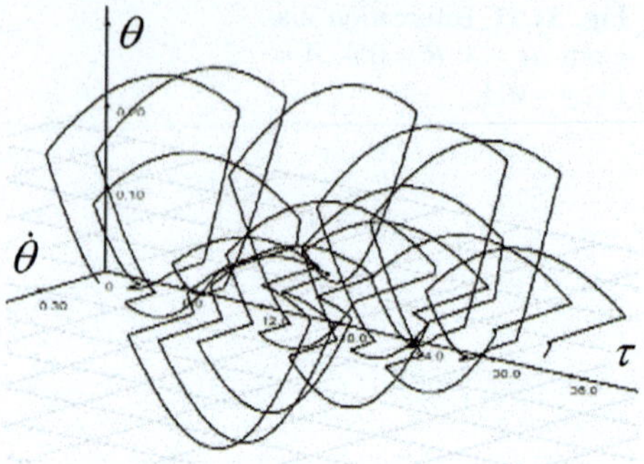

Fig. 31.14 Phase portrait. $\theta_0 = 0.25$, $\omega = 3$, $R = 0.4$, $A = 13$, $\gamma = 0.3$

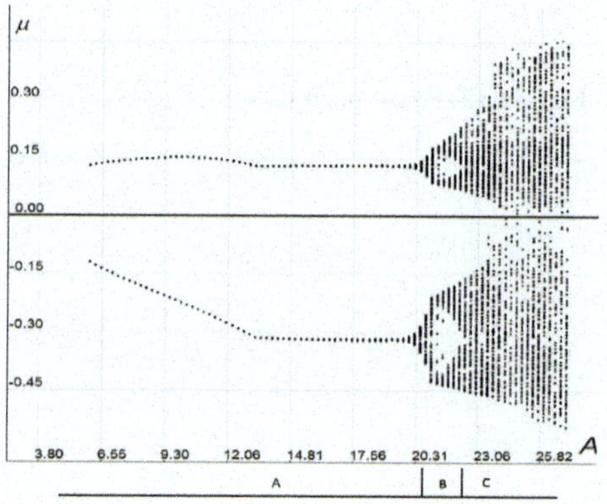

Fig. 31.15 Bifurcation diagram. $\theta_0 = 0.48$, $\omega = 3$, $R = 0.4$, $\gamma = 0.3$.

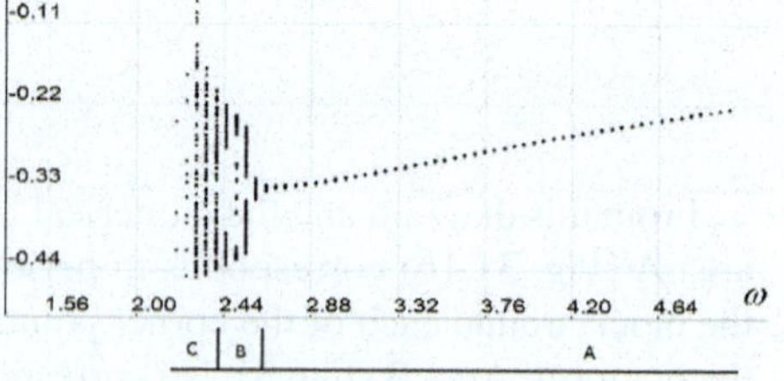

Fig. 31.16 Bifurcation diagram. $\theta_0 = 0.48$, $R = 0.4$, $A = 17$, $\gamma = 0.3$

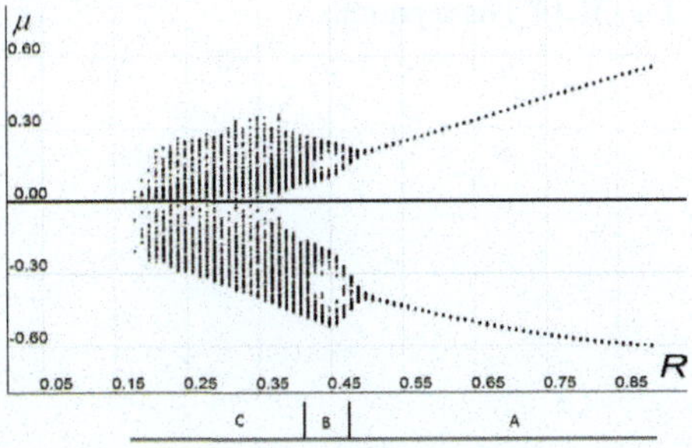

Fig. 31.17 Bifurcation diagram. $\theta_0 = 0.48$, $\omega = 3$, $R \in [0.1, 0.9]$, $A = 23$, $\gamma = 0.3$

31.17) corresponds to two-turn asymmetrical movements with four hits of the block on the platform. Area C (Fig. 31.17) corresponds to the chaotic movement of the block.

And, finally, Fig. 31.18 shows the bifurcation diagram for the off-duty parameter γ for the values of the remaining parameters equal to $\omega = 3$, $R = 0.5$, $A = 23$, $\theta_0 = 0.48$.

From Fig. 31.18 and Fig. 31.19 it follows that for the value of the duty cycle parameter γ, belonging to the area A, there correspond asymmetric one-turn periodic movements of the block with two impacts (Fig. 31.20, Fig. 31.21). The ranges of γ values belonging to the B regions correspond to asymmetric two-turn periodic motions of the block with four hits. Regions C correspond to chaotic movements.

It should be noted that qualitatively, in this case, the single-turn mode of motion differs from the previous ones in that there is an additional rotation of the block relative to one of the corners of the block without impacts (Fig. 31.20 and Fig. 31.21).

31.6 Conclusion

1. A mathematical model has been developed that describes the vibrations of a rigid rectangular body on a horizontally vibrating with acceleration $\ddot{U}(t)$ rigid supporting plane.

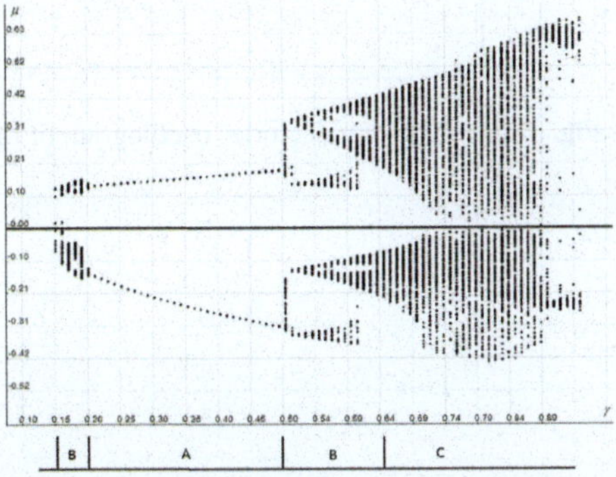

Fig. 31.18 Bifurcation diagram. $\omega = 3$, $R = 0.5$, $A = 23$, $\theta_0 = 0.48$

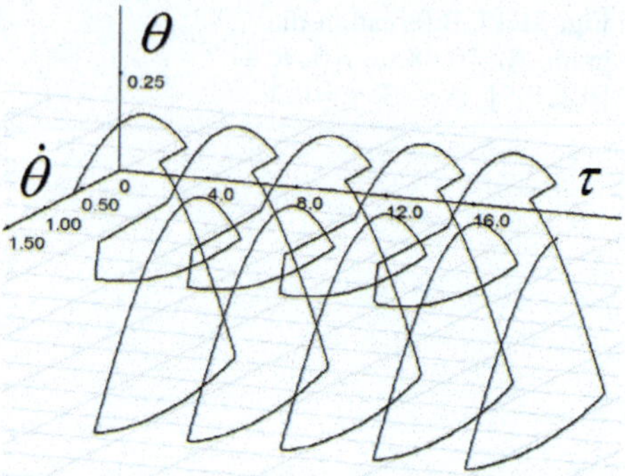

Fig. 31.19 Phase portrait.

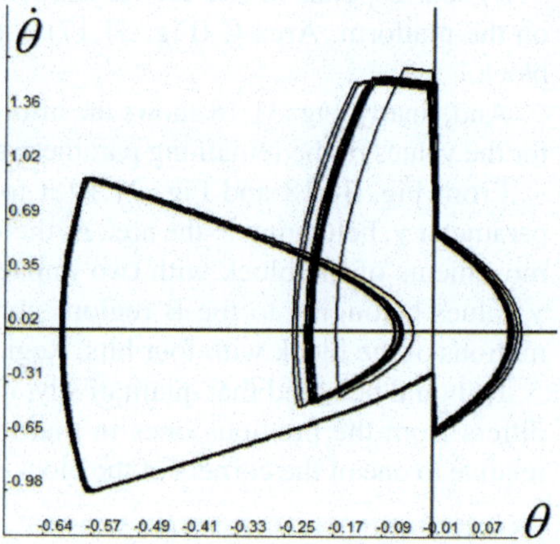

Fig. 31.20 Section of the phase portrait by a plane $\theta = 0$.

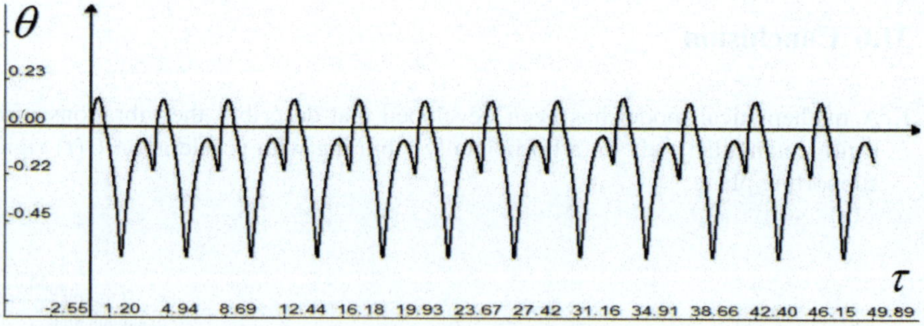

Fig. 31.21: Type of trajectories oscillogram of block trajectories.

2. For an arbitrary set of geometric dimensions of a rigid rectangular body and acceleration parameters $\ddot{U}(t)$ of the supporting plane, point mapping equations of Poincaré surfaces are presented.
3. Bifurcation diagrams are obtained, which made it possible to establish the regions of existence in the multidimensional space of DS parameters of symmetric and asymmetric stable periodic motion modes of a rigid body.
4. A scenario for the occurrence of chaotic rigid body motions under two forms of external influence on the platform is established.

Acknowledgements The work was supported by the Russian Science Foundation grant no. 22-19-00138.

References

[1] Housner GW (1963) The behavior or inverted pendulum structures during earthquakes, Bullet Seismologic Soc Amer **53**(2):403–417.
[2] Pena F, Prieto F, Lourenço PB, Campos Costa A, Lemos JV (2007) On the dynamics of rocking motion of single rigid-block structures, Earthquake Engng Struct Dyn **36**(15):2383–2399.
[3] Prieto F, Lourenço PB (2005) On the rocking behavior of rigid object, Meccanica **40**:121–133.
[4] Andreaus U, Casini P (1999) On the rocking-uplifting motion of a rigid block in free and forced motion: influence of sliding and bouncing, Acta Mech **138**:219–241.
[5] Hogan SG (1992) On the dynamics of digid block motion under harmonic forcing, Earthquake Engineer Tenth World Conf, Balkema, Rotterdam, pp 3973–3977.
[6] Spanos PD, Koh AS (1984) Rocking of rigid blocks due to harmonic shaking, J Engng Mech, ASCE **110**(11):1627–1642.
[7] Metrikin VS, Igumnov LA, Metrikin AV (2020) Application of the theory of controlled dynamic systems to the study of the dynamics of various mechanical objects in seismically active zones, Differential Equations and Control Processes (2):49-65.
[8] Hogan SJ (1990)The many steady state responses of a rigid block, Earthquake Engng Struct Dyn 19(7):1057-1071.
[9] Karapetyan AV, Munitsyna MA (2015) Dynamics of a parallelepiped on a horizontal vibrating plane, In: Automation and Telemechanics, Moscow, Nauka, pp 32-43.
[10] Makris N, Roussos Y (2000) Rocking response of rigid blocks under near-source ground motions, Géotechnique **50**(3)243-262.
[11] Munitdyn A, Munitdyna M (2016) Oscillations of a rigid block on supported base, Vibroengineering PROCEDIA **8**:63-67.

[12] Karapetyan AV, Munitsyna MA (2015) Dynamics of a parallelepiped on a horizontal vibrating plane, Automation and Remote Control **76**(3):394–404.
[13] Anooshehpoor A, Heaton TH, Shi B, Brune JN (1999) Estimates of the ground acceleration at Point Reyes Station during the 1906 San Francisco earthquake, Bull Seismological Soc Amer **89**(4):843–853.
[14] Aslam M, Scalise DT, Godden WG (1980) Earthquake rocking response of rigid bodies, J Struct Div, ASCE **106**(2):377–392.
[15] Campillo M, Gariel JC, Aki K, Sanchez-Sesma FJ (1989) Destructive strong ground motion in Mexico City: Source, path and site effects during the great 1985 Michoacán earthquake, Bull Seismological Soc Amer **79**(6):1718–1735.
[16] Iwan WD, Chen XD (1994) Important near-field ground motion data from the Landers earthquake, Proc 10th Eur Conf Earth-quake Engng, Balkema, Rotterdam, The Netherlands.
[17] Jacobsen LS, Ayre RS (1958) Engineering Vibrations, McGrawHill, New York.
[18] Makris N, Roussos Y (1998) Rocking response and overturning of equipment under horizontal pulse-type motions, Rep. No. PEER98/05, Pacific Earthquake Engng Res Ctr, University of California, Berkeley, CA.
[19] Makris N, Roussos Y (2000) Rocking response of rigid blocks under near-source ground motions, Géotechnique **50**(3):243–262.
[20] Pompei A, Scalia A, Sumbatyan MA (1998) Dynamics of rigid block due to horizontal ground motion, J Engng Mech, ASCE **124**(7):713–717.
[21] Shi B, Anooshehpoor A, Zeng Y, Brune JN (1996) Rocking and overturning of precariously balanced rocks by earthquake, Bull Seismological Soc Amer **86**(5):1364–1371.
[22] Tso WK, Wong CM (1989) Steady state rocking response of rigid blocks, Part 1: Analysis, Earthquake Engng Struct Dyn **18**(1):89–106.
[23] Tso WK, Wong CM (1989) Steady state rocking response of rigid blocks, Part 2: Experiment, Earthquake Engng Struct Dyn **18**(1):107–120.
[24] Yim C-S, Chopra AK, Penzien J (1980) Rocking response of rigid blocks to earthquakes, Earthquake Engng Struct Dyn **8**(6):565–587.
[25] Makris N, Roussos Y (2000) Rocking response of rigid blocks under near-source ground motions, Géotechnique **50**(3):243-262.
[26] Neimark YI (2010) The Point Mapping Method in the Theory of Nonlinear Oscillations (in Russ), Moscow, LIBROKOM.

Chapter 32
Asymptotically Correct Analytical Model for Flexural Response of a Two-Layer Strip with Contrast Elastic Constants

Gennadi Mikhasev and Nguyen Le

Abstract The paper is concerned with the asymptotically consistent models for a two-layer beam-strip with high-contrast elastic properties. In the general case, the strip is acted upon by dynamic body forces varying in the thickness direction, and by variable surface forces. The two dimensionless parameters, the thickness-to-wavelength ratio and the ratio of moduli for soft and stiff layers, are taken as small parameters with the first one being assumed as the main small parameter. A solution of the 2D boundary-value problem is sought in the form of an asymptotic series in the main small parameter. Step-by-step integration of the 2D equations in the thickness direction results in a sequence of relations for stresses and displacements in the form of polynomials in the transverse coordinates with coefficients depending on time and the axial coordinate. The novel asymptotically consistent 1D differential equations, corresponding to the Bernoulli-Euler and Timoshenko-Reissner models, governing vibration (or static deformation) of a beam-strip are derived. As an example, free vibrations of a simply-supported two-layer beam consisting of the soft and stiff layers are analysed relying on both models at different correlation between the geometrical and physical parameters of layers.

Key words: Two-layer strip · Contrast elastic properties · Long-wave asymptotics · Transverse shear deformable theory · Free vibrations

---

Gennadi Mikhasev
Belarusian State University, 4 Nezavisimosti Ave., 220030 Minsk, Belarus,
e-mail: mikhasev@bsu.by

Nguyen Le
Institute of Geophysics, Vietnam Academy of Science and Technology, Hanoi City, Cau Giay District, Bulding A8, 18 Hoang Quoc Viet Street, Vietnam,
e-mail: dinhnguyen081017@gmail.com

H. Altenbach et al. (eds.), *Sixty Shades of Generalized Continua*,
Advanced Structured Materials 170, https://doi.org/10.1007/978-3-031-26186-2_32

32.1 Introduction

Two-layer beams and plates are often met as elements of various thin-walled engineering structures. The use of various materials with high-contrast mechanical properties makes it possible to create the layered structures that have a wide range of properties such as high load-bearing capacity, thermal insulation, sound insulation, vibroprotection and many others. There are different theories and approaches for modelling of mechanical behaviour of laminated beams and plates, e.g., see [1]-[10]. When omitting approaches relying on the accurate but very complicated stress-strain state analysis of elastic bodies (see, among many others, contributions by [11, 12]), these theories can be classified and respectively subdivided into the three main groups [3] which are based on:

- hypotheses for displacements and stresses;
- expansion in series across the plate thickness;
- asymptotic integration of 2D or 3D equations over the beam/plate thickness.

All these approaches reduce 2D/3D equations of elasticity to 1D or 2D equations (for beams and plates, respectively) describing the mechanical behaviour of the reference line or plane. Each of these approaches has both advantages and disadvantages.

The first group of approaches are based on the engineering concept of assuming the kinematic hypotheses for displacements for each layer or/and for the entire packet with boundary conditions at the interfaces. Simplest hypotheses associated with [13] and neglecting the transverse shear strain result in the classical models and, particularly, in the Bernoulli-Euler (BE) type models for layered beams, e.g., see [14]-[20]. Introducing the hypotheses taking into account the transverse shear lead to more accurate and complicated theories. They may be subdivided into the first-order shear deformation theory (FSDT) or Timoshenko type theories (e.g., see [21]-[25]), the higher-order shear deformation theory (HSDT) (among others, see [26]-[33]) and the layer-wise theories (LWT), e.g., see [34]-[39]. Applying the classical theory, or the FSDT and HSDT theories, based on any hypotheses, usually result in the equivalent single layer models (ESL) (e.g., see [40]) for layered beams and plates, which contain a limited number of equations independent of a number of layers, while the LWTs yield the system of differential equations depending on a number of lamina. It is obvious that the accuracy of the LWT, accounting the zig-zag effects, are higher than an accuracy of the ESL theories [41, 42], however for a multi-layered structure the LWTs turns to be extremely complex in the theoretical formulations and computationally costly. The general disadvantage of any theory relying on the kinematic hypotheses is that its accuracy is fixed and can not be improved in the framework of the predetermined hypotheses.

The second approach deals with expansion of required functions (displacements or/and strains and stresses) in power series or in Legendre polynomials with respect to the transverse coordinate [43]-[47]. The consistency of an approach based on expansions in coordinates and its high accuracy is achieved by keeping a large number of terms in series. The advantage of this approach is that it is free of any a priori assumptions, and the drawback is that the construction of more accurate theory inevitably leads to a large number of equations with respect to coefficients in series. As [47] point out, "the gain of accuracy must be counterbalanced by the increase of complexity".

The asymptotic approach is based on expansion of all functions from the 2D/3D equations of elasticity in series in powers of a small parameter [48]. Depending on the class of solutions being constructed and their properties, as a small parameter is taken the ratio of a beam/plate thickness to an external characteristic length, the thickness-to-wavelength ratio, the ratio of moduli of softer and stiffer layers, the ratio of amplitudes of in-plane and transverse displacements, etc. (among others, see [49]-[55]). The problem under consideration is often characterized by two or more small parameters. In this case, the choice of the main small parameter and the asymptotic estimate of others results in different beam and plate theories and, respectively, in different governing equations describing, e.g., low-frequency or high-frequency transverse and tangential vibrations of a thin beam/plate [56] or low-frequency transverse vibrations of a nano-sized plate [55, 57]. The advantage of the asymptotic approach is that its correct application and step-by-step integration allows constructing theories corresponding to different hypotheses. For example, considering the leading and higher approximations in this procedure applied for a multilayered beam/plate, [54] derived equations corresponding to the generalized BE and Timoshenko-Reissner (TR) models, respectively, and [55], using the 3D gradient theory of elasticity for a nanoplate, obtained equations which generalize both the classical equation of Kirchhoff-Love model and equations of the Uflyand-Mindlin theory [24, 58] and take into account the spatial nonlocal effects as well.

The asymptotic approach turns out to be very effective for studying mechanical behaviour of thin-walled layered structures with high-contrast properties of laminates. For the low-frequency vibrations of such structures, the ratio of moduli for soft and stiff layers, together with the thickness-to-wavelength ratio, becomes the main small parameter, in which all functions are expanded into an asymptotic series. To date, there are few works containing the correct asymptotic integration of 3D equations of elasticity for layered structures with high-contrast properties. Basically, these studies are devoted to the asymptotic analysis of the dispersion relations corresponding to the plane anti-symmetric [60, 61] and antiplane shear waves [62, 63, 64] propagating in layered plate taking into account contrast properties of constituent layers. Using the Saint-Venant's principle accounting for the contrast properties of the laminate, [65] and [66] derived asymptotically correct decay boundary conditions. Finally, studying the dynamic behaviour of thin multi-layered structures composed of contrasting "strong" and "weak" layers, [67] developed the asymptotic procedure for analysing the dispersion equations.

In contrast to the mentioned contributions, the current paper is focused on the asymptotic procedure that allows step-by-step integration of 2D elasticity equations for a two-layer beam-strip with high-contrast properties. One of the objectives of the study is the derivation of the hierarchic differential equations corresponding to the BE and TR type models that govern the long-wave dynamics and static deformation of two-layer beam-strips taking into account both contrast in elastic properties and transverse shear.

32.2 Statement of the Problem

We consider a two-layer beam-strip consisting of two different rigidly coupled isotropic elastic layers of thickness h_k, which are characterized by Lamé constants λ_k, μ_k and material densities $\rho_k (k=1,2)$. Two local coordinate systems $O^{(k)}x_1^{(k)}x_2^{(k)}$ are introduced as shown in Fig. 32.1 with $x_1^{(1)} = x_1^{(2)} = x^{(k)}$, where the indices $k=1$ and $k=2$ correspond to the lower and upper layers, respectively. The beam is under action of the external surface load

$$\mathbf{Q}^{(k)} = \left(q_1^{(k)}, q_2^{(k)}\right)$$

and the body forces

$$\mathbf{F}^{(k)} = \left(f_1^{(k)}, f_2^{(k)}\right),$$

where $q_\varsigma^{(k)}(x_1^{(k)}, t)$ are the functions of a longitudinal coordinate $x_1^{(k)}$ and time t, and $f_\varsigma^{(k)}(x_1^{(k)}, x_2^{(k)})$ are functions of longitudinal and transverse coordinates $x_\varsigma^{(k)}$ $(\varsigma=1,2)$.

The 2D equations for both layers read

$$s_{ij,j}^{(k)} - \rho_k \frac{\partial^2 u_i^{(k)}}{\partial t^2} + f_i^{(k)} = 0, \quad i,j,k = 1,2, \tag{32.1}$$

where $s_{ij}^{(k)}$ are components of the stress tensor in both layers under the plane stress state, and $u_i^{(k)}(x_j^{(k)}, t)$ are components of the displacement vector.

For an isotropic materials of both layers, the stresses are given by

$$s_{ij}^{(k)} = \lambda_k e_{ii}^{(k)} \delta_{ij} + 2\mu_k e_{ij}^{(k)}, \tag{32.2}$$

with

$$e_{ij}^{(k)} = \frac{1}{2}\left(\frac{\partial u_i^{(k)}}{\partial x_j^{(k)}} + \frac{\partial u_j^{(k)}}{\partial x_i^{(k)}}\right), \tag{32.3}$$

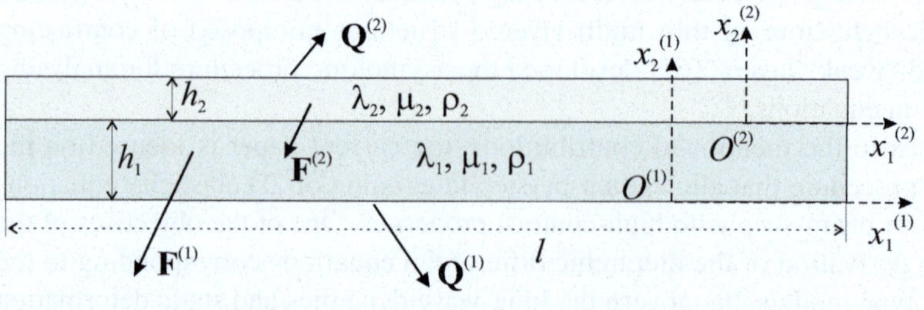

Fig. 32.1: Two-layer beam-strip under surface and body forces. Coordinate systems.

where $e_{ij}^{(k)}$ are linear components of the strain tensor, and δ_{ij} is the Kronecker delta.

The first basic assumption in our study is that the bottom layer is much stiffer than the top one so that $\lambda_2 \ll \lambda_1, \mu_2 \ll \mu_1$. Let l be the strip length or the characterized size of bending deformation. The second assumption introduced here is that $h \ll l$. We introduce two small parameters

$$\varepsilon = \frac{h}{l}, \quad \frac{E_2}{E_1} = \varepsilon_{21} = \varepsilon^{\alpha} \upsilon, \quad \alpha \geq 0, \quad \upsilon \sim 1, \tag{32.4}$$

where ε is a main small parameter, $h = h_1 + h_2$ is the total thickness of the strip, E_k are the Young moduli of layers, and a positive parameter α characterizes the contrast rate of the material constants of the layers.

We transfer to dimensionless variables as follows,

$$\begin{gathered} x_1^{(k)} = lx, \quad x_2^{(k)} = h_k z^{(k)}, \quad \left\{u_1^{(k)}, u_2^{(k)}\right\} = h\left\{u^{(k)}, w^{(k)}\right\}, \\ s_{ij}^{(k)} = E_k \sigma_{ij}^{(k)}, \quad q_2^{(k)} = E_k q^{(k)}, \quad f_2^{(k)} = \frac{E_k}{h_k} \hat{f}_2^{(k)}, \\ q_1^{(k)} = \varepsilon E_k s^{(k)}, \quad f_1^{(k)} = \frac{\varepsilon E_k}{h_k} \hat{f}_1^{(k)}, \quad i, j, k = 1, 2. \end{gathered} \tag{32.5}$$

and assume that all dimensionless forces are values of order $O(1)$ as $\varepsilon \to 0$. Then the latter relations in Eqs. (32.5) mean that the tangential dimensional components $q_1^{(k)}, f_1^{(k)}$ of the surface and body forces are smaller than the transverse ones.

Equation (32.2) for $s_{22}^{(k)}$, together with the strain-displacement relations (32.3), gives the following equation in the dimensionless form:

$$\frac{\partial w^{(k)}}{\partial z^{(k)}} = -\varepsilon \kappa_k c_\nu^{(k)} \frac{\partial u^{(k)}}{\partial x} + \kappa_k c_3^{(k)} \upsilon_{33}^{(k)}, \tag{32.6}$$

where

$$\begin{aligned} \kappa_k &= \frac{h_k}{h}, \\ c_3^{(k)} &= \frac{E_k}{\lambda_k + 2\mu_k}, \\ c_\nu^{(k)} &= \frac{\lambda_k}{\lambda_k + 2\mu_k}. \end{aligned} \tag{32.7}$$

We introduce also additional dimensionless constants which will be used below:

$$\begin{aligned} c_0^{(k)} &= \frac{4\mu_k(\lambda_k+\mu_k)}{E_k(\lambda_k+2\mu_k)}, \\ c_4^{(k)} &= \frac{3\lambda_k+4\mu_k}{\lambda_k+2\mu_k}, \\ c_6^{(k)} &= \frac{3\lambda_k+2\mu_k}{\lambda_k+2\mu_k}, \\ c_g^{(k)} &= \frac{E_k}{\mu_k}. \end{aligned} \tag{32.8}$$

Taking (32.6) into account, from Eq. (32.2) for s_{11} we observe that

$$\sigma_{11}^{(k)} = \varepsilon c_0^{(k)} \frac{\partial u^{(k)}}{\partial x} + c_\nu^{(k)} \sigma_{33}^{(k)}. \tag{32.9}$$

Now governing Eqs. (32.1) can be rewritten as follows,

$$\begin{aligned} \frac{\partial \sigma_{12}^{(k)}}{\partial z^{(k)}} &= -\varepsilon^2 \kappa_k c_0^{(k)} \frac{\partial^2 u^{(k)}}{\partial x^2} - \varepsilon \kappa_k c_\nu^{(k)} \frac{\partial \sigma_{22}^{(k)}}{\partial x} + \varepsilon^4 \kappa_k c_\omega^{(k)} \frac{\partial^2 u^{(k)}}{\partial t^2} - \varepsilon \hat{f}_1^{(k)} \\ \frac{\sigma_{22}^{(k)}}{\partial z^{(k)}} &= -\varepsilon \kappa_k \frac{\partial \sigma_{12}^{(k)}}{\partial x} + \varepsilon^4 \kappa_k c_\omega^{(k)} \frac{\partial^2 w^{(k)}}{\partial t^2} - \hat{f}_3^{(k)}, \end{aligned} \tag{32.10}$$

where $c_\omega^{(k)} = \varepsilon^{-4} h^2 \rho_k / E_k$, and Eq. (32.2) for $s_{12}^{(k)}$ results in the equation

$$\frac{\partial u^{(k)}}{\partial z} = -\varepsilon \kappa_k \frac{\partial w^{(k)}}{\partial x} + \kappa_k c_g^{(k)} \sigma_{12}^{(k)}. \tag{32.11}$$

Here, $\sqrt{c_\omega^{(k)}}$ is the dimensional magnitude which may be treated as the characteristic time for the kth layer.

In what follows, the eight functions $w^{(k)}, u^{(k)}, \sigma_{12}^{(k)}, \sigma_{22}^{(k)}$ will be considered as the main unknowns. To study low-frequency bending vibrations or static flexural response of the strip with a large characteristic dimension of deformations, we assume the following asymptotic estimates [55, 68]

$$w^{(k)} = \varepsilon^{-4} W^{(k)}, \quad u^{(k)} = \varepsilon^{-3} U^{(k)}, \quad \sigma_{12}(k) = \varepsilon^{-1} \tau^{(k)}, \quad \sigma_{22}^{(k)} = \sigma^{(k)}, \tag{32.12}$$

where $W^{(k)}, U^{(k)}, \tau^{(k)}, \sigma^{(k)} \sim 1$. Then Eqs. (32.6), (32.10), (32.11) take the form

$$\begin{aligned}
\frac{\partial W^{(k)}}{\partial z} &= -\varepsilon^2 \kappa_k c_\nu^{(k)} \frac{\partial U^{(k)}}{\partial x} + \varepsilon^4 \kappa_k c_3^{(k)} \sigma^{(k)},\\
\frac{\partial U^{(k)}}{\partial z} &= -\kappa_k \frac{\partial W^{(k)}}{\partial x} + \varepsilon^2 \kappa_k c_g^{(k)} \tau^{(k)},\\
\frac{\partial \tau^{(k)}}{\partial z} &= -\kappa_k c_0^{(k)} \frac{\partial^2 U^{(k)}}{\partial x^2} - \varepsilon^2 \kappa_k c_\nu^{(k)} \frac{\partial \sigma^{(k)}}{\partial x} + \varepsilon^2 \kappa_k c_\omega^{(k)} \frac{\partial^2 U^{(k)}}{\partial \hat{t}^2} - \varepsilon^2 \hat{f}_1^{(k)},\\
\frac{\sigma^{(k)}}{\partial z} &= -\kappa_k \frac{\partial \tau^{(k)}}{\partial x} + \kappa_k c_\omega^{(k)} \frac{\partial^2 W^{(k)}}{\partial \hat{t}^2} - \hat{f}_3^{(k)}, \quad k = 1,2.
\end{aligned} \tag{32.13}$$

In the above equations, omitting the upper index $^{(k)}$ at the transverse coordinate $z^{(k)}$, we assume that consider the kth layer in the local coordinate system $O^{(k)}xz^{(k)}$, where $0 \le z = z^{(k)} \le 1$.

The boundary conditions at the face lines read

$$\begin{aligned}
\tau^{(1)}|_{z=0} &= \varepsilon^2 s^{(1)}(x,t), \quad \sigma^{(1)}|_{z=0} = q^{(1)}(x,t),\\
\tau^{(2)}|_{z=1} &= \varepsilon^2 s^{(2)}(x,t), \quad \sigma^{(2)}|_{z=1} = q^{(2)}(x,t).
\end{aligned} \tag{32.14}$$

When assuming the rigid contact between layers and estimate (32.4) for the material constants, the interface conditions can be written as follows,

$$\begin{aligned}
W^{(1)}|_{z=1} &= W^{(2)}|_{z=0}, \quad U^{(1)}|_{z=1} = U^{(2)}|_{z=0},\\
\tau^{(1)}|_{z=1} &= \varepsilon^\alpha \upsilon \tau^{(2)}|_{z=0}, \quad \sigma^{(1)}|_{z=1} = \varepsilon^\alpha \upsilon \sigma^{(2)}|_{z=0}.
\end{aligned} \tag{32.15}$$

We have the boundary-value problem (32.13)–(32.15) which may be classified depending on a parameter α. Below, we consider only the following two cases: a) $\alpha = 0$, b) $\alpha = 2$. The case a) means that both layer have the elastic constants of the same order, while the case b) corresponds to the high-contrast elastic properties of the layers.

Regardless of the case, a solution of the boundary-value problem (32.13)–(32.15) corresponding to the long-wave deformations can be found in the form of formal asymptotic series

$$\begin{aligned}
W^{(k)} &= w_0^{(k)} + \varepsilon^2 w_2^{(k)} + \ldots, \quad U^{(k)} = u_0^{(k)} + \varepsilon^2 u_2^{(k)} + \ldots,\\
\tau^{(k)} &= \tau_0^{(k)} + \varepsilon^2 \tau_2^{(k)} + \ldots, \quad \sigma^{(k)} = \sigma_0^{(k)} + \varepsilon^2 \sigma_2^{(k)} + \ldots
\end{aligned} \tag{32.16}$$

32.3 Bernoulli-Euler Type Model for Strip Consisting of Layers with Close Material Constants

Let $\alpha = 0$. This case corresponds to a beam-strip consisting of two isotropic layers with different but close elastic properties. We substitute series (32.16) into Eqs. (32.13) and boundary conditions (32.14), (32.15) and consider here only the leading approximation which will be used below for a comparison.

The first two equations of system (32.13) for the transverse displacements with the first interface conditions (32.15) give $w_0^{(1)} = w_0^{(2)} = w_0(x,t)$.

Considering the second pair of Eqs. (32.13) with the second condition (32.15) taking into account, we obtain

$$u_0^{(k)} = -\kappa_k \frac{\partial w_0}{\partial x} z + u_{00}^{(k)}(x,t), \quad k = 1,2, \tag{32.17}$$

where

$$u_{00}^{(2)} = u_{00}^{(1)} - \kappa_1 \frac{\partial w_0}{\partial x}. \tag{32.18}$$

Integrating the third pair of equations from (32.13) and accounting for the homogeneous boundary conditions (32.14) for the shear forces in the leading approximation, we arrive at the following two relations:

$$\begin{aligned} \tau_0^{(1)}(x,z,t) &= \frac{1}{2}\kappa_1^2 c_0^{(1)} \frac{\partial^3 w_0}{\partial x^3} z^2 - \kappa_1 c_0^{(1)} \frac{\partial^2 u_{00}^{(1)}}{\partial x^2} z, \\ \tau_0^{(2)}(x,z,t) &= \frac{1}{2}\kappa_2^2 c_0^{(2)} \frac{\partial^3 w_0}{\partial x^3} (z^2-1) - \kappa_2 c_0^{(2)} \frac{\partial^2 u_{00}^{(2)}}{\partial x^2}(z-1). \end{aligned} \tag{32.19}$$

Applying interface condition (32.15) to Eqs. (32.19) and taking into account formula (32.18), one obtains

$$u_{00}^{(k)}(x,t) = \zeta_k \frac{\partial w_0}{\partial x}, \quad k = 1,2 \tag{32.20}$$

where

$$\zeta_1 = \frac{1}{2} \frac{\kappa_1^2 c_0^{(1)} + \varepsilon_{21}(1-\kappa_1^2) c_0^{(2)}}{\kappa_1 c_0^{(1)} + \varepsilon_{21}\kappa_2 c_0^{(2)}}, \quad \zeta_2 = \frac{1}{2} \frac{\varepsilon_{21}\kappa_2^2 c_0^{(2)} - \kappa_1^2 c_0^{(1)}}{\kappa_1 c_0^{(1)} + \varepsilon_{21}\kappa_2 c_0^{(2)}}. \tag{32.21}$$

Then the longitudinal displacements are

$$u_0^{(k)} = (\zeta_k - \kappa_k z)\frac{\partial w_0}{\partial x}, \quad k = 1,2, \tag{32.22}$$

and the shear stresses read,

$$\tau_0^{(1)}(x,z,t) = \frac{1}{2}\kappa_1 c_0^{(1)}(\kappa_1 z^2 - 2\zeta_1 z)\frac{\partial^3 w_0}{\partial x^3},$$
$$\tau_0^{(2)}(x,z,t) = \frac{1}{2}\kappa_2 c_0^{(2)}[\kappa_2(z^2-1) - 2\zeta_2(z-1)]\frac{\partial^3 w_0}{\partial x^3}. \tag{32.23}$$

Finally, consider the last two equations in system (32.13). Integrating them with respect to z and taking into account inhomogeneous boundary conditions (32.14) at the face lines, one obtains the relations for normal stresses:

$$\sigma_0^{(1)} = -\frac{1}{6}c_0^{(1)}\kappa_1^2(\kappa_1 z^3 - 3\zeta_1 z^2)\frac{\partial^4 w_0}{\partial x^4} + c_\omega^{(1)}\kappa_1 z\frac{\partial^2 w_0}{\partial t^2}$$
$$-\int_0^z \hat{f}_2^{(1)}dz + q^{(1)}(x,t),$$
$$\sigma_0^{(2)} = -\frac{1}{6}c_0^{(2)}\kappa_2^2[\kappa_2(z^3-3z+2) - 3\zeta_2(z^2-2z+1)]\frac{\partial^4 w_0}{\partial x^4}$$
$$+c_\omega^{(2)}\kappa_2(z-1)\frac{\partial^2 w_0}{\partial t^2} + \int_0^z \hat{f}_2^{(2)}dz + q^{(2)}(x,t). \tag{32.24}$$

Applying the last condition from (32.15) at the interface to above stresses, we arrive at the classical form differential equation,

$$I_r\frac{\partial^4 w_0}{\partial x^4} + J_r\frac{\partial^2 w_0}{\partial t^2} = F_r(x,t), \tag{32.25}$$

with

$$I_r = \frac{\left(c_0^{(1)}\kappa_1^2 + \varepsilon_{21}c_0^{(2)}\kappa_2^2\right)^2 + 4\varepsilon_{21}c_0^{(1)}c_0^{(2)}\kappa_1\kappa_2(\kappa_1+\kappa_2)^2}{12\left(c_0^{(1)}\kappa_1 + \varepsilon_{21}c_0^{(2)}\kappa_2\right)},$$
$$J_r = c_\omega^{(1)}\kappa_1 + \varepsilon_{21}c_\omega^{(2)}\kappa_2, \tag{32.26}$$
$$F_r = -q^{(1)}(x,t) + \int_0^1 \hat{f}_2^{(1)}(x,z,t)dz + \varepsilon_{21}\left[q^{(2)}(x,t) + \int_0^1 \hat{f}_2^{(2)}(x,z,t)dz\right].$$

The next approximation leads to very cumbersome transformations and the resulting equations. We restrict ourselves here to the leading approximation, which will be used below to verify the results related to the two-layer strip with high-contrast material constants.

Equation (32.25) can be readily transformed into an equation for the two-layer beam-strip in the dimensional form:

$$\mathcal{I}_r \frac{\partial^4 w^\circ}{\partial x_1^4} + \rho_r h \frac{\partial^2 w^\circ}{\partial t^2} = \mathcal{F}_r(x_1,t). \tag{32.27}$$

Here

$$\mathcal{I}_r = \frac{(K_1 h_1 + K_2 h_2)^2 + 4K_1 K_2 (h_1^2 + h_1 h_2 + h_2^2)}{12(K_1 + K_2)} \tag{32.28}$$

is the reduced flexural rigidity of the two-layer strip with $K_k = E_k h_k/(1-\nu_k^2)$, ν_k is Poisson's ratio of the kth layer, w° is the transverse displacement of the interface line, $\rho_r = (\rho_1 h_1 + \rho_2 h_2)/h$ is the reduced mass density per unit volume, and

$$\mathcal{F}_r = q_2^{(2)}(x_1,t) - q_2^{(1)}(x_1,t) + \sum_{k=1}^{2} \int_0^{h_k} f_2^{(k)}\left(x_1, x_2^{(k)}, t\right) dx_2^{(k)} \tag{32.29}$$

is the reduced external force per unit area.

We note that Eq. (32.25) does not take into account the transverse shear and is related to the Bernoulli-Euler type equation. Regardless of the rate of contrast of elastic parameters of laminae composing a two-layer beam-strip, this equation may be used to predict the long-wave low-frequency flexural vibrations with the error of order $O(\varepsilon^2)$. Accounting for transverse shears in both layers can only be achieved by considering the following approximation, which in the case of peer layers leads to very cumbersome relations.

32.4 Timoshenko-Reissner Type Model for Strip Consisting of Layers with High-Contrast Elastic Properties

Consider the case when $\alpha = 2$ in Eqs. (32.15) and additionally assume that $(q^{(1)})^2 + (\hat{f}_2^{(1)})^2 \neq 0$ at least in one point $x \in (0,1)$. This condition means that the case when the lower rigid layer is free of load is not considered.

We substitute series (32.16) into Eqs. (32.13) and integrate over the segment $[z,0]$. As a result, we arrive at the following sequence of recurrent relations:

$$w_n^{(k)}(x,z,t) = -\kappa_k \int_0^z \left(c_\nu^{(k)} \frac{\partial u_{n-2}^{(k)}}{\partial x} - c_3^{(k)} \sigma_{n-4}^{(k)} \right) dz + w_{n0}^{(k)}(x,t),$$

$$u_n^{(k)}(x,z,t) = -\kappa_k \int_0^z \left(\frac{\partial w_n^{(k)}}{\partial x} - c_g^{(k)} \tau_{n-2}^{(k)} \right) dz + u_{n0}^{(k)}(x,t),$$

$$\tau_n^{(k)}(x,z,t) = -\kappa_k \int_0^z \left(c_0^{(k)} \frac{\partial^2 u_n^{(k)}}{\partial x^2} + c_\nu^{(k)} \frac{\partial \sigma_{n-2}^{(k)}}{\partial x} - c_\omega^{(k)} \frac{\partial^2 u_{n-2}^{(k)}}{\partial t^2} \right) dz$$

$$-H(n-2)H(2-n)\int_0^z \hat{f}_1^{(k)}(x,z,t)dz+\tau_{n0}^{(k)}(x,t), \qquad (32.30)$$

$$\sigma_n^{(k)}(x,z,t)=-\kappa_k\int_0^z\left(\frac{\partial\tau_n^{(k)}}{\partial x}-c_\omega^{(k)}\frac{\partial^2 w_n^{(k)}}{\partial t^2}\right)$$

$$-H(n)H(-n)\int_0^z \hat{f}_3^{(k)}(x,z,t)d+\sigma_{n0}^{(k)}(x,t), \quad n=0,2,4,\ldots,$$

where $H(\xi)$ is the Heaviside function the value of which is zero for negative arguments ξ and one for non-negative ξ, and $w_{n0}^{(k)}, u_{n0}^{(k)}, \tau_{n0}^{(k)}, \sigma_{n0}^{(k)}$ are "constants"of integration with respect to the transverse coordinate z, which are to be determined from boundary conditions. It is obvious that functions with negative subscripts at the right-hand side of relations (32.30) must be set to zero.

The substitution of series (32.15) into boundary conditions (32.14) and (32.15) leads to the sequence of conditions for required functions:

- at the boundaries,

$$\tau_0^{(1)}(x,0,t)=\tau_0^{(2)}(x,1,t)=0,\ \sigma_0^{(1)}(x,0,t)=q^{(1)}(x,t),\ \sigma_0^{(2)}(x,1,t)=q^{(2)}(x,t),$$

$$\tau_2^{(1)}(x,0,t)=s^{(1)}(x,t),\ \tau_2^{(2)}(x,1,t)=s^{(2)}(x,t),\ \sigma_2^{(1)}(x,0,t)=\sigma_2^{(2)}(x,1,t)=0,$$

$$\tau_n^{(1)}(x,0,t)=\tau_n^{(2)}(x,1,t)=\sigma_n^{(1)}(x,0,t)=\sigma_n^{(2)}(x,1,t)=0 \quad \text{for} \quad n=4,6,\ldots, \qquad (32.31)$$

- at the interface,

$$w_n^{(1)}(x,1,t)=w_n^{(2)}(x,0,t), \quad u_n^{(1)}(x,1,t)=u_n^{(2)}(x,0,t) \quad \text{for} \quad n=0,2,4,\ldots,$$

$$\tau_0^{(1)}(x,1,t)=0, \quad \sigma_0^{(1)}(x,1,t)=0,$$

$$\tau_n^{(1)}(x,1,t)=\upsilon\tau_{n-2}^{(2)}(x,0,t), \quad \sigma_n^{(1)}(x,1,t)=\upsilon\sigma_{n-2}^{(2)}(x,0,t) \quad \text{for} \quad n=2,4,\ldots, \qquad (32.32)$$

We have the sequence of boundary-value problems (32.30)–(32.32) which can be solved step-by-step. Note that for the case of high-contrast layers, the boundary conditions at the interface become homogeneous in the leading approximation and inhomogeneous in the first-order and higher approximations.

32.4.1 Leading Approximation

In the leading approximation ($n=0$), Eqs. (32.30) with the corresponding boundary conditions gives

$$
\begin{aligned}
w_0^{(k)} &= w_{00}^{(k)}(x,t) = w_0(x,t) \quad \text{for} \quad k = 1,2, \\
u_0^{(1)} &= -\frac{1}{2}\kappa_1(2z-1)\frac{\partial w_0}{\partial x}, \quad u_0^{(2)} = -\frac{1}{2}\kappa_2\left(2z+\frac{\kappa_1}{\kappa_2}\right)\frac{\partial w_0}{\partial x}, \\
\tau_0^{(1)} &= \frac{1}{2}c_0^{(1)}\kappa_1^2(z^2-z)\frac{\partial^3 w_0}{\partial x^3}, \quad \tau_0^{(2)} = \frac{1}{2}c_0^{(2)}\kappa_2[\kappa_2(z^2-1)+\kappa_1(z-1)]\frac{\partial^3 w_0}{\partial x^3}, \\
\sigma_0^{(1)} &= -\frac{1}{12}c_0^{(1)}\kappa_1^3(2z^3-3z^2)\frac{\partial^4 w_0}{\partial x^4} + c_\omega^{(1)}\kappa_1 z\frac{\partial^2 w_0}{\partial \hat{t}^2} \\
&\quad + q^{(1)}(x,t) - \int_0^z \hat{f}_2^{(1)}(x,z)dz, \\
\sigma_0^{(2)} &= -\frac{1}{12}c_0^{(2)}\kappa_2^2[\kappa_2(2z^3-6z)+\kappa_1(3z^2-6z)+3+\kappa_2]\frac{\partial^4 w_0}{\partial x^4} \\
&\quad + c_\omega^{(2)}\kappa_2(z-1)\frac{\partial^2 w_0}{\partial t^2} + q^{(2)}(x,t) + \int_z^1 \hat{f}_2^{(2)}(x,z)dz
\end{aligned}
\tag{32.33}
$$

where $w_0(x,t)$ is a solution of the differential equation

$$
\boldsymbol{L}^{(1)}w_0 \equiv \frac{1}{12}c_0^{(1)}\kappa_1^3\frac{\partial^4 w_0}{\partial x^4} + c_\omega^{(1)}\kappa_1 z\frac{\partial^2 w_0}{\partial t^2} = F^{(1)}(x,t) \tag{32.34}
$$

with

$$
F^{(1)}(x,t) = -q^{(1)}(x,t) + \int_0^1 \hat{f}_2^{(1)}(x,z)dz. \tag{32.35}
$$

Equation (32.34) describes the transverse displacements of points at the interface line of the two-layer strip. We note that it relates to the Bernoulli-Euler beam model and may be treated as the equation governing transverse response of the stiffer part of the two-layer beam-strip without effect of the softer part. As seen from Eqs. (32.33), in the leading approximation the distribution of longitudinal displacements and stresses through the thickness of the attached upper softer layer depends on the external forces $q^{(2)}, \hat{f}_2^{(2)}$ and on the transverse displacements of the stiffer layer as well.

32.4.2 First-Order Approximation

Consider the next approximation in more detail. The first four equations (32.30) for the displacement components $w_2^{(k)}, u_2^{(k)}$ with conditions at the interface line (32.32) lead to the following relations:

$$
\begin{aligned}
w_2^{(1)} &= \frac{1}{2} c_\nu^{(1)} \kappa_1^2 (z^2 - z) \frac{\partial^2 w_0}{\partial x^2} + w_{20}(x,t), \\
w_2^{(2)} &= -\frac{1}{2} c_\nu^{(2)} \kappa_2^2 \left(z^2 + \frac{\kappa_1}{\kappa_2} z \right) \frac{\partial^2 w_0}{\partial x^2} + w_{20}(x,t), \\
u_2^{(1)} &= \frac{1}{12} c_4^{(1)} \kappa_1^3 (2z^3 - 3z^2) \frac{\partial^3 w_0}{\partial x^3} - \kappa_1 z \frac{\partial w_{20}}{\partial x} z + u_{20}(x,t), \\
u_2^{(2)} &= \frac{1}{12} \left(2c_4^{(2)} \kappa_2^3 z^3 + 3c_4^{(2)} \kappa_1 \kappa_2^2 z^2 - 6c_0^{(2)} c_g^{(2)} \kappa_2^2 z - c_4^{(1)} \kappa_1^3 \right) \frac{\partial^3 w_0}{\partial x^3} \\
&\quad - (\kappa_2 z + \kappa_1) \frac{\partial w_{20}}{\partial x} + u_{20}(x,t),
\end{aligned}
\tag{32.36}
$$

where $u_{20} = u_{20}^{(1)}(x,t)$ is a function independent of z to be defined below.

Consider the next pair of Eqs. (32.30) with respect to $\tau_2^{(k)}$. Taking into account the previous findings and inhomogeneous boundary conditions (32.31) at the lower and upper boundaries, we obtain

$$
\begin{aligned}
\tau_2^{(1)} &= -\frac{1}{12} c_0^{(1)} \kappa_1^4 (z^4 - 2z^3) \frac{\partial^5 w_0}{\partial x^5} - \frac{1}{4} c_\omega^{(1)} \kappa_1^2 \left(c_0^{(1)} c_g^{(1)} z^2 - 2z \right) \frac{\partial^3 w_0}{\partial x \partial t^2} \\
&\quad + \frac{1}{2} c_0^{(1)} \kappa_1^2 z^2 \frac{\partial^3 w_{20}}{\partial x^3} - c_0^{(1)} \kappa_1 z \frac{\partial^2 u_{20}}{\partial x^2} + T^{(1)}(x,z,t), \\
\tau_2^{(2)} &= -\frac{1}{12} c_0^{(2)} \kappa_2 \left[\kappa_2^3 (z^4 - 1) + 2\kappa_1 \kappa_2^2 (z^3 - 1) - 3c_4^{(2)} \kappa_2^2 (z^2 - 1) - c_4^{(1)} \kappa_1^3 (z - 1) \right] \frac{\partial^5 w_0}{\partial x^5} \\
&\quad - \frac{1}{4} c_\omega^{(2)} \kappa_2^2 \left[c_0^{(2)} c_g^{(2)} (z^2 - 1) + 2 \left(\frac{\kappa_1}{\kappa_2} - 2c_\nu^{(2)} \right) (z - 1) \right] \frac{\partial^3 w_0}{\partial x \partial t^2} \\
&\quad + \frac{1}{2} c_0^{(2)} \kappa_2 \left[\kappa_2 (z^2 - 1) + 2\kappa_1 (z - 1) \right] \frac{\partial^3 w_{20}}{\partial x^3} - c_0^{(2)} \kappa_2 (z - 1) \frac{\partial^2 u_{20}}{\partial x^2} + T^{(2)}(x,z,t)
\end{aligned}
\tag{32.37}
$$

where

$$
\begin{aligned}
T^{(1)}(x,z,t) &= s^{(1)}(x,t) - c_\nu^{(1)} \kappa_1 \frac{\partial q^{(1)}}{\partial x} z - \int_0^z \left(\hat{f}_1^{(1)} - c_\nu^{(1)} \kappa_1 \int_0^z \frac{\partial \hat{f}_2^{(1)}}{\partial x} dz \right) dz, \\
T^{(2)}(x,z,t) &= s^{(2)}(x,t) - c_\nu^{(2)} \kappa_2 \frac{\partial q^{(2)}}{\partial x} (z - 1) + \int_z^1 \left(\hat{f}_1^{(2)} + c_\nu^{(2)} \kappa_2 \int_z^1 \frac{\partial \hat{f}_2^{(2)}}{\partial x} dz \right) dz.
\end{aligned}
\tag{32.38}
$$

Substituting formula (32.37) for $\tau_2^{(1)}$ into inhomogeneous condition (32.32) at the interface, we immediately arrive at the differential equation for the function u_{20}:

$$
\begin{aligned}
c_0^{(1)} \kappa_1 \frac{\partial^2 u_{20}}{\partial x^2} &= \frac{1}{12} c_0^{(1)} \kappa_1^4 \frac{\partial^5 w_0}{\partial x^5} + \frac{1}{2} \upsilon c_0^{(2)} \kappa_2 \frac{\partial^3 w_0}{\partial x^3} - \frac{1}{2} c_\omega^{(1)} c_\nu^{(1)} \kappa_1^2 \frac{\partial^3 w_0}{\partial x \partial t^2} \\
&\quad + \frac{1}{2} c_0^{(1)} \kappa_1^2 \frac{\partial^3 w_{20}}{\partial x^3} + T^{(1)}(x,1,t).
\end{aligned}
\tag{32.39}
$$

Now equations for $\tau_2^{(k)}$ can be rewritten. For the lower stiffer layer, the shear stress correction becomes,

$$\tau_2^{(1)} = -\frac{1}{12}c_0^{(1)}\kappa_1^4(z^4-2z^3+z)\frac{\partial^5 w_0}{\partial x^5} - \frac{1}{2}\upsilon c_0^{(2)}\kappa_2\frac{\partial^3 w_0}{\partial x^3} - \frac{1}{4}c_\omega^{(1)}c_0^{(1)}c_g^{(1)}\kappa_1^2(z^2-z)\frac{\partial^3 w_0}{\partial x \partial t^2}$$
$$+\frac{1}{2}c_0^{(1)}\kappa_1^2(z^2-z)\frac{\partial^3 w_{20}}{\partial x^3} - T^{(1)}(x,1,t)z + T^{(1)}(x,z,t), \tag{32.40}$$

and for the upper layer, it is

$$\tau_2^{(2)} = -\frac{1}{12}c_0^{(2)}\kappa_2\Big[\kappa_2^3(z^4-1) + 2\kappa_1\kappa_2^2(z^3-1) - 3c_4^{(2)}\kappa_2^2(z^2-1)$$
$$-\frac{1}{2}c_0^{(1)}c_g^{(1)}\kappa_1^3(z-1)\Big]\frac{\partial^5 w_0}{\partial x^5} - \frac{1}{2}\upsilon\frac{\left(c_0^{(2)}\kappa_2\right)^2}{c_0^{(1)}\kappa_1}(z-1)\frac{\partial^3 w_0}{\partial x^3}$$
$$-\frac{1}{4}c_\omega^{(2)}\kappa_2\Big\{c_0^{(2)}c_g^{(2)}\kappa_2(z^2-1)$$
$$+2\left[\kappa_1 - \frac{2\left(c_\omega^{(2)}c_\nu^{(2)}c_0^{(1)}\kappa_2 + c_\omega^{(1)}c_\nu^{(1)}c_0^{(2)}\kappa_1\right)}{c_\omega^{(2)}c_0^{(1)}}\right](z-1)\Big\}\frac{\partial^3 w_0}{\partial x \partial \hat{t}^2}$$
$$+\frac{1}{2}c_0^{(2)}\kappa_2\left[\kappa_2(z^2-1) + \kappa_1(z-1)\right]\frac{\partial^3 w_{20}}{\partial x^3} + T^{(2)}(x,z,t) - \frac{c_0^{(2)}\kappa_2}{c_0^{(1)}\kappa_1}T^{(1)}(x,1,t). \tag{32.41}$$

Finally, equation for $\sigma_2^{(1)}$ from (32.30) with homogeneous condition (32.31) on the lower boundary results in the following relation:

$$\sigma_2^{(1)} = \frac{1}{120}c_0^{(1)}\kappa_1^5(2z^5-5z^4+5z^2)\frac{\partial^6 w_0}{\partial x^6} + \frac{1}{2}\upsilon c_0^{(2)}\kappa_1\kappa_2 z\frac{\partial^4 w_0}{\partial x^4}$$
$$+\frac{1}{12}c_\omega^{(1)}c_6^{(1)}\kappa_1^3(2z^3-3z^2)\frac{\partial^4 w_0}{\partial x^2 \partial t^2} - \frac{1}{12}c_0^{(1)}\kappa_1^3(2z^3-3z^2)\frac{\partial^4 w_{20}}{\partial x^4} \tag{32.42}$$
$$+c_\omega\kappa_1 z\frac{\partial^2 w_{20}}{\partial t^2} + \frac{1}{2}\kappa_1\left[\frac{\partial T^{(1)}(x,1,t)}{\partial x}z^2 - 2\int_0^z \frac{\partial T^{(1)}(x,z,t)}{\partial x}dz\right].$$

Applying inhomogeneous boundary condition (32.32) at the interface to the stress correction (32.42), we arrive at the following inhomogeneous differential equation

$$\boldsymbol{L}^{(1)}w_{20} = -\upsilon\boldsymbol{L}^{(2)}w_0 - \boldsymbol{L}^{(1,2)}\frac{\partial^2 w_0}{\partial x^2} + F(x,t) \tag{32.43}$$

with the differential operators

$$
\begin{aligned}
\boldsymbol{L}^{(2)} &= \frac{1}{12} c_0^{(2)} \kappa_2^3 \frac{\partial^4}{\partial x^4} + c_\omega^{(2)} \kappa_2 z \frac{\partial^2}{\partial t^2}, \\
\boldsymbol{L}^{(1,2)} &= \frac{1}{60} c_0^{(1)} \kappa_1^5 \frac{\partial^4}{\partial x^4} + \frac{\upsilon}{4} c_0^{(2)} \kappa_2 (1+\kappa_1) \frac{\partial^2}{\partial x^2} - \frac{1}{12} c_\omega^{(1)} c_6^{(1)} \kappa_1^3 \frac{\partial^2}{\partial t^2}
\end{aligned}
\tag{32.44}
$$

and the force function

$$
\begin{aligned}
F(x,t) &= G^{(1)}(x,t) + \upsilon F^{(2)}(x,t), \\
G^{(1)}(x,t) &= -\frac{1}{2}\kappa_1 \left[\frac{\partial T^{(1)}(x,1,t)}{\partial x} - 2\int_0^1 \frac{\partial T^{(1)}(x,z,t)}{\partial x} dz \right], \\
F^{(2)}(x,t) &= q^{(2)}(x,t) + \int_0^1 \hat{f}_2^{(2)}(x,z,t) dz.
\end{aligned}
\tag{32.45}
$$

32.4.3 One-Dimensional Governing Equation

The process of finding unknown functions in series (32.15) can be continued. However, we interrupt it and compose an asymptotic relation for the transverse displacements of the stiffer layer, restricting ourselves to only the first two approximations:

$$
W^{(1)} = w_0(x,t) + \varepsilon^2 \left[\frac{1}{2} c_\nu^{(1)} \kappa_1^2 (z^2 - z) \frac{\partial^2 w_0(x,t)}{\partial x^2} + w_{20}(x,t) \right] + O(\varepsilon^4), \tag{32.46}
$$

where w_{20} is found from Eq. (32.43). We apply the operator $\boldsymbol{L}^{(1)}$ to the both sides of Eq. (32.46) for $z = 1$ and take into account Eq. (32.34). As a result we observe the following differential equation with respect to the displacement $w^\circ = W^{(1)}(x,1,t)$ of the interface:

$$
\left(\boldsymbol{L}^{(1)} + \varepsilon^2 \upsilon \boldsymbol{L}^{(2)} \right) w^\circ + \varepsilon^2 \boldsymbol{D}_{xt} \frac{\partial^2 w^\circ}{\partial x^2} = F^{(1)}(x,t) + \varepsilon^2 \left[G^{(1)}(x,t) + \upsilon F^{(2)}(x,t) \right] + O(\varepsilon^4), \tag{32.47}
$$

where

$$
\boldsymbol{D}_{xt} = \frac{1}{60} c_0^{(1)} \kappa_1^5 \frac{\partial^4}{\partial x^4} + \frac{1}{4} \upsilon c_0^{(2)} \kappa_2 (1+\kappa_1) \frac{\partial^2}{\partial x^2} - \frac{1}{12} c_\omega^{(1)} c_6^{(1)} \kappa_1^3 \frac{\partial^2}{\partial t^2}. \tag{32.48}
$$

Because of the operator $\boldsymbol{D}_{xt}$, the order of differential equation (32.47) is six. This circumstance introduces some uncertainty when setting boundary conditions at the edges of the beam-strip. However, the order of Eq. (32.47) can be reduced. Expressing the derivative $\frac{\partial^4 w^\circ}{\partial x^4}$ from Eq. (32.34) and substituting into (32.48), we arrive at a fourth-order differential equation,

$$
I_{12} \frac{\partial^4 w^\circ}{\partial x^4} + \boldsymbol{J}_{12} \frac{\partial^2 w^\circ}{\partial t^2} = F_{12}(x,t), \tag{32.49}
$$

where

$$I_{12}=\frac{1}{12}\left(c_0^{(1)}\kappa_1^3+\varepsilon^2 \upsilon c_0^{(2)}\kappa_2^3\right),$$

$$\boldsymbol{J}_{12}=c_\omega^{(1)}\kappa_1+\varepsilon^2\upsilon c_\omega^{(2)}\kappa_2-\varepsilon^2 c_\omega^{(1)}\left[3\upsilon\frac{c_0^{(2)}}{c_0^{(1)}}\frac{\kappa_2(1+\kappa_1)}{\kappa_1^2}+\kappa_1^3\left(\frac{1}{5}+\frac{c_6^{(1)}}{12}\right)\frac{\partial^2}{\partial x^2}\right],$$

$$F_{12}=\left\{1-\varepsilon^2\left[3\upsilon\frac{c_0^{(2)}}{c_0^{(1)}}\frac{\kappa_2(1+\kappa_1)}{\kappa_1^3}+\frac{1}{5}\kappa_1^2\frac{\partial^2}{\partial x^2}\right]\right\}F^{(1)}(x,t)$$
$$+\varepsilon^2\left[G^{(1)}(x,t)+\upsilon F^{(2)}(x,t)\right]. \tag{32.50}$$

Equation (32.49) describes the flexural response of the beam-strip with high-contrast elastic constants of layers. Returning to the original variables, it can be rewritten in the dimensional form

$$\mathcal{I}_r\frac{\partial^4 w^\circ}{\partial x_1^4}+\rho_r h\boldsymbol{J}_r\frac{\partial^2 w^\circ}{\partial t^2}=\mathcal{F}_r, \tag{32.51}$$

where

$$\mathcal{I}_r=\frac{1}{12}\left(\frac{E_1h_1^3}{1-\nu_1^2}+\frac{E_2h_2^3}{1-\nu_2^2}\right) \tag{32.52}$$

is the reduced flexural rigidity, ρ_r is the reduced mass density per unit volume introduced above, $\boldsymbol{J}_r$ is the operator of the so-called modified transverse inertia, which is defined as

$$\boldsymbol{J}_r=1-\frac{\rho_1}{\rho_r}\left[\frac{3K_2}{K_1}\left(2+\frac{h_2}{h_1}\right)+\frac{h_1^3(17-7\nu_1)}{60h(1-\nu_1)}\frac{\partial^2}{\partial x_1^2}\right] \tag{32.53}$$

with

$$K_j=\frac{E_jh_j}{(1-\nu_j^2)},$$

and

$$\mathcal{F}_r=E_1F_{12}^*(x_1,t)$$

is the reduced external force per unit length with F_{12}^* calculated by Eq. (32.50) but expressed in terms of dimensional variables. In particular, if

$$q_2^{(1)}(x,t)\neq 0$$

and all other components of the surface and body forces are equal to zero, then

$$\mathcal{F}_r = -\left[1 - 3\frac{K_2}{K_1}\left(1+\frac{h_2}{h_1}\right)\left(2+\frac{h_2}{h_1}\right) - \underline{\frac{h_1^2}{5}\frac{\partial^2}{\partial x_1^2}}\right] q_2^{(1)}(x_1,t). \tag{32.54}$$

In contrast to the model presented in the previous section, Eqs. (32.51)-(32.54) with the underline terms correspond to the TR type model accounting for the shear effects in the stiffer layer. Assuming h_2 (then $K_2 = 0$), we obtain the equation for a single-layer beam-strip, which is the degenerate analogue of the Uflyand–Mindlin plate equation [24, 58, 59] with the shear correction factor $k = 5/(6-\nu_1)$.

32.5 Free Vibrations

First, we compare both models presented above. For the sake of simplicity, we assume that $\nu_1 = \nu_2 = 0$. For the simply-supported beam-strip of the length l, the small deflection is

$$w^\circ = A\sin(\pi n x_1/l)\exp(\mathrm{i}\omega t),$$

where $\mathrm{i} = \sqrt{-1}$, ω is a natural frequency, and n is a number of semi-waves. Then the BE type model gives

$$\frac{\omega^2}{\omega_l^2} = \frac{(1+\epsilon_{21}\kappa_{21}^2)^2 + 4\epsilon_{21}\kappa_{21}(1+\kappa_{21}+\kappa_{21}^2)}{(1+\epsilon_{21}\kappa_{21})(1+\rho_{21}\kappa_{21})} = r_1, \tag{32.55}$$

and the TR type model leads to the following relation

$$\frac{\omega^2}{\omega_l^2} = \frac{1+\epsilon_{21}\kappa_{21}^3}{1+\rho_{21}\kappa_{21} - (1+\kappa_{21})\left[3\upsilon_{21}\kappa_{21}(2+\kappa_{21}) - \frac{17\varepsilon^2(\pi n)^2}{60(1+\kappa_{21})^3}\right]} = r_2, \tag{32.56}$$

where

$$\omega_l = \frac{\pi^2 n^2 h_1}{2l^2}\sqrt{\frac{E_1}{3\rho_1}}, \qquad n = 1,2,\ldots \tag{32.57}$$

are the eigen-frequencies of the BE beam with geometrical and mechanical characteristics of the lower stiffer layer, and all parameters with subscript $_{12}$ are the geometrical and physical parameters ratios of the softer layer to the stiffer one:

$$\kappa_{21} = \frac{h_2}{h_1}, \qquad \epsilon_{21} = \frac{E_2}{E_1}, \qquad \rho_{21} = \frac{\rho_2}{\rho_1}. \tag{32.58}$$

In contrast to Eq. (32.55), relation (32.56) takes into account the transverse shear in the first stiffer layer. Ignoring shear, one can obtain the upper bound for the Young moduli ratio,

$$\epsilon_{21} < \frac{1+\rho_{21}\kappa_{21}}{3\kappa_{21}(1+\kappa_{21})(2+\kappa_{21})}. \tag{32.59}$$

If inequality (32.59) holds, then relation (32.56) can be used formally for any (not very large) number n.

To compare estimates following from Eqs.(32.55) and (32.56), we neglect the shear effects in Eq. (32.56) omitting the term proportional to $(\pi n)^2$. Assuming that $\epsilon_{21} \ll 1$ and expanding both equations into series in powers of ϵ_{21}, it is easy to prove that relations (32.55) and (32.56) give very close results for low-frequency vibrations if

$$\frac{\rho_2}{\rho_1} \sim \rho_{21}^* = \frac{1}{\kappa_{21}(1+\kappa_{21})} = \frac{h_1^2}{hh_2}, \quad h = h_1 + h_2. \tag{32.60}$$

Figure 32.2a displays the ratio $r_{1,2} = \omega^2/\omega_l^2$ calculated for $\epsilon_{21} = 0.01$ on the base of the BE and TR type models versus the thickness ratio κ_{21} at different density ratios ρ_{21}. Figure 32.2b shows the relative deviation of eigenfrequencies

$$\delta = \frac{r_2 - r_1}{r_1} 100\%,$$

found using Eqs. (32.55) and (32.56).

Similar plots of the parameter δ as a function of κ_{21} calculated for more large value of the ratio of the Young moduli, $\epsilon_{21} = 0.05$, are displayed in Fig. 32.3. In all calculations the results of which are presented in Figs. 32.2, 32.3, the density ratio ρ_{21} was taken equal to or close to ρ_{21}^*. It is clearly seen that the divergence of frequencies found by the two models for the assumed parameters ρ_{21}, κ_{21} is small, but it increases together with κ_{21} and ϵ_{21}.

In the next series of calculations we perform the parametric analysis of the lowest natural frequencies accounting for the transverse shear. Figure 32.4 displays the frequency ratio r_2 corresponding to the eigenmodes with $n = 1, 3, 5, 7$ for the two-layer beam-strip at $\varepsilon = 0.1$, $\upsilon = 3$ and with different values of the parameter κ_{21}. Here, in contrast to previous computations, the density ratio was fixed, $\rho_{21} = 0.5$, for any thickness ratio κ_{21}. The dotted line corresponds to the TR type model where, however, the shear effect was ignored (assuming $n = 0$ in Eq. (32.56)). As expected, taking into account the transverse shears in the stiffer part reduces natural frequencies. Behaviour

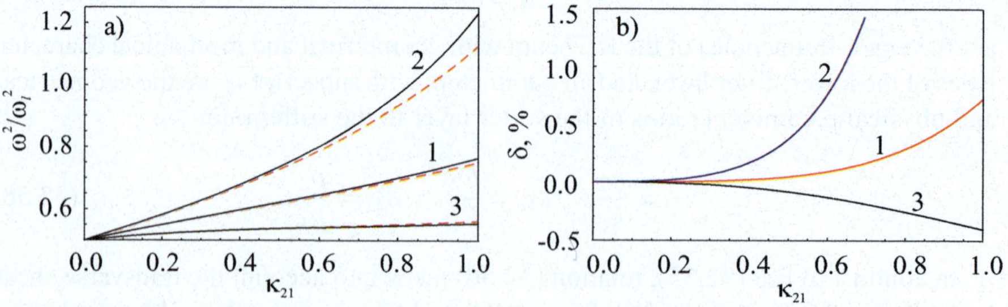

Fig. 32.2: Squared relative eigenfrequency (a) and relative deviation of eigenfrequencies (b) calculated for $\epsilon_{21} = 0.01$ on the base of BE and TR type models (solid and doted lines, respectively) vs. thickness ratio κ_{21} at different density ratios ρ_{21}: $1 - \rho_{21} = \rho_{21}^*$, $2 - \rho_{21} = \rho_{21}^* + 0.5$, $3 - \rho_{21} = \rho_{21}^* - 0.5$. .

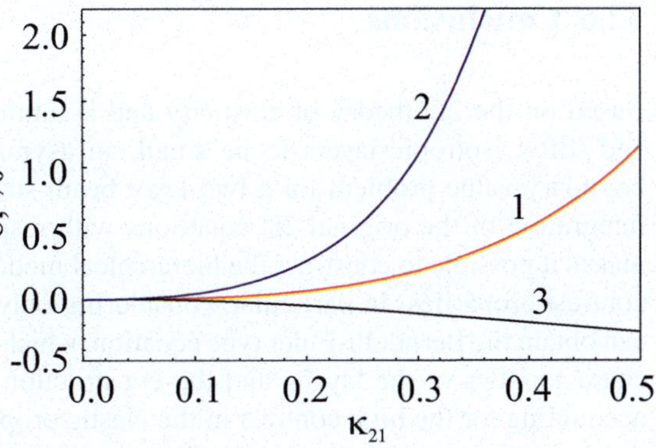

Fig. 32.3 Deviation of eigenfrequencies calculated for $\epsilon_{21} = 0.05$ on the base of BE and TR type models vs. thickness ratio κ_{21} at different density ratios ρ_{21}: 1 – $\rho_{21} = \rho_{21}^*$, 2 – $\rho_{21} = \rho_{21}^* + 0.5$, 3 – $\rho_{21} = \rho_{21}^* - 0.5$.

of the function $r_2(\kappa_{21})$ depends on the mode number. For the lowest frequency ($n = 1$), there is a such ratio of the layers thickness, $\kappa_{21} = \kappa_{21}^\circ$, at which the frequency ratio r_2 is minimal, and for any $n \geq 2$, $r_2(\kappa_{21})$ is the continuously increasing function of κ_{21}.

Finally, Fig. 32.5 displays the squared relative frequency r_2 versus the parameter υ characterizing the elastic moduli ratio of the two strip parts (see Eq. (32.4)). The input parameters were $\varepsilon = 0.1$, $\kappa_{21} = 0.5$, $\rho_{21} = 0.5$. It is seen that all natural frequencies increase together with the parameter υ.

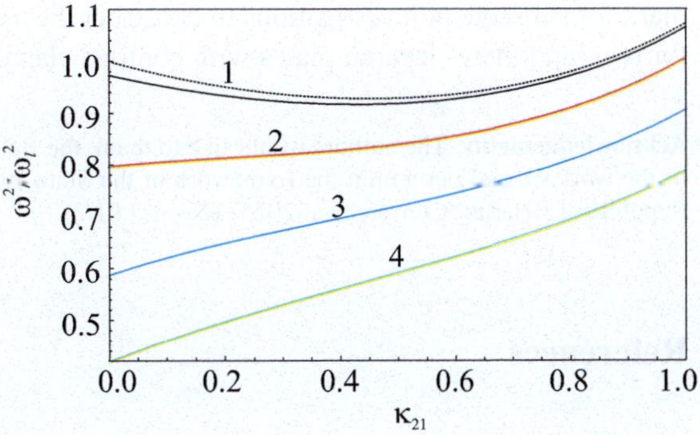

Fig. 32.4 Squared relative eigenfrequency vs. parameter κ_{21} for different wave numbers n: dotted line - $n = 0$, lines 1,2,3, 4 correspond to $n = 1, 3, 5, 7$, respectively.

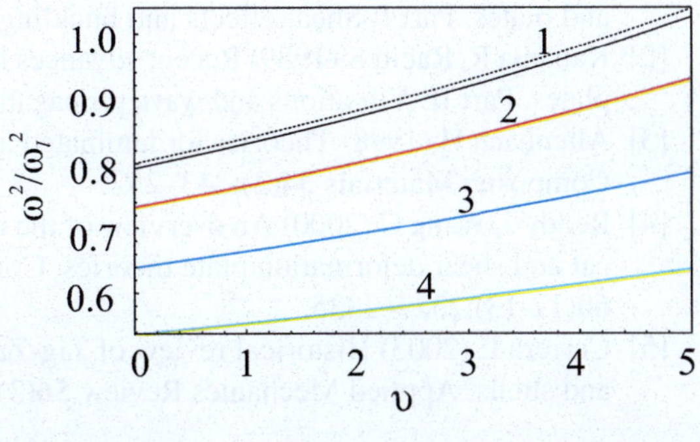

Fig. 32.5 Squared relative eigenfrequency vs. parameter υ for different wave numbers n: dotted line - $n = 0$, lines 1,2,3, 4 correspond to $n = 1, 3, 5, 7$, respectively.

32.6 Conclusions

Based on the 2D theory of elasticity and assuming the ratio of moduli for softer and stiffer isotropic layers to be small, an asymptotic procedure for solving the boundary value problem for a two-layer beam-strip is proposed. The step-by-step integration of the original 2D equations with respect to the transverse coordinate makes it possible to construct the hierarchical models of a two-layer beam-strip with contrast properties. In particular, considering only the leading approximation, one can obtain the Bernoulli-Euler type equation, which is valid for any ratio of the elastic characteristics of the layers, and the construction of the first two approximations accounting for the high contrast in the elastic properties of the two parts, results in the equation corresponding to the Timoshenko-Reissner model. The second model takes into account the shear effects only in the stiffer layer, and to capture this effect in the softer layer, one needs to consider the third-order approximation which generates very cumbersome relations for required functions.

As an example, the analysis of free low-frequency vibration of the simply-supported two-layer beam based on two models was carried out. It was observed that both models give very close results when the density ratio of the layers satisfies a certain relation. The effect of the soft layer attached to the main stiff one on the lowest natural frequencies was studied at different values of the thickness and moduli ratios of layers.

The asymptotic procedure developed in this paper may be considered as a benchmark for subsequent investigations to construct the asymptotically consistent models for two- and three- layered plates with contrast elastic properties.

Acknowledgements The authors would like to thank the Belarusian State University for support of the work carried out within the framework of the State Program of Scientific Research in the Republic of Belarus "Convergence-2025" (No. 1.7.01.2).

References

[1] Kapania R, Raciti S (1989) Recent advances in analysis of laminated beams and plates. Part I: Shear effects and buckling, AIAA Journal **27**(7):923–934.
[2] Kapania R, Raciti S (1989) Recent advances in analysis of laminated beams and plates. Part II: Vibrations and wave propagation, AIAA Journal **27**(7):935–946.
[3] Altenbach H (1998) Theories for laminated and sandwich plates, Mechanics of Composite Materials **34**(3):243–252.
[4] Reddy J, Wang C (2000) An overview of the relationships between of the classical and shear deformation plate theories, Composites Science and Technology **60**(12-13):2327–2335.
[5] Carrera E (2003) Historical review of Zig-Zag theories for multilayered plates and shells, Applied Mechanics Review **56**(3):287–308.

[6] Kreja I (2011) A literature review on computational models for laminated composite and sandwich panels, Central European Journal of Engineering **1**(1):59–80.
[7] Qatu M (2004) Vibration of laminated shells and plates, Elsevier, San Diego.
[8] Reddy J (2004) Mechanics of Laminated Composite Plates and Shells: Theory and Analysis, 2nd edn, CRC Press, Boca Raton.
[9] Mikhasev G, Altenbach H (2019) Thin-walled Laminated Structures. Buckling, Vibrations and Their Suppression, Advanced Structured Materials, Vol 106, Springer, Cham.
[10] Heinisuo M (2022) Classical Beam Theory with Arbitrary Number of Layers, Tampere University.
[11] Malekzadeh P, Fiouz A, Razi H (2009) Three-dimensional dynamic analysis of laminated composite plates subjected to moving load, Composite Structures **90**(2):105–114.
[12] Kulikov G, Plotnikova S (2013) Advanced formulation for laminated composite shells: 3d stress analysis and rigid-body motions, Composite Structures **95**:236–246.
[13] Kirchhoff G (1850) Über das Gleichgewicht und die Bewegung einer elastischen Scheibe, Journal für die reine und angewandte Mathematik (Crelles Journal) **40**:51–88.
[14] Hoff N, Mauther S (1948) Bending and buckling of sandwich beams, Journal of the Aeronautical Sciences **15**(12):707–720.
[15] Reissner E, Stavsky Y (1961) Bending and stretching of certain types of heterogeneous aeolotropic elastic plates, Transactions ASME Journal of Applied Mechanics **28**:402–408.
[16] Stavsky Y (1961) Bending and stretching of laminated aeolotropic plates, Proceedings of the American Society of Cicil Engineers Journal of Engineering Mechanics Division **8**:31–56.
[17] Dong SB, Pister KS, Taylor RL (1962) On the theory of laminated anisotropic shells and plates, Journal of Aerospace Sciences **29**(8):969–975.
[18] Whitney JM, Leissa AW (1969) Analysis of heteregeneous anisotropic plates, Transactions ASME Journal of Applied Mechanics **36**(2):261–266.
[19] Ambartsumyan SA (1970) Theory of Anisotropic Plates, Technomic Publishing, Stamford.
[20] Bert CW (1976) Dynamics of composite and sandwich panels - Parts I and II (corrected title), The Shock and Vibration Digest **8**(11):15–24.
[21] Timoshenko S (1921) LXVI. On the correnction for shear of the differential equation for transverse vibrations of prismatic bar, The London, Edinburgh, and Dublin Philosophical Magazine and Journal of Science **41**(245):744–746.
[22] Reissner E (1944) On the theory of bending of elastic plates, Journal of Mathematics and Physics **23**(1-4):184–191.
[23] Reissner E (1945) The effect of transverse shear deformation on the bending of elastic plates, Transactions ASME Journal of Applied Mechanics **12**(2):A69–A67.

[24] Mindlin R (1951) Influence of rotatory inertia and shear on flexural motions of isotropic elastic plates, Transactions ASME Journal of Applied Mechanics **18**(1):31–38.
[25] Whitney JM, Pagano NJ (1970) Shear deformation in heterogeneous anisotropic plates, Transactions ASME Journal of Applied Mechanics **37**:1031–1036.
[26] Whitney JM, Sun CT (1973) A higher order theory for extensional motion of laminated anisotropic shells and plates, Journal of Sound and Vibration **30**:85–97.
[27] Reddy JN (1984) A refined nonlinear theory of plates with transverse shear deformation, International Journal of Solids and Structures **20**(9/10):881–896.
[28] Librescu L, Khdeir AA, Reddy JN (1987) A comprehensive analysis of state of stress of elastic anisotropic flat plates using refined theories, Acta Mechanica **70**:57–81.
[29] Librescu L, Khdeir (1988) Analysis of symmetric cross-ply laminated elastic plates using a higher-order theory. Part I: Stress and displacement, Composite Structure **9**:189–213.
[30] Karama M, Afaq K, Mistou S (2009) A new theory for laminated composite plates, Proceedings of the Institution of Mechanical Engineering, Part L: Journal of Materials: Design and Applications **223**(2):53-62.
[31] Aydogdu M (2009) A new shear deformation theory for laminated composite plates, Composite Structures **89**:94–101.
[32] Shi P, Dong C, Sun F, Liu W, Hu Q (2018) A new higher order shear deformation theory for static, vibration and buckling responses of laminated plates with the isogeometric analysis, Composite Structures **204**:342–358.
[33] Malikan M, Eremeyev V (2020) A new hyperbolic-polynomial higher-order elasticity theory for mechanics of thick fgm beams with imperfection in the material composition, Composite Structures **249**:112486–16.
[34] Srinivas S (1973) A refined analysis of composite laminates, Journal of Sound and Vibration **30**:495–550.
[35] Sun CT, Whitney JM (1973) On the theories for the dynamic response of laminated plates, AIAA J **11**:372–398.
[36] Bolotin VV, Novichkov YN (1980) Mechanics of Multilayer Structures (in Russ), Mashinostroenie, Moscow
[37] Carrera E (2001) Developments, ideas, and evaluations based upon Reissner's mixed variational theorem in the modeling of multilayered plates and shells, Applied Mechanics Review **54**:301–329.
[38] Sahoo R, Singh B (2014) A new trigonometric zigzag theory for buckling and free vibration analysis of laminated composite and sandwich plates, Composite Structures **117**:316–332.
[39] Naumenko K, Eremeyev VA (2014) A layer-wise theory for laminated glass and photovoltaic panels, Composite Structures **112**:283–291.
[40] Mikhasev G, Altenbach H (2019) Equivalent single layer model for thin laminated cylindrical shells, In: Thin-walled Laminated Structures, Advanced Structured Materials, Vol 106, pp 29–84, Springer, Cham.

[41] Schulze SH, Pander M, Naumenko K, Altenbach H (2012) Analysis of laminated glass beams for photovoltaic applications, International Journal of Solids and Structures **49**(15):2027–2036.

[42] Weps M, Naumenko K, Altenbach H (2013) Unsymmetric three-layer laminate with soft core for photovoltaic modules, Composite Structures **105**:332–339.

[43] Lo K, Christensen R, Wu E (1977) A high-order theory of plate deformation. Part I: Homogeneous plates, Transactions ASME Journal of Applied Mechanics **44**(4):663–668.

[44] Vekua I (1982) Some General Methods of Constructing Different Variants of the Theory of Shallow Shells (in Russ), Nauka, Moscow

[45] Kienzler R (1982) Erweiterung der klassischen Schalentheorie; der Einfluß von Dickenverzerrung und Querschnittverwölbungen, Ingenieur-Archiv **52**:311–322.

[46] Touratier M (1991) An efficient standard plate theory, International Journal of Engineering Science **29**(8):901–916.

[47] Kienzler R, Schneider P (2019) A beam—just a beam in linear plane bending, In: Altenbach H, Chróścielewski J, Eremeyev V, Wiśniewski K (Eds) Recent Developments in the Theory of Shells, Advanced Structured Materials, Vol 106, pp 329–350, Springer, Cham.

[48] Gol'denveiser A (1962) Construction of approximate theory of shells using asymptotic integration of equations of elasticity theories, Prikladnaya Matematika Mekhanika **26**(4):668–686.

[49] Agalovyan L (1997) Asymptotic theory of anisotropic plates and shells (in Russ), Nauka, Moscow.

[50] Babuška I, Lee I, Schwab C (1994) On the a posteriori estimation of the modeling error for the heat conduction in a plate and its use for adaptive hierarchical modeling, Applied Numerical Mathematics **14**(1):5–21.

[51] Schwab C (1996) A-posteriori modeling error estimation for hierarchical plate models, Numerical Mathematics **74**:221–259.

[52] Tovstik P, Tovstik T (2014) A thin-plate bending equation of second-order accuracy, Doklady Physics **59**(8):389–392.

[53] Morozov N, Tovstik P, Tovstik T (2016) Generalized Timoshenko–Reissner model for a multilayer plate, Mechanics of Solids **51**(5):527–537.

[54] Tovstik PE, Tovstik TP (2017) Generalized Timoshenko-Reissner models for beams and plates, strongly heterogeneous in the thickness direction, ZAMM - Journal of Applied Mathematics and Mechanics / Zeitschrift für Angewandte Mathematik und Mechanik **97**(3):296–308.

[55] Mikhasev G (2022) On governing equations for a nanoplate derived from the 3d gradient theory of elasticity, Mathematics and Mechanics of Solids **27**(9):1688–1702.

[56] Kaplunov J, Kossovitch L, Nolde E (1998) Dynamics of Thin Walled Elastic Bodies, Academic Press, New York.

[57] Chebakov R, Kaplunov J, Rogerson G (2017) A non-local asymptotic theory for thin elastic plates, Proceedings of the Royal Society A: Mathematical, Physical and Engineering Sciences **473**(2203):20170249.

[58] Uflyand Y (1948) Wave propagation by transverse vibrations of beams and plates (in Russ), Prikladnaya Matematika i Mekhanika **12**:287–300.
[59] Elishakoff I (2019) Handbook on Timoshenko-Ehrenfest Beam and Uflyand-Mindlin Plate Theories, World Scientific, Singapore.
[60] Kaplunov J, Prikazchikov D, LA P (2017) Dispersion of elastic waves in a strongly inhomogeneous three-layered plate, International Journal of Solids and Structures **113-114**:169–179.
[61] Kaplunov J, Prikazchikov D, Prikazchikova L (2017) Dispersion of elastic waves in laminated glass, Procedia Engineering **199**:1489–1494.
[62] Prikazchikova L, Aydin Y, Erbas B, Kaplunov J (2020) Asymptotic analysis of an anti-plane dynamic problem for a three-layered strongly inhomogeneous laminate, Mathematics and Mechanics of Solids **25**(1):3–16.
[63] Alkinidri M, Kaplunov J, Prikazchikova L (2020) A two mode non-uniform approximation for an elastic asymetric sandwich, In: Papadrakakis M, Fragiadakis M, Papadimitriou C (eds) EURODYN 2020 XI International Conference on Structural Dynamic, pp 528–535.
[64] Alkinidri M, Kaplunov J, Prikazchikova L (2021) Two-mode long-wave low-frequency approximations for anti-plane shear deformation of a high-contrast asymmetric laminate, In: Awrejcewicz J (ed) Perspectives in Dynamical Systems II: Mathematical and Numerical Approaches. DSTA 2019, Springer Proceedings in Mathematics and Statistics, Vol 363, pp 275–285, Springer, Cham.
[65] Kaplunov J, Prikazchikova L, Alkinidri M (2021) Antiplane shear of an asymmetric sandwich plate, Continuum Mechechanics and Thermodynamics **33**:1247–1262.
[66] Prikazchikova L (2022) Decay conditions for antiplane shear of a high-contrast multi-layered semi-infinite elastic strip, Symmetry **2022**(14):1697.
[67] Kaplunov J, Prikazchikova L (2018) Low-frequency vibration modes of strongly inhomogeneous elastic laminates, Izvestiya Saratovskogo Universiteta (N S), Seriya Matematika Mekhanika Informatika **18**(4):447–457.
[68] Tovstik P, Tovstik T (2007) On the 2d models of plates and shells including the transversal shear, ZAMM - Journal of Applied Mathematics and Mechanics / Zeitschrift für Angewandte Mathematik und Mechanik **87**:160–171.

Chapter 33
On Analytical Modeling of Tension-Assisted Winding of Flexible Sheets

Aleksandr Morozov, Wilhelm Rickert, and Sergei Shubin

Abstract This work is devoted to an analytical study of an elastic sheet winding under tension. Treating the model in a 2D axisymmetric formulation, we use kinematic boundary conditions instead of traction ones to introduce the winding tension, which is then equalized by the resulting circumferential stresses. Such an approach allows to avoid any assumptions regarding stress profile where the tension is applied. Based on the theory of linear elasticity a simplified boundary value problem is formulated and its solution is developed in a closed-form. A a comprehensive stress analysis of the winded sheet is performed, based on which the winding tension setup is discussed.

Key words: Winding, Flexible sheet rolling, Winding tension

33.1 Introduction

Winding processes are widely used in various industries as a center-wound roll is the most compact form of storing flat thin sheets of materials such as paper, plastics, fabric, metals etc. Proper winding is under comprehensive analysis by a lot of researchers to avoid typical defects including starring, buckling, air gaps between layers etc. The winding tension is one of the most important parameters to achieve an optimal process without defects. Therefore, a number of analytical, numerical and

Alexandr Morozov · Wilhelm Rickert
Institute of Mechanics, Chair of Continuum Mechanics and Constitutive Theory, Technische Universität Berlin, Einsteinufer 5, 10587 Berlin, Germany
e-mail: morozov@tu-berlin.de, rickert@tu-berlin.de

Sergei Shubin
Institute for Problems in Mechanical Engineering of Russian Academy of Sciences, V.O., Bol'shoy pr. 61, 199178 St. Petersburg, Russian Federation,
e-mail: shubinsn@gmail.com

H. Altenbach et al. (eds.), *Sixty Shades of Generalized Continua*,
Advanced Structured Materials 170, https://doi.org/10.1007/978-3-031-26186-2_33

experimental research has been devoted to find correlations between stresses within roll and winding tension.

The winding treatment by an axisymmetric linear elasticity approach is rooted in a number of classical works. Faupel [1] analyzed the wrapping of an elastically isotropic ribbon (or wire) on a hollow cylinder made of the same material. In the solution he assumed that the coiled ribbon can be simulated by uniform and continuous cylinder and used the Lamé solution as a starting point for the problem. It allowed to obtain the solution for radial and hoop stresses in a closed-form. Another similar solution was obtained by Altmann [2] for the analysis of center-wound rolls.

This analysis was extended by Yagoda [3] adding anisotropy and varying the winding tension. Variable tension during winding is important for the practice as usually two types of tension profiles are used: linear and hyperbolic. The derived solution by Yagoda has an explicit closed-form for the radial and hoop stresses. In particular, it rendered it possible to clarify the previous misunderstanding of sharp stress variations in the vicinity of the rolls core.

These classical works have been followed up by several researchers. In the following the general trends on the topic are briefly investigated. Lin and Wickert [4] analyzed buckling of the inner layers of a wound coil. Hoog et al. [5] developed an inverse solution strategy in order to find the winding tension from the prescribe residual stress. This approach is advantageous for industrial applications, because the residual stress is usually specified (to prevent process defects) and the appropriate winding tension has to be found. Huber et al. [6] provided an analytical model of the winding process taking into account coil winding, mandrel removal, and heating/cooling. The developed solution extends classical works and provides a powerful tool for designers to optimize residual stress. Hakiel [7] investigated the impact of nonlinear material behavior on the stress distribution. Mollamahmutoglu and Good [8] used the theory of large deformations to develop a winding model for highly extensible and compressible materials such as nonwoven fabrics. Lee and Lee [9, 10] developed another model taking into account the bending stress and the stress due to gravity. Lee et al. [11] researched telescopic defect in a wound roll in printing systems. The authors also focused on friction and air films between the layers. Park et al. [12] followed up the work by Sims and Place [13] analysing an assembly mandrel-sleeve-strip. Using numerical simulations, they analyzed the effects of the strip thickness, reaction forces from the mandrel, and gravitational force on the resulting stresses. In particular, they study the impact upon the stress concentration near the coil starting point.

Another approach to the treatment of the winding process of wire-like structures is by means of beam theory. Komodromos et al. [14] used numerical simulations to study plastic strains within the wire winded on a rectangular reel. In the work the clearance between the reel and the wire is analyzed in dependence on various parameters (tension, radii etc). Lee et al. [15] developed a guiding system for a cable winding. Numerical and experimental investigations of wire bonding pull test were analyzed in the work [16] by Schmitz et al. . Ming and Gongliu [17] performed an analysis of a straddle section assuming large displacements and small strains of the fiber. It should be noted, that in winding models based on the beam theory, the interaction between the layers is limited to geometrical conditions, i.e. curvature of the next layer is greater then the curvature of the previous layer by the thickness of the wire meanwhile radial compression is not taken into account.

The current work focuses on an analytical study of a stress analysis of a sheet during a tension assisted winding process. We use kinematic boundary conditions instead of traction ones to introduce the winding tension. Therefore, no assumption on the stress distribution in the region of the applied tension is required, rather only the tension itself is prescribed. Based on linear elasticity we develop a closed-form solution and perform the stress analysis. In Sec. 33.2 the main model assumptions are detailed. Subsequently, in Sec. 33.3 an analytical solution for the first winded layer is presented. Building upon the ideas used for winding the first layer, the extension of the procedure for finding the solution of multiple winded layers is explained in Sec. 33.4 and the solutions are analyzed for various thickness to reel radius ratios. A summary of this work followed by a discussions and a conclusions is given in Sec. 33.5.

33.2 General Assumptions for the Winding Problem

Depending on the sheet-reel dimensions ratio, two possible winding scenarios have to be distinguished. In the first case, if the thickness of the sheet is considerably less than the diameter of the reel, then bending loads in the sheet can be neglected, i.e. the material sheet can be rolled on the reel with a negligible bending moment. To obtain a proper tension setup in a way that the sheet is winded tightly on the reel, a stress analysis has to be performed. In the second scenario, for smaller ratios of the reel diameter to sheet thickness, the bending load cannot be neglected anymore. Moreover, when the sheet is tightly winded onto a small radius reel, it can be damaged by the arising bending stresses. The minimum value of the reel radius at which there is no damage in the cable is the so-called minimum bending radius of a sheet.

In this paper, we study the first of the aforementioned scenarios, i.e. we assume that the sheet of material can be laid stress-free on the reel. This is a reasonable assumption if the reel radius is large enough. We consider linearly elastic behavior of the winded material and an axially symmetric plain-strain formulation. Assuming the stress-free rolling of the sheet onto the reel, a closed form analytical solution can be obtained for the stress state of the sheet winded under constant tension. If the bending moment is of importance, then a similar approach can be used to extended analysis by means of the theory of large displacements with small strains, see, e.g. [18].

If not stated otherwise, the material and geometrical parameters, as well as the tension value from Table 33.1 are used in the calculations. The friction between layers is neglected and subsequent layers assumed to be fully in contact with the previous layers, i.e. there are no gaps between the layers.

Table 33.1: Parameters used in the calculations. .

Parameter	R_i	t	T	λ/μ	μ
Value	100 mm	2 mm	200 N	7/3	37000 N/mm^2

33.3 Winding the First Layer

We consider a layer of material with thickness t and depth d winded with the given tension T onto a cylindrical rigid body cable reel of radius R_i, see Fig. 33.1 (*a*). Furthermore, it is assumed that depth d is large enough to justify the use of plane-strain formulation. In order to fulfill the tension boundary condition, the following trick is made: the length of the rolled material layer is such that there is a gap of angle α_0, see Fig. 33.1 (*b*).

To close the gap, the kinematic condition u_φ^0 is applied to one end of the sheet while the other end is fixed. Subsequently, the parameter α_0 is adjusted to fulfill the winding traction condition. The following semi-inverse *ansatz* in cylindrical coordinates is used,

$$\boldsymbol{u} = u_r(r)\boldsymbol{e}_r + u_\varphi(r,\varphi)\boldsymbol{e}_\varphi, \tag{33.1}$$

where, due to the aforementioned assumptions, the azimuthal displacement has the form

$$u_\varphi = u_\varphi^0 = \varphi r \beta_0\,, \quad \beta_0 = \frac{\alpha_0}{2\pi - \alpha_0}. \tag{33.2}$$

It should be noted that this particular choice leads to strains and therefore to stresses that do not depend upon φ:

$$\boldsymbol{\varepsilon} = \frac{1}{2}(\nabla \otimes \boldsymbol{u} + \boldsymbol{u} \otimes \nabla) = u_r' \,\boldsymbol{e}_r \otimes \boldsymbol{e}_r + \left(\frac{\alpha}{2\pi - \alpha} + \frac{u_r}{r}\right)\boldsymbol{e}_\varphi \otimes \boldsymbol{e}_\varphi. \tag{33.3}$$

The sheet is made of a linear elastic material with Hooke's law as a constitutive relation, where the stiffness tensor is taken in a form

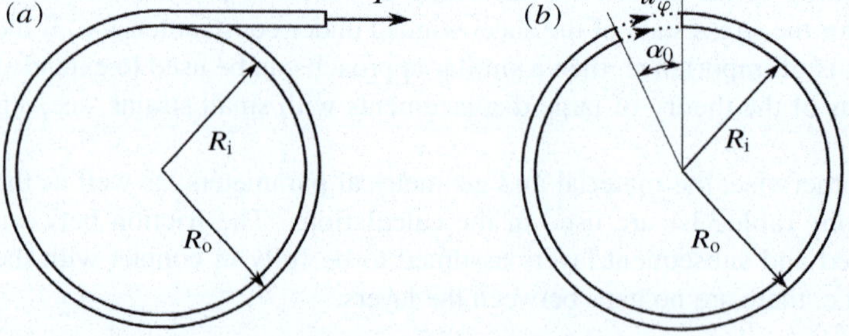

Fig. 33.1: (*a*) Scheme of a tension-assisted winding or rolling process. (*b*) Corresponding model of a winding cable..

$$C = \lambda I \otimes I + 2\mu {}^4 I, \tag{33.4}$$

where λ, μ are the Lamé parameters, and I and ${}^4I$ are the second and fourth order unit tensors, respectively. Insertion into the balance of linear momentum yields the following ordinary differential equation

$$u_r'' + \frac{1}{r} u_r' - \frac{1}{r^2} u_r = \beta_0 \frac{2\mu}{\lambda + 2\mu} \frac{1}{r}, \tag{33.5}$$

where the dash indicates the derivative with respect to r. The general solution to Eq. (33.5) is given by

$$u_r = Ar + \frac{B}{r} + \frac{\mu\beta_0}{\lambda + 2\mu} r\left(1 + \ln(r/R_\mathrm{i})\right), \tag{33.6}$$

where the constants A and B are obtained using the boundary conditions. In particular, for the first layer these read:

$$u_r(r = R_\mathrm{i}) = 0, \qquad \sigma_{rr}(r = R_\mathrm{o}) = 0. \tag{33.7}$$

Even for the single layer problem expressions for the constants A and B are bulky, so they are omitted here. An additional condition for the determination of the closing angle α_0 has to be formulated. To this end, the prescribed tension T is introduced. In a sheet of the depth d, this tension is found as a function of the opening angle α_0 from

$$\frac{T(\alpha_0)}{d} = \int_{r=R_\mathrm{i}}^{R_\mathrm{o}} \sigma_{\varphi\varphi}(r)\,\mathrm{d}r. \tag{33.8}$$

Note that the depth only enters the problem formulation by means of this expression and thus the winding tension always appears together with d, which renders the resulting solutions independent of the depth. The hoop stress depends on the parameter α_0 and can be expressed in terms of the constants A and B

$$\sigma_{\varphi\varphi}(r) = \frac{2\mu}{r^2} B + (\lambda + \mu)\left[2A(\lambda + 2\mu) + \beta_0(\lambda + 6\mu) + 2\beta_0\mu\ln(r/R_\mathrm{i})\right]. \tag{33.9}$$

In view of Eq. (33.5) not α_0 itself, but rather β_0 is required in the displacement solution. Solving the integral in Eq. (33.8) yields

$$\beta_0 = \frac{\lambda + 2\mu}{\lambda + \mu} \frac{TR_\mathrm{i}R_\mathrm{o} - 2(\lambda + \mu)dR_\mathrm{i}R_\mathrm{o}(R_\mathrm{o} - R_\mathrm{i})A - 2\mu d(R_\mathrm{o} - R_\mathrm{i})B}{dR_\mathrm{i}R_\mathrm{o}\left[(R_\mathrm{o} - R_\mathrm{i})(\lambda + 4\mu) + 2\mu R_\mathrm{o}\ln(R_\mathrm{o}/R_\mathrm{i})\right]}. \tag{33.10}$$

Upon obtaining the coefficients A and B by means of the boundary conditions for the first layer, the factor from Eq. (33.10) can be denoted explicitly,

$$\beta_0 = \frac{T/d}{R_\mathrm{i}\mu(\lambda + \mu)} \frac{\mu R_\mathrm{i}^2 + (\lambda + \mu)R_\mathrm{o}^2}{R_\mathrm{o}^2 + 2R_\mathrm{o}^2\ln(R_\mathrm{o}/R_\mathrm{i}) - R_\mathrm{i}^2}. \tag{33.11}$$

For the first layer, β_0 is proportional to the applied tension T and thus the radial displacement and all associated stresses are linear in T as well. The analytical solution for the displacement reads

$$u_r = \frac{\beta_0}{\lambda+2\mu}\left(\mu r \ln(r/R_\mathrm{i}) - \frac{R_\mathrm{o}^2(\lambda+\mu)[\lambda+2\mu+2\mu\ln(R_\mathrm{o}/R_\mathrm{i})]}{2(\mu R_\mathrm{i}^2+(\lambda+\mu)R_\mathrm{o}^2)}\frac{r^2-R_\mathrm{i}^2}{r}\right). \tag{33.12}$$

The distribution of the normalized hoop stress in the layer thickness obtained by Eq. (33.9) with the use of parameters from Table 33.1 is shown in Fig. 33.2 for various thicknesses of the layer. As can be seen from the Fig. 33.2, the distribution of the hoop stress along the cross-section of the layer is almost uniform and the its value is proportional to $\sigma_{\varphi\varphi}/\mu \sim T/(t\,d)$, Fig. 33.3.

From this one can conclude, that another possible *ansatz* to formulate the BVP for the first winded layer would be the following form of the constitutive relation:

$$\boldsymbol{\sigma} = \mathbf{C}\cdot\cdot\,\boldsymbol{\varepsilon} + \sigma^0_{\varphi\varphi}\boldsymbol{e}_\varphi\otimes\boldsymbol{e}_\varphi, \tag{33.13}$$

where $\sigma^0_{\varphi\varphi}$ is a constant and defined by the tension and the cross-section area of the sheet. Relation (33.13) substituted into a balance of linear momentum yields to the ordinary differential equation, which has the form similar to (33.5), but with a different constant the right hand side:

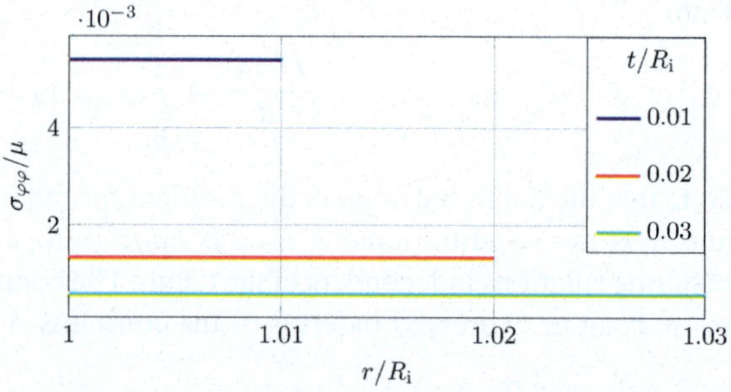

Fig. 33.2 Distribution of the hoop stress in the first layer for different values of the thickness t.

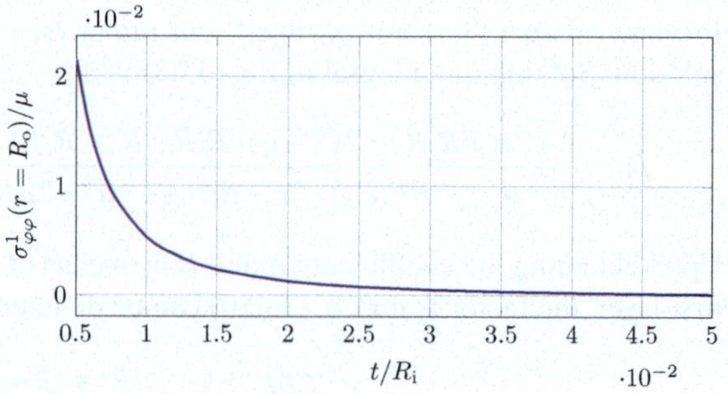

Fig. 33.3 Distribution of the hoop stress calculated at the outer radius of the first layer for different layer thickness t.

$$u_r'' + \frac{1}{r}u_r' - \frac{1}{r^2}u_r = \frac{\sigma^0_{\varphi\varphi}}{\lambda+2\mu}\frac{1}{r}\,, \tag{33.14}$$

and, therefore, has the same solution but with a different parameter. Moreover, such an approach can be extended to the case with bending stress inside the sheet due to the bending load required to lay the layer onto the reel, i.e. the assumption of small bending stiffness is dropped. Then in the case of the linear elasticity, the hoop stress in Eq. (33.13) depends on the radial coordinate and can be decomposed to membrane and bending parts, which to the first order can be approximated as:

$$\sigma^0_{\varphi\varphi}(r) = \sigma^{\text{tens}}_{\varphi\varphi} + \sigma^{\text{bend}}_{\varphi\varphi}(2(r-R_\text{i})/t-1). \tag{33.15}$$

This *ansatz* leads to another type of differential equation for the displacements, which we omit here. However, such an approach requires a constitutive relation to determine parameters $\sigma^{\text{tens}}_{\varphi\varphi}$ and $\sigma^{\text{bend}}_{\varphi\varphi}$. Therefore, the following calculations and analysis is done with the use of our original *ansatz* with kinematic condition (33.2).

33.4 Winding the Subsequent Layers

For the winding of any additional layer consider the configuration depicted in Fig. 33.4. Therein, all previous n layers of thickness t accumulate up to the radius $R_\text{c} = R_\text{i} + nt$, on which the newest layer with its unknown closing angle α is winded upon. It is assumed that the previous displacement solution $\boldsymbol{u}_0$ and its associated stresses $\boldsymbol{\sigma}_0$ are known. The latter can be treated as the residual stress in the already winded layers. The task is then to compute the *additional* displacement $\boldsymbol{w}$ such that the total displacement is given by

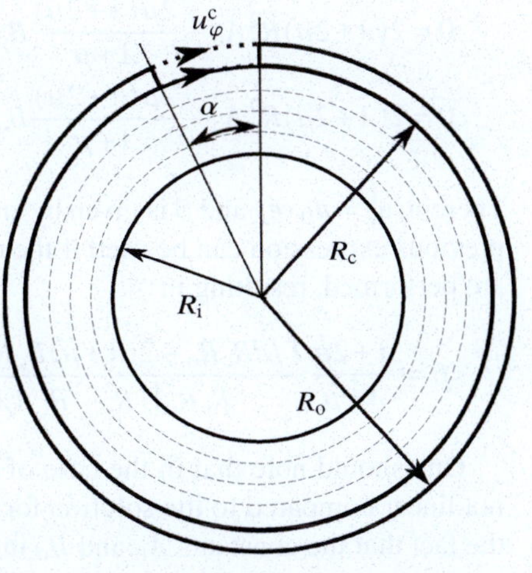

Fig. 33.4 Model of a subsequent layers winding.

$$\boldsymbol{u} = \begin{cases} \boldsymbol{u}_0 + \boldsymbol{w}_\mathrm{i} \, , & r \in [R_\mathrm{i}, R_\mathrm{c}] \, , \\ \boldsymbol{w}_\mathrm{o} \, , & r \in [R_\mathrm{c}, R_\mathrm{o}] \, . \end{cases} \tag{33.16}$$

The balance equation for the outer layer has the same form as in Eq. (33.5), therefore the general solution for $\boldsymbol{w}_\mathrm{o}$ has the form similar to Eq. (33.6). For the already winded layers there is no explicit dependence on any previous closing angles. Since the residual stresses $\boldsymbol{\sigma}_0$ satisfy the balance of linear momentum, solution for $\boldsymbol{w}_\mathrm{i}$ can be obtained in a form of classical solution to the axially symmetric problem. Therefore

$$\begin{aligned} \boldsymbol{w}_\mathrm{i} &= (A_\mathrm{i} r + B_\mathrm{i} r^{-1}) \boldsymbol{e}_r \, , \\ \boldsymbol{w}_\mathrm{o} &= \left(A_\mathrm{o} r + B_\mathrm{o} r^{-1} + \frac{\mu}{\lambda + 2\mu} \beta r \big(1 + \ln(r/R_\mathrm{i})\big) \right) \boldsymbol{e}_r + \beta \varphi r \boldsymbol{e}_\varphi \, . \end{aligned} \tag{33.17}$$

Note that the contained value $\beta = \alpha/(2\pi + \alpha)$ is a new parameter to be determined from a condition similar to the one given in Eq. (33.8), except for the integration domain that is now given by the outer layer, i.e. $r \in [R_\mathrm{c}, R_\mathrm{o}]$. The four constants A_i, B_i, A_o, B_o are determined from the following set of boundary and transition conditions:

$$u_r(r = R_\mathrm{i}) = 0, \qquad [\![u_r]\!] = 0 \, , \quad [\![\sigma_{rr}]\!] = 0 \, , \quad \sigma_{rr}(r = R_\mathrm{o}) = 0, \tag{33.18}$$

where $[\![\cdot]\!]$ denotes the jump of a function evaluated at the interface with $r = R_\mathrm{c}$. Since the solution for winding of the previous layer already satisfies the zero displacement condition at R_i and it is stress free at the interface, the following system of equations arises:

$$\begin{aligned} 0 &= A_\mathrm{i} R_\mathrm{i} + \frac{B_\mathrm{i}}{R_\mathrm{i}} \, , \\ 0 &= u_r^0(r = R_\mathrm{c}) + (A_\mathrm{i} - A_\mathrm{o}) R_\mathrm{c} + \frac{B_\mathrm{i} - B_\mathrm{o}}{R_\mathrm{c}} - \frac{\mu}{\lambda + 2\mu} \beta R_\mathrm{c} \big(1 + \ln(R_\mathrm{c}/R_\mathrm{i})\big) \, , \\ 0 &= 2(\lambda + 2\mu) R_\mathrm{c}^2 A_\mathrm{o} - \frac{2\mu(\lambda + 2\mu)}{\lambda + \mu} B_\mathrm{c} + \beta \big(\lambda + 4\mu + 2\mu \ln(R_\mathrm{c}/R_\mathrm{i})\big) \, , \\ 0 &= 2(\lambda + 2\mu) R_\mathrm{o}^2 A_\mathrm{o} - \frac{2\mu(\lambda + 2\mu)}{\lambda + \mu} B_\mathrm{o} + \beta \big(\lambda + 4\mu + 2\mu \ln(R_\mathrm{o}/R_\mathrm{i})\big) \, . \end{aligned} \tag{33.19}$$

Therein, $u_r^0 = \boldsymbol{u}_0 \cdot \boldsymbol{e}_r$ and β is given by an expression similar to Eq. (33.10). In fact, the previous expression can be used if the replacements $R_\mathrm{i} \to R_\mathrm{c}$, $A \to A_\mathrm{o}$ and $B \to B_\mathrm{o}$ are performed, resulting in

$$\beta = \frac{\lambda + 2\mu}{\lambda + \mu} \frac{T/d R_\mathrm{c} R_\mathrm{o} - 2(\lambda + \mu) R_\mathrm{c} R_\mathrm{o} (R_\mathrm{o} - R_\mathrm{c}) A_\mathrm{o} - 2\mu (R_\mathrm{o} - R_\mathrm{c}) B_\mathrm{o}}{R_\mathrm{c} R_\mathrm{o} \big[(R_\mathrm{o} - R_\mathrm{c})(\lambda + 4\mu) + 2\mu R_\mathrm{o} \ln(R_\mathrm{o}/R_\mathrm{c}) \big]} \, . \tag{33.20}$$

One should note that in the case of multiple layers, relation between β and T is not linear compared to the solution for the first layer, Eq. (33.11). This comes from the fact that the constants A_o and B_o in Eq. (33.20) depend on β.

The solution for n layers is obtained iteratively. The procedure starts with $\boldsymbol{u}_0$ being the first layer solution from Eq. (33.12), which is to be supplemented by the φ-component $\beta_0\varphi r$. Subsequently, system of linear equations (33.19) with $R_c = R_i + t$ is solved in order to obtain the additional displacement $\boldsymbol{w}_1$. The resulting solution is then added to the previous solution, which yields the total displacement for two layers $\boldsymbol{u}_1$. For the next iteration one assigns $\boldsymbol{u}_0 = \boldsymbol{u}_1$ and increases the previous value for R_c by the thickness t. This procedure is then repeated until the desired number of layers is reached. The resulting radial displacement distribution for different layer counts is shown in Fig. 33.5. For the given parameters, the order of magnitude of the radial displacement is small compared to the radius of the reel. The maximum displacement for 150 winded layers is approximately one eighth of the thickness of one layer and thus can be neglected when planning the resultant storage dimensions.

Note that the φ-component of any previous displacement solution does not enter the system in Eq. (33.19). However, as can be seen from the results in Fig. 33.6, it enters the total displacement solution, which is used to calculate the stresses and thus results in the discontinuous hoop stress distribution. In Fig. 33.6, the hoop stress solutions are additionally presented in magnified windows. The hoop stresses are discontinuous and have jumps at the interfaces between the layers. Note that the increase of the number of winded layers decreases the hoop stress in the inner layers.

As can be seen from the magnifications in Fig. 33.6, the level of the hoop stress in the outer layer is approximately the same for every layer count. This is due to the condition imposed by the winding tension. However, the stress distribution in the outer layer changes with the number of winded layers. For few layers the distribution (see, e.g., Fig 33.2 with only one layer or the magnified blue curve for ten layers in Fig. 33.6) is almost uniform and thus the assumption in Eq. (33.13) is also reasonable

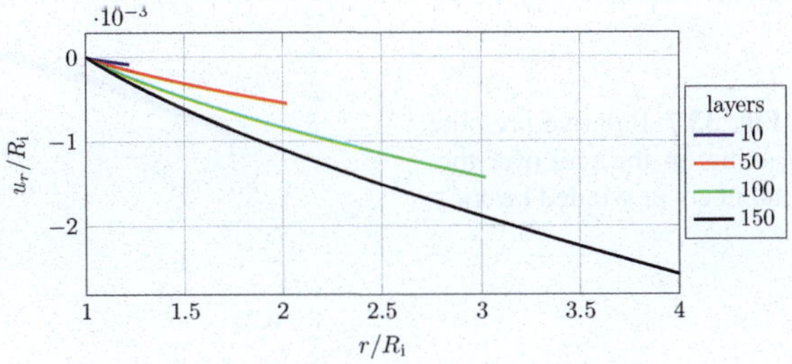

Fig. 33.5 The radial displacement distribution for different number of winded layers.

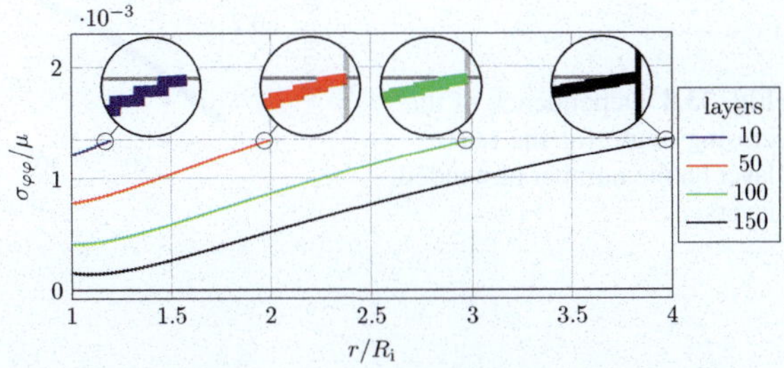

Fig. 33.6 Hoop stress distribution inside the cable assembly for different numbers of winded layers.

for the outer layer. However, with increasing winded layers, the stress distribution in the outer layer (see, e.g., the magnified black curve for 150 layers in Fig. 33.6) should be approximated rather by a linear function in the radial coordinate than a constant, e.g., with an *ansatz* as in Eq. (33.15).

In Fig. 33.7 the pressure acting on the reel normalized with respect to the pressure of the first layer is shown in dependence on the number of winded layers. Note that σ_{rr}^n is the radial stress solution for n winded layers, i.e., σ_{rr}^1 refers to the solution for only one winded layer. For the first winded layers the increase in pressure is almost proportional to the number of layers. However, the increase in pressure decays with further winding. For the analyzed set of parameters the deviation from the linear behavior becomes visible at 25 layers.

For each new layer the closing angle α is a constant value, however, note that this value changes from layer to layer. For a constant winding tension T, the closing angle for the outer layer increases with the number of winded layers, see Fig. 33.8.

Tight winding can be defined by means of the pressure between the two outmost layers, i.e. at $r = R_c$, which is depicted in Fig. 33.9 as a function of the number of winded layers. In contrast to the reel pressure, the pressure between the two outmost layers decreases with an increasing number of winded layers. In the case of 150 layers this pressure is already four times less than the pressure for one layer.

The solution shown in Fig. 33.9 can be used to to determine the appropriate tension profile to obtain a tight winding, e.g. constant torque or hyperbolic profile.

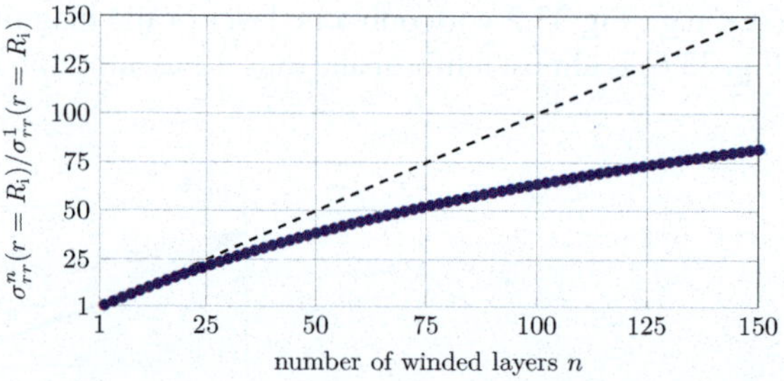

Fig. 33.7 Relative pressure acting on the reel over the numbers of winded layers.

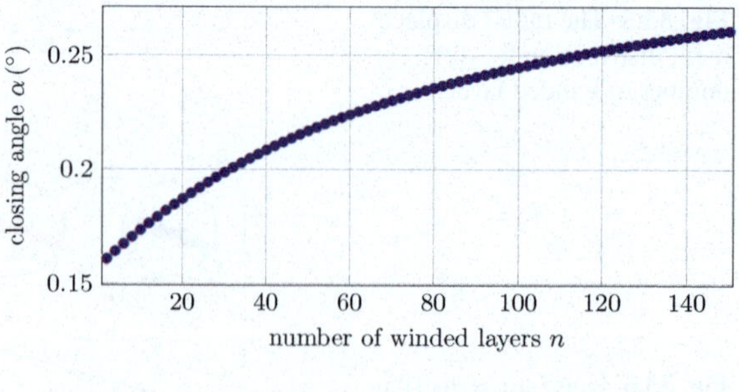

Fig. 33.8 Dependency of the closing angle for the outer layer on the number of winded layers.

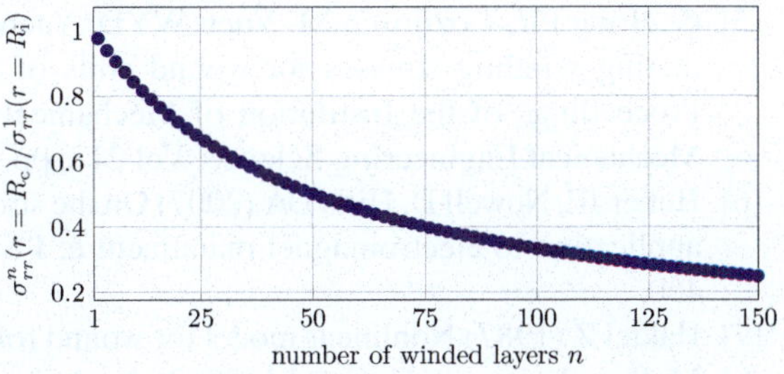

Fig. 33.9 Relative pressure in the interface between the already winded layers and the outer layer.

33.5 Conclusions

In this work an analytical study of a winding problem of an elastic sheet is performed. The winding tension is introduced by means of a kinematic boundary condition, which allowed to avoid any assumptions regarding the stress profile in the sheet cross-section. The modelling is done within the framework of the linear elasticity with the plane-strain formulation.

A solution in a closed form is obtained for the first winded layer and then extended to the case of many layers. The evolution of the pressure between the first layer and the reel, as well as between the outmost layer and the previous one is obtained. It is shown that with an increasing number of layers, at constant tension, the pressure between two outer layers decays. This might affect the tightness of the winding. On the other hand, the pressure between the first layer and the reel increases with the increasing number of layers, which might lead to the reel buckling or damage. In addition, with an increasing number of winded layers, the hoop stress inside some of the inner layers tend to change from tension to compression.

In the current model the reel radius was assumed considerably larger than the thickness of the sheet. It allowed to neglect bending loads needed to roll the sheet onto the reel. A possible *ansatz* to overcome this assumption was introduced for further development of the model.

References

[1] Faupel JH (1964) Engineering design: a synthesis of stress analysis and materials engineering, John Wiley, New York.

[2] Altmann HC (1968) Formulas for computing the stresses in center-wound rolls, J Tech Assoc Pulp Paper Industry **51**(4):176–179.

[3] Yagoda HP (1980) Resolution of a core problem in wound rolls, J Appl Mechanics **47**:847–854.

[4] Lin PM, Wickert JA (2003) Analysis of core buckling defects in sheet metal coil processing, ASME J Manuf Sci Eng **125**(4):771–777.

[5] de Hoog FR, Cozijnsen M, Yuen WYD, Yuen WYD, Huynh HN (2004) Predicting winding stresses for wound coils of linear orthotropic material, In: Proceedings of the Institution of Mechanical Engineers, Part C: Journal of Mechanical Engineering Science, Vol 218, pp 13–25.
[6] Huber JE, Nowell D, Hills DA (2007) On the stress analysis of a wound coil with application to electromagnet manufacture, J Strain Anal Eng Des **42**(6):447–460.
[7] Hakiel Z (1987) Nonlinear model for wound roll stresses, TAPPI J **70**:113–117.
[8] Mollamahmutoglu C, Good J (2013) Analysis of large deformation wound roll models, J Appl Mech **80**:041,016.
[9] Lee J, Lee C (2006) An advanced model for the numerical analysis of the radial stress in center-wound rolls, Int J Mech Sci **105:**360–368.
[10] Lee J, Lee C (2018) Model-based winding tension profile to minimize radial stress in a flexible substrate in a roll-to-roll web transporting system, IEEE/ASME Transactions on Mechatronics **23**(6):2928–2939.
[11] Lee C, Kang H, Kim H, Shin K (2009) Effect of taper tension profile on the telescoping in a winding process of high speed roll to roll printing systems, J Mech Sci Technol **23**:3036–3048.
[12] Park Kt, Park Hc (2018) Effects of design parameters and tension on behavior of a coil using finite element analysis, J Iron Steel Res Int **25**:883–891.
[13] Sims RB, Place JA (1953) The stresses in the reels of cold reduction mills, Br J Appl Phys **4**(7):213-216.
[14] Komodromos A, Tekkaya AE, Hofmann J, Fleischer J (2018) Experimental and numerical investigations of wire bending by linear winding of rectangular tooth coils, AIP Conference Proceedings **1960**:150007.
[15] Lee C, Kang H, Shin K (2008) Development of a guiding system and visual feedback real-time controller for the high-speed self-align optical cable winding, J Mech Sci Technol **22**:2464–2474.
[16] Schmitz S, Kripfgans J, Schneider-Ramelow M, Müller WH, Lang KD (2014) Investigating wire bonding pull testing and its calculation basics, In: Proceedings of the 5th Electronics System-integration Technology Conference (ESTC), pp 1–5, DOI 10.1109/ESTC.2014.6962766.
[17] Ming J, Gongliu Y (2011) Research of optical fiber coil winding model based on large-deformation theory of elasticity and its application, Chinese J Aeronaut **24**:640–647.
[18] Lámer G (1985) Notes on the theory of large displacement with small strain, Periodica Polytechnica Civil Engineering **29**(1-2):53–65.

Chapter 34
On Using Rotations as Primary Variables in the Nonlinear Theory of Thin Elastic Shells

Wojciech Pietraszkiewicz

Abstract We discuss such a formulation of the non-linear theory of thin shells in which finite rotations are among independent fields of the boundary value problem. We show that in the general case of finite shell strains the surface bending tensor and the internal couple resultant tensor are non-symmetric in general, when defined with respect to the rotated base. Transforming the surface principle of virtual work we derive several strong and weak boundary value problems. In particular, we construct global variational principles of the total potential energy, Hu – Washizu and Simmonds – Danielson type. For a homogeneous, isotropic, rubber-like shell it is shown that the simplest approximation to the strain energy function does not contain the skew-symmetric part of the bending tensor.

Key words: Thin shells · Non-linear theory · Finite rotations · Rubber-like elastic shell · Variational principles

34.1 Introduction

In the mechanical model of a thin shell it is usually assumed that deformation of the shell as three-dimensional body can be described with sufficient accuracy by deformation of its reference surface. Equilibrium conditions of such a shell are usually found by postulating the virtual work principle for the reference surface. In the Lagrangean description this leads to a complex boundary value problem for three components of the surface translation field as the only independent field variables of the non-linear boundary value problem (BVP). For numerical analysis by the finite element method (FEM) this formulation requires C^1 interelement continuity

Wojciech Pietraszkiewicz
Faculty of Civil and Environmental Engineering, Gdańsk University of Technology, G. Narutowicza St. 11/12, 80-233 Gdańsk, Poland,
e-mail: wpietrasz@wp.pl

H. Altenbach et al. (eds.), *Sixty Shades of Generalized Continua*,
Advanced Structured Materials 170, https://doi.org/10.1007/978-3-031-26186-2_34

of translation components ot the elements. Thus, such formulation becomes very complex from numerical point of view, see discussion in [1].

In recent two decades the numerical analysis of shell structures by FEM has been based mainly on the so-called resultant (or Reissner type or the six-parameter type) shell formulation, in which six independent components of the translation vector $\boldsymbol{u}$ and of the rotation tensor $\boldsymbol{Q}$ fields are the only independent components of the BVP, see [2, 3]. In this case the corresponding finite elements require only the C^0 interelement continuity of components of $\boldsymbol{u}$ and $\boldsymbol{Q}$, which considerably simplifies the numerical non-linear analysis of shells, see [1].

Here we formulate an alternative BVP of thin shells in terms of the translation vector $\boldsymbol{u}$, the proper rotation tensor $\boldsymbol{R}$ and other surface fields defined on the shell reference surface. The alternative BVP of thin shell theory follows through description of deformation of the reference surface in the nonholonomic rotated surface base following from the polar decomposition of the shell deformation gradient [4]. Alumyae [5] first introduced a similar rotated surface base in a descriptive way and relative to this base defined the surface deformation measures: the symmetric strain tensor $\eta_{\alpha\beta}$ and the nonsymmetric bending tensor $\mu_{\alpha\beta}$. In [5] appropriate surface equilibrium and compatibility conditions were also formulated, [6]. Simmonds and Danielson [7, 8] introduced the rotation explicitly and formulated the BVP in terms of the finite rotation and stress funtion vectors as indepenrent field variables and constructed the corresponding variational principle [8]. The formulation of the non-linear theory of thin shells was somewhat extended in papers [1], [9]-[11] and the results were summarized in monographs [1], [2], [4], [12].

But discussions were limited to small elastic strains when the classical constitutive equations are well known. In such a case only symmetric part of the surface bending tensor explicitly appears in the strain energy density within the consistent first approximation. As a result, in all relations of such a classical shell formulation only symmetric components of the surface couple tensor are present [4].

In this report we extend applicability of thin shell theory given in rotations and other fields into problems of finite strains, arbitrary anisotropy and non-homogeneity of the elastic material. As a result of appropriate transformations of the principle of virtual work we explicitly introduce the non-symmetric components of the bending tensor and non-symmetric components of the surface couples. We formulate the extended principle of virtual work as well as strong and weak BVPs in terms of rotations and other sutface fields as primary variables. We also derive several global variational principles such as the principle of total potential energy, the principle of Hu-Washizu and the principle of Simmonds and Danielson type.

34.2 Geometry and Deformation of the Reference Surface

In the present paper we shall use the notation system applied in [4], [13]-[15]. An arbitrary point M of the shell reference surface $\mathcal{M}$ is defined by the position vector

$$\boldsymbol{r} = \overline{OM} = \boldsymbol{r}(\theta_1, \theta_2) = \sum_{k=1}^{3} r^k(\theta^\alpha)\boldsymbol{i}_k \equiv r^k \boldsymbol{i}_k, \quad k = 1,2,3, \quad \alpha = 1,2,$$

where θ^α are the surface coordinates, and $\boldsymbol{i}_k$ is the orthonormal base of the three-dimensional Euclidean physical space $\mathcal{E}$.

The deformed reference surface in the same base is described by

$$\bar{\boldsymbol{r}} = \chi[\boldsymbol{r}(\theta^\alpha)] = \bar{r}^k(\theta^\alpha)\boldsymbol{i}_k = \boldsymbol{r}(\theta^\alpha) + \boldsymbol{u}(\theta^\alpha), \tag{34.1}$$

where θ^α are convective surface coordinates, χ is the deformation function, $\boldsymbol{u} \in V$ is the translation field, while V is the 3D vector space.

In each point $M \in \mathcal{M}$ the following geometric relations are given:

$$\begin{aligned} &\boldsymbol{a}_\alpha = \boldsymbol{r}_{,\alpha}, \qquad a_{\alpha\beta} = \boldsymbol{a}_\alpha \cdot \boldsymbol{a}_\beta, \quad a = \det(a_{\alpha\beta}) > 0, \\ &\boldsymbol{a}^\alpha \cdot \boldsymbol{a}_\beta = \delta^\alpha_\beta, \quad a^{\alpha\beta} = \boldsymbol{a}^\alpha \cdot \boldsymbol{a}^\beta, \quad \boldsymbol{a}^\beta = a^{\beta\alpha}\boldsymbol{a}_\alpha, \\ &\boldsymbol{n} = \frac{1}{\sqrt{a}}\boldsymbol{a}_1 \times \boldsymbol{a}_2, \; b_{\alpha\beta} = -\boldsymbol{n}_{,\alpha} \cdot \boldsymbol{a}_\beta, \; \epsilon_{\alpha\beta} = (\boldsymbol{a}_\alpha \times \boldsymbol{a}_\beta) \cdot \boldsymbol{n}, \end{aligned} \tag{34.2}$$

where $\boldsymbol{a}_\alpha$ and $\boldsymbol{a}^\alpha$ are natural and dual surface base vectors, $a_{\alpha\beta}$ and $a^{\alpha\beta}$ are components of the surface metric tensor, $\boldsymbol{n}$ is the normal to $\mathcal{M}$ describing its orientation, $b_{\alpha\beta}$ are components of the surface curvature tensor, $\epsilon_{\alpha\beta}$ and $\epsilon^{\alpha\beta}$ are surface permutation symbols for which $\epsilon_{12} = -\epsilon_{21} = \sqrt{a}$, $\epsilon_{11} = \epsilon_{22} = 0$, while δ^β_α is the Kronecker symbol: $\delta^1_1 = \delta^2_2 = 1, \delta^1_2 = \delta^2_1 = 0$.

The surface boundary $\partial\mathcal{M}$ consists of a finite number of closed piecewise smooth curves that do not meet in cusps, each described parametrically by $\boldsymbol{r}(s) = \boldsymbol{r}[\theta^\alpha(s)]$, where s is the arc length along any regular part of $\partial\mathcal{M}$. At each regular point $M \in \partial\mathcal{M}$ we have the unit tangent vector $\boldsymbol{\tau} = d\boldsymbol{r}/ds \equiv \boldsymbol{r}' = \tau^\alpha \boldsymbol{a}_\alpha$ and the unit external normal vector $\boldsymbol{\nu} \equiv \boldsymbol{r}_{,\nu} = \boldsymbol{\tau} \times \boldsymbol{n} = \nu^\alpha \boldsymbol{a}_\alpha$, where $(\ldots)_{,\nu}$ is the external surface derivative orthogonal to $\partial\mathcal{M}$.

Analogous geometric relations are valid also on the deformed surface points of $\overline{\mathcal{M}}$. They are expressed by analogous symbols, which on the deformed reference surface we denote by additional dash: $\bar{\boldsymbol{a}}_\alpha, \bar{\boldsymbol{a}}^\alpha, \bar{a}_{\alpha\beta}, \bar{a}^{\alpha\beta}, \bar{\boldsymbol{n}}, \bar{b}_{\alpha\beta}, \bar{\epsilon}_{\alpha\beta}, \bar{a}, \bar{\boldsymbol{\tau}}, \bar{\boldsymbol{\nu}}$, etc. All quantities for $\overline{\mathcal{M}}$ are expressible in terms of analogous quantities on $\mathcal{M}$ and the translation field $\boldsymbol{u}$ by the relations given in [4, 13].

Components of the Green type surface deformation measures are defined by

$$\gamma_{\alpha\beta}(\boldsymbol{u}) = \frac{1}{2}(\bar{a}_{\alpha\beta} - a_{\alpha\beta}), \quad \kappa_{\alpha\beta}(\boldsymbol{u}) = -(\bar{b}_{\alpha\beta} - b_{\alpha\beta}), \tag{34.3}$$

where $\gamma_{\alpha\beta}$ are quadratic functions of $\boldsymbol{u}$ and $\boldsymbol{u}_{,\alpha}$, while $\kappa_{\alpha\beta}$ are non-rational functions of $\boldsymbol{u}, \boldsymbol{u}_{,\alpha}$ and $\boldsymbol{u}_{,\alpha\beta}$.

The neighbourhoods of $\mathcal{M}$ and $\overline{\mathcal{M}}$ in $\mathcal{E}$ we can parameterise by three convective coordinates θ^α, ζ, where ζ is the distance along the normal vectors $\boldsymbol{n}$ and $\bar{\boldsymbol{n}}$. The 3D deformation gradient at the reference surface is defined by

$$\boldsymbol{F} = \nabla\chi(\boldsymbol{r}+\zeta\boldsymbol{n})|_{\zeta=0} = \bar{\boldsymbol{a}}_\alpha \otimes \boldsymbol{a}^\alpha + \bar{\boldsymbol{n}} \otimes \boldsymbol{n}, \quad \det\boldsymbol{F} = \sqrt{\frac{\bar{a}}{a}} > 0, \tag{34.4}$$

where $\otimes$ is the tensor product.

Polar decomposition of the tensor $\boldsymbol{F}$ takes the form

$$\boldsymbol{F} = \boldsymbol{RU} = \boldsymbol{VR}, \quad \boldsymbol{V} = \boldsymbol{RUR}^{\mathrm{T}}, \quad \boldsymbol{s}_\alpha = \boldsymbol{U}\boldsymbol{a}_\alpha = \boldsymbol{R}^{-1}\bar{\boldsymbol{a}}_\alpha, \quad \boldsymbol{r}_\alpha = \boldsymbol{R}\boldsymbol{a}_\alpha = \boldsymbol{V}^{-1}\bar{\boldsymbol{a}}_\alpha. \tag{34.5}$$

Here $\boldsymbol{R} \in SO(3)$ is the proper orthogonal tensor, $\boldsymbol{U}$ and $\boldsymbol{V}$ are right and left stretch tensors, $\boldsymbol{s}_\alpha$ is a nonholonomic stretched surface base, and $\boldsymbol{r}_\alpha$ is a nonholonomic rotated surface base of the reference surface. The following relations are satisfied:

$$\begin{aligned} &\boldsymbol{U} = \boldsymbol{s}_\alpha \otimes \boldsymbol{a}^\alpha + \boldsymbol{n} \otimes \boldsymbol{n}, \quad \boldsymbol{V} = \bar{\boldsymbol{a}}_\alpha \otimes \boldsymbol{r}^\alpha + \bar{\boldsymbol{n}} \otimes \bar{\boldsymbol{n}}, \\ &\boldsymbol{R} = \bar{\boldsymbol{a}}_\alpha \otimes \boldsymbol{s}^\alpha + \bar{\boldsymbol{n}} \otimes \boldsymbol{n} = \boldsymbol{r}_\alpha \otimes \boldsymbol{a}^\alpha + \bar{\boldsymbol{n}} \otimes \bar{\boldsymbol{n}}, \\ &\boldsymbol{U}^{\mathrm{T}} = \boldsymbol{U}, \quad \boldsymbol{V}^{\mathrm{T}} = \boldsymbol{V}, \quad \boldsymbol{R}^{\mathrm{T}} = \boldsymbol{R}^{-1}, \quad \det\boldsymbol{R} = +1. \end{aligned} \tag{34.6}$$

Let us introduce the following relative strain measures of the surface:

$$\begin{aligned} &\boldsymbol{\eta} = \boldsymbol{U} - \boldsymbol{1} = \eta_{\alpha\beta}\boldsymbol{a}^\alpha \otimes \boldsymbol{a}^\beta, \quad \boldsymbol{\varepsilon} = \boldsymbol{V} - \boldsymbol{1} = \boldsymbol{\varepsilon}_\beta \otimes \boldsymbol{r}^\beta, \quad \boldsymbol{\varepsilon}_\beta = \eta_{\alpha\beta}\boldsymbol{r}^\alpha, \\ &\boldsymbol{\mu} = (\boldsymbol{R}^{\mathrm{T}}\bar{\boldsymbol{n}}_{,\beta} - \boldsymbol{n}_{,\beta} \otimes \boldsymbol{a}^\beta = \mu_{\alpha\beta}\boldsymbol{a}^\alpha \otimes \boldsymbol{a}^\beta, \\ &\boldsymbol{\lambda} = (\bar{\boldsymbol{n}}_{,\beta} - \boldsymbol{R}\boldsymbol{n}_{,\beta} \otimes \boldsymbol{r}^\beta = \boldsymbol{\lambda}_\beta \otimes \boldsymbol{r}^\beta, \quad \boldsymbol{\lambda}_\beta = \mu_{\alpha\beta}\boldsymbol{r}^\alpha, \\ &\eta_{\alpha\beta} = \eta_{\beta\alpha}, \quad \mu_{\alpha\beta} \neq \mu_{\beta\alpha}, \end{aligned} \tag{34.7}$$

where $\boldsymbol{1} = \boldsymbol{a}_\alpha \otimes \boldsymbol{a}^\alpha + \boldsymbol{n} \otimes \boldsymbol{n} = \boldsymbol{r}_\alpha \otimes \boldsymbol{r}^\alpha + \bar{\boldsymbol{n}} \otimes \bar{\boldsymbol{n}}$ is the spatial metric tensor. The measures introduced in (34.7) satisfy several kinematical relations given in [4, 15].

34.3 Equilibrium Conditions

In the non-linear theory of thin shells equilibrium conditions are traditionally introduced by the 2D principle of virtual work. For each kinematically admissible virtual displacement $\delta\boldsymbol{u}$ the principle is postulated at the shell base surface in the Lagrangean description in the form [4, 13, 15]

$$\begin{aligned} &\iint_{\mathcal{M}} [N^{\alpha\beta}\delta\gamma_{\alpha\beta}(\boldsymbol{u}) + M^{\alpha\beta}\delta\kappa_{\alpha\beta}(\boldsymbol{u})]\,dA \\ &= \iint_{\mathcal{M}} [\boldsymbol{p}\cdot\delta\boldsymbol{u} + \boldsymbol{h}\cdot\delta\bar{\boldsymbol{n}}(\boldsymbol{u})]\,dA + \int_{\partial\mathcal{M}_f} [\boldsymbol{N}^*\cdot\delta\boldsymbol{u} + \boldsymbol{H}^*\cdot\delta\bar{\boldsymbol{n}}(\boldsymbol{u})]\,ds. \end{aligned} \tag{34.8}$$

Here $N^{\alpha\beta}$ and $M^{\alpha\beta}$ mean the surface components of the resultant symmetric forces and couples of the Piola-Kirchhoff type, $\delta\gamma_{\alpha\beta}$ and $\delta\kappa_{\alpha\beta}$ are virtual surface deformation measures, the surface distributed forces and couples $\boldsymbol{p}$ and $\boldsymbol{h}$ acting on $\mathcal{M}$,

$\boldsymbol{N}^*$ and $\boldsymbol{H}^*$ the resultant vectors of forces and static moments applied on $\partial\mathcal{M}_f$, but measured at the unit length elements at the undeformed reference surface boundary $\partial\mathcal{M}_f$, while δ means the symbol of variation. Let us note that because $\bar{\boldsymbol{n}}\cdot\delta\bar{\boldsymbol{n}}=0$ only surface components of $\boldsymbol{h}$ and $\boldsymbol{H}^*$ can be taken into account in the theory of thin shells.

When $\boldsymbol{u}$ is the only primary variable of the non-linear thin shells theory it is the only variable subject to independent variation. This allows one to transform the virtual work principle (34.8) to the form allowing to derive three local equilibrium equations and four natural boundary conditions [13].

When the rotations are to be additional primary field variables of the theory of thin shells, the principle (34.8) has to be appropriately modified with the help of relations given in the Sect. 34.2.

The surface deformation measures $\gamma_{\alpha\beta},\kappa_{\alpha\beta}$ and $\eta_{\alpha\beta},\mu_{\alpha\beta}$ are related by (see [4], Eq. (5.11)):

$$\begin{aligned}\gamma_{\alpha\beta}&=\eta_{\alpha\beta}+\frac{1}{2}\eta^{\lambda}_{\alpha}\eta_{\lambda\beta},\\ \kappa_{\alpha\beta}&=\frac{1}{2}[(\delta^{\lambda}_{\alpha}+\eta^{\lambda}_{\alpha})\mu_{\lambda\beta}+(\delta^{\lambda}_{\beta}+\eta^{\lambda}_{\beta})\mu_{\lambda\alpha}]-\frac{1}{2}(b^{\lambda}_{\alpha}\eta_{\lambda\beta}+b^{\lambda}_{\beta}\eta_{\lambda\alpha}).\end{aligned}\tag{34.9}$$

This is the reason why the internal virtual work density in (34.8) can alternatively be expressed in the alternative new variables

$$\begin{aligned}&N^{\alpha\beta}\delta\gamma_{\alpha\beta}+M^{\alpha\beta}\delta\kappa_{\alpha\beta}=S^{\alpha\beta}\delta\eta_{\alpha\beta}+H^{\alpha\beta}\delta\mu_{\alpha\beta},\\ &S^{\alpha\beta}=N^{\alpha\beta}+\frac{1}{2}(\eta^{\alpha}_{\lambda}N^{\lambda\beta}+\eta^{\beta}_{\lambda}N^{\alpha\lambda})-\frac{1}{2}[(b^{\alpha}_{\lambda}-\mu^{\alpha}_{\lambda})M^{\lambda\beta}+(b^{\beta}_{\lambda}-\mu^{\beta}_{\lambda})M^{\alpha\lambda}],\\ &H^{\alpha\beta}=(\delta^{\alpha}_{\lambda}+\eta^{\alpha}_{\lambda})M^{\lambda\beta},\end{aligned}\tag{34.10}$$

where now $S^{\alpha\beta}=S^{\beta\alpha}$, but $H^{\alpha\beta}\neq H^{\beta\alpha}$.

But if rotations are to be independent field variables, there should be satisfied additional geometric constraints on virtual deformation measures relative to the base $\boldsymbol{r}_\alpha,\bar{\boldsymbol{n}}$ formulated in [4]:

$$\epsilon^{\alpha\beta}\boldsymbol{r}_\alpha\cdot\delta\eta_{\lambda\beta}\boldsymbol{r}^{\lambda}=0,\quad \bar{\boldsymbol{n}}\cdot\delta\eta_{\lambda\beta}\boldsymbol{r}^{\lambda}=0.\tag{34.11}$$

These constraints can be introduced into (34.8) with the help of S and Q^β inside $\overline{\mathcal{M}}$ and A and B^β along the boundary $\partial\overline{\mathcal{M}}$. It follows from the analysis given in [4] that always there is $A=B^\beta\nu_\beta$, and the only non-vanishing component is $B\equiv B^\beta\tau_\beta$. Taking this result into account in the relations (34.9) and (34.10), the modified virtual work principle takes now the modified form

$$\begin{aligned}&\iint\limits_{\mathcal{M}}(\boldsymbol{N}^{\beta}\cdot\delta\eta_{\lambda\beta}\boldsymbol{r}^{\lambda}+H^{\alpha\beta}\boldsymbol{r}_{\alpha}\cdot\delta\mu_{\lambda\beta}\boldsymbol{r}^{\lambda})dA=\iint\limits_{\mathcal{M}}(\boldsymbol{p}\cdot\delta\boldsymbol{u}+\boldsymbol{m}\cdot\boldsymbol{w})dA\\ &\quad+\int\limits_{\partial\mathcal{M}_f}(\boldsymbol{N}^*\cdot\delta\boldsymbol{u}+\boldsymbol{M}^*\cdot\boldsymbol{w})ds+\int\limits_{\partial\mathcal{M}_f}B\bar{\boldsymbol{n}}\cdot\delta\eta_{\lambda\beta}\boldsymbol{r}^{\lambda}\tau^{\beta}ds,\end{aligned}\tag{34.12}$$

where we have introduced the following concise notations:

$$
\begin{aligned}
\boldsymbol{N}^{\beta} &= (S^{\alpha\beta}+\epsilon^{\alpha\beta}S)\boldsymbol{r}_{\alpha}+Q^{\beta}\bar{\boldsymbol{n}},\\
\boldsymbol{M}^{\beta} &=\bar{\boldsymbol{n}}\times H^{\alpha\beta}\boldsymbol{r}_{\alpha},\quad \boldsymbol{M}^{*}=\bar{\boldsymbol{n}}\times\boldsymbol{H}^{*},\quad \boldsymbol{m}=\bar{\boldsymbol{n}}\times\boldsymbol{h},\\
\boldsymbol{w} &=\frac{1}{2}(\bar{\boldsymbol{a}}^{\alpha}\times\delta\bar{\boldsymbol{a}}_{\alpha}+\bar{\boldsymbol{n}}\times\delta\bar{\boldsymbol{n}})=\frac{1}{2}(\boldsymbol{r}^{\alpha}\times\delta\boldsymbol{R}\boldsymbol{R}^{\mathrm{T}}\boldsymbol{r}_{\alpha}+\bar{\boldsymbol{n}}\times\delta\boldsymbol{R}\boldsymbol{R}^{\mathrm{T}}\bar{\boldsymbol{n}}),\\
\boldsymbol{w}_{\tau} &=\frac{1}{2}(\bar{\boldsymbol{\nu}}\times\delta\bar{\boldsymbol{\nu}}+\bar{\boldsymbol{\tau}}\times\delta\bar{\boldsymbol{\tau}}+\bar{\boldsymbol{n}}\times\delta\bar{\boldsymbol{n}}).
\end{aligned}
\tag{34.13}
$$

$\boldsymbol{w}$ $\boldsymbol{w}_{\tau}$ are now the respective virtual rotations in the interior of the reference surface and along its boundary. Further transformation of (34.12) according to [4, 15] allows one to represent it in the form

$$
\begin{aligned}
&-\iint_{\mathcal{M}}[(\boldsymbol{N}^{\beta}|_{\beta}+\boldsymbol{p})\cdot\delta\boldsymbol{u}+(\boldsymbol{M}^{\beta}|_{\beta}+\bar{\boldsymbol{a}}_{\beta}\times\boldsymbol{N}^{\beta}+\boldsymbol{m})\cdot\boldsymbol{w}]da\\
&+\int_{\partial\mathcal{M}_f}\{[\boldsymbol{N}^{\beta}\nu_{\beta}+(B\bar{\boldsymbol{n}})'-\boldsymbol{N}^{*}]\cdot\delta\boldsymbol{u}+(\boldsymbol{M}^{\beta}\nu_{\beta}+B\bar{\boldsymbol{a}}_{\nu}-\boldsymbol{M}^{*})\cdot\boldsymbol{w}_{\tau}\}ds\\
&+\int_{\partial\mathcal{M}_d}\{[\boldsymbol{N}^{\beta}\nu_{\beta}+(B\bar{\boldsymbol{n}})']\cdot\delta\boldsymbol{u}+(\boldsymbol{M}^{\beta}\nu_{\beta}+B\bar{\boldsymbol{a}}_{\nu})\cdot\boldsymbol{w}_{\tau}\}ds+\sum_{M_n\in\partial\mathcal{M}}(B\bar{\boldsymbol{n}})_n\cdot\delta\boldsymbol{u}_n=0,
\end{aligned}
\tag{34.14}
$$

where $\bar{\boldsymbol{a}}_{\nu}=\bar{\boldsymbol{r}}'\times\bar{\boldsymbol{n}}$.

Kinematically admissible virtual translations $\delta\boldsymbol{u}$ and rotations $\boldsymbol{w}_{\tau}$ along $\partial\mathcal{M}_d$ are $\delta\boldsymbol{u}=\boldsymbol{0}$ and $\boldsymbol{w}_{\tau}=\boldsymbol{0}$, and the integral along $\partial\mathcal{M}_d$ in (34.14) vanishes identically. Satisfaction of the extended virtual work (34.14) requires that additionally to constraint conditions (34.11) there should be satisfied the following local extended equilibrium equations and extended natural boundary conditions:

$$
\begin{aligned}
&\boldsymbol{N}^{\beta}|_{\beta}+\boldsymbol{p}=\boldsymbol{0},\quad \boldsymbol{M}^{\beta}|_{\beta}+\bar{\boldsymbol{a}}_{\beta}\times\boldsymbol{N}^{\beta}+\boldsymbol{m}=\boldsymbol{0}\quad\text{inside}\quad\mathcal{M},\\
&\boldsymbol{N}^{\beta}\nu_{\beta}+(B\bar{\boldsymbol{n}})'-\boldsymbol{N}^{*}=\boldsymbol{0},\quad \boldsymbol{M}^{\beta}\nu_{\beta}+B\bar{\boldsymbol{a}}_{\nu}-\boldsymbol{M}^{*}=\boldsymbol{0}\quad\text{along}\quad\partial\mathcal{M}_f,\\
&(B\bar{\boldsymbol{n}})_n=\boldsymbol{0}\quad\text{in singular points}\quad M_n\in\partial\mathcal{M}_f.
\end{aligned}
\tag{34.15}
$$

Equilibrium conditions analogous to (34.15) for the geometrically non-linear thin shell theory are given in [4], formulae (5.57)-(5.59). Within the accuracy of consistent first approximation shell theory the non-symmetric part of $H^{\alpha\beta}$ cannot be described from the constitutive relations and still remains indefinite. This was the reason why in [4] it was taken $\boldsymbol{M}^{\beta}=\bar{\boldsymbol{n}}\times H^{(\alpha\beta)}\boldsymbol{r}_{\alpha}$, when $H^{(\alpha\beta)}=\frac{1}{2}(H^{\alpha\beta}+H^{\beta\alpha})$ is the symmetric part of $H^{\alpha\beta}$.

However, in the derivation given above we have not assumed that strains are small and the material is homogeneous or even elastic, as it was assumed in [4]. This is the reason that in further considerations we will use $(34.13)_2$ for the vector of internal surface couples $\boldsymbol{M}^{\beta}$.

34.4 Boundary Value Problem with Independent Rotations

The extended equilibrium conditions (34.15) allow one to choose various forms of the BVPs of the non-linear thin shell theory formulated in rotations.

Let us add to (34.15) the kinematic boundary condition following from the kinematically admissible virtual boundary constraints $\delta \boldsymbol{u} = \boldsymbol{0}$ and $\boldsymbol{w}_\tau \times \bar{\boldsymbol{n}} = \boldsymbol{0}$ along $\partial \mathcal{M}_d$:

$$\boldsymbol{u} = \boldsymbol{u}^*, \quad \boldsymbol{R}_\tau \boldsymbol{n} = \boldsymbol{R}_\tau^* \boldsymbol{n} \quad \text{on} \quad \partial \mathcal{M}_d, \tag{34.16}$$

where by $(\ldots)^*$ we denote the fields known at the boundary.

Inside the internal region of $\mathcal{M}$ the constraints for virtual deformation measures (34.11) require the constraints upon the global deformation measures:

$$\begin{aligned} &\epsilon^{\alpha\beta}\eta_{\alpha\beta} = \epsilon^{\alpha\beta}\boldsymbol{r}_\alpha \cdot \boldsymbol{\varepsilon}_\beta = \epsilon^{\alpha\beta}(\boldsymbol{R}\boldsymbol{a}_\alpha)\cdot(\boldsymbol{a}_\beta + \boldsymbol{u}_{,\beta} - \boldsymbol{R}\boldsymbol{a}_\beta) = 0, \\ &\bar{\boldsymbol{n}} \cdot \boldsymbol{\varepsilon}_\beta = (\boldsymbol{R}\boldsymbol{n})\cdot(\boldsymbol{a}_\beta + \boldsymbol{u}_{,\beta}) = 0. \end{aligned} \tag{34.17}$$

Kinematical relations between the modified deformation measures $\eta_{\alpha\beta}, \mu_{\alpha\beta}$ and fields $\boldsymbol{u}, \boldsymbol{R}$ have the explicit form [4]

$$\boldsymbol{\varepsilon}_\beta = (\boldsymbol{R}^{\mathrm{T}} - \boldsymbol{1})\cdot \boldsymbol{r}_\beta + \boldsymbol{u}_{,\beta}, \quad \boldsymbol{\lambda}_\beta = \frac{1}{2}(\boldsymbol{r}_\alpha \times \boldsymbol{R}_{,\beta}\boldsymbol{R}^{\mathrm{T}}\boldsymbol{r}^\alpha + \bar{\boldsymbol{n}} \times \boldsymbol{R}_{,\beta}\boldsymbol{R}^{\mathrm{T}}\bar{\boldsymbol{n}}) \times \bar{\boldsymbol{n}}. \tag{34.18}$$

For thin shells made of the elastic material we assume that there exists the function of elastic energy $\sum(\gamma_{\alpha\beta}, \kappa_{\alpha\beta}; \boldsymbol{r})$, which variation has the form given in $(34.10)_1$. Explicit forms of $\sum(\gamma_{\alpha\beta}, \kappa_{\alpha\beta}; \boldsymbol{r})$ for various classes of elastic shells, and the corresponding constitutive equations, depend on mechanical properties of the shell material and on geometry of the undeformed reference surface - its thickness, internal structure along the thickness direction etc. This was symbolically marked in $\sum$ by adding $\boldsymbol{r}$ to the set of its arguments. The function $\sum$ is usually constructed by consistent reduction to the surface corresponding known elastic energy W of the 3D solid body or directly by 2D discussion of the body supported by experiments for the reference surface. Several forms of 2D $\sum$, convenient in the analyses of shells have been constructed in [1, 2, 14], where various kinematic hypotheses were postulated, or using consistent reduction to the 2D form of the 3D stress and strain state of the shell.

With the help of (34.9) the function $\sum$ can easily be transformed to the form depended explicitly upon $\eta_{\alpha\beta}, \mu_{\alpha\beta}$:

$$\sum = \sum[\gamma_{\alpha\beta}(\eta_{\lambda\mu}), \kappa_{\alpha\beta}(\eta_{\lambda\mu}, \mu_{\lambda\mu}); \boldsymbol{r}] = \hat{\sum}(\eta_{\lambda\mu}, \mu_{\lambda\mu}; \boldsymbol{r}) \tag{34.19}$$

Differentiating the function $\hat{\sum}(\eta_{\lambda\mu}, \mu_{\lambda\mu}; \boldsymbol{r})$ relative to the 2D deformation measures we obtain the corresponding constitutive equations:

$$S^{\alpha\beta} = \frac{\partial \hat{\Sigma}}{\partial \eta_{\alpha\beta}} = S^{\beta\alpha}, \quad H^{\alpha\beta} = \frac{\partial \hat{\Sigma}}{\partial \mu_{\alpha\beta}} \neq H^{\beta\alpha}. \tag{34.20}$$

Let us introduce (34.18) into (34.20) and then into (34.15). We note that the equilibrium conditions are expressed now in terms of $\boldsymbol{u}, \boldsymbol{R}, S, Q^{\beta}$ and B as the only independent field variables on $\mathcal{M}$. This allows one for the following strong formulation of the BVP of the non-linear theory of thin elastic shells: with known vector fields of surface forces and moments $\boldsymbol{p}(\theta^{\alpha}), \boldsymbol{m}(\theta^{\alpha})$ and $\boldsymbol{N}^{*}(s), \boldsymbol{M}^{*}(s)$ of C^{-1} class it is necessary to calculate the translations $\boldsymbol{u}(\theta^{\alpha})$ and rotations $\boldsymbol{R}(\theta^{\alpha})$ of class C^{0} as well as the Lagrange multipliers $S(\theta^{\alpha}), Q^{\beta}(\theta^{\alpha}), B(s)$ of class C^{0}, which satisfy the geometric boundary conditions (34.16) as well as the equilibrium conditions $(34.15)_{2,3}$.

Each state of the surface $\mathcal{M}$ is now described by fields $\boldsymbol{u} \in V, \boldsymbol{R} \in SO(3), S \in R, Q^{\beta} \in R \times R$ and $B \in R$. The set of all possible solutions $\cup \equiv (\boldsymbol{u}, \boldsymbol{R}, S, Q^{\beta}, B)$ is the configuration space of solutions $C(\mathcal{M}) = V \times SO(3) \times R^{4}$, and virtual change of the solution $\delta\cup$ belongs to the space $T_{\cup}C(\mathcal{M})$ which is tangent to $C(\mathcal{M})$ at $\cup \in C(\mathcal{M})$. In the neighborhood $\cup$ each field $\cup(\alpha)$ along $\mathcal{L}: R \to C(\mathcal{M})$ has the form

$$\cup(\alpha) = \cup + \alpha\delta\cup, \quad \alpha \in R, \quad \cup(0) = 0. \tag{34.21}$$

Let us express the functional (34.12) of the virtual work in the following extended form taking into account the relations for kinematic constraints (34.17):

$$\begin{aligned} G[\cup;\delta\cup] = & \iint_{\mathcal{M}} \delta\hat{\sum}[\eta_{\alpha\beta}(\boldsymbol{u},\boldsymbol{R}), \mu_{\alpha\beta}(\boldsymbol{R})]dA \\ & - \iint_{\mathcal{M}} [\boldsymbol{p}(\boldsymbol{u},\boldsymbol{R}) \cdot \delta\boldsymbol{u} + \boldsymbol{m}(\boldsymbol{u},\boldsymbol{R}) \cdot \boldsymbol{w}]dA \\ & - \int_{\partial\mathcal{M}_f} [\boldsymbol{N}^{*}(\boldsymbol{u},\boldsymbol{R}) \cdot \delta\boldsymbol{u} + \boldsymbol{M}^{*}(\boldsymbol{u},\boldsymbol{R}) \cdot \boldsymbol{w}_{\tau}]ds \\ = + & \iint_{\mathcal{M}} \delta\{[\epsilon^{\alpha\beta} S\boldsymbol{r}_{\alpha}(\boldsymbol{R}) + Q^{\beta}\bar{\boldsymbol{n}}(\boldsymbol{R})] \cdot \eta_{\lambda\beta}(\boldsymbol{u},\boldsymbol{R})\boldsymbol{r}^{\lambda}(\boldsymbol{R})\}dA \\ & - \int_{\partial\mathcal{M}} \delta[B\bar{\boldsymbol{n}}(\boldsymbol{R}) \cdot \eta_{\lambda\beta}(\boldsymbol{u},\boldsymbol{R})\boldsymbol{r}^{\lambda}(\boldsymbol{R})\tau^{\beta}]ds \end{aligned} \tag{34.22}$$

The functional (34.22) is linear relative to S, Q^{β}, B and is nonrational in terms of $\boldsymbol{u}, \boldsymbol{R}$ and their surface derivatives. The BVP of the nonlinear theory of thin shells can now be formulated in the following weak form: with known fields $\boldsymbol{p}, \boldsymbol{m}, \boldsymbol{N}^{*}$ and $\boldsymbol{M}^{*}$ of class C^{-1} it is necessary to find the solution $\cup = (\boldsymbol{u}, \boldsymbol{R}, S, Q^{\beta}, B)$, in which $\boldsymbol{u}, \boldsymbol{R}$ are geometrically admissible of class C^{0}, and S, Q^{β}, B are of class C^{-1} such that for any kinematically admissible $\delta\cup$, in which $\delta\boldsymbol{u} = \boldsymbol{0}$ and $\boldsymbol{w}_{\tau} = \boldsymbol{0}$ on $\partial\mathcal{M}_{d}$, the virtual functional (34.22) becomes null identically:

$$G[\cup;\delta\cup] = 0. \tag{34.23}$$

Based on the principle (34.23) one can construct effective numerical algorithms. We stress here the required lower class of continuity of solution fields than in the analysing of strong BVP. In particular if one applies the finite element method one is allowed to construct the elements of C^0 class with the simplest shape functions.

The formulation (34.23) is the extended virtual work principle of the non-linear theory of thin elastic shells in which the rotations were explicitly taken as independent field variables. The kinematical constraints (34.17) in $\mathcal{M}$ and along $\partial\mathcal{M}$ have been introduced directly to the functional with the help of Lagrange multipliers.

When the external loading has the potential such that

$$G_{\text{ext}}[\boldsymbol{u},\boldsymbol{R};\delta\boldsymbol{u},\boldsymbol{w}] = \delta V[\boldsymbol{u},\boldsymbol{R};\delta\boldsymbol{u},\boldsymbol{w}], \tag{34.24}$$

we can introduce the functional of total potential energy

$$\begin{aligned} I(\mathsf{U}) = & \iint\limits_{\mathcal{M}} \{\hat{\sum}[\eta_{\alpha\beta}(\boldsymbol{u},\boldsymbol{R}),\mu_{\alpha\beta}(\boldsymbol{R})] + \epsilon^{\alpha\beta} S\eta_{\alpha\beta}(\boldsymbol{u},\boldsymbol{R}) + Q^{\beta}\bar{\boldsymbol{n}}(\boldsymbol{R})\cdot\boldsymbol{\varepsilon}_{\beta}(\boldsymbol{u},\boldsymbol{R})\}dA \\ & - \int\limits_{\partial\mathcal{M}} B\bar{\boldsymbol{n}}(\boldsymbol{R})\cdot\boldsymbol{\varepsilon}_{\beta}(\boldsymbol{u},\boldsymbol{R})\tau^{\beta}ds + V(\boldsymbol{u},\boldsymbol{R}), \end{aligned} \tag{34.25}$$

for which the geometric boundary conditions (34.16) are external conditions. Then the extended virtual work principle takes the form of global variational principle

$$\delta I[\mathsf{U};\delta\mathsf{U}] = 0. \tag{34.26}$$

Stationarity conditions of the functional $I[\mathsf{U}]$ are: the equilibrium equations $(34.7)_1$, natural static boundary conditions $(34.15)_{2,3}$ ant the constraint conditions (34.17) in $\mathcal{M}$ and along $\partial\mathcal{M}$. The expression (34.26) represents the modified principle of total potential energy in the non-linear theory of thin elastic shells formulated in terms of rotations and other fields as independent field variables.

The functional (34.25) is linear relative to S, Q^{β}, B as well and rational in terms of $\boldsymbol{u},\boldsymbol{R}$ and their first surface derivatives. The principle (34.26) also allows one to construct approximate solutions of class C^{-1} for S, Q^{β}, B and of class C^0 for $\boldsymbol{u},\boldsymbol{R}$, which in the case of finite element solution method allows one to formulate the simplest finite elements of class C^0 with simplest shape functions.

When formulating and solving some problems of thin shells it may be useful to discuss the general free functional of the Hu – Washizu type

$$\begin{aligned} J = & \iint\limits_{\mathcal{M}} \{\hat{\Sigma}(\eta_{\alpha\beta},\mu_{\alpha\beta}) + \epsilon^{\alpha\beta} S\eta_{\alpha\beta} + Q^{\beta}\bar{\boldsymbol{n}}(\boldsymbol{R})\cdot\eta_{\lambda\beta}\boldsymbol{r}^{\lambda}(\boldsymbol{R}) \\ & - S^{\alpha\beta}[\eta_{\alpha\beta} - \eta_{\alpha\beta}(\boldsymbol{u},\boldsymbol{R})] - H^{\alpha\beta}[\mu_{\alpha\beta} - \mu_{\alpha\beta}(\boldsymbol{u},\boldsymbol{R})]\}\, dA \\ & - \int\limits_{\partial\mathcal{M}} B\bar{\boldsymbol{n}}(\boldsymbol{R})\cdot\eta_{\lambda\beta}\boldsymbol{r}^{\lambda}(\boldsymbol{R})\tau^{\beta}ds \\ & - \int\limits_{\partial\mathcal{M}_d} \langle\{[(S^{\alpha\beta} + \epsilon^{\alpha\beta}S)\boldsymbol{r}_{\beta}(\boldsymbol{R}) + Q^{\beta}\bar{\boldsymbol{n}}(\boldsymbol{R})]\nu_{\beta} + [B\bar{\boldsymbol{n}}(\boldsymbol{R})]'\}\cdot(\bar{\boldsymbol{r}} - \bar{\boldsymbol{r}}^*) \\ & + [H^{\alpha\beta}\boldsymbol{r}_{\alpha}(\boldsymbol{R})\nu_{\beta} - B\bar{\boldsymbol{r}}'(\boldsymbol{u},\boldsymbol{R})]\cdot(\boldsymbol{R}_{\tau}\boldsymbol{n} - \boldsymbol{R}^*_{\tau}\boldsymbol{n})\rangle\, ds + V(\boldsymbol{u},\boldsymbol{R}_{\tau}). \end{aligned} \tag{34.27}$$

The functional J is constructed from the functional I by introducing to it the kinematical relations (18) and geometric boundary conditions (34.16) with the Lagrange multipliers. Then some of stationarity conditions of the functional J allow to identify these multipliers as $S^{\alpha\beta}, H^{\alpha\beta}, \boldsymbol{N}^{\beta}\nu_{\beta}+(B\bar{\boldsymbol{n}})'$ and $H^{\alpha\beta}\boldsymbol{r}_{\alpha}\nu_{\beta}-B\bar{\boldsymbol{r}}'$, respectively. So identified multipliers have been introduced to the functional (34.27). As a result, the functional J is now defined on fields $\boldsymbol{u}, \boldsymbol{R}, \eta_{\alpha\beta}, \nu_{\alpha\beta}, S^{\alpha\beta}, S, Q^{\beta}, H^{\alpha\beta}$ and B which can freely be variated. The variational principle $\delta J = 0$ of the Hu - Washizu type is an alternative weak formulation of the complete set of relations of the non-linear theory of thin elastic shells discussed in this paper. Satisfying à priori some stationarity conditions of the functional J and/or assuming some additional simplifying assumptions this functional can be transformed to various simpler free or conditional functionals.

An interesting special case of such transformations is the functional $J_2(\boldsymbol{\theta}, \boldsymbol{\psi}, B)$ of the Simmonds – Danielson type, which is described by the finite rotation vector $\boldsymbol{\theta}$, stress function vector $\boldsymbol{\psi}$ and a Lagrange multiplier B as the independent variables. The finite rotation vector appearing here is given by the formula $\boldsymbol{\theta} = 2\tan\omega/2\boldsymbol{e}$, where ω is the rotation angle about some axis with a unit vector $\boldsymbol{e}$. Such a functional can be constructed with the following additional assumptions:

1. Equilibrium equations $(34.7)_1$ of surface forces can be satisfied by the expression $\boldsymbol{N}^{\beta} = \epsilon^{\alpha\beta}\boldsymbol{\psi}_{,\alpha} + \boldsymbol{P}^{\beta}$, where $\boldsymbol{P}^{\beta}$ is a particular solution of the equilibrium equations.
2. The surface elastic strain energy density $\hat{\Sigma}$ is an additive function of relative deformation measures $\hat{\Sigma} = \hat{\Sigma}_{\eta}(\eta_{\alpha\beta}) + \hat{\Sigma}_{\mu}(\mu_{\alpha\beta})$, where $\hat{\Sigma}_{\eta}(\eta_{\alpha\beta})$ is the reversible function.
3. Potentials of surface and boundary loads have the forms $\Phi = \boldsymbol{P}^{\beta}|_{\beta}\cdot\bar{\boldsymbol{r}} + f[\bar{\boldsymbol{n}}(\boldsymbol{\theta})]$ and $\Psi = \boldsymbol{N}^{*}\cdot\bar{\boldsymbol{r}} + g[\bar{\boldsymbol{n}}(\boldsymbol{\theta})]$.
4. The shell boundary $\partial\mathcal{M}$ can be divided into separate parts $\partial\mathcal{M}_d$ and $\partial\mathcal{M}_f$ so that there is no boundary parts with mixed boundary conditions.
5. There are satisfied à priori the following stationarity conditions:

$$
\begin{gathered}
\eta_{\alpha\beta} = \frac{\partial\hat{\Sigma}_{\eta}}{\partial S^{\alpha\beta}}, \quad H^{\alpha\beta} = \frac{\partial\hat{\Sigma}_{\mu}}{\partial\mu_{\alpha\beta}}, \quad \mu_{\alpha\beta} = \mu_{\alpha\beta}(\boldsymbol{\theta}), \quad \boldsymbol{N}^{\beta} = \epsilon^{\alpha\beta}\boldsymbol{\psi}_{,\alpha} + \boldsymbol{P}^{\beta} \quad \text{inside } \mathcal{M}, \\
\bar{\boldsymbol{r}} = \bar{\boldsymbol{r}}^{*} \quad \text{along} \quad \partial\mathcal{M}_d, \quad \bar{\boldsymbol{r}}_n = \bar{\boldsymbol{r}}_n^{*} \quad \text{at each} \quad M_n \in \partial\mathcal{M}_d.
\end{gathered}
\tag{34.28}
$$

The construction of the functional $J_2(\boldsymbol{\theta}, \boldsymbol{\psi}, B)$ follows analogically to the construction of similar functional for elastic shells under small strains discussed in detail in [4], Sect. 5.7. Omitting here those complex transformations, we give here the final form of the Simmonds – Danielson type functional which is correct in the case of large elastic strains of thin shells:

$$
\begin{aligned}
J_2(\boldsymbol{\theta},\boldsymbol{\psi},B) = & \iint\limits_{\mathcal{M}} \{\sum\nolimits_{\mu}(\boldsymbol{\theta}) - \hat{\sum\nolimits}^{c}_{S}(\boldsymbol{\theta},\boldsymbol{\psi}) - (\epsilon^{\alpha\beta}\boldsymbol{\psi}_{,\alpha} + \boldsymbol{P}^{\beta})\cdot \boldsymbol{r}_{\beta}(\boldsymbol{\theta}) + f[\bar{\boldsymbol{n}}(\boldsymbol{\theta})]\}dA \\
& + \sum_{n=1,3,\ldots} \langle \int\limits_{s_n}^{s_{n+1}} \{(\boldsymbol{G}_n + \boldsymbol{C}_n - \boldsymbol{\theta})\cdot \bar{\boldsymbol{r}}'(\boldsymbol{\theta},\boldsymbol{\psi}) + f[\bar{\boldsymbol{n}}(\boldsymbol{\theta})]\}ds \\
& - [\boldsymbol{G}_n(s_{n+1})\cdot \bar{\boldsymbol{r}}^*_{n+1} + \boldsymbol{C}_n \cdot (\bar{\boldsymbol{r}}^*_{n+1} - \bar{\boldsymbol{r}}^*_n)]\rangle - \int\limits_{\partial\mathcal{M}} B\bar{\boldsymbol{n}}(\boldsymbol{\theta})\cdot \boldsymbol{\varepsilon}_{\beta}(\boldsymbol{u},\boldsymbol{\theta})\tau^{\beta} ds \\
& - \int\limits_{\partial\mathcal{M}_d} \{\boldsymbol{\theta}\cdot \bar{\boldsymbol{r}}^{*,} - \boldsymbol{P}^{\beta}\nu_{\beta}\cdot \bar{\boldsymbol{r}}^* + [H^{\alpha\beta}(\boldsymbol{\theta})\boldsymbol{r}_{\alpha}(\boldsymbol{\theta})\nu_{\beta} - (B\bar{\boldsymbol{r}}^*)']\cdot[\bar{\boldsymbol{n}}(\boldsymbol{\theta}) - \bar{\boldsymbol{n}}(\boldsymbol{\theta}^*)]\}ds.
\end{aligned}
\tag{34.29}
$$

Here

$$
\boldsymbol{G}_n(s) = \int\limits_{s_n}^{s} (\boldsymbol{N}^* - \boldsymbol{P}^{\beta}\nu_{\beta})ds
$$

are functions on $\partial\mathcal{M}_d$, $\hat{\sum}^{c}_{S}$ is the complementary energy density such that

$$
\hat{\sum\nolimits}^{c}_{S}(S^{\alpha\beta}) = S^{\alpha\beta}\eta_{\alpha\beta} - \sum\nolimits_{\eta}(\eta_{\alpha\beta}),
$$

while $\boldsymbol{C}_n$ are constant vectors, which should assure the vector $\boldsymbol{\psi}$ to be continuous when approaching the boundary $\partial\mathcal{M}_f$. The variational principle

$$
\delta J_2[\boldsymbol{\theta},\boldsymbol{\psi},B,\delta\boldsymbol{\theta},\delta\boldsymbol{\psi},\delta B) = 0 \tag{34.30}
$$

is equivalent to satisfaction of several relations of the non-linear shell theory discussed in [4].

34.5 Constitutive Equations of Rubber-Like Shells

Let us discuss here the special form of the elastic energy density for homogeneous shells made of isotropic incompressible rubber-like material. It was shown in [14] that for such a shell the function

$$
\Sigma \cong hW_{(0)}(\gamma_{\kappa\rho}) + \frac{h^3}{24}\underline{[W^{\alpha\beta}_{(1)}(\gamma_{\kappa\rho})(\kappa^{\lambda}_{\alpha}\kappa_{\lambda\beta} - \kappa^{\lambda}_{\lambda}\kappa_{\alpha\beta})} + W^{\alpha\beta\lambda\mu}_{(2)}(\gamma_{\kappa\rho})\chi_{\alpha\beta}\chi_{\lambda\mu}] \tag{34.31}
$$

is the consistent first approximation to the elastic deformation energy of the shell. In (34.31) the functions $W_{(0)}$, $W^{\alpha\beta}_{(1)}$ and $W^{\alpha\beta\lambda\mu}_{(2)}$ mean, respectively, modified function of 3D elastic energy density and its first and second derivatives relative the tangent strain measure components, all defined at the shell middle surface $\mathcal{M}$, h is the undeformed shell thickness and $\chi_{\alpha\beta} = -(\sqrt{a/\bar{a}}\bar{b}_{\alpha\beta} - b_{\alpha\beta})$.

We have shown in [14] that the underlined terms in (34.31) have only small influence in the elastic deformation energy density and can be omitted in engineering analyses. Additionally, for thin shells the parts of elastic energy from stretches and bendings may have quantities of similar order only in the case when the largest order of the strains $\gamma_{\alpha\beta}$ is at most $O(\theta)$, where θ is the small parameter defined in [14]. In such a case the deformations are large in the sense that $1+\gamma^2 \approx 1$, but not finite in the sense $\gamma = O(1)$, and $\kappa_{\alpha\beta}$ can be approximated by the relation

$$\chi_{\alpha\beta} \cong \kappa_{\alpha\beta} + b_{\alpha\beta} - \kappa_{\alpha\beta}\gamma^{\kappa}_{\kappa}. \tag{34.32}$$

In this case the simplest approximation to the shell strain energy function for rubber-like material takes the form

$$\Sigma \cong hW_{(0)}(\gamma_{\kappa\rho}) + \frac{h^3}{24} W^{\alpha\beta\lambda\mu}_{(2)}(\gamma_{\kappa\rho})[\kappa_{\alpha\beta}\kappa_{\lambda\mu}(1-2\gamma^{\kappa}_{\kappa}) + 2b_{\alpha\beta}\kappa_{\lambda\mu}\gamma^{\kappa}_{\kappa}], \tag{34.33}$$

where we accounted the symmetry of $W^{\alpha\beta\lambda\mu}_{(2)}$ under exchange of indices α and β, and λ and μ, as well as $\alpha\beta$ and $\lambda\mu$.

Let us now transform (34.33) to the form expressed in terms of $\eta_{\alpha\beta}, \mu_{\alpha\beta}$, but with approximation used in the second term of (34.33). Let $\mu_{\alpha\beta} = \mu_{(\alpha\beta)} + \epsilon_{\alpha\beta}\rho$, where $\mu_{(\alpha\beta)} = \frac{1}{2}(\mu_{\alpha\beta} + \mu_{\beta\alpha})$. With this decomposition the tensor $\kappa_{\alpha\beta}$ in $(34.7)_2$ takes the form

$$\kappa_{\alpha\beta} = \mu_{(\alpha\beta)} - \frac{1}{2}[\eta^{\lambda}_{\alpha}(b_{\lambda\beta} - \mu_{(\lambda\beta)}) + \eta^{\lambda}_{\beta}(b_{\alpha\lambda} - \mu_{(\alpha\lambda)})] + \frac{1}{2}(\eta^{\lambda}_{\alpha}\epsilon_{\lambda\beta} + \eta^{\lambda}_{\beta}\epsilon_{\lambda\alpha})\rho. \tag{34.34}$$

From the compatibility of strains [4] we obtain the relation

$$\rho = \frac{1}{2+\eta^{\kappa}_{\kappa}} \epsilon^{\alpha\beta}\eta^{\lambda}_{\alpha}(b_{\lambda\beta} - \mu_{(\lambda\beta)}) \tag{34.35}$$

Estimated value of non-symmetric part $\epsilon_{\alpha\beta}\rho$ of the tensor $\mu_{\alpha\lambda}$ following from (34.35) is of the order $\max(\eta/R, \eta\mu)$, where η is the largest value of the strain tensor $\eta_{\alpha\beta}$, μ is the largest value of symmetric part $\mu_{(\alpha\beta)}$ of the bending tensor, and R is the lowest radius of curvature of the base surface in the discussed point $M \in \mathcal{M}$. It follows from (34.36) that with the relative error $o(\theta^2)$ the expression (34.34) takes the following approximate form:

$$\kappa_{\alpha\beta} \cong \mu_{(\alpha\beta)} - \frac{1}{2}[\eta^{\lambda}_{\alpha}(b_{\lambda\beta} - \mu_{(\lambda\beta)}) + \eta^{\lambda}_{\beta}(b_{\alpha\lambda} - \mu_{(\alpha\lambda)})]. \tag{34.36}$$

Introducing (34.36) and $(34.9)_1$ to the expression (34.33) and omitting small terms with relative error $o(\theta^2)$ we obtain

$$\begin{aligned}\hat{\Sigma} \cong\ & h\hat{W}_{(0)}(\eta_{\kappa\rho}) + \frac{h^3}{24} W^{\alpha\beta\lambda\mu}_{(2)}(\eta_{\kappa\rho})\{\mu_{(\alpha\beta)}\mu_{(\lambda\mu)} + 2(b_{\alpha\beta} - \mu_{(\alpha\beta)})\mu_{(\lambda\mu)}\eta^{\kappa}_{\kappa} \\ & - [\eta^{\sigma}_{\alpha}(b_{\sigma\beta} - \mu_{(\sigma\beta)}) + \eta^{\sigma}_{\beta}(b_{\alpha\sigma} - \mu_{(\alpha\sigma)})]\mu_{(\lambda\mu)}\}.\end{aligned} \tag{34.37}$$

It follows from (34.37) that in the simplest consistent approximation to the strain energy density of thin rubber-like shell the function $\hat{\Sigma}$ contains only the symmetric part $\mu_{(\alpha\beta)}$ of the bending tensor. This is the reason why in this special class of elastic shells only the symmetric part $H^{(\alpha\beta)}$ of surface couple tensor can be described by the constitutive equation (34.20). The non-symmetric part of the moment tensor is still indefinite in such a non-linear theory of rubber-like shells. In this case also all the relations of BVPs formulated in Sect. 34.4 of this paper can be simplified by omitting the non-symmetric parts of the tensors $\mu_{\alpha\beta}$ and $H^{\alpha\beta}$.

References

[1] Chróścielewski J, Makowski J, Pietraszkiewicz W (2004) Statics and Dynamics of Multi-Shells: Non-Linear Theory and Finite Element Method (in Polish), IFTR PASci Press, Warsaw.

[2] Libai A, Simmonds JG (19998) The nonlinear theory of elastic shells, 2nd ed, Cambridge University Press, Cambridge.

[3] Pietraszkiewicz W (2020) Elastic shells, resultant nonlinear theory, In: H Altenbach, A Ochsner (Eds) Encyclopedia of Continuum Mechanics, pp 822-830, Springer-Verlag, Berlin.

[4] Pietraszkiewicz W (1989) Geometrically nonlinear theories of thin elastic shells, Advances in Mechanics, textbf12:51-130.

[5] Alumyae NA (1949) Differential equilibrium equations of thinwalled elastic shells in the postcritical stage (in Russ), Prikladnaja Matematika i Mekhanika **13**:95-106.

[6] Alumyae NA (1956) On expressing basic relations of the nonlinear shell theory (in Russ), Prikladnaja Matematika i Mekhanika **20**:136-139.

[7] Simmonds JG, Danielson DA (1970) Nonlinear shell theory with a finite rotation vector, Proc. Koninkl. Nederl Akademie van Wetenschappen - Amsterdam, Series B, **73**:460-478.

[8] Simmonds JG, Danielson DA (1972) Nonlinear shell theory with finite rotation and stress function vectors, Trans ASME J Applied Mechanics **39**:1085-1090.

[9] Pietraszkiewicz W (1979) Finite rotations and Lagrangean description in the non-linear theory of shells, PWN, Warszawa-Poznań.

[10] Libai A, Simmonds JG (1983): Nonlinear elastic shell theory, In: JW Hutchinson, TY Wu (Eds) Advances in Applied Mechanics **23**:271-371, Academic Press, New York.

[11] Atluri SN (1984) Alternate stress and conjugate strain measures, and mixed variational formulations involving rigid rotations, for computational analyses of finitely deformed solids, with applications to plates and shells – I. Theory, Computers and Structures **18**:93-116.

[12] Valid R (1995) The nonlinear theory of shells through variational principles, John Wiley and Sons, Chicester et al.

[13] Pietraszkiewicz W (1984) Lagrangian description and incremental formulation in the non-linear theory of thin shells, Int J Non-Linear Mechanics **19**:115-140.

[14] Schieck B, Pietraszkiewicz W, Stumpf H (1992) Theory and numerical analysis of shells undergoing large elastic strains, Int J Solids and Structures **29**:689-709.
[15] Pietraszkiewicz W (2020) Thin elastic shells, Lagrangian geometrically nonlinear theory, In: H Altenbach, A Öchsner (Eds) Encyclopedia of Continuum Mechanics, pp 2518-2526, Springer-Verlag, Berlin.
[16] Koiter WT (1960) A consistent first approximation in the general theory of thin elastic shells, In: W.T. Koiter (Ed), Proc. IUTAM Symp on Theory of Thin Shells, pp 12-33, North-Holland PCo, Amsterdam.

Chapter 35
Continuum Description of Extended Mass-in-Mass Metamaterial Models

Alexey V. Porubov

Abstract Various extended mass-in-mass metamaterial models can be studied using their long wavelength continuum limits. It is shown that continuum limit of the classic model describes the band gap at semi-infinite interval of the values of the wave numbers. Generalization of the continuum limit by inclusion of the dispersion of the main chain of the classic model doesn't affect the width of the band gap. Extended models with the additional attached masses reveal appearance of extra band gaps, and the position and the width of the band gap change in comparison with the classic mas–in-mass model. The parameters responsible for the variations of the bang gaps are introduced.

Key words: Metamaterial · Continuum limit · Dispersion relation · Band gap

35.1 Introduction

The study of discrete crystalline lattices with a complex structure [1]-[4] allow us to understand the features the materials with microstructure, in particular, materials with man-made internal structure or metamaterials. Complex structure of the lattices brings additional degrees of freedom, allowing effective modelling of mechanical [2, 3] processes in solids. Among the discrete complex models considerable attention is paid to the acoustic metamaterials [5]-[7]. One of the simplest but instructive metamaterial models is the mas-in-mass lattice model, [5]-[14]. Propagation or non-propagation of acoustic waves in an acoustic metamaterial is caused by the presence of a band gap in the dispersion relation. Experimental realization of the metamaterial model with the internal resonators can be found in [15]-[18].

Alexey V. Porubov
Institute for Problems in Mechanical Engineering, Bolshoy 61, V.O., Saint-Petersburg, Russian Federation,
e-mail: alexey.porubov@gmail.com

H. Altenbach et al. (eds.), *Sixty Shades of Generalized Continua*,
Advanced Structured Materials 170, https://doi.org/10.1007/978-3-031-26186-2_35

These results also point out to the possible extensions of the mass-in -mass metamaterial model. Such extended models have been considered in [10, 16], [19]-[23]. A nonlinearly elastic chain with masses when attached mass with are attached through two main masse has been studied in [10]. The model with additional attached masses has been considered in [16, 19, 20], while the model with additional inner attached mass can be found in [19, 21]. These papers us discrete and mainly linearly-elastic modelling. Among the new features discovered, one can note multiple band gaps.

Discrete modeling based on the study of the difference governing equations is a most favorable method for finding the solution of the metamaterial problems. The continuum modeling is used much less frequently [11]-[14]. The usual transfer from discrete to continuum approach is based on the long wavelength continuum limit giving rise to the partial differential governing equations [1]-[4]. It restricts the study by the long strain waves. At the same time, continuum equations are more likely to study analytically especially in the nonlinearly-elastic springs in the original lattice models.

The plan of the current work is as follows. In Sect. 35.2 the classic linearly-elastic mas-in mass metamaterial model is considered in the continuum approximation, various continuum limits are analyzed using the dispersion relation analysis. Then three extended linearly-elastic models are studied in Sects. 35.3-35.5 using the same continuum limit approach. Of special attention is the band gap description for all models. The Conclusions summarize the relative analysis of all extended models and outline the future directions of their study.

35.2 Classic Mass-in-Mass Chain

The classic mass-in-mass chain is shown in Fig 35.1. Here interactions between the masses, m, are modelled by linearly elastic springs with the stiffness equal to $\beta_0 m$. The attached masses, m_1, are connected to the masses m by the springs with the stiffness equal to $\beta_1 m_1$. The governing equations are [5]:

$$\ddot{u}_n = \beta_0(u_{n-1} - 2u_n + u_{n+1}) + \eta\beta_1(v_n - u_n), \tag{35.1}$$

$$\ddot{v}_n = -\beta_1(v_n - u_n). \tag{35.2}$$

where masses m_1 do not interact between each other. The displacement of the mass m with the number n is denoted by u_n, while that of m_1 is denoted by v_n. and $\eta = m_1/m$ is the mass ratio. The solution of linear discrete equations (35.1) and (35.2) can be found in many papers.

Another description of the dynamic processes can be done using a continuum limit of Eqs. (35.1) and (35.2). The continuum functions $u(x,t)$, $v(x,t)$ are introduced for description of the displacements of the masses m, m_1. The continuum displacements of the neighboring masses are sought using the long-wave approximation [1], based on the Taylor series,

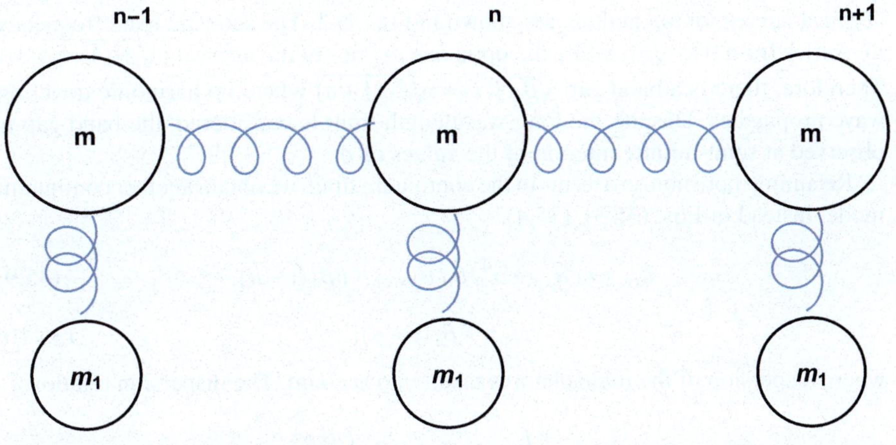

Fig. 35.1: Classic mass-in-mass chain.

$$u_{n\pm 1} = u \pm h\, u_x + \frac{h^2}{2} u_{xx} + \pm \frac{h^3}{6} u_{xxx} + \frac{h^4}{24} u_{xxxx} + \ldots,$$

where the lattice step, h, is considered as an increment of the spatial variable, x. Retaining only the first nonzero term in the expansion, we obtain

$$u_{tt} = a^2 u_{xx} + \eta\beta_1(v-u), \tag{35.3}$$

$$v_{tt} = -\beta_1(v-u), \tag{35.4}$$

where $a = \sqrt{\beta_0}\, h$ is the long-wave velocity for the main lattice.

Dispersion relation is obtained using the solution of the form

$$u = A\exp\Big(\imath\,(px-\omega t - x_0)\Big), \qquad v = B\exp\Big(\imath\,(px-\omega t - x_0)\Big), \tag{35.5}$$

where x_0 accounts for an initial position of the wave, A and B are the wave amplitudes, p is the wave number, ω is the wave frequency, $\imath$ is the imaginary unit. Substitution of Eq. (35.5) into Eqs. (35.3), (35.4) gives rise to

$$A = \frac{B(\beta_1 - \omega^2)}{\beta_1}, \tag{35.6}$$

and the dispersion relation, whose solutions are $\omega = \omega_a$, $\omega = \omega_o$, where

$$\omega_a^2 = \frac{\beta_1(1+\eta) + a^2p^2 - \sqrt{(\beta_1(1+\eta)+a^2p^2)^2 - 4\beta_1 p^2 a^2}}{2}, \tag{35.7}$$

$$\omega_o^2 = \frac{\beta_1(1+\eta) + a^2p^2 + \sqrt{(\beta_1(1+\eta)+a^2p^2)^2 - 4\beta_1 p^2 a^2}}{2}. \tag{35.8}$$

Typical curves for ω_a and ω_o are shown in Fig. 35.2. The acoustic band frequency ω_a varies from 0 to $\sqrt{\beta_1}$, while the optic one ω_o lies in the interval $(\sqrt{\beta_1(1+\eta)}, \infty)$. Therefore, there is a band gap $\sqrt{\beta_1} < \omega < \sqrt{\beta_1(1+\eta)}$ where no harmonic traveling wave propagates. Despite the long wavelength limit is considered, the band gap is observed at semi-infinite interval of the values of p.

Retaining more nonzero terms in the continuum limit, we obtain another continuum model instead of Eqs. (35.3), (35.4),

$$u_{tt} = a^2 u_{xx} + a^4/\beta_0\, u_{xxxx} + \eta\beta_1(v-u), \tag{35.9}$$

$$v_{tt} = -\beta_1(v-u), \tag{35.10}$$

where dispersion of the main chain is taken into account. The dispersion relations,

$$\omega_a^2 = \frac{1}{2}\left(\beta_1(1+\eta) + a^2 p^2 \frac{a^2 p^4}{12\beta_0} - \right.$$

$$\left. \sqrt{\left(\beta_1(1+\eta) + a^2p^2 - a^2p^2\frac{a^2p^4}{12\beta_0}\right)^2 - 4\beta_1\beta_0 p^2 a^2(1-a^2p^2/(12\beta_0))}\right) \tag{35.11}$$

$$\omega_o^2 = \frac{1}{2}\left(\beta_1(1+\eta) + a^2 p^2 \frac{a^2 p^4}{12\beta_0} + \right.$$

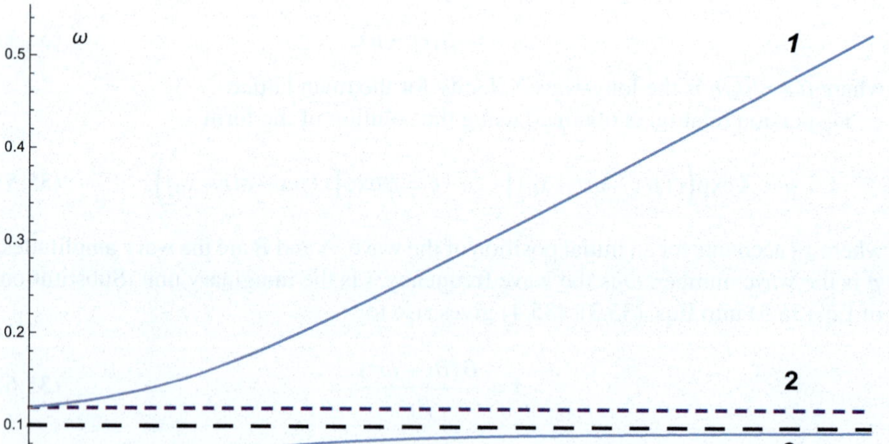

Fig. 35.2: Acoustic ω_a (solid curve 4) and optic ω_o (solid curve 1) branches for the classic mass-in-mass model. Thin dashed horizontal lime cooresponds to the value of ω_o at $p = 0$. The mediumdashed horizontal line 3 corresponds to the value of ω_a at $p \to \infty$. The lines bound a bang gap.

$$\sqrt{\left(\beta_1(1+\eta)+a^2p^2-a^2p^2\frac{a^2p^4}{12\beta_0}\right)^2-4\beta_1\beta_0p^2a^2(1-a^2p^2/(12\beta_0))}\Bigg). \tag{35.12}$$

are different form Eqs. (35.7), (35.8). However, they describe the same bang gap $\sqrt{\beta_1} < \omega < \sqrt{\beta_1(1+\eta)}$ as that obtained from the simpler continuum model Eqs. (35.3), (35.4). The typical dispersion curves are similar to those shown in Fig. 35.2.

35.3 Chain with Extra Attached Masses

The chain with extra attached masses m_1 is shown in Fig. 35.3. We consider different stiffnesses of the springs of the attached masses, that of the masses shown lower the main chain is still equal to $\beta_1 m_1$, while that of the shown upper is equal to $\gamma_1 m_1$, The model equations of motion are

$$\ddot{u}_n = \beta_0(u_{n-1}-2u_n+u_{n+1})+\eta\beta_1(v_n-u_n)+\eta\gamma_1(w_n-u_n), \tag{35.13}$$

$$\ddot{v}_n = -\beta_1(v_n-u_n). \tag{35.14}$$

$$\ddot{w}_n = -\gamma_1(w_n-u_n). \tag{35.15}$$

The continuum functions $u(x,t)$, $v(x,t)$, $w(x,t)$ are introduced for description of the displacements of the masses m, lower m_1. and upper m_1 respectively. Then the

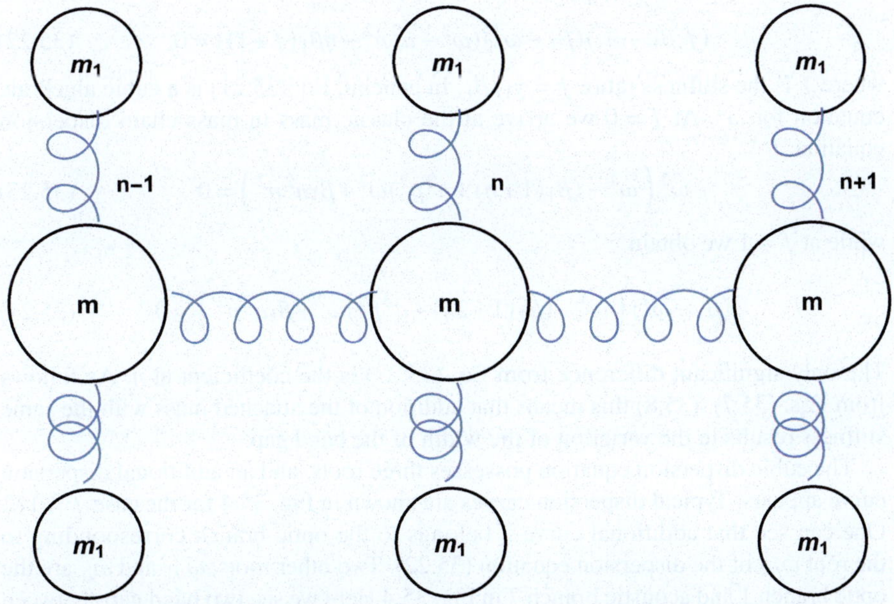

Fig. 35.3: Chain with extra attached masses.

continuum equations of motion are

$$u_{tt} = a^2 u_{xx} + \eta\beta_1(v-u) + \eta\gamma_1(w-u), \tag{35.16}$$

$$v_{tt} = -\beta_1(v-u), \tag{35.17}$$

$$w_{tt} = -\gamma_1(w-u), \tag{35.18}$$

The solution is sought of the form,

$$u = A\exp\Big(\imath\,(px-\omega t - x_0)\Big), \qquad v = B\exp\Big(\imath\,(px-\omega t - x_0)\Big),$$

$$w = C\exp\Big(\imath\,(px-\omega t - x_0)\Big), \tag{35.19}$$

It follows from Eq.(35.17) that

$$B = \frac{\beta_1 A}{\beta_1 - \omega^2}, \tag{35.20}$$

while Eq. (35.18) is resolved for C,

$$C = \frac{\gamma_1 A}{\gamma_1 - \omega^2}. \tag{35.21}$$

Substitution of (35.19) with (35.20), (35.21) into Eq; (35.16) results in the dispersion equation,

$$\eta\beta_1^2(\beta_1 f - \omega^2) + \eta f \beta_1^2(\beta_1 - \omega^2) +$$
$$(f\beta_1 - \omega^2)(\beta_1 - \omega^2)(\omega^2 - a^2p^2 - \eta\beta_1(f+1)) = 0 \tag{35.22}$$

where f is the stiffness ratio, $f = \gamma_1/\beta_1$. In general, Eq. (35.21) is a cubic algebraic equation for ω^2. At $f = 0$ we arrive at the classic mass-in mass chain dispersion equation,

$$\omega^2\Big(\,\omega^4 - (\beta_1(1+\eta) + a^2p^2)\omega^2 + \beta_1 a^2 p^2\,\Big) = 0 \tag{35.23}$$

while at $f = 1$ we obtain

$$(\beta_1 - \omega^2)\Big(\,\omega^4 - (\beta_1(1+2\eta) + a^2p^2)\omega^2 + \beta_1 a^2 p^2\,\Big) = 0 \tag{35.24}$$

The only significant difference from Eq. (35.23) is the coefficient at η. As follows from Eqs. (35.7), (35.8) this means that addition of the attached mass with the same stiffness results in the variation of the width of the band gap.

The cubic dispersion equation possesses three roots, and an additional dispersion curve appears. Typical dispersion curves are shown in Fig. 35.4 for the case $f = 1/2$. One can see that additional curve 3 belongs to the optic branch corresponding to the root ω_{o2} of the dispersion equation (35.22). Two other roots, ω_{o1} and ω_a are the optic branch 1 and acoustic branch 7 in Fig. 35.4 Here we see two band gaps between

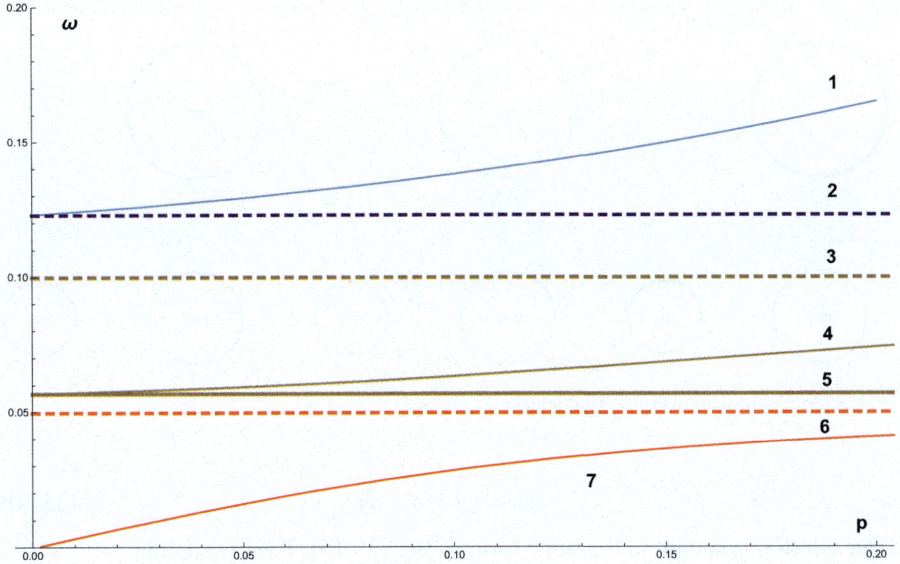

Fig. 35.4: Additional band gap caused by the extra attached mass. 1, solid line: optic curve ω_{o1}; 2, dashed horizontal line: ω_{o1} at $p = 0$; 3, dashed horizontal line: ω_{o2} at $p \to \infty$; 4, solid curve: additional optic curve ω_{o2}; 5, horizontal dashed line: ω_{o2} at $p = 0$; 6, dashed horizontal line: ω_a at $p \to \infty$; 7, solid line: acoustic curve ω_a.

the limiting values of the roots. The dynamics of the band gaps due to the variation of f will be studied elsewhere.

35.4 Chain with Extra Internal Attached Masses

The chain with extra internal attached masses is shown in Fig. 35.5. Now we use the notation w_n for the internal mass displacement, while the notations u_n, v_n has the same meaning as in the classic mass-in-mass model. The equations of motion are

$$\ddot{u}_n = \beta_0(u_{n-1} - 2u_n + u_{n+1}) + \eta\beta_1(v_n - u_n) \tag{35.25}$$

$$\ddot{v}_n = \beta_1(u_n - v_n) + \xi\beta_2(w_n - u_n),. \tag{35.26}$$

$$\ddot{w}_n = -\beta_2(w_n - u_n). \tag{35.27}$$

where ξ is the mass ratio, $\xi = m_2/m_1$, βm_2 is the stiffness of the internal spring.

The continuum limit of the equations of motion gives rise to

$$u_{tt} = a^2 u_{xx} + \eta\beta_1(v - u), \tag{35.28}$$

$$v_{tt} = \beta_1(u - v) + \xi\beta_2(w - u), \tag{35.29}$$

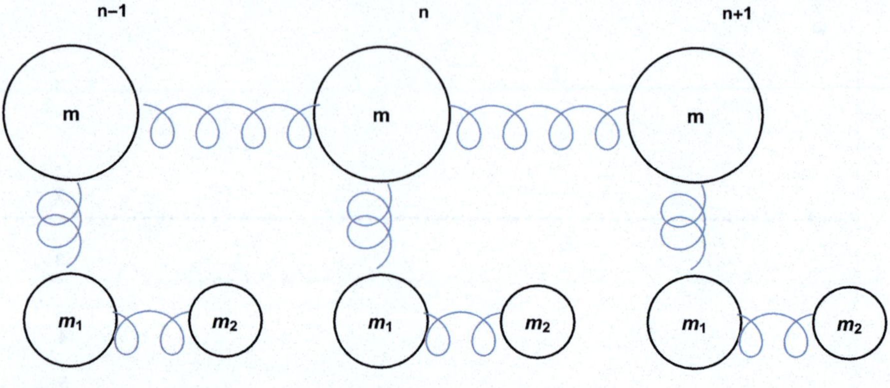

Fig. 35.5: Chain with extra internal attached masses.

$$w_{tt} = -\beta_2(w-u). \tag{35.30}$$

The solution is sought in the same form as Eq. (35.19). Then we obtain

$$C = \frac{\beta_2\, B}{\beta_2 - \omega^2} \tag{35.31}$$

$$A = \frac{\eta\beta_1\, B}{\eta\beta_1 - \omega^2 + a^2\, p^2}. \tag{35.32}$$

The dispersion equation is

$$\eta\beta_1^2(\beta_2-\omega^2) + (\eta\beta_1 - \omega^2 + a^2 p^2)(\xi\beta_2^2 + (\omega^2 - \beta_1 - \beta_2\xi)(\beta_2 - \omega^2)) = 0. \tag{35.33}$$

Now we introduce the stiffness ratio $f = \beta_2/\beta_1$. The dispersion equation is a cubic equation for ω^2. Contrary to the previous model, it reduces to the quadratic equation only in the trivial cases, $\xi = 0$, $f = 0$. Then the solution to Eq. (35.33) always describes two bang gaps with dispersion curves similar to those shown in Fig. 35.4. The detailed description of the dependences of the width and position on the values of ξ and f will be done elsewhere.

35.5 Attached Masses Through one Element of the Main Chain

The model with attached masses through one element of the main chain is shown in Fig. 35.6. The discrete equations of motion are

$$\ddot{u}_n = \beta_0(w_{n-1} - 2u_n + w_{n+1}) + \eta\beta_1(v_n - u_n) \tag{35.34}$$

$$\ddot{v}_n = \beta_1(u_n - v_n),. \tag{35.35}$$

$$\ddot{w}_n = \beta_0(u_{n-1} - 2w_n + u_{n+1}) \tag{35.36}$$

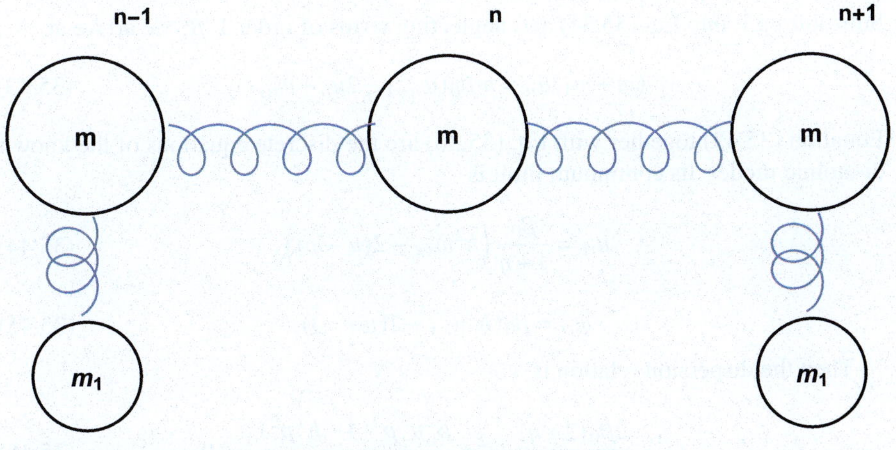

Fig. 35.6: Attached masses through one element of the main chain.

where u is the displacement of the element of the main chain with the attached mass, v is the displacement of the attached mass, w is the displacement of the element without the attached mass.

The continuum limit of Eqs. (35.34)-(35.36) is

$$u_{tt} = \beta_0 h^2 w_{xx} - 2\beta_0(u-w) + \eta\beta_1(v-u), \tag{35.37}$$

$$v_{tt} = \beta_1(u-v), \tag{35.38}$$

$$w_{tt} = \beta_0 h^2 u_{xx} - 2\beta_0(w-u). \tag{35.39}$$

The solution of Eqs. (35.37)-(35.39) is sought in the form of Eq. (35.19). Then we obtain

$$B = \frac{\beta_1 A}{\beta_1 - \omega^2},\ C = \frac{(2-p^2h^2)\beta_0 A}{2\beta_0 - \omega^2}. \tag{35.40}$$

The dispersion relation is

$$\omega^6 - \omega^4\Big(4\beta_0 + \beta_1(1+\eta)\Big) +$$
$$\omega^2\Big(((2\beta_0+\eta\beta_1)(2\beta_0+\beta_1) + \beta_1(2\beta_0-\eta\beta_1) - \beta_0^2(2-h^2p^2)\Big) +$$
$$\beta_0^2\beta_1(2-h^2p^2)^2 - 4\beta_0^2\beta_1 = 0. \tag{35.41}$$

This equation is cubic for ω^2, and we can anticipate an appearance of multiple band gap like shown in Fig. 35.4. However, the model has a limit corresponding to the solid connection with the attached mass, $\beta_1 \gg 1$. Then Eq. (35.35) can be asymptotically resolved for v_n,

$$v_n = u_n - \frac{1}{\beta_1} u_{n,tt} + \frac{1}{\beta_1^2} u_{n,tttt} + \ldots \tag{35.42}$$

Substituting it into Eq. (35.34) and neglecting terms of order $1/\beta_1^2$ we arrive at

$$(m+m_1)\ddot{u}_n = m\beta_0(w_{n-1} - 2u_n + w_{n+1}) \tag{35.43}$$

Equation (35.43) together with Eq. (35.36) are the discrete equations of the known di-atomic model. Its continuum limit is

$$u_{tt} = \frac{\beta_0}{1+\eta}\left(h^2 w_{xx} - 2(u-w)\right), \tag{35.44}$$

$$w_{tt} = \beta_0(h^2 u_{xx} - 2(w-u)). \tag{35.45}$$

Then the dispersion relation is

$$\omega^4 - \frac{2\beta_0(2+\eta)}{1+\eta}\omega^2 + \frac{\beta_0^2 h^2 p^2(4-h^2p^2)}{1+\eta} = 0 \tag{35.46}$$

This is quadratic equation for ω^2 whose solutions describe acoustic and optic modes,

$$\omega_a^2 = \frac{1}{2}\left(\frac{2\beta_0(2+\eta)}{1+\eta} - \sqrt{D}\right) \tag{35.47}$$

$$\omega_o^2 = \frac{1}{2}\left(\frac{2\beta_0(2+\eta)}{1+\eta} - \sqrt{D}\right), \tag{35.48}$$

where

$$D = \frac{4\beta_0^2}{(1+\eta)^2}\left((2+\eta)^2 - (4-h^2p^2)h^2p^2(1+\eta)\right). \tag{35.49}$$

For small values of $h\,p$, the acoustic branch frequency vanishes, $\omega_a \to 0$, while the optic tends to the value ω_{o0},

$$\omega_{o0}^2 = \frac{2\beta_0(2+\eta)}{1+\eta}. \tag{35.50}$$

For large values $h\,p$, the optic frequency tends to infinity. However, the acoustic frequency becomes negative for $hp > 2$. The maximum value of the frequency is achieved at $hp = \sqrt{2}$ and it is equal to ω_{am},

$$\omega_{am}^2 = \frac{1}{2}\frac{2\beta_0(2+\eta)}{1+\eta}\left(1 - \sqrt{1 - \frac{4(1+\eta)}{(2+\eta)^2}}\right). \tag{35.51}$$

Therefore, the continuum di-atomic model predicts the band gap $\omega_{am} < \omega < \omega_{o0}$ but only for the finite interval $0 < hp < 2$.

35.6 Conclusions

Extended metamaterial mass-in-mass models can be analyzed in the long wavelength continuum approximation. It allows us to describe l extra band gaps and variations in the width of the band gap at semi -infinite interval of the wave numbers. Also the parameters responsible for the variations of the bang gasp can br obtained. The presented results are preliminary, more findings are anticipated both in the linear and nonlinear problems.

First, more careful study is needed for description of the band gaps variations in the extended models. We introduced the ratio parameters, like f and ξ in the models with additional attached masses. Dynamics of the width of the additional band gap, appearance or decay of the additional dispersion curve should be studied.

Next problem concerns realization of the bang gaps due to the periodic boundary excitations. Preliminary results have been obtained for the classic mass-in-mass model where numerical simulations of the boundary excitations revealed the appearance of the band gap when the excitation frequency lies inside the interval predicted by Eqs. (35.7), (35.8) [12].

Dispersion curves are obtained on the basis of the periodic solution. Recently no band gap has been found in the study of localized wave dynamics for the classic mass-in-mass model [14]. Similar study should be performed for the extended models including obtaining the solutions for the phase and group velocities.

Of course nonlinear consideration of the extended model deserves special study both analytical and numerical. For the classic mass-in-mass model, some findings can be found in [13].

Acknowledgements The work was supported by the Ministry of Science and Higher Education of the Russian Federation in the framework of the state assignment under contract No. 121112500318-1

References

[1] Born M, Huang K (1954) Dynamical Theory of Crystal Lattices, Clarendon Press, Oxford.

[2] Askar A (1985) Lattice Dynamical Foundations of Continuum Theories, World Scientific, Singapore.

[3] Ostoja-Starzewski M (2002) Lattice models in micromechanics, Appl Mech Rev **55**:35-60.

[4] Andrianov IV, Awrejcewicz J, Weichert, D (2010) Improved continuous models for discrete media, Mathematical Problems in Engineering **2010**:986242.

[5] Huang HH, Sun CT, Huang GL (2009) On the negative effective mass density in acoustic metamaterials, Intern J Eng Sci **47**610-617.

[6] Ma G, Sheng P (2016) Acoustic metamaterials: From local resonances to broad horizons. Sci Adv **2**(2):1-16.

[7] Eremeyev VA, Turco E (2020) Enriched buckling for beam-lattice metamaterials. Mech Res Comm **1**03:103458.

[8] Lazarov BS, Jensen JS (2007) Low-frequency band gaps in chains with attached non-linear oscillators, Intern J Non-Linear Mech **4**2:1186-1193.

[9] Cveticanin L, Cveticanin D (2018) Acoustic Metamaterials: Theory and Application, In: N Herisanu, V Marinca (Eds), Acoustics and Vibration of Mechanical Structures—AVMS-2017, Springer Proceedings in Physics, vol. 198, pp 21-32.

[10] Fang X, Wen J, Bonello B, Yin J, Yu D (2017) Wave propagation in one-dimensional nonlinear acoustic metamaterials, New Journal of Physics **1**9:053007.

[11] Erofeev V, Kolesov D, Malkhanov A (2020) Nonlinear strain waves in a metamaterial defined a mass-to-mass chain, IOP Conf Ser: Mater Sci Eng **7**09:033037.

[12] Porubov AV, Antonov ID (2021) On control of harmonic waves in an acoustic metamaterial, Mechanics Research Communications **1**16:103745.

[13] Porubov AV (2021) Wave modulation in a nonlinear acoustic metamaterial, International Journal of Non-Linear Mechanics **1**37:103788.

[14] Porubov AV, Krivtsov AM (2022) Dispersive propagation of localized waves in a mass-in-mass metamaterial lattice, Continuum Mech Thermodyn **34**:1475–1483.

[15] Yao S, Zhou X, Hu G (2008) Experimental study on negative effective mass in a 1D mass–spring system, New J Phys **1**0:043020.

[16] Zhou J, Cheng Y, Zhang H, Huang G, Hu G (2015) Experimental study on interaction between a positive mass and a negative effective mass through a mass–spring system, Theor Appl Mech Lett **5**:196-199.

[17] Yang T, Song Z-G, Clerkin E, Zhang Y-W, Sun J-H, Su Y-S, Chen L-Q, Hagedorn P (2017) A programmable nonlinear acoustic metamaterial, AIP Advances **7**095323.

[18] Zhou X, Liu X, Hu G (2012) Elastic metamaterials with local resonances: an overview, Theor Appl Mech Lett **2**:041001.

[19] Oyelade AO, Akano TT (2020) Graded hierarchical architecture metamaterial in vibration suppression, UPB Sci Bull, Series D **8**2(3):41-50.

[20] Bukhari M, Barry O (2020) Spectro-spatial analyses of a nonlinear metamaterial with multiple nonlinear local resonators, Nonlinear Dyn **9**9:1539–1560.

[21] Liu C, Reina C (2018) Broadband locally resonant metamaterials with graded hierarchical architecture, Journal of Applied Physics **1**23:095108.

[22] Huang GL, Sun CT (2010) Band gaps in a multiresonator acoustic metamaterial, Journal of Vibration and Acoustics **1**32:031003.

[23] Hu G, Tang L, Das R, Gao S, Liu H (2017) Acoustic metamaterials with coupled local resonators for broadband vibration suppression, AIP Advances **7**:025211.

Chapter 36
Multi-Element Metamaterial's Design Through the Relaxed Micromorphic Model

Leonardo A. Perez Ramirez, Gianluca Rizzi, and Angela Madeo

Abstract Exploring the dynamical response of mechanical metamaterials has gathered increasing attention in the last decades, enabling the design of microstructures exotically interacting with elastic waves (focusing, channeling, band-gaps, negative refraction, cloaking, and many more). Yet, the application and use of such metamaterials in engineering practice is still deficient due to the lack of effective models unveiling metamaterials' interactions with more classical materials at finite scales. In this paper, we show that the relaxed micromorphic model can bring an answer to this open problem and can be effectively used to explore and optimize metamaterials' structures consisting of metamaterials' and classical materials' bricks of finite size.We investigate two examples, namely a double-shield structure that can be used to widen the frequency range for which the internal region can be protected and a multiple-shield structure that optimizes both the screening of the regions internal to the single shields and of the zones exterior to the shields themselves. The exploration of these complex meta-structures has been enabled by the finite element implementation of the relaxed micromorphic model that predicts their response at a fraction of the computational cost when compared to classical simulations.

Key words: Finite-size · Metamaterials · Metastructure · Relaxed micromorphic model · Shield device

---

Leonardo A. Perez Ramirez · Gianluca Rizzi · Angela Madeo
Chair of Continuum Mechanics, Faculty of Architecture and Civil Engineering, TU-Dortmund, August-Schmidt-Str. 8, Dortmund, Germany,
e-mail: leonardo.perez@tu-dortmund.de, gianluca.rizzi@tu-dortmund.de, angela.madeo@tu-dortmund.de

---

The original version of this chapter was revised: Chapter title and the author name have been updated. The correction to this chapter is available at https://doi.org/10.1007/978-3-031-26186-2_46

H. Altenbach et al. (eds.), *Sixty Shades of Generalized Continua*,
Advanced Structured Materials 170, https://doi.org/10.1007/978-3-031-26186-2_36

36.1 Introduction

Metamaterials are architectured materials whose mechanical properties go beyond those of classical materials thanks to local resonance or Bragg scattering phenomena occurring in their heterogeneous microstructure. This micro-heterogeneity allows them to show exceptional mechanical features such as negative Poisson's ratio [1], twist and bend in response to being pushed or pulled [2]-[4], band-gaps [5]-[12], cloaking [13]-[16], focusing [17], [18], channeling [19]-[23], negative refraction [12], [21], [24]-[27], and many others.

In the last two centuries, the advancement of knowledge on finite-size classical materials modeling has enabled the design of engineering structures (buildings, bridges, airplanes, cars, etc.) resisting to static and dynamic loads. Today, while the modeling of infinite-size metamaterials is achievable via reliable homogenization techniques, we must acknowledge that such methods are not optimized to deal with finite-size metamaterials' modeling, mainly due to the lack of well-posed homogenized boundary conditions [28]-[34]. This conceptual gap has prevented us to explore structures made up of both metamaterials and classical materials, and optimize them towards efficient wave control and energy recovery.

Other enriched models, such as strain-gradient, micropolar, Cosserat or classical micromorphic [35]-[46] can be used to describe dispersive behaviours or even higher-frequency modes. However, their use has not been widespread for modeling metamaterials due to limited additional degrees of freedom or to an excessive number of elastic parameters.

In this paper, we show that the mechanical response of finite-size metamaterials can be explored using an elastic- and inertia-augmented micromorphic model (relaxed micromorphic) which is able to describe the main metamaterials' fingerprint characteristics (anisotropy, dispersion, band-gaps, size-effects, etc.), while keeping a reduced structure (free of unnecessary parameters). This model can be linked a posteriori to real metamaterials' microstructures via a fitting approach. The reduced model's structure, coupled with the introduction of well-posed boundary conditions, allows us to unveil the dynamic response of metamaterials' bricks of finite size and complex shapes. Playing LEGO with such bricks can enable the design of surprising structures, combining metamaterials and classical materials, actively controlling noise, vibrations, seismic waves, and many others, while being able to recover energy.

At present, the response of finite-size metamaterials' structures is mostly explored via direct Finite Element (FEM) simulations implementing all microstructures' details (e.g., [47]-[50]). Despite the accurate static and dynamic response that these simulations can provide, they suffer from unsustainable computational costs, already for rather simple structure topoligies. Therefore, the exploration of large-scale structures combining metamaterials' and classical materials' bricks of different type, size and shape, has been until today out of reach.

In this paper, we lay the basis to overcome this stagnation point and we show how the thorough application of the relaxed micromorphic model can open the way to the design of multi-metamaterial's structures that can control elastic wave propagation.

We will focus here on protection devices (shielding), but a similar approach could be adopted for many other applications.

36.2 The Relaxed Micromorphic Model: Constitutive laws, Equilibrium Equations, and Boundary Conditions

In this section, we present the constitutive relations, the equilibrium equations, and the associated boundary conditions for the relaxed micromorphic model [51], [52]. The equilibrium equations and the boundary conditions can be derived by means of a variational approach through the associated Lagrangian[1]

$$\mathcal{L}\left(\dot{u},\nabla\dot{u},\dot{P},\mathrm{Curl}\dot{P},\nabla u,P,\mathrm{Curl}P\right) := K\left(\dot{u},\nabla\dot{u},\dot{P}\right) - W\left(\nabla u,P,\mathrm{Curl}P\right)\,, \tag{36.1}$$

where K and W are the kinetic and strain energy, respectively [52], defined as:

$$\begin{aligned}
K\left(\dot{u},\nabla\dot{u},\dot{P}\right) =& \frac{1}{2}\rho\left\langle\dot{u},\dot{u}\right\rangle + \frac{1}{2}\langle\mathbb{J}_{\mathrm{m}}\,\mathrm{sym}\,\dot{P},\mathrm{sym}\,\dot{P}\rangle + \frac{1}{2}\langle\mathbb{J}_{\mathrm{c}}\,\mathrm{skew}\,\dot{P},\mathrm{skew}\,\dot{P}\rangle \\
&+ \frac{1}{2}\langle\mathbb{T}_{\mathrm{e}}\,\mathrm{sym}\nabla\dot{u},\mathrm{sym}\nabla\dot{u}\rangle + \frac{1}{2}\langle\mathbb{T}_{\mathrm{c}}\,\mathrm{skew}\nabla\dot{u},\mathrm{skew}\nabla\dot{u}\rangle \qquad (36.2)\\
&+ \frac{1}{2}\langle\mathbb{M}_{\mathrm{s}}\,\mathrm{sym}\,\mathrm{Curl}\,\dot{P},\mathrm{sym}\,\mathrm{Curl}\,\dot{P}\rangle \\
&+ \frac{1}{2}\langle\mathbb{M}_{\mathrm{a}}\,\mathrm{skew}\,\mathrm{Curl}\,\dot{P},\mathrm{skew}\,\mathrm{Curl}\,\dot{P}\rangle\,, \\
W\left(\nabla u,P,\mathrm{Curl}P\right) =& \frac{1}{2}\langle\mathbb{C}_{\mathrm{e}}\,\mathrm{sym}\left(\nabla u - P\right),\mathrm{sym}\left(\nabla u - P\right)\rangle \\
&+ \frac{1}{2}\langle\mathbb{C}_{\mathrm{c}}\,\mathrm{skew}\left(\nabla u - P\right),\mathrm{skew}\left(\nabla u - P\right)\rangle \\
&+ \frac{1}{2}\langle\mathbb{C}_{\mathrm{micro}}\,\mathrm{sym}\,P,\mathrm{sym}\,P\rangle \\
&+ \frac{1}{2}\langle\mathbb{L}_{\mathrm{s}}\,\mathrm{sym}\,\mathrm{Curl}P,\mathrm{sym}\,\mathrm{Curl}\,P\rangle \\
&+ \frac{1}{2}\langle\mathbb{L}_{\mathrm{a}}\,\mathrm{skew}\,\mathrm{Curl}P,\mathrm{skew}\,\mathrm{Curl}\,P\rangle\,,
\end{aligned}$$

where $u \in \mathbb{R}^3$ is the macroscopic displacement field, $P \in \mathbb{R}^{3\times3}$ is the non-symmetric micro-distortion tensor, ρ is the macroscopic apparent density, $\mathbb{J}_{\mathrm{m}}$, $\mathbb{J}_{\mathrm{c}}$, $\mathbb{T}_{\mathrm{e}}$, $\mathbb{T}_{\mathrm{c}}$, $\mathbb{M}_{\mathrm{s}}$, $\mathbb{M}_{\mathrm{a}}$, are 4th order micro-inertia tensors, and $\mathbb{C}_{\mathrm{e}}$, $\mathbb{C}_{\mathrm{m}}$, $\mathbb{C}_{\mathrm{c}}$, $\mathbb{L}_{\mathrm{s}}$, $\mathbb{L}_{\mathrm{a}}$ are 4th order elasticity tensors. Further details on the definition of these tensors can be found in [52], [53]. In the tetragonal symmetry case, these tensors in Voigt notation can be expressed as follows

[1] $\langle\cdot,\cdot\rangle$ is the scalar product between tensors of order greater than zero, $(\dot{\cdot})$ is a derivative with respect to time.

$$\mathbb{C}_{\mathrm{e}} = \begin{pmatrix} \lambda_{\mathrm{e}}+2\mu_{\mathrm{e}} & \lambda_{\mathrm{e}} & \star & \dots & 0 \\ \lambda_{\mathrm{e}} & \lambda_{\mathrm{e}}+2\mu_{\mathrm{e}} & \star & \dots & 0 \\ \star & \star & \star & \dots & 0 \\ \vdots & \vdots & \vdots & \ddots & \\ 0 & 0 & 0 & & \mu_{\mathrm{e}}^{*} \end{pmatrix} = \begin{pmatrix} \kappa_{\mathrm{e}}+\mu_{\mathrm{e}} & \kappa_{\mathrm{e}}-\mu_{\mathrm{e}} & \star & \dots & 0 \\ \kappa_{\mathrm{e}}-\mu_{\mathrm{e}} & \kappa_{\mathrm{e}}+\mu_{\mathrm{e}} & \star & \dots & 0 \\ \star & \star & \star & \dots & 0 \\ \vdots & \vdots & \vdots & \ddots & \\ 0 & 0 & 0 & & \mu_{\mathrm{e}}^{*} \end{pmatrix}, \tag{36.3}$$

$$\mathbb{C}_{\mathrm{micro}} = \begin{pmatrix} \lambda_{\mathrm{m}}+2\mu_{\mathrm{m}} & \lambda_{\mathrm{m}} & \star & \dots & 0 \\ \lambda_{\mathrm{m}} & \lambda_{\mathrm{m}}+2\mu_{\mathrm{m}} & \star & \dots & 0 \\ \star & \star & \star & \dots & 0 \\ \vdots & \vdots & \vdots & \ddots & \\ 0 & 0 & 0 & & \mu_{\mathrm{m}}^{*} \end{pmatrix} = \begin{pmatrix} \kappa_{\mathrm{m}}+\mu_{\mathrm{m}} & \kappa_{\mathrm{m}}-\mu_{\mathrm{m}} & \star & \dots & 0 \\ \kappa_{\mathrm{m}}-\mu_{\mathrm{m}} & \kappa_{\mathrm{m}}+\mu_{\mathrm{m}} & \star & \dots & 0 \\ \star & \star & \star & \dots & 0 \\ \vdots & \vdots & \vdots & \ddots & \\ 0 & 0 & 0 & & \mu_{\mathrm{m}}^{*} \end{pmatrix},$$

$$\mathbb{J}_{\mathrm{m}} = \begin{pmatrix} \eta_3+2\eta_1 & \eta_3 & \star & \dots & 0 \\ \eta_3 & \eta_3+2\eta_1 & \star & \dots & 0 \\ \star & \star & \star & \dots & 0 \\ \vdots & \vdots & \vdots & \ddots & \\ 0 & 0 & 0 & & \eta_1^{*} \end{pmatrix} = \rho L_{\mathrm{c}}^2 \begin{pmatrix} \kappa_\gamma+\gamma_1 & \kappa_\gamma-\gamma_1 & \star & \dots & 0 \\ \kappa_\gamma-\gamma_1 & \kappa_\gamma+\gamma_1 & \star & \dots & 0 \\ \star & \star & \star & \dots & 0 \\ \vdots & \vdots & \vdots & \ddots & \\ 0 & 0 & 0 & & \gamma_1^{*} \end{pmatrix},$$

$$\mathbb{T}_{\mathrm{e}} = \begin{pmatrix} \overline{\eta}_3+2\overline{\eta}_1 & \overline{\eta}_3 & \star & \dots & 0 \\ \overline{\eta}_3 & \overline{\eta}_3+2\overline{\eta}_1 & \star & \dots & 0 \\ \star & \star & \star & \dots & 0 \\ \vdots & \vdots & \vdots & \ddots & \\ 0 & 0 & 0 & & \overline{\eta}_1^{*} \end{pmatrix} = \rho L_{\mathrm{c}}^2 \begin{pmatrix} \overline{\kappa}_\gamma+\overline{\gamma}_1 & \overline{\kappa}_\gamma-\overline{\gamma}_1 & \star & \dots & 0 \\ \overline{\kappa}_\gamma-\overline{\gamma}_1 & \overline{\kappa}_\gamma+\overline{\gamma}_1 & \star & \dots & 0 \\ \star & \star & \star & \dots & 0 \\ \vdots & \vdots & \vdots & \ddots & \\ 0 & 0 & 0 & & \overline{\gamma}_1^{*} \end{pmatrix}, \tag{36.4}$$

$$\mathbb{L}_{\mathrm{s}} = L_{\mathrm{c}}^2 \begin{pmatrix} \star & \star & \star & & \dots & 0 \\ \star & \star & \star & & \dots & 0 \\ \star & \star & \star & & \dots & 0 \\ & & & \alpha_1 & 0 & 0 \\ \vdots & \vdots & \vdots & & & \\ & & & 0 & \alpha_1 & 0 \\ 0 & 0 & 0 & 0 & 0 & \star \end{pmatrix}, \qquad \mathbb{M}_{\mathrm{s}} = \rho L_{\mathrm{c}}^4 \begin{pmatrix} \star & \star & \star & & \dots & 0 \\ \star & \star & \star & & \dots & 0 \\ \star & \star & \star & & \dots & 0 \\ & & & \beta_1 & 0 & 0 \\ \vdots & \vdots & \vdots & & & \\ & & & 0 & \beta_1 & 0 \\ 0 & 0 & 0 & 0 & 0 & \star \end{pmatrix},$$

$$\mathbb{J}_{\mathrm{c}} = \begin{pmatrix} \star & 0 & 0 \\ 0 & \star & 0 \\ 0 & 0 & 4\eta_2 \end{pmatrix} = \rho L_{\mathrm{c}}^2 \begin{pmatrix} \star & 0 & 0 \\ 0 & \star & 0 \\ 0 & 0 & 4\gamma_2 \end{pmatrix}, \qquad \mathbb{T}_{\mathrm{c}} = \begin{pmatrix} \star & 0 & 0 \\ 0 & \star & 0 \\ 0 & 0 & 4\overline{\eta}_2 \end{pmatrix} = \rho L_{\mathrm{c}}^2 \begin{pmatrix} \star & 0 & 0 \\ 0 & \star & 0 \\ 0 & 0 & 4\overline{\gamma}_2 \end{pmatrix},$$

$$\mathbb{C}_c = \begin{pmatrix} \star & 0 & 0 \\ 0 & \star & 0 \\ 0 & 0 & 4\mu_c \end{pmatrix}, \qquad \mathbb{L}_a = L_c^2 \begin{pmatrix} 4\alpha_2 & 0 & 0 \\ 0 & 4\alpha_2 & 0 \\ 0 & 0 & \star \end{pmatrix}, \qquad \mathbb{M}_a = \rho\, L_c^4 \begin{pmatrix} 4\beta_2 & 0 & 0 \\ 0 & 4\beta_2 & 0 \\ 0 & 0 & \star \end{pmatrix}.$$

Only the in-plane components are reported since these are the only ones that play a role in the plane-strain simulations presented in the following sections.

36.2.1 Equilibrium Equations

The action functional $\mathcal{A}$ is defined as

$$\mathcal{A} = \iint\limits_{\Omega\times[0,T]} \mathcal{L}\left(\dot{u}, \nabla\dot{u}, \dot{P}, \operatorname{Curl}\dot{P}, \nabla u, P, \operatorname{Curl} P\right)\, dx\, dt\,. \tag{36.5}$$

The first variation $\delta\mathcal{A}$ of the action functional can be related to the virtual work of the internal forces as

$$\delta\mathcal{A} := \int_0^T \mathcal{W}^{\text{int}}\, dt\,. \tag{36.6}$$

Here, the variation operator δ indicates variation with respect to the kinematic fields (u, P). Furthermore, it follows from the least-action principle that $\delta\mathcal{A} = 0$ uniquely defines both the equilibrium equations and the boundary conditions (both Neumann and Dirichlet). Thus, the relaxed micromorphic equilibrium equations in strong form are

$$\rho\,\ddot{u} - \operatorname{Div}\widehat{\sigma} = \operatorname{Div}\widetilde{\sigma}\,, \qquad\qquad \overline{\sigma} + \operatorname{Curl}\overline{m} = \widetilde{\sigma} - s - \operatorname{Curl} m\,, \tag{36.7}$$

where

$$\begin{aligned}
\widetilde{\sigma} &:= \mathbb{C}_e \operatorname{sym}(\nabla u - P) + \mathbb{C}_c \operatorname{skew}(\nabla u - P)\,, & \widehat{\sigma} &:= \mathbb{T}_e \operatorname{sym}\nabla\ddot{u} + \mathbb{T}_c \operatorname{skew}\nabla\ddot{u}\,,\\
m &:= \mathbb{L}_s \operatorname{sym}\operatorname{Curl} P + \mathbb{L}_a \operatorname{skew}\operatorname{Curl} P\,, & \overline{\sigma} &:= \mathbb{J}_m \operatorname{sym}\ddot{P} + \mathbb{J}_c \operatorname{skew}\ddot{P}\,,\\
\overline{m} &:= \mathbb{M}_s \operatorname{sym}\operatorname{Curl}\ddot{P} + \mathbb{M}_a \operatorname{skew}\operatorname{Curl}\ddot{P}\,, & s &:= \mathbb{C}_{\text{micro}} \operatorname{sym} P\,.
\end{aligned} \tag{36.8}$$

The associated boundary conditions are presented in Sect. 36.2.2.

36.2.2 Boundary and Interface Conditions

Well-posed boundary conditions are essential for the analysis of macroscopic metamaterial's samples of finite size. We introduce the work of external surface forces

$\mathcal{W}^{\text{ext}}$ acting on $\partial\Omega$ over the time interval $[0,T]$ for the relaxed micromorphic model as

$$\int_0^T \mathcal{W}^{\text{ext}}\, dt = \iint\limits_{\partial\Omega\times[0,T]} \langle f^{\text{ext}}, \delta u\rangle\, dx\, dt + \iint\limits_{\partial\Omega\times[0,T]} \langle \Phi^{\text{ext}}, \delta P\rangle\, dx\, dt\,, \tag{36.9}$$

where f^{ext} and Φ^{ext} are the external surface forces and double forces, respectively.

According to the Principle of Virtual Work, we can state that the displacement field u and the micro-distortion field P have to satisfy the equality

$$\int_0^T \mathcal{W}^{\text{int}} = \int_0^T \mathcal{W}^{\text{ext}}, \tag{36.10}$$

which implies the following duality condition to hold at a free relaxed micromorphic interface

$$\iint\limits_{\partial\Omega\times[0,T]} \langle t, \delta u\rangle = \iint\limits_{\partial\Omega\times[0,T]} \langle f^{\text{ext}}, \delta u\rangle, \qquad \iint\limits_{\partial\Omega\times[0,T]} \langle \tau, \delta P\rangle = \iint\limits_{\partial\Omega\times[0,T]} \langle \Phi^{\text{ext}}, \delta P\rangle, \tag{36.11}$$

where the generalized traction t and the double traction τ are defined as

$$t = (\widetilde{\sigma} + \widehat{\sigma})\, n\,, \qquad \tau = m \times n\,, \tag{36.12}$$

while n is the normal at the boundary $\partial\Omega$, and the cross product acts row-wise. It is also recalled the expression of the traction for a classical Cauchy material

$$t_{\text{Cauchy}} = \sigma_{\text{Cauchy}}\, n\,, \tag{36.13}$$

where $\sigma_{\text{Cauchy}} = \lambda\, \text{tr}\,(\text{sym}\nabla u)\, \mathbb{1} + 2\mu\, \text{sym}\nabla u$ is the stress tensor for an isotropic linear elastic constitutive model.

The interface conditions between two relaxed micromorphic domains Ω^- and Ω^+ can be similarly derived (see Fig. 36.1) and, in their strong form, they read

$$\begin{cases} t^+ = t^-\,, \\ u^+ = u^-\,, \end{cases} \qquad \begin{cases} \tau^+ = \tau^-\,, \\ (P\times n)^+ = (P\times n)^-\,. \end{cases} \tag{36.14}$$

As a particular case, if Ω^+ is a classical Cauchy continuum, then the interface conditions reduce just to

$$\begin{cases} t^+_{\text{Cauchy}} = t^-\,, \\ u^+ = u^-\,, \end{cases} \qquad \begin{cases} \tau = 0, \\ \text{or} \\ (P\times n) = 0\,. \end{cases} \tag{36.15}$$

Since in a classical Cauchy material neither m nor P are defined, from the perspective of the higher order interface conditions the problem reduces to a free-end boundary

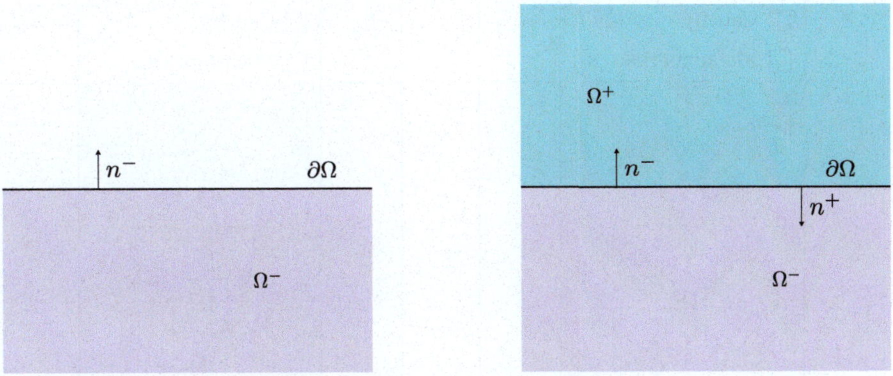

Fig. 36.1: Diagram of the boundary and interface of a relaxed micromorphic continuum: (*left*) boundary with free surface, and (*right*) interface between two different relaxed micromorphic continua.

conditions case. Therefore we can either chose the Neumann boundary conditions $\tau = 0$, which can be interpreted as a "free microstructure" condition, or the Dirichlet boundary conditions $(P \times n) = 0$, which can be interpreted as a "fixed microstructure" condition.

In the reminder of the paper we will set the curvature terms to zero ($\mathbb{L}_s = \mathbb{L}_a = \mathbb{M}_s = \mathbb{M}_a = 0$ in Eq. (36.2)), since their effect is most relevant in statics than in dynamics [53, 54]. It follows that neither the boundary conditions on τ nor the one on P must be assigned in this particular case.

36.2.3 Numerical Results

We perform a time harmonic study to analyze the response of finite size metamaterials using both fully microstructured and relaxed micromorphic models. An incident plane wave u^{I} travels in the domain Ω^1_{C} and acts on the single and double shield devices of Fig. 36.2. The shields are made up of metamaterials, which themselves are made of unit cells of the type shown in Fig. 36.3. For the single-shield configuration, the size of the unit cell is $L_{\mathrm{c}} = 2.5$ mm, this size corresponds to a scaling factor $s_{\mathrm{F}} = 2.5$. On the double-shield, the metamaterial of the outer shield has $s_{\mathrm{F}} = 1$ while the inner shield has $s_{\mathrm{F}} = 2.5$. The properties of the material composing the matrix of the unit cell are reported in Fig. 36.3. The dispersion curves of the two unit cells are presented together in Fig. 36.4. It can be seen the widening of the band-gap effect due to the two unit cell's sizes. The parameters of the relaxed micromorphic model for the unit cell with $s_{\mathrm{F}} = 1$ (see also [53]) and $s_{\mathrm{F}} = 2.5$ can be found in Table 36.1. To obtain the parameters for the unit cell with $s_{\mathrm{F}} = 2.5$ form the ones of the unit cell with $s_{\mathrm{F}} = 1$, it is just necessary to scale L_{c} with the same factor of the size of the unit cell in Eqs. (36.3)-(36.4) (for more details see [52, 55]).

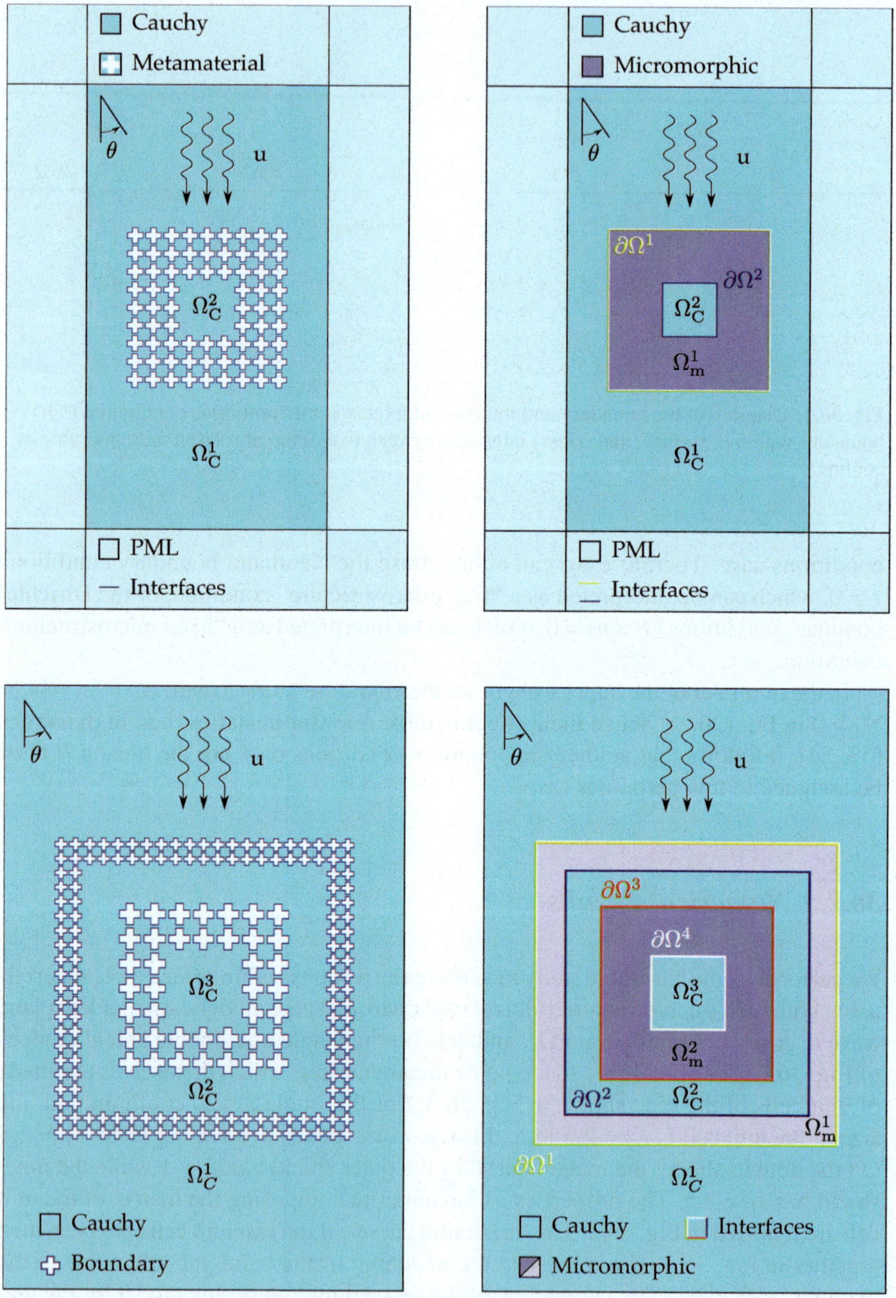

Fig. 36.2: Schematics of metamaterials' shields: (*top*) single shield and (*bottom*) double shield. In the latter, the perfectly matched layer surrounding the outermost Cauchy continuum is still present but it is here omitted for better depiction of the details of the double-shield's configuration.

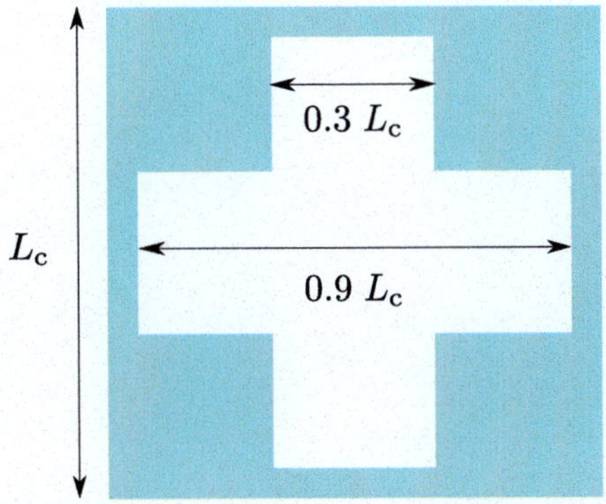

L_c [mm]	ρ_{A} [kg/m^3]	λ_{Al} [GPa]	μ_{Al} [GPa]
1	2700	5.11	2.63

Fig. 36.3: (*left panel*) unit cell geometry; (*right panel*) material and geometrical properties of the aluminum unit cell: size of the unit cell L_c, density ρ_{Al}, and the Lamé constants λ_{Al} and μ_{Al}.

Table 36.1: Parameters of the relaxed micromorphic model for the unit cell of Fig. 36.3 when (*left*) $L_c = 1$ mm and (*right*) $L_c = 2.5$ mm. It is highlighted that in order to obtain the parameters for the unit cell that is just homothetically scaled by a factor $s_F = 2.5$ it is enough to change the value of L_c from 1 to 2.5 mm in the expressions reported in Eqs. (36.3)-(36.4) (see also [52, 55]).

λ_e [GPa]	μ_e [GPa]	μ_e^* [GPa]	μ_c [GPa]
2.33	10.92	0.67	2.28×10^{-3}
λ_m [GPa]	μ_m [GPa]	μ_m^* [GPa]	ρ [kg/m^3]
5.27	12.8	8.33	1485
η_1 [kg/m]	η_2 [kg/m]	η_3 [kg/m]	η_1^* [kg/m]
8.60×10^{-5}	10^{-7}	-2.20×10^{-5}	3.30×10^{-5}
$\overline{\eta}_1$ [kg/m]	$\overline{\eta}_2$ [kg/m]	$\overline{\eta}_3$ [kg/m]	$\overline{\eta}_1^*$ [kg/m]
5.60×10^{-5}	7.30×10^{-4}	1.70×10^{-4}	9×10^{-7}

λ_e [GPa]	μ_e [GPa]	μ_e^* [GPa]	μ_c [GPa]
2.33	10.92	0.67	2.28×10^{-3}
λ_m [GPa]	μ_m [GPa]	μ_m^* [GPa]	ρ [kg/m^3]
5.27	12.8	8.33	1485
η_1 [kg/m]	η_2 [kg/m]	η_3 [kg/m]	η_1^* [kg/m]
5.38×10^{-4}	6.25×10^{-7}	-1.38×10^{-4}	2.06×10^{-4}
$\overline{\eta}_1$ [kg/m]	$\overline{\eta}_2$ [kg/m]	$\overline{\eta}_3$ [kg/m]	$\overline{\eta}_1^*$ [kg/m]
3.50×10^{-4}	4.56×10^{-3}	1.06×10^{-3}	5.63×10^{-6}

All the numerical studies are performed under the plane strain hypothesis. The incident wave u^{I} travels in the external Cauchy continuum and can be defined either as a pressure P or a shear S wave as

$$u^{\mathrm{I}} = a\,\psi^{\mathrm{P/S,I}}\, e^{i\left(k_1^{\mathrm{P/S,I}} x_1 + k_2^{\mathrm{P/S,I}} x_2 - \omega t\right)}, \tag{36.16}$$

with

$$\begin{cases} k_1^{\mathrm{P,I}} = \sqrt{\dfrac{\rho_{\mathrm{Al}}}{\lambda_{\mathrm{Al}}+2\mu_{\mathrm{Al}}}\omega^2 - \left(k_2^{\mathrm{P,I}}\right)^2} \\ k_1^{\mathrm{S,I}} = \sqrt{\dfrac{\rho_{\mathrm{Al}}}{\mu_{\mathrm{Al}}}\omega^2 - \left(k_2^{\mathrm{S,I}}\right)^2} \end{cases}, \tag{36.17}$$

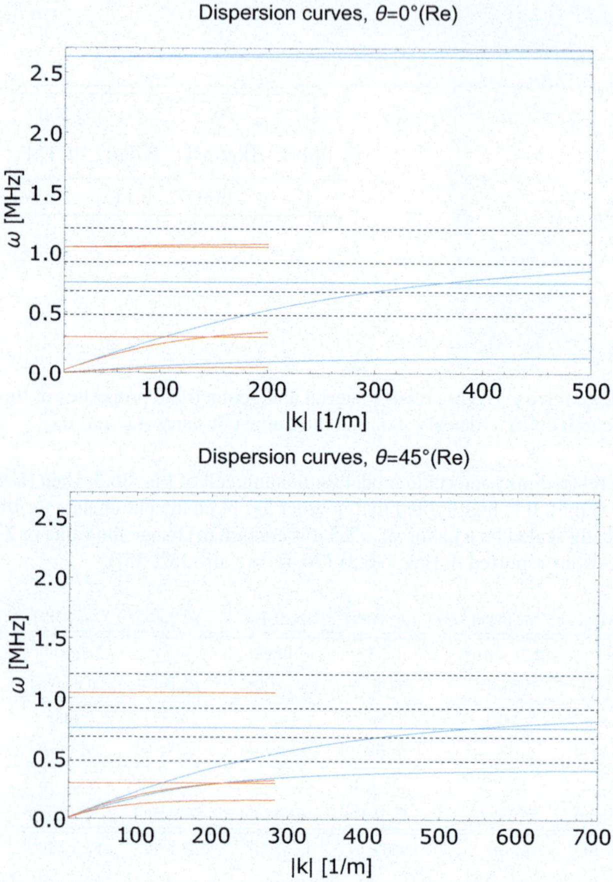

Fig. 36.4: Dispersion curves of the metamaterials issued by the unit cell with $s_F = 2.5$ in red and with $s_F = 1$ in blue. The dashed lines represent the frequencies $\{0.48, 0.68, 0.92, 1.2\}$ MHz that have been used for the simulations presented in the following sections.

where a is the amplitude, $\psi^{P/S,I}$ is the normalized eigenvector associated with the pressure or shear wave, $k_1^{P/S,I}$ and $k_2^{P/S,I}$ are the components of the wave vector, and ω is the frequency.

The perfect contact condition at each interface between the relaxed micromorphic continuum and the classical Cauchy continuum (see Fig. 36.2) are reported in Eq. (36.15). For the unit cell considered in this work, in the dynamic range studied, the curvature terms have a negligible effect in the metamaterial's response [53], [56]. This allows for setting $\mathbb{L}_s = \mathbb{L}_a = \mathbb{M}_s = \mathbb{M}_a = 0$, thus the curvature terms vanish and no double traction has to be defined on the boundaries and the interface conditions reduce to just Eq. $(36.15)_1$.

Let us consider the equilibrium conditions derived from Eq. (36.15) as they act, for example, on the boundary $\partial\Omega^1$. The superposition principle allows for the decomposition of the displacement and traction on the Cauchy side, into an incident and scattered component. The scattered components are indicated by the subscript "sca", and the superscripts refer to their domain. The traction t^{I} is that produced by the incident wave and the conditions can be summarized as

$$\begin{cases} u^- = u^+ & \Longleftrightarrow \quad u^{\Omega^1_{\mathrm{m}}} = u^{\Omega^1_{\mathrm{C}}}_{\mathrm{sca}} + u^{\mathrm{I}} \\ t^- = t^+ & \Longleftrightarrow \quad t^{\Omega^1_{\mathrm{m}}} = t^{\Omega^1_{\mathrm{C}}}_{\mathrm{sca}} + t^{\mathrm{I}} \end{cases} \quad \text{on} \quad \partial\Omega^1 . \tag{36.18}$$

As a consequence, on the boundary $\partial\Omega^1_{\mathrm{C}}$, we can then prescribe

$$u^{\Omega^1_{\mathrm{C}}}_{\mathrm{sca}} = u^{\Omega^1_{\mathrm{m}}} - u^{\mathrm{I}} , \tag{36.19}$$

and the associated tractions $t^{\Omega^1_{\mathrm{C}}}_{\mathrm{sca}}$ and $t^{\Omega^1_{\mathrm{m}}}$ derive from the solution of the correspondent displacement fields. Furthermore, on the boundary, the tractions of Ω^1_{C} differ with respect to the tractions Ω^1_{m} by

$$t^{\Omega^1_{\mathrm{C}}}_{\mathrm{sca}} - t^{\Omega^1_{\mathrm{m}}} = -t^{\mathrm{I}} . \tag{36.20}$$

Thus we must prescribe $-t^{\mathrm{I}}$ on the boundary $\partial\Omega^1$ of Ω^1_{C} to ensure the continuity of tractions reported in Eq. (36.18). For the microstructured material, it is sufficient to prescribe $-t^{\mathrm{I}}$ on the free boundary of the holes. Analogous conditions apply to the other boundaries.

36.3 Parametric Study on the Thickness of a Shielding Device: Capability Limit for the Relaxed Micromorphic Model

In this section, we present the results of a simple shield device modeled both with a full-microstructure model and with the relaxed micromorphic model. The displacement in the subsequent figures is normalized by the amplitude of the incident wave. In Fig. 36.5, we can see that a shield device made up of only one unit cell is not able to ensure the desired shielding effect, and this is true both for the microstructured and the relaxed micromorphic models. We remark that the reflection pattern is nevertheless well captured by the relaxed micromorphic model, even if the shield's thickness is of only one unit cell, while the displacement pattern inside the shield shows a loss of accuracy. This is due to the fact that the relaxed micromorphic model, being a homogenized model, needs a certain domain size to guarantee results with a high level of accuracy, and we show this by performing a parametric study on the shield's thickness (see Figs. 36.5-36.9).

We can infer from Fig. 36.6-36.9 that both the outer and inner displacement fields are well reproduced by the relaxed micromorphic model, as early as the shield is thick enough. Indeed, we established that both the shielding effect and the relaxed micromorphic model accuracy can be considered to be satisfactory already with a 3

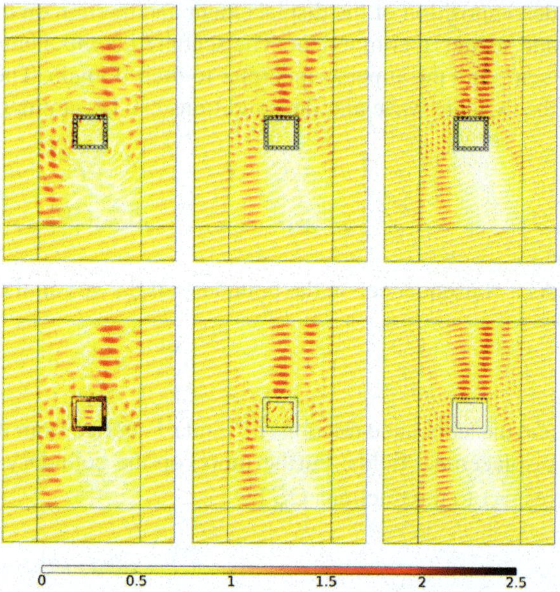

Fig. 36.5 Real part of the normalized displacement field of the microstructured material and its equivalent micromorphic continuum for a shield of 1 unit cell thick, for an incident angle $\phi = 15°$: (*first row*) microstructured material and (*second row*) equivalent micromorphic model for $\omega = 0.48$ MHz, $\omega = 0.68$ MHz, and $\omega = 0.92$ MHz. All frequencies fall in the band-gap range of the considered metamaterial.

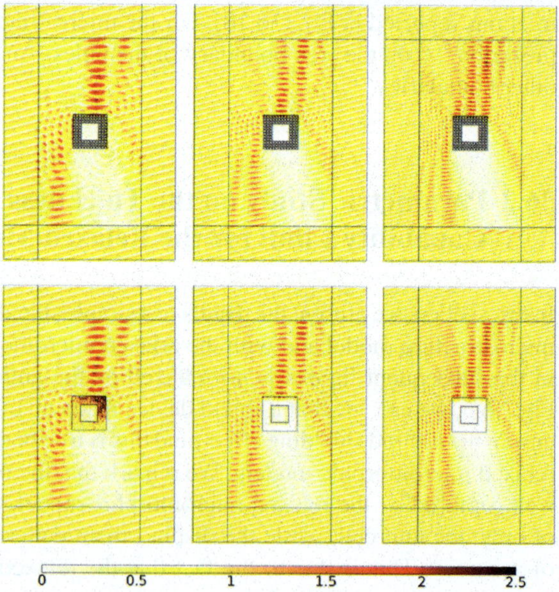

Fig. 36.6 Real part of the normalized displacement field of the microstructured material and its equivalent micromorphic continuum for a shield of 3 unit cells thick, for an incident angle $\phi = 15°$: (*first row*) microstructured material and (*second row*) equivalent micromorphic model for $\omega = 0.48$ MHz, $\omega = 0.68$ MHz, and $\omega = 0.92$ MHz. All frequencies fall in the band-gap region of the considered metamaterial.

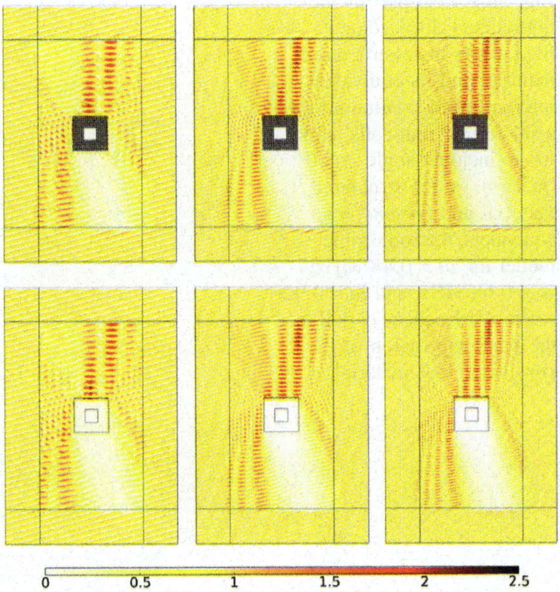

Fig. 36.7 Real part of the normalized displacement field of the microstructured material and its equivalent micromorphic continuum for a shield of 5 unit cells thick, for an incident angle $\phi = 15°$: (*first row*) microstructured material and (*second row*) equivalent micromorphic model for $\omega = 0.48$ MHz, $\omega = 0.68$ MHz, and $\omega = 0.92$ MHz. All frequencies fall in the band-gap region of the considered metamaterial.

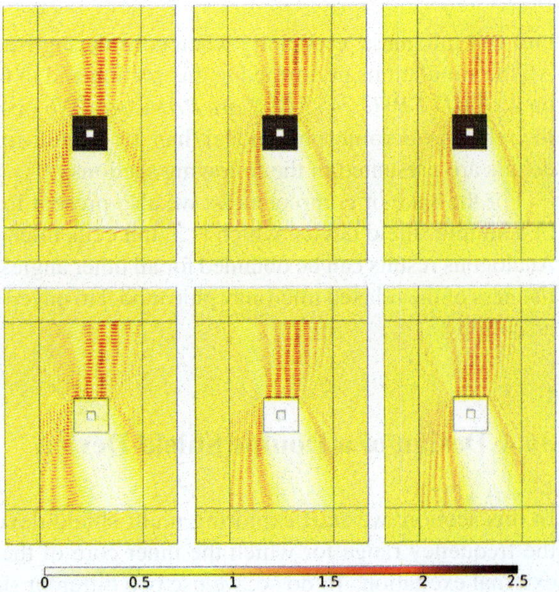

Fig. 36.8 Real part of the normalized displacement field of the microstructured material and its equivalent micromorphic continuum for a shield of 10 unit cells thick, for an incident angle $\phi = 15°$: (*first row*) microstructured material and (*second row*) equivalent micromorphic model for $\omega = 0.48$ MHz, $\omega = 0.68$ MHz, and $\omega = 0.92$ MHz. All frequencies fall in the band-gap region of the considered metamaterial.

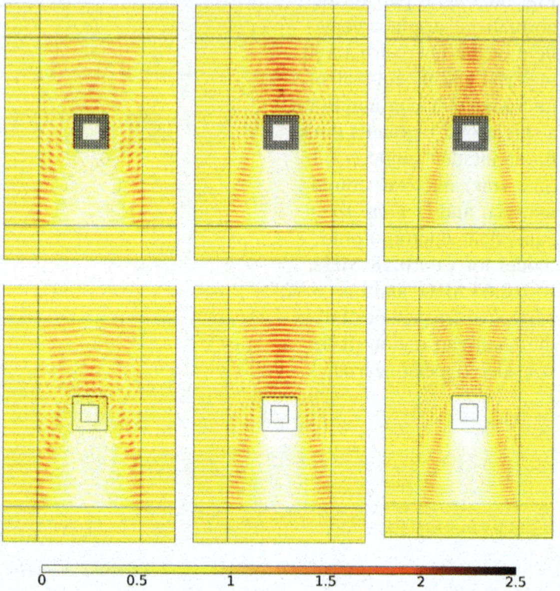

Fig. 36.9 Real part of the normalized displacement field of the microstructured material and its equivalent micromorphic continuum for a shield of 3 unit cells thick, for an incident angle $\phi = 0°$: (*first row*) microstructured material and (*second row*) equivalent micromorphic model for $\omega = 0.48$ MHz, $\omega = 0.68$ MHz, and $\omega = 0.92$ MHz. All frequencies fall in the band-gap region of the considered metamaterial.

unit cell thickness, especially when considering frequencies in the band-gap range. All the simulations performed in this section and in the following ones have been run on 2×*AMD EPYC 7453 28-Core Processor*. We observe faster computing times for the relaxed micromorphic model than for the full-microstructure simulations, more details are presented in the following sections.

For the sake of completeness, we also present in Fig. 36.8 the reflection pattern of a simple shield device which is 3 unit cells thick and for a 0° angle of incidence. Analogous results can be obtained for all other angles of incidence, showing the effectiveness of the relaxed micromorphic model to unveil the macroscopic metamaterial's anisotropy.

36.4 Design of a Double Shield Device

In this section, we start exploring multi-shield devices with the aim of increasing the frequency range for which the inner core of the structure is protected from an external excitation. To do so, we use two different sizes of the unit cell presented in Fig. 36.3 with $s_F = 1$ for the outer and $s_F = 2.5$ for the inner shield. In this way the inner structure's core will be protected in the overlapping intervals [1.11,2.65] MHz and [0.44,1.06] MHz.

We show in Fig. 36.10 a first multi-shield device composed of 3 unit cells for the outer shield and 1 unit cell for the inner shield: the scattering behavior is given for an incidence angle of 15° and for a frequency of $\omega = 0.92$ MHz, which falls in the band-gap of the outer shield. We see that the relaxed micromorphic model recovers well the scattering pattern of the double-shield device at a fraction of the computational cost and with lower memory use. The simulation that uses the relaxed micromorphic model takes 61 s in contrast to the 156 s needed by the geometrically detailed simulation. The gain in computational time, which is already visible for these simple structures, will be more significant when studying larger structures, as those presented in Sect. 36.5.

In Fig. 36.11 we show the same results for a multi-shield device made of two 3x3 cell shields. The results are again convincing, at a fraction of the computational cost. Similar results can be shown for other angles of incidence. In order to complete our exploration, we now choose a frequency $\omega = 1.2$ MHz which falls in the band-gap of the inner shield metamaterial. Figure 36.12 shows the structure's response for $\omega = 1.2$ MHz and an angle of incidence of 0°.

Once again, the relaxed micromorphic model performs really well at a fraction of the computational cost when reproducing the displacement field of the outer Cauchy

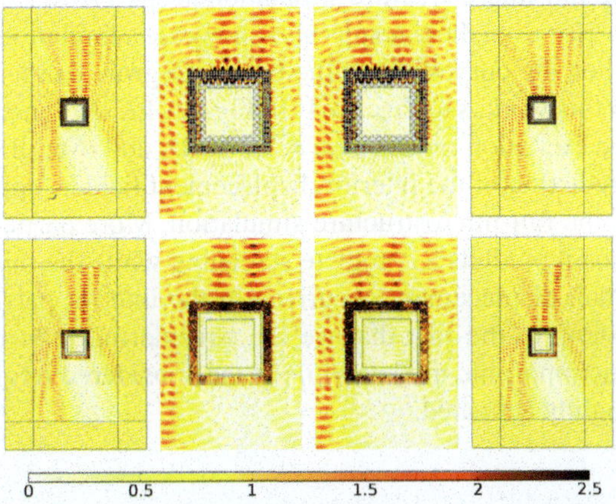

Fig. 36.10 Imaginary (*first two columns*) and real (*last two columns*) part of the normalized displacement field of the microstructured material and its equivalent micromorphic continuum: (*first row*) microstructured material and (*second row*) equivalent micromorphic model for $\omega = 0.92$ MHz. The thickness of the outer shield is 3 unit cells, and the thickness of the inner shield is 1 unit cell.

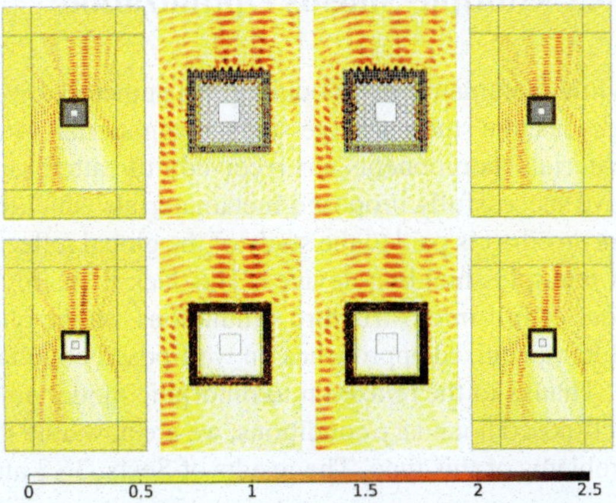

Fig. 36.11 Imaginary (first two columns) and real (last two columns) part of the normalized displacement field of the microstructured material and its equivalent micromorphic continuum: (*first row*) microstructured material and (*second row*) equivalent micromorphic model for $\omega = 0.92$ MHz. The thickness of the outer shield is 3 unit cells, and the thickness of the inner shield is 3 unit cells.

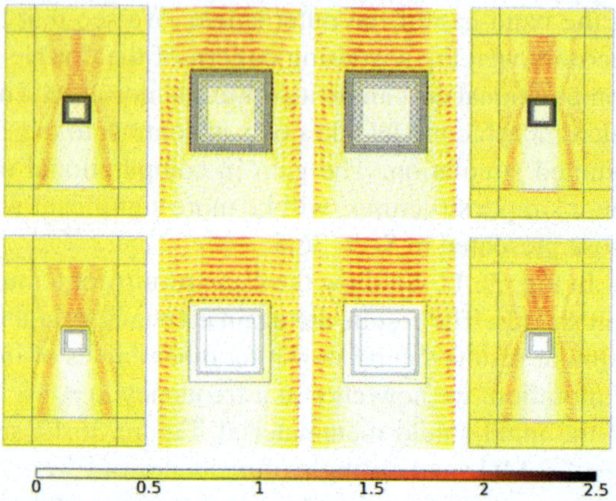

Fig. 36.12 Imaginary (first two columns) and real (last two columns) part of the normalized displacement field of the microstructured material and its equivalent micromorphic continuum: (*first row*) microstructured material and (*second row*) equivalent micromorphic model for $\omega = 1.2$ MHz. The thickness of the outer shield is 3 unit cells, and the thickness of the inner shield is 1 unit cell.

domain. In this case, the solution of the relaxed micromorphic model takes 94 s when the full-microstructure simulation needs 178 s.

However, these results also showcase a limitation of the micromorphic type models when the curves above the band gap grow flat, as is the case for the unit cell used in this work, see Fig. 36.4. In Fig. 36.12, the inner shield does not properly capture the excitation modes at $\omega = 1.2$ MHz, and overall there is a slight inaccuracy in the prediction of the interior displacement field as opposed to the displacement present in the full-microstructure simulation. More particularly, the relaxed micromorphic model predicts a complete screening while the microstructured simulations show a small displacement occurring in the interior part. This difference is due, to a big extent, to the fact that we are considering only the first 6 modes for the relaxed micromorphic model, while at the considered frequency higher order modes can be slightly activated.

36.5 Multiple-Shields Optimization

While metamaterials' modeling has gathered a wealth of research effort in the last decades, the study of large-scale metamaterial's structures consisting of many metamaterials' components has received little attention. For example, many researchers have successfully proposed metamaterials' shielding configurations [48], [57]-[59], but no effort is paid to the study of the global reflection pattern of a system consisting of multiple shielding devices. This lack is mainly due to the computational limitations of full-microstructure simulations, on the one hand, and to lack of knowledge of homogenized approaches on well-posed boundary conditions, on the other.

Thanks to the relaxed micromorphic model presented so far, with the associated well-posed boundary conditions, we are now able to explore large-scale optimization problems of this type. The results of Sects. 36.3 and 36.4 show an additional region with no displacement on the opposite side of the boundary of the shield that is first impacted by the incident wave. We now explore a multiple-shields structure in order to investigate the displacement field in the region situated between and behind a group of single shields. Figure 36.13 shows the global reflection pattern of the multiple-shields systems. It is apparent that the configurations chosen are effective for local shielding of the internal small squares, as well as for the global shielding between the simple shields, as far as a 0° incidence is considered. For other angles of incidence the global shielding effect clearly loses its effectiveness. To optimize the global shielding effect an additional "metamaterial's ring" can be added to the configuration of Fig. 36.14 and the local and global reflection patterns can be explored for different angles of incidence. Such explorations would not be possible, or very time consuming, with simulations with the full-microstructure. The simulation of Fig. 36.14 of the relaxed micromorphic model needed 35 min 8 s, while that of the microstructured case required 2 h 11 min 46 s. The computational gain clearly grows fast as soon as larger structures are investigated.

36.6 Conclusions

In this paper, we showed that the relaxed micromorphic model can be effectively used to describe wave propagation and scattering in complex structures combining metamaterials' and classical materials' bricks of finite size. We successfully designed a double-shield device extending the frequency range for which the screening effect occurs, thanks to the use of unit cells with different size. We also designed a multiple-shield structure allowing to protect both the internal part of each shield and the external region between the shields. Thanks to the model's simplification (homogenized model featuring few frequency-independent parameters), these designs were obtained at a fraction of the computational cost with respect to classical simulations based on Cauchy elasticity. This gain in computational time will be primordial for further works where more and more complex structures will be investigated with the aim of targeting real engineering applications. The results of the present paper open the way

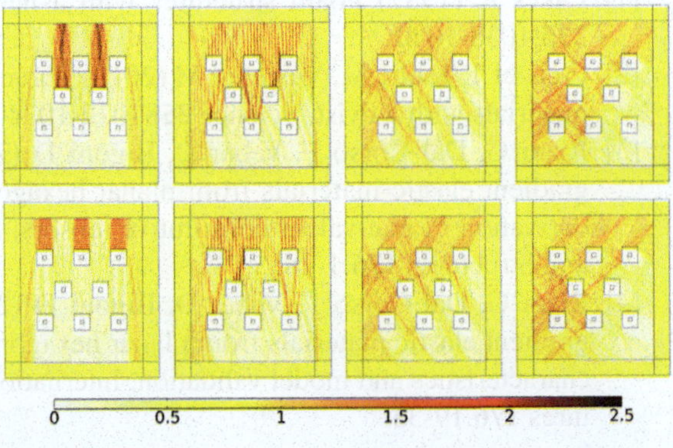

Fig. 36.13 Normalized displacement field of a multiple shield structure for an incident wave of frequency $\omega = 0.92$ MHz: (*first row*) imaginary and (*second row*) real part of the displacement. Form *left* to *right* the angle of the incident wave is $\phi = 0°$, $\phi = 15°$, $\phi = 30°$, and $\phi = 45°$. The thickness of each single shields is 10 unit cells.

Fig. 36.14 Peripheral shield 4 unit cells thick for an incident wave with a frequency of $\omega = 0.92$ MHz and an angle of $\phi = 30°$. Real (*left*) and imaginary (*right*) part of the normalized displacement for the (*first row*) microstructured material and (*second row*) equivalent micromorphic continuum. The thickness of each single shields is 10 unit cells.

to more systematic explorations of engineering large-scale structures that can control elastic waves, thus creating favorable conditions for effective energy conversion and re-use.

Acknowledgements Angela Madeo, Leonardo A. Perez Ramirez, and Gianluca Rizzi acknowledge support from the European Commission through the funding of the ERC Consolidator Grant METALEGO, No. 101001759. The authors also gratefully acknowledge the computing time provided on the Linux HPC cluster at Technical University Dortmund (LiDO3), partially funded in the course of the Large-Scale Equipment Initiative by the German Research Foundation (DFG) as project 271512359.

References

[1] Lakes R (1987) Foam structures with a negative Poisson's ratio, Science **235**:1038-1040.
[2] Frenzel T, Kadic M, Wegener M (2017) Three-dimensional mechanical metamaterials with a twist, Science **358**1072-1074.
[3] Rizzi G, Dal Corso F, Veber D, Bigoni D (2019) Identification of second-gradient elastic materials from planar hexagonal lattices. Part I: Analytical derivation of equivalent constitutive tensors, International Journal of Solids and Structures **176**:1-18.
[4] Rizzi G, Dal Corso F, Veber D, Bigoni D (2019) Identification of second-gradient elastic materials from planar hexagonal lattices. Part II: Mechanical characteristics and model validation, International Journal of Solids and Structures **176**.19-35.

[5] Liu Z, Zhang X, Mao Y, Zhu Y, Yang Z, Chan C, Sheng P (2000) Locally resonant sonic materials, Science **289**:1734-1736.
[6] Wang P, Casadei F, Shan S, Weaver J, Bertoldi K (2014) Harnessing buckling to design tunable locally resonant acoustic metamaterials, Physical Review Letters **113**:014301.
[7] Bilal O, Ballagi D, Daraio C (2018) Architected lattices for simultaneous broadband attenuation of airborne sound and mechanical vibrations in all directions, Physical Review Applied **10**:054060.
[8] Celli P, Yousefzadeh B, Daraio C, Gonella S (2019) Bandgap widening by disorder in rainbow metamaterials, Applied Physics Letters **114**:091903.
[9] El Sherbiny M, Placidi L (2018) Discrete and continuous aspects of some metamaterial elastic structures with band gaps, Archive of Applied Mechanics **88**:1725-1742.
[10] Koutsianitis P, Tairidis G, Drosopoulos G, Stavroulakis G (2019) Conventional and star-shaped auxetic materials for the creation of band gaps, Archive of Applied Mechanics **89**:2545-2562.
[11] Goh H, Kallivokas L (2019) Inverse metamaterial design for controlling band gaps in scalar wave problems, Wave Motion **88**:85-105.
[12] Zhu R, Liu X, Huang, G (2015) Study of anomalous wave propagation and reflection in semi-infinite elastic metamaterials, Wave Motion **55**:73-83.
[13] Bückmann T, Kadic M, Schittny R Wegener M (2015) Mechanical cloak design by direct lattice transformation, Proceedings of the National Academy of Sciences **112**(16):4930-4934.
[14] Misseroni D, Colquitt D, Movchan A, Movchan N, Jones I (2016) Cymatics for the cloaking of flexural vibrations in a structured plate, Scientific Reports **6**:1-11.
[15] Norris A, Amirkulova F, Parnell W (2014) Active elastodynamic cloaking, Mathematics and Mechanics of Solids **19**:603-625.
[16] Misseroni D, Movchan A, Bigoni D (2019) Omnidirectional flexural invisibility of multiple interacting voids in vibrating elastic plates, Proceedings of the Royal Society A **475**:20190283.
[17] Guenneau S, Movchan A, Pétursson G, Ramakrishna S (2007) Acoustic metamaterials for sound focusing and confinement, New Journal of Physics **9**:399.
[18] Cummer S, Christensen J, Alù A (2016) Controlling sound with acoustic metamaterials, Nature Reviews Materials **1**:1-13.
[19] Kaina N, Causier A, Bourlier Y, Fink M, Berthelot T, Lerosey G (2017) Slow waves in locally resonant metamaterials line defect waveguides, Scientific Reports **7**:1-11.
[20] Tallarico D, Trevisan A, Movchan N, Movchan A (2017) Edge waves and localization in lattices containing tilted resonators, Frontiers In Materials **4**:16.
[21] Bordiga G, Cabras L, Piccolroaz A, Bigoni D(2019) Prestress tuning of negative refraction and wave channeling from flexural sources, Applied Physics Letters **114**:041901

[22] Wang Y, Wang T, Liang J, Wang Y, Laude V (2018) Channeled spectrum in the transmission of phononic crystal waveguides, Journal Of Sound And Vibration **437**:410-421.
[23] Miniaci M, Pal R, Manna R, Ruzzene M (2019) Valley-based splitting of topologically protected helical waves in elastic plates, Physical Review B **100**:024304.
[24] Willis J (2016) Negative refraction in a laminate, Journal Of The Mechanics And Physics Of Solids **97**:10-18.
[25] Srivastava A (2016) Metamaterial properties of periodic laminates, Journal of the Mechanics and Physics of Solids **96**:252-263.
[26] Lustig B, Elbaz G, Muhafra A, Shmuel G (2019) Anomalous energy transport in laminates with exceptional points, Journal of the Mechanics and Physics of Solids **133**:103719.
[27] Morini L, Eyzat Y, Gei M (2019) Negative refraction in quasicrystalline multilayered metamaterials, Journal of the Mechanics and Physics of Solids **124**:282-298.
[28] Chen W, Fish J (2001) A dispersive model for wave propagation in periodic heterogeneous media based on homogenization with multiple spatial and temporal scales, J Appl Mech **68**:153-161.
[29] Willis J (2009) Exact effective relations for dynamics of a laminated body, Mechanics of Materials **41**:385-393.
[30] Craster R, Kaplunov J, Pichugin A (466) High-frequency homogenization for periodic media, Proceedings of the Royal Society A: Mathematical, Physical and Engineering Sciences **466**:2341-2362.
[31] Willis J (2011) Effective constitutive relations for waves in composites and metamaterials, Proceedings Of The Royal Society A: Mathematical, Physical And Engineering Sciences **467**:1865-1879.
[32] Willis J (2012) The construction of effective relations for waves in a composite, Comptes Rendus Mécanique **340**:181-192.
[33] Boutin C, Rallu A, Hans S (2014) Large scale modulation of high frequency waves in periodic elastic composites, Journal of the Mechanics and Physics of Solids **70**:362-381.
[34] Sridhar A, Kouznetsova V, Geers M (2018) A general multiscale framework for the emergent effective elastodynamics of metamaterials, Journal of the Mechanics and Physics of Solids **111**:414-433.
[35] Dell'Isola F, Madeo A, Placidi L (2012) Linear plane wave propagation and normal transmission and reflection at discontinuity surfaces in second gradient 3D continua, ZAMM-Journal of Applied Mathematics and Mechanics/Zeitschrift für Angewandte Mathematik und Mechanik **92**:52-71.
[36] Ciallella A, Giorgio I, Eugster S, Rizzi N, Dell'Isola F (2022) Generalized beam model for the analysis of wave propagation with a symmetric pattern of deformation in planar pantographic sheets, Wave Motion **113**:102986.
[37] Dell'Isola F, Giorgio I, Andreaus U (2015) Elastic pantographic 2D lattices: a numerical analysis on the static response and wave propagation, Proceedings of the Estonian Academy of Sciences **64**:219-225.

[38] Altenbach J, Altenbach H, Eremeyev V On generalized Cosserat-type theories of plates and shells: a short review and bibliography, Archive of Applied Mechanics **80**:73-92.

[39] Eremeyev V (2019) Two-and three-dimensional elastic networks with rigid junctions: modeling within the theory of micropolar shells and solids, Acta Mechanica **230**:3875-3887.

[40] Nazarenko L, Glüge R, Altenbach H (2022) On variational principles in coupled strain-gradient elasticity, Mathematics and Mechanics of Solids **27**(11):2256-2274.

[41] Eremeyev V, Alzahrani F, Cazzani A, Dell'Isola F., Hayat T, Turco E, Konopińska-Zmysłowska V On existence and uniqueness of weak solutions for linear pantographic beam lattices models, Continuum Mechanics and Thermodynamics **31**:1843-1861.

[42] Eremeyev V, Dell'Isola F, Boutin C, Steigmann D (2018) Linear pantographic sheets: existence and uniqueness of weak solutions, Journal of Elasticity. **132**:175-196.

[43] Altenbach H, Pouget J, Rousseau M, Collet B, Michelitsch T (Eds.) Generalized Models and Non-classical Approaches in Complex Materials, vol. 2, Advanced Structured Materials, vol. 90, Springer.

[44] Eremeyev V, Altenbach H (2017) Basics of mechanics of micropolar shells, In: Shell-like Structures, CISM International Centre for Mechanical Sciences, vol. 572, pp 63-111.

[45] Mindlin R (1963) Microstructure in linear elasticity, Columbia Univ New York, Dept of Civil Engineering, 1963.

[46] Eringen A (1968) Mechanics of micromorphic continua, In: Mechanics Of Generalized Continua, Ed: Kröner E., pp 18-35.

[47] Krushynska A, Miniaci M, Bosia F, Pugno N (2017) Coupling local resonance with Bragg band gaps in single-phase mechanical metamaterials, xtreme Mechanics Letters **12**:30-36.

[48] Miniaci M, Krushynska A, Bosia F, Pugno N (2016) Large scale mechanical metamaterials as seismic shields, New Journal of Physics **18**:083041.

[49] Baravelli E, Ruzzene M (2013) Internally resonating lattices for bandgap generation and low-frequency vibration control, Journal of Sound and Vibration **332**:6562-6579.

[50] Elford D, Chalmers L, Kusmartsev F, Swallowe G (2011) Matryoshka locally resonant sonic crystal, The Journal of the Acoustical Society Of America **130**:2746-2755.

[51] Neff P, Ghiba I, Madeo A, Placidi L, Rosi G (2014) A unifying perspective: the relaxed linear micromorphic continuum, Continuum Mechanics and Thermodynamics **26**:639-681.

[52] Voss J, Rizzi G, Neff P, Madeo A (2022) Modeling a labyrinthine acoustic metamaterial through an inertia-augmented relaxed micromorphic approach, Mathematics and Mechanics of Solids (accepted).

[53] Rizzi G, D'Agostino M, Neff P, Madeo A (2022) Boundary and interface conditions in the relaxed micromorphic model: Exploring finite-size metastructures for elastic wave control, Mathematics and Mechanics of Solids **27**:1053-1068.
[54] Rizzi G, Neff P, Madeo A (2022) Metamaterial shields for inner protection and outer tuning through a relaxed micromorphic approach, Philosophical Transactions of the Royal Society A **380**:20210400.
[55] Demore F, Rizzi G, Collet M, Neff P, Madeo A (2022) Unfolding engineering metamaterials design: Relaxed micromorphic modeling of large-scale acoustic meta-structures, Journal of the Mechanics and Physics of Solids **168**:104995
[56] D'Agostino M, Barbagallo G, Ghiba I, Eidel B, Neff P, Madeo A Effective description of anisotropic wave dispersion in mechanical band-gap metamaterials via the relaxed micromorphic model, Journal of Elasticity **139**299-329.
[57] Achaoui Y, Antonakakis T, Brûlé S, Craster R, Enoch S, Guenneau S (2017) Clamped seismic metamaterials: Ultra-low frequency stop bands, New Journal of Physics **19**:063022.
[58] Miniaci M, Kherraz N, Cröenne C, Mazzotti M, Morvaridi M, Gliozzi A, Onorato M, Bosia F, Pugno N (2021) Hierarchical large-scale elastic metamaterials for passive seismic wave mitigation, EPJ Applied Metamaterials **8**:14.
[59] Palermo A, Krödel S, Marzani A, Daraio C (2016) Engineered metabarrier as shield from seismic surface waves, Scientific Reports **6**:1-10.

Chapter 37
On Magnetically Induced Motion of Micropolar Ferrofluids

Wilhelm Rickert, Margarita Dementeva, Gregor Ganzosch, Elena N. Vilchevskaya, and Wolfgang H. Müller

Abstract In this paper the effect of external magnetic fields on a ferrofluid is studied numerically. The details of the proposed model are presented and compared to the literature. The governing equations are normalized and the order of magnitude of the dimensionless numbers is analyzed for realistic material parameters. Numerical difficulties arising from the multitude of parameters are discussed. As an exemplary problem a cylindrical container filled with a ferrofluid is considered. Two different magnetic excitations are investigated for different boundary conditions. First, a spherical permanent magnet is placed beneath a closed ferrofluid container and the induced motion is calculated using the finite element method and the software package FEniCS. The difference between slip and no slip boundary conditions are analyzed. Furthermore, the results are compared to a different excitation from a solenoid, which generates a uniform magnetic field inside of the ferrofluid container.

Key words: Micropolar fluids · magnetic relaxation · electromagnetic forces

37.1 Introduction

Ferrofluids are a peculiar type of fluid. Due to their microstructure consisting of magnetic nanoparticles immersed in an oily liquid, they exhibit a magnetic response.

Wihelm Rickert · Margarita Dementeva · Gregor Ganzosch · Wolfgang H. Müller
Institute of Mechanics, Chair of Continuum Mechanics and Constitutive Theory, Technische Universität Berlin, Einsteinufer 5, 10587 Berlin, Germany,
e-mail: rickert@tu-berlin.de,m.dementeva@tu-berlin.de,ganzosch@tu-berlin.de, wolfgang.h.mueller@tu-berlin.de

E. N. Vilchevskaya
Institute for Problems in Mechanical Engineering of the Russian Academy of Sciences, Bol'shoy pr. 61, V.O., 199178 St. Petersburg Russia
e-mail: vilchevska@gmail.com

H. Altenbach et al. (eds.), *Sixty Shades of Generalized Continua*,
Advanced Structured Materials 170, https://doi.org/10.1007/978-3-031-26186-2_37

There are numerous technical applications for ferrofluids, see the review in [1]. In particular, for microchannel flow [2] and ferrofluid flow sensors [3] the quantitative prediction of the flow behavior is important. Although there are attempts to model ferrofluids as ordinary continua, *e.g.*, [4], they are best described by using a generalized continuum theory, such as micropolar theory, see [5, 6]. For a micropolar fluid not only the mass density $\rho(\boldsymbol{x},t)$ and the velocity field $\boldsymbol{v}(\boldsymbol{x},t)$ are to be determined from the mechanical balance laws, but also the angular velocity field $\omega(\boldsymbol{x},t)$. The latter represents the local inner rotation present in the substructure. Of course, by introducing another kinematic variable one also has to consider additional inertia. Motivated by rigid body mechanics, the symmetric microinertia tensor $\boldsymbol{J}$ represents the inner rotational inertia. For a comprehensive introduction into micropolar theory, see [7, 8].

In this paper the coupling to electromagnetism is investigated theoretically. To this end, a cylindrical container shown in Fig. 37.1 filled with ferrofluid is considered. Starting from a state at rest, an external magnetic field is applied, which causes fluid flow. The field activation shall be instantaneous by either abruptly placing a spherical permanent magnet beneath the container or by turning on a solenoid around the container.

Because the ferrofluid is magnetizable, it reacts to the external field, which results in a force as well as a moment distribution. This in turns causes the ferrofluid to flow and also triggers local rotations. The aim is to study the resulting flow fields as well as the force and moment distributions. The structure of the paper is as follows. First, the governing equations are presented and subsequently simplified to the scenario described above. Second, a proper normalization of the resulting system of equations is performed. The obtained dimensionless numbers are discussed and interpreted. Furthermore, by using real material parameters the order of magnitude of each dimensionless number is estimated. Finally, several simplified and normalized initial boundary value problems are solved in cylindrical coordinates with the finite element method and the FEniCS software [9].

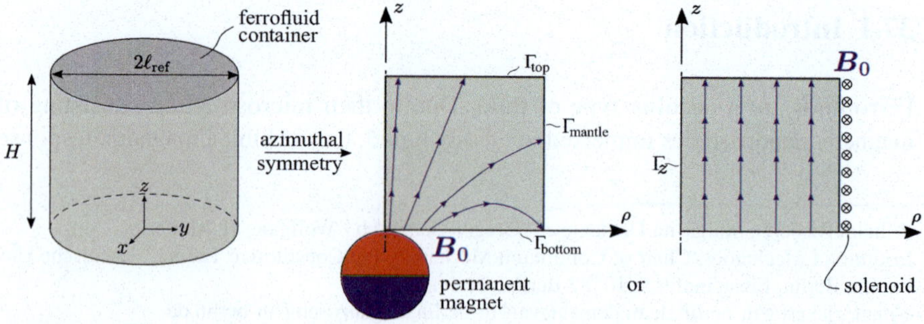

Fig. 37.1: Schematic of the considered ferrofluid container subjected to the magnetic field $\boldsymbol{B}_0$ of a spherical permanent magnet and a solenoid .

37.2 A Flow Problem Coupled to Electromagnetism

The sought fields of a micropolar continuum are given by the mass density $\rho(\boldsymbol{x},t)$, the micro-inertia tensor $\boldsymbol{J}(\boldsymbol{x},t)$, the linear velocity $\boldsymbol{v}(\boldsymbol{x},t)$, the angular velocity $\omega(\boldsymbol{x},t)$ and the temperature $T(\boldsymbol{x},t)$. For each field quantity there exists a balance law. The corresponding balances are given by the mass balance, the microinertia balance, the linear momentum balance, the angular momentum balance and the balance of energy. The last two balances can be simplified by subtracting the linear momentum contributions. As a result the balance of spin and the balance of internal energy are obtained. The corresponding densities are given by

$$\boldsymbol{s} = \boldsymbol{J} \cdot \boldsymbol{\omega}$$

and $u(\boldsymbol{x},t)$, respectively. The system of the balance equations in a spatial description [10] is given by

$$\frac{\mathrm{d}\rho}{\mathrm{d}t} = -\rho(\nabla \cdot \boldsymbol{v}) \,, \tag{37.1}$$

$$\frac{\mathrm{d}\boldsymbol{J}}{\mathrm{d}t} = \boldsymbol{\omega} \times \boldsymbol{J} - \boldsymbol{J} \times \boldsymbol{\omega} + \boldsymbol{\chi}_J \,, \tag{37.2}$$

$$\rho \frac{\mathrm{d}\boldsymbol{v}}{\mathrm{d}t} = \nabla \cdot \boldsymbol{\sigma} + \rho \boldsymbol{g} + \boldsymbol{f}^{(\mathrm{em})} \,, \tag{37.3}$$

$$\rho \frac{\mathrm{d}\boldsymbol{s}}{\mathrm{d}t} = \nabla \cdot \boldsymbol{\mu} + \boldsymbol{\sigma}_\times + \rho \boldsymbol{m} + \boldsymbol{m}^{(\mathrm{em})} + \rho \boldsymbol{\chi}_J \cdot \boldsymbol{\omega} \,, \tag{37.4}$$

$$\rho \frac{\mathrm{d}u}{\mathrm{d}t} = \boldsymbol{\sigma} : (\nabla \otimes \boldsymbol{v} + \boldsymbol{1} \times \boldsymbol{\omega}) + \boldsymbol{\mu} : (\nabla \otimes \boldsymbol{\omega}) - \nabla \cdot \boldsymbol{q} + \rho r + h^{(\mathrm{em})} \,. \tag{37.5}$$

Therein, the material time derivative in spatial description $\mathrm{d}(\cdot)/\mathrm{d}t$ applied to a vector field $\boldsymbol{a}(\boldsymbol{x},t)$ is defined by

$$\frac{\mathrm{d}\boldsymbol{a}}{\mathrm{d}t} = \frac{\partial \boldsymbol{a}}{\partial t} + \boldsymbol{v} \cdot (\nabla \otimes \boldsymbol{a}) \tag{37.6}$$

and the double outer contraction is given by

$$(\boldsymbol{a} \otimes \boldsymbol{b}) : (\boldsymbol{c} \otimes \boldsymbol{d}) := (\boldsymbol{a} \cdot \boldsymbol{c})(\boldsymbol{b} \cdot \boldsymbol{d}).$$

The balances in Eqs. (37.1)–(37.5) contain supply, flux and production terms. The supply terms are given by the gravitational acceleration $\boldsymbol{g}$ as well as the specific moment density $\boldsymbol{m}$ and the specific radiation r. While the supply terms can be controlled from outside, at least theoretically, the production terms are intrinsic to the system and cannot be influenced externally without changing the system. The occurring production terms are the microinertia production $\boldsymbol{\chi}_J$, the electromagnetic force density $\boldsymbol{f}^{(\mathrm{em})}$, the electromagnetic moment density $\boldsymbol{m}^{(\mathrm{em})}$ and the electromagnetic heating $h^{(\mathrm{em})}$. Of course, terms such as the vector invariant of the stress tensor [11]

$$\boldsymbol{\sigma}_\times = (\sigma_{ij} \boldsymbol{e}_i \otimes \boldsymbol{e}_j)_\times = \sigma_{ij} \boldsymbol{e}_i \times \boldsymbol{e}_j \,, \tag{37.7}$$

or $\rho\boldsymbol{\chi}_J \cdot \boldsymbol{\omega}$ act as production terms. Finally, the balances are to be supplemented by constitutive relations. These have to be provided for the internal energy and for the fluxes, namely the stress tensor $\boldsymbol{\sigma}$ as well as the couple stress tensor $\boldsymbol{\mu}$ and the energy flux vector $\boldsymbol{q}$. For simplicity, linear and isotropic constitutive relations are assumed in the following, see [7]:

$$\boldsymbol{\sigma} = (\lambda\nabla\cdot\boldsymbol{v} - p)\mathbf{1} + \mu(\boldsymbol{v}\otimes\nabla + \nabla\otimes\boldsymbol{v}) + 2\tau(\boldsymbol{\omega} - \boldsymbol{w})\times\mathbf{1} , \tag{37.8}$$

$$\boldsymbol{\mu} = \alpha(\nabla\cdot\boldsymbol{\omega})\mathbf{1} + \beta(\boldsymbol{\omega}\otimes\nabla + \nabla\otimes\boldsymbol{\omega}) - \gamma(\boldsymbol{\omega}\otimes\nabla - \nabla\otimes\boldsymbol{\omega}) , \tag{37.9}$$

$$\boldsymbol{q} = -\kappa\nabla T + \zeta(\nabla\otimes\boldsymbol{\omega})_{\times} , \tag{37.10}$$

with the vorticity vector

$$\boldsymbol{w} = \frac{1}{2}\nabla\times\boldsymbol{v}$$

and the pressure field $p(\boldsymbol{x},t)$. The various material parameters are the shear viscosity μ, the second viscosity λ and the rotational shear viscosity τ. Additionally, the generalized viscosities[1] α, β and γ introduce internal length scales, because their dimension is that of of a viscosity multiplied by a square length. Finally, κ is the classical heat conductivity and ζ is a generalized conductivity.

The resulting system of equations cannot be solved in general without providing initial and boundary values. Furthermore, the production terms need to be stated. In particular, following [12], the electromagnetic contributions according to the generalized LORENTZ model in a rest frame are given by:

$$\boldsymbol{f}_{\mathrm{L}}^{(\mathrm{em})} = (q^{\mathrm{f}} - \nabla\cdot\boldsymbol{P})\boldsymbol{E} + \left(q^{\mathrm{f}}\boldsymbol{v} + \boldsymbol{j}^{\mathrm{f}} + \frac{\partial\boldsymbol{P}}{\partial t} + \nabla\times\boldsymbol{M}\right)\times\boldsymbol{B} , \tag{37.11}$$

$$\boldsymbol{m}^{(\mathrm{em})} = \boldsymbol{M}\times\boldsymbol{B} + \boldsymbol{P}\times\boldsymbol{E} , \tag{37.12}$$

$$h^{(\mathrm{em})} = \boldsymbol{j}^{\mathrm{f}}\cdot\boldsymbol{E} . \tag{37.13}$$

Note that this is not the only choice, see the discussion in Sect. 37.4. In the above equations $\boldsymbol{E}$ is the electric field, $\boldsymbol{B}$ is the magnetic flux density, q^{f} and $\boldsymbol{j}^{\mathrm{f}}$ are the free electric charge density and the conductive electric current density, respectively. By means of the electromagnetic source terms the electromagnetic fields are introduced and thus MAXWELL's equations must be solved on top of the mechanical equations from above. Not only that, but for both the magnetization $\boldsymbol{M}$ and the polarization $\boldsymbol{P}$ as well as the conductive current $\boldsymbol{j}^{\mathrm{f}}$ constitutive relations must be prescribed.

In particular, for suspensions of spherical particles in an oily liquid a relaxation equation is to be solved [5, Sect. 8.9],

$$\frac{\mathrm{d}\boldsymbol{M}}{\mathrm{d}t} = \boldsymbol{\omega}\times\boldsymbol{M} - \frac{1}{\tau_{\mathrm{r}}}(\boldsymbol{M} - \boldsymbol{M}_{\mathrm{eq}}) , \tag{37.14}$$

where τ_{r} is the relaxation time and $\boldsymbol{M}_{\mathrm{eq}}$ is the equilibrium magnetization. This equilibrium magnetization points in the direction of the applied magnetic field

[1] Note that the bulk viscosity is given by $\frac{2}{3}\mu + \lambda$.

$\mathfrak{H}_0 = \frac{1}{\mu_0}\boldsymbol{B}_0$. Note that the notation is adapted from [13] where $\mathfrak{H}$ is the free current potential also known as the magnetic field strength. In a rest frame they are connected by means of the aether relations $\boldsymbol{B} = \mu_0(\mathfrak{H} + \boldsymbol{M})$, where μ_0 is the vacuum permeability. The free current potential $\mathfrak{H}$ corresponds to the magnetic field outside of magnetizable matter with $\boldsymbol{M} \equiv \boldsymbol{0}$ via $\mathfrak{H} = \frac{1}{\mu_0}\boldsymbol{B}$, which is why they are sometimes used synonymously. Inside magnetizable matter, however, the material response $\boldsymbol{M}$ is to be included. Note that even outside of magnetic matter the free field $\mathfrak{H}$ is influenced by the magnet. Since MAXWELL's equations are linear the response field of the magnet $\mathfrak{H}'$ is to be added to the applied field, *i.e.*, $\mathfrak{H} = \mathfrak{H}_0 + \mathfrak{H}'$, where $\mathfrak{H}'$ is to be computed from solving MAXWELL's equations. In general, the magnetization is a function of the complete free current potential $\mathfrak{H}$, and not only the applied field $\mathfrak{H}_0$, as is best understood from the example of a linear spherical magnet, see [14, Sect. 5.11].

Turning back to the equilibrium magnetization of a ferrofluid, it is interesting to note that its direction does not depend upon the complete field $\mathfrak{H}$, but rather only on the applied field, $\boldsymbol{M}_{\text{eq}} \sim \mathfrak{H}_0$. The same holds true for the magnitude of the equilibrium magnetization. It is found from a microscopic analysis and a subsequent homogenization procedure. As a result the equilibrium magnetization is given in terms of the LANGEVIN function, [5, Sect. 2.7],

$$\|\boldsymbol{M}_{\text{eq}}\| = M_{\text{s}}\mathcal{L}(\alpha_{\text{m}}) , \quad \mathcal{L}(\alpha_{\text{m}}) = \coth(\alpha_{\text{m}}) - \frac{1}{\alpha_{\text{m}}} , \quad \alpha_{\text{m}} = \frac{\pi\mu_0 M_{\text{s}} H_{\text{m}} D^3}{6\phi k T} , \tag{37.15}$$

where M_{s} is the saturation magnetization, $H_{\text{m}} = \|\mathfrak{H}_0\|$ is the magnitude of the applied magnetic field strength, D is the diameter of the nanoparticles, ϕ is their volume fraction, k is BOLTZMANN's constant and T is the temperature. For strong magnetic fields, the parameter α_{m} becomes large. In that case, the LANGEVIN function can be approximated by $\mathcal{L}(\alpha_{\text{m}}) \approx 1 - \alpha_{\text{m}}^{-1}$.

Some of the various parameters introduced above are given in Table 37.1 for a special ferrofluid. Note that the value for the rotational shear viscosity τ was estimated according to [15]. There exist other estimates, such that $\tau = \frac{3}{2}\phi\mu$ the one given in [16]. This expression would result in an estimate of $\tau \approx \mu/10$, which is much lower than the value used in Table 37.1. Suffice it to say that the values for the rotational shear viscosity are rather uncertain.

37.3 Simplifications and Normalization

In this section a container filled with ferrofluid under the influence of an external field $\boldsymbol{B}_0(\boldsymbol{x}, t)$ is investigated. The question is what types of motion are introduced by the field $\boldsymbol{B}_0(\boldsymbol{x}, t)$ and how can they be calculated numerically. For this purpose a normalization procedure is performed and the obtained dimensionless numbers are analyzed using real material parameters. In order to simplify the analysis, several assumptions about the material itself are made. To this end, isothermal conditions are assumed. Hence, for a known temperature T the mechanical system consisting

Table 37.1: Parameters of ferrofluid (type L5N, parameters provided by the supplier [17]) in the first block of rows. In the second block estimates of material parameters are given. The third block contains information on the spherical permanent magnet and the geometry of the container.

description	symbol	value
magnetic polarization	$\mu_0 M_s$	44×10^{-3} T
volume fraction	ϕ	7.3 %
diameter of nano particles	D	10 nm
shear viscosity	μ	6 mPa s
mass density	ρ_0	1210 kg/m^3
rotational shear viscosity	τ	(1 to 2) μ
relaxation time	τ_r	1×10^{-6} s
remanence of the spherical magnet	$\mu_0 M_0^{\text{sph}}$	1 T
radius of the spherical magnet	R	15 mm
radius of the container	ℓ_{ref}	50 mm
height of the container	H	50 mm

of the balance of mass as well as the balances of linear momentum and angular momentum is already closed and the balance of internal energy does not need to be considered. Similarly, for incompressible matter, the mass density of the mixture as a whole is a constant, $\rho = \rho_0 = \text{const}$. Therefore, the incompressibility constraint $\nabla\cdot\boldsymbol{v} = 0$ follows from the balance of mass. Note that from now on ρ_0 is used for the mass density and ρ is used for the radial distance in cylindrical coordinates. Another simplification regarding the micropolar medium is to consider a material with spherical microparticles. For these, the microinertia tensors becomes a spherical tensor, *i.e.*, $\boldsymbol{J} = J(\boldsymbol{x},t)\mathbf{1}$. Furthermore, if there is no microinertia production, $\chi_J \equiv \mathbf{0}$, it follows from the $\boldsymbol{J}$ balance that $J = J_0 = \text{const}$. Additionally, it is assumed that no electric effects are relevant.

In order not to solve MAXWELL's equations, the external magnetic field $\boldsymbol{B}_0$ is assumed to be known and its effect on the ferrofluid is studied. For a spherical permanent magnet as in Fig. 37.1 (left side) the field is given in cylindrical coordinates $\{\rho,\varphi,z\}$ as follows [18]

$$\boldsymbol{B}_0 = \frac{\mu_0 M_0^{\text{sph}} R^3}{3\sqrt{(\rho^2+z_*^2)^5}}\left(3\rho z_*\boldsymbol{e}_\rho + [2z_*^2-\rho^2]\boldsymbol{e}_z\right), \quad z_* = z+R\,. \tag{37.16}$$

For the approximately homogeneous magnetic field of the solenoid $\boldsymbol{B}_0 = B_{\text{ref}}\boldsymbol{e}_z$ is assumed. By means of the above simplifications the following system of equations results:

$$\rho_0\left(\frac{\partial \boldsymbol{v}}{\partial t}+\boldsymbol{v}\cdot(\nabla\otimes\boldsymbol{v})\right)=-\nabla p+(\mu+\tau)\Delta\boldsymbol{v}+2\tau\nabla\times\boldsymbol{\omega}+ \\ +\rho_0\boldsymbol{g}+(\nabla\times\boldsymbol{M})\times\boldsymbol{B}_0\,, \tag{37.17}$$

$$\rho_0 J_0\left(\frac{\partial \boldsymbol{\omega}}{\partial t}+\boldsymbol{v}\cdot(\nabla\otimes\boldsymbol{\omega})\right)=(\beta+\gamma)\Delta\boldsymbol{\omega}+(\alpha+\beta-\gamma)\nabla(\nabla\cdot\boldsymbol{\omega})+ \\ +2\tau\nabla\times\boldsymbol{v}-4\tau\boldsymbol{\omega}+\boldsymbol{M}\times\boldsymbol{B}_0\,, \tag{37.18}$$

$$\frac{\partial \boldsymbol{M}}{\partial t}+\boldsymbol{v}\cdot(\nabla\otimes\boldsymbol{M})=\boldsymbol{\omega}\times\boldsymbol{M}-\frac{1}{\tau_\mathrm{r}}\left(\boldsymbol{M}-\mathcal{L}(\alpha_\mathrm{m})\frac{\boldsymbol{B}_0}{\|\boldsymbol{B}_0\|}\right). \tag{37.19}$$

This system of equations differs from other investigations, such as in [4] or [19], for two reasons. First, it treats $\boldsymbol{\omega}$ as an independent field variable rather than some quantity related to the vorticity vector. As a consequence, the aforementioned authors introduce a reduced set of parameters, *i.e.*, α, β and γ are not present. Second, the proposed system of equations uses the generalized LORENTZ force model instead of the commonly employed magnetic dipole force model, which is usually applied in context with ferrofluids, see [1]. Hence, a short comparison of both force and moment expressions is performed in Sect. 37.4.

Upon introducing the following normalization,

$$\boldsymbol{x}=\ell_\mathrm{ref}\tilde{\boldsymbol{x}}\,,\qquad t=t_\mathrm{ref}\tilde{t}\,,\qquad \boldsymbol{v}=v_\mathrm{ref}\tilde{\boldsymbol{v}}\,, \tag{37.20}$$

$$\boldsymbol{\omega}=\omega_\mathrm{ref}\tilde{\boldsymbol{\omega}}\,,\qquad \boldsymbol{M}=M_\mathrm{ref}\tilde{\boldsymbol{M}}\,,\qquad p=\rho_0 v_\mathrm{ref}^2\tilde{p}\,, \tag{37.21}$$

with the special choices $v_\mathrm{ref}=\ell_\mathrm{ref}/t_\mathrm{ref}$ and $\omega_\mathrm{ref}=t_\mathrm{ref}^{-1}$ it follows that

$$\frac{\partial \tilde{\boldsymbol{v}}}{\partial \tilde{t}}+\tilde{\boldsymbol{v}}\cdot(\tilde{\nabla}\otimes\tilde{\boldsymbol{v}})=-\tilde{\nabla}\tilde{p}+\frac{1}{Re(1-N^2)}(\tilde{\Delta}\tilde{\boldsymbol{v}}+2N^2\tilde{\nabla}\times\tilde{\boldsymbol{\omega}})+ \\ +\frac{1}{Fr^2}\tilde{\boldsymbol{g}}+\frac{1}{F_0}(\tilde{\nabla}\times\tilde{\boldsymbol{M}})\times\tilde{\boldsymbol{B}}_0\,, \tag{37.22}$$

$$\Theta\left(\frac{\partial \tilde{\boldsymbol{\omega}}}{\partial t}+\tilde{\boldsymbol{v}}\cdot(\tilde{\nabla}\otimes\tilde{\boldsymbol{\omega}})\right)=\frac{1}{\mathrm{Re}L^2}\tilde{\Delta}\tilde{\boldsymbol{\omega}}+\frac{1}{\mathrm{Re}K^2}\tilde{\nabla}(\tilde{\nabla}\cdot\tilde{\boldsymbol{\omega}})+ \\ +\frac{2N^2}{Re(1-N^2)}(\tilde{\nabla}\times\tilde{\boldsymbol{v}}-2\tilde{\boldsymbol{\omega}})+\frac{1}{F_0}\tilde{\boldsymbol{M}}\times\tilde{\boldsymbol{B}}_0\,, \tag{37.23}$$

$$\frac{\partial \tilde{\boldsymbol{M}}}{\partial t}+\tilde{\boldsymbol{v}}\cdot(\tilde{\nabla}\otimes\tilde{\boldsymbol{M}})=\tilde{\boldsymbol{\omega}}\times\tilde{\boldsymbol{M}}-\frac{t_\mathrm{ref}}{\tau_\mathrm{r}}\left(\tilde{\boldsymbol{M}}-\frac{M_\mathrm{s}}{M_\mathrm{ref}}\mathcal{L}(\alpha_\mathrm{m})\frac{\boldsymbol{B}_0}{\|\boldsymbol{B}_0\|}\right). \tag{37.24}$$

The dimensionless numbers that were used are:

$$Re=\frac{\rho_0 v_\mathrm{ref}\ell_\mathrm{ref}}{\mu}\,,\quad Fr=\sqrt{\frac{g_0}{v_\mathrm{ref}\ell_\mathrm{ref}}}\,,\quad F_0=\frac{\rho_0 v_\mathrm{ref}^2}{M_\mathrm{ref}B_0}\,, \tag{37.25}$$

$$N=\sqrt{\frac{\tau}{\mu+\tau}}\,,\quad L=\frac{\ell_\mathrm{ref}}{\ell}\,,\quad K=\frac{\ell_\mathrm{ref}}{k}\,,\quad \Theta=\frac{J_0}{\ell_\mathrm{ref}^2}\,, \tag{37.26}$$

with the internal length scales

$$\ell = \sqrt{\frac{\beta+\gamma}{\mu}}\,, \quad k = \sqrt{\frac{\alpha+\beta-\gamma}{\mu}}\,. \tag{37.27}$$

Parameters that are commonly introduced in classical fluid dynamics are the REYNOLDS number Re as well as the FROUDE number Fr. In context of micropolar theory, the parameters N and L are added in, *e.g.*, [7]. The coupling parameter N assumes values between zero and one, where zero characterizes no coupling between linear momentum and spin and the value one corresponds to a strong influence. The length scale parameter L is different in nature, not only because it is from the interval $L \in [0,\infty)$, but because it offers a microscopic interpretation. For large values of L the internal length scale ℓ is small compared to the system's characteristic length scale. However, for small values of L the internal length scale provided by the microstructure of the material is comparable to the length of the system resulting in a strong influence on the mesoscopic scale. This dependence upon the system's length scales enables the description of both, macroscopic flow applications as well as the analysis of, *e.g.*, microchannel flow [2]. Similarly, the parameters K and Θ represent length scale ratios. The latter, however, is special as it directly relates to the microparticle's size. One can show, that for spherical particles of diameter D the micro inertia is given by $J_0 = \frac{1}{10}D^2$, see [10].

In order to perform any numerical analysis, all occurring numbers must be given and thus the reference parameters need to be specified. The reference length scale is naturally given by the radius of the container. By means of the special choice $v_{\text{ref}} = \ell_{\text{ref}}/t_{\text{ref}}$ either the reference velocity or the reference time scale need to be prescribed. Unfortunately, neither the initial values nor the boundary values in Sect. 37.5 indicate a natural reference velocity. In view of Table 37.1 the relaxation time τ_{r} offers a naturally occurring parameter with the dimension of time. However, a reference time of the order of microseconds seems inadequate for the mechanical motion of fluids on the macroscale. According to numerical tests and other experiments performed with the analyzed ferrofluid another special choice is made. As reference parameters we choose

$$\ell_{\text{ref}} = 5\,\text{cm}\,, \quad t_{\text{ref}} = 4\,\text{s}\,, \quad B_0 = \mu_0 M_{\text{ref}} = \tfrac{2}{3}\mu_0 M_0^{\text{sph}} R^3 \ell_{\text{ref}}^{-3}\,, \tag{37.28}$$

such that with the values given in Table 37.1 one has

$$\begin{gathered} Re = 144\,, \quad Fr = 1.8\times 10^{-2}\,, \quad F_0 = 3.7\times 10^{-8}\,, \\ N = 0.82\,, \quad \Theta = 4\times 10^{-15}\,. \end{gathered} \tag{37.29}$$

Note that in the case of the solenoid the same numerical value $B_{\text{ref}} = B_0$ is used in order to render both problems comparable. Such a magnetic flux density could be achieved by applying an appropriate current to the solenoid. One important observation is that since N is smaller than one there is no complete coupling, *i.e.*, the angular velocity is *not* solely given by the vorticity. Furthermore, the parameter magnetic parameter α_{m} is a function of the position function via $H_{\text{m}}(\boldsymbol{x})$. By means of the assumed parameters we have

$$\frac{\alpha_{\rm m}}{H_{\rm m}} \approx 7.8\times 10^{-5} \,. \tag{37.30}$$

In the vicinity of the spherical magnet a value of $\alpha_{\rm m} \approx 1.7$ results, whereas further away this value drops to $\alpha_{\rm m} \approx 0.5$. Hence, the simplification of the LANGEVIN function is not justified everywhere and its original form from Eq. (37.15) must be used.

Unfortunately, there seems to be no data available for the parameters α, β and γ. As an estimate, the one constant approximation could be employed such that $\alpha = \beta = \gamma$. This results in the relation $L = \sqrt{2}K$ between the remaining dimensionless numbers. In order to estimate the order of magnitude of (say) L one needs to turn to its interpretation. The parameter L represents the ratio between the systems length scale $\ell_{\rm ref}$ and an internal length scale ℓ stemming from the material properties. As discussed above, it is known that the limiting case for a classical fluid is obtained if the coupling parameter tends to zero, $N \to 0$, and the length scale parameter tends to infinity, $L \to \infty$. Since the coupling parameter in Eq. (37.29) is close to one, which indicates strong coupling, it is to be expected that L reflects a similar behavior. Hence, it is assumed that L is small, *i.e.*, $L = 1$.

In fact, if an azimuthally symmetric *ansatz* is made for all fields,

$$\tilde{\boldsymbol{v}} = \tilde{v}_\rho(\tilde{\rho},\tilde{z})\boldsymbol{e}_\rho + \tilde{v}_z(\tilde{\rho},\tilde{z})\boldsymbol{e}_z \,, \quad \tilde{\boldsymbol{\omega}} = \tilde{\omega}(\tilde{\rho},\tilde{z})\boldsymbol{e}_\varphi \,, \tag{37.31}$$

$$\tilde{\boldsymbol{M}} = \tilde{M}_\rho(\tilde{\rho},\tilde{z})\boldsymbol{e}_\rho + \tilde{M}_z(\tilde{\rho},\tilde{z})\boldsymbol{e}_z \,, \tag{37.32}$$

one has $\tilde{\nabla}\cdot\tilde{\boldsymbol{\omega}} = 0$ and K is not relevant anymore. Furthermore, it can be noted that $\boldsymbol{M}_{\rm eq} \sim \boldsymbol{B}_0$. Hence, only the difference $\tilde{\boldsymbol{M}} - \tilde{\boldsymbol{M}}_{\rm eq}$ contributes to the moment density $\boldsymbol{M}\times\boldsymbol{B}_0$. In equilibrium, for which $\tilde{\boldsymbol{M}} = \tilde{\boldsymbol{M}}_{\rm eq}$ the moment density vanishes.

37.4 Electromagnetic Force Density

In this section a short comparison between the proposed electromagnetic force model and the electromagnetic force commonly introduced in the ferrofluid literature is performed. For a in depth discussion of electromagnetic force models, see [20].

In context of ferrofluids, the electromagnetic force density is usually modeled via, see [1],

$$\boldsymbol{f}^{\rm (em)}_{\rm dipole} = \mu_0 \boldsymbol{M}\cdot(\nabla\otimes\boldsymbol{\mathfrak{H}}),$$

which is in contrast to Eq. (37.11). Note that the electromagnetic moment density used in the literature agrees with Eq. (37.12). In order to assess the differences resulting from the different force models assume a fully saturated ferrofluid with $\boldsymbol{M}_{\rm F} = M_{\rm S}\boldsymbol{e}$, where $\boldsymbol{e} = B_0^{-1}\boldsymbol{B}_0$ is the direction of the applied magnetic field with the magnitude $B_0 = \|\boldsymbol{B}_0\|$ from the spherical magnet. It generates a magnetic flux of $\boldsymbol{B}_{\rm F} = \mu_0\boldsymbol{M}_{\rm F}$. Since $\boldsymbol{B} = \boldsymbol{B}_0 + \boldsymbol{B}_{\rm F}$, the free current potential is given by

$$\mathfrak{H} = \frac{1}{\mu_0}\boldsymbol{B} - \boldsymbol{M}_{\mathrm{F}} = \frac{1}{\mu_0}\boldsymbol{B}_0$$

it follows that

$$\begin{aligned} \boldsymbol{f}^{(\mathrm{em})}_{\mathrm{dipole}} &= M_{\mathrm{s}} B_0 \boldsymbol{e}\cdot(\nabla\otimes\boldsymbol{e}) - M_{\mathrm{s}}(\boldsymbol{e}\cdot\nabla B_0)\boldsymbol{e} \\ &= M_{\mathrm{s}} B_0 \left[\boldsymbol{e}\cdot(\nabla\otimes\boldsymbol{e}) - \frac{(\boldsymbol{e}\cdot\nabla B_0)}{B_0}\boldsymbol{e}\right] . \end{aligned} \tag{37.33}$$

In contrast, the generalized LORENTZ model results in

$$\begin{aligned} \boldsymbol{f}^{(\mathrm{em})}_{\mathrm{L}} &= M_{\mathrm{s}}(B_0 + \mu_0 M_{\mathrm{s}})(\nabla\times\boldsymbol{e})\times\boldsymbol{e} \\ &= M_{\mathrm{s}}(B_0 + \mu_0 M_{\mathrm{s}})\boldsymbol{e}\cdot(\nabla\otimes\boldsymbol{e}) . \end{aligned} \tag{37.34}$$

where

$$(\nabla\otimes\boldsymbol{e})\cdot\boldsymbol{e} = \boldsymbol{0}$$

is true for all normalized vector fields $\boldsymbol{e}$. The differences in the leading factor between these expressions may be neglected, because in many scenarios one has $\mu_0 M_{\mathrm{s}} \ll B_0$. However, the directions of the force expressions differ clearly.

Analyzing different impacts of these force models on the flow behavior is beyond the scope of this paper. Suffice it to say that further investigations are required and the corresponding debate regarding the force expression is not settled yet, contrary to what is sometimes claimed in the literature, *e.g.*, [21].

37.5 Initial Values, Boundary Conditions and Implementation

Initially, the fluid is at rest. Therefore, we put

$$\tilde{\boldsymbol{v}}(\tilde{\boldsymbol{x}},\tilde{t}=0) = \boldsymbol{0}\,, \quad \tilde{\omega}(\tilde{\boldsymbol{x}},\tilde{t}=0) = 0\,, \quad \tilde{\boldsymbol{M}}(\tilde{\boldsymbol{x}},\tilde{t}=0) = \boldsymbol{0}\,. \tag{37.35}$$

Furthermore, two different sets of boundary conditions are proposed. First, with the real boundary $\partial\Omega = \Gamma_{\mathrm{bottom}} \cup \Gamma_{\mathrm{mantle}} \cup \Gamma_{\mathrm{top}}$ and the symmetry boundary part Γ_z, the no slip boundary conditions are denoted as follows

$$\mathrm{bc}_{\text{no-slip}} = \begin{cases} \tilde{\omega} = 0\,, & \tilde{\boldsymbol{v}} = \boldsymbol{0}\,, \quad\quad \tilde{\boldsymbol{x}} \in \partial\Omega\,, \\ \tilde{\omega} = 0\,, & \tilde{\boldsymbol{v}}\cdot\boldsymbol{e}_\rho = 0\,, \quad \tilde{\boldsymbol{x}} \in \Gamma_z\,. \end{cases} \tag{37.36}$$

Therein, the condition at Γ_z ensures azimuthal symmetry. These boundary conditions are commonly used as they are formed on a sound physical basis, see the discussion in [22]. However, they are also quite restrictive and result in the formation of thin boundary layers [23]. Therefore, it is also interesting to analyze what happens if there is slip between the fluid and the container's boundaries. In order to simplify the investigation, slip is only allowed at the bottom surface, where the magnetic field from the permanent magnet is strongest. Furthermore, the contact between the ferrofluid and the bottom surface is assumed to be frictionless. Hence, the second set of boundary conditions reads,

$$\mathrm{bc}_{\mathrm{slip}} = \begin{cases} \tilde{\omega} = 0\,, & \tilde{\boldsymbol{v}} = \boldsymbol{0}\,, \quad \tilde{\boldsymbol{x}} \in \Gamma_{\mathrm{mantle}} \cup \Gamma_{\mathrm{top}}\,, \\ \tilde{\omega} = 0\,, & \tilde{\boldsymbol{v}} \cdot \boldsymbol{e}_z = 0\,, \quad \tilde{\boldsymbol{x}} \in \Gamma_{\mathrm{bottom}}\,, \\ \tilde{\omega} = 0\,, & \tilde{\boldsymbol{v}} \cdot \boldsymbol{e}_\rho = 0\,, \quad \tilde{\boldsymbol{x}} \in \Gamma_z\,. \end{cases} \tag{37.37}$$

Note that in the case of the homogeneous excitation

$$\boldsymbol{B}_0 = B_{\mathrm{ref}} \boldsymbol{e}_z$$

only the no slip conditions are used. Furthermore, since the relaxation equation for the magnetization has a diffusive term, *i.e.*, it is a convection equation, no boundary conditions are to be prescribed.

The equations are solved using finite elements for the spatial discretization as well as an implicit EULER scheme for the temporal discretization. In particular, the FEniCS software was used [9] and combined with a newly delevoped cylindrical coordinates method, which will be explained in another publication. The calculations are performed with the assumption of azimuthal symmetry. Hence, the computational domain becomes 2D, which reduces computational costs. However, the obtained solutions still correspond to a 3D problem. In terms of the numerical implementation, some of the operations in Eqs. (37.22)–(37.24) cannot be implemented naturally on a 2D domain in the FEniCS-software. In particular, the cross products need to be evaluated manually, *e.g.*,

$$\begin{aligned} \tilde{\nabla} \times \tilde{\omega} &= -\frac{\partial \tilde{\omega}}{\partial \tilde{z}} \boldsymbol{e}_\rho + \frac{1}{\tilde{\rho}} \frac{\partial}{\partial \tilde{\rho}} (\tilde{\rho} \tilde{\omega}) \boldsymbol{e}_z\,, \\ \tilde{\boldsymbol{M}} \times \tilde{\boldsymbol{B}}_0 &= (\tilde{B}_\rho \tilde{M}_z - \tilde{B}_z \tilde{M}_\rho) \boldsymbol{e}_\varphi\,, \\ (\tilde{\nabla} \times \tilde{\boldsymbol{M}}) \times \tilde{\boldsymbol{B}}_0 &= \left(\frac{\partial \tilde{M}_\rho}{\partial \tilde{z}} - \frac{\partial \tilde{M}_z}{\partial \tilde{\rho}} \right) (\tilde{B}_z \boldsymbol{e}_\rho - \tilde{B}_\rho \boldsymbol{e}_z)\,, \\ \tilde{\omega} \times \tilde{\boldsymbol{M}} &= \tilde{\omega} (\tilde{M}_z \boldsymbol{e}_\rho - \tilde{M}_\rho \boldsymbol{e}_z)\,. \end{aligned} \tag{37.38}$$

In fact, Eq. (37.23) reduces to a scalar equation as it only has a φ-component. Furthermore, the parameters in Eq. (37.29) represent a rather extreme scenario as N is close to its maximum value and Θ is a very small number when compared to all other numerical values in the system of equations. It was found through numerical tests, that Eqs. (37.22)–(37.23) are to be multiplied with $(1 - N^2)$ in order to reduce floating-point errors. Note that the left hand side of Eq. (37.23) can be neglected due to the small parameter Θ. Hence, the spin balance could be solved quasi stationary.

37.6 Results

All simulation are performed until approximately 1 s of real time, *i.e.*, 25% of t_{ref}. After that time, turbulent behavior starts to develop and the time integration with a constant time step and with a fixed grid size fails.

For the setup with the spherical permanent magnet the electromagnetic force and electromagnetic moment distributions look roughly the same independently of the boundary conditions. Additionally, they do not change much over time. Therefore, in Fig. 37.2 only one of each distribution is depicted. From the figure it can be seen, that the presence of the magnet causes a force density near the contact point with the container. This force distribution is localized in a bubble and it generally points in ρ direction, *i.e.*, outward from the center. Similarly, in the same region the electromagnetic moment is localized. As shown in the plot, the moment acts counter clockwise. The order of magnitude of these normalized quantities are small. However, upon dividing them with F_0 from Eq. (37.29), which is of the order of 10^{-8}, their dominating influence becomes apparent. The huge differences in scales between the parameters and the corresponding fields is an indicator for an inadequate normalization. However, in view of the fact, that the success of a finite element analysis of incompressible flow is mostly governed by the REYNOLDS number, the normalization appears to be acceptable. In particular, the successful simulation is in favor of the normalization.

In what follows, the translational velocity field and the angular velocity field will be discussed. The magnetization field looks roughly the same as the external magnetic field $\boldsymbol{B}_0$. This is due to the fact, that $\boldsymbol{M}$ is governed by $\boldsymbol{M}_{\mathrm{eq}}$, which has a limited magnitude of M_{s} and points in direction of the external field. The difference $\boldsymbol{M} - \boldsymbol{M}_{\mathrm{eq}}$ is small, but multiplied with the comparatively large $\boldsymbol{B}_0$ in order to produce a relevant moment density $\boldsymbol{m}^{(\mathrm{em})}$.

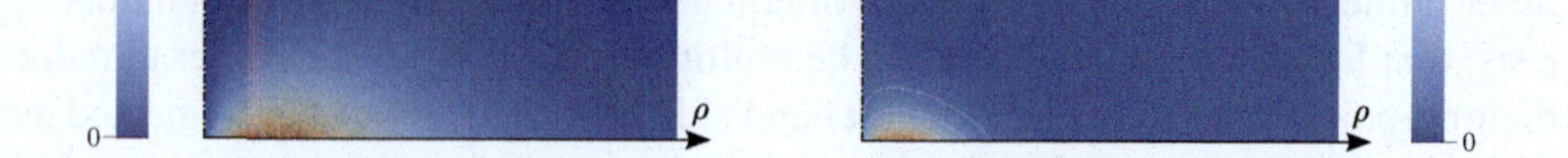

Fig. 37.2: Electromagnetic force distribution (left) and contour plot of the φ component of the electromagnetic moment distribution (right) .

37.6.1 Slip

In Fig. 37.3 it can be seen that for the velocity (bottom row) two vortices arise. As times increases, the big vortex moves in ρ-direction and increases in size. It sweeps across the bottom surface of the container and the velocity accumulates there, as there is no friction with the surface. Similarly, for the angular velocity (bottom row) two different vortices are present. The inner one rotates clockwise[2] and has initially a similar order of magnitude as the outer one in red, which rotates counter clock wise. As time evolves, the inner vortex propagates outwards, grows in size and annihilates the outer one. These results are to be expected, since there is little resistance for the vortices by allowing perfect slip at the bottom of the container. At some point in time the vortices hit the outer wall, *i.e.*, the mantle, where turbulent behavior occurs and the simulation breaks down as the mesh is too coarse to resolve the small vortices.

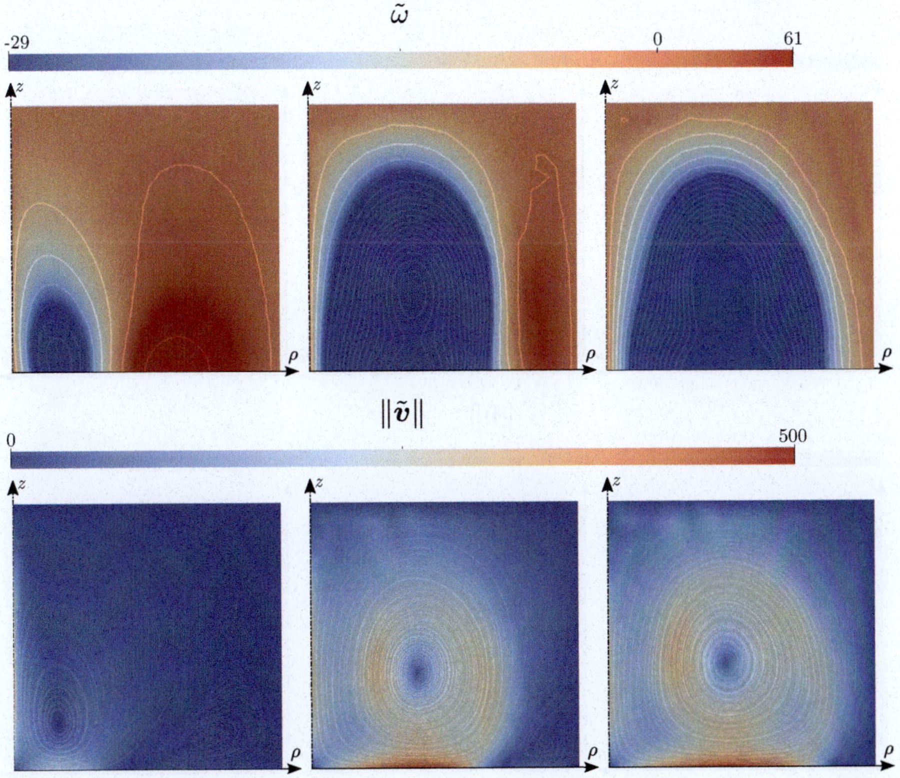

Fig. 37.3: Contourplot of $\tilde{\omega}$ (top) and stream density plot of $\tilde{\boldsymbol{v}}$ (bottom) for increasing time from left ro right using perfect slip conditions .

[2] In the given coordinate system the φ direction points towards the reader and thus a positive value of $\tilde{\omega}$ corresponds to a counter clockwise rotation around that axis.

37.6.2 No Slip

In contrast to the previous example the field evolutions in Fig. 37.4 are obtained for sticky boundary conditions. It can be seen that the initial velocity distribution is similar (second row, first image). However, since the ferrofluid now sticks to the bottom surface, the main vortex cannot travel freely anymore. Instead, a boundary layer develops and the outer vortex grows. Additionally, the outer vortex pushes the bigger one upwards. As a result the flow quickly becomes turbulent and the simulation breaks down due to the coarse discretzation. Interestingly, the angular velocity field looks different initially. As time increases, two domains with opposite directions of rotation develop, which do seem to be able to coexist. From this analysis one can see, that the angular velocity solutions highly depend on the translational velocity field and the boundary conditions imposed on it.

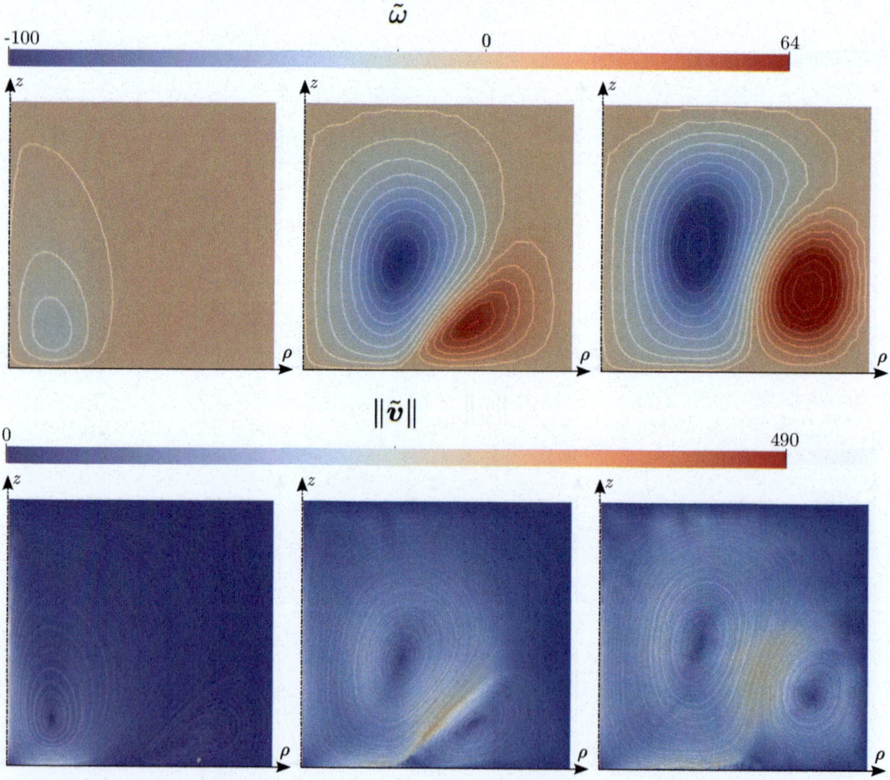

Fig. 37.4: Contourplot of $\tilde{\omega}$ (top) and stream density plot of $\tilde{\boldsymbol{v}}$ (bottom) for increasing time from left ro right using no slip conditions .

37.6.3 Homogeneous Magnetic Field

In this section the resulting flow fields in the ferrofluid are analyzed for the case of a homogeneous external magnetic field. In Fig. 37.5 the electromagnetic force (top row) and electromagnetic moment (bottom row) are depicted. Interestingly, there is no electromagnetic force in the initial state. The electromagnetic moment is small but non-zero and is concentrated near the boundaries of the container (second row, first image). As time increases the electromagnetic force is shearing the fluid as it points in outward ρ-direction near the bottom of the container and inwards near the top. Due to the no-slip boundary conditions it can be expected that this will result in a large vortex in the center of the plot. The depicted vortex is of course actually a vortex ring, as an azimuthally symmetric 3D container is analyzed.

This effect can be seen in Fig. 37.6. In comparison to the previous boundary value problem, the velocity at the z-axis shows the same behavior as for the inhomogeneous magnetic field. However, the overall qualitative behavior is different since only one vortex develops and the magnitude of the velocity field in the bulk is much

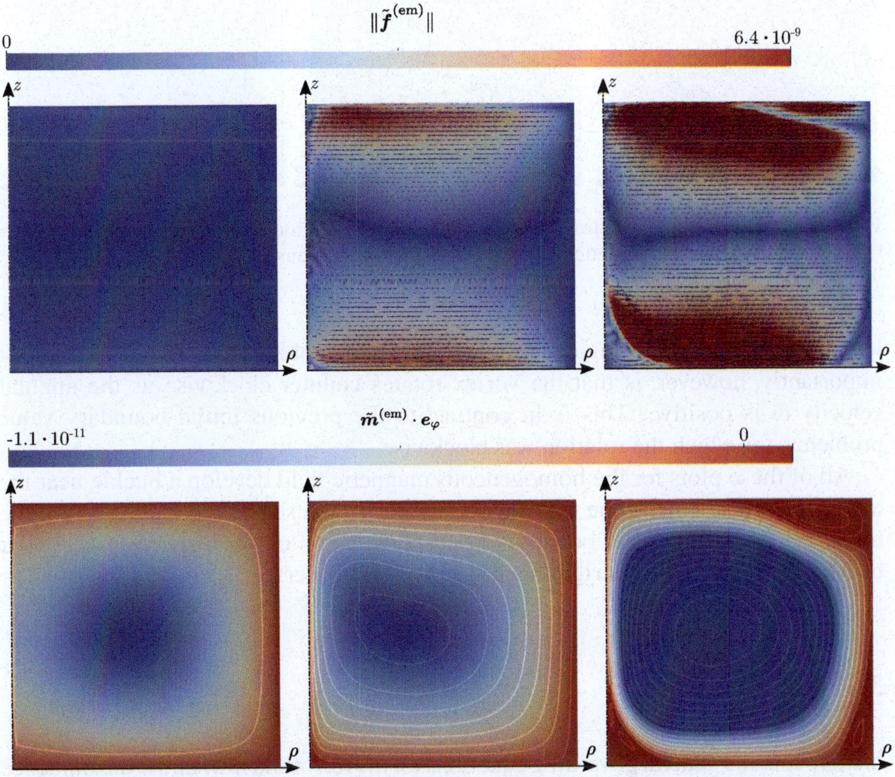

Fig. 37.5: Electromagnetic force distribution (top) and contour plot of the φ component of the electromagnetic moment distribution (bottom) for increasing time from left to right .

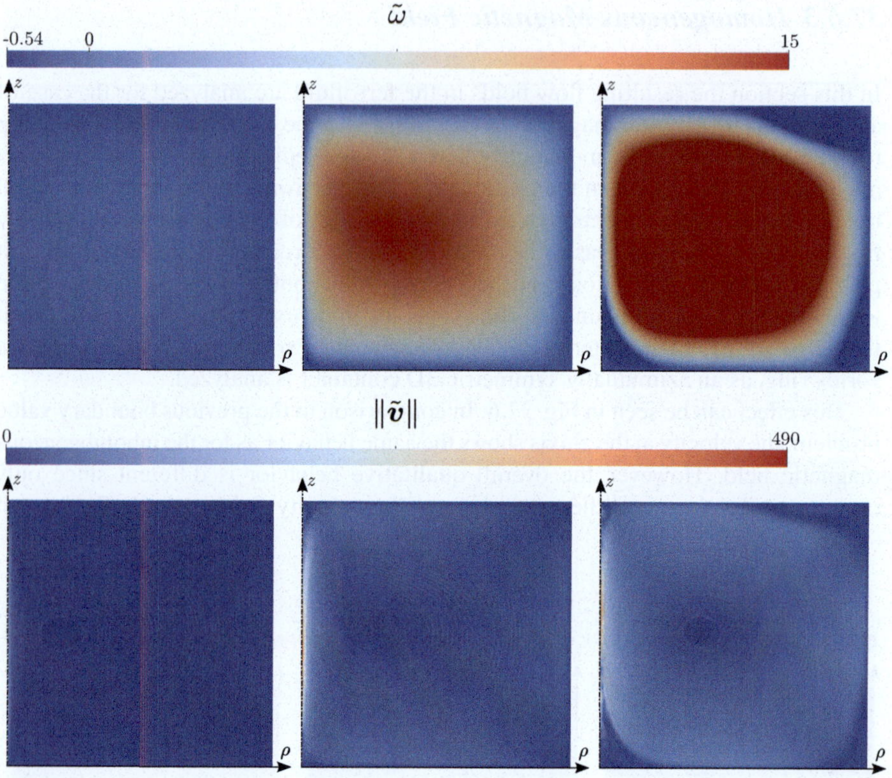

Fig. 37.6: Plot of $\tilde{\omega}$ (top) and stream density plot of $\tilde{\boldsymbol{v}}$ (bottom) for increasing time from left ro right for a homogeneous magnetic field and with no slip conditions .

lower. Furthermore, the angular velocity closely resembles the vorticity field. Most importantly, however, is that the vortex rotates counter clockwise as the angular velocity $\tilde{\omega}$ is positive. This is in contrast to the previous initial boundary value problems, for which the rotation was clockwise.

All of these plots for the homogeneous magnetic field develop a buckle near the top right corner. This is due to a dead water corner caused by a combination of the big vortex and the no-slip boundary conditions. Upon closer inspection, a similar buckle starts to develop in right hand side bottom corner.

37.7 Summary

In this paper the general governing equations for the ferrofluid flow under the influence of an external magnetic field analyzed. They were simplified and subsequently normalized. The resulting dimensionless numbers are discussed and their values are obtained from real material parameters. Several three dimensional initial boundary value problems are solved numerically in cylindrical coordinates using the finite element method.

References

[1] Torres-Diaz I, Rinaldi C (2014) Recent progress in ferrofluids research: Novel applications of magnetically controllable and tunable fluids, Soft Matter **10**(43):8584-8602.

[2] Papautsky I, Brazzle J, Ameel TA, Frazier AB (2008) Microchannel fluid behavior using micropolar fluid theory In: Proceedings MEMS 98. IEEE. Eleventh Annual International Workshop on Micro Electro Mechanical Systems. An Investigation of Micro Structures, Sensors, Actuators, Machines and Systems (Cat. No.98CH36176).

[3] Andò B, Baglio S, Beninato A (2013) A flow sensor exploiting magnetic fluids, Sensors and Actuators A: Physical **189**:17-23.

[4] Schumacher KR, Riley J, Finlayson BA (2011) Turbulence in ferrofluids in channel flow with steady and oscillating magnetic fields, Physical Review E Stat Nonlin Soft Matter Phys **83**(1 Pt 2):016307.

[5] Rosensweig RE (1985) Directions in ferrohydrodynamics, Journal of Applied Physics **57**(8):4259-4264.

[6] Shliomis MI (2001) Ferrohydrodynamics: Testing a third magnetization equation, Physical Review E **64**(6):060501.

[7] Cowin SC (1974) The theory of polar fluids, In C-S Yih (Ed), Advances in Applied Mechanics, vol. 14, pp 279-347, New York, Academic Press.

[8] Eringen AC (1999) Microcontinuum Field Theories Springer, New York.

[9] Alnæs M, Blechta J, Hake J, Johansson A, Kehlet B, Logg A, Richardson C, Ring J, MRognes ME, Wells GN (2013) The FEniCS Project Version 1.5, Archive of Numerical Software **3**(100):9-23.

[10] Vilchevskaya EN, Müller WH, Eremeyev VA (2022) Extended micropolar approach within the framework of 3m theories and variations thereof, Continuum Mechanics and Thermodynamics **34**(2):533-554.

[11] Eremeyev VA, Cloud MJ, Lebedev LP (2018) Applications of Tensor Analysis in Continuum Mechanics, World Scientific.

[12] Müller I (1985) Thermodynamics, Pitman.

[13] Truesdell CA, Toupin R (1960) The classical field theories, In: Handbuch der Physik, Bd. III/1, pp 226-793; appendix, pp 794-858, Springer, Berlin, 1960 (with an appendix on tensor fields by JL Ericksen).

[14] Jackson JD (1999) Classical Electrodynamics, American Association of Physics Teachers.

[15] Kolpashchikov VL, Migun NP, Prokhorenko PP (1983) Experimental determination of material micropolar fluid constants, International Journal of Engineering Science **21**(4):405-411.

[16] Brenner H (1970) Rheology of a dilute suspension of dipolar spherical particles in an external field, Journal of Colloid and Interface Science **32**(1):141-158.
[17] Laborladen online shop, https://www.laborladen.de/Ferrofluid-fluessiges-Eisen-10ml, Accessed: 2022-05-31.
[18] Kovetz A (2000) Electromagnetic Theory, vol. 975, Oxford University Press, Oxford.
[19] Papadopoulos PK, Vafeas P, Hatzikonstantinou PM (2012) Ferrofluid pipe flow under the influence of the magnetic field of a cylindrical coil, Physics of Fluids **24**(12):122002.
[20] Reich FA, Rickert W, Müller WH (2018) An investigation into electromagnetic force models: differences in global and local effects demonstrated by selected problems, Continuum Mechanics and Thermodynamics **30**(2):233-266.
[21] Pfeifer RNC, Nieminen TA, Heckenberg NR, Rubinsztein-Dunlop H (2007) Colloquium: Momentum of an electromagnetic wave in dielectric media, Reviews of Modern Physics **79**(4):1197-1216.
[22] Day MA (1990) The no-slip condition of fluid dynamics, Erkenntnis **33**(3):285-296.
[23] El-Kabeir SMM, El-Zahar ER, Modather M, Gorla RSR, Rashad AM (2019) Unsteady MHD slip flow of a ferrofluid over an impulsively stretched vertical surface, AIP Advances **9**(4):045112.

Chapter 38
Manufacturing Quality Evaluation of Photopolymer Resin 3D-Printed Scaffolds Using Microtomography

Evgeniy V. Sadyrin, Andrey L. Nikolaev, Sergei V. Chapek, Dmitry V. Nazarenko, Sergei M. Aizikovich, and Yun-Che Wang

Abstract In the present paper the approach utilizing X-ray computed microtomography (micro-CT) for evaluation of the manufacturing quality of 3D-printed scaffolds is proposed. The approach is based on matching and comparison the spatial image imported from the stereolithography (STL) files with the real geometry of the scaffold obtained by micro-CT scanning. Using the approach six photopolymer resin scaffolds with varying wall thickness (0.5, 0.7, 0.9, 1.1, 1.3, 1.5 mm) fabricated using digital light processing (DLP) were tested. The manufacturing artifacts were studied on the virtual two-dimensional slices of the internal structure of the scaffolds. A decrease of thickness deviations of the real geometry from the STL model for the scaffolds with increase of the thickness walls was observed and described.

Key words: Scaffold · Microtomography · Photopolymer resin · Digital light processing

Evgeniy V. Sadyrin · Andrey L. Nikolaev · Sergei M. Aizikovich
Laboratory for Mechanics of Biomaterials, Don State Technical University, Rostov-on-Don 344000, Russian Federation,
e-mail: e.sadyrin@sci.donstu.ru, andreynicolaev@eurosites.ru, saizikovich@gmail.com

Sergei V. Chapek
The Smart Materials Research Institute, Southern Federal University, 178/24 Sladkova Street, 344090, Rostov-on-Don, Russian Federation,
e-mail: chapek@sfedu.ru

Dmitry V. Nazarenko
Laboratory "Intelligent Electrical Agricultural Machines and Complexes", Don State Technical University, 1 Gagarin Square, 344000, Rostov-on-Don, Russian Federation,
e-mail: dmnazar@mail.ru

Yun-Che Wang
National Cheng Kung University, 1 University Road, Tainan, Taiwan 70101, R.O.C.,
e-mail: yunche@ncku.edu.tw

H. Altenbach et al. (eds.), *Sixty Shades of Generalized Continua*,
Advanced Structured Materials 170, https://doi.org/10.1007/978-3-031-26186-2_38

38.1 Introduction

Biocompatibility of biological tissue substitutes is determined by the combination of their physical, chemical and mechanical properties, surface morphology, bioactivity, antibacterial properties and resistance to the environment of use. Approaches to the creation of such biological tissue substitutes are being developed in tissue engineering and regenerative medicine. However, over the past three decades, significant advances have been made in the treatment of damaged, pathological or pathologically altered tissues. The tissue scaffold is a structure that enables adhesion of the cells and promotes diffusion of nutrients and waste. Thus, the scaffold skeleton provides the mechanical integrity of the tissue engineering structure. A key feature of scaffold structures is the requirement of sufficient porosity to enable bone and vascular ingrowth but at the same time sufficient strength for their load bearing requirement. Nowadays the various scaffolds receive significant attention from the engineering, mechanical, biological and medical communities. New technologies, in particular 3D printing techniques, are being designed to fabricate more complex scaffolds for hard tissue engineering [1, 2].

Polymer composites are widely used for bone implants and scaffolds [3]. The widespread utilization and uptake of these materials is made possible by rapid advances and development of common methods for 3D printing of thermoplastics [4], such as selective laser sintering (SLS) and fused filament fabrication (FFF) [5]. Stereolithography (SLA) method stands out due to its ability to fabricate details with high precision, surface quality and relatively low expenses [6, 7]. The SLA method, based on light source and imaging technology, is represented by two techniques: digital light processing (DLP) and laser-based stereolithography. In DLP technique, the light ray passes through a digital micro-mirror device and the resulted image is projected onto the resin vat. In the same time, in laser-based stereolithography uses a laser beam that is projected onto a swinging mirror, and reflected on the resin vat to cure the photopolymer resin based on the image of each layer. Overall, the DLP method is much faster of these two due to its ability to cure a whole layer on-the-spot [8] - [10]. Photopolymers are used extensively as coatings and adhesives in a wide range of industries. Photopolymer resins for SLA are systems that polymerize through free radical photopolymerization (acrylates, methacrylates, and acrylamides) and cationic photopolymerization (epoxides, vinyl ethers) [11, 12].

Computed microtomography (micro-CT) is a modern method for non-destructive investigation of inner structure of objects, localization of defects and for other purposes. In micro-CT scanning, the X-rays are generated by the source and emitted to the target sample. Thus, the specimen is divided into a series of virtual 2D slices which are irradiated from the edges with X-rays. Upon transversing through the slice, the X-rays are attenuated and the emergent X-rays with reduced intensities are captured by the detector array. From the detector measurements, the X-ray paths are calculated and the attenuation coefficients are derived [13, 14]. A 2D pixel map is created from these computations and each pixel is denoted by a threshold value which corresponds to the attenuation coefficient measured at a similar location within the specimen. The quality of the maps is dependent on the scanning resolution which ranges from 1 to 50 μm [15] - [18]. Micro-CT characterization comprises three major sequential processes: acquisition, reconstruction, and analysis.

Lin et al. [19] conducted micro CT studies on porous scaffolds formed using porogens in order to optimize the scaffold fabrication process. Baptista et al. [20] conducted experimental and numerical characterization of 3D-printed scaffolds under monotonic compression with the aid of micro-CT volume reconstruction. Salmon and Sasov [21] conducted a study of synthetic bone scaffolds for orthopedic research and the imaging of vascular networks using micro-CT. Hilldore et al [22] proposed a method to determine material volumes from micro-CT data for accounting materials with significantly overlapping peaks and small volumes. The development of in vitro techniques to characterize the behavior of cells in biomedical scaffolds was presented by Shepherd et al [23]. A comparison of cell seeding methods on scaffolds using micro-CT was given by Palmroth et al [24]. Clark et al [25] have performed high resolution X-ray micro-CT imaging of fibrin scaffold using large area single photon counting detector. Seo et al [26] performed *in situ* analysis of the hydration ability of bone graft material using a synchrotron radiation X-ray micro-CT. Mechanical in situ tests using micro-CT tests are conducted as to ascertain the mechanical properties of the scaffold [27] - [29]. In this work we propose an approach utilizing micro-CT for evaluation of the manufacturing quality of photopolymer resin 3D-printed scaffolds, fabricated using DLP technique. The approach is based on matching the spatial image imported from the stereolithography (STL) files with the real geometry of the scaffold obtained by micro-CT scanning with subsequent comparison of their wall thickness values.

38.2 Materials and Methods

For fabricating 3D-printed scaffolds, an Asiga UV MAX (Asiga, Australia, Sydney) with a lateral resolution of 62 μm, a maximum build volume of 119 mm × 67 mm × 75 mm in X-, Y-, and Z-direction, respectively, and a 385 nm light source was used. The three-dimensional models of the scaffolds were designed using the software Autodesk Netfabb (Autodesk Inc., San Rafael, United States). Each scaffold has dimension of 10 mm x 10 mm x 10 mm. The wall thickness of the scaffolds was design as follows: 0.5, 0.7, 0.9, 1.1, 1.3, 1.5 mm. The computer-aided designs were converted into standard tessellation language STL files by the Asiga Composer (Asiga, Australia, Sydney) software for further use in the 3D-printing process.

We have chosen 25 μm layer thickness as a tradeoff between reduced printing time and acceptable printing quality. The resulting voxel size is 62 × 62 × 25 μm^3.

Within the scope of this project, a commercially available resin (FunToDo Nano Clear, Alkmaar, The Netherlands) was used. All 3D printed scaffolds were cleaned by sonication in isopropyl-alcohol (IPA) and compress air blowing, followed by a second sonication in IPA and final drying in air. Finally, the 3D printed scaffolds were post-cured for 280 s with 10 UV-light flashes per second in an Otoflash G171 chamber (NK-Optik GmbH, Baierbrunn, Germany).

Micro-CT scanning was conducted using an Xradia Versa 520 unit (Carl Zeiss X-ray Microscopy, Inc., Pleasanton, USA). The micro-CT scanning parameters for all samples were as follows: X-ray tube voltage 30 kV, power 1.5 V, pixel size 28.8 μm, sample rotation by 360°, frames per image 1, exposure time on each frame 1 s. No X-ray tube filter (air) was used (selected based on the observed transmittance values according to the recommendations of the Xradia Versa 520 User's Guide A003030 Rev. B.). In the process of scanning, 1601 projections of each sample were obtained. Reconstruction of a three-dimensional image was carried out with automatically adjusted values of center shift, Gaussian blur filter 0.5. The beam hardening parameter (0.12 – 0.16) was used to correct the artifacts in reconstructed images in the form of depressions and shaded areas between dense objects, caused by a stronger absorption of X-ray photons in the middle of an object, compared to the edges (associated with an increase in beam hardening). For each scan, the sample was placed at the closest possible distance to the X-ray source. The flat panel detector was used to acquire the images 1536 x 972 pixels. The acquisition was performed with camera binning factor = 2. The exposure time was selected to maintain count (intensity) values 5000 with the selected source parameters. During each tomography procedure, 10 reference (air) X-ray images were acquired with equal time intervals between them. For the reconstruction of the set of virtual slices Scout-and-Scan Reconstructor 12.0.8086.19558 (Carl Zeiss X-ray Microscopy, Inc., USA) software was used.

Matching of the spatial image imported from the STL files with the real geometry of the scaffold obtained by micro-CT with subsequent comparison of their wall thickness values was conducted using VGSTUDIO Max 3.5 software (Volume Graphics, Heidelberg, Germany). To determine the surface of the scaffold the instrument "Surface determination", algorithm: Advanced (classic), starting counter was found by histogram. Particles were removed while starting contour healing. To match the images "Best fit registration" instrument was used. The real geometry served as object to be registered while the STL model served as object to be fitted against. The following options were enabled: "Improved refinement"; "Consider surface orientation". As an example, Fig. 38.1 demonstrates the 3D rendering of the scaffold with the wall thickness of 0.9 mm and its STL model before matching.

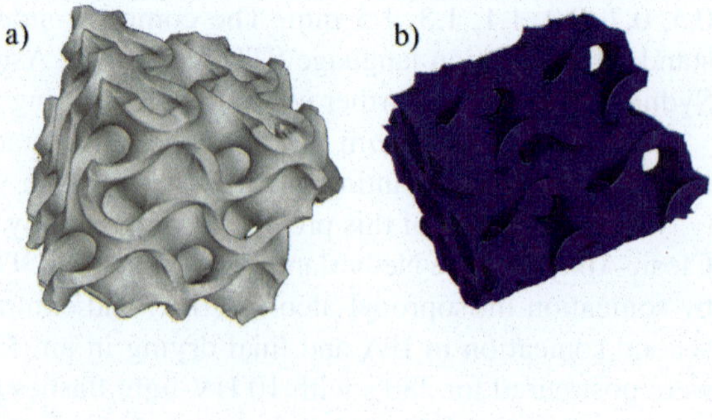

Fig. 38.1 3D rendering of the scaffold with the wall thickness 0.9 mm (a) and its STL model (b).

The difference in the thickness of the scaffold walls was measured on the STL model and micro-CT scan at 7 points on each sample: 6 points correspond to the central position on each of the cube face in the vicinity of the external scaffold surface, one point corresponds to the cube center. For this purpose, instrument "Distance" was used. Additionally, the artifacts of the printing were observed. In case the artifact demonstrated spherical shape, its diameter was measured in the vicinity of the center of the symmetry.

38.3 Results and Discussion

Matching the spatial image imported from the STL with the real geometry of the scaffold obtained by micro-CT enabled to calculate the differences between the wall thicknesses of each pair "STL model – real geometry" and outline some observations regarding the artifacts of printing.

For the scaffold with the wall thickness 0.5 mm (Fig 38.2) the following features were observed: artifacts of internal structure (chips up to 15.4 % thickness, a single pore of irregular shape with dimensions 0.282 x 0.150 mm, material overlap, cracks up to 0.518 mm in depth).

For the scaffold with the wall thickness 0.7 mm (Fig 38.3) the artifacts of internal structure up to 7.7 % thickness were also found. Overall increase of the surface waviness was spotted.

For the scaffold with the wall thickness 0.9 mm the overall decrease of the wall thickness values compared to the STL model (see Table 38.1) was observed (Fig. 38.4). The artifacts in the form of knobs 0.274 mm and 0.373 mm in diameter were

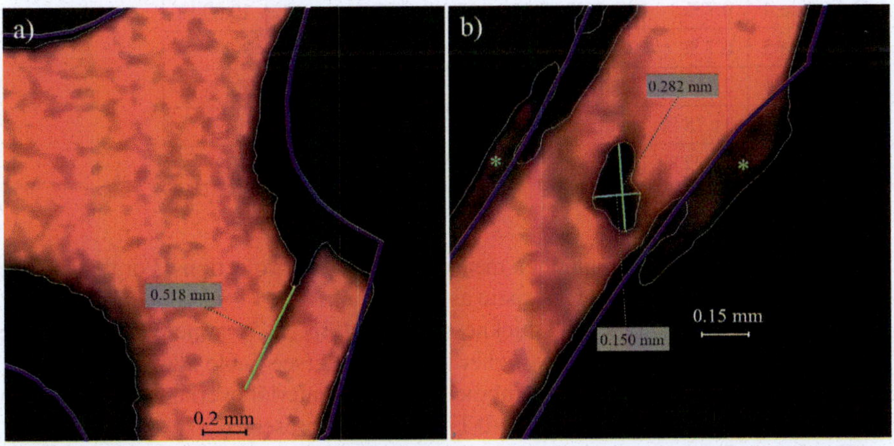

Fig. 38.2: Artifacts of the scaffold with the wall thickness 0.5 mm: (a) crack; (b) a single pore and material overlap (marked with *). Solid blue line – boundaries according to STL model, white line – determined surface of the scanned scaffold.

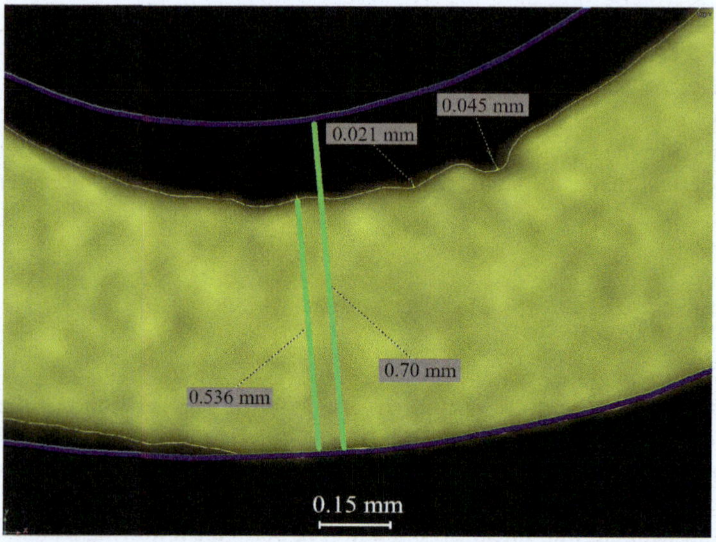

Fig. 38.3: Artifacts of the scaffold with the wall thickness 0.7 mm. Solid blue line – boundaries according to STL model, white line – determined surface of the scanned scaffold.

Table 38.1: The deviations of the thickness of the real geometry from the STL model for the scaffolds with the wall thickness of 0.5, 0.7 and 0.9 mm..

Wall thickness 0.5 mm		Wall thickness 0.7 mm		Wall thickness 0.9 mm	
Cube face 1	33.06 %	Cube face 1	20.13 %	Cube face 1	10.25 %
Cube face 2	32.66 %	Cube face 2	12.35 %	Cube face 2	14.22 %
Cube face 3	32.52 %	Cube face 3	20.23 %	Cube face 3	10.37 %
Cube face 4	0.7 %	Cube face 4	24.79 %	Cube face 4	10.69 %
Cube face 5	6.7 %	Cube face 5	12.55 %	Cube face 5	3.23 %
Cube face 6	2.74 %	Cube face 6	9.22 %	Cube face 6	5.2 %
Cube center	31.78 %	Cube center	13.49 %	Cube center	11.27 %
Average	20.02 %	Average	16.11 %	Average	9.32 %

found on the 3D rendering and then measured on the 2D projections. For this sample the material overlap was characteristic as well.

For the scaffold with the wall thickness 1.1 mm (Fig. 38.5) numerous sites, where actual thickness was surprisingly higher were studied (see Table 38.2, Cube faces 1, 4). Sometimes such sites were combined with the material overlap. Artifacts of in the form of knobs with heights up to 0.177 mm were presented.

For the scaffold with the wall thickness 1.3 mm (Fig. 38.6) a single artifact in the form of knob 0.117 mm high was observed. The sample also revealed some sites,

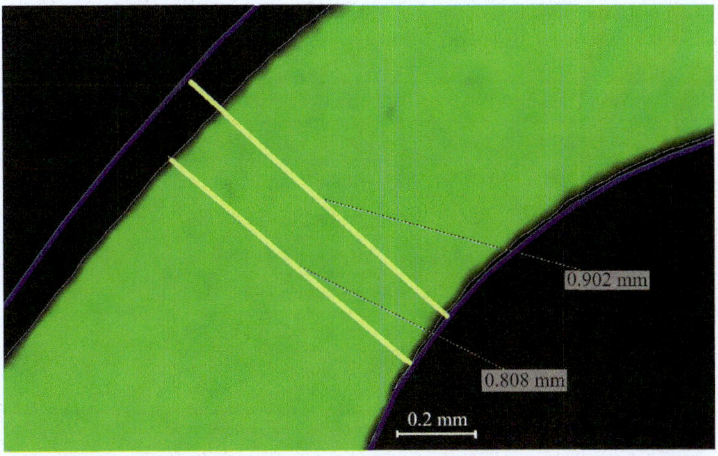

Fig. 38.4: Artifacts of the scaffold with the wall thickness 0.9 mm. Solid blue line – boundaries according to STL model, white line – determined surface of the scanned scaffold.

where actual thickness was higher that it was planned in the STL model (see Table 38.2, Cube face 6).

For the scaffold with the wall thickness 1.5 mm (Fig. 38.7) a paired artifacts in the form of knobs (with sharp edges) up to 0.43 mm was spotted. Minor increase of thickness in some sites (see Table 38.2, Cube face 4, Cube center), surface waviness and a single artifact of internal structure 0.09 mm in diameter were found

The measurements for the deviations of the thickness of the real geometry from the STL models for the each sample studied are listed in the Tables 38.1 and 38.2. The

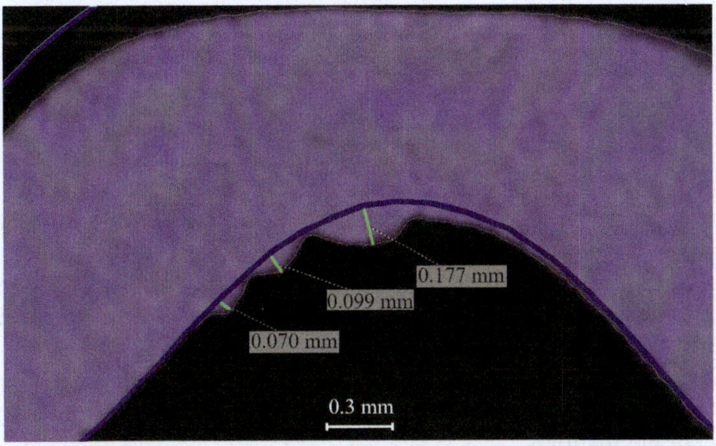

Fig. 38.5: Artifacts of the scaffold with the wall thickness 1.1 mm. Solid blue line – boundaries according to STL model, white line – determined surface of the scanned scaffold.

Table 38.2: The deviations of the thickness of the real geometry from the STL model for the scaffolds with the wall thickness of 1.1, 1.3 and 1.5 mm.

Wall thickness 1.1 mm		Wall thickness 1.3 mm		Wall thickness 1.5 mm	
Cube face 1	6.36 %	Cube face 1	11.28 %	Cube face 1	0.14 %
Cube face 2	6.25 %	Cube face 2	0.3 %	Cube face 2	4.43 %
Cube face 3	10.71 %	Cube face 3	11.03 %	Cube face 3	2.86 %
Cube face 4	10.22 %	Cube face 4	7.31 %	Cube face 4	1.44 %
Cube face 5	9.38 %	Cube face 5	12.11 %	Cube face 5	10.47 %
Cube face 6	10.05 %	Cube face 6	5. 88 %	Cube face 6	11.32 %
Cube center	8.51 %	Cube center	11.05 %	Cube center	2.67 %
Average	8.78 %	Average	8.42 %	Average	4.76 %

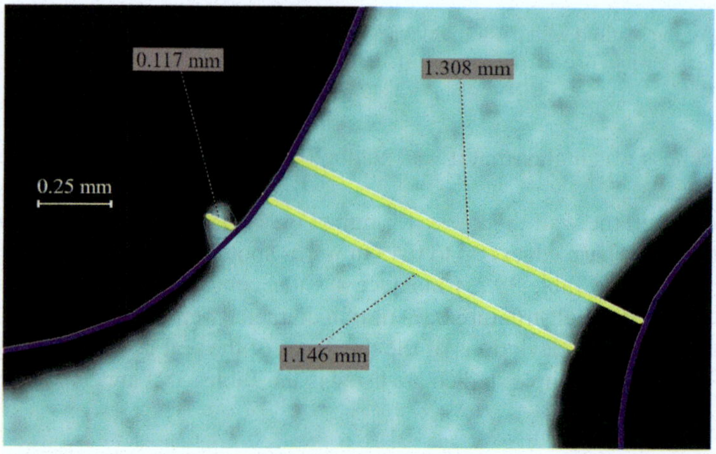

Fig. 38.6: Artifacts of the scaffold with the wall thickness 1.3 mm. Solid blue line – boundaries according to STL model, white line – determined surface of the scanned scaffold.

analysis of the results reveals that the deviations of the thickness of the real geometry from the STL model for the scaffolds decrease with the increase in the wall thickness: from 20.02 % for the wall thickness 0.5 mm to 4.76 % for the wall thickness 1.5 mm. This phenomenon presumably may be caused by shrinkage of the photopolymer resin material. Moreover, the sample with the wall thickness 0.5 mm demonstrated the highest number and variety of the artifacts of inner structure. The cracks should be especially mentioned as the crack tips may serve as the stress concentration sites, thus such areas are susceptible for fracture. In this regard addition of the hydroxyapatite in the scaffold composition may be considered in future to prevent fracture during the cyclic loading of the scaffolds.

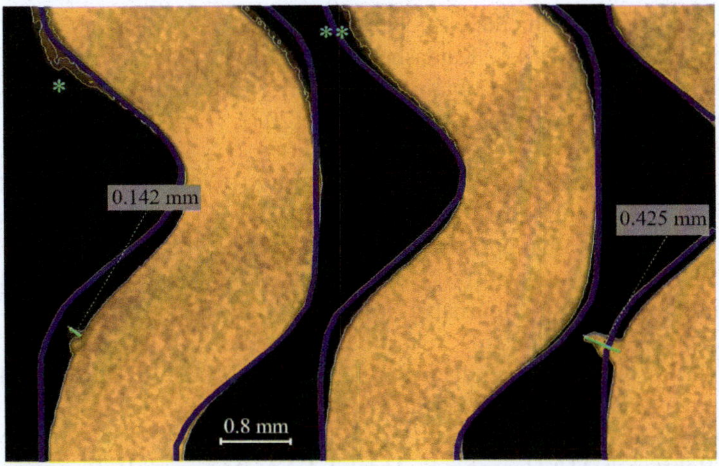

Fig. 38.7: Artifacts of the scaffold with the wall thickness 1.5 mm. Solid blue line – boundaries according to STL model, white line – determined surface of the scanned scaffold. Minor increase of thickness is marked with *. Surface waviness site is marked with **.

38.4 Conclusion

In the present paper the geometry of the 3D-printed scaffolds with the varying wall thickness fabricated using DLP technique was studied. The approach to study the geometry included matching and comparison the spatial image imported from the STL files with the real geometry of the scaffold obtained by micro-CT scanning. The manufacturing artifacts (pores, cracks, material overlap, surface waviness, knobs and pikes) were studied on the virtual two-dimensional slices of the internal structure of the scaffolds. A decrease of thickness deviations of the real geometry from the STL model for the scaffolds with increase of the thickness walls from 20.02 % for the wall thickness 0.5 mm to 4.76 % for the wall thickness 1.5 mm was observed presumably caused by shrinkage of the photopolymer resin material.

Acknowledgements S.M. Aizikovich acknowledges the support of the grant of the Russian Science Foundation, grant number 22-49-08014. Y.C. Wang acknowledges the support of the Russia-Taiwan Bilateral Project, contract No. MOST 111-2923-E-006-003-MY3. All micro-CT experiments were conducted in the Laboratory for Mechanics of Biomaterials, Don State Technical University (http://nano.donstu.ru).

References

[1] Bracaglia LG, Smith BT, Watson E, Arumugasaamy N, Mikos AG, Fisher JP (2017) 3D printing for the design and fabrication of polymer-based gradient

scaffolds, Acta biomaterialia **56**:3-13. DOI 10.1016/j.actbio.2017.03.030
[2] Pushkariov VI, Nikolaev AL, Kaidashev EM (2014) Synthesis and characterization of ZnO nanorods obtained by catalyst-free thermal technique, Journal of Physics: Conference Series **541**:012031. DOI 10.1088/1742-6596/541/1/012031
[3] Ahangar P, Cooke ME, Weber MH, Rosenzweig DH (2019) Current biomedical applications of 3D printing and additive manufacturing. Applied Sciences **2019**(9):1713. DOI 10.3390/app9081713
[4] Aimar A, Palermo A, Innocenti B (2019) The role of 3D printing in medical applications: a state of the art, Journal of Healthcare Engineering **2019**:5340616. DOI 10.1155/2019/5340616
[5] Gardan J (2017) Additive manufacturing technologies: state of the art and trends, International Journal of Production Research **54**:3118-3132. DOI 10.1080/00207543.2015.1115909
[6] Saed AB, Behravesh AH, Hasannia S, Ardebili SAA, Akhoundi B, Pourghayoumi M (2020) Functionalized poly l-lactic acid synthesis and optimization of process parameters for 3D printing of porous scaffolds via digital light processing (DLP) method, Journal of Manufacturing Processes **56**:550-561. DOI 10.1016/j.jmapro.2020.04.076
[7] Naghieh S, Ravari MK, Badrossamay M, Foroozmehr E, Kadkhodaei M (2016) Numerical investigation of the mechanical properties of the additive manufactured bone scaffolds fabricated by FDM: The effect of layer penetration and post-heating, Journal of the Mechanical Behavior of Biomedical Materials **59**:241-250. DOI 10.1016/j.jmbbm.2016.01.031
[8] Zhou C, Chen Y (2012) Additive manufacturing based on optimized mask video projection for improved accuracy and resolution, Journal of Manufacturing Processes **14**:107-118. DOI 10.1016/j.jmapro.2011.10.002
[9] Maines EM, Porwal MK, Ellison CJ, Reineke TM (2021) Sustainable advances in SLA/DLP 3D printing materials and processes, Green Chemistry **23**:6863-6897. DOI 10.1039/D1GC01489G
[10] Xue D, Wang Y, Mei D (2019) Multi-step exposure method for improving structure flatness in digital light processing-based printing, Journal of Manufacturing Processes **39**:106-113. DOI 10.1016/j.jmapro.2019.02.013
[11] Ligon SC, Liska R, Stampfl J, Gurr M, Mülhaupt R (2017) Polymers for 3D printing and customized additive manufacturing, Chemical Reviews **117**:10212-10290. DOI 10.1021/acs.chemrev.7b00074
[12] Gibson I, Rosen DW, Stucker B (2010) Photopolymerization processes, In: Additive Manufacturing Technologies, pp 78-119, Springer, Boston, MA. DOI 10.1007/978-1-4419-1120-9_4
[13] Wright LD, Young RT, Andric T, Freeman JW (2010) Fabrication and mechanical characterization of 3D electrospun scaffolds for tissue engineering, Biomedical Materials **5**:055006. DOI 10.1088/1748-6041/5/5/055006
[14] Cengiz IF, Oliveira JM, Reis RL (2017) Micro-computed tomography characterization of tissue engineering scaffolds: effects of pixel size and rotation

step, Journal of Materials Science: Materials in Medicine: **28**:1-11. DOI 10.1007/s10856-017-5942-3

[15] Sadyrin EV, Yogina DV, Swain MV, Maksyukov SY, Vasiliev AS (2021) Efficacy of dental materials in terms of apparent mineral density restoration: Composite resin, glass ionomer cement and infiltrant, Composites Part C: Open Access **6**:100192. DOI 10.1016/j.jcomc.2021.100192

[16] Sadyrin EV (2022) Correlating the mechanical properties to the mineral density of brown spot lesion in dentine using nanoindentation and X-ray microtomography, In: Advanced Materials Modelling for Mechanical, Medical and Biological Applications, pp 389-398, Springer, Cham. DOI 10.1007/978-3-030-81705-3_21

[17] Tkachev SY, Mitrin BI, Karnaukhov NS, Sadyrin EV, Voloshin MV, Maksimov AY, Goncharova AS, Lukbanova EA, Zaikina EV, Volkova AV, Khodakova DV, Mindar MV, Yengibarian MA, Protasova TP, Kit SO, Ermakov AM, Chapek SV, Tkacheva MS (2021) Visualization of different anatomical parts of the enucleated human eye using X-ray micro-CT imaging, Experimental Eye Research **203**:108394. DOI 10.1016/j.exer.2020.108394

[18] Sadyrin E, Swain M, Mitrin B, Rzhepakovsky I, Nikolaev A, Irkha V, Yogina D, Lyanguzov N, Maksyukov S, Aizikovich S (2020) Characterization of enamel and dentine about a white spot lesion: mechanical properties, mineral density, microstructure and molecular composition, Nanomaterials **10**:1889. DOI 10.3390/nano10091889

[19] Lin AS, Barrows TH, Cartmell SH, Guldberg RE (2003) Microarchitectural and mechanical characterization of oriented porous polymer scaffolds, Biomaterials **24**:481-489. DOI 10.1016/S0142-9612(02)00361-7

[20] Baptista R, Pereira MFC, Mauricio A, Rechena D, Infante V, Guedes M (2021) Experimental and numerical characterization of 3D-printed scaffolds under monotonic compression with the aid of micro-CT volume reconstruction, Bio-Design and Manufacturing **4**:222-242. DOI 10.1007/s42242-020-00122-3

[21] Salmon PL, Sasov AY (2007) Application of nano-CT and high-resolution micro-CT to study bone quality and ultrastructure, scaffold biomaterials and vascular networks, Advanced bioimaging technologies in assessment of the quality of bone and scaffold materials, pp 323-331, Springer, Berlin, Heidelberg. DOI 10.1007/978-3-540-45456-4_19

[22] Hilldore A, Wojtowicz A, Johnson AW (2007) Micro-CT based quantification of non-mineralized tissue on cultured hydroxyapatite scaffolds, Journal of Biomedical Materials Research Part A **82**:1012-1021. DOI 10.1002/jbm.a.31264

[23] Shepherd DV, Shepherd JH, Best SM, Cameron RE (2018) 3D imaging of cells in scaffolds: direct labelling for micro CT, Journal of Materials Science: Materials in Medicine **29**:1-4. DOI 10.1007/s10856-018-6089-6

[24] Palmroth A, Pitkänen S, Hannula M, Paakinaho K, Hyttinen J, Miettinen S, Kellomäki M (2020) Evaluation of scaffold microstructure and comparison of cell seeding methods using micro-computed tomography-based tools, Journal of the Royal Society Interface **17**:20200102. DOI 10.1098/rsif.2020.0102

[25] Clark JN, Tavana S, Heyraud A, Tallia F, Jones JR, Hansen U, Jeffers JR (2020) Quantifying 3D strain in scaffold implants for regenerative medicine, Materials **2020**(13):3890. DOI 10.3390/ma13173890

[26] Seo SJ, Kim YG (2020) In-situ analysis of the hydration ability of bone graft material using a synchrotron radiation X-ray micro-CT. Journal of Applied Biomaterials & Functional Materials **18**:2280800020963476. DOI 10.1177/2280800020963476

[27] Mitrin BI, Chapek SV, Sadyrin EV, Swain MV (2021) Mechanical properties and failure mechanisms of 3D-printed PLA scaffolds: A preliminary study, IOP Conference Series: Materials Science and Engineering **1029**:012074. DOI 10.1088/1757-899X/1029/1/012074

[28] Dewanckele J, Boone MA, Coppens F, Van Loo D, Merkle AP (2020) Innovations in laboratory-based dynamic micro-CT to accelerate in situ research, Journal of Microscopy **277**:197-209. DOI 10.1111/jmi.12879

[29] Farah S, Anderson DG, Langer R (2016) Physical and mechanical properties of PLA, and their functions in widespread applications — A comprehensive review, Advanced Drug Delivery Reviews **107**:367-392. DOI 10.1016/j.addr.2016.06.012

Chapter 39
Comparison of Homogenization Techniques in Strain Gradient Elasticity for Determining Material Parameters

Bekir Cagri Sarar, M. Erden Yildizdag, and Bilen Emek Abali

Abstract Additive manufacturing techniques, especially fused deposition modeling (FDM) based polymer 3-D printers are opening new possibilities in engineering design. Multiscale structures have a macroscale and miscroscale geometry. Composite materials are available as filaments allowing multiphysics applications. Microscale structure or even heterogeneous materials cause deviations from predictions of the materials response by using the classical first-order theory. Hence, we use the second-order modeling in mechanics and study how to determine additional constitutive parameters that arise with the strain gradient theory. Different homogenization methods are demonstrated to result in differing material parameters based on assumptions used in modeling.

Key words: Additive manufacturing · 3-D printing · Homogenization · Strain gradient elasticity

---

Bekir Cagri Sarar
University of L'Aquila, International Research Center on Mathematics and Mechanics of Complex Systems, L'Aquila, Italy,
e-mail: bekircagri.sarar@graduate.univaq.it

M. Erden Yildizdag
University of L'Aquila, International Research Center on Mathematics and Mechanics of Complex Systems, L'Aquila, Italy & Istanbul Technical University, Faculty of Naval Architecture and Ocean Engineering, Istanbul, Turkey,
e-mail: yildizdag@itu.edu.tr

Bilen Emek Abali
Uppsala University, Division of Applied Mechanics, Department of Materials Science and Engineering, Uppsala, Sweden,
e-mail: bilenemek@abali.org

H. Altenbach et al. (eds.), *Sixty Shades of Generalized Continua*,
Advanced Structured Materials 170, https://doi.org/10.1007/978-3-031-26186-2_39

39.1 Introduction

Additive manufacturing (AM) allows to create architectured and composite materials with microstructure at different length scales [1, 2]. Metamaterials are classified according to the main interaction between their complex microstructures. Mechanical interaction between microstructural elements may be listed as negative effective bulk modulus, a negative effective mass density, or negative Poisson's ratio as known from [3]-[5].

3-D printing paves the way for the construction of materials that were deemed impossible to be manufactured with traditional production processes [6]-[9]. With the well-known layer-by-layer manufacturing process, complex geometries emerge in the literature but their accurate models are still in their early beginning. Development in 3-D printers has facilitated the production of metamaterials with exotic properties in their microstructures that are not found in nature. Macroscale structure has a physical response provided by lower-scale microstructures as well as interactions between each scale. The multiscale structure of bone is a most common example biomechanical inspiring from nature to designing metamaterials. Particularly, we know from [10]-[12] tissues of bones have different microscale structures respectively osteon, cortical bone, trabecula, and cancellous bone. Another class of man-made design is known as pantographic structures, which is designed to increase flexibility while minimizing deformation [13]-[16].

Composite materials in 3-D printing are available in the market. For instance, carbon fiber reinforced polyamide (PA) matrix has such a high specific (per mass) strength that it is also called "black aluminum" in AM jargon. The reinforcement is in different sizes regarding material supplier, often in spherical form, and they may have a conductivity as well depending on the agglomeration of carbon particles. Moreover, it has a higher thermal resistance then other off-the-shelf filaments. This composite material is produced in various sizes from powders to grains by means of electrofusion of bauxite, iron oxide and micro-elements in a tilting arc furnace. Composite materials cause a heterogeneous structure such that the microscale structure affects the macroscale materials response as demonstrated in [17, 18]

Multiscale materials response is challenging in terms of modeling. Although Cauchy continuum theories have satisfied researchers until now, higher gradient theories [19] have in fact been the subject of research explained in [20]. Due to metamaterial's complex microstructure, detailed modeling of the multiscale structure fails to be feasible in engineering practice. One particular procedure is increasing the complexity of the material model [21, 22]. The simplest model is incorporating not only strains but also strain gradients in deformation energy [23] and define the parameters by a suitable "homogenization" technique as in [24]. Homogenization is the procedure to circumvent the microstructure and increase the material model's complexity. Ample homogenization approaches are available in the literature for the case of strain gradient elasticity [25]-[31]. The methods are mainly based on the energy equivalence between micro-macro scales, yet they use different assumptions especially for boundary conditions. In this work, we examine and compare two homogenization methods:

- AH – asymptotic homogenization [32, 33] and
- CH – computational homogenization [26, 27].

39.2 Determining Strain Gradient Parameters

We begin with known material models at the microscale and arrive material parameters at the macroscale for an assumed material model by means of a suitable homogenization technique. Effects of the microscale are then incorporated by the macroscale material model. This upscaling may differ in results and a rather straight-forward method is to verify them by computations, for examples we refer to [34, 35]. Herein, we use (first-order) linear elasticity as known as HOOKE's law at the microscale and (second-order) the strain gradient elasticity at the macroscale.

In all homogenization techniques, the main assertion is that the deformation energy at the microscale is equal to the energy at the macroscale. The second-order theory is utilized to define the deformation energy at the macroscale with the higher-order constitutive parameters on a homogenized continuum, while the first-order theory is applied to derive deformation energy at the microscale with the given microstructure. So-called effective parameters are the main objective and they are determined with a given material equation at the microscale. For example, a quadratic deformation energy is used

$$e^{\mathrm{m}} = \frac{1}{2}\varepsilon^{\mathrm{m}}_{ij}C^{\mathrm{m}}_{ijkl}\varepsilon^{\mathrm{m}}_{kl}\,, \tag{39.1}$$

for defining the energy density (per volume), where $\boldsymbol{\varepsilon}^{\mathrm{m}}$ is the linearized strain measure and $\boldsymbol{C}^{\mathrm{m}}$ is the known elasticity tensor of rank 4 defined in each position at the microscale. Strain measure is used to ensure objectivity, in the general sense, a GREEN–LAGRANGE strain measure reduces to linearized symmetric displacement gradients in the linear case, $\varepsilon^{\mathrm{m}}_{ij} = (u^{\mathrm{m}}_{i,j} + u^{\mathrm{m}}_{j,i})/2$, where displacement at the microscale, $\boldsymbol{u}^{\mathrm{m}}$, is the solution of divergence of stress. Stress is energy derivative with respect to strain. A comma notation denotes a partial derivative in space. Elasticity tensor, $\boldsymbol{C}^{\mathrm{m}}$, is positive definite in each coordinate. Indeed $\boldsymbol{C}^{\mathrm{m}}$ is heterogeneous. At the macroscale, we model by using the strain gradient elasticity with the same type of (quadratic) energy density

$$e^{\mathrm{M}} = \frac{1}{2}\varepsilon^{\mathrm{M}}_{ij}C^{\mathrm{M}}_{ijkl}\varepsilon^{\mathrm{M}}_{kl} + \varepsilon^{\mathrm{M}}_{ij}G^{\mathrm{M}}_{ijklm}\varepsilon^{\mathrm{M}}_{kl,m} + \frac{1}{2}\varepsilon^{\mathrm{M}}_{ij,k}D^{\mathrm{M}}_{ijklmn}\varepsilon^{\mathrm{M}}_{lm,n}\,, \tag{39.2}$$

with the usual restrictions as objectivity and positive definiteness [36, 37]. These additional metamaterials tensors, $\boldsymbol{C}^{\mathrm{M}}$, $\boldsymbol{G}^{\mathrm{M}}$, and $\boldsymbol{D}^{\mathrm{M}}$, are sought after. In the case of a repeating microstructure, we characterize one unit cell or a representative volume element (RVE) built by a few unit cells. By using periodic boundary conditions on the RVE boundaries, we may capture the complete response of the body. However, use of boundary conditions may be misleading. In an attempt of using periodic boundaries for displacement, one may realize that strain gradient is immediately vanishing.

Periodicity in the geometry does not mean that the displacement is periodic, as a stark contrast to the classical homogenization schemes.

The deformation energy within a domain, for example within the RVE, Ω, is identical at micro- and macroscales

$$\int_\Omega e^{\mathrm{m}} \, \mathrm{d}V = \int_\Omega e^{\mathrm{M}} \, \mathrm{d}V \, . \tag{39.3}$$

Displacements or stress may differ, but the energy is the same. This axiom is mainly shared by different homogenization techniques.

Asymptotic homogenization technique relies on a TAYLOR expansion of the displacement around the geometric middle. The expansion involves second derivatives leading to additional relations expressed by the periodic variables. If these variables are computed, their values are used to integrate to $\boldsymbol{C}^{\mathrm{M}}$, $\boldsymbol{G}^{\mathrm{M}}$, and $\boldsymbol{D}^{\mathrm{M}}$. Asymptotic homogenization may be applied by solving governing equations for expansion or it may be directly applied in an analysis by [38] by means of Fast Fourier Transformation (FFT).

Computational homogenization may use the whole geometry or again an RVE, but this time, boundaries are given according to a specific assumption. Either the boundaries are deforming the same as at the microscale case. Or there is another assumption for deriving the boundary conditions. Quadratic Boundary Conditions (QBC) are extensively explained in [39].

39.3 Microscale Structure

We consider one of the widely available heterogeneous structure of a composite material of a matrix embedding round inclusions. Aforementioned materials used in 3-D printing are in focus. For example round particle reinforced polymers are available as filaments to be used in FDM 3-D printers. We consider such an ideal material with round inclusions as drawn in Fig. 39.1. By using such an inclusion and a structure in an orthotropic manner, we expect to obtain constitutive parameters in D4 symmetry class.

In order to express rank 4, 5, 6 tensors, $\boldsymbol{C}^{\mathrm{M}}$, $\boldsymbol{G}^{\mathrm{M}}$, $\boldsymbol{D}^{\mathrm{M}}$, respectively, we use a VOIGT-like notation. Namely, we use A, B indicating $\{11, 22, 12\}$ and α, β denoting $\{111, 221, 121, 112, 222, 122\}$, such that we obtain

$$C^{\mathrm{M}}_{AB} = \begin{pmatrix} C^{\mathrm{M}}_{1111} & C^{\mathrm{M}}_{1122} & C^{\mathrm{M}}_{1112} \\ C^{\mathrm{M}}_{2211} & C^{\mathrm{M}}_{2222} & C^{\mathrm{M}}_{2212} \\ C^{\mathrm{M}}_{1211} & C^{\mathrm{M}}_{1222} & C^{\mathrm{M}}_{1212} \end{pmatrix} ,$$

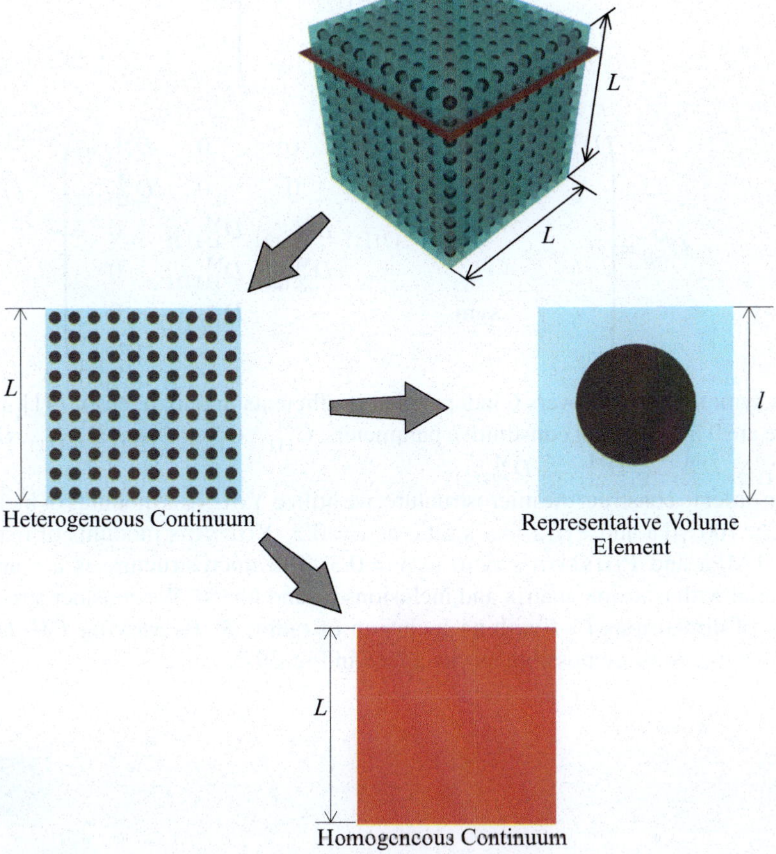

Fig. 39.1: Decomposition of the unit cell from a periodic heterogeneous microstructure and its equivalent homogenized continuum.

$$D^{\mathrm{M}}_{\alpha\beta} = \begin{pmatrix} D^{\mathrm{M}}_{111111} & D^{\mathrm{M}}_{111221} & D^{\mathrm{M}}_{111121} & D^{\mathrm{M}}_{111112} & D^{\mathrm{M}}_{111222} & D^{\mathrm{M}}_{111122} \\ D^{\mathrm{M}}_{221111} & D^{\mathrm{M}}_{221221} & D^{\mathrm{M}}_{221121} & D^{\mathrm{M}}_{221112} & D^{\mathrm{M}}_{221222} & D^{\mathrm{M}}_{221122} \\ D^{\mathrm{M}}_{121111} & D^{\mathrm{M}}_{121221} & D^{\mathrm{M}}_{121121} & D^{\mathrm{M}}_{121112} & D^{\mathrm{M}}_{121222} & D^{\mathrm{M}}_{121122} \\ D^{\mathrm{M}}_{112111} & D^{\mathrm{M}}_{112221} & D^{\mathrm{M}}_{112121} & D^{\mathrm{M}}_{112112} & D^{\mathrm{M}}_{112222} & D^{\mathrm{M}}_{112122} \\ D^{\mathrm{M}}_{222111} & D^{\mathrm{M}}_{222221} & D^{\mathrm{M}}_{222121} & D^{\mathrm{M}}_{222112} & D^{\mathrm{M}}_{222222} & D^{\mathrm{M}}_{222122} \\ D^{\mathrm{M}}_{122111} & D^{\mathrm{M}}_{122221} & D^{\mathrm{M}}_{122121} & D^{\mathrm{M}}_{122112} & D^{\mathrm{M}}_{122222} & D^{\mathrm{M}}_{122122} \end{pmatrix} .$$

Indeed the use of Voigt-like notation is not unique. Different sequences are possible and for some symmetry classes they may create a better representation [40]. For the chosen microstructure herein, D_4 is called the dihedral group, it is the symmetry classification of square mainly expressed as tetragonal system. The rank 5 tensor vanishes because of centro-symmetric configuration and rank 4, 6 tensors read

$$C^{\mathrm{M}}_{AB} = \begin{pmatrix} C^{\mathrm{M}}_{1111} & C^{\mathrm{M}}_{1122} & 0 \\ & C^{\mathrm{M}}_{1111} & 0 \\ \text{sym.} & & C^{\mathrm{M}}_{1212} \end{pmatrix},$$

$$D^{\mathrm{M}}_{\alpha\beta} = \begin{pmatrix} D^{\mathrm{M}}_{111111} & D^{\mathrm{M}}_{111221} & 0 & 0 & 0 & D^{\mathrm{M}}_{111122} \\ & D^{\mathrm{M}}_{221221} & 0 & 0 & 0 & D^{\mathrm{M}}_{221122} \\ & & D^{\mathrm{M}}_{121121} & D^{\mathrm{M}}_{221122} & D^{\mathrm{M}}_{111122} & 0 \\ & & & D^{\mathrm{M}}_{221221} & D^{\mathrm{M}}_{111221} & 0 \\ & \text{sym.} & & & D^{\mathrm{M}}_{111111} & 0 \\ & & & & & D^{\mathrm{M}}_{121121} \end{pmatrix}.$$

This symmetry class covers 6 independent coefficients in rank 6 tensor [41]. In total, there are 9 independent constitutive parameters, C^{M}_{1111}, C^{M}_{1122}, C^{M}_{1212}, D^{M}_{111111}, D^{M}_{111221}, D^{M}_{111122}, D^{M}_{221221}, D^{M}_{221122}, $D^{\mathrm{M}}_{1211211}$.

In order to construct the microstructure, we utilize YOUNG's modulus of inclusions as $E = 100\,\mathrm{MPa}$ and POISSON's ratio as $\nu = 0.3$; YOUNG's modulus of matrix as $E = 1\,\mathrm{MPa}$ and POISSON's ratio as $\nu = 0.3$. The microstructure is a composite material with isotropic matrix and inclusions (round fibers). We consider a rectangle RVE of dimensions $l \times l$ with an inclusion of radius R. By varying R/l different volume fractions are possible as visualized in Fig. 39.2.

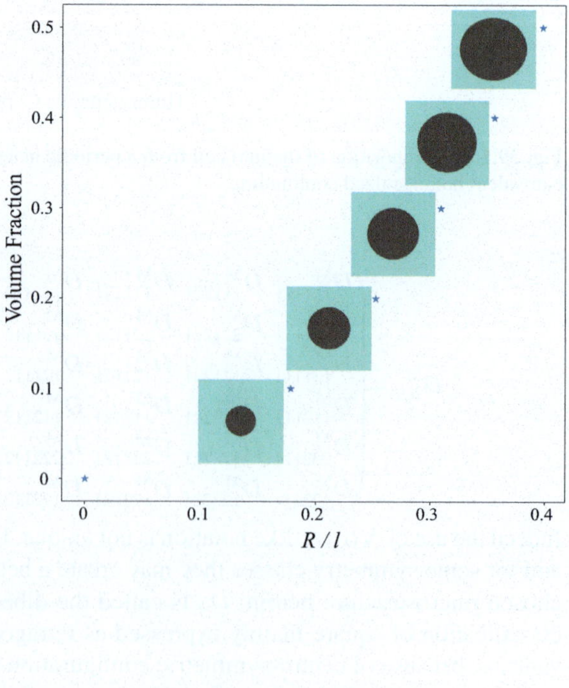

Fig. 39.2 Change of microstructure in the relation between volume fraction and ratio of the radius of inclusion and edge length of the unit cell.

39.4 Results and Discussion

We determine parameters by using asymptotic homogenization with the method as described in [42] and compare with computational homogenization as given by [27]. Specifically, we construct 30% volume ratio and $l = 1$ mm, and the asymptotic homogenization method reads

$$
C^{\mathrm{M}}_{AB} = \begin{pmatrix} 2.2 & 0.8 & 0 \\ 0.8 & 2.2 & 0 \\ 0 & 0 & 0.5 \end{pmatrix} \mathrm{MPa}\,,
$$

$$
D^{\mathrm{M}}_{\alpha\beta} = \begin{pmatrix} 0.122 & 0.022 & 0 & 0 & 0 & -0.017 \\ 0.022 & -0.027 & 0 & 0 & 0 & -0.007 \\ 0 & 0 & 0.016 & -0.007 & -0.017 & 0 \\ 0 & 0 & -0.007 & -0.027 & 0.022 & 0 \\ 0 & 0 & -0.017 & 0.022 & 0.122 & 0 \\ -0.017 & -0.007 & 0 & 0 & 0 & 0.016 \end{pmatrix} \mathrm{N}\,. \tag{39.4}
$$

For a direct comparison, we rewrite the computational homogenization results, in the same digits, as follows:

$$
C^{\mathrm{M}}_{AB} = \begin{pmatrix} 2.2 & 0.8 & 0 \\ 0.8 & 2.2 & 0 \\ 0 & 0 & 0.6 \end{pmatrix} \mathrm{MPa}\,,
$$

$$
D^{\mathrm{M}}_{\alpha\beta} = \begin{pmatrix} 0.018 & -0.006 & 0 & 0 & 0 & 0.002 \\ -0.006 & 0.005 & 0 & 0 & 0 & -0.0002 \\ 0 & 0 & 0.009 & -0.0002 & 0.002 & 0 \\ 0 & 0 & -0.0002 & 0.005 & -0.006 & 0 \\ 0 & 0 & 0.002 & -0.006 & 0.018 & 0 \\ 0.002 & -0.0002 & 0 & 0 & 0 & 0.009 \end{pmatrix} \mathrm{N}\,. \tag{39.5}
$$

Obviously, rank 4 tensors belong to the first order theory and they are identical. However, rank 6 results differ, i.e. the second order results are not matching. Instead of claiming that one result may be error-prone, we consider having several possible results inherently because of the nonlinear energy definition. The common assertion in both methods is that the energy at microscale and macroscale is equal. Therefore, different values are possible. We emphasize that the boundary conditions of the RVE are mismatching in both methodologies. Hence, it is even inadequate to expect the same results from both techniques. Since the analysis herein is only for a specific case, we investigate further for an interval of R/l values. All values are compiled in Table 39.1. The same results from Table 39.1, we demonstrate also in Fig. 39.3, where the calculated values are given as dots and lines are cubic spline between these discrete values. The differences between methods are obvious. The symmetry class is captured correctly in all methods, also they show the same tendency for increasing R/l. Of course, $R/l = 0$ means a homogeneous material without hard (stiff) inclusions, hence all second order parameters vanish. As R/l starts increasing, methods' solutions differ in their behavior as well. All parameters tend to change, AH predicts higher strain gradient contribution than CH. Especially the boundary layer as investigated in [43] occurring in strain gradient elasticity may lead to differing results depending on the used homogenization method's boundary conditions for RVE.

In both AH and FFT, the problem is based on an asymptotic expansion going back to [44] on 1D analysis. The main difference is a so-called correction term discussed in [45] and then applied in [32] and in AH implementation herein. This term contributes for a self-consistency that the size of the RVE (one or more unit cell) becomes irrelevant. FFT uses an efficient solution procedure, possible for rectangular RVE but AH may have irregular shapes owing to the rather general solution method of the finite element method.

Table 39.1: Determined parameters by using different techniques, asymptotic homogenization (AH) and computational homogenization (CH).

Parameter	Method	Value
D^{M}_{111111}	AH	0.122 N
	CH	0.018 N
D^{M}_{111221}	AH	0.022 N
	CH	-0.006 N
D^{M}_{111122}	AH	-0.017 N
	CH	0.002 N
D^{M}_{221221}	AH	-0.027 N
	CH	0.005 N
D^{M}_{221122}	AH	0.007 N
	CH	-0.0002 N
D^{M}_{121121}	AH	0.016 N
	CH	0.009 N

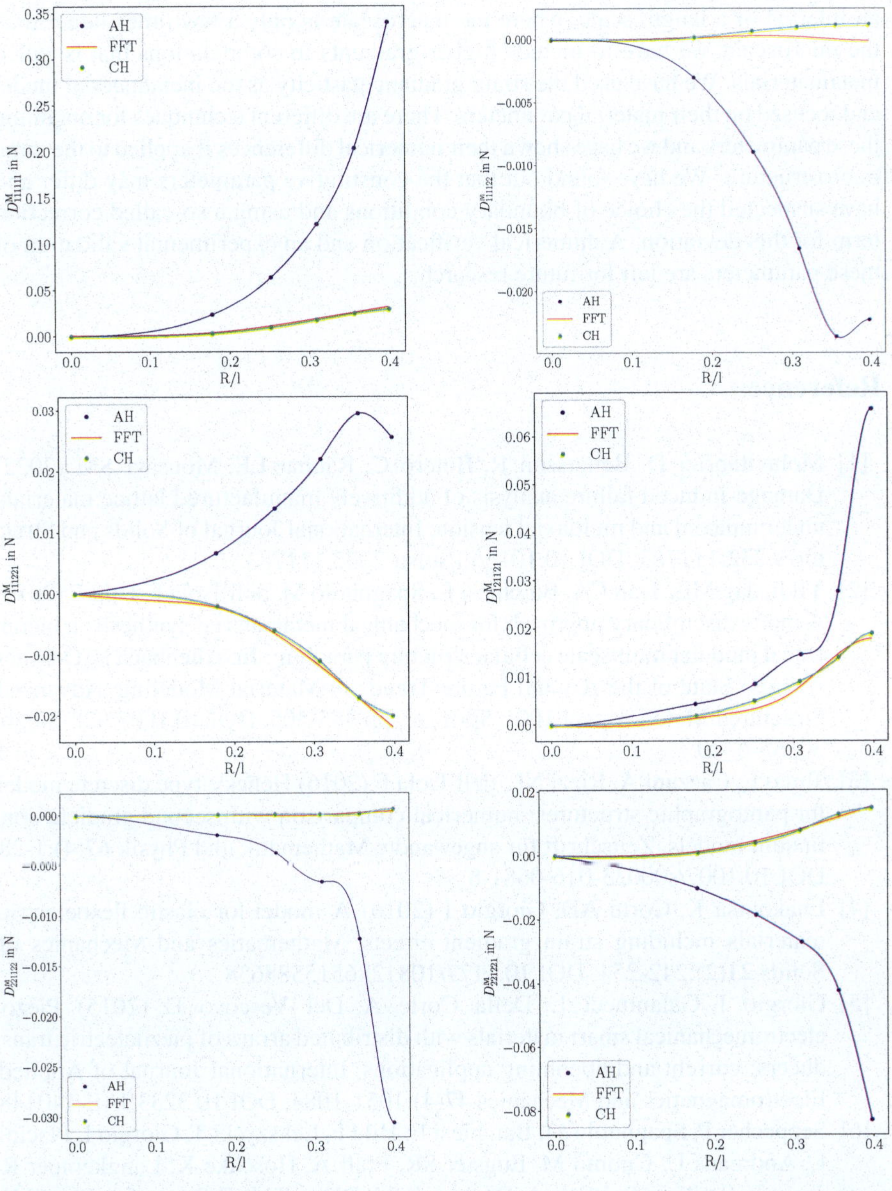

Fig. 39.3: Comparison between asymptotic homogenization (AH), computational homogenization (CH), and asymptotic analysis (FFT).

39.5 Conclusion

We stress that multiscale materials are in use partly because of technological advances in additive manufacturing. With the increasing complexity of the microstructure, for an interval of a length scale, where the macroscale is only a few times larger than the microscale, we have to include higher gradients in computations that is called metamaterials. We have used the strain gradient elasticity as the metamaterial model and focused on their material parameters. There are different techniques for obtaining these parameters and we have shown their numerical differences if applied to the same microstructure. We have found out that the constitutive parameters may differ and have suspected the choice of boundary conditions and using a so-called correction term for this deviation. A numerical verification and an experimental validation of these parameters are left for future research.

References

[1] Molavitabrizi D, Bengtsson R, Botero C, Rännar LE, Mousavi SM (2022) Damage-induced failure analysis of additively manufactured lattice materials under uniaxial and multiaxial tension, International Journal of Solids and Structures **252**:111783. DOI 10.1016/j.ijsolstr.2022.111783

[2] Yildizdag ME, Tran CA, Barchiesi E, Spagnuolo M, dell'Isola F, Hild F (2019) A multi-disciplinary approach for mechanical metamaterial synthesis: a hierarchical modular multiscale cellular structure paradigm. In: Altenbach H, Öchsner A (Eds) State of the Art and Future Trends in Material Modeling, Advanced Structured Materials, vol.100), Springer, pp 485-505. DOI 10.1007/978-3-030-30355-6_20

[3] Turco E, Cazzani A, Rizzi NL, dell'Isola F (2016) Hencky-type discrete model for pantographic structures: numerical comparison with second gradient continuum models, Zeitschrift für angewandte Mathematik und Physik **67**(4):1-28. DOI 10.1007/s00033-016-0681-8

[4] Enakoutsa K, Corte AD, Giorgio I (2016) A model for elastic flexoelectric materials including strain gradient effects, Mathematics and Mechanics of Solids **21**(2):242-254. DOI 10.1177/1081286515588638

[5] Giorgio I, Galantucci L, Della Corte A, Del Vescovo D (2015) Piezo-electromechanical smart materials with distributed arrays of piezoelectric transducers: current and upcoming applications, International Journal of Applied Electromagnetics and Mechanics **47**(4):1051-1084. DOI 10.3233/JAE-140148

[6] Seppecher P, Spagnuolo M, Barchiesi E, Hild F, Lekszycki T, Giorgio I, Placidi L, Andreaus U, Cuomo M, Eugster SR, Pfaff A, Hoschke K, Langkemper R, Turco E, Sarikaya R, Misra A, De Angelo M, D'Annibale F, Bouterf A, Pinelli X, Misra A, Desmorat B, Pawlikowski M, Dupuy C, Scerrato D, Peyre P, Laudato M, Manzari L, Göransson P, Hesch C, Hesch S, Franciosi P, Dirrenberger J, Maurin F, Vangelatos Z, Grigoropoulos C, Melissinaki V, Farsari M, Müller W, Abali BE, Liebold C, Ganzosch G, Harrison P, Drobnicki R, Igumnov L, Alzahrani F, Hayat T (2019) Advances in pantographic structures: design, manufacturing, models, experiments and image analyses, Continuum Mechanics and Thermodynamics **31**(4):1231-1282. DOI 10.1007/s00161-019-00806-x

[7] Golaszewski M, Grygoruk R, Giorgio I, Laudato M, Cosmo FD (2019) Metamaterials with relative displacements in their microstructure: technological challenges in 3d printing, experiments and numerical predictions, Continuum Mechanics and Thermodynamics **31**(4):1015-1034.DOI 10.1007/s00161-018-0692-0

[8] dell'Isola F, Seppecher P, Alibert JJ, Lekszycki T, Grygoruk R, Pawlikowski M, Steigmann D, Giorgio I, Andreaus U, Turco E, Gołaszewski M, Rizzi N, Boutin C, Eremeyev VA, Misra A, Placidi L, Barchiesi E, Greco L, Cuomo M, Cazzani A, Corte AD, Battista A, Scerrato D, Zurba IE, Rahali Y, Ganghoffer JF, Müller W, Ganzosch G, Spagnuolo M, Pfaff A, Barcz K, Hoschke K, Neggers J, Hild F (2019) Pantographic metamaterials: an example of mathematically driven design and of its technological challenges, Continuum Mechanics and Thermodynamics **31**(4):851-884. DOI 10.1007/s00161-018-0689-8

[9] dell'Isola F, Giorgio I, Pawlikowski M, Rizzi N.L. (2016) Large deformations of planar extensible beams and pantographic lattices: heuristic homogenization, experimental and numerical examples of equilibrium, Proceedings of the Royal Society A: Mathematical, Physical and Engineering Sciences **472**(2185):20150,790. DOI 10.1098/rspa.2015.0790

[10] Branecka N, Yildizdag ME, Ciallella A, Giorgio I (2022) Bone remodeling process based on hydrostatic and deviatoric strain mechano-sensing, Biomimetics **7**(2):59. DOI 10.3390/biomimetics7020059

[11] Maggi A, Li H, Greer JR (2017) Three-dimensional nano-architected scaffolds with tunable stiffness for efficient bone tissue growth, Acta biomaterialia **63**:294-305. DOI 10.1016/j.actbio.2017.09.007

[12] Giorgio I, Andreaus U, Lekszycki T, Della Corte A (2017) The influence of different geometries of matrix/scaffold on the remodeling process of a bone and bioresorbable material mixture with voids, Mathematics and Mechanics of Solids **22**(5):969-987. DOI 10.1177/1081286515616052

[13] De Angelo M, Barchiesi E, Giorgio I, Abali BE (2019) Numerical identification of constitutive parameters in reduced-order bi-dimensional models for pantographic structures: application to out-of-plane buckling, Archive of Applied Mechanics **89**(7):1333-1358. DOI 10.1007/s00419-018-01506-90

[14] Barchiesi E, Spagnuolo M, Placidi L (2019) Mechanical metamaterials: a state of the art, Mathematics and Mechanics of Solids **24**(1):212-234. DOI 10.1177/1081286517735695

[15] Del Vescovo D, Giorgio I (2014) Dynamic problems for metamaterials: review of existing models and ideas for further research, International Journal of Engineering Science **80**:153-172. DOI 10.1016/j.ijengsci.2014.02.022 (2014)

[16] dell'Isola F, Lekszycki T, Pawlikowski M, Grygoruk R, Greco L (2015) Designing a light fabric metamaterial being highly macroscopically tough under directional extension: first experimental evidence, Zeitschrift für angewandte Mathematik und Physik **66**(6):3473-3498. DOI 10.1007/s00033-015-0556-4

[17] Vazic B, Abali BE, Yang H, Newell P (2021) Mechanical analysis of heterogeneous materials with higher-order parameters, Engineering with Computers. DOI 10.1007/s00366-021-01555-9

[18] Abali BE, Vazic B, Newell P (2022) Influence of microstructure on size effect for metamaterials applied in composite structures, Mechanics Research Communications **122**:103,877. DOI 10.1016/j.mechrescom.2022.103877

[19] Mindlin RD, Eshel N (1968) On first strain-gradient theories in linear elasticity, International Journal of Solids and Structures **4**(1):109-124. DOI 10.1016/0020-7683(68)90036-X

[20] Yildizdag ME, Placidi L, Turco E (2022) Modeling and numerical investigation of damage behavior in pantographic layers using a hemivariational formulation adapted for a Hencky-type discrete model, Continuum Mechanics and Thermodynamics. DOI 10.1007/s00161-022-01154-z

[21] Eremeyev VA, dell'Isola F (2018) A note on reduced strain gradient elasticity, In: Generalized Models and Non-classical Approaches in Complex Materials 1, pp 301-310, Springer, Cham. DOI 10.1007/978-3-319-72440-9_15

[22] Mandadapu KK, Abali BE, Papadopoulos P (2021) On the polar nature and invariance properties of a thermomechanical theory for continuum-on-continuum homogenization, Mathematics and Mechanics of Solids **26**(11):1581-1598. DOI 10.1177/1081286521994323

[23] dell'Isola F, Della Corte A, Giorgio I (2017) Higher-gradient continua: The legacy of piola, mindlin, sedov and toupin and some future research perspectives, Mathematics and Mechanics of Solids **22**(4):852-872. DOI 10.1177/1081286515616034

[24] dell'Isola F, Giorgio I, Placidi L, Spagnuolo M, Peyre P, Dupuy C, Dirrenberger J, Pawlikowski M, Igumnov LA (2019) Pantographic metamaterials: A view towards applications, Materials Physics and Mechanics **42**:637-645. DOI 10.18720/MPM.4252019_17

[25] Eremeyev VA (2016) On effective properties of materials at the nano-and microscales considering surface effects, Acta Mechanica **227**(1):29-42. DOI 10.1007/s00707-015-1427-y

[26] Abali BE, Yang H, Papadopoulos P (2019) A computational approach for determination of parameters in generalized mechanics, In: Higher Gradient Materials and Related Generalized Continua, pp 1-18, Springer, Cham. DOI 10.1007/978-3-030-30406-5_1

[27] Yvonnet J, Auffray N, Monchiet V (2020) Computational second-order homogenization of materials with effective anisotropic strain-gradient behavior, International Journal of Solids and Structures **191**:434-448. DOI 10.1016/j.ijsolstr.2020.01.006

[28] Altenbach H, Forest S (2016) Generalized Continua as Models for Classical and Advanced Materials, Springer, Cham. DOI 10.1007/978-3-319-31721-2

[29] Dos Reis F, Ganghoffer JF (2012) Construction of micropolar continua from the asymptotic homogenization of beam lattices, Computers & Structures **112**:354-363. DOI 10.1016/j.compstruc.2012.08.006

[30] Solyaev Y (2022) Self-consistent assessments for the effective properties of two-phase composites within strain gradient elasticity, Mechanics of Materials **169**:104,321. DOI 10.1016/j.mechmat.2022.104321

[31] Areias P, Melicio R, Carapau F, Carrilho Lopes J(2022) Finite gradient models with enriched RBF-based interpolation, Mathematics **2022**(10):2876. DOI 10.3390/math10162876
[32] Barboura S, Li J (2018) Establishment of strain gradient constitutive relations by using asymptotic analysis and the finite element method for complex periodic microstructures, International Journal of Solids and Structures **136-137**:60-76. DOI 10.1016/j.ijsolstr.2017.12.003
[33] Abali BE, Barchiesi E (2021) Additive manufacturing introduced substructure and computational determination of metamaterials parameters by means of the asymptotic homogenization, Continuum Mechanics and Thermodynamics **33**(4):993-1009. DOI 10.1007/s00161-020-00941-w
[34] Aydin G, Yildizdag ME, Abali BE (2022) Strain-gradient modeling and computation of 3-D printed metamaterials for verifying constitutive parameters determined by asymptotic homogenization, In: Theoretical Analyses, Computations, and Experiments of Multiscale Materials, pp 343-357. Springer, Cham. DOI 10.1007/978-3-031-04548-6_16
[35] Aydin G, Sarar BC, Yildizdag ME, Abali BE (2022) Investigating infill density and pattern effects in additive manufacturing by characterizing metamaterials along the strain-gradient theory, Mathematics and Mechanics of Solids **27**(10):2002-2016. DOI 10.1177/10812865221100978
[36] Nazarenko L, Glüge R, Altenbach H (2021) Uniqueness theorem in coupled strain gradient elasticity with mixed boundary conditions, Continuum Mechanics and Thermodynamics **34**(1):93-106. DOI 10.1007/s00161-021-01048-6
[37] Nazarenko L, Glüge R, Altenbach H (2021) Positive definiteness in coupled strain gradient elasticity, Continuum Mechanics and Thermodynamics **33**(3):713-725. DOI 10.1007/s00161-020-00949-2
[38] Tran TH, Monchiet V, Bonnet G (2012) A micromechanics-based approach for the derivation of constitutive elastic coefficients of strain-gradient media, International Journal of Solids and Structures **49**(5):783-792. DOI 10.1016/j.ijsolstr.2011.11.017
[39] Auffray N, Bouchet R, Brechet Y (2010) Strain gradient elastic homogenization of bidimensional cellular media, International Journal of Solids and Structures **47**(13):1698-1710. DOI 10.1016/j.ijsolstr.2010.03.011
[40] Auffray N, Bouchet R, Brechet Y (2009) Derivation of anisotropic matrix for bidimensional strain-gradient elasticity behavior, International Journal of Solids and Structures **46**(2):440-454. DOI 10.1016/j.ijsolstr.2008.09.009
[41] Auffray N, Dirrenberger J, Rosi G (2015) A complete description of bidimensional anisotropic strain-gradient elasticity, International Journal of Solids and Structures **69**:195-206. DOI 10.1016/j.ijsolstr.2015.04.036
[42] Yang H, Abali BE, Müller WH, Barboura S, Li J (2022) Verification of asymptotic homogenization method developed for periodic architected materials in strain gradient continuum, International Journal of Solids and Structures **238**:111,386. DOI 10.1016/j.ijsolstr.2021.111386
[43] Barchiesi E, Ciallella A, Giorgio I (2022) On boundary layers observed in some 1d second-gradient theories, In: Theoretical Analyses, Computations,

and Experiments of Multiscale Materials, pp 359-376. Springer, Cham. DOI 10.1007/978-3-031-04548-6_17
[44] Boutin C (1996) Microstructural effects in elastic composites, International Journal of Solids and Structures **33**(7):1023-1051. DOI 10.1016/0020-7683(95)00089-5
[45] Li J (2011) Establishment of strain gradient constitutive relations by homogenization, Comptes Rendus Mécanique **339**(4):235-244. DOI 10.1016/j.ijsolstr.2017.12.003

Chapter 40
Buckling of Rectangular Micropolar Plate with Prestressed Coatings

Denis N. Sheydakov

Abstract Within the framework of the nonlinear Cosserat continuum, the stability of a rectangular plate with coatings (sandwich plate) is studied. It is assumed that the coatings were attached to the plate after preliminary deformations of biaxial tension-compression and contain internal stresses. When determining the subcritical state in the plate and the coatings, writing of constitutive relations with respect to different reference configurations is used. For the physically linear model of a micropolar material, the equations of neutral equilibrium are derived, which describes the behavior of the considered sandwich plate in a perturbed state. Using a special substitution, the linearized boundary value problems are formulated both for a general case of a sandwich plate and for a plate with identical top and bottom coatings. By solving these problems numerically for specific materials, the critical curves and corresponding buckling modes can be found, and the stability regions can be constructed in the plane of loading parameters.

Key words: Nonlinear elasticity · Deformation stability · Micropolar medium · Internal stresses · Rectangular plate · Prestressed coating

40.1 Introduction

Advanced materials with a complex microstructure are widely used in the modern automotive and aircraft industries. However, their behavior quite often cannot be adequately described within the framework of the classical elasticity theory due to the absence of internal size parameters. One approach to account for the microstructure influence is to use the model of a micropolar continuum (Cosserat continuum) [1]-[6],

---

Denis N. Sheydakov
Southern Scientific Centre of the Russian Academy of Sciences (SSC RAS), Chekhov Ave. 41, 344006 Rostov-on-Don, Russian Federation,
e-mail: sheidakov@mail.ru

H. Altenbach et al. (eds.), *Sixty Shades of Generalized Continua*,
Advanced Structured Materials 170, https://doi.org/10.1007/978-3-031-26186-2_40

i.e., medium with couple stresses and rotational degrees of freedom. The Cosserat continuum has been successfully applied for modeling granular and powder-like materials [7, 8], metal and polymer foams [9, 10], polycrystalline and composite materials [11, 12], bones [13, 14], beam-lattice materials [15, 16], masonry [17, 18], etc. It allows, in particular, describing the size effects observed experimentally for these materials [19, 20].

When analyzing the behavior of composites micropolar bodies, it is important to take into account the influence of internal stresses. The prestressed inclusions can appear in the process of assembling a composite body due to the plastic strains, heating, phase transitions, etc., or can be created artificially. A distinctive feature of such bodies with internal stresses is the lack of a unified natural (stress-free) reference configuration. For this reason, when deriving the governing equations for different parts of a composite body, the writing of the constitutive relations with respect to different reference configurations is used [21]-[25].

The present paper is dedicated to studying the stability of composite nonlinearly elastic bodies with a complex microstructure and internal stresses. Some problems on elastic buckling of bodies with prestressed parts and micropolar bodies were considered previously in [15], [26]-[35]. Within the Cosserat continuum model, we studied the bifurcation of equilibrium for a quite common structural element – a rectangular plate with prestressed coatings. The static Euler method was used for the stability analysis. Within this method, the critical values of loading parameters are determined from the condition of existence for non-trivial solutions of a linearized boundary value problem.

40.2 Micropolar Plate with Prestressed Coatings

Within the nonlinear Cosserat continuum model, we consider a rectangular plate whose side lengths in the natural (stress-free) state are b_1 and b_2, and the thickness is $2h$. The plate has top and bottom coatings in the form of rectangular panels. These panels are attached to the plate after their preliminary deformation. We assume that the coatings experience initial deformations of biaxial tension-compression, described by the following relations (hereinafter, the superscripts "+" and "–" will denote the quantities related to the top coating and the bottom coating, respectively) [31, 36]:

$$\begin{aligned} x_k &= a_k^{\pm}\zeta_k^{\pm}, \qquad 0 \le \zeta_k^{\pm} \le b_k^{\pm}, \quad a_k^{\pm}b_k^{\pm} = b_k, \quad k = 1,2, \\ x_3 &= a_3^{\pm}\zeta_3^{\pm} + \theta^{\pm}, \qquad 0 \le \zeta_3^{\pm} \le h_0^{\pm}, \quad \theta^- = -a_3^- h_0^- - h, \quad \theta^+ = h, \end{aligned} \tag{40.1}$$

$$\boldsymbol{C}_0^{\pm} = a_1^{\pm}\boldsymbol{e}_1 \otimes \boldsymbol{e}_1 + a_2^{\pm}\boldsymbol{e}_2 \otimes \boldsymbol{e}_2 + a_3^{\pm}\boldsymbol{e}_3 \otimes \boldsymbol{e}_3, \quad \boldsymbol{H}_0^{\pm} = \boldsymbol{I} = \boldsymbol{e}_1 \otimes \boldsymbol{e}_1 + \boldsymbol{e}_2 \otimes \boldsymbol{e}_2 + \boldsymbol{e}_3 \otimes \boldsymbol{e}_3, \tag{40.2}$$

where $\zeta_1^+, \zeta_2^+, \zeta_3^+$ and $\zeta_1^-, \zeta_2^-, \zeta_3^-$ are the Cartesian coordinates in natural reference configurations χ^+ and χ^- of rectangular panels acting as the top and bottom coatings, b_1^+, b_2^+, h_0^+ and b_1^-, b_2^-, h_0^- are the side lengths and thicknesses of these panels before the deformation; x_1, x_2, x_3 are the Cartesian coordinates in a prestressed state χ;

$\{\boldsymbol{e}_1,\boldsymbol{e}_2,\boldsymbol{e}_3\}$ is the orthonormal vector basis of the Cartesian coordinates; a_1^+,a_2^+,a_3^+ and a_1^-,a_2^-,a_3^- are the coefficients of tension-compression of the coating panels along the coordinate axes; $\boldsymbol{C}_0^+,\boldsymbol{H}_0^+$ and $\boldsymbol{C}_0^-,\boldsymbol{H}_0^-$ are the deformation gradients and the microrotation tensors corresponding to the transition from the natural reference configuration to the prestressed state of the top coating and the bottom coating, respectively ($\chi^{\pm} \to \chi$); $\boldsymbol{I}$ is the unit tensor. The coefficients a_1^+,a_2^+ and a_1^-,a_2^- are assumed to be given, while the coefficients a_3^+ and a_3^- are determined from the absence of loads on the faces of the coating panels.

Next, we study the deformation of biaxial tension-compression for the considered micropolar plate with coatings (sandwich plate), choosing χ as the reference configuration, which is natural for the plate, but prestressed for the coatings. This deformation is described by the relations [31, 36, 37]

$$
\begin{gathered}
X_1 = \alpha_1 x_1, \quad 0 \le x_1 \le b_1; \qquad X_2 = \alpha_2 x_2, \quad 0 \le x_2 \le b_2,\\
X_3 = \begin{cases} f^-(x_3), & -h-h^- \le x_3 \le -h,\\ f(x_3), & |x_3| \le h,\\ f^+(x_3), & h \le x_3 \le h+h^+, \end{cases} \qquad h^{\pm} = a_3^{\pm} h_0^{\pm},
\end{gathered}
\tag{40.3}
$$

$$
\begin{aligned}
\boldsymbol{R} &= \alpha_1 x_1 \boldsymbol{e}_1 + \alpha_2 x_2 \boldsymbol{e}_2 + f(x_3)\,\boldsymbol{e}_3,\\
\boldsymbol{R}^{\pm} &= \alpha_1 x_1 \boldsymbol{e}_1 + \alpha_2 x_2 \boldsymbol{e}_2 + f^{\pm}(x_3)\,\boldsymbol{e}_3,\\
\boldsymbol{C} &= \overset{\circ}{\nabla}\boldsymbol{R} = \alpha_1 \boldsymbol{e}_1 \otimes \boldsymbol{e}_1 + \alpha_2 \boldsymbol{e}_2 \otimes \boldsymbol{e}_2 + \frac{df(x_3)}{dx_3}\boldsymbol{e}_3 \otimes \boldsymbol{e}_3,\\
\boldsymbol{C}^{\pm} &= \overset{\circ}{\nabla}\boldsymbol{R}^{\pm} = \alpha_1 \boldsymbol{e}_1 \otimes \boldsymbol{e}_1 + \alpha_2 \boldsymbol{e}_2 \otimes \boldsymbol{e}_2 + \frac{df^{\pm}(x_3)}{dx_3}\boldsymbol{e}_3 \otimes \boldsymbol{e}_3,\\
\boldsymbol{H} &= \boldsymbol{H}^{\pm} = \boldsymbol{e}_1 \otimes \boldsymbol{e}_1 + \boldsymbol{e}_2 \otimes \boldsymbol{e}_2 + \boldsymbol{e}_3 \otimes \boldsymbol{e}_3.
\end{aligned}
\tag{40.4}
$$

Here X_1,X_2,X_3 are Cartesian coordinates in the actual (deformed) configuration X; α_1 and α_2 are the given tension-compression coefficients along the axes X_1 and X_2, respectively; $f(x_3), f^+(x_3), f^-(x_3)$ are some unknown functions characterizing the thickness deformation of the sandwich plate; h^+ and h^- are the thicknesses of the prestressed coatings; $\overset{\circ}{\nabla}$ is a nabla-operator in the reference configuration χ; $\boldsymbol{R},\boldsymbol{R}^+,\boldsymbol{R}^-$, $\boldsymbol{C},\boldsymbol{C}^+,\boldsymbol{C}^-$ and $\boldsymbol{H},\boldsymbol{H}^+,\boldsymbol{H}^-$ are the radius-vectors, the deformation gradients and the microrotation tensors corresponding to the transition from the chosen reference to the actual configuration ($\chi \to$ X) of a micropolar plate with prestressed coatings.

It follows from relations (40.3) – (40.4) that the wryness tensors $\boldsymbol{L},\boldsymbol{L}^+,\boldsymbol{L}^-$ for the sandwich plate are equal to zero [38, 39, 40]

$$
\begin{aligned}
\boldsymbol{L} &= \frac{1}{2}\operatorname{tr}\left[\boldsymbol{H}\cdot\left(\overset{\circ}{\nabla}\times\boldsymbol{H}\right)^{\mathrm{T}}\right]\boldsymbol{I} - \boldsymbol{H}\cdot\left(\overset{\circ}{\nabla}\times\boldsymbol{H}\right)^{\mathrm{T}} = 0,\\
\boldsymbol{L}^{\pm} &= \frac{1}{2}\operatorname{tr}\left[\boldsymbol{H}^{\pm}\cdot\left(\overset{\circ}{\nabla}\times\boldsymbol{H}^{\pm}\right)^{\mathrm{T}}\right]\boldsymbol{I} - \boldsymbol{H}^{\pm}\cdot\left(\overset{\circ}{\nabla}\times\boldsymbol{H}^{\pm}\right)^{\mathrm{T}} = 0,
\end{aligned}
\tag{40.5}
$$

and the stretch tensors $\boldsymbol{Y},\boldsymbol{Y}^+,\boldsymbol{Y}^-$ are expressed as follows (hereinafter the $'$ denotes the derivative with respect to x_3)

$$\begin{aligned} \boldsymbol{Y} &= \boldsymbol{C}\cdot\boldsymbol{H}^{\mathrm{T}} = \alpha_1 \boldsymbol{e}_1\otimes\boldsymbol{e}_1 + \alpha_2\boldsymbol{e}_2\otimes\boldsymbol{e}_2 + f'\boldsymbol{e}_3\otimes\boldsymbol{e}_3, \\ \boldsymbol{Y}^{\pm} &= \boldsymbol{C}^{\pm}\cdot\boldsymbol{H}^{\pm\mathrm{T}} = \alpha_1 \boldsymbol{e}_1\otimes\boldsymbol{e}_1 + \alpha_2\boldsymbol{e}_2\otimes\boldsymbol{e}_2 + f^{\pm\prime}\boldsymbol{e}_3\otimes\boldsymbol{e}_3. \end{aligned} \tag{40.6}$$

We assume that the elastic properties of the plate ($|x_3| \le h$) and its top ($h \le x_3 \le h + h^+$) and bottom ($-h - h^- \le x_3 \le -h$) coatings are described by the model of a physically linear micropolar material. In this case, the specific strain energy of the plate W is a quadratic form of the tensors $\boldsymbol{Y} - \boldsymbol{I}$ and $\boldsymbol{L}$ [20, 36, 41]

$$\begin{aligned} W = &\frac{1}{2}\lambda\mathrm{tr}^2(\boldsymbol{Y}-\boldsymbol{I}) + \frac{1}{2}(\mu+\kappa)\,\mathrm{tr}\left((\boldsymbol{Y}-\boldsymbol{I})\cdot(\boldsymbol{Y}-\boldsymbol{I})^{\mathrm{T}}\right) + \\ &+\frac{1}{2}\mu\mathrm{tr}(\boldsymbol{Y}-\boldsymbol{I})^2 + \frac{1}{2}\gamma_1\mathrm{tr}^2\boldsymbol{L} + \frac{1}{2}\gamma_2\mathrm{tr}\left(\boldsymbol{L}\cdot\boldsymbol{L}^{\mathrm{T}}\right) + \frac{1}{2}\gamma_3\mathrm{tr}\boldsymbol{L}^2, \end{aligned} \tag{40.7}$$

$$\mu+\kappa>0, \qquad \lambda+2\mu+\kappa>0, \qquad \gamma_2>0, \qquad \gamma_1+\gamma_2+\gamma_3>0,$$

and the constitutive relations for the Piola-type stress and couple stress tensors $\boldsymbol{D}$ and $\boldsymbol{G}$ at $|x_3| \le h$ have the form [31, 32]:

$$\begin{aligned} \boldsymbol{D} &= \frac{\partial W}{\partial \boldsymbol{Y}}\cdot\boldsymbol{H} = \left[\lambda\mathrm{tr}(\boldsymbol{Y}-\boldsymbol{I})\boldsymbol{I} + (\mu+\kappa)(\boldsymbol{Y}-\boldsymbol{I}) + \mu\left(\boldsymbol{Y}^{\mathrm{T}}-\boldsymbol{I}\right)\right]\cdot\boldsymbol{H}, \\ \boldsymbol{G} &= \frac{\partial W}{\partial \boldsymbol{L}}\cdot\boldsymbol{H} = \left[\gamma_1(\mathrm{tr}\boldsymbol{L})\boldsymbol{I} + \gamma_2\boldsymbol{L} + \gamma_3\boldsymbol{L}^{\mathrm{T}}\right]\cdot\boldsymbol{H}, \end{aligned} \tag{40.8}$$

where $\lambda, \mu, \kappa, \gamma_1, \gamma_2, \gamma_3$ are the micropolar elastic parameters of the plate material.

The specific strain energies W^+ and W^- for the top and bottom coatings are written as follows:

$$\begin{aligned} W^{\pm} = &\frac{1}{2}\lambda^{\pm}\mathrm{tr}^2(\boldsymbol{Y}_1^{\pm}-\boldsymbol{I}) + \frac{1}{2}(\mu^{\pm}+\kappa^{\pm})\,\mathrm{tr}\left((\boldsymbol{Y}_1^{\pm}-\boldsymbol{I})\cdot(\boldsymbol{Y}_1^{\pm}-\boldsymbol{I})^{\mathrm{T}}\right) + \\ &+\frac{1}{2}\mu^{\pm}\mathrm{tr}(\boldsymbol{Y}_1^{\pm}-\boldsymbol{I})^2 + \frac{1}{2}\gamma_1^{\pm}\mathrm{tr}^2\boldsymbol{L}_1^{\pm} + \frac{1}{2}\gamma_2^{\pm}\mathrm{tr}\left(\boldsymbol{L}_1^{\pm}\cdot\boldsymbol{L}_1^{\pm\mathrm{T}}\right) + \frac{1}{2}\gamma_3^{\pm}\mathrm{tr}\boldsymbol{L}_1^{\pm 2}, \end{aligned} \tag{40.9}$$

$$\mu^{\pm}+\kappa^{\pm}>0, \quad \lambda^{\pm}+2\mu^{\pm}+\kappa^{\pm}>0, \quad \gamma_2^{\pm}>0, \quad \gamma_1^{\pm}+\gamma_2^{\pm}+\gamma_3^{\pm}>0.$$

Here $\lambda^+, \mu^+, \kappa^+, \gamma_1^+, \gamma_2^+, \gamma_3^+$ and $\lambda^-, \mu^-, \kappa^-, \gamma_1^-, \gamma_2^-, \gamma_3^-$ are the micropolar elastic parameters for the materials of the coatings; $\boldsymbol{Y}_1^+, \boldsymbol{L}_1^+$ and $\boldsymbol{Y}_1^-, \boldsymbol{L}_1^-$ are the stretch tensors and the wryness tensors corresponding to the transition from the natural (stress-free) reference configuration to the actual (deformed) state of the top coating and the bottom coating, respectively ($\mathrm{X}^{\pm} \to \mathrm{X}$). To obtain expressions for these tensors, we use the formulae for transforming the deformation gradient and the microrotation tensor when changing the reference configuration [21, 42, 43, 44]:

$$\begin{aligned} &\boldsymbol{C}_1^{\pm} = \boldsymbol{C}_0^{\pm}\cdot\boldsymbol{C}^{\pm}, \qquad \boldsymbol{H}_1^{\pm} = \boldsymbol{H}_0^{\pm}\cdot\boldsymbol{H}^{\pm}, \\ &\boldsymbol{Y}_1^{\pm} = \boldsymbol{C}_0^{\pm}\cdot\boldsymbol{Y}^{\pm}\cdot\boldsymbol{H}_0^{\pm\mathrm{T}}, \qquad \boldsymbol{L}_1^{\pm} = \boldsymbol{L}_0^{\pm} + \boldsymbol{C}_0^{\pm}\cdot\boldsymbol{L}^{\pm}\cdot\boldsymbol{H}_0^{\pm\mathrm{T}}, \\ &\boldsymbol{L}_0^{\pm} = \frac{1}{2}\mathrm{tr}\left[\boldsymbol{H}_0^{\pm}\cdot\left(\mathring{\nabla}_{\pm}\times\boldsymbol{H}_0^{\pm}\right)^{\mathrm{T}}\right]\boldsymbol{I} - \boldsymbol{H}_0^{\pm}\cdot\left(\mathring{\nabla}_{\pm}\times\boldsymbol{H}_0^{\pm}\right)^{\mathrm{T}}, \end{aligned} \tag{40.10}$$

where $\boldsymbol{C}_1^+, \boldsymbol{H}_1^+$ and $\boldsymbol{C}_1^-, \boldsymbol{H}_1^-$ are the deformation gradients and the microrotation tensors for the transition from the corresponding natural reference configurations to the actual state of the coatings ($\chi^\pm \to X$); $\boldsymbol{L}_0^+$ and $\boldsymbol{L}_0^-$ are the wryness tensors corresponding to the transition from the natural to the prestressed reference configuration of the top coating and the bottom coating, respectively ($\chi^\pm \to \chi$); $\overset{\circ}{\nabla}_+$ and $\overset{\circ}{\nabla}_-$ are the nabla-operators in the natural reference configurations χ^+ and χ^-.

According to representation (40.9) for the coatings, the Piola-type stress and couple stress tensors $\boldsymbol{D}_1^+, \boldsymbol{D}_1^-$ and $\boldsymbol{G}_1^+, \boldsymbol{G}_1^-$ with respect to natural reference configurations χ^+ and χ^- have the form [31, 32]:

$$
\begin{aligned}
\boldsymbol{D}_1^\pm &= \left[\lambda^\pm \mathrm{tr}\,(\boldsymbol{Y}_1^\pm - \boldsymbol{I})\,\boldsymbol{I} + (\mu^\pm + \kappa^\pm)\,(\boldsymbol{Y}_1^\pm - \boldsymbol{I}) + \mu^\pm \left(\boldsymbol{Y}_1^{\pm\mathrm{T}} - \boldsymbol{I}\right)\right] \cdot \boldsymbol{H}_1^\pm, \\
\boldsymbol{G}_1^\pm &= \left[\gamma_1^\pm\,(\mathrm{tr}\boldsymbol{L}_1^\pm)\,\boldsymbol{I} + \gamma_2^\pm \boldsymbol{L}_1^\pm + \gamma_3^\pm \boldsymbol{L}_1^{\pm\mathrm{T}}\right] \cdot \boldsymbol{H}_1^\pm.
\end{aligned}
\tag{40.11}
$$

The constitutive relations for the coatings with respect to the prestressed reference configuration χ are derived using the formulae connecting the Piola-type stress and couple stress tensors in different reference configurations [21, 28, 36]

$$
\boldsymbol{D}^\pm = J_0^{\pm -1} \boldsymbol{C}_0^{\pm\mathrm{T}} \cdot \boldsymbol{D}_1^\pm, \qquad \boldsymbol{G}^\pm = J_0^{\pm -1} \boldsymbol{C}_0^{\pm\mathrm{T}} \cdot \boldsymbol{G}_1^\pm, \qquad J_0^\pm = \det \boldsymbol{C}_0^\pm. \tag{40.12}
$$

Thus, according to equations (40.10) – (40.12) for the top and bottom coatings, the Piola-type stress and couple stress tensors $\boldsymbol{D}^+, \boldsymbol{D}^-$ and $\boldsymbol{G}^+, \boldsymbol{G}^-$ relative to the prestressed reference configuration χ have the form [34, 35]:

$$
\begin{aligned}
\boldsymbol{D}^\pm &= J_0^{\pm -1} \boldsymbol{C}_0^{\pm\mathrm{T}} \cdot \left[\lambda^\pm \mathrm{tr}\left(\boldsymbol{C}_0^\pm \cdot \boldsymbol{Y}^\pm \cdot \boldsymbol{H}_0^{\pm\mathrm{T}} - \boldsymbol{I}\right)\boldsymbol{I} + \mu^\pm \left(\boldsymbol{H}_0^\pm \cdot \boldsymbol{Y}^{\pm\mathrm{T}} \cdot \boldsymbol{C}_0^{\pm\mathrm{T}} - \boldsymbol{I}\right) + \right. \\
&+ \left. (\mu^\pm + \kappa^\pm)\left(\boldsymbol{C}_0^\pm \cdot \boldsymbol{Y}^\pm \cdot \boldsymbol{H}_0^{\pm\mathrm{T}} - \boldsymbol{I}\right)\right] \cdot \boldsymbol{H}_0^\pm \cdot \boldsymbol{H}^\pm, \\
\boldsymbol{G}^\pm &= J_0^{\pm -1} \boldsymbol{C}_0^{\pm\mathrm{T}} \cdot \left[\gamma_1^\pm \mathrm{tr}\left(\boldsymbol{L}_0^\pm + \boldsymbol{C}_0^\pm \cdot \boldsymbol{L}^\pm \cdot \boldsymbol{H}_0^{\pm\mathrm{T}}\right)\boldsymbol{I} + \right. \\
&+ \left. \gamma_2^\pm \left(\boldsymbol{L}_0^\pm + \boldsymbol{C}_0^\pm \cdot \boldsymbol{L}^\pm \cdot \boldsymbol{H}_0^{\pm\mathrm{T}}\right) + \gamma_3^\pm \left(\boldsymbol{L}_0^{\pm\mathrm{T}} + \boldsymbol{H}_0^\pm \cdot \boldsymbol{L}^{\pm\mathrm{T}} \cdot \boldsymbol{C}_0^{\pm\mathrm{T}}\right)\right] \cdot \boldsymbol{H}_0^\pm \cdot \boldsymbol{H}^\pm.
\end{aligned}
\tag{40.13}
$$

It follows from expressions (40.2), (40.4) – (40.6), (40.8), (40.10), (40.13) that the Piola-type couple stress tensors $\boldsymbol{G}, \boldsymbol{G}^+, \boldsymbol{G}^-$ are equal to zero for the considered deformation (40.3) of a micropolar plate with prestressed coatings, and the Piola-type stress tensors $\boldsymbol{D}, \boldsymbol{D}^+, \boldsymbol{D}^-$ are

$$
\begin{aligned}
\boldsymbol{D} &= [\lambda s + \psi\,(\alpha_1 - 1)]\,\boldsymbol{e}_1 \otimes \boldsymbol{e}_1 + [\lambda s + \psi\,(\alpha_2 - 1)]\,\boldsymbol{e}_2 \otimes \boldsymbol{e}_2 + \\
&+ [\lambda s + \psi\,(f' - 1)]\,\boldsymbol{e}_3 \otimes \boldsymbol{e}_3, \qquad \psi = 2\mu + \kappa, \\
\boldsymbol{D}^\pm &= \frac{\lambda^\pm s^\pm + \psi^\pm\,(\alpha_1 a_1^\pm - 1)}{a_2^\pm a_3^\pm}\boldsymbol{e}_1 \otimes \boldsymbol{e}_1 + \frac{\lambda^\pm s^\pm + \psi^\pm\,(\alpha_2 a_2^\pm - 1)}{a_1^\pm a_3^\pm}\boldsymbol{e}_2 \otimes \boldsymbol{e}_2 + \\
&+ \frac{\lambda^\pm s^\pm + \psi^\pm\,(f^{\pm\prime} a_3^\pm - 1)}{a_1^\pm a_2^\pm}\boldsymbol{e}_3 \otimes \boldsymbol{e}_3, \qquad \psi^\pm = 2\mu^\pm + \kappa^\pm,
\end{aligned}
\tag{40.14}
$$

$$s = \alpha_1 + \alpha_2 + f' - 3, \qquad s^{\pm} = \alpha_1 a_1^{\pm} + \alpha_2 a_2^{\pm} + f^{\pm\prime} a_3^{\pm} - 3.$$

The equilibrium equations of the considered sandwich plate in the absence of mass forces and moments are written as follows [21, 37, 41]:

$$\begin{aligned}
&\overset{\circ}{\nabla}\cdot\boldsymbol{D}^- = \boldsymbol{0}, \quad \overset{\circ}{\nabla}\cdot\boldsymbol{G}^- + \left(\boldsymbol{C}^{-\mathrm{T}}\cdot\boldsymbol{D}^-\right)_{\times} = \boldsymbol{0}, \quad -h - h^- \le x_3 \le -h,\\
&\overset{\circ}{\nabla}\cdot\boldsymbol{D} = \boldsymbol{0}, \quad \overset{\circ}{\nabla}\cdot\boldsymbol{G} + \left(\boldsymbol{C}^{\mathrm{T}}\cdot\boldsymbol{D}\right)_{\times} = \boldsymbol{0}, \quad |x_3| \le h,\\
&\overset{\circ}{\nabla}\cdot\boldsymbol{D}^+ = \boldsymbol{0}, \quad \overset{\circ}{\nabla}\cdot\boldsymbol{G}^+ + \left(\boldsymbol{C}^{+\mathrm{T}}\cdot\boldsymbol{D}^+\right)_{\times} = \boldsymbol{0}, \quad h \le x_3 \le h + h^+.
\end{aligned} \tag{40.15}$$

The symbol $_\times$ represents the vector invariant of a second-order tensor. The boundary conditions

$$\begin{aligned}
&\boldsymbol{e}_3\cdot\boldsymbol{D}^-\big|_{x_3=-h-h^-} = \boldsymbol{0}, \qquad \boldsymbol{e}_3\cdot\boldsymbol{D}^+\big|_{x_3=h+h^+} = \boldsymbol{0},\\
&\boldsymbol{e}_3\cdot\boldsymbol{D}\big|_{x_3=-h} = \boldsymbol{e}_3\cdot\boldsymbol{D}^-\big|_{x_3=-h}, \qquad \boldsymbol{e}_3\cdot\boldsymbol{D}\big|_{x_3=h} = \boldsymbol{e}_3\cdot\boldsymbol{D}^+\big|_{x_3=h},\\
&f(-h) = f^-(-h), \qquad f(h) = f^+(h), \qquad f(0) = 0,
\end{aligned} \tag{40.16}$$

express the absence of loads on the surfaces of the coatings ($x_3 = \pm h \pm h^{\pm}$), the rigid coupling of the coatings with the plate ($x_3 = \pm h$), and the absence of vertical displacement at $x_3 = 0$.

By solving boundary value problem (40.15), (40.16) while taking into account relations (40.4), (40.14), we find the unknown functions $f(x_3), f^+(x_3), f^-(x_3)$:

$$\begin{aligned}
&f(x_3) = \alpha_3 x_3, \qquad f^{\pm}(x_3) = \alpha_3^{\pm} x_3 \pm h\left(\alpha_3 - \alpha_3^{\pm}\right),\\
&\alpha_3 = \frac{\lambda(3 - \alpha_1 - \alpha_2) + \psi}{\lambda + \psi}, \qquad \alpha_3^{\pm} = \frac{\lambda^{\pm}\left(3 - \alpha_1 a_1^{\pm} - \alpha_2 a_2^{\pm}\right) + \psi^{\pm}}{a_3^{\pm}\left(\lambda^{\pm} + \psi^{\pm}\right)}.
\end{aligned} \tag{40.17}$$

The coefficients a_3^+ and a_3^- in the case of a physically linear micropolar material (9) are expressed as follows [21]:

$$a_3^{\pm} = 1 - \frac{\lambda^{\pm}\left(a_1^{\pm} + a_2^{\pm} - 2\right)}{\lambda^{\pm} + \psi^{\pm}}.$$

40.3 Equations of Neutral Equilibrium

We assume that in addition to the above-described subcritical state of equilibrium X for the micropolar rectangular plate with prestressed coatings, there is an infinitely close perturbed equilibrium state $\tilde{\mathrm{X}}$ under the same external loads, which is determined by the radius-vectors

$$\tilde{\boldsymbol{R}} = \boldsymbol{R} + \eta\boldsymbol{v}, \quad \tilde{\boldsymbol{R}}^{\pm} = \boldsymbol{R}^{\pm} + \eta\boldsymbol{v}^{\pm}$$

and the microrotation tensors

$$\tilde{\boldsymbol{H}} = \boldsymbol{H} - \eta\boldsymbol{H}\times\boldsymbol{\omega}, \quad \tilde{\boldsymbol{H}}^{\pm} = \boldsymbol{H}^{\pm} - \eta\boldsymbol{H}^{\pm}\times\boldsymbol{\omega}^{\pm}.$$

Here η is a small parameter; $\boldsymbol{v}, \boldsymbol{v}^+, \boldsymbol{v}^-$ are the vectors of additional displacements; $\boldsymbol{\omega}, \boldsymbol{\omega}^+, \boldsymbol{\omega}^-$ are the linear incremental rotation vectors, which characterize the small rotation of the micropolar medium particles, measured from the subcritical state X. This perturbed state is described by the equations [41, 45, 46]:

$$\begin{aligned} &\overset{\circ}{\nabla}\cdot \boldsymbol{D}^{-\bullet} = \mathbf{0}, \quad \overset{\circ}{\nabla}\cdot \boldsymbol{G}^{-\bullet} + \left[\overset{\circ}{\nabla}\boldsymbol{v}^{-\mathrm{T}}\cdot \boldsymbol{D}^{-} + \boldsymbol{C}^{-\mathrm{T}}\cdot \boldsymbol{D}^{-\bullet}\right]_{\times} = \mathbf{0}, \\ &\overset{\circ}{\nabla}\cdot \boldsymbol{D}^{\bullet} = \mathbf{0}, \quad \overset{\circ}{\nabla}\cdot \boldsymbol{G}^{\bullet} + \left[\overset{\circ}{\nabla}\boldsymbol{v}^{\mathrm{T}}\cdot \boldsymbol{D} + \boldsymbol{C}^{\mathrm{T}}\cdot \boldsymbol{D}^{\bullet}\right]_{\times} = \mathbf{0}, \\ &\overset{\circ}{\nabla}\cdot \boldsymbol{D}^{+\bullet} = \mathbf{0}, \quad \overset{\circ}{\nabla}\cdot \boldsymbol{G}^{+\bullet} + \left[\overset{\circ}{\nabla}\boldsymbol{v}^{+\mathrm{T}}\cdot \boldsymbol{D}^{+} + \boldsymbol{C}^{+\mathrm{T}}\cdot \boldsymbol{D}^{+\bullet}\right]_{\times} = \mathbf{0}, \end{aligned} \tag{40.18}$$

where $\boldsymbol{D}^{\bullet}, \boldsymbol{D}^{+\bullet}, \boldsymbol{D}^{-\bullet}$ and $\boldsymbol{G}^{\bullet}, \boldsymbol{G}^{+\bullet}, \boldsymbol{G}^{-\bullet}$ are the linearized Piola-type stress and couple stress tensors for the plate and its top and bottom coatings. By linearizing constitutive relations (40.8), (40.13), we obtain the expressions for these tensors in the case of a physically linear micropolar material (40.7), (40.9) [34, 35]

$$\begin{aligned} \boldsymbol{D}^{\bullet} &= \left[\lambda\,(\mathrm{tr}\boldsymbol{Y}^{\bullet})\boldsymbol{I} + (\mu+\kappa)\boldsymbol{Y}^{\bullet} + \mu\boldsymbol{Y}^{\bullet\mathrm{T}}\right]\cdot \boldsymbol{H} - \\ &- \left[\lambda\mathrm{tr}\,(\boldsymbol{Y}-\boldsymbol{I})\boldsymbol{I} + (\mu+\kappa)\,(\boldsymbol{Y}-\boldsymbol{I}) + \mu\,(\boldsymbol{Y}^{\mathrm{T}}-\boldsymbol{I})\right]\cdot \boldsymbol{H}\times\boldsymbol{\omega}, \\ \boldsymbol{G}^{\bullet} &= \left[\gamma_1\,(\mathrm{tr}\boldsymbol{L}^{\bullet})\boldsymbol{I} + \gamma_2\boldsymbol{L}^{\bullet} + \gamma_3\boldsymbol{L}^{\bullet\mathrm{T}}\right]\cdot \boldsymbol{H} - \\ &- \left[\gamma_1\,(\mathrm{tr}\boldsymbol{L})\boldsymbol{I} + \gamma_2\boldsymbol{L} + \gamma_3\boldsymbol{L}^{\mathrm{T}}\right]\cdot \boldsymbol{H}\times\boldsymbol{\omega}, \end{aligned} \tag{40.19}$$

$$\begin{aligned} \boldsymbol{D}^{\pm\bullet} &= J_0^{\pm-1}\boldsymbol{C}_0^{\pm\mathrm{T}}\cdot\Big\{\Big[\lambda^{\pm}\mathrm{tr}\left(\boldsymbol{C}_0^{\pm}\cdot\boldsymbol{Y}^{\pm\bullet}\cdot\boldsymbol{H}_0^{\pm\mathrm{T}}\right)\boldsymbol{I} + \mu^{\pm}\boldsymbol{H}_0^{\pm}\cdot\boldsymbol{Y}^{\pm\bullet\mathrm{T}}\cdot\boldsymbol{C}_0^{\pm\mathrm{T}} + \\ &+ (\mu^{\pm}+\kappa^{\pm})\boldsymbol{C}_0^{\pm}\cdot\boldsymbol{Y}^{\pm\bullet}\cdot\boldsymbol{H}_0^{\pm\mathrm{T}}\Big]\cdot\boldsymbol{H}_0^{\pm}\cdot\boldsymbol{H}^{\pm} - \\ &- \Big[\lambda^{\pm}\mathrm{tr}\left(\boldsymbol{C}_0^{\pm}\cdot\boldsymbol{Y}^{\pm}\cdot\boldsymbol{H}_0^{\pm\mathrm{T}} - \boldsymbol{I}\right)\boldsymbol{I} + \mu^{\pm}\left(\boldsymbol{H}_0^{\pm}\cdot\boldsymbol{Y}^{\pm\mathrm{T}}\cdot\boldsymbol{C}_0^{\pm\mathrm{T}} - \boldsymbol{I}\right) + \\ &+ (\mu^{\pm}+\kappa^{\pm})\left(\boldsymbol{C}_0^{\pm}\cdot\boldsymbol{Y}^{\pm}\cdot\boldsymbol{H}_0^{\pm\mathrm{T}} - \boldsymbol{I}\right)\Big]\cdot\boldsymbol{H}_0^{\pm}\cdot\boldsymbol{H}^{\pm}\times\boldsymbol{\omega}^{\pm}\Big\}, \\ \boldsymbol{G}^{\pm\bullet} &= J_0^{\pm-1}\boldsymbol{C}_0^{\pm\mathrm{T}}\cdot\Big\{\Big[\gamma_1^{\pm}\mathrm{tr}\left(\boldsymbol{C}_0^{\pm}\cdot\boldsymbol{L}^{\pm\bullet}\cdot\boldsymbol{H}_0^{\pm\mathrm{T}}\right)\boldsymbol{I} + \gamma_2^{\pm}\boldsymbol{C}_0^{\pm}\cdot\boldsymbol{L}^{\pm\bullet}\cdot\boldsymbol{H}_0^{\pm\mathrm{T}} + \\ &+ \gamma_3^{\pm}\boldsymbol{H}_0^{\pm}\cdot\boldsymbol{L}^{\pm\bullet\mathrm{T}}\cdot\boldsymbol{C}_0^{\pm\mathrm{T}}\Big]\cdot\boldsymbol{H}_0^{\pm}\cdot\boldsymbol{H}^{\pm} - \\ &- \Big[\gamma_1^{\pm}\mathrm{tr}\left(\boldsymbol{L}_0^{\pm} + \boldsymbol{C}_0^{\pm}\cdot\boldsymbol{L}^{\pm}\cdot\boldsymbol{H}_0^{\pm\mathrm{T}}\right)\boldsymbol{I} + \gamma_2^{\pm}\left(\boldsymbol{L}_0^{\pm} + \boldsymbol{C}_0^{\pm}\cdot\boldsymbol{L}^{\pm}\cdot\boldsymbol{H}_0^{\pm\mathrm{T}}\right) + \\ &+ \gamma_3^{\pm}\left(\boldsymbol{L}_0^{\pm\mathrm{T}} + \boldsymbol{H}_0^{\pm}\cdot\boldsymbol{L}^{\pm\mathrm{T}}\cdot\boldsymbol{C}_0^{\pm\mathrm{T}}\right)\Big]\cdot\boldsymbol{H}_0^{\pm}\cdot\boldsymbol{H}^{\pm}\times\boldsymbol{\omega}^{\pm}\Big\}, \end{aligned} \tag{40.20}$$

$$\begin{aligned} \boldsymbol{Y}^{\bullet} &= \left(\overset{\circ}{\nabla}\boldsymbol{v} + \boldsymbol{C}\times\boldsymbol{\omega}\right)\cdot\boldsymbol{H}^{\mathrm{T}}, \qquad \boldsymbol{Y}^{\pm\bullet} = \left(\overset{\circ}{\nabla}\boldsymbol{v}^{\pm} + \boldsymbol{C}^{\pm}\times\boldsymbol{\omega}^{\pm}\right)\cdot\boldsymbol{H}^{\pm\mathrm{T}}, \\ \boldsymbol{L}^{\bullet} &= \overset{\circ}{\nabla}\boldsymbol{\omega}\cdot\boldsymbol{H}^{\mathrm{T}}, \qquad \boldsymbol{L}^{\pm\bullet} = \overset{\circ}{\nabla}\boldsymbol{\omega}^{\pm}\cdot\boldsymbol{H}^{\pm\mathrm{T}}. \end{aligned} \tag{40.21}$$

Here $\boldsymbol{Y}^{\bullet}, \boldsymbol{Y}^{+\bullet}, \boldsymbol{Y}^{-\bullet}$ are the linearized stretch tensors, and $\boldsymbol{L}^{\bullet}, \boldsymbol{L}^{+\bullet}, \boldsymbol{L}^{-\bullet}$ are the linearized wryness tensors for the considered sandwich plate.

Linearized boundary conditions on the surfaces of the coatings ($x_3 = \pm h \pm h^{\pm}$) and at the interfaces between the coatings and the plate ($x_3 = \pm h$) are written as follows [31]:

$$\begin{aligned} &\boldsymbol{e}_3 \cdot \boldsymbol{D}^{-\bullet}\big|_{x_3=-h-h^-} = \boldsymbol{0}, \qquad &\boldsymbol{e}_3 \cdot \boldsymbol{G}^{-\bullet}\big|_{x_3=-h-h^-} = \boldsymbol{0}, \\ &\boldsymbol{e}_3 \cdot \boldsymbol{D}^{+\bullet}\big|_{x_3=h+h^+} = \boldsymbol{0}, \qquad &\boldsymbol{e}_3 \cdot \boldsymbol{G}^{+\bullet}\big|_{x_3=h+h^+} = \boldsymbol{0}, \end{aligned} \tag{40.22}$$

$$\begin{aligned} &\boldsymbol{e}_3 \cdot \boldsymbol{D}^{\bullet}\big|_{x_3=-h} = \boldsymbol{e}_3 \cdot \boldsymbol{D}^{-\bullet}\big|_{x_3=-h}, \qquad &\boldsymbol{v}\big|_{x_3=-h} = \boldsymbol{v}^-\big|_{x_3=-h}, \\ &\boldsymbol{e}_3 \cdot \boldsymbol{G}^{\bullet}\big|_{x_3=-h} = \boldsymbol{e}_3 \cdot \boldsymbol{G}^{-\bullet}\big|_{x_3=-h}, \qquad &\boldsymbol{\omega}\big|_{x_3=-h} = \boldsymbol{\omega}^-\big|_{x_3=-h}, \\ &\boldsymbol{e}_3 \cdot \boldsymbol{D}^{\bullet}\big|_{x_3=h} = \boldsymbol{e}_3 \cdot \boldsymbol{D}^{+\bullet}\big|_{x_3=h}, \qquad &\boldsymbol{v}\big|_{x_3=h} = \boldsymbol{v}^+\big|_{x_3=h}, \\ &\boldsymbol{e}_3 \cdot \boldsymbol{G}^{\bullet}\big|_{x_3=h} = \boldsymbol{e}_3 \cdot \boldsymbol{G}^{+\bullet}\big|_{x_3=h}, \qquad &\boldsymbol{\omega}\big|_{x_3=h} = \boldsymbol{\omega}^+\big|_{x_3=h}. \end{aligned} \tag{40.23}$$

We assume that there is no friction at the ends of the considered sandwich plate ($x_1 = 0, b_1; x_2 = 0, b_2$), and constant normal displacements are given. This leads to the following linearized end conditions:

$$\begin{aligned} &\boldsymbol{e}_1 \cdot \boldsymbol{D}^{\bullet} \cdot \boldsymbol{e}_2\big|_{x_1=0,b_1} = \boldsymbol{e}_1 \cdot \boldsymbol{D}^{\bullet} \cdot \boldsymbol{e}_3\big|_{x_1=0,b_1} = \boldsymbol{e}_1 \cdot \boldsymbol{v}\big|_{x_1=0,b_1} = 0, \\ &\boldsymbol{e}_2 \cdot \boldsymbol{D}^{\bullet} \cdot \boldsymbol{e}_1\big|_{x_2=0,b_2} = \boldsymbol{e}_2 \cdot \boldsymbol{D}^{\bullet} \cdot \boldsymbol{e}_3\big|_{x_2=0,b_2} = \boldsymbol{e}_2 \cdot \boldsymbol{v}\big|_{x_2=0,b_2} = 0, \\ &\boldsymbol{e}_1 \cdot \boldsymbol{D}^{\pm\bullet} \cdot \boldsymbol{e}_2\big|_{x_1=0,b_1} = \boldsymbol{e}_1 \cdot \boldsymbol{D}^{\pm\bullet} \cdot \boldsymbol{e}_3\big|_{x_1=0,b_1} = \boldsymbol{e}_1 \cdot \boldsymbol{v}^{\pm}\big|_{x_1=0,b_1} = 0, \\ &\boldsymbol{e}_2 \cdot \boldsymbol{D}^{\pm\bullet} \cdot \boldsymbol{e}_1\big|_{x_2=0,b_2} = \boldsymbol{e}_2 \cdot \boldsymbol{D}^{\pm\bullet} \cdot \boldsymbol{e}_3\big|_{x_2=0,b_2} = \boldsymbol{e}_2 \cdot \boldsymbol{v}^{\pm}\big|_{x_2=0,b_2} = 0, \end{aligned} \tag{40.24}$$

$$\begin{aligned} &\boldsymbol{e}_1 \cdot \boldsymbol{G}^{\bullet} \cdot \boldsymbol{e}_1\big|_{x_1=0,b_1} = \boldsymbol{e}_2 \cdot \boldsymbol{\omega}\big|_{x_1=0,b_1} = \boldsymbol{e}_3 \cdot \boldsymbol{\omega}\big|_{x_1=0,b_1} = 0, \\ &\boldsymbol{e}_2 \cdot \boldsymbol{G}^{\bullet} \cdot \boldsymbol{e}_2\big|_{x_2=0,b_2} = \boldsymbol{e}_1 \cdot \boldsymbol{\omega}\big|_{x_2=0,b_2} = \boldsymbol{e}_3 \cdot \boldsymbol{\omega}\big|_{x_2=0,b_2} = 0, \\ &\boldsymbol{e}_1 \cdot \boldsymbol{G}^{\pm\bullet} \cdot \boldsymbol{e}_1\big|_{x_1=0,b_1} = \boldsymbol{e}_2 \cdot \boldsymbol{\omega}^{\pm}\big|_{x_1=0,b_1} = \boldsymbol{e}_3 \cdot \boldsymbol{\omega}^{\pm}\big|_{x_1=0,b_1} = 0, \\ &\boldsymbol{e}_2 \cdot \boldsymbol{G}^{\pm\bullet} \cdot \boldsymbol{e}_2\big|_{x_2=0,b_2} = \boldsymbol{e}_1 \cdot \boldsymbol{\omega}^{\pm}\big|_{x_2=0,b_2} = \boldsymbol{e}_3 \cdot \boldsymbol{\omega}^{\pm}\big|_{x_2=0,b_2} = 0. \end{aligned} \tag{40.25}$$

The vectors of additional displacements $\boldsymbol{v}, \boldsymbol{v}^+, \boldsymbol{v}^-$ and the vectors of incremental rotation $\boldsymbol{\omega}, \boldsymbol{\omega}^+, \boldsymbol{\omega}^-$ in Cartesian coordinates have the form:

$$\begin{aligned} &\boldsymbol{v} = v_1\boldsymbol{e}_1 + v_2\boldsymbol{e}_2 + v_3\boldsymbol{e}_3, \qquad &\boldsymbol{\omega} = \omega_1\boldsymbol{e}_1 + \omega_2\boldsymbol{e}_2 + \omega_3\boldsymbol{e}_3, \\ &\boldsymbol{v}^{\pm} = v_1^{\pm}\boldsymbol{e}_1 + v_2^{\pm}\boldsymbol{e}_2 + v_3^{\pm}\boldsymbol{e}_3, \qquad &\boldsymbol{\omega}^{\pm} = \omega_1^{\pm}\boldsymbol{e}_1 + \omega_2^{\pm}\boldsymbol{e}_2 + \omega_3^{\pm}\boldsymbol{e}_3. \end{aligned} \tag{40.26}$$

Taking into account relations (40.2), (40.4) – (40.6), (40.17), (40.19) – (40.21), (40.26), the components of the linearized Piola-type stress tensors $\boldsymbol{D}^{\bullet}, \boldsymbol{D}^{+\bullet}, \boldsymbol{D}^{-\bullet}$ are written as [31]:

$$\boldsymbol{D}^{\bullet} = \sum_{i,j=1}^{3} \tilde{D}_{ij}\boldsymbol{e}_i \otimes \boldsymbol{e}_j, \qquad \boldsymbol{D}^{\pm\bullet} = \sum_{i,j=1}^{3} \tilde{D}^{\pm}_{ij}\boldsymbol{e}_i \otimes \boldsymbol{e}_j, \tag{40.27}$$

$$\begin{aligned}
\tilde{D}_{11} &= (\xi+\tau)\frac{\partial v_1}{\partial x_1} + \lambda\frac{\partial v_2}{\partial x_2} + \lambda\frac{\partial v_3}{\partial x_3},\\
\tilde{D}_{22} &= \lambda\frac{\partial v_1}{\partial x_1} + (\xi+\tau)\frac{\partial v_2}{\partial x_2} + \lambda\frac{\partial v_3}{\partial x_3},\\
\tilde{D}_{33} &= \lambda\frac{\partial v_1}{\partial x_1} + \lambda\frac{\partial v_2}{\partial x_2} + (\xi+\tau)\frac{\partial v_3}{\partial x_3},
\end{aligned} \tag{40.28}$$

$$\begin{aligned}
\tilde{D}_{12} &= \tau\frac{\partial v_2}{\partial x_1} + \mu\frac{\partial v_1}{\partial x_2} + A_3\omega_3, \qquad & \tilde{D}_{13} &= \tau\frac{\partial v_3}{\partial x_1} + \mu\frac{\partial v_1}{\partial x_3} - A_2\omega_2,\\
\tilde{D}_{21} &= \tau\frac{\partial v_1}{\partial x_2} + \mu\frac{\partial v_2}{\partial x_1} - A_3\omega_3, \qquad & \tilde{D}_{23} &= \tau\frac{\partial v_3}{\partial x_2} + \mu\frac{\partial v_2}{\partial x_3} + A_1\omega_1,\\
\tilde{D}_{31} &= \tau\frac{\partial v_1}{\partial x_3} + \mu\frac{\partial v_3}{\partial x_1} + A_2\omega_2, \qquad & \tilde{D}_{32} &= \tau\frac{\partial v_2}{\partial x_3} + \mu\frac{\partial v_3}{\partial x_2} - A_1\omega_1
\end{aligned}$$

$$\begin{aligned}
\tilde{D}^{\pm}_{11} &= \frac{a^{\pm}_1(\xi^{\pm}+\tau^{\pm})}{a^{\pm}_2 a^{\pm}_3}\frac{\partial v^{\pm}_1}{\partial x_1} + \frac{\lambda^{\pm}}{a^{\pm}_3}\frac{\partial v^{\pm}_2}{\partial x_2} + \frac{\lambda^{\pm}}{a^{\pm}_2}\frac{\partial v^{\pm}_3}{\partial x_3},\\
\tilde{D}^{\pm}_{22} &= \frac{\lambda^{\pm}}{a^{\pm}_3}\frac{\partial v^{\pm}_1}{\partial x_1} + \frac{a^{\pm}_2(\xi^{\pm}+\tau^{\pm})}{a^{\pm}_1 a^{\pm}_3}\frac{\partial v^{\pm}_2}{\partial x_2} + \frac{\lambda^{\pm}}{a^{\pm}_1}\frac{\partial v^{\pm}_3}{\partial x_3},\\
\tilde{D}^{\pm}_{33} &= \frac{\lambda^{\pm}}{a^{\pm}_2}\frac{\partial v^{\pm}_1}{\partial x_1} + \frac{\lambda^{\pm}}{a^{\pm}_1}\frac{\partial v^{\pm}_2}{\partial x_2} + \frac{a^{\pm}_3(\xi^{\pm}+\tau^{\pm})}{a^{\pm}_1 a^{\pm}_2}\frac{\partial v^{\pm}_3}{\partial x_3},
\end{aligned}$$

$$\begin{aligned}
\tilde{D}^{\pm}_{12} &= \frac{a^{\pm}_1\tau^{\pm}}{a^{\pm}_2 a^{\pm}_3}\frac{\partial v^{\pm}_2}{\partial x_1} + \frac{\mu^{\pm}}{a^{\pm}_3}\frac{\partial v^{\pm}_1}{\partial x_2} + \frac{A^{\pm}_3\omega^{\pm}_3}{a^{\pm}_2 a^{\pm}_3},\\
\tilde{D}^{\pm}_{13} &= \frac{a^{\pm}_1\tau^{\pm}}{a^{\pm}_2 a^{\pm}_3}\frac{\partial v^{\pm}_3}{\partial x_1} + \frac{\mu^{\pm}}{a^{\pm}_2}\frac{\partial v^{\pm}_1}{\partial x_3} - \frac{A^{\pm}_2\omega^{\pm}_2}{a^{\pm}_2 a^{\pm}_3},\\
\tilde{D}^{\pm}_{21} &= \frac{a^{\pm}_2\tau^{\pm}}{a^{\pm}_1 a^{\pm}_3}\frac{\partial v^{\pm}_1}{\partial x_2} + \frac{\mu^{\pm}}{a^{\pm}_3}\frac{\partial v^{\pm}_2}{\partial x_1} - \frac{A^{\pm}_3\omega^{\pm}_3}{a^{\pm}_1 a^{\pm}_3},\\
\tilde{D}^{\pm}_{23} &= \frac{a^{\pm}_2\tau^{\pm}}{a^{\pm}_1 a^{\pm}_3}\frac{\partial v^{\pm}_3}{\partial x_2} + \frac{\mu^{\pm}}{a^{\pm}_1}\frac{\partial v^{\pm}_2}{\partial x_3} + \frac{A^{\pm}_1\omega^{\pm}_1}{a^{\pm}_1 a^{\pm}_3},\\
\tilde{D}^{\pm}_{31} &= \frac{a^{\pm}_3\tau^{\pm}}{a^{\pm}_1 a^{\pm}_2}\frac{\partial v^{\pm}_1}{\partial x_3} + \frac{\mu^{\pm}}{a^{\pm}_2}\frac{\partial v^{\pm}_3}{\partial x_1} + \frac{A^{\pm}_2\omega^{\pm}_2}{a^{\pm}_1 a^{\pm}_2},\\
\tilde{D}^{\pm}_{32} &= \frac{a^{\pm}_3\tau^{\pm}}{a^{\pm}_1 a^{\pm}_2}\frac{\partial v^{\pm}_2}{\partial x_3} + \frac{\mu^{\pm}}{a^{\pm}_1}\frac{\partial v^{\pm}_3}{\partial x_2} - \frac{A^{\pm}_1\omega^{\pm}_1}{a^{\pm}_1 a^{\pm}_2},
\end{aligned} \tag{40.29}$$

and the components of the linearized Piola-type couple stress tensors $\boldsymbol{G}^{\bullet}, \boldsymbol{G}^{+\bullet}, \boldsymbol{G}^{-\bullet}$ are

$$\boldsymbol{G}^{\bullet} = \sum_{i,j=1}^{3} \tilde{G}_{ij} \boldsymbol{e}_i \otimes \boldsymbol{e}_j, \qquad \boldsymbol{G}^{\pm\bullet} = \sum_{i,j=1}^{3} \tilde{G}_{ij}^{\pm} \boldsymbol{e}_i \otimes \boldsymbol{e}_j, \tag{40.30}$$

$$\begin{aligned}
\tilde{G}_{11} &= \gamma \frac{\partial \omega_1}{\partial x_1} + \gamma_1 \frac{\partial \omega_2}{\partial x_2} + \gamma_1 \frac{\partial \omega_3}{\partial x_3}, \\
\tilde{G}_{22} &= \gamma_1 \frac{\partial \omega_1}{\partial x_1} + \gamma \frac{\partial \omega_2}{\partial x_2} + \gamma_1 \frac{\partial \omega_3}{\partial x_3}, \\
\tilde{G}_{33} &= \gamma_1 \frac{\partial \omega_1}{\partial x_1} + \gamma_1 \frac{\partial \omega_2}{\partial x_2} + \gamma \frac{\partial \omega_3}{\partial x_3},
\end{aligned} \tag{40.31}$$

$$\begin{aligned}
\tilde{G}_{12} &= \gamma_2 \frac{\partial \omega_2}{\partial x_1} + \gamma_3 \frac{\partial \omega_1}{\partial x_2}, \qquad & \tilde{G}_{13} &= \gamma_2 \frac{\partial \omega_3}{\partial x_1} + \gamma_3 \frac{\partial \omega_1}{\partial x_3}, \\
\tilde{G}_{21} &= \gamma_2 \frac{\partial \omega_1}{\partial x_2} + \gamma_3 \frac{\partial \omega_2}{\partial x_1}, \qquad & \tilde{G}_{23} &= \gamma_2 \frac{\partial \omega_3}{\partial x_2} + \gamma_3 \frac{\partial \omega_2}{\partial x_3}, \\
\tilde{G}_{31} &= \gamma_2 \frac{\partial \omega_1}{\partial x_3} + \gamma_3 \frac{\partial \omega_3}{\partial x_1}, \qquad & \tilde{G}_{32} &= \gamma_2 \frac{\partial \omega_2}{\partial x_3} + \gamma_3 \frac{\partial \omega_3}{\partial x_2},
\end{aligned}$$

$$\begin{aligned}
\tilde{G}_{11}^{\pm} &= \frac{a_1^{\pm}\gamma^{\pm}}{a_2^{\pm}a_3^{\pm}} \frac{\partial \omega_1^{\pm}}{\partial x_1} + \frac{\gamma_1^{\pm}}{a_3^{\pm}} \frac{\partial \omega_2^{\pm}}{\partial x_2} + \frac{\gamma_1^{\pm}}{a_2^{\pm}} \frac{\partial \omega_3^{\pm}}{\partial x_3}, \\
\tilde{G}_{22}^{\pm} &= \frac{\gamma_1^{\pm}}{a_3^{\pm}} \frac{\partial \omega_1^{\pm}}{\partial x_1} + \frac{a_2^{\pm}\gamma^{\pm}}{a_1^{\pm}a_3^{\pm}} \frac{\partial \omega_2^{\pm}}{\partial x_2} + \frac{\gamma_1^{\pm}}{a_1^{\pm}} \frac{\partial \omega_3^{\pm}}{\partial x_3}, \\
\tilde{G}_{33}^{\pm} &= \frac{\gamma_1^{\pm}}{a_2^{\pm}} \frac{\partial \omega_1^{\pm}}{\partial x_1} + \frac{\gamma_1^{\pm}}{a_1^{\pm}} \frac{\partial \omega_2^{\pm}}{\partial x_2} + \frac{a_3^{\pm}\gamma^{\pm}}{a_1^{\pm}a_2^{\pm}} \frac{\partial \omega_3^{\pm}}{\partial x_3},
\end{aligned} \tag{40.32}$$

$$\begin{aligned}
\tilde{G}_{12}^{\pm} &= \frac{a_1^{\pm}\gamma_2^{\pm}}{a_2^{\pm}a_3^{\pm}} \frac{\partial \omega_2^{\pm}}{\partial x_1} + \frac{\gamma_3^{\pm}}{a_3^{\pm}} \frac{\partial \omega_1^{\pm}}{\partial x_2}, \qquad & \tilde{G}_{13}^{\pm} &= \frac{a_1^{\pm}\gamma_2^{\pm}}{a_2^{\pm}a_3^{\pm}} \frac{\partial \omega_3^{\pm}}{\partial x_1} + \frac{\gamma_3^{\pm}}{a_2^{\pm}} \frac{\partial \omega_1^{\pm}}{\partial x_3}, \\
\tilde{G}_{21}^{\pm} &= \frac{a_2^{\pm}\gamma_2^{\pm}}{a_1^{\pm}a_3^{\pm}} \frac{\partial \omega_1^{\pm}}{\partial x_2} + \frac{\gamma_3^{\pm}}{a_3^{\pm}} \frac{\partial \omega_2^{\pm}}{\partial x_1}, \qquad & \tilde{G}_{23}^{\pm} &= \frac{a_2^{\pm}\gamma_2^{\pm}}{a_1^{\pm}a_3^{\pm}} \frac{\partial \omega_3^{\pm}}{\partial x_2} + \frac{\gamma_3^{\pm}}{a_1^{\pm}} \frac{\partial \omega_2^{\pm}}{\partial x_3}, \\
\tilde{G}_{31}^{\pm} &= \frac{a_3^{\pm}\gamma_2^{\pm}}{a_1^{\pm}a_2^{\pm}} \frac{\partial \omega_1^{\pm}}{\partial x_3} + \frac{\gamma_3^{\pm}}{a_2^{\pm}} \frac{\partial \omega_3^{\pm}}{\partial x_1}, \qquad & \tilde{G}_{32}^{\pm} &= \frac{a_3^{\pm}\gamma_2^{\pm}}{a_1^{\pm}a_2^{\pm}} \frac{\partial \omega_2^{\pm}}{\partial x_3} + \frac{\gamma_3^{\pm}}{a_1^{\pm}} \frac{\partial \omega_3^{\pm}}{\partial x_2}.
\end{aligned}$$

The following notation is adopted here:

$$\begin{aligned}
A_1 &= \lambda s + \mu(\alpha_2 + \alpha_3) - \psi, \qquad & A_1^{\pm} &= \lambda^{\pm} s^{\pm} + \mu^{\pm}\left(\alpha_2 a_2^{\pm} + \alpha_3^{\pm} a_3^{\pm}\right) - \psi^{\pm}, \\
A_2 &= \lambda s + \mu(\alpha_1 + \alpha_3) - \psi, \qquad & A_2^{\pm} &= \lambda^{\pm} s^{\pm} + \mu^{\pm}\left(\alpha_1 a_1^{\pm} + \alpha_3^{\pm} a_3^{\pm}\right) - \psi^{\pm}, \\
A_3 &= \lambda s + \mu(\alpha_1 + \alpha_2) - \psi, \qquad & A_3^{\pm} &= \lambda^{\pm} s^{\pm} + \mu^{\pm}\left(\alpha_1 a_1^{\pm} + \alpha_2 a_2^{\pm}\right) - \psi^{\pm},
\end{aligned}$$

$$\begin{aligned}
\tau &= \mu + \kappa, \qquad & \xi &= \lambda + \mu, \qquad & \gamma &= \gamma_1 + \gamma_2 + \gamma_3, \\
\tau^{\pm} &= \mu^{\pm} + \kappa^{\pm}, \qquad & \xi^{\pm} &= \lambda^{\pm} + \mu^{\pm}, \qquad & \gamma^{\pm} &= \gamma_1^{\pm} + \gamma_2^{\pm} + \gamma_3^{\pm}.
\end{aligned}$$

For solving linearized boundary value problem (40.18), (40.22) – (40.25) for a system of eighteen partial differential equations, we use the substitution [31]

$$\begin{aligned}
&\upsilon_1 = V_1(x_3)\sin\beta_1 x_1\cos\beta_2 x_2, && \upsilon_1^{\pm} = V_1^{\pm}(x_3)\sin\beta_1 x_1\cos\beta_2 x_2,\\
&\upsilon_2 = V_2(x_3)\cos\beta_1 x_1\sin\beta_2 x_2, && \upsilon_2^{\pm} = V_2^{\pm}(x_3)\cos\beta_1 x_1\sin\beta_2 x_2,\\
&\upsilon_3 = V_3(x_3)\cos\beta_1 x_1\cos\beta_2 x_2, && \upsilon_3^{\pm} = V_3^{\pm}(x_3)\cos\beta_1 x_1\cos\beta_2 x_2,\\
&\omega_1 = \Omega_1(x_3)\cos\beta_1 x_1\sin\beta_2 x_2, && \omega_1^{\pm} = \Omega_1^{\pm}(x_3)\cos\beta_1 x_1\sin\beta_2 x_2,\\
&\omega_2 = \Omega_2(x_3)\sin\beta_1 x_1\cos\beta_2 x_2, && \omega_2^{\pm} = \Omega_2^{\pm}(x_3)\sin\beta_1 x_1\cos\beta_2 x_2,\\
&\omega_3 = \Omega_3(x_3)\sin\beta_1 x_1\sin\beta_2 x_2, && \omega_3^{\pm} = \Omega_3^{\pm}(x_3)\sin\beta_1 x_1\sin\beta_2 x_2,\\
&\beta_1 = \pi m_1/b_1, \qquad \beta_2 = \pi m_2/b_2, && m_{1,2} = 1,2,\ldots
\end{aligned} \tag{40.33}$$

which leads to the separation of variables x_1, x_2 and allows satisfying linearized end conditions (40.24), (40.25).

Taking into account relations (40.4), (40.14), (40.26) – (40.33), equations of neutral equilibrium (40.18) are written as follows:

$$\begin{aligned}
&\tau V_1'' - \left[\left(\beta_1^2+\beta_2^2\right)\tau + \beta_1^2\xi\right]V_1 - \beta_1\beta_2\xi V_2 - \beta_1\xi V_3' + A_2\Omega_2' - \beta_2 A_3\Omega_3 = 0,\\
&\tau V_2'' - \left[\left(\beta_1^2+\beta_2^2\right)\tau + \beta_2^2\xi\right]V_2 - \beta_1\beta_2\xi V_1 - \beta_2\xi V_3' - A_1\Omega_1' + \beta_1 A_3\Omega_3 = 0,\\
&(\xi+\tau)V_3'' - \left(\beta_1^2+\beta_2^2\right)\tau V_3 + \beta_1\xi V_1' + \beta_2\xi V_2' + \beta_2 A_1\Omega_1 - \beta_1 A_2\Omega_2 = 0,\\
&\gamma_2\Omega_1'' + \left[(\alpha_2+\alpha_3)A_1 - \beta_1^2\gamma - \beta_2^2\gamma_2\right]\Omega_1 - \beta_1\beta_2(\gamma_1+\gamma_3)\Omega_2 +\\
&\qquad + \beta_1(\gamma_1+\gamma_3)\Omega_3' + A_1V_2' + \beta_2 A_1 V_3 = 0,\\
&\gamma_2\Omega_2'' + \left[(\alpha_1+\alpha_3)A_2 - \beta_1^2\gamma_2 - \beta_2^2\gamma\right]\Omega_2 - \beta_1\beta_2(\gamma_1+\gamma_3)\Omega_1 +\\
&\qquad + \beta_2(\gamma_1+\gamma_3)\Omega_3' - A_2V_1' - \beta_1 A_2 V_3 = 0,\\
&\gamma\Omega_3'' + \left[(\alpha_1+\alpha_2)A_3 - \gamma_2\left(\beta_1^2+\beta_2^2\right)\right]\Omega_3 - \beta_1(\gamma_1+\gamma_3)\Omega_1' -\\
&\qquad - \beta_2(\gamma_1+\gamma_3)\Omega_2' - \beta_2 A_3 V_1 + \beta_1 A_3 V_2 = 0,
\end{aligned} \tag{40.34}$$

$$
\begin{aligned}
&\frac{a_3^\pm \tau^\pm}{a_1^\pm a_2^\pm} V_1^{\pm\prime\prime} - \left[\left(\frac{a_1^\pm \beta_1^2}{a_2^\pm} + \frac{a_2^\pm \beta_2^2}{a_1^\pm}\right)\frac{\tau^\pm}{a_3^\pm} + \frac{a_1^\pm \beta_1^2 \xi^\pm}{a_2^\pm a_3^\pm}\right] V_1^\pm - \\
&\quad - \frac{\beta_1 \beta_2 \xi^\pm}{a_3^\pm} V_2^\pm - \frac{\beta_1 \xi^\pm}{a_2^\pm} V_3^{\pm\prime} + \frac{A_2^\pm \Omega_2^{\pm\prime}}{a_1^\pm a_2^\pm} - \frac{\beta_2 A_3^\pm \Omega_3^\pm}{a_1^\pm a_3^\pm} = 0, \\
&\frac{a_3^\pm \tau^\pm}{a_1^\pm a_2^\pm} V_2^{\pm\prime\prime} - \left[\left(\frac{a_1^\pm \beta_1^2}{a_2^\pm} + \frac{a_2^\pm \beta_2^2}{a_1^\pm}\right)\frac{\tau^\pm}{a_3^\pm} + \frac{a_2^\pm \beta_2^2 \xi^\pm}{a_1^\pm a_3^\pm}\right] V_2^\pm - \\
&\quad - \frac{\beta_1 \beta_2 \xi^\pm}{a_3^\pm} V_1^\pm - \frac{\beta_2 \xi^\pm}{a_1^\pm} V_3^{\pm\prime} - \frac{A_1^\pm \Omega_1^{\pm\prime}}{a_1^\pm a_2^\pm} + \frac{\beta_1 A_3^\pm \Omega_3^\pm}{a_2^\pm a_3^\pm} = 0, \\
&\frac{a_3^\pm \left(\xi^\pm + \tau^\pm\right)}{a_1^\pm a_2^\pm} V_3^{\pm\prime\prime} - \left(\frac{a_1^\pm \beta_1^2}{a_2^\pm} + \frac{a_2^\pm \beta_2^2}{a_1^\pm}\right)\frac{\tau^\pm}{a_3^\pm} V_3^\pm + \frac{\beta_1 \xi^\pm}{a_2^\pm} V_1^{\pm\prime} + \\
&\quad + \frac{\beta_2 \xi^\pm}{a_1^\pm} V_2^{\pm\prime} + \frac{\beta_2 A_1^\pm \Omega_1^\pm}{a_1^\pm a_3^\pm} - \frac{\beta_1 A_2^\pm \Omega_2^\pm}{a_2^\pm a_3^\pm} = 0, \\
&\frac{a_3^\pm \gamma_2^\pm}{a_1^\pm a_2^\pm} \Omega_1^{\pm\prime\prime} + \left[\left(\frac{\alpha_2}{a_3^\pm} + \frac{\alpha_3^\pm}{a_2^\pm}\right)\frac{A_1^\pm}{a_1^\pm} - \frac{a_1^\pm \beta_1^2 \gamma^\pm}{a_2^\pm a_3^\pm} - \frac{a_2^\pm \beta_2^2 \gamma_2^\pm}{a_1^\pm a_3^\pm}\right] \Omega_1^\pm - \\
&\quad - \frac{\beta_1 \beta_2 \left(\gamma_1^\pm + \gamma_3^\pm\right)}{a_3^\pm} \Omega_2^\pm + \frac{\beta_1 \left(\gamma_1^\pm + \gamma_3^\pm\right)}{a_2^\pm} \Omega_3^{\pm\prime} + \frac{A_1^\pm}{a_1^\pm a_2^\pm} V_2^{\pm\prime} + \frac{\beta_2 A_1^\pm}{a_1^\pm a_3^\pm} V_3^\pm = 0, \\
&\frac{a_3^\pm \gamma_2^\pm}{a_1^\pm a_2^\pm} \Omega_2^{\pm\prime\prime} + \left[\left(\frac{\alpha_1}{a_3^\pm} + \frac{\alpha_3^\pm}{a_1^\pm}\right)\frac{A_2^\pm}{a_2^\pm} - \frac{a_1^\pm \beta_1^2 \gamma_2^\pm}{a_2^\pm a_3^\pm} - \frac{a_2^\pm \beta_2^2 \gamma^\pm}{a_1^\pm a_3^\pm}\right] \Omega_2^\pm - \\
&\quad - \frac{\beta_1 \beta_2 \left(\gamma_1^\pm + \gamma_3^\pm\right)}{a_3^\pm} \Omega_1^\pm + \frac{\beta_2 \left(\gamma_1^\pm + \gamma_3^\pm\right)}{a_1^\pm} \Omega_3^{\pm\prime} - \frac{A_2^\pm}{a_1^\pm a_2^\pm} V_1^{\pm\prime} - \frac{\beta_1 A_2^\pm}{a_2^\pm a_3^\pm} V_3^\pm = 0, \\
&\frac{a_3^\pm \gamma^\pm}{a_1^\pm a_2^\pm} \Omega_3^{\pm\prime\prime} + \left[\left(\frac{\alpha_1}{a_2^\pm} + \frac{\alpha_2}{a_1^\pm}\right)\frac{A_3^\pm}{a_3^\pm} - \frac{\gamma_2^\pm}{a_3^\pm}\left(\frac{a_1^\pm \beta_1^2}{a_2^\pm} + \frac{a_2^\pm \beta_2^2}{a_1^\pm}\right)\right] \Omega_3^\pm - \\
&\quad - \frac{\beta_1 \left(\gamma_1^\pm + \gamma_3^\pm\right)}{a_2^\pm} \Omega_1^{\pm\prime} - \frac{\beta_2 \left(\gamma_1^\pm + \gamma_3^\pm\right)}{a_1^\pm} \Omega_2^{\pm\prime} - \frac{\beta_2 A_3^\pm}{a_1^\pm a_3^\pm} V_1^\pm + \frac{\beta_1 A_3^\pm}{a_2^\pm a_3^\pm} V_2^\pm = 0.
\end{aligned}
\tag{40.35}
$$

The linearized boundary conditions (40.22), (40.23) take the form:

1) on the surfaces of the coatings ($x_3 = \pm h \pm h^\pm$)

$$
\begin{gathered}
\frac{a_3^{\pm}\tau^{\pm}}{a_1^{\pm}a_2^{\pm}}V_1^{\pm\prime} - \frac{\beta_1\mu^{\pm}}{a_2^{\pm}}V_3^{\pm} + \frac{A_2^{\pm}}{a_1^{\pm}a_2^{\pm}}\Omega_2^{\pm} = 0, \\
\frac{a_3^{\pm}\tau^{\pm}}{a_1^{\pm}a_2^{\pm}}V_2^{\pm\prime} - \frac{\beta_2\mu^{\pm}}{a_1^{\pm}}V_3^{\pm} - \frac{A_1^{\pm}}{a_1^{\pm}a_2^{\pm}}\Omega_1^{\pm} = 0, \\
\frac{a_3^{\pm}\left(\xi^{\pm}+\tau^{\pm}\right)}{a_1^{\pm}a_2^{\pm}}V_3^{\pm\prime} + \frac{\beta_1\lambda^{\pm}}{a_2^{\pm}}V_1^{\pm} + \frac{\beta_2\lambda^{\pm}}{a_1^{\pm}}V_2^{\pm} = 0, \\
\frac{a_3^{\pm}\gamma_2^{\pm}}{a_1^{\pm}a_2^{\pm}}\Omega_1^{\pm\prime} + \frac{\beta_1\gamma_3^{\pm}}{a_2^{\pm}}\Omega_3^{\pm} = 0, \qquad \frac{a_3^{\pm}\gamma_2^{\pm}}{a_1^{\pm}a_2^{\pm}}\Omega_2^{\pm\prime} + \frac{\beta_2\gamma_3^{\pm}}{a_1^{\pm}}\Omega_3^{\pm} = 0, \\
\frac{a_3^{\pm}\gamma^{\pm}}{a_1^{\pm}a_2^{\pm}}\Omega_3^{\pm\prime} - \frac{\beta_1\gamma_1^{\pm}}{a_2^{\pm}}\Omega_1^{\pm} - \frac{\beta_2\gamma_1^{\pm}}{a_1^{\pm}}\Omega_2^{\pm} = 0
\end{gathered}
\tag{40.36}
$$

2) at the interfaces between the coatings and the plate ($x_3 = \pm h$)

$$
\begin{aligned}
\tau V_1' - \beta_1\mu V_3 + A_2\Omega_2 &= \frac{a_3^{\pm}\tau^{\pm}}{a_1^{\pm}a_2^{\pm}}V_1^{\pm\prime} - \frac{\beta_1\mu^{\pm}}{a_2^{\pm}}V_3^{\pm} + \frac{A_2^{\pm}\Omega_2^{\pm}}{a_1^{\pm}a_2^{\pm}}, \\
\tau V_2' - \beta_2\mu V_3 - A_1\Omega_1 &= \frac{a_3^{\pm}\tau^{\pm}}{a_1^{\pm}a_2^{\pm}}V_2^{\pm\prime} - \frac{\beta_2\mu^{\pm}}{a_1^{\pm}}V_3^{\pm} - \frac{A_1^{\pm}\Omega_1^{\pm}}{a_1^{\pm}a_2^{\pm}}, \\
(\xi+\tau)V_3' + \beta_1\lambda V_1 + \beta_2\lambda V_2 &= \frac{a_3^{\pm}\left(\xi^{\pm}+\tau^{\pm}\right)}{a_1^{\pm}a_2^{\pm}}V_3^{\pm\prime} + \frac{\beta_1\lambda^{\pm}}{a_2^{\pm}}V_1^{\pm} + \frac{\beta_2\lambda^{\pm}}{a_1^{\pm}}V_2^{\pm}, \\
\gamma_2\Omega_1' + \beta_1\gamma_3\Omega_3 &= \frac{a_3^{\pm}\gamma_2^{\pm}}{a_1^{\pm}a_2^{\pm}}\Omega_1^{\pm\prime} + \frac{\beta_1\gamma_3^{\pm}}{a_2^{\pm}}\Omega_3^{\pm}, \\
\gamma_2\Omega_2' + \beta_2\gamma_3\Omega_3 &= \frac{a_3^{\pm}\gamma_2^{\pm}}{a_1^{\pm}a_2^{\pm}}\Omega_2^{\pm\prime} + \frac{\beta_2\gamma_3^{\pm}}{a_1^{\pm}}\Omega_3^{\pm}, \\
\gamma\Omega_3' - \beta_1\gamma_1\Omega_1 - \beta_2\gamma_1\Omega_2 &= \frac{a_3^{\pm}\gamma^{\pm}}{a_1^{\pm}a_2^{\pm}}\Omega_3^{\pm\prime} - \frac{\beta_1\gamma_1^{\pm}}{a_2^{\pm}}\Omega_1^{\pm} - \frac{\beta_2\gamma_1^{\pm}}{a_1^{\pm}}\Omega_2^{\pm}, \\
V_1 = V_1^{\pm}, \qquad V_2 &= V_2^{\pm}, \qquad V_3 = V_3^{\pm}, \\
\Omega_1 = \Omega_1^{\pm}, \qquad \Omega_2 &= \Omega_2^{\pm}, \qquad \Omega_3 = \Omega_3^{\pm}.
\end{aligned}
\tag{40.37}
$$

Thus, the stability analysis of a micropolar plate with prestressed coatings is reduced to solving linear homogeneous boundary value problem (40.34) – (40.37) for a system of eighteen ordinary differential equations.

40.4 Micropolar Plate with Identical Coatings

It is obvious that in the case of identical top and bottom coatings

$$
h^+ = h^-, a_1^+ = a_1^-, a_2^+ = a_2^-, \lambda^+ = \lambda^-, \mu^+ = \mu^-, \kappa^+ = \kappa^-, \gamma_1^+ = \gamma_1^-, \gamma_2^+ = \gamma_2^-, \gamma_3^+ = \gamma_3^-
$$

boundary value problem (40.34) – (40.37) has two independent sets of solutions [31]. The **First set** is formed by solutions for which the deflection of a sandwich plate is an odd function of x_3 (symmetric buckling):

$$
\begin{aligned}
&V_1(x_3) = V_1(-x_3), && \Omega_1(x_3) = -\Omega_1(-x_3), && \\
&V_2(x_3) = V_2(-x_3), && \Omega_2(x_3) = -\Omega_2(-x_3), && 0 \le x_3 \le h, \\
&V_3(x_3) = -V_3(-x_3), && \Omega_3(x_3) = \Omega_3(-x_3), && \\
&V_1^+(x_3) = V_1^-(-x_3), && \Omega_1^+(x_3) = -\Omega_1^-(-x_3), && \\
&V_2^+(x_3) = V_2^-(-x_3), && \Omega_2^+(x_3) = -\Omega_2^-(-x_3), && h \le x_3 \le h+h^+. \\
&V_3^+(x_3) = -V_3^-(-x_3), && \Omega_3^+(x_3) = \Omega_3^-(-x_3), &&
\end{aligned}
$$

For the **Second set** of solutions, the deflection is an even function of x_3 (flexural buckling):

$$
\begin{aligned}
&V_1(x_3) = -V_1(-x_3), && \Omega_1(x_3) = \Omega_1(-x_3), && \\
&V_2(x_3) = -V_2(-x_3), && \Omega_2(x_3) = \Omega_2(-x_3), && 0 \le x_3 \le h, \\
&V_3(x_3) = V_3(-x_3), && \Omega_3(x_3) = -\Omega_3(-x_3), && \\
&V_1^+(x_3) = -V_1^-(-x_3), && \Omega_1^+(x_3) = \Omega_1^-(-x_3), && \\
&V_2^+(x_3) = -V_2^-(-x_3), && \Omega_2^+(x_3) = \Omega_2^-(-x_3), && h \le x_3 \le h+h^+. \\
&V_3^+(x_3) = V_3^-(-x_3), && \Omega_3^+(x_3) = -\Omega_3^-(-x_3), &&
\end{aligned}
$$

Due to this, it is sufficient to consider only half of the sandwich plate ($0 \le x_3 \le h+h^+$) to study its stability. The boundary conditions at $x_3 = 0$ follow from the evenness and oddness of the unknown functions $V_1, V_2, V_3, \Omega_1, \Omega_2, \Omega_3$:

1) for the **First set** of solutions:

$$
\begin{aligned}
&V_1'(0) = V_2'(0) = V_3(0) = 0, \\
&\Omega_1(0) = \Omega_2(0) = \Omega_3'(0) = 0.
\end{aligned}
\tag{40.38}
$$

2) for the **Second set** of solutions:

$$
\begin{aligned}
&V_1(0) = V_2(0) = V_3'(0) = 0, \\
&\Omega_1'(0) = \Omega_2'(0) = \Omega_3(0) = 0.
\end{aligned}
\tag{40.39}
$$

Thus, the stability analysis of a micropolar plate with identical prestressed coatings can be reduced to solving two linear homogeneous boundary value problems for system of twelve ordinary differential equations: (40.34) – (40.38) and (40.34) – (40.37), (40.39).

40.5 Conclusion

Within the nonlinear Cosserat continuum, we have studied the stability of a rectangular plate with top and bottom coatings (sandwich plate) under biaxial tension-compression. The coatings were assumed to be subjected to preliminary deformations and contain internal stresses. When determining the subcritical deformed state in different parts of the sandwich plate, writing of constitutive relations with respect to different reference configurations was used. For the physically linear model of a micropolar material, we have derived the equations of neutral equilibrium, which describe the behavior of a rectangular plate with prestressed coatings in a perturbed state. Using a special substitution, the linearized boundary value problems were formulated both for a general case of a sandwich plate and for a plate with identical top and bottom coatings. In the general case, the stability analysis was reduced to solving linear homogeneous boundary value problem (40.34) – (40.37) for a system of eighteen ordinary differential equations. For the plate with identical top and bottom coatings, the stability analysis was reduced to solving two linear homogeneous boundary value problems – (40.34) – (40.38) and (40.34) – (40.37), (40.39) – for a system of twelve ordinary differential equations. For specific materials of plate and coatings this can be done numerically, using method described in [31].

Acknowledgements This work was carried out partially with the financial support of the Russian Foundation for Basic Research and the Administration of the Krasnodar Territory (grant 19-48-230042-r_a), and as part of the implementation of the State assignment for the Southern Scientific Centre of the Russian Academy of Sciences (SSC RAS), state registration number 122020100343-4.

References

[1] Altenbach J, Altenbach H, Eremeyev VA (2010) On generalized Cosserat-type theories of plates and shells: a short review and bibliography, Archive of Applied Mechanics **80**:73-92.
[2] Cosserat E, Cosserat F (1909)Theorie des Corps Deformables. Hermann et Fils, Paris (1909)
[3] Eringen AC (1999) Microcontinuum Field Theory. Vol I Foundations and Solids, Springer, New York
[4] Kafadar CB, Eringen AC (1971) Micropolar media - I The classical theory. International Journal of Engineeering Scieces **9**(3):271-305.
[5] Maugin GA (1998) On the structure of the theory of polar elasticity Philosophical Transactions of the Royal Society of London. Series A: Mathematical, Physical and Engineering Sciences **356**(1741):1367-1395.
[6] Toupin RA (1964) Theories of elasticity with couple-stress, Archive for Rational Mechanics and Analysis **17**:85-112.
[7] Capriz G, Giovine P, Mariano PM (Eds) (2008) Mathematical Models of Granular Matter, Springer, Berlin.

[8] Pasternak E, Mühlhaus, HB (2005) Generalised homogenisation procedures for granular materials, In: Mathematics and Mechanics of Granular Materials (Ed by JM Hill, JAPS Selvadurai), pp 199-229, Springer Netherlands, Dordrecht.

[9] Diebels S (1999) A micropolar theory of porous media: constitutive modelling, Transport in Porous Media **34**(1-3):193-208.

[10] Lakes RS (1986) Experimental microelasticity of two porous solids, International Journal of Solids and Structures **22**:55-63.

[11] Forest S, Sab K (1998) Cosserat overall modelling of heterogeneous materials, Mechanics Research Communication **25**(4):449-454.

[12] Neff P (2006) A finite-strain elastic-plastic Cosserat theory for polycrystals with grain rotations, Int J Eng Sci **44**(8-9):574-594.

[13] Park HC, Lakes RS (1986) Cosserat micromechanics of human bone: strain redistribution by a hydration-sensitive constituent, J Biomech **19**(5):385-397.

[14] Yang JFC, Lakes RS (1982) Experimental study of micropolar and couple stress elasticity in compact bone in bending, J Biomech **15**(2):91-98.

[15] Chroscielewski J, dell'Isola F, Eremeyev VA, Sabik A (2020) On rotational instability within the nonlinear six-parameter shell theory, International Journal of Solids and Structures **196**:179-189.

[16] Eremeyev VA, Turco E (2020) Enriched buckling for beam-lattice metamaterials. Mechanics Research Communication **103**:103458.

[17] Baraldi D, Cecchi A, Tralli A (2015) Continuous and discrete models for masonry like material: A critical comparative study, Eur J Mech A Solids **50**39-58.

[18] Reccia E, Leonetti L, Trovalusci P, Cecchi A (2018) A multiscale/multidomain model for the failure analysis of masonry walls: A validation with a combined FEM/DEM approach, International Journal for Multiscale Computational Engineering **16**(4):325-343.

[19] Diebels S, Steeb H (2002) The size effect in foams and its theoretical and numerical investigation, Philosophical Transactions of the Royal Society of London. Series A: Mathematical, Physical and Engineering Sciences **458**(3):2869-2883.

[20] Lakes R (1995) Experimental methods for study of Cosserat elastic solids and other generalized elastic continua, In: H Mühlhaus (Ed) Continuum models for materials with micro-structure, pp 1-22, Wiley, New York.

[21] Levin V (2017) Equilibrium of micropolar bodies with predeformed regions. The superposition of large deformations, Journal of Applied Mathematics and Mechanics **81**(3):223-227.

[22] Levin VA, Zubov, LM, Zingerman KM (2018) Multiple joined prestressed orthotropic layers under large strains, International Journal of Engineering Science **133**:47-59

[23] Merodio J, Ogden RW, Rodriguez J (2013) The influence of residual stress on finite deformation elastic response, International Journal of Non-Linear Mechanics **56**:43-49.

[24] Merodio J, Ogden RW (2016) Extension, inflation and torsion of a residually stressed circular cylindrical tube, Continuum Mechanics and Thermodynamics **28**:157-174.

[25] Zubov LM (2019) Universal solution of nonlinear elasticity for a hollow cylinder with prestressed coatings, Acta Mechanica **230**:4137-4143.
[26] Bowden N, Brittain S, Evans AG, Hutchinson JW, Whitesides GM (1998) Spontaneous formation of ordered structures in thin films of metals supported on an elastomeric polymer, Nature **393**(6681):146-149.
[27] Cao Y, Hutchinson JW (2012) Wrinkling phenomena in neo-Hookean film/substrate bilayers, Transactions ASME Journal of Applied Mechanics **79**(3):031019.
[28] Eremeev VV, Zubov LM (2017) Buckling of a two-layered circular plate with a prestressed layer, Mathematics and Mechanics of Solids **22**(4):773-781.
[29] Hutchinson JW (2013) The role of nonlinear substrate elasticity in the wrinkling of thin films, Philosophical Transactions of the Royal Society of London. Series A: Mathematical, Physical and Engineering Sciences **371**(1993):20120422.
[30] Sheydakov DN (2011) Buckling of elastic composite rods made of micropolar material subjected to combined loads, In: H Altenbach, V Erofeev, G Maugin (Eds) Mechanics of Generalized Continua - From Micromechanical Basics to Engineering Applications, Advanced Structured Materials, Vol 7, pp 255-271, Springer, Berlin.
[31] Sheydakov DN (2011) On stability of elastic rectangular sandwich plate subject to biaxial compression, In: H Altenbach, VA Eremeyev (Eds) Shell-like Structures – Non-classical Theories and Applications, Advanced Structured Materials, Vol 15, pp 203-216, Springer, Berlin.
[32] Sheydakov DN (2016) Size effect on buckling of non-uniform circular plate made of foam material, Materials Physics and Mechanics **28**(1-2):26-30
[33] Sheydakov DN, Altenbach H (2016) Stability of inhomogeneous micropolar cylindrical tube subject to combined loads, Mathematics and Mechanics of Solids **21**(9):1082-1094.
[34] Sheydakov DN (2021) Stability of circular micropolar rod with prestressed two-layer coating, Continuum Mechanics and Thermodynamics **33**:1313-1329.
[35] Sheydakov DN (2022) Influence of internal stresses on stability of multilayer micropolar tubes, Mathematics and Mechanics of Solids **27**(10):2239-2255.
[36] Lurie AI (1990) Non-linear Theory of Elasticity, North-Holland, Amsterdam.
[37] Zubov LM (1997) Nonlinear Theory of Dislocations and Disclinations in Elastic Bodies, Springer, Berlin.
[38] Nikitin E, Zubov LM (198) Conservation laws and conjugate solutions in the elasticity of simple materials and materials with couple stress, Journal of Elasticity **51**:1-22.
[39] Pietraszkiewicz W, Eremeyev VA (2009) On natural strain measures of the non-linear micropolar continuum, International Journal of Solids and Structures **46**:774-787.
[40] Zubov LM (2016) Universal deformations of micropolar isotropic elastic solids, Mathematics and Mechanics of Solids **21**(2):152-167.
[41] Eremeyev VA, Zubov LM (1994) On the stability of elastic bodies with couple-stresses. Mechanics of Solids **29**(3):172-181.

[42] Eremeyev VA, Pietraszkiewicz W (2012) Material symmetry group of the non-linear polar-elastic continuum, International Journal of Solids and Structures **49**(14):1993-2005.
[43] Eremeyev VA, Pietraszkiewicz W (2016) Material symmetry group and constitutive equations of micropolar anisotropic elastic solids, Mathematics and Mechanics of Solids **21**(2):210-221.
[44] Truesdell C (1977) A First Course in Rational Continuum Mechanics. Academic Press, New York.
[45] Green AE, Adkins JE (1960) Large Elastic Deformations and Non-Linear Continuum Mechanics, Clarendon Press, Oxford.
[46] Ogden RW, Steigmann DJ, Haughton DM (1997) The effect of elastic surface coating on the finite deformation and bifurcation of a pressurized circular annulus Journal of Elasticity **47**(2):121-145.

Chapter 41
A Cosserat Model for Fiber-Reinforced Elastic Plates

David J. Steigmann, Mircea Bîrsan, and Milad Shirani

Abstract We present a theory for thin elastic plates reinforced by a single family of continuously distributed fibers. The novelty of the model is the assignment of intrinsic flexural and torsional elasticity to the embedded fibers, regarded as material curves. Our derivation is based on a dimension reduction from the three-dimensional theory to obtain a plate model valid to leading order in the small plate thickness.

Key words: Cosserat elasticity · Fiber-reinforced solids · Plate theory · Kirchhoff rods

41.1 Introduction

The theory of fiber-reinforced composite plates is the subject of a vast technical literature, all of which is based on the premise that continuously distributed fibers confer anisotropy to the homogenized continuum but do not otherwise participate in its mechanical response; that is, the intrinsic flexural and torsional resistance of the fibers is not taken into account. Here we extend this classical theory by taking these effects into account explicitly. We model the fibers as continuously distributed extensible, unshearable spatial rods of the Kirchhoff type [1]. The kinematics of these

---

David J. Steigmann
Department of Mechanical Engineering, University of California, Berkeley, CA 94720, USA, e-mail: dsteigmann@berkeley.edu

Mircea Bîrsan
Faculty of Mathematics, University Duisburg-Essen, Thea-Leymann Str. 9, 45127 Essen, Germany & Alexandru Ioan Cuza University of Iaşi, Department of Mathematics, Blvd. Carol I, no. 11, 700506 Iaşi, Romania, e-mail: mircea.birsan@uni-due.de

Milad Shirani
Department of Mechanical Engineering, University of California, Berkeley, CA 94720, USA, e-mail: milad_shirani@berkeley.edu

H. Altenbach et al. (eds.), *Sixty Shades of Generalized Continua*,
Advanced Structured Materials 170, https://doi.org/10.1007/978-3-031-26186-2_41

rods are controlled by a rotation field and its gradient, interacting with the deformation of the underlying continuum. Accordingly the parent three-dimensional theory is a special case of nonlinear Cosserat elasticity theory [2]-[4] in which the underlying deformation and rotation fields are weakly coupled by the requirement that the fibers convect with the continuum deformation as material curves. This theory has been developed and applied in a series of recent papers [5]-[8].

The present work is concerned with the development of the leading-order-in-thickness theory emerging from an asymptotic treatment of the three-dimensional potential energy of a thin plate composed of such material. To simplify the development as far as possible, we confine attention to the practically important case in which the fibers lie parallel to the plate midplane prior to deformation. Nevertheless, the fibers may be curvilinear in this plane.

In Sect. 41.2 we outline the three-dimensional theory for the sake of completeness. Section 41.3 is devoted to the asymptotic development of this theory in the limit of small plate thickness h. This is illustrated by adopting a well-established strain-energy function from three-dimensional Cosserat elasticity [9]-[11]. In strong contrast to existing plate theories derived from conventional three-dimensional nonlinear elasticity, we show that our leading order $O(h)$ model satisfies a crucial coercivity condition that plays a central role in establishing the existence of solutions to equilibrium boundary-value problems. In conventional theories the leading-order model is membrane theory which, as is well known, is ill-posed due to the presence of destabilizing compressive stresses in the range of the relevant constitutive function. Accordingly the conventional theory requires a convexification step to restore well-posedness, or, alternatively, regularization by higher-order expansions in powers of h. The enhanced structure of the three-dimensional parent theory in the present model obviates the need for such special measures, and ensures that the simplest order - h model is well posed. We thus arrive at a new model for fiber-reinforced plates that is suitable for applications.

Some preliminary aspects of the present model are outlined in [12]. However, the present work goes well beyond this to develop a complete theory. We defer illustrative examples to a forthcoming paper concerned with the development of an associated shell theory.

Standard notational conventions are used throughout the chapter. These are explained, as the need arises, in the course of the ensuing development. For instance, the notations $\boldsymbol{A}^t, \mathrm{skw}(\boldsymbol{A}), \det \boldsymbol{A}$ and $\mathrm{tr}(\boldsymbol{A})$ represent respectively the transpose, the skew part, the determinant and the trace of a tensor $\boldsymbol{A}$. Also, $\mathrm{ax}(\boldsymbol{S})$ is the axial vector of any skew-symmetric tensor $\boldsymbol{S}$. Latin and Greek indices take values in $\{1,2,3\}$ and $\{2,3\}$ respectively, and, when repeated, are summed over their ranges. Finally, bold subscripts are used to denote derivatives of scalar functions with respect to their vector or tensor arguments.

41.2 The Three-Dimensional Cosserat Model for Fiber-Reinforced Elastic Solids

We consider elastic bodies with a single family of curved and twisted fibers. The theory of fiber-reinforced elastic solids described as constrained Cosserat continua has been established in [5]-[7]. In this section, we present briefly this model with small modifications, which are suitable to introducing the plate theory in the next section.

41.2.1 Kinematical Variables and Strain Measures in Cosserat Elasticity

Consider the reference configuration $\mathcal{K}$ of a Cosserat body, in which the material points are identified by the position vector $\boldsymbol{X} = X_A \boldsymbol{E}_A$ where $\{\boldsymbol{E}_A\,;\, A = 1,2,3\}$ is a given (fixed) orthonormal vector base. In the deformed configuration, the position vector of a generic point is denoted by $\boldsymbol{x} = x_i \boldsymbol{e}_i$ with respect to another fixed orthonormal vector base $\{\boldsymbol{e}_i\,;\, i = 1,2,3\}$.

The kinematical variables of a Cosserat continuum are the deformation field $\chi(\boldsymbol{X})$ and the rotation field $\boldsymbol{R}(\boldsymbol{X}) \in \mathrm{SO}(3)$. Thus, we have $\boldsymbol{x} = \chi(\boldsymbol{X})$, or with components $x_i = \chi_i(X_A)$, and also $\boldsymbol{R}(\boldsymbol{X}) = R_{iA}\boldsymbol{e}_i \otimes \boldsymbol{E}_A$. Of course, these functions can also depend on time, but since the time dependence is not relevant to our development, it will not be written explicitly.

The rotation field $\boldsymbol{R}$ can also be characterized with help of the directors: Let $\{\boldsymbol{D}_1, \boldsymbol{D}_2, \boldsymbol{D}_3\}$ be the right-handed triad of orthonormal directors in the reference configuration, and $\{\boldsymbol{d}_1, \boldsymbol{d}_2, \boldsymbol{d}_3\}$ the directors in the deformed configuration, such that $\boldsymbol{R} = \boldsymbol{d}_j \otimes \boldsymbol{D}_j$, i.e. $\boldsymbol{d}_j = \boldsymbol{R}\boldsymbol{D}_j$ $(j = 1,2,3)$.

Let ∇ stand for the gradient with respect to $\boldsymbol{X}$, that is

$$\nabla \boldsymbol{f} = \boldsymbol{f}_{,A} \otimes \boldsymbol{E}_A \qquad \text{for any field } \boldsymbol{f},$$

where $(\cdot)_{,A} = \dfrac{\partial(\cdot)}{\partial X_A}$ designate the partial derivatives with respect to X_A. Then, we can write the deformation gradient $\boldsymbol{F}$ and the rotation gradient $\nabla\boldsymbol{R}$ in Cartesian index notations as follows

$$\boldsymbol{F} = \nabla\chi = F_{iA}\boldsymbol{e}_i \otimes \boldsymbol{E}_A = \chi_{i,A}\boldsymbol{e}_i \otimes \boldsymbol{E}_A, \qquad \nabla\boldsymbol{R} = \boldsymbol{R}_{,B} \otimes \boldsymbol{E}_B = R_{iA,B}\boldsymbol{e}_i \otimes \boldsymbol{E}_A \otimes \boldsymbol{E}_B. \tag{41.1}$$

Let $\mathcal{U}(\boldsymbol{F}, \boldsymbol{R}, \nabla\boldsymbol{R}; \boldsymbol{X})$ be the strain energy density per unit reference volume of an elastic Cosserat continuum. Supposing that the strain energy is Galilean-invariant, we can reduce it to the form (see, e.g., [3, 13])

$$\mathcal{U}(\boldsymbol{F}, \boldsymbol{R}, \nabla\boldsymbol{R}; \boldsymbol{X}) = \mathcal{W}(\boldsymbol{E}, \boldsymbol{\Gamma}; \boldsymbol{X}), \tag{41.2}$$

where the *strain tensor* $\boldsymbol{E}$ and the *wryness tensor* $\boldsymbol{\Gamma}$ are given by

$$\boldsymbol{E} = \boldsymbol{R}^t\boldsymbol{F} - \boldsymbol{I} = E_{AB}\boldsymbol{E}_A \otimes \boldsymbol{E}_B \quad \text{with} \quad E_{AB} = R_{iA}F_{iB} - \delta_{AB}, \tag{41.3}$$

$$\boldsymbol{\Gamma} = \text{ax}(\boldsymbol{R}^t\boldsymbol{R}_{,C}) \otimes \boldsymbol{E}_C = \Gamma_{DC}\boldsymbol{E}_D \otimes \boldsymbol{E}_C \quad \text{with } \Gamma_{DC} = \frac{1}{2}e_{BAD}R_{iA}R_{iB,C} \tag{41.4}$$

where δ_{AB} is the Kronecker symbol, e_{ABC} is the permutation symbol, and $\boldsymbol{I}$ is the identity tensor in the 3-space. From (41.4) we see that the wryness tensor $\boldsymbol{\Gamma}$ stands in one-to-one relation to the Cosserat strain measure $\boldsymbol{R}^t\nabla\boldsymbol{R}$ (third order tensor). To write this dependence explicitly, let us introduce the third order permutation tensor $\boldsymbol{e} = e_{ABC}\boldsymbol{E}_A \otimes \boldsymbol{E}_B \otimes \boldsymbol{E}_C$ and denote by ":" the operation (scalar contraction of two indices) defined by

$$\boldsymbol{X} : \boldsymbol{Y} = X_{ABC}Y_{BC}\boldsymbol{E}_A, \qquad \text{respectively} \quad \boldsymbol{X} : \boldsymbol{Z} = X_{ABC}Z_{BCD}\boldsymbol{E}_A \otimes \boldsymbol{E}_D,$$

for any tensors of the form $\boldsymbol{X} = X_{ABC}\boldsymbol{E}_A \otimes \boldsymbol{E}_B \otimes \boldsymbol{E}_C$, $\boldsymbol{Y} = Y_{AB}\boldsymbol{E}_A \otimes \boldsymbol{E}_B$, $\boldsymbol{Z} = Z_{ABC}\boldsymbol{E}_A \otimes \boldsymbol{E}_B \otimes \boldsymbol{E}_C$. Then, the axial vector $\text{ax}(\boldsymbol{S})$ of any skew tensor $\boldsymbol{S} = S_{AB}\boldsymbol{E}_A \otimes \boldsymbol{E}_B$ can be expressed as

$$\text{ax}(\boldsymbol{S}) = -\frac{1}{2}\boldsymbol{e} : \boldsymbol{S} = -\frac{1}{2}e_{ABC}S_{BC}\boldsymbol{E}_A. \tag{41.5}$$

So, the wryness tensor $\boldsymbol{\Gamma}$ is given in terms of $\boldsymbol{R}^t\nabla\boldsymbol{R}$ by the relation

$$\boldsymbol{\Gamma} = -\frac{1}{2}(\boldsymbol{e} : \boldsymbol{R}^t\boldsymbol{R}_{,C}) \otimes \boldsymbol{E}_C = -\frac{1}{2}\boldsymbol{e} : (\boldsymbol{R}^t\boldsymbol{R}_{,C} \otimes \boldsymbol{E}_C) = -\frac{1}{2}\boldsymbol{e} : (\boldsymbol{R}^t\nabla\boldsymbol{R}), \tag{41.6}$$

which will be of later use.

41.2.2 Fiber-Reinforced Materials

We consider a three-dimensional elastic solid with the embedded curved fibers that support bending and twisting moments. We model the fibers as continuously distributed Kirchhoff rods. In this context, Cosserat elasticity arises as a natural framework, in which the deformation field χ describes the matrix elastic material and the rotation field $\boldsymbol{R}$ describes the bending and twisting of fibers.

We employ the model of Kirchhoff rods as it was presented in [1, 14]; a brief account of the model has also been included in [7]. Remark that the model also accommodates a small axial strain along the rod.

Let us choose the initial director $\boldsymbol{D}_1$ to be the unit tangent vector to the centerline of the fiber, in any point. Then, $\boldsymbol{D}_2$ and $\boldsymbol{D}_3$ are orthonormal unit vectors in the fiber cross section. According to the Kirchhoff theory of rods, each cross section deforms as a rigid disc, and the cross sections remain orthogonal to the rod center line during deformation. Thus, the director $\boldsymbol{d}_1 = \boldsymbol{R}\boldsymbol{D}_1$ is the unit tangent to the centerline of fibers in the deformed configuration, while $\boldsymbol{d}_2$ and $\boldsymbol{d}_3$ span the cross-sectional plane. For convenience, let us also denote by $\boldsymbol{D} = \boldsymbol{D}_1$ and $\boldsymbol{d} = \boldsymbol{d}_1$ the unit tangent vectors in fiber direction. Let s be the arclength parameter along the fiber centerline in the reference configuration. Denote by $\boldsymbol{Y}(s)$ the position vector along the curved centerline in the reference configuration and by $\boldsymbol{y}(s)$ the points of the deformed centerline. Then, we have

$$\boldsymbol{Y}'(s) = \boldsymbol{D} \qquad \text{and} \qquad \boldsymbol{y}'(s) = \lambda \boldsymbol{d}, \tag{41.7}$$

where $\lambda = |\boldsymbol{y}'(s)|$ is the stretch of the rod.

We assume that the elastic fibers and matrix elastic material are perfectly bonded. Then, the interface between the matrix and fiber is convected by the deformation as a material surface. Hence, on the basis of Hadamard's compatibility condition one obtains the relation (see [7] for details)

$$\boldsymbol{FD} = \lambda \boldsymbol{d} \qquad \text{with} \qquad \boldsymbol{d} = \boldsymbol{RD}, \tag{41.8}$$

where $\boldsymbol{F}$ is the matrix deformation gradient and $\lambda = |\boldsymbol{FD}|$ is the fiber stretch. Taking into account that $\boldsymbol{d}$ is orthogonal to $\boldsymbol{d}_\gamma = \boldsymbol{RD}_\gamma$ $(\gamma = 2,3)$, the relation (41.8) yields

$$\boldsymbol{FD} \cdot \boldsymbol{RD}_\gamma = 0, \qquad \text{or} \qquad \boldsymbol{D}_\gamma \cdot \boldsymbol{R}^t \boldsymbol{FD} = 0,$$

i.e.

$$\boldsymbol{D}_\gamma \cdot \boldsymbol{ED} = 0 \quad (\gamma = 2,3), \tag{41.9}$$

where $\boldsymbol{E}$ is the strain tensor defined in (41.3). The relations (41.9) represent two constraints for the deformation gradient $\boldsymbol{F}$ and the rotation $\boldsymbol{R}$, which are otherwise independent fields. Thus, this model for fiber-reinforced materials can be regarded as a constrained Cosserat continuum with the constraints (41.9). The two scalar equations (41.9) express that the fibers are convected as material curves relative to the matrix. The directors $\boldsymbol{D}_2$ and $\boldsymbol{D}_3$ are embedded in the fiber cross-section but not in the matrix. Hence, the current directors $\boldsymbol{d}_2$, $\boldsymbol{d}_3$ can shear relative to the matrix continuum, but they remain orthogonal to the fiber centerline.

Let us consider the strain measure associated to the fibers. In the Kirchhoff theory the measure for the relative flexure and twist of the rod due to deformation is given by

$$\boldsymbol{\gamma} = \operatorname{ax}(\boldsymbol{R}^t \boldsymbol{R}'), \tag{41.10}$$

also called the *relative curvature-twist vector*. Throughout this paper, we designate by $(\cdot)' = \frac{\mathrm{d}(\cdot)}{\mathrm{d}s}$ the directional derivative along $\boldsymbol{D}$, for a fiber initially oriented along a unit vector field $\boldsymbol{D}(\boldsymbol{X})$. Remark that we have

$$\boldsymbol{V}' = (\nabla \boldsymbol{V})\boldsymbol{D} \qquad \text{and} \qquad f' = (\nabla f) \cdot \boldsymbol{D}, \tag{41.11}$$

for any vector or tensor field $\boldsymbol{V}$ and any scalar field f. Hence, we can write $\boldsymbol{R}' = (\nabla \boldsymbol{R})\boldsymbol{D}$ and in view of (41.5) and (41.6) we obtain

$$\operatorname{ax}(\boldsymbol{R}^t \boldsymbol{R}') = -\frac{1}{2}\boldsymbol{e} : (\boldsymbol{R}^t \boldsymbol{R}') = -\frac{1}{2}\boldsymbol{e} : [\boldsymbol{R}^t (\nabla \boldsymbol{R})\boldsymbol{D}] = -\frac{1}{2}[\boldsymbol{e} : \boldsymbol{R}^t (\nabla \boldsymbol{R})]\boldsymbol{D} = \boldsymbol{\Gamma}\boldsymbol{D}. \tag{41.12}$$

Thus, from (41.10) and (41.12) we deduce that

$$\gamma = \Gamma D, \tag{41.13}$$

which means that the flexure and twisting of fibers are described by ΓD. Then, the strain energy density (41.2) for the Cosserat model of fiber-reinforced materials can be expressed in the form [7]

$$\mathcal{W}(E, \Gamma; X) = \widetilde{W}(E, \gamma; X), \tag{41.14}$$

where $\widetilde{W}$ stands now for the strain energy per unit reference volume.

41.2.3 Equilibrium and Constitutive Equations

We follow the same lines as in [7, 15] and present the variational derivation of the equilibrium equations, starting with the strain energy density $\widetilde{W}(E, \gamma)$. Equilibrium states are assumed to satisfy the virtual-power statement. Let

$$\mathcal{S} = \int_{\mathcal{K}} \widetilde{W} \, dv$$

be the total strain energy and $\mathcal{P}$ the virtual power of the loads acting on the body (having the form (41.32)). To accommodate the constraints (41.9) in the virtual-power statement, we relax them and consider the extended energy

$$\mathcal{E} = \mathcal{S} + \int_{\mathcal{K}} \Lambda_\gamma D_\gamma \cdot ED \, dv, \tag{41.15}$$

where Λ_γ $(\gamma = 2, 3)$ are Lagrange multipliers corresponding to the constraints (41.9). Then, the extended variational problem is

$$\dot{\mathcal{E}} = \mathcal{P}, \tag{41.16}$$

where superposed dots represent variational derivatives. These are induced by the derivatives with respect to the parameter ϵ, of the one-parameter deformation and rotation fields $\chi(X; \epsilon)$ and $R(X; \epsilon)$ (evaluated at $\epsilon = 0$), where $\chi(X) = \chi(X; 0)$ and $R(X) = R(X; 0)$ are equilibrium fields. Then, we have

$$\dot{\widetilde{W}} = \sigma \cdot \dot{E} + M \cdot \dot{\gamma}, \tag{41.17}$$

where

$$\sigma = \widetilde{W}_E \qquad \text{and} \qquad M = \widetilde{W}_\gamma \tag{41.18}$$

are evaluated at equilibrium. Moreover, we have

$$(\Lambda_\gamma \boldsymbol{D}_\gamma \cdot \boldsymbol{E}\boldsymbol{D})^{\cdot} = \Lambda_\gamma \left[(\boldsymbol{D}_\gamma \otimes \boldsymbol{D}) \cdot \boldsymbol{E} \right]^{\cdot} + \dot{\Lambda}_\gamma \boldsymbol{D}_\gamma \cdot \boldsymbol{E}\boldsymbol{D} = (\boldsymbol{\Lambda} \otimes \boldsymbol{D}) \cdot \dot{\boldsymbol{E}} + \dot{\Lambda}_\gamma \boldsymbol{D}_\gamma \cdot \boldsymbol{E}\boldsymbol{D}. \tag{41.19}$$

Here and henceforth we designate by $\boldsymbol{\Lambda}$ and $\boldsymbol{\lambda}$, respectively, the vectors

$$\boldsymbol{\Lambda} = \Lambda_\gamma \boldsymbol{D}_\gamma \qquad \text{and} \qquad \boldsymbol{\lambda} = \boldsymbol{R}\boldsymbol{\Lambda} = \Lambda_\gamma \boldsymbol{d}_\gamma . \tag{41.20}$$

To express the variational derivatives $\dot{\boldsymbol{E}}$ and $\dot{\boldsymbol{\gamma}}$ we denote by

$$\boldsymbol{v} = \dot{\boldsymbol{\chi}} \qquad \text{and} \qquad \boldsymbol{\Omega} = \dot{\boldsymbol{R}}\boldsymbol{R}^t \quad \text{with} \quad \boldsymbol{\omega} = \text{ax}(\boldsymbol{\Omega}). \tag{41.21}$$

Then, we have $\dot{\boldsymbol{F}} = \nabla \boldsymbol{v}$, $\dot{\boldsymbol{R}} = \boldsymbol{\Omega}\boldsymbol{R}$ and, hence

$$\dot{\boldsymbol{E}} = \left(\boldsymbol{R}^t \boldsymbol{F} - \boldsymbol{I} \right)^{\cdot} = \boldsymbol{R}^t \nabla \boldsymbol{v} + \dot{\boldsymbol{R}}^t \boldsymbol{F} = \boldsymbol{R}^t \left(\nabla \boldsymbol{v} - \boldsymbol{\Omega}\boldsymbol{F} \right). \tag{41.22}$$

On the other hand, from (41.13) we deduce

$$\dot{\boldsymbol{\gamma}} = \dot{\boldsymbol{\Gamma}}\boldsymbol{D} \tag{41.23}$$

and using relations (41.5) and (41.6) we find

$$\begin{aligned}
\dot{\boldsymbol{\Gamma}} &= \text{ax}\left(\boldsymbol{R}^t \boldsymbol{R}_{,C}\right)^{\cdot} \otimes \boldsymbol{E}_C = -\frac{1}{2} \boldsymbol{e} : \left(\boldsymbol{R}^t \boldsymbol{R}_{,C}\right)^{\cdot} \otimes \boldsymbol{E}_C \\
&= -\frac{1}{2} \boldsymbol{e} : \left(\dot{\boldsymbol{R}}^t \boldsymbol{R}_{,C} + \boldsymbol{R}^t \dot{\boldsymbol{R}}_{,C} \right) \otimes \boldsymbol{E}_C \\
&= -\frac{1}{2} \boldsymbol{e} : \left[-\boldsymbol{R}^t \boldsymbol{\Omega} \boldsymbol{R}_{,C} + \boldsymbol{R}^t \left(\boldsymbol{\Omega}_{,C} \boldsymbol{R} + \boldsymbol{\Omega} \boldsymbol{R}_{,C} \right) \right] \otimes \boldsymbol{E}_C \\
&= -\frac{1}{2} \boldsymbol{e} : \left(\boldsymbol{R}^t \boldsymbol{\Omega}_{,C} \boldsymbol{R} \right) \otimes \boldsymbol{E}_C = \text{ax}\left(\boldsymbol{R}^t \boldsymbol{\Omega}_{,C} \boldsymbol{R} \right) \otimes \boldsymbol{E}_C = \boldsymbol{R}^t \text{ax}\left(\boldsymbol{\Omega}_{,C} \right) \otimes \boldsymbol{E}_C \\
&= \boldsymbol{R}^t \boldsymbol{\omega}_{,C} \otimes \boldsymbol{E}_C = \boldsymbol{R}^t \nabla \boldsymbol{\omega},
\end{aligned} \tag{41.24}$$

Here, we have used that the following relation holds

$$\text{ax}\left(\boldsymbol{Q}\boldsymbol{A}\boldsymbol{Q}^t \right) = \boldsymbol{Q}\,\text{ax}(\boldsymbol{A}), \tag{41.25}$$

for any tensors $\boldsymbol{Q} \in SO(3)$ and $\boldsymbol{A}$ skew-symmetric. Then, from (41.23), (41.24) it follows

$$\dot{\boldsymbol{\gamma}} = \boldsymbol{R}^t (\nabla \boldsymbol{\omega}) \boldsymbol{D}, \tag{41.26}$$

and

$$\boldsymbol{M} \cdot \dot{\boldsymbol{\gamma}} = \boldsymbol{R}\boldsymbol{M} \cdot (\nabla \boldsymbol{\omega}) \boldsymbol{D} = (\boldsymbol{m} \otimes \boldsymbol{D}) \cdot \nabla \boldsymbol{\omega}, \tag{41.27}$$

where we have denoted by $\boldsymbol{m}$ the vector

$$\boldsymbol{m} = \boldsymbol{R}\boldsymbol{M}. \tag{41.28}$$

Inserting (41.19), (41.22), (41.27) in equations (41.15) and (41.17), we obtain

$$\dot{\mathcal{E}} = \int_{\mathcal{K}} \left[\boldsymbol{R} \left(\boldsymbol{\sigma} + \boldsymbol{\Lambda} \otimes \boldsymbol{D} \right) \cdot \left(\nabla \boldsymbol{v} - \boldsymbol{\Omega} \boldsymbol{F} \right) + \left(\boldsymbol{m} \otimes \boldsymbol{D} \right) \cdot \nabla \boldsymbol{\omega} + \dot{\boldsymbol{\Lambda}} \cdot \boldsymbol{E}\boldsymbol{D} \right] \mathrm{d}v$$

$$
\begin{aligned}
&= \int_{\mathcal{K}} \Big[(\boldsymbol{R}\boldsymbol{\sigma} + \boldsymbol{\lambda} \otimes \boldsymbol{D}) \cdot \nabla \boldsymbol{v} - \operatorname{skw}\big[(\boldsymbol{R}\boldsymbol{\sigma} + \boldsymbol{\lambda} \otimes \boldsymbol{D})\boldsymbol{F}^t\big] \cdot \boldsymbol{\Omega} + (\boldsymbol{m} \otimes \boldsymbol{D}) \cdot \nabla \boldsymbol{\omega} \\
&+ \dot{\boldsymbol{\Lambda}} \cdot \boldsymbol{E}\boldsymbol{D} \Big] \mathrm{d}v = \int_{\mathcal{K}} \Big[(\boldsymbol{R}\boldsymbol{\sigma} + \boldsymbol{\lambda} \otimes \boldsymbol{D}) \cdot \nabla \boldsymbol{v} \\
&- 2\operatorname{ax}\big\{\operatorname{skw}\big[(\boldsymbol{R}\boldsymbol{\sigma} + \boldsymbol{\lambda} \otimes \boldsymbol{D})\boldsymbol{F}^t\big]\big\} \cdot \boldsymbol{\omega} + (\boldsymbol{m} \otimes \boldsymbol{D}) \cdot \nabla \boldsymbol{\omega} + \dot{\boldsymbol{\Lambda}} \cdot \boldsymbol{E}\boldsymbol{D} \Big] \mathrm{d}v. \qquad (41.29)
\end{aligned}
$$

Here, we have used hat $\boldsymbol{A} \cdot \boldsymbol{S} = 2\operatorname{ax}(\boldsymbol{A}) \cdot \operatorname{ax}(\boldsymbol{S})$ for any two skew-symmetric tensors $\boldsymbol{A}$ and $\boldsymbol{S}$. In relation (41.29) we can apply formulas of the type

$$
\operatorname{Div}(\boldsymbol{T}^t \boldsymbol{u}) = \boldsymbol{T} \cdot \nabla \boldsymbol{u} + \boldsymbol{u} \cdot \operatorname{Div} \boldsymbol{T} \qquad \text{for any tensor } \boldsymbol{T} \text{ and vector } \boldsymbol{u}, \qquad (41.30)
$$

together with the divergence theorem, to obtain

$$
\begin{aligned}
\dot{\mathcal{E}} &= \int_{\mathcal{K}} \dot{\Lambda}_\gamma (\boldsymbol{D}_\gamma \cdot \boldsymbol{E}\boldsymbol{D})\, \mathrm{d}v - \int_{\mathcal{K}} \Big\{ \boldsymbol{v} \cdot \operatorname{Div}(\boldsymbol{R}\boldsymbol{\sigma} + \boldsymbol{\lambda} \otimes \boldsymbol{D}) \\
&+ \boldsymbol{\omega} \cdot \Big[\operatorname{Div}(\boldsymbol{m} \otimes \boldsymbol{D}) + 2\operatorname{ax}\big\{\operatorname{skw}\big[(\boldsymbol{R}\boldsymbol{\sigma} + \boldsymbol{\lambda} \otimes \boldsymbol{D})\boldsymbol{F}^t\big]\big\} \Big] \Big\} \mathrm{d}v \\
&+ \int_{\partial\mathcal{K}} [\boldsymbol{v} \cdot (\boldsymbol{R}\boldsymbol{\sigma} + \boldsymbol{\lambda} \otimes \boldsymbol{D})\boldsymbol{n} + \boldsymbol{\omega} \cdot (\boldsymbol{m} \otimes \boldsymbol{D})\boldsymbol{n}]\, \mathrm{d}a, \qquad (41.31)
\end{aligned}
$$

where $\boldsymbol{n}$ is the outer unit normal to the (piecewise smooth) boundary curve $\partial\mathcal{K}$.

We take into account that the virtual power has the form

$$
\mathcal{P} = \int_{\mathcal{K}} (\boldsymbol{l} \cdot \boldsymbol{v} + \boldsymbol{\pi} \cdot \boldsymbol{\omega})\, \mathrm{d}v + \int_{\partial\mathcal{K}} (\boldsymbol{t} \cdot \boldsymbol{v} + \boldsymbol{c} \cdot \boldsymbol{\omega})\, \mathrm{d}a, \qquad (41.32)
$$

where $\boldsymbol{l}$ and $\boldsymbol{\pi}$ are densities of force and couple acting in $\mathcal{K}$, while $\boldsymbol{t}$ and $\boldsymbol{c}$ are densities of force and couple acting on $\partial\mathcal{K}$. Then, we use the relations (41.31) and (41.32) in the virtual-power statement (41.16). Since the variations $\boldsymbol{v}$ and $\boldsymbol{\omega}$ and $\dot{\Lambda}_\gamma$ are arbitrary, we can apply the fundamental lemma to obtain the constraints (41.9) and the differential equations of equilibrium

$$
\begin{aligned}
&\operatorname{Div}(\boldsymbol{R}\boldsymbol{\sigma} + \boldsymbol{\lambda} \otimes \boldsymbol{D}) + \boldsymbol{l} = \boldsymbol{0} \qquad \text{and} \\
&\operatorname{Div}(\boldsymbol{m} \otimes \boldsymbol{D}) + 2\operatorname{ax}\big\{\operatorname{skw}\big[(\boldsymbol{R}\boldsymbol{\sigma} + \boldsymbol{\lambda} \otimes \boldsymbol{D})\boldsymbol{F}^t\big]\big\} + \boldsymbol{\pi} = \boldsymbol{0} \qquad \text{in } \mathcal{K},
\end{aligned} \qquad (41.33)
$$

together with the natural boundary conditions

$$
(\boldsymbol{R}\boldsymbol{\sigma})\boldsymbol{n} + (\boldsymbol{D} \cdot \boldsymbol{n})\boldsymbol{\lambda} = \boldsymbol{t} \quad \text{on } \partial\mathcal{K}_t \qquad \text{and} \qquad (\boldsymbol{D} \cdot \boldsymbol{n})\boldsymbol{m} = \boldsymbol{c} \quad \text{on } \partial\mathcal{K}_c. \qquad (41.34)
$$

Here $\partial\mathcal{K}_t$ is a part of $\partial\mathcal{K}$ where position is not assigned and $\partial\mathcal{K}_c$ is a part where rotation is not assigned. We assume position to be assigned on $\partial\mathcal{K} \setminus \partial\mathcal{K}_t$ (here $\boldsymbol{v} = \boldsymbol{0}$), and rotation to be assigned on $\partial\mathcal{K} \setminus \partial\mathcal{K}_c$ (here $\boldsymbol{\omega} = \boldsymbol{0}$).

Finally, we can put the equilibrium equations (41.33) in another form, which is similar to the equilibrium equations for Kirchhoff rods, and thus, facilitates the interpretation of various terms appearing in the equations. In view of relations of the type (41.11) and

$$\mathrm{Div}\,(\boldsymbol{a}\otimes\boldsymbol{b}) = (\nabla\boldsymbol{a})\,\boldsymbol{b} + (\mathrm{Div}\,\boldsymbol{b})\,\boldsymbol{a} \qquad \text{for any vectors } \boldsymbol{a},\boldsymbol{b}, \tag{41.35}$$

we can write

$$\begin{aligned} &\mathrm{Div}\,(\boldsymbol{\lambda}\otimes\boldsymbol{D}) = (\nabla\boldsymbol{\lambda})\,\boldsymbol{D} + (\mathrm{Div}\,\boldsymbol{D})\,\boldsymbol{\lambda} = \boldsymbol{\lambda}' + (\mathrm{Div}\,\boldsymbol{D})\,\boldsymbol{\lambda} \quad \text{and} \\ &\mathrm{Div}\,(\boldsymbol{m}\otimes\boldsymbol{D}) = (\nabla\boldsymbol{m})\,\boldsymbol{D} + (\mathrm{Div}\boldsymbol{D})\,\boldsymbol{m} = \boldsymbol{m}' + (\mathrm{Div}\boldsymbol{D})\,\boldsymbol{m}. \end{aligned} \tag{41.36}$$

Also, we have $\boldsymbol{F}\boldsymbol{D} = (\nabla\boldsymbol{\chi})\,\boldsymbol{D} = \boldsymbol{\chi}' = \boldsymbol{x}'$, so we can transform the term

$$2\,\mathrm{ax}\big\{\mathrm{skw}\big[(\boldsymbol{\lambda}\otimes\boldsymbol{D})\boldsymbol{F}^t\big]\big\} = 2\,\mathrm{ax}\big[\mathrm{skw}(\boldsymbol{\lambda}\otimes\boldsymbol{F}\boldsymbol{D})\big] = \mathrm{ax}(\boldsymbol{\lambda}\otimes\boldsymbol{\chi}' - \boldsymbol{\chi}'\otimes\boldsymbol{\lambda}) = \boldsymbol{x}'\times\boldsymbol{\lambda}. \tag{41.37}$$

Substituting (41.36) and (41.37) into (41.33), we obtain the following form of the equilibrium equations

$$\begin{aligned} &\boldsymbol{\lambda}' + (\mathrm{Div}\,\boldsymbol{D})\,\boldsymbol{\lambda} + \mathrm{Div}\,(\boldsymbol{R}\boldsymbol{\sigma}) + \boldsymbol{l} = \boldsymbol{0} \qquad \text{and} \\ &\boldsymbol{m}' + \boldsymbol{x}'\times\boldsymbol{\lambda} + (\mathrm{Div}\,\boldsymbol{D})\,\boldsymbol{m} + 2\,\mathrm{ax}\big[\mathrm{skw}(\boldsymbol{R}\boldsymbol{\sigma}\boldsymbol{F}^t)\big] + \boldsymbol{\pi} = \boldsymbol{0} \qquad \text{in } \mathcal{K}. \end{aligned} \tag{41.38}$$

By comparison with the balance equations of Kirchhoff rods, the last relations reveal the following interpretation of terms [7]: the vectors $\boldsymbol{\lambda}$ and $\boldsymbol{m}$, respectively, can be interpreted as shear force and moment densities acting on fibers "cross sections" (i.e., on surfaces that intersect fibers orthogonally). Further, the term $\mathrm{Div}\,(\boldsymbol{R}\boldsymbol{\sigma})$ can be interpreted as a distributed density of force transmitted to a fiber by the matrix in which it is embedded, whereas $2\,\mathrm{ax}\big[\mathrm{skw}(\boldsymbol{R}\boldsymbol{\sigma}\boldsymbol{F}^t)\big]$ is a density of distributed couple transmitted by the matrix to the fiber.

Remark 41.1. The two terms involving $\mathrm{Div}\,\boldsymbol{D}$ in the equilibrium equations (41.38) are vanishing, provided that the initial arrangement of fibers is such that $\mathrm{Div}\,\boldsymbol{D} = 0$. In this case, the field of unit tangent vectors to the initial fibers is divergence-free. Remind that any divergence-free vector field can be represented as $\mathrm{curl}\,\boldsymbol{U}$, for some potential vector field $\boldsymbol{U}$. Then, since we have $\boldsymbol{D} = \mathrm{curl}\,\boldsymbol{U}$ in this case, we see that $\boldsymbol{U}$ must satisfy the non-trivial condition $\|\mathrm{curl}\,\boldsymbol{U}\| = 1$. □

Following the discussion in [7], we assume in this paper that the strain energy density $\widetilde{W}$ decouples and has the following form

$$\widetilde{W}\,(\boldsymbol{E},\gamma;\boldsymbol{X}) = \widetilde{w}\,(\boldsymbol{E};\boldsymbol{X}) + \widetilde{\varphi}\,(\gamma;\boldsymbol{X}) \tag{41.39}$$

for some homogeneous quadratic function $\widetilde{\varphi}\,(\,\cdot\,;\boldsymbol{X})$.

41.3 Plate Theory for Fiber-Reinforced Laminae

In this section, we derive a plate model for laminae with fibers that resist flexure and twist by starting from the three-dimensional constrained Cosserat model presented in the previous section.

The reference configuration is a cylinder of the form $\mathcal{K} = \Omega \times \left[-\frac{h}{2}, \frac{h}{2}\right]$ where h is the thickness of the plate and Ω is the reference midsurface. Since Ω is flat, we assume that Ω belongs to the plane spanned by the unit vectors $\boldsymbol{E}_1, \boldsymbol{E}_2$. Then, $\boldsymbol{E}_3 = \boldsymbol{k}$ is the unit normal to Ω, see Fig. 41.1. We consider that the fibers are curved, but they are parallel to Ω, i.e., the fiber unit tangent $\boldsymbol{D}$ is orthogonal to $\boldsymbol{k}$.

Let us introduce some notations that are convenient for a plate theory. We decompose the position vector $\boldsymbol{X}$ (in the reference configuration) into its planar and normal parts as follows

$$\boldsymbol{X} = \boldsymbol{u} + \varsigma \boldsymbol{k} \qquad \text{with} \qquad \boldsymbol{u} = X_\alpha \boldsymbol{E}_\alpha \qquad \text{and} \qquad \varsigma = X_3 , \tag{41.40}$$

where ς is the thickness coordinate and $\alpha = 1, 2$. We mark any vector or tensor field which depends on $\boldsymbol{X}$ by a tilde, such as e.g. $\widetilde{\chi}, \widetilde{\boldsymbol{R}}, \widetilde{\boldsymbol{D}}_i, \widetilde{\boldsymbol{F}}, \widetilde{\boldsymbol{E}}, \widetilde{\gamma}$ (these fields are defined in $\mathcal{K}$ as in the previous section).

Thus, the strain measures for the three-dimensional constrained Cosserat model are given by

$$\widetilde{\boldsymbol{E}} = \widetilde{\boldsymbol{R}}^t \widetilde{\boldsymbol{F}} - \boldsymbol{I} \qquad \text{and} \qquad \widetilde{\gamma} = \text{ax}\left(\widetilde{\boldsymbol{R}}^t \widetilde{\boldsymbol{R}}'\right) = \widetilde{\boldsymbol{\Gamma}} \widetilde{\boldsymbol{D}}, \tag{41.41}$$

where $\widetilde{\boldsymbol{F}}$ and $\widetilde{\boldsymbol{\Gamma}}$ are defined by (41.1) and (41.4), respectively, and $(\cdot)'$ is the directional derivative along $\widetilde{\boldsymbol{D}}$. The constraints (41.9) can be written in the form

$$\widetilde{\boldsymbol{D}}_\gamma \cdot \widetilde{\boldsymbol{E}} \widetilde{\boldsymbol{D}} = 0, \qquad \text{or equivalently,} \qquad \left(\widetilde{\boldsymbol{D}}_\gamma \otimes \widetilde{\boldsymbol{D}}\right) \cdot \widetilde{\boldsymbol{E}} = 0 \qquad (\gamma = 2, 3). \tag{41.42}$$

Furthermore, for any field $\widetilde{f}(\boldsymbol{X})$ we denote by $\hat{f}(\boldsymbol{u}, \varsigma)$ the field defined by

$$\widetilde{f}(\boldsymbol{X}) = \widetilde{f}(\boldsymbol{u} + \varsigma \boldsymbol{k}) = \hat{f}(\boldsymbol{u}, \varsigma),$$

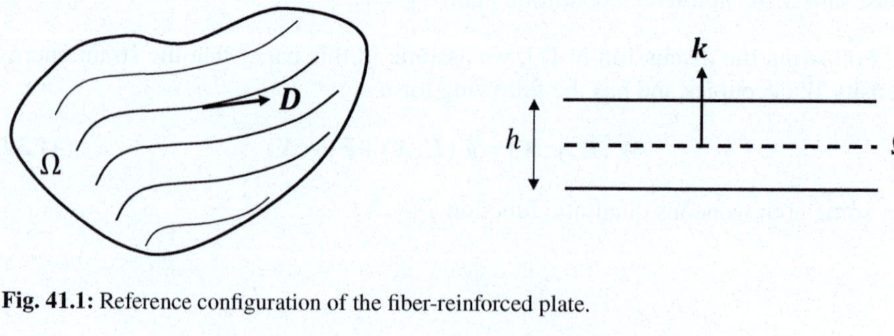

Fig. 41.1: Reference configuration of the fiber-reinforced plate.

in order to show explicitly the dependence on $\boldsymbol{u}$ and ς. Then, we can expand the strain energy density with respect to the small parameter ς in the form

$$\widetilde{W}\left(\widetilde{\boldsymbol{E}},\widetilde{\boldsymbol{\gamma}}\right)=\hat{W}\left(\hat{\boldsymbol{E}},\hat{\boldsymbol{\gamma}}\right)=\left[\hat{W}\left(\hat{\boldsymbol{E}},\hat{\boldsymbol{\gamma}}\right)\right]_{|\varsigma=0}+\varsigma\frac{\mathrm{d}}{\mathrm{d}\varsigma}\left[\hat{W}\left(\hat{\boldsymbol{E}},\hat{\boldsymbol{\gamma}}\right)\right]_{|\varsigma=0}+O\left(\varsigma^2\right) \tag{41.43}$$

and integrating over the thickness we obtain the total strain energy

$$\begin{aligned}\widetilde{\mathcal{S}}&=\int_{\mathcal{K}}\widetilde{W}\left(\widetilde{\boldsymbol{E}},\widetilde{\boldsymbol{\gamma}}\right)\mathrm{d}v=\int_{\Omega}\mathcal{U}\,\mathrm{d}a \qquad \text{with}\\ \mathcal{U}&=\int_{-\frac{h}{2}}^{\frac{h}{2}}\hat{W}\left(\hat{\boldsymbol{E}},\hat{\boldsymbol{\gamma}}\right)d\varsigma=h\,W(\boldsymbol{E},\boldsymbol{\gamma})+o\,(h),\end{aligned} \tag{41.44}$$

where we have put $\boldsymbol{E}=\hat{\boldsymbol{E}}_{|\varsigma=0}$, $\boldsymbol{\gamma}=\hat{\boldsymbol{\gamma}}_{|\varsigma=0}$, and in general we denote by $\boldsymbol{f}=\hat{\boldsymbol{f}}_{|\varsigma=0}$ for any field $\hat{\boldsymbol{f}}(\boldsymbol{u},\varsigma)$.

In order to express $\boldsymbol{E}$ and $\boldsymbol{\gamma}$ in terms of functions which depend only on the planar variables (X_1,X_2), we expand the deformation field $\widetilde{\boldsymbol{\chi}}(\boldsymbol{X})=\hat{\boldsymbol{\chi}}(\boldsymbol{u},\varsigma)$ in Taylor series about $\varsigma=0$ as

$$\hat{\boldsymbol{\chi}}(\boldsymbol{u},\varsigma)\;=\;\boldsymbol{r}(\boldsymbol{u})+\varsigma\,\boldsymbol{g}(\boldsymbol{u})+\cdots, \tag{41.45}$$

where $\boldsymbol{r}(\boldsymbol{u})=\hat{\boldsymbol{\chi}}(\boldsymbol{u},0)$ is the midsurface deformation and $\boldsymbol{g}(\boldsymbol{u})=\left(\frac{\partial\hat{\boldsymbol{\chi}}}{\partial\varsigma}\right)_{|\varsigma=0}$. Then, the deformation gradient is given by (see [16])

$$\hat{\boldsymbol{F}}=\nabla\hat{\boldsymbol{\chi}}+\frac{\partial\hat{\boldsymbol{\chi}}}{\partial\varsigma}\otimes\boldsymbol{k},$$

and for the midsurface $\varsigma=0$ we get

$$\boldsymbol{F}=\nabla\boldsymbol{r}+\boldsymbol{g}\otimes\boldsymbol{k}, \tag{41.46}$$

where ∇ stands now for the two-dimensional gradient with respect to (X_1,X_2). Let us decompose the identity tensor as

$$\boldsymbol{I}=\boldsymbol{1}+\boldsymbol{k}\otimes\boldsymbol{k} \qquad \text{with} \qquad \boldsymbol{1}=\boldsymbol{E}_\alpha\otimes\boldsymbol{E}_\alpha, \tag{41.47}$$

where $\boldsymbol{1}$ is the identity tensor in the midsurface plane ($\alpha=1,2$). From $(41.41)_1$, (41.46) and (41.47) we find the form of the strain tensor

$$\boldsymbol{E}=\boldsymbol{R}^t\boldsymbol{F}-\boldsymbol{I}=\left(\boldsymbol{R}^t\nabla\boldsymbol{r}-\boldsymbol{1}\right)+\left(\boldsymbol{R}^t\boldsymbol{g}-\boldsymbol{k}\right)\otimes\boldsymbol{k}, \tag{41.48}$$

where $\boldsymbol{R}=\hat{\boldsymbol{R}}_{|\varsigma=0}$ and $\boldsymbol{D}=\hat{\boldsymbol{D}}_{|\varsigma=0}$. Also, from $(41.41)_2$ and (41.4) we get

$$\boldsymbol{\gamma}=\boldsymbol{\Gamma}\boldsymbol{D}=\left[\mathrm{ax}\left(\boldsymbol{R}^t\boldsymbol{R}_{,i}\right)\otimes\boldsymbol{E}_i\right]\boldsymbol{D}=\left[\mathrm{ax}\left(\boldsymbol{R}^t\boldsymbol{R}_{,\alpha}\right)\otimes\boldsymbol{E}_\alpha\right]\boldsymbol{D}, \tag{41.49}$$

since $\boldsymbol{E}_3\cdot\boldsymbol{D}=0$, where $\alpha=1,2$. If we denote the tensors

$$\boldsymbol{\varepsilon} = \boldsymbol{R}^t \nabla \boldsymbol{r} - \mathbf{1} \qquad \text{and} \qquad \boldsymbol{\varkappa} = \operatorname{ax}\left(\boldsymbol{R}^t \boldsymbol{R}_{,\alpha}\right) \otimes \boldsymbol{E}_\alpha, \tag{41.50}$$

then the relations (41.48) and (41.49) can be written as

$$\boldsymbol{E} = \boldsymbol{\varepsilon} + \left(\boldsymbol{R}^t \boldsymbol{g} - \boldsymbol{k}\right) \otimes \boldsymbol{k} \qquad \text{and} \qquad \boldsymbol{\gamma} = \boldsymbol{\varkappa} \boldsymbol{D}. \tag{41.51}$$

Note that the strain measures $\boldsymbol{\varepsilon}$ and $\boldsymbol{\varkappa}$ are the strain tensor and bending-curvature tensor, respectively, which are usually employed in the theory of 6-parameter (or Cosserat) shells and plates, see e.g. [17]-[22].

Constraints (41.42) on the midplane $\varsigma = 0$ read

$$\boldsymbol{D}_\gamma \cdot \boldsymbol{E}\boldsymbol{D} = 0 \qquad (\gamma = 2, 3). \tag{41.52}$$

Taking into account (41.44), the strain energy for the plate to the leading order is

$$\mathcal{S} = \lim_{h \to 0} \frac{1}{h} \widetilde{\mathcal{S}} = \int_\Omega W(\boldsymbol{E}, \boldsymbol{\gamma})\, da. \tag{41.53}$$

In view of constraints (41.52), we consider the extended energy

$$\mathcal{E} = \int_\Omega \Big[W(\boldsymbol{E}, \boldsymbol{\gamma}) + \Lambda_\gamma (\boldsymbol{D}_\gamma \cdot \boldsymbol{E}\boldsymbol{D}) \Big] da, \tag{41.54}$$

where Λ_2 and Λ_3 are Lagrange multipliers. In order to deduce the equilibrium equations for fiber-reinforced plates, we employ the same variational procedure as in the three-dimensional theory (see Sect. 41.2.3). To accommodate constraints (41.52) we relax them and consider the extended variational problem corresponding to the virtual power statement

$$\dot{\mathcal{E}} = \mathcal{P}, \tag{41.55}$$

where the superposed dot stands for variational derivative as before, and $\mathcal{P}$ is the virtual power of the loads (41.65). Equilibria of the plate are defined as states that satisfy virtual power equation (41.55).

Let us compute the variational derivatives

$$\dot{W}(\boldsymbol{E}, \boldsymbol{\gamma}) = \boldsymbol{\sigma} \cdot \dot{\boldsymbol{E}} + \boldsymbol{M} \cdot \dot{\boldsymbol{\gamma}} \qquad \text{with} \qquad \boldsymbol{\sigma} = W_{\boldsymbol{E}} \quad \text{and} \quad \boldsymbol{M} = W_{\boldsymbol{\gamma}}. \tag{41.56}$$

Here, we have

$$\dot{\boldsymbol{E}} = (\boldsymbol{R}^t \boldsymbol{F} - \boldsymbol{I})^{\dot{}} = \boldsymbol{R}^t \dot{\boldsymbol{F}} + \dot{\boldsymbol{R}}^t \boldsymbol{F} = \boldsymbol{R}^t \left(\dot{\boldsymbol{F}} - \boldsymbol{\Omega} \boldsymbol{F}\right) \qquad \text{with} \qquad \dot{\boldsymbol{F}} = \nabla \boldsymbol{v} + \dot{\boldsymbol{g}} \otimes \boldsymbol{k}, \tag{41.57}$$

where we use the notations

$$\boldsymbol{v} = \dot{\boldsymbol{r}} \qquad \text{and} \qquad \boldsymbol{\Omega} = \dot{\boldsymbol{R}} \boldsymbol{R}^t, \qquad \boldsymbol{\omega} = \operatorname{ax}(\boldsymbol{\Omega}). \tag{41.58}$$

Moreover, from $(41.51)_2$ we have

$$\dot{\boldsymbol{\gamma}} = \dot{\boldsymbol{\varkappa}} \boldsymbol{D} = \boldsymbol{R}^t (\nabla \boldsymbol{\omega}) \boldsymbol{D}, \tag{41.59}$$

since we can prove, similarly to (41.24), that

$$\dot{\boldsymbol{\varkappa}} = \boldsymbol{R}^t \nabla\boldsymbol{\omega}. \tag{41.60}$$

Then, we can repeat transformations (41.19), (41.27), (41.29) and find

$$\begin{aligned}\dot{\mathcal{E}} = \int_\Omega \Big[&(\boldsymbol{R}\boldsymbol{\sigma} + \boldsymbol{\lambda}\otimes\boldsymbol{D})\cdot\dot{\boldsymbol{F}} - 2\,\mathrm{ax}\{\mathrm{skw}[(\boldsymbol{R}\boldsymbol{\sigma} + \boldsymbol{\lambda}\otimes\boldsymbol{D})\boldsymbol{F}^t]\}\cdot\boldsymbol{\omega} \\ &+(\boldsymbol{m}\otimes\boldsymbol{D})\cdot\nabla\boldsymbol{\omega} + \dot{\boldsymbol{\Lambda}}\cdot\boldsymbol{E}\boldsymbol{D}\Big]\mathrm{d}a,\end{aligned} \tag{41.61}$$

where we have designated by

$$\boldsymbol{m} = \boldsymbol{R}\boldsymbol{M} = \boldsymbol{R}\,\boldsymbol{W}_\gamma\,, \qquad \boldsymbol{\Lambda} = \Lambda_\gamma \boldsymbol{D}_\gamma \qquad \text{and} \qquad \boldsymbol{\lambda} = \boldsymbol{R}\boldsymbol{\Lambda} = \Lambda_\gamma \boldsymbol{d}_\gamma\,. \tag{41.62}$$

Taking into account the expression of $\dot{\boldsymbol{F}}$ in $(41.57)_2$ and the relation (41.30), we deduce

$$\begin{aligned}(\boldsymbol{R}\boldsymbol{\sigma} + \boldsymbol{\lambda}\otimes\boldsymbol{D})\cdot\dot{\boldsymbol{F}} &= (\boldsymbol{R}\boldsymbol{\sigma}\mathbf{1} + \boldsymbol{\lambda}\otimes\boldsymbol{D})\cdot\nabla\boldsymbol{v} + (\boldsymbol{R}\boldsymbol{\sigma} + \boldsymbol{\lambda}\otimes\boldsymbol{D})\cdot(\dot{\boldsymbol{g}}\otimes\boldsymbol{k}) \\ &= \mathrm{Div}\big[(\boldsymbol{R}\boldsymbol{\sigma}\mathbf{1} + \boldsymbol{\lambda}\otimes\boldsymbol{D})^t\boldsymbol{v}\big] - \boldsymbol{v}\cdot\mathrm{Div}(\boldsymbol{R}\boldsymbol{\sigma}\mathbf{1} + \boldsymbol{\lambda}\otimes\boldsymbol{D}) + \dot{\boldsymbol{g}}\cdot(\boldsymbol{R}\boldsymbol{\sigma})\boldsymbol{k}\,.\end{aligned} \tag{41.63}$$

We substitute (41.63) into (41.61) and use again the divergence theorem to obtain

$$\begin{aligned}\dot{\mathcal{E}} = &\int_\Omega \Big[\dot{\Lambda}_\gamma(\boldsymbol{D}_\gamma\cdot\boldsymbol{E}\boldsymbol{D}) + \dot{\boldsymbol{g}}\cdot\boldsymbol{R}\boldsymbol{\sigma}\boldsymbol{k}\Big]\mathrm{d}a - \int_\Omega \Big\{\boldsymbol{v}\cdot\mathrm{Div}\,(\boldsymbol{R}\boldsymbol{\sigma}\mathbf{1} + \boldsymbol{\lambda}\otimes\boldsymbol{D}) \\ &+\boldsymbol{\omega}\cdot\Big[\mathrm{Div}(\boldsymbol{m}\otimes\boldsymbol{D}) + 2\,\mathrm{ax}\{\mathrm{skw}[(\boldsymbol{R}\boldsymbol{\sigma} + \boldsymbol{\lambda}\otimes\boldsymbol{D})\boldsymbol{F}^t]\}\Big]\Big\}\mathrm{d}a \\ &+\int_{\partial\Omega} [\boldsymbol{v}\cdot(\boldsymbol{R}\boldsymbol{\sigma}\mathbf{1} + \boldsymbol{\lambda}\otimes\boldsymbol{D})\boldsymbol{n} + \boldsymbol{\omega}\cdot(\boldsymbol{m}\otimes\boldsymbol{D})\boldsymbol{n}]\,\mathrm{d}l\,.\end{aligned} \tag{41.64}$$

Next, we equate this with the virtual power

$$\mathcal{P} = \int_\Omega (\boldsymbol{l}\cdot\boldsymbol{v} + \boldsymbol{\pi}\cdot\boldsymbol{\omega})\,\mathrm{d}a + \int_{\partial\Omega} (\boldsymbol{t}\cdot\boldsymbol{v} + \boldsymbol{c}\cdot\boldsymbol{\omega})\,\mathrm{d}l\,, \tag{41.65}$$

and we can apply the fundamental lemma, since the variations $\boldsymbol{v}$, $\boldsymbol{\omega}$, $\dot{\boldsymbol{g}}$ and $\dot{\Lambda}_\gamma$ are arbitrary. Thus, the variations $\dot{\Lambda}_\gamma$ $(\gamma = 2,3)$ return constraints (41.52) while the variations $\dot{\boldsymbol{g}}$ yield the equation

$$\boldsymbol{R}\boldsymbol{\sigma}\boldsymbol{k} = \mathbf{0}, \tag{41.66}$$

which expresses that the normal stress $\boldsymbol{\sigma}\boldsymbol{k}$ is zero. Further, the variations $\boldsymbol{v}$ and $\boldsymbol{\omega}$ deliver the equilibrium equations

$$\begin{aligned}&\mathrm{Div}\,(\boldsymbol{R}\boldsymbol{\sigma}\mathbf{1} + \boldsymbol{\lambda}\otimes\boldsymbol{D}) + \boldsymbol{l} = \mathbf{0} \qquad \text{and} \\ &\mathrm{Div}\,(\boldsymbol{m}\otimes\boldsymbol{D}) + 2\,\mathrm{ax}\{\mathrm{skw}[(\boldsymbol{R}\boldsymbol{\sigma} + \boldsymbol{\lambda}\otimes\boldsymbol{D})\boldsymbol{F}^t]\} + \boldsymbol{\pi} = \mathbf{0} \qquad \text{in } \Omega,\end{aligned} \tag{41.67}$$

and the natural boundary conditions

$$(\boldsymbol{R}\boldsymbol{\sigma}\mathbf{1})\boldsymbol{n}+(\boldsymbol{D}\cdot\boldsymbol{n})\boldsymbol{\lambda}=\boldsymbol{t}\quad\text{on } \partial\Omega_t \qquad \text{and} \qquad (\boldsymbol{D}\cdot\boldsymbol{n})\boldsymbol{m}=\boldsymbol{c}\quad \text{on } \partial\Omega_c\,. \tag{41.68}$$

Here, $\boldsymbol{l}$ and $\boldsymbol{\pi}$ are densities of force and couple acting in Ω, while $\boldsymbol{t}$ and $\boldsymbol{c}$ are densities of force and couple acting on $\partial\Omega$. The subset $\partial\Omega_t$ is a part of $\partial\Omega$ where position is not assigned and $\partial\Omega_c$ is a part where rotation is not assigned. We assume position to be assigned on $\partial\Omega\setminus\partial\Omega_t$ and rotation to be assigned on $\partial\Omega\setminus\partial\Omega_c$.

In view of the relation (41.66), we can transform the following term appearing in the second equilibrium equation

$$(\boldsymbol{R}\boldsymbol{\sigma}+\boldsymbol{\lambda}\otimes\boldsymbol{D})\boldsymbol{F}^t=(\boldsymbol{R}\boldsymbol{\sigma}+\boldsymbol{\lambda}\otimes\boldsymbol{D})\big((\nabla\boldsymbol{r})^t+\boldsymbol{k}\otimes\boldsymbol{g}\big)=(\boldsymbol{R}\boldsymbol{\sigma}\mathbf{1}+\boldsymbol{\lambda}\otimes\boldsymbol{D})(\nabla\boldsymbol{r})^t. \tag{41.69}$$

In order to obtain a plate theory, we follow the general ideas from the classical theory of shells and plates (see, e.g. [23, 16]), namely we consider the relation (41.66) as an equation for the determination of the vector $\boldsymbol{g}$. We consider as in (41.39) the decoupling

$$W(\boldsymbol{E},\boldsymbol{\gamma})=w(\boldsymbol{E})+\varphi(\boldsymbol{\gamma}). \tag{41.70}$$

Then, we have $\boldsymbol{\sigma}=w_{\boldsymbol{E}}$ and Eq. (41.66) can be written in the form

$$w_{\boldsymbol{E}}\,\boldsymbol{k}=\mathbf{0}. \tag{41.71}$$

We see that the left-hand side is a function of the tensor $\boldsymbol{E}$, (where $\boldsymbol{E}=\boldsymbol{\varepsilon}+\big(\boldsymbol{R}^t\boldsymbol{g}-\boldsymbol{k}\big)\otimes\boldsymbol{k}$, cf. $(41.51)_1$). Assume that the energy density is such that equation (41.71) can be uniquely solved with respect to $\boldsymbol{g}$. Then, this solution will depend on the plate strain tensor $\boldsymbol{\varepsilon}$; let us denote the unique solution Eq. (41.71) by

$$\bar{\boldsymbol{g}}=\bar{\boldsymbol{g}}\,(\boldsymbol{\varepsilon}). \tag{41.72}$$

In the next section we shall consider a specific energy function $w\,(\boldsymbol{E})$ which satisfies this assumption (i.e., Eq. (41.71) is uniquely solvable) and we shall determine solution (41.72) in explicit form. Inserting (41.72) into $(41.51)_1$ we get the strain tensor

$$\bar{\boldsymbol{E}}=\boldsymbol{\varepsilon}+\big(\boldsymbol{R}^t\bar{\boldsymbol{g}}\,(\boldsymbol{\varepsilon})-\boldsymbol{k}\big)\otimes\boldsymbol{k} \tag{41.73}$$

as a function of $\boldsymbol{\varepsilon}$. Once we have determined the vector $\bar{\boldsymbol{g}}$ as in (41.72), we substitute it in the expression of the energy $w\,(\boldsymbol{E})$ to obtain the new energy function for the plate model $\bar{w}\,(\boldsymbol{\varepsilon})$, which will depend only on $\boldsymbol{\varepsilon}$. This means

$$w(\bar{\boldsymbol{E}})=w(\boldsymbol{\varepsilon}+\big(\boldsymbol{R}^t\bar{\boldsymbol{g}}(\boldsymbol{\varepsilon})-\boldsymbol{k}\big)\otimes\boldsymbol{k})=:\bar{w}(\boldsymbol{\varepsilon}). \tag{41.74}$$

For the whole strain energy we write

$$W(\bar{\boldsymbol{E}},\boldsymbol{\gamma})=w(\bar{\boldsymbol{E}})+\varphi(\boldsymbol{\gamma})=\bar{w}(\boldsymbol{\varepsilon})+\varphi(\boldsymbol{\gamma})=:\bar{W}(\boldsymbol{\varepsilon},\boldsymbol{\gamma}). \tag{41.75}$$

We remark that constraints (41.52) can also be expressed with $\boldsymbol{\varepsilon}$ in the form (since $\boldsymbol{k}\cdot\boldsymbol{D}=0$)

$$\boldsymbol{D}_{\gamma}\cdot\boldsymbol{\varepsilon}\boldsymbol{D}=0 \qquad (\gamma=2,3). \tag{41.76}$$

Corresponding to the new form of the strain energy function $\bar{W}(\boldsymbol{\varepsilon},\boldsymbol{\gamma})$ we define the new stress tensor by

$$\boldsymbol{P} = \bar{W}_{\boldsymbol{\varepsilon}} \qquad \text{and let} \qquad \boldsymbol{p} = \boldsymbol{R}\boldsymbol{P}. \tag{41.77}$$

Then, the relation between $\boldsymbol{P}$ and $\boldsymbol{\sigma}$ is given by

$$\boldsymbol{P} = \boldsymbol{\sigma}\boldsymbol{1}. \tag{41.78}$$

Indeed, we can verify this directly using the components of $\boldsymbol{E} = E_{AB}\boldsymbol{E}_A \otimes \boldsymbol{E}_B$: In view of $(41.51)_1$, we have $\boldsymbol{\varepsilon} = \boldsymbol{E}\boldsymbol{1} = E_{A\beta}\boldsymbol{E}_A \otimes \boldsymbol{E}_\beta$ (with $\beta = 1,2$), so

$$\boldsymbol{P} = \frac{\partial \bar{W}}{\partial \boldsymbol{\varepsilon}} = \frac{\partial W}{\partial E_{A\beta}} \boldsymbol{E}_A \otimes \boldsymbol{E}_\beta = \left(\frac{\partial W}{\partial E_{AB}} \boldsymbol{E}_A \otimes \boldsymbol{E}_B \right) \boldsymbol{1} = \boldsymbol{\sigma}\boldsymbol{1},$$

i.e. the relation (41.78) holds true.

By virtue of (41.69) and (41.78), we can write now equilibrium equations (41.67) and boundary conditions (41.68) in terms of the stress tensor $\boldsymbol{p}$ as follows

$$\begin{aligned} &\operatorname{Div}(\boldsymbol{p} + \boldsymbol{\lambda} \otimes \boldsymbol{D}) + \boldsymbol{l} = \boldsymbol{0} \qquad \text{and} \\ &\operatorname{Div}(\boldsymbol{m} \otimes \boldsymbol{D}) + 2\operatorname{ax}\{\operatorname{skw}[(\boldsymbol{p} + \boldsymbol{\lambda} \otimes \boldsymbol{D})(\nabla \boldsymbol{r})^t]\} + \boldsymbol{\pi} = \boldsymbol{0} \qquad \text{in } \Omega, \end{aligned} \tag{41.79}$$

and

$$\boldsymbol{p}\boldsymbol{n} + (\boldsymbol{D} \cdot \boldsymbol{n})\boldsymbol{\lambda} = \boldsymbol{t} \quad \text{on } \partial\Omega_t \qquad \text{and} \qquad (\boldsymbol{D} \cdot \boldsymbol{n})\,\boldsymbol{m} = \boldsymbol{c} \quad \text{on } \partial\Omega_c. \tag{41.80}$$

Remark 41.2. Of course, field equations (41.79) and (41.80) can also be deduced directly by a variational argument as in (41.54)–(41.68), starting from the virtual-power statement with the strain energy $\bar{W}(\boldsymbol{\varepsilon},\boldsymbol{\gamma})$. □

Finally, using equations of the type (41.36) for the two-dimensional divergence, as well as the relation

$$\begin{aligned} 2\operatorname{ax}\{\operatorname{skw}[(\boldsymbol{\lambda} \otimes \boldsymbol{D})(\nabla \boldsymbol{r})^t]\} &= 2\operatorname{ax}\{\operatorname{skw}[(\boldsymbol{\lambda} \otimes (\nabla \boldsymbol{r})\boldsymbol{D})]\} \\ &= 2\operatorname{ax}[\operatorname{skw}(\boldsymbol{\lambda} \otimes \boldsymbol{r}')] = \operatorname{ax}(\boldsymbol{\lambda} \otimes \boldsymbol{r}' - \boldsymbol{r}' \otimes \boldsymbol{\lambda}) = \boldsymbol{r}' \times \boldsymbol{\lambda}, \end{aligned} \tag{41.81}$$

we can write equilibrium equations (41.79) in the following alternative form

$$\begin{aligned} &\boldsymbol{\lambda}' + (\operatorname{Div} \boldsymbol{D})\boldsymbol{\lambda} + \operatorname{Div} \boldsymbol{p} + \boldsymbol{l} = \boldsymbol{0}, \\ &\boldsymbol{m}' + \boldsymbol{r}' \times \boldsymbol{\lambda} + (\operatorname{Div} \boldsymbol{D})\boldsymbol{m} + 2\operatorname{ax}\{\operatorname{skw}[\boldsymbol{p}(\nabla \boldsymbol{r})^t]\} + \boldsymbol{\pi} = \boldsymbol{0}. \end{aligned} \tag{41.82}$$

This form, which has similarities with the balance equations for Kirchhoff rods, facilitates the interpretation of various terms, like in the three-dimensional theory.

Remark 41.3. If the arrangement of fibers in the reference configuration is such that $\text{Div}\,\boldsymbol{D} = 0$, then the two terms involving $\text{Div}\,\boldsymbol{D}$ vanish, and equilibrium equations (41.82) simplify. In this case, we can represent the vector field $\boldsymbol{D}$ (provided that Ω is simply connected) in the form $\boldsymbol{D} = \boldsymbol{k} \times \nabla\phi$, where $\phi(X_1, X_2)$ is a scalar field. Then, the trajectories of the vector field $\boldsymbol{D}$ are curves on which $\phi(X_1, X_2)$ is constant. However, since $\boldsymbol{D}$ is a unit vector, the field ϕ must satisfy the condition $\|\nabla\phi\| = 1$. We can prove geometrically, with help of the divergence theorem, that the relation $\text{Div}\,\boldsymbol{D} = 0$ holds if and only if the fibers in the reference configuration are "parallel with respect to the normal direction" (see details in the Appendix). However, we do not assume this divergence-free condition in our model. □

To resume, the basic equations of the theory of fiber-reinforced plates are equilibrium equations (41.82), boundary conditions (41.80), together with the constraints (41.76). Here, the vector $\boldsymbol{\lambda}$ is determined by the Lagrange multipliers $(41.62)_2$, while $\boldsymbol{m}$ and $\boldsymbol{p}$ are given by constitutive relations $(41.62)_1$ and (41.77), respectively, where the plate strain energy density has the form (41.75).

41.4 Specific Constitutive Assumptions

In this section, we consider a specific form of the strain energy function which allow us to solve Eq. (41.71) and determine unique solution (41.72). As a result, we are able to write the explicit form of the plate strain energy density (41.74).

Consider that the three-dimensional energy function for the constrained Cosserat continuum is given by (41.39), i.e.

$$\widetilde{W}(\tilde{\boldsymbol{E}}, \tilde{\boldsymbol{\gamma}}) = \tilde{w}(\tilde{\boldsymbol{E}}) + \tilde{\varphi}(\tilde{\boldsymbol{\gamma}}), \tag{41.83}$$

where

$$\tilde{w}(\tilde{\boldsymbol{E}}) = \mu\|\text{sym}\,\tilde{\boldsymbol{E}}\|^2 + \mu_c\|\text{skw}\,\tilde{\boldsymbol{E}}\|^2 + \frac{1}{2}\lambda(\text{tr}\,\tilde{\boldsymbol{E}})^2 \tag{41.84}$$

and

$$\tilde{\varphi}(\tilde{\boldsymbol{\gamma}}) = \frac{1}{2}\tilde{\boldsymbol{\gamma}} \cdot \tilde{\boldsymbol{G}}\tilde{\boldsymbol{\gamma}}. \tag{41.85}$$

Here, $\tilde{\boldsymbol{E}}$ and $\tilde{\boldsymbol{\gamma}}$ are the three-dimensional strain measures given by (41.41), and the second order tensor $\tilde{\boldsymbol{G}}$ is assumed to be positive definite. The coefficients λ, μ are the Lamé constants, while μ_c is the Cosserat couple modulus of the three-dimensional Cosserat material, which are assumed to satisfy the inequalities

$$\mu > 0, \qquad 3\lambda + 2\mu > 0, \qquad \mu_c > 0. \tag{41.86}$$

We mention that the energy (41.84) has been employed previously for isotropic Cosserat media in [4, 24]. Inequalities (41.86) assure that energy function (41.84) is positive definite (in fact coercive), since we can write

$$\tilde{w}(\tilde{\boldsymbol{E}}) = \mu \|\mathrm{dev}_3(\mathrm{sym}\,\tilde{\boldsymbol{E}})\|^2 + \mu_c \|\mathrm{skw}\,\tilde{\boldsymbol{E}}\|^2 + \frac{1}{2}\kappa\big(\mathrm{tr}\,\tilde{\boldsymbol{E}}\big)^2 \geq \min\{\mu, \mu_c, \frac{3}{2}\kappa\}\, \|\tilde{\boldsymbol{E}}\|^2, \tag{41.87}$$

where $\kappa = \frac{1}{3}(3\lambda + 2\mu) > 0$ is the bulk modulus and $\mathrm{dev}_3 \boldsymbol{X} = \boldsymbol{X} - \frac{1}{3}(\mathrm{tr}\,\boldsymbol{X})\boldsymbol{I}$ is the (three-dimensional) deviatoric part of any tensor $\boldsymbol{X}$.

Now, if we reduce the energy function to the midsurface $\varsigma = 0$, according to the procedure presented in the previous section, we have

$$W(\boldsymbol{E}, \boldsymbol{\gamma}) = \widetilde{W}(\tilde{\boldsymbol{E}}, \tilde{\boldsymbol{\gamma}})_{|\varsigma=0} = \tilde{w}(\tilde{\boldsymbol{E}})_{|\varsigma=0} + \tilde{\varphi}(\tilde{\boldsymbol{\gamma}})_{|\varsigma=0}\,, \tag{41.88}$$

i.e.

$$\begin{aligned} W(\boldsymbol{E}, \boldsymbol{\gamma}) &= w(\boldsymbol{E}) + \varphi(\boldsymbol{\gamma}) \qquad \text{with} && (41.89)\\ w(\boldsymbol{E}) &= \mu \|\mathrm{sym}\,\boldsymbol{E}\|^2 + \mu_c \|\mathrm{skw}\,\boldsymbol{E}\|^2 + \frac{1}{2}\lambda\big(\mathrm{tr}\,\boldsymbol{E}\big)^2 \qquad \text{and} && (41.90)\\ \varphi(\boldsymbol{\gamma}) &= \frac{1}{2}\boldsymbol{\gamma}\cdot \boldsymbol{G}\boldsymbol{\gamma}, && (41.91) \end{aligned}$$

where $\boldsymbol{E}$ and $\boldsymbol{\gamma}$ are now given by (41.51). Then, the stress tensor $\boldsymbol{\sigma}$ defined in (41.56) has the form

$$\boldsymbol{\sigma} = W_{\boldsymbol{E}} = w_{\boldsymbol{E}} = 2\mu\,\mathrm{sym}\,\boldsymbol{E} + 2\mu_c \mathrm{skw}\,\boldsymbol{E} + \lambda\big(\mathrm{tr}\,\boldsymbol{E}\big)\boldsymbol{I}. \tag{41.92}$$

In order to solve Eq. (41.66) we use relation $(41.51)_1$ and compute

$$\begin{aligned} \boldsymbol{\sigma k} &= \big[\mu(\boldsymbol{E}+\boldsymbol{E}^t) + \mu_c(\boldsymbol{E}-\boldsymbol{E}^t) + \lambda\big(\mathrm{tr}\,\boldsymbol{E}\big)\boldsymbol{I}\big]\boldsymbol{k}\\ &= (\mu+\mu_c)\boldsymbol{E}\boldsymbol{k} + (\mu-\mu_c)\boldsymbol{E}^t\boldsymbol{k} + \lambda\big(\mathrm{tr}\,\boldsymbol{E}\big)\boldsymbol{k} \;=\; (\mu+\mu_c)(\boldsymbol{R}^t\boldsymbol{g}-\boldsymbol{k})\\ &\quad +(\mu-\mu_c)\big[\boldsymbol{\varepsilon}^t\boldsymbol{k} + ((\boldsymbol{R}^t\boldsymbol{g}-\boldsymbol{k})\cdot\boldsymbol{k})\boldsymbol{k}\big] + \lambda\big[\mathrm{tr}\,\boldsymbol{\varepsilon} + (\boldsymbol{R}^t\boldsymbol{g}-\boldsymbol{k})\cdot\boldsymbol{k}\big]\boldsymbol{k}\\ &= \big[(\mu+\mu_c)\boldsymbol{1} + (\lambda+2\mu)\boldsymbol{k}\otimes\boldsymbol{k}\big](\boldsymbol{R}^t\boldsymbol{g}-\boldsymbol{k}) + \big[(\mu-\mu_c)\boldsymbol{\varepsilon}^t\boldsymbol{k} + \lambda\big(\mathrm{tr}\,\boldsymbol{\varepsilon}\big)\boldsymbol{k}\big].\end{aligned} \tag{41.93}$$

Then, the equation $\boldsymbol{\sigma k} = \boldsymbol{0}$ is equivalent to

$$\boldsymbol{R}^t\boldsymbol{g} - \boldsymbol{k} = -\Big(\frac{1}{\mu+\mu_c}\boldsymbol{1} + \frac{1}{\lambda+2\mu}\boldsymbol{k}\otimes\boldsymbol{k}\Big)\big[(\mu-\mu_c)\boldsymbol{\varepsilon}^t\boldsymbol{k} + \lambda\big(\mathrm{tr}\,\boldsymbol{\varepsilon}\big)\boldsymbol{k}\big],$$

i.e.

$$\boldsymbol{R}^t\boldsymbol{g} - \boldsymbol{k} = -\frac{\mu-\mu_c}{\mu+\mu_c}\,\boldsymbol{\varepsilon}^t\boldsymbol{k} - \frac{\lambda}{\lambda+2\mu}\big(\mathrm{tr}\,\boldsymbol{\varepsilon}\big)\boldsymbol{k}\,, \tag{41.94}$$

which means that the unique solution is

$$\bar{\boldsymbol{g}} = \boldsymbol{R}\Big[\boldsymbol{k} - \frac{\mu-\mu_c}{\mu+\mu_c}\,\boldsymbol{\varepsilon}^t\boldsymbol{k} - \frac{\lambda}{\lambda+2\mu}\big(\mathrm{tr}\,\boldsymbol{\varepsilon}\big)\boldsymbol{k}\Big], \tag{41.95}$$

corresponding to (41.72). Next, we can make transformation (41.74), i.e. we insert the solution (41.95) into the energy function $w(\boldsymbol{E})$ and rewrite it as a function of $\boldsymbol{\varepsilon}$: From (41.73) and (41.95) it follows that

$$\bar{\boldsymbol{E}} = \boldsymbol{\varepsilon} - \frac{\mu - \mu_c}{\mu + \mu_c}\boldsymbol{\varepsilon}^t \boldsymbol{k} \otimes \boldsymbol{k} - \frac{\lambda}{\lambda + 2\mu}(\operatorname{tr}\boldsymbol{\varepsilon})\boldsymbol{k} \otimes \boldsymbol{k} \tag{41.96}$$

and inserting this in relation (41.74) we find the plate energy function

$$\begin{aligned}\bar{w}(\boldsymbol{\varepsilon}) &= \mu \left\| \operatorname{sym}\boldsymbol{\varepsilon} - \frac{\mu - \mu_c}{\mu + \mu_c}\operatorname{sym}(\boldsymbol{\varepsilon}^t \boldsymbol{k} \otimes \boldsymbol{k}) - \frac{\lambda}{\lambda + 2\mu}(\operatorname{tr}\boldsymbol{\varepsilon})\boldsymbol{k} \otimes \boldsymbol{k} \right\|^2 \\ &+ \mu_c \left\| \operatorname{skw}\boldsymbol{\varepsilon} - \frac{\mu - \mu_c}{\mu + \mu_c}\operatorname{skw}(\boldsymbol{\varepsilon}^t \boldsymbol{k} \otimes \boldsymbol{k}) \right\|^2 + \frac{2\lambda\mu^2}{(\lambda + 2\mu)^2}(\operatorname{tr}\boldsymbol{\varepsilon})^2 .\end{aligned} \tag{41.97}$$

After some rather lengthy but straightforward calculations we can write the right-hand side of (41.97) in the following compact form

$$\bar{w}(\boldsymbol{\varepsilon}) = \mu \|\operatorname{sym}\boldsymbol{\varepsilon}\|^2 + \mu_c \|\operatorname{skw}\boldsymbol{\varepsilon}\|^2 + \frac{\lambda\mu}{\lambda + 2\mu}(\operatorname{tr}\boldsymbol{\varepsilon})^2 - \frac{(\mu - \mu_c)^2}{2(\mu + \mu_c)}\|\boldsymbol{\varepsilon}^t \boldsymbol{k}\|^2 . \tag{41.98}$$

To resume, the strain energy density for fiber-reinforced plates is

$$\bar{W}(\boldsymbol{\varepsilon}, \boldsymbol{\gamma}) = h\bar{w}(\boldsymbol{\varepsilon}) + \frac{1}{2}h\boldsymbol{\gamma} \cdot \boldsymbol{G}\boldsymbol{\gamma}, \tag{41.99}$$

where the strain measures $\boldsymbol{\varepsilon}$ and $\boldsymbol{\gamma}$ are defined by $(41.50)_1$, $(41.51)_2$, the energy function $\bar{w}(\boldsymbol{\varepsilon})$ is given by (41.98) and $\boldsymbol{G}$ is a positive definite second order tensor.

Remark 41.4. For instance, in the case of fibers consisting of an isotropic material and having circular cross-sections on the micro-scale, the tensor $\boldsymbol{G}$ is given by

$$\boldsymbol{G} = T\boldsymbol{D} \otimes \boldsymbol{D} + F(\boldsymbol{D}_2 \otimes \boldsymbol{D}_2 + \boldsymbol{D}_3 \otimes \boldsymbol{D}_3), \tag{41.100}$$

where T (torsional stiffness) and F (flexural stiffness) are positive material constants. Thus, the part of the energy function accounting for fibers strain is [14, 8]

$$\varphi(\boldsymbol{\gamma}) = \frac{1}{2}\boldsymbol{\gamma} \cdot \boldsymbol{G}\boldsymbol{\gamma} = \frac{1}{2}\left[T\phi^2 + F(\gamma_2^2 + \gamma_3^2)\right], \tag{41.101}$$

where $\phi = \boldsymbol{\gamma} \cdot \boldsymbol{D}$ is the fiber twist strain and $\gamma_\alpha = \boldsymbol{\gamma} \cdot \boldsymbol{D}_\alpha$ ($\alpha = 2, 3$) are the fiber flexure strains. □

41.4.1 Coercivity

To show that the plate model is well posed, let us prove that energy function (41.98) is coercive.

Lemma 41.1. *Assume that the constitutive coefficients satisfy conditions* (41.86). *Then, the strain energy function* $\bar{w}(\boldsymbol{\varepsilon})$ *given by* (41.98) *is coercive, in the sense that there exists a constant* $c_0 > 0$ *such that*

$$\bar{w}(\boldsymbol{\varepsilon}) \geq c_0 \|\boldsymbol{\varepsilon}\|^2. \tag{41.102}$$

Proof. To show this, let us put the right-hand side of (41.98) in an alternative form. Using the decomposition

$$\boldsymbol{\varepsilon} = (\mathbf{1} + \boldsymbol{k} \otimes \boldsymbol{k})\boldsymbol{\varepsilon} = \mathbf{1}\boldsymbol{\varepsilon} + \boldsymbol{k} \otimes \boldsymbol{\varepsilon}^t \boldsymbol{k},$$

we can transform (41.98) in the equivalent form

$$\bar{w}(\boldsymbol{\varepsilon}) = \mu \|\mathrm{sym}(\mathbf{1}\boldsymbol{\varepsilon})\|^2 + \mu_c \|\mathrm{skw}(\mathbf{1}\boldsymbol{\varepsilon})\|^2 + \frac{\lambda\mu}{\lambda + 2\mu}\big[\mathrm{tr}(\mathbf{1}\boldsymbol{\varepsilon})\big]^2 + \frac{2\mu\,\mu_c}{\mu + \mu_c}\|\boldsymbol{\varepsilon}^t \boldsymbol{k}\|^2. \tag{41.103}$$

Further, let us introduce the two-dimensional deviator operator by

$$\mathrm{dev}_2 \boldsymbol{X} = \boldsymbol{X} - \frac{1}{2}(\mathrm{tr}\,\boldsymbol{X})\mathbf{1}. \tag{41.104}$$

Then, we can decompose any tensor of the type $\boldsymbol{X} = X_{i\alpha}\boldsymbol{E}_i \otimes \boldsymbol{E}_\alpha$ (with $\alpha = 1,2$) as a direct sum (i.e., orthogonal decomposition) in the form

$$\boldsymbol{X} = \mathrm{dev}_2(\mathrm{sym}\,\boldsymbol{X}) + \mathrm{skw}\,\boldsymbol{X} + \frac{1}{2}(\mathrm{tr}\,\boldsymbol{X})\mathbf{1}. \tag{41.105}$$

It follows that

$$\begin{aligned} &\|\boldsymbol{X}\|^2 = \|\mathrm{dev}_2(\mathrm{sym}\,\boldsymbol{X})\|^2 + \|\mathrm{skw}\,\boldsymbol{X}\|^2 + \frac{1}{2}(\mathrm{tr}\,\boldsymbol{X})^2 \qquad \text{and} \\ &\|\mathrm{sym}\,\boldsymbol{X}\|^2 = \|\mathrm{dev}_2(\mathrm{sym}\,\boldsymbol{X})\|^2 + \tfrac{1}{2}(\mathrm{tr}\,\boldsymbol{X})^2 \end{aligned} \tag{41.106}$$

Using the last relation we deduce from (41.103) the representation

$$\bar{w}(\boldsymbol{\varepsilon}) = \mu \|\mathrm{dev}_2\mathrm{sym}(\mathbf{1}\boldsymbol{\varepsilon})\|^2 + \mu_c \|\mathrm{skw}(\mathbf{1}\boldsymbol{\varepsilon})\|^2 + \frac{\mu(3\lambda + 2\mu)}{2(\lambda + 2\mu)}\big[\mathrm{tr}(\mathbf{1}\boldsymbol{\varepsilon})\big]^2 + \frac{2\mu\,\mu_c}{\mu + \mu_c}\|\boldsymbol{\varepsilon}^t \boldsymbol{k}\|^2. \tag{41.107}$$

In view of assumptions (41.86), we see that all coefficients in (41.107) are positive. Hence, the quadratic energy function $\bar{w}(\boldsymbol{\varepsilon})$ is positive definite. To show its coercivity, let us denote by

$$c_0 = \min\left\{\mu, \mu_c, \frac{\mu(3\lambda + 2\mu)}{\lambda + 2\mu}, \frac{2\mu\,\mu_c}{\mu + \mu_c}\right\} > 0.$$

Then, relations $(41.106)_1$ and (41.107) yield

$$\begin{aligned} \bar{w}(\boldsymbol{\varepsilon}) &\geq c_0\Big(\|\mathrm{dev}_2\mathrm{sym}(\mathbf{1}\boldsymbol{\varepsilon})\|^2 + \|\mathrm{skw}(\mathbf{1}\boldsymbol{\varepsilon})\|^2 + \frac{1}{2}\big[\mathrm{tr}(\mathbf{1}\boldsymbol{\varepsilon})\big]^2 + \|\boldsymbol{\varepsilon}^t \boldsymbol{k}\|^2\Big) \\ &= c_0\Big(\|\mathbf{1}\boldsymbol{\varepsilon}\|^2 + \|\boldsymbol{\varepsilon}^t \boldsymbol{k}\|^2\Big) = c_0 \|\boldsymbol{\varepsilon}\|^2, \end{aligned} \tag{41.108}$$

i.e. the strain energy function $\bar{w}(\boldsymbol{\varepsilon})$ is coercive.

Remark 41.5. Since the energy function is quadratic and positive definite, we deduce that $\bar{w}(\boldsymbol{\varepsilon})$ is a convex function of $\boldsymbol{\varepsilon}$. On the basis of these properties, one can show the existence of minimizers for the corresponding minimization problem (see. e.g. [21, 22]). Thus, the plate model is well-posed. □

41.4.2 Minimum Property

Let us show now that we are justified to insert the vector $\bar{\boldsymbol{g}}(\boldsymbol{\varepsilon})$ in the energy function (as we did in (41.74)), since from the energetical viewpoint we obtain the optimal model.

Lemma 41.2. *With the above notations, the vector* $\bar{\boldsymbol{g}} = \bar{\boldsymbol{g}}(\boldsymbol{\varepsilon})$ *given by* (41.95) *minimizes the strain energy function* $w(\boldsymbol{E})$ *for given fields* $\boldsymbol{R}$ *and* $\boldsymbol{r}$, *i.e. it holds*

$$w\big(\boldsymbol{\varepsilon}+(\boldsymbol{R}^t\boldsymbol{g}-\boldsymbol{k})\otimes\boldsymbol{k}\big) > w\big(\boldsymbol{\varepsilon}+(\boldsymbol{R}^t\bar{\boldsymbol{g}}(\boldsymbol{\varepsilon})-\boldsymbol{k})\otimes\boldsymbol{k}\big), \qquad \textit{for all} \quad \boldsymbol{g}\neq\bar{\boldsymbol{g}}(\boldsymbol{\varepsilon}). \tag{41.109}$$

Proof. Let us define the function

$$F(\boldsymbol{g}) := w\big(\boldsymbol{\varepsilon}+(\boldsymbol{R}^t\boldsymbol{g}-\boldsymbol{k})\otimes\boldsymbol{k}\big) \quad \text{at fixed} \quad \boldsymbol{R} \text{ and } \boldsymbol{r}. \tag{41.110}$$

Consider a path $\boldsymbol{g}(x)$ and let $f(x) := F\big(\boldsymbol{g}(x)\big)$. The derivatives of $f(x)$ are

$$\begin{aligned} f'(x) &= \frac{\mathrm{d}}{\mathrm{d}x}F\big(\boldsymbol{g}(x)\big) = w_{\boldsymbol{E}}\big(\boldsymbol{\varepsilon}+(\boldsymbol{R}^t\boldsymbol{g}-\boldsymbol{k})\otimes\boldsymbol{k}\big)\cdot\big(\boldsymbol{R}^t\boldsymbol{g}'(x)\otimes\boldsymbol{k}\big) \\ &= \boldsymbol{\sigma}(x)\cdot\boldsymbol{R}^t\big(\boldsymbol{g}'(x)\otimes\boldsymbol{k}\big) = \boldsymbol{R}\boldsymbol{\sigma}(x)\cdot\big(\boldsymbol{g}'(x)\otimes\boldsymbol{k}\big) = \big(\boldsymbol{R}\boldsymbol{\sigma}(x)\boldsymbol{k}\big)\cdot\boldsymbol{g}'(x) \end{aligned} \tag{41.111}$$

and

$$f''(x) = \big(\boldsymbol{R}\boldsymbol{\sigma}(x)\boldsymbol{k}\big)\cdot\boldsymbol{g}''(x) + \big(\boldsymbol{R}\boldsymbol{\sigma}'(x)\boldsymbol{k}\big)\cdot\boldsymbol{g}'(x). \tag{41.112}$$

Using (41.92) and $(41.51)_1$ we compute

$$\begin{aligned} \boldsymbol{\sigma}'(x)\boldsymbol{k} &= \big[(\mu+\mu_c)(\boldsymbol{R}^t\boldsymbol{g}'(x)\otimes\boldsymbol{k}) + (\mu-\mu_c)(\boldsymbol{k}\otimes\boldsymbol{R}^t\boldsymbol{g}'(x)) + \lambda(\boldsymbol{R}^t\boldsymbol{g}'(x)\cdot\boldsymbol{k})\boldsymbol{I}\big]\boldsymbol{k} \\ &= (\mu+\mu_c)\boldsymbol{R}^t\boldsymbol{g}'(x) + (\mu-\mu_c)(\boldsymbol{R}^t\boldsymbol{g}'(x)\cdot\boldsymbol{k})\boldsymbol{k} + \lambda(\boldsymbol{R}^t\boldsymbol{g}'(x)\cdot\boldsymbol{k})\boldsymbol{k} \\ &= \big[(\mu+\mu_c)\boldsymbol{1} + (\lambda+2\mu)\boldsymbol{k}\otimes\boldsymbol{k}\big]\boldsymbol{R}^t\boldsymbol{g}'(x). \end{aligned} \tag{41.113}$$

Also, the derivative of $F(\boldsymbol{g})$ with respect to the vector $\boldsymbol{g} = g_A\boldsymbol{E}_A$ is

$$\begin{aligned} F_{\boldsymbol{g}} &= \frac{\partial F}{\partial g_A}\boldsymbol{E}_A = \Big[w_{\boldsymbol{E}}\cdot\frac{\partial}{\partial g_A}\big(\boldsymbol{\varepsilon}+(\boldsymbol{R}^t\boldsymbol{g}-\boldsymbol{k})\otimes\boldsymbol{k}\big)\Big]\boldsymbol{E}_A \\ &= \big[\boldsymbol{\sigma}\cdot(\boldsymbol{R}^t\boldsymbol{E}_A\otimes\boldsymbol{k})\big]\boldsymbol{E}_A = \big[(\boldsymbol{R}\boldsymbol{\sigma}\boldsymbol{k})\cdot\boldsymbol{E}_A\big]\boldsymbol{E}_A = \boldsymbol{R}\boldsymbol{\sigma}\boldsymbol{k}\,. \end{aligned} \tag{41.114}$$

Using (41.113), (41.114), we can express derivatives (41.111), (41.112) as follows

$$f'(x) = F_{\boldsymbol{g}}\big(\boldsymbol{g}(x)\big)\cdot\boldsymbol{g}'(x) \qquad \text{and} \tag{41.115}$$

$$f''(x) = F_{\boldsymbol{g}}\big(\boldsymbol{g}(x)\big)\cdot\boldsymbol{g}''(x) + \boldsymbol{R}^t\boldsymbol{g}'(x)\cdot\big[(\mu+\mu_c)\boldsymbol{1} + (\lambda+2\mu)\boldsymbol{k}\otimes\boldsymbol{k}\big]\boldsymbol{R}^t\boldsymbol{g}'(x). \tag{41.116}$$

Consider now the special path of the form $\boldsymbol{g}(x) = (1-x)\boldsymbol{g}_1 + x\boldsymbol{g}_2$ with $x \in [0,1]$, where $\boldsymbol{g}_1 \neq \boldsymbol{g}_2$ are two fixed vectors. Then, we have

$$\boldsymbol{g}'(x) = \boldsymbol{g}_2 - \boldsymbol{g}_1 \neq \boldsymbol{0} \qquad \text{and} \qquad \boldsymbol{g}''(x) = \boldsymbol{0}. \tag{41.117}$$

Inserting this in (41.116) we obtain

$$\begin{aligned} f''(x) &= \boldsymbol{R}^t(\boldsymbol{g}_2-\boldsymbol{g}_1)\cdot\big[(\mu+\mu_c)\boldsymbol{1} + (\lambda+2\mu)\boldsymbol{k}\otimes\boldsymbol{k}\big]\boldsymbol{R}^t(\boldsymbol{g}_2-\boldsymbol{g}_1) \qquad (41.118)\\ &= (\mu+\mu_c)\sum_{\alpha=1}^{2}|\boldsymbol{R}^t(\boldsymbol{g}_2-\boldsymbol{g}_1)\cdot\boldsymbol{E}_\alpha|^2 + (\lambda+2\mu)|\boldsymbol{R}^t(\boldsymbol{g}_2-\boldsymbol{g}_1)\cdot\boldsymbol{k}|^2 > 0, \end{aligned}$$

since $\mu+\mu_c > 0$ and $\lambda+2\mu > 0$. Thus, the function $f(x)$ is convex on $[0,1]$. Hence, we deduce

$$\begin{aligned} f'(x) &= f'(0) + \int_0^x f''(s)\,\mathrm{d}s > f'(0) \qquad \text{and}\\ f(1)-f(0) &= \int_0^1 f'(x)\,\mathrm{d}x > f'(0). \end{aligned} \tag{41.119}$$

Since $f(1) = F(\boldsymbol{g}_2)$ and $f(0) = F(\boldsymbol{g}_1)$, we obtain

$$F(\boldsymbol{g}_2) - F(\boldsymbol{g}_1) > f'(0) = F_{\boldsymbol{g}}(\boldsymbol{g}_1)\cdot\boldsymbol{g}'(0) = F_{\boldsymbol{g}}(\boldsymbol{g}_1)\cdot(\boldsymbol{g}_2-\boldsymbol{g}_1), \tag{41.120}$$

which holds for all vectors $\boldsymbol{g}_1 \neq \boldsymbol{g}_2$. The last relation expresses the convexity of the function $F(\boldsymbol{g})$ with respect to $\boldsymbol{g}$.

Suppose now that $\boldsymbol{g}_1 = \bar{\boldsymbol{g}}(\boldsymbol{\varepsilon})$ given by (41.95) and $\boldsymbol{g}_2 = \boldsymbol{g}$ is arbitrary. Then, the inequality (41.120) reads

$$F(\boldsymbol{g}) - F(\bar{\boldsymbol{g}}) > F_{\boldsymbol{g}}(\bar{\boldsymbol{g}})\cdot(\boldsymbol{g}-\bar{\boldsymbol{g}}) = \boldsymbol{R}\boldsymbol{\sigma}(\bar{\boldsymbol{g}})\boldsymbol{k}\cdot(\boldsymbol{g}-\bar{\boldsymbol{g}}) = 0, \tag{41.121}$$

since $\boldsymbol{\sigma}(\bar{\boldsymbol{g}})\boldsymbol{k} = \boldsymbol{0}$ by construction. Thus, we obtain $F(\boldsymbol{g}) > F(\bar{\boldsymbol{g}})$ for any $\boldsymbol{g} \neq \bar{\boldsymbol{g}}$, i.e. inequality (41.109) holds. The proof is complete.

In conclusion, we are justified to choose the vector $\bar{\boldsymbol{g}}(\boldsymbol{\varepsilon})$ (given by (41.95)) in the expression of the energy, since it minimizes the strain energy function $W(\boldsymbol{E},\boldsymbol{\gamma})$ among all other choices of $\boldsymbol{g}$ (for fixed $\boldsymbol{R}$ and $\boldsymbol{r}$).

Acknowledgements The research of MB has been funded by the Deutsche Forschungsgemeinschaft (DFG, German Research Foundation) – Project no. 415894848: BI 1965/2-1.

Appendix

In this Appendix we want to characterize geometrically the case $\operatorname{Div}\boldsymbol{D} = 0$, where $\boldsymbol{D}$ is the unit tangent vector field to the fibers in $\Omega \subset \mathbb{R}^2$. Let us introduce the unit normal vector field in Ω given by the relation $\boldsymbol{N} = \boldsymbol{D} \times \boldsymbol{k}$.

Definition 41.1. Let $\varphi \subset \Omega$ be a fiber and $P \in \Omega$ be a point which is not on the fiber. We define the *distance in normal direction* from P to the fiber φ as the length of the curve segment $\widetilde{PQ}$, where $Q \in \varphi$ is a point on the fiber and the curve $\widetilde{PQ}$ has unit tangent vector $\boldsymbol{N}$ (i.e., the curve $\widetilde{PQ}$ is in normal direction, see Fig. 41.2a).

Definition 41.2. We say that two fibers $\varphi, \psi \subset \Omega$ are *parallel with respect to the normal direction* if the distance in normal direction from any point $P \in \varphi$ to the fiber ψ is constant along φ.

Example 41.1. If Ω is a circular domain and the fibers are concentric circle lines, then the fibers are parallel with respect to the normal direction (see Fig. 41.2b).

Lemma 41.3. *Let* $\boldsymbol{D}$ *be the unit tangent vector to the family of fibers in* Ω*. Then, we have* $\operatorname{Div}\boldsymbol{D} = 0$ *if and only if any two fibers* φ *and* ψ *are parallel with respect to the normal direction.*

Proof. Consider two fibers φ and ψ and two points P, R on the fiber φ (see Figure 41.2a). Consider the curve in normal direction (with unit tangent $\boldsymbol{N}$) passing through P and intersecting the fiber ψ in Q. Also, consider the curve in normal direction passing through R and intersecting the fiber ψ in S. We want to show that the lengths of the curve segments $\widetilde{PQ}$ and $\widetilde{RS}$ are equal. To this aim, we apply the divergence theorem on the domain Δ, which is bounded by the fibers φ, ψ and the curves $\widetilde{PQ}$ and $\widetilde{RS}$. We get

$$\int_{\Delta} \operatorname{Div}\boldsymbol{D}\, da = \int_{\partial\Delta} \boldsymbol{D} \cdot \boldsymbol{\nu}\, dl\,, \tag{41.122}$$

where $\boldsymbol{\nu}$ is the outer unit normal to $\partial\Delta$. Note that on the fibers $\varphi, \psi \subset \partial\Delta$ we have $\boldsymbol{\nu} = \pm\boldsymbol{N}$, so it holds $\boldsymbol{D} \cdot \boldsymbol{\nu} = 0$. On the other hand, on the boundary curves $\widetilde{PQ}$ and $\widetilde{RS}$

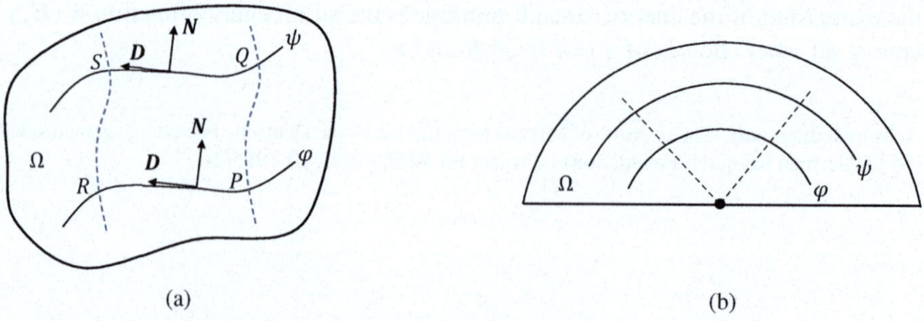

Fig. 41.2: (a) Two fibers φ and ψ which are parallel with respect to the normal direction; (b) Circular domain Ω with concentric fibers.

the unit normal is $\boldsymbol{\nu} = \pm\boldsymbol{D}$, so we have $\boldsymbol{D}\cdot\boldsymbol{\nu} = \pm 1$. Hence, taking the absolute value of equation (41.122) we obtain

$$\left|\int_{\Delta} \mathrm{Div}\,\boldsymbol{D}\,\mathrm{d}a\right| = \left|\int_{\widetilde{PQ}} 1\,\mathrm{d}l - \int_{\widetilde{RS}} 1\,\mathrm{d}l\right|. \tag{41.123}$$

Let us assume now that $\mathrm{Div}\,\boldsymbol{D} = 0$. Then, from relation (41.123) we get

$$\int_{\widetilde{PQ}} 1\,\mathrm{d}l = \int_{\widetilde{RS}} 1\,\mathrm{d}l, \qquad \text{i.e.} \qquad \text{length}(\widetilde{PQ}) = \text{length}(\widetilde{RS}).$$

Since P, R are arbitrary points on φ, we conclude by definition that the fibers φ and ψ are parallel with respect to the normal direction.

Conversely, let us suppose that any two fibers φ and ψ in Ω are parallel with respect to the normal direction. We can construct the domain Δ as above, and from the relation (41.123) we deduce

$$\int_{\Delta} \mathrm{Div}\,\boldsymbol{D}\,\mathrm{d}a = 0.$$

Then, using the continuity of $\mathrm{Div}\,\boldsymbol{D}$ and letting the distances between the fibers φ, ψ and between the points P, R tend to zero, we obtain that $\mathrm{Div}\,\boldsymbol{D} = 0$ in any point of Ω.

References

[1] Dill E (1992) Kirchhoff's theory of rods, Arch. Hist. Exact Sci. **44**:1-23.
[2] Cosserat E, Cosserat F (1909) Théorie des corps déformables, Hermann et Fils (reprint 2009), Paris.
[3] Pietraszkiewicz W, Eremeyev V (2009) On natural strain measures of the non-linear micropolar continuum, Int J Solids Struct **46**:774-787.
[4] Neff P, Bîrsan M, Osterbrink F (2015) Existence theorem for geometrically nonlinear Cosserat micropolar model under uniform convexity requirements, J Elasticity **121**:119-141.
[5] Steigmann D (2012) Theory of elastic solids reinforced with fibers resistant to extension, flexure and twist, Int J Non-Linear Mech **47**:734-742.
[6] Steigmann D (2015) Effects of fiber bending and twisting resistance on the mechanics of fiber-reinforced elastomers, In: L Dorfmann, R Ogden (Eds) Nonlinear Mechanics of Soft Fibrous Materials, CISM Courses and Lectures 559, pp 269-305, Springer, Wien.
[7] Shirani M, Steigmann D (2020) A Cosserat model of elastic solids reinforced by a family of curved and twisted fibers, Symmetry **12**:1133.
[8] McAvoy R, Steigmann D (2022) Cosserat elasticity of helically wound cylinders, J Elasticity, DOI 10.1007/s10659-022-09934-z online first.

[9] Neff P, Jeong J, Münch I, Ramezani H (2010) Linear Cosserat Elasticity, Conformal Curvature and Bounded Stiffness, In: G Maugin, V Metrikine (Eds) Mechanics of Generalized Continua. One hundred years after the Cosserats, Advances in Mechanics and Mathematics, vol 21, pp 55-63, Springer, Berlin.

[10] Neff P (2004) A geometrically exact Cosserat-shell model including size effects, avoiding degeneracy in the thin shell limit. Part I: Formal dimensional reduction for elastic plates and existence of minimizers for positive Cosserat couple modulus, Cont Mech Thermodynamics **16**:577-628.

[11] Bîrsan M, Ghiba I, Martin R, Neff P (2019) Refined dimensional reduction for isotropic elastic Cosserat shells with initial curvature, Math Mech Solids **24**:4000-4019.

[12] Krishnan J (2016) A generalized Cosserat theory with applications to fiber-reinforced plate, Dissertation, University of California at Berkeley.

[13] Eremeyev V, Pietraszkiewicz W (2012) Material symmetry group of the non-linear polar-elastic continuum, Int J Solids Struct **49**:1993-2005.

[14] Landau L, Lifshitz E (1986) Theory of Elasticity, 3rd edn, Pergamon, Oxford.

[15] Steigmann D (1996) The variational structure of a nonlinear theory for spatial lattices, Meccanica **31**:441-455.

[16] Steigmann D, Bîrsan M, Shirani M (2023) Lecture Notes on the Theory of Plates and Shells, Series Solid Mechanics and Its Applications, Springer, to appear.

[17] Libai A, Simmonds J (1998) The Nonlinear Theory of Elastic Shells., 2nd edn, Cambridge University Press, Cambridge.

[18] Chróścielewski J, Makowski J, Pietraszkiewicz W (2004) Statics and Dynamics of Multifold Shells: Nonlinear Theory and Finite Element Method (in Polish), Wydawnictwo IPPT PAN, Warsaw.

[19] Eremeyev V, Pietraszkiewicz W (2004) The nonlinear theory of elastic shells with phase transitions, J Elasticity **74**:67-86.

[20] Eremeyev V, Pietraszkiewicz W (2006) Local symmetry group in the general theory of elastic shells, J Elasticity **85**:125-152.

[21] Bîrsan M, Neff P (2013) Existence theorems in the geometrically non-linear 6-parameter theory of elastic plates, J Elasticity **112**:185-198.

[22] Bîrsan M, Neff P (2014) Existence of minimizers in the geometrically non-linear 6-parameter resultant shell theory with drilling rotations, Math Mech Solids **19**:376-397.

[23] Steigmann D (2013) Koiter's shell theory from the perspective of three-dimensional nonlinear elasticity, J Elasticity **111**:91-107.

[24] Sander O, Neff P, Bîrsan M (2016) Numerical treatment of a geometrically nonlinear planar Cosserat shell model, Computational Mechanics **57**:817-841.

Chapter 42
On the Structure of Solutions in the Vicinity of Discontinuity of Boundary Conditions for Gradient Models

Alexander O. Vatulyan, Sergey A. Nesterov, Victor O. Yurov, and Oksana V. Yavruyan

Abstract A study is made of the influence of gradient effects on the structure of singular solutions in mixed problems of the theory of elasticity and electroelasticity in the vicinity of a discontinuity of boundary conditions. Antiplanar problems for a strip with delamination and the problem of deformation of a polarized strip when an electric potential is applied to an electrode of finite length located on its upper boundary are considered. The study of the problems under consideration was carried out within the framework of gradient models of the theory of elasticity and electroelasticity, in particular, a simplified one-parameter Aifantis model was used. Problem solving is reduced to solving and analyzing boundary integral equations (BIEs) with irregular kernels. For the problem of a strip with delamination within the framework of the gradient theory of elasticity (GTE), an integral equation with a cubic singularity is obtained. It is shown that the use of the gradient model significantly changes the structure of the BIE solution in the vicinity of the stress concentrator - the peaks of delamination.

The study of the problem for an electroelastic strip is aimed at analyzing the electrostatic field in the vicinity of the electrode boundaries and analyzing the normal component of the electric induction vector on the electrode. After applying the Fourier transform, the boundary value problem with mixed boundary conditions is reduced to a boundary integral equation (BIE). A comparative analysis of the results obtained on the basis of classical and gradient electroelasticity models depending on the values of the gradient parameter is carried out. The analysis shows that gradient

---

Alexander O. Vatulyan · Victor O. Yurov · Oksana V. Yavruyan
Institute of Mathematics, Mechanics and Computer Sciences named after I.I. Vorovich, Southern Federal University, 8a Mil'chakova Str., 344090 Rostov-on-Don & Southern Mathematical Institute, Vladikavkaz Scientific Center of Russian Academy of Sciences, 22 Markusa Str., 362027 Vladikavkaz, Russian Federation,
e-mail: aovatulyan@sfedu.ru,vyurov@sfedu.ru,oyavruyan@sfedu.ru

Sergey A. Nesterov
Southern Mathematical Institute, Vladikavkaz Scientific Center of Russian Academy of Sciences, 22 Markusa Str., 362027 Vladikavkaz, Russian Federation,
e-mail: 1079@list.ru

H. Altenbach et al. (eds.), *Sixty Shades of Generalized Continua*,
Advanced Structured Materials 170, https://doi.org/10.1007/978-3-031-26186-2_42

effects do not appear when considering the one-parameter Aifantis model within the framework of the formulation under consideration.

Key words: Stripe · Electrode · Delamination · Gradient elasticity/electroelasticity theory · Boundary integral equation · Singular solutions

42.1 Introduction

Mixed problems are problems of solid mechanics for a region whose boundary is divided into sections, each of which has its own boundary conditions, i.e. there are sections with discontinuity of boundary conditions. Mixed problems include contact problems - problems about a stamp, problems of the theory of cracks, and problems of electroelasticity, when there are one or more electrodes of finite length at the boundary of the body under study, which are included in the circuit.

The classical solution of such problems, obtained using standard models of the linear theory of elasticity and electroelasticity, is well known and has a singular character in the vicinity of characteristic points - stress concentrators, for example, the edge of a stamp, the top of a crack, the top of a wedge, the edge of the electrode. A more detailed study of the behavior of mechanical and electric fields in the vicinity of concentrators is possible using non-classical models of the theory of elasticity, for example, models of the gradient theory of elasticity, thanks to which it is possible to quantify singular solutions, and in some problems to obtain regular solutions.

Piezoelectric transducers (sensors and actuators) are widely used in modern industry. At the same time, piezoceramics is widely used as a material for piezoelectric transducers due to the high coefficient of electromechanical coupling. The practice of designing piezoelectric elements indicates that by cutting both surface and internal electrodes, it is possible to achieve greater energy transformation [1]. The electroelastic interaction of piezoceramics with surface electrodes of finite length is also important in various devices for generating surface waves. However, according to the experimental data [2, 3], a large concentration of the electric field and stresses arise at the edges of the electrodes, which can lead to the formation of cracks and failure of the piezoelectric transducers. Therefore, a theoretical study of the effects created by split electrodes is necessary.

To calculate elements with surface electrodes of finite length, it is required to solve a planar or antiplanar electroelasticity problem with mixed boundary conditions. However, due to the mixed conjugation conditions at the edges of the electrodes, the normal component of the electric induction vector suffers a jump. From the point of view of the classical theory of electroelasticity, plane and antiplane problems of deformation of a half-space and a strip with surface and internal split electrodes have been studied quite fully [4]-[10]. So in [8], using the Fourier transform, the antiplane problem of electroelasticity is reduced to solving integral equations by the Wiener-Hopf method. Expressions are obtained for the coefficients of stress intensity and induction. Calculations have shown that the normal component of the electric induction vector has a root singularity and increases indefinitely when approaching the edge of the electrode. However, the experimental results indicate the final value of the electric and mechanical fields around the split electrodes [3].

At present, to refine the behavior of physical fields in the vicinity of mating surfaces and cracks, non-classical mechanics is used, based on gradient models, which take into account the gradients of deformation and physical fields in constitutive relations. The first models of the gradient theory of elasticity were proposed by Tupin and Mindlin in the 1960s [11, 12]. The problems of gradient elasticity, thermoelasticity, and electroelasticity for layered bodies, strips, and half-planes were studied in [13]-[17].

Studies in the field of gradient electroelasticity can be divided into two groups: one, within the framework of the theory of gradient elasticity, takes into account the effect of the strain gradient, but does not take into account the electric field gradient [18], the other takes into account the effect of the electric field gradient, but neglects the effect of the strain gradient [19].

Antiplane problems on the deformation of bodies with cracks within the framework of models of the gradient theory of elasticity and electroelasticity are presented in [20]-[30]. Boundary integral equations (BIEs) are obtained for the jumps of displacement fields on the crack faces, and the stress intensity factors at the crack tips are calculated.

In this paper, we consider model mixed problems of the gradient theory of elasticity for a strip with a crack and problems of electroelasticity for a strip with a surface electrode in the framework of the Aifantis gradient model. The aim of this work is to analyze the influence of gradient effects on the nature of singular solutions in the vicinity of the discontinuity of the boundary conditions.

42.2 Constitutive Relations of the Gradient Theory of Elasticity and Electroelasticity

In the gradient model proposed by Mindlin [11], the function of the strain energy density is presented as [17, 26]

$$W = \frac{1}{2} c_{ijkl} \varepsilon_{ij} \varepsilon_{kl} + \frac{1}{2} A_{ijklmn} \varepsilon_{ij,k} \varepsilon_{lm,n} \tag{42.1}$$

$$T = \frac{1}{2} \rho \dot{u}_i \dot{u}_i + \frac{1}{2} \rho l_m^2 \dot{u}_{i,j} \dot{u}_{i,j} \tag{42.2}$$

where c_{ijkl}, A_{ijklmn} are the components of the tensor of elastic moduli of the linear theory of elasticity and components of the tensor of gradient moduli, l_m is the gradient parameter of the dynamic problem. In the gradient model, the following mechanical fields are introduced

$$m_{ijk} = \frac{\partial W}{\partial \varepsilon_{ij,k}}$$

are moment stresses,

$$\sigma_{ij} = \tau_{ij} - m_{ijk,k}$$

are total stresses,

$$\tau_{ij} = \frac{\partial W}{\partial \varepsilon_{ij}}$$

are the components of the Cauchy stress tensor.

The most popular GTE model is the simplified Aifantis model [26] used for static and dynamic problems. Thus, for an anisotropic medium, there are constitutive relations and equations of motion of the form [17]

$$\sigma_{ij} = c_{ijkl}(1 - l_s^2 \nabla^2)\varepsilon_{kl} \tag{42.3}$$

$$c_{ijkl}\left(1 - l_s^2 \nabla^2\right) u_{k,lj} = \rho\left(1 - l_m^2 \nabla^2\right) \ddot{u}_i$$

The equations of motion within the framework of the gradient theory of elasticity for an isotropic medium in terms of displacements are written in the form

$$(\lambda + \mu)\left(1 - l_2^2 \nabla^2\right) u_{j,ij} + \mu\left(1 - l_3^2 \nabla^2\right) u_{i,jj} = \rho\left(1 - l_1^2 \nabla^2\right) \ddot{u}_i \tag{42.4}$$

here parameters l_1, l_2, l_3 are dimensional parameters determined by gradient effects.

Let a piezoelectric body occupies a volume V with a surface S. Limited only by the gradient effects of the electric field, the electric enthalpy density can be represented as [16]:

$$H(\varepsilon_{ij}, \varepsilon_{ij,k}, E_i, E_{i,j}) = \frac{1}{2} c_{ijkl}\varepsilon_{ij}\varepsilon_{kl} - e_{kij}\varepsilon_{ij}E_k - \frac{1}{2}\ni_{ij}E_iE_j - \frac{1}{2}\alpha_{ijkl}E_{i,j}E_{k,l} \tag{42.5}$$

Here e_{kij} are the third-order components of the piezoelectric modulus tensor, $\ni_{ij}$ and α_{ijkl} are components of tensors of permittivity coefficients of the 2nd and 4th order, respectively.

The mathematical formulation of the problem of gradient electroelasticity has the form:

$$\begin{aligned} \tau_{ij,j} &= 0 \\ P_{i,i} &= 0 \end{aligned} \tag{42.6}$$

Boundary conditions can be specified relative to mechanical and electric fields, as well as additional boundary conditions due to the consideration of the gradient model. On the part of the boundary, on which there is an electrode, mixed boundary conditions are set

$$\tau_{ij}n_j = t_i, u_i = v_i, \tag{42.7}$$

$$Q_{ij}n_in_j = 0, (D_j - Q_{jk,k})n_j - (Q_{jk}n_k)_{,j} + (Q_{jk,k}n_kn_l)_{,l}n_j = q,$$

$$\varphi = V_0, E_in_i = 0 \tag{42.8}$$

Here

$$\tau_{ij} = \frac{\partial H}{\partial \varepsilon_{ij}} = c_{ijkl}\varepsilon_{kl} - e_{kij}E_k$$

are the components of the Cauchy stress tensor,

$$D_i = -\frac{\partial H}{\partial E_i} = e_{ijk}\varepsilon_{jk} + \ni_{ij}E_j$$

are the components of the electric induction vector,

$$Q_{ij} = -\frac{\partial H}{\partial E_{i,j}} = \alpha_{ijkl}E_{k,l}$$

are the quadrupole moment tensor components,

$$P_i = D_i - Q_{ij,j}$$

are the components of the total electric induction vector, t_i are the components of the vector of given forces on the surface of the body, q is the surface charge density, n_i are the components of the unit vector of the normal to the surface of the body at the considered point. Next, we consider two model problems within the framework of the gradient models described above the problem of delamination of an elastic strip and the antiplanar problem of electroelasticity with a finite electrode.

42.3 Problem for a Strip with Delamination

Lets consider the anti-plane steady-state oscillation of an isotropic elastic strip with thickness h, fixed on the lower boundary, on which there is a tunnel delamination of length $2a$. Let x_1-axis be directed at the lower boundary of the strip, x_3 - axis be directed perpendicularly to x_1, and x_2-axis be aligned with the delamination axis. Taking into account steady state oscillations

$$u_2(x_1, x_3, t) = w(x_1, x_3)e^{-i\omega t},$$

when the applied load is represented in the form of

$$\sigma_{23} = p_0\delta(x_1 - L)e^{-i\omega t}.$$

After separating the time factor the boundary value problem takes the form

$$\left(1 - l^2\nabla^2\right)\left(\nabla^2 + k^2\right)w = 0, \;\; k^2 = \rho\omega^2/\mu, \tag{42.9}$$

$$\begin{aligned} &\sigma_{23}(x_1, h) = p_0\delta(x_1 - L), \quad m_{332}(x_1, h) = 0, \quad m_{332}(x_1, 0) = 0 \\ &w(x_1, 0) = 0, \; x_1 \notin [-a, a], \; \sigma_{32}(x_1, 0) = 0, x_1 \in [-a, a], \\ &m_{332} = \mu l^2 w_{,33}, \; \sigma_{23} = \mu(1 - l^2\nabla^2)w_{,3} \end{aligned} \tag{42.10}$$

A simplified Aifantis model is considered here, in which the assumption is ($l_1 = l_2 = l_3 = l$).

Discontinuous boundary conditions are given at the lower boundary of the strip, at the delamination the condition of no stress on the banks is given and outside the delamination the condition of rigid fixation of the strip is given. The problem statement is closed by the conditions of wave propagation at infinity, formulation of which is based on the principle of limiting absorption [31].

The standard research scheme for solving of crack theory problems is to reduce the initial problem to a boundary integral equation (BIE) with reference to the function $\varphi(\eta) = w_{,1}(\eta,0)$. In the case of the classical problem of linear theory of elasticity we have a singular boundary integral equation, where the corresponding solution and the Cauchy stresses have a root singularity in the area of delamination tip

$$\varphi(\eta) \approx (1-\eta^2)^{-1/2}, \ \eta \in [-1,1] \tag{42.11}$$

when using GTE, the boundary integral equation is characterised not only by an operator with a singular kernel, but also by an integral with a cubic singularity, which are interpreted in the meanings of the Cauchy general value

$$-\frac{2s_1^2}{\pi}\int_{-1}^{1}\frac{\varphi(\eta)}{(\eta-\tau)^3}d\eta+\frac{(1-s_2^2)}{\pi}\int_{-1}^{1}\frac{\varphi(\eta)}{\eta-\tau}d\eta+\int_{-1}^{1}N(\eta,\tau)\varphi(\eta)d\eta = F(\tau), \tau \in [-1,1]$$
$$s_1^2 = \left(\frac{l}{a}\right)^2, \ s_2^2 = \frac{l^2k^2}{2} \tag{42.12}$$

where $N(\eta,\tau)$ is the regular part of the core.

The asymptotic solution of the BIE (42.12) in the vicinity of the tip is found in the form

$$\varphi(\eta) = \left(1-\eta^2\right)^{1/2} g(t), \ t \in [-1,1], g(t) \in C[-1,1] \tag{42.13}$$

Numerical investigation of solution (42.13) constructed by means of the Chebyshev polynomial expansion was carried out. An asymptotic analysis of the problem in the framework of GTE for delamination of small relative size, the solution of BIE (42.12) is investigated in depending on the input parameters of the problem - the frequency parameter, the gradient parameter and the relative crack size, and a simplified analytical expression for the swap function is obtained as a result of asymptotic analysis. A numerical study of the problem was carried out and the working ranges of the asymptotic approach were revealed [25].

42.4 Problem for an Electroelastic Strip with a Surface Electrode

Lets consider the anti-plane problem of deformation of a transversally isotropic piezoelastic strip of thickness h when an electrode of finite length $2a$, located on its stress-free upper bound, is supplied with an electric potential $\varphi = V_0$. The other part of the upper bound is unelectroded. The lower bound of the strip is rigidly stiff and grounded. The coordinate system is chosen similar to the previous problem. The nonzero component of the mechanical displacement field is $u_2 = w(x_1,x_3), u_1 = u_3 = 0$, the electric field potential is $\varphi = \varphi(x_1,x_2)$.

The mathematical formulation of the problem can be presented as follows

$$\begin{aligned} &P_{1,1} + P_{3,3} = 0 \\ &\tau_{21,1} + \tau_{23,3} = 0 \end{aligned} \tag{42.14}$$

with the constitutive relations

$$\begin{aligned} &D_1 = e_{15} w_{,1} - \ni_{11} \varphi_{,1}, D_3 = e_{15} w_{,3} - \ni_{11} \varphi_{,3} \\ &P_1 = D_1 - Q_{11,1} - Q_{13,3}, P_3 = D_3 - Q_{31,1} - Q_{33,3}, \\ &Q_{ij} = -l^2 \ni_{11} \varphi_{,ij} \\ &\tau_{13} = c_{44} w_{,1} + e_{15} \varphi_{,1}, \tau_{23} = c_{44} w_{,3} + e_{15} \varphi_{,3}, \end{aligned} \tag{42.15}$$

and the boundary conditions

$$\varphi(x_1,0) = Q_{33}(x_1,0) = w(x_1,0) = 0 \tag{42.16}$$

$$\begin{aligned} &Q_{33}(x_1,h) = \tau_{23}(x_1,h) = 0 \\ &\varphi(x_1,h) = V_0, x_1 \in [-a,a], P_3(x_1,h) = 0, x_1 \notin [-a,a]. \end{aligned} \tag{42.17}$$

where l - is the gradient parameters. Thus, we have discontinuous boundary conditions at the upper boundary of the strip and the study of the behavior of the solution in the area of the discontinuity - the edges of the electrode - is of the greatest interest.

Let us consider an auxiliary boundary problem

$$\begin{aligned} &c_{44} \nabla^2 w + e_{15} \nabla^2 \varphi = 0 \\ &e_{15} \nabla^2 w - \ni_{11} \left(1 - l^2 \nabla^2\right) \nabla^2 \varphi = 0 \\ &w(x_1,0) = \varphi(x_1,0) = \varphi_{,33}(x_1,0) = 0 \end{aligned} \tag{42.18}$$

$$\begin{aligned} &\varphi_{,33}(x_1,h) = w_{,33}(x_1,h) = 0 \\ &P_3(x_1,h) = \chi(x_1),\ \chi(x_1) = 0,\ x_1 \notin [-a,\ a],\ \chi(x_1) \neq 0,\ x_1 \in [-a,\ a] \\ &P_3 = e_{15} w_{,3} - \ni_{11} \varphi_{,3} + \ni_{11} l^2 (\varphi_{,311} + \varphi_{,333}) \end{aligned} \tag{42.19}$$

Let us exclude the function w from the system of differential equations, which gives us a fourth-order differential equation with respect to the function describing the electric field potential

$$\left(1 + k_0^2 - l^2 \nabla^2\right) \nabla^2 \varphi = 0,$$

where

$$k_0^2 = \frac{e_{15}^2}{c_{44} \ni_{11}}$$

is the electromechanical coefficient.

Applying the Fourier integral transformation by the x_1 coordinate to the resulting differential equation, taking the following term

$$D=\frac{\partial}{\partial x_3},\tilde{\varphi}(\alpha,x_3)=\int\limits_{-\infty}^{+\infty}\varphi(x_1,x_3)e^{i\alpha x_1}dx_1,\ \varphi(x_1,x_3)=\frac{1}{2\pi}\int\limits_{-\infty}^{+\infty}\tilde{\varphi}(\alpha,x_3)e^{-i\alpha x_1}d\alpha$$

As a result, we obtain a differential equation of the form

$$(1+k_0^2-l^2(D^2-\alpha^2))(D^2-\alpha^2)\tilde{\varphi}=0 \tag{42.20}$$

The characteristic equation for (42.20) is

$$(1+k_0^2-l^2(\lambda^2-\alpha^2))(\lambda^2-\alpha^2)=0 \tag{42.21}$$

and has the roots

$$\lambda_1=|\alpha|,\lambda_2=-\lambda_1,\quad \lambda_3=\sqrt{\alpha^2+\frac{1+k_0^2}{l^2}},\ \lambda_4=-\lambda_3 \tag{42.22}$$

Then the solution of (42.20) can be presented as

$$\tilde{\varphi}(\alpha,x_3)=\sum_{j=1}^{4}C_j(\alpha)e^{\lambda_j(\alpha)x_3} \tag{42.23}$$

By satisfying general solution (42.23) with boundary conditions, we find

$$C_2=-C_1,C_3=-C_1\frac{\operatorname{sh}(|\alpha|h)}{|\alpha|\operatorname{sh}(\lambda_3 h)},\quad C_1(\alpha)=-\frac{\tilde{\chi}(\alpha)}{2\ni_{11}(1+k_0^2)|\alpha|\operatorname{ch}(|\alpha|h)}$$

then

$$\tilde{\varphi}(\alpha,x_3)=2C_1(\alpha)\left(\operatorname{sh}(|\alpha|x_3)-\frac{\alpha^2}{\lambda_3^2}\frac{\operatorname{sh}(|\alpha|h)}{\operatorname{sh}(\lambda_3 h)}\operatorname{sh}(\lambda_3 x_3)\right) \tag{42.24}$$

Applying the inverse Fourier transform, taking into account the condition at the electrode, $\varphi(x_1,h)=V_0$, $x_1\in[-a,a]$, we obtain a boundary integral equation with a difference kernel

$$\int\limits_{-a}^{a}\chi(\xi)K(\xi-x_1)d\xi=-V_0\ni_{11},\ x_1\in[-a,a]$$

$$K(\xi-x_1)=\frac{1}{2\pi}\int\limits_{-\infty}^{\infty}L(|\alpha|)\frac{e^{i\alpha(\xi-x_1)}}{|\alpha|\,(1+k_0^2+l^2\alpha^2)}d\alpha \tag{42.25}$$

$$L(|\alpha|)=\frac{\operatorname{sh}(|\alpha|h)}{\operatorname{ch}(|\alpha|h)}$$

Introducing dimensionless variables and functions

$$y = x_1/a,\ \eta = \xi/a,\ \beta = a\alpha,\ s = l/a,\ t = h/a,\ f_0 = -V_0 \ni_{11} a \tag{42.26}$$

we obtain the boundary integral equation

$$\begin{gathered}
A[\chi^*] = \int_{-1}^{1} \chi^*(\eta) K^*(\eta - y) d\eta = f_0,\ y \in [-1, 1] \\
K^*(\eta - y) = \frac{1}{2\pi} \int_{-\infty}^{\infty} L^*(|\beta|) \frac{e^{i\beta(\eta - y)}}{|\beta| \left(1 + k_0^2 + s^2\beta^2\right)} d\beta \\
L^*(|\beta|) = \frac{\mathrm{sh}(|\beta| t)}{\mathrm{ch}(|\beta| t)}
\end{gathered} \tag{42.27}$$

Let us consider an integral operator

$$\begin{gathered}
A_0[\chi^*] = \int_{-1}^{1} \chi^*(\eta) K_0^*(\eta - y) d\eta,\ y \in [-1, 1] \\
K_0^*(\eta - y) = \frac{1}{2\pi\left(1 + k_0^2\right)} \int_{-\infty}^{\infty} L^*(|\beta|) \frac{e^{i\beta(\eta - y)}}{|\beta|} d\beta
\end{gathered} \tag{42.28}$$

Note the following relationship between the integral operators

$$\left(1 + k_0^2\right) A_0[\chi^*] = \left(1 + k_0^2 - s^2 \frac{d^2}{dy^2}\right) A[\chi^*] = f_0 \left(1 + k_0^2\right)$$

Thus, we obtain the next integral equation

$$A_0[\chi^*] = f_0 \tag{42.29}$$

Note that BIE (42.29) corresponds to the BIE of the classical problem of linear theory of elasticity (at $s = 0$), the asymptotic solution of which for $t \gg 1$ given $L^*(|\beta|) = 1$, is reduced to the analysis of the known equation with logarithmic kernel [32], it has the root singularity at the electrode boundary and is represented in the form

$$\chi^*(y) = \left(1 - y^2\right)^{-1/2},\ y \in [-1, 1] \tag{42.30}$$

the constant C can be found by substituting (42.30) into the original integral equation (42.29).

42.5 Conclusion

A special property of gradient models in the investigation of crack theory problems is the possibility to obtain regular solutions of BIE, which makes it possible to give an estimation of the stresses in the area of delamination tip. The stresses in the area of delamination tip have an explicit extremum, the estimation of which gives a possibility to formulate a new strength criterion.

Meanwhile, the results of the anti-plane problem for the electroelastic strip with a surface electrode show that the degree and the character of singularity of the total electric induction does not change while using one-parameter gradient model of Aifantis, and the gradient model containing only gradients of electric fields does not contribute to the solution and any gradient effects does not exhibit. In order to clarify the behaviour of the corresponding solution in the vicinity of the electrode boundary it is necessary to consider more complex gradient models or to take into account the gradient effects in the mechanical fields.

Acknowledgements The research was supported by RSF No. 22-11-00265.

References

[1] Vatul'yan AO, Rynkova AA (2007) A Model of Bending Vibrations of Piezoelectric Bimorphs with Split Electrodes and Its Applications, Mechanics of Solids **42**:595-602. DOI 10.3103/S0025654407040127

[2] Zuo JZ, Sih GC (2000) Energy density theory formulation and interpretation of cracking behavior for piezoelectric ceramics, Theoretical and Applied Fracture Mechanics **34**(1):17-33. DOI /10.1016/S0167-8442(00)00021-5

[3] Shindo Y, Yoshida M, Narita F et al (2004) Electroelastic field concentrations ahead of electrodes in multilayer piezoelectric actuators: Experiment and finite element simulation, Journal of the Mechanics and Physics of Solids **52**:1109-1124. DOI 10.1016/j.jmps.2003.09.017

[4] Parton VZ, Kudryavtsev BA (1988) Elektromagnitouprugost' p'yezoelektricheskikh i elektroprovodnykh tel [Electromagnetic resistance of piezoelectric and electrically conductive bodies], Nauka, Moscow.

[5] Shindo Y, Narita F, Sosa H (1998) Electroelastic analysis of piezoelectric ceramics with surface electrodes, International Journal of Engineering Science **36**:1001-1009. DOI 10.1016/S0020-7225(98)00007-X

[6] He LH, Ye RQ (2000) Concentration of electric field near electrodes on piezoelectric layer, Theoretical and Applied Fracture Mechanics **33**:101-106. DOI 10.1016/S0167-8442(00)00005-7

[7] Li XF, Duan XY (2001) Electroelastic analysis of a piezoelectric layer with electrodes, International Journal of Fracture **111**:73-78. DOI 10.1023/A:1012425817745

[8] Chen CD, Chue CH (2003) Fracture mechanics analysis of a composite piezoelectric strip with an internal semi-infinite electrode, Theoretical and Applied Fracture Mechanics **39**:291-314. DOI 10.1016/S0167-8442(03)00022-3
[9] Li XF, Tang GJ (2003) Electroelastic analysis for a piezoelectric layer with surface electrodes, Mechanics Research Communications **30**:345-351. DOI 10.1016/S0093-6413(03)00036-3
[10] Wang BL, Han JC, Du SY (2011) Anti-plane mechanical and in-plane electrical fields for a finite electrode/punch on a finite piezoelectric layer, Fatigue & Fracture of Engineering Materials & Structures **34**:139-148. DOI 10.1111/j.1460-2695.2010.01501.x
[11] Toupin RA (1962) Elastic materials with couple-stresses, Archive for Rational Mechanics and Analysis **11**:385-414. DOI 10.1007/BF00253945
[12] Mindlin RD (1964) Micro-structure in linear elasticity, Archive for Rational Mechanics and Analysis **16**:51-78. DOI 10.1007/BF00248490
[13] Vatulyan AO, Nesterov SA, Yurov VO (2021) Investigation of the stress-strain state of a hollow cylinder with a coating based on the gradient model of thermoelasticity, PNRPU Mechanics Bulletin **4**:60-70. DOI 10.15593/perm.mech/2021.4.07
[14] Vatulyan AO, Nesterov SA (2022) Modeling of thermoelastic deformation of a thin-layer "coating-substrate" system, Journal of Physics: Conference Series. DOI 10.1088/1742-6596/2317/1/012012
[15] Solyaev Y, Lurie S (2021) Electric field, strain and inertia gradient effects on anti-plane wave propagation in piezoelectric materials, Journal of Sound and Vibration **494**:115898. DOI 10.1016/j.jsv.2020.115898
[16] Eremeyev VA, Ganghoffer JF, Konopińska-Zmysłowska V, Uglov NS (2020) Flexoelectricity and apparent piezoelectricity of a pantographic micro-bar, International Journal of Engineering Science **149**:103213. DOI 10.1016/j.ijengsci.2020.103213
[17] Eremeyev VA, Rosi G, Naili S (2019) Comparison of anti-plane surface waves in strain-gradient materials and materials with surface stresses, Mathematics and Mechanics of Solids **24**:2526-2535. DOI 10.1177/1081286518769
[18] Shodja HM, Ghazisaeidi M (2007) Effects of couple stress on anti-plane problems of piezoelectric media with inhomogeneities, European Journal of Mechanics - A/Solids **26**:647-658. DOI 10.1016/j.euromechsol.2006.09.001
[19] Yang XM, Hu YT, Yang JS (2004) Electric field gradient effects in anti-plane problems of polarized ceramics, International Journal of Solids and Structures **41**:6801-6811. DOI 10.1016/j.ijsolstr.2004.05.018
[20] Yue Y, Xu K, Aifantis E (2014) Microscale size effects on the electromechanical coupling in piezoelectric material for anti-plane problem, Smart Materials and Structures **23**(12):125043. DOI 10.1088/0964-1726/23/12/125043
[21] Sladek J, Sladek V, Wünsche M, Zhang C (2018) Effects of electric field and strain gradients on cracks in piezoelectric solids, European Journal of Mechanics - A/Solids **71**:187-198. DOI 10.1016/j.euromechsol.2018.03.018

[22] Solyaev Y, Lurie S (2019) Pure bending of the piezoelectric layer in second gradient electroelasticity theory, Acta Mechanica **230**:4197-4211. DOI 10.1007/s00707-019-02484-x

[23] Li J, Wang B, Zhang C (2019) Anti-plane fracture mechanics analysis of a piezoelectric material layer with strain and electric field gradient effects, Mechanics Research Communications **102**:103439. DOI 10.1016/j.mechrescom.2019.103439

[24] Fannjiang AC, Paulino GH, Chan YS (2002) Strain gradient elasticity for antiplane shear cracks: a hypersingular integrodifferential equation approach, SIAM Journal on Applied Mathematics **62**:1066-1091. DOI 10.1137/S0036139900380487

[25] Vatul'yan AO, Yavruyan OV (2021) Vibrations of a layer with delamination in the framework of the gradient elasticity theory, Russian Journal of Nondestructive Testing **57**:825-837. DOI 10.1134/S1061830921100090

[26] Askes H, Aifantis K (2011) Gradient elasticity in static and dynamics: An overview of formulations, length scale identification procedures, finite element implementations and new results, International Journal of Solids and Structures **48**:1962-1990. DOI 10.1016/j.ijsolstr.2011.03.006

[27] Vatulyan AO, Plotnikov DK, Yurov VO (2020) On the deformation of coated functionally graded structures with delamination, Acta Mechanica **232**:1863–1873. DOI 10.1007/s00707-020-02846-w

[28] Yavruyan O, Yavruyan K (2022) On the Specifics of Investigation for the Dynamic Problems of Cracked Layer by the Gradient Elasticity Theory, Lecture Notes in Networks and Systems **510**:1681–1688. DOI 10.1007/978-3-031-11051-1_173

[29] Vasiliev V, Lurie S, Solyaev Y (2021) New approach to failure of pre-cracked brittle materials based on regularized solutions of strain gradient elasticity, Engineering Fracture Mechanics **258**(10):108080. DOI 10.1016/j.engfracmech.2021.108080

[30] Askes H, Bennett T, Aifantis EC (2007) A new formulation and C0–implementation of dynamically consistent gradient elasticity, International Journal for Numerical Methods in Engineering **72**(1):111–126. DOI 10.1002/nme.2017

[31] Vorovich II, Babeshko VV (1989) Dinamicheskie smeshannie zadachi teorii uprugosti dlya neklassicheskih oblastey [Dynamic mixed problems of elasticity theory for nonclassical domains], Nauka, Moscow.

[32] Vorovich II, Aleksandrov VM, Babeshko VV (1974) Neklassicheskiye smeshannye zadachi teorii uprugosti [Non-classical Mixed Problems of Elastic Theory]. Nauka, Moscow.

Chapter 43
A Damaged Medium Model for Assessing Life Characteristics of Polycrystalline Structural Alloys with Joint Mechanical Fatigue and Long-Term Strength of Material

Ivan A. Volkov, Leonid A. Igumnov, Aleksandr A. Belov, and Andrey I. Volkov

Abstract The processes of initial strength degradation properties of structural materials (metals and their alloys) under combined low-cycle fatigue and long-term strength of material are considered. A mathematical model describing the processes of cyclic viscoplastic deformation and damage accumulation in structural alloys under multiaxial disproportionate modes of combined thermomechanical loading has been developed from the standpoint of mechanics of damaged medium (MDM). The version of constitutive relations of viscoplasticity based on the idea of plasticity surfaces and creep and gradientality of velocity vectors of plastic and creep deformations to the relevant surface in the loading point. This version of state equation reflects the main effects of cyclic viscoplastic deformation of material for arbitrary folded loading paths. The version of kinetic equations for damage accumulation is based on introduction of a scalar damage parameter, on energy principles and takes into account the main effects of nucleation, growth and merging of microdefects under arbitrary complex modes of cyclic thermomechanical loading. The condition when damage degree reaches its critical value is used as a strength damage criterion. The results of numerical simulation of the processes of deformation and damage accumulation in structural alloys under combined fatigue and creep are presented. The results of comparison of calculated and experimental data showed that the proposed MDM model qualitatively and accurately enough for practical calculations quantitatively describes the durability of materials and structures under combined low-cycle fatigue and long-term strength.

Key words: Non-stationary creep · Low-cycle fatigue · Modeling · Mechanics of damaged medium · Stress-strain state · Damage · Service life · Numerical and full-scale experiment

---

Ivan A. Volkov · Leonid A. Igumnov · Aleksandr A. Belov · Andrey I. Volkov
National Research Lobachevsky State University of Nizhny Novgorod, 23 Gagarin Avenue, Nizhny Novgorod 603022, Russian Federation,
e-mail: pmptmvgavt@yandex.ru, igumnov@mech.unn.ru, belov_a2@mech.unn.ru, pmptmvgavt@yandex.ru

H. Altenbach et al. (eds.), *Sixty Shades of Generalized Continua*,
Advanced Structured Materials 170, https://doi.org/10.1007/978-3-031-26186-2_43

43.1 Introduction

One of the main tasks of modern mechanical engineering is substantiation of service life of equipment and systems of critical engineering objects (CEO) at the stage of their design, estimation of the depletion life and forecast of the residual life of in-service structural units and service life extension after the objects have worked out their standard service life. These tasks are especially relevant for facilities with a service life of several decades (nuclear power plants, petrochemical equipment, storage tanks for gaseous and liquefied chemical products, main gas and oil pipelines, etc.). The working conditions are characterized by multi-parametric non-stationary thermomechanical effects, the effects of external fields of various nature leading to development of various degradation mechanisms of the initial strength properties of structural materials and the initial service life exhaustion of structural units [1].

There are a large number of mechanisms that can affect the processes of service life exhaustion of a particular object depending on its working conditions: high-cycle fatigue (HCF), low-cycle fatigue (LCF), long-term strength (LS) etc. (taking into account their interaction). For these mechanisms, the formation of a macroscopic crack is the result of the successive action of a certain number of very complex (from a physical point of view) transformation processes of initial structure of a structural material, including the nucleation, growth and interaction of various defects in the crystal lattice in metals and interaction of hierarchical structural components of various levels.

The process of destruction under the mutual influence of long-term strength and low-cycle fatigue occurs as a result of cyclic load application at elevated temperature, load stand and activation of time-dependent deformation mechanisms. Creep damage (long-term strength) and non-stationary plastic deformation (low-cycle fatigue) of a material has a different physical character. From a microscopic point of view, creep damage is considered to be of an intergranular nature, while fatigue damage corresponds to the micro and macro cracks accumulation along the grain body. When, in accordance with the loading conditions (cyclic deformation at high temperature), both processes develop simultaneously, there is a possibility for their interaction, which manifests itself in the form of highly non-linear effects that are unfavorable for the overall durability of the material [2]-[8]. At the microscopic level, this is explained by the fact that during creep, intergranular microcracks tend to form and develop at a very early stage of the life of an elementary volume of a structural material. The presence of such defects due to creep contributes to initiation of transcritical cracks during fatigue.

To date, a system has been developed for classifying the criteria for low-cycle fracture during creep at high temperatures, depending on the accepted measure of the intensity of fatigue damage accumulation and the type of equations that determine the damage parameter [9]. The most representative are the groups of criteria of the following types: deformation (including deformation or its components that characterize the deformation cycle as a whole); kinetic (determining the rate of damage and its critical value); energy (associating the accumulation of damage with the dissipation of the energy of inelastic deformation); criteria based on long-term strength characteristics [10]-[13].

It should be noted that, while satisfactorily describing the patterns of low-cycle fracture under relatively simple loading programs, the above criteria lose this property when it is required to reflect the effect of the features of a particular cycle, for example, the nature (creep, relaxation) of exposure, the alternation order of the plastic and viscous deformation stages, as in a half-cycle tension and compression, the law of temperature change, etc.

Thus, neither the process of damage accumulation as a result of creep nor the process of damage accumulation as a result of MCF has been properly studied so far. Therefore, it is not surprising that the process in which both types of damage accumulate and interact simultaneously has also not been fully studied, nevertheless in a number of practically important cases, these processes proceed simultaneously, interacting with each other, significantly affecting the service life of structural element. Circumstances are getting more complicated by the fact that during operation such elements are subjected to alternating stresses at constant temperatures, alternating temperatures at constant stresses or the stress-strain state and temperature (in most cases) change simultaneously with different periods. In addition, experimental facts indicate that the interaction between the processes of low-cycle fatigue and long-term strength is synergistic. At present, there is no general theory that would make it possible to predict long-term strength during a cyclic change in temperature under the action of a constant stress, a cyclic change in stress exposure with various durations at a constant temperature and under conditions when both stress and temperature change according to different laws. Under actual operating conditions of structural elements of power equipment, the duration of thermal-force loading cycles can range from several seconds to tens and hundreds of hours, and sign-variable viscoplastic deformation of the material in the region of structural or technological stress concentrators is accompanied by temperature-time effects and creep. The initiated creep deformation and damage process can develop both in the process of exposure under load and during a monotonous loading or unloading process occurring at a rather low rate. Simultaneous development of damage processes in the load change cycle as a result of low-cycle fatigue and long-term strength has a noticeable effect on the mechanism of low-cycle fracture and the durability of the material, where the shape of loading cycle, the asymmetry and sign of stress-es, the type of stress state against which creep deformations develop are important factors.

The above results lead to the conclusion that the linear law of damage summation (Palmgren-Miner hypothesis) on the degradation mechanism of low-cycle fatigue and creep is not applicable, since these are completely different processes that exhibit a nonlinear interaction.

In the present work, from the modern standpoint of the MDM, a mathematical model is developed that describes the processes of cyclic viscoplastic deformation and damage accumulation in structural materials (metals and their alloys) during the degradation of the initial strength properties of materials through the joint mechanisms of low-cycle fatigue and long-term strength. In order to qualitatively and quantitatively evaluate the model, a study was made on the influence of the laws of change in mechanical deformation and temperature (the shape of the loading cycle) on the durability of structural steels under various loading cycles.

43.2 Constitutive Relations of the Mathematical Model of Mechanics of Damaged Medium

The model of a damaged medium under degradation mechanisms that combine low-cycle fatigue and long-term strength consists of three interrelated components [17]-[20]:

- constitutive relations that describe the viscoplastic behavior of material, taking into account the dependence on the failure process;
- evolutionary equations describing the kinetics of damage accumulation;
- strength criterion of the damaged material.

43.2.1 Constitutive Relations of Thermoviscoplasticity

The constitutive relations of thermoviscoplasticity are based on the following basic provisions:

- the material of the medium is initially isotropic and there are no damages in it (only the anisotropy caused by the processes of plasticity and creep is taken into account; the anisotropy of the elastic properties caused by the processes of damage to the material is not taken into account);
- strain and strain rate tensors e_{ij}, $\dot{e}_{ij}$, represent the sum of the "instantaneous" and "temporal" components. The "instantaneous" component of the strain tensors and strain rates includes the elastic component e^e_{ij}, $\dot{e}^e_{ij}$, (independent of the loading history and determined by the final state of the process) and the plastic component – e^p_{ij}, $\dot{e}^p_{ij}$, (depending on the history of the loading process). The time component of the strain tensors and strain rates e^c_{ij}, $\dot{e}^c_{ij}$ is related to the evolution of the creep process;
- the initial yield surface is described by a surface in the form of a Mises hypersphere. The evolution of the change in the plastic loading surface is de-scribed by the change in its radius C_p and the displacement of its center $\dot{\rho}^{pl}_{ij}$;
- in the stress space there is a family of equipotential creep surfaces of radius C_c and having a common center ρ^c_{ij};
- the principle of gradientality is valid;
- is the change in the volume of the body elastically $e^{irr}_{ij} = 0$, where $e^{irr}_{ij} = e^p_{ij} + e^c_{ij}$ is the inelastic deformation, which includes the "instantaneous" and "temporary" components;
- deformation processes characterized by small deformations are considered;

- the only structural parameter that characterizes the degree of material damage at the macrolevel is the scalar parameter ω - damage ($\omega_0 \leq \omega \leq \omega_f$);
- the influence of the level of accumulated damage on the processes of deformation of the material is taken into account by introducing effective stresses.

It is assumed that in the elastic region the deviatoric σ'_{ij}, e'_{ij} and spherical σ, e components of the stress σ_{ij} and strain tensors e_{ij}, as well as their velocities are related by generalized Hooke's law:

$$\sigma = 3K\left[e - \alpha\left(T - T_0\right)\right], \sigma_{ij} = 2Ge'^{e}_{ij}, \dot{\sigma} = 3K\left[\dot{e} - \dot{\alpha}T - \alpha\dot{T}\right)\right] + \frac{\dot{K}}{K}\sigma,$$
$$\dot{\sigma}_{ij} = 2G\dot{e}'^{e}_{ij} + \frac{\dot{G}}{G}\sigma'_{ij}, \tag{43.1}$$

where $\alpha(t)$ is the linear thermal expansion coefficient of the material, $K(t)$ is the bulk modulus, $G(t)$ is the shear modulus, T_0 is the base temperature, and T is the current temperature. To describe the effects of monotonic and cyclic deformation in the stress space, a yield surface is introduced, which is described by the von Mises hypersphere [21, 22]:

$$F_p = S_{ij}S_{ij} - C_p^2 = 0, \; S_{ij} = \sigma'_{ij} - \rho^{pl}_{ij} \tag{43.2}$$

where C_p is the radius of the yield surface, and ρ^{pl}_{ij} are the coordinates of its center. To describe complex deformation modes in the deformation space, the surface of cyclic "memory" is introduced [23]:

$$a_e = \left(e^p_{ij} - \xi_{ij}\right)\left(e^p_{ij} - \xi_{ij}\right) - a^2_{max} = 0, \tag{43.3}$$

where a_{max} is the maximum value of the intensity of the plastic strain tensor e^p_{ij}, and ξ_{ij} is the tensor of unilaterally accumulated plastic strains.

When modeling the kinetics of the stress-strain state (SSS) during cyclic plastic deformation of a material, it is necessary to describe the processes of hardening and softening of structural materials as reliably as possible, since these mechanisms play a decisive role in the accuracy of estimates of the resource characteristics of the material. Under fatigue loading, there is a competition be-tween the processes of hardening and softening caused in the material by the mechanisms of changing the phase composition, microstructural state, changing the density of dislocations, as well as the effect of changing temperature [4, 5]. The physical mechanisms that occur in polycrystalline metals and alloys and affect the processes of hardening and softening depend on the operating temperature, the type of deformation trajectory, the length of the plastic deformation path, the effective amplitudes of the plastic deformation intensity, and the degree of stabilization of the deformation process.

We postulate that in the temperature range T, at which annealing effects can be neglected and specifying the parameters in relation (43.5), the isotropic hardening (softening) of the material can be represented as:

$$\dot{C}_p(\chi,T) = q_\chi \dot{\chi}_p^{mon} + a_p\left(Q_p^p - C_p^{cyc(p)}\right)\dot{\chi}_p^{cyc} + a_c\left(Q_p^c - C_p^{cyc(c)}\right)\dot{\chi}_c^{cyc} + q_T\dot{T} \tag{43.4}$$

$$C_p = C_p^0 + \int_0^t \dot{C}_p dt, \tag{43.5}$$

$$q_\chi = q_2 A + (1-A)\, q_1,\ Q_p^p = Q_2^p A + (1-A)\, Q_1^p,\ Q_p^c = Q_2^c A + (1-A)\, Q_1^c$$

$$A = 1-\cos^2\Theta,\ \cos\Theta = n_{ij}^e n_{ij}^s,\ n_{ij}^e = \frac{\dot{e}'_{ij}}{\left(\dot{e}'_{ij}\dot{e}'_{ij}\right)^{1/2}},\ n_{ij}^s = \frac{S_{ij}}{\left(S_{ij}S_{ij}\right)^{1/2}} \tag{43.6}$$

$$\dot{\chi}_P = \left(\frac{2}{3}\dot{e}_{ij}^p \dot{e}_{ij}^p\right)^{1/2},\ \chi_P = \int_0^t \dot{\chi}_P dt, \tag{43.7}$$

$$\dot{\chi}_p^{mon} = \begin{cases} 0 & \text{at} \quad a_e < 0 \vee e_{ij}^p \dot{e}_{ij}^p \le 0, \\ \dot{\chi}_P & \text{at} \quad a_e = 0 \wedge e_{ij}^p \dot{e}_{ij}^p \ge 0., \end{cases} \tag{43.8}$$

$$\dot{\chi}_p^{cyc} = \begin{cases} \dot{\chi}_P & \text{at} \quad a_e < 0 \vee e_{ij}^p \dot{e}_{ij}^p \le 0, \\ 0 & \text{at} \quad a_e = 0 \wedge e_{ij}^p \dot{e}_{ij}^p \ge 0., \end{cases} \tag{43.9}$$

In formulas (43.4) - (43.9) the following notations are introduced: C_p^0 is the value of the initial radius of the material yield surface; q_1, q_2, q_T are the modules of monotonic isotropic hardening (softening) under loading along ray loading paths, with a break in the deformation path by 90° and a change in temperature T, respectively; Q_1^p, Q_2^p, Q_1^c, Q_2^c are modules of cyclic isotropic hardening (softening); χ_p^{mon}, χ_p^{cyc}, χ_c^{cyc} are the length of the trajectory of plastic deformation of the material in monotonous sections, in sections of regular and irregular cyclic loading, in sections of load holding, respectively.

The first term of Eq. (43.1) describes isotropic hardening as a result of monotonic plastic deformation of the material, the second term - in the areas of cyclic deformation, the third term - under the action of the creep process, and the fourth - with a change in temperature T. Equations (43.1) - (43.9) describe the isotropic hardening of the material depending on the loading disproportionality parameter A.

We postulate that the rate of change of the microstress tensor $\dot{\rho}_{ij}$ is taken as:

$$\dot{\rho}_{ij} = \dot{\rho}_{ij}^{pl} + \dot{\rho}_{ij}^c \tag{43.10}$$

$$\dot{\rho}_{ij}^{pl} = f(\chi)\left(\dot{\rho}_{ij}^p + \dot{\rho}_{ij}^r\right),\ \rho_{ij} = \int_0^t \dot{\rho}_{ij} dt, \tag{43.11}$$

$$\dot{\rho}_{ij}^p = g_1^p \dot{e}_{ij}^p - g_2^p \rho_{ij}^p \dot{\chi} + g_T^p \rho_{ij}^p \dot{T}, \tag{43.12}$$

$$\dot{\rho}_{ij}^r = g_1^r \dot{e}_{ij}^p - g_2^r \left(\frac{\rho_{min} - \rho_u^r}{\rho_u^r}\right) \rho_{ij}^r \dot{\chi} \cos\gamma + g_T^r \rho_{ij}^r \dot{T}, \tag{43.13}$$

$$\cos\gamma = \begin{cases} \left| \dfrac{\dot{\rho}_{ij}\rho^r_{ij}}{\left(\dot{\rho}_{ij}\dot{\rho}_{ij}\right)^{1/2}\left(\rho^r_{ij}\rho^r_{ij}\right)^{1/2}} \right| & \text{at } \cos\gamma < 0, \\ 0 & \text{at } \cos\gamma > 0, \end{cases} \quad \rho^r_u = \left(\rho^r_{ij}\rho^r_{ij}\right)^{1/2}, \tag{43.14}$$

$$g^{p,r}_T = \frac{1}{g^{p,r}_1}\left[\frac{\partial g^{p,r}_1}{\partial T}\right],$$

$$f(\chi) = f\left(\chi^{mon}_p\right) + f\left(\chi^{cyc}_p\right) + f\left(\chi_\xi\right), \tag{43.15}$$

$$f\left(\chi_{c,\xi}\right) = 1 + k^{c,\xi}_1\left(1 - e^{-k^{c,\xi}_2 \chi_{c,\xi}}\right) \tag{43.16}$$

$$\dot{\chi} = \dot{\chi}_p + \dot{\chi}_c,$$

where $g^{p,r}_1, g^{p,r}_2, k^{c,\xi}_1, k^{c,\xi}_2$ are material parameters determined experimentally. Here and below, for any value B enclosed in $<>$, the following condition is satisfied:

$$< B >= \begin{cases} B, & B > 0, \\ 0, & B \le 0. \end{cases}$$

In formula (43.11), the first term in parentheses describes the evolution of ρ^{pl}_{ij}, associated with the formation of macroscopic plastic deformations, and the second term, with one-sidedly accumulated plastic deformations ξ_{ij}. Dependence (43.11) makes it possible to describe the main effects of anisotropy caused by inelastic deformation of a material under alternating loading, as well as the effects arising from the implementation of hard ("landing" of the plastic hysteresis loop) and soft ("stepping out" of the plastic hysteresis loop) loading modes. The function $f(\chi)$ takes into account the change ρ^{pl}_{ij} in cyclic block asymmetrical modes of cyclic plastic loading of the material. The components of the plastic strain rate tensor $\dot{e}^p_{ij}$ are determined based on the associated flow law:

$$\dot{e}^p_{ij} = \lambda_p S_{ij}.$$

If the magnitude of stresses, temperature and loading rate are such that creep effects are significant, the parameters of the material deformation process must be determined at the loading stage, taking into account the creep process. To describe creep processes in the stress space, a family of equipotential creep surfaces F_C with a common center ρ^c_{ij} and different radii C_c determined by the current stress state is introduced [22]:

$$F^{(i)}_C = S^c_{ij}S^c_{ij} - C^2_c, \; i = 0, 1, 2, \ldots \tag{43.17}$$

According to associative law

$$\dot{e}^c_{ij} = \lambda_c \frac{\partial F^{(i)}_c}{\partial S^c_{ij}} = \lambda_c S^c_{ij}, \tag{43.18}$$

where λ_c corresponds to the current surface $F_c^{(i)}$, which determines the current stress state S_{ij}^c.

Among these equipotential surfaces, one can single out a surface with a radius $\overline{C}_c$ corresponding to zero creep velocity:

$$F_c^{(0)} = \overline{S}_{ij}^c \overline{S}_{ij}^c - \overline{C}_c^2 = 0, \qquad \overline{S}_{ij}^c = \overline{\sigma}'_{ij} - \rho_{ij}^c,$$

where $\overline{S}_{ij}^c$ and $\overline{\sigma}'_{ij}$ are the set of stress states corresponding (with a certain tolerance) to zero creep velocity. Taking into account the mutual influence of the plasticity and creep processes, we postulate that,

$$\overline{C}_c = \overline{C}_c(\chi_c, T) + C_c^{cyc}(\chi_c^{cyc}, T)$$

$$\dot{C}_c^{cyc} = a_c\left(Q_C^p - C_c^{cyc}\right)\dot{\chi}_c^{cyc},\ Q_c^p = Q_2^{cr} A + (1-A) Q_1^c,$$

$$\dot{\chi}_c = \left(\frac{2}{3}\dot{e}_{ij}^c, \dot{e}_{ij}^c\right)^{1/2}, \chi^c = \int_0^t \dot{\chi}^c dt,$$

$$\dot{\chi}_c^{mon} = \begin{cases} 0, \text{ at } a_e < 0 \vee e_{ij}^c \dot{e}_{ij}^c \leq 0, \\ \dot{\chi}_c, \text{ at } a_e = 0 \wedge e_{ij}^c \dot{e}_{ij}^c \geq 0, \end{cases} \qquad \dot{\chi}_c^{cyc} = \begin{cases} \dot{\chi}_c \text{ at } a_e < 0 \vee e_{ij}^c \dot{e}_{ij}^c \leq 0, \\ 0 \text{ at } a_e = 0 \wedge e_{ij}^c \dot{e}_{ij}^c \geq 0, \end{cases}$$

$$\lambda_c = \lambda_c(\psi_c, T),\ \psi_c = \left[\frac{\left(S_{ij}^c S_{ij}^s\right)^{1/2} - \overline{C}_c}{C_c}\right],\ \lambda_c = \begin{cases} 0, \ \psi_c \leq 0, \\ \lambda_c, \ \psi_c > 0, \end{cases} \tag{43.19}$$

where $\overline{C}_c$ and λ_c are experimentally determined functions of temperature T.

The evolutionary equation for changing the center coordinates of the creep surface will be taken as [17]:

$$\dot{\rho}_{ij}^c = g_1^c \dot{e}_{ij}^c - g_2^c \rho_{ij}^c \dot{\chi} + g_T^c \rho_{ij}^c \dot{T},$$

where g_1^c and $g_2^c > 0$ are experimentally determined material parameters.

Concretizing relation (43.18), the gradient law can be represented as:

$$\dot{e}_{ij}^c = \lambda_c(\psi_c, T) S_{ij}^c = \lambda_c \psi_c S_{ij}^c = \lambda_c \left(\frac{\sqrt{S_{ij}^c S_{ij}^c} - \overline{C}_c}{C_c}\right) S_{ij}^c.$$

The intensity of the creep strain rate tensor has the form:

$$\dot{e}_u^c = \lambda_c \left(\frac{\sqrt{S_{ij}^c S_{ij}^c} - \overline{C}_c}{C_c}\right) S_u^c \tag{43.20}$$

Taking into account (43.19), the expression for the length of the trajectory of creep deformations takes the form:

$$\dot{\chi}_c = \sqrt{\frac{2}{3}}\dot{e}_u^c = \sqrt{\frac{2}{3}}\lambda_c\left(\sqrt{S_{ij}^c S_{ij}^c} - \overline{C}_c\right),$$

where:

$$\lambda_c = \frac{\dot{\chi}_c}{\sqrt{\frac{2}{3}}\left(\sqrt{S_{ij}^c S_{ij}^c} - \overline{C}_c\right)}. \tag{43.21}$$

The time dependence χ_c of the process t for $S_u^c = \text{const}$ multiaxial deformation along a ray trajectory has the form shown in Fig. 43.1. On the curve $\chi_c \sim t$ (Fig. 43.1), with a certain degree of conditionality, three sections can be distinguished:

1. section of unsteady creep $\left(0 - \chi_c^{(1)}\right)$ - the creep deformation rate $\dot{\chi}_c$ decreases;
2. section of unsteady creep $\left(\chi_c^{(1)} - \chi_c^{(2)}\right)$ – creep strain rate $\dot{\chi}_c$ is approximately constant $\dot{\chi}_c \cong \text{const}$;
3. section of unsteady creep $\left(\chi_c > \chi_c^{(2)}\right)$ – creep strains grow rapidly (precedes failure) and $\dot{\chi}_c$ sharply increases.

The sections lengths essentially depend on the value of $S_u^c = \text{const}$.

Equations (43.19) - (43.23) describe the unsteady and steady sections of the creep curve at different stress levels and the main effects of the creep process under al-ternating stress. Creep equations (43.19) - (43.23) are connected with equations (43.1) - (43.18), which describe "instant" plastic deformations at the stage of loading through the stress deviator σ'_{ij} and the corresponding algorithm for determining $\dot{e}_{ij}^c$ and $\dot{e}_{ij}^p$ at the stage of loading by means of certain relationships between "time" and "instantaneous" scalar and tensor quantities.

At the stage of development of damages scattered over the volume, the influ-ence of damage on the physical and mechanical characteristics of the material is observed. This effect can be taken into account by introducing effective stresses [17, 18]:

$$\widetilde{\sigma}'_{ij} = F_1(\omega)\,\sigma'_{ij} = \frac{G}{\overline{\overline{G}}}\sigma'_{ij} = \frac{\sigma'_{ij}}{(1-\omega)\left[1 - \frac{(6K+12G)}{(9K+8G)}\omega\right]}, \tag{43.22}$$

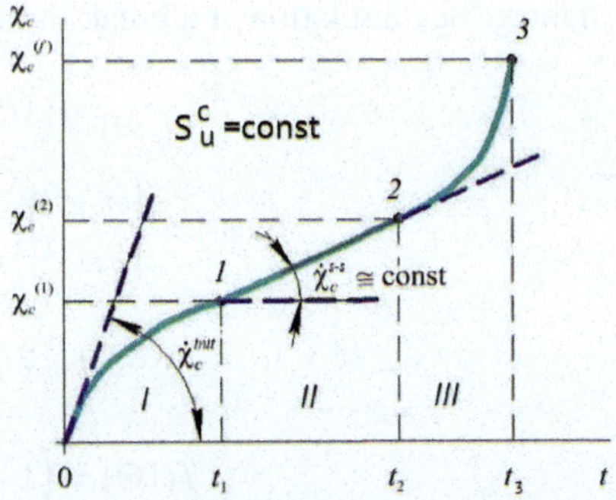

Fig. 43.1 Creep curve under multiaxial deformation.

$$\widetilde{\sigma} = F_2(\omega)\sigma = \frac{K}{\widetilde{K}}\sigma = \frac{\sigma}{4GK(1-\omega)/(4G+3K\omega)}, \tag{43.23}$$

where $\widetilde{G}$ and $\widetilde{K}$ the effective moduli of elasticity defined by Mackenzie formulas (43.24). The effective variable is defined similarly $\widetilde{\rho}_{ij}$:

$$\widetilde{\rho}_{ij} = F_1(\omega)\rho_{ij}.$$

43.2.2 Evolutionary Equations for Damage Accumulation

Damage accumulation equations are based on the relationship between the damage value and internal macroscopic parameters, the critical value of which determines the moment of failure (macroscopic crack nucleation in an elementary volume of a material). The dependence of the dissipated energy spent on the formation of defects under fatigue loading is determined through the work of microstress tensor $\widetilde{\rho}_{ij}$ on inelastic deformations $\dot{e}_{ij}^{irr}$:

$$W_0 = \widetilde{\rho}_{ij}\dot{e}_{ij}^{irr}.$$

In problems of evaluating the service life characteristics, it is necessary to take into account the influence of multiaxial loading, the presence of which significantly reduces the service life due to both an increase in the effective components of the strain and stress tensor under proportional loading, and due to the rotation of the main areas of stress and strain tensors under nonproportional loading.

Taking into account the considered effects that affect the service life characteristics, the equation for the rate of accumulation of fatigue damage under conditions of low-cycle and high-cycle loading can be represented as:

$$\dot{\omega} = f_1(\beta) f_2(\omega) f_3(W_0) f_4(\Theta) W_0. \tag{43.24}$$

In (43.24), the following notation is introduced for functions: $f_1(\beta)$ - taking into account the effect of the "volume" of the stress state; $f_2(\omega)$ – taking into account the effect of the accumulated level of damage on the rate of damage accumulation; $f_3(W_0)$ – taking into account the current relative level of dissipated energy used for the formation of microdefects; $f_4(\Theta)$ – taking into account the change in the rate of damage accumulation at a break in the deformation trajectory.

$$f_1(\beta) = \exp(k\beta),$$

$$f_2(\omega) = \begin{cases} 0, \text{ at } W_0 \le W_a, \\ c\omega^{-1/3}(1-\omega)^{-2/3}, \text{ at } W_0 > W_a, \end{cases}$$

$$f_3(W_p) = \frac{W_0 - W_a}{W_f - W_a},$$

$$f_4(\Theta) = \left(1-\cos^2\theta\right)\beta + \left(\cos^2\theta\right),$$

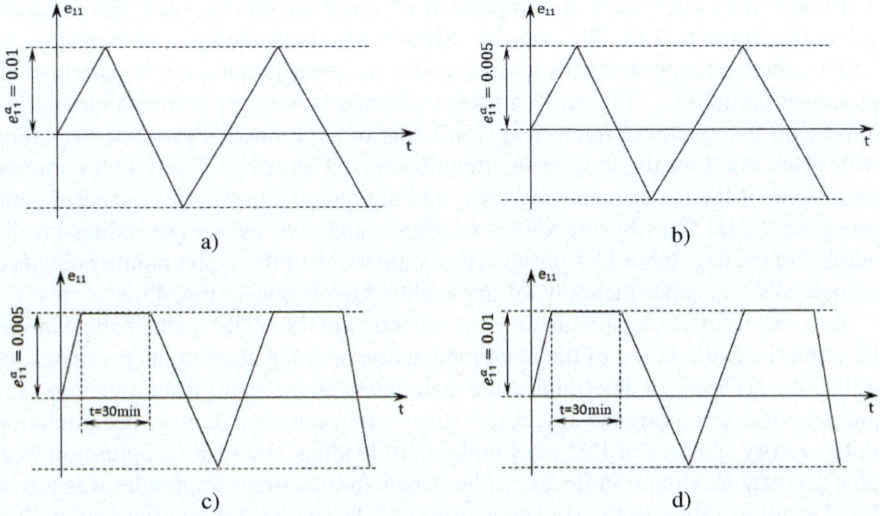

Fig. 43.2: Loading law for laboratory specimens.

c, W_a – are material parameters,

$$W_f = W_{pf}^{cyc} \frac{\dot{\chi}_p^{cyc}}{\dot{\chi}} + W_{cf}^{cyc} \frac{\dot{\chi}_c^{cyc}}{\dot{\chi}} + W_{pf}^{mon} \frac{\dot{\chi}_p^{mon}}{\dot{\chi}} + W_{cf}^{mon} \frac{\dot{\chi}_c^{mon}}{\dot{\chi}}$$

where $W_{fp}^{cyc,mon}(T)$ and $W_{fc}^{cyc,mon}$ – are material parameters. As a criterion for failure, the condition for achieving a damage value of a critical value ω is taken:

$$\omega = \omega_f \leq 1.$$

43.3 Numerical Results

The problem of determining the kinetics of stress-strain state and damage accumulation in laboratory samples of steel 316 under hard loading - tension-compression was solved as an example of practical implementation of the MDM mathematical model for numerical simulation of the combined action of creep and fatigue processes, as well as the accuracy of determining the material parameters and scalar functions with time delays at maximum strain and without them for two strain amplitudes [7]. The temperature in the studies was $650\,^{\circ}C$. In Fig. 43.2 shows the loading laws implemented in tests. The physical and mechanical characteristics of steel 316 [7] and the material parameters at a temperature of $650\,^{\circ}C$ of the damaged medium model [7]-[23] are given in Tables 43.1–43.4.

Figures 43.3-43.5 show a comparison of creep curves for steel 316 at stress levels σ_{11} equal to 183, 207 and 241 MPa, respectively, and at a temperature of $650\,^{\circ}C$, obtained experimentally (points) and by numerical simulation of experimental processes (solid line). Figure 43.6 shows a comparison of experimental data (light dots) with the results of numerical simulation of experimental processes (marked with black dots) for the long-term strength curve. Figures 43.7 and 43.8 compare the results of the experiment (markers) and numerical simulation of experimental processes (solid line) by the values of stress ranges in cycles depending on the number of cycles. Table 43.5 presents the comparison of the cycles number obtained numerically and experimentally for the loading laws shown in Fig. 43.2.

As a second example of evaluating the reliability of the MDM mathematical model for numerical simulation of the combined action of creep and fatigue processes, as well as the accuracy of determining the material parameters and scalar functions, the problem of determining the kinetics of stress-strain state and damage accumulation in laboratory samples of P92 steel under hard loading - tension-compression with time delays at maximum strain and without them for two strain amplitudes was solved [8]. The temperature in the studies was 600 °C. Figure 43.9 shows the loading law implemented in the experiments.

The physical and mechanical characteristics of R92 steel [20] and the material parameters at a temperature of $650\,^{\circ}C$ of the damaged medium model [17]–[23] are given in Tables 43.6–43.8. Figures 43.10 and 43.11 show cyclic hysteresis loops for the first loading cycle and for a loading cycle equal to half the life ($0,5N_f = 335$),

Table 43.1: Physical and mechanical characteristics and material parameters of the MDM model of steel 316 at $T = 650$ °C.

K (MPa)	G (MPa)	C_p^0 (MPa)	g_1^p (MPa)	g_2^p	g_1^c (MPa)	g_2^c
153450	58450	74	6858	290,4	5100	260
a_p	a_c	W_{pf}^{cyc}	W_{cf}^{cyc}	W_{cf}^{mon}	W_{pf}^{mon}	
3,2	75	820	10	50	10	

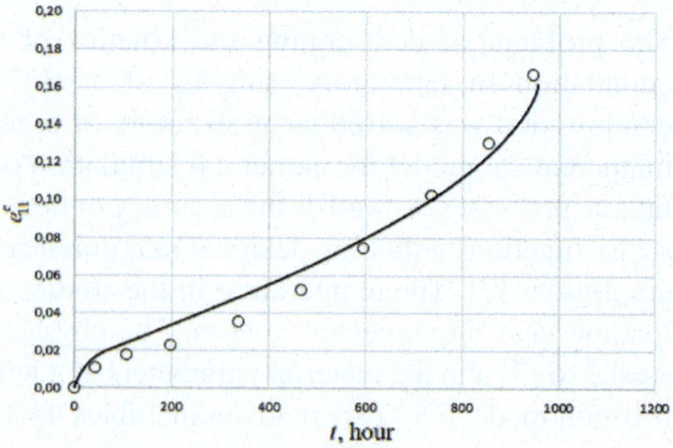

Fig. 43.3 Creep curve at 183 MPa.

Table 43.2: The radius of the loading surface corresponding to the zero creep rate $\overline{C}_c$ (MPa) of steel 316 at $T = 650°$C.

χ_c	0	0,01	0,05	0,08	0,1	0,12	0,14	0,16	0,17
$\overline{C}_c$	40	40	40	35	27	14	6	3	2

Table 43.3: Modulus of cyclic hardening $Q_1^p\,(a_{emax})$ and $Q_1^{cr}\,(a_{emax})$ (MPa) of steel 316 at $T = 650°$C.

a_{emax}	0	0,001	0,003	0,004	0,006	0,01
$Q_1^p\,(a_{emax})$	0	50	80	90	95	110
$Q_1^c\,(a_{emax})$	0	-30	-33	-33	-33	-33

Table 43.4: Dependence of proportionality coefficients λ_c^0 and $\lambda_c^{II}\lambda$ on stress intensity σ_u for steel 316 at $T = 650°$C.

σ_u, MPa	0	50	100	150	170	200
λ_c^0, 1/MPa·h, 10^{-6}	0	9,0	9,0	9,0	21	72
λ_c^{II}, 1/MPa·h, 10^{-6}	0	1,455	1,455	1,455	3,6	1,854

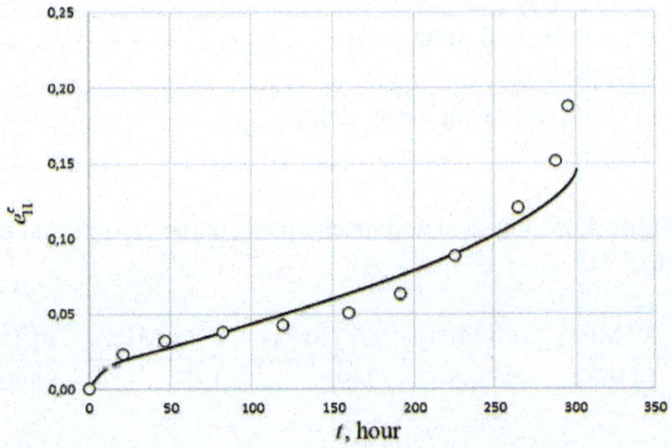

Fig. 43.4 Creep curve at 207 MPa.

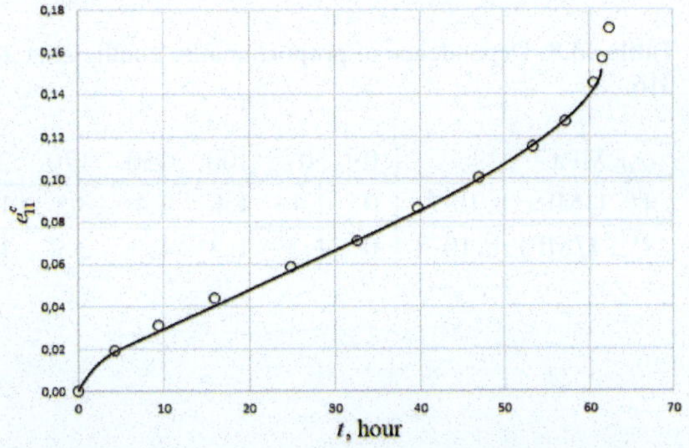

Fig. 43.5 Creep curve at 241 MPa.

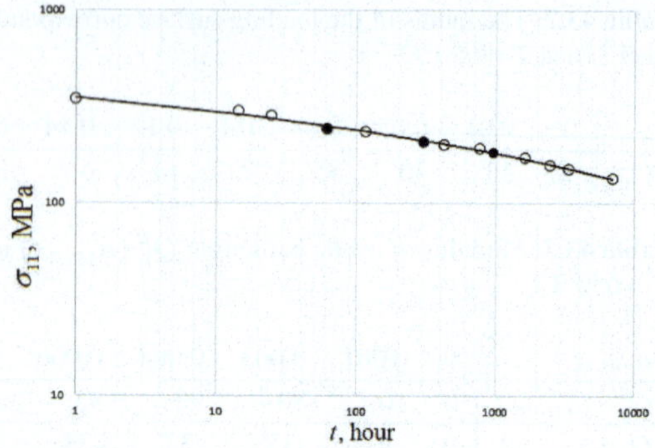

Fig. 43.6 316 steel tensile curve.

Table 43.5: Comparison of calculated and experimental values of cycles to failure.

Loading law	Experimental number of cycles to failure	Estimated number of cycles before failure
$e_{11}^{a} = 0,01$ without time delay (Fig. 43.2 a)	863	964
$e_{11}^{a} = 0,005$ without time delay (Fig. 43.2 b)	4000	3639
$e_{11}^{a} = 0,01$ with time delay (Fig. 43.2 c)	338	385
$e_{11}^{a} = 0,005$ with time delay (Fig. 43.2 d)	985	950

Table 43.6: Physical and mechanical characteristics and material parameters of the MDM model of steel 316.

K(MPa)	G(MPa)	C_p^0(MPa)	$\overline{C}_{c0}$(MPa)	g_1^p(MPa)	g_2^p	g_1^c(MPa)	g_2^c	a
141667	65385	249	125	38900	660	5100	260	4

Table 43.7: Modulus of cyclic hardening $Q_1^{pl}(a_{emax})$ and $Q_1^{cr}(a_{emax})$ (MPa) of steel 316.

a_{emax}	0	0,001	0,003	0,004	0,006	0,01
$Q_1^{pl}(a_{emax})$	0	-92	-92	-92	-92	-92
$Q_1^{cr}(a_{emax})$	0	-50	-50	-50	-50	-50

Table 43.8: Dependence of proportionality coefficients λ_c^0 and λ_c^{II} on stress intensity σ_u for steel 316.

σ_u, MPa	0	50	100	150	170	250
λ_c^0, 1/MPa · h, 10^{-4}	0	1,4	1,4	1,4	3,5	15
λ_c^{II}, 1/MPa · h, 10^{-6}	0	1,3	1,3	1,3	3,2	17,5

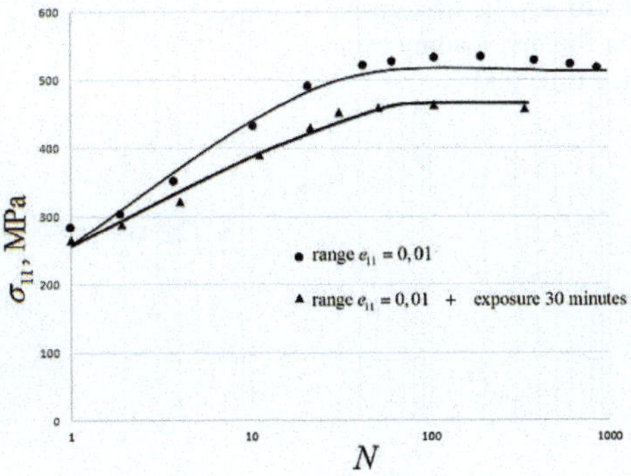

Fig. 43.7 Dependence of the stress range in the loading cycle on the number of cycles.

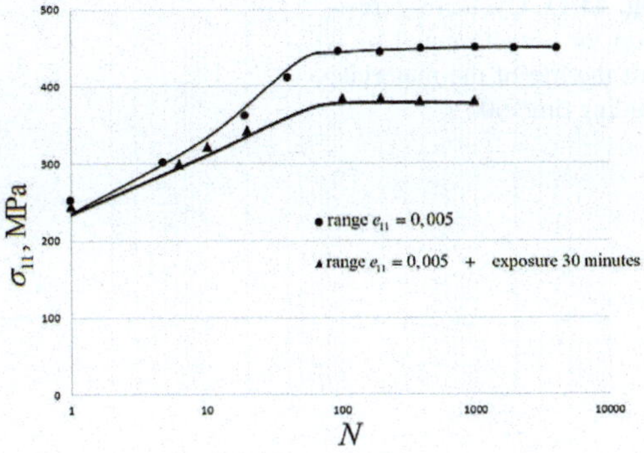

Fig. 43.8 Dependence of the stress range in the loading cycle on the number of cycles.

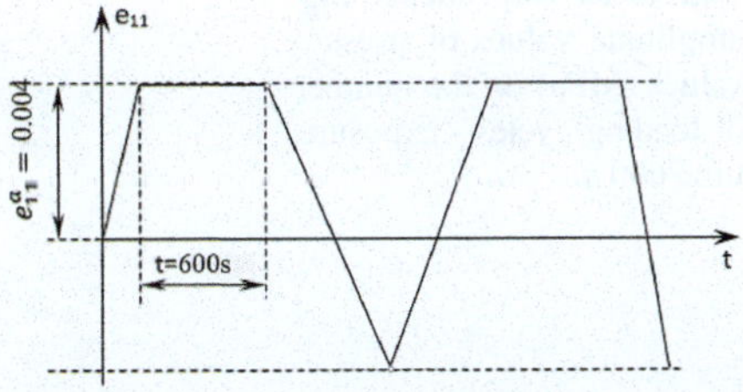

Fig. 43.9 Loading law for laboratory specimens.

respectively (dots mark the experimental data, the solid line shows the numerical results).

Figure 43.12 shows the numerical results of the dependence of the maximum and minimum stresses in a cycle on the number of cycles, where the experimental data are marked with dots. Figures, tables and service life characteristics of materials given in them demonstrate the qualitative and quantitative coincidence of numerical and experimental data and testify to the reliability of the constitutive relations of the MDM and the developed methodology for determining the material parameters included in these relations.

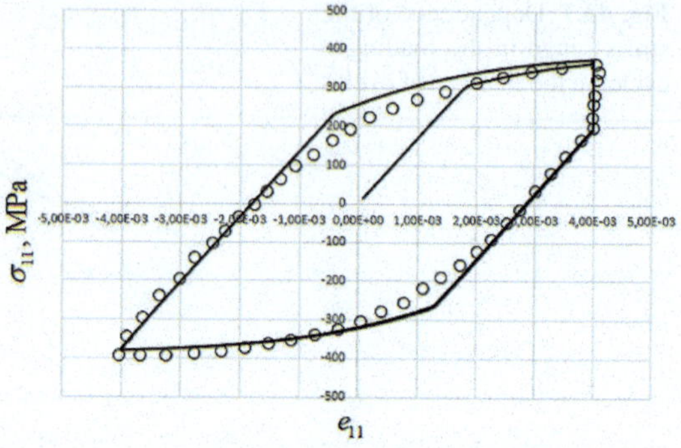

Fig. 43.10 Cyclic hysteresis loop for the first loading cycle - holding time 600 s.

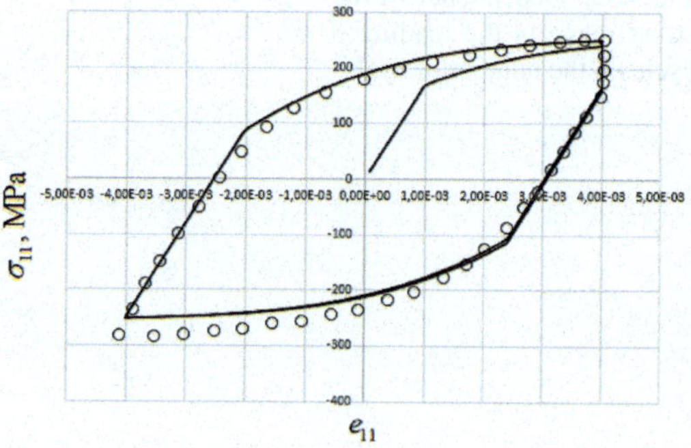

Fig. 43.11 Cyclic hysteresis loop for a cycle equal to half the life of the material - holding time 600 s.

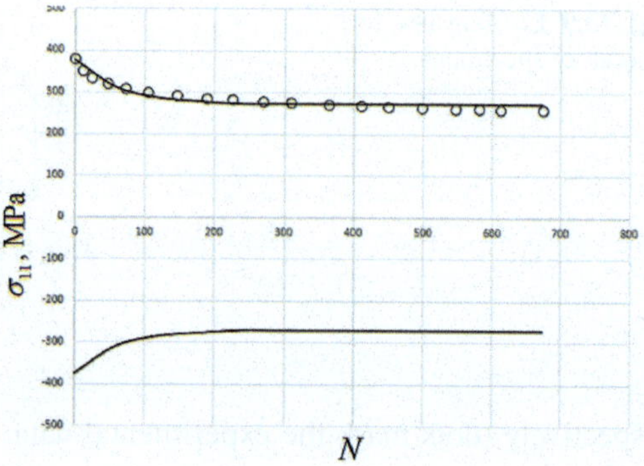

Fig. 43.12 Dependence of amplitude values of stress values (MPa) on the number of loading cycles - exposure time 600 s.

43.4 Conclusion

A mathematical model of MDM has been developed that describes the pro-cesses of deformation and damage accumulation under fatigue and long-term strength, based on the energy approach and a unified form of representation of the damage accumulation process. For the mathematical model of the MDM, the mathematical parameters and scalar functions included in the constitutive relations are determined.

For the purpose of evaluating the reliability of the developed mathematical model of the mechanics of damaged medium, numerical studies were carried out to determine the kinetics of stress-strain state and damage accumulation in la-boratory samples of steels 316 and P92 under rigid loading - tension-compression with and without time delays at maximum deformation for two de-formation amplitudes.

The presented in this paper analysis of the equations of mechanics of damaged medium and the results of assessing their reliability make it possible, in the authors' opinion, to recommend them at this stage for calculating the durability of structural units of machine-building objects operating under conditions of non-stationary thermomechanical loading. At the same time, further experimental and theoretical verification of their reliability is necessary for the cases of non-stationary multiaxial stress-strain states and non-isothermal processes accompanied by a significant rotation of the main areas of stress and strain tensors (for various complex loading trajectories) [26].

Acknowledgements The work was supported by the Russian Science Foundation grant no. 22-19-00138.

References

[1] Mitenkov AM, Kaidalov VB, Korotkikh YG et al., Methods for substantiating the resource of nuclear power plants. - Moscow, Mashinostroenie, 2007.

[2] Woodford DA (1979) Creep damage and residual life concept, Journal of Theoretical Foundations of Engineering Calculations, **101**(4):1-8.

[3] Majumar, M (1980) Time dependent mechanical fatigue, Journal of Theoretical Foundations of Engineering Calculations, **102**(1):111 - 121.

[4] Gomyuk. The study of the behavior of steel 316 under loading schemes of fatigue, creep and combined action of fatigue and creep / Gomyuk, Bew Quoc, Biron. Journal of modern engineering. – 1991. - (1). – pp 14 - 23.

[5] Gomyuk. Calculation of durability of stainless steel 304 under conditions of fatigue and creep interaction using the theory of continuous damage / Gomyuk, Beau Quoc. Journal of Theoretical Foundations of Engineering Calculations. – 1986. - (3):111 - 136.

[6] Zamrik S, Davies D (1991) Application of the plasticity exhaustion method to fatigue under uniaxial loading - creep damage assessment for stainless steel 316, Modern Engineering, Ser B, (8).

[7] Zamrik S (1990) Calculation of damage due to the mutual influence of creep and fatigue of stainless steel 316 under axial loading, Modern Engineering, Ser E, (7).

[8] Zhang T, Wang X, Zhang W, Hassan T, Gong J (2020) Fatigue–creep interaction of P92 steel and modified constitutive modelling for simulation of the responses, Metals **10**:307.

[9] Dulnev RA, Kotov PI (1980) Thermal Fatigue of Metals (in Russ), Moscow, Mashinostroenie.
[10] Gusenkov A.P., Kotov P.I. Low-cycle fatigue under nonisothermal loading (in Russ.). Moscow: Mashinostroenie (1983).
[11] Sorensen CB (1974) Low-cycle resistance at elevated temperatures and bearing capacity of structural elements, In: Proceedings of the All-Union Symposium on Low Cycle Fatigue at High Temperatures. Chelyabinsk, issue 4, pp 3-46.
[12] Shesterikov SA, Arshakuni AL, Cheredeeva LY (1989) Method for the temperature-force prediction of the long-term strength of metals, Strength of Materials **21**(9):6–9.
[13] Boyle J, Spence J (1983) Stress Analysis for Creep, Butterworth-Heinemann.
[14] Kazantsev AG (1985) Interaction of low-cycle fatigue and creep in nonisothermal loading, Strength of Materials **17**(5):25-31.
[15] Bernard - Connolly, Beau Quoc, Biron. Fatigue of corrosion-resistant steel 304 during tests under conditions of multistage controlled deformation. Theoretical foundations of engineering calculations. 1983, (3):47-53.
[16] Volkov IA, Egunov VV, Igumnov LA, Kazakov DA, Korotkikh YG, Mitenkov FM (2015) Assessment of the service life of structural steels by using degradation models with allowance for fatigue and creep of the material, Journal of Applied Mechanics and Technical Physics **56**(6):995–1006.
[17] Volkov IA, Korotkikh YG (2008) Equations of state of viscoelastic-plastic media with damage, Moscow, Fizmatlit.
[18] Volkov IA, Igumnov LA (2017) Introduction to the continuum mechanics of a damaged medium, Moscow, Fizmatlit, 2017.
[19] Volkov IA, Korotkikh YG (2014) Modeling of fatigue life of materials and structures under low-cycle loading, Mechanics of Solids **49**(3):290-301.
[20] Bolshukhin MA, Lebedev VV, Kozin AV, Korotkikh YG, Panov VA, Pakhomov VA (2014) Modeling damage accumulation processes resulting from thermal pulsation (in Russ), Problems of Strength and Plasticity **76**(2):134-143.
[21] Mitenkov FM, Volkov IA, Igumnov LA, Korotkikh YG and others (2015) Applied Theory of Plasticity. Moscow: Fizmatlit.
[22] Volkov IA, Igumnov LA, Korotkikh YG (2015) Applied theory of viscoplasticity, Nizhny Novgorod: UNN Publishing House.
[23] Volkov IA, Igumnov LA, Tarasov IS, Shishulin DN, Pichkov SN, Markova MT (2019) Modeling plastic deformation of polycristalline structural alloys under block-type nonsymmetrical regimes of softlow-cycle loading (in Russ), Problems of Strength and Plasticity **81**(1):63–76 (2019).
[24] MacKenzie JK (1959) The elastic constants of a solids containing spherical holes, Proc Phys Soc Section B **63**(1):2–11 (1950).
[25] Troshchenko VT (2006) Nonlocalized fatigue damage of metals and alloys. Part 3. Strain and energy criteria. Strength of Materials **38**(1):1-19.
[26] Xu L, Kojima T, Itoh T (2021) Creep–fatigue life evaluation of type 304 stainless steel under non-proportional loading, International Journal of Pressure Vessels and Piping **194**(Part A):104515.

Chapter 44
Bandgap Properties of a Class of Chiral and Achiral Metamaterials

Yun-Che Wang, Tse-Chun Liao, Kai-Wen Tan, and Sergey M. Aizikovich

Abstract Bandgaps in the dispersion curves of acoustic metamaterials are of great scientific and engineering importance for developing novel materials in vibration-isolation and noise-reduction applications. In this work, two dimensional achiral and chiral composite metamaterials are studied for their bandgap properties in relation to microstructures and physical properties of constituents. In particular, parametric studies have been conducted to investigate the effects of the Young's modulus and density mismatches, as well as thickness of ligaments on bandgaps. When the ratio of matrix density to skeleton density ρ_m/ρ_s less than 1, largest bandgaps can be found in the modulus mismatch $E_m/E_s = 1$ chiral case, not in the $E_m/E_s = 100$ cases, indicating the interplay among chirality, modulus and density on bandgap structures in the metamaterials. When chirality, modulus and density are given, the thickness of ligaments provides additional degrees of freedom to design bandgaps.

Key words: Bandgap · Dispersion curve · Chiral · Metamaterials · Finite element analysis

44.1 Introduction

Metamateirals, which are materials with artificial microstructure, exhibit internal degrees of freedom, and can be analysed as the generalized continua [1]-[3]. Chirality in metamateirals provides additional degrees of freedom for material designers,

Yun-Che Wang · Tse-Chun Liao · Kai-Wen Tan
National Cheng Kung University, 1 University Road, Tainan, Taiwan 70101, R.O.C.,
e-mail: yunche@ncku.edu.tw, 3951242a@gmail.com, n66091427@gs.ncku.edu.tw

Sergey M. Aizikovich
Research and Education Center "Materials", Don State Technical University, Rostov-on-Don 344000, Russian Federation,
e-mail: saizikovich@gmail.com

H. Altenbach et al. (eds.), *Sixty Shades of Generalized Continua*,
Advanced Structured Materials 170, https://doi.org/10.1007/978-3-031-26186-2_44

and can give rise to negative Poisson's ratio [4] and deformation mode coupling phenomena. The deformation mode coupling phenomena entail that the materials can twist or bend under purely axial deformation, or vice versa [5]-[8]. A non-centrosymmetric Cosserat mathematical model has been used to analyze the chiral materials [9]. Negative Poisson's ratio materials have been demonstrated by using the re-entrant cell shape [10]-[16]. Some of other approaches to obtain negative Poisson's ratio are utilizing the rotating rigid blocks [17] or phase transitions [18, 19]. Negative Poisson's ratio can be related to negative moduli through the inter-relationships among elastic constants. However, in most cases, negative moduli raise a concern of stability [20]-[24].

Bandgaps in metamaterials have a significant role in the applications of noise reduction and vibration isolation [25]-[28]. The forbidden frequency bands shown in the dispersion curves as gaps do not allow the existence of acoustic or elastic waves with corresponding frequency and wave vectors. Therefore, in addition to filtering out waves with specific frequencies, one can use bandgaps to reduce total transmission amplitudes if the incoming waves contain all frequencies. Study of bandgaps has been a tradition in solid state physics for the lattice dynamics of atoms [29]. Through the design of metamaterials, phononic bandgap structures can be fabricated through additive manufacture methods [30]. In [31] is presented the theoretical foundation of the anisotropic strain gradient elasticity to analyze chiral metamaterials.

In this work, the finite element analysis [32] is adopted to calculate the dispersion curves of a class of chiral and achiral composite metamaterials under the plane strain assumption. Effects of microstructrual geometry, modulus ratio, density ratio on the bandgap properties of the metamaterials are investigated.

44.2 Theoretical Considerations

The dispersion relationship between frequency ω and wave vector k can be analytically studied in some discrete or continuum systems. For a monoatomic chain with each atom mass M and each spring constant K, its dispersion relationship is as follows [29].

$$\omega^2 = \frac{4K}{M}\sin^2(ka/2) \tag{44.1}$$

The distance between two nearest atoms is a. Only does one branch exist. However, for a diatomic chain with M_1, M_2, K_1, K_2 being masses and spring constants, and alternatively positioned, its dispersion relationship is as follows.

$$\omega^2 = \frac{(M_1+M_2)(K_1+K_2)}{2M_1M_2} \pm \frac{\sqrt{(M_1+M_2)^2(K_1+K_2)^2 - 16M_1M_2K_1K_2\sin^2(ka/2)}}{2M_1M_2}$$

The $\pm$ sign indicates the existence of two branches. If wave vector k approaches zero, two branches can be obtained from the above equation. To the first order approximation, the optic branch shows no dependence on k, as follows.

$$\left(\omega^2\right)_{\text{optic}} = \frac{(M_1+M_2)(K_1+K_2)}{M_1 M_2} \tag{44.2}$$

The acoustic branch is proportional to k^2, as below.

$$\left(\omega^2\right)_{\text{acoustic}} = \frac{K_1 K_2 a^2 k^2}{(M_1 M_2)(K_1+K_2)} \tag{44.3}$$

When k is small, $(\omega^2)_{\text{optic}} > (\omega^2)_{\text{acoustic}}$. At the Brillouin zone boundary, which is $k = \pi/a$, $(\omega^2)_{\text{optic}} = 2K_1/M_1$ and $(\omega^2)_{\text{acoustic}} = 2K_1/M_2$ if $K_1 = K_2$. The terms of optic and acoustic branch or mode have the roots in lattice dynamics [29]. The gap between the two branches, i.e. Eqs. (44.2) and (44.3), indicates the forbidden frequency bands. The formation of the gap is due to resonant phenomena, and empty space in the models is not required. The above dispersion relationship can be used to determine the displacements of all atoms, as follows.

$$u = \sum_k \tilde{u}_k e^{i(\boldsymbol{k}\cdot\boldsymbol{x} - \omega_k t)} \tag{44.4}$$

In the long wave limit, the dispersion relationship of the three dimensional, isotropic and homogeneous elastic solid is as follows.

$$\rho\omega^2 \tilde{u}_m = c_{mnpq} k_n k_p \tilde{u}_q \tag{44.5}$$

Theory of elasticity is insufficient to describe the material with microstructure. Under the framework of the strain gradient elasticity (SGE) [31], the dispersion relationship of the microstructured solid can be estimated by

$$\rho V^2 \tilde{u}_i = Q_{il} \tilde{u}_l, \tag{44.6}$$

where

$$Q_{il} = c_{ijlm} - \omega^2 J_{ijlm} \hat{\xi}_j \hat{\xi}_m + \frac{\omega^2}{V^2} A_{ijklmn} \hat{\xi}_j \hat{\xi}_k \hat{\xi}_m \hat{\xi}_n. \tag{44.7}$$

The second order inertia tensor is denoted by J_{ijlm}, and A_{ijklmn} is the sixth order tensor relating hyperstress and strain gradient tensors. The phase velocity is denoted by V, hence $\hat{\xi}$ is related to the wave vector, as follows.

$$\hat{\xi}_i = \frac{V}{\omega} k_i. \tag{44.8}$$

Eqs. (44.6), (44.7) and (44.8) can be used to study the bandgap properties of SGE materials, but the homogenized material parameters are required as input. It is noted that although the SGE framework is widely accepted for homogenized lattice materials containing empty space between ligaments, the composite lattice materials studied in this work require the homogenization of composite materials between the matrix and the lattice skeleton.

In the next sections, we adopt the finite element analysis to study three composite models with microstructure. Our numerical results may be used as input data, along with experimental data, to determine the parameters used in above homogenized SGE framework.

44.3 Numerical Considerations

In this work, the finite element method is adopted to numerical analyze three metamaterial models shown in Figs. 44.1 (a), (c) and (e). Their corresponding finite element meshes are shown in (b), (d) and (f). Denser mesh is applied to the regions with sharp angles to obtain more accurate results from the stress concentration regions. The models are composed of two materials, namely ligament and matrix. Young's modulus is assumed to be $E_f = 2$ and $E_s = 200$ GPa for the $E_s/E_f = 100$ case, and $E_f = E_s = 200$ GPa for the $E_s/E_f = 1$ case. For Poisson's ratio, $\nu_f = 0.45$ and $\nu_s = 0.34$ are assumed for all cases. The subscripts f and s indicate the filler (i.e. matrix) and skeleton, respectively. The width of ligament is 4 mm, and the edge length of the square specimen is 36 mm. In (a) the model is achiral, hence possesses centro-symmetry. Models in (b) and (c) are chiral with different arm geometry. Periodic boundary conditions

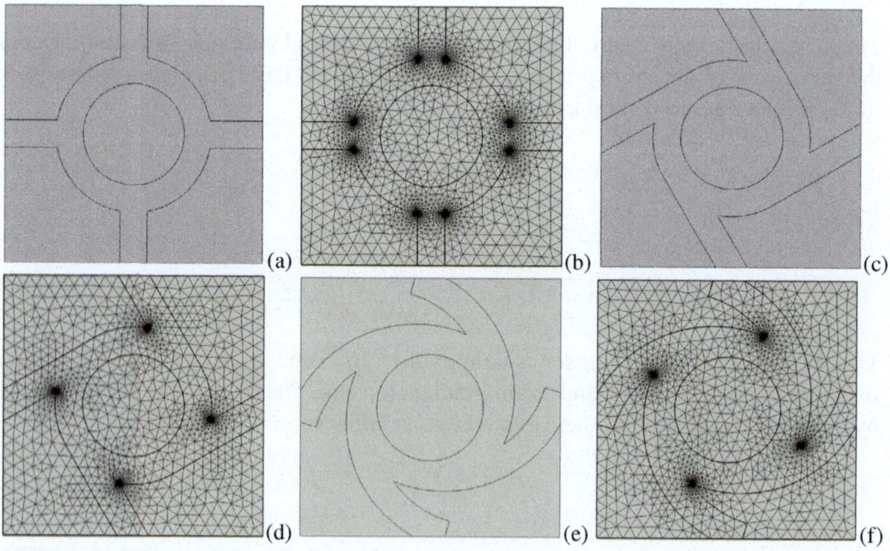

Fig. 44.1: Geometry of the (a) achiral model, (c) chiral model with straight arms and (e) chiral model with curved arms. Corresponding finite element meshes are shown in (b), (d) and (f). The edge length of the square is 36 mm, and the thickness of the ligament is 4 mm.

are applied to all edges of the models. The COMSOL (2022) software package was used in this study to carry out the finite element numerical procedure. The shape function is 5th-order Lagrange interpolation function. All constituents are isotropic, homogeneous, linearly elastic materials. Affine deformation is assumed. The path to sweep wave vector from Point 1 to 2, 2 to 3, and 3 to 1 in the irreducible first Brillouin zone is shown in Figure 44.2. The tuning parameter for the wave vector is k from 0 to 3. Let the edge length of the square be L. When $k < 1$, $k_x = \pi k/L$ and $k_y = 0$. When $1 \leq k < 2$, $k_x = \pi/L$ and $k_y = \pi(k-1)/L$. When $k \geq 2$, $k_x = \pi(3-k)/L$ and $k_y = \pi(3-k)/L$. Because of the symmetry in the squared geometry, the chosen path represents all possible wave vectors in the system.

44.4 Results and Discussion

Figure 44.3 shows the dispersion curves of achiral model with ligament thickness $t = 4$ mm. Five different density ratios are shown, and it can be seen that when skeleton ligaments have higher density, more bandgaps can be found. The interplay between resonating ligament and matrix determines the bandgap patterns.

When varying the ligament thickness, Fig. 44.4 shows that the ligament thickness can be used as a strong tuning parameter to control bandgap patterns, as they are very sensitive to the thickness. When the ligament thickness $t = 3$ mm, the total bandgap in all bands below 40 kHz is the largest. Since the modulus and density ratios are fixed, varying the thickness of ligament effectively changes the structural rigidity of the ligament, and at the same time, reduces the amount of surrounding matrix.

Selected deformation mode shapes at four frequencies are shown in Figure 44.5. The dark blue color indicates the the zero displacement magnitude, and the red color the largest displacement, which is about 50 nm. Mode shapes are the eigenvector solutions, hence their magnitudes are obtained through suitable normalization. In this particular case, ligament thickness $t = 3$ mm gives rise to the widest bandgap in total.

For a given ligament thickness, the comparisons of modulus and density ratios are shown in Fig. 44.6 for the lowest frequency bands. Both bandgap start frequency

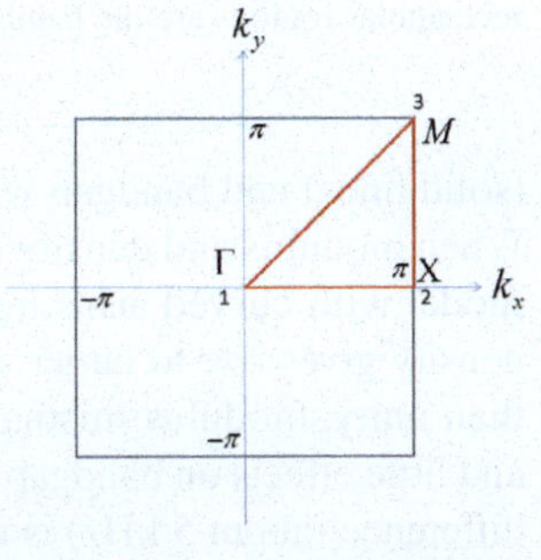

Fig. 44.2 The sweeping path $1 \rightarrow 2 \rightarrow 3 \rightarrow 1$ for the wave vector in the two-dimensional square unit cell.

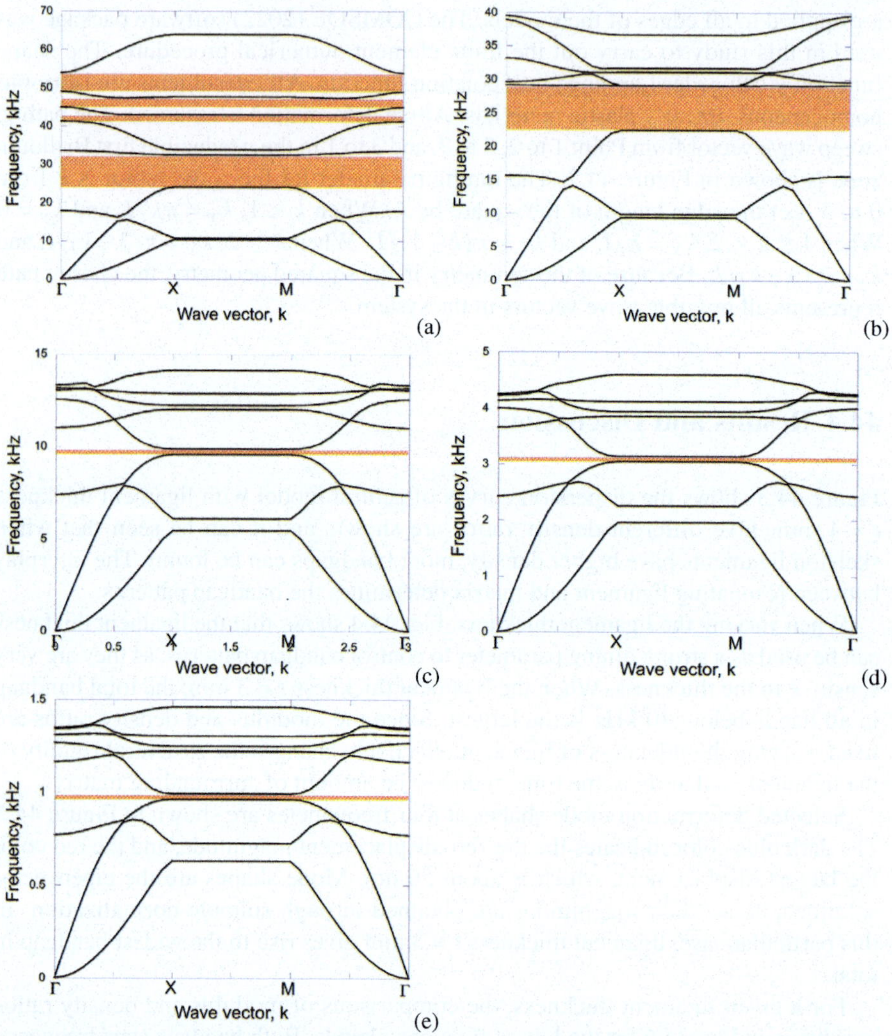

Fig. 44.3: Numerically calculated dispersion curves of the achiral model when $E_s/E_f = 100$ for (a) $\rho_s/\rho_f = 100$, (b) $\rho_s/\rho_f = 10$, (c) $\rho_s/\rho_f = 1$, (d) $\rho_s/\rho_f = 0.1$, (e) $\rho_s/\rho_f = 0.01$. The colored rectangular regions are the bandgaps.

(solid lines) and bandgap width (dashed lines) are shown on the double y-axis plots. When modulus and density ratio equals to unity, bandgaps are not found in the chiral model with curved arms, regardless of modulus ratios. In general, larger ligament density gives rise to larger start frequency and gap width. When density ratio is less than unity, modulus mismatch shows some effects on the bandgap start frequency, and little effects on bandgap width. When density ratio equals to 10, largest gap width difference (about 5 kHz) is found in the achiral model for the studied modulus ratios.

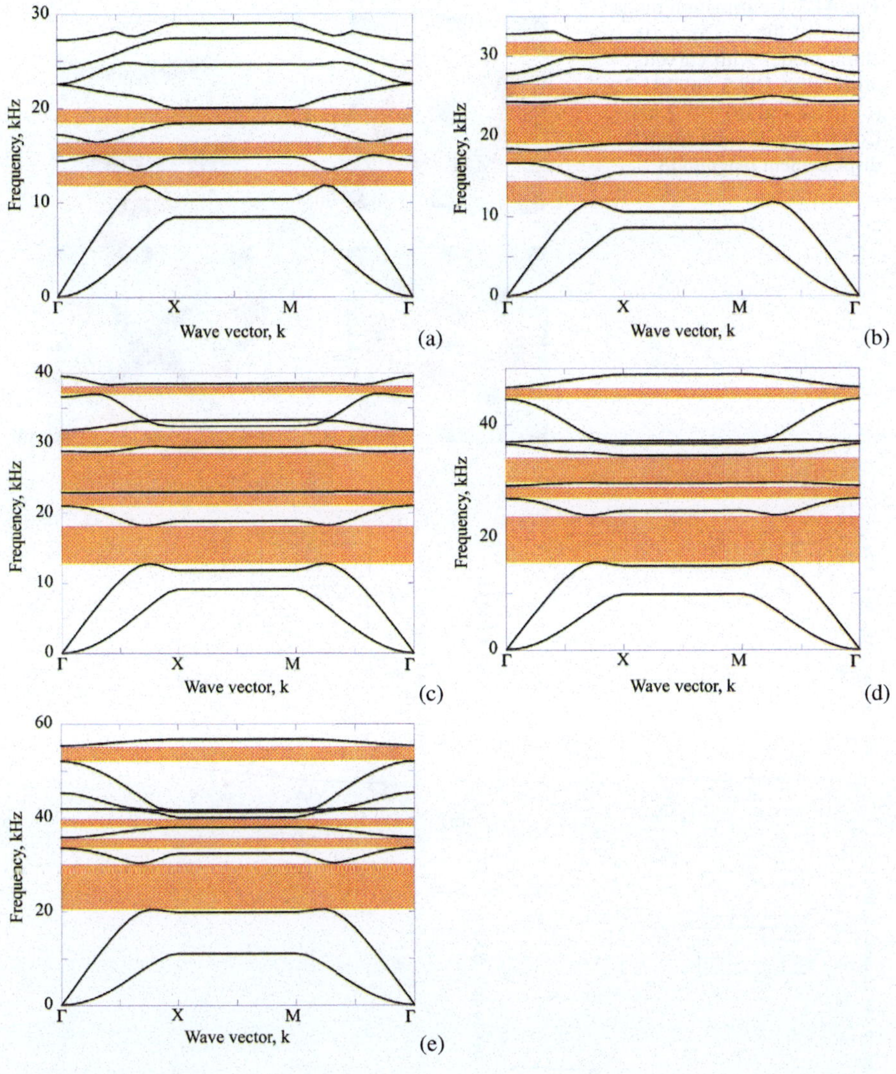

Fig. 44.4: Dispersion curves of the chiral models with straight arms for various ligament thickness when $E_s/E_f = 100$ and $\rho_s/\rho_f = 8$. (a) t=1, (b) t=2, (c) t=3, (d) t=4, (e) t=5 mm. The colored rectangular regions are the bandgaps.

44.5 Conclusions

Three metamaterial models are studied for their bandgap properties under the plane strain assumption. Effects of chirality, density ratio, modulus ratio, and ligament thickness are investigated. When the density ρ_s/ρ_f is small bandgap start frequency and width are small in general. The largest bandgap start frequency (about 130 kHz) is

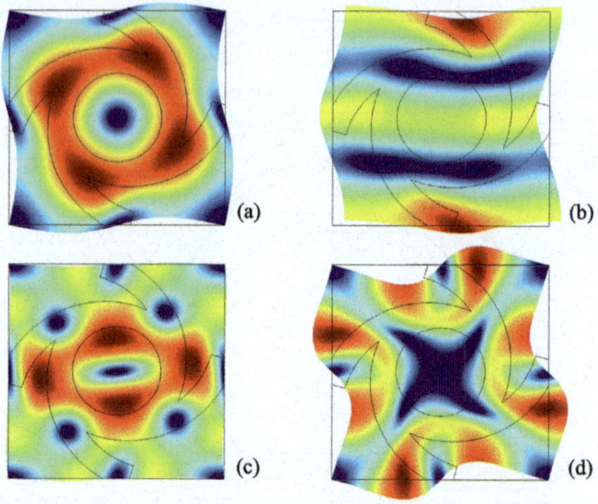

Fig. 44.5 Deformation mode shapes of the $\rho_s/\rho_f = 10$ chiral model with curved arms at (a) 109.5, (b) 131.2, (c) 171.2 and (d) 185.2 kHz. Color indicates the total displacement magnitude

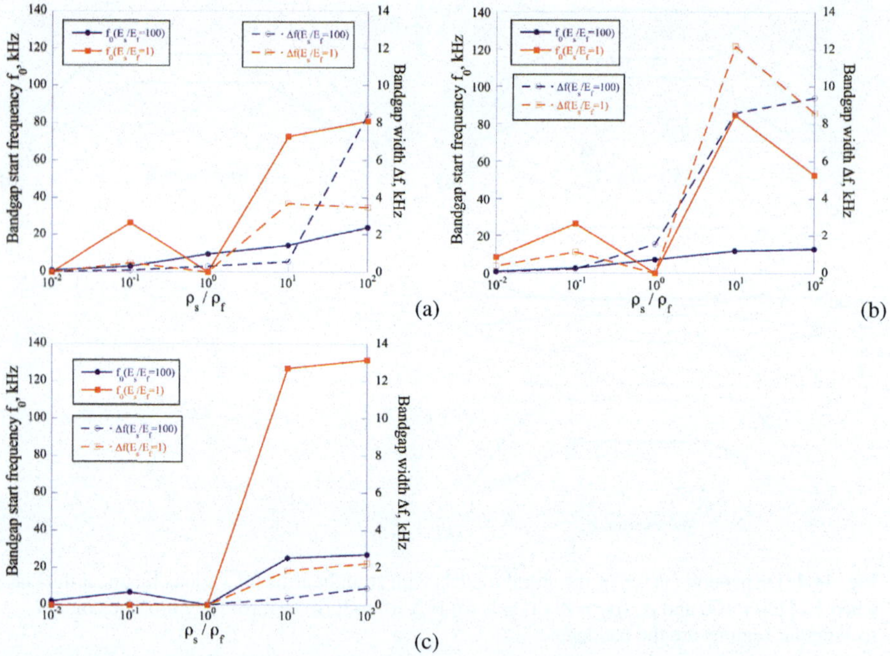

Fig. 44.6: Bandgap start frequency and width of the lowest frequency mode for the (a) achiral model, (b) chiral model with inclined straight arms and (c) chiral model with curved arms.

found in the chiral model with curved arms when $\rho_s/\rho_f = 100$. The largest bandgap width (about 12 kHz) is found in the chiral model with inclined straight arms when $\rho_s/\rho_f = 100$. When the ligament thickness $t = 3$ mm, the total bandgap in all bands below 40 kHz is the largest. Although the models are in two dimensions, generalization to three dimensions is straightforward in the numerical approaches. Our results from two-dimensional analysis may shed light on future studies on three-dimensional metamaterials.

Acknowledgements This research was, in part, supported by Taiwan Ministry of Science and Technology, under the contract number MOST 111-2923-E-006-003-MY3 and MOST 109-2221-E-006-016-MY3. Sergei Aizikovich acknowledges the support of the grant of the Russian Science Foundation, grant number 22-49-08014.

References

[1] Altenbach H, Eremeyev VA (2013) Generalized Continua from the Theory to Engineering Applications, CISM International Centre for Mechanical Sciences Vol 541, Springer, Wien.

[2] Eremeyev VA, Lebedev LP, Altenbach H (2013) Foundations of Micropolar Mechanics, Springer, Heidelberg.

[3] Eremeyev VA (2019) Two- and three-dimensional elastic networks with rigid junctions: modeling within the theory of micropolar shells and solids, Acta Mech **230**:3875–3887.

[4] Prall D, Lakes RS (1997) Properties of a chiral honeycomb with a Poisson's ratio of -1, International Journal of Mechanical Sciences **39**:305-314.

[5] Frenzel T, Kadic M, Wegener M (2017 Three-dimensional mechanical metamaterials with a twist, Science **358**:1072-1074

[6] Wang YC, Ko TW, Ren X (2020) Effective mechanical responses of a class of 2D chiral materials, Phys. Status Solidi B **257**:2000277.

[7] Wang YC, Ko TW, Tan KW (2022) Effective viscoelastic properties of chiral structures with hierarchy. In: Altenbach H, Eremeyev VA, Galybin A, Vasiliev A (eds), Advanced Materials Modelling for Mechanical, Medical and Biological Applications, Advanced Structured Materials, vol. 155, Springer, Cham.

[8] Wang YC, Tan KW, Ko TW, Liao TC (2022) Numerical study of simultaneously negative thermal expansion and Poisson's ratio in a class of auxetic structures, Phys. Status Solidi B **2022**:2200418.

[9] Lakes RS, Benedict RL (1982) Noncentrosymmetry in micropolar elasticity, International Journal of Engineering Science **20**:1161-1167.

[10] Lakes R (1987) Foam structures with a negative Poisson's ratio, Science **235**:1038-1040.

[11] Wojciechowski KW (1989) Two-dimensional isotropic system with a negative Poisson ratio, Physics Letters A **137**:60-64.

[12] Evans KE, Nkansah MA, Hutchinson IJ, Rogers SC (1991) Molecular network design, Nature 353:124.

[13] Lim T-C (2015) Auxetic Materials and Structures, Springer.

[14] Wang YC, Shen MW, Liao SM (2017) Microstructural effects on the Poisson's ratio of star-shaped two-dimensional systems, Phys. Status Solidi B **254**:1700024.

[15] Wang YC (2020) Negative Poisson's ratio in plane elasticity, In: H. Altenbach, A. Öchsner (eds) Encyclopedia of Continuum Mechanics, Springer, Berlin, Heidelberg, pp 1800–1806.

[16] Wang YC, Lai HW, Ren XJ (2020) Enhanced auxetic and viscoelastic properties of filled reentrant honeycomb, Phys. Status Solidi B **257**:1900184.

[17] Grima JN, Manicaro E, Attard D (2011) Auxetic behaviour from connected different-sized squares and rectangles, Proceedings of the Royal Society A **467**:439-458.

[18] Lakes RS (2017) Negative-Poisson's-ratio materials: auxetic solids, Annual Review of Materials Research **47**:63-81.

[19] Lakes R (2020) Composites and Metamaterials, World Scientific, Singapore.

[20] Lakes RS, Lee T, Bersie A, Wang YC (2001) Extreme damping in composite materials with negative-stiffness inclusions, Nature **410**:565-567.

[21] Jaglinski T, Kochmann D, Stone D, Lakes RS (2017) Composite materials with viscoelastic stiffness greater than diamond, Science **315**:620-622.

[22] Lakes RS, Drugan W (2002) Dramatically stiffer elastic composite materials due to a negative stiffness phase?, Journal of the Mechanics and the Physics of Solids **50**:979-1009.

[23] Wang YC, Lakes RS (2005) Composites with inclusions of negative bulk modulus: extreme damping and negative Poisson's ratio, Journal of Composite Materials **39**:1645-1657.

[24] Wang YC, Ko CC, Chang KW, Ko TW (2021) Negative-stiffness composite systems and their coupled-field properties, Continuum Mech. Thermodyn. **33**:1857-1872.

[25] Meng J, Deng Z, Zhang K, Xu X, Wen F, Körner C (2015) Band gap analysis of star-shaped honeycombs with varied Poisson's ratio, Smart Mater. Struct. **24**:095011.

[26] Krushynska AO, Miniaci M, Bosia F, Pugno NM (2017) Coupling local resonance with Bragg band gaps in single-phase mechanical metamaterials, Extreme Mechanics Letters **12**:30-36.

[27] Frazier MJ, Kochmann DM (2917) Band gap transmission in periodic bistable mechanical systems, Journal of Sound and Vibration **388**:315-326.

[28] Hou J, Zhang Z, Li D (2022) Study on bandgap and directional wave propagation of a two-dimensional lattice with a nested core, J. Phys. D: Appl. Phys. **55**:205302.

[29] Dove MT (1993) Introduction to Lattice Dynamics, Cambridge University Press, Cambridge, UK.

[30] Wormser M, Wein F, Stingl M, Körner C (2017) Design and additive manufacturing of 3D phononic band gap structures based on gradient based optimization, Materials **2017**(10):1125.

[31] Rosi G, Auffray N (2016) Anisotropic and dispersive wave propagation within strain-gradient framework, Wave Motion **63**:120-134.

[32] Cook RD, Malkus DS, Plesha ME, Witt RJ (2001) Concepts and Applications of Finite Element Analysis, 4th ed, John Wiley & Sons, Hoboken, NJ, USA.

[33] COMSOL website. http://www.comsol.com (Nov. 2022)

Chapter 45
Large Strains of a Spherical Shell with Distributed Dislocations and Disclinations

Leonid M. Zubov and Mikhail I. Karyakin

Abstract A theory of nonlinear deformation of the elastic Cosserat shells with continuously distributed dislocations and disclinations is formulated. Displacements, rotations, and strains are considered to be arbitrarily large, and the rotation field is kinematically independent of the displacement field. A system of nonlinear differential equations is derived that describes the stress state of an elastic shell with given external loads and given dislocation and disclination densities. This system consists of equilibrium equations and incompatibility equations and contains, as unknown functions, the tensor fields of metric and flexural strains of the elastic shell. The general theory is illustrated by solving a nonlinear problem of the equilibrium of a spherical shell with a spherically symmetric distribution of dislocations and disclinations.

Key words: 6-parameter shell theory · Incompatibility equations · Dislocation and disclination densities · Spherical symmetry

45.1 Introduction

Shell structures of completely different scale are actively used in various industries. Examples include containers for storing and transporting liquid and bulk cargo, and capsules for medicines and cosmetics. Shell-type, or two-dimensional, objects naturally arise in biological systems, as well as in technological processes of various nature. In this connection, crystalline and glassy colloidosomes [1], multielectron

Leonid Zubov
Southern Federal University, Russian Federation,
e-mail: lmzubov@sfedu.ru

Mikhail Karyakin
South Mathematical Institute, Vlkadikavkaz & Southern Federal University, Rostov-on-Don, Russian Federation,
e-mail: karyakin@sfedu.ru

H. Altenbach et al. (eds.), *Sixty Shades of Generalized Continua*,
Advanced Structured Materials 170, https://doi.org/10.1007/978-3-031-26186-2_45

bubbles on helium films [2], capsules of Janus and patchy particles [3], nematic vesicles [4], and viral capsids [5] deserve mention.

Two-dimensional crystals can be successfully used in nano-design, development of submicron devices and the latest drugs. They are ideal candidates for a range of applications, such as nanocontainers for storing and delivering drugs and therapeutic genes to cells.

Note that many of the two-dimensional objects mentioned above are closed shells, topologically equivalent to a sphere. In this case, new crystallographic patterns arise that are absent in ordinary flat and bulk crystals. One of these features is the mandatory presence of structural defects. It is the presence of certain types of defects that determines the curvature of two-dimensional structures.

Topological defects such as dislocations and disclinations can be found in thin shell structures at different length scales: from the world of carbon allotropes, as in fullerenes, nanotubes, and graphene, to biological systems such as in lipid membranes, and in synthetic structures such as colloidosomes, colloidal particle shells lying at the interface between two fluids [6]-[9].

Most experimental studies have shown that these defects in 2D crystals significantly affect their physical, chemical, and mechanical properties [10]. Dislocation-type defects play an essential role in the mechanical behavior of surface crystals, nanotubes, nanofilms, and other two-dimensional physical systems [11]. Simulation results of the mechanical deformation of colloidal crystalline shells illustrate the role played by geometrically necessary topological defects in controlling plastic yielding and failure and provide general guiding principles to optimize the structural and mechanical stability of curved colloidal crystals [12]. Defects formed on the surface of spherical crystals are a kind of «scaffolding» for the formation of mesomolecules and the occurrence of various chemical reactions [13]. Deeper theoretical understanding of the dislocations influence upon the properties of 2D systems can create background that will open a new direction in science and technology — defect engineering [14].

Many common approaches to studying the effect of a given distribution of structural defects in two-dimensional systems on stress and strain fields are based on non-Euclidean geometry and related methods and concepts [15]. The geometry and energy of disclinated graphene configurations were analyzed with the help of molecular dynamics simulation technique and in the framework of the theory of defects in elastic continuum [14]. Modeling the most topologically regular two-dimensional nanocrystals including various possible polymorphic forms of the HIV viral capsid is performed by minimizing the shell elastic energy in [16]. The studies of viral capsids within the framework of the nonlinear theory of elastic shells [17]-[19] demonstrate good fits to experimentally determined virus shapes and explain many experimental observations in viruses.

This work applies the nonlinear theory of micropolar shells, also called the 6-parameter model. In this theory, the shell is treated as a two-dimensional continuum, i.e., a material surface, each point of which has six degrees of freedom of an absolutely rigid body. Based on the concept of isolated Volterra dislo- cations in a multiply connected two-dimensional elastic body, we build a model of a nonlinearly elastic micropolar shell with distributed dislocations by passing to the limit from a discrete set of dislocations to their continuous distribution. The system of equations describing the nonlinear deformation of a shell with distributed dislocations contains the tensor fields of distortion and rotation as unknown functions. To extend the nonlinear theory of shells to the case of the presence of isolated and distributed disclinations, we transform the indicated system of equations in such a way that the tensor fields of metric and bending strains become unknown functions. To model continuously distributed disclinations, their density function is introduced by analogy with geometrically linear theory as a nonzero right-hand side of the nonlinear compatibility equations for bending strains.

Previously, the theory of continuously distributed dislocations and disclinations was developed in [20]-[23] within the framework of the classical Kirchhoff-Love model of plates and shells (both linear and non-linear). The nonlinear theory of isolated and continuously distributed dislocations in three-dimensional micropolar media is presented in [24]-[27].

45.2 The Model of a Nonlinear Elastic Micropolar Shell

According to the mathematical model of micropolar shells, also called the Cosserat-type theory or the 6-parameter theory, the shell is a two-dimensional continuum, that is, a material surface. Each point (particle) of this continuum has six degrees of freedom of an absolutely rigid body [28]-[35].

Let us denote by σ the surface of the shell in the reference configuration, i. e., in the undeformed state, and define the equation of the surface σ in the form $\boldsymbol{r} = \boldsymbol{r}(q^1, q^2)$, where $\boldsymbol{r}$ is a radius-vector of the particle of σ, and q^1, q^2 are some Gaussian coordinates. The surface Σ of the shell in the current configuration is also referred to the coordinates q^γ ($\gamma = 1,2$), and the radius-vector $\boldsymbol{R}(q^1, q^2)$ specifies the position of the material point on Σ. The field of rotations, or proper orthogonal tensor field $\boldsymbol{H}(q^1, q^2)$, characterizes the orientation of the shell particles. The tensor field $\boldsymbol{H}$ is kinematically independent of the displacement field of the shell $\boldsymbol{u} = \boldsymbol{R} - \boldsymbol{r}$. The strained state of the Cosserat-type shell is determined by the vector and tensor fields $\boldsymbol{R}(q^1, q^2)$ and $\boldsymbol{H}(q^1, q^2)$, at that $\boldsymbol{H} \cdot \boldsymbol{H}^{\mathrm{T}} = \boldsymbol{I}$, $\det \boldsymbol{H} = 1$, $\boldsymbol{I}$ is three-dimensional identity tensor. The elastic properties of the shell are determined by the function of the specific (per surface σ area unit) potential energy $W(\boldsymbol{E}, \boldsymbol{L})$.

$$\boldsymbol{E} = \boldsymbol{F} \cdot \boldsymbol{H}^{\mathrm{T}}, \; \boldsymbol{L} = \frac{1}{2} \boldsymbol{r}^\gamma \otimes \left(\frac{\partial \boldsymbol{H}}{\partial q^\gamma} \cdot \boldsymbol{H}^{\mathrm{T}} \right)_\times , \tag{45.1}$$

$$\boldsymbol{F} = \operatorname{grad} \boldsymbol{R}. \tag{45.2}$$

Here $\boldsymbol{F}$ is the shell distortion tensor, $\boldsymbol{E}$ is the strain measure, $\boldsymbol{L}$ is the bending strain tensor. The symbol $\boldsymbol{A}_\times = A_{sk} \boldsymbol{e}^s \times \boldsymbol{e}^k$ denotes the vector invariant of the second-rank tensor $\boldsymbol{A} = A_{sk} \boldsymbol{e}^s \otimes \boldsymbol{e}^k$. Gradient and divergence operations on the surface are introduced by the relations [36]:

$$\operatorname{grad}\boldsymbol{\Lambda} = \boldsymbol{r}^{\gamma} \otimes \frac{\partial \boldsymbol{\Lambda}}{\partial q^{\gamma}}, \ \operatorname{div}\boldsymbol{\Lambda} = \boldsymbol{r}^{\gamma} \cdot \frac{\partial \boldsymbol{\Lambda}}{\partial q^{\gamma}},$$
$$\boldsymbol{r}_{\varkappa} = \frac{\partial \boldsymbol{r}}{\partial q^{\varkappa}}, \ \boldsymbol{r}^{\gamma} \cdot \boldsymbol{r}_{\varkappa} = \delta^{\gamma}_{\varkappa}, \ \boldsymbol{r}^{\gamma} \cdot \boldsymbol{n} = 0, \ \gamma, \varkappa = 1,2 \tag{45.3}$$

In (45.3) $\boldsymbol{r}_{\varkappa}$ and $\boldsymbol{r}^{\gamma}$ are main and reciprocal vector bases on σ, $\boldsymbol{n}$ is unit normal to the surface σ, $\delta^{\gamma}_{\varkappa}$ is the Kronecker delta, $\boldsymbol{\Lambda}$ is an arbitrary differentiable tensor field of any rank.

The stress state of the shell is characterized by the surface stress $\boldsymbol{P}$ and couple stress $\boldsymbol{\Pi}$ tensors of Kirchhoff type. These tensors are expressed in terms of strain tensors $\boldsymbol{E}$ and $\boldsymbol{L}$ by using constitutive relations

$$\boldsymbol{P}(\boldsymbol{E},\boldsymbol{L}) = \frac{\partial W}{\partial \boldsymbol{E}}, \ \boldsymbol{\Pi}(\boldsymbol{E},\boldsymbol{L}) = \frac{\partial W}{\partial \boldsymbol{L}}. \tag{45.4}$$

Note that the tensors participating in (45.4) have the property

$$\boldsymbol{n} \cdot \boldsymbol{E} = \boldsymbol{n} \cdot \boldsymbol{L} = \boldsymbol{n} \cdot \boldsymbol{P} = \boldsymbol{n} \cdot \boldsymbol{\Pi} = 0.$$

If the elastic shell is in equilibrium under the given force and moment loads, then the stress and couple stress tensors obey the equations of statics

$$\operatorname{div}(\boldsymbol{P} \cdot \boldsymbol{H}) + \boldsymbol{f} = 0, \ \operatorname{div}(\boldsymbol{\Pi} \cdot \boldsymbol{H}) + \left(\boldsymbol{F}^{\mathrm{T}} \cdot \boldsymbol{P} \cdot \boldsymbol{H}\right)_{\times} + \boldsymbol{l} = 0, \tag{45.5}$$

where $\boldsymbol{f}$ and $\boldsymbol{l}$ are intensities (per surface σ area unit) of external distributed forces and moments.

45.3 Continuously Distributed Dislocations in an Elastic Micropolar Shell

The concept of dislocations in an elastic shell arises naturally when solving the problem of determining the displacement field from a given single-valued differentiable field of the distortion tensor $\boldsymbol{F}(\boldsymbol{r})$. This problem is obviously equivalent to the problem of determining the vector field $\boldsymbol{R}(\boldsymbol{r})$ characterizing the positions of shell particles in the deformed state. Based on (45.2) we have

$$\boldsymbol{R}(\boldsymbol{r}) = \int_{\boldsymbol{r}_0}^{\boldsymbol{r}} d\boldsymbol{r} \cdot \boldsymbol{F} + \boldsymbol{R}(\boldsymbol{r}_0). \tag{45.6}$$

If the initial value $\boldsymbol{R}(\boldsymbol{r}_0)$ is given at some point $\boldsymbol{r}_0$, then under the condition

$$\operatorname{div}(\boldsymbol{d} \cdot \boldsymbol{F}) = 0, \ \boldsymbol{d} = -\boldsymbol{I} \times \boldsymbol{n} \tag{45.7}$$

the expression (45.6)in a simply connected domain defines a single-valued function $\boldsymbol{R}(\boldsymbol{r})$. Tensor $\boldsymbol{d}$ in (45.7) is the surface discriminant tensor [36, 37].

Let us now consider some section σ_0 of the shell surface and assume that the domain σ_0 is multiply-connected and homeomorphic to a circle with N circular holes, and the function $\boldsymbol{F}(\boldsymbol{r})$ is single-valued in the multiply-connected domain. In this case, the property of uniqueness of displacements is, generally speaking, lost. The multivaluedness can be eliminated by making the domain σ_0 simple connected by drawing the required number of cuts (partitions). In this case, the values of the function $\boldsymbol{R}(\boldsymbol{r})$ will differ on different sides of the cut. It follows from the relation(45.6) that the displacement jump at the intersection of each cut is described by the formula

$$\boldsymbol{u}_+ - \boldsymbol{u}_- = \boldsymbol{R}_+ - \boldsymbol{R}_- = \boldsymbol{B}_k, \tag{45.8}$$

where $\boldsymbol{B}_k$ $(k = 1, 2, \ldots N)$ are vectors constant for the given cut, called Burgers vectors. These vectors do not depend on the choice of the cuts system and, according to (45.6), are expressed in terms of the distortion field by the formulas

$$\boldsymbol{B}_k = \oint_{\Gamma_k} d\boldsymbol{r} \cdot \boldsymbol{F}, \tag{45.9}$$

where Γ_k is any closed contour enclosing only one k-th hole. A non-zero value of $\boldsymbol{B}_k$ means that the shell contains isolated dislocations.

If the number of dislocations in a limited part of the shell is considerably high, switching to a continuous distribution of defects is advisable. The total Burgers vector of a discrete set of N isolated dislocations contained in the subregion σ_0 according to (45.9) is given by

$$\boldsymbol{B} = \sum_{k=1}^{N} \boldsymbol{B}_k = \sum_{k=1}^{N} \oint_{\Gamma_k} d\boldsymbol{r} \cdot \boldsymbol{F}. \tag{45.10}$$

Due to the well-known properties of curvilinear integrals and the single-valuedness of the tensor field $\boldsymbol{F}$, the sum of integrals in (45.10) can be replaced by a single integral over the closed contour Γ_0 covering all holes in the domain σ_0:

$$\boldsymbol{B} = \oint_{\Gamma_0} d\boldsymbol{r} \cdot \boldsymbol{F}. \tag{45.11}$$

To pass from a discrete set of dislocations to their continuous distribution, we tend to the hole diameters to zero and transform the contour integral (45.11) by the known [36, 37] formula

$$\iint_{\sigma_0} \operatorname{div}(\boldsymbol{d} \cdot \boldsymbol{\Lambda})\, d\sigma = \oint_{\Gamma_0} d\boldsymbol{r} \cdot \boldsymbol{\Lambda}$$

into a surface integral over the region σ_0 bounded by the contour Γ_0:

$$B = \iint_{\sigma_0} \alpha \, d\sigma, \quad \alpha = \operatorname{div}(d \cdot F). \tag{45.12}$$

Since B is the total Burgers vector of all dislocations con- tained in an arbitrary region σ_0, the vector field $\alpha(r)$ should be called the dislocation density. In what follows, we will consider the dislocation density to be a given function of Gaussian coordinates, similar to the external loads f and l.

In the presence of continuously distributed dislocations in the shell, the vector fields $u(r)$ and $R(r)$ do not exist, so Eq. (45.2) does not have sense. It is replaced according to (45.12) by the incompatibility equation

$$\operatorname{div}(d \cdot F) = \alpha. \tag{45.13}$$

The complete system of equations describing the equilibrium state of an elastic shell with distributed dislocations consists of relationships (45.1), (45.4), (45.5) and (45.13). This system is easily reduced to equations for two unknown tensors: the distortion tensor $F(r)$ and the rotation tensor $H(r)$.

45.4 Transformation of Incompatibility Equations and Equilibrium Equations

If the elastic shell in addition to distributed and concentrated dislocations contains isolated disclinations then the distortion F and rotation H tensor fields will be multivalued functions of the coordinates q^1, q^2 whereas in contrast the strain measure E and the bending strain tensor L are single-valued tensor fields.

To analyze the stress state of a shell with isolated disclinations, it is expedient to compose a system of resolving equations in which the multivalued functions F and H are excluded, and the unknowns are the tensor functions E and L. Thus, system of equations (45.1), (45.4), (45.5), (45.13) should be transformed by excluding tensors F and H as unknown functions.

First, we transform appropriately expression for the bending strain tensor (45.1). Introducing vectors $L_\varkappa = r_\varkappa \cdot L$, we have a decomposition $L = r^\varkappa \otimes L_\varkappa$. Then, based on (45.1), the following equalities hold ($\varkappa, \gamma = 1, 2$)

$$\frac{\partial H}{\partial q^\varkappa} \cdot H^{\mathrm{T}} = -L_\varkappa \times I, \quad \frac{\partial H}{\partial q^\gamma} \cdot H^{\mathrm{T}} = -L_\gamma \times I, \tag{45.14}$$

or

$$\frac{\partial H}{\partial q^\varkappa} = -L_\varkappa \times H, \quad \frac{\partial H}{\partial q^\gamma} = -L_\gamma \times H. \tag{45.15}$$

Differentiating the first relation (45.15) with respect to q^γ, and the second one with respect to $q^\varkappa$ we get

$$\frac{\partial^2 \boldsymbol{H}}{\partial q^\gamma \partial q^\varkappa} = -\frac{\partial \boldsymbol{L}_\varkappa}{\partial q^\gamma} \times \boldsymbol{H} - \boldsymbol{L}_\varkappa \times \frac{\partial \boldsymbol{H}}{\partial q^\gamma}, \tag{45.16}$$

$$\frac{\partial^2 \boldsymbol{H}}{\partial q^\varkappa \partial q^\gamma} = -\frac{\partial \boldsymbol{L}_\gamma}{\partial q^\varkappa} \times \boldsymbol{H} - \boldsymbol{L}_\gamma \times \frac{\partial \boldsymbol{H}}{\partial q^\varkappa}. \tag{45.17}$$

Since the left parts of expressions (45.16) and (45.17) are the same, we have

$$\frac{\partial \boldsymbol{L}_\varkappa}{\partial q^\gamma} \times \boldsymbol{H} + \boldsymbol{L}_\varkappa \times \frac{\partial \boldsymbol{H}}{\partial q^\gamma} = \frac{\partial \boldsymbol{L}_\gamma}{\partial q^\varkappa} \times \boldsymbol{H} + \boldsymbol{L}_\gamma \times \frac{\partial \boldsymbol{H}}{\partial q^\varkappa}. \tag{45.18}$$

Multiplying (45.18) by $\boldsymbol{H}^{\mathrm{T}}$ from the right and accounting (45.14) we get

$$\left(\frac{\partial \boldsymbol{L}_\varkappa}{\partial q^\gamma} - \frac{\partial \boldsymbol{L}_\gamma}{\partial q^\varkappa} \right) \times \boldsymbol{I} = \boldsymbol{L}_\varkappa \times \boldsymbol{I} \times \boldsymbol{L}_\gamma - \boldsymbol{L}_\gamma \times \boldsymbol{I} \times \boldsymbol{L}_\varkappa. \tag{45.19}$$

An easily verifiable identity $\boldsymbol{a} \times \boldsymbol{I} \times \boldsymbol{b} = \boldsymbol{b} \otimes \boldsymbol{a} - (\boldsymbol{a} \cdot \boldsymbol{b})\boldsymbol{I}$ and (45.19) implies the equality

$$\left(\frac{\partial \boldsymbol{L}_\varkappa}{\partial q^\gamma} - \frac{\partial \boldsymbol{L}_\gamma}{\partial q^\varkappa} \right) \times \boldsymbol{I} = \boldsymbol{L}_\gamma \otimes \boldsymbol{L}_\varkappa - \boldsymbol{L}_\varkappa \otimes \boldsymbol{L}_\gamma. \tag{45.20}$$

The left and the right parts of expression (45.20) are antisymmetric tensors of the second rank. Their equality is equivalent to the equality of their vector invariants:

$$\frac{\partial \boldsymbol{L}_\varkappa}{\partial q^\gamma} - \frac{\partial \boldsymbol{L}_\gamma}{\partial q^\varkappa} = \boldsymbol{L}_\varkappa \times \boldsymbol{L}_\gamma. \tag{45.21}$$

System (45.21) can be called the compatibility equations for bending deformations. They are the necessary and sufficient conditions for the unique solvability of Eq. (45.14) with respect to the rotation tensor $\boldsymbol{H}$ in a simply connected domain for a given value of this tensor at some point.

For further transformation of system (45.21), we use the directly verifiable identity

$$\operatorname{div}(\boldsymbol{d} \cdot \boldsymbol{L}) = d^{\varkappa\gamma} \frac{\partial \boldsymbol{L}_\gamma}{\partial q^\varkappa}; \; d^{\varkappa\gamma} \triangleq \boldsymbol{r}^\varkappa \cdot \boldsymbol{d} \cdot \boldsymbol{r}^\gamma \tag{45.22}$$

and introduce the tensor $\boldsymbol{L}^\perp$, called the adjugate tensor of the $\boldsymbol{L}$ [36]

$$\boldsymbol{L}^\perp \triangleq \boldsymbol{L}^2 - (\operatorname{tr} \boldsymbol{L})\boldsymbol{L} + \frac{1}{2}\left(\operatorname{tr}^2 \boldsymbol{L} - \operatorname{tr} \boldsymbol{L}^2\right)\boldsymbol{I}. \tag{45.23}$$

Equation (45.23) together with representation $\boldsymbol{L} = \boldsymbol{r}^\varkappa \otimes \boldsymbol{L}_\varkappa$ implies the equality

$$\boldsymbol{L}^\perp = \frac{1}{2}\left(\boldsymbol{L}_\gamma \times \boldsymbol{L}_\varkappa\right) \otimes \left(\boldsymbol{r}^\gamma \times \boldsymbol{r}^\varkappa\right). \tag{45.24}$$

Using the following formula for the unit normal $\boldsymbol{n}$ to the surface σ

$$\boldsymbol{n} d^{\varkappa\gamma} = \boldsymbol{r}^\varkappa \times \boldsymbol{r}^\gamma,$$

and considering (45.24), we get

$$\boldsymbol{L}^{\perp} \cdot \boldsymbol{n} = \frac{1}{2} \left(\boldsymbol{L}_\gamma \times \boldsymbol{L}_\varkappa\right) d^{\gamma\varkappa}. \tag{45.25}$$

Based on (45.22) and (45.25), the system of compatibility equations for bending strains (45.21) is written as a vector equation

$$\operatorname{div}(\boldsymbol{d} \cdot \boldsymbol{L}) + \boldsymbol{L}^{\perp} \cdot \boldsymbol{n} = 0. \tag{45.26}$$

Turn to the transformation of incompatibility equation (45.13). Given (45.1), we rewrite it as

$$\operatorname{div}(\boldsymbol{d} \cdot \boldsymbol{E} \cdot \boldsymbol{H}) = \boldsymbol{\alpha}. \tag{45.27}$$

Multiplying (45.27) by $\boldsymbol{H}^{\mathrm{T}}$ and bearing in mind that $\boldsymbol{H} \cdot \boldsymbol{H}^{\mathrm{T}} = \boldsymbol{I}$, we get

$$\operatorname{div}(\boldsymbol{d} \cdot \boldsymbol{E}) + \boldsymbol{r}^{\varkappa} \cdot \boldsymbol{d} \cdot \boldsymbol{E} \cdot \frac{\partial \boldsymbol{H}}{\partial q^{\varkappa}} \cdot \boldsymbol{H}^{\mathrm{T}} = \boldsymbol{\alpha}^{*}. \tag{45.28}$$

The vector $\boldsymbol{\alpha}^{*} = \boldsymbol{\alpha} \cdot \boldsymbol{H}^{\mathrm{T}}$ is called the modified dislocation density vector. Referring to (45.14), instead of (45.28) we will have the equation

$$\operatorname{div}(\boldsymbol{d} \cdot \boldsymbol{E}) - \boldsymbol{r}^{\varkappa} \cdot \boldsymbol{d} \cdot \boldsymbol{E} \times \boldsymbol{L}_\varkappa = \boldsymbol{\alpha}^{*}. \tag{45.29}$$

Let us prove the identity

$$-\boldsymbol{r}^{\varkappa} \cdot \boldsymbol{d} \cdot \boldsymbol{E} \times \boldsymbol{L}_\varkappa = \left(\boldsymbol{E}^{\mathrm{T}} \cdot \boldsymbol{d} \cdot \boldsymbol{L}\right)_{\times}. \tag{45.30}$$

Indeed, since $\boldsymbol{E}^{\mathrm{T}} \cdot \boldsymbol{d} \cdot \boldsymbol{r}^{\varkappa} = -\boldsymbol{r}^{\varkappa} \cdot \boldsymbol{d} \cdot \boldsymbol{E}$, we have

$$\begin{aligned}\left(\boldsymbol{E}^{\mathrm{T}} \cdot \boldsymbol{d} \cdot \boldsymbol{L}\right)_{\times} &= \left(\boldsymbol{E}^{\mathrm{T}} \cdot \boldsymbol{d} \cdot \boldsymbol{r}^{\varkappa} \otimes \boldsymbol{L}_\varkappa\right)_{\times} \\ &= \left(\boldsymbol{E}^{\mathrm{T}} \cdot \boldsymbol{d} \cdot \boldsymbol{r}^{\varkappa}\right) \times \boldsymbol{L}_\varkappa = -\boldsymbol{r}^{\varkappa} \cdot \boldsymbol{d} \cdot \boldsymbol{E} \times \boldsymbol{L}_\varkappa.\end{aligned}$$

So, incompatibility equation (45.13) with respect to the distortion tensor has been transformed into an equation for metric and bending strains, i.e. regarding tensors $\boldsymbol{E}$ and $\boldsymbol{L}$

$$\operatorname{div}(\boldsymbol{d} \cdot \boldsymbol{E}) + \left(\boldsymbol{E}^{\mathrm{T}} \cdot \boldsymbol{d} \cdot \boldsymbol{L}\right)_{\times} = \boldsymbol{\alpha}^{*}. \tag{45.31}$$

Let us now transform equilibrium equations (45.5) in order to exclude the distortion and rotation tensors, $\boldsymbol{F}$ and $\boldsymbol{H}$, as unknown functions. Multiplying the force balance equation by $\boldsymbol{H}^{\mathrm{T}}$ and taking (45.14) into account, we obtain

$$\operatorname{div}\boldsymbol{P} - \boldsymbol{r}^{\varkappa} \cdot \boldsymbol{P} \times \boldsymbol{L}_\varkappa + \boldsymbol{f}^{*} = 0; \; \boldsymbol{f}^{*} = \boldsymbol{f} \cdot \boldsymbol{H}^{\mathrm{T}}. \tag{45.32}$$

Then we get

$$\boldsymbol{r}^{\varkappa} \cdot \boldsymbol{P} \times \boldsymbol{L}_{\varkappa} = \left(\boldsymbol{P}^{\mathrm{T}} \cdot \boldsymbol{r}^{\varkappa}\right) \times \boldsymbol{L}_{\varkappa} = \left[\left(\boldsymbol{P}^{\mathrm{T}} \cdot \boldsymbol{r}^{\varkappa}\right) \otimes \boldsymbol{L}_{\varkappa}\right]_{\times} = \left(\boldsymbol{P}^{\mathrm{T}} \cdot \boldsymbol{L}\right)_{\times},$$

and Eq. (45.32) is written as

$$\operatorname{div} \boldsymbol{P} - \left(\boldsymbol{P}^{\mathrm{T}} \cdot \boldsymbol{L}\right)_{\times} + \boldsymbol{f}^{*} = 0. \tag{45.33}$$

Moment equilibrium equation (45.5) is transformed in a similar way and takes the form

$$\operatorname{div} \boldsymbol{\Pi} - \left(\boldsymbol{\Pi}^{\mathrm{T}} \cdot \boldsymbol{L} + \boldsymbol{P}^{\mathrm{T}} \cdot \boldsymbol{E}\right)_{\times} + \boldsymbol{l}^{*} = 0; \; \boldsymbol{l}^{*} = \boldsymbol{l} \cdot \boldsymbol{H}^{\mathrm{T}}. \tag{45.34}$$

If we assume that the modified intensities of the force and moment loads $\boldsymbol{f}^{*}$ and $\boldsymbol{l}^{*}$ are known functions, then, by virtue of the constitutive relations (45.4), we see that the Eqs. (45.33) and (45.34) contain as unknown functions the tensor fields of metric and bending strains $\boldsymbol{E}$ and $\boldsymbol{L}$.

Using system of Eqs. (45.26), (45.31), (45.33) and (45.34), one can consider the problem of equilibrium of an elastic shell containing distributed dislocations and isolated disclinations. While the quantitative characteristic of an isolated dislocation is the Burgers vector, then the power of an isolated disclination is determined by the Frank vector [24, 38, 39]. The Frank vector in the nonlinear case, that is at large rotations and strains of the shell, is expressed in terms of a multiplicative curvilinear integral over a closed contour enveloping an isolated disclination [40]. The complex properties of curvilinear multiplicative integrals [24], due to the non-commutativity of finite rotations, do not allow one to explicitly express the total Frank vector of the set of disclinations in terms of the tensor field of bending strains and to carry out the limit transition from a discrete set of isolated disclinations to their continuous distribution. This makes it difficult to rigorously formulate the concept of disclination density in a nonlinear elastic shell. The mentioned transition is possible in the geometrically linear theory of the Cosserat-type shells [41], i.e., for small strains and small rotations. In this case, the Frank vector is expressed in terms of the usual curvilinear integral, and the disclination density is defined as a function of coordinates, the integral of which over the area of an arbitrary region is equal to the total Frank vector of the disclinations contained in this region. The incompatibility equation for bending strains in the geometrically linear case was derived in [41] and has the form

$$\operatorname{div}(\boldsymbol{d} \cdot \boldsymbol{L}) = \boldsymbol{\beta}, \tag{45.35}$$

where $\boldsymbol{\beta}$ is disclination density vector. The left side of (45.35) equation differs from the left side of the (45.26) compatibility equation in that it does not contain the term $\boldsymbol{L}^{\perp} \cdot \boldsymbol{n}$. According to (45.23), this term is nonlinear and is a homogeneous function of the second degree with respect to the tensor $\boldsymbol{L}$.

To study continuously distributed disclinations in nonlinearly elastic shells, we formally, by analogy with geometrically linear theory, introduce the disclination density as a vector function $\boldsymbol{\beta}(q^1, q^2)$ that replaces zero in the right part of compatibility equation (45.26):

$$\operatorname{div}(\boldsymbol{d} \cdot \boldsymbol{L}) + \boldsymbol{L}^{\perp} \cdot \boldsymbol{n} = \boldsymbol{\beta}. \tag{45.36}$$

Equation (45.36) will be called the incompatibility equation for bending strains in the nonlinear theory of elastic shells of the Cosserat type.

45.5 Equilibrium of a Closed Spherical Shell with Distributed Dislocations and Disclinations

In a closed spherical shell of radius r_0, we take the geographic coordinates $0 \leqslant \varphi \leqslant 2\pi$ (longitude) and $-\pi/2 \leqslant \theta \leqslant \pi/2$ (latitude) as Gaussian coordinates. The unit normal to the sphere will be denoted by $\boldsymbol{e}_r$, and the unit vectors tangent to the coordinate lines by $\boldsymbol{e}_\varphi$ and $\boldsymbol{e}_\theta$. The metric $\boldsymbol{g}$ and discriminant $\boldsymbol{d}$ tensors on the sphere have the form

$$\begin{aligned} \boldsymbol{g} &= \boldsymbol{I} - \boldsymbol{e}_r \otimes \boldsymbol{e}_r = \boldsymbol{e}_\varphi \otimes \boldsymbol{e}_\varphi + \boldsymbol{e}_\theta \otimes \boldsymbol{e}_\theta, \\ \boldsymbol{d} &= -\boldsymbol{I} \times \boldsymbol{e}_r = \boldsymbol{e}_\varphi \otimes \boldsymbol{e}_\theta - \boldsymbol{e}_\theta \otimes \boldsymbol{e}_\varphi, \end{aligned} \tag{45.37}$$

and the gradient operator is given by the relation

$$\operatorname{grad} \boldsymbol{\Lambda} = \frac{1}{r_0 \cos\theta} \boldsymbol{e}_\varphi \otimes \frac{\partial \boldsymbol{\Lambda}}{\partial \varphi} + \frac{1}{r_0} \boldsymbol{e}_\theta \otimes \frac{\partial \boldsymbol{\Lambda}}{\partial \theta}. \tag{45.38}$$

Let us assume that the external loads and the densities of dislocations and disclinations are vectors directed along the normal to the sphere:

$$\boldsymbol{f}^* = f_0 \boldsymbol{e}_r, \ \boldsymbol{l}^* = l_0 \boldsymbol{e}_r, \ \boldsymbol{\alpha}^* = \alpha_0 \boldsymbol{e}_r, \ \boldsymbol{\beta} = \beta_0 \boldsymbol{e}_r, \tag{45.39}$$

where f_0, l_0, α_0, β_0 are constant values. Metric and bending strains of the shell will be sought in the form of spherically symmetric tensor fields [42]

$$\mathbf{E} = E_1 \boldsymbol{g} + E_2 \boldsymbol{d}, \ \mathbf{L} = L_1 \boldsymbol{g} + L_2 \boldsymbol{d}, \tag{45.40}$$

$$E_\mu, L_\mu = \text{const}, \ \mu = 1,2.$$

We assume that the shell material is isotropic and is described by the following constitutive relations [31]

$$\begin{aligned} \boldsymbol{P} &= a_1 (\operatorname{tr} \boldsymbol{U}) \boldsymbol{g} + a_2 (\boldsymbol{U} \cdot \boldsymbol{g})^{\mathrm{T}} + a_3 (\boldsymbol{U} \cdot \boldsymbol{g}) + a_4 \boldsymbol{U} \cdot \boldsymbol{n} \otimes \boldsymbol{n}, \ \boldsymbol{U} = \boldsymbol{E} - \boldsymbol{g}; \\ \boldsymbol{\Pi} &= b_1 (\operatorname{tr} \boldsymbol{L}) \boldsymbol{g} + b_2 (\boldsymbol{L} \cdot \boldsymbol{g})^{\mathrm{T}} + b_3 (\boldsymbol{L} \cdot \boldsymbol{g}) + b_4 \boldsymbol{L} \cdot \boldsymbol{n} \otimes \boldsymbol{n}, \end{aligned} \tag{45.41}$$

where a_k, b_k $(k = 1,2,3,4)$ are material constants. The relations (45.40) and (45.41) imply that the tensors $\boldsymbol{P}$ and $\boldsymbol{\Pi}$ will also be spherically symmetric

$$\boldsymbol{P} = P_1 \boldsymbol{g} + P_2 \boldsymbol{d}, \ \boldsymbol{\Pi} = \Pi_1 \boldsymbol{g} + \Pi_2 \boldsymbol{d}. \tag{45.42}$$

The external force load in (45.39) is expressed in terms of the given hydrostatic pressure p using (45.40) and the transformation formula for the surface element area under deformation [37] as follows

$$f_0 = \left(E_1^2 + E_2^2\right) p. \tag{45.43}$$

The pressure p is positive if it is internal and negative if it is external.

In view of (45.38)–(45.40) and (45.42) equilibrium equations (45.32), (45.34) and incompatibility equation (45.31), (45.36) in a spherically symmetric problem are reduced to a nonlinear algebraic system of equations for unknown constants E_1, E_2, L_1, L_2

$$P_1 + r_0 \left(P_1 L_2 - P_2 L_1\right) = \frac{r_0}{2}\left(E_1^2 + E_2^2\right) p, \tag{45.44}$$

$$\Pi_1 + r_0 \left(\Pi_1 L_2 - \Pi_2 L_1 + P_1 E_2 - P_2 E_1\right) = \frac{r_0}{2} l_0, \tag{45.45}$$

$$E_2 + r_0 \left(E_1 L_1 + E_2 L_2\right) = \frac{r_0}{2}\alpha_0, \tag{45.46}$$

$$L_2 + \frac{r_0}{2}\left(L_1^2 + L_2^2\right) = \frac{r_0}{2}\beta_0. \tag{45.47}$$

Stress and couple stress components in (45.44), (45.45) are expressed in terms of strain components using constitutive relations (45.41) and have the form

$$\begin{gathered} P_1 = (2a_1 + a_2 + a_3)(E_1 - 1), \; P_2 = (a_3 - a_2) E_2, \\ \Pi_1 = (2b_1 + b_2 + b_3) L_1, \; \Pi_2 = (b_3 - b_2) L_2. \end{gathered} \tag{45.48}$$

We present two simple solutions to system (45.44)–(45.47) in a particular case $\alpha_0 = 0$, $\beta_0 = 0$, $l_0 = 0$, $p \neq 0$.

- First solution:

$$E_1 = \frac{A \pm \sqrt{A^2 - 4Ap}}{2p} \quad \text{or} \quad A\left(E_1 - 1\right) = E_1^2 p, \quad A = \frac{2}{r_0}\left(2a_1 + a_2 + a_3\right),$$

$$E_2 = 0, \; L_1 = 0, \; L_2 = 0$$

The solution exists only for $p < A/4$. The corresponding loading diagram $p(E_1)$ is schematically shown in Fig. 45.1.

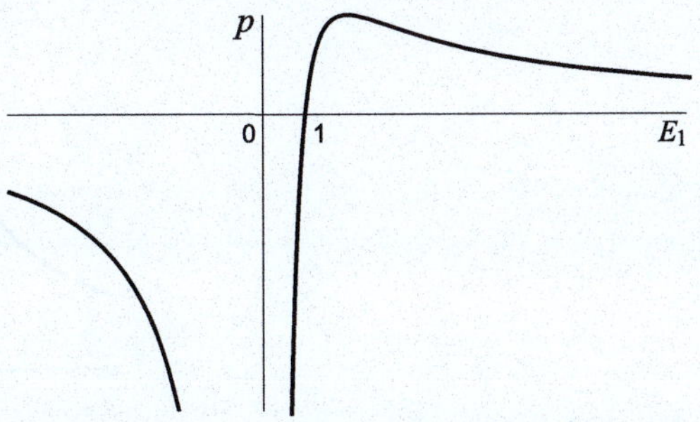

Fig. 45.1 Loading diagram for the shell in the absence of dislocations and disclinations

- Second solution:

$$E_1 = \frac{-A \pm \sqrt{A^2+4Ap}}{2p} \quad \text{or} \quad A(1-E_1) = E_1^2 p,$$

$$E_2 = 0, \; L_1 = 0, \; L_2 = -2r_0^{-1}.$$

This solution exists for $p > -A/4$. The loading diagram corresponding to it differs from the previous case only by the sign (Fig. 45.2)

Note the non-uniqueness of the solution of the considered problem even in this simple case. One value of pressure p corresponds to up to four different values of the deformation parameter E_1. Even at zero pressure, system (45.44)–(45.47), in addition to the trivial solution corresponding to the undeformed state, also admits a second solution describing the equilibrium of a spherical shell turned inside out.

45.6 Numerical Results

The quantities r_0, p, l_0, α_0, β_0 and the material shell constants a_k, b_k $(k = 1,2,3)$ are considered to be the given parameters when solving system (45.44)–(45.47). To obtain some numerical results, as the values of the material constants of the shell, we will choose the following values consistent with [30]

$$a_1 = \frac{2\mu h \nu}{1-\nu}, \qquad a_2 = 0, \quad a_3 = 2\mu h,$$
$$b_1 = \frac{\mu h^3 \nu}{6(1-\nu)}, \quad b_2 = 0, \quad b_3 = \frac{\mu h^3}{6}. \tag{45.49}$$

Here μ is the shell material shear modulus, ν is the Poisson's ratio, h is the shell thickness. So six parameters a_k, b_k are replaced by three material parameters. Below, the division by μ will be used to non-dimensionalize the forces, moments, energy, and stress tensors components. Therefore, the parameter μ will not enter the final dimensionless system of equations. Thus the thickness h and Poisson ratio ν of the shell were added to the list of the problem parameters.

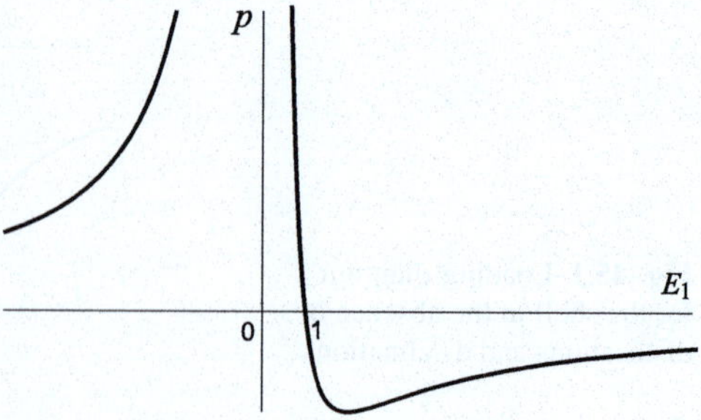

Fig. 45.2 Loading diagram for the everted shell

Let us turn the parameters of the problem, which have the dimension of length, into dimensionless ones, referring them to the radius of the shell. So, for example, the dimensionless shell thickness will take the form

$$\tilde{h} = \frac{h}{r_0}.$$

For non-dimensionalization of the dislocation density α_0 and the disclination density β_0 we also use the value of the radius by the following formulas

$$\tilde{\alpha_0} = \alpha_0 r_0, \ \tilde{\beta_0} = \beta_0 r_0^2.$$

The quantities E_1, E_2 are dimensionless by definition. The non-dimensionalization of the parameters of the bending strain tensor will be carried out according to the scheme

$$\tilde{L}_\gamma = L_\gamma r_0 \ (\gamma = 1,2).$$

Finally, we use the following schemes to make the force and moment quantities included in the equations dimensionless:

$$\tilde{P}_\gamma = \frac{1}{\mu r_0} P_\gamma, \ \tilde{\Pi}_\gamma = \frac{1}{\mu r_0^2} \Pi_\gamma, \ (\gamma = 1,2)$$

$$\tilde{p} = \frac{1}{\mu} p, \ \tilde{l}_0 = \frac{1}{\mu r_0} l_0.$$

In what follows, the tilde over the dimensionless parameters will be omitted. The examples below consider a shell whose thickness is 5% of the radius, i. e. the ratio $h/r_0 = 1/20$. Poisson's ratio was always taken 1/3. It was noted above that the essentially nonlinear system (45.44)–(45.47) can have a number of solutions depending on the values of the parameters. In the examples given, we confine ourselves to the case of a regular, that is, non-inverted shell. For this, in the calculations, the zero value was chosen as the initial approximation for the parameter L_2.

Figures 45.3-45.5 show the dependences of the dimensionless characteristics E_1, E_2 and L_2 of the deformed state of the unloaded shell ($p = 0, l_0 = 0$) on the dislocation density for different values of the disclination density. As for the L_1 parameter, a direct analysis of system (45.44)–(45.47), taking into account (45.48), (45.44), shows that for any values of the disclination parameter, the following ratio holds up to h^2

$$L_1 = \frac{1}{2}\alpha_0.$$

The important characteristic of the stress-strain state of the shell is its elastic energy. To calculate it, we note that constitutive relations (45.41) correspond to the following form of the function of specific potential energy [34]

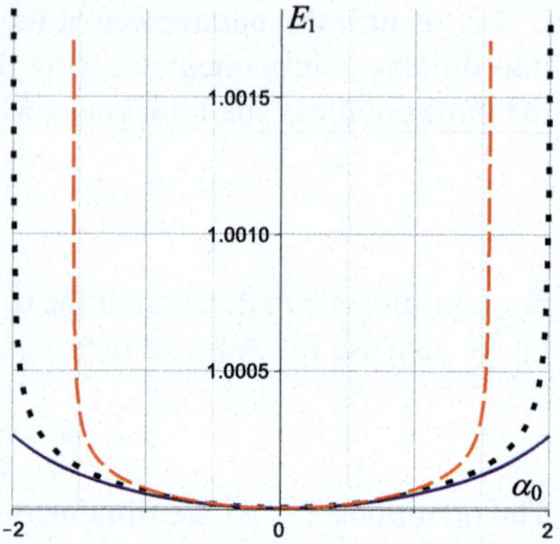

Fig. 45.3 Dependence of the strain parameter E_1 on the dislocation density α_0: $\beta_0 = 0$ — dotted line, $\beta_0 = 0.4$ — solid line, $\beta_0 = -0.4$ — dashed line.

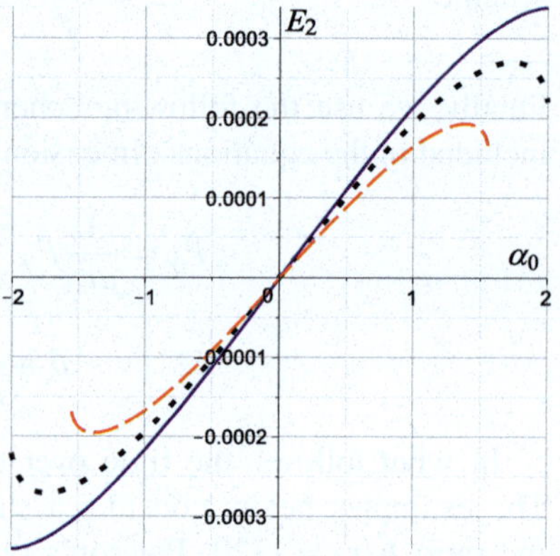

Fig. 45.4 Dependence of the strain parameter E_2 on the dislocation density α_0: $\beta_0 = 0$ — dotted line, $\beta_0 = 0.4$ — solid line, $\beta_0 = -0.4$ — dashed line.

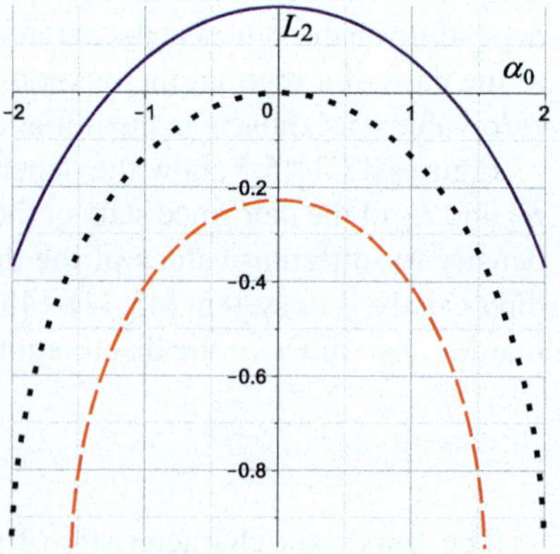

Fig. 45.5 Dependence of the bending strain parameter L_2 on the dislocation density α_0: $\beta_0 = 0$ — dotted line, $\beta_0 = 0.4$ — solid line, $\beta_0 = -0.4$ — dashed line.

$$
\begin{aligned}
W = \frac{1}{2} \Big(& a_1 \mathrm{tr}^2 \boldsymbol{U} + a_2 \mathrm{tr} \boldsymbol{U}^2 + a_3 \mathrm{tr} \left(\boldsymbol{U} \cdot \boldsymbol{U}^{\mathrm{T}} \right) + a_4 \boldsymbol{n} \cdot \boldsymbol{E}^{\mathrm{T}} \cdot \boldsymbol{E} \cdot \boldsymbol{n} \\
& + b_1 \mathrm{tr}^2 \boldsymbol{L} + b_2 \mathrm{tr} \boldsymbol{L}^2 + b_3 \mathrm{tr} \left(\boldsymbol{L} \cdot \boldsymbol{L}^{\mathrm{T}} \right) + b_4 \boldsymbol{n} \cdot \boldsymbol{L}^{\mathrm{T}} \cdot \boldsymbol{L} \cdot \boldsymbol{n} \Big).
\end{aligned} \tag{45.50}
$$

Taking (45.40) into account, the specific energy W is a constant value, i.e., it does not depend on the position of the point on the sphere. This means that the total strain energy of the shell S can be found by multiplying W by the surface area of the sphere:

$$S = 4\pi r_0^2 W.$$

The quantity $S_0 = S/(4\pi\mu r_0^3)$ will be further taken as a dimensionless characteristic of the strain potential energy of the sphere. Considering (45.40), (45.49), (45.50) we arrive at the following expression

$$S_0 = (1+\nu)(E_1 - 1)^2 + (1-\nu)E_2^2 + \frac{\eta^2}{12}\left[(1+\nu)L_1^2 + (1-\nu)L_2^2\right].$$

Figures 45.6 and 45.7 show the dependence of elastic energy on the defect density for an unloaded shell. The dependence of the energy S_0 on the dislocation density α_0 for various disclination densities is shown in Fig. 45.6. It can be seen that the energy is an even function of the parameter α_0. Fig. 45.7 shows the dependence of the energy S_0 on the disclination density β_0 for various dislocation densities. Due to the mentioned above symmetry of the energy as a function of the dislocation density, the graphs are plotted only for zero and positive values of the parameter α_0.

The effect of internal pressure on the energy of the shell is shown in Fig. 45.8, where the solid lines correspond to the level lines of the unloaded shell on the plane of parameters α_0, β_0. The dotted lines are the energy level lines of the shell under internal pressure $p = 0.001$. It can be seen from the figure that the pattern of level lines is not symmetric with respect to the disclination density: as the pressure increases, the surface of the energy function rises and shifts in the direction of positive values of the parameter β_0.

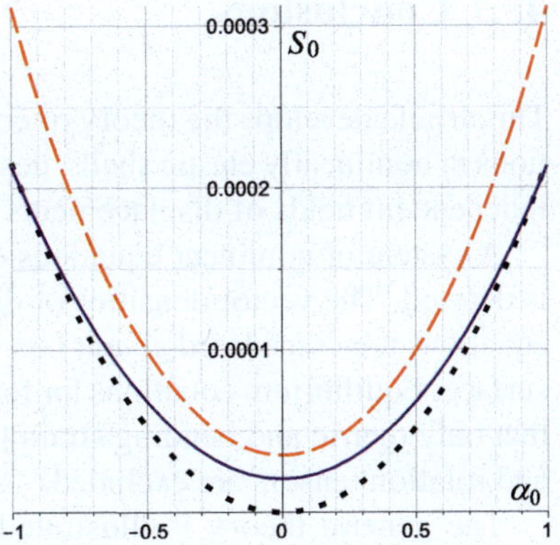

Fig. 45.6 Dependence of the energy on the dislocation density α_0: $\beta_0 = 0$ — dotted line, $\beta_0 = 0.5$ — solid line, $\beta_0 = -0.5$ — dashed line.

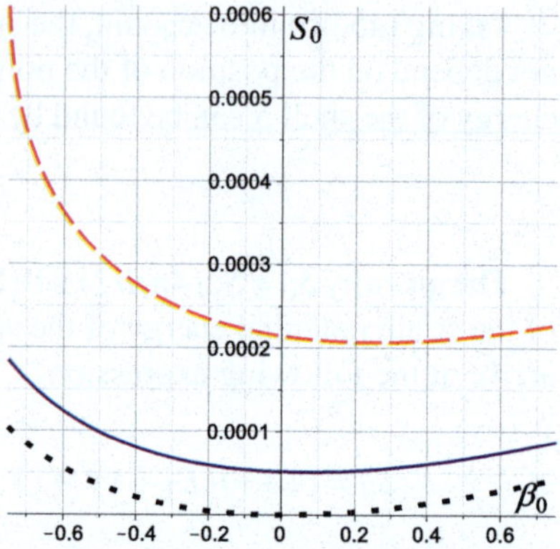

Fig. 45.7 Dependence of the energy on the disclination density β_0: $\alpha_0 = 0$ — dotted line, $\alpha_0 = 0.5$ — solid line, $\alpha_0 = 1$ — dashed line.

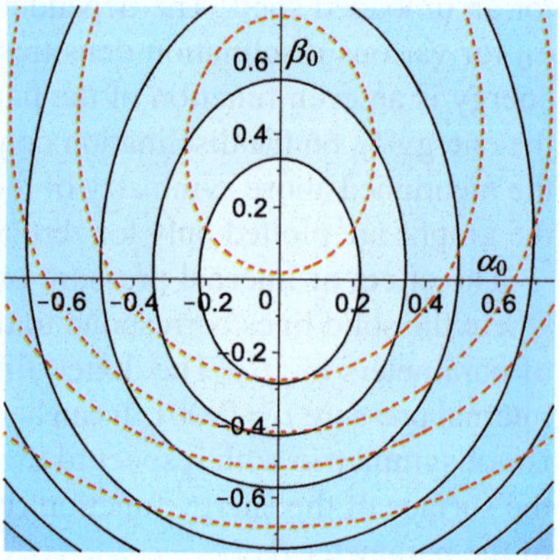

Fig. 45.8 The shell energy level lines as functions of dislocation and disclination density: solid lines — the unloaded shell ($p = 0$), dotted lines — the shell under internal pressure $p_0 = 0.001$.

45.7 Conclusions

This article develops the theory of continuously distributed dislocations and disclinations in nonlinearly elastic shells, described by the Cosserat model with kinematically independent fields of displacements and rotations.

A system of nonlinear equations of incompatibility of metric and bending strains is derived. The vector densities of dislocations and disclinations contained in these equations are considered given, i.e., known functions of the coordinates on the shell surface. Equilibrium equations for forces and moments are transformed in such a way that only metric and bending strain tensors are unknown functions, while distortion and rotation tensors are excluded.

The general theory is illustrated by solving the problem of large strains of a closed spherical shell with a spherically symmetric distribution of dislocations and disclinations. This problem is reduced to a nonlinear algebraic system of equations, which is solved numerically. The effect of dislocations and disclinations on the stress-strain state of a shell loaded with internal pressure is analyzed.

References

[1] Shilpi S, Jain A, Gupta Y, Jain SK (2007) Colloidosomes: An emerging vesicular system in drug delivery, Critical Reviews in Therapeutic Drug Carrier Systems **24**(4):361-391. DOI 10.1615/critrevtherdrugcarriersyst.v24.i4.20

[2] Leiderer P (1995) Ions at helium interfaces, Zeitschrift für Physik B Condensed Matter **98**(3):303-308. DOI 10.1007/BF01338394

[3] Rozynek Z, Mikkelsen A, Dommersnes P, Fossum JO (2014) Electroformation of Janus and patchy capsules, Nature Communications **5**(1):3945. DOI 10.1038/ncomms4945

[4] Keber FC, Loiseau E, Sanchez T, DeCamp SJ, Giomi L, Bowick MJ, Marchetti MC, Dogic Z, Bausch AR (2014) Topology and dynamics of active nematic vesicles, Science **345**(6201):1135-1139. DOI 10.1126/science.1254784

[5] Zandi R, Dragnea B, Travesset A, Podgornik R (2020) On virus growth and form, Physics Reports **847**:1-102. DOI 10.1016/j.physrep.2019.12.005

[6] Azizi A, Zou X, Ercius P, Zhang Z, Elías AL, Perea-López N, Stone G, Terrones M, Yakobson BI, Alem N (2014) Dislocation motion and grain boundary migration in two-dimensional tungsten disulphide, Nature Communications **5**(1):4867. DOI 10.1038/ncomms5867

[7] Butz B, Dolle C, Niekiel F, Weber K, Waldmann D, Weber HB, Meyer B, Spiecker E (2014) Dislocations in bilayer graphene, Nature **505**(7484):533-537. DOI 10.1038/nature12780

[8] Kim T, Park JY, Hwang J, Seo G, Kim Y (2020) Supramolecular two-dimensional systems and their biological applications, Advanced Materials **32**(51):2002405. DOI 10.1002/adma.202002405

[9] Negri C, Sellerio AL, Zapperi S, Miguel MC (2015) Deformation and failure of curved colloidal crystal shells, Proceedings of the National Academy of Sciences of the United States of America **112**(47):14545–14550. DOI 10.1073/pnas.151825811

[10] Xiong Z, Zhong L, Wang H, Li X (2021) Structural defects, mechanical behaviors, and properties of two-dimensional materials, Materials **14**(5):1192. DOI 10.3390/ma14051192

[11] Gutkin MY, Ovid'ko IA (2004) Plastic Deformation in Nanocrystalline Materials. Springer, Berlin, Heidelberg. DOI 10.1007/978-3-662-09374-0

[12] Qi W-K, Zhu T, Chen Y, Ren J-R (2009) Topological aspect of disclinations in two-dimensional crystals, Chinese Physics B **18**(3):1002-1008. DOI 10.1088/1674-1056/18/3/026

[13] Petrov CY, Shevchenko IA (2013) Topological defects in the two-dimensional spherical crystals, Engineering Journal of Don **26**(3):103.

[14] Romanov A, Rozhkov M, Kolesnikova A (2018) Disclinations in polycrystalline graphene and pseudo-graphenes. Review, Letters on Materials **8**(4):384-400. DOI 10.22226/2410-3535-2018-4-384-400

[15] Roychowdhury A, Gupta A (2018) On structured surfaces with defects: Geometry, strain incompatibility, stress field, and natural shapes, Journal of Elasticity **131**(2):239-276. DOI 10.1007/s10659-017-9654-1

[16] Roshal DS, Konevtsova OV, Myasnikova AE, Rochal SB (2016) Assembly of the most topologically regular two-dimensional micro and nanocrystals with spherical, conical, and tubular shapes, Phys. Rev. E **94**:052605. DOI 10.1103/PhysRevE.94.052605

[17] Lidmar J, Mirny L, Nelson DR (2003) Virus shapes and buckling transitions in spherical shells. Phys. Rev. E **68**:051910. DOI 10.1103/PhysRevE.68.051910

[18] Šiber A (2006) Buckling transition in icosahedral shells subjected to volume conservation constraint and pressure: Relations to virus maturation, Phys. Rev. E **73**:061915. DOI 10.1103/PhysRevE.73.061915

[19] Perotti LE, Aggarwal A, Rudnick J, Bruinsma R, Klug WS (2015) Elasticity theory of the maturation of viral capsids, Journal of the Mechanics and Physics of Solids **77**:86-108. DOI 10.1016/j.jmps.2015.01.006

[20] Zubov LM (2007) Von Kármán equations for an elastic plate with dislocations and disclinations, Doklady Physics **52**(1):67-70. DOI https://doi.org/10.1134/S102833580701017X

[21] Zubov L, Stolpovskii A (2008) A theory of dislocations and disclinations in elastic plates, Journal of Applied Mathematics and Mechanics **72**(6):724-737. DOI https://doi.org/10.1016/j.jappmathmech.2009.01.005

[22] Zubov L (2010) The linear theory of dislocations and disclinations in elastic shells, Journal of Applied Mathematics and Mechanics **74**(6):663-672. DOI 10.1016/j.jappmathmech.2011.01.006

[23] Zubov LM (2012) Large deformations of elastic shells with distributed dislocations, Doklady Physics **57**(6):254-257. DOI https://doi.org/10.1134/S1028335812060092

[24] Zubov LM (1997) Nonlinear Theory of Dislocations and Disclinations in Elastic Bodies, Springer, Berlin. DOI https://doi.org/10.1007/978-3-540-68430-5

[25] Zubov LM (2011) Continuum theory of dislocations and disclinations in nonlinearly elastic micropolar media, Mechanics of Solids **46**(3):348-356. DOI https://doi.org/10.3103/S0025654411030022

[26] Karyakin MI, Zubov LM (2011) Theory of isolated and continuously distributed disclinations and dislocations in micropolar media, In: H Altenbach, GA Maugin, V Erofeev (eds.) Mechanics of Generalized Continua, pp 275-290. Springer, Berlin, Heidelberg. DOI https://doi.org/10.1007/978-3-642-19219-7_14

[27] Zelenina AA, Zubov LM (2018) Spherically symmetric deformations of micropolar elastic medium with distributed dislocations and disclinations, In: F dell'Isola, VA Eremeyev, A Porubov (eds.) Advances in Mechanics of Microstructured Media and Structures, pp 357-369. Springer International Publishing, Cham. DOI https://doi.org/10.1007/978-3-319-73694-5_19

[28] Zhilin PA (1976) Mechanics of deformable directed surfaces, International Journal of Solids and Structures **12**(9):635-648. DOI 10.1016/0020-7683(76)90010-X

[29] Altenbach J, Altenbach H, Eremeyev VA (2010) On generalized Cosserat-type theories of plates and shells: a short review and bibliography, Archive of Applied Mechanics **80**(1):73-92. DOI 10.1007/s00419-009-0365-3

[30] Chróscielewski J, Makowski J, Pietraszkiewicz W (2004) Statics and dynamics of multyfolded shells. Nonlinear theory and finite element method (in Polish), Wydawnictwo IPPT PAN, Warszawa.

[31] Eremeyev VA, Zubov LM (2008) Mechanics of Elastic Shells (in Russ), Nauka, Moscow.

[32] Libai A, Simmonds JG (1998) The Nonlinear Theory of Elastic Shells (2nd edn.), Cambridge University Press.

[33] Zubov LM, Eremeyev VA (2003) Mechanics of elastic micropolar shells, Far Eastern Math. J. **4**(1):182-225.

[34] Altenbach H, Eremeyev VA (2014) Vibration analysis of non-linear 6-parameter prestressed shells, Meccanica **49**(8):1751-1761. DOI 10.1007/s11012-013-9845-1

[35] Eremeyev VA, Lebedev LP, Cloud MJ (2015) The Rayleigh and Courant variational principles in the six-parameter shell theory, Mathematics and Mechanics of Solids **20**(7):806-822. DOI 10.1177/1081286514553

[36] Zubov LM, Karyakin MI (2006) Tensor Calculus (in Russ), Vuzovskaya Kniga, Moscow.

[37] Zubov LM (1982) Methods of Nonlinear Theory of Elasticity in the Theory of Shells (in Russ), Rostov Univ. Press., Rostov-on-Don.

[38] deWit R (1973) Theory of disclinations: II. Continuous and discrete disclinations in anisotropic elasticity, J Res Natl Bur Stand A Phys Chem **77A**(1):49-100. DOI 10.6028/jres.077A.003

[39] deWit R (1973) Theory of disclinations: III. Continuous and discrete disclinations in isotropic elasticity, J Res Natl Bur Stand A Phys Chem **77A**(3):359-368. DOI 10.6028/jres.077A.024

[40] Zubov LM (1989) Nonlinear theory of isolated dislocations and disclinations in elastic shells, Izv. AN SSSR. MTT pp 139–145.

[41] Zubov LM, Karyakin MI (2023) Theory of Cosserat-type elastic shells with distributed dislocations and disclinations, In: H Altenbach, SM Bauer, AK Belyaev, DA Indeitsev, VP Matveenko, YV Petrov (eds.) Advances in Solid and Fracture Mechanics – A Liber Amicorum to Celebrate the Birthday of Nikita Morozov, Advanced Structured Materials. vol. 180, pp 259-278. Springer, Cham. DOI 10.1007/978-3-031-18393-5_17

[42] Goloveshkina EV, Zubov LM (2021) Spherically symmetric tensor fields and their application in nonlinear theory of dislocations, Symmetry **2021**(13):830. DOI 10.3390/sym13050830

Correction to: Multi-Element Metamaterial's Design Through the Relaxed Micromorphic Model

Leonardo A. Perez Ramirez, Gianluca Rizzi, and Angela Madeo

Correction to:
Chapter 36 in: H. Altenbach et al. (eds.), *Sixty Shades of Generalized Continua*, Advanced Structured Materials 170, https://doi.org/10.1007/978-3-031-26186-2_36

In the original version of this chapter, the title of Chapter has been changed from "Multi-element MEtamaterial's Design Through the Relaxed Micromorphic Model" to "Multi-element Metamaterial's Design Through the Relaxed Micromorphic Model" and the author name "Ramirez" has been changed to "Perez Ramirez". The Correction chapter has been updated with the changes.

The updated original version of this chapter can be found at https://doi.org/10.1007/978-3-031-26186-2_36

H. Altenbach et al. (eds.), *Sixty Shades of Generalized Continua*,
Advanced Structured Materials 170, https://doi.org/10.1007/978-3-031-26186-2_46

The manufacturer's authorised representative in the EU is Springer Nature Customer Service Centre GmbH, Europaplatz 3, 69115 Heidelberg, Germany. If you have any concerns regarding our products, please contact ProductSafety@springernature.com

Printed and bound by CPI Group (UK) Ltd, Croydon, CR0 4YY
01/10/2026
02225872-0019